U0162514

海岸海洋科学研究与实践

Coastal Ocean Science *Study and Practice*

（下 册）

王 颖 著

南京大学出版社

图书在版编目(CIP)数据

海岸海洋科学研究与实践:上下册 / 王颖著.
-- 南京:南京大学出版社,2021.4
ISBN 978-7-305-24300-4

Ⅰ.①海… Ⅱ.①王… Ⅲ.①海岸—海洋学—文集
Ⅳ.①P7-53

中国版本图书馆 CIP 数据核字(2021)第 051072 号

出版发行 南京大学出版社
社　　址 南京市汉口路 22 号　　　　邮　编 210093
出 版 人 金鑫荣

书　　名 海岸海洋科学研究与实践(上下册)
著　　者 王　颖
责任编辑 蔡文彬　　　　编辑热线 025-83686531

照　　排 南京南琳图文制作有限公司
印　　刷 徐州绪权印刷有限公司
开　　本 787×1092　1/16　印张 125.75　字数 3050 千
版　　次 2021 年 4 月第 1 版　2021 年 4 月第 1 次印刷
ISBN 978-7-305-24300-4
总 定 价 498.00 元

网址:http://www.njupco.com
官方微博:http://weibo.com/njupco
官方微信号:njupress
销售咨询热线:(025) 83594756

目　录

第三篇　海岸带资源环境管理与海洋经济

第四篇　海疆权益研究

第五篇　全球海岸海洋综合研究

第十三章 海岸海洋与全球变化研究 …… 371

第十四章 区域对比研究 …… 485

第六篇　学科发展与教学科普

第七篇 科学交流活动与考察研究

第三篇
海岸带资源环境管理与海洋经济

据海岸带环境与资源禀赋探讨其发展规划与多方面结合的管理途径。

中国在平原海岸带建设深水港并成功应用的突出实例:渤海曹妃甸深水港、南黄海洋口深水港建设,为工、农业发达的平原建设与成功应用的石油能源深水大港。

发展建设“数字海洋”。

Coastal Management in China*

INTRODUCTION

China is located in the eastern part of the Euro-Asian continent adjacent to the Pacific Ocean including its margin seas: Bohai Sea, Yellow Sea, East China Sea and South China Sea. The total coastline is 32 000 km long including the coastlines of 6 500 islands. The coastline along mainland is 18 000 km long from the Yalu River mouth on the China-Korea border in the north, to the Beilun River mouth on the China-Viet Nam border in the south (Figure 1). The coastlines extend across three climatic zones:

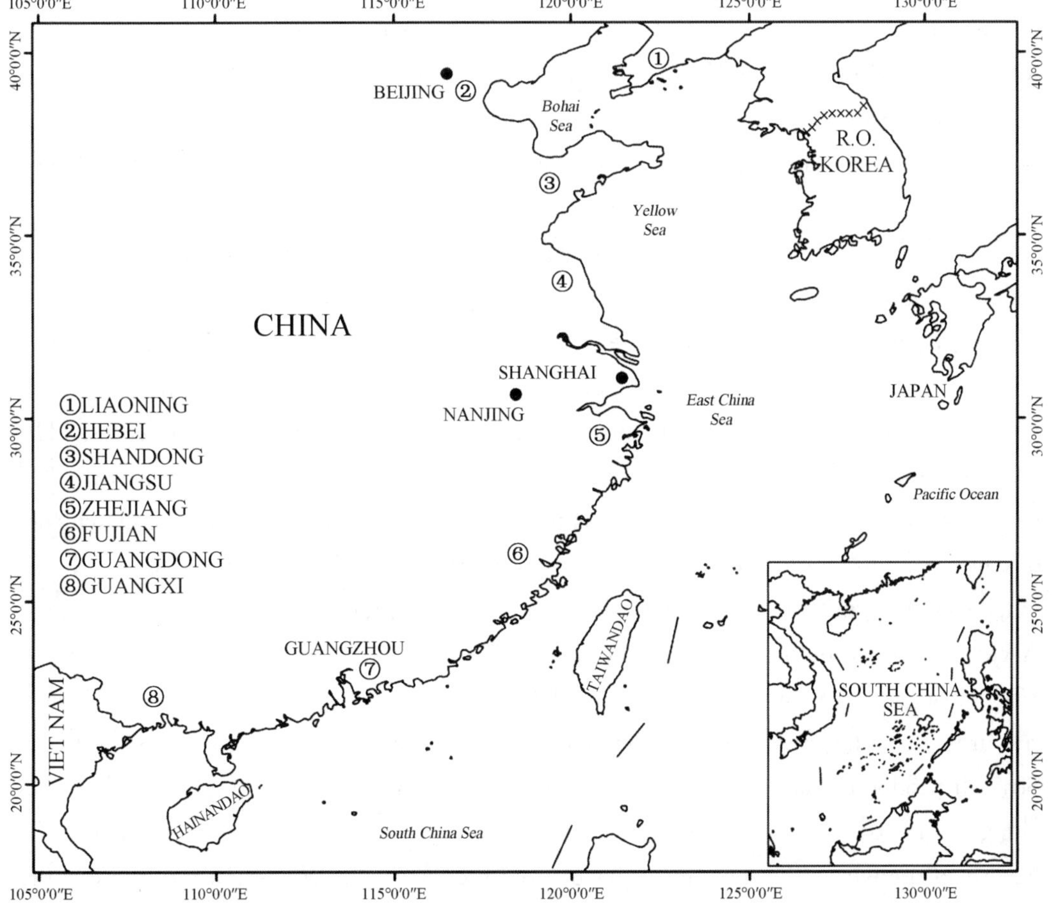

Figure 1　The location map of China coastal zone

* Ying Wang; Paolo Fabbri, *Ocean Management in Global Change*, pp. 469 – 479. London and New York: ELSEVIER APPLIED SCIENCE, 1992.

temperate, subtropical and tropical. The coastline features are affected by monsoon winds, the Pacific tidal waves and currents, and also with influence from several large rivers, such as Huanghe (Yellow River), Changjiang (Yangtze River) and Zhujiang (Pearl River). Thus, there is a variety of coastal types, such as the vast area of flat coasts either with sandy beaches and barriers, or with tidal flats in the North China; indented coasts in the mountain area or hilly land distributed mainly in the South China; river-mouth coasts including deltas and estuaries; coral reef coasts and mangrove coast, etc.[(1)]

The definition of the coastal zone limits seems uncertain for about 133 maritime countries in the world. It depends on each country's own needs of coastal zone development and administration. China carried out a comprehensive investigation of coastal and tidal flat resources from 1980 to 1986: the limit of coastal zone has been defined that from sea shore extended 10 Km landward as its inner boundary and from sea shore down to 15～20 m bathymetric contours as its outer boundary. The boundary may be extended landward or seaward in the area of steeper mountain coast, larger estuary, offshore island and submarine sandy ridge field[(2)]. The definition of coastal range fits the physical geographical implication of coastal zone as the area of land-sea interaction. It is also suitable for the course of regional changing, economic development and administration.

There are eight provinces, two cities and an autonomous region along China coastal zone, including: Liaoning, Hebei, Tianjin City, Shandong, Jiangsu, Shanghai City, Zhejiang, Fujian, Guangdong, Hainan and Guangxi Autonomous Region. The coastal area is 14 per cent of China's total area, and it concentrates 41 per cent of China's population, 56 per cent of total output value, and 70 per cent of larger cities. The coastal region is characterized by developed economy and heavy population, with 386 person/sq. km as its average density; hence, the wild land, tidal flats and embayments of the coastal region are valuable land resources for China. Thus, to divide an independent region of coastal zone in China has its practical economic significance. According to the definition of coastal zone, the total land area of the coast is about 350 000 sq. km including 6 per cent of tidal flats. The flats are an important new land resource in the heavy populated economic zone, because of their prograding shorelines. Sea salt production and saline chemical industry are the important traditional activities along China coast, the annual production of sea salt is 15 million tons, probably the largest in the world. Fishing and aquaculture products in China are 4. 5 million tons per year, and the aquaculture part is about 1/3 of the total. The production of kelp, laver prawn, mussel, sea eel and scallop has been developed as a large scale aquaculture industry, which has improved the Chinese food composition notably. The number of harbours has been increased from 61 to 253, and these can handle up to 5 million tons of cargo a year. By the end of 1990, China has 20 million tons of oceanic commercial ships, the annual transportation reaches 12 million tons, which is the eighth commercial fleet in the world. The estimated reserves of offshore petroleum are about 200 billion tons, and the annual crude oil production is 5 million tons. The natural gas production will be 30 billion cubic metres per year in 1994. The systematical exploitation of offshore oil and natural gas makes a notable impact on the coastal environment, economic development and other resources utilization. Even though, China still lacks of fresh water and energy resources along its economic developed coastal region, and the construction of nuclear power stations along the coastal zone has brought a series of new conditions recently. These circumstances arouse public and government attention to the importance of coastal management.

THE DEVELOPMENT OF COASTAL MANAGEMENT

Some projects have been conducted in China in order to enhance the exploitation, utilization and management of the coastal resources and environments.

1. The nation-wide comprehensive investigation along tidal flats and the whole coastal zone of mainland was carried out during 1980 to 1986. The investigation included natural environmental elements, resources and social and economic conditions. More than 500 units and 1 500 people in all included coastal provinces have been involved to organize multidisciplinary survey teams including 30 per cent of senior scientists from universities, institutes and governmental departments. Specific working items: Hydraulics, Meteorology, Geology, Geomorphology, Marine Biology, Marine Chemistry, Environment Protection, Vegetation and Forests, Soil, Land Use and Social Economics. Resources invents included land resources, biological resources, salt and saline chemical resources, mineral resources, marine energy, harbour, tourism industry, etc. The methods of coastal zone investigation were including: monitory observation and survey on the specific spots and selected sections combined with regional reconnaissance and grid line controlled survey. Historic records and previous works have been collected and summarized for review; remote sensing images, air photos, computer data and calculation have been used mostly in modern equipped laboratories. Through the investigation, a great deal of first-hand data has been gathered, and all of the results have been published in a series of scientific works. It includes 130 volumes of "Data collections of China Coastal Zone Investigation", "Atlas of China Coastal Zone" including 21 kinds of 1 218 pieces of map in the 1 : 20 000 scale, 10 volumes of "Regional Reports on the Comprehensive Investigation of Coastal Zone", 13 volumes of "Specialistic Reports of China Coastal Zone", and one volume book of "The Reports on Comprehensive Investigation of China Coastal Zone".

These achievements offer material and scientific information for coastal basic research, resources exploitation environment protection, and coastal management. They also stimulate public interest and better understanding of the coastal environment and its exploitation.

Since 1988, China has carried out another nation-wide project: Sea Islands Investigation with emphasis on larger islands which have population of more than 1 000 people. The investigation contains both natural environment, natural resources and social economic states, and the whole project will be finished in 1993.

Above two projects are important fundamental construction for coastal zone economic development and coastal management.

2. China has drawn up the plan for regional coast exploitation, from small scale of systematic exploiting, gradually to the stage of regional coast exploitation.

The experiment was started in 1982 and was limited to agriculture at first, then including medium and small harbour renovation.

There is a total of 42 experimental stations throughout the Country. The experimental items include:

1) Productive sea water cultivation;

2) Marine farming or pasture of prawns;

3) Littoral sandy wasteland and saline soil transformation;

4) Selection of sustainable plants;

5) Forest and fruit trees plantation;

6) Livestock and poultry husbandry;

7) Harbour renovation and coastal protection.

A number of good results were achieved through the experiments. For example, Jiangsu Province has larger area of tidal flats, about 400 000 hectares, to develop aquaculture and plantation. About 40 species of medicinal herbs and economic crops have been successfully planted on the tidal flats, such as: Lycium Chinense, Astragalus Membranaceus, Asparagus, and other 40 species of salty enduring economic plants. There are also 147 species of trees and flowers which have been planted and are growing luxuriantly. Second example, a modal tidal flat farm has been set up in the Rudong County Jiangsu Province with the scientific and technological support from Nanjing Agricultural University. The farm carries out systematic experiments on coast utilization. It operates aquaculture on bare flat; cattle, sheep, hare husbandry on marsh and grassland; fishing pond and poultry on superi-flat; farming wasteland or mature tidal flat to grow cotton, corn, mulberry trees, fruit trees, and vegetables. Rice can be planted in the area with fresh water irrigation and the salt marshes have been changed with mature soil after ten years farmed as cotton or corn fields. The farm also produce frozen storage and exports of fruits and vegetables. Since 1985, the farm has got profits of more than 40 million yuan per year.

Third example from an embayed coast in Zhuanghe County, Liaoning Province, where coastal zone is systematically exploited to aquaculture scallops in the intertidal zone, inside of coastal dike to divert sea water breeding prawns; and using littoral wastes and bay head area as rice fruit fields. The three-dimension operating of coastal zone benefits the local people, even during 1980 summer suffered from heavy storm of typhoon, the 670 hectares of farmed tidal flats still gained more than 2 million yuan net income. The economic development improves local living standards and changes the traditional conception of coast as "a bitter water saline soil" to a three dimension operation of new land for aquaculture, agriculture, forest, animal husbandry, and commercial trade. Even several units who joined the coastal survey before, now reorganize as the Consultant Company or Technical Service Centre. They form a new pattern of coastal development and management.

3. *Administrative Organisation and regional Division.* The State Oceanic Administration is an Agency of the Central Government of China, and there are three branches: the North Sea Subbureau in Qingdao, the East Sea Subbureau in Shanghai and the South Sea Subbureau in Guangzhou. Since 1986, after the coastal survey project, each province has set up the provincial Oceanic Administration for managing, researching, planning and exploiting coastal and oceanic area. There is also an Oceanic Environmental Monitory Organisation for protecting the coast and the sea nation-wide.

Since the coastal development in the 1980's, China has set up a series of special economic zones and high technology developing zones along coastal region, and 24 harbours open to the world. In 1988, the second larger island of China, renamed as Hainan Province, was also opened to the world as a new special zone. All of the economic special zones are managed in preferential trade and tariff rights.

In 1963, China set up the first natural preserve of Snake Island for protecting Pallas pit viper, the island is located off Dalian Coast. Now, there are 45 National Natural Preserves with a total area of 13 500 sq. km and covering most of coastal environments and ecosystems. Such as Changli Golden Beach Natural Preserves (Hebei Province), Laotieshan Island Migratory Birds Natural Preserves (Liaoning Province), Qidong Spartina Pasture Preserves, Yancheng Red-Crowned Crane Reed Flat Protection, Dafeng Davis Deer Natural Preserves (Jiangsu Province), Nanji Islands Oceanic Natural Protection (Zhejiang Province), Shankou Mangrove Coast Ecosystem Natural Preserves (Guangxi), Dazhou Island Oceanic Ecosystem Protection, Sanya Coral Reef Protection, Lingshui Macaque Island Protection, Dongzhai Mangrove Forests Preserves, etc. The Government supports these protected areas with special organization staffs and regulation for protecting wild life resources and natural environment.

A successful experiment is being carried for hibernated birds. There are about

1 000 red-crowned cranes in the world, and during cold season, from October to March, there are more than 600 red-crowned cranes dwelling in the Yancheng Reed Flat Natural Preserves. The wild vast wetland full of fish, fresh or brackish water has offered an ideal habitat for the precious birds and other kinds of wild ducks.

Among Scientific Association, there are Coastal Exploitation and Management Research Group and Marine Law Society under Chinese Society of Oceanography, Special Committee of Marine Geography under Chinese Society of Geography, and also the state Pilot Laboratory of Coast and Island Exploitation has been set up recently in Nanjing University. It has carried out a series of international symposia to improve coastal zone research and coastal management. All aspects mentioned above indicate that coastal utilization and management are now in the ascendant.

4. *Legislation.* The progress of coastal investigation and utilization activities, gradually creates contradictions between professions, departments and localities. Special laws of coastal management urgently need to be set up. In the summer of 1982, the formulation of the coast management law was started. After nine revisions, the "Regulations Regarding the Coastal Management of China" has been drawn up and submitted to the State Council for examination and approval.

Jiangsu Province, where Nanjing University is located, was the first to finish the coastal investigation, and since November 19th, 1985, the Province has issued formally "Coastal Zone Management Provisional Regulations of Jiangsu Province". It is the first province to put the coastal management regulations into effect. The other provinces are following the regulations. This situation indicates that coastal management in China progresses from a single administrative control to the legal management and then towards the comprehensive, systematical regulation and legislation management.

By the end of 1990, the laws related to the coastal management are as follows[3]: "Regulations Regarding Exploitation Regarding Environment Protection on Offshore Oil Exploration and Exploitation", 1983; "Regulations on Sea Water Management for Preventing Ships Pollution", 1983; "Regulations of Waste Dumping into Sea", 1985; "The Law of Fishery", 1986; "The Law of Marine Traffic Safety", 1983; "The Law of Mineral Resources", 1986; "The Law of Land Management", 1986; "The Law of Water", 1987.

The establishment of a series of Ocean and Coastal Natural Preserves is a result of enactment of the law of marine environment protection. The law enforcement agency pays more and more attention to the coastal zone, by using systematic theory and methods, concerning the relationship among the various resources, technical possibility and economic value, to solve the coastal developing problems.

While our country studies to promulgate the detailed rules of coastal management, we draw lots of lessons and experiences on coastal management both theoretically and practically from other countries and international organisations. We still need to continue the international academic exchanges in this field; it will enable us to conduct the regulations that are really suitable to the condition of China. Any successful experiences or unsuccessful lessons from China may be also the mirror for other countries who are paying efforts on their own coastal zone development.

PROBLEMS AND PERSPECTIVES

Coastal zone development in China started only in the early part of this century. The coast was wildlands and forbidden area during Qing Dynasty. Because the feudal kingdom followed the "blocked country" policy, it was the government to control fishing and salt products. Thus, the coastal zone, except several river mouths, was a depopulated area.

Development of China coastlands began in the late nineteen seventies as a result of the government carrying out the open policy. The coastal management work is in the early developing stage, thus, it involves new problems or contradictions.

1. *The scope of the coastal zone and the limitation of jurisdiction*. The problems involve mainly local counties, because there is no certain definition of the boundaries of the coastal zone. Most of coastal counties use 15 contours which have applied since the national coastal survey, but some others use either the base line of territorial waters, or the fishery administrative line. Thus, several foreshore islands have the ownership argument. Other example is the boundary of tidal flats, according to No. 14 article of the coastal regulation of Jiangsu Province. The flats are divided by land administrative boundary or by the centre line of natural rivers. The definition is suitable for longitude direction coastline of Jiangsu Province, but contradictions appear in the coast zone along arch-shipped outline of the Bohai Sea. The problems are also with the seaward extension of tidal flats.

It is clear that the jurisdictional limitation for each coastal province and county need to be made certain, and the landward boundary of coast can be divided both by physical geographical standard and by administrative limits[5,6].

2. *Multisource investment and plural departments administration*. The Ministry of Agriculture is responsible for aquatic products including the resources exploitation and protection, fishing harbour monitoring and pollution prevention. Tidal flat and wetland reclamation, and coastal farm and pasture land operation, etc. All the rights are authorized by "The law of fishery", "The Law of Land Management" and "The Law of Animal Husbandry". Ministry of Transportation has authority over harbor construction and management, harbour water monitoring and pollution prevention. Ministry of Hydraulic and Electric Power undertakes preventing tidal flooding, diversion, dyke,

sluice, coal and nuclear power station construction and management. Ministry of Urban Rural Area Construction and Environment Protection is in charge of coastal city planning management and construction; it examines and approves urban land utilization and the prevention of environmental pollution. Ministry of Light Industry is in charge of sea salt production, reed fields products and other light industry products. Ministry of Forestry is responsible for construction of coastal forest belts and its management, based on the law of forestry. Thus, each department exploits the coastal zone from respective point of view and in the respective style. There is no an authoritative agency to coordinate all the departments by comprehensive overall consideration and with long-term plans for coastal zone development. As a result, it is often difficult to overcome various natural disaster along coastal zone. There are multiple-use problems, such as: the struggle between salt fields and agricultural farmland. The extension of silt field for light industry has deteriorated the brackish water environment of reed ponds and grass flats as the favourite habitat for red-crown crane and other birds. The damming in river mouth areas for preventing saline water intruding the irrigation system has caused the underdam siltation and affected water depth of navigation channels fish and crab spawning migration, which decreases the crab crop. Beach sand and coast bedrock mining destroy the natural balance of sediment budget, with resulting coastal erosion. Overfishing of crockers in the Lusi fishing ground destroy the fishery resources; the yellow croakers crop decreased from several hundred thousand tons to 30～50 thousand tons per year. On the other hand, disobeying the natural processes to reclaim the lower tidal flats near tidal inlet with a low dyke construction, only after one storm by the typhoon No. 8114, the meandering tidal inlet washed away the whole dyke system and the 1 500 hectares reclaimed land, it lost several million yuan. As a natural and economic complex, the coastal zone especially needs overall planning and comprehensive management, taking it as a systematical project and improving local government to play major role in the coastal development [(4)]. It can be expected high profit.

3. The problems between resources utilization and environment protection become more and more serious especially in the area of estuaries, coastal bays, harbours and sea shore cities. The development of coastal industry and agriculture results in serious pollution, deteriorating the excellent coastal environment quality. During May of 1982, in a small bay, waste water discharge polluted the bay water with rich nutrients of hydrogen and phosphorus. Suddenly, sea algae generated 3 000 times more than before, which consumed large amount of dissolved oxygen, transparency clarity of sea water dropped from 3m deep to 0. 5m and the plankton increased 80 times. Fishes and crabs choked to death gradually as their grills over filled with fragments of algae and plankton. Thus, sea water colour changed and the cultivated prawns died. Red tide exploded frequently nowadays, it caught public close attention. Over exploitation of

underground waters in the coastal zone for irrigation caused subsidence of water table, and the intrusion of sea water and large areas of farmland had to be abandoned. Reclamation of tidal inlet embayment has decreased tidal prism and changed the natural flushing pattern, increasing siltation in navigation channels and decreasing the value of deep-water harbours.

Coastal environment which is sensitively changeable and natural factors are related closely. It requires more attention to a systematic management.

4. Immigrants from inner land to the coastal zone have some new problems resulted from living customs, management style and economic profits. It needs careful plan of immigration, and the relationship between new immigrants and the local people should be treated properly and harmoniously.

Experiences and lessons enhance public consciousness of the nature and importance of the coastal zone. It plays a pivotal role to improve domestic and international trade. It has great developing potentially and economic value. Coastal zone development can be taken as a turning point for large scale oceanic exploitation in the XX Century. The key point is to enhance the nation's consciousness of ocean territory. Marine protection and exploitation should be studied as important policy of our nation and to improve the total capacity of marine exploitation. Attention should be paid to marine legislation, resource exploitation and utilization, protection of eco-environment, prevention of disaster and other major tasks. Management work coordinates the development for all kinds of professions. The high level management of ocean and sea coast will provide a more rational and durable use of all resources.

REFERENCES

[1] Wang, Ying and Aubrey, David. 1987. The Characteristics of the China Coastline. *Continental Shelf Research*, 7 (4): 329 - 349.

[2] Su, Shen Jien. 1990. The Comprehensive Review on the Seven Years of National Coasts and Tidal Flats Investigation. In: Comprehensive Management of the Ocean. Science and Technology Press of Guangxi. 218 - 224.

[3] Li, Dechau. 1990. A Comparative Study on Coastal Zone Management: the traditional way and the modern way. In: Comprehensive Management of the Ocean. Science and Technology Press of Guangxi. 135 - 144.

[4] Li, Dechau. 1990. Preliminary Study on Coastal Management. In: Comprehensive Management of the Ocean. Science and Technology Press of Guangxi. 237 - 245.

[5] Ren, Mei-e. 1990. The problems on Coastal Management of China. In: Comprehensive Management of the Ocean. Science and Technology Press of Guangxi. 124 - 134.

[6] Zhao, Enpao. 1990. Administrative Divisions of Coasts and Adjacent Seas. In: Comprehensive Management of the Ocean. Science and Technology Press of Guangxi. 187 - 195.

China—*Coastal Management in the Asia-Pacific Region: Issues and Approaches* Selected Chapter*

Environment, Resources, Utilization

The coastal zone of China is located in the eastern part of the Euro-Asian continent, adjacent to the Pacific Ocean. The marginal seas surrounding China include the Bohai Sea, the Yellow Sea, the East China Sea and the South China Sea. The total coastline is 32 000 km, including 18 000 km along the mainland, with 14 000 km encompassing the coastlines of 6 500 islands. Presently, the area of the territorial sea is 300 000 sq. km, with a larger area of adjacent continental shelves. The Exclusive Economic Zone (EEZ), and the total area of the marine territory of China is more than 3 000 000 sq. km, i. e. about a third of its terrestrial territory.

Along mainland China, the coastline stretches from the Yalu River mouth on the China-Korea border in the north, to the Beilun River mouth on the China-Vietnam border in the south. The coastlines, including the islands, extend across three climatic zones: temperate; subtropical; and tropical. The coastal landforms reflect geological structures, either bedrock embayed coasts along mountain ranges in emergent regions, or coastal plains and river mouths in subsiding regions. The coastal features also are affected by monsoonal wind waves, Pacific tides, and sediment discharges from several large rivers, such as Yellow River, Changjiang (Yangtze River) and Zhujiang (Pearl River). Thus, there is a variety of coastal environmental types: vast areas of coastal plains with either sandy beaches and barriers, or tidal flats in North China; bedrock embayed coasts, mainly in South China; river mouths, including deltas and estuaries; and coral reefs and mangroves in the tropical and subtropical areas.

Even though developing scientific trends indicate that the coastal ocean has a broad definition, including the surf zone, continental shelf, continental slope and continental rise, the definition of coastal zone limits seems uncertain for about 133 maritime countries in the world. Coastal limits depend upon each country's own needs for coastal zone development and administration.

---

* Ying Wang: *COASTAL MANAGEMENT IN THE ASIA-PACIFIC REGION: ISSUES AND APPROACHES* PART Ⅲ—NATIONAL STATUS AND APPROACHES TO COASTAL MANAGEMENT, Chapter 12, pp. 177 – 186. Published by Tokyo: Japan International Marine Science and Technology Federation (ISBN 1 875865 09 2) 1995.

China carried out a comprehensive investigation on the environments and resources of coasts and tidal flats during 1980—1986. The limit of the coastal zone has been defined as extending 10 km landward, from the seashore, and seaward down to the 15～20 metre bathymetric contours. The boundary may be extended landward or seaward in areas of steep mountainous coasts, larger estuaries, offshore islands and submarine sand ridge fields (Su, 1990). This definition of the coastal range fits the physical geographical implications of the coastal zone as the area of active land and sea interaction. It is suitable also for the course of changing regional economic development and administration.

There are eight provinces, two cities and an autonomous region along China's coastal zone. From north to south, they are: Liaoning Province, Hebei Province, Tianjin City, Shandong Province, Jiangsu Province, Shanghai City, Zhejiang Province, Fujian Province, Guangdong Province, Hainan Province, and the Guangxi Autonomous region (Figure 1). The coastal region makes up 14% of China's total area, and has a concentration of

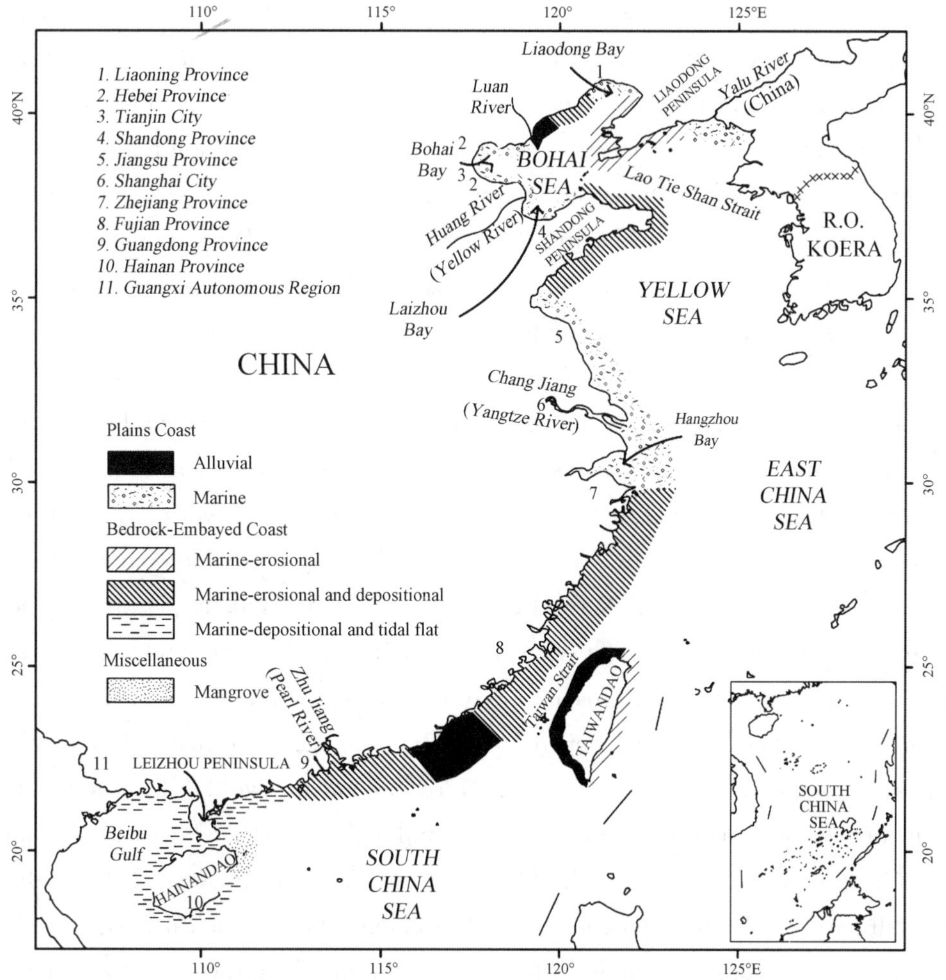

Figure 1　Distribution of Coastal Environment Types and Provinces of China

(Source: Wang & Aubrey, 1987)

41% of China's population, with an average of 386 people per sq km. This region contains 70% of the country's urban areas and accounts for 66% of total output value.

According to the definition of the coastal zone, the total land area of the coast is about 350 000 sq. km, 6% of which is tidal flats. The prograding tidal flats, wild lands and embayments of the coastal zone are valuable land resources for China, especially in the heavily populated economic zone.

Coastal zone development in China only began in the early part of 20 century. The coast had been a wild and forbidding area during the Qing Dynasty (from 1644—1911), with the government controlling fishing and salt production. Therefore, except for some river mouths, the coastal zone was a sparsely populated area separating the land and the sea. Large-scale development began in the late 1970s as a result of the open policy of the present Chinese Government. By the end of 1979, the total value of the marine estate of China was 7. 4 billion yuan. The annual increasing rate was 19% during 1980—1990 (not including the ship building industry), and 28% during 1990—1993. The total sum of the marine estate of China was 43. 8 billion yuan in 1990, and 90 billion yuan in 1993. Sea salt production and the saline chemical industry are important traditional activities along China's coast. The annual production of sea salt is 19. 8 million tonnes, probably the largest in the world. Fishing and aquaculture in China produces 6. 1 million tonnes per year, with aquaculture providing 2. 42 million tonnes. The production of kelp, laver, prawns, mussels, sea eel and scallops has been developed as a large-scale aquaculture industry, which has notably improved the Chinese food composition.

The number of harbours have increased from 61 to 260, with 373 sites for deep water berths, handling up to 720 million tonnes of cargo a year. China has 22 million tonnes of commercial oceanic ships. Annual transportation reaches 11 908 billion tonnes/km, which is the eighth largest commercial fleet in the world. The estimated reserves of offshore petroleum are about 200 billion tonnes, and annual crude oil production is 5 million tonnes. Natural gas production will annually provide 3. 4 billion cubic metres from early 1996. The systematic exploitation of offshore oil and natural gas makes a notable impact on the coastal environment, economic development and other resource usage. The estimated sum of coastal placer (i. e. valuable heavy mineral deposits in the coastal zone, such as zircon, rutile, garnet, gold, etc.) is about 430 million tonnes, and marine energy is 1 000 million kilowatts. Several tidal power stations have been established in the Shandong and Zhejiang Provinces. Even though China still lacks freshwater and energy resources along the economically well developed coastal region, the construction of nuclear power stations along the coastal zone recently has changed these conditions. These circumstances attract the attention of the public and the government and aid in understanding of the importance of coastal management.

The Development of Coastal Management

Projects have been carried out with a view to increasing exploitation and improving the protection and management of coastal resources and environments.

Survey

Between 1980—1986, a comprehensive nationwide survey was made of the tidal flats and entire coastal zone of mainland China. Environmental factors, natural resources and socio-economic conditions were considered. More than 500 agency units and 10 000 people were involved in the organization of multidisciplinary survey teams. One third of team members were senior scientists from universities, institutes, and government departments. Specific disciplines involved included oceanography, meteorology, geology, geomorphology, marine biology, marine chemistry, environmental protection, vegetation and forests, soils, land use and social economics. Areas studied included land resources, biological resources, salt and salt-chemical resources, mineral resources, marine energy, harbours, the tourism industry, etc.

As a result of this survey, a large amount of data has been gathered, providing a scientific foundation for further research on coastal areas, exploitation of resources, and environmental protection and coastal management. The information can also be used to stimulate public interest in the coastal zone environment and its exploitation.

During 1988—1993, another nationwide project was carried out: the Sea Island Survey. It assessed the environment, natural resources and socio-economic conditions of offshore islands.

These two projects provide an important basis for economic development, environmental protection, and management of the coastal zone.

Experimental Exploitation

China has drawn up a plan for regional exploitation of coastal areas, including a systematic strategy for small scale and regional developments. This experiment began in 1982 and was initially restricted to agriculture. Then, gradually, multi-purpose experiments were made, for example the renovation of small and medium sized harbours.

At present, there are 42 coastal experiment stations throughout China. The experimental issues include:

1) sea water cultivation;

2) marine farming (cultivation of prawns);

3) amelioration of littoral sandy wasteland and saline soil;

4) selection of sustainable plants;

5) planting of forest and fruit trees;

6) raising of livestock and poultry; and

7) harbour renovation and coastal protection.

A number of positive results have been achieved through experimentation. For example, Jiangsu Province has developed a large area of tidal flats (about 400 000 ha) for agricultural and forestry development. In Liaoning Province the coastal zone is being exploited systematically for farming scallops in the intertidal zone, and for using littoral waste areas and the bay head area to grow rice and fruit.

Economic development improves the standard of living for the local population and changes the traditional conception of the coast as an environment of 'bitter water and saline soil' to a three dimensional perspective encompassing the claiming of new land for aquaculture, agriculture, forestry, animal husbandry and trade. Several units that were part of the coastal survey have since been reorganized as consultancy firms or technical service centres, as part of the new pattern of coastal development and management.

The first step, by the State Oceanic Administration (SOA) to achieving integrative coastal management, has been dividing China's whole coastal area into functional zones, which are:

1) Exploitation Zone. Includes harbour areas, natural gas and petroleum areas, tourist areas, salt field areas, place mine areas, coal mine areas, ground halogen water areas, plantation areas, reedy marshes, pastoral land, aquacultural areas, fishing grounds, transportation areas, city construction areas, etc. ;

2) Remediation and Governance Zone. Areas, such as restored aquatic productive areas, important economic fish breeding areas, preserved fishing grounds, sheltered forest zones, restoring polluted areas, protected underground water areas, etc. ;

3) Natural Preserved Zone;

4) Special Function Zone. Includes military areas, waste dumping areas, pollutant drain off areas, etc. ; and

5) Conservation Zone. Possibly provides opportunities for future development, etc.

Administrative Organization

In China there is a State Oceanic Administration Bureau under the central government, with three branches, the North Sea, the East China Sea, and the South China Sea sub-bureaus. Since 1986, following completion of the coastal survey project, each province has established a provincial oceanic administration to develop strategies for the management, research, planning and exploitation of coastal and shallow sea areas. There is also an Oceanic Environmental Monitoring Organization for protecting the coasts and seas of the whole country.

Following the coastal development of the 1980s, China has set up a series of special economic zones and highly technical development zones along the entire coastal region. Twenty four harbour cities are accessible and open to the world. In 1988 the second largest island of China, Hainan Island, was designated as Hainan Province in order to accelerate its economic development. It is also globally accessible, as a new special zone. All of the special economic zones have preferential trade and traffic rights.

In 1963 China set up its first natural reserve, Snake Island, for the protection of the Pattas pit viper. The island is located off the Dalian coast in Liaoning Province. There are currently 45 national natural reserves in coastal areas, with a total area of 13 500 sq km, covering most coastal environments and ecosystems. These include Changli Golden Beach Natural Reserve (in Hebei Province), Laotieshan Island Migrating Birds Natural Reserve (in Liaoning Province), Qidong Spartina Pasture Natural Reserve, Yancheng Red-Crowned Crane Reed Flat Natural Reserve, Dafeng Davis Deer Nature Reserve (in Jiangsu Province), Nanji Island Oceanic Natural Reserve (in Zhejiang Province), Shankou Mangrove Coast Ecosystem Natural Reserve (in Guangxi Autonomous region), Dazhou Island Oceanic Ecosystems Natural Reserve, Sanya Coral Reef Natural Reserve, Lingshui Macaque Island Natural Reserve, Dongzhai Mangrove Forests Natural Reserve in Hainan Province, etc. The Central Government supports these reserves by providing staff and regulations to protect wildlife resources and the natural environment.

A successful conservation experiment has been carried out for hibernating birds. There are only about 1 000 red-crowned cranes in the world. During the cold season, October-March, more than 600 of them dwell in the Yancheng Reed Flat Natural Reserve of Jiangsu Province. This vast and wild wetland is full of fish which live in fresh or brackish water, and provides an ideal habitat for these and other types of wild birds.

Among scientific associations are the Coastal Exploitation and Management Research Group and the Marine Law Society, which are part of the Chinese Society for Oceanography. A Special Committee of Marine Geography under the Chinese Geographical Society, as well as a State Pilot Laboratory of Coast and Island Exploitation, has been established at Nanjing University. The university has sponsored a series of international symposiums to improve the knowledge of coastal zone research and coastal management.

By the end of 1993, there were more than 100 marine education and research institutes in China. These institutes have developed international academic co-operation links with more than 40 countries in the world.

Legislation

The development of coastal investigation and utilization activities has gradually created more and more conflicts of interest between professions, departments and localities. Therefore a special law on coastal management was urgently needed. In the summer of 1982 such legislation began to be formulated. After nine revisions, the *Regulations Regarding the Coastal Management of China* were drawn up and submitted to the State Council for approval.

Jiangsu Province, where Nanjing University is located, is the first province to have put the coastal management regulations into effect, and was the first to complete its coastal survey. As early as 19th November, 1985, this province issued the *Provincial Coastal Zone Management Regulations of Jiangsu Province*. Many other provinces, such as Liaoning, Shandong, Guangdong, Hainan and the City of Tianjin also have formulated regulations since 1986 (Ren, 1990; Zhao, 1990). This situation indicates that progress is being made towards the development of comprehensive, systematic regulations and legislation for the management of coastal areas in China.

The laws relating to coastal management include:

1) *Regulations Regarding Environment Protection on Offshore Oil Exploration and Exploitation*, 1983;

2) *Regulations on Sea Water Management for Prevention of Ship Pollution*, 1983;

3) *Regulation of Waste Dumping into the Sea*, 1985;

4) *The Law of Fishery*, 1986;

5) *The Law of Marine Traffic Safety*, 1983;

6) *The Law of Mineral Resources*, 1986;

7) *The Law of Land Management*, 1986; *and*

8) *The Law of Water*, 1987.

These laws contain provisions for the management of coastal and ocean areas. However, to establish a series of coastal and ocean reserves has required the enactment of a specific law on marine environmental protection. Increasing attention is being given to the legal aspects of protecting and managing the coastal zone, using systematic theory and methods concerning the relationship among the various resources, technical possibilities and economic values to solve coastal development problems.

New Approach

Coastal management in China is at an early development stage. Management approaches are now being introduced which address the following key issues.

1) Definition of the scope of the coastal zone and the limitation of jurisdiction. This problem involves mainly local governments, because there is no specific definition of the boundaries of the coastal zone. Several foreshore islands are subject to competing ownership claims. It is clear that the jurisdiction limitation of each coastal province needs to be clarified, and the landward boundary of coast needs to take into account both

physical geographical standards and administrative limits (Ren, 1990; Zhao, 1990).

2) Multiple sources of investment and a number of administration departments. Aquatic products, land, water, forest, harbour and energy resources are the concern of different departments, such as the Ministry of Agriculture, the Ministry of Forestry, the Ministry of Light Industry, the Ministry of Urban Rural Area Construction and Environment Protection, etc. There are many conflicting problems, but there is not an authoritative agency to co-ordinate all the departments and provide comprehensive holistic consideration and long-term plans for coastal zone development. Therefore, it is often difficult to overcome various natural disasters along the coastal zone. As a naturally and economically complex system, the coastal zone needs integrated planning and comprehensive management. A systematic approach may improve the ability of local government to play a major role in coastal development.

3) The problems between resource utilization and environmental protection become increasingly serious, especially when associated with estuaries, coastal bays, harbours and coastal cities. The development of coastal industry and agriculture results in serious pollution, which deteriorates the excellent coastal environment quality. Reclamation of tidal inlet embayments has decreased the tidal prism and changed the natural flushing patterns. Increasing siltation in navigation channels decreases the value of deep water harbours. The coastal environment is sensitive to change, and natural factors are closely related, therefore more attention to integrated management approaches is required.

4) Immigrants to the coastal zone from inland areas have encountered several new problems associated with living customs, management styles and economic profits. Careful planning of immigration, and the relationship between new immigrants and the local people should be treated properly and harmoniously.

Conclusion

Experiences and lessons enhance public consciousness of the nature and importance of the coastal zone, which plays a pivotal role in improving domestic and international trade. Coastal zone development is at a turning point for large-scale oceanic exploitation in the 21st Century. The key point is to enhance the nation's consciousness of the oceanic territory. Marine protection and exploitation should be studied as an important part of the nation's policy, with an aim to improve the total capacity of marine exploitation. Attention should be paid to marine legislation, resource exploitation and utilization, protection of ecological environment, and prevention of disaster. Management co-ordinates the work and development of a wide range of professions. High levels of management of the ocean and coast will provide more rational and durable usage of all resources for sustainable development.

REFERENCES

[1] Li D. 1990. A Comparative Study on Coastal Zone Management: the Traditional Way and the Modern Way. In: *Comprehensive Management of the Ocean, Science and Technology*. Press of Guangxi. 135 - 144.

[2] Li D. 1990. Preliminary Study on Coastal Management. In: *Comprehensive Management of the Ocean, Science and Technology*. Press of Guangxi. 237 - 245.

[3] Ren, M.-E. 1990. The Problems on Coastal Management of China. In: *Comprehensive Management of the Ocean, Science and Technology*. Press of Guangxi. 124 - 134.

[4] Su, S. 1990. The Comprehensive Review on the Seven Years of National Coasts and Tidal Flats Investigation. In: *Comprehensive Management of the Ocean, Science and Technology*. Press of Guangxi. 218 - 224.

[5] Wang, Y. and Aubrey, D. 1987. The Characteristics of the China Coastline. *Continental Shelf Research*, 7 (4): 329 - 349.

[6] Zhao, E. 1990. Administrative Divisions of Coasts and Adjacent Seas. *Comprehensive Management of the Ocean, Science and Technology*. Press of Guangxi. 187 - 195.

Economic Development and Integrated Management Issues in Coastal China*

China's coastal zone is located along its continental margin bordering the Pacific Ocean and includes three marginal seas: the Yellow Sea, the East China Sea and the South China Sea. It has a coast line that extends 32 000 km and includes some 6 500 islands. The coastline of the islands measures *ca.* 14 000 km (Figure 1). This coastal zone stretches across several climate zones, ranging from the temperate, to subtropical and tropical. The entire area is humid and receives plenty of rain fall, although there are differences between the North and the South. In general the coastal climate is controlled by the monsoon. In the winter it is dominated by the Mongolian high pressure system and the climate is dry and cold, with prevailing winds from the north. In the summer the area is dominated by a low pressure system, rich in rainfall, and prevailing winds are from the south and southeast.

Chinese rivers are rich in sand- and silt-sized sediments and their large volume and fresh-water discharges have strongly influenced the coastal environment. The annual sediment discharge of the Yellow River is 1.19×10^9 tons, of which 64 % is sequestered in the delta and tidal flats, only 36% making it to the sea (Pang, 1979). In the last 4 000 years, Yellow River had changed its course eight times, delivering a huge volume of sediment that formed the Northern China and the Jiangsu Plains, and a wide range of coastal mud flats. This sediment load has also made a significant contribution to the bottom sediments of the Bohai and Yellow Seas. The Changjiang (Yangtze) River is the largest river in China with an annual water discharge of 925 billion cubic meters, which carries a total amount of 0.49×10^9 tons of sediment to the sea (Wang et al., 1983).

The Changjiang River sediments deposited on the continental shelf of the East China Sea have formed a submarine delta whose outer edge can reach 25～30 m of water depth. The delta started to form around 7 000 years ago when the sea level reached the modern highest and the point-sourced deltaic shoreline began to regress seaward. Sediments off the mouth of Changjiang River are distributed in a more than 50 km wide area towards the east. Main longshore transportation is to the south from the river mouth and fine-grained sediments can be transported to the northern coast of Fujian province. This is the main source of silt and clay deposited in the embayments along

---

* Ying Wang, Zhewen Luo, Dakui Zhu: Bilal U. Hap et al, *Coastal Zone Management Imperative for Maritime Developing Nations*, pp. 371 - 384, Chapter12, Kluwer Academic Publishers, 1997.

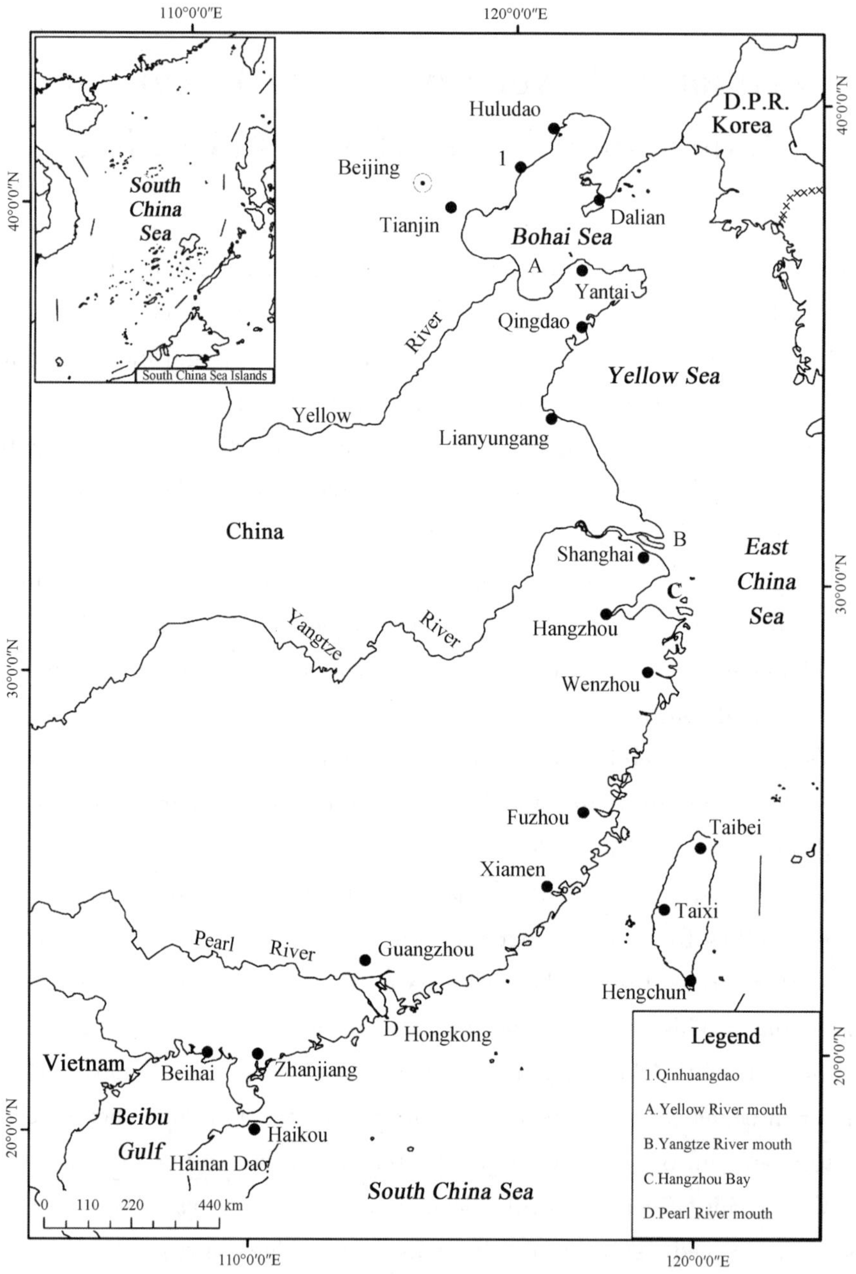

Figure 1 China's Coastal Zone

Zhejiang coast, as well as in the inner continental shelf (Wang, 1986).

In the coastal zone the population density is high and the economic development is more rapid than the hinterland. The Chinese coastal zone has experienced the "forbidden" time during the feudal dynasties, the "frontier exploitation" at the beginning of this century, a nationwide investigation, planning and development phase since the 1980's. In the 1990s it is perceived as an economic golden zone. Thus, 41 % of China's population and over 60 % of its total gross domestic product (GDP) is concentrated in this zone which has become the vanguard of the economic development of China. However, the dense population and rapid development has raised problems of its own. For example, in the area of Changjiang River Delta, 3.9% of total population of the country with 13.5% of the national GDP output resides in about 0.6 % of China's total area, i.e. in less than 0.06 hectare/person, the lowest ratio in the country. The present population increase rate in the delta area is about 1% per year, while the arable land is decreasing at the rate of 0.5% each year. Agricultural output value is only about 10 % of the total value of both industry and agriculture GDP (Cui, 1994). These pressures make it imperative that the relationship between human activities and land use in the coastal zone is better planned and integrated.

The coastal zone of China may have suffered great pressure from environmental pollution. Contaminants are carried to the sea by rivers, and more than 70% of this material may come to be concentrated in the coastal zone. The area of Changjiang River Delta, e.g., receives 9 % of the waste water, 5 % of the waste gas and 4 % of other residuals of the Chinese industry, in a land area that is only 0.6 % of the nations total. The flat-lying nature of the delta with its extensive river network also makes it susceptible to flooding, which seriously impacts the local economy and endangers human life. According to the statistical data, the Chinese coast suffers from about seven typhoons each year, causing economic loss exceeding 10 billion RMB Yuan (*ca.* 8 Yuan = 1 US $ in 1996). Flooding in the coastal lowlands is estimated to cost at least 10 billion RMB a year, often exceeding these figures. The property loss from storm surges in 1994 totaled 19.2 billion RMB, accompanied by a loss of 1 240 human lives. The flood disaster in 1991 took 4 500 human lives in the Changjiang River Delta alone, and the economic loss exceeded 50 billion RMB.

Sustainable development of the coastal environment is perhaps the biggest challenge facing China. The coastal zone is in an area where population pressure, resource use and environmental preservation all come into conflict. Thus, integrated management of this precious zone, with particular attention to strategic planning, policy making and legislation that will suit the special coastal needs of China, become a high priority.

China's Coastal Regions, their Development and Management

The coastal zone of China can be subdivided into four regions based on the area's natural condition and present economic status. These factors and other developmental trends of these regions are considered below:

1. Yellow and Bohai Seas Coastal Area

This coastal region extends from the Yalu River mouth on the China-Korea border to the south of Haizhou Bay and includes the coasts of three provinces: Liaoning, Hebei and Shandong, as well as the Tianjin District. The total length of the coastline is 5 668 km. The two peninsulas of Liaodong and Shandong in this area are characteristic embayment coasts of China. The land along the coast is largely hills of granite and metamorphic rocks, providing steep slopes, deep water and curved coastline. Many of the better Chinese harbors are located in the region. The granite and metamorphic weathering produces sandy beaches inside the bays, providing scenic spots, ideal for recreation. Along the coasts of Bohai Sea, bordering the Northern China and Liao River plains, mud flats have developed through the huge supply of sediment from the Yellow and Liao Rivers.

Several metropolitan areas, including Beijing, Tianjin, Shenyang, Dalian and Qingdao, are located in the coastal region, making it the most important economic zone in North China. Tianjin acts as the harbor city for the capital city of Beijing. The Bohai Sea offshore oil fields produce about half of the national output. China's major coal fields are in the inland provinces of Shanxi (Taiyuan) and Shaanxi (Xi'an), which is transported to the south through the harbors along the Bohai Sea. The harbors of Dalian, Qinhuangdao, Tianjin, and Qingdao, together have an annual capacity of transportation of over 50 million tons. These harbors are a hub both for the communication industry and overall economy of the metropolitan cities. In addition, there are a series of middle-sized harbors, such as Yingkou, Tangshan, Huanghua, Longkou, Yantai, and Rizhao, giving this region the highest density of harbors and transportation capacity in China.

The coastal region of the Yellow Bohai Seas, linking North China to the ocean, is also the national hub for oil, energy resources, chemical engineering and fishery industries. Major issues for the coastal zone management in the region can be summarized as the need to:

- produce a regional plan of the coastal zone development based on the characteristics of natural coastal evolution, i. e., the embayments and mud-flats coasts, as well as their developmental trends;
- prevent oil and coal pollution during the processing and transportation, and to strengthen the protection of the marine environment of the Bohai Sea coastal areas. On and offshore Bohai Sea is the major oil production area, while Qinhuangdao and Huanghua harbors are the biggest coal exporting harbors in China;
- strengthen the protection and management of fishing resources against over exploitation in the Yellow Sea region. The Yellow Sea is surrounded by the land and fishing is the traditional base of the local economy; and
- protect, during regional development, ecologically sensitive areas. Natural,

ecological and historical preservation zones or natural parks, such as, reed wetlands at Liao River mouth for bird breeding, poisonous snake island, shell beach ridges, early man's remnants, east end of the Great Wall, etc.

2. Changjiang River Delta Region

This region includes the coastal zone of Jiangsu and Zhejiang provinces and the Shanghai District. The length of coastline is 2 965 km. The coastal zone in this region is strongly influenced by the Changjiang River, and is economically closely linked to the delta area.

Three types of coasts are characteristic of this area:

1) The Yellow Sea coast north of Jiangsu province is a mud-flat plain coast that evolved a flat morphology due to flooding by ancient Yellow and Changjiang River. The intertidal zone is about 10 km wide, and the coastal slope in the order of 0. 1‰～0. 2‰. The abandoned Yellow River Delta in the north lost its sediment supply when the Yellow River shifted from the Yellow Sea side to the Bohai Sea in 1855. The coast of the abandoned delta is being eroded landward at a rate of 20～25 m/year. Other parts of the Yellow Sea coast are protected by a group of large submarine sand ridges. This is also aided by sediment supply from the offshore sand ridges which makes the coastal mud flats accrete rapidly as the coastline regresses seaward at a rate of 40～100 m per year.

2) The Changjiang River mouth and the Hangzhou Bay in the middle are characterized by deltaic and estuarine landforms.

3) A series of northeast tending folded mountains create a bedrock-embayed coast in the southern part. Fine sediments of the Changjiang and Qiantang Rivers have been deposited into these long narrow bays to form mud-flat features within bedrock-embayed coasts.

The ChangJiang River Delta region is the most advanced economic zone of China that continues develop at a rapid pace. The hinterland of Shanghai harbor includes the Changjiang River basin and a part of the north and the west of China. The regional import and export trade is amounts to about 40% of the whole country. The central harbor of Shanghai is surrounded by Ningbo and Wenzhou harbors in the south, and Nantong and Lianyun harbors in the north which are unable to satisfy to the actual needs of the region. Several new harbors are therefore under construction. New and creative ideas are being employed to use deep tidal channels within the submarine sand ridges and mud flat coast to build deep-water harbors of 20 000 tons capacity.

This region, including the southern part of the Yellow Sea and the East China Sea, will be one of China's major area of offshore oil and gas production in the future. The estimated reserves of offshore petroleum are about 5. 9 billion tons and natural gas is estimated to be 2 630 billion cubic meters. Exploitation of these resources has begun which will be of enormous help to the economic development of the coastal region.

The tidal flats are another special resource of the region. There are 5 100 km^2 of tidal flats in the Jiangsu, 380 km^2 in Shanghai, and 2 800 km^2 in Zhejiang, constituting about 40% of China's total tidal flats area. While the population of the Changjiang River Delta region is high, it is land poor. Tidal flats therefore represent potential new land, which is of obvious value for future regional development (Zhu, 1986).

There are 1 934 islands off this coast of China. The Zhoushan Islands and Lusi offshore are two major fisheries areas in China. The numerous ports and tidal flats has led this region to become a base of marine aquaculture in China.

Because it faces the open sea, this coastal region has often suffered from natural disasters, including typhoons and strong tidal bore and storm surges, which have caused serious erosion in the abandoned Yellow River Delta, Hangzhou estuary and other parts the of Zhejiang coast.

Key issues for the coastal zone management include:

- A comprehensive and systematic management of the Changjiang River mouth navigation channel, including coastal engineering constructions, water resources and environment, waste disposal in and out of the mouth of the river, and pollution control;
- Management of fisheries in Zhoushan and Lusi fishing fields;
- Comprehensive management of the whole Hangzhou estuary with due consideration to local economic development patterns, as well as the protection of natural tidal bore environment, the prediction and prevention of the effects of coastal engineering on the deep channel, sandy shoals and bottom morphology, which are all related to harbor development;
- setting aside of national protective zones, for such activities as red crown, cranes and Davis deer breeding in the wetlands of North Jiangsu coast, etc.

3. Embayment Coastal Region of Southern China

This region includes the Fujian and Guangdong provinces, as well as the islands offshore Taiwan. The length of coast line along the mainland is some 6 419 km. The main trend of the coast follows the northeastery direction of the granitic mountain ranges, but is cut by a group of northwest tending faults to form a series of long narrow bays, i. e. tidal channel embayments. Some of indented bays have been filled by sediment forming sandbars and lagoons and a narrow coastal plain. The Pearl River is the third largest river in China with an annual discharge of 331. 9 billion m^3. The sediment discharge, however, is low, with an annual average sediment load of about 87. 5 million tons. Several small deltas have developed inside of the Pearl River estuary, forming a transitional pattern from the estuary to an open depositional delta. This has led to an estuary full of channels, tidal flats, and rock islands. The sediments of the Pearl River entering the sea have been transported by longshore drifting towards to the southwest, developing mud flats also over western part of the Guangdong coasts inside the bedrock bays.

The coasts of Fujian and Guangdong provinces are also one of the fastest developing areas in China. Trans-shipment trade via Hongkong encourages production and export trade in the area which is about half of China's total export. The processing industry in the area is also expending rapidly. In the last ten years 60% of Hongkong's processing industry has been transferred to these two provinces of South China. On the continental shelf outside of the Pearl River mouth, the offshore oil reserves are estimated to be about 6.03 billion tons and those of natural gas *ca.* 730 billion m^3. Since the South China Sea may have richest hydrocarbon resources in China, the area is fast becoming one of the main centers of marine petroleum industry.

The region's tropical location and ample fresh-water resources also make it a production base for the tropical plantation and hard-currency earning agriculture. It is also important as a tourist area with many visitors, especially the overseas Chinese, which is an important factor in its economic development. Nevertheless, the region is still short of energy, raw materials and basic industries, and communication with northern industrial areas is in need of improvement.

The main focal points for the area's coastal management include:

- The integrated management of the coasts along the Taiwan Straits;
- The management of the area around the mouth of the Pearl River and the issues related to Hongkong, Macao and other offshore islands; and
- Planning and management of the tropical plantations and eco-agriculture of the Pearl River Delta, the integration of tourism industry and the preservation of historically important sites, as well as the natural beauties of the coastal landforms.

4. The Region of Beibu Bay and Hainan Island

The region covers the coasts of Guangxi Autonomous Region and Hainan Island, with a total coastline is about 2 750 km. The Beibu Bay has a characteristically erosional-depositional embayed coast. Sand barriers and lagoons enclose the inner coastline, coastal platforms, headlands and large-scale mangrove swamps. The northern part of the Hainan Island is also largely erosional-depositional embayed coast despite the existence of some river delta. The southern shoreline of the Hainan Island is a depositional embayed coast and a transition from the embayment coast to a plain coast. The erosional process has moved the headland backwards, so that huge sandbars block the bays and the coastline tends to be smooth. Due to its tropical climate, Hainan Island contains China's major coral reef and mangrove areas.

This is a relatively less developed region of China. The region's economy started to develop only in the last decade and holds great potential for the future. The Beibu Bay of Guangxi is the natural outlet to the ocean for the inland provinces of the South-West China. Several deep-water harbors are being constructed, such as the Fangcheng, Beihai and Qinzhou, to provide links with the railway of South-West China. The Beibu Bay and the adjacent region of Hainan Island are also potential offshore hydrocarbon areas. It may contain reserves of up to 8.9 billion tons of oil and 9 380 billion m^3 of gas. The natural gas supply to mainland started in 1996. There are also numerous coastal placer deposits containing titanium, zircon, rutile, etc. Hainan Island is also known as the Chinese Hawaii for its tropical tourist facilities.

The issues for coastal management in this region include:

- Definition of the international boundaries in the Beibu Bay and the islands of Hainan province;
- Development of transportation system into the region is of highest priority;
- Environmental protection in the course of exploitation of offshore oil and gas resources;
- Integrated planning and coordination of all coastal zone activities, such as harbor construction aquaculture, production of sea algae, and public recreation;
- Protection and rational utilization of the coral reef and mangroves resources; and
- The protection and proper administration of tourism in the tropical zone.

Status and Problems of Coastal Zone Management in China

The coastal zone is the transitional area linking the land and the sea, and is at the forefront of Chinese economic development, of great value both environmentally and economically. Along with developments in science- and technology-led industrial production, its value is bound to increase. Thus, it is crucial to treat the coastal zone as area of great long-term value.

Present States

The coastal zone has become a hot center of activity, but its management needs to be integrated: Both Chinese authorities and the public have realized the value of resources of the coastal zone. It is now commonly understood that before exploitation of coastal resources for economic development can begin, a thorough investigation of the coastal zone is necessary. Thus, a nationwide comprehensive investigation of the tidal flats and the entire coastal zone of the mainland was carried out in the years 1980 to 1986. The investigation focused on natural environmental elements, resources and social and economic conditions. Another, nationwide sea-islands investigation, with emphasis on larger islands, was carried out in the years 1988 to 1993. In addition, there were a number of detailed site investigations for the larger construction projects along the coastal area. Yet, in practice, coastal development seems to suffer from a lack of coordinated management. The governmental agencies, such as the ministry of communication, agriculture, hydraulics, health environment, public safety and the Navy have all imposed regulations, often covering the same area of concerns in the coastal zone, but with a special focus of their own missions. There has been a lack of integrated management.

Unbalanced use of resources and wastage is common: The present coastal exploitation has been haphazard and spontaneous. A lack of coordinated planning and management has led to very uneven utilization, some areas remaining unaffected while other are crowded and over exploited. Improper activities include: dredging on the shoals to construct navigational channels when many deep water coasts remain idle, and industries with no need for the sea water occupying large waterfront areas, while those activities that could use sea water remain miles away from the coast.

Some ocean engineering and coastal construction activities have seriously changed the natural evolution of the coastal areas, disrupting natural balance, and leading to accelerated coastal erosion and siltation. Some of these effects are irreversible.

Deterioration of the coastal ecology is occurring at a rapid pace: In the unplanned and uncoordinated economic development of the coasts, resources have often been squandered. For example, irresponsible reclamation, beach-sand mining, mangrove felling, coral-reef mining, over-pumping of underground water, overfishing, and dumping of contaminated water and waste material, have all led to disastrous results that include deterioration of ecology, salt-water intrusion of aquifers, coastal erosion and siltation of navigational channels. Thus, the setting up of proper legislative infrastructure for the proper management of the coasts is of utmost importance in China.

Problems

Pressures of accelerated population and industrial growth: The coastal region of China is characterized by developed economy and heavy population. This very limited land resource has to provide the land for industry, agriculture and civilian use. The average personal arable land area in the coast zone is about 30%～70% of the national level. Most of national industries are already located in the coastal cities, and new ones are moving in continuously, exerting ever greater pressure on the very limited land and fresh water resources.

Unsustainable exploitation: Present coastal zone development practices are based on subjective elements, rather than on objective environmental and resource availability criteria, leading to misuse and wrongful exploitation of the coastal resources.

China has more than twenty harbor cities along its coast, their population amounts to 20% of the nation with an industrial output that is about 1/3 of the country. Nevertheless, the capacity of the harbors remain slow for national needs. Over 200 potential sites for the construction of large to medium scale harbors have been identified. There are 89 000 hectares of tidal flat and shallows suitable for aquatic activities, of which only 1/5 are in use so far (Wang, 1992).

Poorly conceived coastal-engineering construction works are responsible for much damage to the coastal environment, while the pollution and overfishing have led heavy losses in coastal aquatic resources. Reclamation is often initiated without consideration to the nature of local environment and water resource. This has resulted in reclaimed land that is not suitable for cultivation for long periods of time.

The damage to coral reef and mangroves has caused intense coastal erosion and environmental deterioration.

The practice of damming in the areas around the river mouth for prevention of saline-water intrusion into the irrigation systems has caused the behind-the-dam siltation and affected water depth of the navigational channels.

Decentralized administration by plural departments: The management of the coastal zone is carried out in an uncoordinated manner by several administrative departments independent of each other and without an overall objective plan. Plans abound and planning may be undertaken separately for various missions, e. g. for the coastal-city urban plan, land application plan, harbor construction plan, fisheries plan, etc. Legislation is often enacted to ensure the development of a single economic activity. Because of this lack of coordination among relevant departments, the issue of environmental protection is often neglected. In addition, local governments tend to provide rather weak and ineffective management, making it difficult to ensure a balance between exploitation and protection.

Objectives and Requirements of Coastal Zone Management

Since the coastal zone is an area of special interest for both land and water resources, its management must be integrated to maximize the utilization of resources and environment and to ensure sustainable development of the region. This should be the major objective of all coastal zone management schemes.

Important aspects for the management of Chinese coast include: enforcing the existing regulations for the coastal zone and its resources; establishment of a comprehensive coastal legislation system; setting up of an coordinated administrative system that includes all levels of the governments and representatives of the local stakeholders; strengthening of scientific research; development of a coastal database and management system; a strategic plan for the coastal development; monitoring of coastal practices; protection of the environment and resources, and training of specialists who will direct future coastal management and development. The State Oceanic Administration of China has just completed the "Oceanic Agenda 21" of China. It is expected to further the cause of the integrated coastal zone management within a scheduled time scale.

The system of integrated coastal management in China can be summarized in the framework structure shown in Figure 2.

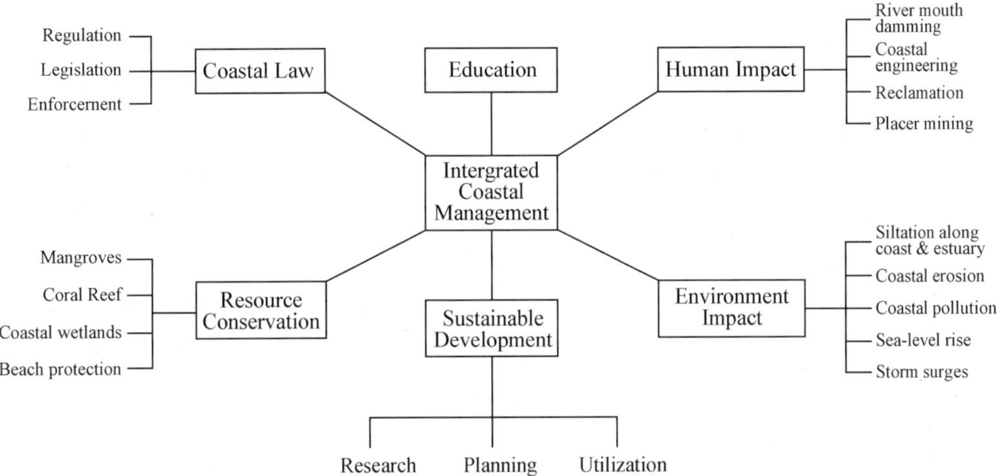

Figure 2　A System of Coastal Zone Management for China

Integrated management of Tidal Flats in China—A Case in Point: Distributed along the 4 000 km stretch of the muddy coastal plain and larger river deltas are tidal flats that represent a potential new land resource for China. Sediment discharge from the rivers has supplied nitrite-rich soil to the flats that accumulates rapidly as the flats area prograde toward the sea. The area of tidal flats above the base level is estimated to be about 2. 07 million hectares, and the area between the base level and 10m depth is about 8 million hectares. Thus, some 20 000 to 30 000 hectares of potential new land are added to the Chinese coastline each year. The tidal flats are mainly distributed along the Bohai Bay, the Yellow Sea Delta and the Jiangsu coastal zone. Their significance cannot be underestimated, especially when one considers the fact that China has lost some 500 000 hectares of cultivated land each year since 1979.

As a special resource, the tidal flats have a huge potential and many functions that include, salt- and fresh-water resources, aquatic products, terrestrial and marine plants, tourism, harbors and tidal energy. Integrated models for the development of farming, animal husbandry and fisheries are needed in order to practice coastal agriculture efficiently and in an ecologically sound manner.

During the different stages of tidal flat evolution different utilitarian models and activities need to be adopted:

- Study of the environment to determine the proper use of all resources, e. g. , use the deeper water section of tidal channels for harbor construction, while aquaculture (seaweed, laver, sea horse and marine peal) can be practiced in the shallower water;
- Different aquaculture models in accordance with the ecological characteristics of intertidal zone, e. g. , sandy flats for raising scallops, mussels and oysters; silty flats for arks and clams; muddy flats for cockles, marshes for bird breeding and sand worms production;
- Three dimensional development models that are suitable for the supratidal zone: aquaculture in the depressed ponds; duck and goose raising in the fishing ponds; grasslands for animal husbandry, etc. Different cultivation models should be adopted on the basis of different maturation stages of the flat soil and irrigation conditions, e. g. , a combination of cotton-mulberry-wheat or corn-rice fields, flower, herb and forest plantation, etc. , and
- Integrated development models should be established for production, transportation and marketing in farming, animal husbandry and fisheries sectors. If hydrocarbons are extracted and transported in the flats they should be exploited harmoniously with the environment.

With a lot of tidal flats remaining unused, the current models are rather rudimentary considering economic development only in single sectors with short-term utility. No integrated, three-dimensional, developmental model exists at the present time that can be deployed for sustainable, longer-term development. Many of the existing practices are counter productive, for example, salt ponds built behind the red-crown crane habitat destroy the brackish-water environment with a loss of protection to these rare birds. Upstream damming of rivers has decreased river sediment loads, while the threat of potential global warming and sea-level rise storm surges and flooding has increased. Tidal flats are often threatened by erosion and redeposition processes and terrestrial pollutants, especially liquid waste discharged from the rivers often cause red tide and mass fish mortality. It is, therefore, imperative that integrated coastal zone management should be strengthened. Research and exploitation, as well as utilization and protection should go hand in hand, which will smooth the way towards sustainable development in coastal zone.

REFERENCES

[1] Cui, Gonghao. 1994. Social and economic situation and prospect in Changjiang River Delta (In Chinese). In: Deltas in China. Senior Education Press of Beijing. 157 - 218.

[2] Pang, J. and Si, Shuheng. 1979. The estuary change of Hunaghe River. *Oceanologia et Limnologia*

Sinica, 10: 136－141.

[3] Ren, Mei-e. 1990. The problem on coastal management of China. In: Comprehensive Management of the Ocean. Science and Technology Press of Guangxi. 124－134.

[4] Wang, Kangshan, Su, Jilan, Dong, Lixian. 1983. Hydrographic features of the Changjiang estuary (In Chinese). In: Proceedings of International Symposium on Sedimentation on the Continental Shelf, with special reference to the East China Sea 1. 137－147.

[5] Wang, Ying, Ren, Mei-e, Zhu, Dakui. 1986. Sediment supply to the continental shelf by the major rivers of China. *Journal of Geological Society*, *London*, 143: 935－944.

[6] Wang, Ying and Aubrey, D. G. 1987. The characteristics of the China coastline. *Continental Shelf Research*, 7(4): 329－349.

[7] Wang, Ying. 1992. Coastal management in China Ocean Management in Global Change. Elsevier Applied Science. 460－469.

[8] Zhu, D. K. 1986. The utilization of coastal mudflat of China. *Scientia Geographica Sinica*, 6(1): 34－40.

The Utilization of Coastal Tidal Flats: A Case Study on Integrated Coastal Area Management from China*

1. INTRODUCTION

Integrated coastal area management (ICAM) is a continuous and dynamic process by which decisions are taken for sustainable use, development and protection of coastal environments and resources. It consists of a legal and institutional framework which ensures that development and management plans for coastal areas are integrated with environmental (including social) goals. Furthermore, such developments should not be made without the participation of all those affected. The purpose of ICAM is to maximize the benefits provided by the coastal area, and to minimize conflicts and harmful effects of such activities upon each other. It starts with an analytical process in the course of which objectives for the development and management of the coastal area are defined. What ICAM should ensure is that the process of setting the objectives, planning their implementation, and executing them involves as broad a spectrum of interest groups as possible, that the best possible compromises between the different interests are found, and that a balance is achieved in the overall use of the coastal zone (Anonymous 1993a).

In the context of this paper, the coastal zone is the interface where the land meets the sea. It is a transitional zone dominated by interactive terrestrial and marine processes. It should cover an area which includes the lower river basins up to 10 km or more inland, the beach and surf zone, the continental shelf, the continental slope and the continental rise.

2. THE COASTAL SETTING OF CHINA

In China the coastal zone is located along the eastern Euro-Asian continental margin adjacent to the Pacific Ocean. It includes the Yellow Sea, the East China Sea and the

---

* Ying Wang, Xinqing Zou and Dakui Zhu: B. W. Flemming, M. T. Delafontaine & G. Liebezeit eds., *Muddy Coast Dynamics and Resources Management*, pp. 287－294, ELSVIER, 2000.

摘要载于：*International Conference of Muddy Coast*. pp. 73, Wilhelmshaven, Germany, 1997.

South China Sea (Figure 1). The total length of the coastline is 32 000 km. Of this, 18 400 km encompass the mainland coast between the Yalu River in the north, which forms the border between China and Korea, and the Beilun River in the south which forms the border between China and Vietnam. The remaining are 13 600 km of coastline and of 6 500 islands (Wang & Aubrey 1987). Four coastal types can be recognised: 1) bedrock-embayed coast, 2) coastal plain coasts, 3) river mouth coasts, and 4) biogenic coasts comprising coral reefs and mangroves. Tidal flats are widely distributed along the coastal plain and river mouth coasts, including large parts of the Bohai Sea, Yellow Sea, and East China Sea, as well as the mangrove coasts in the South China Sea (Wang 1983; Figure 1).

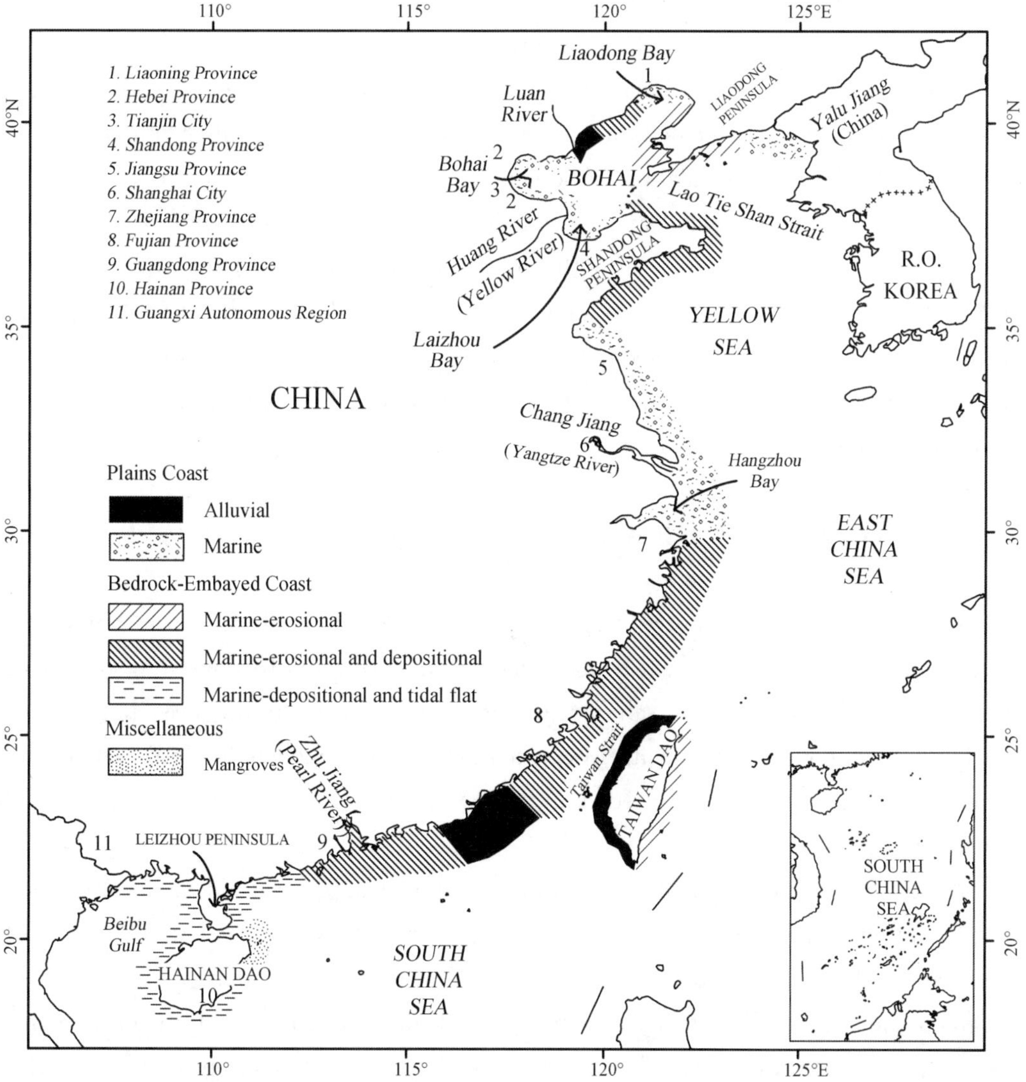

Figure 1　Map showing the coastal types of China (extracted from Wang & Aubrey 1987)

Muddy tidal flats, which line some 4 000 km of the coastal plains and large river deltas, form a valuable land resource, especially in the heavily populated coastal areas of China. Fluvial sediments are discharged by several large rivers in China, supplying nutrients and fertilized soils to the tidal flats. Because of the huge volume of sediment (ca. 1.0×10^9 tons per year; Wang et al. 1998; Zhu et al. 1998) supplied by the rivers, tidal flats have accreted rapidly and are still prograding gradually towards the sea. Muddy tidal flats are thus a rich endowment contributed by land-sea interactions. The tidal-flat area above the theoretical base level (for safe navigation, China has adopted a practical base level, called zero depth, which is equal to the lowest low-tide level) amounts to about 2.07 million hectares, the total area between the base level and the −10 m bathymetric contour contributing another million hectares (Anonymous 1995).

About 20 000 to 30 000 hectares of new land are created every year. This new land is of enormous significance in the eastern, heavily populated coastal plain region of China where it forms the basis of advanced economic development. Taking the Changjiang (Yangtze) River delta as an example, more than 73 million inhabitants or 6.09% of China's total population (Anonymous 1993b) live on an area of 99 500 km^2 (1.04% of the territorial land area), contributing 15.4% to the total GDP in 1995 (Anonymous 1995). In recent years, the population is increasing at the rate of 1% per year, while the arable land is decreasing at the rate of 0.05% each year. Agriculture output value in the area is only about 10% of the local GDP. In the light of this, there is an urgent need to harmonize the relationship between the population and the land resources in this region.

3. TIDAL-FLAT RESOURCES

As a kind of spatial resource, tidal flats have a huge potential value. In China, they have a multifunctional character, being utilised as a marine and freshwater resource, for the production of tidal energy, salt, aquatic products and terrestrial plants, but also for tourism and harbour development. Integrated development models for farming, animal husbandry, and fisheries are needed in order to sustain an ecologically sound and efficient agriculture.

Along the Jiangsu coast of the Yellow Sea, for example, the tidal flats are about 10~13 km wide and, with an area of 600 000 hectares, they represent roughly 14% of the total farmland in Jiangsu province (Wang & Zhu 1994). The tidal flats are currently prograding at a rate of 1 400 hectares per year, being a valuable land resource in this densely populated province which is characterized by a highly developed economy with very limited farmland. Although farmland makes up only 0.08 hectare per capita, it is nevertheless decreasing at a rate of 26 700 hectares per year due to industrial expansion and urbanization.

The increasing exploitation of the tidal-flat resources will inevitably intensify social and economic conflicts. Problems created by multiple jurisdictions and the competition between users of resources without the benefit of a conflict resolution mechanism, inadequate regulations for protecting resources, and the lack of nationally or locally adapted coastal policies for informed decision making will all translate into a loss of capability for future sustainable development. As a resource is depleted, conflicts may develop to the point of threatening human life and public order. Tidal flats currently offer physical and biological opportunities for human use, and ICAM tries to find an optimum balance between such uses based on a given set of objectives. In particular, there is growing concern about the destruction of the ecosystem by the demands of population growth and economic expansion. The considerable value of the natural ecosystem for the supply of sustainable extractive and non-extractive products is often underated in comparison with other often non-sustainable uses.

4. INTEGRATED COASTAL AREA MANAGEMENT (ICAM)

Based on the ecological character, there should be different exploitation models for the different types of tidal flats:

1) Deep-water sections could be used for harbour construction, whereas aquaculture (in particular seaweed, lavers, sea horse, and pearl farming) could be developed in shallow-water sections.

2) In accordance with the different ecological types found in the intertidal zone (sandy flats, silt flats, mud flats and marshes), different types of aquaculture models should be developed (Figure 2).

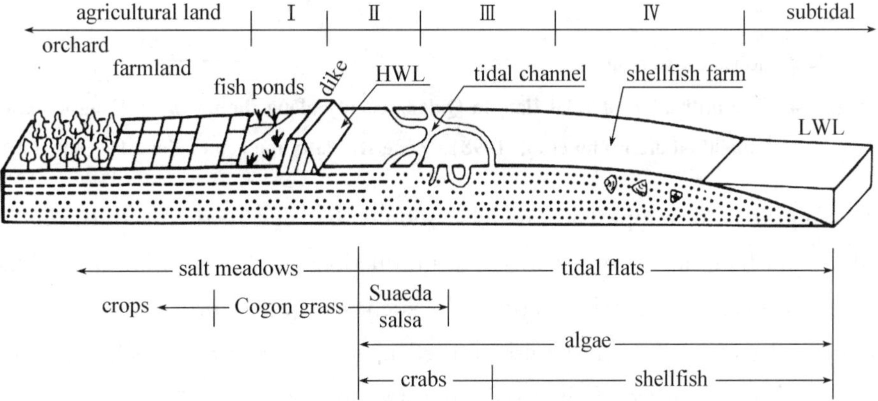

Figure 2 The utilization of tidal flats in Rudong county along the northern Jiangsu coast (modified after Zhu et al. 1998). Zone I: salt meadow; Zone II: mud flat; Zone III: silty mud flat; Zone IV: silt flat.

3) Three-dimensional models are particularly suitable for the development of supratidal zones. Thus, aquaculture should focus on topographic depressions, whereas animal husbandry and protected reserves for rare wild life (e. g., for the Davis deer (Milu) and the red-crown crane) should utilise salt marshes and grassland, etc. (Figure 3). Different cultivation models should also be considered for the various maturation stages in tidal-flat evolution, for example, cotton-mulberry-wheat or corn-rice fields (Figure 4).

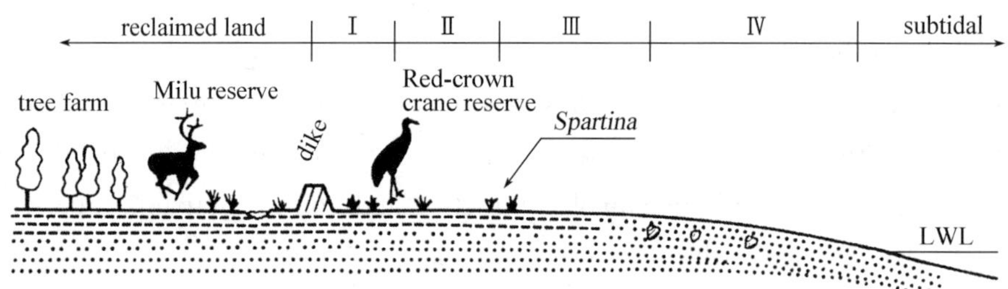

Figure 3 The utilization of tidal fiats in Dafeng county along the northern Jiangsu coast (modified after Zhu et al. 1998). Zone I: grass flat; Zone II: mud flat; Zone III: silty mud flat; Zone IV: silt flat.

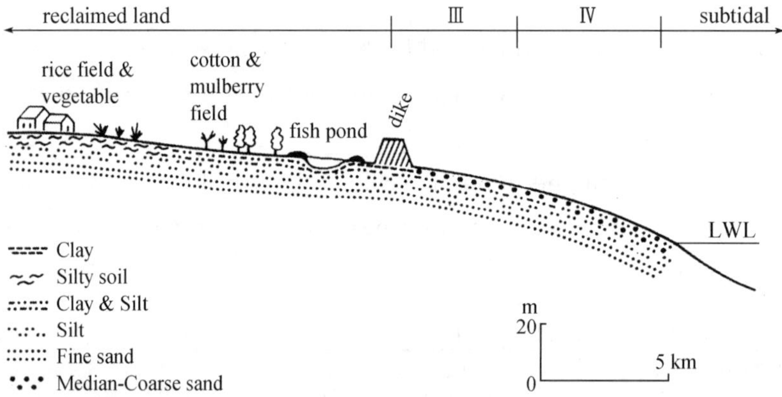

Figure 4 The utilization of tidal flats in Qidong county along the northern Jiangsu coast (modified after Zhu et al. 1998). Zone III: silty mud flat; Zone IV: silt flat.

4) Integrated development models should be designed for farming, animal husbandry, and fisheries in order to make production, transportation and marketing a coordinated process. Industrial activities, such as goods processing, oil and gas extraction, and transportation should proceed in such a way as to achieve a harmonic relationship with the nature of the tidal environment.

5) Model farms are especially useful in demonstrating the successful application of comprehensive development schemes.

With large tracts of tidal flats being still unused, tidal-flat development currently takes place on a rather small scale, following the old pattern of isolated economic development which is only concerned with immediate benefits. This results in sharp contradiction between short-term resource consumption or utilization and long-term supply. Since it is not a type of integrated and three-dimensional development, it is unsuitable for a sustainable development with higher benefit. As a result, it will produce conflicts between activities such as farming, forestry, animal husbandry, fisheries, industry, harbour development, tourism, salt production, reed planting, and chemical extractions. For example, salt ponds constructed behind the crane habitat will destroy the brackish-water environment and thereby threaten the protection of this rare bird species. Due to global warming and sea-level rise, river sediment discharge is decreasing while storm surges, floods and other marine calamities are increasing. Because of the shallow slope and numerous tidal channels, the tidal flats are often threatened by erosional and depositional processes, droughts, and flood disasters which can attain a frequency up to 70% in the region.

Taking typhoon no. 11 of 18—20 August 1997 as an example, this cyclone hit the south coast of Zhejiang province with wind speeds of 12 Beaufort, i. e. more than 32. 7 m/sec. Since it coincided with the highest annual astronomical high tide, the tide reached a level of 6 m, causing a major disaster on the coastal lowlands (2～4 m above sea level). Even after it landed, the water level of the lower reaches of the Changjiang River was still 5 m high, this being the highest water level on record. Land-based pollutants, especially liquid wastes discharged by rivers, often cause red tides, killing fish and shrimps in the process. In China, more than 60% of anthropogenic contaminants are concentrated in the coastal zone. Thus, the Changjiang River delta, which makes up only about 1% of the national territorial area, has taken up 9% of national waste waters, 5% of waste gas, and 4% of other industrial residues (Fu & Li 1996).

5. CONCLUSIONS

Integrated coastal area management should be emphasized not only by decision makers but also by stockholders and by the interested public. The combination of research and exploitation, utilization and protection, calamity prevention and benefits will certainly smoothen the way for sustainable development in coastal agriculture. The scheme shown in Figure 5 is suggested as a systematic approach to integrated coastal area management.

It clearly shows the links between the natural environment and human impacts in the form of coastal law (regulation), education and ICAM to protect and conserve the natural environment and its resources (cf. Wang et al. 1997; Healy & Wang 2000).

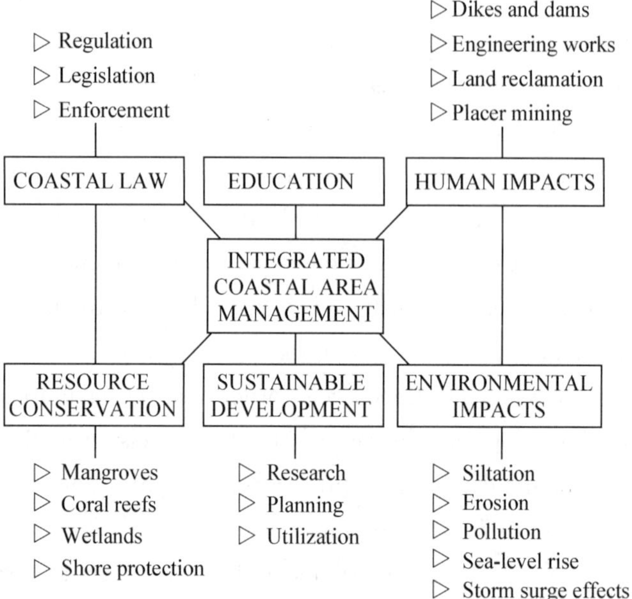

Figure 5 Proposed scheme for systematic integrated coastal area management (modified after Wang et al. 1997).

ACKNOWLEDGEMENTS

This study was supported by the State Laboratory of Coast and Island Exploitation, Nanjing University (Contribution No. SCIEL21198118).

REFERENCES

[1] Anonymous. 1993a. The Noordwijk guidelines for integrated coastal zone management. In: The World Bank Environment Department, Land, Water and Natural Habitats Division. World Coast Conference 1993, November 1993, Noordwijk. 1-6.

[2] Anonymous. 1993b. China Population Statistics Yearbook 1993. State Statistical Bureau P. R. China, China Statistical Publishing House.

[3] Anonymous. 1995. Statistical Yearbook of China 1995. State Statistical Bureau P. R. China, China Statistical Publishing House.

[4] Fu, W. X. and Li, G. T. 1996. Ocean pollution and ocean environment protection. In: Marine Geography of China (Chap. 23). Ocean Science Press, Beijing. 474-517 (in Chinese).

[5] Healy, T. and Wang, Y. eds. 2000. Muddy Coasts of the World: Processes, Deposits and Function. Elsevier, Amsterdam (in press).

[6] Wang, Y. 1983. The mudflat system of China. *Can. J. Fish. Aquat. Sci.*, 40 (Suppl. 1): 160-171.

[7] Wang, Y. and Aubrey, D. G. 1987. The characteristics of the China coastline. *Cont. Shelf Res.*,

7(4)：329－349.

[8] Wang, Y. and Zhu, D. K. 1994. Tidal flats in China. In: Zhou Di, Liang Yuan-bo, Zeng Cheng-Kui (C. K. Tseng) eds., Oceanology of China Seas (Vol. 2). The Netherlands: Kluwer Academic Publishers. 445－456.

[9] Wang, Y., Luo, Z., Zhu, D. 1997. Economic development and integrated management issues in coastal China. In: Haq, B. U., Haq, S. M., Kullenberg, G., Stel, J. H. eds., Coastal Zone Management Imperative for Maritime Developing Nations. The Netherlands: Kluwer Academic Publishers. 371－384.

[10] Wang, Y., Ren, M.-E., Syvitski, J. 1998. Sediment transport and terrigenous fluxes. In: The Sea (Vol. 10). John Wiley & Sons, New York, pp. 253－292.

[11] Zhu, D. K., Martini, I. P., Brookfield, M. E. 1998. Morphology and land-use of the coastal zone of the North Jiangsu Plain Jiangsu Province, Eastern China. *J. Coast. Res.*, 14: 591－599.

关于开发海南岛港口的几点建议*

二十世纪六十年代我曾多次参加海南岛港口选建工作。二十年后再来，感到海南岛的面貌变化很大。在海南经济开始起飞之际，愿就自己一隅之见提几点建议：

一、海南岛有五大资源，是独具特色的热带海岛

一是矿产资源。石碌的富铁矿与南部的钛砂矿是重要的矿产资源。西部浅海的石油、天然气，是海南岛经济起飞的动力资源。新的工业布局与经济规划，应考虑以天然气与石油作为能源。同时，海南岛是莺歌海海域油气资源开发的主要依托和最近的陆上基地，油气资源开发事业发展迅猛，必能对海南岛开发带来新的生命力与巨大的推动。对此，要有足够的估计并积极做好迎接准备，必须从新的起点考虑港口、工业以及城市之布局。

二是热带作物。海南岛位于北回归线以南，岛北端的纬度比中国台湾的南端还低2°，而且地形条件好，中间高山，四周平坦。台湾地形陡峻多山，仅西部有些平原。从热带作物分析，海南岛更有其特长。台湾盛产稻米、蔗糖和热带水果，但海南岛有橡胶、咖啡、胡椒，并且宜于发展茶叶与蔗糖工业。海南岛亦应发展热带水果（香蕉、菠萝、芒果、柑桔、木瓜）、蔬菜与林木，以供应我国大陆广大市场之需要。尤其是反季蔬菜与水果可发展专线空运以解决大城市冬季蔬菜之不足。仅这项经济发展潜力就很大。海南岛气候适宜发展硬材林，近年来虽然种植了木麻黄、桉树、与台湾相思等林带，但原始的热带雨林及贵重的硬材林遭到相当严重的破坏，至今仍见“刀耕火种”式开荒。由于无林木蓄水，以致雨季时暴雨成灾，旱季时干旱缺水。日本是岛国，没有大的河流，但是日本诸岛无山不绿，无水不清。因为他们仍注意保护生态环境，不砍伐本国林木，保护森林以调节气候与保持水土。海南岛要注意到造林与护林，因地制宜地发展不同林种，尤其是我国迫切需要的硬材林。成林后，要严格执行有计划地砍伐。西海岸台地由于气候干燥，宜保留草场发展养牛，使之成为我国肉食牛的基地。这一项经济效益亦大。

三是海洋水产资源。由于四面环海，且具有不同的海岸类型，故海洋水产品种丰富，宜于发展养殖。岩石海岸可以发展石斑鱼与龙虾；砂质海岸养殖珍珠或牧养对虾与鱿鱼等；珊瑚礁上可以生产琼胶，将它与热带水果生产结合起来，可以发展果冻粉的生产。要进行科学研究来发展海产养殖，发挥海南岛特色。

四是旅游资源。日本、百慕大、波多黎各与夏威夷，皆发展海岛旅游事业。尤其是夏威夷被誉为旅游事业的天然博物馆。海南岛与夏威夷类似，四季温暖，火山玄武岩风化的肥沃土壤与碧绿的农田，港湾碧海，椰林海滩，平坦的基岩台地发展公路交通方便。火山

---

* 王颖：《中国港口参考资料》，1984年第3期，第44－47页。

口，温泉，黎族寨子与手工纺织品，历史古迹与丰富的热带食品等，如果配以现代化的交通通信设施与别具风格的不同等级的旅馆、村舍，则可以吸引日本、东南亚华侨、欧洲以及国内的旅客。日本每年有 60 万人去台湾岛旅游。海南岛比台湾纬度低，冬季休养的条件更好，发展旅游事业大有潜力。建议海南岛派出得力的干部到夏威夷、日本与新加坡等处学习考察，提高科学管理水平，发展具有海南特色的旅游事业。

五是人力资源。海南岛现有人口 580 万，不少人来自华南、华北与东北等省。当地土著与黎苗少数民族朴实友好，自古以来，与外来人员和睦共处，不断吸收岛外文化，同时也传播文化至岛外，并且有大量移民侨居海外。多元文化与侨乡的背景利于吸收岛外资金促进发展。

充分考虑这五项资源的特点，制定出切实可行又具远见的经济发展规划，必会加速海南岛的发展。

二、经济发展交通先行

海南岛四面环海，海岸线长达 1 369 km（一说为 1 528 km），有 60 个港湾，全岛面积为 33 920 km^2（南北端点间 258 km，东西长 180 km）。对外联系首先是海上航运，其次是空运。岛内交通首要为公路，其次是西岸的铁路专线。在各种交通工具中，要大力发展海上运输，因此港口建设就显得很重要。

概言之，海南岛的海岸是港湾式海岸，具有优良的建港条件。北部海岸系沿玄武岩流与湛江系地层所组成的台地侵蚀而成，南部与沿湛江系台地与相间分布的花岗岩低山丘陵所发育的。海岸的特点是：岸线曲折，波浪作用活跃，以砂砾质沉积为主，在一些狭长海湾中，潮流作用亦盛，并有细颗粒的悬移质泥沙；地处热带海岛，因此岩岸发育珊瑚礁，砂质海岸带风力作用亦强。大体上，海岛北部以东北向风浪为强常向，东岸以北东与南东向盛行，南岸与西南岸以西南向风浪为主导。

根据海岸的成因与作用过程差异，可将本岛的港湾海岸划分为下列三类：

（一）潮汐汊道（Tidal Inlet）

它是沿着构造断裂带和岩层交界的软弱带或沉溺的谷地发育而成。其特点是：狭长的海湾，口门有小岛或沙坝束狭、水深流急，海湾上游没有河流或者河流的影响很小，主要是纳潮海湾，潮流作用盛。砂质沉积为主，但湾内有细颗粒的淤泥物质。在潮汐汊道口门深水通道两侧，常常发育着拦门沙浅滩。内拦门沙地形平坦，泥沙颗粒细，系涨潮流所堆积成的。一般地说，内拦门沙对航行危害不大。位于口门外的外拦门沙，颗粒粗，系落潮流携带湾内的泥沙于口外开阔处堆积，有的是沿岸纵向来源的泥沙。即使落潮流速度大，但由于过水断面增大及口门外激浪的顶托，会形成浅滩，影响船只进出港湾。保持潮汐汊道港湾与航道的水深，主要取决于两个因素：① 纳潮量。它取决于潮差与涨落潮延时之对比。潮差大，纳潮量大，或涨潮延时大于落潮延时利于维持航道天然水深。为保持大的纳潮量，要尽量扩大或维持湾内水域面积。② 弄清湾内与沿岸泥沙来源状况，采取措施减少泥沙来量，并造成泥沙天然越过之条件。

海南岛沿岸为不正规全日潮，除个别地区（如后水湾）外，潮差不大，但涨落潮延时条件有利，7～8 h 的涨潮，湾内蓄积了大量海水，5～6 h 的落潮，流速加大有利于维持口门外的航道水深。海南岛的榆林港、三亚港、洋浦湾、清澜港、新村港、铁炉港等都是潮汐汊道。其中榆林与三亚具有建成万吨级深水港的优良条件。三亚港北有山地屏障，四季温暖，并不受东北向强常风浪影响；湾东有鹿回头岭屏障，不受东南向风浪与泥沙影响；西侧湾外有东、西瑁州罗列，减小了西南向风浪作用的强度。港区泥沙来源少，自 1966 年始挖的航道至今无回淤。目前已沿西侧沙坝建成两个 5 000 t 级泊位，潟湖内湾停泊渔船，万吨级轮需在白排东的锚地驳船转作业。三亚港是海南岛与东南亚各国联系的门户，亦为南海诸岛的后方基地；它西连莺歌海地区的油气与铁矿产地，东通旅游海岸，北部与海口相接通往大陆。三亚地处枢纽，必须充分发挥它的门户作用才能加强海口的“咽喉”地位。三亚港泥沙来量很少，港域水深条件可供万吨级船舶出入。如果自白排礁向岸，沿珊瑚礁尾这一天然基础，筑成 1.5 km 长的突堤式码头，既可防南西向风浪，又可供火车上堤直达白排*，白排礁附近水深 7～9 m，13 m 等深线靠近岸边，适当疏浚，该码头可供 5 万 t 级的大船使用。三亚港具有建设大中小港的优良自然条件，需加强规划安排。要合理使用码头岸线，同时，三亚港内之潟湖不宜开辟盐田，不能缩小该水域面积，以维持航道天然水深。内陆的大坡水与月川水上游，可适当改造，以拦蓄淡水供港市发展所需。

洋浦湾是海南岛北部与西部自然条件最优越的深水港湾，它是沿火山玄武岩流与湛江系地层之交界带发育的。湾内风浪隐蔽，水稳条件好，具有天然的基岩码头岸线多处。可以建成万吨级以上的深水大港。开发洋浦港必然对海南岛的经济发展有很大的促进。要预见到西部石油工业的发展以及分担八所港的散矿，需要及早发展洋浦港。该港已积累了大量的勘测资料，建港条件比较成熟。虽然目前该处无城市依托，但港口之兴建会带动工业与城市之发展。国内外的大都市，如纽约、东京、天津、上海、广州等，最初都是随着港口的建立而兴起的。洋浦湾建港存在的开挖拦门沙航道、部分水域较狭窄及对地震的预防等问题，我认为采取适当的工程措施是可以解决的，关键是经济规划与港口布局。

清澜港水深条件宜发展为千吨级的港口，港外沿岸原无大量泥沙流，本可开挖口门航道，但是，由于岸外开挖珊瑚礁，湾内砍伐了红树林，造成了泥沙来量增多。主要考虑到布局与经济力量，北岸究竟发展几个港口？海口港地位重要，而洋浦港自然条件好，但清澜是东岸的中型港口，地方上有力量可以发展，如经济力量有限，此港宜靠后。

新村港湾内水深条件好，但口外有较活跃的泥沙运动，开挖航道必须配以相应的工程措施拦截沿岸泥沙向港口口门运移，否则仍会淤积。

（二）沙坝潟湖海岸

如港北港，沿岸有大量的泥沙来源（海浪冲刷花岗岩风化壳或湛江系所造成），南东向波浪垂直向岸传播，泥沙横向运动活跃，海底泥沙被推积向岸成为长大的沙坝。沿沙坝岸线，泥沙的纵向运动与风沙运动也很活跃，随季节不同泥沙沿岸运动方向亦有所改变。由于横向泥沙运动活跃，港北港口门内外有数列砂质浅滩，妨碍船只航行。沙坝后侧的潟湖

* 南京大学地理系：三亚港泥沙来源与海岸发展趋势研究报告，1973 年。

水深一般较潮汐汉道小，并且处于逐渐淤浅的过程中。因此，港北港可作为小型渔港或供水产养殖，需要考虑对口门做专门处理。

（三）河口海湾

河口三角洲向海伸展突出所形成的宽坦砂质海湾岸，是海南岛北岸与西岸的特色。南渡江三角洲向海伸出，泥沙向海与向两侧扩散不远，多被东北向常强风浪推回，堆积成海岸沙坝。数列沙坝，自河口向两侧规模逐渐减小，反映着泥沙数量之变化。琼州海峡风浪强，潮流作用亦强，海峡两端的海流皆有规模较大的潮流三角洲堆积。建港时需考虑这些因素。

海南岛北岸需建深水港。海口市为海南的政治、经济中心，扼守海峡与大陆联系的咽喉，并有老港之基础。个人认为白沙门条件更适宜进行深水港选址调查工作，可以有几个不同的工程方案设想，如：① 白沙门东侧更接近 10 m 等深线处，修建防波堤挡东北向风浪并起导沙至深水使天然越过，防波堤以西可依沙坝为基础修建数列突堤式码头；② 白沙门；③ 白沙门以西，风浪小，泥沙来量少，但水浅。总之，白沙门一带水深条件好，但建港需要有防波堤掩护，堤头宜至深水并注意突堤与海岸之交角，以达到导沙的作用。

八所港位于一小河三角洲突出岬的南岸，该港浪大，水域小，全靠防波堤掩护，它是这类海岸建港的实例模型，所存在的一些问题，可供进行新的港口（如白沙门）设计时加以借鉴。八所港在原有码头与设备的基础上已发挥了很好的作用。可以进行适当的改造与扩建。根据自然条件与现有技术以及经济力量，似不宜大规模扩建。从长远发展看，海南岛西岸工业区宜由洋浦港为运输主力并分担八所港部分任务。

总之，发展海南岛的经济需加强海运，重点抓三个深水港的建设：① 尽可能地发挥北部海口港与南部三亚港的作用。海口港的深水码头宜选建新址，以白沙门一带更适于多做些工作；三亚港具有发展五万吨级大港的自然条件与支援莺歌海油气田开发的战略基地作用，目前要加强扶植与规划，使之发挥更大作用；② 从长远利益考虑西岸应建立洋浦港，要从现在抓起以迎接西部浅海油气资源开发之冲击，并分担八所港部分任务。

其他是中小型港口，地方经济发展则会逐步解决资金与设备，目前可进行先期调查工作以积累资料。

三、海港是联系陆地与海洋之桥梁，但不是终点

海南岛港口条件优越，海洋运输方便，而陆上宜大力发展公路，及沟通南北航空。海陆衔接好，转运快，可减少仓库和堆场的数量与面积。海南岛陆上直线距离不超过 300 km，无大规模客流，而多新鲜蔬菜与分散的货源，宜发展公路网，便于及时而又灵活地运输。矿区可应用西岸专线铁路，但集装箱货运与旅游事业宜发展公路交通。海南岛四周陆地平坦，多荒地，加上砂质与基岩的地基，便于开拓公路，宜及早发展。除加宽改善现有三条公路干线外，尚需连通东西干线，修建支线网络，并有计划地修建港市间的高速公路。油气资源开发会促进公路运输，后者又可促进发展旅游事业。海南岛南岸三亚一带冬泳海滨及海岛景象可吸引欧亚游客，游客乘船或乘机至三亚登岛后，即可利用方便的公路沿

岛旅行，便宜而有趣，会增加对旅客的吸引力，而公路旅游事业的发展，又可促进岛内村镇的繁荣。

海港加公路，岛内短途运输灵活机动；岛外长途运输费用低廉。再加上空运方便迅速，必会大大改善海南交通运输面貌与增加经济效益。

四、规划、实施与教育

发展交通要加强经济规划，安排好联运体系与抓紧施工；热带资源与海洋水产发展要加强科学研究与普及推广；发展海南岛旅游事业宜学习日、美经验；加速发展海南岛的经济，要重视开发智力资源与提高管理水平。针对本岛特色与经济建设的需要，办新型的大学，培养海洋地质、水产、港口、石油化工、民用建筑、计算机与英语等七方面的高水平人才，使学校成为建设海南的智力中心。

中国海岸及其建港条件的分析*

一、中国海岸环境和类型

中国海岸从辽宁的鸭绿江口至广西的北仑河口，海岸线全长 18 400 km，包括了 6 000 余个岛屿，则岸线总长 32 000 余 km。海岸的基本轮廓受到地质构造及大河沉积作用的控制，同时亦受季风波浪及潮汐作用的影响。

中国沿海陆地及大陆架，有一系列 NE 及 NNE 走向的隆起及沉降带，它们与海岸斜交，从西向东是渤海凹陷、胶辽隆起、南黄海凹陷、浙闽隆起、东海大陆架沉降带及台湾断块带等。基岩港湾海岸及基岩岛屿即沿着隆起带发育，而平原海岸常伴有大河，通常形成在沉降带。在华南海岸 NE 向的块状断裂决定了港湾海岸的主要方向，而沿着 NW 向的断裂常发育为狭长的海湾、潮汐汊道。

中国海岸受河流影响非常深刻，黄河、长江、珠江三条大河及许多次一级河流将大量泥沙淡水带入海洋(表 1)，使中国海岸的泥沙特别丰富。

表 1　中国主要河流的水沙资料

河　流	年径流量(m^3)	年输沙量(吨)	平均含沙量	平均潮差(m)
鸭绿江	27.8×10^9	4.75×10^6	0.33～0.42	4.48
滦　河	38.9×10^8	20.08×10^6	3.94	1.50
黄　河	48.5×10^9	11.9×10^8	37.7	0.80
长　江	9.25×10^{11}	4.86×10^8	0.54	2.77
珠　江	3.7×10^{11}	0.85×10^8～1.0×10^8	0.12～0.334	0.86～1.63

来自河流的沉积物极大部分沉积在河口及邻近的内陆架，沿岸搬运测得的最大距离为 150 km(渤海湾、黄河)，自河口向大陆架深水区搬运的最大距离为 50 km(东海、长江)，珠江沉积物向海主要搬运到水深 20～30 m，距岸 35 km，而砂质沉积物仅在河口、沿岸堆积，至水深 10 m 距岸仅 4 km(渤海、滦河)。因此，河流入海泥沙控制了河流作用区海岸的发育及各类海岸地貌的形成。

中国海岸的潮汐形成于太平洋的潮波，它同海岸及大陆架海底相互作用，南海主要是全日潮，潮差小(1～2 m)，渤海、黄海及东海主要是半日潮，有较大的潮差(3 m 或 3 m 以上)，产生强的潮流，在一些河口湾产生了涌潮。潮流能侵蚀海岸海底，搬运海岸沉积物，

---

* 王颖，朱大奎：(全国海洋开发工程技术交流会秘书处编)《海洋开发工程技术论文集(上)》，106－110 页，海洋出版社，1987 年 12 月。

转载于：《科技进步与经济建设——中国科学技术协会 1988 年学术年会论文集》，240－243 页，学术期刊出版社，1988 年 9 月。

形成潮流侵蚀与堆积地貌。潮汐作用展宽了海岸带动力作用的范围。

波浪是塑造海岸的主要营力，表面波主要是风波，随季节而变动。冬季盛行北向波，9月影响到中国台湾以北海域，10月到达10°N，11月遍及整个海域。1月份波浪方向沿着蒙古高气压边缘顺时针方向变化，渤海西北向波浪，在东海及南海通常变动在北至东北向波浪。

夏季主要是南向浪，从南海向北运动，最初出现于2月，5月南向浪控制了5°N以南，7月控制了整个海域。波浪方向是沿着印度低压系统作反时针方向旋转，在15°N以南的南海区呈西南向，向北到东海及北黄海、渤海转为南向东南向。夏季与冬季之间是个过渡时间，其风浪方向是变动的，没有优势的浪向。

波浪方向、强度及季节变化的研究，将为确定海岸带侵蚀、堆积的动态，海岸泥沙运动的方向、数量，及海岸演变过程提供重要的依据。

中国海岸主要有4个类型，而其中基岩港湾海岸与粉砂淤泥质平原海岸是最基本的(表2)。

表2 中国海岸类型

海岸类型	亚 类	分布地点
Ⅰ. 基岩港湾海岸	Ⅰ-1. 海蚀港湾岸	辽东、山东半岛
	Ⅰ-2. 海蚀堆积港湾岸	胶、辽、冀东、浙闽粤
	Ⅰ-3. 海积港湾岸	辽西、莱州湾东、北部湾沿岸
	Ⅰ-4. 潮汐汊道港湾岸	华南、海南岛
Ⅱ. 粉砂淤泥质平原海岸	Ⅱ-1. 冲积平原岸	冀东、江苏北部等
	Ⅱ-2. 海积平原岸	渤海湾、江苏沿岸
Ⅲ. 河口海岸	Ⅲ-1. 三角洲岸	黄河、长江等三角洲
	Ⅲ-2. 河口湾岸	杭州湾、蓟运河口
Ⅳ. 珊瑚礁与红树林海岸	Ⅳ-1. 珊瑚礁岸	南海诸岛、海南岛、台湾等
	Ⅳ-2. 红树林岸	27°N以南大陆沿岸
		海南岛、台湾等岛屿

二、基岩港湾海岸

这类海岸发育于山地直接临海，以港湾曲折及有基岩岛屿为特征。水下岸坡较陡(坡度0.01～0.005)波浪能量较高，受到侵蚀与沉积两种作用。海岸沉积物是砂或砾石，海岸地貌的细节与波浪作用有关，同时取决于海岸的开敞性、水下岸坡、岩石类型及沉积物的供应。按发育阶段及特征，有4个亚类。

海蚀港湾岸：发育于坚硬的结晶岩区域，侵蚀速率缓慢，自冰后期海面上升至今日位置以来，海岸很少变化。沿岸河流或海岸侵蚀只供应小量的沉积物，只在很小的范围内有沉积作用，如小湾的湾顶、湾口及沙咀等。海岸地貌主要是侵蚀的，而缺乏堆积地貌。这类海岸有利于建设海港，如辽东半岛的大连港、营口新港。

海蚀堆积港湾岸：这类广泛发育于中生代花岗地区，盖有薄层风化碎屑沉积物，例如山东半岛。这些物质易受波浪侵蚀，或由小河搬运，将大量砂粒物质带到海里。因而，广泛发育了各种海蚀的与海积的地貌，如岩滩、海蚀崖、湾坝、沙嘴、连岛坝等。这类海岸常可找到合适的港址。

海积港湾岸：通常基岩是软弱的，海蚀作用使港湾的岬角受侵蚀，港湾内堆积，常在向着盛行风浪方向形成了平直的堡岛岸线，在一个港湾曲折的原始基岩线前形成一条很窄的海岸平原，表明从港湾岸发展到平原海岸的过渡。

潮汐汊道港湾岸：广泛分布于华南，该处主要是 NE 向的岸线，被一些 NW 向的断裂形成的狭长的海湾所分割，有一些小岛、沙嘴保护了这些湾口，形成潮汐汊道。汊道底部受落潮流冲刷，很少沉积，在汊道口外形成落潮流三角洲。这类海岸亦有利于建港。

作为一个合适的港址，必须具备两个条件：一是要有一定的水深供船只航行进出港口；二是港区须风浪平静，便于船只靠岸作业。要同时满足这两个条件常会遇到困难。因为水深条件好的岸段通常波浪能量较大，而风浪平静岸段常发生泥沙淤积，在建港进行地貌分析时，就是要查明海岸各自然因素以解决这对矛盾。对港湾海岸主要有三方面。

(1) 分析海岸动态，要采用综合的方法查明海岸的演变过程与目前的动态，处于侵蚀的或堆积的岸段，显然是不利于建港。港湾海岸各岸段通常侵蚀岸段与堆积岸段交替分布，在侵蚀与堆积的过渡区，常是比较稳定的岸段，在弧形港湾的两侧，或开敞的缓弧的湾顶，常是比较稳定的岸段，可用以建港。秦皇岛油港即选建于开敞的弧形港湾，建港后十几年来岸线一直比较稳定。

(2) 分析海岸带泥沙来源，泥沙流方向及强度。在天然海岸有沿岸泥沙流通过，而海岸并不表现出明显的侵蚀与堆积，但建港后，纵向泥沙流被拦截，天然的动态平衡被破坏，则在港口上游一侧堆积，而下游沿岸受侵蚀。上游的堆积浅滩扩大，泥沙可进入港口阻塞进港航道。综合分析海岸地貌可查明泥沙来源、方向、强度。

如秦皇岛—辽宁兴城海岸，沿岸以岬角河口为界，可分为六个泥沙活动区，各区段之间泥沙互不交换。从岸向海亦可分出三个泥沙运动带，Ⅰ岸边泥沙运动带，主要是砂质沉积，其外界相当于波浪破碎带；Ⅱ泥沙扩散带，粉砂及砂，为过渡区；Ⅲ泥沙基本稳定带，泥质沉积，是波场外的沉积。若Ⅰ、Ⅱ带较宽，反映泥沙来源丰富，有沿岸输沙，一般要避免在该处建港；Ⅰ、Ⅱ带较窄，通常无明显沿岸输沙，若有也易于工程处理。

山海关—秦皇岛海岸分布着一系列砂砾堤。由沿岸堤的分析得出，砂砾沉积物来自沙河石河，而沙河以西的砂砾物质是过去的，现在没有再向海供应。综合上述分析，查明秦皇岛南李庄岸段，无泥沙流通过，岸线稳定。该处建成油港、煤港后，港区海岸稳定，没发生港口上方淤积，下方侵蚀，危及港口工程的现象。

在岸边泥沙运动带内开挖进港航道，常会在破波带处淤积，严重影响港口使用，如秦皇岛新开河渔港。对这类现象，若查明无明显沿岸泥沙流，则可在航道二侧修筑突堤，堤头水深越过泥沙活动带，使进港航道中不产生破波，阻止泥沙横向运动，新开河渔港，在修筑双突堤(至－20 m)后，防淤效果很好。

(3) 分析水下地形，利用港湾中心深槽。港湾中通常有潮流作用形成的深槽。有涨落潮流共用的深槽，也有涨潮流落潮流不同流路的。通常，主深槽与海岸垂直，主要为落潮流所流经，而涨潮流成片状，分成几股涨潮流深槽，在两条深槽相交汇处潮流冲刷力最强，如浙江乐清湾，平均潮差 4.5 m，流速 2.5～3 m/s，中心深槽深度大于 10 m。又如山东龙口港是个向西南开敞的浅水海湾，平均潮差 0.9 m，流速仅 0.2 m/s，湾中南北有两个湾坝(浅滩)，进出海湾的水量在沙坝之间通过.形成一水深 3～4 m 的深槽，龙口港沿此深槽开挖航道。已浚深至 7 m，多年来均回淤量不大，因此，综合分析水下地形，选择稳定的中心深槽用作通海航道，在港湾海岸建港中具有重要意义。

三、平原海岸

冲积平原岸，大部分位于山地区域向海一侧平原由来自山区河流冲积物所组成。沿岸冲积平原及水下岸坡度为 1∶100～1∶1 000，岸外有很宽的破浪带，使沉积物发生沿岸的及向岸的运动。因此，堆积地貌如沙咀、堡岛、海底沙坝及宽广的海滩均甚发育，海滩大多在发展前进的，沿岸有一系列滩脊、潟湖及潟湖后的沙坝。

海积平原岸，这类海岸非常宽广平坦，分布于构造上沉降的大河的低平区域，如苏北平原、华北平原、辽河下游平原，海岸坡度 1∶1 000～1∶5 000，波浪作用仅在岸外远处，海岸塑造中潮流起了主要的作用，沉积物主要是河流带来的粉砂淤泥，这些细粒物质不能沉积在浅水的近岸带，而是堆积在深水区或在潮滩上部。

典型的平原海岸剖面有盐沼湿地，潮间带泥质浅滩(潮滩)及水下岸坡。盐沼湿地的陆地一侧分布着贝壳堤，贝壳堤代表了当时的岸线，贝壳堤向陆一侧，有一些残留的潟湖凹地，渤海湾潮滩宽 4～6 km，坡度 1∶1 000～1∶3 000。苏北潮滩宽 10～13 km，坡度 1∶1 000～1∶5 000，潮间带内可划分出 4 个地貌沉积带。水下岸坡坡度 1∶500～1∶3 000，沉积物在低潮位最粗，是粉砂极细砂，向深水区及向潮滩粒度减少，是粉砂泥质沉积物。

平原海岸反映了松散沉积物侵蚀与沉积作用之间的平衡。关键是沉积物的供应，特别是来自河流的泥沙，若河流的供应量大于侵蚀量海岸是前进的，沉积物供应减少或停止，则将引起海岸线后退。华北及苏北海岸均受黄河控制，随着黄河变迁，华北及苏北平原海岸发生大规模的侵蚀与堆积过程的交替。

通常，平原海岸建港要比港湾岸困难。主要平原岸坡度小，深水区距岸远，开挖通海航道的距离过长；平原岸物质是粉砂淤泥，受水流搬运，易在港内淤积。但多年来中国建港实践与科学研究，表明在淤泥质平原海岸亦能建成深水大港(天津新港)。另外，平原海岸亦有些有利的海岸环境，可供建港所用。

(1) 在淤泥质平原海岸建港，要研究大河入海泥沙对港口的影响，避免大河直接影响，将泥沙带入港口航道。港口淤积的原因，主要是港外浅水海域风浪掀沙，随潮流进入港内淤积。为减少淤积，要尽量减少港内水域面积，增加港内落潮流流速。

(2) 平原海岸的岸外常有潮流深槽，是重要的海岸资源。深槽是潮流与泥沙平衡条件下形成的。有的长期处稳定状态，可利用为港口航道。如，江苏岸外分布着一片辐射状沙洲，沙洲区南北长 200 km，东西宽 90 km，水深 0～25 m，由 70 多条巨大的沙脊及沙脊间的深槽组成。其中如东县岸外的黄沙洋(小洋口)及启东县岸外的小庙洪(吕四)主槽稳定,可用于建港。

黄沙洋距岸 8 km,水深大于 10 m,最大水深 27 m,在蒋家沙与太阳沙之间,深槽多年来为潮流进出沙洲区的主要通道,而稳定不变。小庙洪距岸 5 km,水深大于 9 m,最大水深 20 m,南北有腰沙冷家沙等沙洲掩护,风浪较小。小庙洪是潮流及长江径流入海后沿岸向北的主要通道,水流在深槽中呈往复流最大流速大于 1 m/s,经多年的图件卫片比较,小庙洪及腰沙(浅滩)都相当稳定,在浅平的江苏海岸,岸外深槽是很可利用的。同样,渤海湾曹妃甸深槽是渤海北支潮流的主要通道,是可开发为深水港的重要的海岸资源。

开发江苏岸外海洋石油*

占地球表面70%的海洋，蕴藏着丰富的食物蛋白质、石油矿产和能源等资源。20世纪80年代以来，海洋工程即海洋资源的开发，与电子、宇航、信息及生物工程等，都已成为新技术革命的重要组成部分，而海洋石油资源是其中最引人注目的。

我国近海有宽阔的大陆架，海洋石油开发处在年轻期。经20多年普查勘探，发现了具有良好生储油条件的六大沉积盆地：渤海盆地、南黄海盆地、东海盆地、南海珠江口盆地、北部湾盆地与莺歌海盆地。发现油气田10多个。世界舆论普遍认为，"中国大陆架的海上油田是世界上屈指可数的原油宝库"。

南黄海盆地位于江苏岸外，是陆上苏北油田向海洋的延伸部分，盆地面积10万km^2。我国石油部估计，可采储量为2.9亿t。应该看到，这是与苏北油田衔接的我国的一项巨大的海洋石油资源，江苏拥有分享这份资源的天然权利。

在南黄海石油开发中，我省能参与哪些工作呢？

(1) 海洋石油开发需要巨额投资与先进技术，我国采取与外国合作方式，勘探开发中钻井、采油平台、海底管线等工程，需要海洋水文、气象、海底动力地貌与沉积、工程地质、地震灾害与水下工程等方面的科学资料。前些年，通过向外商反承包，已从外国投资的22亿美元中收回5亿美元外汇，并从工作中提高了我方技术水平。江苏有众多的大学、研究院(所)，有雄厚的海洋科技力量，应认真考虑参与这项开发海洋石油的海洋环境、地质研究工作，分享这项投资对我省技术、经济发展带来的巨大利益。

(2) 南黄海石油的开发勘探，需要大量的后勤供应，如作业船的停泊维修，各种器材供应、修理，中外作业人员的生活娱乐设施等。目前，初期的勘探基地在上海，但南黄海盆地近在江苏岸外，当大规模勘探开发时，再由上海为基地显然是不合理的、不经济的。江苏应为建立南黄海石油基地作准备，为引导南黄海油气资源从苏北登陆创造条件。

(3) 海洋石油开发是工艺复杂、技术密集的现代化系统工程，而我省经济发达，人口密集，能源缺乏，南黄海油气开发，可就近供应解决部分能源，并推动一批新型企业的发展。海洋石油及石化工业的兴起，将促进我省钢铁、冶金、土木建筑、造船、机械、运输、深海工程以及海洋调查勘探，海上救护、环保、海洋预报等一系列企业及技术的发展。这将为改变我省的产业结构和交通布局，推动沿海港口、城市发展，有效地摆脱苏北经济不够发达的状况，提供一个极为有利的契机。

为在江苏沿海建立南黄海石油基地及苏北沿海油气贮运、加工、石油化工基地，我建议：组织江苏科技力量，进行前期可行性研究。

建立海洋石油勘探开发基地、油气贮运加工及化工工业基地，需要解决深水港口及后方供应(城市)。根据南京大学多年的工作，我认为连云港及南通是合适的。连云港的长

---

* 王颖：《江苏政协》，1991年6月第7期，第22页。

处是已有港口及城市依托，条件较好，只是距油田距离稍远。南通有两处，如东的洋口港，利用辐射沙洲中最大的深水道——黄沙洋，可建立10万t级深水港。这是我国沿海可以利用天然水深建立大型深水港的八个深水港址之一，是十分难得可贵的深水资源。近年来南京大学与河海大学合作已为这一宝贵的深水资源做了许多勘探研究，并经国内专家论证，如东的黄沙洋最宜于开发大型石油港，沿岸有大片滩涂荒地，适宜建立化工基地及城镇。南通另一宝贵水域是启东县吕四岸外的小庙洪水道，可建设为5万t级深水港。南京水利科学院亦已做了许多研究论证工作。由此可见，江苏沿海是有条件为南黄海油气登陆、贮运、加工的。希望省及有关部门对开发岸外海洋石油能引起重视，并做好必要的前期准备工作。

开发南黄海海洋石油的建议*

江苏省是一个经济发达的大省，也是个海洋大省。有 1 000 km 的海岸线，众多的河口、水道与港湾，600 多万亩海涂（占全国 1/5），岸外有 2 万多 km 的浅海沙洲区。这巨大的沙洲及深水道是国内唯一、世界罕见的宝贵的浅海自然资源。江苏省岸外南黄海海底还蕴藏着丰富的海洋石油资源。这些海洋石油的开发，将会促进江苏经济进一步发展，特别会促进苏北经济腾飞。这里，我愿就开发江苏岸外海洋石油与振兴江苏经济问题提供一些情况与建议。

一、国内情况

占地球表面 70％的海洋，蕴藏着丰富的食物蛋白质、石油矿产、海水化学与能源等资源。20 世纪 80 年代以来，人们已确认，海洋工程、电子、宇航、信息与生物工程等成为世界新技术革命的重要组成部分。

所谓“海洋工程”是海洋开发利用的总称。海洋石油资源是其中最引人注目的。据法国石油研究所估计全球石油储量为 3 000 亿 t，海洋石油为 1 350 亿 t，占 45％。随着海洋勘探的增加，海洋石油储量的比例还会上升。自 1887 年北美加利福尼亚岸外数米水深处打了第一口海洋油井钻探开始，至 1946 年美国在路易斯安那州外水深十几米处成功地开采海洋石油。钻井数已达 3 000 多个，至今已有 100 多个国家勘探开采海洋石油。钻井水深已达 2 286 m（1983 年），海洋石油产量近 7 亿 t，约占世界石油产量的 30％。

海洋石油业的兴起与迅猛发展，不仅推动了传统的航运与渔业，并且带动一批新兴的产业，推动经济的发展。在一些国家已成为振兴国家的经济支柱。例如，挪威是传统的海洋渔业和海洋运输国家，自从北海石油开发兴起，政府收入的 55％～85％来自北海油田。英国在 20 世纪 70 年代以前是石油进口国，自开发北海油田后，现已每日出口石油百万桶，1982 年海洋油气产值为 241 亿美元，占当年英国总产值的 4％，在国际石油市场具有一定影响。正如英国前财政大臣佛里·豪爵士所说“北海石油帮我们偿还了外债，取消了外汇管制，并减小了税收和公债。总之，北海石油带来的好处是增加了可以自由使用的收入，帮助降低了利率以及缩小了通货膨胀率。使英国走上繁荣与振兴的大道中，北海石油正起着重要作用”。

二、南黄海石油资源

我国近海有宽阔的大陆架，海洋石油开发事业尚年轻。20 多年来的普查勘探，发现

* 王颖：《南京大学学报（地理学专辑）》，1992 年总第 13 期，第 188－190 页。

了具有良好生储油条件的六大沉积盆地：渤海盆地、南黄海盆地、东海盆地、南海珠江口盆地、北部湾盆地与莺歌海盆地，发现油气田 10 多个。1982 年美国总统能源顾问估计，中国近海石油可采储量为 70 亿～105 亿 t，国内估计，海洋石油地质储量为 100 亿～300 亿 t，可采储量 42 亿～100 亿 t。世界普遍认为"中国大陆架的海上油田是世界上屈指可数的原油宝库"。

南黄海盆地位于江苏岸外，是陆上苏北油田向海的延伸部分，盆地面积 10 万 km^2。地质部与石油部已对南黄海进行了一系列的地球物理与勘探工作。该盆地主要发育在早第三纪的古新世及始新世，巨厚的沉积层有良好的生油条件。在下第三系与上古生界地层中已见到油气显示，此外有上古生界与下三叠系海相生储油地层。南黄海盆地中多潜山并有上古生界良好的背斜构造显示，找油范围很广。据美国估计，南黄海石油可采储量为 2.2 亿 t(低值)、5.7 亿 t(中值)、21.6 亿 t(高值)，次于东海盆地，居全国第二位，据石油部门估计为 2.9 亿 t。这是与苏北油田衔接的我国的一项巨大的海洋石油资源。江苏省拥有分享这份资源的天然权利。

三、江苏省面临的任务主要有三项

(1) 海洋石油开发需要巨额投资与先进技术，我国与外国合作勘探开发中，钻井、采油平台、海底管线等工程需要海洋水文、气象、海底的动力地貌与沉积、工程地质、地震灾害与水下工程等方面的科学资料。前些年，通过我国向外商反承包，已从外国投资的 22 亿美元中收回了 5 亿美元，并从工作中提高了我方技术水平。今后若干年的勘探开发中，我国将耗资数百亿元。江苏有众多的大学、研究院所，有雄厚的海洋科技力量，应认真考虑参与这项开发海洋石油的海洋环境、地质等研究工作，分享这项投资将对江苏省技术、经济发展带来巨大利益。

(2) 南黄海石油的开发勘探，需要大量的后勤供应、作业船的停泊维修、各种器材的供应机修、中外作业人员的生活娱乐设施等等。目前，初期的勘探基地在上海，当扩大勘探及开采生产时，这后方陆上基地将形成庞大的工业城镇。南海石油的勘探开发已给广州及湛江两个基地带来繁荣，原先荒芜的湛江郊区，正形成新兴的城镇。南黄海盆地近在江苏岸外，当大规模勘探开发时，再以上海为基地显然是不合理的，也是极不经济的。江苏省应为建立"南黄海石油基地"作准备，着手于基地的选址，做好港口、道路、通信系统及生活设施等前期准备工作，为引导南黄海油气资源从苏北登陆创造条件。

(3) 海洋石油开发是项工艺复杂、技术密集的现代化系统工程。江苏省经济发达，农业产值高，但人口密集，能源缺乏，制约着经济的进一步腾飞。南黄海油气开发，可就近供应部分能源，并推动一批新型企业发展。海洋石油及石油化工工业的需要，将促进钢铁、冶金、土木建筑、造船、机械、运输、海洋工程以及海洋调查勘探、海上救捞、环保、海洋预报等一系列企业及技术的发展。这将改变江苏省的产业结构、交通布局，推动沿海港口、城市发展，有效地摆脱苏北经济不发达的状况，提供一个极为有利的契机。

四、建议：为建立海洋石油基地组织可行性研究

南黄海石油开发将是促进苏北经济发展的良机，不要把它看作仅仅是国家石油部门的事，要与江苏经济联系起来，积极参与南黄海石油的开发。我认为，在江苏沿海建立“南黄海石油基地”及“苏北沿海油气贮运、加工、石油化工基地”，已经到了可以组织江苏科技力量进行前期可行性研究的时候。

要建立海洋石油勘探开发基地和油气贮运加工及化工基地，这要取决于深水港口及后方供应（城市）。根据在南京大学多年的工作，我认为连云港和南通是合适的选择。连云港的长处是已有港口及城市依托，能满足作业船停泊及大吨位输油的需要，条件较好，只是距油田稍远。南通有两处可供选择：如东县的洋口港，利用辐射状沙洲区中最大的深水道——黄沙洋，可建 15 万 t 级深水港，这是我国沿海利用天然水深，可建 15 万 t 级深水港的仅有几个深水港址之一，是十分难得的深水资源。近年来南京大学与河海大学合作已为这宝贵的深水资源做了许多勘探研究工作，并经国内专家论证，如东县的黄沙洋最宜于开发大型石油港、海洋石油基地，沿岸有大片滩涂荒地，适宜于建立化工基地及城镇。南通另一宝贵水城是启东县吕四岸外小庙洪水道，可建设 5 万 t 级的深水港。南京大学 20 世纪 80 年代初在该区域做过勘测研究，南京水利科学研究院做过许多研究论证。因此，江苏沿海是有条件为南黄海油气登陆、贮运、加工建设良好基地的。

我认为，科技人员提出一些尚未被人们了解和重视的、但对国计民生有重大意义的问题，以引起政府领导与有关部门的理解、重视，这也是科技人员的责任与作用。科学建议也是科研活动的重要组成部分。

论发展中的深水大港
——连云港*

连云港地处黄海之滨，是江苏省长达1000多km海岸带唯一的基岩港湾海岸。港阔水深，避风与泊稳条件优良。多年的调查研究、重复观测、模拟试验与扩建港口实践，充分说明连云港是具有巨大潜力的深水良港。它既为中原大地与黄土高原架起了通向海洋的桥梁，亦为贯通欧亚大陆、联通太平洋与大西洋奠定了新起点。其地理位置、历史文化渊源、资源环境特点在政治与经济发展、国内外贸易交流等方面皆具有重大的价值与优越性。

通过短期的专题考察并结合既往的工作实践，拟就港域环境及城市发展进行研讨。

一、港域边界的稳定性

连云港位于华北板块南部的苏鲁隆起，是长期上升剥蚀区。后云台山与东西连岛由上元古界的云台组构成，系一套浅变质的白云二长变质岩、白云石英片岩等，局部有薄层绿泥石云母片岩。前震旦纪的基底岩层经多次变质作用与构造运动，表现为各类断裂、褶皱、片理、片麻理的相互结合、叠加与改造。港区周边的山地呈块状岩体，岩石抗压强度大（>1000 kg/cm^2），工程地质条件优良[1]。地区构造以NE向构造（断裂、褶皱、片理等）为主。但从后云台山与东西连岛的构造地貌分析，连云港海峡系北西西向的地堑，与北西向及东西向的构造运动关系密切。估计系印度洋板块向欧亚板块俯冲时，促使中国板块呈块状向东滑动时所形成[2]。总之，时代较新，控制着地貌的基本格局。后云台山与东西连岛为断块上升部分，仍保留着一系列抬升的古海岸遗迹[3,4]，但是，以数千万年的地质时期变动与人类历史时期及当前港口的建设发展相比较，构造抬升量是微乎其微的。历史记录未发现连云港地区发生过Ⅶ度以上的地震烈度现象。

连云港背依后云台山，面向东西连岛，底辖海峡通道（见图1），港域边界的稳定性可从三方面分析。

（1）后云台山北坡，系一断崖陡坡以狭窄的基座与海港相接。地形表现为危崖耸立，但地貌结构是稳定的。因为陡崖至今仍保留着断层崖的三角面状态，亦未出现阶梯状的滑坡体；坚硬的结晶变质岩，岩性均一致密，呈块状，不具有形成巨大滑动面的厚层软弱岩层；山地岩层的倾向SE15°与陡崖方向相反，向陆地倾斜而山坡植被茂密，不可能形成反向滑坡而影响海港；崖麓无现代水流掏蚀作用或人工挖入，基底是稳定的；冰后期海侵以来，该海岸陡崖几乎无浪蚀后退之表现。所以这种海岸陡崖或者说断层海岸带对于人类历史时期来说，是稳定的[5][6]。该处不具备形成垮山或大型崖崩而毁及海港的自然条件。

---

* 王颖：《南京大学学报（连云港市环境评价和海港发展专辑）》，1992年12月，第31－38页。

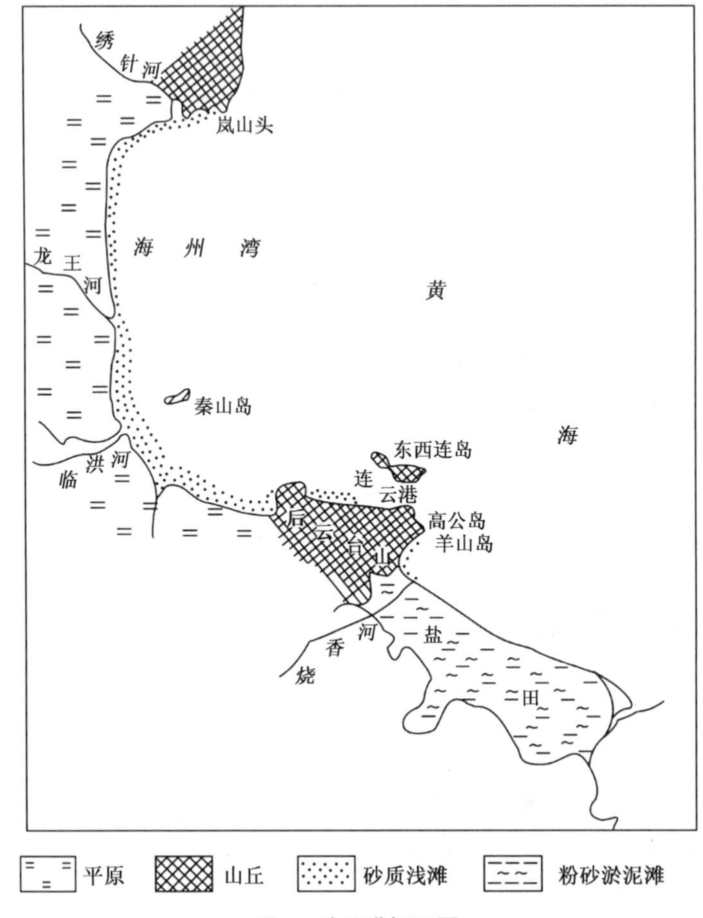

图 1　连云港概况图

其次，泥石流发育于气温较差、降水强度变率大，具有大量松散堆积层的地区，如黄土堆积、火山灰堆积、冰川外围融冻堆积区及厚层风化岩屑堆积区，当暴雨洪流时，会形成泥石流呼啸流下，在泥石流频繁爆发地常形成泥石流冲沟。连云港气候温和、湿润，山地植被茂密而坚岩难以形成大量风化堆积物，北坡陡崖区未发育大型沟道，因此，没有形成灾难性泥石流的条件。

后云台山下的荷花尖一带，有一坡积裙发育于山麓基岩平台上。锥状地形平缓，已有建筑物，锥体两侧有天然流水沟槽，该处需注意排水通畅。但是，即使该堆积体滑动，其活动范围亦很有限。

所以，后云台山系稳定的石山，不致形成巨大的、灾害性的垮山滑坡。只是由于片麻岩地区，岩层节理丰富，需注意对局部风化危岩的处理，防止其崩落对道路建筑的局部伤害。

(2) 东西连岛南坡，植被茂密，除东、西两端有海蚀崖发育外，无大量碎屑堆积于坡麓。仅航修站处由于挖坡筑路，形成松散堆积的坡积体。由于系砂砾质，透水性强，而底部系坚硬的片麻岩，上下层岩性基本一致，且堆积体范围不大，不致形成危害性滑坡。连云港外侧的东西连岛南坡，现为稳定的基岩山坡，今后进行连岛开发时，需注意保持植被并防止在坡麓开挖土方。不良的施工，会影响山坡的稳定性，而形成新的泥沙源而增加港口的淤积量。

(3) 港域海底分布着厚约 10～30 m 的淤泥层，施工时，需注意清基与处理好软土地基。20 世纪 30 年代中期，荷兰人修建 1 号码头，没注意清淤坚基。后由于上部建筑的重力曾发生滑动，经过打钢板桩与加固护坡，自 70 年代至今，1 号码头未发生蠕动而保证了使用。连云港后期施建的码头与堆场均注意爆破、清淤与加固软基。因此，港域海底局部的蠕动与滑坡是可通过相应的工程措施加以防止的。

以上分析了几种不同类型的边坡形变、危害程度及其在连云港区发生的可能性。所有的情况均表明，连云港港域的边界是稳定的，具有良好的工程地质与海岸地貌条件。

二、港口回淤分析

连云港为一向海突出的基岩山地与海岛，形同一巨大的岬角，分隔了北部岬湾曲折的砂质海岸与南部的淤泥质平原海岸；连岛外侧水深浪大、岬礁丛生，因此，港口所在的基岩山地与岩岛实为该区海岸的泥沙分流点，分流点意味着泥沙的越过少，其两侧泥沙运动分别属两个体系。

海岸带泥沙来源主要有三种：河流供沙、海蚀岸供沙与海底供沙。连云港区由河流直接供沙的数量不大，主要是海蚀岸段与海底供沙。

(1) 海峡东侧，泥沙主要来自起源于废黄河口冲刷岸段的海底掀沙。主要是悬移质的粉砂质黏土，呈沿岸泥沙浑水流状自东南向西北方向移动。这段泥沙流主要部分被高公岛与羊山岛组成的岬角岸段所阻隔，泥沙堆积于烧香河口及其东南的海湾中，形成粉砂、黏土质潮滩，构成淮北盐田的物质基础；部分外滨漂沙被涨潮流搬运至连云港区。据估计，在 0～10 km 的范围内，从废黄河口北岸向海输沙总量为 3 500 万 t/a[7]。华东师大河口海岸研究所工作成果认为废黄河口岸段冲刷量为 2 500 万 m^3/a，连云港泥沙主要来自废黄河口北岸冲刷岸段。估计海峡内东向来沙每年约 400 万 t[7][8][9]。这些资料与上述的海岸地貌结构的分析是一致的。

(2) 海峡西部供沙主要是粉砂与细砂，质地较东侧来沙粗，含较多云母矿物，物质来源于连岛西岬海蚀岸段、竹岛与棺材山海蚀岸段，以及岸外海底浅滩。风浪水流与潮流将泥沙自西向东，向海峡港域输送。西侧供砂量据连云港建港指挥部测计为 900 万 t/a[7]。

(3) 东西连岛冲刷岸段及冲沟泥沙汇入海峡的泥沙，质地粗，主要为粗砂与中砂，堆积于凹入岸段形成新月形海滩。东西连岛北岸的小海湾中现代海滩以上，5 m 高处的细砂质高海滩，抬升的古海岸，泥沙来源于当时的海蚀岸段与海底两个方面。

现代连云港泥沙源以悬移质泥沙为主，底移质泥沙为辅。悬移质泥沙的沙量是不大的。据江苏省海岸调查资料：连云港表层海水含沙量为 30～50 mg/L，底层海水含沙量为 30～100 mg/L；连云港以南到长江口的淤泥质平原海岸，其表层海水含沙量为 100 mg/L，底层海水含沙量＞200 mg/L[1]，连云港海水中悬沙含量较南部海水含沙量至少少一倍！

近期卫星照片图相清楚地反映出：漂沙浑水主要分布于连云港外围两侧的凹入湾岸内。所以，连云港位于突出岩岬岸段，居泥沙分流点，海岸自然来沙量少，港域有东西连岛天然屏障，挡风避浪，形成“水深、浪静”的停泊条件。自然界中，海深处，浪多大。这两个矛盾因素在一定条件下结合起来，才能形成优良港湾，连云港是具有此良好自然条件的，具有广阔发展前景的优良港口。

建设中的西大堤将阻断海峡，使陆地与东西连岛相连，形成弯臂式巨型防浪掩护堤，其作用明显：

(1) 挡住 NE 向强浪，突出了东西连岛的泥沙分流点作用。

(2) 扩大港域成为水域达 30 km^2 的港湾，可供各种类型船舶停泊。在东侧深水岸段可建 10 万 t 级码头。

(3) 改变了东西两侧潮流在海峡港域形成会聚之条件，减少了由于水流汇聚而形成的泥沙落淤。

东侧的泥沙以黏土质为主，其粒径小于 0.001 mm 或大于 5ϕ，而海湾内向东的落潮流速约 0.5～0.75 m/s。大于 1 kn 的流速是有足够力量搬运粉砂黏土质泥沙于港外[6][7][8]，所以筑堤与连岛合围屏蔽港口是个好办法。我国有此类港口的例证，如秦皇岛老港建于南山岬矶头上，以弯臂式防波堤掩护，阻挡了东北向强风浪、而并不阻挡沿岸水流、同时强化了矶头的泥沙分流点作用，以两侧海湾拦积泥沙、确保港口经久不淤。秦皇岛港是成功地利用海岸特点的范例，缺点是港域面积太小(图 2)。

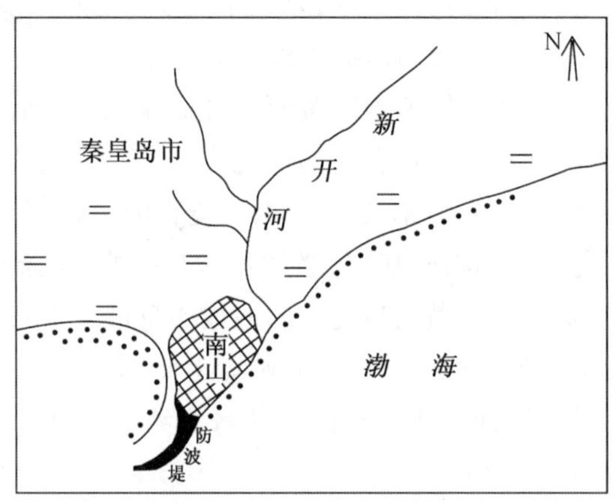

图 2　秦皇岛老港图

连云港港口研究成果认为，港口所在的海岸段是处于冲淤动态平衡而略有刷深，与南部废黄河口海岸冲刷趋向渐缓(岸坡塑造成熟)有关[7]。基于港口已趋蚀积动态平衡的结论，笔者认为：人工建筑改变了原港口环境特点后，必会引起海岸剖面的重新调整。及时了解掌握海岸自然调节趋势，有针对性地布置港口工程，则将有利于促进港口环境呈良性发展。例如：自 1980 年修建西大堤以来，在东西连岛南岸的西大堤根部两侧，已形成新的“入射角填充”堆积(即人工堤上游方向的泥沙被阻而于堤根处堆积成浅滩)，堤东侧的浅滩宽约数百米，物质较细构成潮滩；堤西侧浅滩略窄，而物质颗粒比较粗(表 1)。浅滩质地的不同，不仅反映堤内外动力环境之差异，而且反映出港域东西两侧的不同物质来源。这结果与前节所述的泥沙来源分析相一致。

表 1 西大堤北端两侧浅滩沉积物粒度分析

样 号		L9201	L9202	备注
取样位置		连云港西大堤北端东侧	连云港西大堤北端西侧	1992 年 7 月 22 日下午低潮时采样 采样人：P. Martini，王颖 分析人：王雪瑜，许叶华
各粒级含量%	2～1 mm			
	1～0.5 mm		1.0	
	0.5～0.25 mm		10.3	
	0.25～0.125 mm	4.2	38.9	
	0.125～0.063 mm	6.2	22.6	
	0.063～0.032 mm	7.2	7.2	
	0.032～0.016 mm	2.0	2.0	
	0.016～0.008 mm	11.2	1.0	
	0.008～0.004 mm	13.2	3.0	
	0.004～0.002 mm	2.0	14.0	
	0.002～0.001 mm	8.5		
	<0.001 mm	45.5		
中值粒径 Md_{Φ}		9.5	3.1	
分选系数 QD_{Φ}		2.05	1.30	
密度 SK_{Φ}		−1.0	0.7	
粒级含量	砂	10.4	72.8	
	粉砂	33.6	13.2	
	黏土	56.0	14.0	
定名		粉砂质黏土	细砂	

今后西大堤两侧浅滩还会有所淤长，但是淤积速度会减缓。当西大堤建成后，虽留有通道，但港口环境从海峡转为潮汐汊道型海湾，港口的动力条件与自然发展模式将会改变。需注意维护港口的水域面积以保证相当数量的纳潮量，利用落潮流的天然冲刷力维持航道的水深；港内原有的防波堤的掩护功能改变了，需注意码头与建筑物的方向，避免成为阻滞水流而促淤的丁坝，建筑型式宜随着变化的环境条件而相应地调整；防止港口水域与底土的污染。例如，海头湾、望海楼沙滩泳场的环境保护等。据 1992 年南京大学于港区海底取样的初步结果，海域底上的重金属含量仍低于或接近于北太平洋海域的背景值，但 DDT 类的含量有明显的富集。因此，需重视开展港域水、沙、底土与浅滩沉积速率的重复观测，使海港的扩建工程适应海岸环境的自然特性，以保证港口发展的良性循环。

三、对城市发展的建议

连云港市位居中国海岸的中部，境内岛、山、海、河风光优美且资源丰富，历史文化源远流长并具有发展交通的巨大潜力：海、陆、空结合。因此，在新的改革开放形势下，宜发挥市域特点制定出新的经济与社会发展规划。20 世纪内，连云港市可望发展成为以交通为动脉，以经贸、农工和旅游业为特色的国际港市。规划重要，人才急需，应重视引进国内外资金，同时，要做好投资环境与资源优势的宣传，以唤起国内的共识。

连云港市的海岸类型丰富，在一日之内可访遍北部的沙滩、丛林、碧水激浪的砂质海岸；中部的连岛、岩岬、海湾新月海滩及深水悬崖海岸；南部的盐田、苇滩、野禽、虾塘以及水天一色的潮滩海岸。赏心悦目、陶冶于清新空气与矿泉沐浴，品赏海珍、河鲜、稻粟、生果之味，集博览、调剂身心于一体，充分达到休息与焕发精神之目的。连云港市山地原为海岛，直至近万年来才成为陆地，山地与岛屿上保存着一系列的古海岸遗迹，分布于海拔高度的 5 m、10 m、40 m、100 m 或 120 m，200 m、320 m 至 600 m 高处[3,4]，古海蚀崖、海蚀龛与古岩滩之组合，保存完好，无疑为天然的地质博物馆。加之，新旧石器时代的遗址、4 000 年前的岩画等，集史、地珍迹于一地，为旅游增加了科学内容，具有吸引国际会议集团性旅游的宝贵资源。因此，连云港的旅游事业更具特色：具有田园风光的海滨港市；具有集学术考察、海洋文体活动与疗养休息于一体的度假胜地。必须重视市域全面规划，进一步加强文化、科学教育，提高现代化的服务设施，赋予更多的经济与政策活力，以期待这颗海岸明珠熠熠生辉，与其他港市相连，组成群星灿烂的锦绣中华海岸，在二十一世纪的太平洋地区发挥举足轻重的作用。

参考文献

[1] 任美锷主编. 1986. 江苏省海岸带和海涂资源综合调查(报告). 北京：海洋出版社.

[2] Zhang Yuchang, Wei Zhili, Xu Weiling, Tao Ruiming, and Cheng Ruigen. 1989. The North Jiangsu—South Yellow Sea Basin. In: Chinese Sedimentary Basins (Chapter10). Elsevier. 108 - 123.

[3] 朱大奎. 1984. 江苏海岛的初步研究. 海洋通报，3(2)：36 - 46.

[4] 王颖. 1983. 关于海岸升降标志问题. 南京大学学报(自然科学版)，(4)：745 - 752.

[5] O. K. 列昂节夫. 1965. 海岸与海底地貌学. 北京：中国工业出版社. 92 - 104.

[6] Ying Wang and David G. Aubrey. 1987. The Characteristics of the China Coastline. *Continentel Shelf Research*, 7(4): 329 - 349.

[7] 连云港建港指挥部. 1978. 连云港基地资料汇编.

[8] 朱大奎，许达官. 1982. 江苏中部海岸发育和开发利用问题. 南京大学学报(自然科学版)，18(3)：799 - 818.

[9] 朱大奎，高抒. 1988. 江苏淤泥质海岸剖面研究. 南京大学学报(自然科学报)，24(1)：66 - 84.

Coastal Development and Environmental Protection in China*

Introduction

Mainland China has a 18 400 km coastline. The total length of the country's coasts, including those of its 6 500 islands, is 32 000 km. The physical features along the Chinese coastline were formed by monsoons, Pacific tidal waves and currents, and by rivers including the Yellow (Huanghe), Yangtze (Changjiang) and Pearl (Zhujiang). At the mouths of rivers are deltas and estuaries.

The coastal areas are of various types. In northern China is a vast region of coastal plains, with sandy beaches and barriers and tidal flats. Southern China has rocky coasts with many bays, as well as coral reefs, mangroves, etc. (Wang and Aubrey, 1987).

The ways to define coastal zones seem to vary among different countries, depending on their development and administrative needs. In China the coastal zone is considered to extend 10 km inland, and from the seashore 15 - 20 metres underwater. The boundary may be extended landwards or seawards in areas where there are steep rocky coasts, large estuaries, off-shore islands, or fields of sandy underwater ridges (Su, 1990). Such a definition fits the physical-geographical designation of a coastal zone as an area of land-sea interaction. It is also appropriate in relation to regional historical changes, economic development and administration.

There are eight provinces, two cities and an autonomous region in the Chinese coastal zone. From north to south, they are: Liaoning Province, Hebei Province, Tianjin City, Shandong Province, Jiangsu Province, Shanghai City, Zhejiang Province, Fujian Province, Guangdong Province, Hainan (Island) Province, and the Guangxi Autonomous Region (Figure 1). The coastal zone makes up 14 per cent of China's total area. It has a developed economy and dense population, with an average of 386 persons per sq km. Forty-one per cent of China's population lives here. This zone contains 70 per cent of the country's urban areas and accounts for 56 per cent of total output value.

---

* Ying Wang: *UNEP Industry and Environment*, 1992, Vol. 15, pp. 7 - 10。由陈定茂翻译后以中文刊登于该期刊中文版《产业与环境》, 1993 年 15 卷 1~2 期, 7 - 10 页。

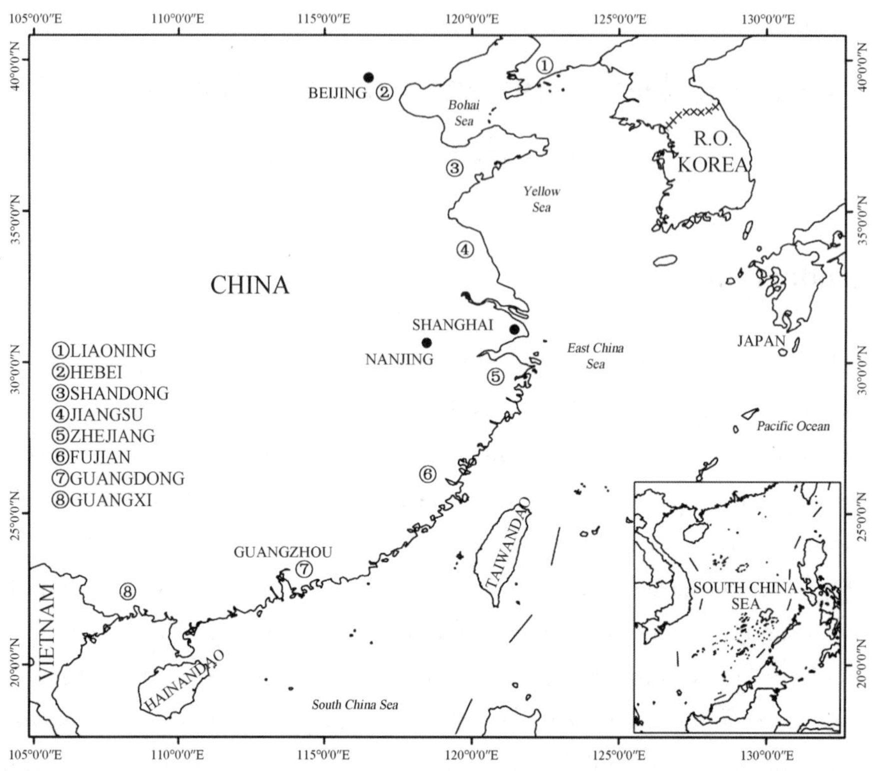

Figure 1　The coast of China

According to this delimitation of the coastal zone, its total land area is about 350 000 sq km, of which 6 per cent is tidal flats. The flats are an important new land resource. Sea salt production and the salt-chemical industry are the important traditional means of exploitation. Annual production of sea salt, perhaps the largest in the world, is 15 million tons. Marine fishing and mariculture in China yield 4.5 million tons of products per year, with mariculture representing about one third of the total.

The number of coastal harbours has increased to 253. These harbours can handle 5 billion tons of cargo a year. By the end of 1990, China had a fleet of ocean-going commercial ships with a capacity of 9.6 million tons.

Proven offshore petroleum reserves are about 20 billion tons, and annual crude oil production is 5 million tons.

Economic development of the coastal region has brought a series of new environmental problems, which have directed the attention of the public and the government to the importance of coastal management and environmental protection.

Coastal zone development and environmental pollution

Coastal zone development in China only began in the early part of this century. The coast had been a wild and forbidding area during the Qing Dynasty.[1] The government controlled fishing and salt production. Thus, the coastal zone except at some river mouths was a sparsely populated area separating the land and the sea.

Large-scale development began in the late 1970s as a result of the open policy of the Chinese government. As this was at the early stage of coastal management and environmental protection work, many new problems and/or contradictions emerged from the development of the coastal region:

a) Generally speaking, the sea along the Chinese coast is not very polluted. Recently, however, pollution (mainly in the form of oil, Hg, Cd, Pb, Zn, As, volatile matter, cyanides, organic matter, etc.) has been increasing year by year. Pollutants come largely from industries, agriculture, urban domestic wastes, harbour and ship wastes, and offshore oilfields. Pollutants from industries make up about 80 per cent of the total, urban domestic wastes about 18 per cent, and other sources about 2 per cent. Every year over 4 billion tons of polluted water enters the China Seas. The Bohai, Yellow and South China seas each receive 20 per cent of the total amount, while the East China Sea receives 40 per cent. The pollutants may concentrate in estuarine areas, and in some cases they cause eutrophication and red tide. Much attention has been paid to the increasing oil pollution in harbours, from offshore oilfields, and near coastal cities.

b) Dams have been constructed at many river mouths to prevent salt water intrusion and protect fresh water resources. These dams cause serious siltation in the lower part of the river mouth. They also prevent fish, prawns and crabs breeding upstream from reaching the river mouth. As a result, local populations lose income from the decrease in silverfish (Jiyunhe, on Bohai Bay), crabs (Jiangsu, along the Yellow Sea), etc. and may also be flooded, as in the summer of 1991 (Zhu, 1986).

c) Due to overexploitation of underground water in the coastal zone, as at Laizhou (莱州) and Longkou(龙口) in Shandong Province, the water table has been lowered and salt water intrusion has occurred more than 100 km inland. There is thus a lack of good quality water for drinking, irrigation and industrial use, so that large areas of cultivated land have been abandoned.

d) Reclamation of tidal inlets has decreased tidal prism,[2] altered the natural flushing pattern, and resulted in increasing siltation of the navigational channels of deepwater harbours. For example, at the harbour of Sanya, in the southern part of Hainan Island, the total area of the tidal inlet has decreased two-thirds owing to reclamation, from 4 140 000 to 1 390 000 square metres, and the tidal prism has decreased almost 50 per cent from 4 900 000 to 2 530 000 cubic metres. Consequently the harbour, which was free from channel siltation around ten years ago, must be dredged annually (Wang et al., 1987).

The conflicting claims of resource utilization and environmental protection are becoming more and more evident, especially in the vicinity of estuaries, coastal bays, harbours and coastal cities. The development of coastal industry and agriculture causes pollution, and the once excellent quality of the coastal environment deteriorates. More attention should therefore be paid to systematic environmental protection and management.

Figure 2　Ten years ago, this commerical centre in Haikou City was a wasteland

Protection and management of the coastal zone

Projects have been carried out with a view to increasing exploitation, and improving protection and management, of coastal resources and environments.

1. Surveys

Figure 3　The new harbour in Hainkou City, Hainan Province

Between 1980 and 1986, a comprehensive nationwide survey was made of tidalflats and of the entire coastal zone of the mainland. Environmental factors, natural resources and socio-economic conditions were considered. More than 500 units and 10 000 people were involved in the organization of multi-disciplinary survey teams. Thirty percent were senior scientists from universities, institutes and governmental departments. Specific disciplines involved included oceanography, meteorology, geology, geomorphology, marine biology, marine chemistry, environmental protection, vegetation and forests, soil, land use, and social economics. Areas studied include land resources, biological resources, salt and salt-chemical resources, mineral resources, marine energy, harbours, the tourist industry, etc.

As a result of this survey, a large amount of data have been gathered. These provide a scientific foundation for further research on coastal areas, exploitation of resources, and environmental protection and coastal management. They can also stimulate public interest in the coastal zone's environment and its exploitation.

Since 1988 another nation-wide project has been carried out: the Sea Island Survey. It covers the environment, natural resources and socio-economic conditions. The project will be completed by 1993.

Figure 4 Dongzhaigang Mangrove Natural Reserve

The above two projects will be an important basis for economic development, environmental protection, and management of the coastal zone.

2. Experiment and exploitation

China has drawn up a plan for regional exploitation of coastal areas, ranging from systematic small-scale to regional. This experiment began in 1982 and was confined at first to agriculture. Then, gradually, multi-purpose experiments were made including the renovation of small and medium-sized harbours.

At present, there are 42 coastal experiment stations throughout China. The experiment items include:

1) sea water cultivation;

2) marine farming (cultivation of prawns),

3) amelioration of littoral sandy wasteland and saline soil;

4) selection of sustainable plants;

5) planting of forest and fruit trees;

6) raising of livestock and poultry; and

7) harbour renovation and coastal protection.

A number of good results have been achieved through experimentation. For example, Jiangsu Province has developed a large area of tidal flats (about 400 000 ha) for agriculture and forests. In Liaoning Province the coastal zone is being exploited systematically for farming scallops in the intertidal zone, and for using littoral waste areas and the bay head area to grow rice and fruit.

Economic development improves the standard of living of the local population, and changes the traditional conception of the coast as "bitter water and saline soil" to that of a three-dimensional operation encompassing the claiming of new land for aquaculture, agriculture, forests, animal husbandry and trade. Several units that were part of the coastal survey have since been reorganized as consultancy firms or technical service centres: this is part of a new pattern of coastal development and management.

3. Administrative organization

In China there is a State Oceanic Administration (SOA, equivalent to the NOAA in the United States) under the central government, with three branches of Sub Bureaus (the North Sea, the East China Sea, and the South China Sea sub-bureaus). Since 1986, following completion of the coastal survey project, each province has set up a provincial oceanic administration for the management, research, planning and exploitation of coastal and shallow sea areas. There is also an Oceanic Environmental Monitoring Organization for protecting the coasts and seas of the whole country.

Following the coastal development of the 1980s, China has set up a series of special economic zones and high tech development zones all along the coastal region. Twenty-four harbor cities are open to the world. In 1988 the second largest island of China, Hainan, was designated as Hainan Province in order to accelerate its economic development too. It is open to the world as a new special zone. All of the special economic zones have preferential trade and traffic rights.

In 1963 China set up its first natural reserve, Snake Island, for the protection of the Pattas pit viper. The island is located off the Dalian coast in Liaoning Province. Today there are 45 national natural reserves in coastal areas, with a total area of 13 500 sq km, covering most coastal environments and ecosystems. These include Changli (昌黎) Golden Beach Natural Reserve (in Hebei Province), Laotieshan (老铁山) Island Migrating Birds Natural Reserve (in Liaoning Province), Qidong (启东) Spartina Pasture Natural Reserve, Yancheng (盐城) Red-Crowned Crane Reed Flat Natural Reserve, Dafeng(大丰) Davis Deer Natural Reserve (in Jiangsu Province), Nanji(南麂) Island Oceanic Natural Reserve (in Zhejiang Province), Shankou(山口) Mangrove Coast Ecosystem Natural Reserve (in Guangxi), Dazhou (大洲) Island Oceanic Ecosystem Natural Reserve, Sanya(三亚) Coral Reef Natural Reserve, Lingshui(陵水) Macaque Island Natural Reserve, Dongzhai(东寨) Mangrove Forests Natural Reserve, etc. The Central Government supports these reserves with special staffs and regulations for protecting wildlife resources and the natural environment.

A successful experiment has been carried out for hibernating birds. There are only around a thousand red-crowned cranes in the world. During the cold season, from October to March, more than 600 of them dwell in the Yancheng Reed Flat Natural Reserve. This vast wild wetland, full of fish who live in fresh or brackish water, is an ideal habitat for these and other types of wild birds.

Among scientific associations, there are the Coastal Exploitation and Management

Research Group and Marine Law Society, under the Chinese Society of Oceanography. A Special Committee of Marine Geography under the Chinese Geographical Society, as well as a State Pilot Laboratory of Coast and Island Exploitation, have been set up at Nanjing University. The university has sponsored a series of international symposiums to improve knowledge of coastal zone research and coastal management.

4. Legislation

The development of coastal investigation and utilization activities has gradually created more and more conflicts of interest between professions, departments and localities. Thus a special law on coastal management was urgently needed. In the summer of 1982 such legislation began to be formulated. After nine revisions, the *Regulations Regarding the Coastal Management of China* were drawn up and submitted to the State Council for approval.

Jiangsu Province, where Nanjing University is located, was the first province to complete its coastal survey. As early as 19 November 1985 this province issued the *Provincial Coastal Zone Management Regulations of Jiangsu Province*. Jiangsu is the first province to have put the coastal management regulations into effect. Many other provinces have also formulated such regulations since 1986 (Ren, 1990; Zhao, 1990). This situation indicates that in China progress is being made towards the development of comprehensive, systematic regulations and legislation for management of coastal areas.

By the end of 1990, the laws related to coastal management were as follows (Li. 1990):

Regulations Regarding Environment Protection on Offshore Oil Exploration and Exploitation, 1983;

Regulations on Sea Water Management for Prevention of Ship Pollution, 1983;

Regulation of Waste Dumping into the Sea, 1985;

The Law of Fishery, 1986;

The Law of Marine Traffic Safety, 1983;

7he Law of Mineral Resources, 1986;

The Law of Land Management, 1986; *and*

The Law of Water, 1987.

These laws contain provisions for the management of coastal and ocean areas. To set up a series of coastal and ocean reserves has required the enactment of a law on marine environmental protection. Increasing attention is being given to the legal aspects of protecting and managing the coastal zone, using systematic theory and methods concerning the relationship among the various resources, technical possibilities and economic values to solve coastal development problems.

We have benefited from international academic exchanges on coastal management.

These will enable us to manage our coastal areas in a way that is suitable to Chinese conditions. Knowledge of any successes or failures in China may also be useful for other countries making efforts to manage the development of their coastal zones.

Notes

1. The Qing Dynasty's 296 - year rule ended in 1912.

2. "Tidal prism" is the amount of water that flows in or out of a coastal inlet as the tide rises and falls, not including entering freshwater.

References

[1] Li, Dechau. 1990. A comprehensive study on coastal zone management: the traditional way and the modern way. In: Comprehensive Management of the Ocean. Science and Technology Press of Guangxi. 135 - 144.

[2] Ren, Mei-e. 1990. The problems on coastal management of China. In: Comprehensive Management of the Ocean. Science and Technology Press of Guangxi. 124 - 134.

[3] Su, Shenjien. 1990. The comprehensive review on the seven years or national coasts and tidal flats investigation. In: Comprehensive Management of the Ocean. Science and Technology Press of Guangxi. 218 - 224.

[4] Wang, Ying, and Aubrey, D. 1987. The characteristics of the China coastline. *Continental Shelf Research*, 7(4): 329 - 349.

[5] Wang, Ying, Schafer, C. T. and Smith, J. 1987. Characteristics of tidal inlet designated for deep water harbour development, Hainan Island. In: Proceedings of Coastal and Port Engineering in Developing Countries, Vol. 1. China Ocean Press. 363 - 369.

[6] Zhao, Enpao. 1990. Administrative divisions of coasts and the adjacent seas. In: Comprehensive Management of the Ocean. Science and Technology Press of Guangxi. 187 - 195.

[7] Zhu, Dakui. 1986. The utilization of coastal mud-flat of China. *Scientia Geographia Sinics*, 6(1): 34 - 40.

Effect of Military Activities on Environment in Eastern and Southeastern China*

Several unique historical examples in China relate to actual impact of the military activities on the environment.

The Yellow River (Huanghe), one of the largest rivers in China, is unique for its extremely heavy sediment load (11.9×10^8 tons · y^{-1}), rather small water discharge (485×10^8 $m^3 y^{-1}$) and migrated river channel in the lower reaches (Y. Wang et al. 1986). The river has experienced eight major changes in its lower course since 2278 B.C., discharging either into the Bohai Sea, or into the Yellow Sea via the Huai River. The shifted distances between North and South were more than 600 km long (Figure 1). Two major migration were created artificially as a result of military activities.

(1) In A. D. 1128, November 15th, Du chuen, the Capital (Kai Feng) official of the Southern Sung dynasty, excavated the Yellow River bank in order to use flood water as a weapon to prevent an invasion of Jin dynasty soldiers from the north (Sung history (475), Jin history). Even though this effort was unsuccessful, initiated a change that caused the Yellow River to flow through the Si River to the south. Since then, the river shifted often in the Huai River area as a disaster for about 250 years.

(2) May 1938, the military committee of Nationalist ordered the Chinese army to excavated the Yellow River bank between Zhenzhou city and Zhongmu county to force the river flowed to the southeast, in an effort to stop the Japanese army from advancing into the western and southern parts of China. After twice explosion during June 2nd to 9th of 1938, the river rushed out of channel in the place named Hua Yuan Kou of Henan province, and resulting flood developed rapidly. It flooded 44 counties in three provinces (Figure 2), 12.5 million people died as a consequence of that operation (The Water Conservancy Committee of the Yellow River, 1982). Even though, the river returned to its original channel in 1947. The flooded area (54 000 km^2) was completely changed from one of fertilized farm land to wild badland with sand dune field and saline alkali soil depressions. The natural river system of the Huai River was destroyed and the whole area suffered from disasters of sand storm, mobile dunes, drought and waterlogging often. Local people lost their houses and land to be refugees to the neighbor provinces for several decades. It has taken 40 years to reestablish conditions suitable for farming.

---

* Ying Wang: *Annual Science Report—Supplement of Journal of Nanjing University*, 1994, Vol. 30(English Series 2), pp. 43 - 46.

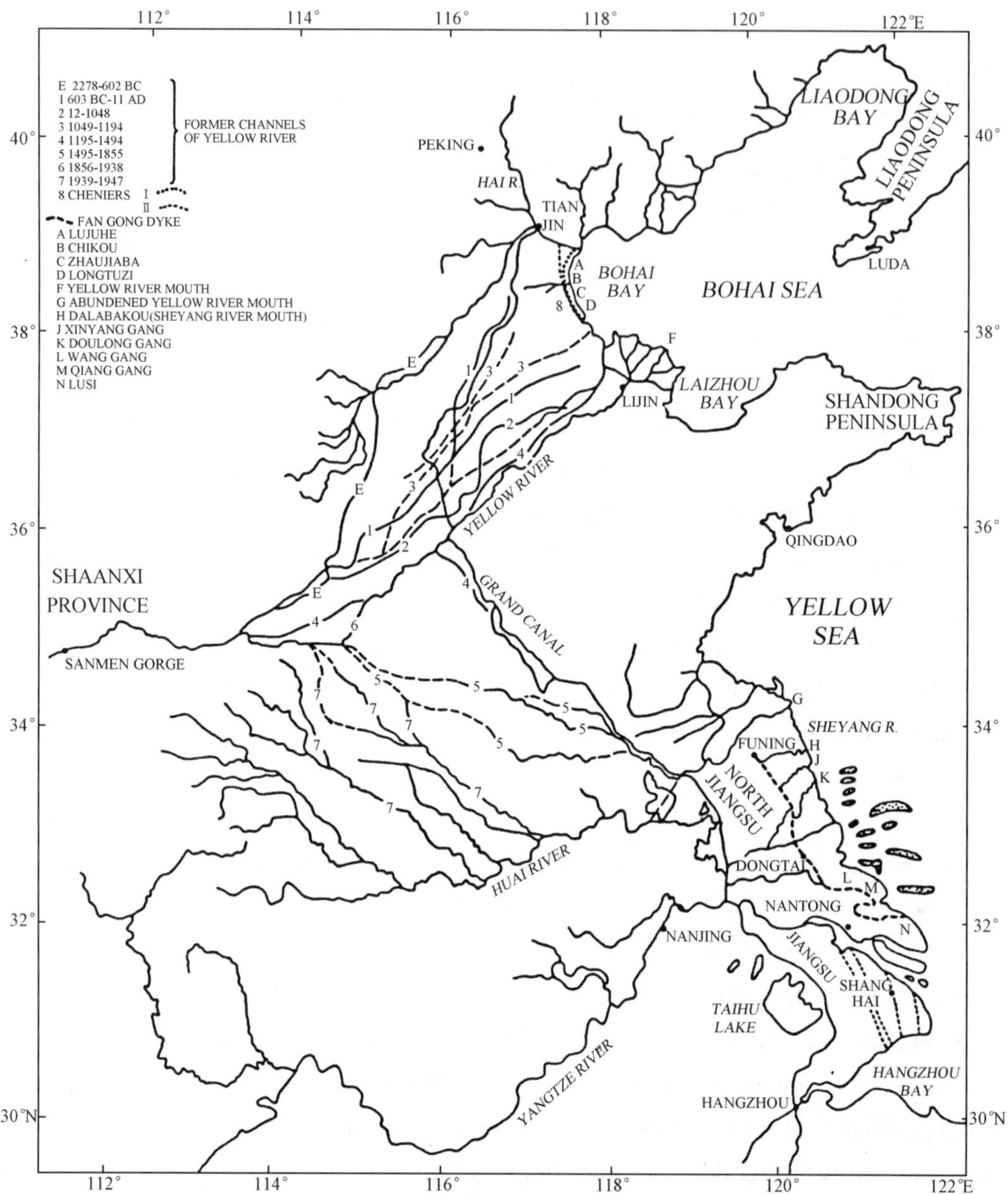

Figure 1　The map of migrated courses of the Yellow River

Such as, to set up drainage system and irrigation network, to plant forests and grasses for fixing the dune sands etc. Now, the abandoned Yellow River area becomes one of the most productive wine fields and fruit base in north China. However, the natural environment has been changed completely since the military operation in 1938. Using flood Yellow River to repel an invasion had been repeated twice in Chinese history, and had also been failed twice, but have had influences over environment lasting for long time plus the loss of huge amount of property and lives of the native. People should be smart to learn from past mistakes and avoid future ones.

North Jiangsu plain had been previously divided by numerous canals and channel

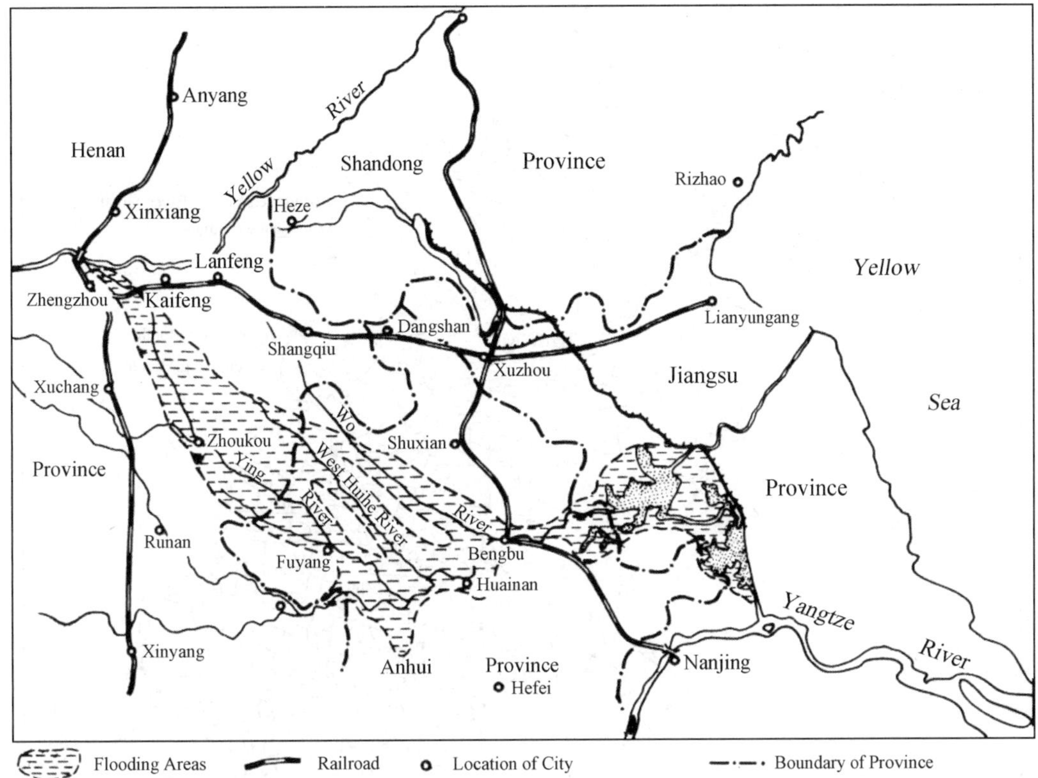

Figure 2　The flooded area of the Yellow River in 1938 diversion

networks in the early 1950's as a strategy for frontier defense. This project changed the natural environment (Figure 3) but had the positive effect of bringing more farmland under irrigation. Today, the north Jiangsu plain is one of the major grain producing area in china as a result of more than 40 years peaceful period since then. However, the military activity related to this event occurred over a very short period, but the environment impact of this activity would last for long time still.

Hainan Island and other small islands in China were deliberately left undeveloped for more than 30 years because it was believed that these areas would be the frontline of the "Third World War". Until the 1980's the islands were in a state of relatively low economic development. However, some improvements were made by the army in setting up transportation facilities such as roads and harbours. Some of the harbours constructed by the military have changed local environments. For example, coastal erosion on the leeward side of jetties, or increased in harbour siltation. All of these local effects are of minor importance compared to the large river diversion projects discussed earlier. Under a 40 year interval of peace in China, most islands have retained their natural, environments and ecosystems. In addition, most surplus army facilities were converted for civilian use during 1980's and have provided some benefits to local populations.

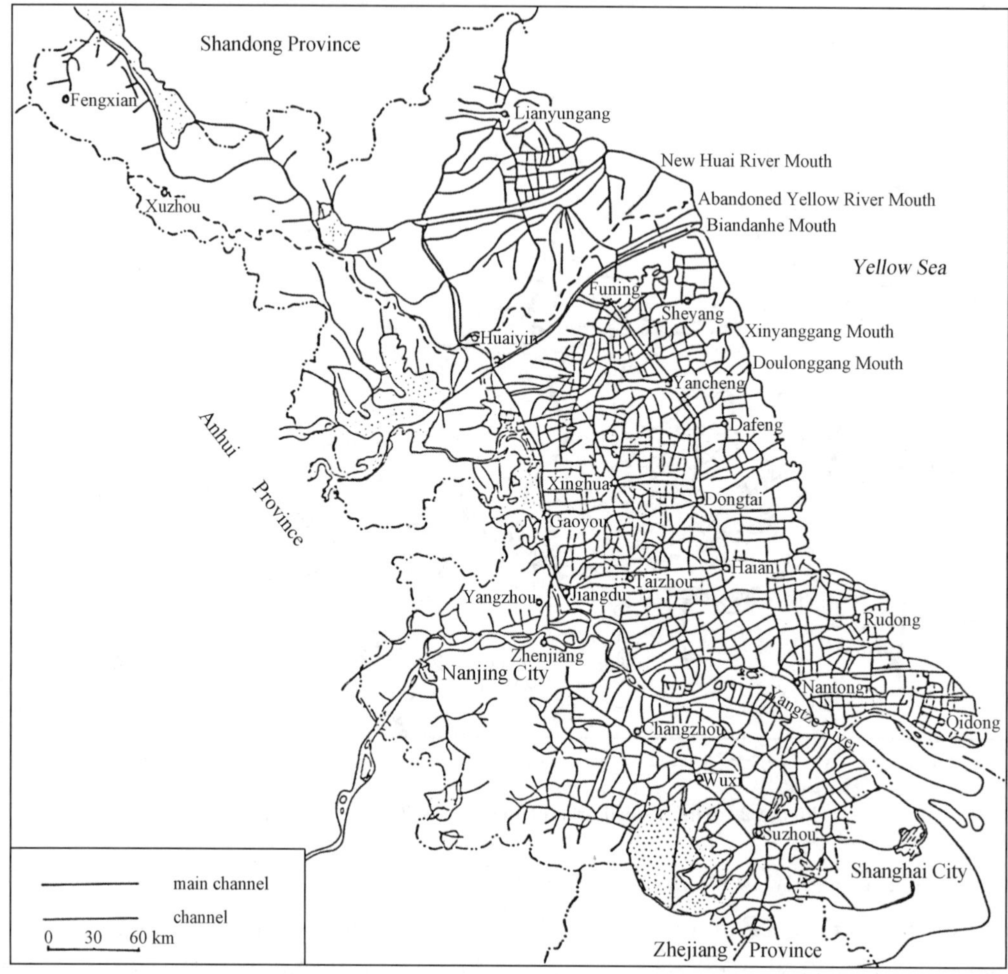

Figure 3 The map of canals and channel networks of north Jiangsu plain

References

[1] Wang Y, Ren M E, Zhu Dakui. 1986. Sediment supply to the continental shelf by the major rivers of China. *Journal of the Geological society*, *London*, 143: 935 - 944.

[2] Sung History. Original records of Gao Zhuong.

[3] Jin History. Geographical records.

[4] The Water Conservancy Committee of Yellow River. 1982. The summary of water conservancy history of Yellow River. Water Conservancy Press. 374 - 381.

湛江市区位优势与海岸海洋旅游发展*

1　海岸海洋发展与湛江市区位特点

1.1　海岸海洋

海岸海洋是一个独立的环境体系，是地球表层海、陆、气、生物各圈层相互作用、质能转移与变化活跃的地带。海岸海洋地带，大城市众多，人口密集，经济发达。人与海洋相互联系紧密，互相影响突出，其发展变化与人类生存密切相关。

海岸海洋成为沿海国关注的热点，缘于1994年11月16日正式生效的《联合国海洋法公约》(The 1982 United Nations Law of Sea Convention)。公约将自陆地延伸到海洋中的大陆架、大陆坡等面积约1.3亿 km^2 的海域，从公海中划分出来，归由各沿海国管辖。管辖海域涉及对环境的监测、利用与资源开发等权益[1]，其范围基本上在海岸海洋地区。据此公约，我国拥有管辖海域面积约300万 km^2。海岸海洋包括整个海陆过渡带：海岸带、大陆架、大陆坡与坡麓的大陆隆(图1)。这一地带具有自然因素变化迅速、生物量巨大、生态系统丰富、缓冲作用突出、人类活动密集等环境特性，既不同于陆地，又不同于海洋[2]。面对资源枯竭、环境恶化与人口压力的严峻挑战，各沿海国已将开发研究重心转向海岸海洋，以期取得高效利用与迅速获益。1998年国际海洋权威论著"The Sea"出版了其系列中的第10卷"The Global Coast Ocean: Processes and Method"，即以全球海岸海洋命名[3]。2000年国际地学家联合会(IGU)发表"海洋地理国际宪章"，将全球海洋正式划分为"海岸海洋"与"深海海洋"两部分[4]。

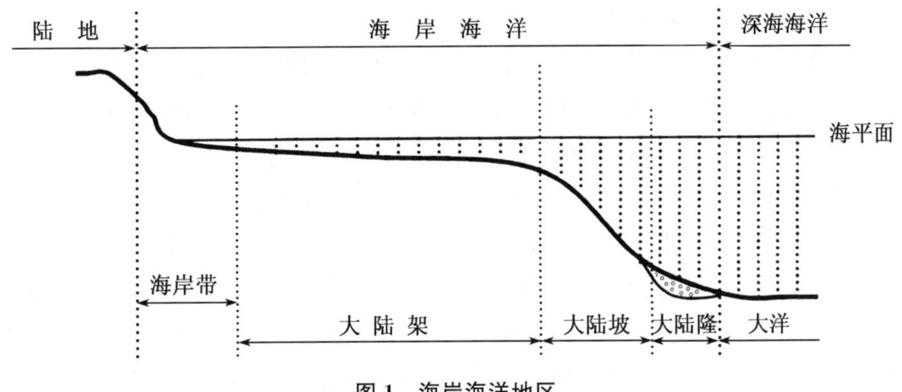

图1　海岸海洋地区

---

* 王颖，张永战，葛晨东：周镇宏，胡日章主编，《21世纪中国海洋开发战略》，471－478页，北京：海洋出版社，2001。

2000年载于：《南海海洋资源综合开发战略高级研讨会论文集》，280－284页，中国湛江，2000年11月。

1.2　湛江市区位

从“海岸海洋”的观点来审视，湛江市的地位非常重要(图 2)。它位于祖国大陆南缘，是与南海诸多岛屿和广阔海疆联系的中心城市端点。湛江市居粤桂琼海岸海洋地区的中心，是南部地区物资集运的海上通道。作为我国沿海首批 14 个开放城市之一，与东、西部相连，是我国与东南亚、欧洲、非洲、大洋洲海上航运的重要口岸之一。

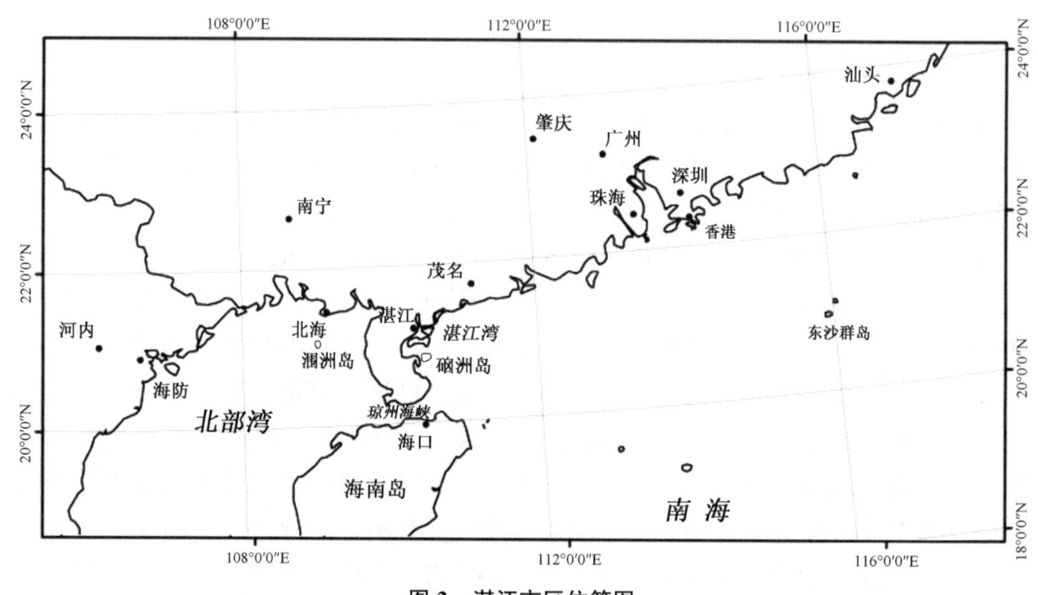

图 2　湛江市区位简图

湛江湾是典型的潮汐汊道型港湾，湾阔水深，港口设施完善，是我国 8 大港口之一。港湾中有东海岛、南三岛等 4 个岛屿环绕，形成天然屏障，水深浪静。东海岛通过东北大堤与霞山相连，北岸蔚律至龙腾 6.5 km 岸线水深达 26～40 m，可能建 30 万 t 级码头，接待国际第五、第六代集装箱船(吃水 16m)。目前，湛江港 5 个终端中拥有 31 个泊位，包括 24 个万 t 级泊位，并建有集装箱专用通道，5 万 t 级船舶可直接靠泊作业。湛江港东行 290 n mile 可达广州，与港、澳、深圳及珠江三角洲城市群联系；西去 202 n mile 抵北海，242 n mile 至钦州湾龙门港，自北海又有航线直通越南海防港(160 n mile)；南面隔琼州海峡与海南岛相望，经 83 n mile 可达海口，有直达客轮，并早有公路与轮渡联运。

湛江市北依广阔的大陆，经由铁路干线与初具规模的高速公路网与内陆相连，东通广州，北达长沙、武汉、郑州、北京，西通南宁、贵阳、昆明，与整个西南地区相连，并进一步与西北、东北、长江三角洲及华南地区相连。待驶过琼州海峡的粤琼铁路贯通后，与海南之间又增加一条大宗运输干线，从而使湛江成为大陆与南海诸岛紧密联系的重要“中继站”与必经的“桥梁”。

湛江市航空港已与北、东、南、西建立航线联系，有直飞香港、广州、上海、北京等城市的 11 条航线，交通便捷。以湛江为中心，已建立了一定规模的海陆空交运网络，扩展了湛江的腹地，有效地增加了湛江的辐射范围。

21 世纪是“海洋世纪”，亚洲太平洋将成为全球经济发展的中心，对海岸海洋地区的权益要求、综合研究和大规模开发利用更将成为一股强有力的浪潮，锐不可当。在这一背景下，从促进海岸海洋发展的观点审视，湛江市的区位与基础设施，使她位于陆海、东西相交的十字形枢纽点上。这一独特的区域位置，应成为湛江在21世纪发展规划的重要出发点。

2　发展海岸海洋旅游，积累资金，提高知名度

2.1　旅游资源与市场

作者于20世纪60年代从事过湛江港调顺码头（“652”工程）选址研究工作，70年代与80年代中期访问过湛江。印象中的湛江十分美丽，具有明媚的热带亚热带风光：阳光明亮，天空湛蓝，阵雨于阳光下飘洒；大道在红沙土台地上天然筑成，高大的椰树与木麻黄林绿荫丛丛；碧波轻浪的海湾中，海牛、翻车鱼悠闲戏水，洁白的海鸥自由地嬉戏在海军士兵飞扬的蓝色飘带中；白色的法式住宅，穹窗屋廊，阴凉清新，令人陶醉。同时，湛江有着特色的生态农业与文物古迹（表1）。

表1　湛江市旅游资源分析

自然生态景观	历史文化景观	民俗风情	交通情况
热带、亚热带海洋风光：三面环海，岸线长达1 566 km，港湾岛屿与细砂海滩众多，轻浪软沙，为海滨圣地。已开发的如：东海岛龙海天、南三岛、吴川吉兆湾和海安白沙湾等，沿岸多白蝶珍珠贝等养殖海水南珠。 湖光岩—独特的火山口玛珥湖：水清倒影，含光异彩，称为“镜湖”。 硇洲岛—火山岛：岩滩浴场，火山岩风化沃土，盛产香蕉，成为“蕉岛”，沿岛海湾养殖九孔鲍鱼。 南亚热带植物园：植有热带、亚热带30多个国家和地区的300多种植物。 国家级红树林自然保护区：位于高桥镇，沿北部湾滩涂蜿蜒20 km。 珊瑚礁：徐闻西海岸有连绵10 km的珊瑚礁群，是我国目前面积最大、保护最好的珊瑚礁区之一。	雷州古城：国家级历史文化名城，文物古迹有雷祖祠、三元塔、十贤祠、天宁寺、西湖公园等。 鹤地水库：我国大型人工水库之一。 法租界残痕：市区内尚有法国式楼房和教堂等建筑与遍布的法国枇杷树、寸金桥、抗法纪念像等；硇洲岛上屹立高16.05 m的法式灯塔，是世界仅有的2座水晶棱镜灯塔之一，与伦敦国际灯塔齐名。 西汉古徐闻港遗址：闻名世界的“海上丝绸之路”始发港。 灯楼角：中国大陆最南点，琼州海峡与北部湾的分界点，建有太阳能灯塔。 明登云古塔与汤显祖贵生书院。	雷剧、雷歌、雷音乐；吴川飘色、泥塑、花桥、单人木偶戏；东海岛的人龙舞；廉江的舞鹰雄；乌石镇的舞蜈蚣；安铺镇的会八音等。	海陆空立体交通网络： 海港：湛江港已同世界上100多个国家和地区通航，与香港、海口有快速、豪华直达客轮往返。可接待国际第四代集装箱船停泊。世界第十大、亚洲最大的豪华旅游邮轮“狮子星”号每周一抵湛，国际游客众多。 路网：公路交通四通八达，是全国45个公路交通枢纽之一；铁路已将西南及中部地区连接起来，同广州、武汉、长沙、南宁、贵阳、昆明及北京均有直达客运列车，至海南的粤海铁路大通道建成后，湛江将为南部路网枢纽端点。 空港：湛江机场可供波音757等大型客机日夜起降，已开通至国内主要城市18条航线，每周班机110班次。

据湛江市的自然条件、生态环境特点、城市发展的历史渊源、民族、文化与宗教特点、经济发展现状及持续繁荣的需要等考虑，当前，宜全面发展旅游事业，以旅游业作为新起点。重视国内客源，首先积极发展国内旅游，积累资金，提高湛江市的知名度，再进一步扩

大向海外发展。

全面发展旅游事业，需要立足于湛江市，并外延与广西、海南等省市连接，突出其区位优势，进行区域旅游规划，确定发展目标与发展步骤。同时，需考虑湛江旅游的内涵，如何充分利用自身资源特点，合理规划与设计，深层次开发，提高其追加性旅游价值，形成具有鲜明特色和极富吸引力的旅游项目，提高旅游质量，以利经久不衰地发展等。

当前，随着我国经济的快速发展，人民生活水平的不断提高，以及长假期的明显增多，国内旅游业出现了蓬勃发展之势。今年五一劳动节及国庆前夕，故宫、黄山、兵马俑、九寨沟等地均已出现旅游高峰，游客数纷纷超过当地的旅游接待能力，故宫日接待游客超过12万人。但国庆节长假却未能保持全面高涨，原因在于基础设施不足，旅游质量与价格规律不适应。为迎接假期旅游高峰，保证旅游质量，一些著名旅游区已陆续提出每日限额的不同方案。许多著名旅游区都在不断地建设、更新、修复、开发新的景点，提高交通、服务与管理水平，充实旅游内容，提高与完善接待能力与旅游质量。另一方面，人民需要新的旅游点，需要具有趣味性、探险性、知识性与休闲性的旅游新内容，这是一个不断增长的新的市场需求。因此，湛江市发展旅游需注意从多方面吸引游客，尤其是扩大吸收华北、华东地区的客源，尽量设计使游人以湛江为起止点的旅游线路，增加游客在湛江的停留时间，有效增加旅游收入。

2.2　海岸海洋旅游路线规划与网络设想

2.2.1　海滨休闲与运动旅游路线

以湛江湾与雷州湾、东海岛、南三岛和硇洲岛为主。充分发挥“3S”(沙滩 Sand beach、阳光 Sun light 和海鲜 Sea food)的优越条件，吸引就近游客。发展海湾滑水、游泳、竞舟、游艇、垂钓与海兽观赏等，可将体育活动、运动员集训、休憩与海岛小憩等集于一体。湛江湾水阔浪轻，东海岛沙滩宽阔，绵延数十千米，为最佳的海滨泳场，这些都是其他海湾无法比拟的。

2.2.2　学术会议与地质观光旅游

设立湛江→广西→越南的旅游大路线。如湛江的湖光岩是沿火山口发育的玛珥湖，学术价值高，湖区宁静。乘船经硇洲火山岛，小憩尝海鲜，香蕉林采蕉，攀灯塔远眺。登舟再向西航行至北部湾中的涠洲岛，它是沿火山口发育的海湾与岛屿，与陆上的湖光岩火山口遥相对应，异曲同工，学术价值很高。继续西行经北海或径直至越南下龙湾，观赏世界珍稀的石灰岩喀斯特海岸。这条学术观光旅游路线可以海上航行为主，海船为游动的旅馆，沿岸停泊，登陆观光，方便旅客，费用低廉。

2.2.3　海岛观光与探险旅游

湛江湾→硇洲岛→海南岛→西沙。以舟船为据点，登岛览胜、休息购物。远航西沙，海天一色，在珊瑚礁岛体验白云悠悠、礁湖潜泳、鲣鸟盘旋、七彩礁盘共同组成的胜景，增长知识，陶冶心情，锻炼体魄，其乐无穷。

2.2.4　文化历史与生态旅游

以汉代古徐闻港和灯角楼与“海上丝绸之路”为内容的文化历史旅游，结合海洋养殖、热带特色农作物种植、红树林与珊瑚礁保护、海洋环境教育与保育等的生态旅游。

各条路线要有引人入胜的景点与丰富的旅游内容，形成区域特点后，进一步发展与广西的喀斯特山水景点、西北黄土高原的深厚文化蕴藏、大漠的飞沙落雁与古遗址、塞北风光、首都北京的历史文化名胜、华东的山水与小桥流水民居以及三峡、川江的胜景相衔接，湛江将会成为中国旅游网点上的一个重镇。同时，经营管理十分重要。如导游的素质、服务质量、交通通信与安全保险等，都应保持高水平、严要求、持久不懈。绝不能言而无信、不实现旅游路线上的景点保证，例如在景点上求近舍远、时间安排在采购点上的多于观光点等，带队或导游获利，而游客受损，这种旅游业不能持久。要设法吸引旅客来过之后，还想再来。

在各旅游点，应有不同档次的旅馆，吸引不同经济能力的游客。同时，相同路线上，可设计不同的内容与时间安排，方便不同游客的不同选择与旅游要求。在车船交替、餐饮、娱乐等方面，尽量做到方便、高效、安全、卫生，这样才能吸引游客，做到有口皆碑，有效地提高知名度，才能长久经营，财源不断。旅游收入积累了城市发展资金，提供了就业机会，其意义未可限量。

旅游业已成为世界第一大产业，而海滨旅游业又居榜首。湛江市宜充分发挥此独特条件，重视以旅游业吸引投资与提高城市知名度，从吸引国内大陆居民旅游，逐渐发展为东北亚与东南亚旅游线路中的重要旅游点。

参考文献

[1] 联合国第三次海洋法会议. 1983. 联合国海洋法公约(中文版)[S]. 北京：海洋出版社，3－328.

[2] 张永战，王颖. 2000. 面向21世纪的海岸海洋科学[J]. 南京大学学报，36(6)：717－726

[3] Wang Ying, Ren Mei-e, James Syvitski. 1998. Sediment Transport and Terrigenous Fluxes[M]. In: K. H. Brink, A. R. Robinsion, eds., The Sea, Volume 10. John Wiley and Sons. 252－292.

[4] 王颖. 1999. 海洋地理国际宪章[J]. 地理学报，54(3)：284－286.

充分利用天然潮流通道
建设江苏洋口深水港临海工业基地*

1　建设洋口深水海港与发展江苏海洋经济的相关性

江苏省大陆海岸线分布于长江口以北，其总长度为 953.875 9 km（岛屿岸线长 26.964 1 km），其中 883.561 km 岸线属于粉砂淤泥质平原海岸；北部苏鲁交界地带有 40.248 3 km 基岩港湾海岸，30.062 5 km 砂质海岸及 19 座基岩岛屿。

平原海岸坡缓水浅，波浪动力衰减，缺少供大船停泊的天然良港，这是苏北长期处于封闭型农业经济的客观环境不利因素。

“九五”期间，我国海洋经济发展速度快于同期 GDP 发展速度，其增长率为 10.9%，如 1999 年为 447.4 亿元，2000 年为 4 133.5 亿元，亦快于同期世界总体发展速度。江苏省地居江海之交，海陆交通理应高度发达，虽提出发展“海上苏东”等战略口号，但现实发展尚不理想，与相邻省市比较，江苏省的海洋经济发展滞后，见表 1[1]，这可能与 800 多km 平原海岸未建成深水良港，未形成临海工农业带有关。传统的海洋经济为海洋渔业、盐业与海洋交通运输业。自 20 世纪 80 年代以来，新兴的海洋产业发展较快，如海水养殖、海洋油气、滨海与海岛旅游、海水利用、海洋医药、海洋食品、海底通信与信息产业。21 世纪的海洋产业为：海洋人工岛、海域再生能源、深海与大洋采矿业、天然气水合物（可燃冰）新能源、海水与海洋生物资源综合利用等。我国海洋经济持续增长，其中又以第二产业发展迅速（表 2）[2]。

表 1　2000 年中国沿海省（区/市）海陆域经济对比　　亿元

省区	海洋经济产值	排序	国内生产总值	排序
广东	1 114.57	1	9 662.23	1
山东	737.76	2	8 542.44	3
上海	601.37	3	4 451.15	7
福建	419.15	4	3 920.07	8
浙江	399.53	5	6 036.34	4
辽宁	326.58	6	4 669.06	6
江苏	146.04	7	8 582.73	2
天津	138.63	8	1 639.36	10
广西	110.45	9	2 050.14	9
海南	70.23	10	518.48	11
河北	69.19	11	5 088.96	5

* 王颖：《水资源保护》，2003 年第 6 期，第 1－5 页。同年载于：第二届江海论坛论文集（论坛主题：建设洋口深水港，促进江苏经济可持续发展），1－12 页。

表 2 中国海洋产业结构发展比较 亿元

年份	总产值	第一产业	第二产业	第三产业	比重
1995	2 463.85	1 176.87	185.03	946.53	48∶8∶38
2000	4 133.50	2 084.34	639.86	1 355.30	51∶16∶33

深入分析,苏北(本文指长江以北部分)的土地面积与人口约占江苏省的 2/3,但人均国内生产总值仅占全省的 3/5,而苏南的人均国内产值为全省的 1.89 倍(表 3),产值与资源相比较,江苏省经济发展的潜力在苏北,苏北经济的发展应高起点,迈出新步伐,发展海洋经济,其关键是建设洋口深水港临海工业。因为江苏北部苏鲁交界带有连云港贯通中原与西北,具有联通欧亚海路通道的巨大潜力。但是,该港口的腹地区域经济尚欠发达,尚未形成与苏南经济发达区物流输运的主渠道;南通港既可沿江联系,又可转口海上,但长江口有拦门沙碍航,航道浚深后水深不足 13 m,制约了大型集装箱货轮畅通。从江苏省社会经济持续保持高水平发展所需,宜加强海港建设,形成海陆江交通动脉网络。其关键是在长江三角洲接近经济发达的苏南并与经济起飞的苏北衔接段,建设深水海港,成为江苏中部的海上门户,发展临港工业,形成新型的海港工贸基地,增强对外进出口联系,带动苏北沿海经济发展,促进"长江三角洲北翼经济带"的建设。

表 3 2001 年江苏省各项指标及比例[3]

地区及比例	年末总人口/万人	土地面积/km^2	耕地面积/m^2	国内生产总值/亿元	人均国内生产总值/元	粮食产量/万 t	棉花产量/万 t	油料产量/万 t
苏 南	2 177.57	27 954	1 166 730	5 446.25	24 969	626.00	2.16	56.35
苏 中	1 737.15	20 429	1 114 200	1 764.73	10 160	784.86	8.64	63.88
苏 北	3 182.28	52 312	2 693 200	2 186.94	6 889	1 593.87	37.52	112.29
全 省	7 097.00	100 695	4 974 130	9 397.92	13 242	3 004.73	48.32	232.52
苏北各项指标在全省的比例/%	44.84	51.95	54.14	23.27	52.02	53.04	77.65	48.29
长江以北地区各项指标在全省的比例/%	69.32	72.24	76.54	42.05	60.77	79.17	95.53	75.77

2 洋口深水潮流通道发育过程与环境特点

苏北 800 多 km 平原海岸唯一具有天然深水航道的地区,位于南黄海辐射沙脊群区。其中最佳的港址是洋口港,该处烂沙洋潮流通道,水深 16～23 m,为天然深水航道,并有众多的沙脊,为发展临港工业与建设人工岛提供土地资源,可发展为 10 万～20 万 t 级大型船舶停泊的海港;其次为小庙泓潮流通道,及兴建中的王港区,海域条件可以发展 5 万 t 级海港。上述论点是南京大学海岸与海岛实验室于 20 世纪 80 年代以来,先后 5 次对辐射沙脊群进行系统深入地调查研究后获得的科学认识。

2.1　基础研究工作

（1）1979—1985 年，南京大学作为主要组织单位与考察队领导，参加了江苏海岸带资源与海涂综合调查，这是对辐射沙脊群的初次规模化调查，积累了地形、地貌、气象水流、泥沙生物等系统资料，初步阐明了辐射沙脊群的动力系统——潮波、流、浪与含沙等特点与作用过程，但对沙脊群内部结构了解尚少。

（2）1988—1989 年，南京大学与河海大学共同开展了对辐射沙脊群区的洋口港与王港建港可行性研究，系统地调查研究了烂沙洋、黄沙洋及西洋 3 条主要潮流通道的水动力、泥沙与海底地貌冲淤变化等。

（3）1992—1993 年，南京大学在辐射沙脊群区研究了洋口港水道的稳定性。地震剖面初次揭示出黄沙洋—烂沙洋潮流通道是承袭一条古河道发展而来的，其通道处于微冲微淤的动态平衡状态。

（4）1993—1996 年，南京大学与河海大学、同济大学及中国科学院海洋研究所共同完成了国家自然科学基金“八五”重点项目——辐射沙脊群演化与沉积特征研究。在此工作中完成了 600 km 海底地层剖面测量，完成了 10 个钻孔，2 000 个沉积样品分析及多点水文观测，完成 1∶250 000 海底地形图（图 1）。

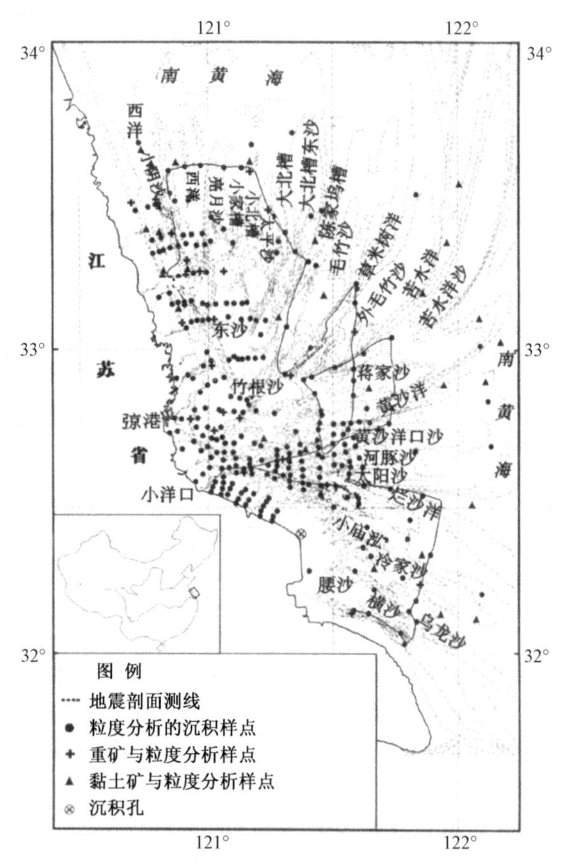

图 1　工作路线和采样点图

（5）1998—2001 年，开展了国家“九五”攻关项目子课题——海岸潮汐水道冲淤动态预测预报研究，南京大学对辐射沙脊群区遥感与实测图像进行了冲淤对比分析，建立了变化动态分析预报技术。

2.2　辐射沙脊群动力地貌

长时期系统深入研究的成果所获得的科学结论是本文的主要依据，该项研究成果证明[4]：

辐射沙脊群位于苏北平原外侧南黄海海域，以弶港为中心，呈扇形展开。它是我国最大的浅海潮流沙脊群，南北范围介于 32°00′N～33°48′N，长达 199.6 km，东西范围介于 120°40′E～122°10′E，宽达 140 km，由 70 多条沙脊与其间的潮流通道组成（见图 2）。它介于海陆交互作用带，水深为 0 m～－25 m 之间，全部面积 22 470 km^2，其中 3 782 km^2 出露于水面之上。南黄海海域为正规半日潮型，涨潮时，潮流自北、东北和东南方向涌向弶港方向，落潮时，以弶港为中心，呈 150°扇面向外逸出，形成辐射状潮流场。潮波的辐聚与辐散，形成潮差大，平均潮差 4.18 m，大潮潮差 6.5 m，最大潮差 9.28 m 于沙脊群核心区黄沙洋测得，是我国潮差最大记录，潮差自黄沙洋—烂沙洋，即辐射沙脊群核心区向南北两侧减小。辐射沙脊群区潮流作用强，涨潮时潮流辐聚，平均流速为 1.2～1.3 m/s；落潮时潮流辐散，流速急，平均流速为 1.4～1.8 m/s。持续上升的海平面使潮流动力与潮流系统加强，形成全球最具代表性的辐合——散射潮流系统海域，沙脊群的基本轮廓与潮流场相符，强大的潮流是维持潮流通道水深的天然动力。辐射沙脊群掩护着其后侧的海岸，自射阳河口至小庙泓段发育了广阔的潮滩。沙脊群外侧受波浪冲刷，泥沙向岸输运，至潮滩处才能停积使潮滩不断地淤涨，而沙脊群范围以外的南北两段平原海岸，却受浪流冲刷后退。

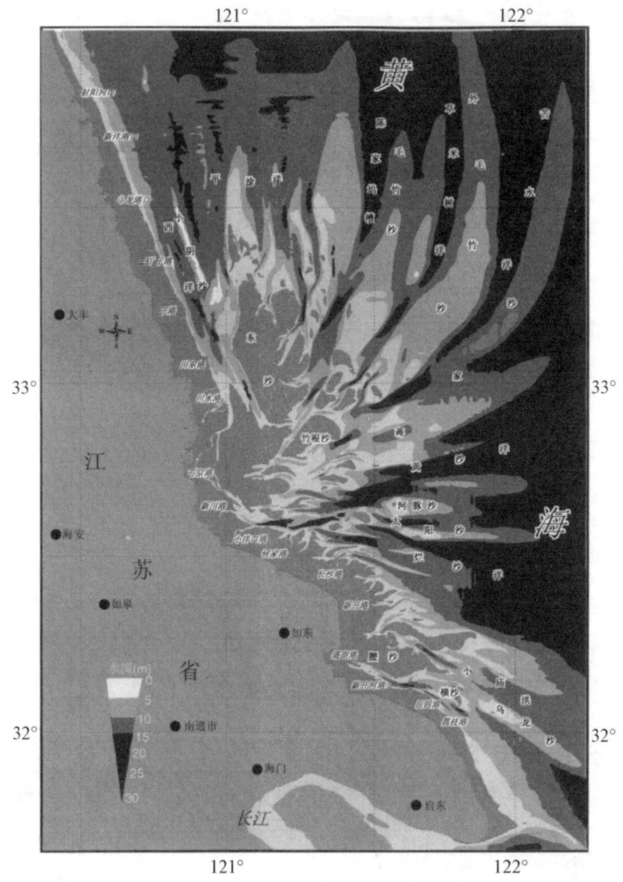

图 2　南黄海辐射沙脊群(附彩图)

辐射沙脊群由细砂与粉砂组成，是在地质历史的新近时期——约 3 万年前，古长江口自弶港入海时堆积的细砂物质，当时的海平面位于现今的－20 m。在近 1 万年来的全新世海面上升过程中，尤其在距今 6 000 年及 1 000 年两个明显的海侵冲刷期，由潮流冲刷改造古长江口泥沙形成长条形沙脊。经过漫长的历史时期，黄河夺淮入黄海后，向辐射沙脊群补充供给了淤泥物质。所以,辐射沙脊群是以晚更新世末期的古长江三角洲体为基础而于全新世(近 10 000 年)海侵过程中发展成型。

烂沙洋与黄沙洋是位于辐射沙脊群的枢纽部的主潮流通道,其中烂沙洋自西太阳沙深槽(−29 m)至口门外海,53 km 的长度内水深介于 16～23 m 之间,它是在古长江干道基础上发育而成的主潮流通道[5]。

2.2.1 烂沙洋、黄沙洋中段地震剖面反映其发展过程

(1) 晚更新世末期古河谷埋藏于现代海底 40 m 厚的沉积层下,该处谷地约宽 10.42 km,深度为 20 m(北岸)～33 m(南岸),钻孔剖面亦反映出厚层的细砂层埋藏于 41 m 深处(图 3)。

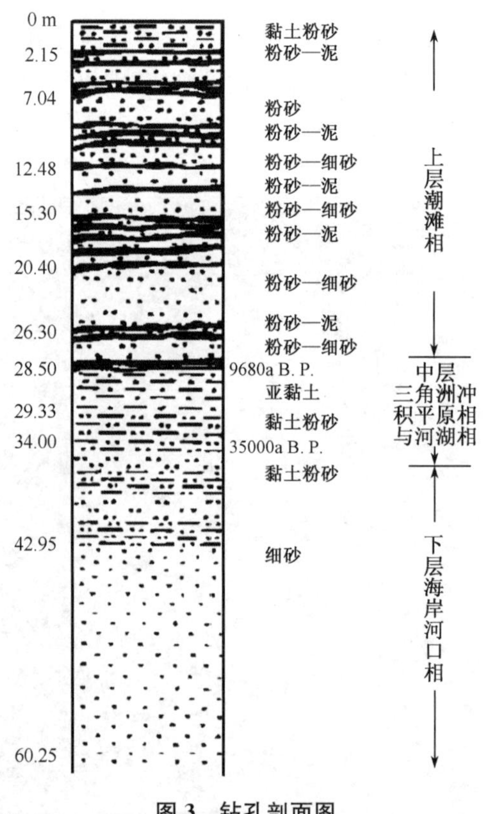

图 3 钻孔剖面图

(2) 在老河谷之上,一个 2.2 km 宽的 V 形谷地埋藏于现代海底 35 m 深处。

(3) 全新世早期(公元前 10000—8000 年)的谷地为 2.2 km 宽的 V 形谷地埋藏于现代海底 20 m 厚的沉积层下。烂沙洋、黄沙洋潮流通道至今基本承袭了老河谷的形态与位置。

2.2.2 烂沙洋、黄沙洋潮流通道外侧(32°5′34″N,121°8′17″E)处的地震剖面分析

(1) 在 38 m 厚的沉积层下,显示出 81 km 的宽谷,以及高达 10 m、宽约 1.8 km 的沙脊。

(2) 6 km 宽的全新世谷地埋藏于 20 m 厚沉积层之下。

地震剖面揭示出烂沙洋与黄沙洋为古长江从苏北入海主干道,时代相当于晚更新世末约 3 万年前,卫星照片仍隐约显现此埋藏谷地(见图 4)。辐射沙脊群北部与南部的脊槽形成时代晚于枢纽部的烂沙洋与黄沙洋通道,是在全新世(公元前 10000 年)以来,长江

从李堡或遥望港入海时形成，大北槽地震剖面亦反映此情况(图 5)。烂沙洋与黄沙洋两条潮流通道的长度、水深及位置所在，亦反映出现代潮流主通道的位置，仍在弶港以东的沙脊枢纽区(表 4)。

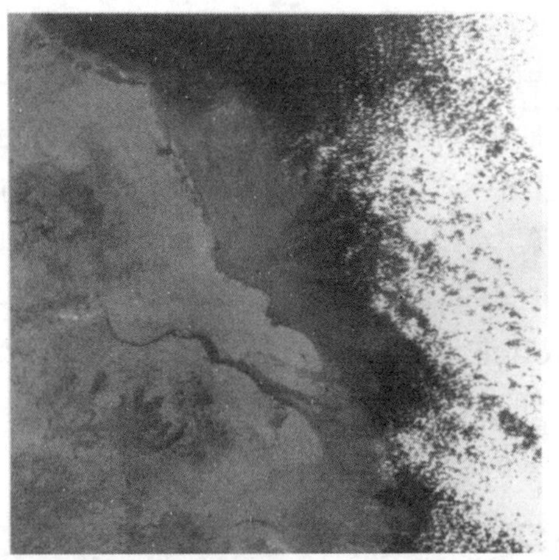

图 4　卫星照片显示的辐射沙洲区古河道(附彩图)

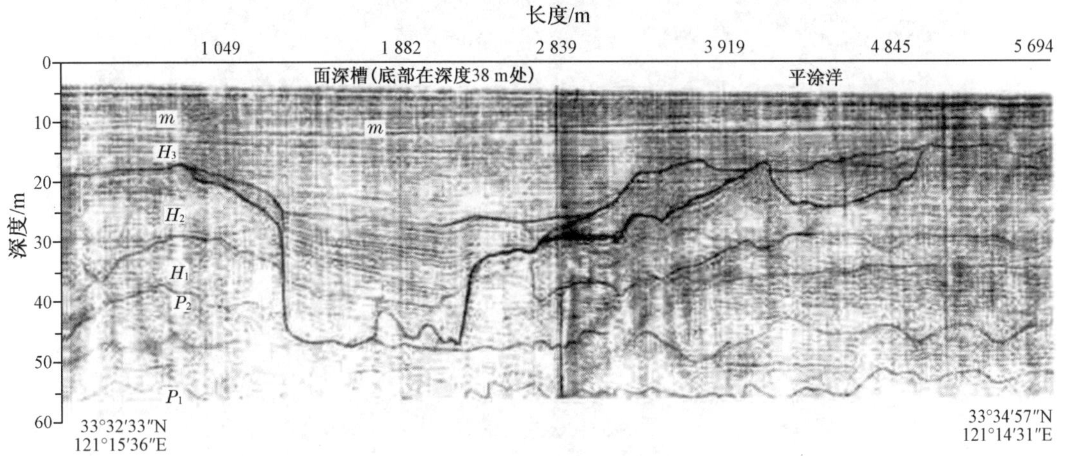

图 5　大北槽地震地层剖面图

表 4　潮流通道长度、水深

部位	潮流通道	潮流通道长度/km	水深/m
枢纽部	烂沙洋	53	16～23(局部 29)
	黄沙洋	45	11～20
南部	小庙泓	28	8～12
北部	西洋(王港)	50	10～13(局部 28.5)

烂沙洋与黄沙洋海域潮流作用强，潮流通道中为往复型潮流，该处实测最大潮差达 9.28 m，平均潮差 4.18 m，涨潮流流速 1～1.5 m/s，每潮涨潮流量 19.5 亿 m^3；落潮流流速为 1.2～2.0 m/s，每潮落潮流量达 21 亿 m^3，落潮流大于涨潮流。巨大的潮流动力，使得烂沙洋多年保持水深达 16～23 m，此水道伸入陆地，是江苏省最宝贵的天然深水航道

资源，也是全国沿海最深长的天然良港港域。

烂沙洋内陆侧有一西太阳沙沙脊，5 m 水深以内，呈东西向延长 1.375 km；10 m 水深以内长 4.25 km，宽 250 km，由粉砂与细砂组成，脊顶水深 0.5 m，落潮时，主干部分出露（见图 6），该沙脊位置稳定。西太阳沙由厚达 90 m 的粉砂、粉砂黏土与粉细砂沉积层组成（据江苏省电力设计院钻孔资料，1995 年 1 月），为在深水航道侧修建码头与扩展建设人工岛的理想基地。太阳沙毗邻的水深为 17.2～20 m，这样的沙脊与深水航道组合，在我国淤泥海岸段仅有 2 处，另一处是位于渤海湾的曹妃甸沙岛与其外侧潮流深槽，该岛向陆侧有大片潮滩（南堡滩涂），海岸环境与辐射沙脊群类似，但沙岛与潮流深水槽的规模小。南京大学亦曾于 1997—1998 年在曹妃甸沙岛为深水港建设从事潮流深槽稳定性研究。2002 年夏该处已完成建港设计，并通过环境评价，将于曹妃甸建深水码头，成为唐山、王滩港与天津港新港间的一个大型船舶通航的现代化深水港，为京津唐钢铁基地的组成部分。曹妃甸建港为江苏树立了可效法的实例。

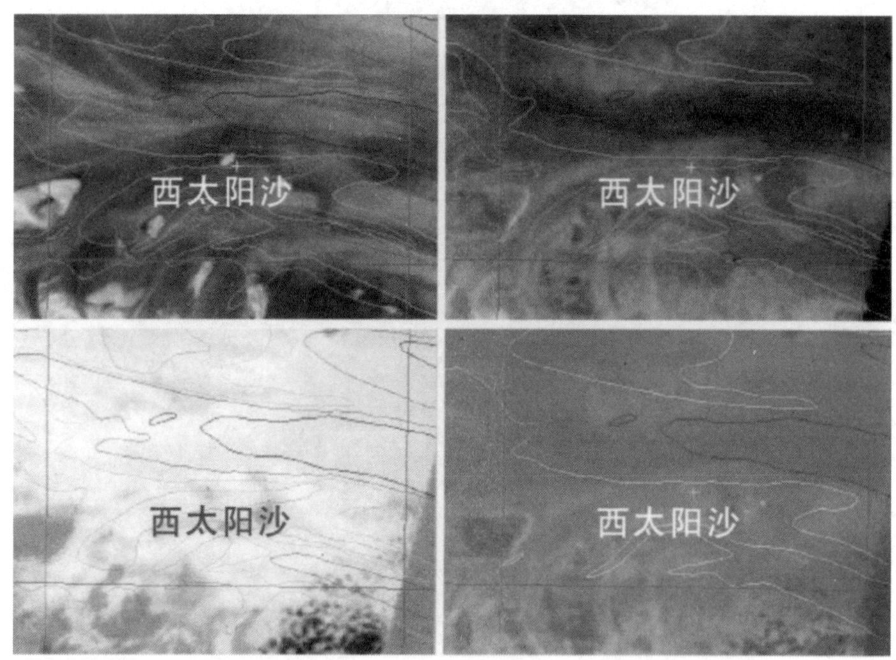

图 6　西太阳沙与烂沙洋遥感图像(附彩图)

3　建设洋口深水港，促进经济飞跃增长

建设洋口深水港是促进苏北发展外向型经济的关键一步。辐射沙脊群枢纽区洋口深水港的建设，将使 883.6 km 的苏北平原海岸，利用出海口直接联通海外世界，使几千年来传统的农业经济发展为现代开放型的临海工农贸经济带。

洋口港位于长江三角洲北端，与南侧的镇海港、建设中的大小洋山港遥相对应，港口建成后可通过平原的河道水网与区内乡镇联通，集输物产发展海陆联运；通过纵横的公路网络，可经苏通大桥、江阴大桥、润扬大桥及南京二桥等，跨过长江，与苏南经济发达区沟通，与长江航道联通；进一步，可通过东部沿海公路干道，延展至东北与海南；苏北平原西部可通过大运河南北联系，使苏北中部平原具有东西两侧公路与运河双通道，组成发达的运输动脉；此举必将大大促进苏北地区的繁荣，成为 21 世纪江苏省经济持续发展的潜力源泉。平原海岸的广袤滩涂与土地资源，易于开辟建立沿海机场，发展空运，形成海陆空立体交通网络，必将加速通、盐、扬地区的农、经、工、贸发展，在古长江活动影响的地区，建成长江三角洲北翼经济发达区，将大大提高长江三角洲的经济实力与辐射范围。

新时代、新形势要求对洋口港交通基地的建设，需从高起点与多元发展思路进行总体规划。西太阳沙与深水航道不仅可建设深水枢纽港，同时，还应发展临港工业，发展进出口物产的加工工业，提高产品价值；建设天然气发电、潮流、风能与太阳能发电的新能源产业群；建设绿色农、林、渔、药、花卉产品的生态农业基地；建设棉、毛、化纤轻纺工业带和乡镇结合的新型城市群以及集自然、文化、历史风情于一体的休闲、度假与体育活动的旅游基地等。关键在于做好洋口港发展的总体规划，以港口为龙头，带动江苏东部地区经济新发展，为江苏省率先实现全面小康与现代化做出重要贡献。

参考文献

[1] 杨荫凯. 2002. 21 世纪初我国海洋经济发展的基本思路[J]. 宏观经济研究，(2)：35 - 38.

[2] 杨文鹤. 2001. 中国海洋统计年鉴(2001)[M]. 北京：海洋出版社.

[3] 江苏省统计局. 2002. 江苏省统计年鉴(2002)[M]. 北京：中国统计出版社.

[4] 王颖. 2002. 黄海陆架辐射沙脊群[M]. 北京：中国环境科学出版社.

[5] Wang Ying, Zhu Da-kui, You Kun-yuan, et al. 1999. Evolution of radiative sand ridge field of the South Yellow Sea and its sedimentary characteristics[J]. *Science in China* (*Series D*), 42(1): 97 - 112.

中国海岸湿地环境特点与开发利用问题*

一、中国海岸湿地环境概况

中国湿地总面积约 6 600 万公顷，占世界湿地总面积的 10%，居全球第四，亚洲第一。我国海岸带跨 39 个纬度，自然环境以季风、潮汐动力与大河影响的河海交互作用为特色[1]。其中海岸湿地所占的比例较大，含 966 万公顷（14 497 万亩）的沿海潮间带滩涂及 200 万公顷海拔在 2 m 以下的低地[2][3]。海岸湿地从东北至海南均有分布，以低地、滩涂与生物群落组合的海岸环境为特征。

（一）中国海岸湿地按其成因与表相分类

1. 河口湿地

以松辽平原盘锦的双台子河口、黄河口、长江口沿岸为代表，系淡水影响较大的湿地（wetlands），特点是多芦苇滩、草滩（三棱草、蒲草、碱蓬、大米草）及泥滩为特征，是丹顶鹤、白头鹤、黑鹳等候鸟栖息地，多半咸水淡水、咸淡水、咸水鱼类以及白鱀豚、中华鲟、江豚等濒危洄游水生动物。

2. 平原海岸湿地

渤海湾、苏北沿岸为我国最大的海岸湿地所在，海拔 0～4 m。陆地平原外侧为芦苇沼泽；近海侧位于平均高潮位与平均大潮位之间为大米草沼泽与碱蓬沼泽，以草滩与盐沼（salt marsh）为特征。苏北海岸湿地近海部分为丹顶鹤越冬区，靠陆侧为麋鹿保护区。

3. 隐蔽的海湾顶部湿地

主要分布在浙、闽、粤、桂、琼，以海南与广西的红树林沼泽海岸为代表（mangrove swamps）；其他港湾多已开发垦殖，改变了天然特性，或留存着一些人工湿地。

（二）海岸湿地的地理环境特征

（1）分布于有淡水汇入及潮流影响的咸淡水交汇河口或平原缓流区。具有较强激浪带的海岸不适于海岸湿地发育。

（2）底质透气性良好，以粉砂质最宜，或含砂之淤泥海岸带。完全为淤泥或黏土则不利于植物繁殖。红树林尚可繁殖于具有大量气孔与裂隙，并经热带气候风化强烈的玄武岩底质上。一旦红树林幼苗繁殖后，在基岩上亦可生长，如海南东寨港为陷落谷与玄武岩基底，印尼及印度孟买象岛等处玄武岩岛屿均有红树林生长。

---

* 全文由王颖，朱大奎撰写于 2005 年 4 月，以中英文摘要载于：《中德长江流域湿地生态功能区划分研讨会论文摘要》，中国长沙，2005 年，6－7 页。

（3）红树林沼泽主要分布于南北回归线之间，是热带与亚热带的丛林沼泽海岸，但在有暖流影响的海岸，分布可达 32°N。但纬度增高，气温稍低，红树林植物种属减少，红树林低矮稀疏。

（4）南方是以红树丛林泥沼为特色，北方以草滩-盐沼为主，在淡水汇集充沛处则为芦苇沼泽。但是，无论南北方，海岸湿地均具有分带性特点[2]：

① 潮上带（平均高潮线以上至特大高潮线之间）与沿海低地；

② 潮间带（平均高、低潮线之间）；

③ 潮下带（平均低潮线至海岸带下界即水深相当于 1/2～1/3 平均波长处），但湿地下界较浅，止于水深 6 m 处。

（三）南、北方海岸湿地分带性实例[3][4][5][6]

1. *海岸红树林沼泽岸*（mangrove swamps）（*表* 1，*图* 1）

表 1　红树林海岸分带性

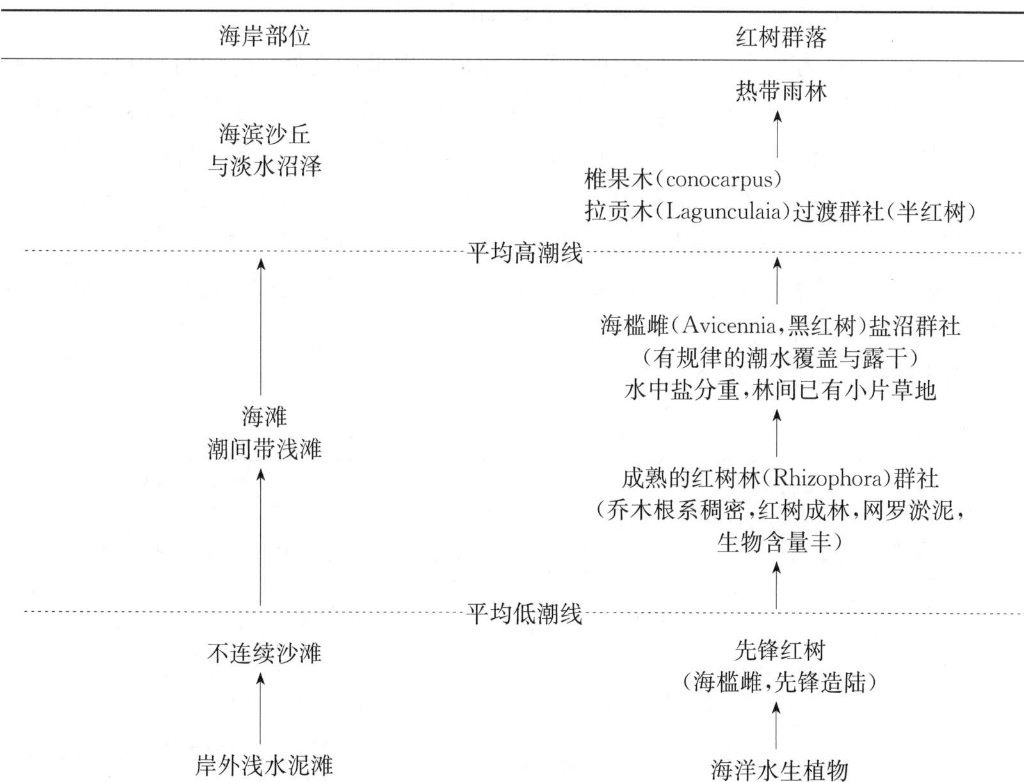

海岸部位	红树群落
海滨沙丘 与淡水沼泽	热带雨林 ↑ 椎果木（conocarpus） 拉贡木（Laguncularia）过渡群社（半红树）
平均高潮线	
海滩 潮间带浅滩 ↑	↑ 海榄雌（Avicennia，黑红树）盐沼群社 （有规律的潮水覆盖与露干） 水中盐分重，林间已有小片草地 ↑ 成熟的红树林（Rhizophora）群社 （乔木根系稠密，红树成林，网罗淤泥， 生物含量丰） ↑
平均低潮线	
不连续沙滩 ↑ 岸外浅水泥滩	先锋红树 （海榄雌，先锋造陆） ↑ 海洋水生植物

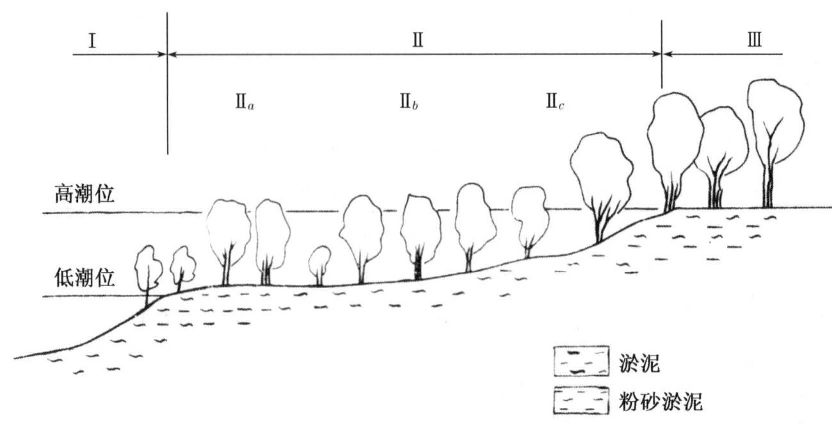

图 1 红树林沼泽湿地的综合剖面

Ⅰ. 潮下带泥滩 Ⅱ. 潮间带红树林 Ⅲ. 潮上带红树林(黄槿-海漆群落)

Ⅱa. 低潮位,白骨壤-海桑群落带 Ⅱb. 中潮位,红树-秋茄群落带 Ⅱc. 高潮位,角果木-海莲群落带

2. 黄海、东海平原潮滩——湿地岸(salt marsh)

发育于华北—苏北平原外缘,是我国海岸湿地范围最广之处。是河海交互作用堆积的大平原,因目前是否拥有黄河或长江泥沙补给,而海岸冲、淤状态与生物群落均有不同。大体上,废黄河与老黄河口岸段受冲刷,被黄海辐射沙脊群掩护岸段,滩涂淤长。例如:1975 年测得辐射沙脊群在 0 m 等深线以上面积为 2 125 km^2;1995 年为 3 782 km^2,平均每年增加 80 km^2;现在每年以 2 万亩的速度自然增长。黄海辐射沙脊群是我国最大的海岸沙体湿地,是 3 万年前古长江在弶港入海时堆积的泥沙,经冰后期海侵潮流冲刷而成,总面积 2.3 万 km^2,相当于江苏面积的 1/5[7][8][9]。

我国平原海岸湿地分带性[8],在全球具有代表性(表 2)。

表 2 平原海岸湿地分带性

部位	分带特征	物质组成	生物与现况
沿岸低地 — 潮上带	盐土平原与低洼地 沿岸贝壳堤	黏土质淤泥 贝壳、壳屑、粉砂,滞存淡水	贝壳堤上有人居,历史文物
	草滩 盐沼湿地(大潮浸淹)	芦苇沼泽、苔草沼泽 大米草盐沼	麋鹿保护区牧场或人工围垦为稻、麦、桑、麻地,鱼塘
		碱蓬、盐沼	禽鸟栖息,沙蚕
潮间带	高潮泥滩、泥沼	黏土质粉砂与粉砂质黏土	草类根茎残留或大米草斑块,毛蚶、弹涂鱼
	中潮位混合滩,滩面具冲刷体与潮水沟	黏土质粉砂与细粉砂互层,光滩	贝类(毛蚶、文蛤等) 禽鸟栖息
	低潮粉砂滩	粉砂与极细砂,具波痕	光滩 蛤、贝、鱼、螺生长 地网捕鱼 人工养殖紫菜

（续表）

部位	分带特征	物质组成	生物与现况
潮下带	潮下带淤泥滩涂	极细砂 粉砂 黏土质粉砂	沉积物粒径向海减小，悬移质沉降 鱼虾繁殖

丹顶鹤越冬场所集中于河口淡水汇入的淤长段，淤泥质粉砂滩涂上，有水草、鱼虾，但又非可陷入的泥沼！麋鹿喜繁衍于盐土平原，但均远离人居之旷野处。

总之，不同的气候条件，不同的底质与不同的海岸部位，其生物群落与组成海岸湿地的形态与环境皆有区别。

二、海岸湿地开发利用问题

（一）海岸湿地面临着人类大范围开发的挑战

（1）华南港湾湿地因降水丰富多已开发为农田与村镇。

（2）红树林盐沼与保护区开发旅游，伐林养殖、鱼虾、伐木为薪炭，疏干为土地造房舍。

（3）浙闽港湾湿地成条块分租，改海湾为鱼塘，养殖贝类或牡蛎，潮上带辟为柑、柚、花卉园地。

（4）平原潮滩湿地[7][10]

① 华北：泥滩不透水，少雨，早已辟为最大的盐场。当代迅速的发展是开采油气田与兴建城镇。

② 苏北：194 万公顷滩涂（约 2 907 万亩），雨水充沛，宜于发展新型的绿色农业与循环经济（图 2）。

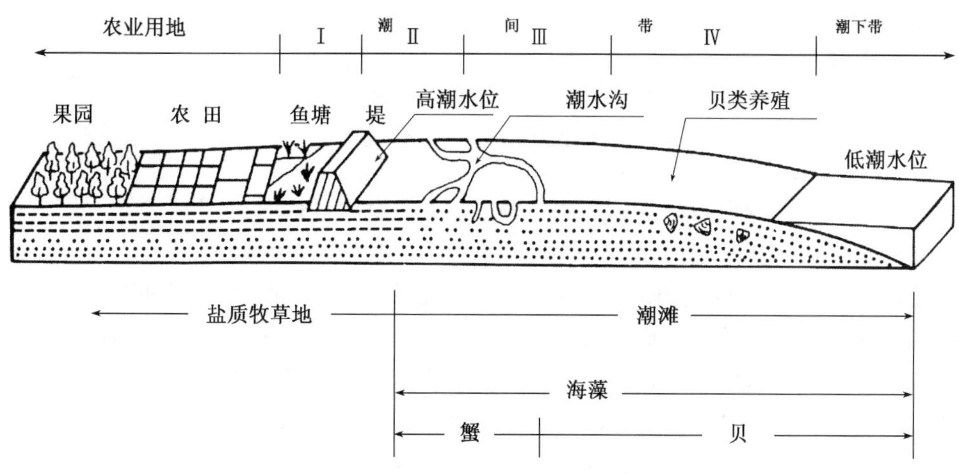

图 2　苏北如东海岸滩涂开发利用[7][10]

Ⅰ. 盐沼　Ⅱ. 泥滩　Ⅲ. 粉砂、淤泥滩　Ⅳ. 粉砂滩

- 潮下带：捕捞与放养鱼、虾，水产繁殖（海马、鳗仔、虾、蟹）。

• 潮间带：光滩养殖。粉砂滩：紫菜、文蛤；泥滩：蛤、沙蚕。

• 潮上带：芦苇田、鱼塘；草滩，放牧、养殖鳗、蟹、虾；随着土地成熟程度可种植桑、棉、麦、稻（淡水充沛处）或开辟为药、花、林、果种植园。

苏北海岸不同地段，因海岸带坡度、泥沙供应、动力作用效应等不同，潮滩的蚀、淤状况有区别，开发利用状况具有差异（图 3 和图 4）。

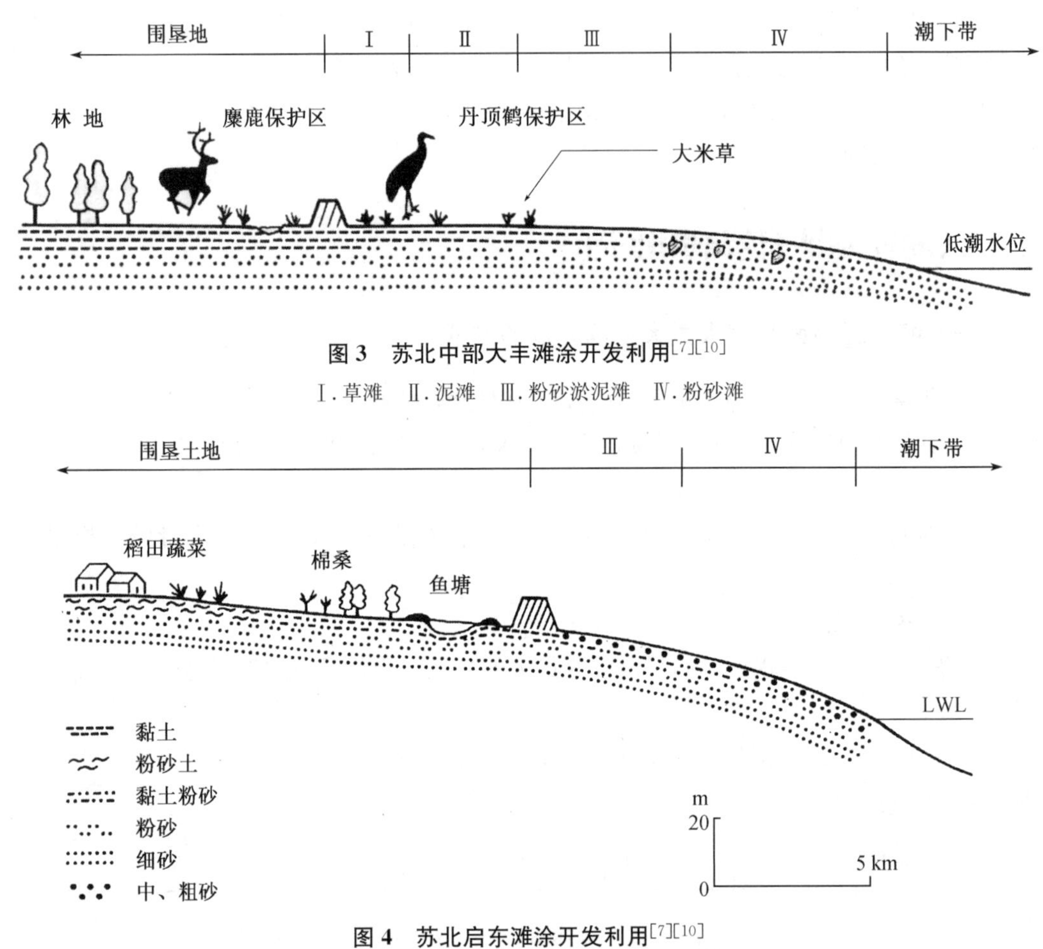

图 3　苏北中部大丰滩涂开发利用[7][10]

Ⅰ.草滩　Ⅱ.泥滩　Ⅲ.粉砂淤泥滩　Ⅳ.粉砂滩

图 4　苏北启东滩涂开发利用[7][10]

Ⅲ.粉砂淤泥滩　Ⅳ.粉砂滩

人口增加，海岸湿地为新生的土地资源，必会为内地移民及沿海经济发展所开发利用。但是，大面积砍伐红树林，疏干草滩，围垦土地，使天然海岸湿地面积减少，环境退化；20 世纪 60 年代人为引进英美的大米草，期望为促淤固滩，但是，大米草在湿热的浙闽沿海疯长，形成外来品种扼杀当地群落之危害。如此等等，向人们提出一个新任务：如何解决湿地保护与人民生计这一对新矛盾？作者认为：出路可能是系统调查研究，制定整体规划，在两者之间找个度——临界值（点、区），保护最佳处，开发最宜点。

（二）海岸湿地的冲淤动态问题

平原与河口湿地海岸由松散沉积物组成，它存在一个内在规律，即海岸动力与泥沙供应间的动态平衡问题[1]。陆源泥沙供应量大于浪、流冲刷携运量，海岸淤长；泥沙补给量与海岸动力冲刷量相当，海岸持有动态平衡；陆源泥沙补给量小于海岸冲刷量，则海岸后退。

实例比较：

(1) 苏北原沿阜宁、盐城、东台的公路大体上是沿北宋时修建的挡海土堤——范公堤修建而成，范公堤是利用了 6 000 年前的贝壳堤古海岸为基础。自 1128 年，黄河夺淮入黄海，每年携运汇入高达 1.4 亿 t 的泥沙，500 多年间使海岸向海推进了近百公里，位于苏北中部的范公堤距现代海堤 40 km。1855 年黄河北归返回渤海，泥沙补给中断，废黄河口遭受显著侵蚀，新海堤修建于老海堤之内，仍节节后退。但是，相对照的是：位于辐射沙脊群后侧受到保护的海岸，潮滩宽度达 30 km，因有来自水下沙脊的泥沙补给，目前仍在淤长(图 5)[11]。

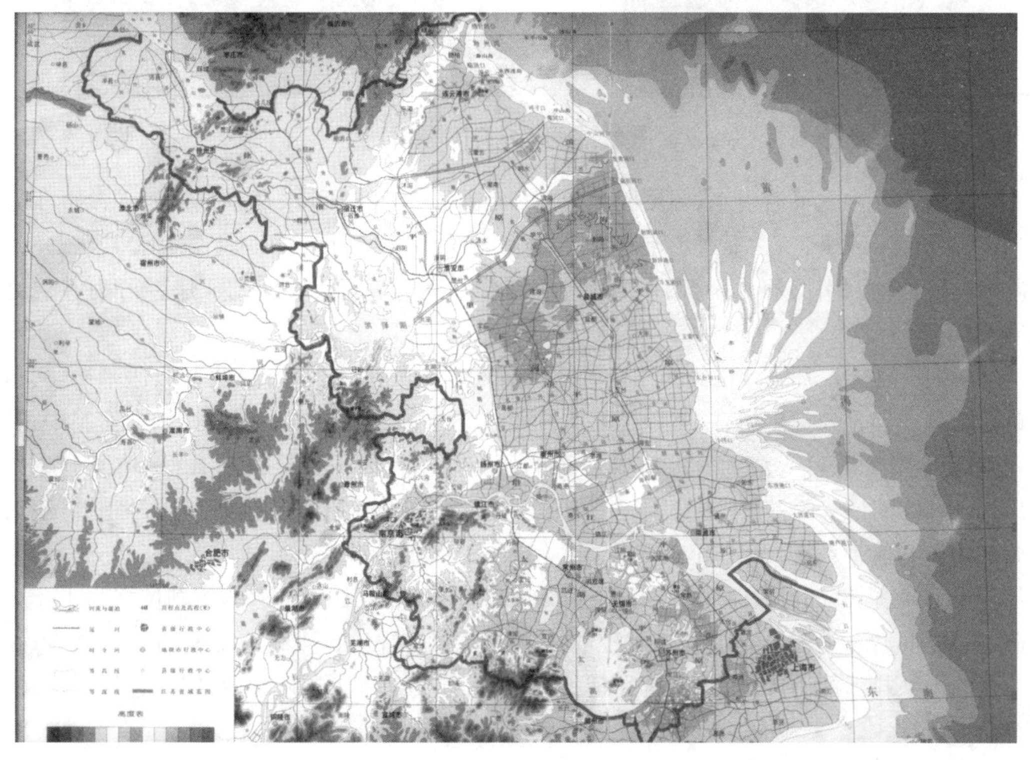

图 5　苏北海岸图(附彩图〈上册〉)

(2) 在渤海湾，自 1855 年黄河返回以来，已发育成为巨大的扇形三角洲以及在入海口外深度达－15 m 的水下三角洲(图 6)。但是，由于河床淤高加积迅速，原黄河三角洲河道迁移，大约每 6～9 年，河道从高河床向低处搬迁，因此河口也是摆动的，呈现自 N→NE→E→SE→S→N 的顺时针摆动。每当河口改道，新河口两侧会形成向海伸出的指状沙咀，沙咀两侧形成烂泥湾，泥沙在河口及两侧淤积，而废弃的老河口则遭受冲刷，海岸后退。由于胜利油田及东营市的发展，人为限制其改道，现代黄河口沙咀逐年南偏，此情况能维持多久？下一步又如何变化？这又是个新问题[12]。

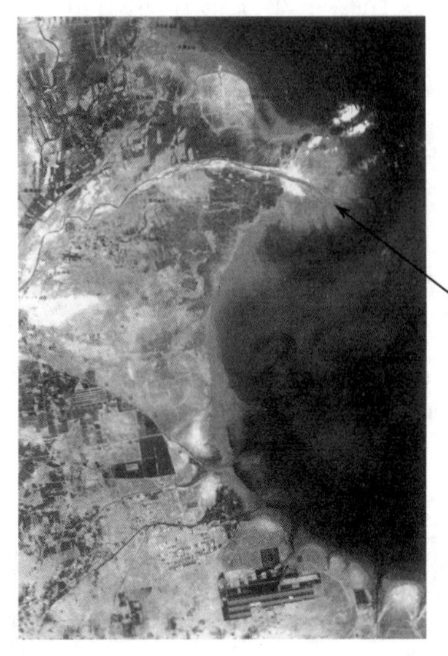

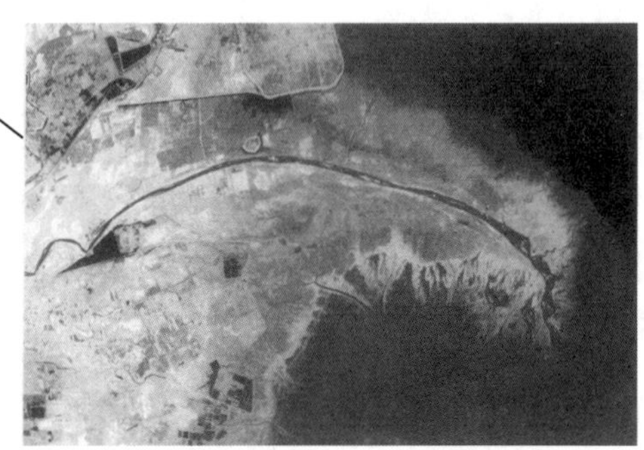

图 6 黄河三角洲(附彩图)

黄河因上、中、下游引水,屡屡断流,1996 年断流达 226 天,水量拦截入海泥沙逐年减少,从 11 亿 t 到 9 亿 t。现人工调水入海,水量有限,泥沙多被拦积陆上,黄河口及三角洲湿地发展如何?引滦河水济天津,河口泥沙锐减,三角洲已侵蚀后退,是否可为黄河之借鉴!

(3) 长江口入海泥沙量约为 4 亿~5 亿 t、入海流量 9 322 亿 m^3,两者似达平衡,河口略有淤积。但是长江上游在建坝;三峡大坝已拦蓄水沙;中线、东线引水计划在实施中;长江到底以多少水、沙入海为宜?应做系统的研究。当代海平面持续上升(平均上升率约 1.4 mm/a),加之,长三角地区过量开采地下水与建筑重载招致的地面沉降,吴淞口相对海平面上升速率更大:2010 年较 1990 年上升 15~20 cm,2030 年上升 25~35 cm,2050 年将上升 40~70 cm*。未来海平面上升 50 cm 时,黄浦江百年一遇的高潮位将为十年一遇,风暴潮频繁;河流入海水量与泥沙量减少,招致海岸侵蚀;滩涂与湿地面积将减少;长三角的水系排泄能力降低,内涝时间延长;海水入侵使地下水质咸化等,灾患连锁反应,形势会很严峻。

总之,从大河三角洲与沿海湿地发展趋势分析,深感到大型河流水利工程、海岸带建设发展,必须研究自然环境特点与其内在演变规律,应考虑流域的系统效应,海陆结合地研究与整体规划,可能是实践人地和谐相关发展的正确途径。

参考文献

[1] Ying Wang. 1980. The coast of China. *Geoscience Canada*, 7(3): 109 - 113.

[2] 王颖,朱大奎. 1990. 中国的潮滩. 第四纪研究,(4):291 - 300.

[3] 王颖. 1963. 红树林海岸. 地理,(3):110 - 138.

* 中国海洋报 2005. 1. 21 第二版,据任美锷与上海台风研究所预测之综合。

[4] 林鹏. 1981. 中国东南部海岸红树林的类群及其分布. 生态学报,(1):283-290.

[5] 毛树珍,黄金森. 1987. 我国的红树林海岸. 热带地貌,8(2):4-12.

[6] Zhu, Dakui. 1988. Mangrove Coast of Hainan Islands, China. *Galaxes*, (7): 251-255, Japan.

[7] Y. Wang, X. Zou and D. Zhu. 2001. The utilization of coastal tidal flats: a case study on integrated coastal area management from China. In: *Muddy Coast Dynamics and Resources Management*, ELSVIER. 287-294.

[8] 王颖,朱大奎,曹桂云. 2003. 潮滩沉积环境与岩相对比研究. 沉积学报,21(4):539-546.

[9] 王颖. 2003. 黄海陆架辐射沙脊群. 北京:中国环境科学出版社,418-427.

[10] Zhu Dakui, et al. 1988. Morphology and Land use of Coastal Zone of the Jiangsu Plain, China. *Journal of Coastal Research*, 14(2): 591-599.

[11] Ying Wang. 1983. The mudflat coast of China. *Canadian Journal of Fisheries and Aquatic Sciences*, 40(Supplement No. 1): 160-171.

[12] Ying Wang, Gustaf Arrhenius and Yongzhan Zhang. 2001. Drought in the Yellow River—an Environment Threat to the Coastal Zone. *Journal of Coastal Research*, ICS 2000 Proceedings, 503-515.

Characteristics and Exploitation of Coastal Wetland of China*

1 Environment

The total area of China's wetland is 66 million hectares, which is about 10% of the total wetland area of the world, i. e. it ranks number four in the world and the number one in Asia[1].

Coastal wetland of China is consisted of 2 million hectares of less than 2 meter's low land, 9.7 million hectares of tidal flats[1,3], and associated biological community. The total area covers 1/5 of China's wetland and runs across 39 latitudes in the northern hemisphere. The nature of coastal wetland is characterized by monsoon wave action, tidal dynamics and large river influence, i. e. the land sea inter action processes[2].

1.1 Three types of coastal wetland of China according to the genetic features (Figure 1)

(1) Estuary wetland. Taking the Yellow River, the Changjiang River and the Shuangtaizi River mouth of the Songliao Plain for example, the wetlands have been influenced greatly by fresh water. With reeds flat, muddy flat and grass flat as its major feature, the major grasses are consisted of *Scirpusmangueta*(三棱草), *Typhaangustifolia*(蒲草), *Suaeda*(碱蓬) and *Spartina Spp*(大米草). The wetland is also the habitat of several migratory birds—such as *Grusjaponensis*(丹顶鹤), *Grus monachirs*(白头鹤), *Ciconia nigra*(黑鹳), and fresh-salt water animals, some of which are in dangerous state—such as *Lipotes vexillifer*(白豚), *Acipenser sinensis*(中华鲟) and *Neophocaena phocaenoides*(江豚), and so forth.

(2) Plain coastal wetland. It is distributed widely along the coasts of Bohai Bay and North Jiangsu coastal plain with an altitude of 0 to 4 meters. There are brackish reeds swamps located in the outer boundary of land, Spatina and Suaeda Swamps distributed between the area of average spring high tidal and average high tidal levels, and part of Subtidal zone Grass flat and salt marsh are the major features. The wetlands along the North Jiangsu coast are the winter habitat of *Grusjaponensis* in the outer part of tidal flat, and the David's deer protected zone in the inner tidal flat.

---

* Ying Wang, Da-kui Zhu:《长江流域资源与环境》,2006年第15卷第5期,第553-559页。

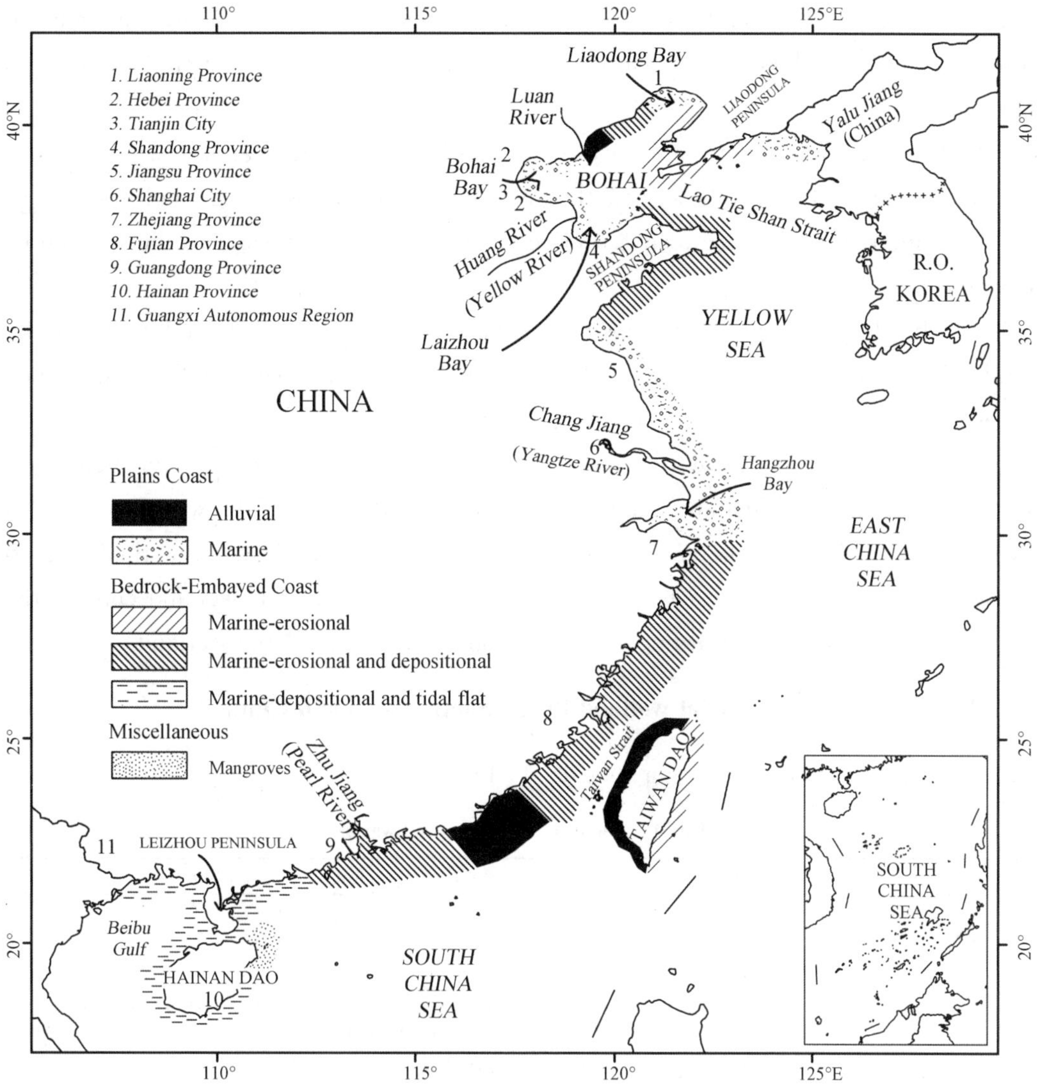

Figure 1 Coastal Types of China

(3) Wetland of sheltered bay head. This kind of coastal wetland is developed in the coastal bays of Zhejiang, Fujian, Guangdong, Guangxi, and Hainan provinces. There are mainly mangrove swamps and coastal coral reefs. Several coastal bays have been used as farm land or aquaculture ponds.

1.2 Geographic environment of coastal wetland

(1) Most wetlands are developed in the river mouth and plain coast with tidal current influenced fresh water and salt water interactive environment. Coastal wetlands are not developed in the area with strong wave action or wide surf zone.

(2) Silts or sandy mud are the optimum bottom sediments, while thick and sticky mud or clay bottom are not suitable for marsh plants growth. Even though, mangrove forests can grow on the basalt rocky coast, such as the mangrove forests in Hainan of

China, Bombay of India and several islands of Indonesia, because there are quite a lot of joints, fissures and holes in the basalt lava, wherein it is easy for air and water to go through.

(3) Mangrove swamps are mainly developed in the tropical area between the Capricorn and the Cancer, however, they also can be developed in the high latitude as 32°N. While there exist warm current effects, only rare, lower mangrove bushes can be developed.

(4) In China, mangrove swamps are the major type of coastal wetland in the south, while the grassed salt marshes are dominant in the north, and reeds ponds are developed widely in the fresh water influenced area.

All of three types have the zonation features[2]. There are superior tidal zone between average high tidal and maximum high tidal levels. Inter tidal zone lies between average high and low tidal levels. Sub tidal zone extends from average low tidal level to the water depth equivalent to 1/2～1/3 of local wave length. But wetland ends in the water depth of 6 meters according to international definition of "the wetland".

1.3 Examples of coastal wetland in the South and North China[3-5]①

(1) Coastal mangrove swamps (Tab. 1, Figure 2).

Tab. 1 Zonation of Mangrove Swamps

Coast location	mangrove assembly
Coastal dune & fresh water swamps	Tropic forests ↑ *Conocarpus*(椎里木) *Laguncularia*(拉贡木) → Transitional assembly (semi-mangrove 半红树)
Average high tide level	
↑ Beach of shoal patches of inter tidal zone ↑	*Avicennia salt swamps assembly*(海榄雌、黑红树) (Regular flood over and dry out by tides, Salty water with several grass islets in the forests) ↑ Mature mangrove forest of *Rhizophora* assembly (Dense roots system of arbor mangrove forests trapping silt to form abounded biomass content)
Average low tide level	
Uncontinuing sand ↑ Outer shallow water muddy patches	Pioneer mangroves (*Avicennia*, pioneer land producer) ↑ marine aquatic plant

① Mao Shuzhen, Huang Jinsen. Mangrove coast of China. Tropic Coast Geomorphology, 1987 (in Chinese). (毛树珍,黄金森. 我国的红树林海岸. 热带地貌,1987.)

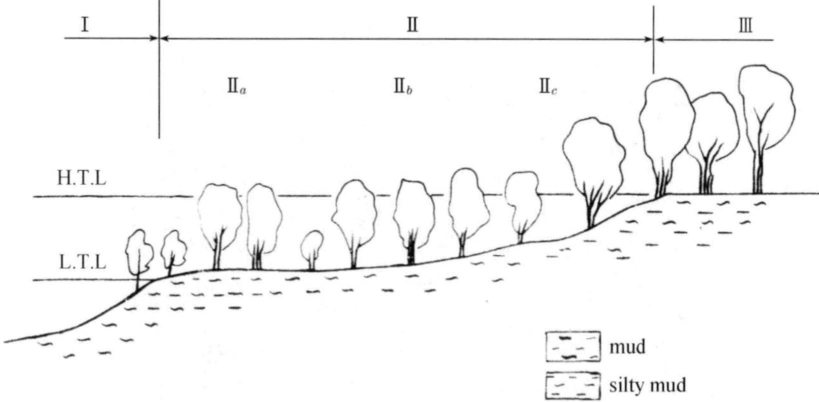

Figure 2 Wetland Profile of Mangrove Swamp

Ⅰ. Sub tidal zone mudflat, Ⅱ. Inter tidal zone mangrove forest, Ⅲ. Superior tidal zone mangrove forests of *Hibiscustiliaceus*-Excoecaria agallocha assembly (黄槿—海漆群落) and *Avicennia marina-Sonneratia* Caseolaris assembly (白骨—海桑群落), Ⅱ a. Low-tidal level of *Avicennia marina-Sonneratia* Caseolaris assembly (白骨—海桑群落), Ⅱ b. Mid-tidal level of Rhizophore-Kandelia candel assembly, Ⅱ c. High-tidal level of *Conocarpus*-Bruguiera sexangula assembly

(2) Plain wetland of tidal flat and salt marshes along the Yellow Sea, and the East China Sea. They are widely developed along the fringe zone of the North China and north Jiangsu plains, the wetland coasts are formed by river-sea interactive processes, and wetlands developing trend relies on fluvial sediment supply to form coastal deposition or erosion, and biological assembly characteristics. Such as, the abandoned Yellow River mouth, in the north Jiangsu, suffers from coast erosion since the Yellow River shifted back to the Bohai Sea in 1855, but the coast flat in the south is accumulated and advanced to sea as it not only sheltered by offshore sandy ridges to avoid wave erosion, but also with sediment supply from outer submarine sandy ridge field.

In 1975, the total area above 0 m counter of the radioactive sandy ridge field were 2 125 km^2, and reached 3 782 km^2 in 1995. It was 80 km^2 increasing annually. Thus, the radioactive sandy ridge field of the Yellow Sea is the largest one of coastal wetlands in China. The sandy ridge field was consisted of old Changjiang River sediments. While the river entered the Yellow Sea 30 000 years ago from the Qianggang area, the sandy bodies have been reformed by tidal current processes since the post glacial period. The total area of the sandy field is 23 000 km^2, which is about 1/5 of the total area of Jiangsu Province (Figure 3)[6-8].

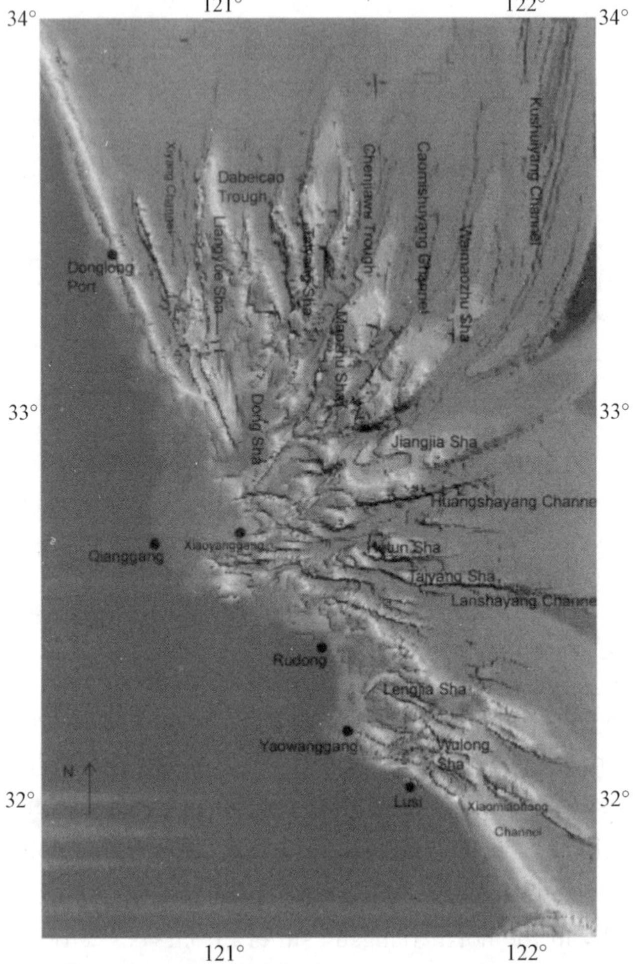

Figure 3 Largest Coastal Wetland of Radioactive Sandy Ridge Field of the Yellow Sea(附彩图)

The plain coastal wetland has zonation features, which can be compared with the world types (Tab. 2).

Tab. 2 Zonation of Plain Coastal Wetland[8, 11]

Location	Zonation	Sediment	Biological assemblage
Coastal lowland	salt earth plain and depressions shell beach ridges	clayey mud, silt shell, shell fragments fresh water pools	settlement villages, historical material
Superior tidal flat	grass flat salt marsh (flood over during spring high tide)	reed and sedge swamps *Spartina* saltmarsh Suaeda salt marshes	David's deer protect area pasture land, reclaim farm land to plant rice, wheat, mulberry field, fish pond migratory birds, perch sandy worm
Inter tidal zone	high tide mud flat muddy poor	clayey silt, silty clay	grass roots or *Spartina* remainders, Ark Spp, Scapharca subcrenata and Periophthaimus

(Continued)

Location	Zonation	Sediment	Biological assemblage
	mid-tide level of mixed flat surface erosional channels and bodies	clayey silt and fine sand interval, bare flat	shells migratory birds
	low tide silt flat	silt and very fine sand ripples	bare flat, shell, fish, aquaculture of seaweed
Low tidal zone	subtide muddy flat	very fine sand silt clayey silt	sediment size decreased towards sea, suspend sediment settle down processes, fish shrimps and prawn

Grusjaponensis always gather in the depositional muddy silt flats of estuary area, where exist fresh water input, feedings of fish and shrimps, and without danger to sink in the thicker mud. But David's deer prefer to multiply on wild salty plain far away from human settlements.

Thus, there are different coastal wetland environments dependent on different climate conditions, geographical locations, coastal dynamics, sediment situations, and biological assemblages.

2 Exploitation of the Coastal Wetland

2.1 Challenge of large scale exploitation by human beings

(1) Wetlands have been exploited as farm lands, villages and towns while the area has plenty of precipitation.

(2) Mangrove forests have been cut down to produce charcoal, or developed as tourist parks, aquaculture field and farm land.

(3) In the bay head of Zhejiang and Fujian provinces, wetlands of muddy flat have been rented to local people, used as fish ponds, aquaculture field for oyster or shell fish production, and the superior flats have been used as orange fruit field and flower nurseries.

(4) In respect to tidal flat wetland of plain coast[7,10], the situation is as following:

① In North China, under dry climate and muddy flat condition, most coastal wetlands have been reformed as largest salt field, petroleum field and new towns.

② In north Jiangsu, plenty precipitation about 77×10^4 hm^2 of tidal flat are ideal site to develop green agriculture and cyclical economy. Nowadays, practice is carried out as following:

- Sub tidal zone can be used for fishing, aquaculture of fish, shrimp, sea horse, and eel.
- Inter tidal zone uses bare flat for aquaculture.

• Silt flat for sea weed and meretrix aquaculture.

• Mud flat for sandy worm and Scapharca subcrenata growing.

• Superior tidal zone is used as reeds or/and fishing ponds; grass flat used as pasture land or aquaculture of eels, crabs and prawns; according the situation of land nature, it can be used as mulberry, cotton, wheat, rice or hurb, flower, and fruit cultivation (Figure 4, Figure 5, and Figure 6).

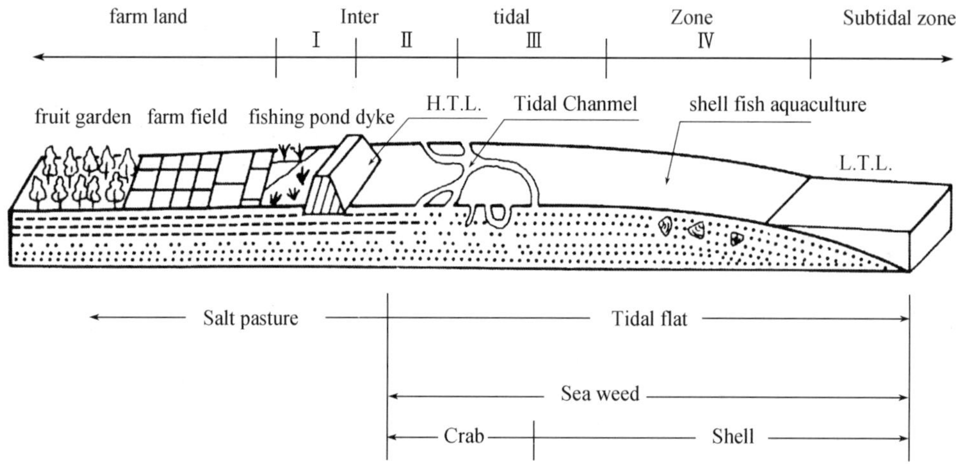

Figure 4　Coastal Tidal Flat Utilization of Rudong County, Southern Part of North Jiangsu Coast[6,9]

Ⅰ. salt swamps　Ⅱ. mud flat　Ⅲ. silty mud flat　Ⅳ. silt flat

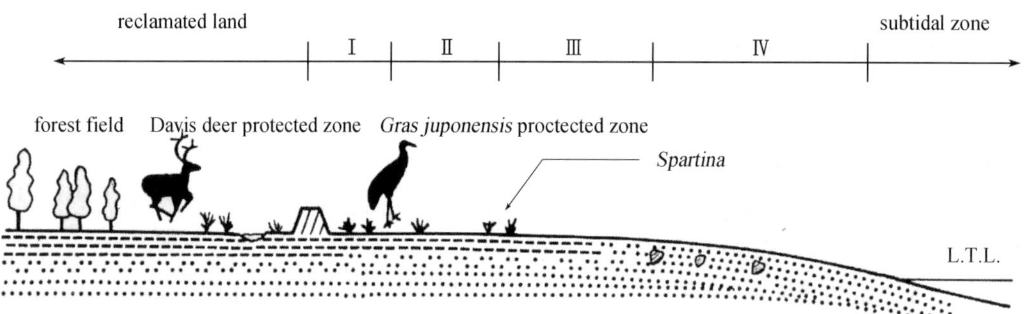

Figure 5　Tidal Flat Utilization of Dafeng City, Middle Part of North Jiangsu Coast[6,9]

Ⅰ. marsh　Ⅱ. mud flat　Ⅲ. silty mud flat　Ⅳ. silt flat

With the population increasing, the coastal wetland will be more and more used for inland settlers and coastal economic development. However, lumbering mangrove forest, dried grass flat, reclaimed coastal wetland have caused the diminution of wetland area and environmental deterioration rapidly.

Spartina was introduced into China during the 1960's for improving siltation and stability of tidal flat as a new land. However, *Spartina* grows crazily in the humid subtropic climate area. As a result, it becomes an invader to harm local species of plant. Thus, people faces a challenge of how to solve the confliction between wetland

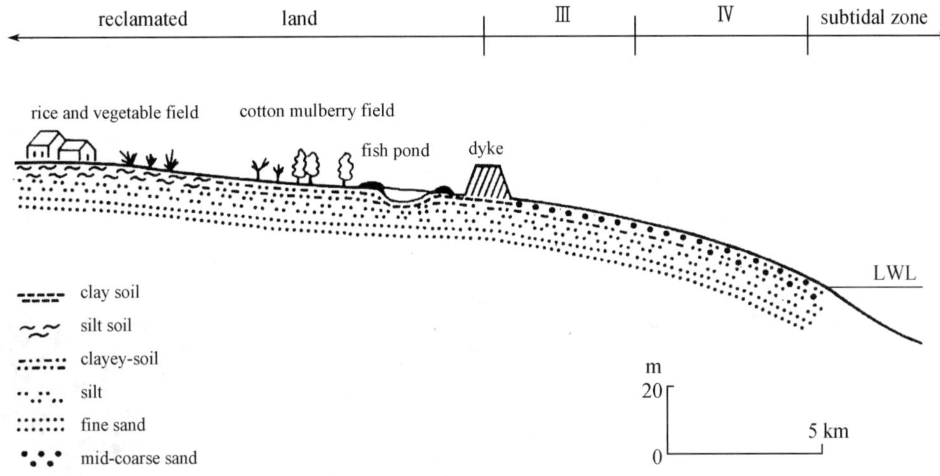

Figure 6　Tidal Flat Utilization of Qidong County, North Jiangsu Coast[7,10]

Ⅲ. silty mud flat　Ⅴ. silt flat

ecological protection and basic living requirement of local residents. We must carry out systematic investigation and comparative study to summarize experiences of successful and unsuccessful development practice, and to find out the critical value (point or range), for best protection of wetland's nature and suitable way of wetland utilization, then, to work and practice a master plan for regional or local development.

2.2　Dynamic Movement of Coastal Wetland

Plain and estuary wetlands consist of loose sediment, and thus, there is a natural regulation of their dynamic trends of the wetland development: the balance between coastal dynamic energy and the quantity of terrigenous sediment supply[1]. The coast will be accumulated and advanced towards the sea, if the quantity of terrigenous sediment supply is larger than the transportation capacity of wave and current energy; vis-a-vis, the coast would suffer from erosion and retreat back landward; if the quantity of sediment supply is equal to the erosional energy of wave and current dynamics, the coast would be stable as in equilibrium status.

The following examples indicate the natural regulation:

(1) The Fangong Dike, from Funing to Dongtai through the Yancheng in Jiangsu Province, was built up on the foundation of a 6 000 y. B. P's natural shell beach ridges during the North Song Dynasty (960 – 1127 A. D.). Since 1128 A. D., the Yellow River had migrated through the Huai River and entered into the Yellow Sea. Because there were 14×10^8 tons sediment being carried out into the Yellow Sea by the river, sediments were redistributed along the adjacent coastal zone. As a result, the river mouth coast advanced towards the sea almost 100 km during the period of about 700 years, and now the Fangong Dike was located inland about 60 km away from modern

coast. After the Yellow River returned back to Bohai Sea in 1855, there has been no sediment supply from the large river, as a result, the abandoned Yellow River mouth of the Yellow Sea suffered from erosion, original villages were submerged under sea water, and the new dykes were built up landward one by one in the abandoned river mouth area. On the contrary, the coast sheltered by offshore submarine sandy ridges received sediment supply from the eroded outer ridges, thus, the tidal flat has accumulated to be 30 km wide and is advancing towards the sea (Figure 7).

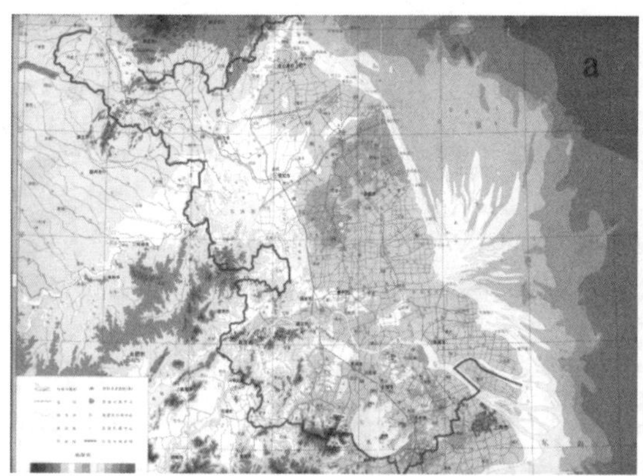

Figure 7 Map of North Jiangsu Coast

a. Map(附彩图〈上册〉) b. Remote sensing images

(2) In the Bohai Sea, since the Yellow River returned back in 1855, it has formed a huge Yellow River delta reaching 15m in water depth. Because of the rapid depositions, river bed elevates annually. Thus, whenever after 6 to 9 years, the river channel on the delta would be shifted to lower plain, to form a new channel, and the shifted flow route will almost follow a clockwise pattern—N →NE→E→SE→ S→ N. Whenever a new channel is formed, it will deposit two finger-shaped bars along both sides of the river mouth. On both sides of coasts, protected by bars, the deposited mud would form muddy pool and it can calm down the wave energy in the bay. But the abandoned former channel parts would suffer from coast erosion as sediment supply suddenly stops. Because of the Victoria Petroleum field and Dongying City have been set up at the lower reach of the Yellow River delta since 1970's, it is necessary to stop the river channel migration artificially, only allowing the Yellow River to enter into the Laizhou Bay. Now, the river mouth shifts gradually in a curve shape to the south (Figure 8). While human changes a river fixed in one spot, how long the river mouth will retain? What will happen in the future? It has raised a kind of new problems [11].

On the other hand, the lower reach, especially in the delta area of the Yellow River, had stopped flowing for 226 days in 1996, because it had diverted water volume beyond river capacity in the upper and middle reaches[9] At present, in order to control run off diverting, definite volume of water should be distributed for delta reaches. What

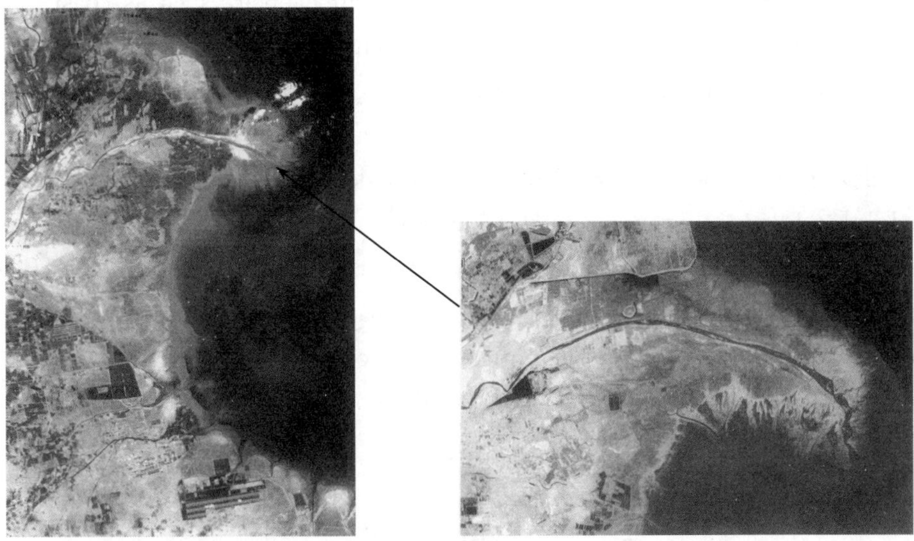

Figure 8　Curved modern delta of the Yellow River(附彩图)

would happen next, when the dam of Sanmenxia Gorge blocked the sediment from running down to lower reaches? It has decreased the sediment input to the Bohai Sea from 11 billion tons to 9.0, and then 6.0 billion tons, and it will continue to decrease. What will be the impact on the natural development of the delta? It notes that since the project of diverting water volume of Luanhe River to supply Tianjin City in 1970's, the delta of Luanhe River has eroded and retreated apparently. Can't it be the lessons for the Yellow River delta?

(3) The annual sediment of the Changjiang River entered the East China Sea was 4～5 billion tons in 1990's, and the volume of water discharge 9 322 billion m^3. It seems to reach a dynamic equilibrium with a little deposition of the delta. However, the Three Gorges Dam will block the down-flowing sediment completely, and there is also a plan for water diverting from its middle and lower reaches to supply fresh water to North China and the capital city. To face these changes, it really needs a systematic planning to figure out how many billion tons of water and silt should be kept for the estuary area and the East China Sea. Besides, the delta also faces the problem of sea level rising (annual rate 1.4mm), with additional land subsidence as a result of over pumping of ground water, and the compacted ground sediment. Taking the Wusongkou Tidal Gauge Station as an example, it has shown a rapid rising of local sea level: compared with 1990, in 2010 it will be 15～20 cm, 25～30 cm in 2030, and 40～70 cm in 2050①. While the sea level rising is 50 cm, the hundred year's highest tidal level of Huangpu River will be once per 10 years. As a comprehensive result of frequent storm surge, water and

① China Ocean News Paper, 21. Jan, 2005. 2nd edition al page.

sediment discharge decrease, a series of disaster will happen, such as coast erosion, tidal flat and wetland diminishing, drainage capacity decrease, longer period of delta land water logging and fresh water becoming saltish. People will suffer from a series of disasters.

Analyzing the developing trend of large river delta and coastal wetland, it can be concluded that before a large river engineering project or coastal construction development starts, the environment nature and its evolutional regulation must be studied. First, it is necessary to concern the effect of whole river basin, and land-sea interaction, as a whole system, then, to work out an entirety plan for practice. It may be the right way to avoid disaster, and to achieve sustainable development so as to set up a harmonic relationship between nature and human beings.

References

[1] Wang Ying, Zhu Dakui. 1990. The tidal flat of China[J]. *Quaternary Sciences*, (4): 291 - 301.

[2] Ying Wang. 1980. The coast of China[J]. *Geoscience Canada*, 7(3): 109 - 113.

[3] Wang Ying. 1963. Mangrove coast[J]. *Geography*, (3): 110 - 138.

[4] Lin Peng. 1981. Coastal mangrove assembly and distribution of southeast China[J]. *Acta Ecological Sinica*, (1): 283 - 290.

[5] Zhu Dakui. 1988. Mangrove coast of Hainan Islands, China[J]. *Galaxes*, (7): 251 - 255.

[6] Y Wang, X Zou, D Zhu. 2001. The utilization of coastal tidal flats: a case study on integrated coastal area management from China [A]. In: Muddy Coast Dynamics and Resources Management [C]. ELSVIER. 287 - 294.

[7] Wang Ying, Zhu Dakui, Cao Guiyun. 2003. Study on tidal flat environment and it's sedimentary comparative[J]. *Acta Sedimentologica Sinica*, 21(4): 539 - 546.

[8] Wang Ying. 2003. The radioactive sandy ridge field of continental shelf of Yellow Sea[M]. Beijing: Environmental Science Press of China. 418 - 427.

[9] Zhu Dakui. 1988. Morphology and land use of coastal zone of the Jiangsu plain, China[J]. *Journal of Coastal Research*, 14(2): 591 - 599.

[10] Ying Wang. 1983. The mudflat coast of China[J]. *Canadian Journal of Fisheries and Aquatic Sciences*, 40 (Supplement No. 1): 160 - 171.

[11] Ying Wang, Gustaf Arrhenius, Yongzhan Zhang. 2001. Drought in the Yellow River and environment threat to the coastal zone[J]. *Journal of Coastal Research* (ICS 2000 Proceedings): 503 - 515.

海洋文化特征与江苏海岸海洋文化发展*

一、海洋文化特征

海洋文化与所在区域的海洋环境、资源特点及经济发展水平密切相关。随着科学与经济的发展进步，海洋环境与资源状况的变化，就会影响反映于海洋文化，并具有不同的内涵。海洋文化具有时代的特征、区域的特征，以及当代全球一体化发展的特征。

1. 广阔的海洋是全球联通的，海洋文化的发展具有开放、传播以至全球影响流动的特点

回顾世界海洋文化的发展，均反映出海洋与人类生存密切相关，但因生产力的发展水平不同，而具有明显的时代特色。

(1) 原始社会-石器时代，已是海洋文化的始萌期。人类(包括最初的猿人)从海岸地带捕捉鱼、虾、贝、蟹，以鱼骨为箭弩猎取禽兽为食，进而豢养家畜与种植稻粟等。在太平洋两岸发现砖石质网坠、岩浆岩质石臼等；在我国周口店、辽宁、河北、浙江河姆渡以及海南岛北部湾等处的古海岸阶地上均发现有绳纹瓦器皿的残片及早期以渔、猎、耕、稼活动为特点的古文化遗迹。上述事例反映出人类生存活动的最初阶段即与海洋密切相关，是海洋文化的初期反映。

(2) 新大陆发现、航海事业发展，对非洲掠夺及贩奴热、殖民地占领与土地分割，贸易与军事争夺，是欧洲封建社会末期与资本主义发展早期与海洋相关的活动；《鲁滨逊漂流记》《基督山恩仇记》等一系列著作，部分地反映当时海洋文化的特点。15 世纪以后，葡萄牙、西班牙、英国、荷兰的崛起，都是靠海上争霸发家的。

(3) 英国 18 世纪中叶，工业革命兴起，开始了开拓海外市场、发展英国海洋经济的历程。经过大约 1 个世纪，英国成为世界贸易大国。借助海洋，其势力、文化与宗教传播至世界各地。19 世纪后，美国实现工业化，主要表现在形成沿海与五大湖区工业化城市带：大西洋海岸的波士顿、纽约、巴尔的摩、华盛顿，太平洋沿岸的西雅图、旧金山、洛杉矶与圣地亚哥，海外贸易带动海洋经济，带动区域发展，以致美国文化、美国生活方式，尤其是在 20 世纪 40 年代以后在亚洲、欧洲获得广泛的传播。

(4) 20 世纪 50 年代后，亚太经济发展之路，同英、美经历相同。日本发展海岸工业，以海外贸易和海运事业为纽带，促进日本经济腾飞，形成东京—大阪—神户和名古屋为主体的深水港群与大城市群。日本平均每 30 km 海岸线就建有一个深水港，兴起海港城市促进海外贸易发展，利用国外的广泛资源发展“两头在外”的临海工业，推动日本经济高度发展。以海洋经济为基础，加上临海工业，相关产值占日本总产值的 1/2。随着日本经济复苏，电器、汽车制造与化肥等商品交易，活跃日本的影视、语言、文学、科技以至餐饮等，

* 王颖：国家海洋局直属机关党委办公室主编，《中国海洋文化论文选编》，231 - 240 页，北京：海洋出版社，2008 年 8 月。

亦伴随经贸而传播，又一次兴起“东洋”留学之热潮。

(5) 20 世纪 60～70 年代亚洲“四小龙”(新加坡、韩国、中国的香港与台湾地区)兴起，是借助于发展海洋经济。台湾西岸以高雄港为中心，建加工区，发展海外全球化经济。海洋经济的发展、商品的交流，有力地促进文化传播与海洋文化的发展。与新加坡的人才交流，韩国以及中国港台的科技、影视、文学著作、饮食风俗等在亚太区形成传播热潮。

(6) 21 世纪，海洋经济高度发展，2000 年世界海洋总产值超过 15 000 亿美元。当代全球化战略，与能源、水资源、矿产资源的争夺，及海疆与海峡通道的纷争等有关。北海油气资源开发，促进了荷兰、丹麦等北欧诸国经济的又一轮繁荣。海洋经济发展的同时，欧洲文化再次向海外传播。近期，俄、加、美对极地与北冰洋的争论，核心是对导弹、潜艇活动基地的争夺，反映出全球化下的海洋资源、环境、疆域纷争是当代海洋文化的突出特点。

2. 海洋文化的第二个特点，是具有区域性特点

即使是在全球一体化的发展趋势中，区域海洋的文化烙印仍然显著。原因在于，在人类活动的历史中，广阔的海洋水域是徒步迁徙与交通往返的障碍。例如史传：太平洋两岸人类迁徙始于第四纪大冰期，当白令海峡成为陆桥时，从亚洲大陆生存的古人向美洲迁移。美洲东岸土人，具有黑发、黄肤、高颧骨、狭长眼的外貌特征，与蒙古人种有关，该处一些农舍布局、摇篮、大蒜挂等生活习俗与亚洲民俗相类似。太平洋东岸具有亚裔、印第安裔和欧裔人种之复式结构与多元文化的结合，是该区域的特点。大洋的阻隔，在漫长的历史时期，形成了明显的区域海洋文化特征：

(1) 大西洋文化

渔猎、航海、15～16 世纪地理大发现、宗教传播、移民与港口城镇建设、海洋贸易、现代科技与油气开发等，组成大西洋海洋文化的特色内容。

(2) 地中海文化

航海发现、掠夺与海外开拓，多语种与殖民地文化、宗教、文艺复兴，多民族结构、农耕、田园与酿酒文化构成其独特之处。

(3) 太平洋文化

太平洋东西两侧经济、文化发展不平衡。

早期是西部的亚太文化——秦、汉、唐、明之儒家文化与佛教文化向日本、韩国、东南亚诸国传播；东南亚地居太平洋与印度洋交汇处，具有周边移民带来的多元文化(语言、文学、习俗)及佛教、回教与印度教交汇之特色(新加坡牛车水小区为代表)。20 世纪后期兴起的亚洲“四小龙”海洋经济文化，以及 21 世纪中国的海外贸易影响加大所兴起的以京、沪、穗为中心的中国文化效应。

太平洋东部是亚欧移民与美洲土著文化之结合。海洋文化以渔业(如：大马哈鱼与金枪鱼捕获、加工、外销)，牧业(牛、羊、驼畜牧)，肉、毛、皮加工、制造与贸易，海啸、地震灾害与宗教祈福等结合，成为具有特色的南美太平洋文化。

海岛文化是太平洋海洋文化的重要特色，从北向南众多的海岛跨越不同的气候带，经受不同陆地国家的政治经济影响，在人种、语言、文化、宗教与艺术活动等方面各有特色。虽然大洋是连通的，但限于生产力发展水平与海上交往能力以及长时期隔绝的历史，形成大和族文化、鲜族文化、汉族闽粤文化，夏威夷太平洋群岛与澳洲、新西兰等海岛移民文

化，决然有别。这可能是太平洋海岛文化研究的一个中心课题。海岛文化具有依海繁衍、安居、自力更生的特色，是蓝色海洋文化之代表。

3. 中国海洋文化隶属于亚洲—太平洋海洋文化，但更具有边缘海文化的特点

先民沿海聚居，开发早，但发展缓慢，主要是陆地一统文化为主。先秦统一中国，相传派遣 3 000 童男、童女赴东瀛海上，采取长生药，促成了海上移民与汉文化东传。盛唐与明朝的丝绸、瓷器、香料、药材贸易与移民，经海路至东南亚及东非，伴随贸易传播了语言、文字、汉医、宗教、艺术以及耕作、工艺技术，但主要限于第一岛弧以西的渤海、黄海、东海、南海等海域。

20 世纪以来，中国经海上与欧美交往增加。浙江温州的移民通过对西欧的商贸、餐饮与文化传播的世纪性延续已成为海商贸易的劲旅。当代中国海洋贸易，以能源、原材料、机械制造、轻工业、小商品以及餐饮等为多，船舶吨位居世界前列，贸易范围遍及欧、美、非、拉等大洲。中国海洋文化另一特色，是在 20 世纪 80 年代初以来，贯彻邓小平同志提出的改革开放的大政方针；中国派遣留学生至海外学习，一批批的海归学者促进了与海外的交流。国家海洋局负责组织我国科技界完成查清中国海的使命，同时，已进军三大洋、南极洲与北冰洋，建立科考站，持续不断地进行对海洋、大气、地质地貌、矿产、能源与水产等的考察活动。海洋文化内涵日臻丰富，科学著作、报告文学、影视、歌舞艺术已与世界海洋文化发展潮流一致，具有环球交流的特点。但是，历史上闭关锁国，以陆地聚居和以农耕为中心的传统思想影响仍深，至今，国人的海洋意识，我国的海洋科学、文化及国防力量仍需极大的增强。这些，可视为当代我国海洋文化发展的新起点。

二、江苏海洋文化之发展

江苏省地处我国东部沿海之中枢，是以由江、海交互作用形成的大河三角洲与滨海平原为主体，位于长江以北的海岸线长达 953.875 9 km（19 座岛屿岸线为 29.964 1 km）。江苏省经济、教育发达，但是海洋文化仍居于初级阶段，虽始于徐福东渡、汉军戍边，但经长期封滩防海，仍以渔、盐、围垦等经济活动为主，缺乏大规模的现代海洋经济活动，未形成海洋文化体系。但是，江苏的海岸海洋独具特色，海洋科技力量强，海洋文化与海洋经济发展潜力极大。

1. 海岸类型丰富

北部苏鲁交界地的赣榆地区，是在准平原基础上发育的砂砾质海岸（图 1），岸线长 30.062 5 km；连云港一带具有 40.248 3 km 的基岩港湾海岸，是海侵基岩山地发育，谷地被河-海交互沉积所淤填发育成海湾或滨海平原；主体的海岸类型为粉砂淤泥质平原海岸，岸线长达 883.561 km。平原海岸外貌单调，但由于系江-海交互作用、堆积形成，具有在我国沿海地带最广阔、“肥沃”的滩涂与土地资源，以及特有的深水航道资源（图 2、图 3、图 4、图 5、图 6、图 7、图 8），工农业进一步发展和与海外交往潜力大。

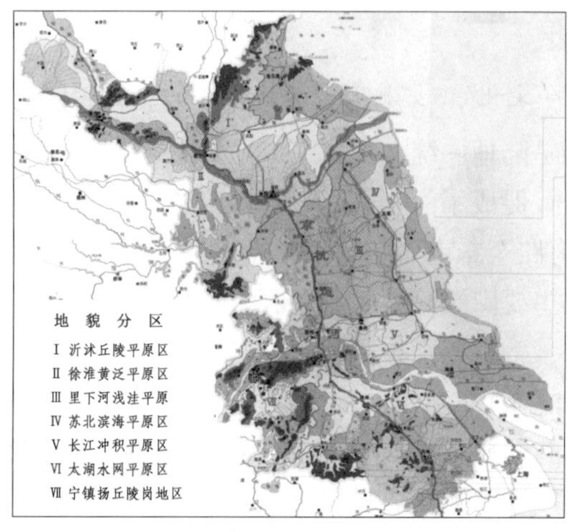

图 1－1　江苏省地貌图(附彩图〈上册〉)

据《江苏省地图集》中国地图出版社，2004.

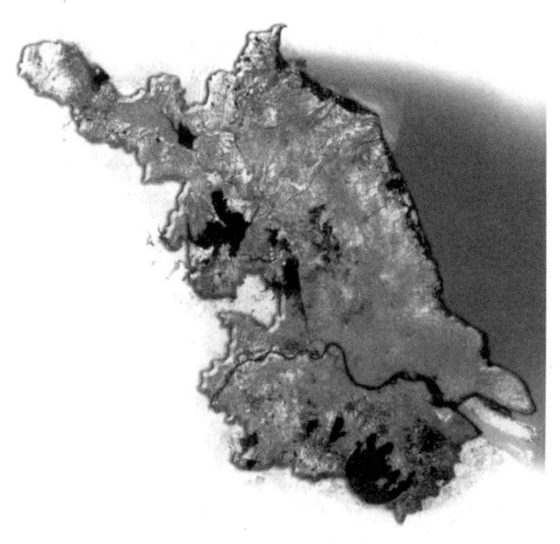

图 1－2　江苏省卫星图(附彩图)

图 2　淤泥草滩(附彩图〈上册〉)

图 3　高潮潮滩淤积带

图 4　中潮冲刷带

图 5　低潮粉砂淤积带

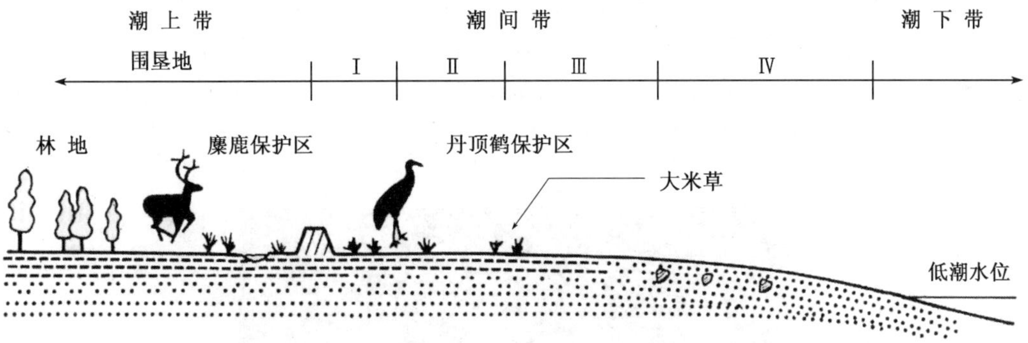

潮 上 带
潮 间 带
潮 下 带
围垦地
Ⅰ
Ⅱ
Ⅲ
Ⅳ
林 地
麋鹿保护区
丹顶鹤保护区
大米草
低潮水位

图 6　苏北潮滩自然保护区

潮 上 带
潮 间 带
潮 下 带
农业用地
Ⅰ
Ⅱ
Ⅲ
Ⅳ
果园
农 田
鱼塘
堤
高潮水位
潮水沟
贝类养殖
低潮水位
盐质牧草地
潮滩
海藻
蟹
贝

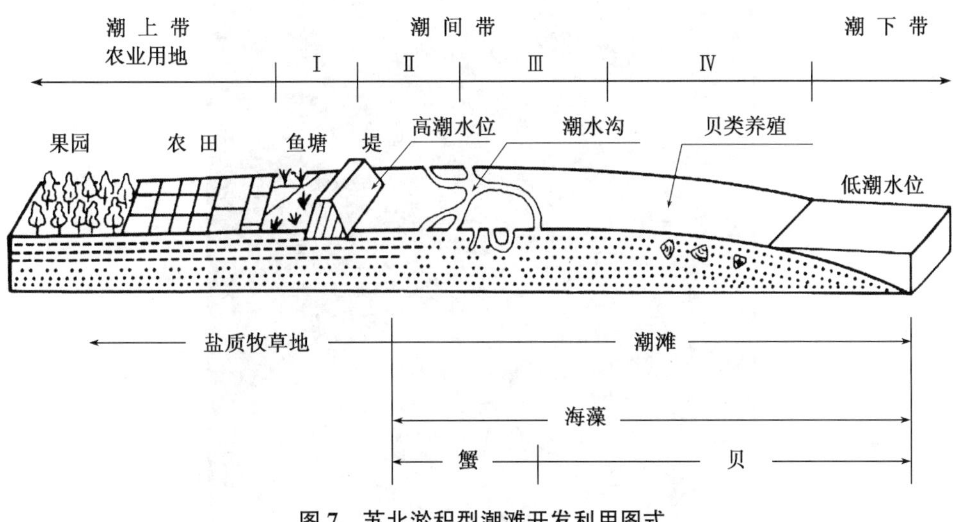

图 7　苏北淤积型潮滩开发利用图式

潮 上 带
潮 间 带
潮 下 带
围垦土地
Ⅲ
Ⅳ
稻田蔬菜
棉桑
鱼塘
LWL
黏土
粉砂土
黏土粉砂
黏砂
细砂
中、粗砂
m
20
0
5 km

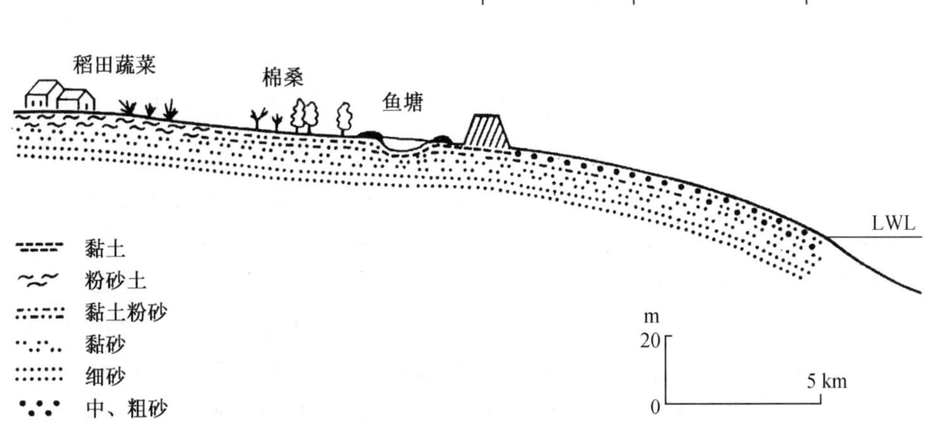

图 8　苏北冲刷型潮滩开发利用图式

江苏岸外的浅海内陆架，在射阳河以南至吕泗北部之间，以弶港为中心，分布着面积达 22 470 km^2 的辐射沙脊群（图 9）。它由 70 多条沙脊及脊间水道所构成，自低潮线向海达水深 20～25 m 处。其中出露于海面以上的沙洲面积达 3 782 km^2，大部分与海岸相连成陆，即东沙与条子泥宝贵的土地资源。

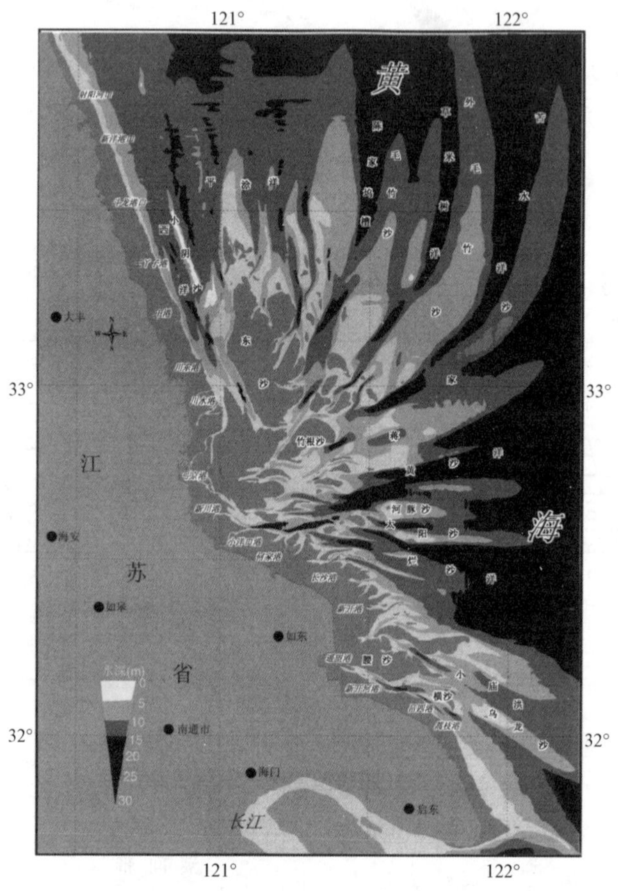

图 9　苏北黄海辐射沙脊群（附彩图）

2. 江苏海岸十分独特，也十分重要

江苏海岸其岸坡平缓，海水浅浑，未若岛屿港湾具有蓝天碧海、浪涛海鸥之引人景观，但它孕育着新生的土地与滩涂资源（3 700 km^2），如渔、虾、蟹、贝、海藻水产与药物资源，以及尚未被探明的石油天然气能源。众所周知，油气主要聚藏于埋藏的巨大沙体中，辐射沙脊群是由长江、黄河、淮河的泥沙所堆积。全新世以来，海平面上升，由于强劲的潮流冲刷形成辐射状沙脊群。沙脊群与毗连海域必定是油气富聚区。更为独特的是，一般的平原海岸与港湾海岸不同，缺乏深水海港。苏北平原海岸由于辐射沙脊群的存在，在沙脊之间沿古长江河谷发育的深水潮流通道形成波浪隐蔽之航道。该区为不正规半日潮型，中枢的弶港为南黄海无潮点，涨潮时潮流向中枢辐聚，而落潮时向海外辐散，形成我国最大的潮差达 9.28 m，平均潮差 4.16 m（南京大学、河海大学，1996），强劲的涨潮流（1～1.5 m/s）与落潮流（1.2～2.0 m/s）往返冲刷，大潮涨潮通量为 20.8 亿 m^3，落潮流通量达 26 亿 m^3（小潮落潮通量为 7.3 亿 m^3，涨潮为 8.7 亿 m^3），维持着沙脊群间通道之顺畅；该处蓝沙洋-黄沙洋水道为 3 万年前古长江河谷，全新世海平面上升淹没了古河谷，由于有强大的潮流往复冲刷，使该处绝大部分保持为水深 16～25 m 的通道，是水深浪静的天然海港。2005 年 9 月，已有空载的 10 万 t 船试航通行。辐射沙脊群数条主要的潮流通道，均赋有天然的大中型深水海港条件(图 10、表 1)。

表 1　黄海辐射沙脊群潮流通道水深条件比较

部位	潮流通道	长度(km)	水深(m)	适建港口(万 t)
枢纽部	蓝沙洋	53	16～23(局部 29)	11～20
	黄沙洋	45	11～20	
南部	小庙泓	28	8～12	5
北部	西洋(王港、大丰港)	50	10～13	5

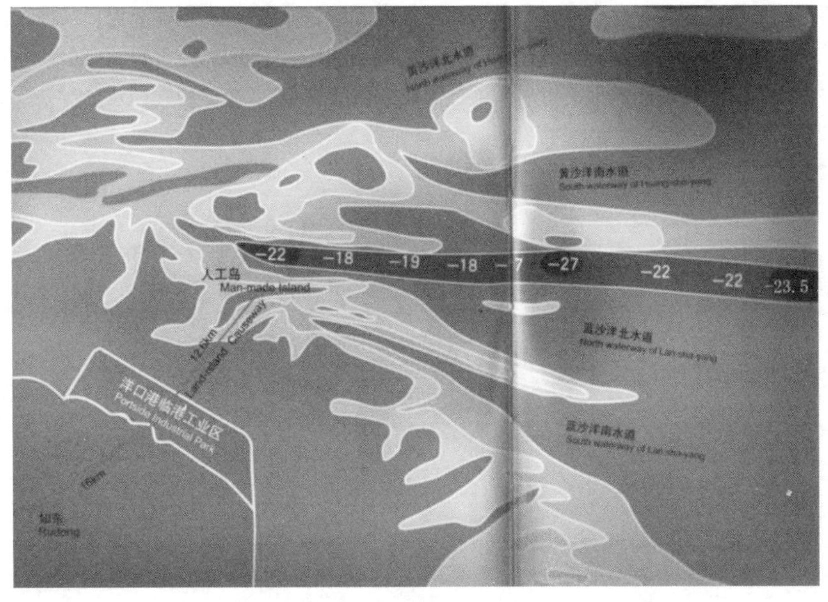

图 10　长三角北翼洋口港蓝沙洋深水通道(附彩图)

深水潮流通道、沙脊群的掩护与天然码头基础条件，与淤涨的滩涂土地资源结合，构成了不需移民，不需深挖的深水大港基础。加之苏北大地发达的稻麦、桑棉、菜蔬农业，与高素质的劳动力结合，使兰沙洋可发展为 20 万 t 级深水港和临海工业-绿色农产品与产品加工业、制造业、新能源清洁工业(油气、潮能、风能、太阳能)及对外贸易基地。洋口港已于江苏省“十一五”期间投建，它的建成将促进江苏海洋经济发展，改变长三角北翼无大港的缺陷，是提升长三角经济新一轮高速发展的制动闸。

江苏海岸从北到南，连云港与洋口港可建设 10 万～20 万 t 级大型海港；大丰港与吕泗港为 5 万 t 级海港；沿河口之陈家港、射阳河口港等，可建万吨级以下港口。以群策群力的科学智慧建设江苏省大、中、小一系列海港(表 1)，是改变传统的陆地农耕经济为海洋外向型经济的关键，将带动利用苏北广阔的滩涂、充沛的太阳能、风能、潮能及淡水资源，发达的农业基础和知识型密集的劳动力；在江苏省发达的科技与经济力量支持下，发展临海工业、海洋交通与经贸，形成现代化的海洋经济实力，促进江苏工农业产值新增长。21 世纪中叶，江苏将以油、气与新能源为动力，构建农产品绿色工业基地，推动海内外联通的商、贸、游憩经营活动，以发展河海交汇的城镇与宜人的民居为特色的现代海洋文化而光芒四射。

参考文献

[1] 王颖主编. 1996. 中国海洋地理. 北京：科学出版社.
[2] 王颖主编. 2002. 黄海陆架辐射沙脊群. 北京：中国环境科学出版社.
[3] 佘之祥. 2003. 洋口港的开发与长江三角洲北翼的发展. 水资源保护，(19)6：10 - 11.
[4] 孙月平. 2003. 洋口港的开发建设与江苏"江海联动"战略研究. 水资源保护，(19)6：12 - 18.
[5] 朱大奎. 2003. 江苏海洋环境与沿海经济发展. 水资源保护，(19)6：19 - 21.
[6] 田汝耕. 2003. 试论洋口可建成江海直通运河工业带式的世界级海港. 水资源保护，(19)6：25 - 28.
[7] 王颖. 2003. 充分利用天然潮流通道建设江苏洋口深水港临海工业基地. 水资源保护，(19)6：1 - 4.
[8] 罗一民. 2003. 打造洋口港深水大海港实施江海联运新战略. 水资源保护，(19)6：32 - 33.

曹妃甸深水港
——从理想到现实*

一、“游荡”的滦河

曹妃甸，当年叫沙垒田岛，是位于南堡东南方约19 km，呈NE-SW向延伸的沙岛。沙岛长度约1 500 m，宽度据1945年测量为750 m，而至1959年时，其宽度为450 m。沙岛高出海面2～4 m，最高处为风成沙丘，顶部筑有灯塔。1959年我参加天津新港泥沙回淤研究，担任北队队长，从滦河口调查至新港，探求滦河泥沙是否影响到新港。通过实地调查滦河三角洲海岸，了解到滦河是条多沙河，受偏东向盛行季风风浪影响，入海泥沙被推移向岸，形成一系列围绕河口的沙坝，使三角洲海岸线具有双重性：原始的平原海岸线，潟湖与沙坝外侧的海岸线。滦河在下游平原区为游荡型河流，其最早的入海口可追溯到南堡一带。南堡以西，平原为淤泥质地，与滦河三角洲平原的砂质土壤迥然不同；以后，滦河曾迁移到西河（溯河）—大庄河（小清河）一带入海，当时的海岸线较现今偏南，曹妃甸、腰坨、草坨、蛤坨是当时河口的岸外沙坝；后来，滦河又向东迁移至大清河蜓湖林口入海，遗留下石臼坨、月坨、打网岗、灯笼铺等沙坝；嗣后，滦河继续向东北方迁移至老米沟、稻子沟入海，遗留下蛇岗等系列沙坝；最后于1915年滦河迁至现代河口入海，又形成了一系列环绕河口的新沙坝。沙坝发育与滦河供沙有关，同时表明，该地海向风力促成盛行的偏东向风浪动力（冬春季以NE向风力强劲，夏秋季盛行E、ES风）始终未变。1959年调查时，曾选登一些大型沙坝调查，了解到曹妃甸为棕黄色石英、长石质细砂组成的岸外沙坝，顶部叠加的风成沙丘为粉砂细砂，海滩上出现中砂，砂中含普通角闪石、辉石、绿帘石、钛石等铁矿与锆石等重矿物以及贝壳屑。打网岗砂坝为浅黄色细砂与中砂，矿物成分与曹妃甸沙岛类似，均为滦河供砂。曹妃甸沙岛南侧相伴以渤海潮流北上之通道，水深条件好，是渤海最优越的深水航道。

二、海岸沙坝与潮流通道的稳定组合

1997—2000年，南京大学海岸与海岛开发教育部重点实验室在曹妃甸海域重复观测，进行海港选址预可行性研究时，看到曹妃甸沙岛范围虽有所减小，但历经50年后，它仍然屹立海中。这是十分奇妙的地貌组合，盖因渤海的潮流是自黄海经渤海海峡传入的，初始主流向西到达蓟运河口海外，促使蓟运河口呈喇叭口形。然后，渤海潮流转向，北支沿海湾走势，大体上在南堡岸外转向东北，流向辽东湾。渤海北上潮流的主通道紧临曹妃甸沙岛外侧流过，强劲的潮流流速达1～2 kn，潮流通道与小沙岛相邻分布，历经半个世

* 王颖：《纵横——唐山曹妃甸专刊》，18－22页，2008年。

纪曹妃甸沙岛依然存在，说明该岛海岸稳定。因此提出，利用沙坝为天然的码头基础，与相邻的潮流深槽组建深水海港。更何况，沙岛后侧岸陆方向有大片浅滩可吹填造陆供港口利用。科学研究之成果与经 50 年先后实地查勘的经历，有力地支持利用曹妃甸建深水海港的设想。这一建议获得唐山市领导、计经委与以后发改委的支持，再经交通部、规划部门及建港专家的反复论证，最终列入建港规划，开展了相关的水文、泥沙、工程、设计规划工作。

港口投建，进一步提出了曹妃甸建港对潮流通道稳定性与该区生态环境的影响。2005—2006 年，南京大学第三次组队调查研究。

构成曹妃甸岛的砂层厚度超过 40 m，其基础是宽度达 1.94 km 的大沙坝。近一万年来在全新世海平面上升过程中，约在 3 400 年前，海水浸淹到该沙岛，虽经历长期的海浪冲刷作用，仅余沙丘的基部出露在海面上，但是宽大层厚的沙坝基础却依然存在。地震剖面表明，曹妃甸沙岛南侧为渤海潮流主通道深槽，其宽度达 6 000～7 000 m，水深超过 20 m，最深处超过 30 m。该深槽具有承袭性的特点，现代深槽位置重叠于老谷之上，老谷底有沉积层淤积，但谷地位置不变，水深大，反映出潮流主通道深槽是稳定的。在深槽中打钻至海底 70 m 深处，了解到深槽中沉积物是以黏土与粉砂为主，含少量细砂，是潮水落淤的细颗粒泥沙，而非海浪侵蚀沙坝之产物。事实上，紧临曹妃甸沙坝南侧的渤海主潮流通道的深槽吸收了外海传来海浪之波能，减缓了海岸带波浪变形，一定程度上保护了沙坝免遭风浪侵袭，这是曹妃甸沙岛长期屹立于海上的原因之一。正是潮流深槽与沙坝的稳定组合，使它成为最佳的天然深水海港港址。所以，深入的科学研究成果，进一步肯定了曹妃甸沙坝基础与潮流深槽的稳定性。因而，可以发挥该海岸环境优越性，建设冀东深水大港工业基地。

曹妃甸沙坝孤悬海外，长久不消失的另一个重要原因，是它有深厚的砂层基础。沙坝雏形始现于海底 60 m 深处，砂基厚度超过 50 m，说明当时滦河于西河口及大庄河一带出海时期长，入海沙量巨大。岛上的砂层呈棕黄色，似经历过较长时期与湿热气候环境的沉积过程。这些现象发人深省，启迪我们研究探索曹妃甸沙岛形成的历史过程。

三、曹妃甸沙岛发育与滦河三角洲变迁

滦河是一条多沙性河流，发源于河北省巴颜图古尔山麓，流经内蒙古高原、燕山山脉，至乐亭县南部的兜网铺入海，全长 877 km。流域面积 44 900 km^2，多年平均径流量 47.210 8 m^3，输沙量 2 219 104 t，1983 年引滦入津及 1984 年引滦入唐后，滦河年平均径流量降为 18.4×10^8 m^3，比工程前减少 61%，冬、春季滦河断流，入海泥沙锐减，河口三角洲淤进停滞，并遭受海浪侵蚀。入海泥沙为花岗岩与变质岩类风化形成，并为河流搬运的中细砂。滦河口沿岸潮差较小，仅 1～1.5 m，在南堡的东西两侧湾岸中形成两个小型的顺时针方向的旋转流。南堡呈现向海突出的三角滩地，成为分流分沙点，它两侧的泥沙特性迥然不同，东侧以滦河的细砂、粉砂为主，西侧为黏土淤泥质。滦河三角洲沿岸常年盛行偏东风，波浪作用强。东东南风与东东北风强盛，波浪最大，波高达 3～4 m，波长平均 30～31 m，最长达 41 m。波浪作用可影响到沿岸海底，掀沙力强，形成自海向岸横向地推

移泥沙作用，发育了海岸沙坝围封原始河口海岸的格局。当偏东向波浪与海岸斜交时，可掀带泥沙顺风顺流地沿着海岸搬运，其中以沿海岸向西南方搬运的泥沙量大于向东北侧搬运的。泥沙横向搬运形成围封河口的沙坝，而沿岸纵向运移的泥沙，促成了 NE-SW 向长大沙坝的形成，打网岗、曹妃甸沙坝是属于这一类型。

滦河发育了两个大型三角洲体：① 时代较新的三角洲体呈现向海展开的扇形，它以滦州以南的马城为顶点，西起西河口（溯河口），东至现代滦河三角洲，中轴大体在大清河与王滩港之间，石臼坨、月坨、打网岗为其河口区海岸沙坝；② 西部较老的三角洲界于涧河口与西河口（溯河口）之间，当时的古滦河可能曾于高尚堡、南堡、沙河口及涧河口之间出海，曹妃甸、腰坨、草坨、蛤坨应形成于老三角洲发育时。

南堡以北的陆地上，仍有一系列断流的干涸河道及小型的河流，如沙河、小戟门河、嘴东河（双龙河）、小青河等分流入海。这些河流已不成天然河系，一些废河谷被改造为输、排水渠道或平原水库，体现了唐山人民对自然环境的恰当应用。

自两个三角洲平原的内陆方向追溯至迁西（兴城）与旧城之间，观察到滦河主河道呈直角拐弯的非正常改道，以及被构造抬升而废弃的老河道。滦河改道始发点现象如下：

（1）滦河自山地流出，至唐山境内时为自北向南流，但在旧城与迁西之间河流突呈直角拐弯改向东流。经忍字口与尹庄河间返向南流，先后经黑沿子河口、嘴东河河口及西河（溯河）河口入海。嗣后，又在尹庄改道再向东流，至罗家屯再向南经大庄河、大清河入海。最后又改道向东移，曾经湖林口、老米沟、浪窝口以及 1915 年改从现代滦河口入海。河道辗转向东北迁移表明地面有自东向西的掀升，河水循低地而流。始自迁西的突然转折向东，从卫星图片中判断，在迁西南部因地震形成原河道成断块隆起，阻挡河流南流。

（2）在滦河直角拐弯向东流的白龙山村，仍保留一段具有天然河曲形式被废弃的老河道，保留着 NE30°走向，但已成为高出当地地面 12～14 m 的阶地。此废河谷阶地叠置在基岩河谷中，并有两级基岩陡坎，上部基岩陡坎比废河道高 20～22m，呈 NE20°方向延伸；下部的基岩陡坎比废河道阶地面高 2.4 m。组成阶地的物质，底部是 3 m 厚的具透镜体的、由河床堆积的砂砾层，上覆 1 m 厚的河漫滩相的细砂与粗粉砂层，这是典型的河流堆积物。

据上述现象判断是突发的断裂抬升事件使白龙山村的一段古河道抬升成小丘，堵挡河流南下，迫使滦河改道向东。两级基岩陡坎岸，反映出突发型的断裂抬升具有间歇性特点：第一次抬升 9 m 后，河流力图仍保持原来自 NE 向 SW 的流向；但第二次又抬升了 3 m，彻底结束了河流向南之流路，而直转向东。白龙山村地居山地向平原过渡带，属唐山地震多发带，判断为突发的断裂抬升系地震活动造成。地震构造掀升的后续影响，造成地面自西向东的倾斜，河流顺势沿低地流，故而滦河下游不断地向东北方向迁移。

判断古滦河在迁西改道的时期，应是在曹妃甸沙坝发育成型的后期。

迁西及罗家屯以南的广袤平原，虽经开辟为农田，但是根据平原地貌组合的残留遗迹及沉积物特性，仍可追溯平原成因如下：

（1）平原上有断续的河道与干涸洼地，尚可辨识出自迁西向南，自忍字口与尹庄间向南，以及自罗家屯向南，古、今滦河的三股主要河道。大体上是在滦县以南，河道分流增多，呈现出两个三角洲体的水流分布形式。

（2）滦河三角洲平原地势由北部山地外围向南部海域倾斜，组成平原的泥沙亦由北向南变细，从北部的砂砾层与中、细砂，渐为细砂、极细砂与粗粉砂，再次为粉砂与粉尘沉积，至南堡已为黏土质淤泥沉积。沉积物的矿物成分均以石英、长石为主，深色矿物及稀有矿物成分均与滦河泥砂相同。

（3）迁安市以南，平原上残留着 9 列近东西向排列的沙坨（沙岗），是古海岸沙坝之遗迹，系列沙坨代表着滦河三角洲平原发展过程中不同时期的古海岸线位置，其中第 1 列与第 7 列沙坨，是由海岸沙坝并陆后的残留地貌。其中第 1 列与第 2 列沙坨，是滦河流出山地丘陵区后，所发育的岸外沙坝及海湾湾口沙坝之残留，当时泥沙充分，堆积的地貌高大；沉积层为黄棕色，是因经过湿热的天气影响，估计可能是更新世中期的堆积。第 3 列到第 7 列沙坨原为由滦河泥沙供给，经风浪与岸流堆积的海岸沙坝，它环绕着原始的平原海岸，拦阻了滦河泥沙，于坝后的浅海湾或潟湖中逐步淤填为冲积平原，并改变原海岸沙坝成为陆上沙坨，平原上河渠多，经充足淡水灌溉成为绿色田野。自第 7 列沙坝向南，景色迥异，因缺乏淡水与林草，是淤泥质、盐土质平原。第 8 列与第 9 列沙坨蜒高尚堡、南堡和尖坨子，均不是海岸沙坝，而是盐土平原潮滩受激浪冲刷掏蚀出的贝壳物质被浪流堆积于海岸上的沿岸堤。分析当时滦河已改道，无泥沙供应该处，故而海岸遭受冲刷发育贝壳堤。

据曹妃甸沙岛分布位置，其形成时代是在滦河改道向东流之前，以及初次经忍字口与尹庄间向南流，经过第 2 列塔坨、第 3 列大茨榆坨、第 4 列青坨营及第 7 列东黄坨，形成于海岸淤进阶段，当滦河自高尚堡一带出海时，曹妃甸则是在岸外沙坝基础上加积沙丘而成沙岛，由于滦河曾长期稳定地向南入海，故而沙坝基础砂层厚。对比冀东海岸沙丘、沙滩广泛发育的时段，曹妃甸沙岛形成的地质年代相当于晚更新世低海面后期与全新世海侵初期。

四、河海交互作用平原潮汐汊道与深水港建设

曹妃甸沙岛具有深厚的沙坝基础，岛外濒临渤海潮流主通道深槽，被选建成为停泊 30 万 t 巨轮的深水海港；岛后波影区浅滩被吹填成广阔的港域陆地，两侧海岸的潮汐汊道已无河沙下淤，潮水畅通，规划建成一系列中、小型辅助港区，形成曹妃甸港口群。这是中国海岸动力地貌研究的一项重要成果，研究体现于海港工程，意义重大。21 世纪曹妃甸深水港建设，改变了平原区缺乏良港的现状，推动了渤海区钢铁、能源、化工、农贸等临海工业及唐山滨海新城的飞跃发展，促进京、津、唐海洋经济联合基地经济新高涨。

中国海陆过渡带
——海岸海洋环境特征与变化研究*

1　海岸海洋概念的兴起与发展动态

人们对地球环境与资源的关注是从陆地逐渐走向海洋的，20 世纪后期认识到在陆海之间存在着一个相互交界的过渡地带。这个地带被称为海岸海洋(Coastal Ocean)。它涵盖了陆地延伸至海洋的全部地带。据 1994 年 UNESCO 在比利时 Liege 大学会议定义：其范围包括海岸带、大陆架、大陆坡及坡麓的堆积体——大陆隆(图 1)。海岸海洋是一个独立的环境体系，该地带因有海水覆盖而区别于陆地，又因有陆地延伸为海底而区别于深海大洋。地球表层的气、水、岩、生物在此地带交互作用变化频繁，风力、波浪、潮流均在海陆交互带加强，其生物量巨大，生态系统丰富，缓冲作用突出，与人类生存活动关系密切。

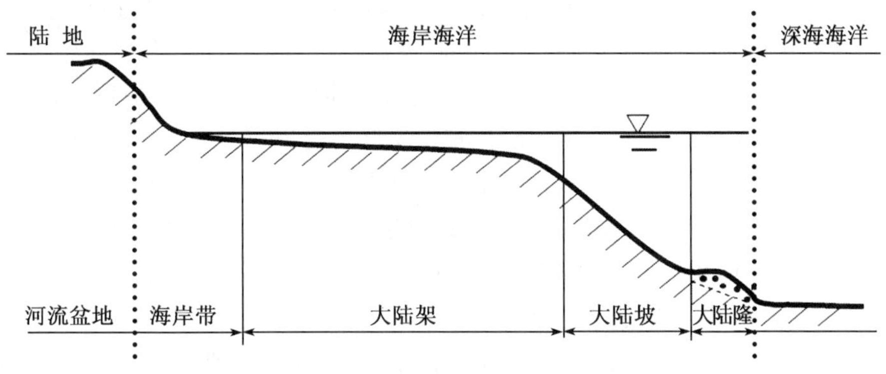

图 1　海陆过渡带——海岸海洋

20 世纪后，人们认识到浅海与深海之区别，但是未将海陆过渡带作为统一的体系而加以研究。1994 年“联合国海洋法”的实施，将大陆架与 200 海里专属经济区划归沿海国管辖，这些均在海陆过渡带的范围之内。全球有 370 多处海域具有权益纷争①，中国的管辖海域约 300×10^4 km²，在东海岛屿与油气田开发、南海岛礁归属方面与周边国均存在着海疆权益之纷争。海洋国土权益推动了对海岸海洋之关注，在国际上兴起了对海岸海洋科学研究之热潮。国际海洋学界也关注到海岸带、岛屿与大陆架，才逐渐地认识到海陆过渡带是有别于陆地，又区别于海洋的一个独立的环境体系。

20 世纪 90 年代后期，国际科学联合会(ICSU)将海岸带海陆相互作用(LOICZ)作为国际岩石圈-生物圈计划项目(IGBP)的重要组成部分，开展研究“河流—河口—海岸—大陆架”物质输移过程与通量，被各国奉为前沿研究课题。LOICZ 研究重点承袭英国的“陆

* 王颖，季小梅：《地理科学》，2011 年第 31 卷第 2 期，第 129－134 页。

① 据《中国海洋报》1996 年 5 月 17 日文章与国家海洋资料中心 2005 年资料综合。

海相互作用研究"项目(1992—1998 LOIS)的特色,但未包含大陆坡与大陆隆,未完全概括海陆过渡带整个一体化环境。但是,LOICZ项目却又阐明了海岸带与大陆架海域的重要性:占全球面积的6%,占海洋面积的8%,但集聚着全世界60%人口,70%大城市,提供25%的初级生产力与90%的捕鱼量,50%的全球反硝化作用,容纳全球80%的有机残体,75%~90%的悬移质与污染物。中国海洋学界的年轻一代仍沿用LOICZ以"Coastal Zone"(海岸带)来概括海岸与大陆架的研究内容。

1994年5月由联合国教科文组织的政府间海洋委员会(IOC),在比利时列日大学举行了第一次"海岸海洋先进科学技术研究国际学术大会(COASTS, Coastal Ocean Advanced Science and Technology Study)",具有重要里程碑意义的科学进步。该会获得IOC欧洲委员会、SCOR(Scientific Committee on Ocean Research)、美国NSF(National Science Foundation)及ONR(Office Of Navel Research)支持。参加会议代表100多名,中国代表4名:国家海洋局在UNESCO IOC的代表李海清,海洋局二所的苏纪兰,中国科学院海洋研究所的胡敦欣及南京大学王颖(王颖的论文被选为大会报告:"河海交互作用与陆源通量")。这次会议正式提出Global Coastal Ocean(全球海岸海洋),以此为名在国际专门报道海洋研究新理念与调查研究进展的权威学术专著期刊"The Sea"出版"The Global Coastal Ocean"[1]。两部巨著内容包括:海岸海洋的作用过程与研究方法(Vol. 10),综合性的区域海岸海洋:西部的、东部的、极地海洋与半封闭的海与群岛(Vol. 11),论著中对Coastal Ocean明确定义:"它是居于地球的陆地与深海之间的客体",包括"陆架海与陆坡海以及海岸带"。1999年国际地理学家联合会以英、法、中等5国文字出版的"国际海洋地理宪章"[2]明确地将全球海洋区分为"海岸海洋"(Coastal Ocean)及"深海大洋"(Deep Ocean)。这是国际海洋科学与地理科学学界对海岸海洋的正式确定。

20世纪60年代,中国已重视海岸带的"海陆两栖性"的特点,适应生产建设(如选港和建港)的实际需要,将海岸陆上部分与水下岸坡结合,研究泥沙运动与海岸演变,推动了海岸科学研究。80年代,先后进行全国海岸带与海涂资源综合调查、不同海区的大陆架考察及全国海岛资源综合调查等,获得了大量基础资料。90年代,开展了全国海岸带功能区划研究,范围逐步扩展至内陆架地区,开始了海陆结合、注重海陆相互作用进行研究的新阶段。21世纪初,"海域法"通过并开始实施。2006年以来,由国家海洋局统一组织、领导开展的"中国近海海洋综合调查与评价"项目(908项目),将海岸带、大陆架与陆坡海结合为一体进行调查。同时,中国已积极地参加到众多的海岸海洋国际合作研究计划中,包括PAGES、JGOFS、LOICZ等,地区间的合作研究也已展开,如中韩合作进行黄海大陆架的对比研究等。有所不足的是,在中国将海岸带区别由环境部门管理陆上部分,海洋局管理水边线以下部分,分割了海岸带之整体,未能形成协同的一体化管理。

2　中国海陆过渡带之环境特征

中国地处最大的大陆与最大的大洋之交,毗邻黄、渤、东、南陆缘海与太平洋,大陆岸线长达18 000多km,6 500多个岛屿岸线长达14 000 km,海洋国土面积达300×10^4 km^2[3]。渤海为内海,黄、渤海是陆架海;东海具有世界最宽阔的大陆架与拉张的大陆坡;南海海底地貌类型丰富:堆积型的大陆架与古海岸带、断裂沉降的阶梯状大陆坡,以及拉张有洋壳涌入,具有海底火山的深海平原;台湾以东濒临太平洋,悬崖陡坡海岸直插大洋底部[3,4]。

中国海岸类型丰富,海岸海洋环境与作用过程独具特色,河海交互作用与人类活动影响之特点鲜明:

(1) 南北跨越39个纬度带,具有寒、温、热3个气候带的不同环境特点[4]。黄海和渤海处在北温带海的边缘,东海属亚热带性质,南海地处热带与亚热带。沿海环境生态有明显的地带性差异:华北与东南沿海以大河平原海岸为特色,而在华南热带气候环境下,海岸繁殖着红树林与珊瑚礁,形成典型的高生产力、高生物多样性的生态系统。

(2) 地质构造具隆起与凹陷相间、构造运动活跃、区域变异大等特征。海岸海洋受一系列与海岸斜交或平行的NE或NNE向的隆起带与沉降带所控制[5]。地质构造格局,控制了海岸带的基本类型[6]:基岩港湾海岸及岩石岛屿多发育于褶皱隆起带,台湾东海岸带发育有断层海岸;而平原海岸多发育于大河分布的沉降带:辽东湾松辽平原、黄河与长江等河口与渤海、黄东海平原。南海,北东向的断块构造控制着基岩港湾海岸线的主要走向,而海湾与潮汐汊道多沿北西向断裂带发育[3]。大体上杭州湾以北的海岸是隆升的基岩港湾海岸与沉降的平原海岸交错分布,而以南的海岸基本是基岩港湾海岸为主体[4],间伴着沉溺河口(湾)岸。

(3) 季风风浪是主导的波浪作用,尤以冬季东北季风与夏季东南季风的影响突出[7],在很大程度上控制了沿岸流的作用。每年夏、秋季节,中国海岸带从南到北部可能受到热带气旋和台风的直接侵袭或影响;冬半年,中国海岸带从北到南均程度不同地受到寒潮或强冷空气的影响。台风和冷空气大风是形成灾害性风浪的主要原因。

(4) 具有显著的潮流作用,中国海以潮汐与潮流作用为特色,南海以不正规半日潮与全日潮为主,黄海、渤海、东海为半日潮。潮波自太平洋经相关海峡传入中国海过程中,受陆架海底地形及海岸轮廓的影响变化较大,其中以南黄海与东海北部海域潮流作用强烈。长江口以北的江苏沿海海域位于潮波系统的波腹区,潮差大(平均潮差4.0～6.0 m,最大潮差9.28 m)[8],以潮流作用(大潮平均流速2.0 m/s,最大流速2.5 m/s)为主导,发育了典型的独具特色的淤泥质海岸和一系列潮流沙脊群。江苏滨海平原是全球最具代表性的粉砂淤泥质海岸与潮滩湿地集中分布区,在世界淤泥质海岸中占有重要地位。浙、闽沿岸的一些港湾内,由于海岸曲折等地形集能作用,不仅出现较大的潮差(3～8m),并且汇集源自长江口的悬移泥质,发育形成基岩港湾海岸内的淤泥质滩涂如乐清湾、象山港、三门湾、沙埕港和三都澳等。

(5) 发源于"世界屋脊"青藏高原的8条大河,有5条汇入中国海,带来巨大的物质、能量通量,影响着海域的动力条件、海岸及陆架地质地貌、海洋生物等各方面,尤以对海岸塑造、陆架沉积作用影响最为重要。未兴建规模化的水利工程前,中国河流——大河及上千条中小型河流——每年输送超过1.8×10^{12} m^3的径流,近2×10^9 t的悬移质泥沙和3×10^8 t左右的溶解物质入海[9,10],其影响不仅限于河口区,而且波及河口两侧岸线及大陆架毗邻区,在太平洋西岸岛弧-海沟系的弧后盆地堆积形成广阔的大陆架。不同类型的河口动力型式各异,入海泥沙的输移与沉积模式亦相差较大。王颖等根据区域基础地质、河流特性、河口及浅海环境与动力因素差异,以5条不同类型的河流为代表,分析研究了其不同的泥沙运动与河口沉积的特性以及对相邻陆架之影响[10]。

3　关注中国海岸海洋的开发利用

中国海除台湾岛以东濒临太平洋沿岸有因菲律宾板块俯冲被推挤上来的海洋沉积层所组成的特殊地貌:独特的海蚀柱(女王头)、蜡烛台盘等壮丽的大洋海岸景观。渤、黄、东、南海均为被岛弧围拦的陆缘海(marginal sea),海上防卫受制于海峡通道。各边缘海区域特点突出:地质历史时期以来,黄河、长江巨量泥沙汇入发育了黄渤海平原海岸与大河三角洲。人口密集,沿海城市发达,新生的土地资源,为西部荒漠区与水库淹没区提供了移民新家园;东海与南海的岛屿既是国土的前卫,又是开发海洋的据点;先民开拓的历史遗迹,亚热带浅海丰富的生物资源与油气藏,既是海洋开发的宝贵资源,又是海洋旅游、休憩与教育基地;沿海风能、阳光及中国潮汐动力丰富的特点,是绿色能源,是发展低碳经济的重要基础。历史上中国以农立国,众多的人口仍以农业为发展的根本,东部平原是中国人口密集的经济发展带,应是绿色大农业基地。本文以平原海岸为例,阐述其开发利用的特点与发展中的问题。

3.1　因地制宜,根据区域特点,进行最佳利用

例如,黄海平原位于长江三角洲北翼与苏北农业发达区,但长达 884 km 的海岸无深水海港。传统的观念是平原海岸水浅沙多,不宜建港。自 20 世纪 80 年以来,海岸带调查,尤其是国家自然基金重点项目研究发现南黄海陆架辐射沙脊群中的烂沙洋—黄沙洋潮流通道,是承袭晚更新世末(3.4×10^4 a 前)古长江河谷而发育的,古长江南迁,此处已无河沙下泄,却具有强大潮流动力的往复冲刷,使水道稳定,水深 16～25 m,可通行($10\sim20)\times10^4$ t 级海轮。利用承袭古长江谷地的潮流通道建深水港,是平原海岸与陆架地貌组合建深水港的重大突破[11]。有力地促进长江三角洲北翼能源、工农、商贸经济带发展。

“桑基鱼塘”于明代后期兴起于珠江三角洲地区。该三角洲是由东、西、北三江汇合于港湾内发育的充填式三角洲,地势低洼,洪涝灾害频繁。人们据自然环境特点,筑堤围拦三角洲水洼地为鱼塘。围堤上植桑,桑叶饲蚕,桑葚、蚕沙喂鱼,鱼塘逐渐淤涨为田地。桑基鱼塘妥善地解决了拦蓄洪水、低洼地积水内涝的问题,同时创建了独特的海陆交互带资源相互作用的人工生态系统,世界上罕见[12]。当代,城镇扩建与工业发展消失了农田利用的生态模式,助长了洪、涝、风暴潮灾害的频发机率。

苏北海岸有 194×10^4 hm^2 滩涂,雨水充沛,宜发展绿色农业与循环经济:潮下带,捕捞与放养鱼、虾,水产繁殖(海马、鳗仔、虾、蟹);潮间带,进行光滩养殖,粉砂滩适宜于养殖紫菜、文蛤,泥滩则养殖蛆、沙蚕;潮上带,发展苇田和鱼塘;在草滩放牧、养殖鳗、蟹、虾;随着土地成熟程度可种植桑、棉、麦、稻(淡水充沛处)或开辟为药、花、林、果种植园[13-15]。

3.2　以苏北平原海岸为例剖析开发利用中的问题

主要是不了解河海环境体系的特点,人为地改变体系间水沙动态平衡而招致灾害。地处淮河流域入海尾闾的里下河地区,涉及扬州、盐城、淮安与南通四区,面积约 22×

10^4 km^2，其地势四周高，中间低洼，降雨后“四水投塘”，水位上涨，虽经湖荡调蓄下泄，但排水受海潮顶托，易因洪致涝，被称为“锅底洼”。新中国成立后，以射阳河、新洋港、斗龙港、黄沙港为里下河地区排水入海的主要通道[16]。20 世纪 60 年代，为拦蓄利用河流淡水，在入海河道的河口纷纷建立挡潮闸，以期防止盐水入侵。但河口建闸后，改变了河口地段的动力条件，涨潮流受阻，使动能转变为势能，潮位和潮流过程出现相位差，涨潮历时缩短而涨潮流速加大，落潮历时加长而落潮流速减小，使涨潮带进的泥沙量远大于落潮带出的泥沙量，改变了原有天然情况下的输沙平衡，引起闸下河道淤积，闸上水质恶化，以致闸门难于打开[17]。1985 年统计，江苏沿海修建的 58 座挡潮闸，除位于侵蚀岸段的 18 座淤积较小外，大部分发生淤积，其中，一般淤积的 20 座（占总数 34%），严重淤积的 15 座（占 26%），基本淤死的 5 座（占 9%）[18]。这是开发建设，不调查海陆过渡带动力环境、潮汐河道双向流的特点，人为地按陆地河流仅有下行水流设计的严重教训。

又如：1977 年，东台市在梁垛潮流通道建闸，筑堤拟围垦滩涂约 6 667 hm^2。由于不了解滩涂上潮水沟摆动规律及风暴浪与大潮叠加的严峻动力，1981 年 14 号台风与农历八月大潮叠加，水位猛增突破历史最高潮位，退潮顶冲区最大流速达 298 cm/s，风暴潮的风浪高出堤顶 0.5～1.0 m，一次风暴造成潮水沟摆动与冲破围堤[19,20]，使闸门废弃孤立于滩上。实例说明，要依据海岸地貌特点，掌握海洋动力与泥沙条件，相应地规划设计堤坝位置、起围高程及合宜的围淤面积才能达到预期效果。

3.3　人类活动与海平面上升交互作用的影响

历史悠久而频繁的人类活动直接或通过河流作用间接地对海岸海洋环境的发展变化产生重大影响，加速了海岸海洋环境和资源的变化。中国的海岸海洋面临着人类大范围开发的挑战：三峡工程、南水北调工程、河口深水航道工程、河口围垦工程等重大工程的实施和海平面的上升都对海岸海洋的水动力产生显著影响。

以长江为例，长江潮流界、潮区界的位置随上游径流大小而变化，枯季潮流界一般在镇江，而洪季潮流界则下移到江阴；枯季潮区界上界一般在大通，而洪季潮区界则下移到芜湖；枯季咸水界一般认为在徐六径[21-23]。三峡工程改变了上游的径流量，海平面上升改变了河口入射波的特性，这两方面水动力因素的变化都对长江潮区界和潮流界的位置产生影响。在平水年，三峡工程 1 月份增加 1 300 m^3/s 的流量，潮区界平均下移约 5 km，潮流界平均下移约 13～18 km；10 月份流量减小 8 500 m^3/s，潮区界平均上移约 28 km，潮流界上移约 35～48 km[24]。海平面上升通过抬高潮位使潮流界、潮区界位置向上游移动，当徐六径处平均水位升高 50 cm 时，7 月潮区界平均上移 18 km，大潮期潮流界上移约 17 km；而 1 月份潮区界平均上移 47 km，大潮期潮流界 31 km[24]。

流域开发招致河流入海水沙量骤减，据戴仕宝等对中国 9 条主要河流（松花江、辽河、海河、黄河、淮河、长江、钱塘江、闽江、珠江）入海泥沙的研究：近 50 年来其输沙率均值为 13×10^8 t/a 左右，并从 20 世纪 60 年代开始呈显著的下降趋势，1994—2003 年的年均入海泥沙 6.6×10^8 t/a 仅为 1964—1973 年的 37%[25]（表 1）。长江流量$(3\sim4)\times10^8$ m^3/s，洪水期流量$(8\sim9)\times10^8$ m^3/s，宜昌站最大洪水量为 11×10^8 m^3/s，三峡大坝建立与中、东线引水约占径流量的 22%；原入海泥沙量 4.7×10^8 t/a，三峡水库拦储泥沙约占总输沙量的 1/3～1/2，加上沿江数百个水库拦截泥沙，使入海泥沙量锐减达总量的 2/3，影响到河口、三角洲海岸冲刷后退，湿地退化。初始为外冲内淤（海底三角洲冲蚀，部分泥沙搬运向岸，使岸滩暂时性淤涨），随着入海泥沙量持续减少，水下三角洲受蚀降低，而持续上升的海平面，频繁的风暴潮灾害，导致三角洲海岸全面侵蚀。黄河因沿途引水，入海泥沙量锐减，三角洲海岸已在侵蚀、湿地退化。珠江三角洲已明显招致盐水入侵。

表 1　中国主要河流径流量与输沙量（根据文献[26]整理）

河流（水文站）		多年平均	2001	2002	2003	2004	2005	2006
长江（大通）	年径流量（10^8 m^3）	9 051	8 250	9 926	9 248	7 884	9 015	6 886
	年输沙量（10^8 t）	4.33	2.76	2.75	2.06	1.47	2.16	0.848
黄河（利津）	年径流量（10^8 m^3）	331.2	46.53	41.90	192.6	198.8	206.8	191.7
	年输沙量（10^8 t）	8.392	0.197	0.543	3.69	2.58	1.91	1.49
淮河（蚌埠＋临沂）	年径流量（10^8 m^3）	285.44	92.12	228.92	671.56	241.75	481.63	245.53
	年输沙量（10^8 t）	1 241	35.9	493	809.4	417.7	847.4	223.44
海河（石匣里＋响水堡＋下会＋张家坟）	年径流量（10^8 m^3）	16.90	3.51	2.01	3.37	5.46	4.85	4.62
	年输沙量（10^8 t）	2 060	57.32	103.12	60.38	46.07	6.16	32.46
珠江（高要＋石角＋博罗）	年径流量（10^8 m^3）	2 866.1	3 384.7	3 131.3	2 369.9	2 135	2 501.8	2 889.1
	年输沙量（10^8 t）	7 980.5	6 531	5 738.5	1 877	2 690.8	3 632	4 525
松花江（佳木斯）	年径流量（10^8 m^3）	675.2	—	381.6	536.1	413.8	596.5	425.5
	年输沙量（10^8 t）	1 270	—	1 030	1 270	832	2 430	1 590
辽河（铁岭＋新民）	年径流量（10^8 m^3）	35.02	4.3	5.194	4.66	13.77	33.80	11.50
	年输沙量（10^8 t）	1 868	68.86	60.98	91.84	73.8	261	62
钱塘江（兰溪＋花山＋诸暨）	年径流量（10^8 m^3）	202.7	—	284.24	172.53	113.54	201.21	177.65
	年输沙量（10^8 t）	292.3	—	285.4	95.1	73.96	170.74	144.43
闽江（竹岐＋永泰）	年径流量（10^8 m^3）	577.1	—	651.27	399.33	322.07	683.7	715.1
	年输沙量（10^8 t）	695.7	—	326.1	86.2	53.35	737	716.8

历史与现实实例表明，人们应调查研究区域环境的特点，探索其动力过程机制，阐明其发展变化趋势，依据自然规律，在阈值极限的范围内规划设计工程与开发利用，才可达到事半功倍与避免灾害的效果。

4 结 论

海陆过渡带是客观存在的独立环境体系，其特点在于地球表层大气、海水、岩石与生物圈层体系在这一地带相互作用、交叉渗透、变化频繁，环境变异灵敏度高、耐受性低，人类活动在这一地带的反应效应显著。东部沿海久为中国生息繁衍之地，尚未对海陆交互带一体化的环境体系特点，有明确的认识，当代大规模开发更引发了一系列问题，负面效应如海水污染已招致降水与淡水质量下降，转而危及人生。因此，需重视研究海陆过渡带的环境特性，了解其环境因素多端的变异性，正确地规范人类开发活动是当务之急，十分重要。

参考文献

[1] Brink K H, Robinson A R(ed.). 1998. The Global Coastal Ocean[M]. John Willey & Sons Inr.

[2] 王颖译，任美锷，吴传钧校. 1999. 国际海洋地理宪章(中文版)[J]. 地理学报，54(3):284-286.

[3] 王颖，朱大奎. 1994. 海岸地貌学[M]. 北京：高等教育出版社.

[4] 王颖(主编). 1996. 中国海洋地理[M]. 北京：科学出版社.

[5] Wang Y. 1980. The coast of China[J]. *Geoscience Canada*, 7(3): 109-113.

[6] Wang, Ying, D. G. Aubrey. 1987. The characteristics of China coastline[J]. *Continental Shelf Research*, 7(4): 329-349.

[7] 苏纪兰. 2005. 中国近海水文[M]. 北京：海洋出版社.

[8] 王颖. 2002. 黄海陆架辐射沙脊群[M]. 北京：中国环境科学出版社.

[9] 程天文，赵楚年. 1984. 我国沿岸入海河川径流量与输沙量的估算[J]. 地理学报，39(4):418-427.

[10] 王颖，傅光翮，张永战. 2007. 第四纪研究，27(5):674-689.

[11] 王颖. 2003. 充分利用天然潮流通道　建设江苏洋口深水港临海工业基地[J]. 水资源保护，19(6):1-4,63.

[12] 钟功甫，蔡国雄. 1987. 我国基(田)塘系统生态经济模式——以珠江三角洲和长江三角洲为例[J]. 生态经济，(3):15-20.

[13] Zhu Dakui, L. P. Martini, M. E. Brookfield. 1998. Morphology and Land-Use of Coastal Zone of the North Jiangsu Plain Jiangsu Province, Eastern China[J]. *Journal of Coastal Research*, 14(2): 591-599.

[14] Wang Y, Zou X, Zhu D. 2000. The utilization of coastal tidal flats: a case study on integrated coastal area management from China[C] In: Flemming B W, Delafontaine M T, Liebezeit G. Muddy Coast Dynamics and Resources Management. Elsevier Science, 287-294.

[15] Wang Y, Zhu D. 2006. Characteristics and exploitation of coastal wetland of China[J]. *Resources and Environment in the Yangtze Basin*, 15(5): 553-559.

[16] 闵凤阳，汪亚平. 2008. 江苏淤泥质海岸入海河道闸下淤积研究[J]. 海洋科学，32(12):87-91.

[17] 许朋柱. 1994. 海平面上升对里下河地区洪涝灾害的影响[J]. 地理科学，14(4):315-323,339.

[18] 任美锷. 1986. 江苏省海岸带和海涂资源综合调查报告[M]. 北京：海洋出版社，2-18.

[19] 陈才俊. 1991. 江苏沿海特大风暴潮灾研究[J]. 海洋通报，10(6):19-24.

[20] 陈宏友. 1993. 死生港潮沟活动及其对巴斗围垦的影响[J]. 海洋湖沼通报，15(3):35-39.

[21] 宋兰兰. 2002. 长江潮流界位置探讨[J]. 水文，22(5):25-26,34.

[22] 左书华. 2006. 长江河口典型河段水动力、泥沙特征及影响因素分析[D]. 华东师范大学硕士论文.

[23] 陈沈良，胡方西，胡辉，谷国传. 2009. 长江口区河海划界自然条件及方案探讨[J]. 海洋学研究，27(增刊):1-9.

[24] 李佳. 2004. 长江河口潮区界和潮流界及其对重大工程的响应[D]. 华东师范大学硕士论文.

[25] 戴仕宝，杨世伦，郜昂，刘哲，李鹏，李明. 2007. 近50年来中国主要河流入海泥沙变化[J]. 泥沙研究，(2):49-58.

[26] 中华人民共和国水利部. 中国河流泥沙公报(2001—2006年)[R]. 北京：中国水利水电出版社.

我国海港建设发展初议*

一、总况

中国位居亚洲与太平洋之间，拥有 960 万 km^2 的陆域，还有 32 000 km 长的海岸线与约 300 万 km^2 的海疆。陆地与海洋文化并举是中国历史传统光辉的一页。18 世纪后，清政府长期闭关锁国使现代科学、文化与经济发展滞后，以致甲午海战一败涂地，扼杀了新生的海洋军力，国人沦为“东亚病夫”。20 世纪初期，军阀混战，经建一统民国，又遭日寇侵略，国土沦丧，人民流离失所。直至 20 世纪中期，中国共产党统一领导，经过 50 年的耕耘发展，中国已以亚洲强国在崛起。21 世纪，再启对外开放，振兴经济，发展科学文化而展现辉煌，进入重振海上雄风的新时代。

海港建设是发展海洋交通、贸易，提高海洋生产值的先行与关键。20 世纪 60 年代之初，周恩来正式提出“三年改变港口面貌”的号召，开启了我国海岸带开发建设与人才培养的新阶段。仅经数年，就遭受近 10 年的“文化大革命”动乱、破坏，海洋经济发展深受政治运动影响，几经起落，而发展迟缓，但却不能抑制在建港实践中成长的人员才干与沿海经济发展势头。

纵观近百年来我国海港选建与发展，大致具有三个阶段的发展特点。

第一阶段：20 世纪早、中期，港口选建特点是利用河口段主泓道与基岩港湾深水域，以岬角蔽风，船只在海湾内碇泊，或在岬角延建防浪堤，在湾内或防波堤背侧开挖深水港池，构成湾内或堤内码头与港池组合的泊地，供船只停靠。

第二阶段：20 世纪末期与 21 世纪初期，选建港是在经济发达的三角洲与平原海岸地带，利用岸外沙坝、水下沙脊或岩礁为基础，与潮流通道组合建设岸外人工岛深水港域，选建港口科学水平具有飞跃式进步。

第三阶段：海港建设的发展与全球化经济发展的需求密切结合，建设海陆结合，江海联运的港口群，发挥不同规模港口组合的功能优势，是当代港口建设发展的新趋势。

试以代表性海港建设实例加以分析讨论。

二、我国海港选建发展与实例

海港是供船只畅通进出，安全碇泊，具有客货上、落设施，以及获取燃料、食物与淡水补给的地方。建设港口的重要自然条件是水深与浪静，不冻不淤，以利船只停泊。但是，在深水海域风浪频繁，而浪静处易致泥沙淤积，是相互对立的一对矛盾。因此，应对需建港区进行海洋动力地貌调查研究，选港调研与工程实验分析十分重要。通过调查确定：区

* 王颖：《中国学科发展战略 · 海岸海洋科学》，第 222 - 235 页，第九章. 北京：科学出版社，2016 年。

域河、海动力环境——风、浪、水流动力特点与水深条件；海域泥沙组成特性、泥沙来源与泥沙量；海岸与海底岩层、结构、地质构造、地震活动以及海岸与港口区域的侵蚀、淤积动态与稳定性分析。选择无大量或突发性泥沙来源，深水与动力条件适宜处，建港后可稳定地使用，不需时常地疏浚维护，以期达到事半功倍的效果。

我国建港发展的三个阶段，是适应当时社会经济条件的结果，与当时的船舰吨位、航海能力、对海洋运输的需求量以及海洋工程技术水平有关。

（一）第一阶段

建港在20世纪初至30年代，继之60～80年代，当时的大船主要是1万～5万t级的，利用天然港湾或辅建防波堤的港口主要有：旅顺港、大连港、秦皇岛港、青岛港、连云港、北仑港、厦门港、湛江港、北海港、榆林港、基隆港等，无论是老港址或新码头，均沿海岸基岩港湾而建。沿平原河口建港，最具代表性的是天津新港、上海港及广州港。丹东港、宁波港、温州港、福州港、泉州港以及洋浦港、八所港等处则兼具河口与基岩海湾的双重特性。沿长江河口段及珠江下游具有优良的水深，但所建的港口，多具有河口深水段迁移、航道不规则、水深多变的特点。尤其是因江潮顶托，长江口拦门沙淤浅碍航，后经多次疏浚，拦门沙段航道水深达−12.5 m，5万t级轮船可全天候通行，但10万t级大船需趁潮通行。加之，没有系统规划利用长江天然航道，下游多处大桥跨江，阻碍大型海轮沿江深入。多数基岩海湾港口却得以持续稳定的利用，至20世纪末，除北仑港具金塘深水航道外，早期建港的原设计水深已不敷10万t级以上的巨轮进出与碇泊，需待扩建。

（二）第二阶段

伴随着中国改革开放的进展，海外联系与海洋经济发展，尤其是对石油、矿砂的大量需求，海港建设迫切需要适应10万～20万t级大、中型油轮船只的停靠与碇泊。因此，在我国尤其是经济发达的渤海湾与长江三角洲海域，发展了人工岛大型深水港建设时期，是我国建港第二阶段的特点。先后建成30万t级的唐山曹妃甸深水港，长江三角洲北翼天然10万t级洋口西太阳沙人工岛深水港及15万t级洋山岛深水港等，标志着20世纪90年代至21世纪初建港的新阶段。

1. 曹妃甸

或叫沙壘田岛，原为古滦河入海所堆积的海岸沙坝，由于滦河向东迁移，成为遗留于距岸23 km的沙岛（图1），南京大学海岸地貌与沉积研究室通过在该区的动力地貌勘测了解到，当年古滦河水沙量大，曹妃甸沙坝具有深厚的沙基（图2），故而虽孤悬外海，却经久未蚀尽。沙岛前500 m处水深达−25 m，是处于渤海主潮流自西向东北方向流动的潮流通道−30 m深槽上。该深槽东西向长达6 km，南北向宽5 km，最深处为−36 m。因而可利用具有深厚沙基的岸外沙坝岛与渤海主潮流通道最佳的天然组合设计建设30万t级深水港。历经评审，却缘于为2008年奥林匹克运动会在北京举办，石景山钢铁厂需搬迁，才促成此最佳深水大港的投建。它以曹妃甸沙坝为基础，扩建为人工岛式码头，岛后浅海围填成陆，是新型的钢铁石油基地所在，而曹妃甸沙坝—人工岛成为屹立在渤海最大的30万t级大型泊位深水海港（图3）。自开港后稳定使用，曹妃甸深水码头的建成印证了建港前期海岸海洋动力地貌调研成果的科学准确性。据发展规划，曹妃甸港可建设 30 万～70 万 t 级深水泊位 263 个，可发展成为年吞吐能力 5 亿 t 的现代化的国际贸易大港。自 2003 年港口建设以来，未形成骤淤或严重回淤，确保了深水港的发展利用。

图 1　曹妃甸沙岛(附彩图)

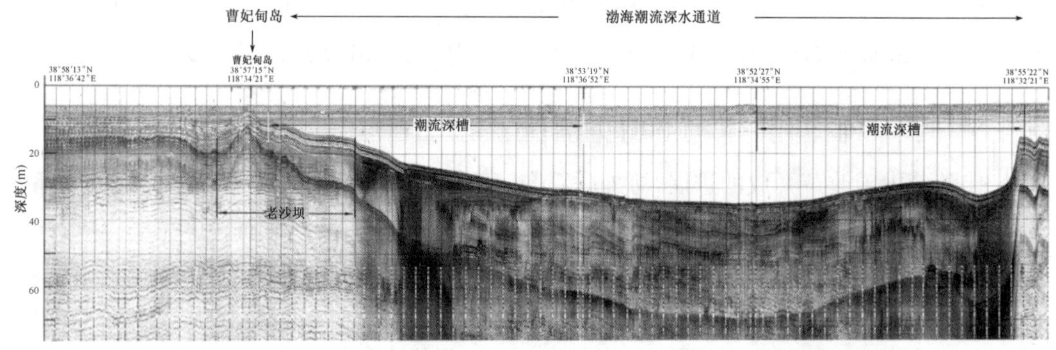

图 2　曹妃甸沙岛与相邻深槽地层剖面

2. 洋口港

洋口港位于我国沿海绵延面积最大的苏北平原岸外。在南黄海内陆架分布着巨大的黄海辐射沙脊群，南北长达 199.6 km(介于 32°00′N～33°48′N)，东西宽约 140 km(介于 120°40′E～122°10′E)，以弶港为中心，自陆向海呈扇形展开。由 70 多条沙脊与脊间的潮流通道组成，水深范围介于－25～0 m，全部面积为 22 470 km^2，是世界上最大的浅海潮流沙脊群，其中约 3 782 km^2 已出露为沙洲，是新生的土地资源(图 4)(王颖，2003)。

(a)

(b)

图 3 曹妃甸港人工岛深水码头(附彩图)

资料来源:(a) 据 2006 年中国科学院北京遥感卫星地面站遥感影像;(b) 据 2007 年遥感影像

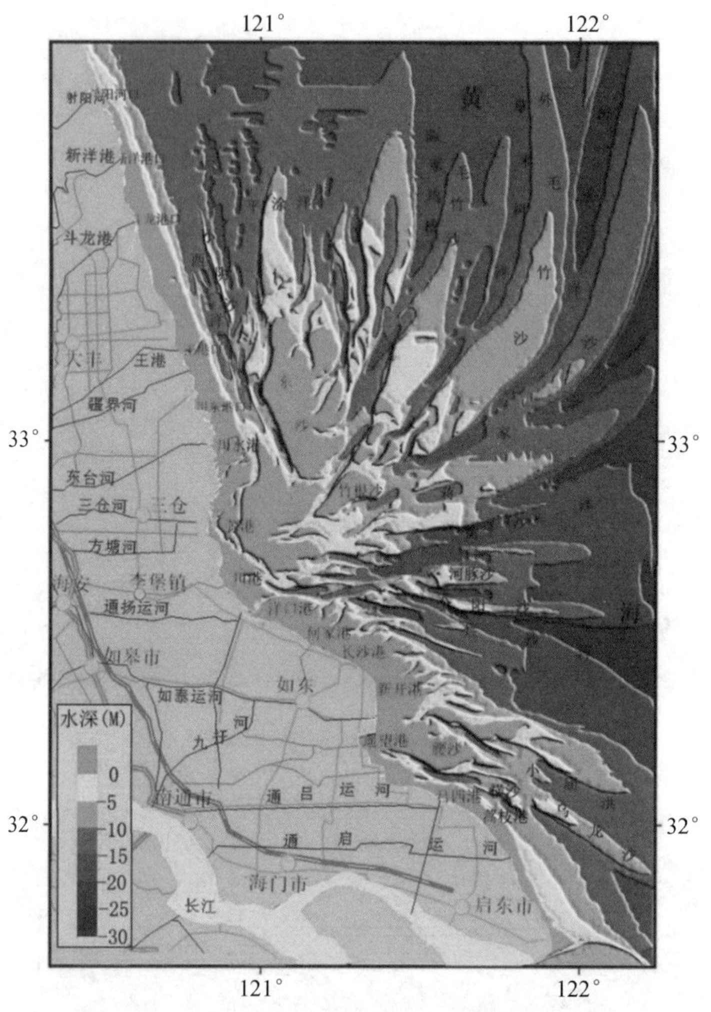

图 4　南黄海辐射沙脊群全貌(据 1979 年海图)(附彩图)

沙脊群是 3.5 万～4.3 万年前冰期低海面时，古长江由弶港一带(黄沙洋与烂沙洋水道)出口时堆积的三角洲，在全新世近 1 万年的海侵过程中，经潮浪冲刷改造成辐射状沙脊与潮流通道的组合地貌。从东海向黄海传播的涨潮波，与来自山东半岛的反射潮波相交，在弶港岸外辐聚，落潮时呈 150°扇面向外散射形成辐射状潮流场，平均潮差 4.18 m，大潮潮差 6.5 m，在黄沙洋形成我国最大的潮差 9.28 m；涨潮流平均流速 1.2～1.3 m/s，落潮流急，平均流速 1.4～1.8 m/s，沙脊群为半日潮型，每湾潮通量达 21×10^8 m^3 水量，每涨潮流量 19.5×10^8 m^3，形成全球最具代表性的辐合-散射潮流系统。潮流冲刷改造古长江三角洲为辐射状沙脊群，潮流是维持通道水深的强大动力，形成天然深水航道，如黄沙洋、烂沙洋、小庙泓均是沿古长江河谷发育的潮流通道(图 5)。

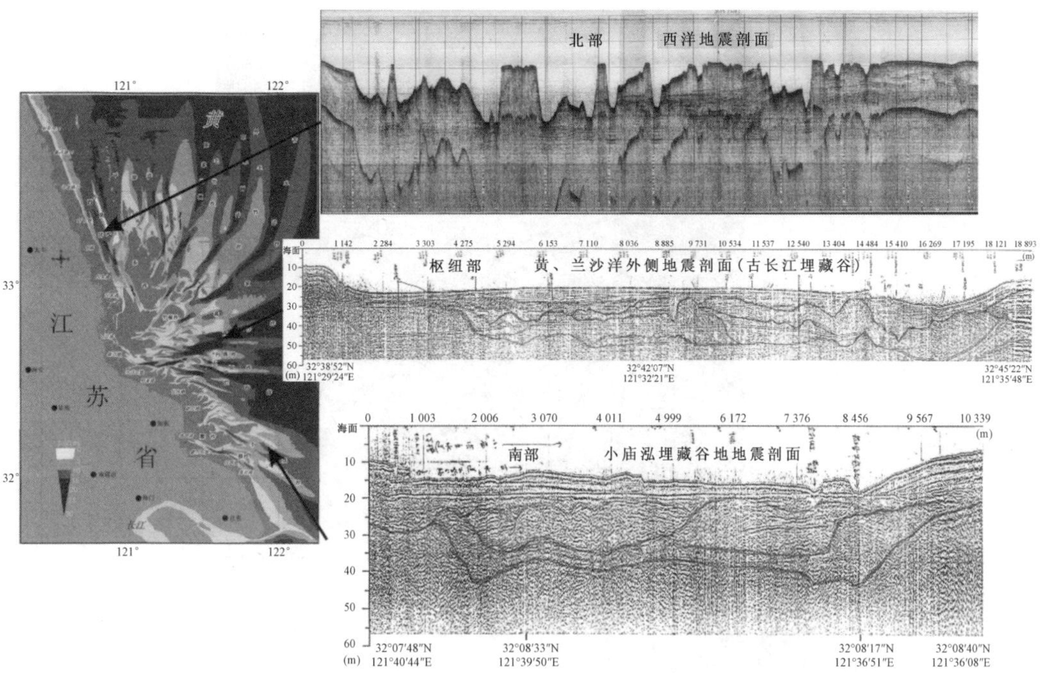

图 5 潮流通道地震剖面(左图附彩图)

由于古长江南迁,古河谷已无泥沙下泄,却有强劲的潮流冲刷,使黄沙洋-烂沙洋潮流通道保持−25～−15 m 水深;潮流通道两侧之沙脊可蔽风浪,是建造人工岛码头的基础;二者结合,为平原海岸提供了建设深水港的优异条件,利用烂沙洋与黄沙洋潮流通道具建成深水港的优异条件。利用烂沙洋与黄沙洋潮流通道建成深水港航道,海底标高取−19 m,宽度 350 m,1∶5 边坡,可满足 25 万 t 级,并兼 30 万 t 级原油巨轮乘潮通航。通过对烂沙洋北水道之西太阳沙(水下沙脊)的地质基础稳定性,细砂与粉砂质沉积组成,及周边水流动力的深入调查探测,洋口港建设采用桩基围护与填沙工程结合,扩建成 1.25 km^2 的人工岛,稳定地屹立于如东外海,作为码头堆场与场房用地,以及管道的依托。以 12.5 km 长的黄海大桥与陆地相连,利用深水潮流通道开辟为不经疏深的 10 万 t 级优异海港,并有潮滩滩涂供临港开发后备用地。洋口港建设开辟了利用辐射沙脊群与大型潮流通道组合建港的范例,突破了潮滩平原海岸无深水港资源之禁例。沙脊群北部、东台县所属的苦水洋,与洋口港相似,可发展为人工岛与潮流通道组合的深水海港。洋口港屹立于南黄海展现新型大港:油、气基地与新型能源示范点的新风貌(图 6)。洋口港因尚未获得国家投资,目前仍保留为单一人工岛与输油管道栈桥组合的形式。另外,也是具有扩展建设大型港口群的潜力所在。作为长江三角洲北翼深水大港,宜开挖洋口—长江运河以加强江海联运优势——以长江三角洲顶点南京为枢纽,内联华中与西南广大腹地;以上海为中心连接长三角南北两翼的深水港,构造直通国际的商贸枢纽港口群,可极大地增强我国海洋经济实力。

(a) 人工岛式洋口港

(b) 洋口港连接人工岛的输油、输气管道(栈桥)

图 6　人工岛式洋口港

资料来源:(a) 来自 Google Earth 2014 年遥感影像;(b) 来自 http://www.ntykg.com/

3. 洋山港

位于长江口以南,修建 31 km 长的东海大桥联结陆地与海上的大小洋山岛,利用岛间的−15 m 水深的海域建成 1.6 km 长码头,5 个深水泊位,年处理约 200 万标准集装箱,至 2020 年,将具有 52 个泊位,码头水深 14.5～15.0 m,年处理 2 500 万个标准箱能力(图 7),成为上海港的组成部分。

上述三个人工岛建造,为深水大港之实例,均以外海岛礁或沙脊为基础,人工围栅、填土、夯实而成。但是,仅洋口港仍为沙岛形式,岛上建有液化天然气管道枢纽及临港工业建筑设施。曹妃甸沙坝已并陆,洋山港具连岛坝形式。三处人工岛海港均稳固屹立于海上发挥作用。

向外海转移,建设人工岛深水港,是国际港口发展的趋势性动向,我国尤其明显,钢铁企业纷纷向深水港转移,建设临港工业基地:首钢整体搬迁至曹妃甸,矿石与钢铁成品运输以海港为基地,减少成本,有足够的滩涂空间可扩展为现代化的临港工业与新居民区;

鞍钢向鲅鱼圈深水港转移；上海宝钢向湛江东海岛转移；武钢向广西防城企沙港转移；等等。因为可利用 25 万 t 级深水港，进口矿石与输出产品，码头后方利用海岸滩涂围地，供钢铁企业使用。由于我国石油能源短缺，海上油气输入更需要深水海港，因此，外海人工岛建设将会继续发展。

(a) 洋山港三连岛规划图示

(b) 洋山港-人工陆连岛鸟瞰图(Google Earth2013 年遥感影像)(附彩图)

图 7　洋山港

（三）第三阶段

发展海陆结合、江海联运组合的港口群，这是当代海港建设发展的又一新阶段，而且是大、中、小结合，发挥港尽其利，并促进海运向陆地延伸的新举措。

实际上，人工岛式海港已渐发展为大、中、小港口结合的、现代化综合的港口群，发展中的曹妃甸港是一明显的例证。2010 年以来，截至 2014 年夏季，曹妃甸港以陆海结合，外滨与陆岸并举，扩建了大中型码头 68 个，在建码头 22 个，已从单一的人工岛深水港转变为港口群(图 8)，具有直联日本、韩国、巴西等国的直航功能，又成为“北煤南运”的基地，支持三北地区的海上通道。并成为新兴的临海工业基地，海水淡化基地与海港城镇。

图 8　曹妃甸港口组合遥感影像(附彩图)

资料来源：据 2010 年中国科学院北京遥感卫星地面站遥感影像

深水岸段是珍贵的资源，而具人工岛建设基础之海域又更加珍稀。我国港口目前的分布局势，实际是成簇分布：在渤海如旅顺—大连港口，营口—鲅鱼圈港，京唐港—曹妃甸港—天津新港，黄骅港—东营港，葫芦岛港—山海关渔港—南李庄军港—秦皇岛新港与老港，龙口港；在黄海如青岛新、老港口，日照港—连云港—滨海港，洋口港—吕四港—南通港；在东海如上海港—芦潮港—洋山港，宁波港—镇海港—北仑港—舟山港，厦门港—湄洲港；在南海如广州港—香港—赤湾港—盐田港，北海港—钦州港—防城港，海口港—八所港—洋浦港，榆林港—三亚港—莺歌海港等。大体上均是具有相邻的、群组特性，但却分别经营，缺少统一规划与分工发展，影响了港口持续地发挥作用，抑制了经济效益。关键在于统一规划与领导管理，发挥区域分工与港口组合特性的功能。面对 21 世纪全球化经济与海洋世纪发展的时代需求，宜从全国规模地规划与建设海陆结合、江海联运、大中小结合的港口群，其中以经济发达的辽东湾旅大海区，京津唐沿岸，青岛港城，长江三角洲南北翼，珠江口海域，以及国际前沿的南海岛礁海域等，组建港口群十分迫切。可以采取中央统一领导部署，地区与部门结合的办法进行。以经济发达的长三角为例，上海是国际大都市，通江联海，商贸经济向全球辐射是发展中的突出特点。上海港南翼为基岩港湾岸，分布着一系列大港，如洋山港、宁波—北仑港，港阔水深，但腹背为山地丘陵，港区陆域与发展公路运输系统有一定的限制，因此，宜重视港湾河道的开发利用，在海港的陆地侧组建水运与陆运结合的运输网络系统，疏通开扩河港航道与公路建设并举，海运与陆运结合，必促进物资流通与区域经济繁荣，更重要的是增强人民海洋意识的直接观感。上海港以北，南通地区是地理上的长三角北翼，其特点是：气候温、湿相宜，平原土地开阔，有江淮与运河水系与陆地沟通，粮、棉、果、药及水产品丰富，经济发达，人民受教育水平高，多长寿之乡。长三角北翼这些特点不仅是环渤海平原海岸带，也是浙、闽、粤或珠三角地区所不能匹配的，后者虽是雨量充沛，农业品出产丰富，但没有长三角北翼那么广阔的平原与成长中的潮滩、海涂为新生的、后备土地资源。但长三角北翼的资源环境特点尚未得到相宜地发展与利用，尤其是海洋经济尚处于发展的初期阶段。

江苏的工农业生产值在沿海十一个省市中位居第二，但是海洋经济产值至 2010 年尚位居第五。而江苏的沿海地市主要在南黄海，海洋经济产值主要于长三角北翼地区。经济发达的江苏省，有责任致力发展与提高海洋经济水平，海洋经济低于南、北两侧省市，正说明了江苏应为长三角北翼的重点，致力于海洋经济发展，成为上海国际大都市发展中的后援基地：南通港已形成先进的造船业与海洋工程机械制造业基地；吕四港渔业基地发展着远洋捕捞与近岸生产相结合，具备了优良的发展基础；与吕四港相伴为邻，通州湾为南通市沿海新兴的城镇中心；洋口港为长三角北翼最佳的天然深水大港，已展现为新能源与工农业产品的基地。进一步，宜统一规划江海联运，由内及外，通江达海，组建南通、吕四-通州湾与洋口港三处集结的港口群，大、中、小港口结合，分工互补，共赢发展。关键是构建江海运河，其步骤如下：① 着手于顺直、浚深通吕运河，扩充为 1 000～3 000 t 级航道，使吕四港内经南通港通长江，北联通州湾港，成为中心型双港。② 规划与开挖洋口港-南通港的江海运河。古长江自北向南迁移过程中，在洋口地区出口的古河道遗迹有三条(图 9)，其中以 b、c 两处条件适宜：距离短，淤浅时代稍晚，江海运河可为第二长江口通道，可避免河口拦门沙淤积阻航。河道开挖宜顺直、浚深，以通行 5 000～10 000 t 船只为目的，必要时可采用船闸提升水位，北欧鹿特丹港、阿姆斯特丹港及安特卫普港均为河海联运的工程先例，可为借鉴。发展两条通江运河，使洋口港直通内陆，组合以通州湾为中心港的沿海港口，构建长三角北翼港口群，将是我国海港建设发展新阶段的里程碑与海运发展的新跃点。促进长三角北翼海洋经济持续进步，构成以上海为中心的国际海港大都市群，其经济发展必日新月异。

图 9 卫星图片显示古长江河道(附彩图)

三、结语

我国海港建设三个阶段的特点，反映着社会经济发展和人民生活水平提高的需求。20 世纪中后期以来，航空运输快速便捷，取代了海洋客运与部分货运，但是人类生活需求的工业原料、食粮、燃料及大宗散杂货仍需船只运输。人类生存离不开清洁的空气与淡水，海洋是气、水源泉。海岸与海底资源开发与工程、海洋军事活动、航海探险与科学调查、海洋旅游与休闲度假、海洋环境保护与防灾、海洋救助与海洋管理等，仍需要船舰活动与港口上、落，人类生产与生存活动离不开海洋。尤其是当今人口增长，陆地环境利用与资源开发压力倍增的状况下，开发海洋资源、利用与保护海洋环境是必然的出路，航海与港口建设具有贯彻始终的重要作用。

我国位居亚太边缘海之内侧，陆、海国土兼备，东部沿海是经济发达与人口密集的地带，开辟海上通途，开拓海洋，发展海洋经济，提高海洋科技、文化对中华民族世代延续发展至关重要。致力研究与总结海港建设实践，具有开发海洋起点的现实意义与长远的科学价值。

参考文献

[1] 王颖主编. 2003. 黄海陆架辐射沙脊群. 北京：中国环境科学出版社.

黄海西平原海岸发育与生态环境*

一

黄海西平原界于 31°33′～35°37′N 之间，大陆岸线全长为 935.9 km。粉砂淤泥质平原-潮滩海岸为主要类型：基岩港湾海岸分布于连云港市，岸线长约 40 km，但水下岸坡仍有淤泥沉积；平原-沙滩海岸岸线长约 30 km，主要分布于海州湾以北的赣榆县。黄海西平原海岸的发育过程中，河流作用突出，赋有较多的人类活动影响。

通过对地貌特点、沉积类型与海岸发育动态遗证的综合分析，认识到该海岸经历了港湾海岸、沙坝潟湖海岸和淤泥质平原潮滩海岸三种海岸类型的发展阶段。

早期的海岸位于沿海平原西部，呈现为一系列低缓山丘所环绕的海湾，范围北起苏鲁交界的岚山头，向西南至塔山、石门、王庄集、盱眙、桂五、六合、汤山、茅山、宜溧山地、太湖洞庭山至昆山一线，这一系列断续相间的低缓山丘可能是第三纪古海岸线的遗迹。其东侧的湖泊洼地中，可能有老海湾沉积的残留。平原东部在南北两端的一些低山，是古海岛的遗迹。

长江与淮河分别在海湾中部偏南、偏北处横穿入海，据沿河岗丘分布的形式，两河“入海”口，均具有感潮河口的迹象。长江在扬州至仪征为北岸与镇江至栖霞山为南岸之间，呈现明显的喇叭口形。隋唐以来开挖大运河是利用了古沿海洼地与江淮水系河道之利。长江、淮河的泥沙填充着海湾，并为沙坝潟湖海岸的发育奠定了物质基础。龙岗、大岗是贝壳沙堤，反映出 7 000～5 000 年前海岸波浪动力与底质特性。嗣后沙坝逐渐与岸陆并接，范公堤是沿唐、宋时海岸沙坝或沿岸沙堤基础上加筑而成。

黄河自 1195—1494 年第五次及 1855—1938 第六次大改道夺淮入黄海，沿河道两侧发育了溢流扇，并向黄海汇入大量细颗粒泥沙，促进了海岸向海淤进，并发育了宽广的淤泥质潮滩海岸。

二

海岸生态系受环境发育与人类垦殖活动控制而具有明显的分区、分带现象。

(1) 现代海岸：低潮滩为光滩或养殖贝、藻，辟为丹顶鹤珍禽养殖；高潮滩为草滩，家禽或牛羊放牧，芦苇田塘或为鱼、虾养殖塘。

(2) 沿范公堤两侧平地种植桑、麻、花卉药材或麦、棉。

(3) 沿岸沙堤或废黄河河道种植果树或林木。

---

* 王颖：《第六届海峡两岸国家公园暨保护区研讨会》，55－56 页，2002 年 3 月 20－21 日，中国台北，国家公园学会主办。

(4) 平原河网区,种植稻、麦、棉、桑。

(5) 海州湾沿岸养殖虾、蟹,平原沙地种植薯、粟、花生等。

目前大规模开发,打乱了多年人类经营对环境相适应的经验,生态环境遭受一定干扰破坏,如何正确处理人、地关系,建设永续利用的海岸平原生态系,已提上议事日程。

“数字海洋”的发展与实践*

南京大学从事海岸海洋研究始自1958年，为解决天津新港回淤的泥沙来源研究。在长达51年期间，将海岸海洋动力、地貌与沉积研究相结合，进行海陆交互带的基础理论与海岸工程应用研究。在科学研究与实践中，自1990年以来，依据生产实践的需要与科学进步，逐渐地发展建设“数字海洋”科学与应用新方向。

【实例1】

1994—1995年，为深圳湾人工岛工程探寻海底沙源：在50 km^2 水域设计86条剖面，以GPS实时动态差分定位，以Geopulse作浅地层剖面探测，应用GIS技术做出深圳湾三维地层模型，标出海底基岩与其上55m厚的沉积层与砂层界面层深，获得河床相层的埋深及砂层储量 4.66×10^8 m^3，为人工岛建设提供沙源保证。

【实例2】

南黄海海底辐射沙脊群是分布于内陆架0～25 m水深的大型地貌，由70多条沙脊与相间的潮流通道组成，系3万年前古长江由弶港入海时的泥沙堆积，在全新世海面上升过程中，由潮流冲刷而成。承袭古长江发育的潮流通道与沙脊组合是长三角北翼平原海岸可建设离岸岛型深水海港的最佳港址。继之，应用GPS、Geopulse、GIS及RS技术相结合，利用信息系统强大的空间数据获取、管理与分析功能，使研究范围覆盖沙脊群22 470 km^2 的整个海域，建立了平坦海底立体监测的新方法，获取了辐射沙脊群全区自1963年至1979年的冲淤变化，以及主要潮流通道至2006年的冲淤变化富有价值的成果，又为江苏省滩涂与土地资源开发提供重要的科学依据。

【实例3】

海陆交互带——渤海曹妃甸深水海港选址与信息系统研究。曹妃甸沙岛是老滦河口海外沙坝，濒临30 m深的渤海主潮流通道，自1959年初勘，历经20世纪70年代、80年代及90年代多次重复调查，证实沙岛与潮流通道的稳定性，以及海洋动力的恒定特点。因此，选定以曹妃甸沙岛为码头依据，建设30万t级深水海港。2003年应用GPS定位，Geopulse探测海底地震剖面，证明：曹妃甸沙岛有深厚的沙基，沙岛南侧是渤海主潮流进入辽东湾的深水通道，水深36 m。沙岛后侧潮滩可填筑为港口陆域。应用海洋-地理信息系统结合与多方论证，设计曹妃甸可建成260个泊位海港，其中16个为 30×10^4 t大型泊位。进一步采用ArcGIS软件二次开发，借用大型商用关系型数据库软件及可视化系统整合开发，建立了现代化的数字曹妃甸信息系统，应用于曹妃甸新区以东海陆交互带的建设规划与管理，扩大了海港功能、临海工业区与新城镇体系建设。

【实例4】“数字南海”边缘海大型集成项目，与中国南海研究院合作完成，包括：① 数字南海软件设计：空间数据库、引擎及客户终端3D显示查询系统；② 数字南海硬件

---

* 王颖，朱大奎，傅光翮：《中国地理学会百年庆典学术论文摘要集》，56页，2009年10月，北京。

系统，建成海口基地实验室；③ 多要素、多尺度、多时态的南海及周边区域海洋环境资源、人文、经济及军事力量信息。经过集存、整编、分析、研究，完成海底地形三维模型，地质、地貌、沉积、水文、气象、矿产、水产、渔业、岛礁、港口、沿岸国经济、军事等专题成果，作为数字南海的数据源进入系统并建立网页。结合数字地球三维空间可视化技术，将不同的空间信息与研究成果，通过计算机综合地表现出来，进行相应的空间分析，获得对复杂南海问题的技术解决方案，具有重要意义。

实践项目鼓舞着研究与探索不断地前进！新技术应用与实地调查研究相继结合，赋予地球科学以强大的生命力。

The Development of China's Coastal Zone*

Introduction

The coast of China is 18 400 kilometres long extending from the mouth of the Yalu River on the China-Korea border in the north to the mouth of the Beilun River on the China-Vietnam border in the south. Including the islands—more than 6 000 of them—the total length of its coastline is approximately 32 000 km. The general outline of the coast is controlled mainly by geological structure and deposits from a number of large rivers. The continental shelf and adjacent mainland of China are composed of a series of uplifted and depressed belts, with a NE or NNE tendency intersecting obliquely with the coastline (Figure 1). A bedrock-embayed coast with rock islands is formed along the uplifted belts, and a plain coast, usually with a large river, is always formed along the depressed belts. In the South China Sea, block faults running NE define the main trend of the bedrock embayed coast, with inlets developed along faults running NW.

The sedimentary processes along the coast of China are greatly affected by the large rivers which deliver enormous quantities of sediment (Table 1).

Table 1 Annual Water and Sediment Discharge of the Three Major Rivers of China

River	Annual water discharge 10^8 m^3	Annual sediment discharge 10^8 tons
Yellow River	485. 65	12. 0
Chang Jiang (Yangtze River)	9 793. 5	4. 0～5. 0
Pearl River	3 700. 00	0. 85～1. 0

* Ying Wang: Derek R. Diamond, Karlheinz Hottes, Wu Chuanchun, eds, *Regional Planning in Different Political Systems-The Chinese Setting*-(Proceedings of the International Meeting on Regional Development Planning-Theory and Practice-. Beijing, 29 March - 5 April, 1983), 93 - 99, Bochum W. Germany, 1984.

1983 年曾由 Department of Geography, University of Nanjing 以"The Seacoast and Coastal Exploitation in China"为题刊印。

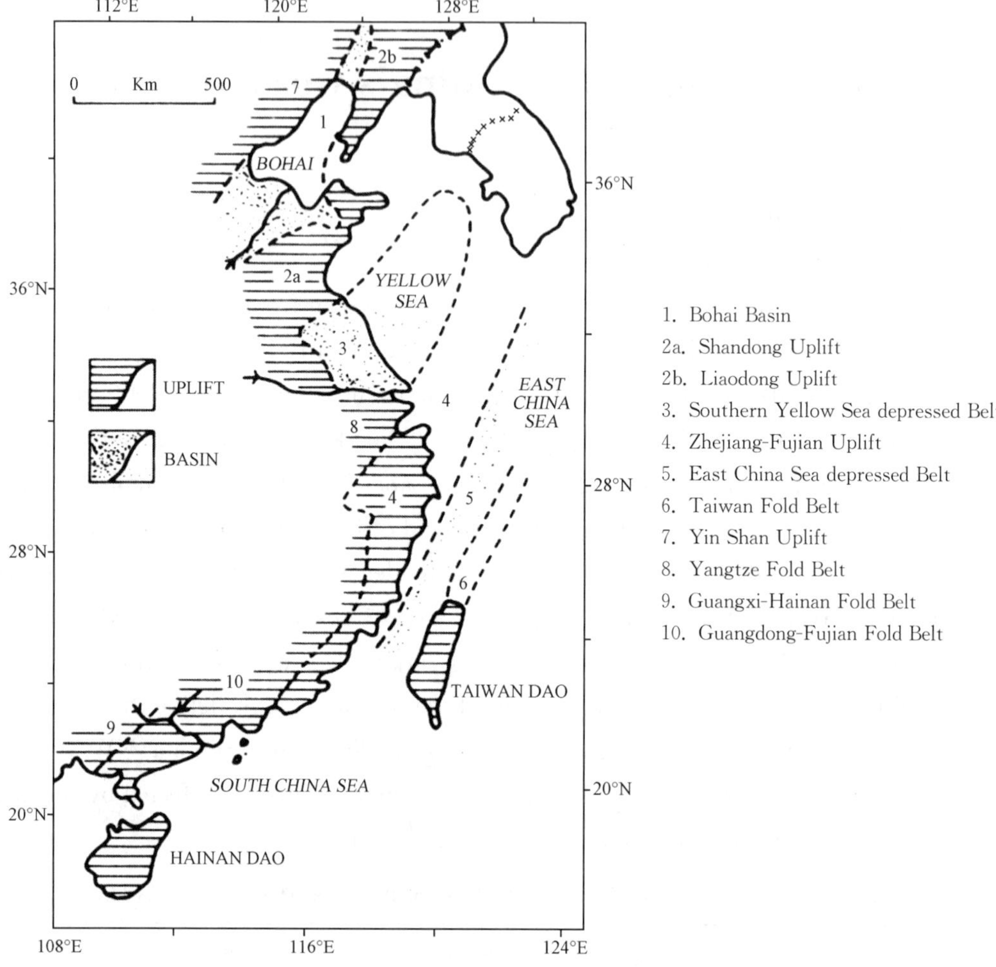

Figure 1　Tectonic Map of the Mainland Coast

China's coastal tides have great regional variations. Those of the South China Sea are predominantly diurnal, with a small range (1 or 2 m). The Bohai, Yellow and East China Seas have a dominant semidiurnal tide with a large range (3 m. or more), producing strong tidal currents. Tidal bores occur in many estuaries. Over one-third of China's coastline is made up of tidal flats.

The action of surface waves, principally wind waves, is the main mechanism in shaping most of the coast. Wave patterns are dominated by the monsoon. Northerly waves prevail in the winter, moving southwards to reach the north of Taiwan Province in September, 10° N in October, and covering the entire seaboard in November. During January the wave directions change clockwise along the edge of the Mongolian high pressure system. This change is related to the monsoon, with a northwesterly wave direction in the Bohai Sea gradually veering north to northeast in the southern East China Sea and South China Sea.

Southerly waves prevail in the summer, veering northward from the South China

Sea. They appear first in February, and by May they dominate a wide area to the south of 5° N. In July southerly waves prevail along the entire seaboard. The wave directions change anticlockwise along the edge of the Indian Ocean low pressure system, being southwesterly in the South China Sea to the south of 15° N. These change northwards through southerly to south-easterly waves in the East China, the northern Yellow and Bohai Seas.

There is a transitional period between summer and winter, during which the wind directions fluctuate, and there are no prevailing waves.

The Major Coastal Types and the Exploitation

The sea coast of China can be classified into two major types: the bedrock embayed coast and the plain coast. Both may include estuarial coasts, which are important because of the large population they support. In the south, under the tropical and subtropical climate, the coastline is also modified by the growth of coral reefs and mangrove swamps.

1. Bedrock-embayed coast

The coast is characterized by irregular headlands, small bays and islands. Wave energy is high, influencing both erosion and deposition. This type of coast is found along the Liaodong Peninsula, Shandong Peninsula, the Zhejiang, Fujian, Guangdong and Guangxi Provinces.

The bedrock embayed coast can be divided into four subtypes according to the stage of development.

(1) Marine-erosion embayed coast

This is developed on hard crystalline rock where the rate of erosion is slow with little sediment supply to the coast, for example, the coast of Liaodong Peninsula. The deep water bays along the coast are favourable to harbour development, for instance the Luda Port in the Dalian Bay and a new oil harbour in the Nianyu Bay. Many small bays with gentle waves and good water quality have always been ideal breeding grounds for various fish and aquatic plants; seven kelp and sea food farms were established and about ten thousand mu of tidal flats in the bays of Dalian area were turned into breeding grounds for shellfish in the early 1950's. The kelp, mussels, sea cucumbers, scallops, clams and especially oysters have always been outstandingly popular with Chinese people throughout the country[(1)].

(2) The marine erosion deposition type

This is most commonly developed where there are mesozoic granites covered with a thick layer of weathered deposits, such as those on the Shandong Peninsula. This material is readily eroded by wave action and by the many small rivers that carry large amounts of sandy material to the sea. Therefore, erosion features such as rock benches and terraces and deposition features, such as bay-bars, sand spits and tombolos are developed. The sediments contain valuable local deposits, including zircon, rutile, monozite and gold. Many fishing harbours have been established in the small bays behind sand spits. There is a tidal power station at Baisha, one of the small bays at the southern side of Shandong Peninsula, but it is subject to silting up from the sand carried past it into the bay.

(3) Marine deposition type

The bedrock of this type of coast is relatively soft. Marine erosion has supplied a large amount of sediment, forming a straight, protruding sand barrier protecting the bay or lagoon. This type of coast is very common in south China. The bays are famous for pearl cultivation, such as Hepu county in the Guangxi Province. Some sand barriers have been developed as narrow coastal plains in front of an irregular line that represents the original bedrock coastline. The plains are valued by local people to develop for farming, especially in areas with a high population density, as in some parts in the Fujian and Guangdong provinces.

(4) The tidal inlet-embayed coast

This type of coast is also common in south China. Here the major NE-SW lie of the coastline is cut by long, narrow bays following NW faults. Small islands or sand spits protect the mouths of the bays, forming tidal inlets which are swept clean by the powerful ebb tide currents. By using the high tidal range of 8.9 metres in the Yueqing Bay of this type of coast, China has set up the first bilateral—tidal power plant in Jiangxia, Zhejiang Province, where three of six planned generators have been set up with an output of 300 kilowatts each per generator. This type of inlet is also suitable for harbours but reclamation wherever it alters the tidal configuration of the bay will cause silting problems, as is the situation in Shantou Harbour, Guangdong Province. This harbour can handle two million tons of shipping annually, and has passenger services to Hong Kong and Singapore, but has suffered from silting of the navigation channels since the bay was reclaimed during the 1970's.

Xiamen (Amoy) was an island until 1949 when it was joined to the mainland at Jimei by a two kilometre causeway along which the railway runs. Within this kind of tidal inlet, Xiamen has a fine deep-water port at Dongdu, capable of handling 50 000 ton container ships. The harbour is being developed under a central government plan. Combined with the railway and a new airport, the Xiamen Special Economic Zone, northwest of the city, is begining to take shape. It is the fourth SEZ so far in China. By 1985 the SEZ will be provided with 40 000 tons of water and 24 000 kilowatts of electricity per day. The plan envisages 100 to 120 factories, providing jobs for 20 000 people. Raw materials and goods to meet the needs of the undertakings and their foreign personnel will enter duty free. After a three to five-year start-up period, a 15 percent profits tax will be levied on undertakings by the province. The average annual land rent will be about RMB four yuan per square metre, ranging from three to RMB 20 yuan and contracts will be for 30 to 50 years. This is a new type of coastal development in China.

2. Plains Coast

On the basis of origin, the plain coast can be subdivided into two types.

(1) The alluvial plain coast

Most of this type of coast is located seaward of mountain ranges, such as the east coast of Hebei Province, the Hangjiang delta plain in Guangdong Province, and the west coast of Taiwan Province. There is a very wide break zone, with active sediment movement both alongshore and enshore. Barrier bars, sand spits, submarine bars and extensive sandy beaches are the typical features of this coast. The coastal plains with natural irrigation systems are traditional farm land or fruit forests, and the sandy beaches provide ideal recreation areas for tourists. However, in some places local people have removed sand and gravel from beaches for construction, causing problems of coastal erosion.

(2) The marine deposition plains coast

This type of coast is flat and very extensive, and is located on the lower reaches of large rivers in areas of subsiding basement, such as the North Jiangsu Plain, North China Plain and the Liaohe Plain. The coastal slopes are very gentle, with gradients of the order of 1 : 1 000 to 1 : 5 000. As a result, effective wave action is well offshore, tidal currents thus play a major role in shaping the coast. The sediment consists mainly of silt and mud to form the extensive intertidal mud flats. For example, along the west coast of Bohai Bay, tidal flats are 4 to 6 kilometres wide and much wider in the southern part of the North Jiangsu Plain.

Along this coast, there are salt fields and coastal fishing on a fairly large scale for a considerable distance, as well as rich offshore oil and gas resources. In the southern part of the North Jiangsu Plain, where the annual precipitation is over 1 000 milimetres, the salt marsh and mud flats above mean high tide level can be reclaimed for farming. The lower part of tidal flat is better for shellfish farming or for ground net fishing. However, the mudflat coast lacks suitable sites for harbours because the small depth of water and the huge quantity of sediment along the coastal zone could cause serious silting. For example, Tianjin's new port was built on the mudflats of Bohai Bay but, shortly after its completion, the harbour had to be deepened due to extensive silting. Six million m^3 of silt were removed annually. This example has since prompted extensive coastal research programmes.

3. The River Mouth Coast

(1) The Yellow River has the highest rates of sediment discharge of any Chinese

river, and carries the sediments to form the turbid mud plume into the sea. During historic time, there have been eight major shifts in the course of the Yellow River, which once flowed into Bohai Bay and at another time to the North Jiangsu coast of the Yellow Sea. Its latest change of course from the Yellow Sea back to Bohai Bay was in 1855. Within a period of 120 years a fan-shaped delta protruded seawards from Lijin with a distal margin in 15 metres of water (Figure 2). The delta has been developed as new farmland and pasture. The mud bays along both sides beside the river mouth finger bars are fishing grounds especially suitable for prawn spawning. Prawns are a favourite food with the Chinese people and a valuable export. In addition to the agriculture, the delta and adjacent shallow sea is one of the major fields of petroleum and natural gas.

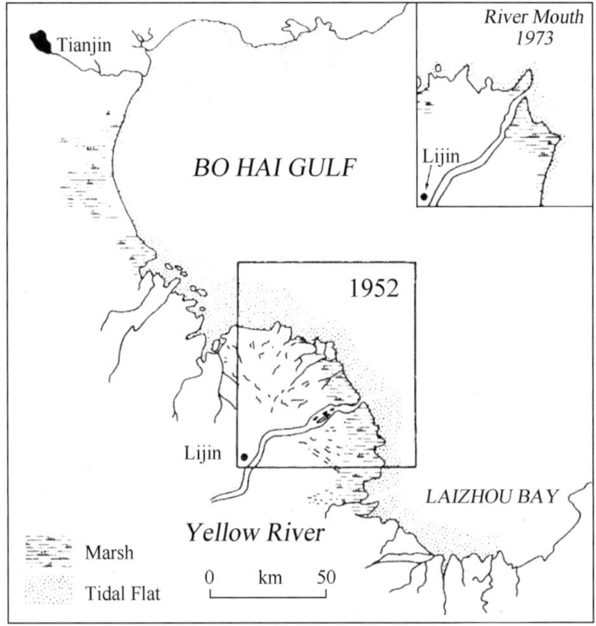

Figure 2　Yellow River Delta

(2) The Chang Jiang (Yangtze) estuary faces the open East China Sea. Wave action at its mouth is stronger than at the Yellow River. Most of the sediment accumulates in the inner part of the river mouth, and as a result a series of sand islands and rivermouth banks are formed. It causes dredging problems for navigation, and the modern Chang Jiang river delta is relatively small, and is growing towards the southeast. Shanghai is located at the river mouth, it is the largest port of China and also the largest city with well-developed industry, agriculture, science, technology, trade and finance. A new economic zone, with Shanghai as its centre has been set up in the Yangtze Delta area, to unify the ten cities and 50 counties of Suzhou, Wuxi, Changzhou, Hangzhou etc. It will promote the national economy and create a powerful centre for the nation's foreign trade. The Yangtze Delta is known as the 'land of fish and rice' because of its thriving

fisheries and agriculture. With a population of 50 million, it covers an area of 8 400 square kilometres, only 0.6 per cent of China's total territory. However, annual industrial and agricultural output value accounts for nearly one fifth of the national total.

The delta is the only area in China where the combined annual output value of industry and agriculture averages $1 000 per capita, the national target for the end of the century.

(3) The Pearl River delta includes the huge alluvial plain lying in the lower reaches of the river which cross central and southern Guangdong Province. It is transitional between the estuarine and delta types of coastline, with sediment accumulating in a bedrock bay sheltered by islands (Figure 3). Huangpu port at Guangzhou in the Pearl River delta, is a junction for water and land transport and the biggest foreign trade port in south China. The construction of a new container terminal, designed to handle 200 000 containers a year will make Huangpu port one of the country's largest international container terminals.

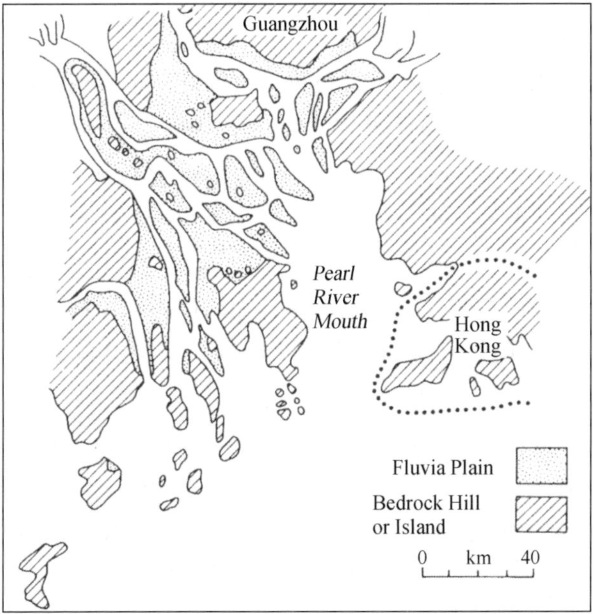

Figure 3 Pearl River Delta

The Pearl River delta is one of the most crowded, fertile and beautiful areas in China. It is highly developed, especially in agriculture. A few examples will give a brief view of comprehensive planning of the region. There are more than 100 granite hills in the delta. Due to the chemical weathering processes in the tropical climate zone, the rock was readily weathered into kaolin, the essential material for Shiwans porcelain industry. Shiwan is a porcelain centre with a history dating back almost 1 000 years.

In a region of rivers and lakes where Nanhai and Shunde counties meet, the landscape is wonderful with numerous interconnected fish ponds. Mulberry and banana trees and sugar cane grow along the embankments of the ponds. The local people have been conscientious in maintaining an ecological balance. They grow mulberry trees to raise silkworms, then feed the fish with silkworm droppings and pupae. The plants surrounding the ponds help create a favourable microclimate for growth of the fish and the pond sludge fertilizes the plants. Quite a unique ecosystem.

There is a sandy area in the southern part of the delta, with numerous power-operated irrigation and drainage pumping stations. They represent a significant achievement in the delta's farm land capital construction.

Land has been reclaimed from the sea in the eight estuaries of the Pearl River. Most of the silt carried into the sea is held back by the tide, consequently the land is rapidly inching its way into the sea—at an annual rate of about 80～130 metres.

(4) The estuaries, such as Hangzhou Bay just south of Shanghai, carry very little sediment. The tital range is high, up to 9 metres, depending on the shape of the coastline, and there is a distinct tidal bore. During the historical period, the Hangzhou estuary has tended to shoal and as a result, the tidal ranges have decreased. The estuary has great potential for tidal power energy.

Coral reefs and mangrove coasts (Biogenic coasts) are important to comprehensive exploitation in south China. Most of the South China Sea islands are atolls, rich in fishing resources and are also wonderful scenic places for tourism and scientific study. Fringing reefs are found surrounding Hainan Island, Taiwan and many other small islands. At present, the coral growth is poor. Many small bays, with fringing reefs, around Hainan Island are good for cultivating aquatic products. The workers there use the white-lip oyster to cultivate pearls. In 1981 they cultivated a huge pearl, 1. 55 cm in diameter.

In China, the mangrove can grow south of latitude 27° N. Along the mainland coast the mangrove swamps grew intermittently at river mouths, bays or lagoons, but the growth is stunted with only small bushes (mainly *Avicennia*). Mangrove jungles with 24 varieties grow along the northeast of Hainan Island.

Summary

China's coast and adjacent continental shelf have great potential and are important to the development of the country's economy. At present, comprehensive surveys along the whole coastal zone are being carried out, and plan for full scale exploitation of marine resources drawn up. It is an important part of China's programme to quadruple industrial and agricultural production in the next 20 years. For example, sea ports have been the weakest link in the transport buildup, and in the Sixth Five-Year Plan, 132 deep-water berths will be constructed at 15 coastal ports, including Luda, Qinhuangdao, Tianjin New Port, Qingdao, Shijiusuo, Lianyungang, Shanghai, Huangpu, and Zhanjiang. Of these, 54 will come into operation by the end of 1985, increasing loading a unloading capacity by 100 million tons to 317 million tons. By the end of the century, the nation's total coastal berths are expected to reach 1 000 from 330 in 1981. The number of deep-water berths that can be used by ships of the 10 000 ton class will rise from 143 to around 600. Also, the handling capacity of the ports is expected to climb to about 700 million tons, 2. 2 times the 1980 figure, or an average growth rate of 6 per cent per year. Petroleum resources have been widely surveyed, especially several old submarine deltas and accumulated basins in the depressed belts of the continental shelf for offshore oil and gas exploitation. In an area of one million square kilometres, more than 100 wells have been sunk and oil-bearing zones have been identified. Oil production has started in the Bohai and South China Seas. The potential for tidal, wave and thermal energy from sea water is also being investigated. Marine biologists have developed techniques for breeding kelp, laver, and prawns. Seafood production increased from 88 000 tons in 1954 to 458 000 tons in 1981. In addition to salt production, potassium, bromine, iodine and other chemicals are being extracted from sea water.

However, the growth of industries and agriculture along the coastal region, especially in the estuaries, harbours and some coastal areas, is threatening the marine environment with pollution. The toxic content of marine creatures has increased, and this has pushed fishing grounds further out. Some coastal resorts have been damaged by pollution and reclamation projects in several bay areas have caused silting of the harbours or navigation routes nearby. The daming of the lower reaches of some rivers has destroyed the natural environmental balance in the river mouth areas and caused rapid siling in front of the dams. To solve such problems and encourage the comprehensive use of coastal zones, more regional planning and marine legislation are needed to protect the ecological environment and enhance coastal management.

References

[1] Chai, I-chi and Xian-yan Lee. 1964. Some characters of coral reef of Southern Coast of Hainan Island. *Oceanology and Limnology*, 6(2): 205 - 218.

[2] Chao, Qing-ying, Hao-ming He and Xien-luan Cheng. 1977. The Huang He Delta and shoreline development. Geomorphological Symposium of the Geographical Society of China, Tianjin.

[3] Cheng, Chi-yu. 1957. Note on the Development of the Yangtze Estuary. *Acta Geographical Scinica*, 23(3): 241 - 253.

[4] Jin, Xiang-lung. 1977. Submarine Geology of the Yellow Sea and East China Sea. *Ocean*, 5.

[5] Tseng, Chiu-suen. 1957. Note on the Hankiang Delta. *Acta Geographica Sinica*, 23 (3): 255 - 278.

[6] Tseng, Chiu-suen. 1982. Type Division of Atolls in China. *Marine Science Bulletin*, (4): 43 - 50.

[7] Wang, Ying. 1976. The Sea Bottom of the South China Sea. *Ocean*, 2.

[8] Wang, Ying. 1977. The Submarine Geomorphology of Bohai Sea. *Ocean*, 6: 5 - 8.

[9] Wang, Ying. 1980. The Coast of China. *Geoscience Canada*, 7(3):109 - 113.

[10] Wang, Ying. 1983. The Mudflat Coast of China. *Canadian Journal of Fisheries and Aquatic Science*, 40, (Supplement 1): 160 - 171.

建设长三角北翼洋口深水港推动海洋经济发展*

一、长江三角洲经济社会发展需要扩大深水港建设

长江三角洲地处我国海岸带中枢，内含15个城市、13个地区，组成特大中心城市6个：上海、南京、苏州、无锡、杭州与宁波。全部面积99 610 km²，约占全国陆地面积1%，人口7 404.71万，约占全国人口的6%，创造的工农业产值占全国生产总值(GDP)的19.5%，贡献22%的中央财政收入(据佘之祥，2004)，其经济实力强，文化科技底蕴深厚，经济发展速度快。在全球经济一体化发展的新形势下，长三角经济的进一步发展，必须发挥其江海联运的优势：以三角洲西顶点的南京为枢纽，通过长江黄金水道内联华中与西南广大腹地；外以上海为中心，联合三角洲南北两端深水港，构建直通海外的临海工业与国际商贸枢纽港口群。

纵观长江三角洲港口现状，有以下两点不足：

(1) 深水港主要分布于长江三角洲南部(图1)，如：上海港、洋山港、宁波港(镇海、北

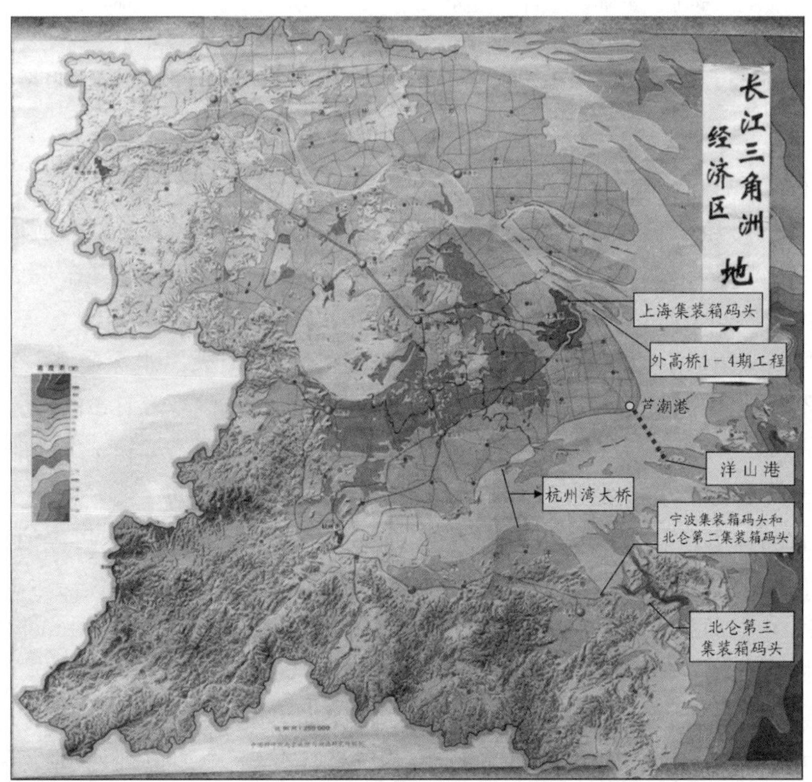

图1　长江三角洲深水港分布图

* 王颖撰写全文，以摘要载于：《2004年海洋经济地理论坛论文集》，大连，2004年9月。

仑与宁波老港)，长江三角洲北翼没有海港，形似跛足巨人，没有发挥长三角的海、陆、空立体发展优势。

随着长三角经济发展，2002 年该区所有港口的集装箱总吞吐量为 1 050 万个标准箱，其中上海港与宁波港合计吞吐量占 90%以上。2002 年全国 3 700 万标准箱的 23%为上海所吞吐，达到 860 万标准箱，已超过高雄港而成为世界第四大集装箱港口。宁波港于 2003 年货物吞吐量首次突破 1.85 亿 t，保持全国大陆港口第二位，集装箱吞吐量突破 277 万标准箱，同比增长 48.6%，2004 年可增长 30%，从而进入全球前 20 个大港口。宁波北仑港现有两个集装箱码头，设计年吞吐量为 50 万个标准箱，2004 年将完成第三个码头建设。北仑港天然深水区多，码头水深 13.5 m，可停泊 8 万 t 巴拿马级集装箱，航道最大水深 17～24 m，宁波港发展势头已超过上海(图 2 和图 3)[1]。

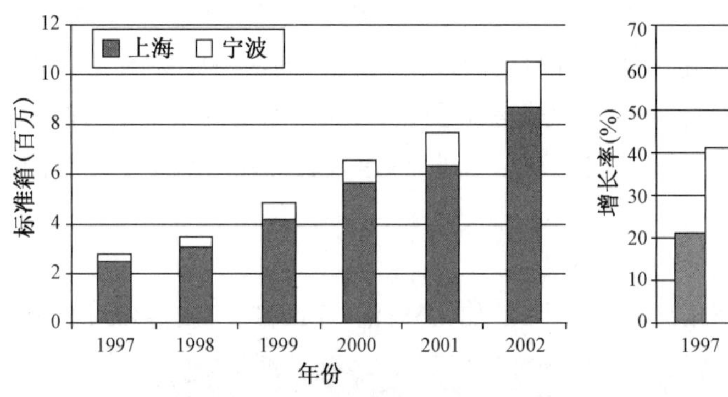

图 2　上海和宁波集装箱吞吐量比较

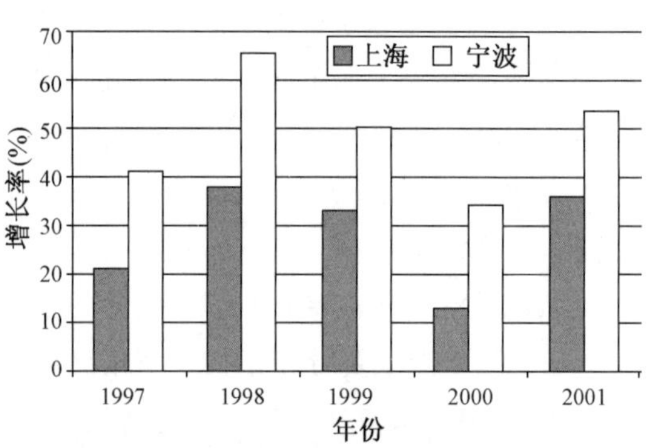

图 3　上海和宁波集装箱吞吐增长情况

* 据 Kevin Cullinane 2004.5. 疏浚与港口建设，DPC

上海港发展有口门碍航与码头岸线不足之问题。长江口由于咸淡水交汇、河水径流与潮流顶托，口门有拦门沙浅滩淤积，天然水深仅 6.5 m，成为 164 km 长江口航道的主要障碍。沙洲与浅滩堆积，使长江口分支成南北两支，南支分成三个水道，口门总宽度为 90 km，年平均水流量为 29 600 m^3/s，1999—2000 年年平均淤积量为 3.42×10^8 t[2]。长江口航道主要利用南支南通道的北槽(图 4)，该处为水流主通道，但天然水深受涨、落潮流及径流变化而不断地变化，自 1998 年至 2001 年，已将北槽水深疏浚至 8.5 m；计划至 2005 年初，将航道挖深至 10 m；最终航道挖至 12.5 m，不需趁潮即可通行 5 万 t 级的船只[2]。

为改善其吞吐能力，上海市政府采取两项措施：

① 外高桥深水方案自 1993 年实施：创建 1.63 km^2，水深 13 m 港区，可停泊 4 艘 4 000 标准箱的集装箱船只，但水深 13 m 仍难维持，作为国际深水港，上海港水深需达 16 m。

② 在洋山群岛利用大小洋山间平均水深为 15 m 的海域建港，以 31 km 长的东海大桥与市区相连，预计 20 年完成，总投资 120 亿美元。从 2002 年开始，投资 14.5 亿美元，至 2005 年完工，将建成 1600 m 长码头，包括 5 个深水泊位，具有年处理 180 万至 200 万标准箱能力(图 5)。至 2020 年全部完工时，港口将具有 52 个泊位，码头水深达 14.5～15 m，年处理 2 500 万标准箱能力[1]。上述分析可知，洋山港水深 15 m，与国际深水港标准有差距！

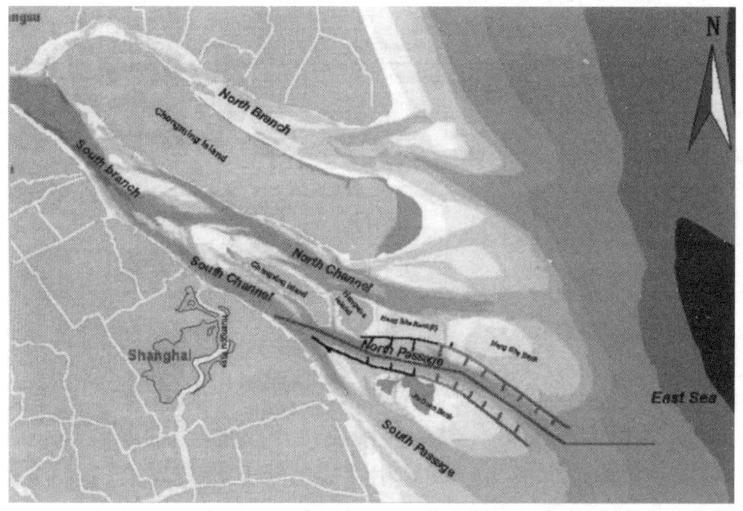

图 4　长江口航道示意图(据李文正,2004)

图 5　洋山港集装箱码头效果图(据 Kevin Cullinane,2004)

(2) 沿江航运主要是小吨位船只,以上海港为中心,沿长江内联太仓、南通、张家港、镇江至南京为万吨级。长江口拦门沙碍航,浚深至 8.5 m 时,6 万 t 级大船需趁潮进入长江口,江阴与镇江间航道浅滩需浚深、或大船减载,才能通行到南京港,由于长江大桥跨江阻拦,5 万 t 级海轮沿长江内航尚不能越过南京港。按目前国际航运发展,20 万 t 级集装箱船输运成本低,运输价值高,10 万 t 级集装箱船已属中小型船只。沿江发展受长江口拦门沙横卡咽喉,不能发挥长江水道的天然运输能力。长江河港运输难以与海港输运能力所比对(表 1)[3]。

表1　长江三角洲地区各主要港口集装箱吞吐量预测表[3]

单位：万 TEU(标准箱)

港口	长江下游各主要港口								上海港		宁波港		其他	合计
	南京	镇江	张家港	南通	太仓	其他	小计	占比例(%)	吞吐量	占比例(%)	吞吐量	占比例(%)		
2005	45	12	30	45	26	20	178	11.87	1035	69	260	17.33	27	1 500
2010	80	20	40	80	50	38	308	13.22	1510	64.80	470	20.17	42	2 330

* 据严以新、许长新，2003。

所以，必须建设长江三角洲北翼通海港口，洋口港天然水深 16～23 m，原为古长江河道，长江南徙，古河谷无河沙下泄，却有天然潮流维持水深，具有建设深水海港的最佳条件。建设洋口深水港使南通形成与宁波遥相呼应的长三角两翼重镇，使扬州—泰州—南通一线，直接通海，发挥苏北土地广袤、劳动力多、素质高、农业发达的优势，为长三角的经济可持续增长增添巨大活力。

二、江苏发展海洋经济，必须建设深水海港

江苏省大陆海岸线分布于长江口以北，其总长度为 953.875 9 km，其中 883.561 km 岸线属于粉砂淤泥质平原海岸；北部苏鲁交界地带有 40.248 3 km 基岩港湾海岸，30.062 5 km 砂质海岸及 19 座基岩岛屿，岛屿岸线长 26.964 1 km。

平原海岸坡缓水浅，波浪动力衰减，缺少供大船停泊的天然良港，这是苏北长期处于封闭型农业经济的客观环境不利因素。

我国在“九五”期间海洋经济发展速度快于同期 GDP 发展速度，其增长率为 10.9%，如：1999 年为 3 651 亿元，2000 年为 4 133.5 亿元，快于同期世界总体发展速度，海洋经济在国民经济中总量不断提高，2003 年我国海洋产业总值达到 10 077.7 亿元(突破 1 万亿元)，海洋产业增加值为 4 455.54 亿元，比上年增长 9.4%，继续保持高于同期国民经济增长速度，相当于全国国内生产总值的 3.8%。环渤海、长三角、珠三角三大海洋经济区基本形成，2003 年其海洋产值占全国海洋产值的 82.3%。

江苏省地居江海之交，海陆交通理应高度发达，虽提出发展“海上苏东”等战略口号，但现实发展尚不理想，与相邻省市比较，江苏省的海洋经济发展滞后(表 2)[4]，可能与 800 多 km 平原海岸未建成深水良港，未形成现代化的临海工农业带有关。传统的海洋经济为海洋渔业、盐业与海洋交通运输业。20 世纪 80 年代以来，新兴的海洋产业发展快：如海水养殖业、海洋油气业、滨海与海岛旅游业、海水利用、海洋医药、海洋食品、海底通讯与信息产业。21 世纪的海洋产业为：海洋人工岛、海域再生能源、深海与大洋采矿业、天然气水合物(可燃冰)新能源、海水与海洋生物资源综合利用等。我国海洋经济持续增长，其中又以第二产业发展迅速(表 3)[5]。长江三角洲北翼港口，应发展 21 世纪的海洋产业，以人工岛、海洋油气与新能源带动制造业，发展以农业、土地及人力资源为特色的临港工业——循环经济体系。

表 2　2000 年中国沿海省(区/市)海陆域经济对比*　　单位:亿元

省区	海洋经济产值(排序)	国内生产总值(排序)
广东	1 114.57(1)	9 662.23(1)
山东	737.76(2)	8 542.44(3)
上海	601.37(3)	4 451.15(7)
福建	419.15(4)	3 920.07(8)
浙江	399.53(5)	6 036.34(4)
辽宁	326.58(6)	4 669.06(6)
江苏	146.04(7)	8 582.73(2)
天津	138.63(8)	1 639.36(10)
广西	110.45(9)	2 050.14(9)
海南	70.23(10)	518.48(11)
河北	69.19(11)	5 088.96(5)

* 据杨荫凯,"宏观经济研究",2002.2

表 3　中国海洋产业结构发展比较　　单位:亿元

年份	总产值	第一产业	第二产业	第三产业	三产比重
1995	2 463.85*	1 176.87	185.03	946.53	48∶8∶38
2000	4 133.50*	2 084.34	639.86	1 355.30	51∶16∶33
2003	10 077.71*	2 821.76	2 922.54	4 333.41	28∶29∶43

* 据中国海洋年鉴 2001 及 2003 年中国海洋经济统计公报

深入分析,江苏省长江以北部分土地面积与人口约占全省的 2/3,但人均国内生产总值仅占全省人均国内生产总值的 3/5,而苏南的人均国内产值为全省人均产值的 1.89 倍(表 4),产值与资源相比较,江苏省经济发展的潜力在苏北*。

表 4　2001 年江苏省长江以北部分各项指标在全省中的比例[6]

指　标	苏南	苏中	苏北	全省	苏北在全省的比例(%)	长江以北地区在全省的比例(%)
年末总人口(万人)	2 177.57	1 737.15	3 182.28	7 097	44.84	69.32
土地面积(km^2)	27 954	20 429	52 312	100 695	51.95	72.24
耕地面积(千公顷)	1 166.73	1 114.20	2 693.20	4 974.13	54.14	76.54
国内生产总值(亿元)	5 446.25	1 764.73	2 186.94	9 397.92	23.27	42.05
人均国内生产总值(元)	24 969	10 160	6 889	13 242	52.02	60.77
粮食产量(万 t)	626.00	784.86	1 593.87	3 004.73	53.04	79.17
棉花产量(万 t)	2.16	8.64	37.52	48.32	77.65	95.53
油料产量(万 t)	56.35	63.88	112.29	232.52	48.29	75.77

* 以下均指长江以北的江苏地区。

苏北经济的发展应以发展海洋经济为起点，迈出新步伐，其关键是建设洋口深水港临海工业。江苏北部苏鲁交界带虽有连云港贯通中原与西北，虽然腹地区域之经济目前尚欠发达，但连云港仍具有联通欧亚海路通道的巨大潜力。在长达 953.8 km 的苏北海岸，不能仅仅有北部一个港口，尤其是长三角北翼迫切需要建深水港联通海外，何况连云港尚未形成与江苏南部经济发达区物流运输的主渠道；位于长江北岸的南通港，既可沿江联系，又可转口海上，但是，长江口航道制约了大型集装箱货轮畅通，南通港制造了 30 万 t 级海轮，也只能空载驶出长江口。从江苏省社会经济持续保持高水平发展所需，宜加强海港建设，形成海陆江交通动脉网络。其关键是在长江三角洲接近经济发达的苏南并与经济起飞的苏北衔接段，建设深水海港，成为江苏中部的海上门户，发展临港工业，形成新型的海港工贸基地，增强对外进出口联系，带动苏北沿海经济发展，促进"长江三角洲北翼经济带"的建设。

三、洋口港——长三角北翼最佳天然深水航道与深水码头组合

苏北 800 多 km 平原海岸唯一具有天然深水航道的地区，位于南黄海辐射沙脊群区。其中最佳的港址是洋口港，该处烂沙洋潮流通道，水深 16～23 m，为天然深水航道，西太阳沙濒临深水航道，是建人工岛码头的天然基础。相近的众多沙脊掩蔽消浪，为发展临港工业与建设人工岛的土地资源，深水航道与码头之组合可发展为 10 万～20 万 t 级大型船只停泊的海港；其次为吕泗港小庙泓潮流通道，及兴建中的王港区，海域条件可以发展 5 万 t 级海港。上述论点是通过南京大学海岸与海岛实验室于 20 世纪 80 年代以来，先后 5 次对辐射沙脊群进行系统深入的调查研究后获得的科学成果，此处再加以概括地论述。

辐射沙脊群位于苏北平原外侧南黄海海域。以弶港为中心，呈扇形展开，它是我国最大的浅海潮流沙脊群，南北范围界于 32°00′N～33°48′N，长达 199.6 km，东西范围介于 120°40′E～122°10′E，宽达 140 km，由 70 多条沙脊与其间的潮流通道组成(图 6)。它界于海陆交互作用带，位于 0 m 至水深－25 m 之间，全部面积 22 470 km^2，其中 3 782 km^2 出露于水面之上。南黄海海域为正规半日潮型，涨潮时潮流自北、东北和东南方向涌向弶港方向，落潮时以弶港为中心，呈 150°扇面向外逸出，形成辐射状潮流场。潮波的辐聚与辐散，形成潮差大，平均潮差 4.18 m，大潮潮差 6.5 m，最大潮差 9.28 m 于沙脊群核心区黄沙洋测得，是我国潮差最大记录，潮差自黄沙洋-烂沙洋，即辐射沙脊群核心区向南北两侧减小。辐射沙脊群区潮流作用强，涨潮时潮流辐聚，平均流速 1.2～1.3 m/s，落潮时段潮流辐散，流速急，平均流速 1.4～1.8 m/s。持续上升的海平面使潮流动力与潮流系统加强，形成全球最具代表性的辐合-散射潮流系统海域，沙脊群的基本轮廓与潮流场相符。强大的潮流是维持潮流通道水深的天然动力。辐射沙脊群掩护着其后侧的海岸，沿射阳河口至小庙泓段发育了广阔的潮滩。沙脊群外侧受波浪冲刷，泥沙向岸输运，至潮滩处停积使潮滩不断地淤涨，而沙脊群范围以外的南北两段平原海岸，却受到浪流冲刷后退。

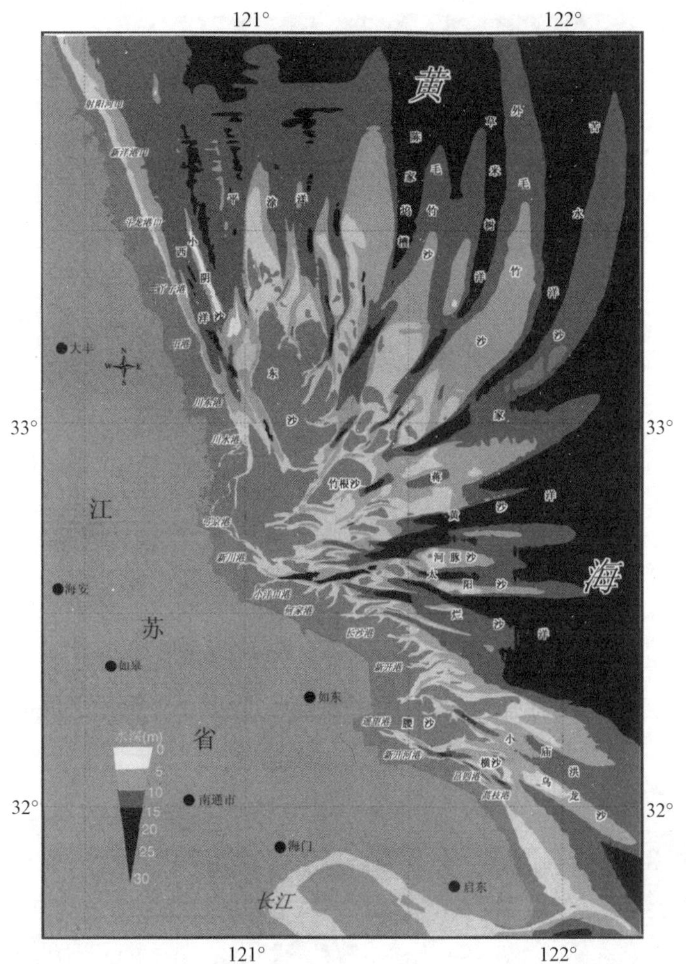

图 6　辐射沙脊群图(附彩图)

辐射沙脊群由细砂与粉砂组成，是在地质历史的新近时期——约 3 万年前，古长江口自弶港入海时堆积的细砂物质，当时海平面位于现今－20 m。在近 1 万年来的全新世海面上升过程中，尤其以 6 000 年及 1 000 年两个明显的海侵冲刷期，由潮流冲刷改造古长江口泥沙形成长条形沙脊。历史时期，黄河夺淮入黄海后，向辐射沙脊群补充供给了淤泥物质。所以辐射沙脊群是以晚更新世末期的古长江三角洲体为基础而于全新世(近 10 000 年)海侵过程中发展成型。

烂沙洋与黄沙洋是位于辐射沙脊群的枢纽部的主潮流通道，其中烂沙洋自西太阳沙深槽(－29 m)至口门外海，53 km 的长度内水深介于 16～23 m 之间，它是在古长江干道的基础上发育而成的主潮流通道[7]。

地震剖面揭示出烂沙洋与黄沙洋为古长江从苏北入海主干道，时代相当于晚更新世末约 30ka 前，卫星照片仍隐约显现此埋藏谷地(图 7)。

晚更新世末期古河谷埋藏于现代海底 40 m 厚的沉积层下，该处谷地约宽 10.42 km，深度为 20 m(北岸)～33 m(南岸)(图 8 和图 9)。钻孔剖面亦反映出厚层的细砂层埋藏于 41 m 深处(图 10)[8]。

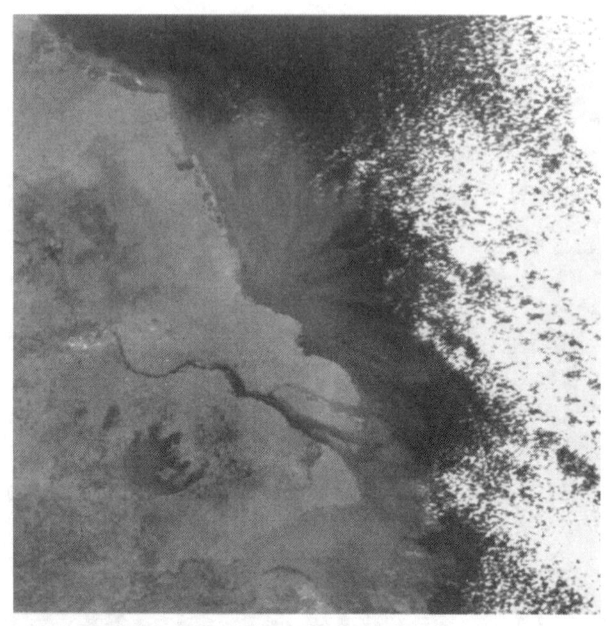

图7　辐射沙洲区卫星图片显示古河道示意图(附彩图)

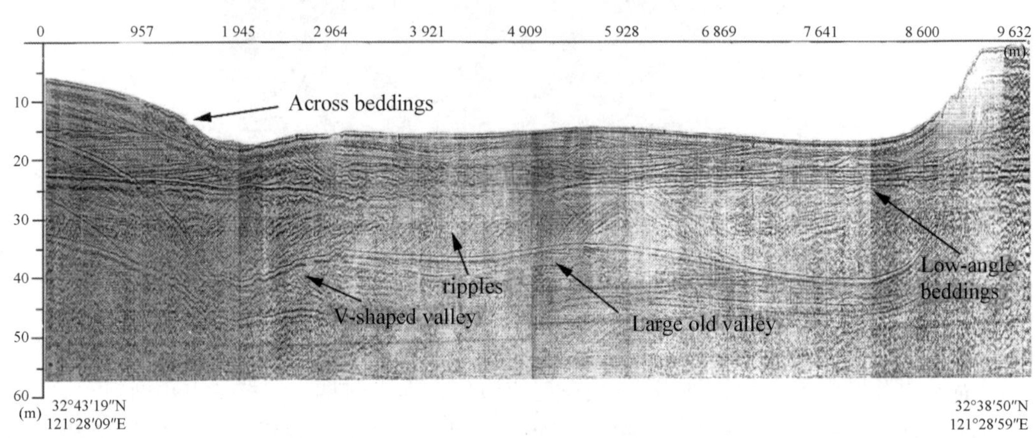

图8　黄沙洋中段地震地层剖面图

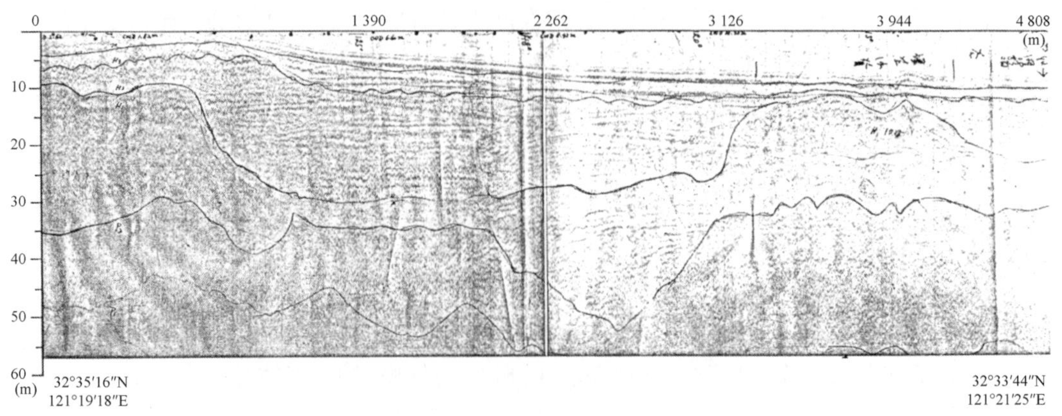

图9　谷中谷与改造的河流沙体

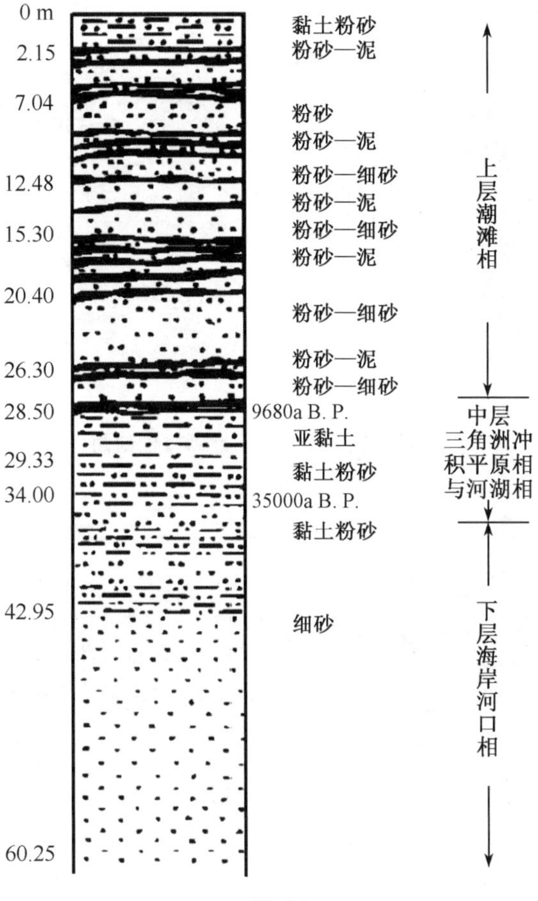

图 10　钻孔剖面图

辐射沙脊群北部与南部的脊槽形成时代晚于枢纽部的烂沙洋与黄沙洋通道，是在全新世(公元前 10 000 年)以来，长江从李堡或遥望港入海时形成(图 11)。大北槽地震剖面亦反映此情况(图 12 和图 13)。烂沙洋与黄沙洋两条潮流通道的长度、水深及位置所在，亦反映出现代潮流主通道的位置，仍在弶港以东的沙脊枢纽区(表 5)。

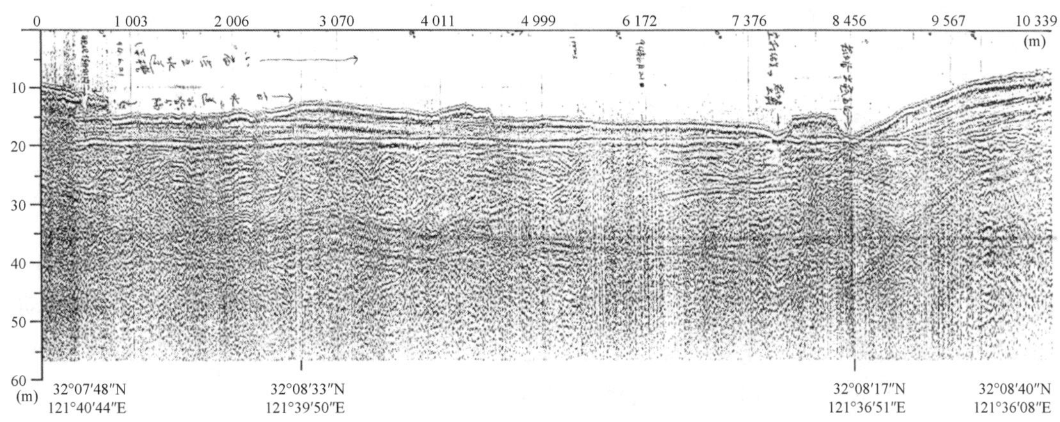

图 11　辐射沙脊群南部小庙泓水道地震剖面

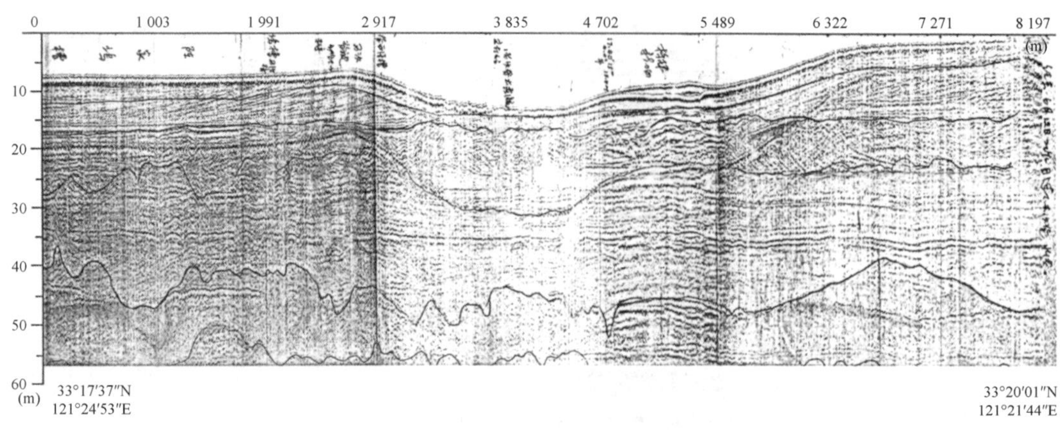

图 12　辐射沙脊群东北部脊槽剖面

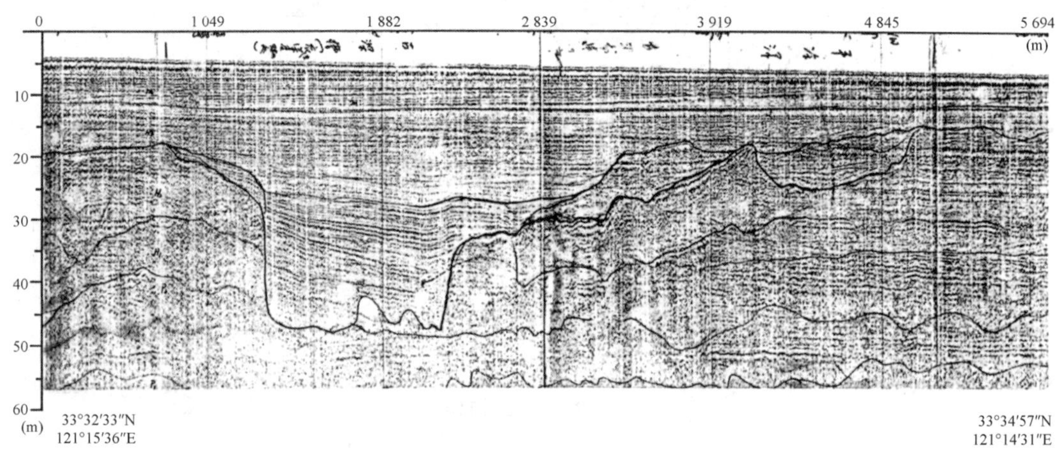

图 13　辐射沙脊群北部大北槽承袭谷剖面

表 5　潮流通道长度、水深表

部位	潮流通道名称	潮流通道长度/km	水深/m
枢纽部	烂沙洋 黄沙洋	53 45	16～23(局部 29) 11～20
南部	小庙泓	28	8～12
北部	西洋(王港)	50	10～13(局部 28.5)

烂沙洋与黄沙洋海域潮流作用强，潮流通道中为往复型潮流。该处实测最大潮差达 9.28 m，平均潮差 4.18 m，涨潮流流速 1～1.5 m/s，每潮涨潮流量 19.5×10^8 m^3；落潮流速度为 1.2～2.0 m/s，每潮落潮流量达 21×10^8 m^3，落潮流大于涨潮流。巨大的潮流动力，使得烂沙洋多年保持水深达 16～23 m，此水道伸入陆地，是江苏省最宝贵的天然深水航道资源，也是全国沿海最深长的天然良港港域。

烂沙洋内陆侧有一西太阳沙沙脊，5 m 水深以内，呈东西向延长 1.375 km，10 m 水深以内长 4.25 km，宽 250 km，由粉砂与细砂组成，脊顶水深 0.5 m，落潮时，主干部分出露

(图 14),该沙脊位置稳定。西太阳沙由厚达 90 m 的粉砂、粉砂黏土与粉细砂沉积层组成(据江苏省电力设计院钻孔资料,1995 年 1 月),为在深水航道侧修建码头与扩展建设人工岛的理想基地。太阳沙毗邻的水深为 17.2～20 m,这样的沙脊与深水航道组合,在我国淤泥海岸段仅有 2 处,另一处是位于渤海湾的曹妃甸沙岛与其外侧潮流深槽,该岛向陆侧有大片潮滩(南堡滩涂),海岸环境与辐射沙脊群类似,但沙岛与潮流深水槽的规模小。南京大学亦曾于 1997—1998 年在曹妃甸沙岛为深水港建设,从事潮流深槽稳定性研究。2002 年夏该处已完成建港设计,并通过环境评价,将于曹妃甸建深水码头,成为唐山、王滩港与天津港新港间的一个大型船舶通航的现代化深水港,为京津唐钢铁基地的组成部分。曹妃甸建港为江苏树立了可效法的实例。

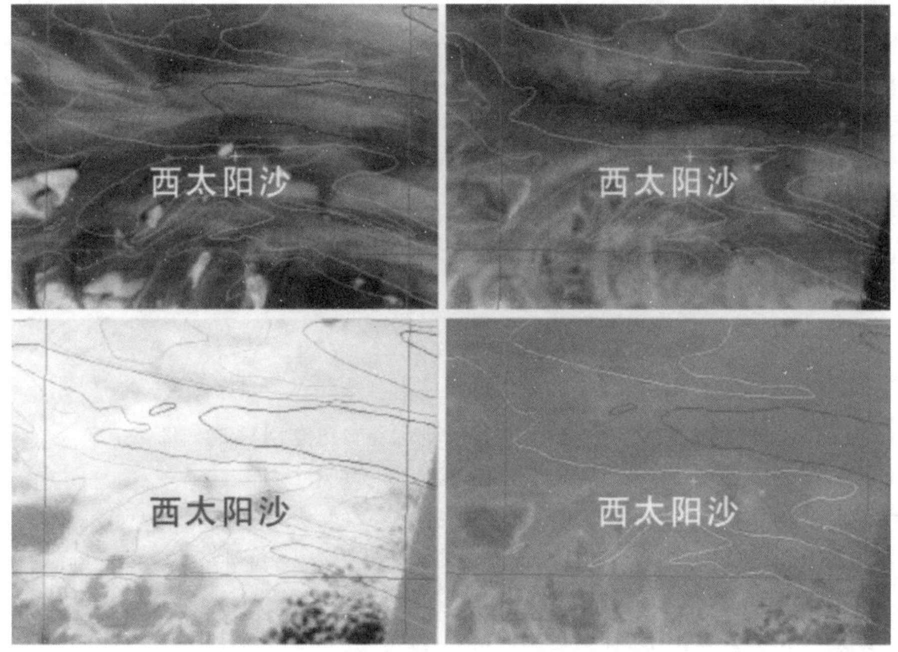

图 14 西太阳沙与烂沙洋遥感图像(附彩图)

四、展望

新时代、新形势要求对洋口港交通基地的建设,需从高起点与多元发展思路进行总体规划。西太阳沙人工岛与烂沙洋深水航道不仅建设深水枢纽港,同时,应发展临港工业,发展进出口物产的加工工业,提高产品价值;建设天然气发电、潮流、风能与太阳能发电的新能源产业群;建设绿色农、林、渔、药、花卉产品的生态农业基地;建设棉、毛、化纤轻纺工业带和乡镇结合的新型城市群以及集自然、文化、历史风情于一体的休闲、度假与体育活动的旅游基地等。关键在于做好洋口港发展的总体规划,以港口为龙头,带动江苏北部地区经济大发展。

利用辐射沙脊群烂沙洋深水通道建设洋口深水港,是促进苏北发展外向型经济的重要一步,关键是做好深水海港发展的总体规划,充分利用深水岸线与深水航道的宝贵资源。洋口港宜首先发展为石油、天然气集输与新型能源基地，解决我国能源短缺与需求量增长之严重不足，这对三角洲经济持续高速发展以及所面临的海上疆域纷争之挑战，具有战略意义[9]。洋口港宜发展临港工业，成为农产品的加工与外贸区，促进几千年以来的传统农业与渔盐业发展为21世纪新型的江苏海洋经济区。

平原海岸的广袤滩涂与土地资源，易于发展海陆空立体交通网络：港口通过平原的河道水网与区内乡镇联通，集疏物资；联通纵横的公路，经苏通大桥、江阴大桥、润扬大桥及南京二桥等，跨过长江，与苏南发达区沟通；还可通过东部沿海公路干道，延展至东北与海南；进一步，可通过开挖洋通运河实现江海直通联运。江海联运推动长江三角洲北翼的建设，不仅促进苏北地区的繁荣，成为江苏省经济持续发展的动力源泉，而且发挥在我国东部的中枢杠杆作用。

洋口港位于长江三角洲北部，与三角洲南翼的宁波港、建设中的大小洋山港遥相对应，形成以上海为中心的长三角深水港口群。南北呼应、东西互补加强大上海国内外联系的重要地位，发展成为东亚经济圈最大的枢纽中心，提高我国在亚洲与太平洋地区的战略地位。

21世纪赋予江苏南通发展的新前景，抓住机遇、迎接挑战，以科学发展观为指导，区域团结、同心协力，为建设繁荣昌盛的长江三角洲做出贡献。

参考文献

[1] Kevin Cullinane. 2004. 决战长江——上海宁波争霸战. 港口疏浚 DPC：30－32.

[2] 李文正. 2004. 当疏浚不足以满足要求时. 港口疏浚 DPC：16－17.

[3] 严以新，许长新. 2003. 论21世纪初洋口港的发展环境与对策[J]. 水资源保护，(6)：5－9.

[4] 杨荫凯. 2002. 21世纪初我国海洋经济发展的基本思路[J]. 宏观经济研究，(2)：35－38.

[5] 杨文鹤. 2001. 中国海洋统计年鉴2001[M]. 北京：海洋出版社.

[6] 江苏省统计局. 2002. 江苏省统计年鉴—2002[M]. 北京：中国统计出版社.

[7] 王颖主编. 2002. 黄海陆架辐射沙脊群[M]. 北京：中国环境出版社.

[8] Wang Ying, Zhu Dakui, You Kunyuan, et al. 1999. Evolution of radiative sand ridge field of the South Yellow Sea and its sedimentary characteristics[J]. *Science in China* (*Series D*), 42(1): 97－112.

[9] 高渊. 2004. 应对石油运输情势、维护国家能源安全，中国组建超级船队. 环球时报，2017年7月14日844期16版.

发挥“江海”优势
力争5年内江苏海洋经济跃居全国前列*
——大力发展江苏海洋经济的建议

• 从数据上看，江苏海洋经济产值若能追平广东，则GDP总量会超出广东300亿元。

• 高起点地规划好、建设好洋口海港，对苏中、苏北发展的带动意义十分巨大。

依江临海、江海交汇是江苏最显著的自然地貌特点，也是江苏得天独厚的资源优势。数万年来，长江、黄河泥沙不断堆积形成了江苏广阔的沿海平原，沿3万年前古长江在弶港入海处发育形成世界上罕见的大型辐射沙脊群。沿古河道发育的烂沙洋、黄沙洋潮流通道，水深16～25 m，拥有强劲的潮流往返冲刷，却无河流泥沙下泄，加之通道两侧有天然沙脊掩护，构成了“深水浪静”的优越建港条件，是我国平原海岸上最优越的天然深水海港港址，也是长江口以北除渤海湾曹妃甸外唯一可建设20万t级海轮泊位的优良港址。利用烂沙洋潮流通道与沙脊组合建设洋口深水港将会加快长江三角洲北翼（江苏地区）海洋经济发展，促进以上海为中心的长三角经济大幅增长。江苏海岸线全长954 km，管辖海域3.75万km^2，浅海面积2.44万km^2，占全国沿海总面积的20%；不断淤涨的滩涂面积现约6 533 km^2为全国之最；海洋资源综合指标位列全国第4。据研究，江苏如能做好“江海”文章，充分发挥优越的港口条件和海洋资源优势，大力发展海洋经济，可望5年时间使全省海洋经济跃居全国前列。

一、发展海洋经济的战略意义

近半个世纪以来，随着世界各大海港的开发，港口城市的建设，海洋经济发展十分迅猛。沿海各国、各地区无不把发展海港、发展海洋经济作为顺应经济全球化的重要战略举措：日本平均每30 km海岸线就建有一座深水港，利用全球资源，发展临海工业；美国在东海岸建设了波士顿—纽约—巴尔的摩—华盛顿的沿海工业带，在西海岸建设了西雅图—旧金山—洛杉矶—圣地亚哥沿海工业带。我国沿海各省市纷纷把建设临海工业基地、发展海洋经济作为本地区经济发展的战略举措：上海加大了对洋山深水港的建设；浙江通过建设杭州湾跨海大桥提升北仑港优势；福建重点开发厦门与泉州港区；广东实施南海石化、广州大南沙计划；山东倾力发展青岛。而江苏的核心经济区远离深水海港，在全球经济一体化加速的形势下，面对国内外的挑战与竞争，江苏要保持在全国的领先地位，就不

* 王颖：《江苏科技信息》，2006年第3期，第1-2页。

能不重视深水海港的开发，就不能不大力发展海洋经济。

二、江苏海洋经济发展滞后的现实和成因

江苏港口条件优越，海洋资源丰富，是全国最早完成海岸、滩涂及海岛海洋综合调查的省份。但江苏海洋经济发展滞缓，落后于全国。据近几年统计数据分析，江苏省海洋经济产值只处于沿海省、市中低水平，如 2003 年处于 11 个沿海省市的第 8 位（见表 1），比广东少 1 482.66 亿元；而当年江苏国内生产总值只比广东少 1 165.04 亿元。从数据上看，若江苏海洋经济产值能追平广东，则 GDP 总量会超出广东 300 亿元。

表 1　2003 年中国沿海省（区、市）海陆域经济对比（亿元）

省区	海洋经济产值	排序	GDP 总量	排序
广东	1 936.07	1	13 625.87	1
山东	1 477.64	2	12 435.93	3
福建	1 344.96	3	5 232.17	8
浙江	1 177.75	4	9 395.00	4
上海	845.91	5	6 250.81	6
天津	568.07	6	2 447.66	10
辽宁	543.41	7	6 002.54	7
江苏	453.41	8	12 460.83	2
河北	182.52	9	7 098.56	5
海南	145.93	10	670.93	11
广西	57.66	11	2 735.13	9

江苏海洋经济发展不快的原因：一是传统观念认为、淤泥质平原海岸"苦海盐边"缺乏深水良港，未能在苏北 800 多 km 平原海岸上建成深水良港，对洋口港建设重视不够；二是缺少大工业项目带动，未能招商引资以形成临海产业带；三是现有海洋产业附加值不高，滩涂垦殖、海洋渔业、盐业等传统产业所占比例过重，海洋医药、海洋生物、海洋工程、海洋能源等高附加值产业发展不足，这与江苏省新兴海洋技术研究开发应用不足有关系。

三、大力发展海洋经济的几点建议

1. 高起点地规划好、建设好洋口海港

无论是沿江开发、苏中发展，还是苏北振兴，开发江苏自己深水海港的战略意义十分重大。长江口天然水道水深不足 7 m，经过建国以来几十年的治理疏浚，至今长江口航道水深才增至 9 m。目前投资 150 亿元的航道修建工程，也只能使水深达到 12.5 m，实现 5 万 t 级海轮在不乘潮不减载的情况下全天候通行。但 5 万 t 级轮运还远不能满足江苏省发展现代服务业、现代物流业的需要。当前国际上主要是发展 15 万～20 万 t 集装箱海

运，要求航海水深达到 16 m 以上，而洋口港航道水深一般为 16～25 m，可确保 15 万～20 万 t 级海轮顺利通航。因此，开发建设洋口港可以直接连通世界各地，带来全球资源，进而会在沿海地区形成新的产业集聚带，为苏中、苏北等腹部地带发展外向型经济提供重要依托和“辐射源”，促进苏中、苏北在新的平台上加速发展。建议将洋口港建成世界级的具有资源配置功能的现代型海港和区域型现代物流中心，按此目标科学地规划洋口港的开发建设，并纳为江苏“十一五”期间重大建设项目。

2. 以洋口港开发建设为契机，优化和提升沿海地区产业结构

建设能源和重化工基地，大力发展临海加工业与外贸经济，发展海洋生物工程、海洋医药工业等新兴产业，发展天然气发电、潮汐、风能和太阳能发电等新能源产业，发展绿色农、林、渔、药、花卉等生态农业基地，发展棉、毛、化纤等轻纺工业，努力形成新的经济增长带。

3. 加大对海岸海洋科学研究的投入

充分发挥南京大学、河海大学等高校的海岸海洋科研优势，重点对海洋医药、海洋生物、海洋食品、海域再生能源、天然气水合物(可燃冰)新能源等海洋技术实施攻关，并积极推进产业化，发展新兴海洋产业。

4. 发挥海岸海洋科学在滨海城市规划建设中的重要作用

深入研究和把握海岸海洋的生态环境与滨海城市发展的变化规律，将人、地、水关系和谐发展的思想渗透贯穿到滨海城市的建设之中，充分发挥海岸海洋科学在滨海城市规划建设中的重要作用。

江苏海岸海洋环境特点与海洋经济发展*

一、江苏海岸环境特点

江苏海岸地处长江三角洲北翼的黄海之滨，自苏鲁边界的绣针河口至苏沪边界的浏河口，大陆岸线长 888.9 km（“江苏近海海洋综合调查与评价”，即 908 专项，2010.5）。海岸线原长为 953.875 km（八・五调查），因人工围海裁弯取直而岸线长度减少。沿海 19 座岛屿岸线长达 26.941 km。

海岸带以粉砂淤泥质平原海岸为主，全长达 818.837 km（图 1、图 2 和图 3）；仅赣榆有 30.062 5 km 砂质海岸（图 4）；连云港一带基岩港湾海岸，岸线长 40 km（图 5 和图 6）。

图 1　粉砂淤泥质平原海岸

* 王颖：2013 年 2 月 2 日在江苏省在宁院士座谈会上的发言，会后以摘要形式作为院士建议，由中科院南京分院院士联络处提交省委省政府。

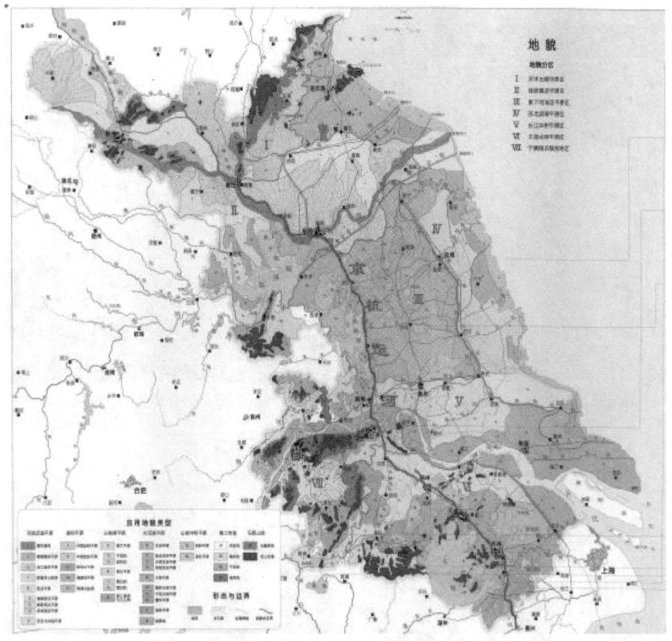

图 2　江苏省地貌图(附彩图〈上册〉)

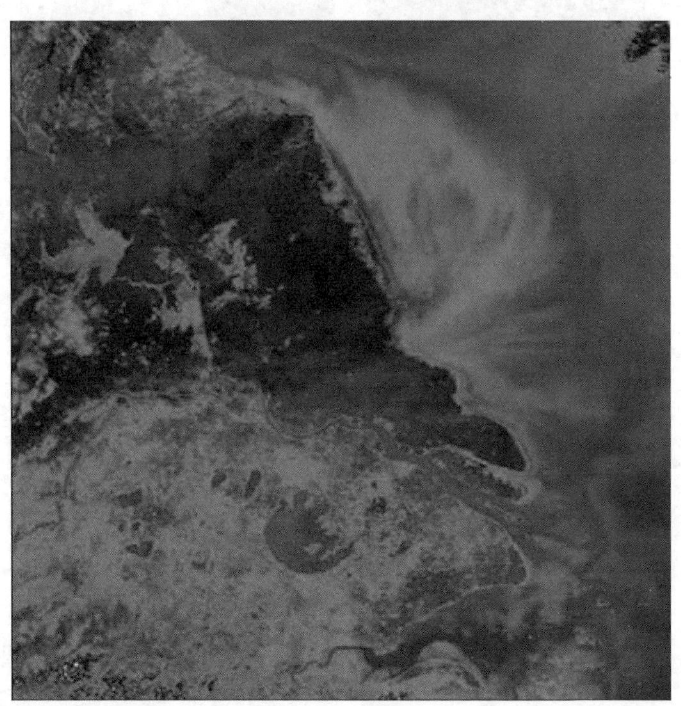

图 3　平原海岸遥感影像(从沿古海湾低地开挖的大运河至岸外辐射沙脊群)(附彩图)

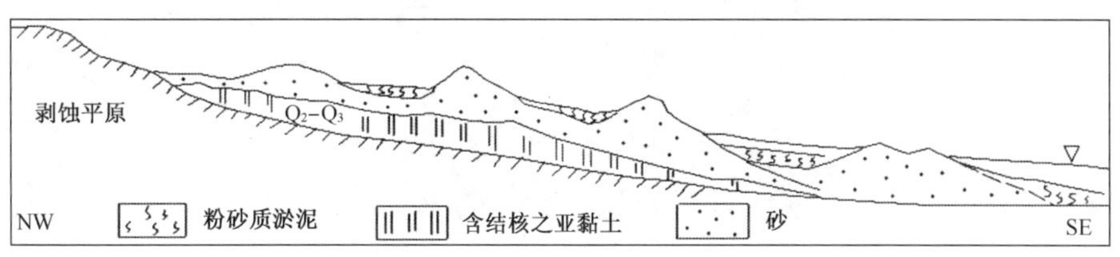

图 4　赣榆砂质海岸剖面(以基岩剥蚀面为基底)

图 5　连云港基岩港湾海岸照片(1)

图 6　连云港基岩港湾海岸照片(2)

沿黄海低地平原是由长江与黄河入海的泥沙被海浪堆积充填古海湾而成。原始的古海岸基本上沿基岩山地、丘陵外缘分布，高邮湖、洪泽湖、骆马湖以及太湖基本上沿古海湾的内缘分布；大运河开挖亦利用了此低地。兴化曾是海湾最低部分；龙冈、上冈、大冈为古海岸线的遗迹（王颖等，2006）。

苏北平原海岸外缘坡度平缓，一般小于 1‰，潮汐作用强，潮差大，平均潮差 4.18 m，大潮潮差 6.5 m，最大潮差 9.28 m，潮滩宽度约 4～5 km，最宽 14 km（条子泥），坡度约为 0.21‰，潮滩宽广、岸坡平缓是苏北平原海岸的突出特点。潮滩具有上部（沿陆）为盐蒿泥滩与泥沼滩（高潮位），中部为冲刷的淤泥质与粉砂混合滩，外部为粉砂滩的分带性，应据滩涂特性而加以不同的利用（图 7）。

(1) 高潮海岸贝壳滩

(2) 岸堤与高潮粉砂滩

(3) 沿岸堤修建的防风林带及草滩

(4) 草滩湿地(附彩图〈上册〉)

(5) 泥沼潮滩(潮滩上、中部)

(6) 潮滩中部冲刷带

(7) 冲刷带与外侧潮滩外部粉砂波痕带

(8) 潮水沟

(9) 潮滩外淤积带

(10) 潮滩外淤积带(波痕)

图 7 粉砂淤泥质平原海岸的潮滩分带相

低潮线以下,海岸坡度平缓与黄海大陆架衔接,是河海交互作用形成的古扬子大三角洲的一部分。主体是南黄海辐射沙脊群,自射阳河口向南至长江口北部的蒿枝港,南北范围界于 32°00′N～33°48′N,长达 199.6 km;东西范围是自海岸带向外分布至水深 25 m 处,界于 120°40′E 至 122°41′E 之间,宽度 140 km。大致以弶港为枢纽,沙脊呈褶扇状向海辐射,脊槽相间,由 25 条大沙脊与其间的潮流通道组成(图 8),是全新世海平面上升过程中,由东海传来的前进潮波与受阻于山东半岛形成的反射潮波相交辐聚后再形成的辐射状落潮流改造古三角洲所形成(图 9)。

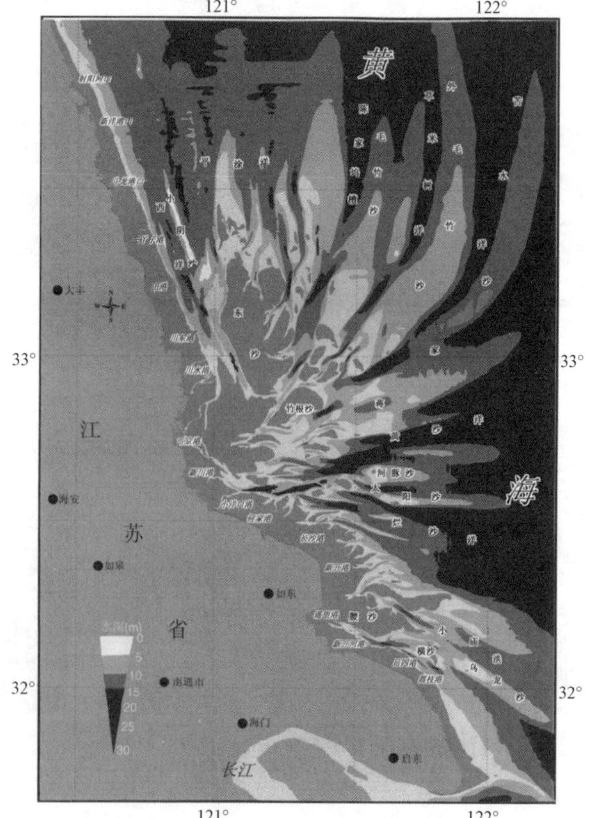

图 8 南黄海辐射沙脊群(附彩图)

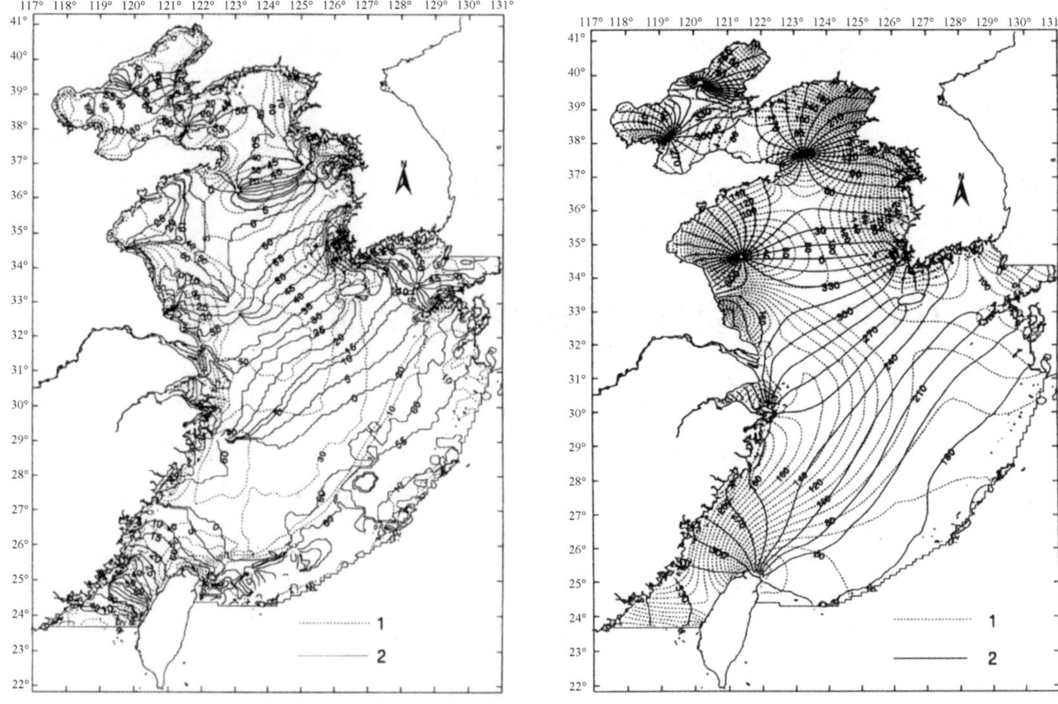

图 9　潮流辐聚与辐散

辐射沙脊群外侧，海底出露古江河三角洲（王颖主编，2012），延展分布于－30 m（北部）及－50 m 水深处（图 10），其范围较南黄海辐射沙脊群的面积大出三倍，为细砂、粉砂、

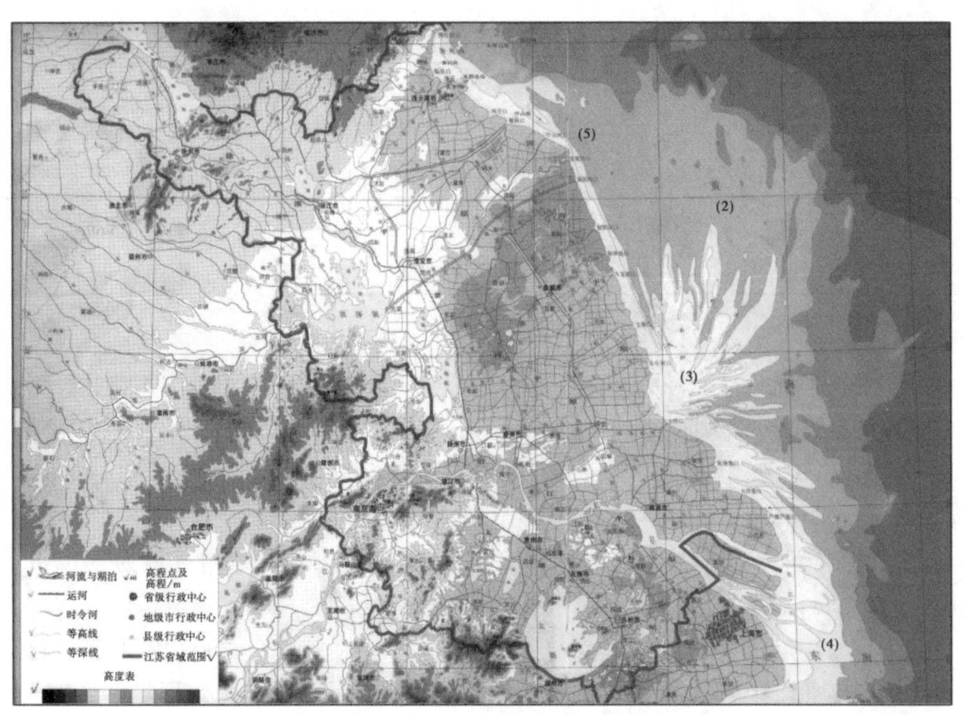

图 10　南黄海古江、河大三角洲(2)(附彩图〈上册〉)

（资料来源：江苏地图集，2004）

图中：(3) 辐射沙脊群　(4) 全新世—现代长江三角洲　(5) 废黄河三角洲

砂质粉砂及粉砂质砂底质，北部为黏土质粉砂，富含结核，系黄河携运之泥沙，细砂之重矿组成表明主要源自长江，故定名为“古江河大三角洲”，其形成时代估计为晚更新世，较辐射沙脊群（＞43000 年）更老。

在大三角洲北部叠置着废黄河三角洲，是 1128 年人工在开封掘开黄河大堤，至 1855 年黄河北归渤海前的 727 年间堆积形成。黄河北归后该三角洲遭受侵蚀后退，目前已形成新的岸坡，适应海域动力环境而冲刷减缓。

二、江苏海岸海洋具有在我国沿海最珍贵的自然资源

河海交互作用的淤泥质平原海岸海洋，地貌单一。但是：广阔滩涂新生的土地资源丰富；淡水资源与海水动能资源丰富；港口资源丰富，这三项结合在全国独居首位。因为渤海平原沿岸滩涂广阔，但缺乏淡水资源；东海与南海有丰沛的降水，但缺少滩涂新生的土地资源，因此江苏得天独厚。

此外，更有辐射沙脊群深水潮流通道，解决了平原海岸水浅、无港口的问题，加之，南黄海堆积型大陆架与沙脊群构成形成丰富的油气资源与储存条件。

江苏省管辖的海岸海洋尚未涉及陆架浅海全部，仍面临着国内外在海洋权益方面的纷争。

发展江苏省海岸海洋经济所面临的关键问题是：依据自然环境资源制定出合理（符合自然规律）的全面一体化的发展规划。各市分别依据区域优势的特点发展，不要重复类似的竞争发展。

（一）滩涂——新生的土地资源

目前江苏省管辖海域面积为 18.842 6 km^2（表 1）。

表 1　江苏海域的范围和面积

类型	内水	领海	毗连区	专属经济区
面积（$\times 10^4$ km^2）	2.175 9	0.982 9	0.999 1	14.688 3

（内水包括沿海、长江口内岸线与理论深度基准面之间 5 000 km^2 之滩涂）

内水、领海和毗连区水域为 4.16 万 km^2，占全省陆地面积 40%（图 11）。

1985 年全国海岸带调查：陆上以 10 km 宽为界，水下至－15 m 深处：

• 江苏省海岸带面积 35 000 km^2，其中：沿岸陆地，5 000 km^2；潮间带浅滩，5 000 km^2；浅海海域，25 000 km^2。

辐射沙脊群至 0 m 等深线水域面积：1975 年实测为 2 125 km^2，1995 年实测为 3 782 km^2，后者相当 567 万亩，约每年增加 80 km^2（12 万亩/年）。

——“908 项目”调查之面积

• 按多年平均计，苏北沿海潮间带滩涂约 700 万亩，每年增加 12 万亩。

• 按现代工程技术水平与围填的沙土源，辟作农田的以起围高程在平均高潮线以上为宜，现代高潮滩大约有 120 万亩可供农业利用。

图 11　中国内水、领海和毗连区示意图

其余约 600 万亩滩涂，在平均高潮线以下，可辟围养殖紫菜、海带滩涂及工业用地，后者填土量大。

农田或城镇利用围填的土地需注意避开潮流通道。20 世纪 70 年代梁垛河口围垦 10 万亩，一次风暴潮形成潮汐摆动而冲破围堤，原梁垛河闸至今残立海中；后再修堤围成 3 万亩。此教训应吸取。

• 有科学依据的稳妥规划：自 20 世纪 80 年代至今，30 年可开发 3 000 000 亩是适宜的(江苏省沿海开发总体规划，2007)。据 2008 年 8 月，国务院《全国土地利用总体规划纲要》要求江苏补充耕地，2006—2010 年指标为 60 000 公顷(90 万亩)是恰当的。

建议：

• 分期开发，从 2010 年的 120 万亩逐步发展至 2020 年为 200 万亩为宜。通过实践总结实施。

• 围滩应避开深水港域，不能减少纳潮量而形成回淤。

• 滩涂应根据其自然发育阶段适宜性地开发(图 12～图 14)。

低潮滩：养殖贝藻

高潮滩：鱼塘、牧场、芦苇、蚕桑、麻、棉、稻(淡水充足处发展水田)、麦(旱作)、果(桃、梨、西瓜……)、药(白菊、丹参、黄芪、白术、白芷、板蓝根)、花(鲜花栽培等，如：玫瑰可以荷兰为榜样发展为国内大市场)。

• 城镇集约用地。

• 临海工业与航运用地。

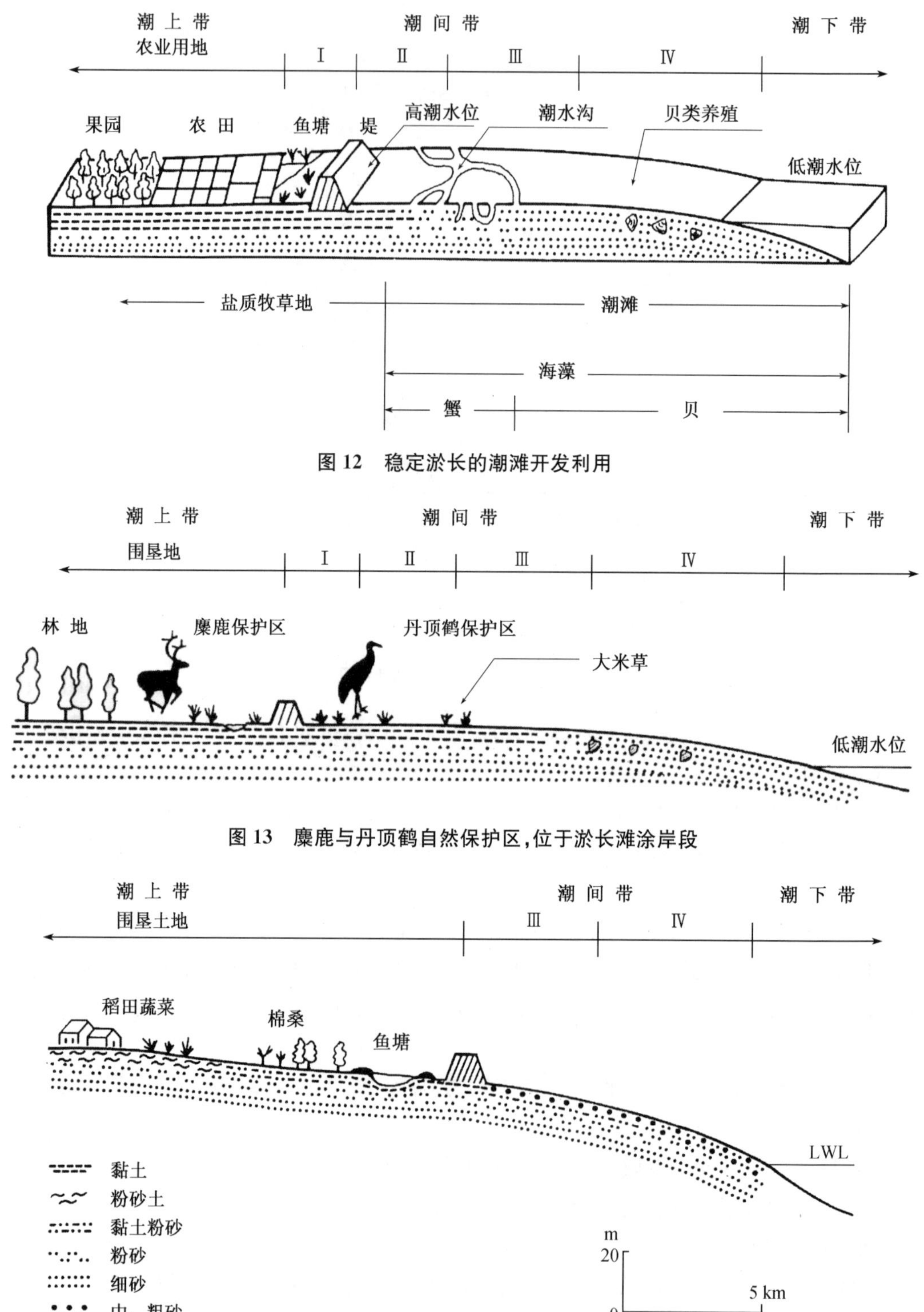

图 12　稳定淤长的潮滩开发利用

图 13　麋鹿与丹顶鹤自然保护区，位于淤长滩涂岸段

图 14　冲刷的南部海岸滩涂利用状况

（二）淡水资源丰富

苏北平原为江海作用形成，土地平坦开阔，有众多的河流和充沛的降水，比渤海平原

海岸淡水资源丰富，但分布不均匀。

水土资源条件启示：江苏沿海适宜于发展大农业，不应仅是粮棉生产基地，而且应是农、林、药、花卉、牧副及加工成品业——新型的现代化大农业基地。

表 2　近 50 年来黄、渤海平原海岸年降水

年代 地区	江苏平原海岸北部 赣榆、连云港 (1957—2000)	江苏平原海岸南部 盐城、东台 (1953—2000)	渤海平原海岸北部 唐山地区 (1957—2000)	渤海平原 天津地区 (1954—2000)
多年平均	926.2 mm	1 076 mm	601.7 mm	553.7 mm
最多	1 482.7 mm (1974 年)	1 978.2 mm (1991 年)	1 162.96 mm (1964 年)	976.2 mm (1977 年)
最少	480.9 mm (1988 年)	462.3 mm (1978 年)	286.01 mm (1957 年)	269.5 mm (1968 年)

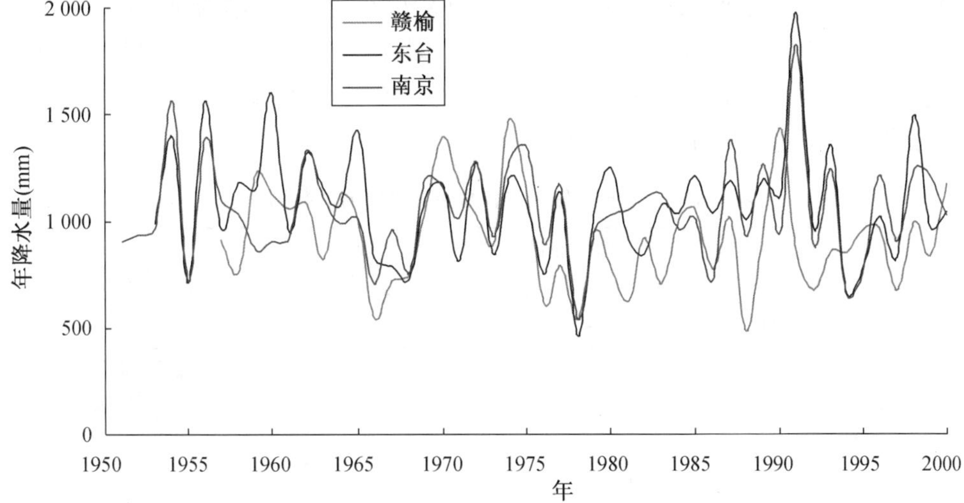

图 15　江苏沿海近 50 年来降水变化趋势

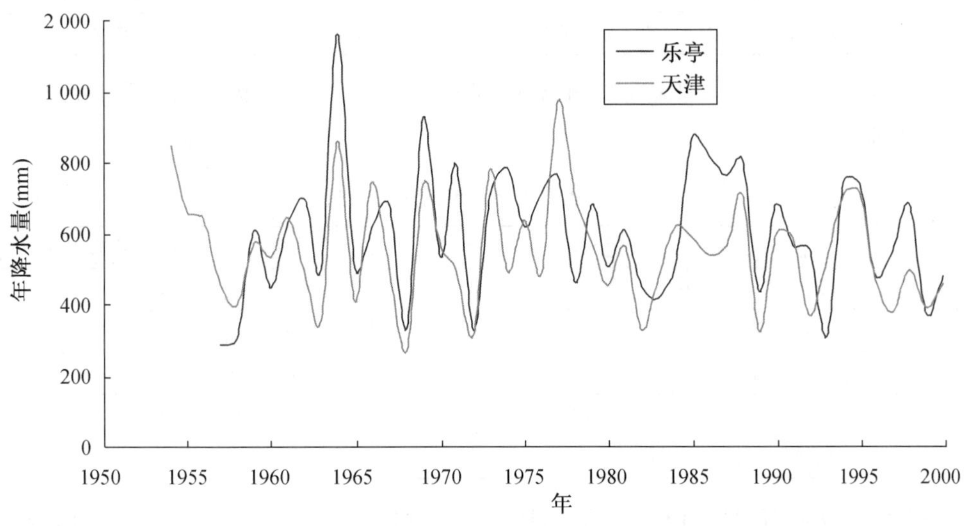

图 16　渤海湾地区近 50 年来降水变化趋势

表 3　2006 年沿海地区水资源量统计

省别	总量(亿 m^3)	地表水(亿 m^3)	地下水(亿 m^3)	人均(m^3/人)
河北省	107.3	42.1	94.3	156.1
江苏省	404.4	314.7	110.7	538.3

中国沿海 11 个省市中，没有哪个省有如此广阔的平原与淡水丰富的水土资源。目前，江苏沿海水质尚佳于渤海湾与珠三角区。平原海岸新能源丰富，太阳能、潮汐能、风能资源丰富，是发展农工业动力。

(三) 港口资源

南黄海平原海岸，坡缓水浅，原不适建港，但是因岸外分布着辐射沙脊群，其潮流通道或为强潮冲刷通道(如：王港、苦水洋、草米树洋、大北槽等)，或为沿古长江河谷发育的强潮通道(如：兰沙洋、黄沙洋、小庙泓)形成天然的深水"航道"，间隔潮流通道的水下沙脊可蔽风消浪，是建造人工岛或防浪堤的基础；广阔的潮滩又提供了丰富的港口与临港工业建设用地。三者结合为江苏省平原海岸提供了最佳的建造深水海港条件——可谓得天独厚，江苏省需认识此优越的自然环境资源(表 4)。

表 4　南黄海辐射沙脊群主要潮流通道数据

地区	潮流通道	潮流通道		潮流通道水深		可建港口(T)	备注
		长度(km)	宽度(km)	口门(m)	内端(m)		
北部	西洋(西通道)	65.0	2.0	13.6	5.0	50 000	冬春 N 向浪大，可建贸易与旅游港
东北部	苦水洋(北水道)	50.0	2～5	24.0	13.5	50 000～200 000	大、中、小港口并举
枢纽部	黄沙洋	45.0	4.0	20.0	11.0	100 000～200 000	可建为中枢区航母基地
	兰沙洋	53.0	2～4	23.0	10.0	100 000～200 000	原油与临港工业基地
南部	小庙泓	28.0	1～2	12.0	6.0	50 000	渔港

上述辐射沙脊区主要潮流通道均为优良的港口资源(表 5)，除苦水洋与黄沙洋外均已初步建港。

表 5　世界著名海港的航道水深

序号	港口	航道水深(m)	序号	港口	航道水深(m)
1	新加坡	>20	6	洛杉矶(美)	15.5
2	西雅图(美)	20	7	上海洋山港	15.5
3	鹿特丹(荷)	18.3	8	中国香港	15
4	高雄	16	9	上海港	8.5
5	南通洋口港	16			

在苏北海岸尚有河口可利用建港，如燕尾港与射阳河口港，目前均已开发为渔港，但均在万吨级下。滨海利用海岸潮汐汊道的流通深水域，已投入建设 10 万 t 级港口。

上述港口与北部已有的连云港及南部长江口内的南通港相结合，南黄海将可发展建设大、中、小并举的河口与海港结合的港口群网络。在长达 900 多 km 的海岸线上，每隔 100 km 建一港口是适宜的，尤其是对苏北农产品丰富、加工业与动力已具一定规模情况下，港口建设是必需的，也是发展海洋经济的必备条件。日本列岛港口罗列，几乎为每 30 km 有一港口。

海港建设可大、中、小结合，建设港口群，成为江苏沿海经济区的门户，发展外向型国际贸易经济带。

北部连云港，为贯通中西部连接欧亚大陆的海上门户。

东北部苦水洋东台港是辐射潮流形成的深水港。

南部洋口港，是长江三角洲北翼最佳的天然深水港址，理想的 20 万 t 级港口，可建成 30 万 t 级众多泊位的深水大港口，集输钢铁、石油与液化天然气、船舶制造业、农业与轻工业产品，将有力促进长三角北翼海洋经济发展。

吕四、小庙泓、大丰王港西洋通道，为 5 万 t 级的港口，可发展为 10 万 t 级港口，分别为海洋渔业与制造业基地，可与洋口港组成长江三角北翼港口群。

南北之间滨海新港可建为 5 万～10 万 t 级煤炭与钢铁海港；燕尾港、射阳河口可建 3 000～10 000 t 级港口。

苏北经济的持续发展，必须发展海洋经济、外向型与世界相通的经济。通过海洋走向世界市场的全球化经济，是通过发展海港与海运事业，发展临港工业、临港工业与城镇体系，扩大海外贸易促进地区经济整体的高速发展。沿海环境资源之优势与江苏已有的教育程度高的人力资源结合，必然会促进江苏省经济持续地、稳定地高速发展，争取国民经济产值位居榜首，为我国做出更大贡献。

三、建议

在省委、省政府大力领导与关怀支持下，制定着具有前瞻性、与生态环境相和谐的江苏省海洋经济发展规划：

(1) 发展建设港口群，推动发展海洋交通运输、油气、海洋渔业、能源、制造业与外贸为特色的临港产业带。

(2) 开发利用可再生能源，建设绿色有机的现代化大农业。

(3) 保护与相宜地开发海岛与滩涂，巩固地保障丹顶鹤与麋鹿自然保护区。

(4) 建设现代化、生态型的海滨城镇与组建自然灾害预警减防灾体系，发展江苏海洋经济，促进 GDP 跃居全国第一。

江苏是中国的经济大省，GDP 在全国居第二位，但江苏内陆经济发达，海洋经济发展滞后。当前正值中国大力发展海洋经济与发展海军保卫海疆之时，海洋发展势不可挡，此时要注意江苏省的海洋经济发展需有序进行，与生态环境保护、疆界纷争放在一起综合考虑，做好规划。建议以海港建设为龙头，发挥江海联运的优势，建设临港产业带，推动海洋经济发展；建设绿色现代大农业，开发可再生能源；发挥人力资源素质高优势，院士和院校人才与政府同心协力，促建人与自然和谐发展的江、海经济带，使 GDP 跃居全国首位。

发展海洋经济　建设海洋强国*

一、引子

中国位于亚洲大陆与太平洋边缘海之交，既是陆地大国，也是海洋大国。但自清政府封建锁国的政策及1894年甲午海战战败以来，中国海洋势力薄弱，较长时期处于被外海岛链围封之态势(图1)。当代，党中央发出"海洋强国"的号召，振奋人心，鼓舞我们发展海洋科学、教育、文化与工程技术，发展海洋经济、加强海洋国防力量、卫护海疆权益，促进亚太地区的和平稳定、国际交流与合作。海洋科学教育、海洋经济发展、海洋防卫力量增强这三方面至关重要。我是从学习海洋地质与动力地貌入门的海洋工作者，以海港港址选建与防止港口回淤为专长，讲海洋经济，真如盲人摸象……，但是，我知道海洋经济很重要。纵观世界发达国家和地区的经济发展，多经历过大陆经济向海洋经济之转变，海洋经济促进其国家经济之飞跃。18世纪英国工业化借助于海外开拓、扩展殖民地原料、劳动力与市场，经历近100年的历程成为世界贸易大国。美国经济发达区主要是面向海洋，

图1　亚太边缘海海底

* 王颖：2016年1月20日在"江苏省海洋经济协会成立暨第一次会员大会"上的发言，南京。

如:大西洋岸波士顿、纽约、费城,太平洋岸的西雅图、旧金山、洛杉矶,墨西哥湾及五大湖区。20世纪日本、韩国及我国台湾地区在近代发展过程中,均是以发展海洋经济为基础,推进经济飞跃提高。现在它们又面临发展新阶段的挑战。

二、对海洋经济的理解

什么是海洋经济?当代海洋经济的概念宽泛,涉海的、与海洋有关的,以海洋为主的开发、利用或服务的经济活动,均可列入海洋经济。据孙世芳等[1]:“在海洋及其空间上进行的一切经济性开发活动和直接利用海洋资源进行生产加工,以及为海洋开发、利用、保护和服务而形成的各种经济形态统称为海洋经济。”我理解:现代海洋经济是指为开发海洋资源和依赖海洋空间而进行的生产活动,是直接或间接为开发海洋资源、利用海洋空间提供相关服务而形成的经济集合。海洋经济涉及的范围广,包括海洋渔业与水产业,海洋油气与矿产业,航海、港口运输与船舶修造业,海洋工程建筑业,海洋盐业、化工、海水利用业(冷却、淡化),海洋生物、医药业,海洋能源与电力业,海洋旅游与探险业,海底考古与打捞业,以及海洋研究教育管理服务业等,可综合概括为十大方面。由于各国海域资源环境的不同,海洋产业发展的历史与特长不同,而各有侧重。但是,现代海洋经济显现三大特征:一是以深水港开发为中心,形成原材料与市场在外,而以现代工业技术为骨干,形成临港工业群:石化、钢铁、电力、盐化工、棉毛与化纤纺织以及水产与食品加工等产业群;二是以信息、电子及海洋生物等高新技术为主干,形成海水淡化、海洋风、浪与潮汐的能源利用以及海洋生物医药等产业;三是以港口为“桥樑”或集散点,形成港口与腹地,港口与城镇间的联系与互动发展。我国以上海港为中心的长三角海陆交互带经济区,以香港、广州为中心的珠三角经济区,以及以天津港-曹妃甸港口群为基础的京、津、冀沿海经济带均为例证。

进入21世纪以来,我国重视沿海经济带的开发,海洋产值增长很快。据统计[2]:

2001年,我国海洋生产总值为9 518.4亿元,占国内生产总值8.68%。

2005年,我国海洋生产总值为17 655.6亿元,占国内生产总值9.55%。

2009年,我国海洋生产总值为32 277.6亿元,占国内生产总值9.47%。

2013年,我国海洋生产总值为54 313.2亿元,占国内生产总值9.55%。

2014年,我国海洋生产总值达到59 936.0亿元,比2013年增长7.7%,对全国GDP贡献为9.4%(表1)。2006年最高达9.98%,2010年为9.86%。

表1 全国海洋生产总值[2]

年份	海洋生产总值(亿元)	产业			海洋生产总值占国内生产总值比重(%)	海洋生产总值增长速度(%)
		第一产业	第二产业	第三产业		
2001	9 518.4	646.3	4 152.1	4 720.1	8.68	
2002	11 270.5	730.0	4 866.2	5 674.3	9.37	19.8
2003	11 952.3	766.2	5 367.6	5 818.5	8.80	4.2

（续表）

年份	海洋生产总值（亿元）	产　业			海洋生产总值占国内生产总值比重（%）	海洋生产总值增长速度（%）
		第一产业	第二产业	第三产业		
2004	14 662.0	851.0	6 662.8	7 148.2	9.17	16.9
2005	17 655.6	1 008.9	8 046.9	8 599.8	9.55	16.3
2006	21 592.4	1 228.8	10 217.8	10 145.7	9.98	18.0
2007	25 618.7	1 395.4	12 011.0	12 212.3	9.64	14.8
2008	29 718.0	1 694.3	13 735.3	14 288.4	9.46	9.9
2009	32 277.6	1 857.7	14 980.3	15 439.5	9.47	9.2
2010	39 572.7	2 008.0	18 935.0	18 629.8	9.86	14.7
2011	45 496.0	2 381.9	21 685.6	21 428.5	9.62	9.9
2012	50 045.2	2 670.6	23 469.8	23 904.8	9.64	8.1
2013	54 313.2	2 918.0	24 909.0	26 486.2	9.55	7.6

表1数据反映出以第三产业增长迅速，其次为第二产业发展亦快，而第一产业慢，是发展的潜力所在（表2、图2）。

表2　2013年海洋生产总值情况表

（据国家海洋局网站，2014.3.12）

项　目	总量（亿元）	增速（%）
海洋生产总值	54 313	7.6
海洋产业	31 969	6.9
主要海洋产业	22 681	6.7
海洋渔业	3 872	5.5
海洋油气业	1 648	0.1
海洋矿业	49	13.7
海洋盐业	56	−8.1
海洋化工业	908	11.4
海洋生物医药业	224	20.7
海洋电力业	87	11.9
海水利用业	12	9.9
海洋船舶工业	1 183	−7.7
海洋工程建筑业	1 680	9.4
海洋交通运输业	5 111	4.6
滨海旅游	7 851	11.7
海洋科研教育管理服务业	9 288	7.3
海洋相关产业	22 344	—

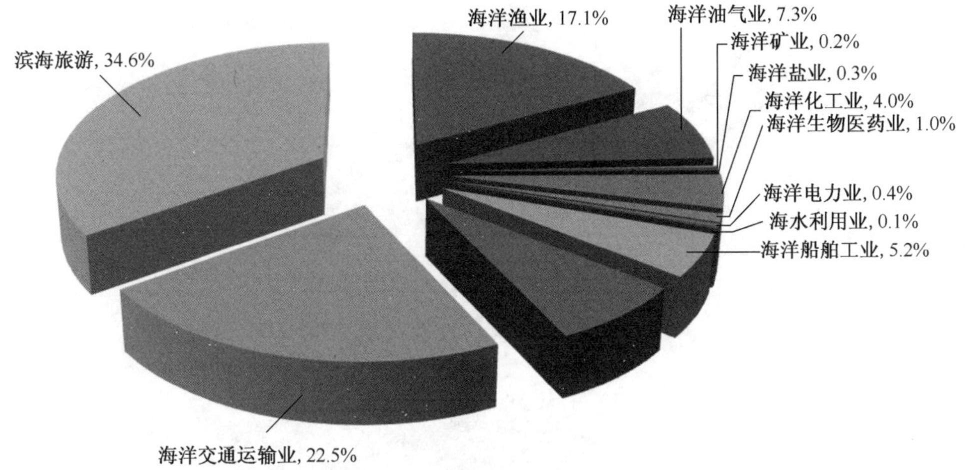

图 2　2013 年主要海洋产业增加值构成图[3]

图 2 表明：滨海旅游业发展快，且稳步增长，亦反映出国人生活水平的提高与对自然界的探求。海洋交通运输业反映出我国外向性经济之增长，而船舶制造业我国居世界前列，是钢铁产量高、高科技设计制造以及劳动力水平高的反映。

三、浅谈江苏海洋经济

（一）江苏海岸很具特点

（1）主要是平原海岸，由江、河汇入海中的泥沙，经浪、潮搬运堆积于大海湾的平原海岸（图 3），外缘有南黄海辐射沙脊群与古扬子大三角洲（图 4）。目前，长江口以北的江苏省岸线长达 888.945 km：基岩岸线为 8.36 km，分布于北部连云港地区；砂质海岸线 1.775 km，主要於赣榆地区；余下的 878.81 km 岸线是粉砂淤泥质平原海岸。[4]不同性质的岩层或沉积物组成的海岸，其抗风浪与潮流的侵蚀程度不同，开发利用的途径也不同。由于辐射沙脊群堆积于海岸带外围，其内侧潮滩受泥沙补给而淤长，所以，江苏海岸地形有一特点是：岸上高（筑堤），岸外高（沙脊、沙岛），而中间与近岸处潮滩低洼。涨潮水位升高迅速，加之，沙脊与潮滩上潮流通道与潮水沟多，所以，江苏潮滩涨潮时水势急湍，如不及早躲避，会溺水身亡。旧时江苏海岸带多建筑避潮墩，为出海捕捞的渔民提供救助。在渤海湾，潮滩是从岸向海坡度降低，渔民在涨潮时可沿潮滩走在潮头前安全抵岸！

（2）河海交互作用的平原海岸的第二个特点是潮滩广阔；辐射沙脊群中出露海面的部分形成沙洲、沙岛，数量众多；沙脊群波影区滩涂广阔。滩涂是新生的土地资源，可有多种用途，海水养殖鱼、虾、贝、蟹、海藻、紫菜，滩洲、沙岛与沿岸陆地可供海鸟与麋鹿栖息繁殖，同时，滩涂亦是新能源建设的基底。

淤涨的滩涂，是海岸带外围削减浪、潮动能的天然防护带，而当河流泥沙供应中断，如黄河从苏北北归回渤海湾后，废黄河三角洲从淤积转变为侵蚀降低，进而海岸受蚀后退。因此，苏北粉砂淤泥质平原海岸大体上以射阳河口为界，以南至长江口为淤长段、新堤建

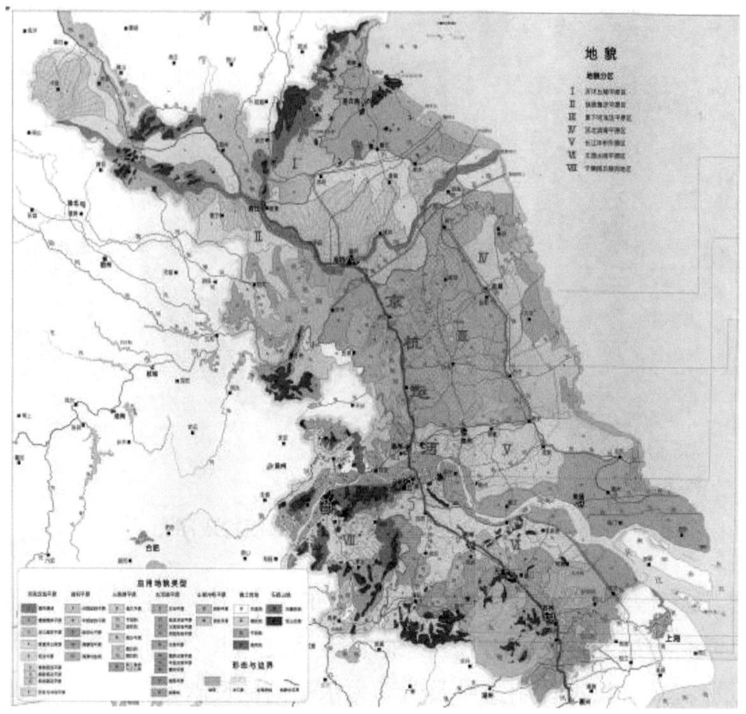

图 3　海湾填积成平原海岸(附彩图〈上册〉)

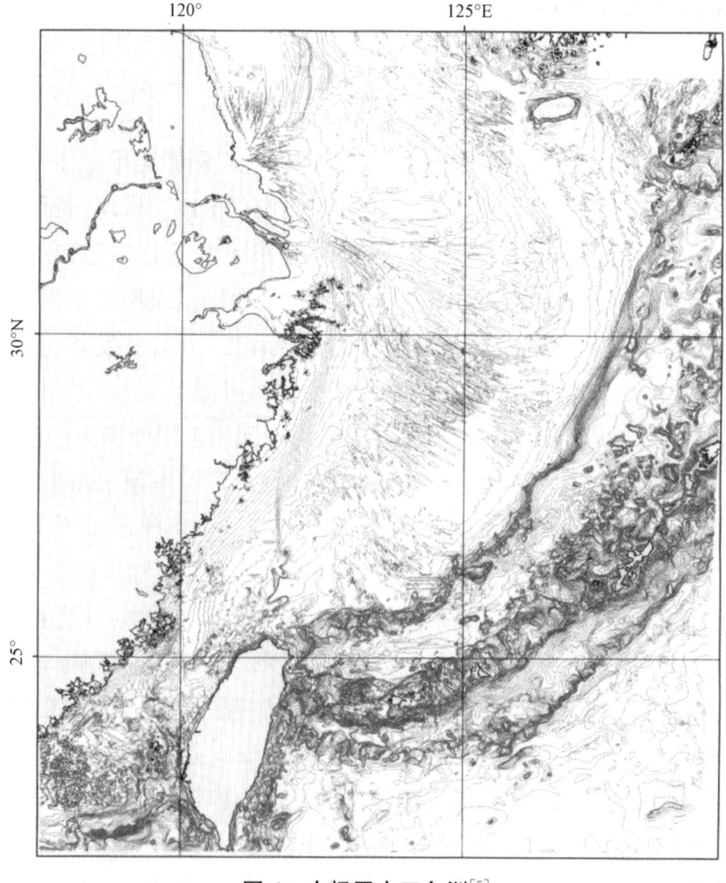

图 4　古扬子大三角洲[5]

在老堤外；而以北海岸受蚀，新堤建在老堤的向陆侧。但是，经过了近 50 年的冲淤变化，已形成新的海岸水下剖面，废黄河口沿岸侵蚀速率已减小，而长江以北的江苏南部海岸在风暴潮亦受冲蚀需防护。因此，在发展海洋经济活动时，需了解海岸潮滩发育的蚀、积动态，还需要了解潮水沟与海岸带河道摆动的趋势。例如，1977 年在苏北弶港以北的新曹农场外滩涂潮水沟上建闸，围垦了十万亩土地。1978—1979 年堤被冲，王家槽潮水沟河曲摆动位移，闸坝废弃，孤立于海上，围垦失败。后在潮滩上部另围造地 3 万亩。这是不了解潮水沟发育自然规律而造成的失败教训。

需要了解潮滩具有分带性，从海向陆，从低潮滩粉砂带变为中潮滩的细粉砂与淤泥互层带，再向陆为淤泥与黏土质淤泥高潮滩带。不同的底质有不同的底栖生物，如文蛤在粉砂滩、毛蚶在淤泥滩、弹跳鱼在黏土质淤泥中，陆上土地亦是，砂质层渗水，需淤泥层隔水才可种水稻等。

（二）江苏海洋经济发展，海港建设先行

平原海岸是河海交互作用形成，潮滩间的大型潮水通道往往是沿古河道发育而成。根据辐射沙脊群中烂沙洋潮流通道是 35 000 年前古长江的河道，古长江南移了，近 10 000 年来，海平面上升，苏北的强劲潮流贯通古河道，水深条件达 25 m，而且已无河沙下泄，因此选定它建深水大港，未经疏浚已建成十万吨级洋口港。北面相邻的黄沙洋是另一股古河道，亦可建港。我们分析，如经适当疏浚打通浅滩，则洋口港可通 30 万 t 级船只。海港建设是发展海洋经济的根本，它是沟通海陆、联系国际的海上“虹桥”，也是维护海洋权益的基地。江苏从建设洋口深水港后，打破了平原潮滩海岸不宜建港的传统桎梏，兴起了纷纷建港之热潮。实际上，全省应进一步规划，根据自然条件与经济发展的需求，将大小规模不一的港口，组建为港口群：宜将洋口港浚深为 30 万 t 级深水港，开发相邻的黄沙洋为 20 万 t 级的集装箱或航空母舰基地，共同组合成江苏深水港中心；向北经大丰港、滨海港与连云港相联，经连云港西通欧亚大陆；向南经吕四港、通州湾与长江南通港结合，通过长江向西，组成江苏第二条东西大通道；而东西相联的两侧以洋口-黄沙洋港为中心，形成我国在黄海的港口群。国内北联青岛、天津、曹妃甸、秦皇岛、营口与大连港；南通上海、舟山、闽、粤、香港，联合发展沿海经济带。更为重要的是，以江苏发达的科技与工农业为基础，着力发展外向型经济：向北方与日本、韩国及俄罗斯相通；南达东盟与新、澳，向东与美、加、南美直联。江苏经济地位与影响加强则非一日而语。

航海事业带动临港工业发展，尤其是造船与海洋装备制造业，在江苏已形成基础，必将促进海洋经贸往来与增强海上防卫能力，这是发展江苏海洋经济，为建设海洋强国尽力的关键。

其次，发挥江苏平原海岸海洋的优势，发展海洋能源产业——风能、潮能、浪能、太阳能与潜在的油气资源发现与开发，是发展海洋经济解决民生之困的重要支柱。平坦的海岸与滩涂新生土地，适宜建大型光电板组合，可发展成为大型太阳能光电场中心。风能在海陆交互的岸、滩上，比内陆与外海增强 30%，在缺煤少油、耕地面积小的江苏海岸，发展风力发电具天然优势！已建的风机场空气振动影响到候鸟迁徙，可能要将风机支柱位置适当向海转移。

辐射沙脊群中的黄沙洋潮流通道尾部测到最大潮差 9.28 m，涨、落潮流急，具有建设潮汐发电站的巨大潜力，利用波浪能提供航标灯电力等。例如：2016 年 1 月，在舟山群岛的岱山县秀山岛南部海域，建设 3.4 MW 潮流能发电机组长 70 m，宽 30 m，平均高 20 m，重达 2 500 t，可抵抗 16 级台风和 4 m 巨浪。安装与运行稳定后，年发电量可达 600 万 kW。发展新能源电力，更需要科技力量支撑，关键是做好海洋能开发规划，分重点、有序地进行，必须将风能、潮能与太阳能联网运行，形成新能源场，可分时、分区连续供电。既解民生用电，又可供工、农业对电力的所需，重要的是利用清洁的天然能源，源源不断，而无污染。

江苏海洋经济的第三方面是与人民食用和医药健康密切相关的。其中，渔业水产在捕捞与养殖贝、虾、蟹、藻已形成规模；海水化学元素提取与海洋生物产品制药已形成采集、处理与制成的系列，但是品种与医、食疗效仍具研发潜力，食品与医药产品发展应关注成分、用途与成品化。紫菜已从海水养殖、采集、晒干发展成多种形式的食品，拓宽了市场，丰富了人民海产品的应用途径，是效法的先例。为鱼类的生息繁衍，我国实行了休渔期，建议：利用休渔期，组织身强力壮、有经验的渔民，给予优良的渔船装备，进行远洋捕捞，到外海与远洋，尤其是南大洋与孟加拉湾富产海区进行深海捕捞，形成远近接合、全年性的渔捞作业，增加国人的海产食品比重。

海岸海洋旅游，已是海洋经济中的重头所在，并且是具巨大发展潜力的海洋经济组成部分。改革开放与人民生活水平提高，走出家门去国内外旅游度假，尤其是去滨海与海岛地区，以致赴极地探险已成为国人关注的热点，发展迅猛。2013 年滨海旅游产业占到海洋主要产业总值的 34.6%(图 2)，滨海旅游还应向海洋旅游、极地与海底探险结合起来。在国内，海底探险应结合考古发现工作，丰富海洋旅游之内涵。江苏海岸海洋特色突出：江、河、海三水交汇，物产丰富，民俗、艺术与饮食特色多样；秦皇汉武征讨、屯田与出使东瀛留下越千年的文化遗迹；当代海洋开发又焕发了新风貌等，组合自然与历史景观，赋寓文化内涵，知识与休闲养生、与体育锻炼、与探险结合，与长周末与年假结合，国内外并重地规划与组织经营，海岸海洋旅游发展前景广阔，对海洋经济会持续贡献加大。

江苏沿海平原的农业与种植业很重要。盐碱土与丰沛的降水条件，稻米、麦、棉、麻、花卉、菓药生长繁茂，形成沿海经济的组成部分。稻、麦需提高品种，棉、麻与化纤结合，花卉与油料作物与医药生材结合发展，均为人民衣、食所需，同时，需加工为多种形式成品，提高产值，内销与出口。

以上是我认识到的有关江苏发展海洋经济的主要产业方面，我是从自然环境资源说明其发展潜力，以历史发展背景表明其基础，大体上的先后顺序表达其发展的紧迫性与重要性。但也不尽然，对发展已具坚实基础的，如：滨海旅游业与农业，虽排序在后，却是最具发展动力的部分。在天时地利条件下，开发海洋经济需要有高科技手段支撑。而关键是人的作用，调查了解情况，总结历史经验，进行新一轮的开发规划、建设与管理，总结再前进。所以在发展海洋经济改变以传统的陆地经营开发为主的经营方式之际，选拔人才与培养人才很重要。江苏需要建设以海洋为重点、为特色，多学科综合的高等学校，南京大学正在与国外海洋重点院校联合，拟在江苏与上海交界的长三角海岸举办大学。愿在省、市领导下，与兄弟院校合作，为发展海洋科技、发展海洋经济，建设海洋强国贡献力量。

参考文献

[1] 孙世芳,余宝林等著.2009.河北省海洋经济可持续发展研究.石家庄:河北人民出版社.

[2] 国家海洋局.2015.中国海洋统计年鉴2014.北京:海洋出版社.

[3] 国家海洋局.2014.2013年海洋生产总值情况表.2013年中国海洋经济统计公报.

[4] 江苏省908专项办公室,张长宽主编.2012.江苏近海海洋综合调查与评价总报告.北京:科学出版社.

[5] 李家彪主编.2008.东海区域地质.北京:海洋出版社.

第四篇
海疆权益研究

中国海海底环境研究与海岸带研究相结合，以区域海洋自然特点与历史归属为依据，论证捍卫中国海海疆权益。

南海海底特征、资源区位与疆界断续线*

1　南海海底特征

通过建立南海海底数字地形模型，可知南海海底地势是西北高、东南低，自中国大陆边缘向南海中心部分呈阶梯状下降[1]，地貌成因组合特征如图1所示。

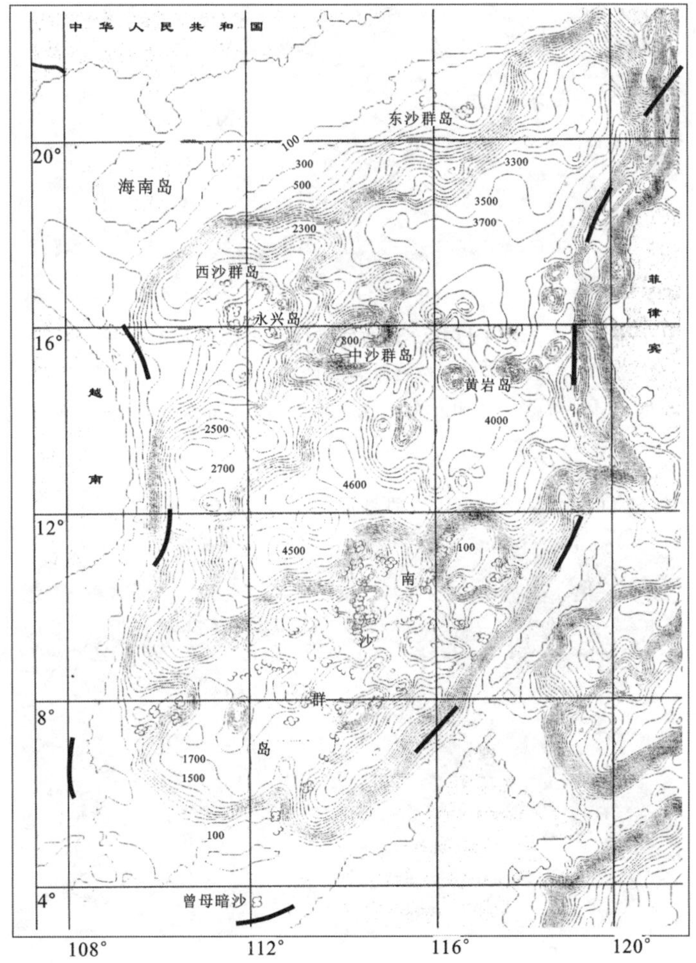

图1　南海海底数字地形模型插值200 m间距等深线图(单位:m)

(1) 南海中央部分是呈北东-南西向延长的菱形深海盆地，其纵长1 500 km，最宽处为820 km。它是由晚第三纪的NE向断裂拉张形成。中央海盆水深3 400 m(北部)至

* 王颖，马劲松:《南京大学学报(自然科学)》，2003年第39卷第6期，第797－805页。
2002年载于:海南南海研究中心，《南海问题研讨会论文集(2002)》，10－18页。

4 200 m（南部），有些地方水深超过 4 400 m。深海盆地中有由孤立的海底山组成的高度达 3 400～3 900 m 的山群，有 27 座相对高度超过 1 000 m 的海山及 20 多座 400～1 000 m 的海丘[2]，多为火山喷发的玄武岩山地，上覆珊瑚礁及沉积层。深海盆地底部平坦，坡度 0.3‰～0.4‰[3]。盆地东北端与西南端有断裂谷形成的深水谷地，内有沉积物充填，谷地终端有海底扇沉积体，已受到日后隆起成为小山脊（图 2 和图 3）。

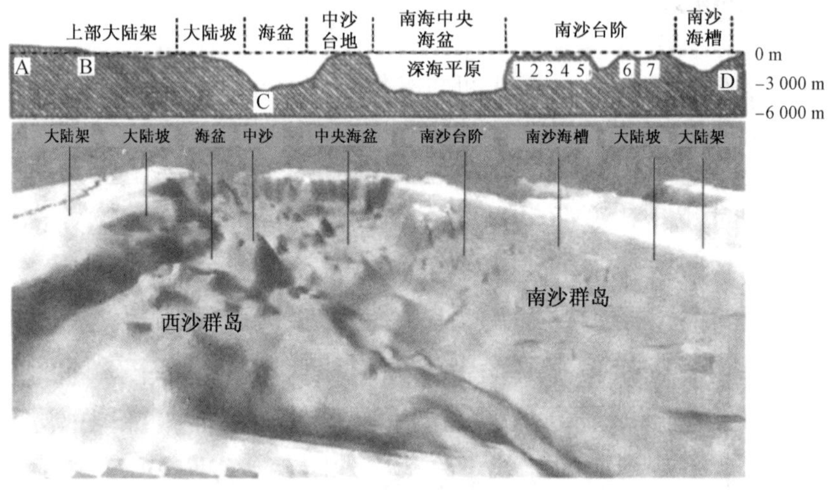

图 2　南海 114°20′E 海底数字地形模型及剖面图

A：香港东侧；B：一统暗沙东侧；C：西沙海槽东侧；D：文莱

1：双子群礁；2：中业群礁；3：道明群礁；4：郑和群礁；5：九章群礁；6：南华礁东侧；7：安渡滩

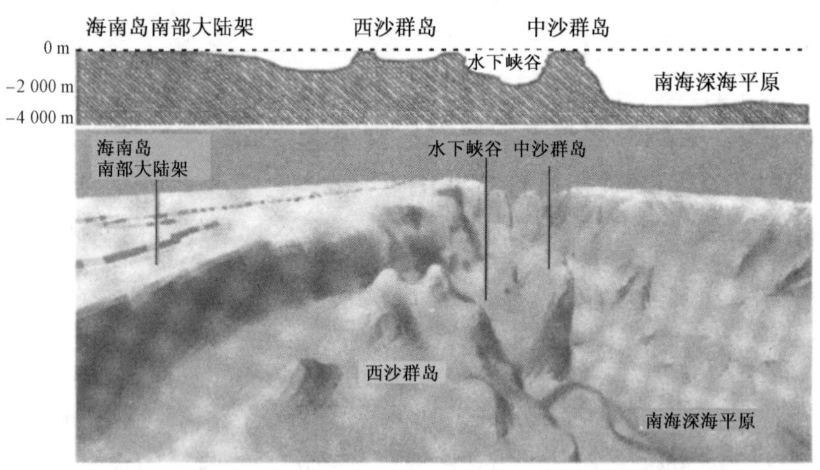

图 3　12°N，109°15′E 至 12°N，120°E 海底数字地形模型及剖面图

（2）深海盆地的南北两侧是块状断裂下沉形成的阶梯状大陆坡。南海大陆坡面积 1 200 000 km²，占海区总面积 49%，水深界于 1 500～3 500 m，其中海盆东北坡及北坡，终止深度为 3 200～3 500 m，其他处终止深度达 3 800～4 000 m，此阶梯状大陆坡系由块断下沉的古大陆架形成[1]。相对高出海底数千米的高差，使其表现为海底高原。

① 位于深海盆地西北侧的是块断下沉的西沙-中沙大陆坡，沿走向北东的断裂延伸

到东沙、台湾浅滩及澎湖列岛。西沙阶梯面水深 900～1 100 m。其上发育着 30 条珊瑚礁、岛、暗礁及火山岛。

珊瑚礁群岛内的各环礁水深不大，界于数十米到 200 m。在中沙群岛与西沙群岛之间，相隔着近 1 100 m 的深水道。在西沙群岛的西北部和北部，还有一条近东西向海槽，长 420 km，宽 8～10 km，槽底平，槽壁陡，水深 1 500(西部)～3 400 m(东部)。

② 位于中央海盆南侧与东南侧，与西沙大陆坡遥相对应，是相对高出海底 2 400 m 的南沙大陆坡。由于边缘两侧均受到北东与北西两组断裂隔断，使南沙大陆坡亦呈现为海底高原形式。顶面水深约 1 800 m，有沟谷、海山、海丘起伏，并发育着 200 多个珊瑚礁岛、暗沙与浅滩，大致均呈北东向排列。南沙基底岩性与西沙、与广东沿海大陆相同，也是由晚第三纪以来沉降、断裂的大陆架形成的。

③ 中央海盆北坡是以东沙群岛为中心的阶梯状凸形坡(图 4)，上部水深约 700 m，坡度为 3.4‰，超过 700 m 水深的大陆坡坡度为 27.7‰。本区陆坡上发育有海底高原，如东沙海底高原。它紧靠北部大陆架外缘，似正三角形向深海突出，面积约 1.2 万 km^2。台阶状地形明显，水深一般在 300～400 m 之间。

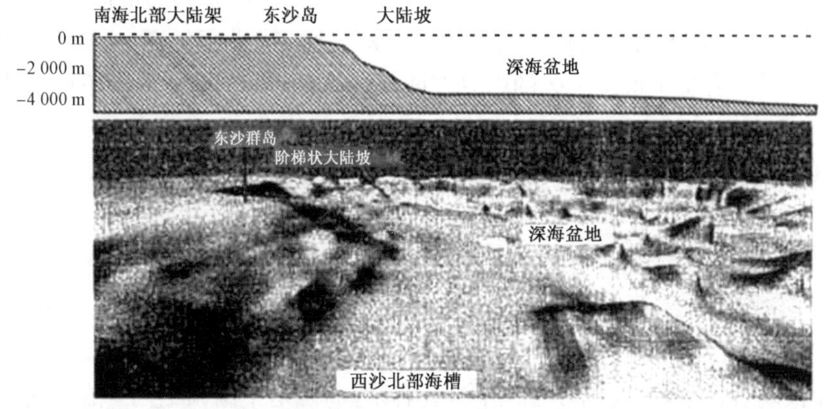

图 4　广东沿海(22°15′N,116°E)至南海东北部(19°10′N,119°10′E)海底数字地形模型及剖面图

④ 中央海盆东坡是沿菲律宾群岛外的窄陡陆坡，宽度很少超过 70 km(约 38 nmile)，坡度达 170.8‰～119.3‰(吕宋海槽与马尼拉海沟)，最缓处在巴拉望岛外缘，坡度为 17.7‰。大陆坡多受水下峡谷切割，坡麓有扇形堆积体。

(3) 南海大陆架主要分布于海区的北、西、南三面，是亚洲大陆向海缓缓延伸的地带。

① 在珠江口以东，陆架水深小于 200 m，珠江口以西陆架水深度超过 200 m，最深达 379 m(图 5)。大陆架宽度各处不等，沿中国大陆约 280 km(约 150 nmile)，岛屿外缘在台湾南部为 14 km(约 7.5 nmile)，海南岛南部为 93 km(约 50 nmile)。

② 南海南部大陆架为巽它陆架，范围界于 150 m 水深以内，宽度超过 300 km，南沙群岛的南屏礁、南康暗沙、立地暗沙、八仙暗沙和曾母暗沙等，是位于该大陆架北部，在水深 8～9 m、20～30 m 间的珊瑚礁、滩。

③ 南海东部陆架分布于吕宋岛、民都洛岛和巴拉望岛边缘，呈南北向或 NE - SW 向的狭窄带状分布，是为岛架，外缘水深 150～200 m，岛架被沟谷切割并延伸至巴拉望海槽。

了解南海海底的地貌特点，有助于对海洋权益的维护[4]，以提出有理与有力的依据。

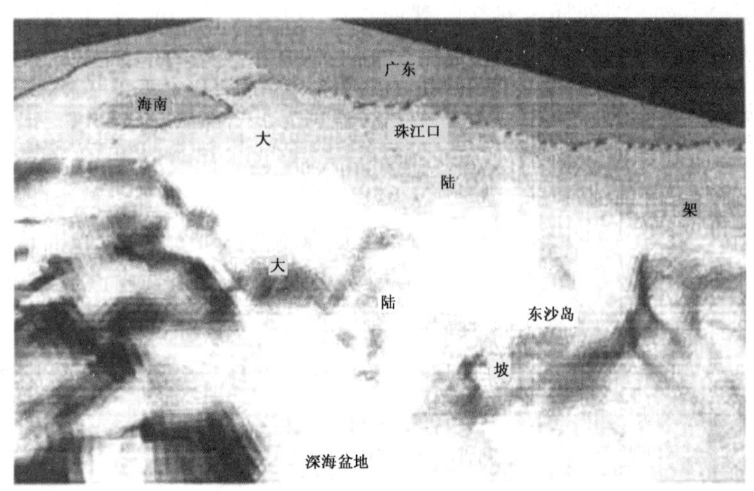

图 5 珠江口外(22°20′N,114°03′E)至(18°30′N,115°15′E)海底数字地形模型

需清楚地认识到,西沙群岛不是位于现代大陆架海域,而是位于由古大陆架下沉折断形成的大陆坡顶。南沙群岛位于巽它陆架上。岛屿是确定该海域归属的依据,因此,岛屿的归属权至关重要。历史的渊源与传统的疆界,国际公认的或沿用的图、文资料是重要的证据,断续线正是具有半个世纪之久的疆界直证,已被国际多方多次引用,不能退让。

2 南海疆界断续线的形成历史及位置

从历史上看,断续线是我国在南海的传统疆界线,国际上对其承认已达半个多世纪之久。1883 年,德国曾在南沙和西沙进行测量,广东省政府提出抗议,德国停止了作业;1911 年辛亥革命后,国际组织多次要求中国政府在南海诸岛上建立气象台、灯塔等航海设施,以保证航行安全,表明国际公认南沙、西沙等岛礁是中国领土。二战期间,南海诸岛被日本侵占,1945 年波茨坦公告明确规定日本侵占中国领土(含南海诸岛)必须归还中国;抗战胜利后的 1946 年,中国政府接收西沙、南沙,在岛上升国旗、立主权碑,昭告世界,并派兵驻守南沙群岛的最大岛屿——太平岛,国际上对此皆予承认;1947 年,中国政府正式公布了《南海诸岛位置图》[5],图上画有断续线形式的国界线 11 段(原件存于南京,中国第二历史档案馆),以后出版的地图均遵照政府规定画有这些断续线,中国对南海诸岛的主权得到国际公认;1951 年,英、美等国对日签订和约,条约也规定日本放弃对台湾、澎湖列岛、西沙、南沙群岛的权利,会上南越代表提出要西沙、南沙主权,这一无理要求遭与会各国摈弃,无人理睬;1952 年,台湾当局以中国名义和日本签订和约,和约第二条为:“兹承认依照公历 1951 年 9 月 8 日在美利坚合众国旧金山市签订之对日和平条约第二条,日本国业已放弃对于台湾及澎湖列岛以及南沙群岛、西沙群岛之一切权利,权利名义与要求。”上述这些会议的参与国、签字国已承认西沙、南沙群岛是中国领土,日本也已将这些地方交还中国政府;1953 年,经我国政府批准去掉北部湾 2 段断续线,改成 9 段,沿用至今,已历时半个多世纪。

20 世纪 70 年代后期,随着对南海丰富自然资源认识的深入,越南、菲律宾、马来西亚和印度尼西亚等国纷纷提出对南海主权的无理要求，抛出多种南海划分界线，造成南海国界问题的矛盾日益突出。由此，有必要对我国南海国界断续线给予确定性的认识。

由于对 9 条断续线的位置需进一步确定其经纬度，因此，我们根据 1947 年民国内政部《南海诸岛位置图》的原图复印件（南京大学杨怀仁教授参与该图工作，复印图存南京大学大地海洋科学系）和 1983 年我国政府正式出版的《南海诸岛》图[6]，采用数字化仿射变换，利用地理信息系统分别量测出新旧各 9 条断续线的经纬度坐标，以度为单位，舍入到小数点后 3 位，具有较高的精确度，具体数据编号列表于后（表 1 和表 2）。然后，将各新旧断续线按照经纬度制成南海新旧国界断续线示意图（图 6）。

表 1　民国内政部制 1∶650 万《南海诸岛位置图》中的断续线位置（1947 年）

断续线编号	位置	端点位置（度）		长度，走向	起止范围（度）	
		起点	终点		东经	北纬
西 1（旧）	西沙群岛以西，平行于越南东岸，近土伦（顺化）	109.452 15.958	110.289 14.288	210.4 km SSE	109.452 110.289	15.958 14.288
西 2（旧）	南沙群岛以西，平行于越南南岸，西北为金兰湾	110.232 10.528	109.399 9.080	185.2 km SSW	110.232 109.399	10.528 9.080
南 1（旧）	南沙群岛西南，接近纳土纳群岛	108.895 6.053	109.686 4.585	186.5 km SSE	108.895 109.686	6.053 4.585
南 2（旧）	曾母暗沙以南，平行于婆罗洲（加里曼丹岛）北岸	111.134 3.725	113.007 4.232	230.8 km E，NEE	111.134 113.007	3.613 4.232
南 3（旧）	平行于婆罗洲北岸，拉布安岛西北	114.201 5.280	115.005 6.214	137.0 km NE	114.201 115.005	5.280 6.214
东 1（旧）	巴拉望岛西南，巴拉巴克海峡以西	116.076 7.447	116.906 8.414	141.2 km NE	116.076 116.906	7.447 8.414
东 2（旧）	南沙群岛东北，菲律宾布桑加岛、龟良岛以西	118.427 11.314	118.866 12.596	152.1 km NNE	118.427 118.866	11.314 12.596
东 3（旧）	黄岩岛东北，平行于吕宋岛西海岸	119.008 15.469	119.260 17.027	175.7 km N	119.008 119.260	15.469 17.027
东北 1（旧）	巴时海峡（巴士海峡）内，台湾七星岩与吕宋岛之间	119.845 19.690	120.975 20.184	131.5 km NEE	119.845 120.975	19.690 20.184

表 2　地图出版社 1∶600 万《南海诸岛》图中的断续线位置（1983 年）

断续线编号	位置	端点位置（度）		长度，走向	起止范围（度）	
		起点	终点		东经	北纬
西 1（新）	西沙群岛以西，平行于越南东海岸，近越南广东群岛	109.284 16.148	109.869 15.163	127.7 km SSE	109.284 109.869	15.163 16.148
西 2（新）	南沙群岛西北，平行于越南东南海岸，西北为越南槟绘湾	110.328 12.278	110.046 11.248	120.5 km S，SSW	110.046 110.328	11.248 12.278
南 1（新）	南沙群岛西南，纳土纳群岛正北方	108.247 7.108	108.322 6.003	124.2 km S，SSE	108.220 108.322	6.003 7.108

（续表）

断续线编号	位置	端点位置（度）		长度，走向	起止范围（度）	
		起点	终点		东经	北纬
南 2（新）	曾母暗沙以南，亚西暗沙西北，平行于加里曼丹岛北岸	111.892 3.434	113.015 3.847	134.4 km NEE	111.892 113.015	3.434 3.847
南 3（新）	巴拉望岛西南，巴拉巴克海峡以西	115.623 7.171	116.326 7.993	119.9 km NE	115.623 116.326	7.171 7.993
东 1（新）	南沙群岛东北，菲律宾布桑加岛、龟良岛西南	118.572 10.918	119.029 12.013	132.0 km NNE	118.572 119.029	10.918 12.013
东 2（新）	黄岩岛东、东北，接近吕宋岛西海岸	119.079 14.939	119.074 16.059	124.7 km N	119.054 119.079	14.939 16.059
东 3（新）	平行吕宋岛西北海岸，近北吕宋海槽	119.474 17.956	119.984 18.971	126.3 km NNE	119.474 119.984	17.956 18.971
东北 1（新）	巴士海峡内，台湾兰屿与吕宋岛之间	121.126 20.817	121.860 21.776	131.1 km NEE	121.126 121.860	20.817 21.776

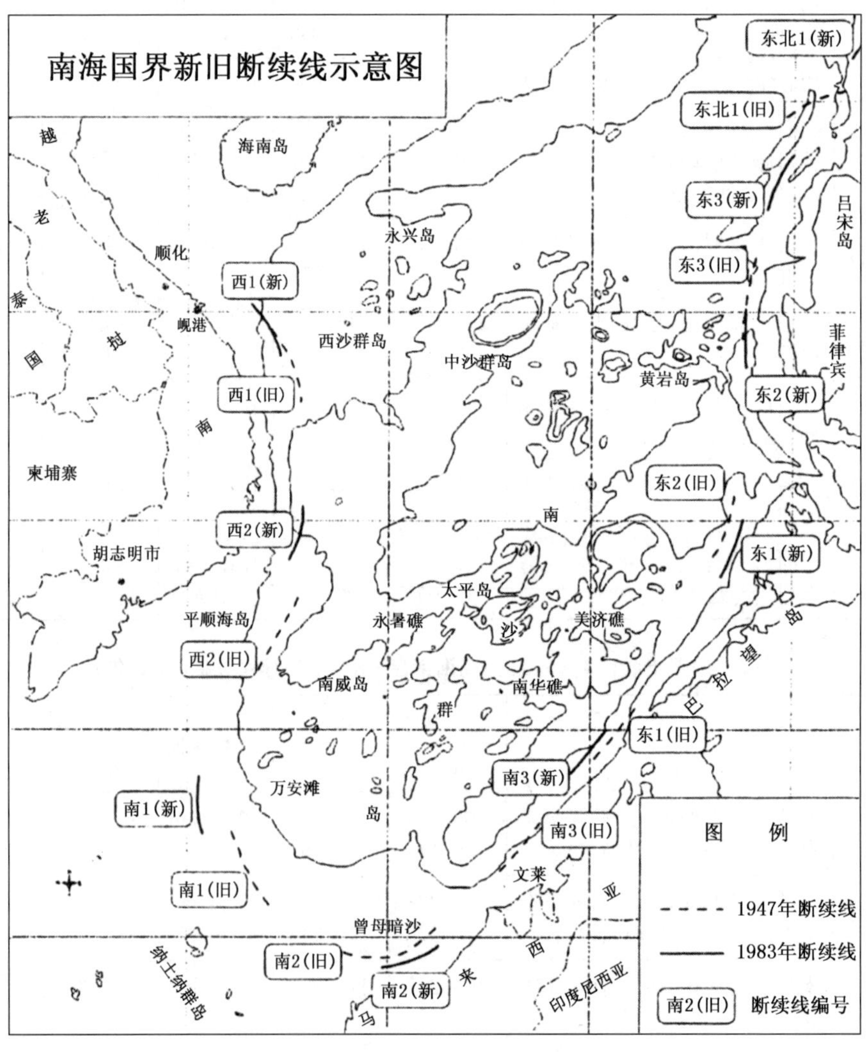

图 6　南海国界新旧断续线对比示意图

通过对比发现，新旧各 9 条断续线的位置并不完全重合，表现出多种位置关系：新旧线是相互衔接，如西 1(旧)与西 1(新)，东 3(旧)与东 2(新)；新旧线是相互平行，有外展，如东 2(旧)与东 1(新)，南 2(新)与南 2(旧)；有的新线在旧线的延伸线的位置上，如南 3(旧)与南 3(新)；一般而言，旧断续线长度较长，通常在 130～210 km 之间，且长度变化较大，不规则；新断续线长度较短，一般在 120～130 km 之间，长度变化不大，较为规则。对应编号的新旧断续线之间的距离从最接近的 18 km 左右(南 2 新旧之间)，到相距最远的 333 km 左右(东 1 新旧之间)。

虽然各条新旧断续线之间存在空间位置上的差异，但新旧断续线所共同界定的领土范围是基本一致的，经计算得出，旧断续线以内的南海海域面积约为 192.7 万 km^2，新断续线以内的南海海域面积约为 202.5 万 km^2，误差小于 5%，显示出中国南海领土主权范围的历史延续性[7]。我们认为作为疆界依据，应明确采用一致的断续线画法，且应该是以 1947 年旧图为准，更有历史依据。

3　南海区位资源与重要性

3.1　最重要的是国土资源，海域疆界划分主要以陆地为依据

海洋法的贯彻，在群岛海域形成“小岛大海洋”的新概念。一个人居的小岛，它可据有内水、12 海里领海、24 海里毗连区及 200 海里专属经济区。因此，南海的岛、礁、滩、沙十分重要，即使暗沙，如果一年中可出露一次，也可为划界依据，如黄海辐射沙洲的外磕角据此定为我国领海基线点。因此，对南海群岛应由国家规划立项，逐年调查，确定方位、面积、海况，渐次移民开发，纳入防卫体系，是根本大计。

3.2　海洋资源

3.2.1　能源

潮汐与波浪能发电；风能发电，在 3 级风力，波高 0.2 m 的情况下，即可发电达 60 W/m^2；利用南海表层水与 1 000 多 m 深的海底的温差在 20 ℃以上，海底地形起伏大，在南海诸岛周围海域形成海底上升流，它不仅将海底丰富的氮磷营养盐带至上部海水形成渔场，而且可利用上升流冷水温差发电，供给南海诸岛电力，与常年的热带日照提供了丰富的太阳能发电相结合，为海岛提供动能，尤其是用以淡化海水，意义重大。

3.2.2　油气与矿产资源

南海油气藏资源被誉为“第二个波斯湾”。分布于珠江口盆地的原油资源估计 40 亿～50 亿 t；莺歌海盆地原油约 4 亿～5 亿 t，天然气已进入开发；北部湾盆地油气资源总量为 4 亿～5 亿 t；曾母盆地第三纪沉积层厚度 4 000～9 000 m，面积约 25 万 km^2，推测油气资源储量约 137 亿 t；万安滩油气资源估计超过 40 亿 t。南海海底还蕴藏着天然气水合物(可燃冰)等。因此，南沙群岛曾母暗沙区是我国必须力保，寸水不让之区。但是，它位于南海南部大陆架上，曾母暗沙未曾出露水面，巽它陆架国必为此起争端。需从历史传统已为国际认同的我国疆界线，以及地理环境界限，做出解决问题的依据，进行有步骤的防卫

实施。

海岸带的砂矿资源:锆英石、独居石、铌钽铁矿、磷钇矿及石英砂等,主要砂矿为航空航天硬合金原料,主要在沿海岛沿岸海滩与近海水域分布。我国已进行开发,争执少。

3.2.3 渔业水产与药物资源

珍稀的热带海洋药物资源值得研究与开发保护。

南海渔业资源丰富,有近 400 万 t。我国渔民去南海的捕鱼活动由来已久,但近来受到越南的非法干涉,海匪骚扰亦甚。我国应考虑对国民与资源保护的有效措施。

3.3 发展海洋经济是 21 世纪增强国力的重要方面,宜重点投资

大力加强海空防卫力量,保卫海疆,是支持我国社会经济可持续发展,人民安居、权益保障的关键。另一方面,亦需深入实地研究南海资源环境的具体特点、现实状况,有理有据地通过国际谈判及外交途径逐步解决。

参考文献

[1] Wang Y. 1996. Marine geography of China. Beijing: Science Press. 45－57(王颖. 1996. 中国海洋地理. 北京:科学出版社. 45－57).

[2] Zeng C K. 1987. The original formation of the seamounts in the basin of the South China Sea. *East Sea Ocean*, 5(1～2): 1－9(曾成开. 1987. 南海海盆中的海山海丘及其成因. 东海海洋, 5(1～2): 1－9).

[3] Feng W K. 1988. Late quaternary geological environment of the north side of the South China Sea. Guangdong Science and Technology Press(冯文科. 1988. 南海北部晚第四纪地质环境. 广东科技出版社).

[4] Zhang Y Z, Wang Y. 2000. Forecasting on coastal ocean science of the 21st centary. *Journal of Nanjing University(Natural Sciences)*, 36(6): 717－726(张永战,王颖. 2000. 面向 21 世纪的海岸海洋科学. 南京大学学报(自然科学),36(6):717－726).

[5] Ministry of Interior of China. 1947. The Location map of the South China Sea Islands(民国内政部方域司. 1947. 南海诸岛位置图).

[6] China map Press. 1983. Map of Islands in South China Sea(中国地图出版社. 1983. 南海诸岛图).

[7] Xu S A. 2001. The intension of the 9 intermittent national boundaries in the South China Sea. In: Proceedings of Seminar on the Problems and Predictions of South China Sea in the 21st Century. 77－83(许森安. 2001. 南海断续国界线的内涵. 见:21 世纪的南海:问题与前瞻学术研讨会论文选. 77－83).

南海海底特征、疆域与数字南海*

一、南海海底特征

南海为西北太平洋一系列边缘海中面积最大、海底类型齐全、结构复杂的海洋。南海面积 350 万 km^2，平均海深 1 212 m，北接我国大陆，东面和南面分别隔以菲律宾岛和大巽它群岛与太平洋为邻，西邻中南半岛和马来半岛，为一半封闭的海盆。其东北有台湾海峡与东海相接，东面有巴士海峡、巴林塘海峡、巴拉巴克海峡与太平洋及苏禄海相通，南面有卡里马塔海峡连接爪哇海，西南有马六甲海峡沟通印度洋。

南海海底是受晚第三纪 NE 向大规模断裂与晚更新世-全新世的 SN 向的断裂所控制形成的拉张盆地，是长轴为北东-南西方向的菱形盆地。海盆中央有地幔物质补偿性上涌，热流量达 1.48～2.84 HFO(钱翼鹏，1982)，海盆基地岩层为大洋型玄武岩、橄榄岩与安山岩，是中国海中唯一具有洋壳型底层的深海盆地(王颖等，1996)。海底地貌复杂，海底隆起与洼陷相间，有海槽与海沟发育，岛屿及珊瑚礁滩广布。海底地势自西北向东南减低，自海盆边缘向中央部分呈阶梯状下降。菱形盆地的四周边缘分布着大陆架；大陆架外侧分布着阶梯状大陆坡，即西沙-中沙群岛大陆坡与南沙群岛大陆坡。大陆坡终止处，是南海海盆的中央部分——深度超过 4 000 m 的深海平原(图 1 和图 2)。

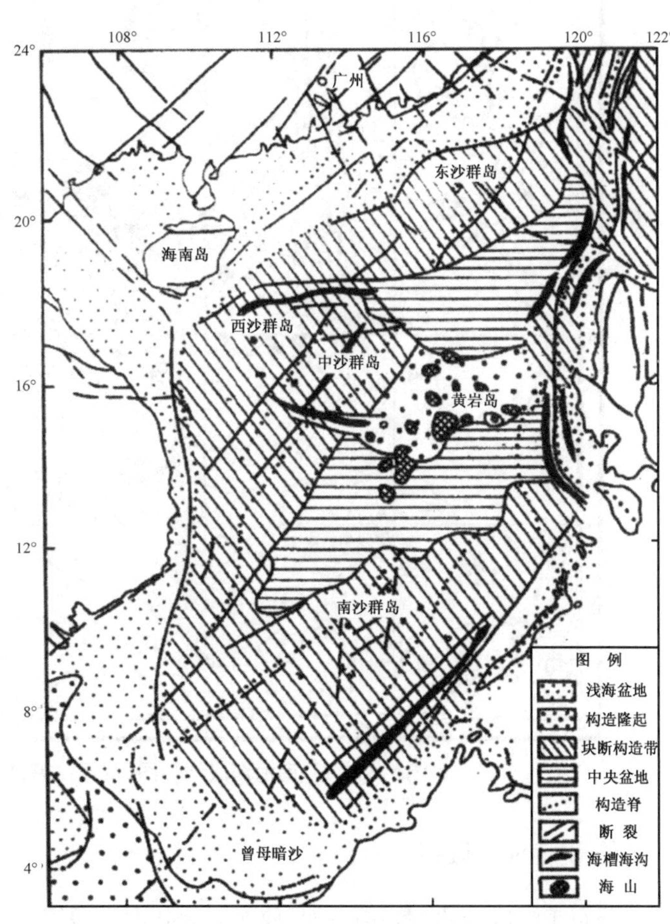

图 1 南海海底构造略图(中国科学院海洋研究所，1982)

* 王颖，马劲松：苏纪兰主编，《郑和下西洋的回顾与思考》，177－193 页，科学出版社，2006 年。

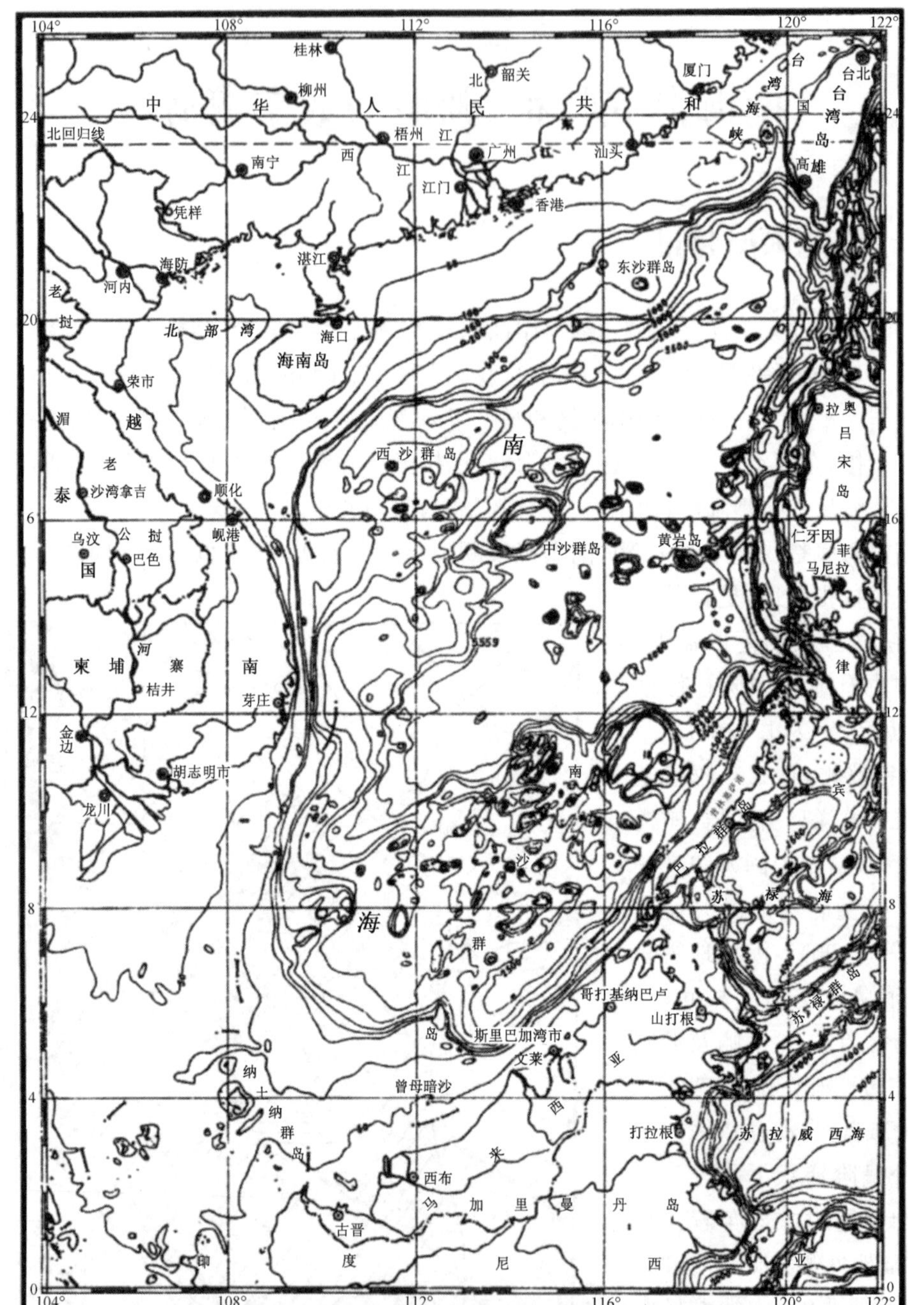

图 2　南海海底地形图(罗章仁等,1996)

(一) 南海大陆架

分布于海区的北、西、南三面,使亚洲大陆向海缓缓延伸的地带。陆架坡度平缓,尚有沉溺的海岸阶地、水下三角洲、珊瑚礁与岩礁等地貌残留。陆架外缘水深一般小于200 m,珠江口以西,陆架外围水深随陆架宽度而增加,一般超过 200 m,最深可达 379 m。大陆架面积约 190 万 km^2,其宽度是西北部与西南部长,西部宽度相对减小,东部陆架狭窄,是岛缘陆架(王颖等,1996)(表 1)。

表 1　南海大陆架宽度

区　域	地　名	宽度/km	坡度/‰
南海西北部（珠江口以东陆架水深200 m，最深处达379 m）	台湾南部(高雄)	13.9	14.3
	汕头(南沃岛)	196.3	0.77
	珠江口	253.7	0.54
	电白	274.1	0.53
	海南岛南部	92.6	2
南海南部（巽它陆架，水深界于150 m以内，南沙群岛的礁滩、暗沙，位于巽它陆架北部，水深8 m、9 m、20～30 m内）	沙捞越	300	
	纳土纳群岛	300	
	昆仑岛	300	
	卢帕河口外	405	0.4～0.5
	巽它陆架	300	0.4～0.5
	(湄公河三角洲—文莱)		0.4～0.5
	(邦加岛—勿里洞岛)		
南海西部（外缘水深150～200 m）	中南半岛东岸(北部湾出口—向南—湄公河三角洲以北)	40～50	3.0～4.0
南海东部岛架（外缘水深150～200 m）	吕宋岛		
	民都洛岛	66	
	巴拉望岛	57	

(二) 南海大陆坡

分布于大陆架外缘，是大陆架向海的延伸部分，面积约120万km^2，占海域面积的49%，是南海最大面积的地貌单元，也是南海疆域纷争的焦点。南海大陆架由于受海盆张裂影响程度不同而具有地貌差异，可划分为四类，但均具有因海盆张裂而造成的断阶特点。

1. 中央盆地北坡

位于我国台湾岛南端与珠江口之间，水深界于150～3 600 m之间。以东沙群岛为中心，呈现为阶梯状的凸形坡，向中央海盆倾斜，在水深700 m以浅，坡度为3.4‰；超过700 m深，坡度为27.2‰，大陆坡上残存海山、海丘及切割水道；台湾岛南端大陆坡坡度为26.5‰(王颖等，1996)(图3和图4)。

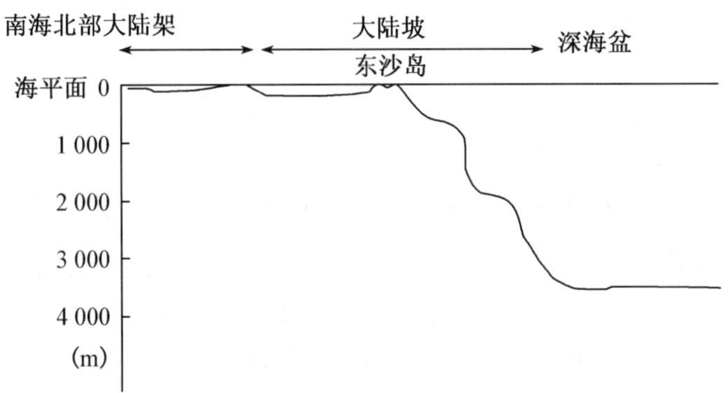

图3　广东沿海(22°15′N，116°E)至南海东北部(19°10′N，119°10′E)海底地形剖面图(王颖等，1996)

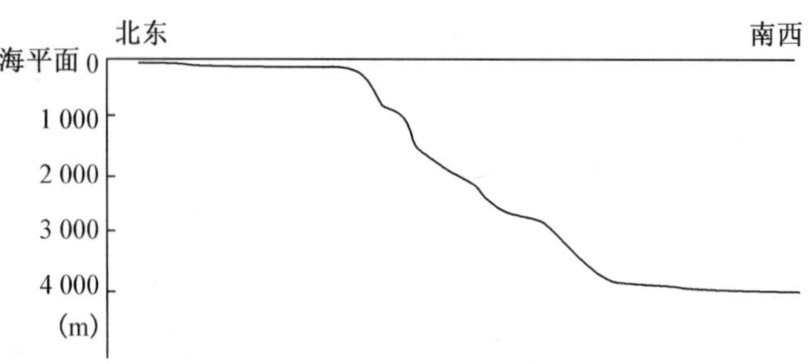

图 4　珠江口外(22°20′N,114°03′E 至 18°30′N,115°15′E)海底地形剖面图(王颖等,1996)

2. 海南岛南部大陆坡

范围介于珠江口外深沟与越南南部之间,水深界于 150 m～1 000 m～3 600 m 之间,与大陆架相接部分为水深在 1 000 m 以内的陆坡,坡度为 27.5‰,1 000 m 以下或有海沟相隔,或直接相连地为一宽度达 494 km 的海底台阶,其外侧以 28.4‰～111.2‰的陡坡直降到 3 600 m 的深海平原。这种阶梯状大陆坡的宽大台阶形成耸立于海底高达 2 000 多 m 的海底高原(图 5 和图 6)。

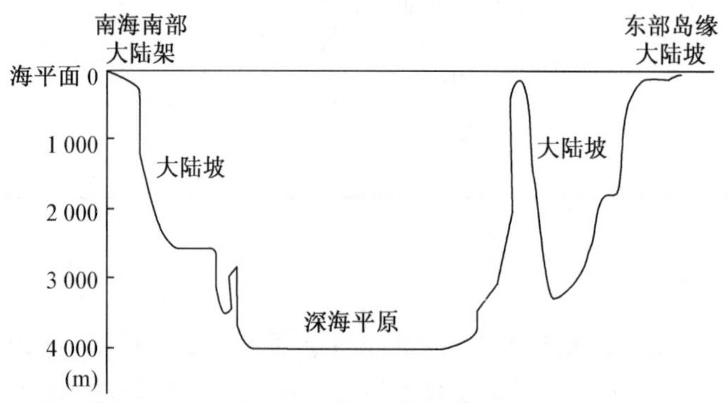

图 5　海南岛南部(18°11′N,109°48′E)至南海东南部(12°32′N,118°40′E)深海盆地海底地形剖面图(王颖等,1996)

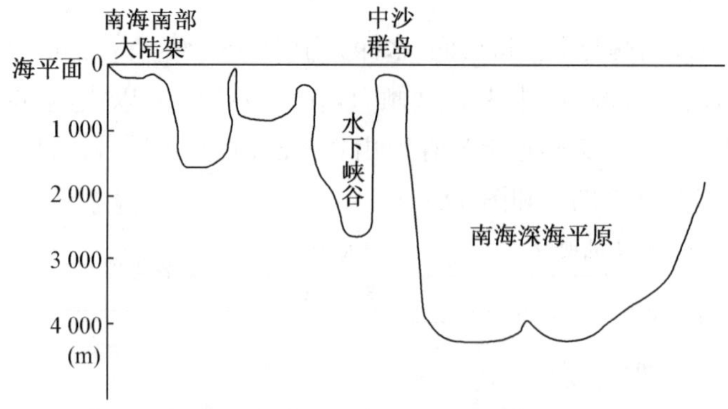

图 6　12°N,109°15′E 至 12°N,120°E 海底地形剖面图(王颖等,1996)

组成阶梯状大陆坡的基底岩层与大陆架基底岩层一致,是前寒武纪古生代至中生代的变质岩和花岗岩,属大陆型地壳,地壳厚度 14～26 km,莫霍界面深度在大陆坡为 14～26 km,在大陆架为 26～32 km(罗章仁等,1996)。基底岩层上为由数千米厚的沉积岩或火山岩所组成的海山(曾成开,1987),海山顶部又发育着厚达千余米的珊瑚礁。西沙群岛、中沙群岛均为位于海底高原上的珊瑚礁群岛。地质地貌结构反映出海底高原是由巨大的东北断裂分割、呈块断下沉淹没的古陆架所形成，时代发生于晚第三纪。经过多次构造活动与海平面的变化，一些珊瑚礁曾在第四纪时抬升至海面上，近期东西向的断裂发生，使海底高原及其上的构造脊均被错动断裂与分割。

3. 中央海盆南—东南坡

以深海平原相隔与海南岛南部大陆坡遥相对峙，也是阶梯状大陆坡，南沙群岛位于其上。在 1 800 m 深处亦为呈 NE－SW 向的台阶式海底高原，即南沙海底高原，高出海底 2 400 m，边缘受 NE 向与 NW 向两组断裂控制均为陡坡，并以海槽与巽它陆架以及与北婆罗洲陆架相接。高原面上沟谷纵横，海山、海丘众多，南沙群岛是由发育于海底高原上的 200 多个呈 NE 向排列的珊瑚礁岛、浅滩与暗沙组成，此处可称为南沙阶梯状大陆坡。

4. 中央海盆东坡

指沿西太平洋岛弧之一的菲律宾群岛发育而成的岛屿，其特点是：狭窄陡峭的阶梯状下降陆坡，宽度很少超过 70 km，巴拉望岛屿外缘的陆坡坡度为 17.7‰；坡麓有海槽或海沟分布，陆坡坡度很陡，如临吕宋海槽的陆坡为 170.8‰，马尼拉海沟处为 119.3‰。大陆坡多受水下峡谷切割穿越而成海峡通道，坡麓峡谷出口处有海底扇堆积体分布（图 5）。

（三）南海中央部分

呈 NE－NW 向延长的菱形深海盆地，其纵长 1 500 m，最宽处为 820 km。它是由晚第三纪的 NE 向断裂拉张形成。中央海盆水深 3 400 m（北部）至 4 200 m（南部），有些地方水深超过 4 400 m。深海盆地中有由孤立的海底山组成的高度达 3 400～3 900 m 的山群，有 27 座相对高度超过 1 000 m 的海山及 20 多座 400～1 000 m 的海丘，多为火山喷发的玄武岩山地，上覆珊瑚礁及沉积层。深海盆地底部平坦，坡度 0.3‰～0.4‰（中国科学院南海海洋研究所，1982）。盆地东北端与西南端有断裂谷形成的深水谷底，内有沉积物充填，谷底终端有海底沉积体，已受到日后隆起成为小山脊（图 7 和图 8）。

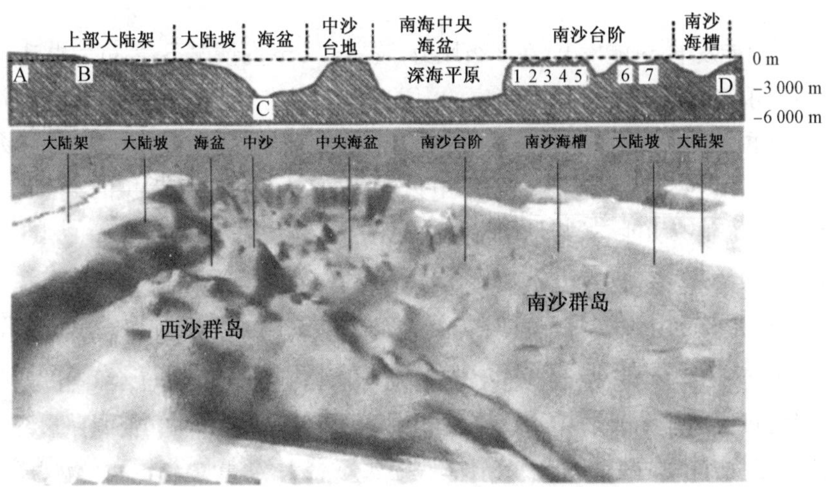

图 7　南海 114°20′E 海底数字地形模型及剖面图

A：香港东侧；B：一统暗沙东侧；C：西沙海槽东侧；D：文莱

1：双子群礁；2：中业群礁；3：道明群礁；4：郑和群礁；5：九章群礁；6：南华礁东侧；7：安渡礁

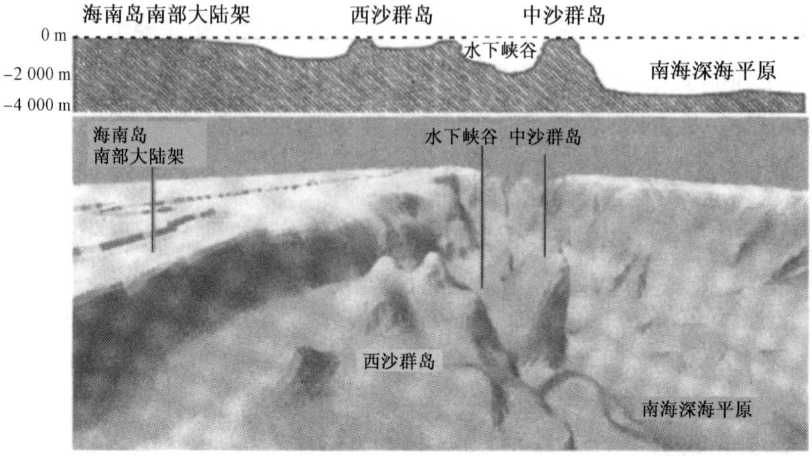

图 8　12°N,109°15′E 至 12°N,120°E 海底数字地形模型及剖面图

二、南海区位资源与重要性

(一) 最重要的是空间国土资源,海域疆界划分主要以陆地为依据

联合国海洋法的实施,在群岛海域形成“小岛大海洋”的新概念。一个人居的小岛,它可据有内水、12 海里领海、24 海里毗邻区及 200 海里专属经济区。因此,海洋中的岛、礁、滩、沙十分重要,即使暗沙,如果一年中可出露一次,也可作为基点依据,如黄海辐射沙脊的外磕角据此定为我国领海基线点。南海群岛应由国家立项,调查规划,渐次开发并纳入防卫体系,是根本大计。

(二) 海洋资源

1. 能源

潮汐与波浪能发电;风能发电,在 3 级风力,波高 0.2 m 的情况下,即可发电达 60 W/m^2;利用南海表层水与 1 000 多 m 深的海底的温差在 20°以上,海底地形起伏,在南海诸岛周围海域形成海底上升流,它不仅将海底丰富的氮磷营养盐带至上部海水形成渔场,而且可利用上升流冷水温差发电,供给南海诸岛电力,与常年的热带日照提供的丰富的太阳能发电相结合,为海岛提供动能,尤其是用以淡化海水,意义重大。

2. 油气与矿产资源

南海油气藏资源丰富:分布于珠江口盆地的原油资源估计约 40 亿～50 亿 t;莺歌海盆地原油约 4 亿～5 亿 t,天然气已进入开发;北部湾盆地油气资源总量 4 亿～5 亿 t;曾母盆地第三纪沉积层厚度 4 000～9 000 m,面积约 25 万 km^2,推测油气资源储量约 137 亿 t;万安滩油气资源估计超过 40 亿 t。南海海底还蕴藏着天然气水合物(可燃冰)等。

海岸带的砂矿资源:锆英石、独居石、铌钽铁矿、磷钇矿及石英砂等,在沿海岛沿岸海滩与近海水域分布,已进行开发。

3．渔业水产与药物资源

南海渔业资源近400万t，珍稀的热带海洋药物资源值得研究与开发保护。我国渔民在南海的捕鱼活动达2 000年之久，近来收到非法干涉与海匪骚扰，应对我国国民与资源保护采取有效措施。

发展海洋经济是我国在21世纪经济持续快速发展与增强国力的重要方面，应深入调查研究南海资源环境特点与开发途径；大力加强海空防卫力量，保卫海疆权益与人民安居。

三、中国的南海疆界

了解南海海底的地貌特点，有助于对海洋权益的维护（冯文科，1988），以提出有理、有力的依据。需清楚地认识到，西沙群岛不是位于现代大陆架海域，而是位于由古大陆架下沉折断形成的大陆坡顶；南沙群岛部分岛屿是位于巽它陆架上。岛屿是确定该海域归属的依据，因此，岛屿的归属权至关重要。历史的渊源与传统的疆界，国际公认的或沿用的图、文资料是重要的疆界直证，不能退让。

历史事件记载：1883年，德国曾在南沙和西沙进行测量，广东省政府抗议，德国停止了作业。1911年辛亥革命后，国际组织多次要求中国政府在南海诸岛上建立气象台、灯塔等航海设施，以保证航海安全，表明国际公认南沙、西沙等岛礁是中国领土。第二次世界大战期间，南海诸岛被日本侵占，1945年波茨坦公告明确规定日本侵占的中国领土（含南海诸岛）必须归还中国；战后1946年，中国政府接收西沙、南沙，在岛上升国旗、立主权碑，昭告世界，并派兵驻守南沙群岛的最大岛屿——太平岛，国际上对此皆予承认。1947年，中国政府正式公布了由内政部方域司勘定的《南海诸岛位置图》（民国内政部方域司，1947），图上画有11段以断续线形式标注的海域国界线，原件存于南京市的中国第二历史档案馆。以后出版的地图均遵照政府规定画有这些断续线，中国对南海诸岛的主权得到国际公认。1951年，英、美等国对日签订合约，条约明确规定日本放弃对台湾、澎湖列岛、西沙、南沙群岛的权利；会上南越代表提出要西沙、南沙的主权，这一无理要求遭到与会各国的摈弃。1952年，台湾当局以中国名义和日本签订和约，和约第二条为："兹承认依照公历1951年9月8日在美利坚合众国旧金山市签订之对日和平条约第二条，日本国业已放弃对台湾及澎湖列岛以及南沙群岛、西沙群岛之一切权利，权利名义与要求。"上述这些会议的参与国、签字国已承认西沙、南沙群岛是中国领土，日本也已将这些地方交还给中国政府。1953年，我国政府批准去掉北部湾2段断续线，改成9段，沿用至今，已历时半个多世纪。

20世纪70年代后期，随着对南海丰富自然资源认识的深入，越南、菲律宾、马来西亚和印度尼西亚等国纷纷提出对南海的主权的无理要求，抛出多种南海的分界线，造成南海国界问题的矛盾日益突出。由此，有必要对我国南海疆域给予确定性的认识。

对9段断续线的地理经纬度位置进一步核实工作是根据1947年民国政府颁布的《南海诸岛位置图》的原图复印件①和1983年我国正式出版的《南海诸岛》图（中国地图出版

① 南京大学杨怀仁教授参加当年内政部方域司的该项工作。

社，1983），采用数字化仿射交换，利用地理信息系统分别测量出来与订正后的各 9 条断续线的经纬度坐标，以度为单位，舍入到小数点后 3 位，获得具有较高的精确度的位置，具体数据与编号列表于后（表 2 和表 3）。

表 2　民国内政部制 1∶650 万《南海诸岛位置图》中的断续线位置（1947）

断续线编号	位置	端点位置(°)		长度，走向	起止范围(°)	
		起点	终点		东经	北纬
西 1(旧)	西沙群岛以西，平行于越南东岸，近土伦(顺化)	109.452 15.958	110.289 14.288	210.4 km SSE	109.452 110.289	15.958 14.288
西 2(旧)	南沙群岛以西，平行于越南南岸，西北为金兰湾	110.232 10.528	109.399 9.080	185.2 km SSW	110.232 109.399	10.528 9.080
南 1(旧)	南沙群岛西南，接近纳土纳群岛	108.895 6.053	109.686 4.585	186.5 km SSE	108.895 109.686	6.053 4.585
南 2(旧)	曾母暗沙以南，平行于婆罗洲(加里曼丹岛)北岸	111.134 3.725	113.007 4.232	230.8 km E，NEE	111.134 113.007	3.613 4.232
南 3(旧)	平行于婆罗洲北岸，拉布安岛西北	114.201 5.280	115.005 6.214	137.0 km NE	114.201 115.005	5.280 6.214
东 1(旧)	巴拉望岛西南，巴拉巴克海峡以西	116.076 7.447	116.906 8.414	141.2 km NE	116.076 116.906	7.447 8.414
东 2(旧)	南沙群岛东北，菲律宾布桑加岛、龟良岛以西	118.427 11.314	118.866 12.596	152.1 km NNE	118.427 118.866	11.314 12.596
东 3(旧)	黄岩岛东北，平行于吕宋岛西海岸	119.008 15.469	119.260 17.027	175.7 km N	119.008 119.260	15.469 17.027
东北 1(旧)	巴时海峡(巴士海峡)内，台湾七星岩与吕宋岛之间	119.845 19.690	120.975 20.184	131.5 km NEE	119.845 120.975	19.690 20.184

表 3　地图出版社 1∶600 万《南海诸岛》图中的断续线位置（1983 年）

断续线编号	位置	端点位置(°)		长度，走向	起止范围(°)	
		起点	终点		东经	北纬
西 1(新)	西沙群岛以西，平行于越南东海岸，近越南广东群岛	109.284 16.148	109.869 15.163	127.7 km SSE	109.284 109.869	15.163 16.148
西 2(新)	南沙群岛西北，平行于越南东南海岸，西北为越南槟绘湾	110.328 12.278	110.046 11.248	120.5 km S，SSW	110.046 110.328	11.248 12.278
南 1(新)	南沙群岛西南，纳土纳群岛正北方	108.247 7.108	108.322 6.003	124.2 km S，SSE	108.220 108.322	6.003 7.108
南 2(新)	曾母暗沙以南，亚西暗沙西北，平行于加里曼丹岛北岸	111.892 3.434	113.015 3.847	134.4 km NEE	111.892 113.015	3.434 3.847
南 3(新)	巴拉望岛西南，巴拉巴克海峡以西	115.623 7.171	116.326 7.993	119.9 km NE	115.623 116.326	7.171 7.993
东 1(新)	南沙群岛东北，菲律宾布桑加岛、龟良岛西南	118.572 10.918	119.029 12.013	132.0 km NNE	118.572 119.029	10.918 12.013

(续表)

断续线编号	位置	端点位置(°)		长度,走向	起止范围(°)	
		起点	终点		东经	北纬
东 2(新)	黄岩岛东、东北,接近吕宋岛西海岸	119.079 14.939	119.074 16.059	124.7 km N	119.054 119.079	14.939 16.059
东 3(新)	平行吕宋岛西北海岸,近北吕宋海槽	119.474 17.956	119.984 18.971	126.3 km NNE	119.474 119.984	17.956 18.971
东北 1(新)	巴士海峡内,台湾兰屿与吕宋岛之间	121.126 20.817	121.860 21.776	131.1 km NEE	121.126 121.860	20.817 21.776

对比测定的新、旧断续线经纬度示意图(图 9),发现新旧 9 条断续线的位置不完全重合,表现出三种位置相关:新、旧线是相互衔接,如西 1(旧)与西 1(新),东 3(旧)与东 2(新);新、旧线是相互平行,有外展,如东 2(旧)与东 1(新),南 2(新)与南 2(旧);有的新线在旧线的延伸线的位置上,如南 3(旧)与南 3(新)。总体上,旧断续线长度变化较大,在 130～210 km 之间,不规则;新断续线长度较短,约在 120～130 km 之间,长度变化不大,较为规则。对应编号的新旧断续线之间的距离从最接近的 18 km 左右(南 2 新旧之间),到相距最远的 333 km 左右(东 1 新旧之间)。

虽然各条新旧断续线之间存在空间位置上的少量差异,但重要的是,新旧断续线所界定的领土范围是基本一致的,订正后的断续线所表示的海域疆界更为合理。经计算得出,旧断续线以内的南海海域面积约为 192.7 万 km^2,新断续线以内的南海海域面积约为 202.5 万 km^2,误差小于 5%,显示出中国南海领土主权范围的历史延续性(许森安,2001)。

四、"数字南海"工作

随着对南海问题的重要性认识的不断深化,2004 年,由南京大学海洋研究中心、教育部南京大学海岸与海岛开发重点实验室与中国南海研究院合作,共同开展了"数字南海"项目的研究与建设工作,获得有关部门的大力支持,进展顺利。

"数字南海"是采用"数字地球"的信息化技术,将整个南海及其周边国家和地区的水深、海底地形、地质地貌结构,海洋波浪潮汐海流时空变化信息,资源矿产分布信息,以及历史、政治、经济、军事等信息,以数字化的三维空间数据库形式进行数据存储,以虚拟现实的方式通过计算机综合地表现出来,并进行相应的空间分析研究,以获得针对复杂南海问题的技术解决方案。

"数字南海"是我国目前正在大力发展的中国数字地球——"数字中国"的重要组成部分。"数字地球"是美国前副总统戈尔在 1998 年最先提出的国家信息化发展目标,我国对"数字地球"十分重视,江泽民同志自 1998 年以来曾两次在重要会议上提到数字地球问题,指出:"数字地球是知识经济时代,国家信息的基础设施。"提出建设"数字中国"的战略发展目标。"数字南海"正是实现这一目标的重要组成部分。"数字南海"系统包括大型的基于因特网的计算机网络硬件系统、大型空间数据库管理系统,以及三维空间虚拟现实系统等。

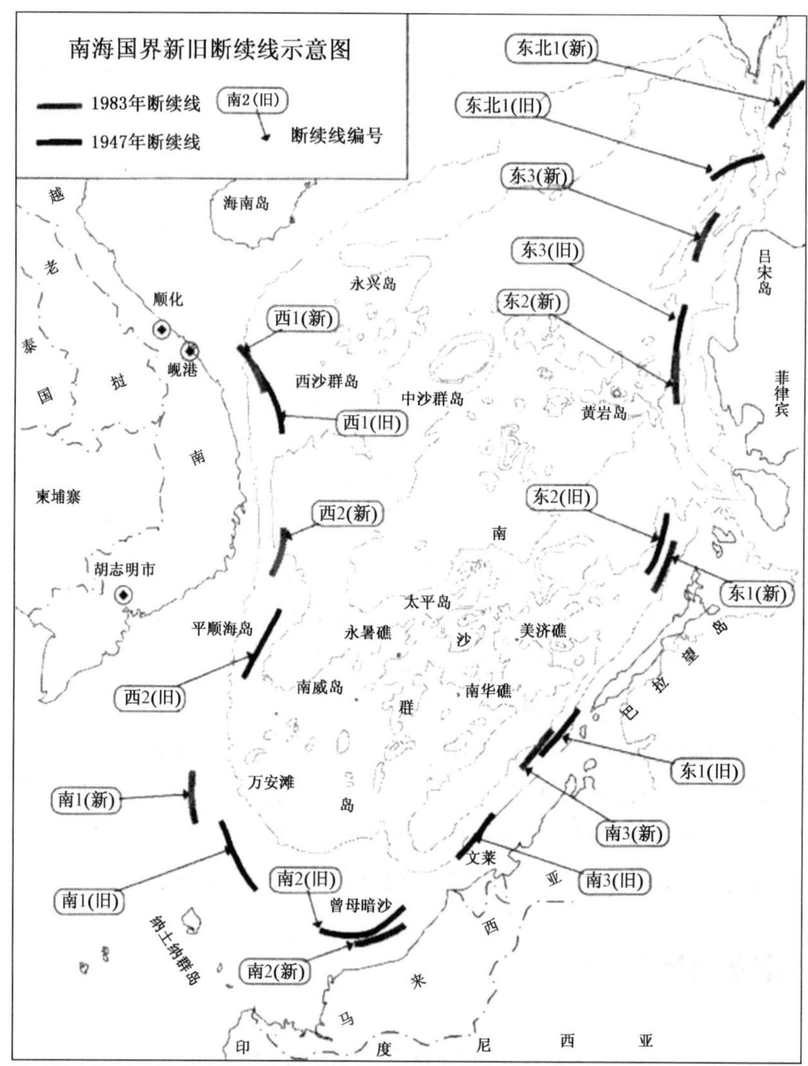

图 9　南海国界新旧断续线对比示意图

"数字南海"数字化存储的数据有 50 多种数据专题，几乎涵盖了南海的所有相关信息，其组成见表 4。

表 4　"数字南海"数据库包含的专题信息

自然概况	地质地貌数据库	南海数字地形数据库，南海地球物理数据库，南海岛礁与地理数据库
	气象气候数据库	南海气压带数据库，南海季风数据库，南海台风气旋活动数据库，南海气温数据库，南海相对湿度数据库，南海气压数据库，南海地表温度数据库，南海总辐射数据库，南海反射辐射数据库，南海光合有效辐射数据库，南海紫外辐射数据库，南海风速数据库，南海风向数据库，南海海域海面风场数据库，南海降雨量分布数据库
	海水理化环境数据库	南海海平面数字化高度数据库，南海海水温度数据库，南海海水颜色数据库，南海海水盐度数据库，南海海水浮力数据库，南海海水冰点数据库，南海海水声音数据库，南海海洋热量交换数据库，南海海水透明度数据库，南海海水密度数据库，南海水质数据库（含 SAL、pH、DO、DO_SAT、PO_4、SiO_2、CHL_A、TIN、$TINoPO_4$）

（续表）

	自然地理数据库	南海岛礁土壤分布数据库，南海岛礁植被分布数据库，南海灾害分布数据库（南海台风灾害数据库，南海地震灾害分布数据库，南海火山灾害分布数据库，南海风暴潮灾害数据库，南海台风浪灾害分布数据库，南海赤潮灾害分布数据库），南海潮汐类型分布数据库
	资源数据库	南海国土资源数据库，南海生物资源数据库，南海油气资源数据库，南海矿产资源数据库，南海太阳能资源数据库，南海风能资源数据库，旅游资源数据库，南海周边浅海滩涂数据库，南海波能数据库，南海温差能数据库，南海海洋化学资源数据库，南海地下水资源数据库
	环境保护数据库	南海海洋生态圈保护数据库，南海海岸生态圈保护数据库
社会经济	交通状况数据库	南海位置数据库，南海水域范围数据库，南海港口数据库，南海交通航线航道数据库，南海航空通道数据库
	岛礁名称数据库	我国南海古代历史地名数据库，现代南海地名数据库，我国南海地名与国外地名对照数据库
政治历史库	相邻国家和地区数据库	国家地区概况资料数据库，双边往来资料数据库，就南海问题我国发言人谈话资料数据库，两国声明、公报资料数据库，我驻该国外交机构、该国驻我国外交机构情况数据库，其他问题资料数据库
	历史资料数据库	南海史地古籍与史料，南海历史地图数据库
	南海问题资料数据库	南海问题的主权论述资料数据库，相关法律资料数据库

“数字南海”系统将整个南海数字化搬进政府决策部门及专门科研机构的实验室，供实地分析研究决策，不仅可以提供静态、动态立体地理景观、卫星遥感影像、虚拟现实仿真图像、数字地图等数字化产品，还可以通过网络直接提供联机信息服务，对我国南海问题的研究、南海的开放管理和南海主权的维护等将起到十分积极的作用。

参考文献

[1] 冯文科. 1988. 南海北部晚第四纪地质环境. 广州：广东科技出版社. 46 - 68.

[2] 罗章仁，陈史坚，郑天祥. 1996. 南海. 见：王颖主编. 中国海洋地理. 北京：科学出版社. 411 - 452.

[3] 民国内政部方域司. 1947. 南海诸岛位置图.

[4] 钱翼鹏. 1982. 南海北部热流测量及其成果. 海洋地质研究，2(4)：102 - 107.

[5] 王颖，蔡明理. 1996. 中国海海底地质与地貌. 见：王颖主编. 中国海洋地理. 北京：科学出版社. 45 - 52.

[6] 许森安. 2001. 南海断续国界线的内涵. 见：“21 世纪的南海：问题与前瞻”学术研讨会论文选. 77 - 83.

[7] 曾成开. 1987. 南海海盆中的海山海丘及其成因. 东海海洋，5(1～2)：1 - 9.

[8] 中国地图出版社. 1983. 南海诸岛. 北京：中国地图出版社.

[9] 中国科学院南海海洋研究所. 1982. 南海海区综合调查研究报告(一)、(二). 北京：科学出版社.

论证南海海疆国界线*

1　引　言

长期以来，中国南海诸岛外围海域之界限被称为断续线，但是对它所代表的内涵，却有着不同的理解。不同的理解，事实上是对这一界线在历史上所反映的疆域权属内涵认识的不一致，而这些不一致的解释和理解，对海洋划界造成了很大的困惑。为探索这一界线的起源，厘清其在历史上的客观定位，以便了解其初始的本质含义，作者对第二次世界大战以后的历史地图及文献记载进行了系统研究，以明确断续线的真实历史含义，以期为南海划界提供科学依据。

2　历史图件数据

为探索这一界线的本质含义，作者查阅了在台北、南京所保存的中华民国时期的图文原件档案《南海诸岛位置图》(6 幅)[1](表 1)，这组图是抗日战争胜利后，国民政府派军队完成对南海诸岛接收并宣示主权后由内政部方域司绘制的，于 1946 年 12 月完成测绘。据民国档案文[(卅六)四内 30844]：是于民国卅六年(1947 年)8 月 7 日经行政院院长张群批准签发。1947 年 11 月上旬《大公报》报道了"南海诸岛内部名称核定"；1947 年 12 月 1 日，内政部方域司公布了《南海诸岛位置图》，在《中央日报》上刊登了南海诸岛名称。《南海诸岛位置图》收录在《南海诸岛地理志略》一书中[2]，并作为《中华民国行政区域图》附图于 1948 年 2 月公开出版发行[3]。

表 1　南海诸岛位置图(6 幅)[1]

图名	比例尺	绘制单位和时间	备注
南海诸岛位置	1∶4 000 000	国民政府内政部方域司，1946	
南沙群岛	1∶2 000 000	国民政府内政部方域司，1946	
中沙群岛	1∶350 000	国民政府内政部方域司，1946	
西沙群岛	1∶350 000	国民政府内政部方域司，1946	
西沙群岛永兴岛及石岛	1∶10 000	国民政府内政部方域司，1946	实测图
南沙群岛太平岛	1∶5 000	国民政府内政部方域司，1946	实测图

* 王颖，葛晨东，邹欣庆：《海洋学报》，2014 年第 36 卷第 10 期，第 1－11 页。

3 分析结果与讨论

在南海诸岛位置图中，明确地标注出与周边各国的界限。符号是“—〈■〉—〈■〉—〈■〉—〈■〉—”，它从图的西北端开始，划分出我国广西省与越南之间的边界，再经中南半岛东侧与西沙、南沙群岛之间，南沙群岛与加里曼丹岛及菲律宾吕宋岛之间以及台湾岛以南与巴士海峡间的分界线。在 1∶4 000 000 地图上标出 10 段分界线(见图 1)作为我国南

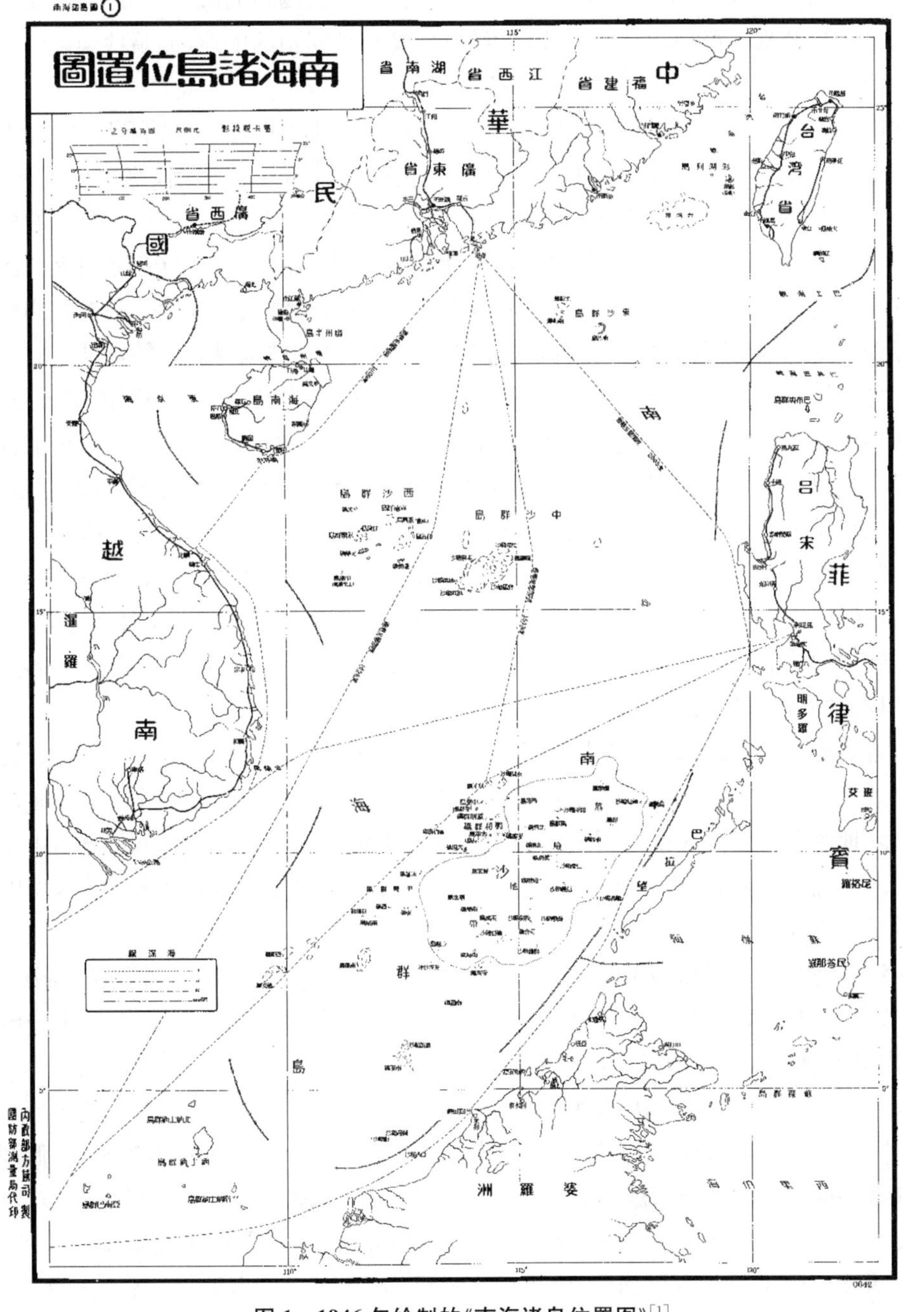

图 1 1946 年绘制的《南海诸岛位置图》[1]

海疆界线。该图对界线的划法、数量与位置均有严格的规定，采用了海域中间线原则划线。图 1 中，未附全部图例的诠释，但从该幅图中界线的划分与连续位置，可明确其为海域国界线。最有力的证据是在图幅的西北端，海域界线与陆域国界相衔接成为一段两国间之界限，该线段东北侧是我国的广西，而该线段以南是越南的同登与谅山，明确标出该符号线段所代表的为国界线（见图 2）。证据之二，在菲律宾巴拉望岛南端与马来西亚所属的婆罗洲（今加里曼丹岛）之间是以与中、越间界限相同的符号明确地划分出两国界线（见图 3）。而且，上述两段线均居于该处两国间之中线部位，至今不变。

除上述两处之外，其他部分也以该线段符号标注出我国与周边国之分界：越南东岸与我国西沙群岛及南沙群岛间距之中线位置划分，岛与岛间距的中线，或暗沙与陆地外缘之中间线等，界线的划分均是一致地，遵循明确的原则。

在南京大学地理与海洋科学学院图书馆查到 1948 年由商务印书馆刊印发行的《中华民国行政区域图》[3]（比例尺为 1：4 200 000），由内政部方域司编制，傅角今主编（见图 4），该图的右下角附有《南海诸岛位置图》（比例尺为 1：9 100 000，图 5）。图中的疆界线是以“—〈■〉—〈■〉”标出，在“图例”中明确地注释为国界（见图 6），并与未定国界的图例的点线（……）有明显的区别。因此，可以清楚无误地判定，南海周边线段是我国在南海的海疆国界线，海疆国界线与陆域国界线是连接在一起的，是陆域国界向海的延伸。它是我国南海诸岛与周边诸国在海域的国界线，区分了我国岛屿海疆与周边各国间的疆界范围。

在表 1 南沙群岛图幅中，包含着南沙全部岛屿，比例尺为 1：2 000 000，其东、西、南三边界限也以直线间夹括弧与内之方块点“—〈■〉—〈■〉—”组成，也有十段线（见图 7）。不同比例尺图上的海疆范围虽有大小不同，但在图上的我国南海海域疆界却均以十段表示（见图 1、图 7），在正式公布的 1：6 500 000《南海诸岛位置略图》[2]和 1：9 100 000《南海诸岛位置图》[3]（见图 5）上，与菲律宾西侧相毗邻的国界线分为两段，则总数为十一段。因此，南海海疆国界线的称谓不宜以数字表示。目前南海周围之界线在去掉北部湾界线后常被称为“九段线”，似不恰当。

作者认识到，南海海疆国界线以不连接的线段（断续线段）标注在图上，海水是波动的，以“线段”表示海域疆界比连续的实线更符合实际，而且具有允许船只无害通过之属性。国际地图上海疆界线也以断续线段表示，如：The New Canadian Oxford Atlas[4]中，在地中海中，意大利与西班牙岛屿的海疆界线，土耳其与希腊之间的海疆界线，均以断续线段表示。因此，海疆国界线以断续线段表示是国际惯例。

西沙群岛与中沙群岛两图幅，比例尺均为 1：350 000，西沙群岛图幅中有图例表明珊瑚礁、沙滨、阔叶林，等深线，水深点与破船遗迹。中沙群岛周边以 20 m 等深线分出岛周轮廓，内有实测水深点以及各暗沙与礁滩位置。

永兴岛及石岛图幅，比例尺均为 1：10 000，为实测地图，标注：“中华民国卅五年（1946）年 11 月 30 日测绘”，岛屿与码头清晰绘出，标出永兴岛与石岛位于同一珊瑚礁上，低潮时礁滩出露，高潮时礁边有浪花（激浪）。该图明确标出当年“永兴岛面积为 1 851 000 方公尺（1 方公尺为 1 m^2），合 7.40 方市里（1 方市里为 0.25 km^2），地势平坦，西南隅岸边有沙丘、拔海最高达 8.3 公尺”。“石岛面积 78 000 方公尺，合 0.312 方市里，最高处拔海 12.4 公尺”。当 2013 年 5 月 29 日我们登岛观察时，因林木丛生，已不见沙丘原貌。再以 GPS 定

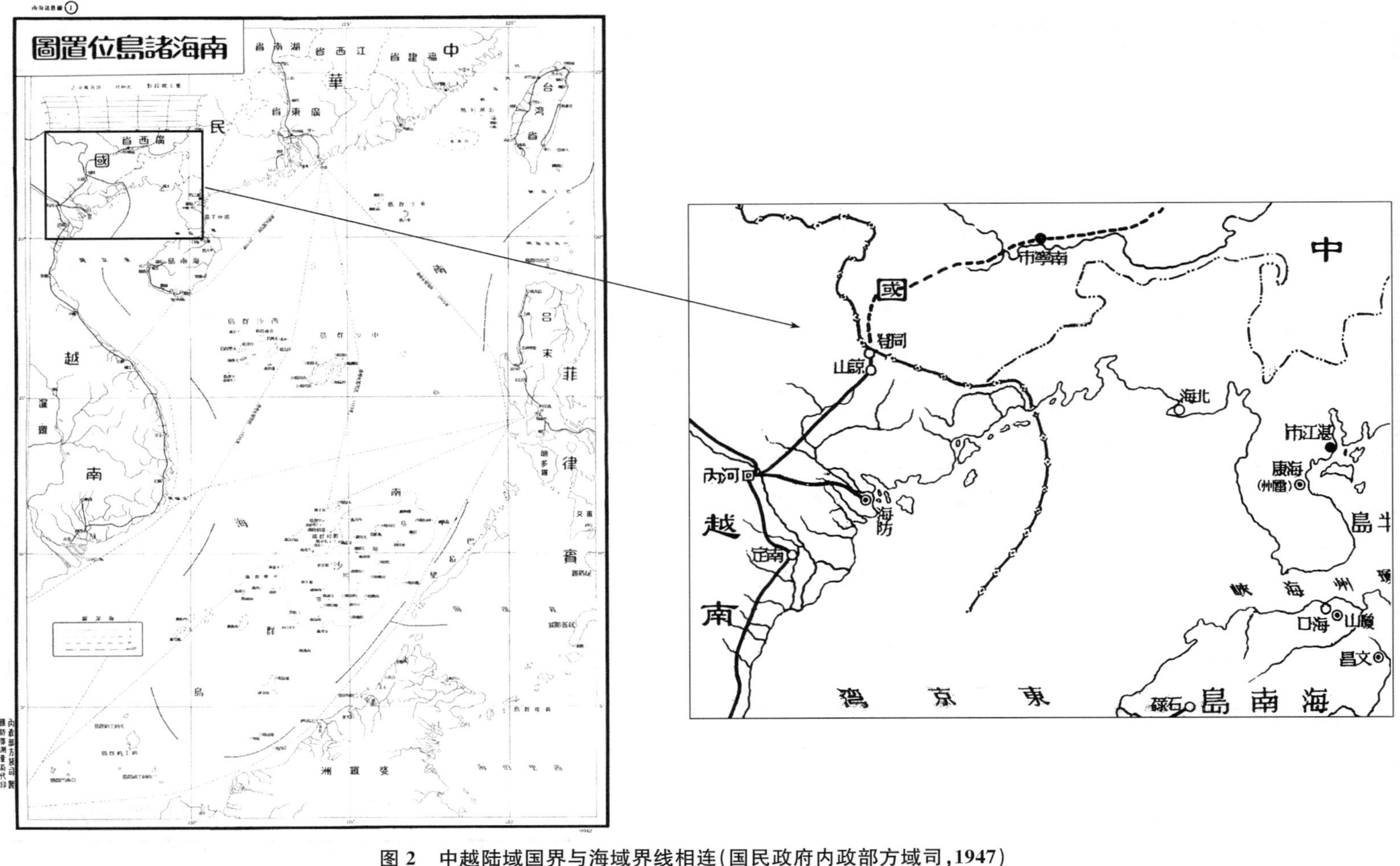

图 2　中越陆域国界与海域界线相连(国民政府内政部方域司,1947)

(海域界线符号与陆域国界符号相同)

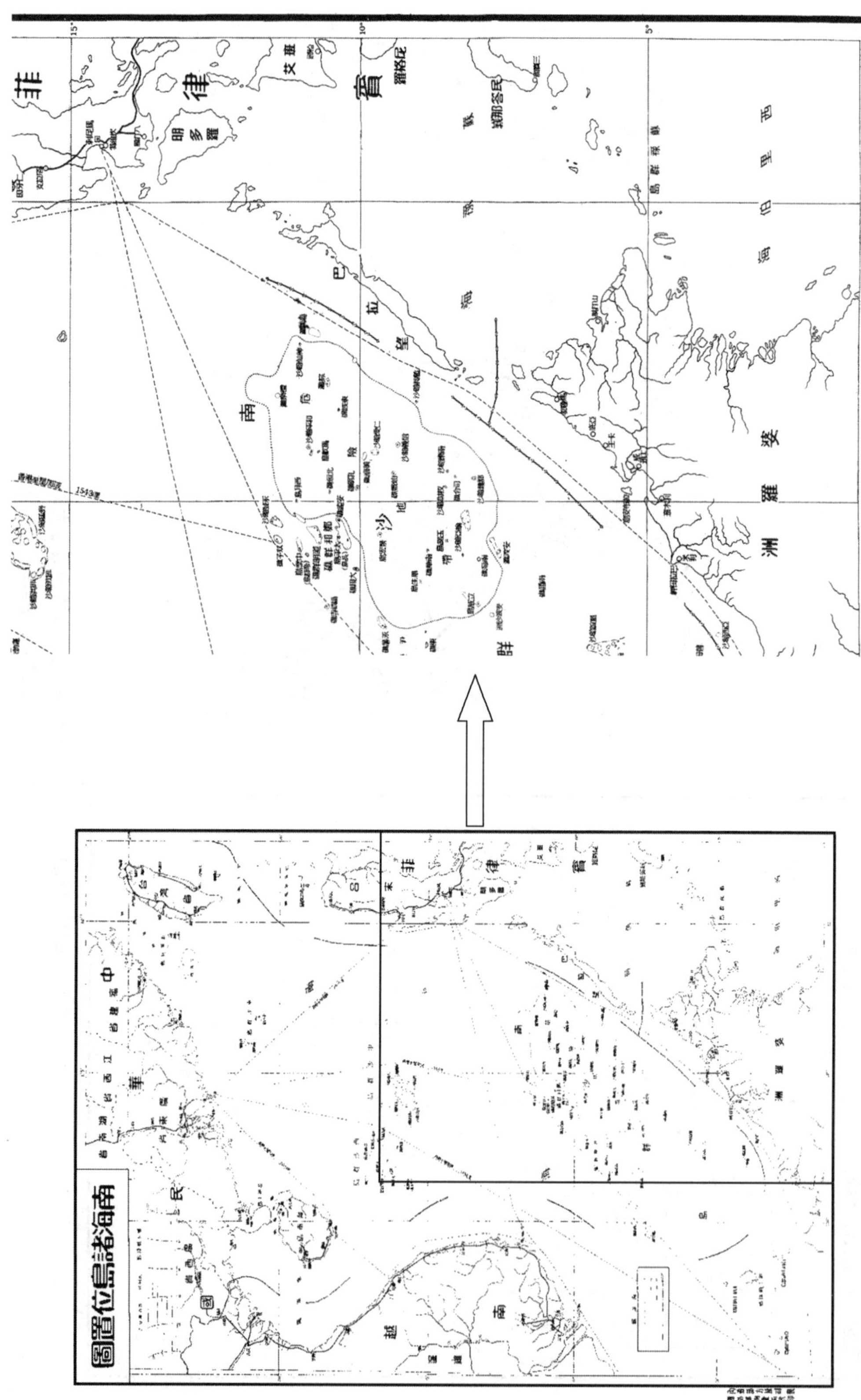

图 3　巴拉望岛与加里曼丹岛间的国界线

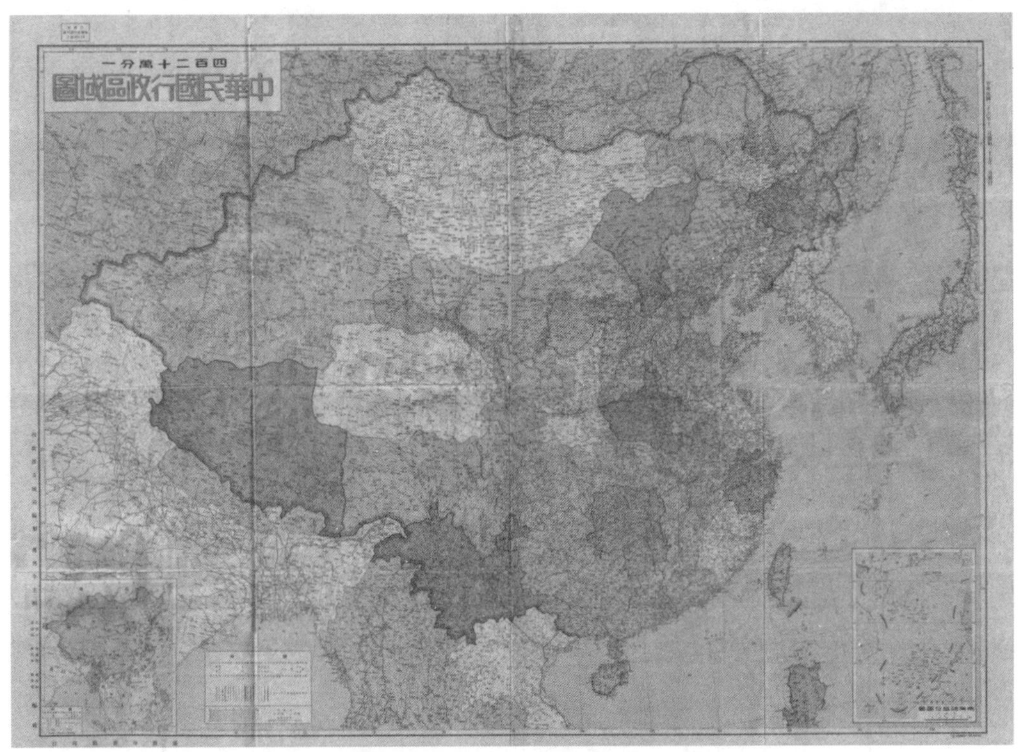

图 4　1948 年出版的《中华民国行政区域图》[3]

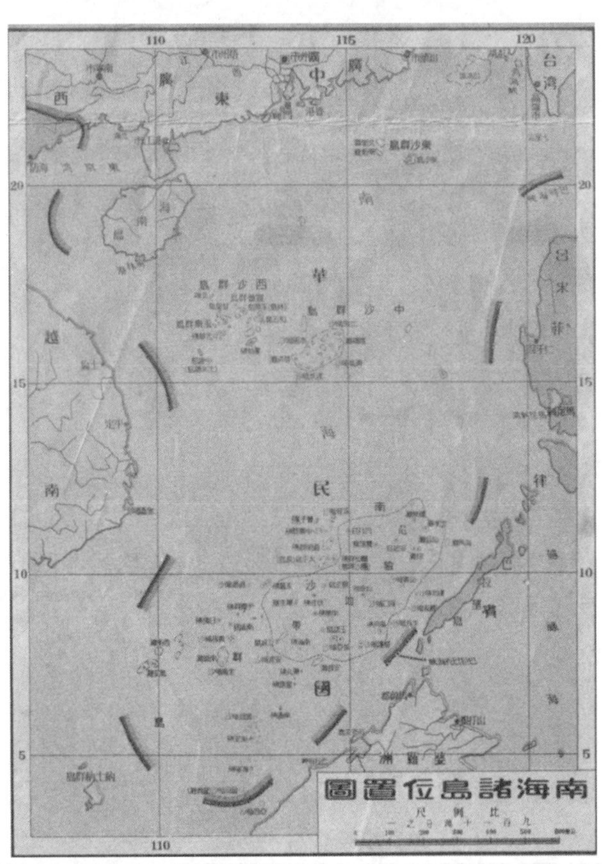

图 5　《中华民国行政区域图》中的《南海诸岛位置图》[3]（原图比例尺为 1∶9 100 000）

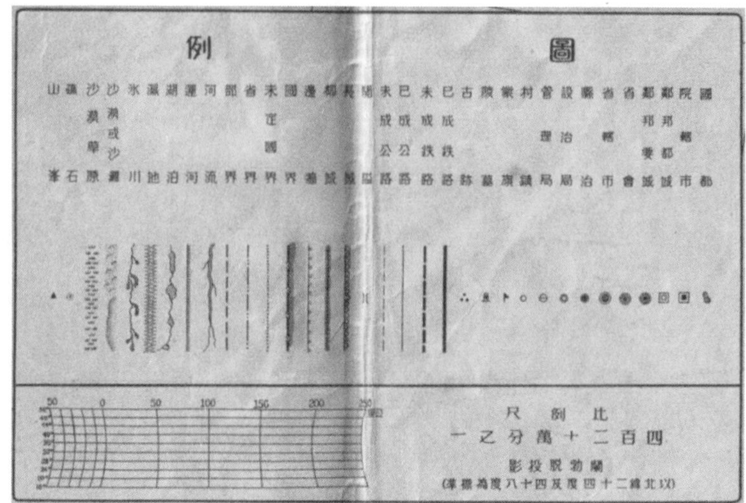

图 6　《中华民国行政区域图》中的“图例”[3]

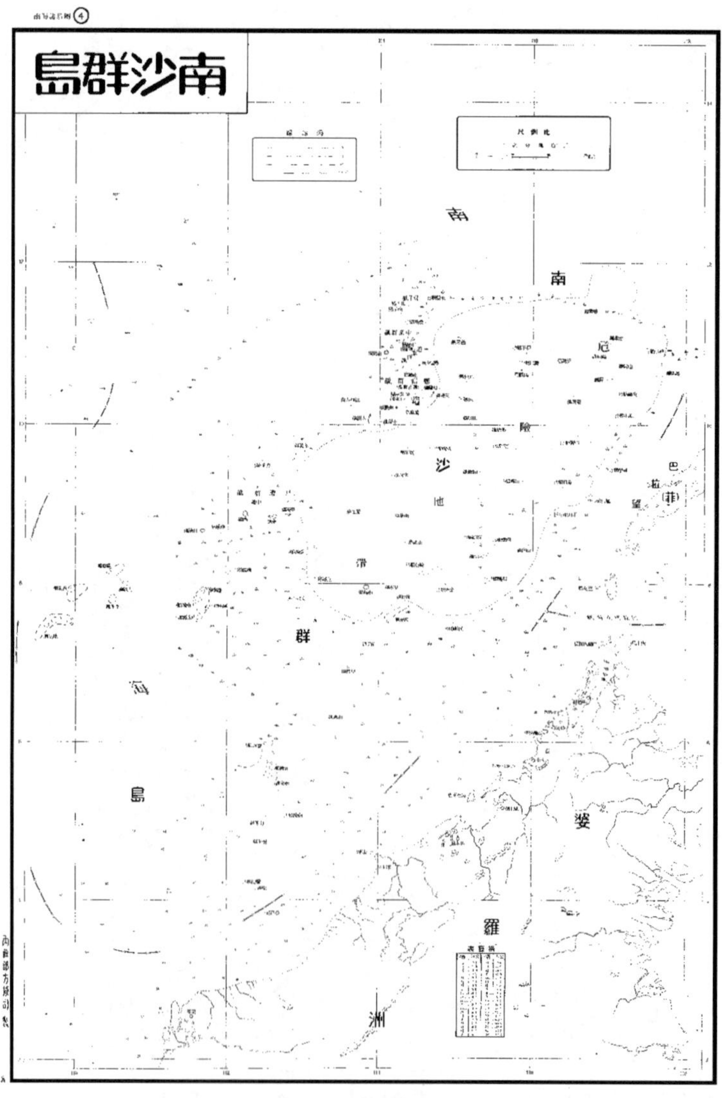

图 7　南沙群岛[1]

位，测得石岛最高处为 15.9 m，而现代文献中标出为 13.8 m。若有现代高度的精确计量，则可知其近 60 多年来地貌发展的变化。原图件中用“拔海”术语形容高出海面之高度，引起我们关注，认为中文之“拔海”比“海拔”更贴切其内涵。后人为何将“拔海”改称为“海拔”不得其解?!

太平岛图比例尺为 1∶5 000，是实测的，图上附注：“一、太平岛面积为 432 000 方公尺合 173 方市里地势平坦拔海 4 公尺(1 公尺为 1 m)；二、岛内建筑物俱已毁坏。”图例有：“房屋(钢骨水泥、木屋、屋基)、碑、电台、旗杆、水井、蓄水池、坟墓、防空洞、树木(椰树、阔叶树林、独立树)、垒石圈、大路、小径、轻便铁路、沙滨、沉船、岩石、珊瑚礁”。图例中还注出“该岛曾有驻军及居民，但测图时建筑已毁”。附注三是：“沿岛周边有珊瑚礁低潮时显露，高潮时岩礁边见浪花”。图框外左上角注明“中华民国三十五(1946)年十二月十二日测绘”。

1949 年中华人民共和国成立后，中国出版的各种地图中有关南海海疆国界线的画法，基本上仍然沿袭了 1948 年当时中国政府公布的地图画法，由华夏史地学社编制、亚光舆地学社 1951 年发行的《中华人民共和国新地图》[5](见图 8)，在其右下角《南海诸岛图》中，所划出的疆界仍同于 1948 年出版的图(见图 9)，而且，在左下角“图例”中注明该图例代表国界(见图 10)。1948 年出版的图[3]中图例的国界线符号与 1979 年由地图出版社出版的《中华人民共和国地图集》[6]中图例的国界线符号相似，与 1991 年由地图出版社和瑞典埃塞尔特地图公司编绘出版的《埃塞尔特世界地图集》[7]所采用的中国地图疆界符号也相似。但是，在这两图册中均以—[■]—[■]—[■]，即括弧中有一方块点表示中国疆界。与 1946 年图中括弧形式稍有区别，但均为括弧。作者对民国政府内政部方域司绘制

图 8　《中华人民共和国新地图》[5](1951 年，原图比例尺为 1∶8 500 000)

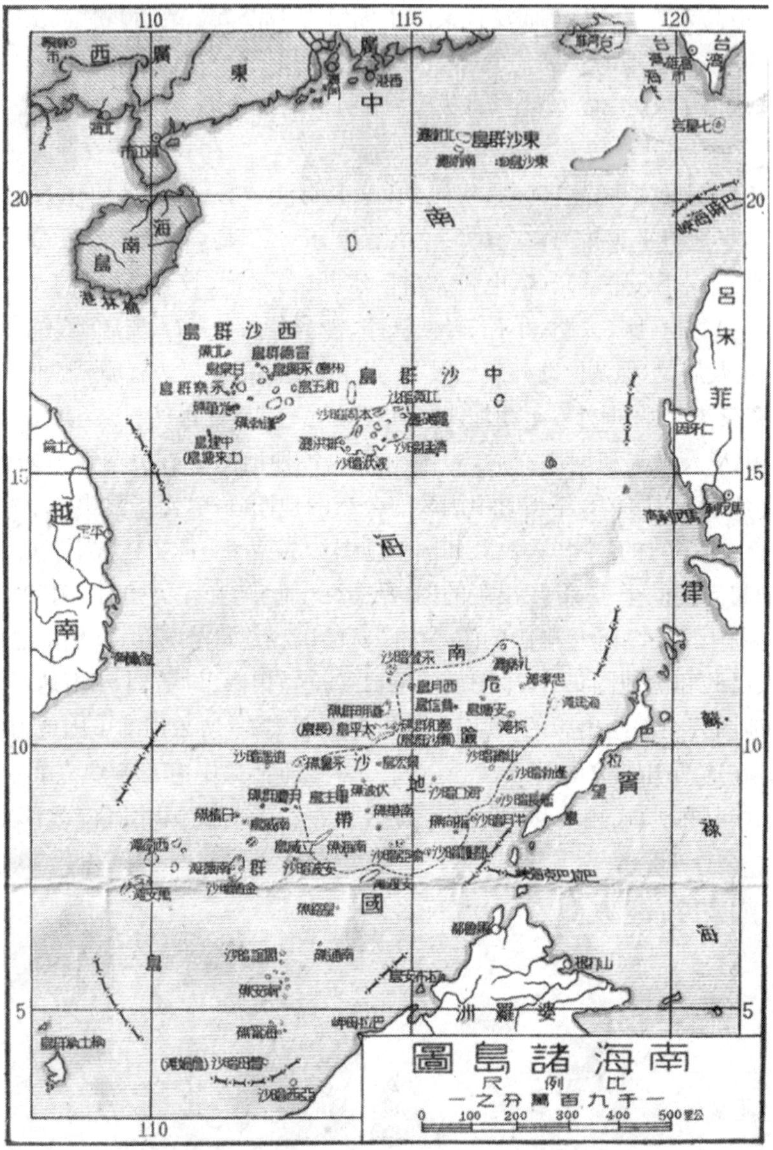

图 9　《中华人民共和国新地图》中的《南海诸岛图》[5]（1951 年，原图比例尺为 1∶19 000 000）

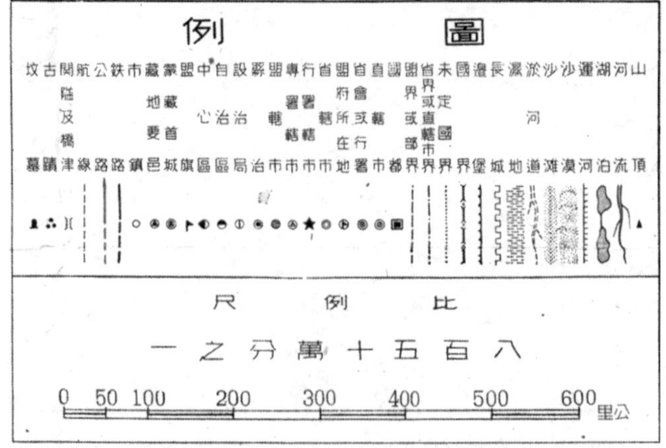

图 10　《中华人民共和国新地图》中的“图例”[5]（1951 年）

的《南海诸岛位置图》中南海海疆国界线的地理位置与空间范围进行了详细的量计，并与1982年地图出版社出版的《南海诸岛图》进行对比，所界定的海域范围基本一致[8]，认识到界限的位置具有严格的依据。

同时，在查阅南海诸岛的相关档案文献时，检索到在1930年1月至1931年10月期间，当时的民国政府已制定了《规定领海界线草案》[9]，开宗明义地指出："领海与领土同为国家主权与法律之所及在对外关系上有关于战时中立权利及平时渔业权利之处颇钜在军事上与海军区区域之划分其相联系之处亦复不少我国对于领海尚无明确规定故事不可缓"（原文中文句均不含标点符号，照录，余同），可获知当年划定南海海疆国界线时所遵循的原则。

在文件的第二部分包含了分析各国领海之例："英美法德及日本为三海里斯堪的那维亚诸国（瑞典挪威）为四海里葡萄牙为六海里苏俄为十二海里"。

该文件第三部分分析领海大小之利害（略）。

第四部分明确提出：我国领海宜制定为十二海里。

第五部分是计算领海之准绳：

（一）计算领海以十二海里，自沿岸之低潮点往外起算

（二）一海湾全在我国海岸线范围以内之领海界

（甲）凡海湾湾口宽不逾24海里者则于湾口对直引以直线由此直线往外起算

（乙）凡海湾湾口宽逾24海里者则择靠湾口最近之一点构成宽口不逾十二海里之处于此对直引线往外起算

（丙）凡海湾得视为我国完全所有时则不问该海湾湾口宽度为何概于湾口封直引线由此线往外起算

（三）一海湾虽在我国海岸线惟兼与他国海岸交界时之领海

凡以海湾湾口或一河之河口宽不逾十二海里为我国与他国交界时计算领海应彼此各由已方海岸之低潮点起算按半而平分之。

（四）港湾　自该湾靠外设备之最终点起对引直线起算该处领海海里数

（五）岛　我国沿海岛屿有单独之领海

（甲）凡众岛成群全属我国如其外周所有各岛相去之距离不逾我领海海里数之一倍者则该处领海自其岛群最靠外之岛往外起算

（乙）凡括于岛群之内之海皆为我国领海

（丙）各岛之距大陆海岸或群岛之两端之岛与大陆之海岸相距其距离不逾领海海里数之一倍者则所包括之海皆为我国领海

（六）海峡　凡海峡之沿岸皆属我国且其入口宽不逾我领海海里数之一倍者则峡内全部为我国领海

凡海峡之沿岸为我国与他国海峡宽不逾领海海里数之一倍者则各以自由（有）国自辖之中线为彼我国之领海范围

凡通联我国内水之海峡其起算领海得准用前述关于海湾领海起算法之规定

凡河川以其直接入海之处为其内水终点并不问其入海之处所有内水面积之宽若干其领海起算概准用前述关于海湾领海起算法之规定

若一河川出口系在一哈夫 Half 者(哈夫系一形狭长而积水不深之浅湾其外面隔堤中穿缺口以与外海相连)其领海起算概准用前述关于海湾领海起算法之规定”。

《规定领海界线草案》是附于行政院决议案(行字第 3606 号)之后,议案文为:“行政院为国务会议决议领海界限拟定为十二海里勘界事宜交海军部办理事给国民政府呈及给内政外交财政海军实业部指令(指令第六八三号),经行政院院长蒋中正副院长宋子文会签批准于民国廿年二月廿六日(1931 年 2 月 26 日)发文。”

4 结 论

综上所述,历史图文档案表明中国对海疆界限早在 1930—1931 年间已有研究与建议,并在 1931 年 2 月经国民政府行政院批准与实施勘测。日本侵华战争使海域勘界中断,但是在第二次世界大战后,民国政府在 1946—1947 年间在完成对南海诸岛接收时依据法规实地勘定南海海疆界限,并正式对外公布。南海周边线段是我国在南海的海疆国界线,海疆国界线与陆域国界线是连接在一起的,是陆域国界向海的延伸。它是中国南海诸岛与周边诸国在海域的国界线,区分了中国岛屿海疆与周边各国间的疆界范围,并以国际惯例的断续线段标明在南海的中国海疆国界线上。在国际上的众多条约、外交会谈、出版的图集及其他正式出版物中,均获认可。南海海疆国界线的称谓不宜以数字表示,目前南海周围之界线在去掉北部湾界线后常被称为九段线,似不恰当。厘清这一事实,对于南海划界具有科学意义。

参考文献

[1] 国民政府内政部方域司. 南海诸岛位置图(六幅). 南京第二档案馆,1947.
[2] 郑资约. 南海诸岛地理志略[M]. 上海:商务印书馆,1947.
[3] 傅角今. 中华民国行政区域图[M]. 上海:商务印书馆,1948.
[4] The New Canadian Oxford Atlas[M]. Canada:Oxford University Press, 1977.
[5] 华夏史地学社编制. 中华人民共和国新地图(1951 年 6 月增订再版)[M]. 亚光舆地学社,1951.
[6] 中华人民共和国地图集[M]. 北京:地图出版社,1979.
[7] 瑞典埃塞尔特地图公司编制. 埃塞尔特世界地图集[M]. 北京:中国地图出版社翻译出版,1991.
[8] 王颖,马劲松. 南海海底特征、疆域与数字南海[M]. //苏纪兰主编. 郑和下西洋的回顾与思考. 北京:科学出版社,2006:177 - 193.
[9] 规定领海界线草案. 中华民国行政院决议案(行字第 3606 号),1931.

我国海疆权益争端辑要*

中国海界于亚洲与太平洋之间，大陆岸线 18 000 km，加上 7 000 多个岛屿，岸线长达 34 000 km。渤海是内海，濒临黄海、东海、南海与台湾以东太平洋岸的海域面积超过 3 000 000 km^2。除内海外，各海域均存在着海疆纷争问题。

一、黄海

黄海是“海湾式”半封闭的大陆架海，西、北、东三侧分别临中国、韩国及朝鲜陆域，海域岛界分明。与朝鲜在鸭绿江河口(图 1)界限及岛屿有主权争议，江口几乎全归朝鲜，我国忍让和解使纷争矛盾未突出。

图 1　鸭绿江口(来源:Google Earth)(附彩图)

苏岩礁(Suyan Islet，125°10′56.81″E，32°07′22.63″N)，位于东海大陆架上，在中国专属经济区内，距海面最浅处 4.6m。韩国从 2000 年下半年到 2003 年 6 月，以岩礁为基础建立了全钢结构塔楼，高 76m(水上 36m，水下 40 m)，由 15 人连续观测。2001 年韩国地质学院将苏岩礁非法命名为“离於岛(Leodo)”，人工塔台命名为“综合海洋科学基地”(图 2 和图 3)，实属违权。按:1963 年 4 月 30 日，我国 15 930 t 巨轮跃进号自青岛首航前往名

* 王颖:2013 年 5 月 17 日南京大学地理与海洋科学学院举办的南京大学“5・20”南京大校庆学术报告会专题报告。

王颖，傅光翩:沈固朝主编《南海经纬》，009 - 010 页，2016 年，南京大学出版社。经作者校改用原稿。

古屋，于 5 月 1 日凌晨误经苏岩礁，在 32°6′N，125°11′42″E 处触礁，曾力图撤离，却使船底“破膛”沉没。当年周恩来总理亲自指挥现场勘察，对该海域测量与海底制图，应是归属的明证。我国于 2000 年和 2002 年两次抗议，反对韩方在两国专属经济区重叠海域单方活动。我国现已加强全面推进海域遥感监视与监测，包括黄岩岛、钓鱼岛、苏岩礁以及西沙、中沙和南沙群岛附近海域在内的中国全部管辖海域的综合管控。

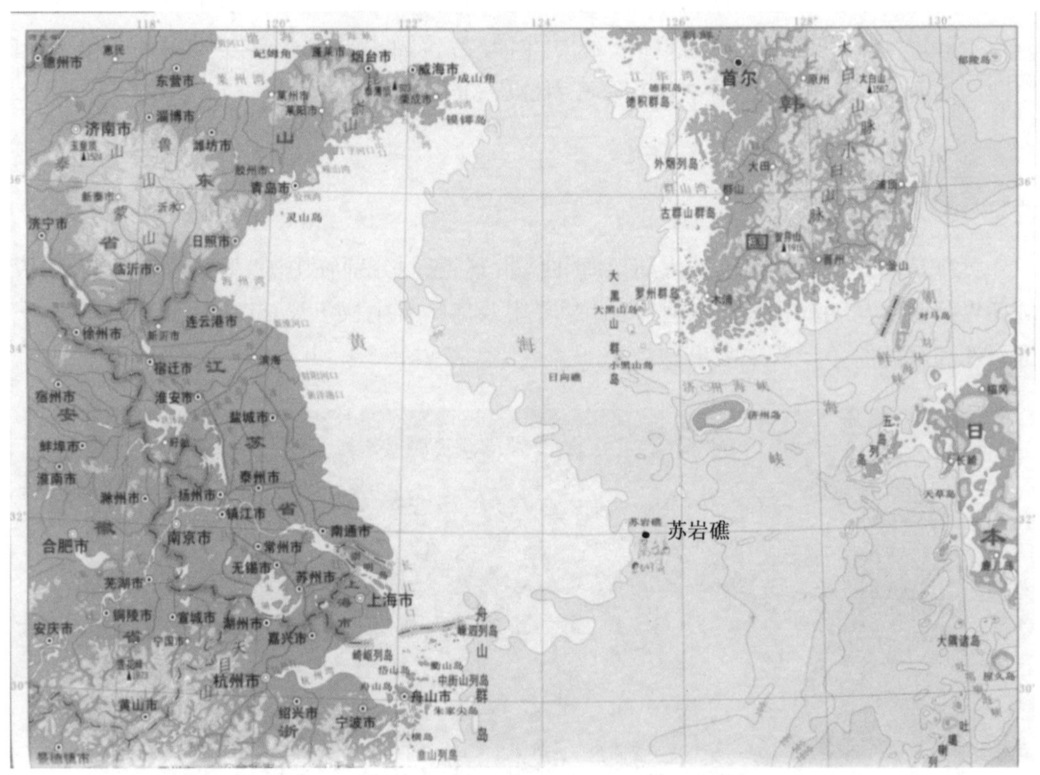

图 2　苏岩礁位置图(附彩图)

(图片来源：国家海洋局，中国海岛(礁)名录，海洋出版社，2012)

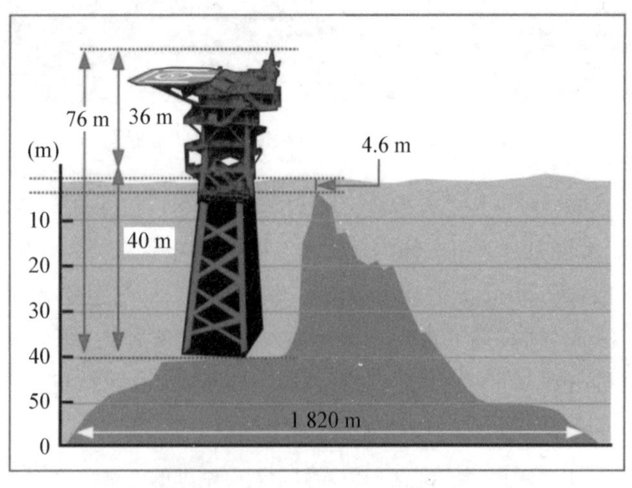

图 3　韩国在苏岩礁违建的“离於岛”人工塔台

(图片来源：图片全球网，http://www. tpqq. com/zhongguotupian/20121007/3346_3. html)

二、东海

东海绝大部分为陆架海，从我国大陆外延，陆架东临冲绳海槽，其东南侧为钓鱼岛主岛。历史传统与依据国际法大陆架条例，东海陆架应归属我国，但陆架前缘钓鱼岛归属，却存在着中、日双边争论。

(1) 钓鱼岛及其附属岛屿(钓鱼台列岛)位于东海大陆架东南边缘，由黄尾屿、赤尾屿、钓鱼岛、飞濑岛、大北小岛、北小岛、大南小岛与南小岛，以及数处暗礁所组成。范围界于 124°35′E 至 123°20′E，25°44′N 至 25°56′N，全部面积约 6.3 km²。钓鱼列岛位置在台湾岛北端的东北方向，距基隆市 120 nmile。与日本冲绳那霸市相距 250 nmile，与琉球群岛以冲绳海槽分隔(图 4)。

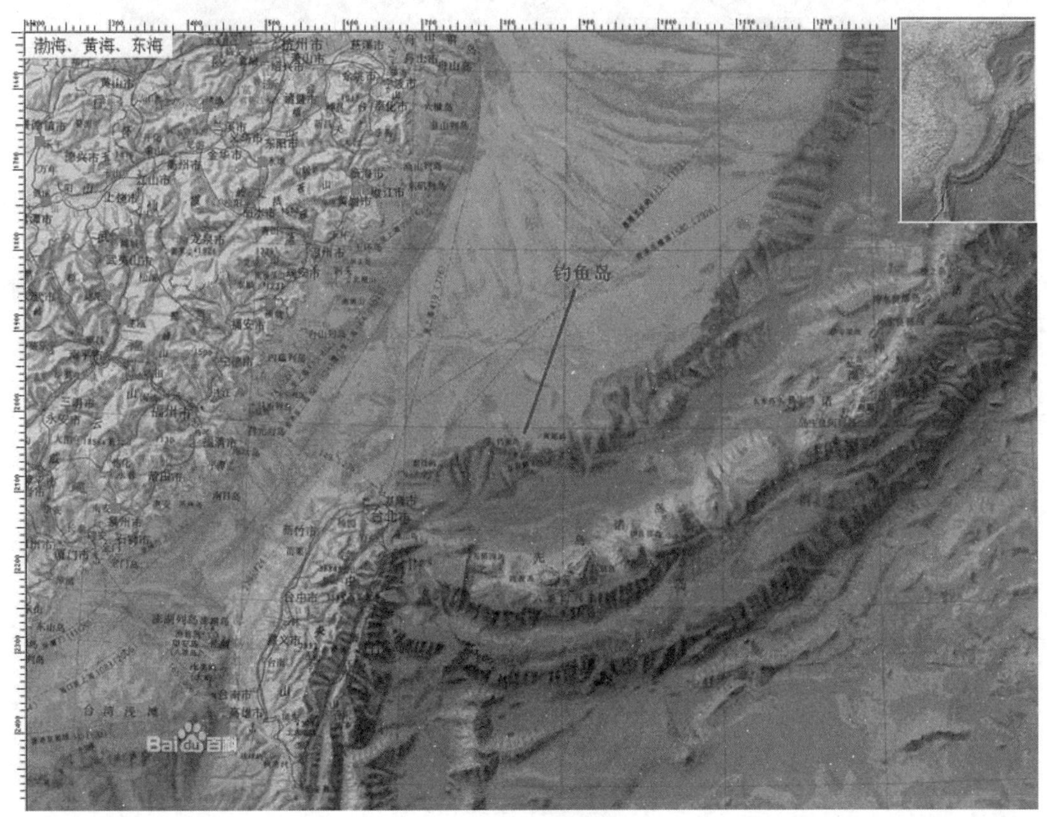

图 4　钓鱼岛位置图

(图片来源：百度百科，http://baike.baidu.com/subview/2876/6475776.htm? fr=aladdin)

钓鱼岛主岛 NW－SE 向延长，面积 4.3km²，南向为悬崖陡坡，北侧为倾斜坡(图 5)，主体为砂岩，并有晚第三纪—第四纪、与台湾大屯火山相当的安山岩分布，与南、北小岛、黄尾屿、赤尾屿的岩层组成相同[1]。

a. 由近及远依次为南小岛、北小岛、钓鱼岛主岛

（图片来源：新浪微博，http://weibo.com/208165559）

b. 钓鱼岛本岛

（图片来源：八一军事，http://www.junshi81.com/ttpl/junshi/2013/0517/12910.html）

图5　钓鱼岛及其周边附属岛屿

（2）中国历史文献记载对钓鱼台列岛发现与开发最早①：

1403 年明永乐元年始称“钓鱼屿”，并形容船只至该海域是“顺风相送”；

1534 年即明嘉靖十三年，第 12 次琉球册封使陈侃《使琉球录》记有钓鱼屿；

1561 年郭汝霖《使琉球录》说明钓鱼屿是与琉球的分界地；

1562 年明嘉靖四十一年，郑若曾《筹海图篇》；

① 苏纪兰，李家彪．“钓鱼台列屿”演示文稿，增订版．杭州：国家海洋局第二研究所，2012.10.

1600年明万历二十八年，夏子阳《使琉球录》注明钓鱼屿为中国之界；

1683年清康熙二十二年，汪楫《使琉球杂录》言及赤尾屿为中国界内；

1767年清乾隆三十二年，《坤舆全图》均用“钓鱼屿”，并注明在中国界内，沿用至今称钓鱼岛或钓鱼台列岛。

(3) 1894—1895年，清“甲午战争”败后签订《马关条约》。1895年1月14日，日本宣称钓鱼屿纳入冲绳县管辖，称其为“尖阁群岛”。

1941年日据时期，台北州与冲绳县政府对上述小岛归属问题进行诉讼，东京法院审判“尖阁群岛”纳入台北州管辖。另一说是1941年台北州与宜兰州争执，东京法院判给宜兰管辖。总之，归属于台湾管辖。

(4) 1943年12月1日，中、美、英三大盟国宣布《开罗宣言》(图6)：第二次世界大战的宗旨之一在于使日本所窃取于中国的领土，包括满洲、台湾、澎湖列岛等归还中国。1945年美国罗斯福总统提出将琉球归还中国，蒋介石犹豫不决，提出中、美共管而搁置。

图6　中、美、英三巨头及蒋宋美龄

(5) 1945年6月23日，第二次世界大战美日血战后，美军占领冲绳岛。

1972年美国政府将琉球群岛行政权移交日本。日本在该海域巡视，抓捕中国渔民、登岛立标、租借与售岛动议等系列活动是对中国领土之挑衅。

中国外交部长杨洁篪2013年3月9日在十二届人大一次会议答记者问时，明确指出：“钓鱼岛及其附属岛屿自古以来就是中国的领土。钓鱼岛问题的根源是日本对中国领土的非法窃取和占据。……日方所作所为严重侵犯中国领土主权，是对二战胜利成果和战后国际秩序的挑战，严重损害中日关系，也损害了本地区的稳定。”

中国采取海监船定期巡航，维护渔民作业及一系列严正声明，是“采取的坚定措施，显示了中国政府和人民维护国家领土主权的坚强意志和决心。中方认为，日方应正视现实，以实际行动纠正错误，同中方一道，通过对话磋商妥善处理和解决有关问题，防止事态升级失控。……日本军国主义发动的侵略战争，给包括中国在内的亚洲受害国人民带来深重灾难……日本只有正视和深刻反省历史，才能搞好同亚洲邻国的关系，才能赢得未来”(杨洁篪，2013年3月9日)。

三、南海

1994 年《联合国海洋法》开始实施，在群岛海域形成"小岛大海洋"的新概念：一个人居的小岛，它可据有内水、12 海里领海、24 海里毗连区及 200 海里专属经济区。因此，海洋中的岛、礁、滩、沙十分重要，即使暗沙，如果一年中可出露一次，也可作为基点依据，如黄海辐射沙脊群的外磕角。

了解南海海底的地貌特点，有助于对海洋权益的维护[2]，以提出有理、有力的依据。需认识到：南海问题与东海不同，南海存在多边纷争；西沙与南沙群岛并非位于现代大陆架海域，而是位于由古大陆架下沉折断形成的大陆坡顶；部分南沙岛礁位于巽它大陆架上（图 7）。岛屿是确定该海域归属的依据，因此，岛屿的归属权至关重要。历史的渊源与传统的疆界，国际公认的或沿用的图、文资料是重要的证据，断续线正是被国际承认半个世纪之久的疆界直证，不能退让[3]。

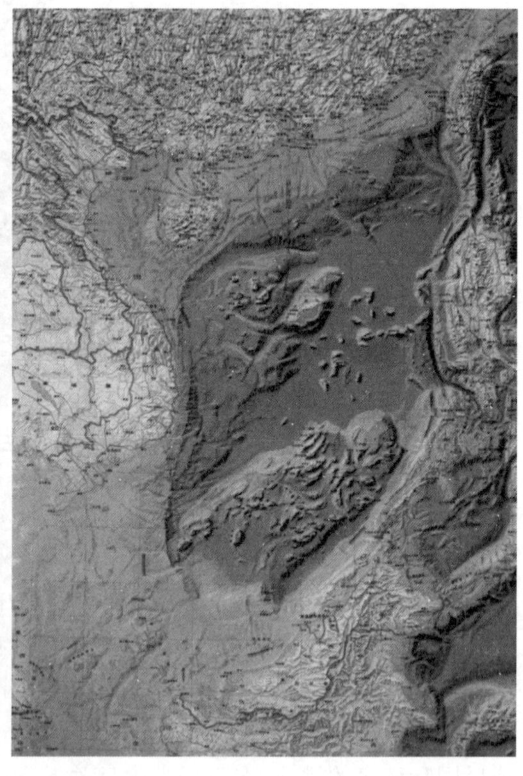

图 7 南海海底图(附彩图)

（图片来源：(左) 南海研究论坛，http://www.nhjd.net/thread-2635-1-1.html。
(右) 互动百科，http://www.baike.com/wiki/%E5%8D%97%E4%B8%AD%E5%9B%BD%E6%B5%B7）

(1) 中国对南海诸岛发现与开发最早[3]：

- 早在东汉(25—220 年)期间，杨孚所著的《异物志》中，对西沙、南沙有记述；
- 三国时期(222—280 年)，万震所著的《南国异物志》对南沙记述；
- 225—230 年，孙权派康泰出使扶南国(今柬埔寨)，归撰《扶南传》，记述南海"涨海中珊瑚礁底有盘石”；

• 梁代(502—557 年)《梁书》中记“扶南东界即大涨海，海中有大洲，洲上有诸薄国，国东有马五洲……”，即指印尼一带；

• 隋唐时期与南海诸国航路增多，对诸岛记录明确；

• 《初学记》(唐徐坚撰)、《旧唐书・地理志》、《琼州府志》对南海诸岛均有记载；

• 唐朝贞观元年(627 年)，设立崖州，首次将南海列入中国管辖范围；

• 宋朝对南海诸岛行使管辖，巡视南海海疆，宋以后将南海诸岛归海南万州管辖；

• 元朝时有关南海记载更加详细，1350 年刊印的汪大渊《岛夷志略》以“万里石塘”记述今东沙、西沙、中沙与南沙；明朝大学士宋濂撰《元史》记述越南东南海岸至印尼的卡里马塔群岛一带地区及西沙、中沙和南沙群岛部分海域；

• 明郑和下西洋(1405—1433 年)，至南海与印度洋，茅元仪《武备志》中有郑和航海路线图成于 1425—1430 年间；

• 清陈伦炯《海国闻见录》(1730 年)中所绘“四海总图”列出南海诸岛；

• 1883 年，德国曾在南沙和西沙进行测量，广东省政府提出抗议，德国停止了作业，1911 年辛亥革命后，国际组织多次要求中国政府在南海诸岛上建立气象台、灯塔等航海设施，以保证航行安全，表明国际公认南沙、西沙等岛礁是中国领土[4]，1921 年版《中国地理沿革图》明确标出南海诸岛(图 8)。

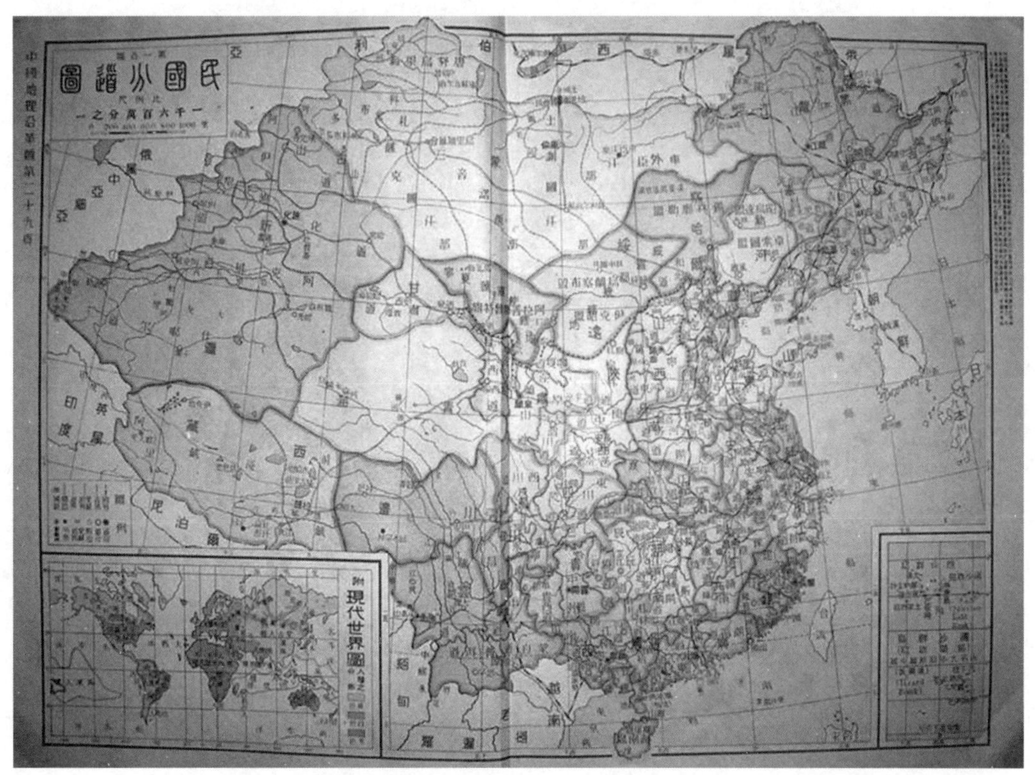

图 8　《中国地理沿革图》的民国全图(1921 年)

(图片来源：数字南海成果报告，2007)

(2) 第二次世界大战期间，南海诸岛被日本侵占。

(3) 第二次世界大战后，国际社会对中国南海主权的承认：

• 1945 年，《波茨坦公告》明确规定日本侵占的中国领土(含南海诸岛)必须归还中国；

• 1946 年，中华民国国民政府接收西沙、南沙，在岛上升国旗、立主权碑，昭告世界，并派兵驻守南沙群岛的最大岛屿——太平岛，国际上对此皆予承认(图 9 和图 10)；

图 9　1946 年 11 月 24 日中国海军收复西沙群岛纪念碑(附彩图)

(图片来源：数字南海成果报告，2007)

图 10　升旗太平岛

(图片来源：数字南海成果报告，2007)

• 1947 年，中华民国国民政府正式公布了由内政部方域司勘定的 1∶650 万《南海诸岛位置图》[5]，图上画有 11 段以断续线形式标注的海域国界线(原件存于南京市的中国第二历史档案馆)，以后出版的地图均遵照政府规定画有这些断续线。

• 1948 年申报馆出版的《中国分省新图》第 5 版，在广东省分幅中标出南海诸岛（图 11），中国对南海诸岛的主权得到国际公认；

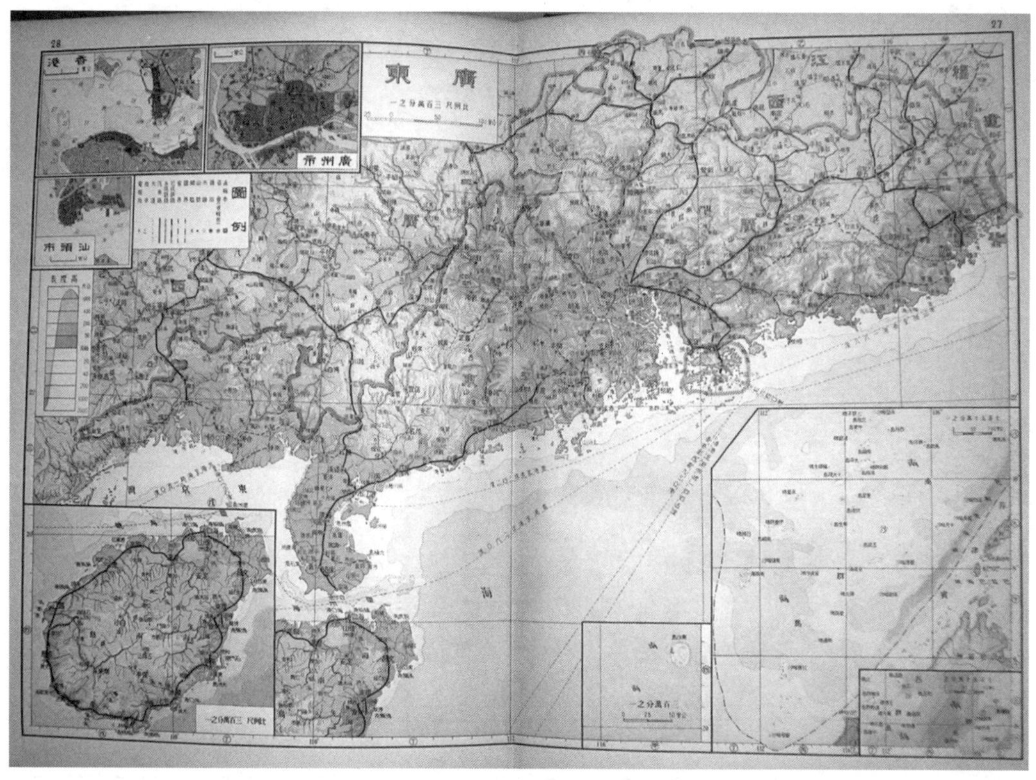

图 11　1948 年申报馆广东省分省图

（图片来源：数字南海成果报告，2007）

• 1951 年 9 月 8 日，英、美等国对日签订《旧金山和约》，第三条规定："将琉球群岛与大东群岛在内的西南群岛管辖权由美军负责，而不交联合国托管"。时任美国国务卿杜勒斯（J. F. Dullus）宣称日本对这些群岛享有"剩余主权"，即美在结束管理后，可将这些岛屿归于日本。《旧金山和约》期间，中国政府和台湾当局未参加。

• 1951 年 8 月 15 日，中国外交部长周恩来发表声明："东沙、西沙、中沙和南沙向为中国领土，中国对群岛的主权，无论对日和约草案有无规定及如何规定，均不受任何影响。"旧金山会议上，南越代表提出要西沙、南沙主权，这一无理要求遭到与会各国摈弃。

• 1952 年 4 月 28 日，台湾当局与日本在台北签订《"中华民国"与日本国间和平条约》，8 月 5 日互换批准书后生效，其中第一条"'中华民国'与日本国间之战争状态，自本约发生效力之日起，即告终止"。第二条"兹承认依照公历 1951 年 9 月 8 日在美利坚合众国金山市签订之对日和平条约第二条，日本国业已放弃对于台湾及澎湖群岛以及南沙群岛、西沙群岛之一切权利、权利名义与要求"。

上述这些会议的参与国、签字国已承认西沙、南沙群岛是中国领土，日本也已将这些地方交还中国政府。1952 年日本全国教育图书公司出版的《标准地图集》标注南海诸岛属于中华人民共和国（图 12）。1953 年，我国政府批准去掉北部湾 2 段断续线，改成 9 段，沿用至今，已历时半个多世纪。

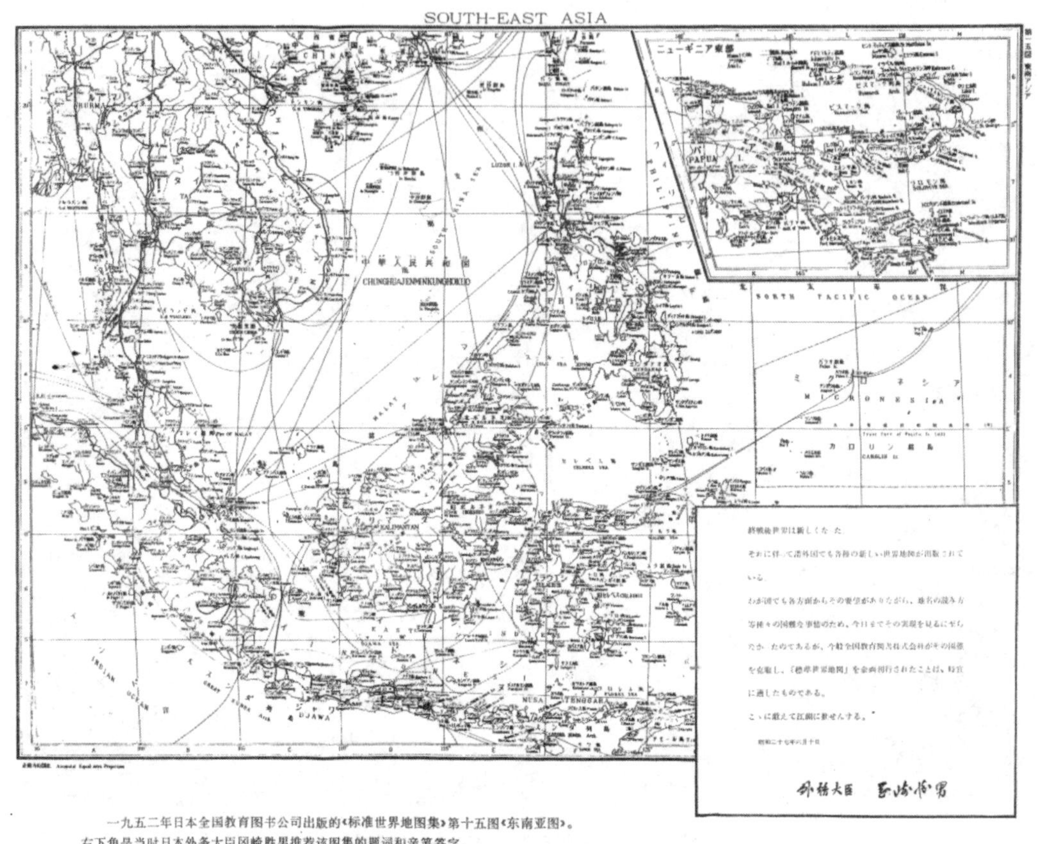

图 12　1952 年日本全国教育图书公司出版的《标准世界地图集》标注的南海诸岛属于中国

（图片来源：数字南海成果报告，2007）

（4）20 世纪 70 年代后期，随着对南海丰富自然资源的调查了解，越南、菲律宾、马来西亚和印度尼西亚等国纷纷提出对南海主权的无理要求，抛出多种南海划分界线，造成南海国界问题的矛盾日益突出。为捍卫主权和国家利益，有必要对我国南海疆域给予确定性的认识。

根据 1947 年国民政府颁布的《南海诸岛位置图》的原图复印件①和 1983 年我国正式出版的《南海诸岛》地图（中国地图出版社，1983），南京大学对 9 条断续线的地理经纬位置作了核实：采用数字化仿射变换，利用地理信息系统分别测量出原来与订正后的各 9 条断续线的经纬度坐标，以度为单位，舍入到小数点后 3 位，获得具有较高的精确度的位置，具体数据与编号列表于后（表 1 和表 2）。

① 南京大学杨怀仁教授参加了当年内政部方域司的该项工作。

表 1 民国内政部制 1∶650 万《南海诸岛位置图》中的断续线位置(1947)

断续线编号	位置	端点位置(度)		长度,走向(km)	起止范围(度)	
		起点	终点		东经	北纬
西 1(旧)	西沙群岛以西,平行于越南东岸,近土伦(顺化)	109.452 15.958	110.289 14.288	210.4 SSE	109.452 110.289	15.958 14.288
西 2(旧)	南沙群岛以西,平行于越南南岸,西北为金兰湾	110.232 10.528	109.399 9.080	185.2 SSW	110.232 109.399	10.528 9.080
南 1(旧)	南沙群岛西南,接近纳土纳群岛	108.895 6.053	109.686 4.585	186.5 SSE	108.895 109.686	6.053 4.585
南 2(旧)	曾母暗沙以南,平行婆罗洲(加里曼丹岛)北岸	111.134 3.725	l13.007 4.232	230.8 E,NEE	111.134 113.007	3.613 4.232
南 3(旧)	平行于婆罗洲北岸,拉布安岛西北	114.201 5.280	115.005 6.214	137.0 NE	114.201 115.005	5.280 6.214
东 1(旧)	巴拉望岛西南,巴拉巴克海峡以西	l16.076 7.447	116.906 8.414	141.2 NE	116.076 116.906	7.447 8.414
东 2(旧)	南沙群岛东北,菲律宾布桑加岛、龟良岛以西	118.427 11.314	118.866 12.596	152.1 NNE	118.427 118.866	11.314 12.596
东 3(旧)	黄岩岛东北,平行于吕宋岛西海岸	119.008 15.469	119.260 17.027	175.7 N	119.008 119.260	15.469 17.027
东北 1(旧)	巴时海峡(巴士海峡)内,台湾七星岩与吕宋岛之间	119.845 19.690	120.975 20.184	131.5 NEE	119.845 120.975	19.690 20.184

表 2 地图出版社 1∶600 万《南海诸岛》图中的断续线位置(1983)

断续线编号	位置	端点位置(度)		长度,走向(km)	起止范围(度)	
		起点	终点		东经	北纬
西 1(新)	西沙群岛以西,平行于越南东岸,近越南广东群岛	109.284 16.148	109.869 15.163	127.7 SSE	109.284 109.869	15.163 16.148
西 2(新)	南沙群岛西北,平行越东南海岸,西北为越南槟绘湾	110.328 12.278	110.046 11.248	120.5 S,SSW	110.046 110.328	11.248 12.278
南 1(新)	南沙群岛西南,纳土纳群岛正北方	108.247 7.108	108.322 6.003	124.2 S,SSE	108.220 108.322	6.003 7.108
南 2(新)	曾母暗沙以南,亚西暗沙西北,平行加里曼丹岛北岸	111.892 3.434	113.015 3.847	134.4 NEE	111.892 113.015	3.434 3.847
南 3(新)	巴拉望岛西南,巴拉巴克海峡以西	115.623 7.171	116.236 7.993	119.9 NE	115.623 116.326	7.171 7.993
东 1(新)	南沙群岛东北,菲律宾布桑加岛、龟良岛西南	118.572 10.918	119.029 12.013	132.0 NNE	118.572 119.029	10.918 12.013
东 2(新)	黄岩岛东、东北,接近吕宋岛西海岸	119.079 14.939	119.074 16.059	124.7 N	119.054 119.079	14.939 16.059
东 3(新)	平行于吕宋岛西北海岸,近北吕宋海槽	119.474 17.956	119.984 18.971	126.3 NNE	119.474 119.984	17.956 18.971
东北 1(新)	巴士海峡内,台湾兰屿与吕宋岛之间	121.126 20.817	121.860 21.776	131.1 NEE	121.126 121.860	20.817 21.776

对比测定的新、旧断续线经纬度示意图(图 13),发现新旧 9 条断续线的位置不完全重合,表现出三种位置相关:新、旧线是相互衔接,如西 1(旧)与西 1(新),东 3(旧)与东 2(新);新、旧线是相互平行,有外展,如东 2(旧)与东 1(新),南 2(新)与南 2(旧);有的新线在旧线的延伸线的位置上,如南 3(旧)与南 3(新)。总体上,旧断续线长度变化较大,在 130～210 km 之间,不规则;而新断续线长度较短,约在 120～130 km 之间,长度变化不大,较为规则。对应编号的新旧断续线之间的距离从最接近的 18 km 左右(南 2 新旧之间),到相距最远的 333 km 左右(东 1 新旧之间)。

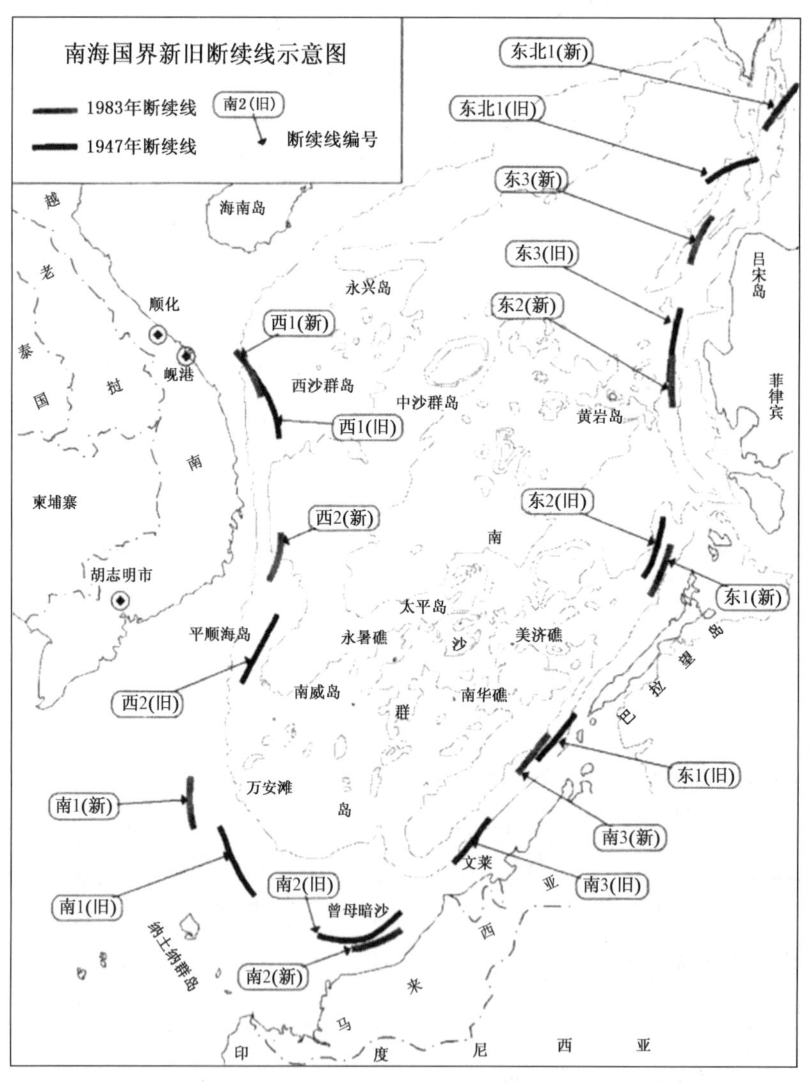

图 13　南海国界新旧断续线对比示意图[3]

虽然各条新旧断续线之间存在空间位置上的少量差异,但重要的是,新旧断续线所界定的领土范围是基本一致的,订正后的断续线所表示的海域疆界更为合理。经计算得出,旧断续线以内的南海海域面积约为 192.7 万 km^2,新断续线以内的南海海域面积约为 202.5 万 km^2,误差小于 5%,显示出中国南海领土主权范围的历史延续性[6]。

(5) 20 世纪 70 年代,西沙海战挫败了越南当局侵占我国岛礁的嚣张气焰。1994 年

《联合国海洋法》实施后，南海周边国家又纷纷非法侵占南沙岛礁①：

• 越南侵占 29 个岛礁（南威岛、鸿庥岛、南子岛、敦谦沙洲、景宏岛、安波沙洲、染青沙洲、中礁、毕生礁、柏礁、西礁、无乜礁、日积礁、大现礁、六门礁、南华礁、舶兰礁、奈罗礁、鬼喊礁、蓬勃堡礁、广雅礁、万安滩、人骏滩、西卫滩、琼礁、金盾暗沙、奥南暗沙、李准滩、单柱石），几乎侵占我国南海大部分疆域，可造成损失油气资源 190 亿 t；

• 菲律宾侵占 9 个岛礁（北子岛、西月岛、双黄沙洲、司令礁、中业岛、马欢岛、南钥岛、费信岛、仁爱礁），失去油气资源量可达 38 亿 t。

• 马来西亚侵占 8 个岛礁（弹丸礁、光星仔礁、南海礁、簸箕礁、榆亚暗沙、北康暗沙、南康暗沙、中康暗沙），会失去 150 亿 t 油气资源。

• 文莱声称对南通礁的主权。

上述四国几乎瓜分了我国南海疆域，形势严峻，不能容忍。

（6）我国目前在南海驻守 9 个岛礁（渚碧礁、南薰礁、东门礁、美济礁、赤瓜礁、永暑礁、华阳礁，台湾地区驻守太平岛与中洲礁）。我国巡航可控制岛礁 13 个（南钥岛、双黄沙洲、杨信沙洲、库归礁、安乐礁、西门礁、舰长礁、信义礁、仁爱礁、蓬勃暗沙、奥援暗沙、康泰滩及曾母暗沙）。定期巡航，维护南海疆域立场坚定！

中国对南海东沙、西沙、中沙与南沙群岛发现最早，均正式列入版图；渔民作业、船舰航行巡视、岛礁与海域的开发经营与建设由来已久；国际公认中国对南海诸岛之权限，海疆权益绝不容侵犯。但是，我们须清醒地认识到，南海海程遥远，缺乏大型船只将难以进行充足的后勤补给，而军力维护国家制海权更为重要，必须大力发展航母与海、空军力，保卫祖国神圣海疆！同时，诚望海峡两岸炎黄子孙携手合力，维护中国海洋权益。

致谢：王敏京同志协助文稿的打印、制图与排版；傅光翮同志提供苏岩礁资料。

参考文献

[1] 章雨旭，乔秀夫. 中国钓鱼岛地质概况[J]. 地质论评，2012，58(6)：1144.

[2] 冯文科. 南海北部晚第四纪地质环境[M]. 广州：广东科技出版社，1988.

[3] 王颖，马劲松. 南海海底特征、疆域与数字南海[M]. //苏纪兰主编. 郑和下西洋的回顾与思考. 北京：科学出版社，2006：177 - 193.

[4] 罗章仁，陈史坚，郑天祥. 南海[M]. //王颖主编. 中国海洋地理. 北京：科学出版社，1996：411 - 452.

[5] 民国内政部方域司编制. 南海诸岛位置图[M]. 民国内政部方域司印制，1947.

[6] 许森安. 南海断续国界线的内涵[C]. //海南南海研究中心. "21 世纪的南海：问题与前瞻"学术研讨会论文选. 海口：海南出版社，2000：80 - 81.

① 马劲松，殷勇：《南沙岛礁现况》，2013 年 3 月 15 日发言稿。

1947 年中国南海断续线精准划定的地形依据*

1 引　言

2009 年，马来西亚和越南提交了对南海外大陆架划界方案联合申请之后，中国在向联合国秘书长潘基文提交照会时首次使用了 1：400 万《南海诸岛位置图》予以回击，并声明中国对南海诸岛及其附近海域拥有无可争辩的主权，并对相关海域及其海床和底土享有主权权利和管辖权，中国政府这一一贯明确的立场为国际社会所周知。但国内一些学者对于《南海诸岛位置图》中断续线的属性存在不同的认识，例如“岛礁归属线”[1]、“历史性权利线”[2]、“群岛水域线”[3-4]和“国界线”[5-6]等。王颖等[6]根据《南海诸岛位置图》里中越陆地界线的海中延伸及加里曼丹岛与婆罗洲界线均与陆地国界线符号相同，明确了南海断续线的性质就是海疆国界线。国际上，在地中海各国的岛屿之间亦是以断续线段表示海域分界线。因此，《南海诸岛位置图》具有维护中国南海主权的重要作用。

在疆域划界工作中，地形变化与地貌特征是划界的重要参考依据。陆域划界通常采用山脊线（如印度东北部与缅甸以顺延阿拉干山脉进行划界）、河流中线进行划分（中国东北部与朝鲜北部以鸭绿江中线进行划界）；海域常以海峡中线（俄罗斯东部与美国阿拉斯加州西部以白令海峡为界）、海洋法意义上的大陆坡及大陆架进行划分（德国、丹麦、荷兰在北海以大陆架进行划界）。南海海域地貌具有弧后盆地的典型特点，海域中部存在深度超过 4 000 m 的深水盆地，而四周则被岛弧及大陆坡、大陆架所包围，南海断续线的独特外形与海域四周环绕的岛弧、大陆坡、大陆架分布位置及走向保持了高度一致，显示通过地貌特征进行划界的可能性。但是，目前结合海底地貌探究南海断续线段形态及划分原则的工作尚处于起步阶段[6-7]。因此，在前人的研究基础上，本文利用地理信息系统的数字化手段，对于《南海诸岛位置图》中（图 1）断续线进行了精确定位，获取了断续线地理位置信息。并且，通过将断续线与海底地形叠合分析，制作地形横剖面，获得断续线下伏地形特征，以此探究当时断续线划定所遵循的地形依据。

2 南海断续线的地理位置及形态特点

依据 1947 年中国政府内政部方域司印制的《南海诸岛位置图》[8]（原件存于南京的中国第二历史档案馆），图上共有 11 段以断续形式划定的断续线，往后出版的地图均遵照政府规定，画有这些断续线，标示了我国最南端疆界，图中注明了西沙群岛、东沙群岛、中沙群岛、南沙群岛的位置和名称，系中国南海地图的重要蓝本。

---

* 唐盟，马劲松，王颖，夏非：《地理学报》，2016 年第 71 卷第 6 期，第 914－927 页。

基金项目：中国科学院学部咨询评议项目（2016ZWH005A－005）。

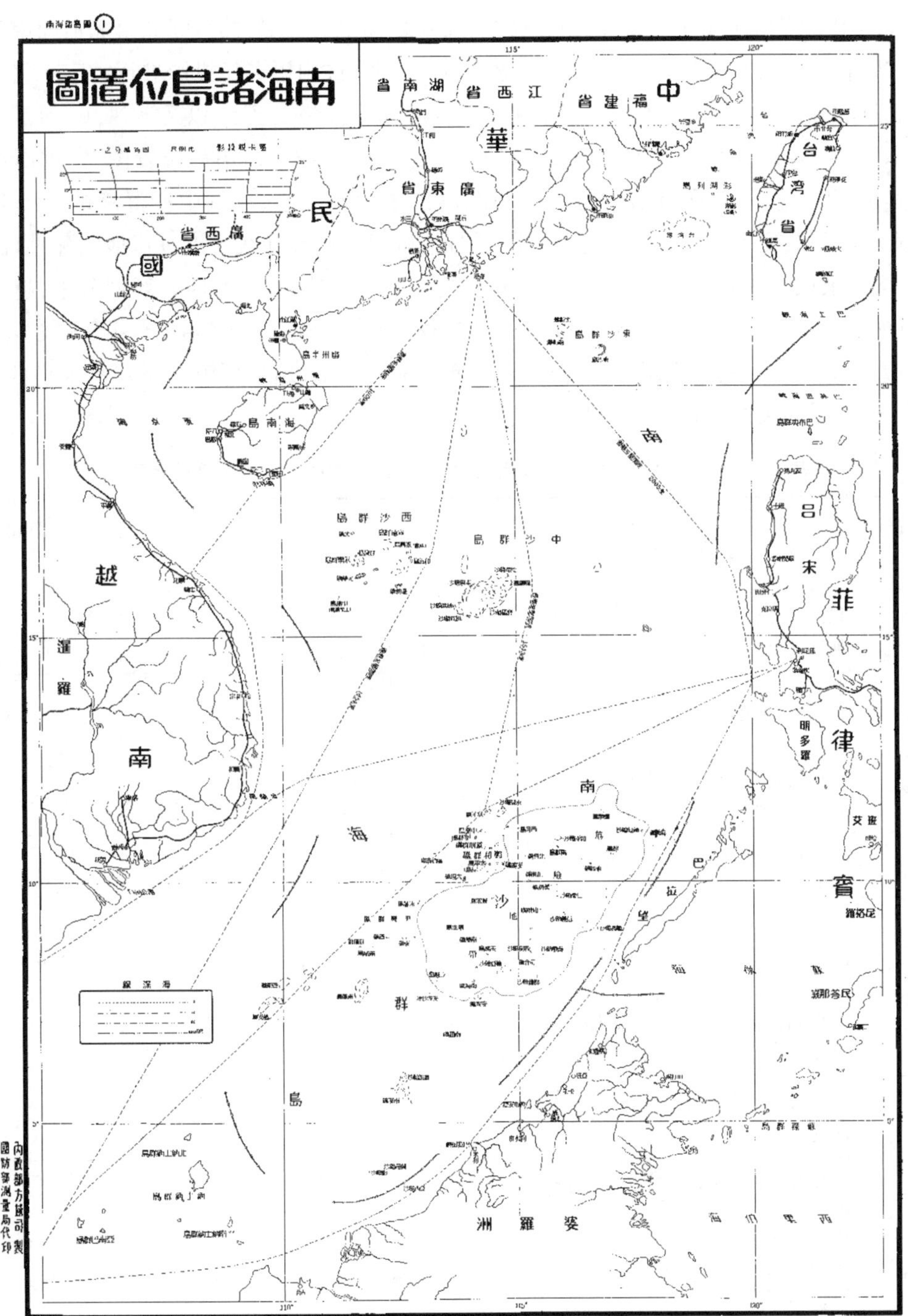

图 1　1947 年 1∶400 万中国《南海诸岛位置图》复印件

注:《南海诸岛位置图》于 1947 年中国内政部印制

王颖等[7]2003 年发表“南海海底特征、资源区位与疆界断续线”一文,分析了 1947 年中国政府内政部印制的 1∶650 万《南海诸岛位置图》上所划出断续线的具体位置,并对于断续线的性质进行了研究和讨论。在此基础上,本文根据 1947 年中国政府内政部印制的 1∶400 万《南海诸岛位置图》的原图复印件(图 1),使用 ArcGIS 10.1 软件进行地图配准及数据处理分析。此图左上角区域标明了“墨卡托投影;比例尺四百万分之一”的繁体字样，且图中经线与纬线互相垂直，相邻经线、纬线之间平行，相邻纬线间隔由赤道向南北两极逐渐增加，以上特征均符合墨卡托投影性质。因此，选用墨卡托投影进行数字仿射变换来完成地图配准。本图图幅上注明了 110°E、115°E 和 120°E 经线以及 5°N、10°N、15°N、20°N、25°N 纬线，除经纬线在 110°E，25°N 的交点被图名"南海诸岛位置图"字框所遮挡外，其余 14 处经纬线交点清晰可见。配准以位于图幅中心、靠近中沙群岛的 115°E 与 15°N 的经纬线交点为坐标系原点，采用 14 处清晰可见的经纬线交点进行地图配准，配准的总均方根误差为 0.095 97 个像元，具有很高精度。此外，以南海西北部划分中、越边界的断续线为第 1 段，并按逆时针顺序，自西向东将断续线编号命名（第 1 至第 11 段），进而利用 ArcGIS 10.1 精准测量出北部湾 2 段，越南东南部 3 段，婆罗洲北岸及吕宋岛西岸 6 段等共 11 段断续线所在的经纬度，其具体位置如表 1 所示。

表 1　1947 年中国 1∶400 万《南海诸岛位置图》中的断续线地理位置数据

断续线编号	地理位置	端点经纬度位置		长度(km) 走向	断续线间距(km)
		起点	终点		
1	北部湾以北，中越陆上界线的延伸	105.015 23.146	107.500 19.964	701.0 SE,SW	103.1
2	海南岛西侧，平行于海南岛西岸	107.246 19.071	108.127 17.547	195.6 SE	297.5
3	西沙群岛以西，平行于越南东岸，近土伦(顺化)	110.000 15.555	110.812 14.248	167.0 SE	369.7
4	西沙群岛以西，平行于越南南岸，西北为金兰湾	110.755 10.911	109.779 9.257	208.0 SW	406.0
5	南沙群岛西南，接近纳土纳群岛	108.778 5.723	109.594 4.083	200.2 SE	173.8
6	曾母暗沙以南，平行于婆罗洲(加里曼丹岛)北岸	111.050 3.505	113.313 4.485	275.2 E, NE	204.8
7	平行于婆罗洲北岸，拉布安岛西北	114.551 5.860	116.729 8.395	361.5 NE	170.4
8	巴拉望岛西南，巴拉巴克海峡以西	117.643 9.632	118.697 11.527	235.1 NE	532.0
9	南沙群岛东北，菲律宾布桑加岛、龟良岛以西	119.246 16.306	119.519 18.755	267.0 NS	115.9
10	巴士海峡内，台湾七星岩与吕宋岛之间	119.876 19.744	121.423 21.200	230.3 NE	61.4
11	台湾台东以东南	121.948 21.451	122.612 22.241	118.5 NE	

注：断续线端点的经纬度坐标单位为东经(°E)，北纬(°N)。

精准测量后发现，断续线并非按照等长、等间距的原则"平均"分布于南海海域。其长短具有较大的差异(表 1)，最长段为第 1 段，长度为 701.0 km，最短段为第 11 段，长度为

118.5 km，两者长度相差 582.5 km，其平均长度为 296.3 km，具有西部长，东部短的特点。测算断续线前段终点与后段起点的直线距离也表明断续线间距亦有较大差异，间距最大的是第 8 段与第 9 段间，为 532.0 km，间距最小的是第 10 段与第 11 段间，为 61.4 km，其平均间距为 243.3 km，总体表现出东西两侧间距大，南北两端间距小的态势。

南海断续线有长短之分，间距有大小之别，导致此结果是由于划定断续线时考虑了多种因素所致。收录本图 1∶850 万比例尺版本的《南海诸岛地理志略》第二章“地质地形”中(第 7 页)亦收录了 1∶1 800 万的南海海底地形图:《南海地形鸟瞰图》[9]，由此可见在划定断续线之前，中国政府已经拥有一定的南海海底地形资料。因此，本文试图从断续线下伏地形地貌特征对断续线划定原则进行探究。

3 南海断续线下伏海底地形特征分析

3.1 南海海底地形地貌概况

南海位于中国南部，是西北太平洋内大型的边缘海，海域面积 356×10^4 km²。南海海底地形西北高，东南低，自周边陆架向南海中央下降，中央为呈 NE－SW 延长的菱形盆地，纵长 1 500 km，最宽处 820 km。其于欧亚大陆、太平洋和印度洋三大板块的交界处，属于大洋型和大陆型地壳构造域之间的过渡型构造域，其构造运动较为复杂，大规模水平运动伴随着大规模的垂直运动，剧烈的陆海扩张伴随着剧烈的陆缘挤压[10-13]，即陆壳在北缘离散阶梯状下降，在南缘拼贴增生。洋壳在中央海盆新生，在东侧马尼拉海沟消减。陆缘地堑在扩张中形成，岛弧-海沟在挤压中发育[14]。

南海大陆架北部及西北部属于水深小于 200 m 的陆架海域，在地质构造和沉积物源特征上与我国华南大陆关系密切，其中北部湾为半封闭的浅海盆地(北部湾盆地)，水深在 20～50 m；南海南部为宽广的巽他大陆架，其宽度超过 300 km，坡度平缓，水深小于 150 m[15]；南海西部为中南半岛大陆架，大陆架沿越南东海岸呈带状分布[16]，宽度约 40～50 km，大陆架外缘水深小于 200 m[17-18]；南海东部为带状分布的狭窄岛架，分布于吕宋岛、巴拉望岛边缘，以急剧变化的地形坡度向吕宋海槽和巴拉望海槽过渡。

南海大陆坡位于大陆架的周边外缘，面积达 120×10^4 km²，系南海区域分布最广阔的大地单元，深度介于 150～3 600 m 之间[19]。深海盆地南北两侧大陆坡是块状断裂下沉形成的阶梯状大陆坡，相对高出深海盆地数千米，表现为海底高原；深海盆地东西两侧陆坡狭窄而坡度陡峭，西、北吕宋海槽、马尼拉海沟均分布于东侧陆坡，其具有阶梯状下降的特点。

南海海盆位于南海中部偏东，面积达 40×10^4 km²[20]，中央水深超过 4 000 m，盆地中分布着由孤立的海底山组成的高出海盆底部 3 400～3 900 m 的海底山群，以及 27 座高度超过 1 000 m 的海山和多座 400～1 000 m 的海丘[21]。海盆北部边缘为华南陆架，其内部有一系列阶梯状正断层以及地堑和地垒交错分布，盆地中填充了巨厚的沉积，是拉张型边缘；南海海盆南部为南沙海槽，南沙海槽是特提斯的残留海，自燕山时期开始向南消减，在加里曼丹岛北部形成褶皱带及冲断层带，为挤压型边缘；南海海盆西部为狭窄的中南半岛大陆架，与海岸线大致平行，陆架上有一系列平直的阶梯状断层，系剪切拉张所形成；南海海盆东部有近似 S－N 向的马尼拉海沟与中国台湾-菲律宾岛弧相分隔，海沟位于向陆一侧，有别于西太平洋边缘其他海沟-岛弧体系，具有挤压型边缘特征。因此，南海海盆自西北向东南裂离蠕散，后缘阶梯状拉张，前缘挤压，两侧剪切，这是南海海盆边缘的基本构造特征[15,22]。

为了较高分辨率地提取南海断续线下伏海底地形数据，本文选用中国人民解放军海军司令部航海保证部提供的相关海域 28 幅海图（表 2），完成配准后提取水深点信息 44 022 个，采用最近邻域法插值，提取出 200 m 间距的等深线，并通过等深线编制出南海海底分层设色数字地形图（图 2），图上可明显辨识大陆架、大陆坡、深海盆地、海沟等地形单元。

表 2　选用的 28 幅中国南海海域海图明细表

编号	图名	投影系	比例尺（万）	深度点数量	编号	图名	投影系	比例尺（万）	深度点数量
10015	福州至广州	墨卡托	1：100	2 221	17600	中南暗沙附近	墨卡托	1：25	853
10016	香港至海防	墨卡托	1：100	1 699	17900	双子群礁北部	墨卡托	1：25	814
10017	中沙群岛至巴士海峡	墨卡托	1：100	1 538	18050	南沙群岛北、中部	墨卡托	1：80	3 441
10018	西沙群岛至南沙群岛	墨卡托	1：100	1 889	18100	双子群礁至郑和群礁	墨卡托	1：25	1 099
10019	黄岩岛至巴拉巴克海峡	墨卡托	1：100	2 648	18102	万安滩至广雅滩	墨卡托	1：25	1 255
10020	金瓯角至纳土纳群岛	墨卡托	1：100	2 417	18103	皇路礁至南安礁	墨卡托	1：25	1 503
10021	纳土纳群岛至巴拉巴克海峡	墨卡托	1：100	1 810	18104	曾母暗沙及附近	墨卡托	1：25	1 773
17010	西沙、中沙群岛	墨卡托	1：50	2 221	18200	礼乐滩	墨卡托	1：25	2 581
17020	中建岛、惹岛至莹角	墨卡托	1：50	1 710	18300	永暑礁至尹庆群礁	墨卡托	1：25	1 532
17030	中沙群岛至南沙群岛	墨卡托	1：50	1 022	18400	郑和群礁至永暑礁	墨卡托	1：25	1 840
17040	莹角至湄公河口、日积礁	墨卡托	1：50	2686	18500	南方浅滩至海口礁	墨卡托	1：25	1 694
17100	西沙群岛	墨卡托	1：25	828	18600	尹庆群礁至南薇滩	墨卡托	1：25	1 551
17101	富贵岛附近	墨卡托	1：25	927	18700	无乜礁至皇路礁	墨卡托	1：25	1 666
17310	黄岩岛	墨卡托	1：10	899	18800	海口礁至榆亚暗沙	墨卡托	1：25	1 249
深度点数量总计									44 022

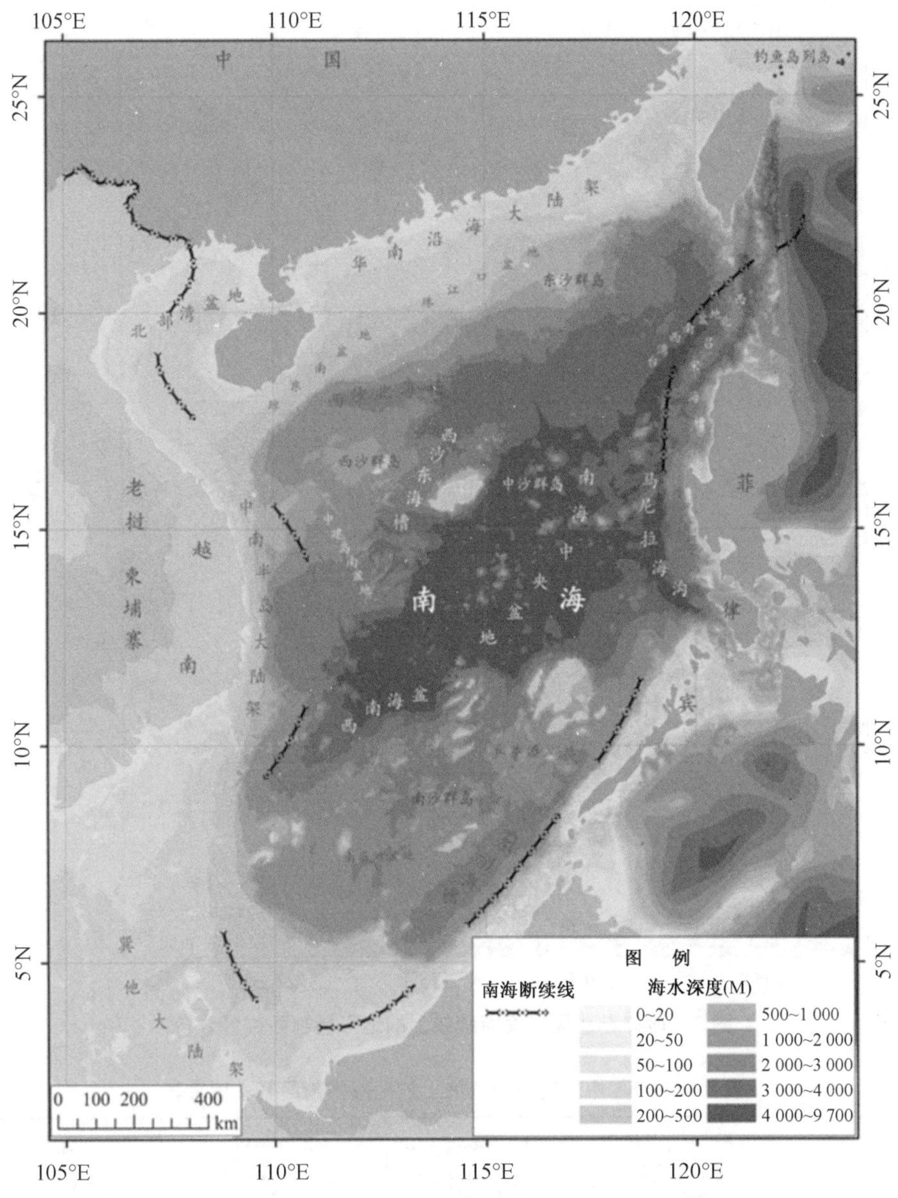

图 2　中国南海海底数字地形图(附彩图)

注:图中"南海断续线"取自 1947 年中国政府内政部制 1∶400 万《南海诸岛位置图》

3.2　南海断续线与海底地形叠合分析

为了进一步探索断续线划定的地形依据,基于 1947 年中国内政部印制《南海诸岛位置图》中的断续线,按照逆时针顺序,自南海西北开始对每一段断续线的起点(A)、中点(B)和终点(C)分别做垂直于各点切线的地形横剖面 32 条(具体位置和基本信息见图 3 和表 3,其中第 1 断续线仅有入海部分的中点和终点)。测算显示,最长的为第 6 断续线 A 点剖面,长 443.7 km;最短的为第 11 断续线 C 点剖面,长 117.6 km。剖面走向以 SW－NE 及 NW－SE 为多,主要分布于南海中央盆地东、西、南 3 个方向。以下分别就南海西部(第 1～4)、南部(第 5～7)和东部(第 8～11)三组断续线与下伏地形关系进行叠合分析。

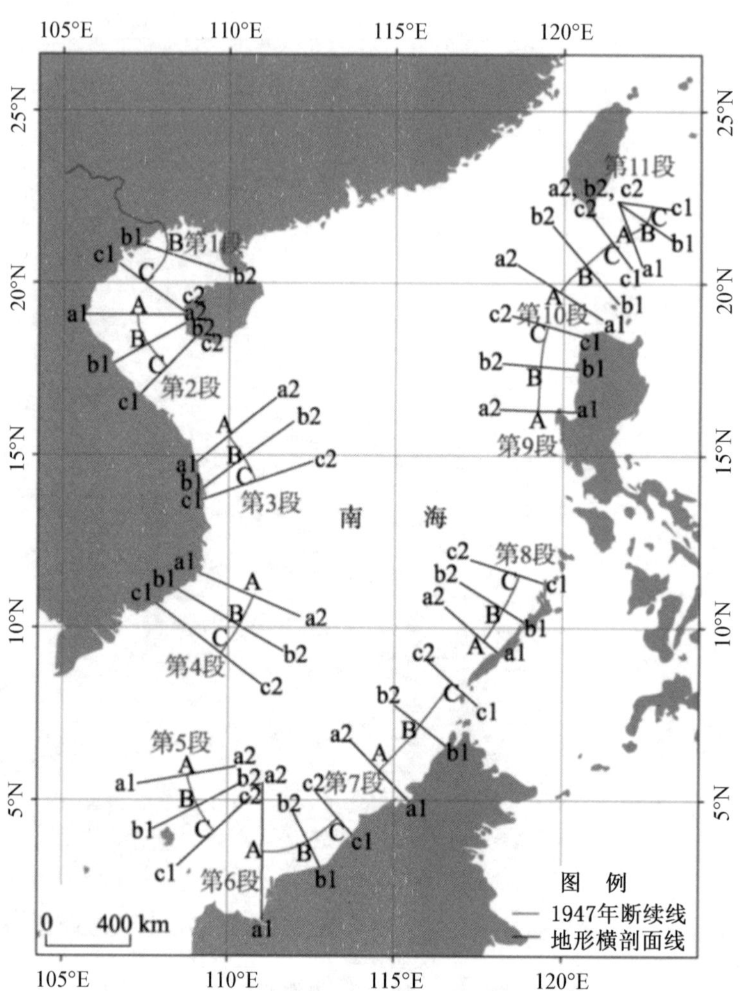

图3　1947年中国南海断续线上的地形横剖面位置分布

表3　1947年中国南海断续线上的地形横剖面基本信息表

名称/编号	端点位置		与断续线相交点位置	范围	长度(km)	走向
	起点	终点				
第1断续线B剖面	107.366 21.178	109.916 20.238	108.083 20.866	越南东兴市到海南岛东北	295.2	NW-SE
第1断续线C剖面	106.658 20.557	108.617 19.166	107.500 19.964	越南海防市至海南东方市	264.8	NW-SE
第2断续线A剖面	105.655 19.063	108.633 19.034	107.246 19.071	越南荣市至海南东方市	326.2	W-E
第2断续线B剖面	106.501 17.667	108.683 18.233	107.534 18.234	越南洞海市至海南莺歌咀	270.3	SW-NE
第2断续线C剖面	107.351 16.766	108.953 18.383	108.127 17.547	越南洞海南至海南岛西南	252.0	SW-NE

（续表）

名称/编号	端点位置		与断续线相交点位置	范围	长度(km)	走向
	起点	终点				
第 3 断续线 A 剖面	109.016 14.766	111.416 16.653	110.000 15.555	越南岘港至永乐群岛	338.5	SW-NE
第 3 断续线 B 剖面	109.216 14.053	111.933 15.966	110.466 14.916	越南归仁至永乐群岛	363.8	SW-NE
第 3 断续线 C 剖面	109.287 13.755	112.516 14.816	110.812 14.248	越南归仁至中建岛以南	383.1	SW-NE
第 4 断续线 A 剖面	109.116 11.553	112.133 10.316	110.755 10.911	越南大叻至南沙群岛西北	362.1	NW-SE
第 4 断续线 B 剖面	108.466 11.083	111.653 9.355	110.333 10.051	越南大叻至南沙群岛西北	399.1	NW-SE
第 4 断续线 C 剖面	107.816 10.702	110.983 8.354	109.779 9.257	越南胡志明市至 南沙群岛西北	432.5	NW-SE
第 5 断续线 A 剖面	107.266 5.466	111.258 5.983	108.778 5.723	大纳土纳岛北侧至 万安滩南	333.5	W-E
第 5 断续线 B 剖面	107.716 4.155	110.433 5.537	109.066 4.816	大纳土纳岛北侧至 万安滩南	331.7	SW-NE
第 5 断续线 C 剖面	108.483 3.083	110.716 5.083	109.594 4.083	大纳土纳岛北侧至 万安滩南侧	326.6	SW-NE
第 6 断续线 A 剖面	111.033 1.544	111.033 5.516	111.050 3.505	马来西亚古晋至广雅滩	443.7	S-N
第 6 断续线 B 剖面	112.752 3.083	115.916 4.683	122.366 3.816	马来西亚北岸至曾母暗沙	201.9	SE-NW
第 6 断续线 C 剖面	113.753 4.010	128.633 5.258	113.313 4.485	马来西亚北岸至北康暗沙	179.0	SE-NW
第 7 断续线 A 剖面	115.433 4.966	113.655 6.752	114.551 5.860	文莱至南薇滩	227.3	SE-NW
第 7 断续线 B 剖面	116.516 6.583	114.966 7.766	115.733 7.166	马来西亚古达至安渡滩	217.2	SE-NW
第 7 断续线 C 剖面	117.466 7.733	115.953 9.063	116.729 8.395	巴拉阿克岛至仙娥礁	218.9	SE-NW
第 8 断续线 A 剖面	118.766 8.667	116.483 9.067	117.643 9.632	巴拉望岛至礼乐滩	331.3	SE-NW
第 8 断续线 B 剖面	119.516 9.766	116.916 11.316	118.216 10.552	巴拉望岛至礼乐滩	328.6	SE-NW
第 8 断续线 C 剖面	120.155 11.083	117.233 11.983	118.697 11.527	巴拉望岛至南海海盆	331.2	SE-NW

（续表）

名称/编号	端点位置		与断续线相交点位置	范围	长度(km)	走向
	起点	终点				
第9断续线A剖面	120.423 16.283	118.083 16.333	119.246 16.306	吕宋岛至中沙群岛	254.7	E-W
第9断续线B剖面	120.433 17.502	118.154 17.543	119.350 17.583	吕宋岛至中沙群岛	248.9	E-W
第9断续线C剖面	120.583 18.454	118.433 19.067	119.519 18.755	吕宋岛至东沙群岛南侧	243.2	E-W
第10断续线A剖面	121.151 18.951	118.583 20.553	119.876 19.744	达卢皮里岛至东沙岛	328.9	SE-NW
第10断续线B剖面	121.633 19.427	119.680 21.710	120.642 20.567	巴林塘海峡至台湾浅滩	332.6	SE-NW
第10断续线C剖面	122.016 20.459	120.852 21.998	121.423 21.200	巴林塘海峡至台湾岛南部	211.1	SE-NW
第11断续线A剖面	128.527 20.539	121.581 22.447	121.948 21.451	太平洋至台湾岛中部	222.1	SE-NW
第11断续线B剖面	123.174 21.312	121.581 22.447	122.461 21.787	太平洋至台湾岛中部	222.4	SE-NW
第11断续线C剖面	123.174 22.167	121.581 22.447	122.612 22.241	太平洋至台湾岛中部	117.6	E-W

注：海底地形横剖面的端点及其与断续线相交点的经纬度坐标单位为东经(°E)、北纬(°N)。

3.2.1　西部断续线

第1断续线陆域部分沿东南走向，经越南东兴市入海后改为西南向，所示范围大致为越南防城市至中国海南岛西北部。

从其地形横剖面形态上看(图4)，两条剖面自陆向海水深先增大后减小，最深处不超过80 m，B点水深40 m，C点水深28 m。沿NE－SW方向，自B点到C点的断续线下伏水深有所减少，地形起伏变化和缓。第1断续线下伏地形均为水深变化较小的北部湾浅海盆地，是中越陆上疆界的海上延续。

第2断续线位于北部湾西南部浅海盆地内，呈弧形，东南走向，朝越南一侧凸出，与中国海南岛西岸及越南北部海岸的岸线形状类似，所示范围大致为越南荣市、洞海市到中国海南岛西部。

从其地形横剖面形态上看(图4)，3条剖面形态相似，均自西向东水深逐步加深至60～95 m后开始减少，最大水深不超过95 m。以断续线所在位置为中心的地形横剖面形态基本对称，说明断续线上A、B、C3点均接近浅海盆地的中线位置，第2断续线是以等距离中线的方式划分了中国与越南在北部湾盆地南部的界线。

第3断续线位于越南东部中南半岛大陆坡上，中建海台西南侧，北端靠近越南岘港，南端靠近越南怀仁，形状呈弧形，向东北方向稍微凸起，沿越南东部海岸线呈东南走向，所

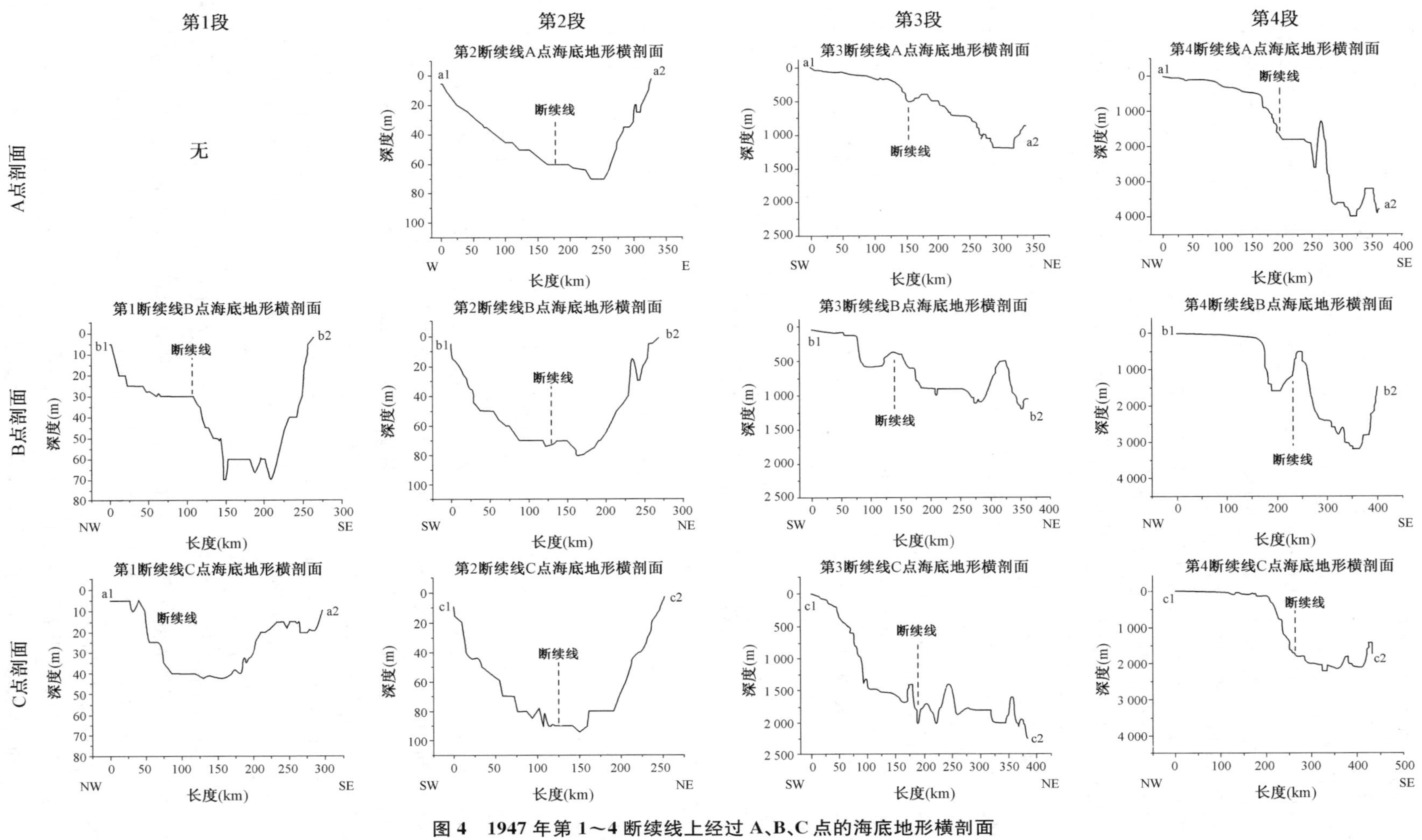

图4　1947年第1～4断续线上经过A、B、C点的海底地形横剖面

示范围大致为越南岘港市至中国永乐群岛。

从其地形横剖面形态上看(图 4),3 条剖面起伏变化较大,下伏地形呈阶梯状,水深自陆向海逐渐增加,距岸 300 km 后水深有所减小,说明此区域内拉张作用明显,大陆坡呈阶梯状破碎下降,海槽、海洼与海岭相间分布,在地形横剖面上表现为凹陷与凸起相互交错分布,系中南半岛陆架与西沙海台的过渡地带。断续线上 A 点,下伏水深约 450 m,位于越南东部陆架的海洼之中,海洼比周围水深低出约 50 m;断续线上 B 点,下伏水深约 380 m,位于越南东部陆架的海山顶部,其西南侧有一个深约 600 m 的海底峡谷,东北侧存在数个因拉张断裂所产生的海沟;断续线上 C 点,下伏水深约 2 400 m,位于海山之间的海沟内。因此,第 3 断续线下伏地形为越南东北部陆架与西沙海台的交界处,将陆架与海台这两个地形单元相分隔,其划分了越南东北部陆坡与中国西沙群岛的界线。

第 4 断续线位于越南东南部,西南走向,北部位于中南半岛大陆坡,南部靠近巽他大陆坡,走向与陆架边缘保持一致,略微朝东南方向凸出,所示范围大致为越南大叻市至中国南沙群岛西北。

从其地形横剖面形态上来看(图 4),3 条剖面起伏变化较大,大陆架、大陆坡、海槽可明显识别。经 A、B 点的横剖面贯穿了相对高差达 800 m 的海岭后,在距岸 350 km 左右处水深急剧加大,由于抵达南海海盆扩张所产生的西南海盆,水深迅速下降至 3 000 m 左右。经过 C 点的横剖面穿过陆架、陆坡和西南海盆南缘之后,抵达南沙群岛广雅滩北部区域。第 4 断续线 A、B、C 3 点下伏中南半岛大陆架东南部与西南海盆南缘水深 1 000～2 000 m 的过渡地带,其划分了越南东南部大陆坡与中国南沙群岛的界线。

3.2.2　南部断续线

第 5 断续线呈弧形,东南走向,略朝纳土纳群岛(今印度尼西亚廖内群岛)方向凸出,形状与纳土纳群岛岸线相似,西距该群岛 112 km,所示范围大致为纳土纳大岛至中国南沙群岛西南。

从其地形横剖面形态上来看(图 5),3 条剖面形态较为接近,均沿西南-东北向水深逐渐增加,在距西侧端点 300 km 处水深明显变化,反映出巽他陆架外侧与陆坡如下特点:① 断续线下伏的巽他陆架平台中部水深在 80～140 m 之间,陆架陆坡转折带水深在 140～150 m 之间,单一陆坡的坡脚水深超过 610 m。② 经过 A 点剖面反映出巽他陆架宽广而平坦的特点,经过 B、C 点剖面反映出该处陆架处于折断下沉的初期,海台与海洼相间分布,系陆架折断下降中的残留遗迹。第 5 断续线 A、B、C 三点下伏水深均不超过 200 m,为变化平缓的巽他大陆架上,其划分了印度尼西亚廖内群岛与中国南沙群岛间的界线。

第 6 断续线走向平行于婆罗洲(加里曼丹岛)西北岸,先朝东后转向北东,呈弧形,朝婆罗洲一侧凸出,所示范围大致为加里曼丹岛以北至中国南沙群岛南部。

从其地形横剖面形态上来看(图 5),3 条剖面形态变化和缓,经过 A 点的横剖面在离岸约 400 km 处出现大陆架与大陆坡的转折点,反映出巽他大陆架宽广而平缓的特点。第 6 断续线上 A、B、C 3 点下伏水深均不超过 100 m,系巽他大陆架北部,其划分了婆罗洲与中国南沙群岛之间的界线。

第 7 断续线平行于婆罗洲东北岸,拉布安岛西北,与岸线形状相似,呈直线状,东北走向,与南沙海槽东缘平行,所示范围大致为文莱至南薇滩。

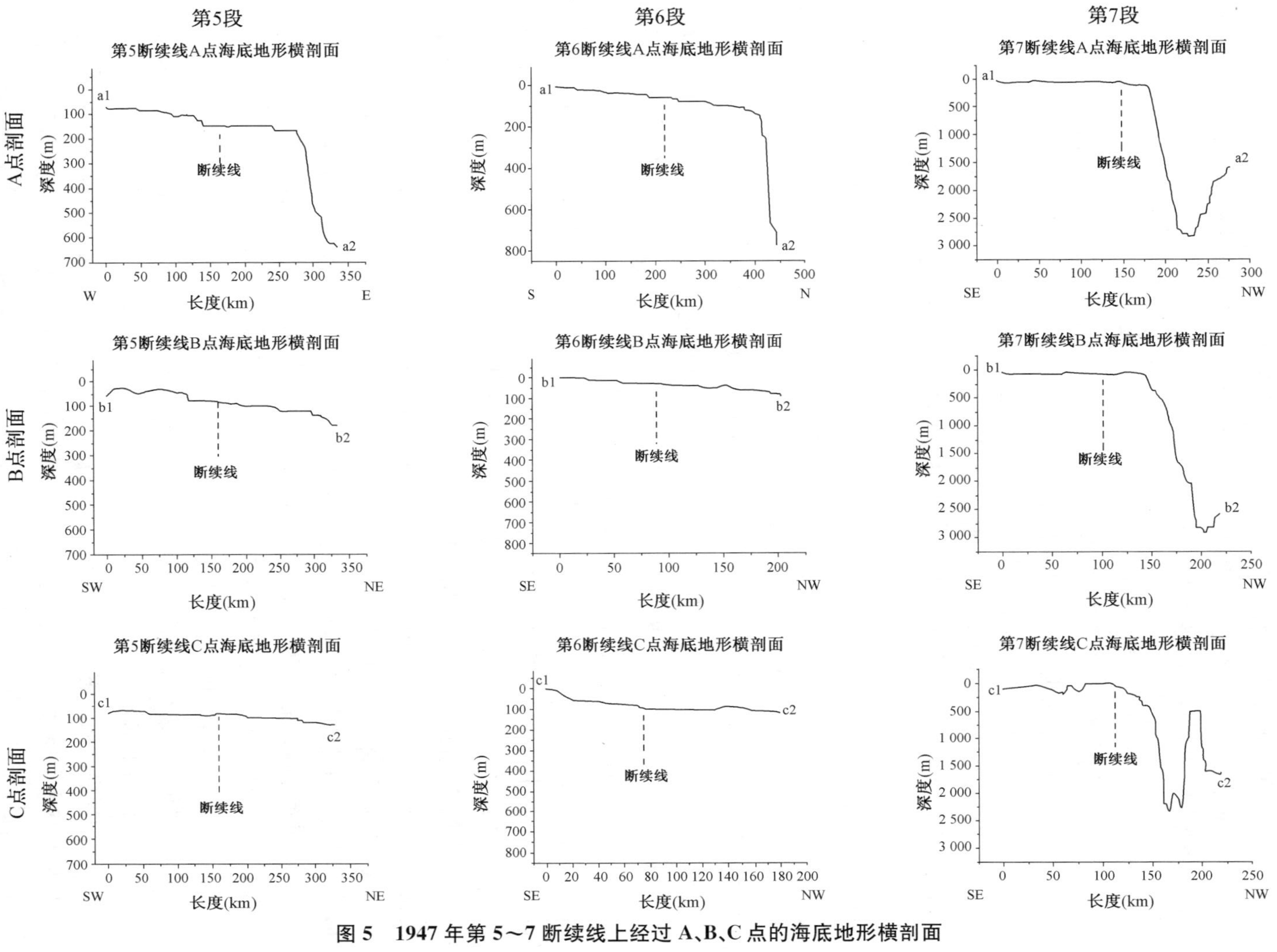

图5　1947年第5～7断续线上经过A、B、C点的海底地形横剖面

从其地形横剖面形态上来看(图 5),3 条剖面形态较为相似,大陆架、大陆坡、海槽可清晰识别。自陆向海距端点 150 km 左右水深陡然加大,在 200 km 处达到最大值 2 800 m,直至 220 km 处水深开始减小。横剖面上显示的深槽系南海板块向加里曼丹岛俯冲所形成南沙海槽。第 7 断续线 A、B、C 3 点下伏地形均为水深小于 200 m 的加里曼丹北部岛架,且 3 点间下伏地形变化较小。因此,它经过加里曼丹岛架与南沙海槽这两个地形单元之间将婆罗洲与我国南沙群岛予以划分。

3.2.3 南海东部断续线

第 8 断续线位于巴拉望岛西侧大陆坡上,巴拉巴克海峡西南,礼乐海台东南,呈直线状,东北走向,与大陆架走向平行,所示范围大致为巴拉望岛至礼乐滩。

从其地形横剖面形态上来看(图 6),3 条剖面地形较大起伏,距东侧端点 150 km 处水深急剧加大,而 275 km 处水深又急剧减小。剖面均贯穿了位于礼乐滩东南部水深约 1 800 m 的海槽,此海槽将南沙海台与巴拉望岛架两个地形单元分隔开来。第 8 断续线上 A、B、C 3 点下伏水深 200～500 m,系在巴拉望岛西侧的狭窄岛坡上,其划分了菲律宾巴拉望岛与中国南沙群岛之间的界线。

第 9 断续线位于菲律宾吕宋岛西、北吕宋海槽内,呈直线状,南北走向,大致与吕宋岛北部岸线平行,所示范围大致为吕宋岛西侧至中沙群岛。

从其地形横剖面形态上来看(图 6),3 条剖面地形起伏较大,经过 B、C 两点的横剖面中出现了两处较深的凹槽,近岸的为西、北吕宋海槽,水深达 3 000 m;远岸的为马尼拉海沟,水深大于 4 000 m,系由于南海深海盆地扩张,大洋板块向下俯冲插入吕宋岛弧之下而形成。第 9 断续线上 A 点之下为下降岛坡中上部,水深 1 700 m,B 点之下为折断下降的西、北吕宋海槽中,水深约 3 000 m,C 点之下为水深大于 2 500 m 的海槽东侧坡麓上,其位于菲律宾吕宋岛西侧岛坡、海槽内,通过西北吕宋海槽将吕宋岛与中沙海台两个不同的地形单元予以区分,划分了菲律宾吕宋岛与中国中沙群岛之间的界线。

第 10 断续线位于巴士海峡西南部,台湾七星岩与吕宋岛之间,呈直线状,东北走向,南端位于马尼拉海沟内,北端位于巴士海峡中,所示范围大致为巴士海峡至台湾浅滩。

其地形横剖面形态特征明显(图 6),具有较大的起伏,经过 A、C 两点的剖面呈“W 形”,存在两处明显的凹谷,此区域内地形变化复杂,主要受挤压作用影响,从而形成了多条 NS 的海底山脉。第 10 断续线之下为菲律宾吕宋岛-中国台湾岛之间多座高出周围海底约 1 500 m 的海山、海台,其上 A、B、C 3 点下伏水深为 1 700～2 500 m 海峡中,其主要划分了菲律宾吕宋岛与我国台湾岛之间的界线。

第 11 段位于台湾东南部之外,呈弧形向太平洋一侧凸出,东北走向,所示范围大致为太平洋至台湾岛中部。

从其地形横剖面形态上来看(图 6),经过 A 点的剖面具有海沟、海岭相间分布的特点,地形起伏大。经过 B、C 点的剖面近岸一侧均贯穿了相对高差达 2 000 m 的海山,随后向 NW 水深逐步加大,最深处达到 5 500 m,在剖面上表现起伏较大。第 11 断续线上 A、B 两点之下为水深 2 600～3 700 m 的海底峡谷内,C 点之下为自台湾向大洋隆起、深度 3 200 m 的斜坡上,下伏地形为台湾东南侧、平均水深约 3 200 m 的“岛弧-海沟”体系的海沟之中,并且穿过巴士海峡而达太平洋,水深大,海底起伏变化大,是中国台湾岛与菲律宾

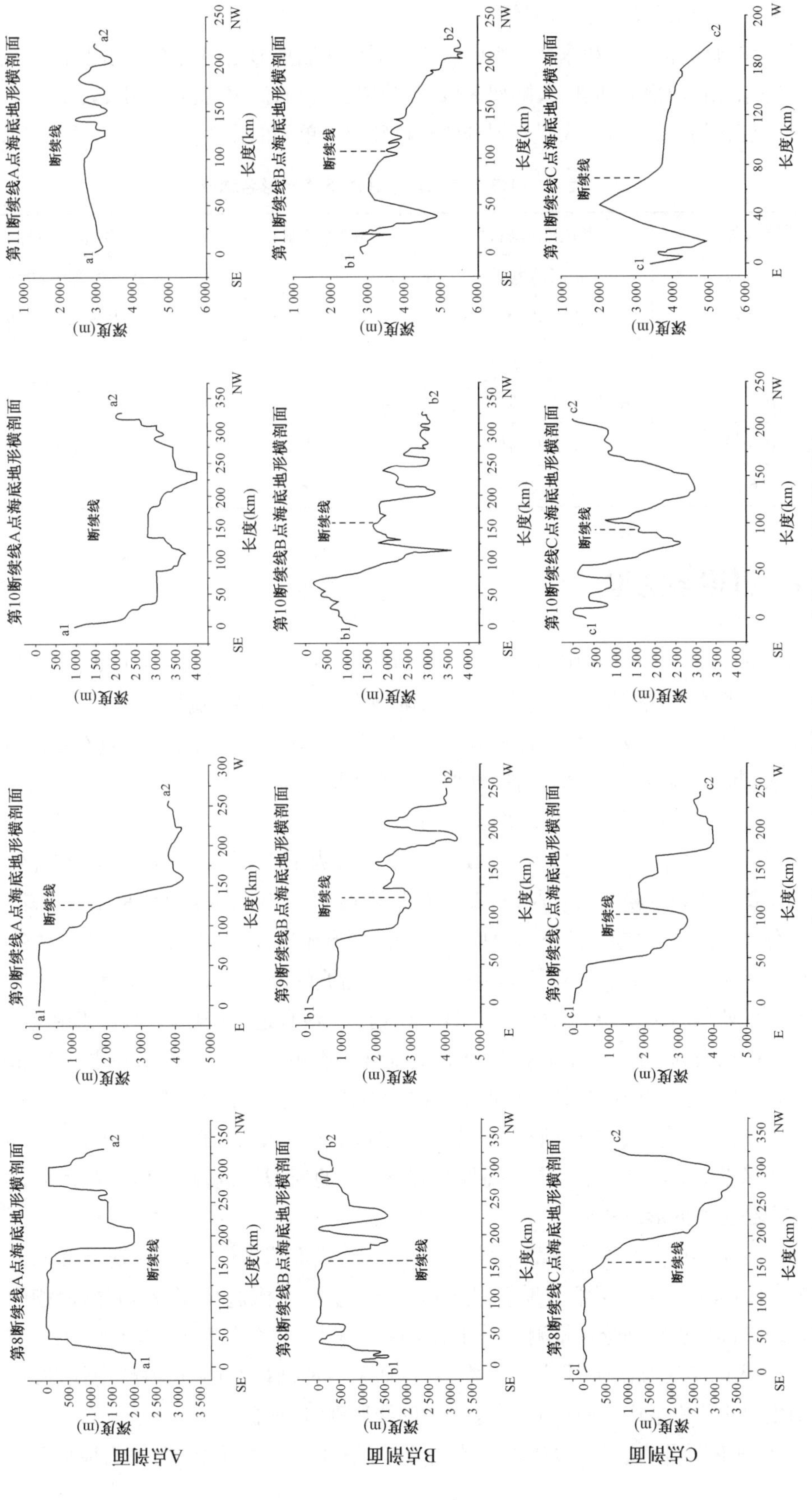

图 6　1947 年第 8～11 断续线上经过 A、B、C 点的海底地形横剖面

吕宋岛和太平洋之间的天然分界线。

综上南海西部、南部和东部三组断续线与下伏海底地形之间的关系可见，断续线主要位于南海东、南、西部大陆坡、大陆架区域，部分位于海沟、海槽、海峡之中（表 4）。在具有较大地形变化的区域，断续线起着划分两种特殊地形单元的作用。

表 4 1947 年南海断续线的下伏地形名称

断续线编号	下伏地形名称	断续线编号	下伏地形名称
第 1 段	北部湾盆地	第 7 段	北婆罗洲大陆坡
第 2 段	北部湾盆地	第 8 段	巴拉望岛西侧大陆架
第 3 段	中南半岛大陆坡北部	第 9 段	西、北吕宋海槽
第 4 段	中南半岛大陆坡南部	第 10 段	巴士海峡
第 5 段	巽他大陆架	第 11 段	台湾岛东侧海沟
第 6 段	巽他大陆架		

4 讨论与结论

（1）研究表明，1947 年中国政府内政部印制的 1∶400 万《南海诸岛位置图》中划定的南海断续线具有明显依据中国的大陆架、大陆坡海底地形划界的特征。断续线主要位于南海东、南、西部大陆坡、大陆架区域，部分位于海沟、海槽、海峡之中。在具有较大地形变化的区域，断续线起着划分两种特殊地形单元的作用。

（2）在 1947 年中国政府内政部印制的 1∶400 万《南海诸岛位置图》上，断续线有长短之分，间距有大小之别。最长段为第 1 段，最短段为第 11 段，具有西部长，东部短的特征。间距最大为第 8 段与第 9 段，间距最小为第 10 段与第 11 段，总体呈东西稀疏，南北紧密的分布特点，推测断续线下伏海底地形特征是形成这种分布特点的重要影响因素。

（3）断续线形态及分布与南海周围大陆架大陆坡的位置保持了较高的一致性。位于北部湾内的两段断续线处于大陆架堆积平原之中，水浅且地形变化和缓；位于中南半岛大陆架东侧的两段断续线，位于陆坡上，走向与陆坡边缘相平行，断续线所在水域水深较大；位于巽他大陆架上的两段断续线靠近纳土纳群岛及婆罗洲，所处位置水深较浅，地形变化平缓；位于南海东南侧的五段断续线，沿台湾岛南部至吕宋岛，直至巴拉望岛、巴士海峡，于岛架边缘近似 NS 向分布。由于岛坡受到 NS 向的断裂控制，断续线所处海底地形具有槽沟交替更迭的形态特征。

（4）1947 年中国政府内政部印制的 1∶400 万《南海诸岛位置图》在划定第 2、3、4、7、9 段断续线时，充分考虑了下伏地形，分别沿海台边缘、海槽边缘、海盆边缘、海槽中线等标志性地形部位进行划分。推测当时在划定这 11 段断续线时遵循了以下原则：

① 在大陆架浅海盆地（北部湾）、海峡处，通常沿两国岸线中点、海峡中线进行划分，与目前划界工作中最广泛使用的“等距离中间线原则”相吻合。

② 在具有明显地形变化的部位，例如南沙海槽、西北吕宋海槽等，实为地形单元分界

带所在，通常沿海槽坡麓或海槽槽沟中线进行划分。

③ 在中南半岛东侧陆坡区及南侧巽他大陆架区，通常结合岸线走向及下伏地形特征进行划分。断续线一般落在同一种地形之上，使之具有完整性、连续性。

参考文献

[1] 李金明. 2011. 南海断续线的法律地位：历史性水域、疆域线、抑或岛屿归属线？南海问题研究，4：22 - 29.

[2] 贾宇. 2005. 南海"断续线"的法律地位. 中国边疆地史研究，2：112 - 120.

[3] Sun Kuan-Ming. 1995. Policy of the Republic of China towards the South China Sea: Recent developments. Marine Policy, 19(5): 403 - 404.

[4] Song Yann-huei and Yu Peter Kien-hong. 1994. China's "historic waters" in the South China Sea: an analysis from Taiwan, R. O. C., American Asian Review, 12(4): 86 - 89.

[5] 姜丽，李令华. 2008. 南海传统九段线与海洋划界问题. 中国海洋大学学报(社会科学版)，6：7 - 8.

[6] 王颖，葛晨东，邹欣庆. 2014. 论证南海海疆国界线. 海洋学报，36(10)：1 - 11.

[7] 王颖，马劲松. 2003. 南海海底特征、资源区位与疆界断续线. 南京大学学报，39(6)：797 - 805.

[8] 民国内政部方域司制. 1947. 南海诸岛位置图.

[9] 傅角今，郑资约. 1947. 南海诸岛地理志略. 商务印书馆，3 - 33.

[10] 姚伯初. 1995. 中南-礼乐断裂的特征及其构造意义[J]. 南海地质研究，7：1 - 14.

[11] 冯文科. 鲍才旺. 1982. 南海地形地貌特征. 海洋地质与第四纪地质，4：80 - 93.

[12] 中国科学院南沙综合科学考察队. 1989. 南沙群岛及其邻近海区综合调查研究报告(上卷). 北京：科学出版社，11 - 106.

[13] 中国科学院南沙综合科学考察队. 1989. 南沙群岛及其邻近海区综合调查研究报告(下卷). 北京：科学出版社，15 - 53.

[14] 姚伯初，万玲，刘振湖. 2004. 南海海域新生代沉积盆地构造演化的动力学特征及其油气资源. 地球科学，5：543 - 549.

[15] 赵焕庭. 南沙群岛自然地理. 1996. 北京：科学出版社，35 - 60.

[16] 刘昭蜀，赵焕庭，范时清等. 2002. 南海地质. 北京：科学出版社，22 - 102.

[17] 陈史坚. 1982. 南沙群岛的自然概况. 海洋通报，1：52 - 58.

[18] 赵焕庭. 1996. 南沙群岛自然区划. 热带地理，16(4)：304 - 309.

[19] 王颖. 1996. 中国海洋地理. 北京：科学出版社，74 - 87.

[20] 冯文科. 1982. 南海地形地貌特征. 海洋地质研究，2(2)：82 - 85.

[21] 曾成开，王小波. 1987. 南海海盆中的海山海丘及其成因. 东海海洋，5(1/2)：1 - 9.

[22] 刘昭蜀. 2000. 南海地质构造与油气资源. 第四纪研究，20(1)：70 - 77.

Evidence of China's Sea Boundary in the South China Sea*

1 Introduction

Over the years, the boundary surrounding the South China Sea Islands has been called a dashed line on conventional maps. There are different understandings about this boundary. In fact, these different interpretations reflect the different understanding of the territorial ownership embodied in the history. Such inconsistent interpretations and understanding have resulted in great confusion in maritime delimitations. To explore the origins of this line usage and objectively clarify its position in maritime history, and also to understand the essence of its initial meaning, the authors have systematically studied historical maps and archives after World War II, with the expectation of clarifying the true meaning of this dashed line and providing a scientific basis for the delimitation of the South China Sea.

2 Materials

To explore the essential meaning of this boundary, the authors reviewed the original archives of the *Location Map of the South China Sea Islands* (six pieces) from the Republic of China era stored in Taipei and Nanjing①(Table 1). This series of maps was drawn by the Territorial Administration Division of the Ministry of Interior after the National Government sent troops to the South China Sea Islands to declare China's sovereignty following the victory of the Anti-Japanese War (World War II). They were completed in December of the 35th year of the Republic of China (1946). According to the archived document [(36) Ministry of Interior-IV 30844] of the Republic of China, they were approved on 7th August in the 36th year of the Republic of China (1947) by Zhang Qun, then the President of the National Executive Administration. In early November 1947, Ta Kung Newspaper reported *the Internal Names of the South China*

---

* Ying Wang, Chendong Ge, Xinqing Zou: *Acta Oceanologica Sinica*, 2017, Vol. 36, No. 4: pp. 1 – 12

① Territorial Administration Division of the Ministry of Interior, the Republic of China. 1947. Location Map of the South China Sea Islands (six pieces) (in Chinese). The Nanjing Second Institute of Archives.

Sea Islands Verified; on 1st December 1947, the Territorial Administration Division of the Ministry of Interior released *the Location Map of the South China Sea Islands*, and published the names of these islands in the Central Daily News. *The Location Map of the South China Sea Islands* was included in the book *Short Records of Geography of the South China Sea Islands* (Zheng, 1947) and was then published in February 1948 as an attached map of *the Administrative District Map of the Republic of China* (Fu, 1948).

Table 1 Location maps of the South China Sea Islands (six pieces)

Map title	Scale	Department and date	Comments
Location Map of the South China Sea Islands	1 : 4 000 000	Territorial Administration Division of the Ministry of Interior, 1946	
The Nansha Islands	1 : 2 000 000	Territorial Administration Division of the Ministry of Interior, 1946	
The Zhongsha Islands	1 : 350 000	Territorial Administration Division of the Ministry of Interior, 1946	
The Xisha Islands	1 : 350 000	Territorial Administration Division of the Ministry of Interior, 1946	
The Yongxing Island and Shidao Island in the Xisha Islands	1 : 10 000	Territorial Administration Division of the Ministry of Interior, 1946	surveyed map
The Taiping Island in the Nansha Islands	1 : 5 000	Territorial Administration Division of the Ministry of Interior, 1946	surveyed map

3 Results and discussion

Boundaries with neighboring countries are clearly marked on the *Location Map of the South China Sea Islands*. The symbols are — <■> — <■> — <■> — <■>— from the northwest end of the map, first denoting the boundaries between Guangxi Province of China and Vietnam, and then the boundaries between the Indo-China Peninsula and Xisha and Nansha Islands, between the Nansha Islands and Borneo, and the Philippines, and then through the Bashi Strait. The 1:4 000 000 map marks out the boundaries of China in ten segments (Fig. 1). Here, strict rules are established governing the delimitation, number and location of boundaries. This map draws lines in the middle of two countries and uses the symbol <■> between lines for the purposes of signification. Although the illustration in Fig. 1 is not accompanied by interpretations of all the legends, it clearly defines the country's land and sea boundary. The most convincing evidence is that, at the northwest end of the map, sea and land boundaries converge into one, thus defining the borders of the two adjacent countries. To the northeastern part of the boundary is Guangxi Province of China, while to the south of the boundary is Tong Deng and Liangshan of Vietnam. The nature of this line with its symbols clearly identifies its status as a border between two countries (Fig. 2). Additional evidence for this is that, between the southern tip of the Palawan Islands of the Philippines and the Borneo of Malaysia, the same symbol that is used to demarcate the boundary between China and Vietnam is here used to delimit the two countries' boundaries (Fig. 3). It is worth noting that these two lines marking the boundaries between two adjacent countries still remain in service.

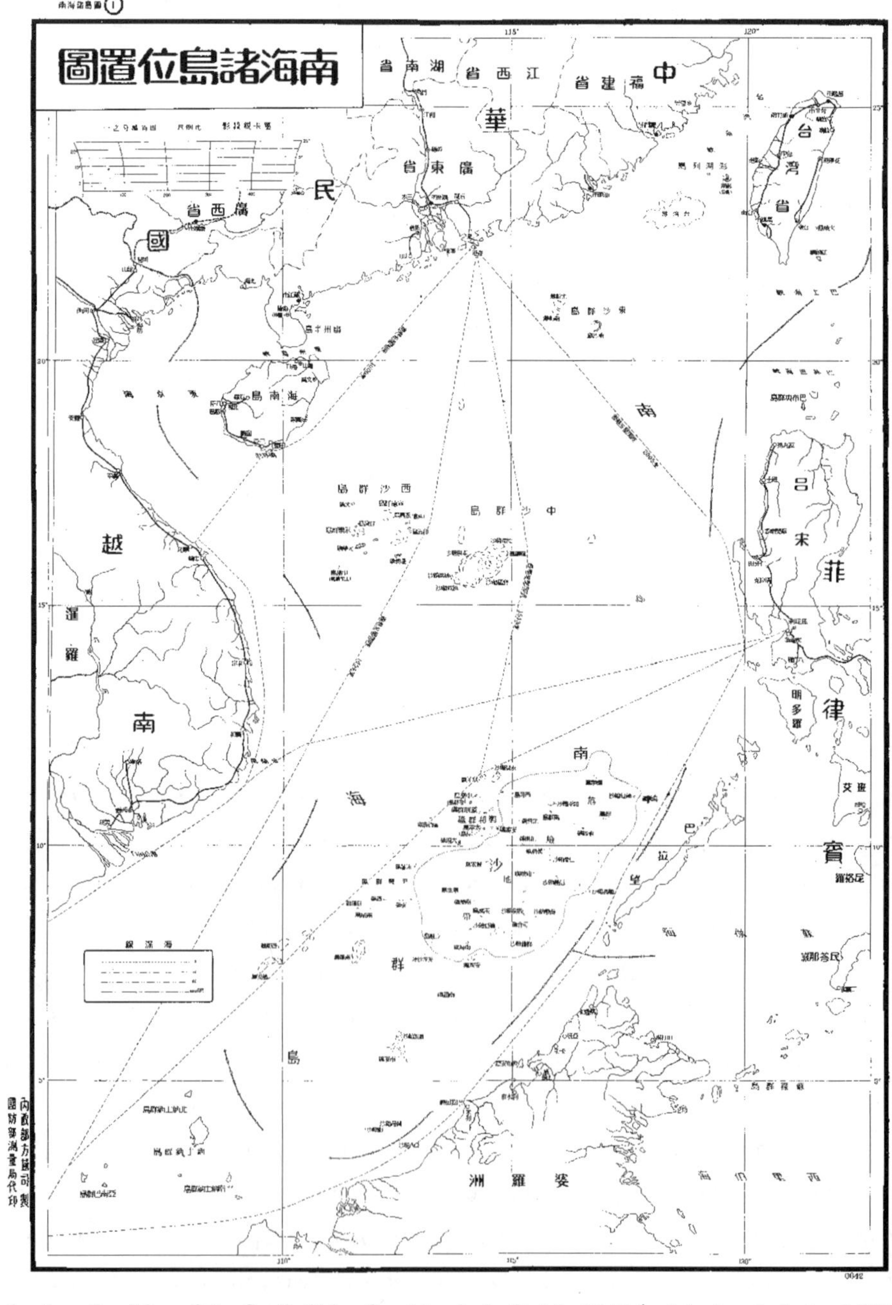

Fig. 1 Location Map of the South China Sea Islands drafted in 1946 (original scale is 1 ∶ 4 000 000)

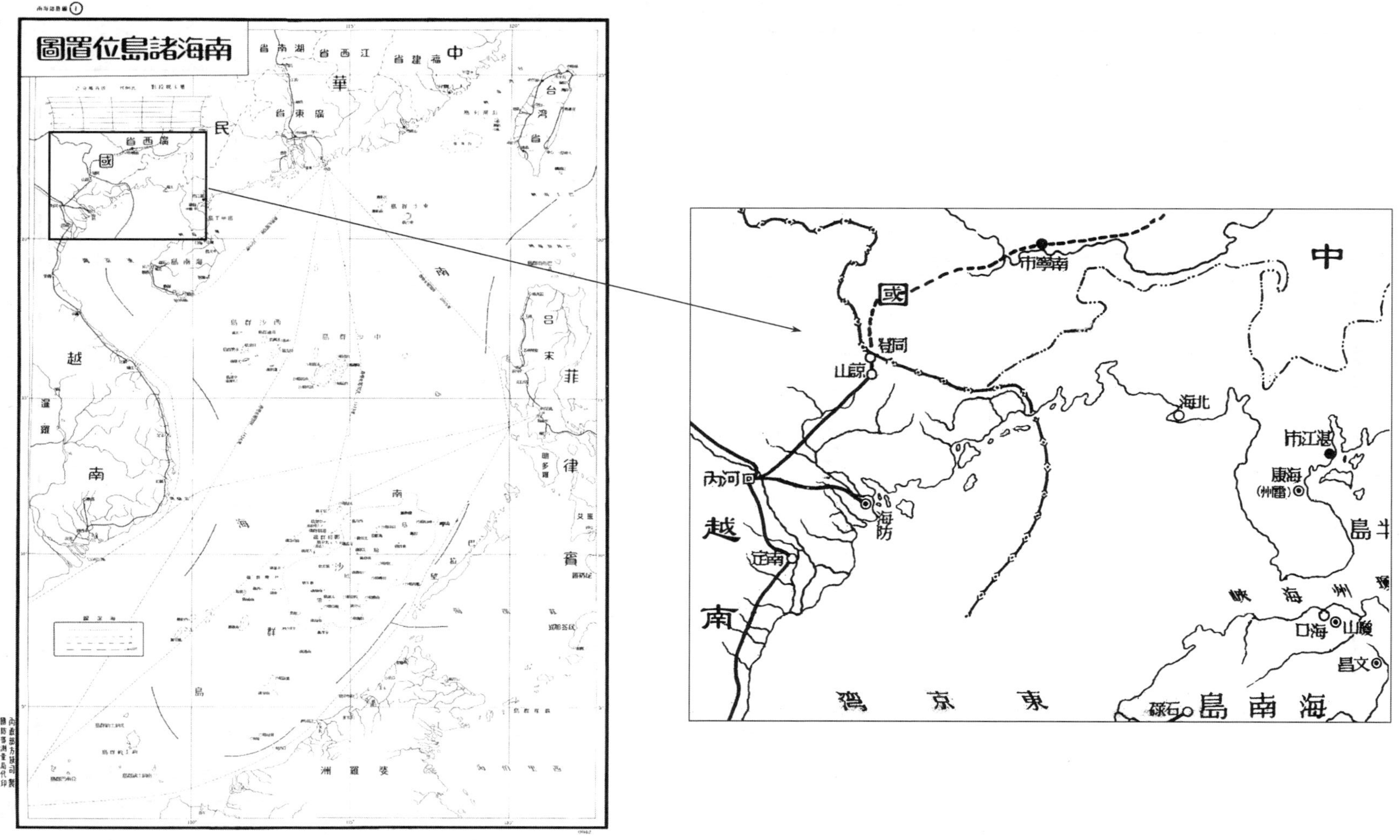

Fig. 2 The terrestrial boundary between China and Vietnam being extended as the Sea Boundary in the South China Sea

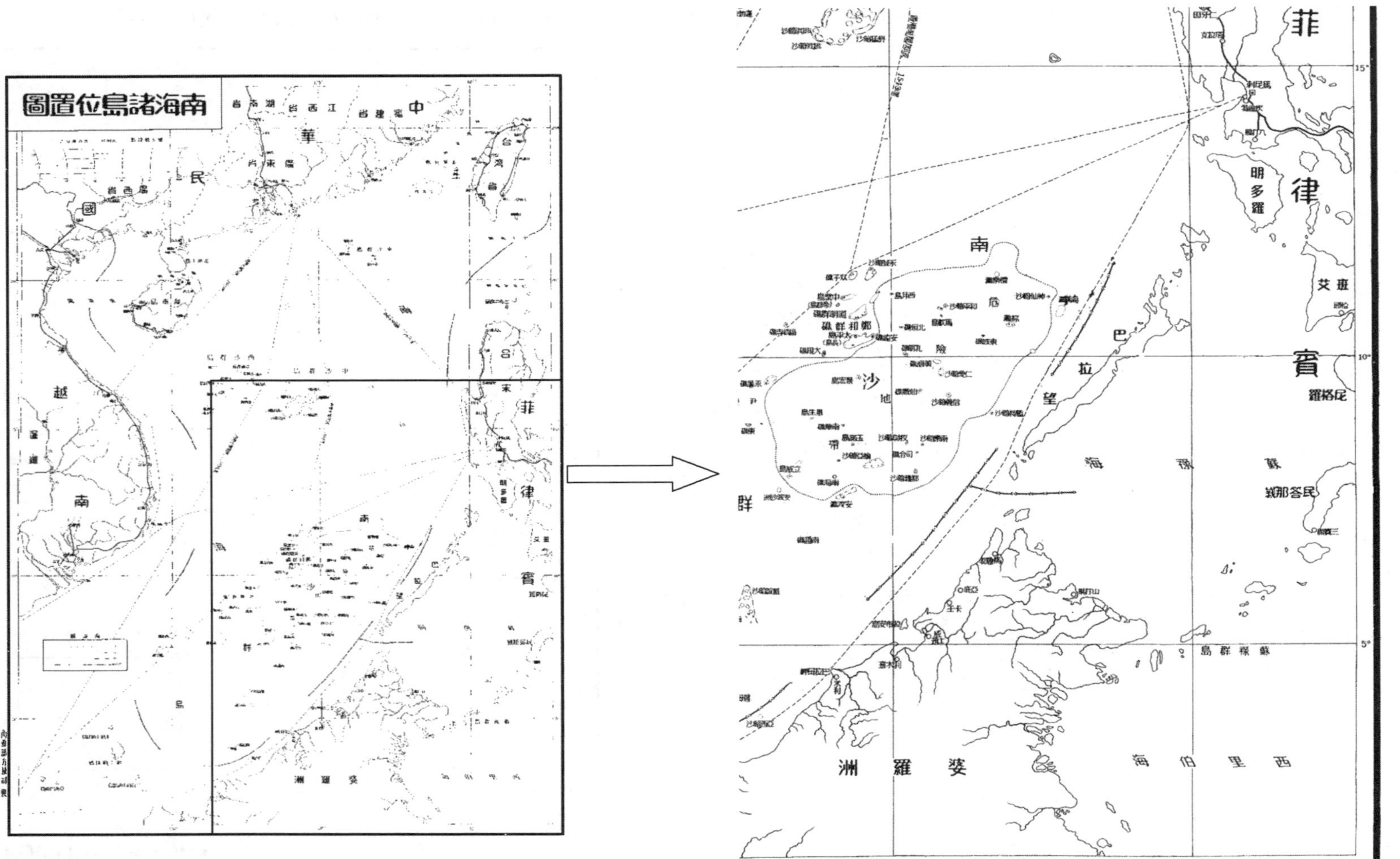

Fig. 3 The boundary between Palawan of Philippine and Borneo

In addition to the above two parts, the other parts of the map also use such disconnected symbols to mark the boundaries between China and its surrounding countries; namely, the middle line between the east shore of Vietnam and China's Xisha Islands and the Nansha Islands. The middle line between islands, the middle line between sandbank and coastal land, and so forth, all follow the same and clear principle.

We saw the *Administrative District Map of the Republic of China* (Fu, 1948) published by the Commercial Press in 1948 in the library of School of Geographic and Oceanographic Sciences, Nanjing University (scaled at 1 : 4 200 000). It was compiled by the Territorial Administration Division of the Ministry of Interior, and Fu Jiaojin was the chief editor (Fig. 4). The bottom right corner of the map provides the *Location Map of the South China Sea Islands* (scaled at 1 : 9 100 000, Fig. 5). The border in the map is marked with — <■> — < ■>, and is explicitly annotated in the legends as a national boundary (Fig. 6). It is also apparently differentiated from the undefined national boundary, represented by dotted lines. Therefore, it can be judged that the disconnected lines surrounding the South China Sea denoted China's national border. Likewise, it can be said that the sea boundary is both connected with and an extension of the land border. That is to say, it is the national boundary that separates the South China Sea Islands from neighboring Asian countries in the sea, demarcating both the boundaries and scope of China in South China Sea and those of other countries. And

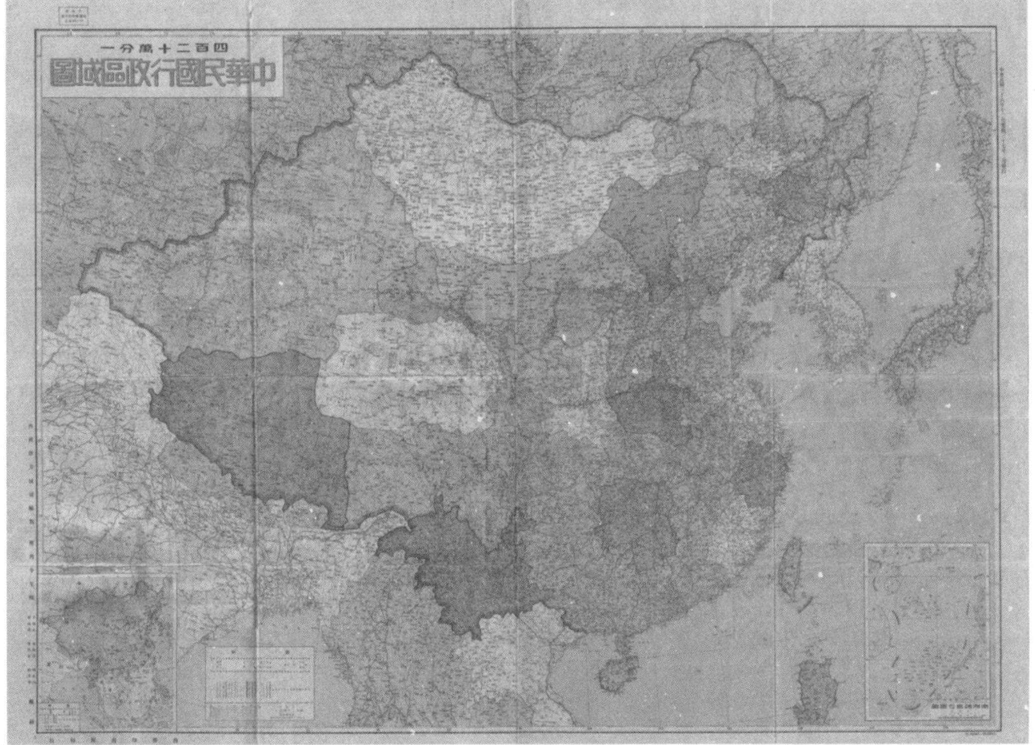

Fig. 4　Administration District Map of Republic of China published in 1948 (original scale is 1 : 4 200 000)

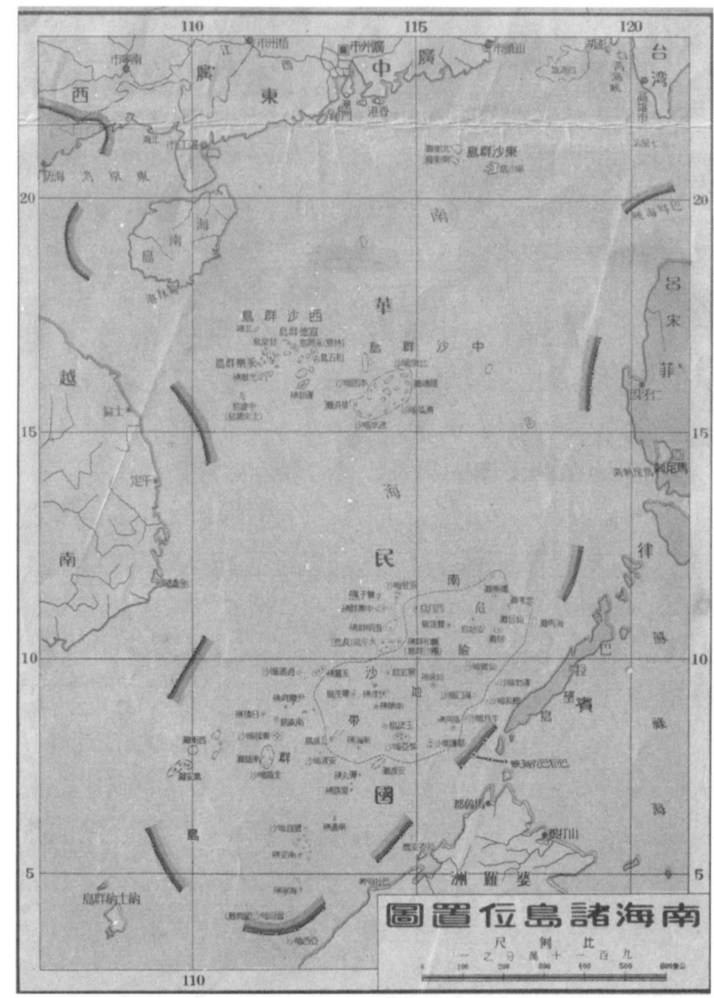

Fig. 5　Location Map of the South China Sea Islands attached to the Administration District Map of Republic of China (original scale is 1∶9 100 000)

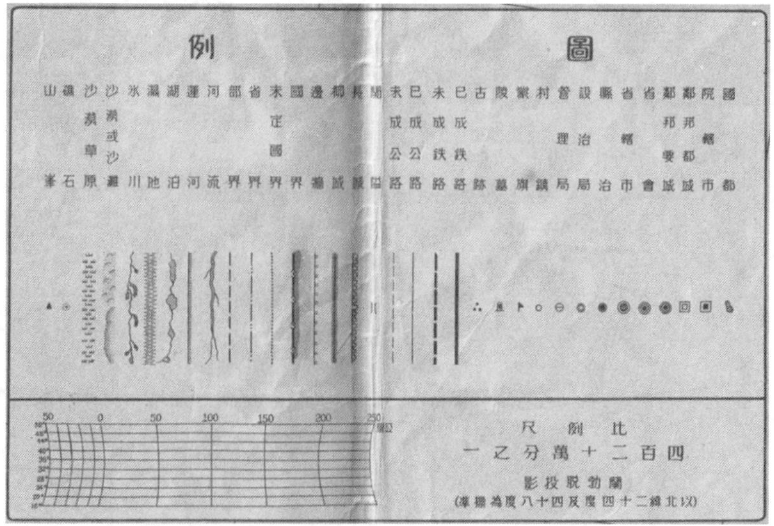

Fig. 6　Legends of the Administration District Map of Republic of China

Chinese Character "中华民国"(The Republic of China) is clearly marked in the area of the South China Sea surrounded by the sea boundary, which indicates undisputedly the area belongs to China.

The Nansha Islands map among the six above, including all islands in Nansha, is scaled at 1 : 2 000 000. Its east, west and south boundaries also use the symbol — <■> — < ■> —, i. e., squares between brackets and within dashed lines; there are also ten segments (Fig. 7). In fact, in this set of maps, even though the scope may differ across maps of different scales, all sea borders are drawn with ten segmented lines (Figs 1 and 7). In the officially published *Sketch Location Map of the South China Sea Islands* in the scale of 1 : 6 500 000 (Zheng, 1947) and the *Location Map of the South China Sea Islands* in the scale of 1 : 9 100 000 (Fu, 1948) (Fig. 5), the country's boundary,

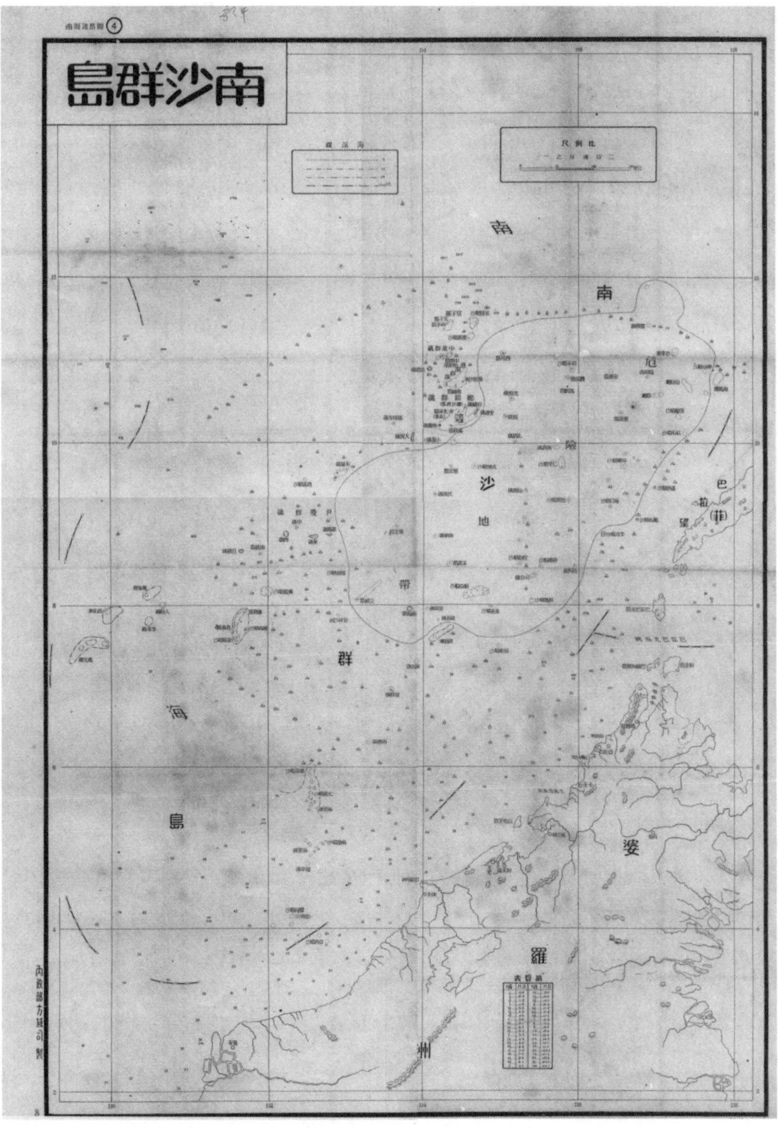

Fig. 7　The Nansha Islands (original scale is 1 : 2 000 000)

which adjoins the west of the Philippines, is divided into two sections, resulting in a total of 11 segments. Therefore, it can be argued that the South China Sea territorial boundaries should not be expressed in numbers. Currently, it seems to be inappropriate to refer to the boundaries around the South China Sea as "nine segmented lines" after taking out two segments of the boundary in the Beibu Gulf. It should be called China's sea boundary in the South China Sea.

We recognized that the sea boundaries in the South China Sea were marked with disconnected lines. These are arguably more realistic owing to the natural fluctuations of sea and tides, and the attribute that vessels can pass without harm. On international maps, sea border lines are also represented with such disconnected lines. For example, in the *New Canadian Oxford Atlas* (Oxford University Press, 1977), the boundaries between Italy and the islands in Spain, and between Turkey and Greece, are both represented with disconnected lines. Therefore, using disconnected lines to represent sea boundary between countries can be seen as an international standard practice.

The maps of Xisha and Zhongsha Islands are drawn to a scale of 1 : 350 000 (Figs 8 and 9). The legends in the map of Xisha Islands include coral reefs, sandy shores, broadleaved forests, contour lines, depth points and wreck sites. The Zhongsha Islands map uses 20 m bathymetric contour plots to outline the Zhongsha Islands, including measured depth points and the locations of various shoals and reefs on the map.

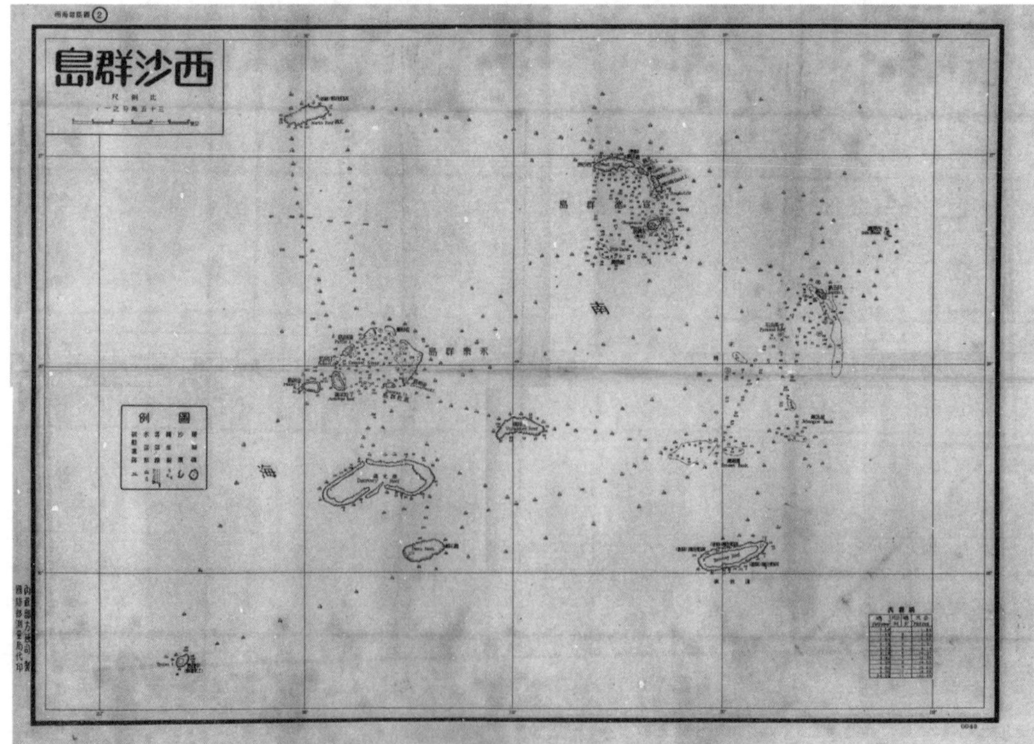

Fig. 8 The Xisha Islands (original scale is 1 : 350 000)

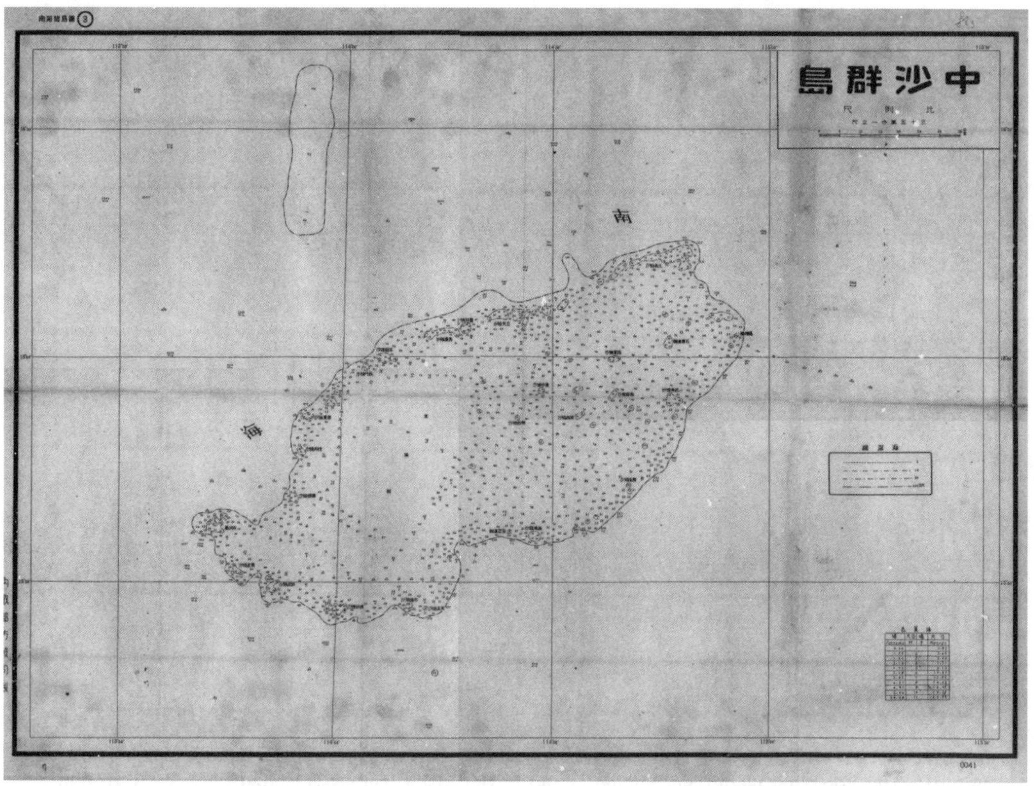

Fig. 9　The Zhongsha Islands (original scale is 1 : 350 000)

The measured maps of Yongxing Island and Shidao Island are drawn at a scale of 1 : 10 000, and marked "mapped on November 30 in the 35th year of the Republic of China" (Fig. 10). The islands and piers are clearly drawn. The maps also delineate that Yongxing Island and Shidao Island are located on the same reef, which is exposed at low tide, with waves forming during high tide (in the surf zone). The map clearly marks that, at the time, the Yongxing Island occupies an area of 1 851 000 m^2, or 7.4 li^2 (Chinese measure, 1 li^2 = 0.25 km^2). It is flat, and there are dunes at the southwest shore, with the highest altitude reaching 8.3 m. In addition: "Stone Island occupies an area of 78 000 m^2, or a total of 0.312 li^2. The highest point is 12.4 m in altitude." When we visited the islands on May 20, 2013, the entire island was covered with forest and we were not able to see the original dunes. GPS positioning determined that the highest point was 15.9 m, while the most up to date literature reports 13.8 m. If there had been a highly accurate measurement in the past, we could have determined the change over the past 60 years. The fact that the original map pieces introduced "拔海" (meaning "pulling the sea") to describe the height above sea level drew our attention; we deem it to be a more suitable term to the connotation here than "海拔" (a term presently used), although we are unsure as to why and how it has evolved historically.

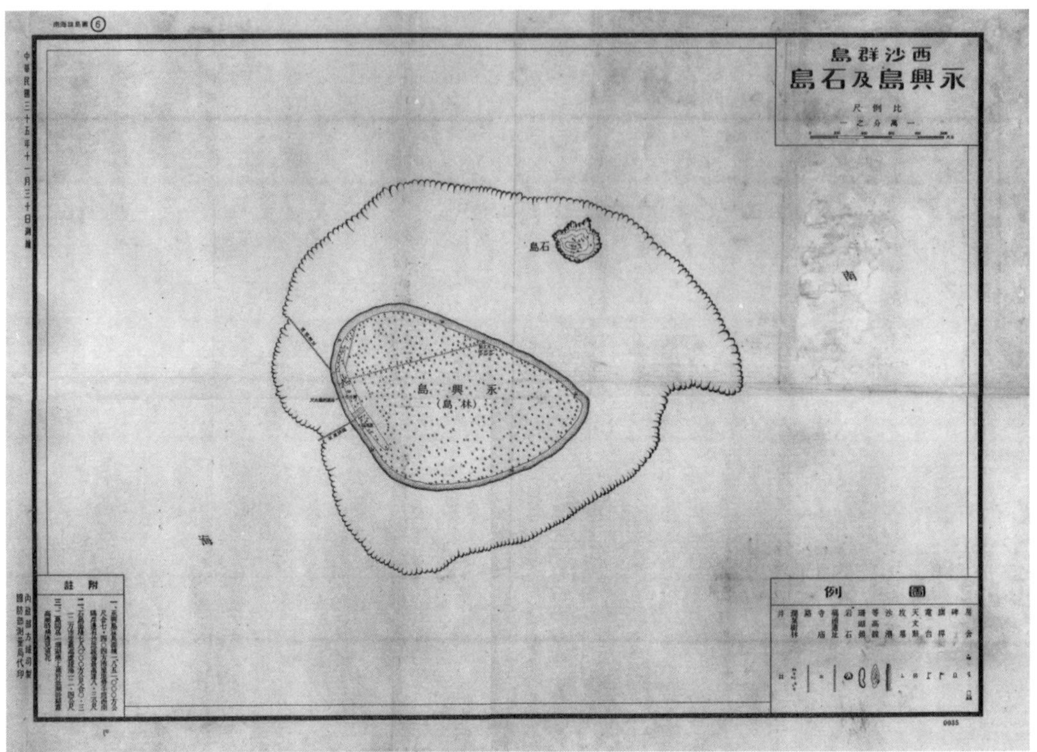

Fig. 10　The Yongxing Island and Shidao Island in the Xisha Islands (original scale is 1 ∶ 10 000)

The measured Taiping (meaning Peace) Island map is scaled at 1 ∶ 5 000 (Fig. 11). There are notes on the map, describing the island as having an area of 432 000 m^2 or 173 li^2 (Chinese measure), and an elevation of 4 m; the structures on the island are said to have been destroyed. The legends include "houses (reinforced concrete, wood house, house base), monument, radio, flagpoles, wells, reservoirs, graves, shelters, trees (coconut trees, broadleaved forests, independent trees), base stone circles, roads, trails, light rails, sand beaches, shipwrecks, rocks, coral reefs," and also indicate that there used to be military forces and residents; however, by the time of mapping, "all buildings were destroyed". The notes also state that "coral reefs around the islands will emerge during low tides, while waves will splash up along the edge of coral reefs during high tides." The upper left corner of the frame additionally states that the island was "mapped on December 12 in the 35th year of the Republic of China."

After the People's Republic of China was founded in 1949, China published a variety of maps relating to the sea boundary in the South China Sea. The drawing method followed the mapping conventions published by China's government in 1948. These new maps of China were compiled by the Huaxia Society of History and Geoscience and published by Yaguang Geography Club in 1951 (Huaxia Society of History and Geoscience, 1951) (Fig. 12). The boundaries on the South China Sea islands map shown on its lower right corner are the same as those on the map published

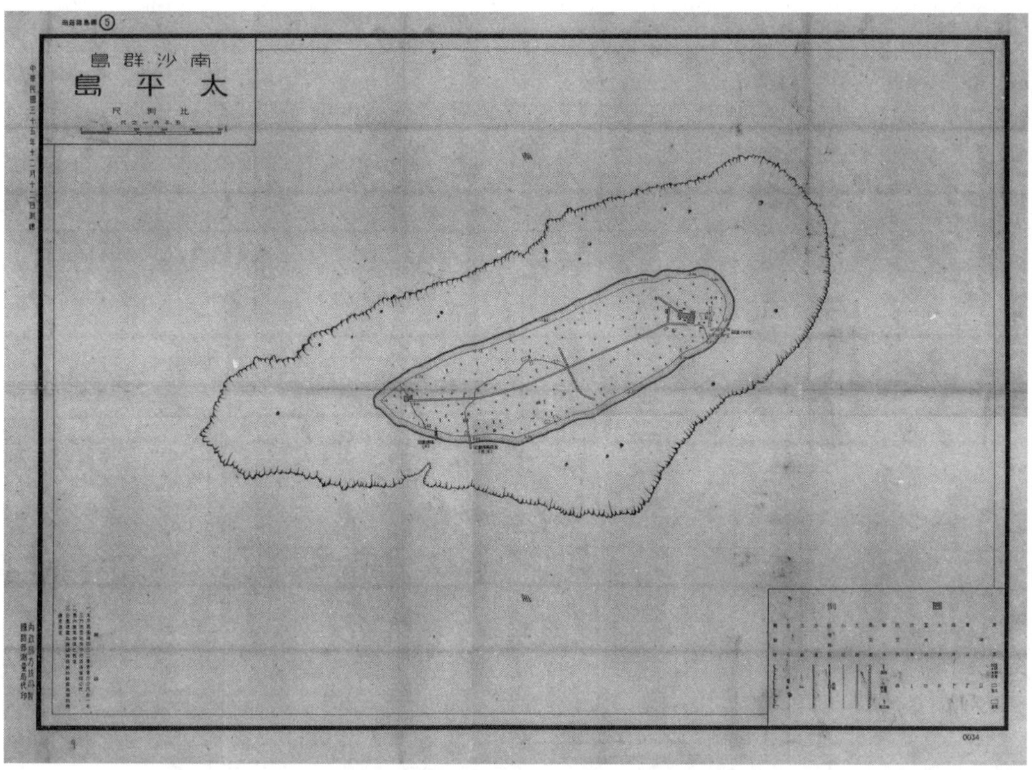

Fig. 11　The Taiping Island in the Nansha Islands (original scale is 1∶5 000)

Fig. 12　The　New Map of the People's Republic of China (1951, original scale is 1∶8 500 000)

in 1948 (Fig. 13). In addition, they represent the national border, as indicated by the legends on the lower left corner (Fig. 14).

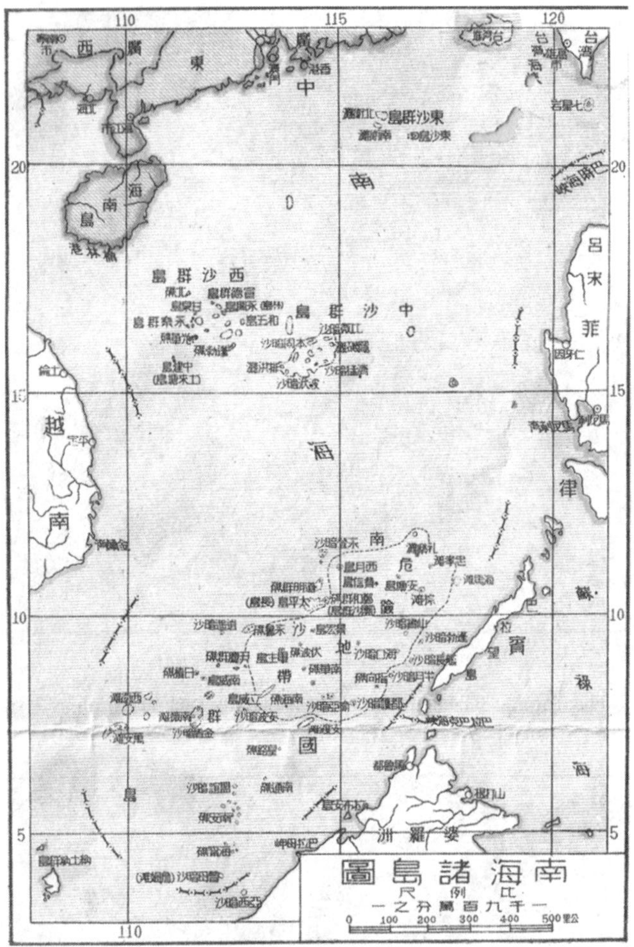

Fig. 13　Location Map of the South China Sea Islands attached to the New Map of the People's Republic of China (1951, original scale is 1 : 19 000 000)

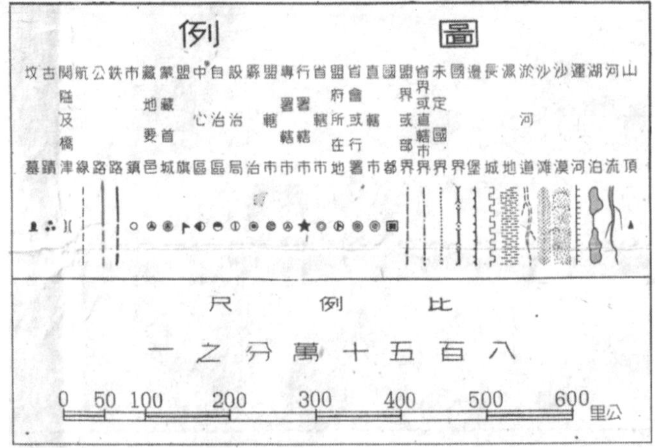

Fig. 14　Legends of the New Map of the People's Republic of China (1951)

Symbols for national borders on the map published in 1948 (Fu, 1948) are similar to those on the *People's Republic of China Atlas* (SinoMaps Press, 1979). The latter is similar to the *Esselte World Atlas* (in Chinese) (Esselte Map Service, 1991) translated and published by SinoMaps Press in 1991. However, on those maps, the symbols, —[■]—[■]—[■], a series of squares in parentheses connected by lines which represent China's border, are slightly different from the 1946 map, although they are both parentheses. The authors studied the locations and spatial details of the China's sea boundary in the South China Sea on the original *Location Map of the South China Sea Islands*, compared it with the *Map of South China Sea Islands* published by SinoMaps Press in 1982, and found the defined sea boundary to be almost the same (Wang and Ma, 2006), implying that the boundary's position came from a strict origin and had been carefully demarcated.

When looking into archives related to the South China Sea Islands from January of 1930 to October of 1931, we found that the government of the Republic of China, at that time had already drafted the *Provisions of Territorial Sea Boundary Draft*①. This document clearly points out that "territorial waters and territorial lands are both involved in a nation's sovereignty and law. In many cases, it owns neutrality right in wartime and fishing right in peacetime in foreign relations, as well as relates to the division of its associated areas in naval districts in military affairs. To date, there has been no regulation in this area, and we should not hesitate in establishing it". From this text, we can derive the principles of how the sea boundary in the South China Sea was designated.

The second part of the document contains examples of each country's territorial sea. The extension of territorial waters is three nautical miles in Britain, USA, France, Germany, Japan, four nautical miles in Scandinavian countries (Sweden, Norway), six nautical miles in Portugal, and 12 nautical miles in Russia.

The third part of the file analyzes the importance of the size of the territorial sea (not discussed in this article).

The fourth part clearly proposes that China territorial sea should be demarcated as 12 nautical miles.

The fifth part of the document contains the criteria for calculating the territorial sea:

(1) Calculate the territorial sea by 12 nautical miles, starting from the lowest low tide point, and then going forth.

(2) When the bay is completely within our sea water border line.

① The National Executive Administration, the Republic of China, 1931. Resolution No. 3606. Provisions of Territorial Sea Boundary Draft.

(a) If the mouth of the bay is no wider than 24 nautical miles, draw a straight line across the bay mouth, and calculate from that line.

(b) If the bay mouth is wider than 24 nautical miles, select a point where the mouth is less than 12 nautical miles and calculate from that line.

(c) If the bay belongs wholly to our nation, draw a straight line across the bay mouth and calculate from that line, without considering the width of the bay mouth.

(3) If the bay is not only inside our sea boundary but also inside another country's sea boundary, the territorial sea should be divided in half between our baseline and the other country's baseline.

(4) Bay: From the farthest point in the bay, draw a straight line to calculate the distance.

(5) Island: the coastal islands have their own territorial sea.

(a) If islands are included in our territory, but the distance between them and other islands is less than twice as much as that of our territorial sea region, the extension should be calculated from the farthest island of the islands group.

(b) Sea waters within the islands should be considered as our territorial sea.

(c) If the distance between the islands and continent, or between two terminal islands of the islands group and the coastal line is no more than twice the territorial waters, the waters included should be considered as territorial waters.

(6) Channels: If coasts of straits all belong to our nation, the channel entrance is less than twice as long as the extent of our territorial waters, the water within the channels should be considered our territorial waters. If the coast of the strait between our country and other countries is less than twice as long as the extent of our territorial waters, it should be divided down the midpoint of the owned area. Whenever a strait is connected with our inland water, the way to calculate the territorial waters should follow the aforementioned rules.

If the entrance points of rivers are terminal of their inner water, there is no need to consider the width of their inner water. The calculation of its territorial water should follow the aforementioned rules.

If the exit of a river is a Half (a Half system refers to a narrow and shallow-water bay, which has dyke outside and gap in the middle that connect the water in the bay with the sea waters beyond), the way to calculate the extent of the territorial water should follow the aforementioned rules.

Provisions of Territorial Sea Boundary Draft is attached to the resolution of the National Executive Administration (No. 3606), which states: "The National Executive Administration made its resolution on the national conference that the limit of the territorial sea should be set at 12 nautical miles, that the issue of boundary settlement should be handled by the Navy, and that a directive should be given to the nationalist government and the Ministry of Interior, the Ministry of Foreign Affairs, the Ministry of Finance, the Navy and the Ministry of Industry (Directive 683)". The resolution was jointly signed and approved for release on February 26th in the 20th year of the Republic of China (1931) by Chiang Kai-shek, the President of the National Executive Administration, and Tsu-wen Soong, the Vice President of the National Executive Administration.

4 Conclusions

The extension of Chinese territorial waters was discussed and proposed as early as 1930—1931, and was approved to proceed with measurement by the National Executive Administration of the government in February, 1931. The Japanese War of Aggression against China interrupted the exploration work. After World War II, the government of the Republic of China completed the reception of the South China Sea Islands and announced this to the world. The disconnected lines surrounding the South China Sea were China's sea boundary, which were connected with and an extension of the land border. In other words, thi was the national boundary that separates the South China Sea Islands and neighboring Asian countries in the waters, and demarcated the boundaries and scope of China's territorial waters around the islands from those of other countries. These boundaries have been marked on the South China Sea as dashed lines in accordance with international practice, and have been approved by numerous treaties, diplomatic talks, published atlases, and other official publications internationally. The boundary of the South China Sea is not suitable to be represented by using numbers and, as such, it is inappropriate at present to refer to the boundary as Nine Segmented Lines after taking out the boundary of Beibu Gulf. It should be called China's sea boundary in the South China Sea. To clarify the veracity of these lines is of great scientific significance to boundary demarcations in the South China Sea.

References

[1] Esselte Map Service. 1991. Esselte Atlas of the World (in Chinese). Beijing: SinoMaps Press.

[2] Fu Jiaojin. 1948. The Administrative District Map of the Republic of China (in Chinese). Shanghai: The Commercial Press.

[3] Oxford University Press. 1977. The New Canadian Oxford Atlas. Don Mills: Oxford University Press (Canada).

[4] SinoMaps Press. 1979. The People's Republic of China Atlas (in Chinese). Beijing: SinoMaps Press.

[5] Wang Ying, Ma Jinsong. 2006. Characteristics of sea bottoms in the South China Sea, territory domain and digital South China Sea. In: Su Jilan, ed. Review and Thinking of Zheng He's Voyage to

the West (in Chinese). Beijing: Science Press, 177 - 193.

[6] Huaxia Society of History and Geoscience. 1951. The New Map of the People's Republic of China (in Chinese). Shanghai: Yaguang Geography Club.

[7] Zheng Ziyue. 1947. Short Records of Geography of the South China Sea Islands (in Chinese). Shanghai: The Commercial Press.

黄岩岛海域环境与战略地位分析*

一、岛礁基础与海域环境

黄岩岛(副名:民主礁)归属中沙群岛。中沙群岛北起宪法暗沙(16°20′N,116°44′E),南至中南暗沙(13°57′N,115°24′E),东起黄岩岛(15°08′N,117°45′E),西至中沙环礁的排洪滩(15°38′N,113°43′E)。中沙群岛海域内有中沙环礁和黄岩环礁,以及一统暗沙、神狐暗沙、宪法暗沙和中南暗沙,黄岩岛位于黄岩环礁中的北端。中沙群岛位于南海中央深海盆以西,处于水深 2 000 m 的断降大陆坡阶面上,四周为明显的断裂槽,东临水深大于 4 000 m 的中央深海盆,西临水深 2 000 m 的中沙海槽。中沙环礁为水下大陆坡环礁,礁体处于海面以下。黄岩岛在中央深海盆区的东侧,位于中沙群岛东南部,距中沙环礁 315 km 处,是在海盆中的东西向海山带上耸立为岛礁的最高海山,其基底岩层为大洋玄武岩,该处因海底被拉张开裂,地球深部地幔物质上涌,构成南海洋壳型海底(图 1)。中央深海盆区中部有宪法暗沙及中南暗沙。深海盆北面的北部大陆坡上有一统、神狐等暗沙。

黄岩环礁及其上的黄岩岛是中沙群岛中唯一出露于海面以上的基岩礁岛,包括相距约 10 海里的南、北二岩(图 2)。黄岩环礁是在 3 500 m 深处的海底火山上部发育的环礁,是我国南海中唯一的大洋型环礁,环礁轮廓似等腰三角形,周长 55 km,腰长 15 km,包括潟湖,全部面积 150 km^2(图 3)[1]。环礁宽约 1 km,环围着中间的三角形潟湖,面积约 130 km^2,水深 10～20 m,内多点礁与喜礁植物繁殖。潟湖出口在东南侧,宽 360～400 m,中间水深 6～11 m,边缘水深约 3 m,小船可进入潟湖避风。环礁四周有大型礁块出露,第二次大战期间,美军舰官兵曾以巨石为目标,进行打靶。环礁顶部之巨石,是经暴风浪冲蚀与堆积之遗迹,反映出该海域经历过罕有的巨浪作用。英文称该处为“Scarborough Shoal”,亦称“Maroona Reef”,缘于 1748 年英国东印度公司的商船“Scarborough 号”在该处触礁沉没[2]。1935 年中国水陆地名委员会公布的地名是“Scarborough Reef”,标明位于中国领土范围。1938 年—1945 年时期,中国出版的地图上用“南石”、“South Rock”标志黄岩岛。1947 年民国政府公布的地名中,称该处为民主礁。1983 年中华人民共和国政府定名为黄岩岛,正式对外公布。“黄岩”之名反映了该岛泛黄色巨石的特征。

黄岩岛的基底是南海深海盆张裂过程中喷发的海底火山,是在呈东西走向的一系列火山锥组成的火山链上。该系列火山锥多为水下海山,仅在火山顶部发育的黄岩环礁与黄岩岛达到海面附近(图 4)。火山链的南北两侧均为深度超过 4 200 m 的海底洼地,是海底扩张的遗迹。东西向的火山链其位置——15°N 附近(14°55′～15°27′N,116°55′～118°05′E),可能是南海深海盆扩张时的轴部,该处地壳厚度 12～14 km,而外围海盆地壳

* 王颖:《亚太安全与海洋研究》,2017 年第 3 期,第 18－28 页。

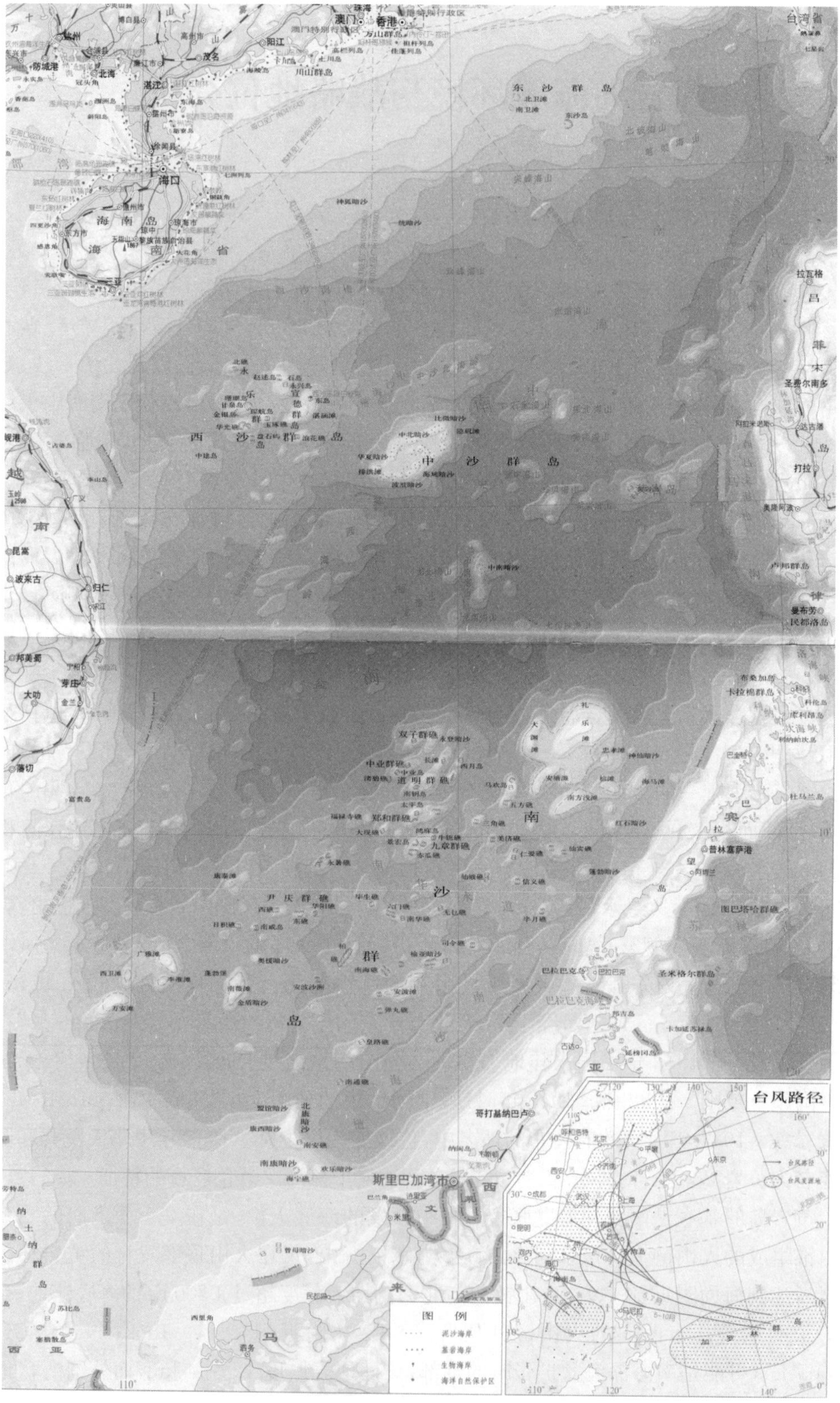

图1 南海海域的中沙群岛与黄岩岛(附彩图)

(据:总参谋部测绘局编制.中华人民共和国地图集新世纪版.北京:星球地图出版社,2011,24-25)

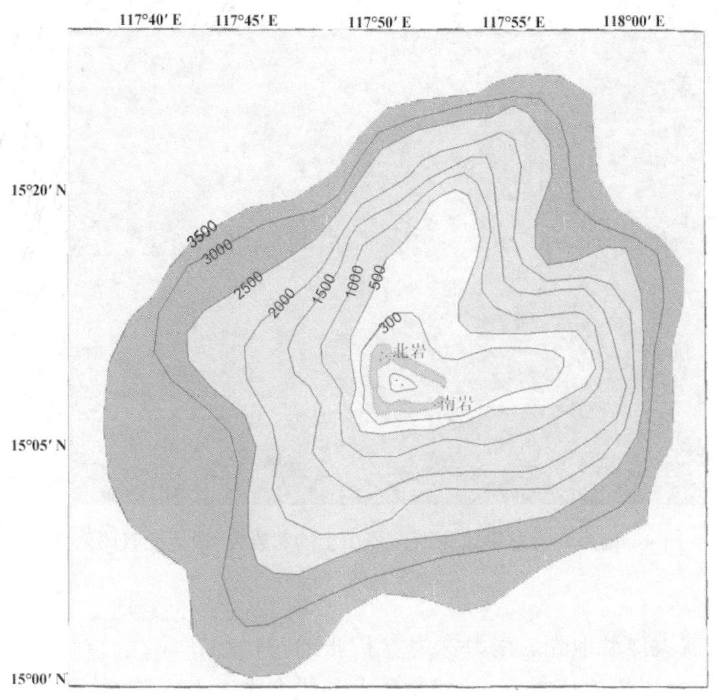

图 2　黄岩岛地形简图

（中心橙色部分为黄岩环礁，数字为等深线值，单位为 m）

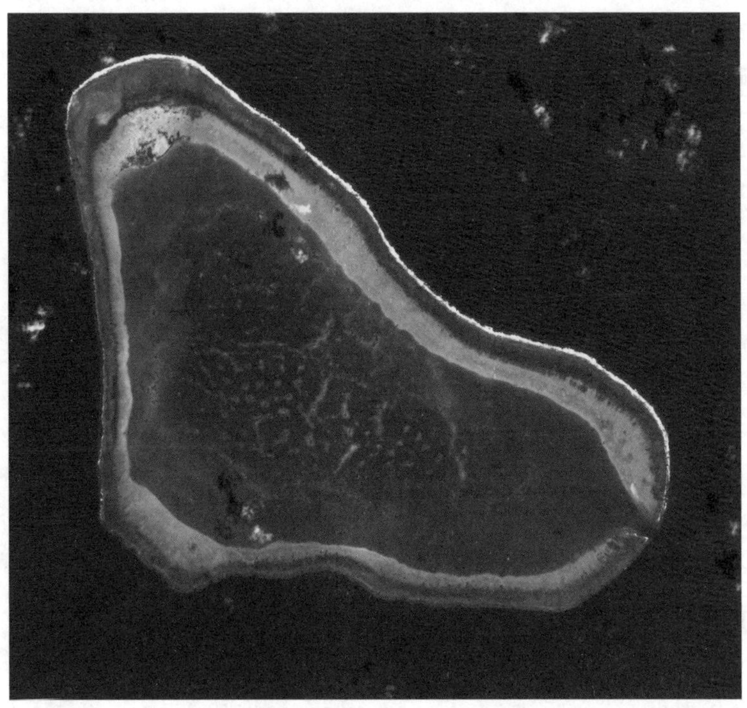

图 3　黄岩环礁卫星图

（蓝色为环礁，环围的中部为潟湖）

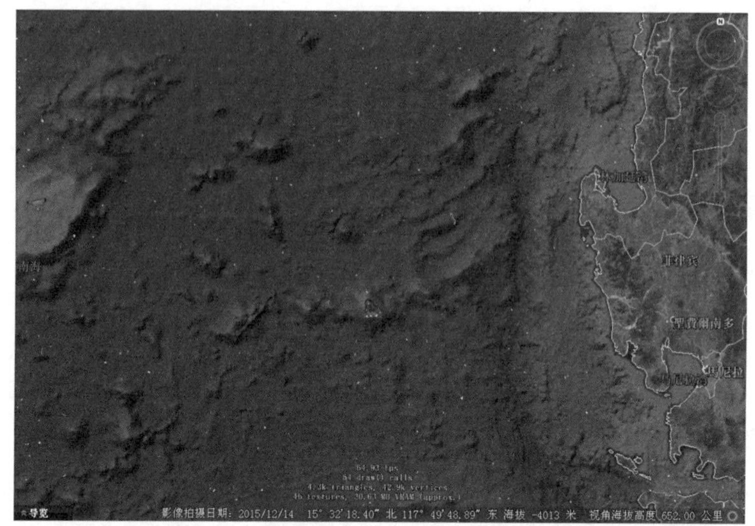

图 4　南海深海盆中近东西向海山上的黄岩岛(白点处)(附彩图)

(据:Google Earth)

厚度 10 km,均属大洋型地壳。南海深海盆扩张的时代约在(32～17)百万年前的中渐新世及早中新世。伴随着海底扩张,地幔物质上涌成多次火山喷发,形成高陡的玄武岩火山锥,当火山堆积高度临近海面时,形成围绕火山顶周围发育珊瑚礁的基础。据黄岩火山链西侧碱性玄武岩定年为(9.9±1.4)百万年,推断该处系中新世末期火山喷发的堆积层[3]。珊瑚环礁发育是在火山形成相当长时间以后,当更新世末期亚洲大陆冰川融化、海平面上升过程中逐渐发育的。

黄岩珊瑚环礁地貌发育完整,仅东南部有出口通道,环礁体宽度 300～1 200 m,其中,西侧宽度 1 000～1 200 m,南侧宽 300～500 m,东北侧宽 600～900 m。礁体的外侧坡度 15°～18°,直下 4 000 m 深海,环礁外坡由于波浪与潮流作用活跃,在 20 m 浅水区内造礁珊瑚及喜礁生物繁殖。20 m 以深造礁珊瑚减少,却以珊瑚块体与碎屑堆积为主。2 000 m以深珊瑚碎块消失,沉积为软泥。环礁顶部礁坪崎岖不平,有数百只巨型礁块突立于深水海域(图 5 为低潮时的黄岩岛——礁块环列),是海水冲刷、溶蚀珊瑚礁坪形成的特征地貌:北岩,是暗黄色浅水巨礁,出露海面高达 1.5 m,出露的整块面积为 4 m^2,是命名为黄岩岛的缘由(图 6 为黄岩岛最大礁石);与北岩隔潟湖相对的南岩,周长 8 m,出露海面高 1.8 m,面

图 5　低潮时的黄岩岛——礁块环列

(据:地图看世界山寨版的博客,"【中沙群岛】黄岩岛",新浪博客,2012.7.5)

积为 3 m^2，南岩表面已发育了溶蚀沟与孔洞，反映其屹立海面已多年（图 7 为黄岩岛第二大礁石，表面具溶蚀孔沟）；另外，还有一些礁头出水高度 0.3～1.5 m。巨大岩块出露是黄岩岛的特殊地貌与海域环境的反映：该处海底山仍具有缓慢地上升活动，使珊瑚礁岩体矗立海面以上；开阔海域海蚀作用活跃，既有海水溶蚀钙质礁的作用，更有滔天巨浪的冲刷打击礁基作用，使整体岩礁裂为巨岩块，亦可将岩块推至礁坪台上。该处罕有的海洋风暴巨浪或海底火山地震引发的海啸巨浪强烈，可从巨岩块体积推算其冲击推移动力。但是，发生巨浪作用的周期间隔很长，实不频繁，所以巨岩块可停滞堆积、被再胶结，并经海水的岁月侵蚀，在南岩面上形成冲刷、溶蚀沟。

图 6　黄岩岛最大礁石

（据：地图看世界山寨版的博客，“【中沙群岛】黄岩岛”，新浪博客，2012.7.5）

图 7　黄岩岛第二大礁石，表面具溶蚀孔沟

（据：地图看世界山寨版的博客，“【中沙群岛】黄岩岛”，新浪博客，2012.7.5）

据黄金森在黄岩海域研究：从地震 P 波等传播波速为 6.5～6.6 km/s，分析黄岩环礁基底为火山喷发的玄武岩；玄武岩之上是为火山喷出物及珊瑚礁碎屑层沉积层，P 波速度为 4.3～4.5 km/s；再上为 P 波 2.1 km/s 的松散沉积。黄岩环礁是在火山基底上发育，形成年代更晚，因第四纪初期仍有火山喷发活动。黄金森认为：火山受蚀发育平顶山，然后再建珊瑚环礁地貌，环礁主体是在全新世发育的。北岩（黄岩）与南岩的堆积时期只有 470±90 年[3]。

二、菲律宾对中国黄岩岛主权之挑衅

菲律宾妄图黄岩岛及周边海域主权，始于 1994 年国际海洋法正式实施以后。海洋法虽明确规定领海、毗连海区及 200 海里专属经济区等范围，但未涉及与阐明海洋法界定的范围与历史传统海疆间的关系，这一缺陷被菲律宾等国利用为觊觎中国南海疆域的借口。但是，直至 1997 年 5 月之前，菲律宾尚未对黄岩岛与其邻接海域提出主权要求。

1997 年 5 月初，当由 6 位中国人、3 位美国人与 2 位日本人组成的无线电业余爱好者登上黄岩岛进行科普活动时，菲律宾军用侦察机与炮艇进行了骚扰，国家海洋局送人到岛的两艘船只的船员无意对抗，在声明黄岩岛是中国领土之后，提前将这组无线电爱好者送回[4]。5 月中旬，两位菲律宾国会议员乘军舰登上黄岩岛，并插上菲律宾国旗与拍照。5

月20日，菲律宾的巡逻艇在距离黄岩岛11 km处拘捕了中国21名渔民，扣留了渔船。5月底，菲律宾总统Fidel Ramos声称："菲律宾有勘探和开发黄岩岛资源的主权，它是在菲律宾的专属经济区之内……"公然要求黄岩岛的主权。6月5日，菲律宾外长Domimgo Siazon在菲众议院外交与国防委员会听证会上声明："我们坚持黄岩岛是我们领土一部分。……已对黄岩岛及其周围水域行使主权和有效的管辖权，……渔民的传统渔区和天气恶劣时的庇护所。……进行了海洋科学研究……建立一座灯塔……"加剧了菲律宾对黄岩岛的侵犯与争端活动。

1999年5月23日和7月19日，菲律宾海军巡逻艇追逐扣捕我国潭门港渔民及撞沉渔船(琼-03091号、琼-03061号)，扣压渔民经月，中国驻菲大使馆及时严正抗议，要求无条件放人与赔偿损失，菲方坚持是意外事故，不予赔偿。11月2日，菲炮艇又在黄岩岛海域追逐我国3艘渔船而导致失踪。2000年1月6日，菲军舰又驱逐我国6艘渔船，我国外交部发言人声明："黄岩岛是中国领土，中国渔民在该海域从事正常捕鱼作业，菲律宾行径严重干扰了中国渔民和平生产活动，中国对此予以强烈关注。"1月31日，菲方再次驱逐中国渔船，开炮威胁，菲海军司令声称开炮是"声响效应"以避免船只相撞……3月25日，菲海军两艘炮舰赶走在黄岩岛避风的我国8艘渔船。嗣后，菲海军舰艇与空军加紧在黄岩岛海域巡逻，登船没收捕鱼工具，掠夺渔获物，驱逐中国渔船，骚扰频繁，行径极其恶劣。我国外交部发言人多次声明："要求菲方尊重中国主权，维护南海和平稳定大局……""黄岩岛是中国固有领土。黄岩岛海域是中国渔民的传统渔场，对此中国有充分的历史和法理依据""黄岩岛从来不在菲律宾领土范围之内，一系列条约对界定菲律宾领土有明确规定，菲律宾领土的西部界限在118°E，而黄岩岛在118°E以西，是中国中沙群岛的组成部分。"我国一系列声明与外交交涉，菲律宾政府均置若罔闻，并进一步加强对黄岩岛海域之控制。

三、中国黄岩岛主权不容侵犯

(一) 黄岩岛海域具有非常重要的战略地位

黄岩岛是我国在南海东北部唯一出露海上的岛礁，是南海东北部的门户要塞；是我国在半封闭型的亚太边缘海域与世界大洋海运交往的枢纽控制点；它扼守台湾海峡南端以及由太平洋进入南海之通道；自我国西沙群岛驶向菲律宾吕宋岛，从我国南沙群岛通向台湾岛，以及出巴士海峡驶向太平洋的中途岛，战略地位重要。

黄岩岛隶属我国中沙群岛，是在南海诸岛中唯一的大洋型环礁上、唯一出露于海面上的礁岛，也是我国在南海深海盆地东部唯一的濒临于马尼拉海沟西侧的火山峰顶上发育的珊瑚环礁，其海洋地质结构具有重要的科学研究价值。其海洋环境位置，具有监测南海边缘海域海洋气候、海洋水流动态与海水循环变化，海洋生物特征与资源量，以及海洋环境特征与发展演变趋势研究的重要地位，是具有代表性的科学观测基地，科学价值独特。

黄岩岛海域历来是我国南海与东海渔民捕捞作业的基地，海洋生物、新能源与矿产资源丰富，发展前景广阔，自古以来即为我国先民开发经营，世代传承至今，我们必须护卫与持续经营。

（二）黄岩岛是中国领土，其历史可追溯到元朝初期

元世祖为统一中国疆域内历法，设太史院改治新历，1279 年，太史院主持郭守敬提出："治历在于测验"，获准而进行"四海测验"，在中国全境选 6 点进行。其中南海点，即在今日之黄岩岛。据《元史・天文志・四海测验》中记载："南海，北极出地一十五度，夏至景在表南，长一尺一寸六分，昼五十四刻，夜四十六刻"，按照夏夜晷景和昼夜时刻，李金明推算出当时南海测点位于 15°12′N，116.7°E，接近现今的黄岩岛（15°08′～15°14′N，117°44′～117°48′E），是元时"四海测验"的端点之一，是在中国南海疆域之内[5]。明清时延续元朝行政建制，南海诸岛划归琼州府的万州管辖。在《郑和航海图》、《清绘府州县厅总图》等多幅舆图上，南海诸岛均在中国版图之内[6]。历代渔民下南海的《更路簿》有关古籍及南海周边地方志上，均有中国渔民赴黄岩岛海域作业的记录与航海路线图。

第二次世界大战前后，中国又多次重申了对黄岩岛拥有主权：1935 年 1 月，民国政府水陆地图审查委员会出版的第一期会刊，刊登了《中国沿海各岛屿华英名对照表》，其中包括黄岩岛（Scarborough Reef），4 月出版《中国南海岛屿图》，黄岩岛标在中国领土范围内；1936 年白眉初编《中华建设新图》中第二图中，黄岩岛位于中国南海疆域线内；1947 年 11 月，内政部方域司专门委员郑资约著《南海诸岛地理志略》出版，在南海诸岛新、旧名称对照表中，中沙群岛包括民主礁（Scarborough Reef）；同时，内政部方域司印制出版了《南海诸岛位置图》，在南海地域的中沙群岛也标有黄岩岛。

20 世纪 80 年代初，中国政府组织对南海诸岛地名普查，中国地名委员会于 1983 年 4 月 24 日公布了《中国南海诸岛部分标准地名》，其中，中沙群岛包括黄岩岛（民主礁）[7]。中国科学院南海海洋研究所于 20 世纪 70—80 年代多次对黄岩岛进行了地质、地貌调查，总结与出版了科学研究成果。在民国时期和新中国建立后的一段时期内，黄岩岛是由广东省管辖。1988 年后，黄岩岛改由海南省实施行政管辖[8]。1992 年颁布的《中华人民共和国领海及其毗连区法》中的"领土条款"明确规定黄岩岛等岛屿是中国领土，中国对黄岩岛拥有无可争辩的主权[9]。事实说明，黄岩岛自古以来为中国领土，先后曾多次测量、调查，延续实施巡视与管理，过去从未遇到过挑战，而今为国际广泛承认。

（三）黄岩岛不属菲律宾领土范围

黄岩岛位于我国海疆界线以内，是我国领土，无可置疑。在地理位置上，它与菲律宾吕宋岛之间，隔以西吕宋海槽与北吕宋海槽。在吕宋海槽以西，又有深达 4 000 m、近南北向的马尼拉海沟将西吕宋海槽与黄岩岛分隔开来。黄岩岛位于近东西向的黄岩海山之顶部（参见图 4）。黄岩岛与东部近南北向的吕宋岛的地质构造单元不同，而且平行于吕宋岛外的吕宋海槽与马尼拉海沟两列深海结构，更将黄岩岛与菲律宾列岛区别开来。

据菲律宾历史，在 1898 年 12 月 10 日美国与西班牙巴黎条约第三条规定："西班牙现将被称为菲律宾列岛的群岛，其中包括位于下列各线内的诸岛屿割让给美国：一条从东经 118°到东经 127°沿着或靠近北纬 20°由西向东穿过巴士海峡航道中间，然后沿着东经 127°到北纬 4°45′，然后沿北纬 4°45′到与东经 119°35′的交叉点，然后从东经 119°35′到北纬

7°40′，然后沿北纬 7°40′到东经 116°的交叉点，然后再以一条直线到达北纬 10°与东经 118°的交叉点，然后再沿东经 118°到达起点。”1900 年 11 月 7 日，美、西华盛顿条约重审上述巴黎条约的内容并把卡加延、苏禄和锡布图及其属岛割让给美国。这些内容中有关菲律宾“条约界线”的西部界线是 118°E，而黄岩岛位于 117°51′E，在该界线以西，明确地不属菲律宾的领土范围。

菲律宾在 1961 年 6 月 17 日第 3046 号法案《关于确定菲律宾领海基线的法案》中宣布，“条约界线”内的“全部水域始终被视为菲律宾群岛的组成部分”。这种主张违背了巴黎条约缔结时的历史事实，不符合现行国际海洋法的有关规定：

(1) 菲律宾划定领海基线采用的是群岛基线，即以直线划定与连接群岛最外缘各点的基线。《联合国海洋法公约》规定，群岛国的领海测算应从划定的群岛基线量起，其宽度不超过 12 海里的界线。然而，菲律宾在第 3046 号法案中却宣布，位于群岛外缘岛屿之外而在“条约界线”之内的全部水域为菲律宾的领海。如此主张的领海不是海洋法规定的具有一定宽度的海域，而是由围绕各岛的经纬线构成的多边形界线，其西南端最宽的领海是 0.5～2 海里，在南中国海一边是 147～284 海里，在太平洋一边是 270 海里，甚至包括了印度尼西亚的米安加斯(帕尔马)岛。显然违反《联合国海洋法公约》有关 12 海里领海宽度的规定，是不合法的。

(2) 菲律宾的“条约界线”使用的是地理速记的简单方法，以经纬度为界线，省却列举割让的众多岛屿。条约割让的仅是界线内的岛屿，没有包括整个海域，故菲律宾不能以此为依据把“条约界线”内的全部水域看成是菲律宾的领海，菲律宾所谓的“条约依据”完全违背了条约缔结时的历史事实。

(3) 菲律宾依据的三个历史条约的主要缔结国是美国与西班牙，这两个当事国的看法是解释条约内容的关键。西班牙方面对此没有表达过看法，而美国政府则对菲律宾的做法持反对意见。菲律宾把“条约界线”内的水域宣布为领海的做法，既违反了《联合国海洋法公约》有关 12 海里领海的规定，又与美国、西班牙 1898 年缔结《巴黎条约》时的事实不符，菲律宾的领海主张是没有根据，是不合法的。菲律宾以“黄岩岛与菲律宾相邻近”“黄岩岛在菲律宾专属经济区内”为借口，声称对黄岩岛拥有主权是毫无根据的。由于黄岩岛的主权向来属于中国，这就决定了菲律宾不可能再把黄岩岛声称在自己的专属经济区之内而拥有主权。

(四) 黄岩岛争端的历程及结果

“黄岩岛事件”促使了中菲黄岩岛争端的激化，2012 年 4 月 10 日，12 艘中国渔船在中国黄岩岛潟湖内正常作业时，被一艘菲律宾军舰干扰，菲军舰一度企图抓扣被其堵在潟湖内的中国渔民，却被赶来的中国两艘海监船所阻止。随后，中国渔政 310 船赶往事发地黄岩岛海域维权，菲方亦派多艘舰船增援，双方持续对峙至 6 月 5 日。6 月 3 日，菲方将其海岸警卫队船撤离黄岩岛，我国两艘公务船清理现场后，于 6 月 5 日驶出潟湖，而把守口门继续在黄岩岛海域执行公务。至此，我国捍卫控制了黄岩岛及周边海域。

菲方虽失去了黄岩岛及周边海域的控制权，却仍觊觎我国黄岩岛海域。自 2012 年 6 月 16 日至 2013 年 1 月 20 日之间，菲律宾却以诬赖中国船只撞沉菲律宾渔船、更改教科书内容，将“南中国海”更名为“西菲律宾海”(包括黄岩岛及周边水域)，扬言以袭击中方在黄岩岛海域的无人机来威胁中国等，挑战我国在中沙群岛的主权，但终告失败。

至 2013 年 1 月 21 日，菲律宾外交部长承认中国已“实质上控制”黄岩岛，菲船不能进驻该海域，黄岩岛争端才正式落下帷幕，我国实际控制了黄岩岛及周边海域。黄岩岛争端是我国南海维权之中的重要事件之一，通过外交手段完成了对岛礁的控制，对于我国维护南海其他岛礁的权益具有借鉴意义。

结　语

我国中沙群岛的黄岩岛位于南海深海海盆东北端，是扼守我国南海海疆东部咽喉通道的要塞，是我国南沙群岛与台湾岛之间的中途岛，也是从我国南海诸岛向东出巴士海峡通往太平洋的中心枢纽与补给站，地位重要，必须巩固捍卫与建设确保。从发展前景看，黄岩海山区，具有高硬度宝石级矿产的巨大潜力。其海域开敞、水深浪大，黄岩岛礁出露海面不高，巨岩耸立是经历过暴风巨浪冲击，裂礁堆石的自然灾害遗迹。因此，岛礁上不适宜大范围建筑与大量常住人口。武装驻守，经常性的军舰巡逻保护岛礁及其水域，维护安全的海上作业与船只航行，在目前仍是因地、因势制宜的良策。

参考文献

[1] 黄金森. 1980. 南海黄岩岛的一些地质特征. 海洋学报，2(2)：112 - 123.
[2] 李孝聪. 2015. 从古地图看黄岩岛的归属. 南京大学学报(哲学、人文与社会科学)，4：76 - 87.
[3] 曾昭璇等著. 1997. 中国珊瑚礁地貌研究. 广州：广东人民出版社，319 - 322.
[4] 李金明. 2003. 近年来菲律宾在黄岩岛的活动评析. 南洋问题研究，3：40 - 47.
[5] 李金明. 1996. 元代“四海测验”中的南海. 中国边疆史地研究，4：35 - 42.
[6] 张磊. 2012. 加强黄岩岛有效控制的国际法依据. 法学，8：67 - 75.
[7] 李金明. 2001. 从历史与国际海洋法看黄岩岛的主权归属. 中国边疆史地研究，11(4)71 - 77.
[8] 吴士存. 中国黄岩岛主权无可争辩. 海南日报，2012 年 5 月 7 日，第 A03 版.
[9] 张磊. 2012. 加强黄岩岛有效控制的国际法依据. 法学，8：67 - 75.

南海研究成果实例*

中国南海研究协同创新中心是属于教育部2011大项目建立的多学科研究中心，海岸与海岛开发教育部重点实验室承担南海资源环境与海疆权益工作，愿藉此次，汇报我在2014年负责完成的两项成果：① 论证南海海疆断续线；② 全球海平面变化与南海典型港湾效应。一大一小两个项目反映我们研究的关注点，是以专业工作为基础，为国家服务的实例。

1 论证南海海疆国界线①

在南海，以断续线段标注的国界线，自1947年以来，明确地标注于我国公开出版的图集与文献中。但是，自1994年“国际海洋法公约”实施以来，周边诸国却对我国南海国界线称谓多样、争论不断：九段线，U形线，袋状线，岛屿归属线，岛礁归属线，群岛水域，历史性水域，历史性权利，传统海疆线等。前列三种名称不能反映名称的内涵，后列六种名称不能提出符合断续线段位置与形式的合理解释。

东沙、西沙、中沙与南沙群岛，的确是中国最早发现并具有历史传统的海域。远自两汉时期，渔民已在南海海域捕鱼与航行作业。唐朝贞观年间划分了南海地域的州、县域以实施管辖，历代延续。

当代南海国界线的划定，是20世纪40年代由同盟国与国际各国明确的。第二次大战后，根据1943年11月22日～26日中、美、英三国首脑在开罗会议讨论与制定，于1943年12月1日正式实效的“开罗宣言”中明确提出：“被日本所窃取于中国之领土，特别是满洲和台湾，应归还中华民国。”1945年7月17日～8月2日，美、英、苏三国首脑举行波茨坦会议，发布了敦促日本投降的“波茨坦公告”再次重申：“开罗宣言之条件必将实施，而日本之主权必将限于本州、北海道、九州、四国及吾人所决定其他小岛之内。”1945年8月15日正午，日本天皇向全国广播宣布接受波茨坦公告，实行无条件投降的诏书。8月21日，日本今井武夫飞抵中国芷江请降。9月2日上午9时，在停泊于东京湾的密苏里战舰上，举行了日本向美、英、中、苏等9个同盟国投降仪式，日本外相重光葵代表日本天皇和政府，日本陆军参谋长梅津美治郎代表日本军大本营在投降书上签字，9时8分，麦克阿瑟以最高司令官的身份签字，接受日本投降，然后是受降的各盟国代表签字。签字结束后，成千架的美军飞机从东京上空呼啸而过，宣告了日本军国主义的彻底失败和世界反法西斯战争的最后胜利。嗣后，驻海外日军陆续向盟国投降，中国战区的投降仪式于9月9日

* 王颖：中国工程院环境与轻纺工程学部、钦州市人民政府与钦州学院，《2015广西海上丝绸之路建设“钦州论坛”论文摘要集》，第6-8页，2015年12月18～20日，钦州。

① 发表于《海洋学报》36卷10期1-11页，2014年10月。

在南京陆军司令部礼堂举行，日本在中国的派遣军总司令官冈村宁次在投降书上签字，并交出随身佩刀以示正式缴械投降，中国陆军司令何应钦接受了日本侵华军参谋长小林浅三递交的投降书。此后，中国开始接收日本归还我国的领土包括南海诸岛。

1946 年 10 月，中国政府组织了一个由军、政官员，地理地质专家组成的接收团，乘坐几艘军舰前往南海，分赴西沙与南沙群岛，从事收复，并对主要岛屿进行实地测量和制图工作，标志着二次大战后，中国政府收复南海诸岛并实施了正式管辖。南京大学杨怀仁教授，当年参加了南海诸岛实测与岛礁名称订正工作，将岛礁的当地土名与渔民俗称，外航与日本名称及我国历史名称等理清与划一。1946 年的勘测工作完成了 6 幅地图："西沙群岛图""中沙群岛图""南沙群岛图""永兴岛-石岛图""太平岛图"及集成的"南海诸岛位置图"。六幅图于 1947 年 8 月 7 日经国民政府行政院院长张群批准签发。1947 年 12 月 1 日，国民政府内政部方域司公布了"南海诸岛位置图"，并在"中央日报"上刊登了南海诸岛名称。1947 年，参加当年接收南海诸岛行动的地理学家郑资约先生出版了专著《南海诸岛地理志略》，其中包含了《南海诸岛位置略图》。1948 年该图亦公布于《中华民国行政区域图》上，并被多次在正式出版的中国地图册内刊印。

在《南海诸岛位置略图》上，正式标志出起自中国与越南交界的陆域向海连续延伸、环绕我国南海疆域的 11 段不连续的界限，图例注明为国界线。1953 年时，去掉了北部湾中的两条界线，余下九条，后人称其为"九段线"。但是，失去了在南海北部湾与陆界连续下延的海域界线，造成了日后对断续线段的理解不一与曲解。应重视国家的疆域的历史由来与整体延续性，任何个人无权任意更改，是我们应吸取的教训。

为什么是断续线？究其划分的依据，是介于我国岛礁与周边相邻诸国的中间线上，对各方均合理；而且，线段所在的海底多是在陆架、陆坡或海槽中的深水部位；由于海面是波动的，大海中波浪为外形传播，而水质点是在其平衡点位置上的振荡运动，所以，最合理的海面上界线，应是不连续的，即断续线段。不仅是我国，国际上也有采用断续线段的实例：如在多岛屿的地中海诸国间的国界，在地图上均是以不连续的线段标注。我国在南海采用不连续线段标注国界，也是国际人道主义的体现，即：在这较广的群岛海域中，提供了国际间无害航行与海上救助之便。但是，本文最关键之意在于，要遵守二次大战后国际认定的宣言与公告，是遵循建设和平安全、繁荣的海洋新秩序的基本原则。

2　当代海平面上升与海南海滩侵蚀①

海平面持续上升，近百年来全球海平面上升 19 cm，上升速率约为 2 mm/a。中国沿海海平面亦呈上升趋势，自 1980 年至 2014 年上升速率为 3 mm/a。当代，风暴潮与人类活动影响加促海平面变化效应。海平面上升在海岸带反应为海滩侵蚀与海岸沙坝向岸后退位移。对比我国在黄海、渤海与南海基岩港湾海岸的海平面上升效应：20 世纪 90 年代中期，南海三亚地区海滩侵蚀与海岸后退均较黄海、渤海、东海为小，亚龙湾海滩侵蚀与滨线后退最小；21 世纪以来，南海海平面持续上升与海南海平面上升处于高值水平，海南东

① 中国科学院学部咨询评议项目："南海资源环境与海疆权益"，王颖主持，2012—2014 年。

部海平面于近期连续居高，夏秋季海平面接近 400 mm，西部海岸海平面夏秋季亦高，但较东部为低，反映着东部海岸海域开阔受季风与风暴潮浪影响更为显著；海南岛南岸海滩侵蚀加剧，亚龙湾海滩侵蚀尤为突出，引起我们关注，提出海湾内人工建筑必须先有预后效应研究，否则促成海湾环境恶化的负面效应是不可逆的。海平面上升招致海滩侵蚀与海滨线后退趋势需予以重视，加强管理并采取有效防治措施。

第五篇
全球海岸海洋综合研究

将海岸海洋研究扩展至深海大洋：

* 完成加拿大大西洋鼓丘海岸形成发展理论总结；
* 总结大西洋拉布拉多大陆架沉积的石英砂表面结构特点，分析其沉积作用过程；
* 总结大西洋深海浊流沉积特征与表面结构；
* 对大西洋有关深海海域的动力环境与核废物埋藏安危之论述。

关注全球变化效应对海陆过渡带环境资源的变化趋势研究；综合研究特征海岸带发展的实例：红树林、珊瑚礁等生物作用海岸带，以及火山地震构造活动海岸带的自然发展与环境资源特点。

进行区域特征海岸带环境与沉积特点研究：沙丘海岸、火山海岸系列；红岩地貌特点与发育；江河湖海与城市发展相关。

Surface Texture of Quartz Grains and Sedimentary Processes on the Southeastern Labrador Shelf*

Introduction

Surface textures of quartz grains as seen by the scanning electron microscope have been used to distinguish various sedimentary environments by Krinsley and Donahue (1968), Krinsley and Doornkamp (1973), and many others. The markings are linked with the main dynamic processes that act upon the grains, such as, transport by turbulent water, wind, glacial ice and diagenesis. For example, the high energy surf zones that are associated with beaches increase the roundness of the grains, produce mechanical V-shaped depressions and small collision marks. The action of wind produces meandering ridges, pitted grain surfaces, and many other less diagnostic markings. The glacial cycle is characteristically reflected by grains that exhibit very irregular surfaces dominated by large or small conchoidal fracture patterns, imbricated breakage blocks and many other markings due to the high energy mechanical action on grains. Diagenesis may produce surface etchings or silica precipitation or both.

Grains usually contain markings that can be linked to more than one process or individual environment, reflecting a complex history of sedimentation. Many markings can be produced in more than one particular environment. Nevertheless, it is possible to determine the most likely sedimentary environment represented by a sample with a sufficiently large number of observations.

This paper describes the surface features of quartz grains found in a late Wisconsinan to Holocene sediment core from the inner Labrador Shelf. The dominant markings from each of four intervals sampled are used to infer the sequence of sedimentary environments represented in the core.

---

* Gustavs Vilks, Ying Wang: *Geological Survey of Canada*, Current Research, Part B, Paper 81-1B, pp. 55-61, 1981.

Methods

Sediments from core 95 (Fig. 1) have been analyzed at 25 cm intervals for foraminiferal content and grain size distribution; data are summarized in Fig. 2. At the intervals 0 to 5 cm, 200 to 205 cm, 450 to 455 cm, and 775 to 780 cm, quartz grains between 1Φ and −0.5Φ in diameter were selected for the study of surface microstructures (Fig. 2).

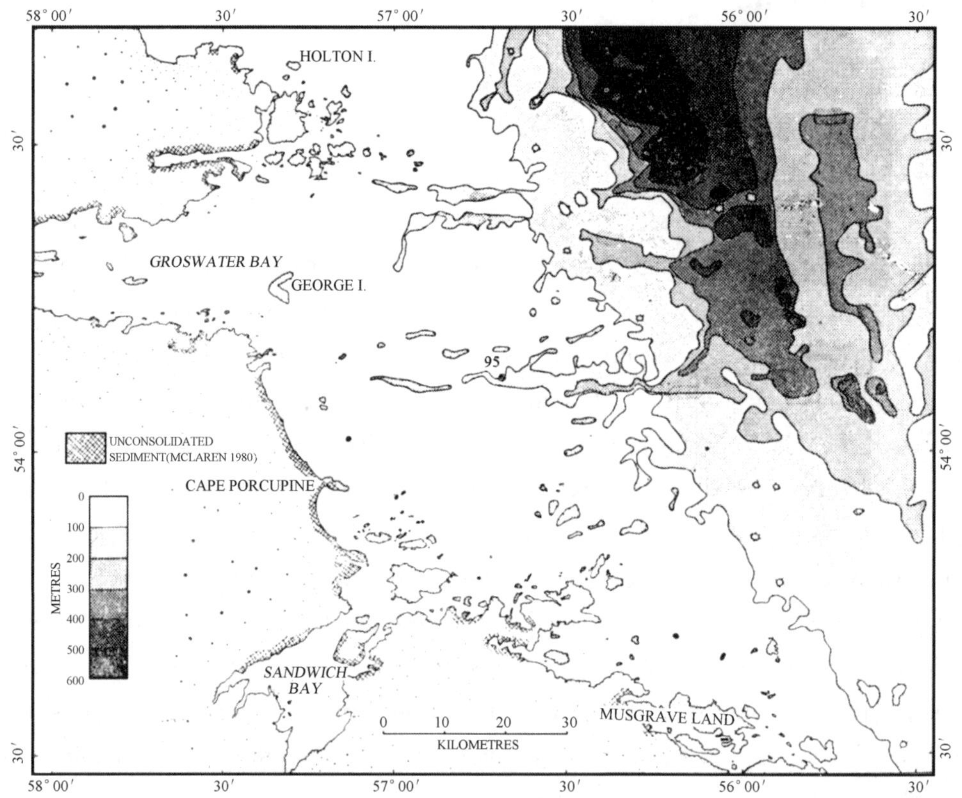

Fig. 1 Bathymetry of inner continental shelf and location of Core 95

The grains were boiled in concentrated hydrochloric acid, stannous chloride, and 30% hydrogen peroxide (Krinsley and Doornkamp, 1973). Between each boiling, each grain was washed thoroughly in distilled water. Fifteen grains from each core sample were coated with an alloy of gold-palladium for SEM analysis on a Cambridge 180 Stereoscan.

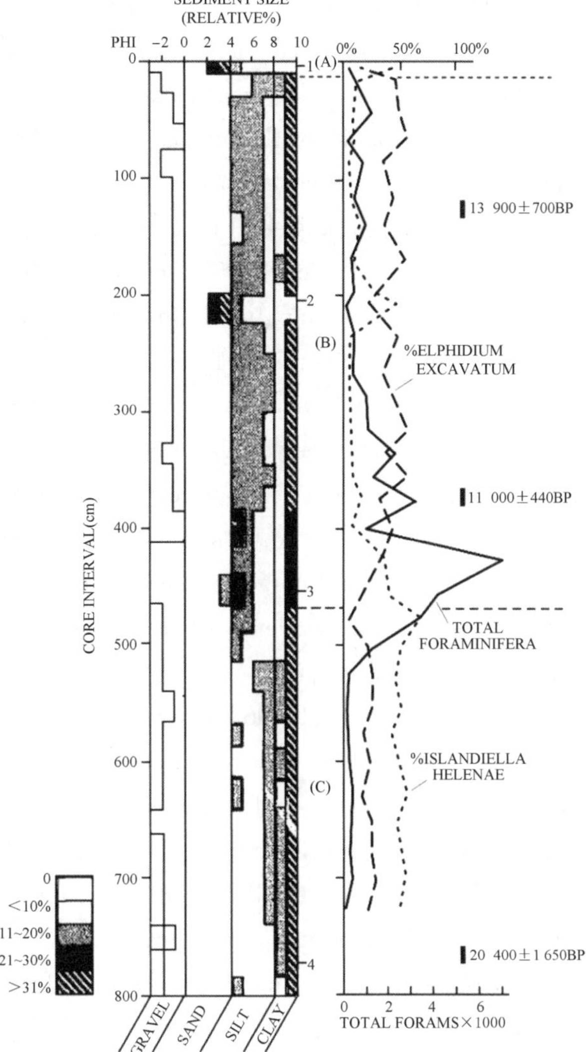

Fig. 2 Sediment texture and foraminifera in Core 95

Numbers with arrows indicate subsample locations for SEM analysis.

Environment

The coastline nearest to the sampling site has been described as unconsolidated lowland by Reinson et al. (1979). It consists of a broad, sandy beach with a well developed back beach area and sand dunes. The few offshore islands consist of bedrock (McLaren, 1980).

The inner Labrador Shelf in Hamilton Inlet area is extremely uneven (Fig. 1). Between the coastline and the depth of 100 metres, the shelf contains minor shoals, depressions and transverse channels across the inner shelf. Core 95 was taken in one of the major transverse channels.

Preliminary study of high resolution seismics and surface samples indicate that the distribution of the bottom sediment on the inner shelf is as variable as its topography (Reinson, 1979). Sand is present in the flat areas between the bedrock highs. Cobble and boulders are present along the flanks of the highs and in saddles between banks. In many depressions sand is found as a thin veneer overlying clay.

Bottom current velocities recorded between July 24 and August 12, 1979, 60 km to the north had peaks in the vicinity of 12cm/s. These maxima had a setting towards west-southwest. Surface currents in the Hamilton Inlet area were estimated to be in the order of 20 cm/s and southeasterly (Dunbar, 1951).

In addition to bottom currents, sediments are also modified by the grounding of icebergs. Approximately 1 000 icebergs per year have been sighted off the northern tip of Newfoundland via the Labrador Current (Dinsmore, 1972). Extreme concentrations of icebergs (20 icebergs per 12 000 km^2) occur in spring (February-April) along the Marginal Channel of the southeastern Labrador Shelf (Gustajtis, 1979). Sidescan sonograms of the seafloor show ample evidence of iceberg scours on the banks and saddles (e. g. Van der Linden et al. , 1976).

Between December and July the Labrador inner shelf is covered with sea ice (Dinsmore, 1972). The ice is carried to the southeast by the Labrador Current and occasionally driven offshore by westerly winds. Major disintegration of the pack ice takes place in the warmer offshore waters, where it also presumably drops most of its sediment load. Sediment characteristics (Vilks, 1980) would suggest that in the marginal channel sea ice is not an important agent for transport. The inner shelf and the nearshore zone, however, are protected by the cover of sea ice from the effect of large storm waves during the winter.

Stratigraphic Sequence in Core 95

Microfossils and sediment textures suggest three sediment units within core 95:

A—0～5 cm: surface veneer of very fine sand with **Islandiella helenae** dominant.

B—5～460 cm: Clayey silt, with lenses of silt and sand, and at 200 cm a 10 cm fine sand bed, **Elphidium excavatum** f. **clavata** dominant.

C—460～800 cm: Silty clay with scattered gravel, very low foraminiferal abundance and **I. helenae** predominating.

Quartz grains for SEM analysis were taken from each of the units and from the fine sand bed (see Fig. 2). The grain size frequency histograms of the sediment from the two silty sand layers are remarkably similar (Fig. 3A, B) with a distinct peak at the 4 phi size interval. The sorting is only moderate because of the trace of gravel in each sample. The clayey silt frequency histogram (Fig. 3C) shows a mode at the 5 phi size interval and a moderate amount of sand. The silt is by far less sorted in the silty clay sample (Fig. 3D), even though sand and gravel are almost lacking.

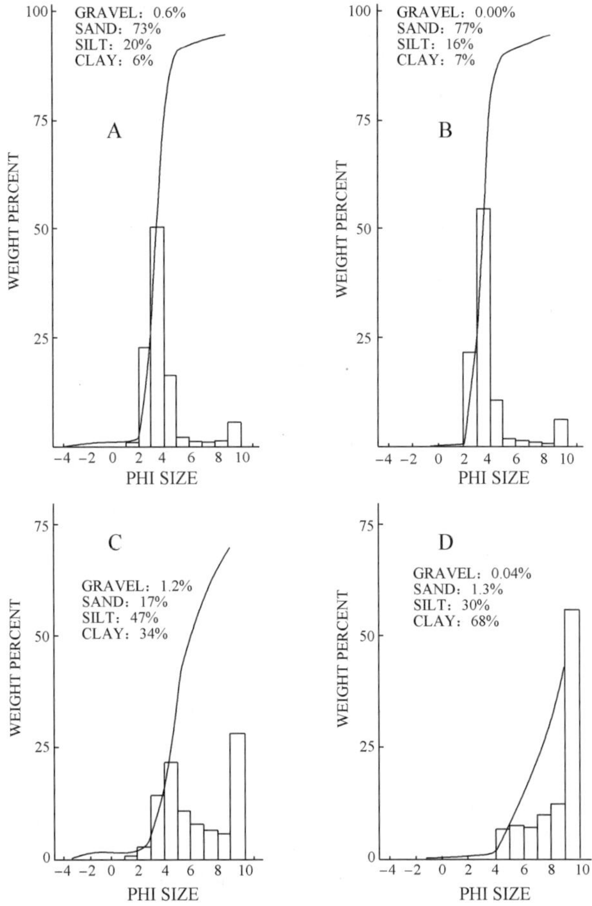

Fig. 3　Grain size distribution of subsamples

In comparison to the coastal sediments of this area, the three textural groups in core 95 are typical continental shelf deposits. They are coarser towards the top with lenses of silty and sandy sediment in the upper half of the core. The trend probably indicates changing hydrodynamic conditions from lower to higher energy, in combination with reduced sediment supply towards the recent times.

The foraminiferal number per 35 ml of sediment is relatively low (Fig. 2). The two layers of silty sand are almost barren of tests, but the numbers increase to 7 000 in the zone of silty clay between 400～500 cm. Below this layer, only trace amounts of tests were found and below 750 cm, the sediment was barren of foraminifera.

The two major species are **Elphidium excavatum** forma **clavata** and **Islandiella** helenae, both almost mutually exclusive. Elsewhere on Labrador Shelf (Vilks, 1980) **E. excavatum** f. **clavata** is abundant in late glacial sediments and **I. helenae** in surface sediments and isolated horizons in Lake Melville cores (Vilks et al., in press).

Total organic carbon ^{14}C dates (Table 1) give an age of 20 400 years at 8 m, 11 000 years at 4 m, and 13 900 years at 1.5 m. These inconsistencies suggest dilution by dead carbon, but the dates may indicate that the core consists mostly of glacial or late glacial sediment. This is confirmed by the predominance of **Elphidium excavatum** forma **clavata** throughout the upper part of the core (unit B), which elsewhere on the Labrador Shelf (Vilks, 1980) characterizes late glacial sediment. This implies that Holocene sedimentation is represented by the surface sand veneer (Unit A). Unit C may represent sedimentation in a proglacial environment with ice-rafting of gravel.

Table 1 ^{14}C dates of Core 95

Core interval cm	^{14}C Date	Laboratory Number
118～133	13 900±700a	GSC - 2993
365～380	11 000±440a	GSC - 3014
760～775	20 400±1 650a	GSC - 2977

Surface Markings of Quartz Grains

Twelve surface marking types and grain shapes were recognized (Table 2). Irregular grain shapes (Plate 1, fig. 1) with angular and subangular edges (Plate 2, fig. 1) indicate a glacial environment. The glacial environment is also characterized by conchoidal fracture patterns (Plate 1, fig. 2; Plate 4, fig. 2), and cleavage faces (Plate 4, fig. 3). Subrounded (Plate 2, fig. 6) to rounded edges, V-shaped pattern (Plate 3, fig. 4) and collision marks (Plate 3, fig. 2), superimposed on glacial markings, may indicate residence in the beach zone. The continental shelf environment that involves sediment reworking and winnowing can be identified with quartz also containing V-shaped patterns and collision marks (Table 2). Diagenesis following deposition results in solution holes (Plate 3, fig. 6), silica deposits (Plate 1, fig. 5, 6), and weathering cracks (Plate 2, fig. 3).

Each of the twelve types of surface markings, except for the rounded edges, were found in all four samples analyzed, (Table 2). The major characteristics of the grains in the layer of 0～5 cm (Plate 2) are: irregular shape with angular to subangular edges and conchoidal fracture patterns. These markings indicate a glacial origin for the grains. Superimposed on these markings, however, are V-shaped depressions, collision marks, solution holes, and silica deposits. This suggests the grains resided for a subsequent period of time on the continental shelf.

Table 2　Major surface markings and shapes of quartz grains

Numbers indicate observations made in each surface texture class

Sample Number	1	2	3	4
Interval (cm) / Surface Markings	0～5	200～205	450～455	775～780
Irregular Shape	8	6	7	10
Angular Edges	3		1	2
Angular to Subangular Edges		6	1	2
Subangular Edges	7		3	4
Subangular to Subrounded Edges		3	4	2
Subrounded Edges	4	4	4	4
Subrounded to Rounded Edges			1	
Rounded Edges	1			
Conchoidal Fractures	8	5	7	9
Cleavage Faces	8	5	2	3
Weathering Cracks	3	5	2	1
V-Shaped Depressions	9	13	10	5
Collision Marks	2	9	10	5
Solution Holes	12	8	11	13
Silica Deposits	8	2	5	13
Number of Grains	15	13	14	14

All grains in the 200～205 cm sample (Plate 2) contain V-shaped depressions indicating grain movement either in a beach zone or continental shelf. Many grain surfaces also contain solution holes, and therefore, surface markings that signify the continental shelf environment are most abundant in this interval (Table 2). The 450 to 455 cm interval (Plate 3) is also dominated by grains with V-shaped depressions, collision marks and solution holes. The grains are less angular and the surface markings due to glacial action are not as fresh as in the other samples.

Most grains of the 775 to 780 cm interval contain glacial induced surface markings, such as, irregular shape and conchoidal fractures (see Plate 4). The initial glacial markings are coated with silica deposits and have superimposed solution holes. The number of grains with V-shaped depressions and collision marks is low but the diagenetic effect is high, suggesting that the sample represents sediments that have been transported initially by ice and subsequently buried with a minimal intervening period of shelf transport.

Conclusions

Although the interpretation of surface textures on quartz grains is to some extent ambiguous, a sufficiently large sample in addition to the study of other parameters, such as grain size, age and fossils can provide sufficient material for a paleosedimentary model. In our study the sediment size analysis indicates a changing hydrodynamic setting from relatively constant and low energy in late glacial time to fluctuating and high energy towards the present time. Surface textures on quartz grains support this trend.

In the 775 to 780 cm interval the markings are dominated by solution holes and silica deposits that supercede the initial glacial effects. The abundance of the diagenetic surface markings suggest considerable age of the grains. The mechanically induced markings, such as, V-shaped depressions and collision marks occur in small numbers indicating a relatively continuous burial in the fine clay.

The clayey silt of the 450 to 455 cm and 200 to 255 cm intervals contain grains with few glacial effects but many mechanically induced markings. These are mainly V-shaped depressions and collision marks, suggesting relatively high-energy hydrodynamic events, such as, storm waves. The material may have migrated from the coastal zone during periods of heavy storms. Unit B (Fig. 2) is a typical continental shelf deposit containing coarse-layered storm deposits (Reineck and Singh, 1973, p. 308).

The surface sandy layer is a relict sediment of glacial marine deposit on the inner shelf. Glacial markings are overlain by V-shaped depressions, collision marks, solution holes and silica deposits, all indicating a prolonged exposure on the seafloor. Thus, Unit A (Fig. 2) may represent Holocene deposits which are only a few centimetres thick because of the low postglacial supply of debris from the land and high inner shelf hydrodynamics.

Grain surface texture, sediment facies, and foraminifera suggest a sequence of three sedimentary regimes for the continental shelf. Unit C at the bottom of Core 95 was deposited in a proglacial environment and in a continuous sequence. Unit B was dominated by storm wave deposition, and Unit A by sediment reworking and the slow rate of sedimentation during postglacial time.

At core station 95 during the late glacial-postglacial period the depth of water did not change considerably, including the time when the eustatic sealevel change was at its lowest (18 000 years BP, Milliman and Emery, 1968). The inner shelf must have been depressed with the load of ice and remained low during the marine maximum when the Lake Melville area was inundated with marine waters. Since deglaciation, the Lake Melville area has been uplifted by between 120 and 180 m (Andrews, 1973) and our evidence suggests that the inner shelf has remained below sea level while following the general trend of isostatic rebound inland.

Explanation to Plates 1 to 4

a—V-shaped marking
b—collision mark (impact pit)
c—solution hole
d—silica deposit
e—subparallel steps

Plate 1.
Quartz grains from interval 0～5 cm.
Fig. 1—irregular grain;
Fig. 2—conchoidal fracture pattern;
Fig. 3—weathering cracks;
Fig. 4—subangular edges and collision marks;
Fig. 5—solution holes and silica deposits;
Fig. 6—silica flowers.

Plate 3.
Quartz grains from interval 450 to 455 cm.
Figs. 1, 3—irregular grain with subrounded edges;
Figs. 2, 4, 6—V-shaped patterns, collision marks and solution holes;
Fig. 5—silica deposits on edges.

Plate 2.
Quartz grains from interval 200 to 205 cm.
Fig. 1—irregular grain with angular to subangular edges;
Fig. 2—conchoidal fracture pattern;
Fig. 3—weathering cracks;
Fig. 4—V-shaped pattern;
Figs. 5, 6—subrounded edges, V-shaped pattern and collision marks.

Plate 4.
Quartz grains from interval 775 to 780 cm.
Fig. 1—irregular grain with subangular to subrounded edges;
Fig. 2—relict conchoidal fracture pattern with silica coating;
Fig. 3—cleavage-controlled grain, subrounded edges;
Fig. 4—solution holes and silica deposits;
Fig. 5—relict conchoidal fracture pattern and silica deposit;
Fig. 6—collision marks, solution holes and subparallel steps.

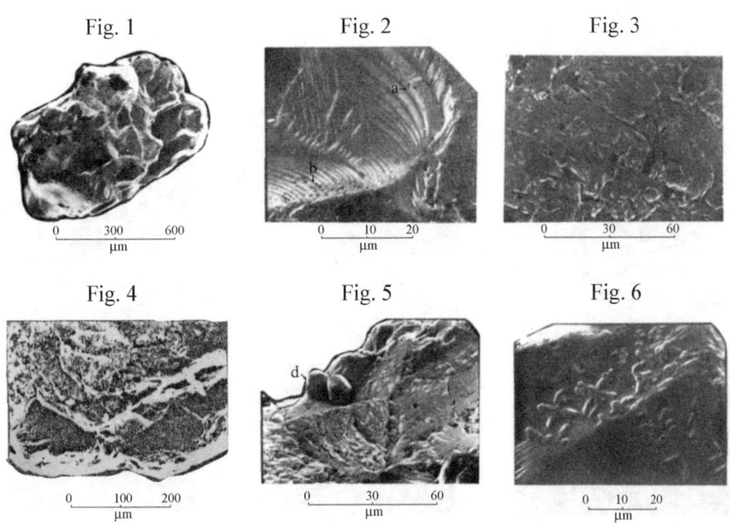

Plate 1

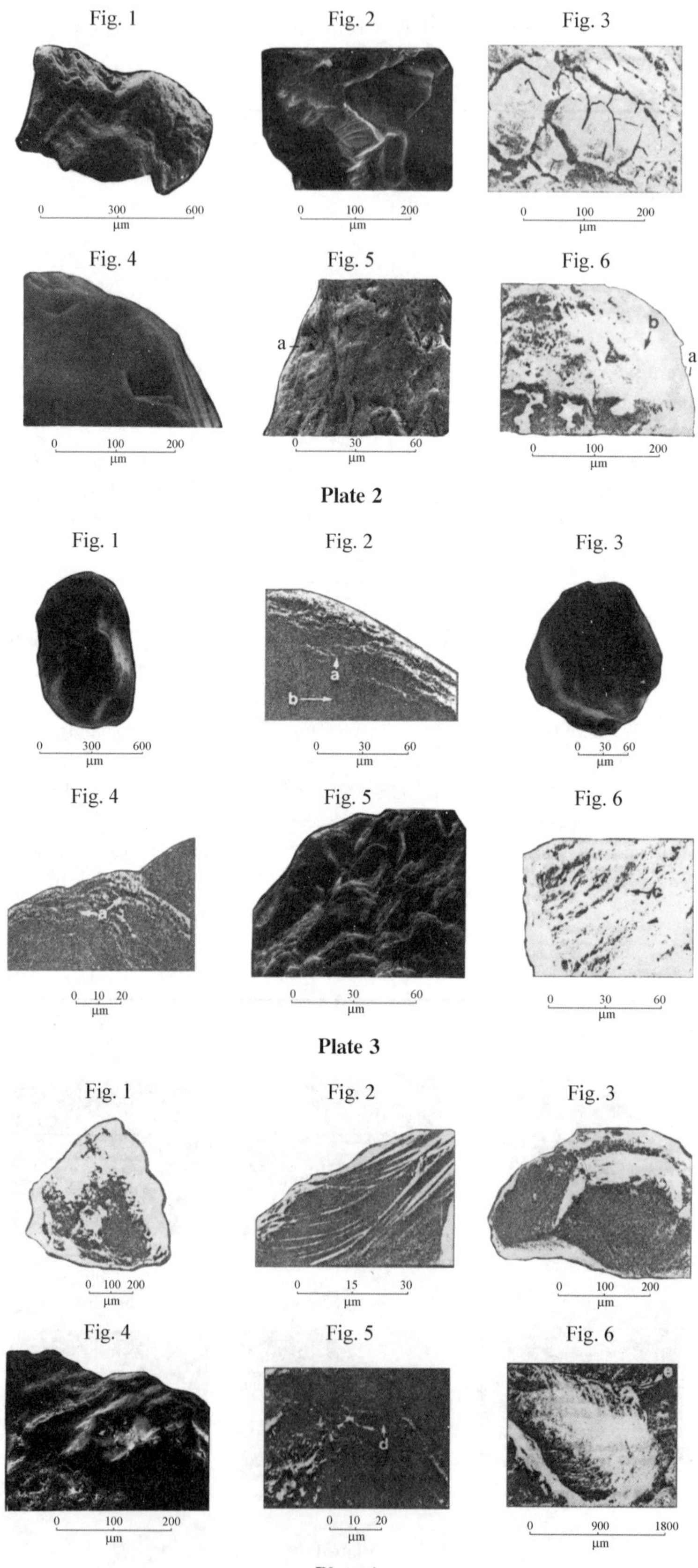

Plate 2

Plate 3

Plate 4

Acknowledgments

We would like to thank B. Deonarine for making all the Scanning Electron Microscope pictures of the paper. D. J. W. Piper and C. F. M. Lewis provided helpful comments in review of the manuscript.

References

[1] Andrews, J. T. 1973. Maps of the maximum postglacial marine limit and rebound for the former Laurentide Ice Sheet (the National Atlas of Canada). *Arctic and Alpine Research*, 5(1): 41 - 48.

[2] Dinsmore, R. P. 1972. Ice and its drift into the North Atlantic Ocean. *ICNAF*, *Special Publication*, (8): 89 - 128.

[3] Dunbar, M. J. 1951. Eastern Arctic Waters. *Fisheries Research Board of Canada*, *Bulletin*, (88): 1 - 131.

[4] Gustajtis, A. K. 1979. Iceberg population distribution study in the Labrador Sea. Data Report, C-CORE (Centre for Cold Ocean Resources Engineering) Publication 79 - 8, 41.

[5] Krinsley, D. H. and Donahue, J. 1968. Environmental interpretation of sand grain surface textures by electron microscopy. *Geological Society of America Bulletin*, 79: 743 - 748.

[6] Krinsley, D. H. and Doornkamp, J. C. 1973. Atlas of quartz sand surface textures. Cambridge University Press, 91.

[7] McLaren, P. 1980. The coastal morphology and sedimentology of Labrador: A study of shoreline sensitivity to a potential oil spill. Geological Survey of Canada, Paper 79 - 28, 41.

[8] Milliman, J. D. and Emery, K. O. 1968. Sea levels during the past 35,000 years. *Science*, 162: 1121 - 1123.

[9] Reineck, H. E. and Singh, I. B. 1973. Depositional Sedimentary Environments. Springer-Verlag, New York, 439.

[10] Reinson, G. 1979. Inner shelf sediment dynamics. Cruise Report 79 - 018. Bedford Institute of Oceanography, Dartmouth, N. S. 4 - 7.

[11] Reinson, G. E., Frobel, D., and Rosen, P. S. 1979. Physical environments of the Groswater Bay-Lake Melville coastal region. Proceedings of the Symposium on Research m the Labrador Coastal and Offshore region, Memorial University of Newfoundland, St. John's, 291.

[12] Van der Linden, W. J., Fillon, R. H., and Monahan, D. 1976. Hamilton Bank, Labrador margin: origin and evolution of a glaciated shelf. Geological Survey of Canada, Paper 75 - 40, 31.

[13] Vilks, G. 1980. Postglacial Basin Sedimentation on Labrador Shelf. Geological Survey of Canada, Paper 78 - 28, 28.

[14] Vilks, G., Deonarine, B., Wagner, F. J., and Winters, G. V. Foraminifera and Mollusca in Surface Sediments of Southeastern Labrador Shelf. Geological Society of America Bulletin. (In Press)

Dynamic Geomorphology of the Drumlin Coast of Southeast Cape Breton Island*

INTRODUCTION

Drumlin coasts are widely distributed in Nova Scotia. The early work of Johnson (1925) and Goldthwait (1924) established Nova Scotia as a type area for this class of coast. This study examines 50 km of the southeastern coast of Cape Breton Island, from Gabarus Bay to Red Cape (Fig. 1). The area has attractive drumlin-coast scenery: there are many lakes and ponds along the shore, protected by gravel beaches and vegetated drumlin islands. Rocky reefs are overlain by reddish clay till cliffs. We have studied the characteristics of this drumlin coast to develop a model describing the major processes of post-glacial development of the coastline. No comparable study of coastal geomorphology

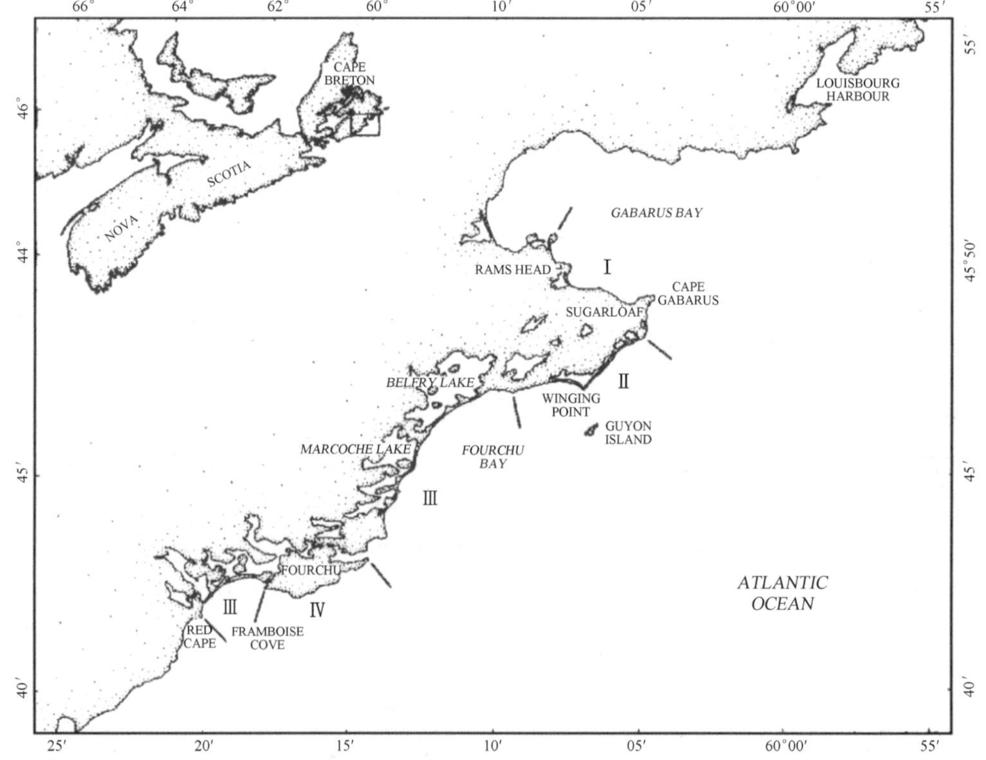

Fig. 1　Map showing location of study area and division into four coastal segments

* Ying Wang, D. J. W. Piper: *Maritime Sediments and Atlantic Geology*, 1982, Vol. 18, No. 1. pp. 1－27. 后附中文译文。

has been made since the classic work of the early twentieth century. During the Holocene transgression, different parts of the coast were exposed to varying wave and current regimes, and were also under the control of the geological structure, lithological character, and original morphology of the land surface. As a result of these factors, different parts of the drumlin coast have different geomorphological features, reflecting different processes and evolution. These particularities exist even within short distances, such as from Cape Gabarus to Framboise Cove.

METHODS

A reconnaissance survey of the area was started in the spring of 1979 when observations were made at one-kilometre intervals along the coastline. On the basis of these observations, a coastal geomorphological map was made and the locations for 16 shore profiles were selected (Figs. 2 and 3). During November of 1979, the whole area was studied and photographed from a helicopter in cooperation with R. Taylor and D. Frobel from the Atlantic Geoscience Centre, Bedford Institute of Oceanography. Systematic field work was carried out between the summer of 1979 and the fall of 1980 in cooperation with R. Taylor and D. Frobel. Profiles were surveyed using Emery poles

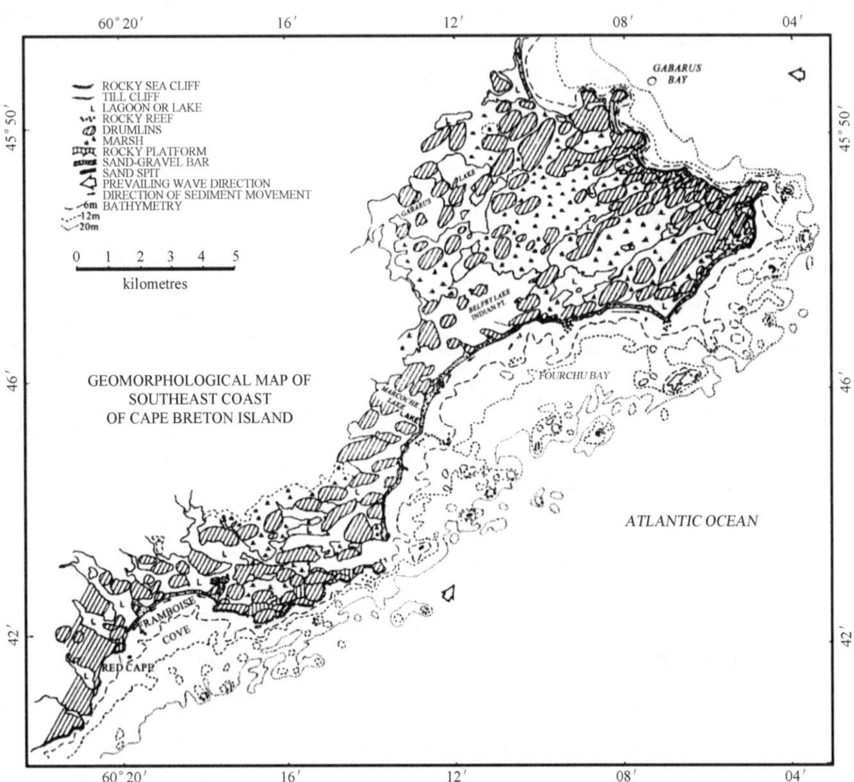

Fig. 2　Detailed geomorphological map of the southeast coast of Cape Breton Island

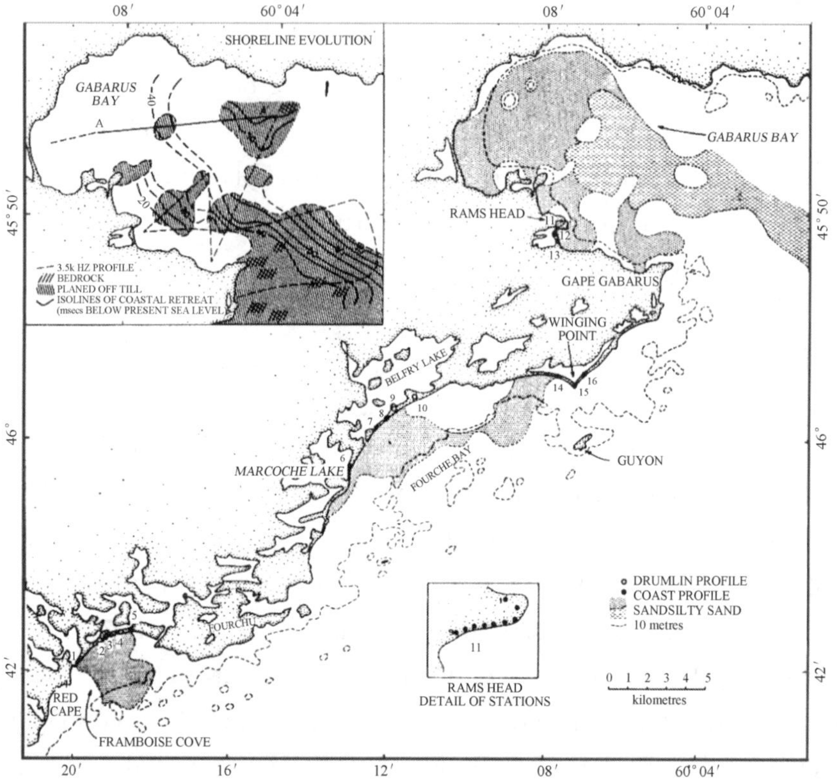

Fig. 3 Southeast Cape Breton Island, showing (a) location of coast and drumlin profiles 1～16 (c. f. Fig. 7 and Table 6), (b) nearshore sediment distribution, based on 200 grab samples and 100 km 3. 5kHz profile. Inset shows inferred development of Gabarus Bay with rising sea level: for further explanation, see text.

(Emery, 1961) with the horizon for reference. Nearshore profiles were obtained from a 3. 6-m ZODIAC, using a Raytheon DE－719 echo sounder. Offshore surveys were conducted from CSL TUDLIK using a Mini Ranger IV navigational system with Raytheon echo sounder and Klein side-scan sonar. Part of the sea floor was surveyed from a chartered fishing vessel, LONE ANGEL, with an over-the-side Ocean Research Equipment (ORE) acoustic profiling unit with a 3. 5 kHz pinger-type source operating at 10 kW power, recorded on an EPC 48-cm analogue recorder with a 0. 25-s sweep.

Sediment samples collected from sea shore and bottom were analyzed by standard methods (Carver 1971), according to the size of the sediment. The size distributions of sandy sediments were determined by sieve analysis and of finer sediment by pipette analysis. The mean size, standard deviation, skewness and kurtosis were computed for each sample. Folk's sediment nomenclature is used. Coarse samples of beach cobbles and pebbles were measured in the field by using a circle scale of Wadell (Krumbein and Pettijohn 1938). One hundred clasts per 0. 25 m^2 area were measured to determine mean size, psephicity (roundness), flatness and sphericity of each sample (these parameters are defined in Table 1).

Table 1 DEFINITION OF CLAST PARAMETERS

a	Length
b	Breadth
c	Thickness
r	The radius of curvature of the most convex direction of the flattest developed face.
Mean Size	$\sqrt[3]{abc}$
Flatness	$\frac{a+b}{2c}$
Psephicity	$\frac{2r}{a}$
Sphericity	$\sqrt[3]{\frac{abc}{a}}$

GEOGRAPHICAL AND GEOLOGICAL SETTING

The study area is located on the central part of the southeastern coast of Cape Breton Island. It extends from the large embayment of Gabarus Bay in the east to Red Cape, 50 km to the west-southwest. Relief at the coastline is relatively low, approximately 2～5 m; inland hills rise 60 m above sea level.

The bedrock belongs to the late Precambrian Fourchu Group (Weeks 1954; Cooke 1973), which is mainly volcanic in origin with rare interbedded sedimentary rocks. Dark green volcanic breccia or tuff is most common. Sedimentary rocks include shale, siltstone and rare sandstone or greywacke. The rocks have been sheared and metamorphosed to a low grade in most areas. The Fourchu Group is cut by intrusive igneous rocks, such as granite, especially around Gabarus Bay.

As a rule, it is difficult to determine the strike and dip of bedding of the Fourchu rocks. It is assumed that an anticlinal structure trends south-westerly across the Louisburg peninsula and Fourchu may be close to the anticlinal axis of these pyroclastic rocks. Shale and siltstone, although common, are not abundant. Narrow shale bands found in the Louisburg peninsula are also common extending northeastward from the mouth of Framboise River and to the north of Fourchu. At the mouth of Framboise River these rocks are associated with fine volcanic breccia or coarse tuffs.

The landforms of the exposed Precambrian rocks in the study area are variable. The north shore of Gabarus Bay is steep, the hills rising abruptly from the sea to a height of 60 m. From the south side of Gabarus Bay to Framboise, the land is much lower and irregular, with rocky lowland, fields of drumlins mostly 20 to 30 m high, and numerous bogs, ponds and lakes. The general trend of the land surface is probably a continuation of the peneplain of the northern tableland of Cape Breton, which dips southwards beneath the sea (Goldthwait 1924), but this has been substan dips southwards beneath the sea (Goldthwait 1924), but this has been substantially modified by Quaternary glacial erosion (King 1972).

During the Wisconsinan glaciation, ice moved in three widely divergent directions (Grant 1972): first toward the east, then toward the northeast, and finally southwards in an offshore direction. Any weathered mantle of rock waste or soil was almost completely removed by the ice, in addition to a great deal of solid rock. The ice rounded or smoothed the hills and ridges, gliding over the obstructing summits or filling small depressions. In consequence, the hills took the form of arched mounds of glacial drift, known as drumlins (Goldthwait 1924). They commonly occur in groups or fields and may display a striking parallelism of arrangement.

The drumlin belt of Nova Scotia is one of the four major such belts in North America (Thornbury 1960). Drumlins are common in several areas along the Atlantic coast of Nova Scotia, but are most abundant and have been most studied in northern Queens and Lunenburg Counties (Flint 1971). East of Halifax drumlins are found at several places along the coast, and the southeastern coast of Cape Breton is one such drumlin district. The trend of the drumlin series is mainly northeasterly, parallel to one of the directions of the ice flow and also to the strike of the bedrock.

There are two kinds of drumlins in the study area: true drumlins that consist mainly of glacial debris and are oriented in the direction of ice movement; and rock-core drumlins, comprising bedrock with a veneer of glacial till on top, that are more or less oriented in the strike of the bedrock. Both are considered as drumlins in this report, because both are of glacial origin, occur in the same area, and are probably related to a single glacial event.

COASTAL PROCESSES

The classic work of Goldthwait (1924) and Johnson (1925), and more recent local studies by Bowen and Piper and their students (Bowen 1975; Piper 1980), have defined some of the general processes by which the drumlin coast of Nova Scotia has developed.

The Atlantic coast of Nova Scotia is a coastline of submergence, on which sea level has risen by some 40 m in the last 10 000 years (Quinlan and Beaumont 1981). The earlier history of relative sea level is unclear, but there is no evidence of stands of sea level above the present level since late Wisconsinan deglaciation (Quinlan and Beaumont 1981). Tide gauge records (Grant 1970) suggest a present sea level rise as great as 4 mm per year.

Wave erosion causes retreat of drumlin cliffs, at rates in excess of one metre per year in exposed areas. Some of the eroded sediment is redeposited in barrier beaches of sand or pebbles between drumlin headlands. Rising sea level allows continuing coastal erosion, as the wave-cut platform submerges and ceases to significantly dissipate wave energy. Sediment supply is generally insufficient to permit beach progradation, and, as sea level rises, catastrophic breaching and landward beach retreat may occur (Bowen 1975). Substantial washover may be a precursor of actual breaching.

Coastal dynamic processes, especially those processes related to waves and currents, are very important factors in shoreline development. These processes control the movement of material along the coast. Because local climatic and oceanographic data are lacking observations from climatological stations and ships are used.

Winds and Waves

During the winter months there are frequent storms over southeastern Cape Breton. These storms bring cold air from the west, northwest, or southwest, i. e. basically from the continent. During summer months winds over the area are generally moderate, mainly from the south or southwest (Tables 2 and 3). The maximum hourly mean speed for this area is between 110 and 125 km hr^{-1} in winter and spring but in summer it is somewhat less (Thomas 1953).

Table 2 1980 WIND FRENQENCY BY DAILY PREVAILING DIRECTION, Sydney (46°10′N, 60°03′W)

Wind Direction	Monthly Frequency (percent)											
	Jan.	Feb.	Mar.	Apr.	May	June	July	Aug.	Sept.	Oct.	Nov.	Dec.
N	12.9		19.35		22.58	23.3		12.9	13.33		10.0	6.45
NNE		3.45	9.67		3.22		3.22			3.22		3.22
NE		3.45	6.45		6.44		3.22	3.22		6.44	10.0	
ENE			3.22	3.22	6.44			3.22			6.66	
E		3.45		6.66							3.33	3.22
ESE	6.45		3.22	3.32	3.22		9.67	3.22	3.33	3.22	3.33	9.67
SE	3.22		3.22	6.66	3.22	3.22	6.45			3.22		
SSE	3.22		3.22	10.0			6.45	9.678	3.33	3.22	3.33	6.45
S	6.45	10.34	6.45	13.3	6.44	13.33	19.35	3.22	23.3	19.35	10.0	
SSW	6.45		6.45	3.32	9.67	23.3		9.67	3.33	6.44	6.66	9.67
SW		17.24	6.45	10.0	3.22	20.0	29.03	16.12	13.33	22.58	6.66	6.45
WSW	9.67	6.89	3.22	3.32	6.44			25.8	13.33	3.22	3.33	16.12
W	45.2	24.13	16.12	3.32	16.13	16.6	12.9	3.22	16.6	19.35	20.0	32.25
WNW	3.22	3.45	3.22		3.22		6.45		3.33			3.22
NW	3.22	3.45	6.45	6.66	6.44			9.67	6.66	3.22	6.66	
NNW		6.89	3.22		3.22		3.22			6.44	10.0	3.22

Table 3　1980 AVERAGE DAILY WIND SPEED FREQUENCY, SYDNEY(46°10′N,60°03′W)

Monthly Frequency	Wind Speed (km/h)					
	7～12	13～19	20～30	31～40	41～50	51～60
January	6.45	22.58	38.7	25.8	6.45	
February	6.89	31.03	48.27	10.34	3.45	
March	6.45	19.35	48.38	16.12	9.67	
April	6.66	46.66	33.33	10.0	3.33	
May	9.67	32.25	51.61	3.22	3.22	
June	10.0	40.0	46.6	3.33		
July	22.58	61.29	16.12			
August	3.22	32.25	45.16	16.12	3.22	
September	6.66	40.00	53.33			
October	3.22	51.61	41.93	3.22		
November	3.33	6.66	20.0	53.33	6.66	10.0
December	12.90	32.45	41.93	12.9		

These wind conditions have a great effect upon wave generation in the coastal area (Tables 4 and 5). Southwesterly and southerly waves are predominant from May to August. Northerly, westerly, and northeasterly waves prevail during October and November. During the other months the waves are mainly from the west, but northerly and southwesterly waves are also significant. Because of the orientation of the coastline, waves from the west and northwest do not have a significant effect on shore processes. Therefore, for the whole year the effective predominant wave and wind are southwesterly. More than two-thirds of storms occur in the winter months (December to February) and the remainder during the transition periods in spring and autumn. The modal wave height offshore is between 1.5 to 2.5 m from November to April and lower during the remainder of the year (Table 5).

Table 4　WAVE DIRECTION FREQUENCY (1970—1972) FOR THE AREA OFFSHORE FROM S. E. CAPE BRETON ISLAND (Neu 1971, 1976)

Direction	Monthly Frequency (percent)											
	Jan.	Feb.	Mar.	Apr.	May	June	July	Aug.	Sept.	Oct.	Nov.	Dec.
N	14.59	12.94	12.69	15.08	5.9	5.11	6.45	3.76	10.0	20.96	18.3	13.97
NE	8.64	6.47	6.35	14.52	11.29	3.4	7.52	2.68	8.3	12.36	17.7	6.98
E	1.62	10.0	9.52	11.73	10.75	5.11	3.22	5.37	14.4	15.05	11.1	8.6
SE	2.16	5.88	3.70	6.70	11.29	5.68	5.37	9.67	8.3	2.68	6.66	5.3

(Continued)

Direction	Monthly Frequency (percent)											
	Jan.	Feb.	Mar.	Apr.	May	June	July	Aug.	Sept.	Oct.	Nov.	Dec.
S	5.94	14.1	8.46	6.70	20.43	30.11	26.88	23.12	124.4	9.14	9.44	5.9
SW	14.60	14.7	17.98	13.96	23.10	30.11	28.49	27.90	13.3	12.36	8.88	10.2
W	27.5	23.5	31.74	20.67	10.20	15.9	18.28	20.40	21.1	15.05	16.1	31.7
NW	24.86	12.3	9.52	10.61	6.98	4.54	3.76	6.99	10.0	12.36	11.6	17.2

Table 5　SIGNIFICANT WAVE HEIGHT* FREQUENCY (1970—1972) FOR THE AREA OFFSHORE FROM SE CAPE BRETON ISLAND (Neu 1971, 1976)

Wind Height (m)	Monthly frequency of Wave Height (percent)											
	Jan.	Feb.	Mar.	Apr.	May	June	July	Aug.	Sept.	Oct.	Nov.	Dec.
0~0.5												
0.5~1.5	20.54	15.88	17.9	27.37	54.3	61.36	72.0	65.6	69.4	50.0	17.7	12.9
1.5~2.5	37.29	38.82	38.62	43.0	30.1	30.11	23.6	26.3	23.88	33.3	35.0	42.47
2.5~3.5	20.0	19.4	23.28	17.3	9.14	7.95	3.22	5.9	3.8	10.75	24.4	30.64
3.5~4.5	14.59	12.35	10.05	9.49	5.38	0.57	1.07	1.07	2.2	3.22	9.44	8.6
4.5~5.5	3.78	5.29	7.93	2.23	1.07					1.07	7.77	4.8
5.5~6.5	3.24	4.7		0.55				1.07	0.55	1.61	2.22	3.2
6.5~7.5	0.54	2.35	1.06								0.5	0.53
7.5~8.5		1.17	0.53								2.7	0.53
8.5~9.5			0.53									0.53
9.5~10.5												

* Where the 'significant wave height' is the mean height of the highest third of all the waves in a wave train.

Tropical hurricanes occur during late summer and early autumn. Although the annual occurrence rate in the northwest Atlantic Ocean is between 6 and 10, only 10 are recorded to have passed across Cape Breton Island between 1871 and 1977 (U. S. Department of Commerce, 1978). The wind speed in hurricanes can be in excess of 180 km hr^{-1}. Under these rare conditions, waves could become extremely high. Based on 1970 data, the 10-year maximum deep-water wave height for the coast of Nova Scotia, including the southeastern coast of Cape Breton, is 16 to 18 m (Neu 1971, 1976). The 100-year wave is between 22 and 24 m.

Waves are refracted upon moving to the coast. The speed of waves is reduced as a result of the decreasing water depth, so that wave fronts are bent while approaching the shoreline. Because of wave refraction, wave energy is concentrated around headlands, while wave action is dispersed in the bays.

Refraction diagrams for the 6-, 10-, and 14-s waves approaching Gabarus Bay from northeast, east and southeast are plotted in Fig. 5. All diagrams show that the wave energy gradually decreases in Gabarus Bay, but is concentrated around the headland of Cape Gabarus.

Refraction diagrams for 10-s waves approaching the Fourchu area, from the southwest, south, southeast, and east and for 6-s waves from the southwest are plotted in Fig. 6.

Tide and Currents

Along the southeastern coast of Cape Breton the tide is mixed and propagates northwestwards from the open sea. The tidal range is not great: the mean range in Gabarus Cove is about 1. 2 m, and spring range is about 1. 5 m. In general, tidal currents are not important geomorphological agents in this area.

Currents along the coast are quite dependent upon the winds, and consequently are variable in speed and direction, but are constrained by local coastal morphology and bathymetry. The prevailing surface current often experienced about 5 km off this coast is 0. 3 to 0. 5 m/s to the west-southwest.

COASTAL TYPES

Drumlins dominate the morphology of the 50 km coastline from Gabarus to Red Cape along the southeastern coast of Cape Breton Island. The coast is transgressive, with embayments formed by flooding of valleys between drumlins. Sand and gravel deposits, derived mainly from the drumlins, accumulate in the embayments and enclose lagoons. Rounded or oval-shaped drumlins, either single or in a group, are drowned, forming dome-shaped islands or promontories.

Drumlin types, the relationships between the direction of the coastline and the drumlins, and topographical characteristics of the submarine coastal slopes all vary along the shore. Thus the action of different directions and intensities of waves and currents forms a variety of coastal morphologies, which can be considered as different developmental stages of the drumlin coast. The coast of the research area is divided into four sectors or subtypes, from north to south, each with a different developmental stage (Fig. 1). These four sectors are:

(1) South coast of Gabarus Bay (Rouse Point—Cape Gabarus—Sugar Loaf). This area is characterized by active marine erosion of drumlins, and the development of small embayments, some with barrier beaches, between headlands of bedrock or drumlins (Fig. 4A).

(2) The Winging Point coast (Sugar Loaf-Winging Point-Indian Point). This is a coastline of marine deposition, with tombolos and barrier beaches backed by eroding drumlins and lagoons (Fig. 4B).

(3) Framboise Cove and Fourchu Bay (Indian Point—Fourchu Head). This is a coastline of marine deposition with an extensive coastal barrier (Fig. 4C).

(4) Fourchu Head to Red Cape. This is a low rocky coastline of marine erosion, from which drumlin tills have been largely removed by marine erosion (Fig. 4D).

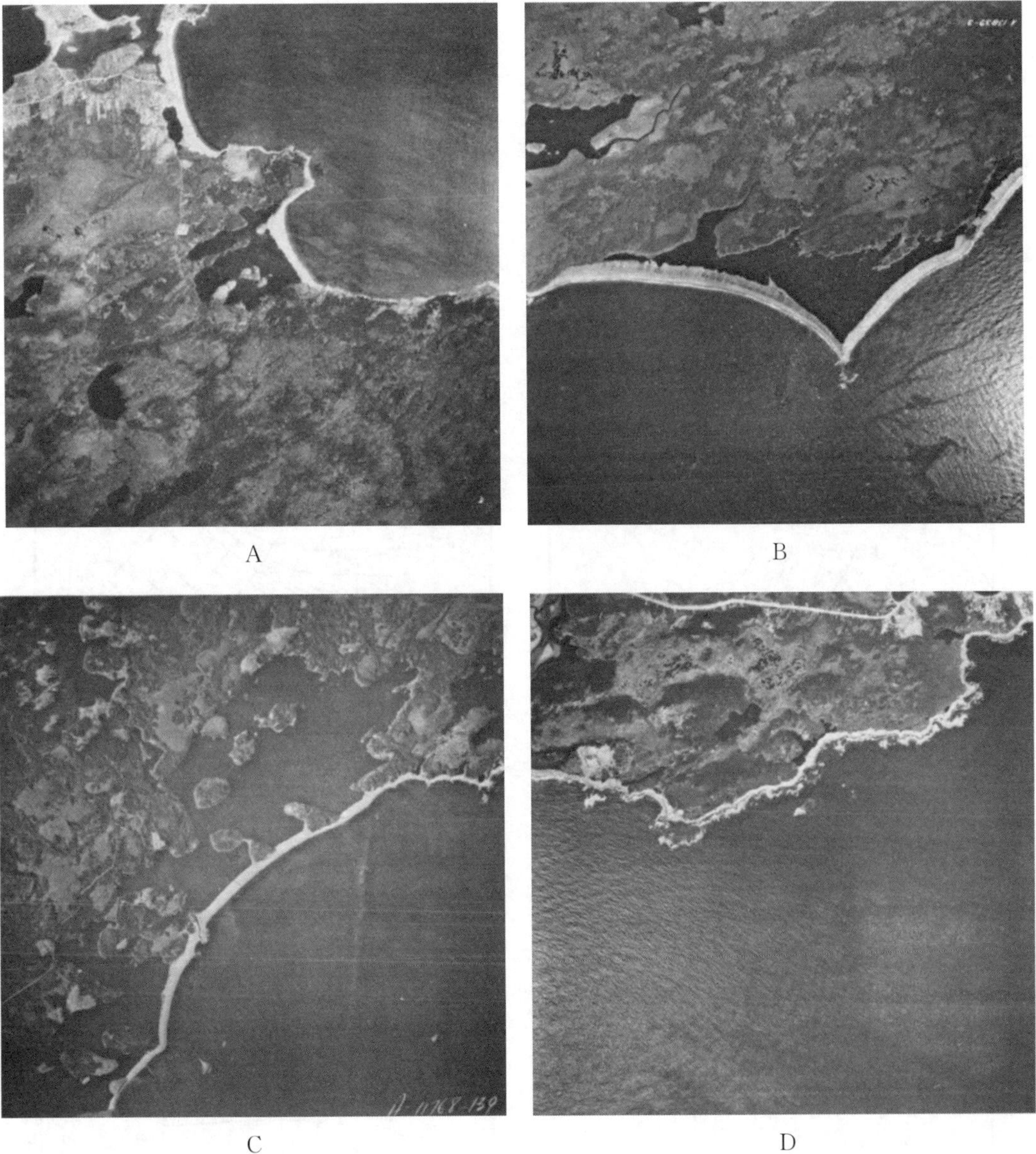

Fig. 4 Representative air photographs illustrating the four types of coastal sediment (c. f. Fig. 1)

A—Rock-cored drumlin coast. Segment I. South coast of Gabarus Bay; B—Double-tombolo drumlin coast. Segment II. Winging Point; C—Valley drumlin coast. Segment III. Coast of Fourchu Bay; D—Eroded low rocky coast. Segment IV. Fourchu Head.

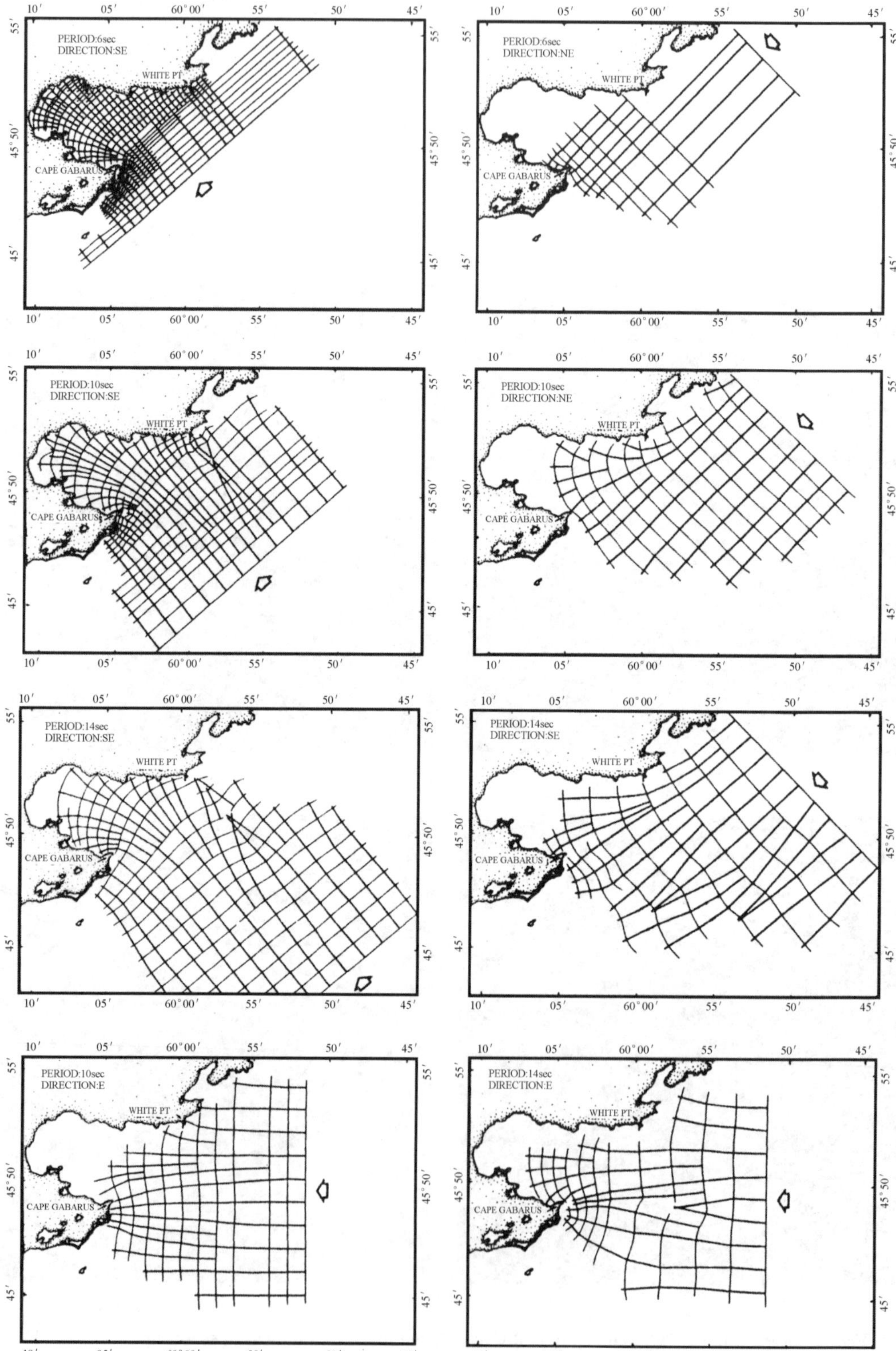

Fig. 5　Wave refraction diagrams for Gabarus Bay, showing effects of varying direction and period of waves.

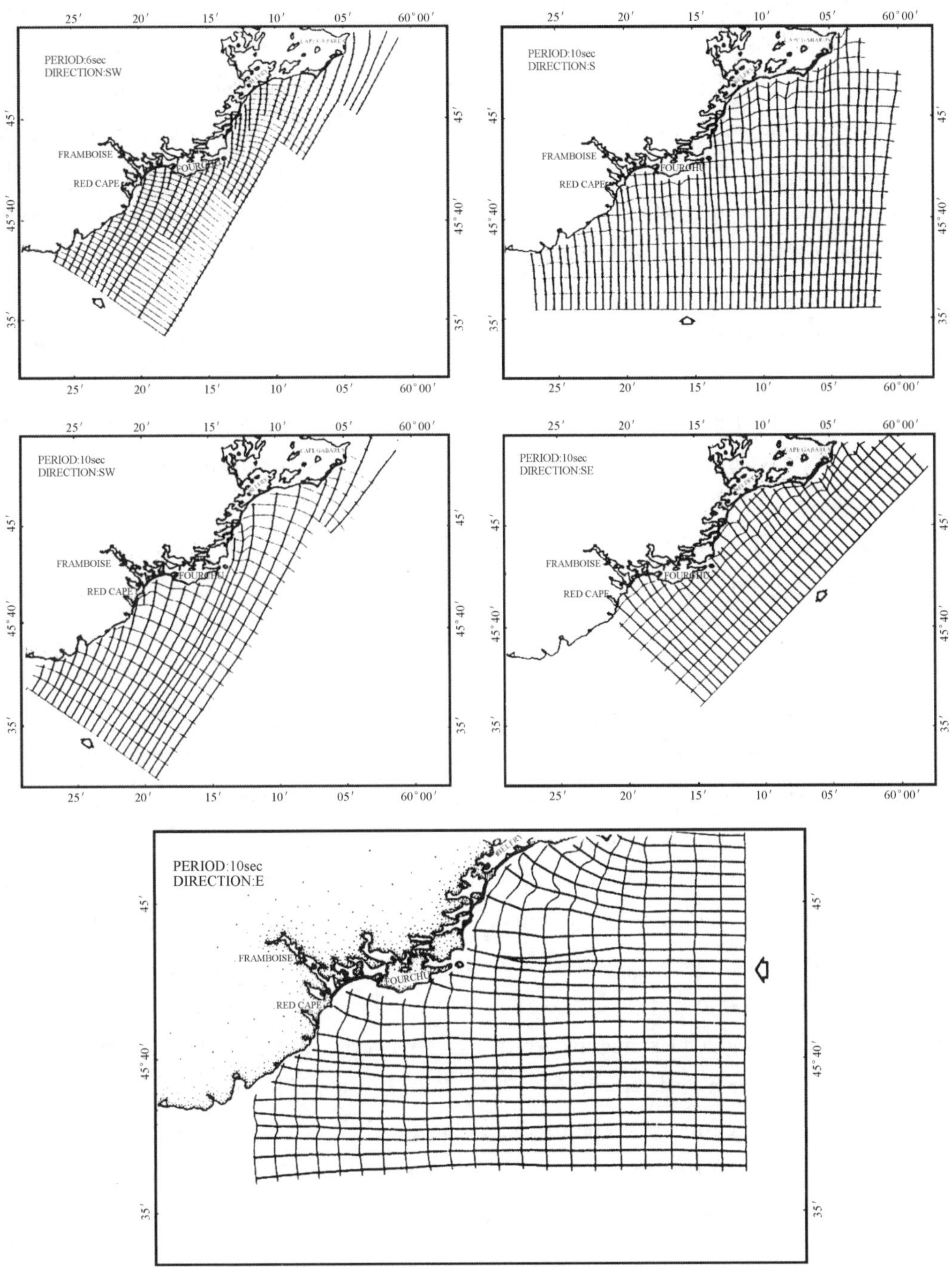

Fig. 6　Wave refraction diagrams for Fourchu and Framboise Bays, showing effects of varying direction and period of waves.

1. South Coast of Gabarus Bay

a. General setting

This coastal type is located along the southern part of Gabarus Bay extending from

Rouse Point to Cape Gabarus and then to Sugar Loaf, a total distance of about 11 km. The basic trend of the coastline is northwest to southeast, except between Gabarus Point and Sugar Loaf where it is north to south. Louisburg Peninsula and Scatarie Island protect the coastline from strong northerly waves. In this area, easterly and northeasterly waves with a large fetch prevail. The bedrock of this part of coast is mainly greyish, siliceous sandstone, shale, siltstone, or limestone. There are granite outcrops in coastal cliffs between Harris Lake and Gull Cove.

b. Drumlin cliffs

The drumlins along this coast are smaller than elsewhere in the study area but are closely spaced and trend northeasterly (060°to 080°). Along the coast they are seen to rest on an irregular rock platform 5 to 8 m above sea level, suggesting that this drumlin field originally developed on rocky, hilly ground; therefore, the drumlins are rock-cored and are covered with only a thin sheet of glacial till. For the most part, the till is 2 to 4 m thick, and is a yellowish sandy gravel. The till is only 0. 5 to 1. 0 m thick on the granite hills west of Gull Cove, but more than 10 m thick between Gull Cove and Cape Gabarus.

The northeast-trending drumlins meet the coastline obliquely, and marine transgression has produced an eroded rocky embayment coast. The drumlin hills enter the sea to become headlands or islands. A 3- to 5-m high sea cliff forms an uneven rocky bench beneath the cliff, facing the sea obliquely. There are several pocket beaches and cliffs 10 m high in the area of granite bedrock.

Cape Gabarus projects seawards in a northeasterly direction. It consists of more than 10 m of greyish-yellow till that thins eastwards as the underlying polished bedrock surface rises to 8 m above sea level. Wave erosion at the base of the till has cut a cliff from which material frequently slumps, supplying sediment to the seashore. Despite the strong wave action, glacial striae are quite well-preserved on the bedrock surface. The top of the headland is a rocky platform 100 to 200 m wide. The platform appears to be wave-cut, because it cuts across the regional glaciated bedrock surface, leaving low cliffs and stacks. The platform marks the extent of erosional retreat of the headland at approximately present sea level, and this can be used as an index of the intensity of wave erosion.

Rams Head has a more sheltered location on the south side of Gabarus Bay. Like Cape Gabarus, it consists of till overlying a bedrock terrace a few metres above sea level. Retreat of the till cliff is more or less 0. 1 m in a year according to the measurement of stakes on the top of the Rams Head drumlin (Table 6). The wave-cut platform is 40 m wide on the north, 75 m on the east, and 20 m on the south, suggesting much less rapid erosion than at Cape Gabarus.

Table 6　MEASUREMENT OF DRUMLIN RETREAT BY USE OF STAKES

Stake No.	Location	1979. 9. 25 (m)	1979. 11. 21 (m)	1980. 5. 12 (m)	1980. 8. 6 (m)
4～100*	Framboise Cove area on the top of sand bar, 100 m east of profile 4	1. 48		1. 00	0. 96
4～200**	200 m east of profile 4	2. 00		1. 74	1. 74
4～300	300 m east of profile 4	2. 30		2. 27	2. 27
4～400	400 m east of profile 4	2. 44		2. 44	2. 44
2*	Framboise Cove area on the top of No. 2 drmlin	2. 00		0. 50	eroded
201***	30 m west of stake 2	2. 30		2. 00	2. 2
202	30 m west of stake 2	2. 05		1. 40	1. 40
203	Landward of stake 2	8. 10		6. 47	6. 40
				1980. 5. 14	
900	Fourchu Bay area on the top of No. 9 drumlin	3. 75		3. 70	3. 70
9002	30 m west of stake 9	2. 55		2. 45	2. 45
9003	30 m west of stake 9	2. 40		2. 35	2. 30
9004	Landward of stake 9	13. 74		13. 70	13. 70
				1980. 5. 12	
10	Fourchu Bay area on top of No. 10 drumlin	1. 30		1. 0	1. 0
1001	Landward of stake 10	6. 30		6. 0	6. 0
		1979. 9. 16		1980. 5. 14	1980. 8. 10
1101	Harris Lake area on the top of Rams Head drumlin. 1101 is on the northern edge of the drumlin. 1102 is on the east side, and 1103 to 1109 are on the southern side of the drumlin (see Fig. 3)	2. 0	1. 95	1. 90	1. 90
1102		2. 0	2. 00	1. 98	1. 98
1103		2. 0	2. 00	1. 98	1. 95
1104		2. 0	2. 00	2. 00	1. 95
1105		2. 0	1. 95	1. 90	1. 82
1106		2. 0		1. 90	1. 90
1107		2. 0		1. 90	1. 90
1108		2. 0		2. 00	2. 00
1109		2. 0		2. 00	lost stake

* Framboise Cove, ** the distance from stake to the edge of dune scarp

*** all results show the distance between stake and the edge of drumlin cliff

c. Barrier beaches

Along this section of coast, small bays have developed between rocky head-lands or drumlins, and have transgressed landward as sea level has risen. Some of these bays are less than a hundred metres wide. Clasts from the headlands have been deposited in the bays to form steep, narrow cobble or pebble bars which enclose lagoons.

Harris Lake is one of the largest embayments along this type of coast, with the total area of the bay being about 1. 5 km^2. Four parallel strings of drumlins extend southwesterly through this area, with the outer two strings marking the edge of the lagoon, and the inner smaller strings forming small islands in the lagoon. The bar at Harris Lake is 750 m long. Its northern end is protected by Rams Head, and in the wave shadow of northeast waves the bar is low and wide (135 m) (Fig. 7, profile 12). It consists of pebbles and coarse sand, the mean size of the pebbles increasing from 25 mm at the bar crest in the north to 35 mm in the middle and 50 mm at the southern end as exposure to waves driven by the northeast winds increases. The bar profile at the southern end is narrower (70 m), higher, and steeper, with a storm beach berm (Fig. 7, Profile 13). The material of the bar appears to have come from the eroded headlands along both sides of the bay. Cobbles, pebbles, and coarse sand accumulate on the bar, while fine sand and silt are deposited on the sea floor offshore (Fig. 3). Mean pebble psephicity at the drumlins on Rams Head is 0. 2, which increases to 0. 3 to 0. 4 on the bar, still a low value reflecting the local sources of the material and the short distance of sediment transport (Fig. 8).

The variation in morphology and grain size of the bar within a single bay results from differences in wave intensity related to the varying coastal exposure. This morphological analysis indicates that waves from the northeast and east are most effective in modifying the coast along Harris Lake.

Two shore profiles (Fig. 7) at the northern (12) and southern (13) ends of the bar illustratethese effects. Profile 12 is located along the northern part of the bay protected by Rams Head. The sediment is coarse sand and pebbles. In 1980, seasonal changes in the beach profile were mainly around mean high tide level but below extreme high tide level. Sediment accumulated in the spring and summer during times of low wave activity, but eroded during the time of strong northeasterly and easterly winter waves. The backshore adjacent to the lagoon receives little sediment.

Profile 13 is located on the narrow, higher southern part of the bar. The cobbles at this end of the bar are moved mainly during storms; average waves do not change the profile significantly. The upper foreshore consists of cobbles and pebbles that accumulated during winter and late fall, when the northeast waves predominated. Only the lower part of the foreshore profile changed during the summer, with the accumulation of finer sediment (mainly coarse sand).

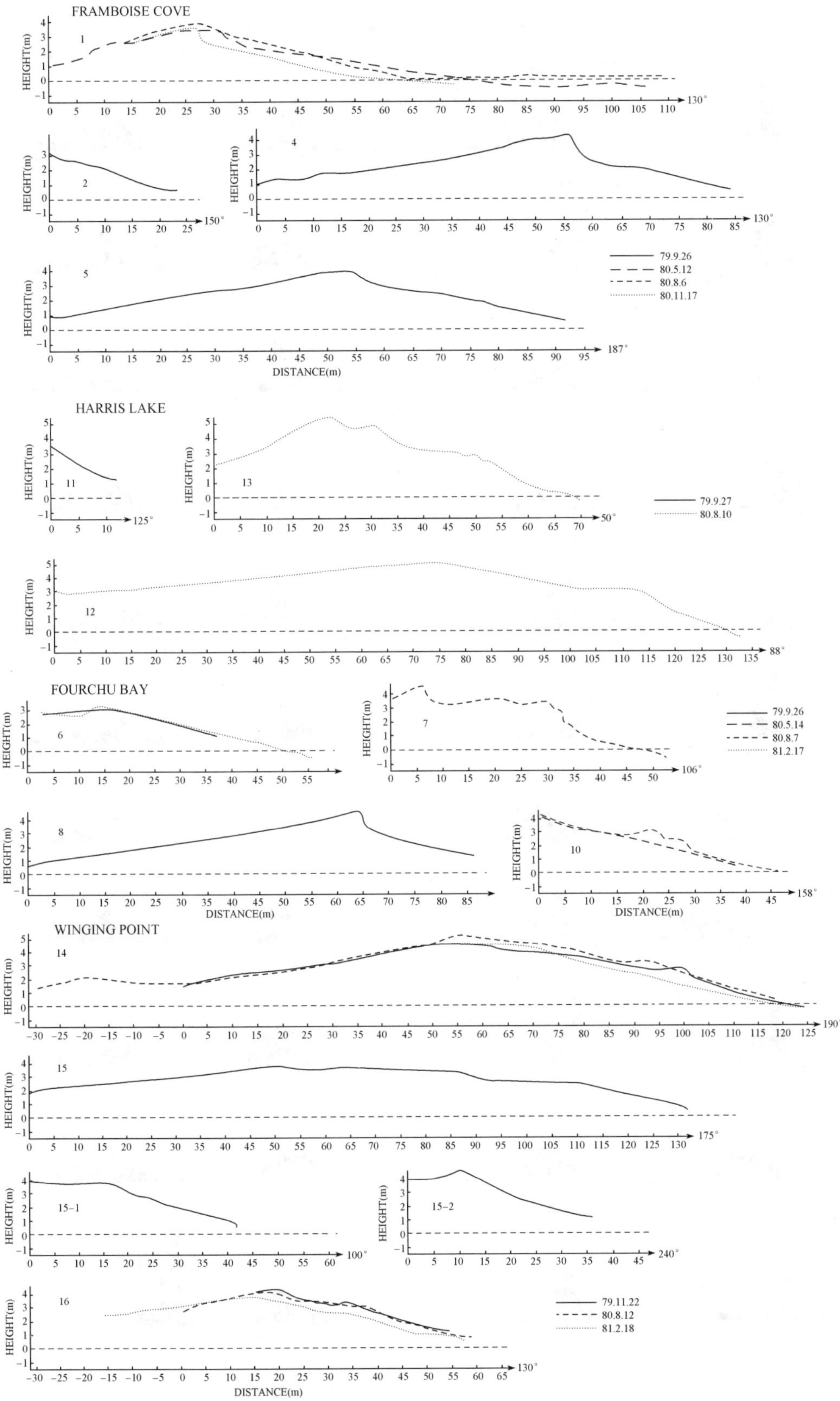

FRAMBOISE COVE

1　130°

2　150°

4　130°

5　187°

DISTANCE(m)

HEIGHT(m)

79.9.26
80.5.12
80.8.6
80.11.17

HARRIS LAKE

11　125°

13　50°

79.9.27
80.8.10

12　88°

FOURCHU BAY

6

7　106°

79.9.26
80.5.14
80.8.7
81.2.17

8

10　158°

DISTANCE(m)

WINGING POINT

14　190°

15　175°

15-1　100°

15-2　240°

16　130°

79.11.22
80.8.12
81.2.18

DISTANCE(m)

Fig. 7　Representative coast profiles of southeast Cape Breton; profiles located in Fig. 3.

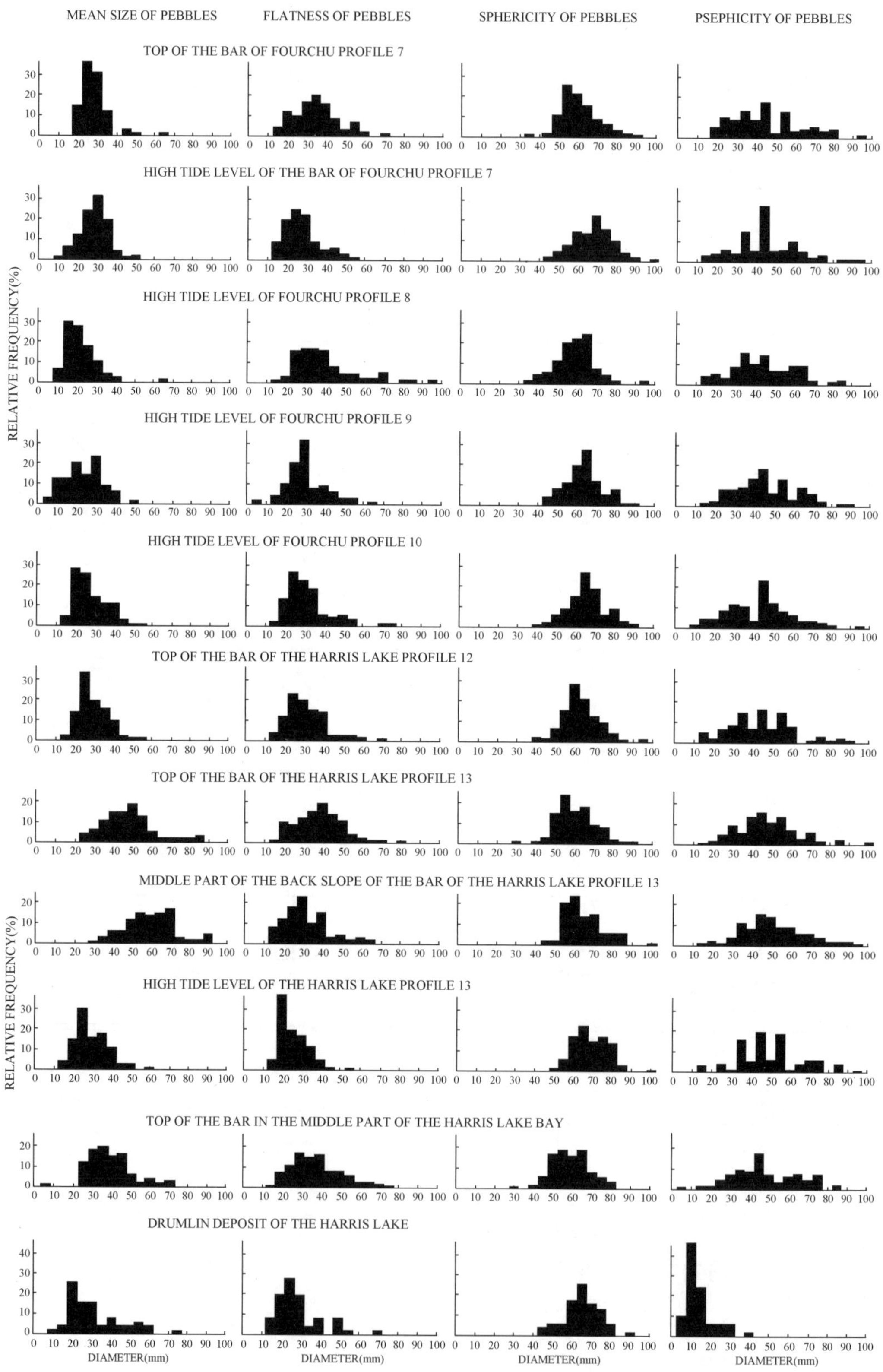
MEAN SIZE OF PEBBLES
FLATNESS OF PEBBLES
SPHERICITY OF PEBBLES
PSEPHICITY OF PEBBLES
TOP OF THE BAR OF FOURCHU PROFILE 7
HIGH TIDE LEVEL OF THE BAR OF FOURCHU PROFILE 7
HIGH TIDE LEVEL OF FOURCHU PROFILE 8
HIGH TIDE LEVEL OF FOURCHU PROFILE 9
HIGH TIDE LEVEL OF FOURCHU PROFILE 10
TOP OF THE BAR OF THE HARRIS LAKE PROFILE 12
TOP OF THE BAR OF THE HARRIS LAKE PROFILE 13
MIDDLE PART OF THE BACK SLOPE OF THE BAR OF THE HARRIS LAKE PROFILE 13
HIGH TIDE LEVEL OF THE HARRIS LAKE PROFILE 13
TOP OF THE BAR IN THE MIDDLE PART OF THE HARRIS LAKE BAY
DRUMLIN DEPOSIT OF THE HARRIS LAKE
RELATIVE FREQUENCY(%)
DIAMETER(mm)

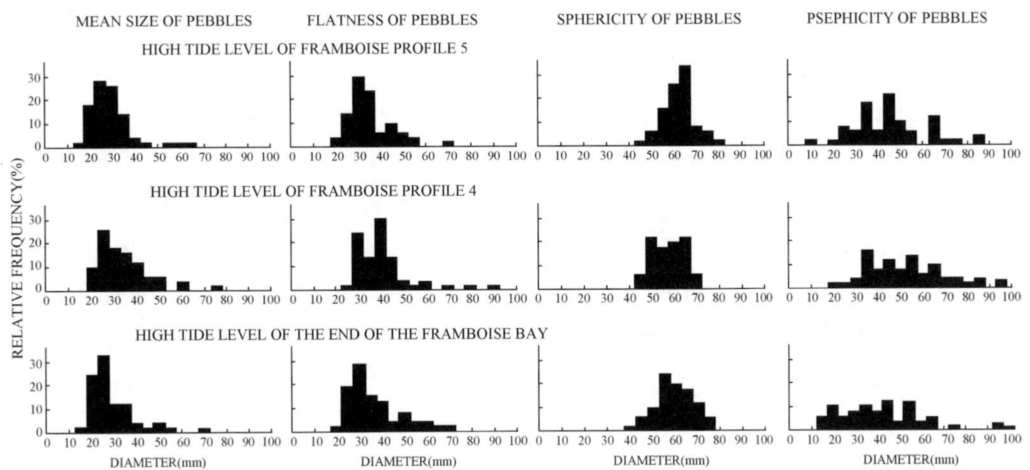

Fig. 8 Histograms showing properties of pebbles at Harris Lake, Fourchu and Framboise; for locations see Figs. 3 and 7. Definition of parameters in Table 1.

2. The Winging Point Coast

a. General setting

A depositional tombolo coast is found from Sugar Loaf southward to Winging Point, then westward to Indian Point. The total length of coastline is 9.5 km. The drumlins in this region are closely spaced, have rocky cores, and at the coast rest on rocky platforms. The drumlins trend 060° to 080°, almost parallel with the coastline. Offshore there are several small islands and reefs following the same trend, which are probably remains of old drumlins and protect the coastline from waves. The main feature of this coast is well developed double tombolo or bi-tie-bar islands (Zenkovich 1967) enclosing lagoons. Most of the drumlins are separated from the sea by lagoons and barrier beaches. Thus the scenery of this type of coast is quite different from the one previously described.

This coast is exposed to the open sea, including long-fetch southerly waves. The prevailing waves are southwesterly for much of the year, except in winter and spring when the northeasterly waves are strongest. Along this coast wave current and wind action are also stronger than in southern Gabarus Bay. Strong marine erosion that accompanied the post-glacial sea level rise has caused erosion and submergence of former drumlins, which become a series of oval-shaped islands or reefs offshore. These eroded drumlins have supplied sufficient sediment to the coast to develop extensive tie bars behind islands or reefs.

b. Winging Point

Winging Point, located at the centre of this segment of the coast, is a typical symmetrical or double tombolo. It is therefore used as an example to describe the coastal development. Winging Point is a rocky reef tied to the coast to the east and west by pebble bars. The reef is the remnant of a rock-cored drumlin, and is exposed during extreme low tide. It consists of Cambrian meta-sediments striking 055° with a 50°NW dip. The trend of this drumlin was northeast-southwest, and other rock ledges nearby also show this orientation. A second reef is located 100 m offshore from Winging Point, and Guyon Island is located 1.5 km off these two reefs. The reefs and islands protect the coast from wave erosion by refracting waves (Fig. 6) and causing the development of the two bars from east and west that connect Winging Point reef to the coast.

The eastern bar is 1.5 km long, and the western bar 2.2 km, the latter being 0.5 to 1.0 m higher than the shorter eastern bar. The entire bar consists of sand and gravel, with coarse sand deposited at the apex immediately north of the reef. Sand dunes are stabilized by grass and occur along the entire western bar, but only on part of the more exposed eastern bar.

Longshore sediment transport controlled by prevailing winds moves sediment toward the bar apex without any noticeable material exchange between the two bays. The bars prevent coastal erosion of drumlins from occurring and supplying significant amounts of sediment. The main direction of sediment movement along this coast is thus normal to the shoreline, reworking old coastal deposits or glacial till.

The east bar at Winging Point (Fig. 7, Profile 16) appears to undergo periodic washover and is getting lower. Waves have eroded the top of the bar and moved the material from the top to the lower slopes of both sides of the bar. This situation reflects the lack of sediment supply to this part of the coastal zone. Washover has resulted in little vegetation being established. In contrast the seaward progradation of dune grass and the evidence of sequential beach profiles show that the western bar is slowly accumulating sediment.

During winter and spring, under the prevailing northeast winds, the east bar is affected by wave erosion and the height decreases (Fig. 7, Profiles 16 and 15-1). The western bar is protected from the northwest waves by Winging Point so that, during the winter, this bar tends to grow (Fig. 7, Profile 14). During the summer southwest waves prevail so that the eastern bar is accumulating while the western bar undergoes wave erosion. During the autumn, both western and eastern bars are eroded because of the changing direction of the approaching waves.

Because of different coastline orientation and exposure, waves cause different seasonal changes of the beach profile. The sediments show no apparent seasonal changes, and on the whole both bars are relatively stable, showing only slight accumulation in the west and erosion in the east.

c. The southwest part of Winging Point coast

The trend of coastline is perpendicular to the trend of the drumlins, so that the shoreline consists of headlands and small embayments. The rock cores of drumlins consist of gneiss and slightly metamorphosed sandstone and slate (strike 040°, dip 20° to 60°SE). Drumlin rock cores form a platform 4 to 5 m above sea level, such as at Rock Point, which is affected by storm waves. Rock benches are covered by cobbles and pebbles forming beaches with crests reaching 6 to 7 m above sea level. Sediment has been deposited in small bays or behind the shoals to form bay-head bars of coarse sand and fine pebbles. There are also several tie-bars (Fig. 4C), such as at Rock Point, where two bars consisting of coarse sand and fine pebbles enclose a small lagoon. The west bar is larger than the east bar, but the east bar contains storm beach berms. Accumulation forms and flag-shaped pine trees both show that southwesterly winds and waves prevail.

Little material appears to be supplied to this coast by coastal erosion or longshore transport. The bars appear to have a stable morphology. Submerged drumlins lying offshore refract the approaching waves, leading to the formation of tie-bars. This part of the coast is in a relatively stable state, without either severe erosion or fast deposition.

3. Fourchu Bay and Framboise Cove

a. General setting

The coast of Fourchu Bay and Framboise Cove consists of barrier bars, backed by lagoons and series of drumlins. The total length of coastline is about 12. 5 km and follows the same general northeasterly (065° to 080°) trend as most of the drumlins. A few drumlins trending 105° were probably formed by an early easterly ice flow. Along this type of coast drumlins were developed in the original depressions or valleys, and formed elliptic or dome-shaped till hills. The till in the drumlins is 10 to 15 m thick and consists of yellowish-brown silty and sandy gravel. The drumlins are widely spaced, lack rock cores and bedrock occurs only at their base. They are thus quite different from the thin rock-cored drumlins developed on the hilly ground farther east. Since the retreat of the ice-sheet, the drumlins exposed to the sea have been completely cut away, leaving elliptical shoals offshore. There are at least four northeast-trending lines of drumlin remnants visible on bathymetric charts and seismic profiles (Fig. 3, Fig. 9) of this area. Along the shoreline, exposed parts of drumlins have been eroded, forming 10- to 15-m high bare till cliffs. Spits trail from their submerged ends and sand bars tie them together in parts. In Fourchu Bay and Framboise Cove, sandbars have united to form an extensive barrier bar that has, in places, built out in front of the drumlin cliffs, thereby reducing the effectiveness of wave attack on the cliffs. Thus, on this drumlin coast, complete, unmodified drumlins are found in the lagoons, eroding drumlins at the coastline, and relict drumlins occur offshore, showing all stages of the coastal development.

The nearshore sediment of Fourchu Bay consists of pebbles near Winging Point, the drumlin shoals, and around the headlands; and fine sand occurs in deeper water.

b. Drumlin cliffs

There are 11 drumlins in this area that have been truncated and form coastal cliffs of till (Fig. 2). Measurements on the top of the drumlin cliffs show that most of these cliffs are still being modified and retreat slightly during storms. A few typical examples are described.

Number 2 drumlin is a 5-m high and 75-m long drumlin located near the middle part of Framboise Bay. The slope of the drumlin cliff is from 54° to 60°. The till is greyish-yellow, sandy gravel. Between September 1979 and September 1980 the cliff retreated 0.3 to 2.0 m with the middle part retreating most rapidly (Table 6). Drumlin No. 10 is a 10-m high accumulation of silty gravel located in the middle part of Fourchu Bay bar. There is a pine forest on the top of the drumlin but only dead trees on the edge of the cliff. The cliff retreated 0.3 m in one year.

Drumlin cliffs appear to retreat faster in Framboise Cove than in Gabarus Bay. The rate of retreat depends on the fetch of the prevailing waves. However, the sections of the coast that are being eroded are far shorter than the depositional sections, so that only a small portion of the sand-bar sediment can have been derived from recently eroded parts of the coast.

c. Fourchu Bay

In Fourchu Bay the sand bar is 3.9 km long and 140 m wide, extending from Indian Point in the east to Belfry Head toward the southwest. Belfry Head is a 10-m high, dome-shaped headland of sandstone and slate, with three groups of nearly vertical joints (striking 160°, 260°, 290°) that control the morphology of the cliff and uneven rocky beach. At the south side of the headland is a small tidal inlet. A sand causeway has been built several times by local people, but this has been breached in storms and kept open by strong currents. Across the Belfry tidal inlet, the bar extends southward for 2.2 km to a new inlet at Marcoche Lake; from here it continues south with small bay bars toward Fourchu Harbour.

The sand bar of Fourchu Bay consists of coarse sand and pebbles. Pebble or cobble bars appear at the base of till cliffs or near rock headlands. Fourchu Bay bar is a bayhead bar consisting of a series of bayhead beaches: the formation of this united large bar enclosed the lagoon and small bays, changing what was originally a curved and embayed coastline to a flat and straight depositional sand-bar coast. The lithology of the beach pebbles is similar to those in the drumlin tills, consisting mainly of sandstone, slate and limestone, with little granite and gneiss. In the process of transport from the drumlins, the pebbles have been changed to form rounded or flat beach pebbles. The psephicity of the bar pebbles is 0.4, considerably higher than drumlin tills (0.11) (Fig. 8) suggesting that the pebbles are stored and reworked in the nearshore zone.

The morphology of the sand bar varies along different parts of the coast:

(i) Symmetric foreshore and backshore—This occurs where the bar is between the drumlins and in front of a lagoon (e. g. Indian Point, and Profile 8, Fig. 7). Near Indian Point the bar consists of coarse sand and pebbles, ranging in diameter from 40～100 mm at the upper part of the bar to 20 mm on the lower part. The bar suffers only slight erosion and both sides of the slope are still symmetrical.

At Profile 8, the sediment size ranges from medium sand at low tide level to gravelly sand at high tide level. The sediments are poorly sorted because the material is supplied locally without extensive transport. The bar faces the open sea and its top retreated 2 m in the winter of 1979—1980. Thus the profile is slightly asymmetric, with a one-metre high scarp along the top edge of the bar (Fig. 7, Profile 8). This part of the bar consists of coarse sand with fine wind-blown sand on the top.

(ii) Asymmetric foreshore and backshore—Profile No. 7 is an example of this morphology. The bar here is narrow, steep and with several beach berms (Fig. 7, Profile 7). Because it is located near Belfry Head, the bar consists mainly of sandstone pebbles, 20 to 40 mm in diameter, derived from the adjacent headland. Roundness and flatness are low, only about 0. 45 and 0. 3 (Fig. 8). Sediment at low tide level is poorly sorted coarse sand. There are pronounced seasonal changes in Profile 7. Pebbles accumulate during winter storms, but are removed from the middle part of the beach face in summer and are deposited on the lower part of the beach.

(iii) A single foreshore—Profile 10 is an example of an inclined beach with a single slope (Zenkovich 1962) at the base of the drumlin cliff (Fig. 7, Profile 10). The swash reaches the foot of the cliff, and all of the water thrown up is incorporated in the backwash. There are mainly pebbles on the upper beach, and granules and medium sand on the lower beach. The profile is accumulating as the drumlin cliff retreats. Because there are no systematic changes in sediment grain size along the beach, there is unlikely to be major longshore movement of the littoral sediment.

As the beach sediment is largely fine sand and most of the pebbles are wellrounded, it is unlikely that much of the sediment is derived directly from drumlin erosion. A long period of sediment reworking is indicated in the nearshore, where sediment from erosion of former drumlin tills accumulates. Only a part of the beach sediments is supplied by modern marine erosion of drumlin cliffs or headlands.

In this part of Fourchu Bay, both drumlin cliffs and pebble bars show limited erosional retreat, so that there is only a limited supply of sediment on this coast. However, wave energy is decreased considerably because the approaching waves are refracted by relict drumlin islands and shoals. Under these circumstances, the coastline is relatively stable.

To the south of Belfry, along the outer coast of Marcoche Lake, the coastline stretches from north to south, and is exposed to the open sea to the east. Because of the large fetch, the easterly waves are quite strong. As a result, extensive overwash transport of sediment occurs, resulting in a relatively flat bar profile (Fig. 7, Profile 6). Waves transport gravel from nearshore landwards and because the sediment supply is plentiful and the bar is growing, the material is excavated for construction by local people. The bar sediment is well sorted fine sand mixed with granules. Because of the artificial influence of the excavation, the beach profile does not show the natural state.

From Marcoche Lake to the south there are two small drumlin cliffs, about 100 m long and 5 m high. At the foot of the cliffs there are almost bare clayey beaches, partly covered with a thin layer of coarse sand. The beach slope is 7° at the clay beach and 4° at the sandy part. Farther to the south a small river mouth separates the two drumlins from a third drumlin, which also has a trend of 060°. This river mouth or tidal inlet is new. A bar here has been washed out by strong currents, probably by flood currents, because the tails of sand bars at both sides of the inlet point landward. The abandoned inlet was at the south side of the third drumlin, now occupied by a new crescentic pebble bar. The bar is asymmetric with two slopes, and about 3m above sea level. The bar is very new and bare of vegetative cover. Beneath the crescentic bar well packed boulders make a smooth bed, formed by strong and fast-running currents, apparently the result of an abandoned tidal channel.

d. Framboise Cove

Framboise Cove bar is 3. 5 km long, extending westward from the east drumlin headland of the cove. It can be divided into three parts: (1) end of the bar, (2) base of the drumlin cliff, and (3) normal beach. A tidal inlet to Bagnell Lake separated the bar from the west drumlin headland of Framboise Cove. The Framboise River discharges into Bagnell Lake 10 km inland, so there are no apparent river mouth features at the bar.

Profile 1 (Fig. 7) is located at the west end of the Framboise Cove bar, and shows a complete profile with two slopes. The material consists of medium sand mixed with 20 to 30 mm pebbles (Fig. 8). The bar is low and flat. Beach profile measurement does not show seasonal changes, perhaps because of the excavation of the sand material.

Profile 2 (Fig. 7) is an inclined beach at the foot of No. 2 drumlin cliff. The beach sediment is mainly sandy gravel with positive skewness and poor sorting, because of mixing with a large amount of slumped mass from the retreating cliff. Profile measurements show that material accretes here, especially during the winter, presumably because winter storms have eroded the cliff and supplied the material to the beach. During the summer moderate waves remove beach sediments to the sea, so that the beach face is slightly decreased.

Profiles 4 and 5 (Fig. 7) are located in the middle part of the Framboise Cove bar.

Here the bar faces the open sea and, because it is far from a drumlin cliff, it should result from sediment processes acting on the bar alone. The bar consists of a complete profile with two slopes. The backshore slope, adjacent to the lagoon, is gentle and has a cover of grass: the foreshore slope is shorter with a 0. 5 m scarp on the top, resulting from wave erosion.

Along this part of the bar the sediment below high tide level consists of fine sand. Around the high tide level there are pebbles in Profile 5, but pebbles with medium sand in Profile 4. Coarse sand with pebbles occurs at the top of the bar. This type of sediment distribution with coarse sediment on the upper part of the bar, gradually becoming finer downslope, indicates that the sediments were transported by waves. Very fine sand is deposited on the sea bottom in the bay. Littoral sediment distribution, especially the pebble size distribution, suggests that there is a longshore transport from both headlands toward the centre of the bay. The mean size of sediments decreases from both ends of the bar toward the middle, so that sediment at Profile 4 is finer than at any of the other profiles. The psephicity and flatness of the pebbles in Profile No. 4 are also higher and they are mixed with well sorted sand (Fig. 8, Table 7). Although the top of the bar is eroded by waves to form scarps, shore profile measurements show that only slight seasonal changes take place. The bar seems to be stable or slightly accumulative.

e. Summary

Fourchu Bay and Framboise Cove have a double coastline. The curved inner coastline of drumlins is enclosed by barrier bars of sand, gravel and pebbles which form an outer straight coastline. Between these two coastlines are large lagoons containing small drumlin islands. There is no significant river sediment supply, neither is very much sediment supplied by the coastal erosion. The coastal sediment originates mainly from the submarine coastal slope of old deposits of the residual material of eroded drumlins. This source of sediment is limited. The coast faces the open sea, but several groups of relict drumlin reefs influence the waves approaching the coastline. Under this condition, the coastline of Fourchu Bay and Framboise Cove is considered dynamically stable over periods for which sea level change is negligible.

4. Fourchu Head to Red Cape

This type of coast dominates from Fourchu Head to Red Cape. The original land was a rocky platform or a series of small hills that developed small rock-core drumlins. The thin till of those drumlins rests on the rock platform some 5 or 10 m above sea level. The trend of drumlins is 055° to 070°, and the coastline is nearly parallel with the drumlin axis. Here the coast faces the open sea to the southeast. Due to the large fetch and wave refraction on this projecting stretch of coast, wave energy is high. The powerful waves cut the sandstone and slate to form a very irregularly eroded rocky coast. Uneven rocky beaches with frequent cobbles are common along the protected part of the coastline. Wave-cut beaches in some parts are more than 400 m wide. For example, the width of the submarine wave-cut beach from Fourchu Head to Framboise east headland is 700 m. Along this coast strong waves have swept away the drumlin tills and inhibited the growth of coastal vegetation.

GEOLOGY OF NEARSHORE AREAS

a. Gabarus Bay

The 3.5 kHz seismic profiles of Gabarus Bay show three types of sea floor (Fig. 9):

(1) small areas of bedrock (with a rough, highly irregular reflective surface) ;

(2) relatively flat areas of till, locally with thin sand or gravel patches; and

(3) areas of thick (up to 30 msec) sediments overlying an undulating hard surface, presumed to betill.

Fig. 9　3.5 kHz profile in northern Gabarus Bay, showing planed-off till drumlins (a); buried drumlins (b); and overlying lagoonal and shallow marine sediments (c); profile located A～A′ in Fig. 3 inset.

The bedrock outcrops east of Cape Gabarus and in northeasterly Gabarus Bay are a continuation of prominent lines of isolated rocky islands or reefs. In both trend and distribution they appear to be the cores of rock-cored drumlins from which the till cover has been stripped during marine transgression.

The flat till surfaces are most probably the result of coastal retreat of till cliffs, planed-off to a wave-cut platform. Isobaths of this till surface thus represent successive positions of the drumlin cliff coastline during the Holocene transgression (Fig. 3). The deepest flat till surfaces are about 50 m below present sea level.

In contrast, the undulating till surfaces preserved beneath sediment have not experienced wave erosion, and have maintained their original form. They occur in water depths greater than 50 m, suggesting that the maximum early Holocene lowering of sea level was about －50 m. They are also widespread in many shallower areas of southwestern Gabarus Bay. In these areas the drumlins were probably submerged in lagoons (as is at present happening in Harris Lake), protected from the open sea by barrier beaches. These beaches presumably linked drumlin headlands, so that their former positions can also be approximately located (Fig. 3).

Since the early Holocene, till cliffs at Cape Gabarus have retreated about 5km in a westerly direction. The isolines of retreat suggest that perhaps half of this sediment would have been transported by longshore drift toward Gabarus Bay and half southwestward toward Fourchu Bay. Assuming the till cliffs were 5m high, some 1×10^8 m^3 till has been eroded from this area, and a further 0.2×10^8 m^3 from within Gabarus Bay.

b. Fourchu Bay and Framboise Cove

The geological history of the coast off Fourchu Bay and Framboise Cove is less easily interpreted. Rocky shoals, perhaps the cores of drumlins, are common. Between these shoals, the seabed sediment comprises sand and pebbles, and thick, accoustically transparent sediment sequences are not seen.

If the drumlins that are widespread on land originally extended at least 4km seawards, then during Holocene transgression large amounts of till would have been eroded and supplied to the coastal zone.

COASTAL DEVELOPMENT

Most drumlin coasts are in areas of submergence in mid to high latitudes. The origin of a drumlin coast is related closely to glacial processes. Drumlin fields were formed during glaciation and the drumlin coast developed during the ensuing Holocene sea level rise. Drumlin coasts typically consist of small bays and myriads of islands. The sediment size in till varies from boulders to clay, and bars consist mainly of sand and gravel. As a result of the different coastal processes, the drumlin coast eventually becomes flat and straight.

The dynamic processes of a drumlin coast depend on the original types of drumlin, the morphology of the bedrock surface on which they rest, coastline orientation and exposure, and topographic features of the submarine coastal slope. These factors influence both the coastal sediment supply and the intensity and characteristics of waves.

Sediment supply from rivers is minor. River processes in high latitude, glaciated uplands are not as important as in middle and lower latitudes, because numerous glaciated lakes have trapped sediment in the drainage area. Because of the moist climate, the ground is covered with a luxuriant growth of grass and bushes, thereby retarding erosion. In mainland Nova Scotia and Cape Breton Island very little sediment is supplied by rivers (Piper and Keen 1976).

The sediment of a drumlin coast is supplied directly or indirectly by drumlin till, either from till cliffs or old coastal sediment on the sea bottom that originally came from drumlin till. Drumlin till is poorly sorted and drumlin coasts also include all sizes of sediment. Along the southeast coast of Cape Breton, the sediment comprises mainly cobbles, pebbles, sand and coarse silt.

Within the study area there are two contrasting developments of till. In irregular hilly areas, till is thin, and occurs in rock-cored drumlins. In valley areas bedrock is low-lying and thick till drumlins occur. These two types of till morphology have very different effects on coastal geomorphology. We can examine this by studying the stages by which the drumlin coast evolves (Fig. 10). In this analysis, we assume that a slow rate of marine transgression took place.

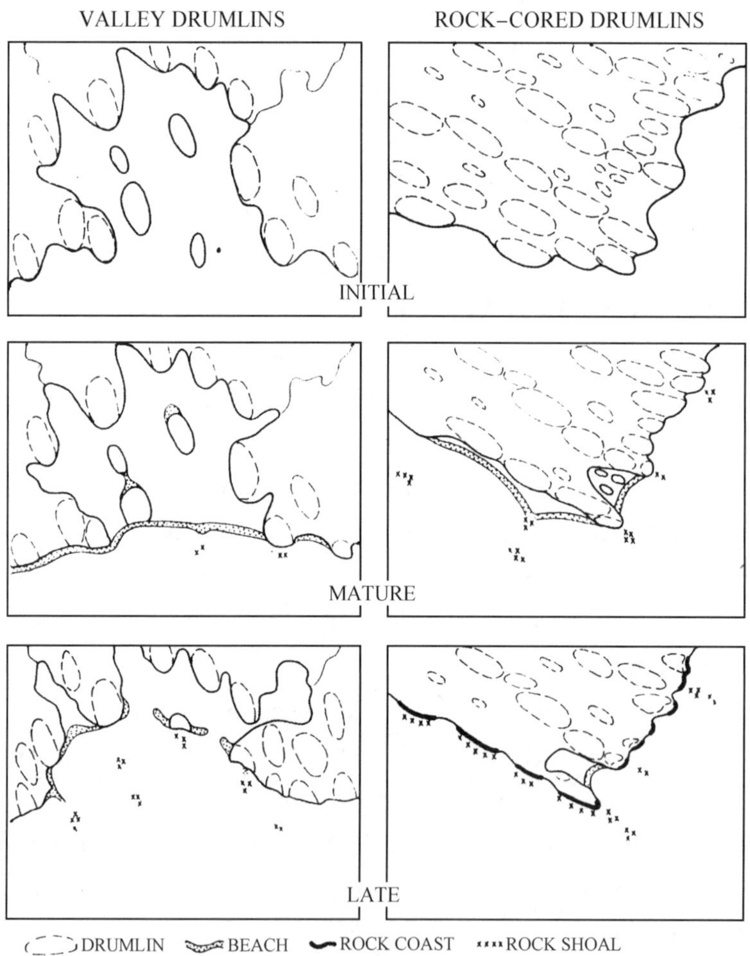

Fig. 10　Cartoons illustrating stages of coastal geomorphic evolution of (a) hilly rock-cored drumlin areas, and (b) valley thick drumlin till areas

1. Hilly coast with rock-cored drumlins

(i) Initial stage—this is where rising sea level in the early Holocene begins to drown and erode the drumlin coastline. At first there are a series of drumlin headlands and small embayments. With time, rocky headlands and residual till cliffs form on each side of the bays. Cobble and pebble bars gradually enclose the bays to form small lagoons. During this stage marine erosion is the leading process and the sediments originate from headlands. The inferred early Holocene coast east of Cape Gabarus would have shown this type of development, and the present coastline from Cape Gabarus south toward Sugar Loaf is an example of this stage.

(ii) Mature stage—As marine erosion progresses, where the trend of the drumlin is oblique or normal to the coastline, two or more small bays may combine to form a large bay. The single bar becomes longer but is still protected by two rocky headlands at the sides of the bay. The sediments come both directly from the headland and through the nearshore, as for example in Harris Lake Bay.

Where the coastline is parallel with the general direction of drumlins, marine erosion and sea level rise caused the coastline to retreat and form a series of small islands or relict drumlins. Sediments are deposited in the wave shadow area of islands or reefs to form well-developed tombolos or tie bars. Thus, the original coast is enclosed by tie bars or tombolo systems. At this stage, strong marine erosion is not a leading process, because waves are refracted by eroded island sand reefs. The coast has reached a relative dynamic equilibrium, except for the rise of sea level. The coast of Winging Point is an example of the stage.

(iii) Late stage—This stage of drumlin coast is developed only where it faces open ocean and where there is a very steep coastal slope. Drumlins have nearly disappeared due to marine erosion, and have formed an irregular curved and eroded rocky lowland coast; for example, the coast westward of Fourchu Head.

2. Valley areas with thick till drumlins

(i) Initial stage—Valley areas, because they are low-lying, form deep bays and inlets during early transgression. They are thus relatively sheltered from open ocean waves. The flooded drumlins form dome-shaped forested islands.

With the start of marine erosion, pine trees are removed, exposing a glacial till cliff. Gravel and coarser material, derived from the eroded till, is carried along the coast and deposited mainly at the landward side of drumlin islands to form a tie bar or a small spit. Finer materials however are carried to the sea. This stage of the drumlin coast type is very short-lived, because marine erosion is rapid on coasts of unconsolidated sediment. This type of coast can be seen in the inner parts of lagoons or where there are long narrow bays, such as in the lagoon of Fourchu Bay and Framboise Cove.

(ii) Mature stage—As the coastline irregularities retreat and more sediment accumulates because of drumlin erosion, extensive bars develop, which enclose a series of lagoons and drumlin till cliffs. There are several islands and drumlin shoals offshore.

Most sediment is reworked from the nearshore and very little is supplied by contemporary marine erosion. Thus, the stage of coastal evolution such as in Fourchu and Framboise Bays is also in a state of dynamic equilibrium.

During this stage, both erosional and depositional processes transform the coastal geomorphology. Either process may dominate at any given time, or the two may proceed in cycles that are related to seasons or longer-term dynamics, including rates of sea level change.

(iii) Late stage—This stage of development is seen only where the coast faces open ocean. Drumlin tills are eroded and swept off, leaving only the bedrock base of the drumlins close to the water surface. In extreme cases, all the drumlin islands and reefs have been eroded and have disappeared, leaving fragments of tie bars, which always meet the coastline at right angles. As the erosion continues by wave action, these residual bars also disappear. Only boulders are left along the coast at random locations. At this stage also, there may be crescentic till cliffs and well-developed tie bars or tombolos in the inner parts of the more sheltered bays. The southeastern coast of Cape Breton Island has not reached this stage. However, analogies exist in the Yarmouth area, at Shag Harbour, and Three Fathom Harbour, on the mainland coast of Nova Scotia.

SUMMARY

Two fundamental types of drumlin coast are distinguished: hilly coasts with rock-cored drumlins, and valley coasts with thick till drumlins. Their development can be described in terms of three stages. These stages do not express different coastal age, because all three stages maybe seen in the same area, and coastal processes of all drumlin coasts here started at much the same time, at the beginning of the Holocene transgression. The stage reflects differences in the speed of coastal development, determined by the intensity of wave action and availability of sediment supply. However, these factors are governed by the original landform and coastline configuration.

ACKNOWLEDGEMENTS

We would like to express our thanks to Drs. A. J. Bowen, H. B. S. Cooke, D. L. Forbes and R. B. Taylor for critical reading of the manuscript and very helpful remarks. We would also like to extend our gratitude to R. B. Taylor, D. Frobel, M. Gorveatt and Tom Duffett for their assistance in the field.

REFERENCES

[1] BOWEN, A. J. (*Editor*) 1975. Maintenance of Beaches. Technical Report, Institute of

Environmental Studies, Dalhousie University, Halifax, N. S. 582.

[2] CARVER, R. E. 1971. Procedures in Sedimentary Petrology. Wiley-Interscience, New York, 653.

[3] COOKE. H. B. S. 1973. Outline of the Geology of Nava Scotia. Department of Geology, Dalhousie University, 26.

[4] EMERY, K. O. 1961. A simple method of measuring beach profiles. Oceanography and Limnology, 6: 90 - 93.

[5] FLINT, R. F. 1971. Glacial and Quaternary. Wiley, New York, 892.

[6] GOLDTHWAIT, J. W. 1924. Physiography of Nova Scotia. Geological Survey of Canada, Memoir 140, 179.

[7] GRANT, D. R. 1970. Recent coastal submergence of the Maritime Provinces, Canada. Canadian Journal of Earth Sciences, 7: 676 - 689.

[8] GRANT, D. R. 1972. Surface geology of southeast Cape Breton Island, Nova Scotia. Geological Survey of Canada, Paper 72 - 1, Part A, 160 - 163.

[9] JOHNSON, D. W. 1925. The New England-Acadian Shoreline. Hafner, New York, 608.

[10] KING, L. H. 1972. Relation of plate tectonics to the geomorphic evolution of the Canadian Atlantic Provinces. Geological Society of America Bulletin, 83: 3082 - 3090.

[11] KRUMBEIN, W. C. and PETTIJOHN, F. J. 1938. Manual of Sedimentary Petrography. Appleton-Century-Crofts, 549.

[12] NEU, H. J. A. 1971. Wave climate of the Canadian Atlantic coast and continental shelf—1970. Bedford Institute of Oceanography, Report 1971 - 10, 103.

[13] NEU, H. J. A. 1976. Wave climate of the North Atlantic—1970. Bedford Institute of Oceanography, Report Series BI - R - 76 - 10.

[14] PIPER, D. J. W. and KEEN, M. J. 1976. Geological Studies in St. Margaret's Bay, Nova Scotia. Geological Survey of Canada, Paper 76 - 18, 18.

[15] PIPER, D. J. W. 1980. Holocene coastal zone and budget, South Shore, Nova Scotia. In: The Coastline of Canada, S. B. McCann, *editor*. Geological Survey of Canada, Paper 80 - 10, 195 - 198.

[16] QUINLAN, G. and BEAUMONT, C. 1981. A comparison of observed and theoretical postglacial relative sea level in Atlantic Canada. Canadian Journal of Earth Sciences, 18: 1146 - 1163.

[17] THOMAS, M. K. 1953. Climatological Atlas of Canada. National Research Council of Canada, 256.

[18] THORNBURY, W. D. 1960. Principles of Geomorphology. John Wiley and Sons, Inc. , New York, London, 390.

[19] U. S. Department of Commerce. 1978. Tropical cyclones of the North Atlantic Ocean, 1871—1977. National Oceanic and Atmospheric Administration.

[20] WEEKS, L. J. 1958. South Cape Breton Island, Nova Scotia. Geological Survey of Canada, Memoir 277, 112.

[21] ZENKOVICH, V. P. 1967. Processes of coastal development. Interscience, New York, 738.

附:(中文译文)

布雷顿岛东南岬鼓丘海岸的动力地貌学*

引 言

鼓丘海岸广泛分布于加拿大新斯科舍省,早期 Johnson(1925)和 Goldthwait(1924)的著作把新斯科舍作为这一类型海岸的典型地区。这次研究调查了布雷顿岛东南岬 50 km 海岸,从 Gabarus 湾到红角(red cape)(图 1)。这一地区有着引人注目的鼓丘海岸风光:海滨有许多湖泊和池沼,外侧环护着砾石海滩和植被葱郁的鼓丘岛屿。岩礁上是覆盖了偏红色冰碛泥海崖,我们通过研究这种鼓丘海岸的特征而发展了一个模式阐明冰后期以来该海岸发育的主要过程。自二十世纪早期以来还没有人做过对该类海岸地貌的对比性研究。经历全新世海侵,该地海岸不同部分经历不同的波浪和海流作用过程,同时又受地质构造、岩石性质和原始陆地地形的控制,这些因素的综合结果表明:不同鼓丘海岸部位有不同的地貌特征,反映不同的作用与发育过程。这些综合作用效应在很短的距离内都能有所反映。例如从 Gabarus 岬到 Framboise 湾的岸段。

方 法

对该地区的踏勘测量工作始于 1979 年春,当时确定沿海岸每隔 1 km 作定点观察。在这些观察的基础上,制成了海岸地貌图,并选定 16 条海岸定位观测剖面(图 2、图 3)。在 1979 年 11 月间笔者和贝福德海洋研究所大西洋地质研究中心的 R. Taylor 和 D. Frobel 一起研究了整个地区并用直升飞机对海岸带拍照。系统性的野外工作是在 1979 年夏和 1980 年秋间进行的,并与 Taylor 和 Frobel 合作,应用 Emery 杆(Emery, 1961)并参考水准点,进行海岸剖面测量,近滨剖面(Near Shore)是应用一个 36m ZODIAC 橡皮艇及一台 Raytheon DE - 719 型回声测深仪测量的。外滨测量(offshore)是以 CSL TUDLIK 应用一小型 Ranger 电视导航系统,配以 Raytheon 回声测声仪和 Klein 旁侧声纳进行。部分海底测量通过包租渔船(LONE 孤独天使号),配以测视海洋研究设备(ORE)声学测量系统单位是在 10 kW 功率下 Pinger 型声源发出的 3. 5 kHz 声音,以 0. 25 s 的扫描幅记录在 EPC 48 cm 的模拟记录器上。

从海岸和海底采集的泥沙样品按泥沙粒径经标准规范方法分析(Carver, 1971),采样泥沙的粒级分布用筛分确定,细颗粒用移液管分析确定。参数如平均粒径、标准偏差、偏度和峰态用计算机程序处理,采用福克斯(Folk's)沉积专门术语,海滩砾石的粗粒级样

* 蔡明理译,王颖校。

品在野外施测，用 Wadell 同心圆盘测计（Krumbein 和 Pettijohn，1983），每 0.25m^2 测 100 个砾石来确定平均粒径、圆度、扁平度和每个样品的球度。

（上述参数定义见表 1）

地质地理背景

研究区域位于布雷顿（Breton）岛东南端海岸的中部，从 Gabarus 大湾的东端向西西南延伸 50 km 至红岬，海岸地势相对低缓，在 2～5 m 间，内陆山峰多在海拔 60 m 左右。基岩为前寒武晚期 Fourchu 组（Weeks，1954；Cooke，1973），主要是火山岩和少量沉积岩夹层。黑绿色火山角砾岩或凝灰岩最普遍，沉积岩有页岩、粉砂岩和少量砂岩或杂砂岩。许多地方岩石受机械剪切和变质程度低。Fourchu 组受侵入火成岩如花岗岩分割，在 Gabarus 湾尤为明显（表 1）。

按理，确定 Fourchu 组岩层的倾向、走向很不容易，可以假设跨越 Louisburg 半岛可能为接近一西南向延伸的背斜构造，而 Fourchu 湾靠近由火山碎屑岩组成的背斜轴部。页岩、粉砂岩虽普遍，但并不丰富，在 Louisburg 半岛从 Framboise 河口到 Fourchu 湾北部，可见窄条带状页岩沿东北向延展，在 Framboise 河口，上述岩石与细颗粒火山角砾岩和粗颗粒火山凝灰岩汇合。

研究地区出露的寒武纪岩层形成的地貌多姿多样，Gabarus 湾北岸陡峭，小山峰突立于海上，高达 60 m。从 Gabarus 湾南到 Framboise 地势低缓而不规则，基岩低地和鼓丘原野多高于海面 20～30 m，并有数不清的泥沼、池塘和湖泊，地面的大致倾向与布雷顿岛北面桌状高地的准平原一致，向南倾斜入海（Goldthwait，1924），但是，本处曾经过第四纪冰川侵蚀修饰改造（King，1972）。

威斯康星冰期期间，冰川分岔向三个方向流动（Grant，1972）：先向东，然后折向东北，最后沿外滨方向向南流去。任何风化岩屑层或土壤包括很多坚岩全被冰川蚀去。冰川还磨圆或平滑了山峰和岭脊，铲去阻挡的山峰，充填了低陷的小洼地。结果，山峰变成了冰碛物组成的穹形丘包，被称为鼓丘（Goldthwait，1924）。它们通常成组或成鼓丘原野状出现，并有相似的平行排列。

新斯科舍地区的鼓丘带是北美四个主要鼓丘带之一（Thornbury，1960）。在沿新斯科舍的大西洋海岸的一些地方鼓丘地形很常见，但最丰富、研究最深入的是在北 Queens 和 Lunenburg 乡间（Flint，1971）。在哈利法克斯以东的一些海岸地段也有鼓丘海岸，布雷顿岬的东南岸是该类鼓丘海岸之一，鼓丘系列的走向主要是东北向，与冰流的运动方向平行，也和基岩的走向一致。

研究区有两种类型的鼓丘：标准鼓丘主要由冰川碎屑组成，排列方向与冰流的方向一致；另一种岩核鼓丘，是在基岩核体表面上，或多或少的有一冰川泥砾盖层，排列方向主要与基岩倾向一致。本文中将这两种皆定为鼓丘，因为都是冰川作用形成，在同一地形成，相对而言，很可能是同一冰川事件的结果。

海岸作用过程

Goldthwait(1924)和Johnson(1925)的早期经典工作,以及近期Bowen和Piper及他们学生的区域研究工作曾解释过有关新斯科舍鼓丘海岸发育的一般过程。

新斯科舍的大西洋海岸是一沉降岸段,该地近10 000年来海面上升了40 m(Quinlan and Beaumont, 1981)。虽然对相对海平面的早期变化史不很清楚,但没证据表明自威斯康星冰盖融化后,海面高于现代海平面多少(Quinlan & Beaumont, 1981)。潮位记录(Grant, 1970)表明现代海平面升高速率达到4 mm/a。

浪蚀导致鼓丘海崖的后退,在开敞区域后退率超过1 m/a,一些被浪蚀的泥沙被再沉积为鼓丘前部的砂质或砾石质沙坝或海滩,海面上升导致连续的海岸侵蚀,因为浪蚀平台下沉而终止了波浪侵蚀力的重要作用。沉积供给一般不足以使海滩的前进发展。而且,因为海平面上升,灾难性的破坏和海滩向岸蚀退现象会发生(Bowen, 1975)。大量的越流冲刷可能是沙坝破口的前兆。

海岸动力过程,尤其和波浪与浪流有关的作用,是海滨发展的重要因素,这些作用过程控制了物质的沿岸运移。由于缺乏台站关于地区性的天气与海洋资料,因而多采用船只对天气观察的记录。

(一) 风和波浪

在冬季的几个月里,频繁的风暴袭击布雷顿的东南端,这些风暴为西风、西北风或西南风,基本来自西部大陆的冷空气。夏季的几个月,和风吹拂,主要是来自南或西南的气流(表2和表3),这一地区在冬春季最大风平均时速在110 km/h和125 km/h。但在夏季,风速和缓得多(Thomas, 1953)。

海岸地区的风情在波浪的生成中起着重要作用(表4和表5)。五月至八月,西南风和南风生波起主导作用,而在10月和11月,北向风、西向风与东北风起主导作用,在其他的月份里,波浪主要来自西方,但是北风和西北风浪也很旺盛,由于岸线延伸的不同方向,西风和西北风向的波浪对海岸发育无明显作用。因此,全年内,最主要的风及风成浪是西南向的。2/3的风暴发生在冬季月份(12月至次年2月),其余的发生在春、秋季的过渡季节。外滨的典型波高为1.5～2.5 m,发生在11月到来年4月,其余月份的波高较低(图5)。

热带飓风多发生的晚夏和早秋,在西北大西洋虽年发生率界于6次和10次之间,仅在1871年和1977年间,曾有经过布雷顿岛屿的飓风记录,风速可达180 km/h(U. S. Department of Commerce, 1978)。飓风风速可超过180 km/h,在这罕见的情况下,波浪可达极高,据1970年数据,新斯科舍海岸的10年一遇的最大深水波高(包括Breton岬东南海岸)达16～18 m(1971,1976)。百年一遇的波高为22～24 m。

波浪在向海岸移动过程中发生折射,随水深减小、波速降低,因此接近海岸时波前端波峰区弯折。由于波浪折射,波能集中在海岬突出端,而在海湾则波能辐散。

趋进Grabarus湾的6 s、10 s、14 s的周期波,多来自东北、东、东南标注在图5的折射图谱上。这些图谱表明在Gabarus湾波能渐消，但在Gabarus岬角的突出地区波能辐聚。

趋进Fourchu地区的10s周期波来自西南、南、东南、东方，而6 s周期波来自西南，它们在图6折射图上标明。

（二）潮汐和流

沿布雷顿的东南海岸潮汐是混合型的，从外海向西北推进。潮差并不很大：Gabarus湾的平均潮差大约1.2 m，大潮时潮差约1.5 m。一般而言，该地潮流不是塑造地貌的主要动力。

沿岸流与风紧密有关，因此流速与流向多变，但受区域海岸地貌和水深控制，主导表面流向常出现在离岸约5 km处，流速约0.3～0.5 m/s，向西—西南向流动。

海岸类型

鼓丘是从Gabarus湾到红岬的50 km岸线上的主导地貌类型。海岸线不断向海推进，鼓丘之间的谷地因海侵发育成海湾，砂与砾石堆积物主要从鼓丘上侵蚀下来，堆积在海湾和封闭的潟湖内，圆形或椭圆形鼓丘，或单或成群被淹没，形成穹形岛或沙岬。

鼓丘类型，鼓丘与岸线方向间相关状况以及水下岸坡的地形特征等，沿海岸均有不同，因而不同方向和强度的波浪和流形成多种海岸地貌，可被视作鼓丘海岸发育的不同阶段。研究区域的海岸可分成四部分或四个亚类，从北向南，每种都有不同的发育阶段（图1）。

该海岸的四个类型如下：

（1）Gabarus湾南岸（Rouse岬角—Cape Gabarus—Sugar Loaf）地区以活跃海蚀作用的鼓丘以及小湾为特征，在鼓丘岬角之间尚有些沙坝海滩（图4A）；

（2）Winging头海岸（Sugar Loaf—Winging Point—Indian point）是海积岸段，发育有连岛坝、岸外坝滩，其后侧有蚀退的鼓丘与潟湖（图4B）；

（3）Framboise湾和Fourchu湾（Indian point—Fourchu head）是一海积岸段，具有开阔的海岸沙坝（图4C）；

（4）Fourchu高地到红岬，是海蚀性低岩岸段，这些地区的鼓丘冰碛已被大量蚀去（图4D）。

（一）Gabarus湾南岸

1. 概况

该海岸类型是处于Gabarus湾南部从Rouse岬角到Cape Gabarus再到Sugar Loaf，全长约11 km。海岸的基本走向从西北向东南，除了在Gabarus Point头和Sugar Loaf间有一段为南北走向。Louisburg半岛和Scatarie岛掩蔽了海岸使不受强烈的北向来浪侵蚀。在这一地区，东和东北向波浪具有很大的吹程。这段海岸的基岩主要是偏灰色的硅质砂岩、页岩、粉砂岩或石灰岩。在Haris湖和Gull湾间有花岗岩露头形成的海崖。

2. 鼓丘海蚀崖

这段海岸的鼓丘比研究区域其他地方的要小，但紧密分布，呈东北走向（060°到

080°)。沿岸似位于一个高出海面 5～8 m 的不规则基岩平台上，反映出该鼓丘田野初始发育在基岩山地上。因此，鼓丘具有基岩核心，上覆薄层冰碛物。大体上，冰碛层厚 2～4 m，呈黄色的砂质砾石沉积。

在 Gull Cove 以西的花岗岩山地区，冰碛物有 0.5～1.0 m 厚，但在 Gull 湾和 Cape Gabarus 间有 10 m 厚。

东北向延伸的鼓丘系列与海岸斜交，海进造成了海蚀型的基岩港湾海岸。鼓丘山地延伸入海形成岬角或岛屿，向海成为 3～5 m 高的海蚀崖及崖下能崎岖不平的岩滩，倾斜入海。

在花岗岩地区，有一些袋状海滩和约 10 m 高的海崖。

Cape Gabarus 沿东北向延伸入海，它由超过 10 m 厚的、而向东减薄的灰黄色冰碛物组成，基底为冰川磨蚀过基岩面，已抬升至海拔 8 m 高。

海浪掏蚀冰碛层底部，造成频繁崩塌与海蚀陡崖，崩塌亦给海崖带来足量沉积物。尽管有强烈的波浪侵蚀，在基底岩面上，仍保留着冰川擦痕。岬角的顶部是一个宽约 100～200 m 的浪蚀基岩平台，它与区域的浪蚀基岩平台相连，基岩平台面上保留着低崖和海蚀柱。海蚀平台标志着岬角在现代海面时广泛地后退，它可以用作浪蚀强度的一个指标。

像 Cape Gabarus 一样，在 Gabarus 湾南侧，Rams 岬角有个更掩蔽区域。它是由冰碛物覆盖在海拔几米高的基岩阶地面上，根据 Rams 岬鼓丘顶部的标志桩的测量，冰碛物悬崖的后退速率每年约±0.1 m 左右(表 6)，浪蚀平台在北部有 40 m 宽，在东部有 75 m 宽，南部 20 m 宽，假设 Gabarus 湾侵蚀量速率相当慢。

3. 沿岸沙坝海滩

沿这段海岸，基岩岬角或鼓丘系列之间发育了小海湾，并因海面上升，而海湾不断向陆伸展。其中一些海湾不足 100 m 宽，从岬角侵蚀下的物质沉积在海湾里，形成陡峭、狭窄的砾石砂坝，环围着内侧的潟湖。

Harris 湖是沿这类海岸最大的海湾之一。全湾面积约有 1.5 km^2，四条平行的鼓丘系列以西南向延伸于本区，外侧两列鼓丘成为潟湖的边缘，内侧的小鼓丘形成湖中的小岛。Harris 湾的沙坝长 750 m，它的北端为 Rams 岬角所掩护，在东北向波浪的波影区沙坝低而宽(135 m)(见图 7 剖面 12)，它由砾石和粗砂组成，砾石的平均粒径从北端的坝峰 25 mm 增强至中部的 35 mm 及南端的 50 mm，因为它朝向东北风增强的波浪方向。坝的南端较狭窄(70 m)且较高、较陡，有一风暴形成的滩肩(图 7 剖面 13)，坝的组成物质来于海湾两端的遭受海蚀的岬角。卵石、砾石和粗砂堆积在坝上，而细砂和粉砂沉积在岸外海底(图 3)。在 Rams 岬角鼓丘上平均砾石的磨圆度是 0.2，在沙坝处的砾石平均磨圆度增至 0.3～0.4，依然相对是个低值，反映了当地物源与短距离的物质搬移(图 8)。

在同一海湾内沙坝地貌与组成粒级的不同是由于波浪强度的不同，与海岸朝向的开敞度有关。地貌分析表明来自东北和东向波浪对 Harris 湖沿岸地貌的塑造最为有效。

两种海岸剖面(图 7)中，在北端(12)和南端(13)的沙坝展现出波浪作用效果。剖面 12 位于湾的北端受到 Rams 岬角的掩护，沉积物为粗砂和砾石。1980 年，海滩剖面的季节性变化主要分布在平均高潮位附近，但低于极端高潮位，泥沙积累多在春夏季低波能时，而侵蚀发生在冬季强烈的东北向与东向波浪作用时。邻接潟湖的后滨地带，接受的泥沙较少。

（二）Winging 角海岸

1. 概况

堆积型的陆连岛海岸分布在自 Sugar Loaf 向南到 Winging 点之间，再向西延到印第安角，整个岸线长 9.5 km。这一区域的鼓丘紧密分布，鼓丘具有基岩核心，海岸鼓丘发育在基岩平台上。鼓丘排列方向呈 060°到 080°，基本平行于岸线走向，外滨也有些小岛与礁石沿同样走向排列，也许是些老鼓丘的残余，可保护岸线不受波浪侵蚀。这段岸线的主要特征是发育着双陆连岛或双沙坝连岛（曾柯维奇，1967）。大多数的鼓丘被潟湖和沙坝与海隔开，因而这种类型的海岸景观与前述有相当大的差异。

海岸向外海开敞，除冬季与春季外，多为具有很长吹程的南向风浪。常年主导来波是西南向的，冬、春季东北向风浪强劲。沿海岸波流和风能也强于南 Gabarus 湾。

伴随着冰后期海面上升强烈的海洋侵蚀导致了侵蚀和早期鼓丘的下降，形成一系列的椭圆形岛屿或岸外岛礁，这些被蚀的鼓丘为海岸提供充足的泥沙，形成在岛礁后侧广泛发育的连岛坝。

2. Winging 岬角

位于这段海岸的中心部位的 Winging 头是个典型的对称型的或双陆连岛，因此，它可作为海岸发育的范例。Winging 是个基岩岬角，以东西两条砾石坝与海岸相连。该基岩礁岬是岩核鼓丘的残留，在极低潮时出露水面。基岩礁由寒武纪的变质岩组成，走向 055°，倾向为 NW50°。此列鼓丘呈东北—西南，附近其他岩石露头也为同一走向。第二列岛礁位于 Winging 岬角岸外 100 m 处，Guyon 岛位于这两岛礁以外的 1.5 km。这些岛礁保护海岸不受折射波侵蚀，从而导致沙坝从东西两个方向发育（图 6），把 Winging 头岛礁和海岸连接起来。

东坝长 1.5 km，西坝长 2.2 km，后者比短的东坝高 0.5～1.0 m。整个坝体由砂砾组成，粗砂沉积在岛礁北部的顶点附近。草丛固定的沙丘沿整个西坝分布，但在开敞的东坝只有部分地区发育沙丘。盛行风控制泥沙沿岸运移，并推动泥沙移向坝的顶点，而在两个海湾间无明显的物质交换。沙坝防止了鼓丘海岸被侵蚀，并补给了大量泥沙，泥沙运移的主导方向是朝向海滨线，泥沙源于改造老的海岸沉积或冰碛物。

在 Winging 岬角的东坝（图 7 剖面 16）出现有越流冲刷，而使坝减低。波浪已蚀去坝顶并将泥沙移向坝的两侧斜坡上。这种情况表明该段海岸缺少泥沙供给。冲浪使得植被难生。相反地，草丛沿沙丘向海繁殖，一序列的沙滩剖面证据表明西坝在缓慢地堆积泥沙。

冬春季，在盛行东北风下，东坝受波浪侵蚀而坝高减低（图 7 剖面 16 和 15－1）。西坝被 Winging 岬保护不受西北波浪侵袭，以至在冬季，坝体不断生长（图 7 剖面 14）。在夏季，西南来波浪起主导作用，以至东坝处堆积，西坝处遭波浪侵袭。秋季，西、东坝皆受侵袭，那是由于来波方向的改变。

由于不同的岸线朝向和敞开度，波浪导致了海滩剖面不同的季节性变化。沉积物没表现出明显的季节性变化，总言之，双坝相对稳定，表现出西部微弱的积累和东部的侵蚀。

3. Winging 岬角海岸的西南部

岸线的走向与鼓丘的走向相垂直，因此，岸线由岬角和小海湾组成。鼓丘的基岩核由片麻岩和微弱变质的砂岩和板岩组成（图 7）（走向 040°，倾向 20°到 60°SE）。鼓丘的基岩核芯形成一高于海面 4～5 m 的平台。比如基岩头岬角部位易受风暴浪侵袭，岩滩被卵石和砾石覆盖，形成海滩，滩顶达高出海面 6～7 m。泥沙沉积在小海湾或浅滩后面，形成粗砂细砾的湾顶坝包围着个小潟湖。还有一些连岛坝（图 4c），例如在 Rock Point，该处两列粗砂、细砾沙坝，围封了小潟湖，西坝比东坝大，但东坝包括风暴成因滩肩。堆积形态和旗形松林，两者都表明该处西南风和波浪盛行。

海岸侵蚀和沿岸输沙对该段海岸供应少，沙坝具有较稳定形态，离岸的沉溺鼓丘，折射着来波，导致形成连岛双坝。这部分海岸也呈现相对稳定状态，没有快速突变的蚀积变化。

（三）Fourchu 湾和 Framboise 湾

1. 概况

Fourchu 湾和 Framboise 湾由拦湾坝组成，背后是潟湖和一系列的鼓丘，岸线总长是 12.5 km，和大部分的鼓丘一致，同样是走向东北（065°至 080°）。

少量鼓丘走向 105°，可能是由早期东部冰流形成。沿这类海岸，鼓丘发育在原始洼地或谷地里，形成椭圆形或穹形冰碛山丘。鼓丘的冰碛层厚约 10～15 m，由黄棕色的粉砂、砂砾组成。此类鼓丘广泛分布，缺少基岩核芯，基岩仅出现在鼓丘的基部。它们与发育在更东边的山地的薄层岩核鼓丘不同。由于冰流的后退，鼓丘暴露向海的部分被完全蚀去，留下椭圆形的岸外浅滩，至少还有四列东北向的鼓丘残余物，在这一地区的等深线图和地震剖面图上很显著（图 3、图 9）。沿海岸线，鼓丘的暴露部分已被蚀去，形成 10～15 m 高光秃秃的冰碛物崖，沙嘴尾迹从沉溺鼓丘和沙坝尾端可逐步连接追溯。在 Fourchu 湾和 Framboise 湾，沙坝已连接形成一个宽广的沿岸沙坝，在一些地方沙坝已并岸位于鼓丘崖前，从而降低了波浪袭击悬崖的有效性。因此，在这儿鼓丘海岸，完全未经改造的鼓丘仅可在潟湖中存在，被侵蚀的鼓丘可在海岸线上发现，而残余鼓丘可在外滨发现，展现出该海岸带发育的各个阶段。

Fourchu 湾的近滨砾石堆积于接近 Winging 岬处、鼓丘浅滩处与岬角外围，细砂堆积在较深水中。

2. 鼓丘崖

该地区有 11 列鼓丘，每一个鼓丘被海蚀截去山嘴，形成冰碛物海蚀崖（图 2），在鼓丘崖顶部进行测量，结果表明：大多数的陡崖仍在被改造，在风暴潮期间仍在慢慢后退。一些典型例子阐述如下：

No. 2 鼓丘 5 m 高，75 m 长位于 Framboise 湾的中部附近，鼓丘崖坡度约 54°～60°，冰碛物为灰黄色砂砾层。测量获知自 1979 年 9 月和 1980 年 9 月间崖后退 0.3～0.2 m，其中端后退尤速。

No. 10 鼓丘是一 10 m 高由粉砂质砾石堆积物组成，位于 Fourchu 湾沙坝中部，鼓丘上部有片松林，仅在崖边有枯死的松树，是因浪蚀毁崖之故，崖退 0.3 m/a，Framboise 湾比 Gabarus 湾鼓丘崖后退迅速。后退速率与盛行浪的吹程有关。但这部分海滨，侵蚀岸段少于堆积岸段。因此，只有小部分的沙嘴沉积物能被带离近期被侵蚀的海滨部分。

3. Fourchu 湾

在 Fourchu 湾，沙坝 3.9 km 长、140 m 宽，从东边的印第安角向外延伸至西南边的 Belfry 头。Belfry 头是个 10 m 高、由砂岩和页岩组成的穹形岬角，有三组近垂直的节理(走向 160°，260°，290°)控制着海崖地貌形态和起伏不平的岩滩。岬角的南边是一潮流通道小湾，当地人曾几次建造了一条人工沙堤，但这条沙堤由被风暴潮浪流冲开而通海。沙堤跨过 Belfry 潮流通道向南延伸 2.2 km，达到 Marcoche 湖的一个新潮流通道；继之，以湾坝形式向南延续至 Fourchu 港。

Fourchu 湾的沙坝由粗砂和砾石组成，砾石坝可在冰碛物崖底或在基岩头地处发现，Fourchu 湾坝是个由一系列湾头滩组成的湾头坝：这个复合大坝建造封闭了潟湖和许多小湾，改变着最初的曲折岸线和海湾，成为平直的堆积型的沙坝海岸。

海滩砾石的岩性同鼓丘冰碛类似，主要为砂岩、页岩和石灰岩，并有少量花岗岩和片麻岩质的。在鼓丘侵蚀物搬运过程中，砾石被磨圆与磨平为海滩砾石。沙坝砾石磨圆度为 0.4，比鼓丘冰碛物砾石的磨圆度(0.11)高出许多。这表明：砾石在近岸带被存积和改造(图 8)。

沙坝的形态，因海岸的不同部位而异。

(1) 对称型的前滨和后滨——这种情况出现在当沙坝位于鼓丘之间和潟湖之前时(如印第安角剖面 8，图 7)。

在印第安角附近，坝由粗砂细砾组成，砾石直径在坝的上部为 40～100 mm，坝下部的砾石粒径为 20 mm。该沙砾坝只经历少许侵蚀，因此坝的两坡依然对称。剖面 8 的泥沙粒径从低潮水平线处的中砂变到高潮位线处的砾质砂，由于物质就近提供，未经长距离运移。朝向外海的沙坝，其顶部在 1979—1980 年的冬天后退约 2 m，因而剖面有点不对称，沿沙坝顶脊有约 1 m 高的陡坎(图 7 剖面 8)。这部分沙坝由粗砂组成，顶部是细粒的风吹沙。

(2) 不对称的前坡后坡剖面，No. 7 是这类形态的典型。这里的沙坝狭窄、陡峻，有几个滩肩(图 7 剖面 7)，由于坐落在 Belfry 头附近，沙坝主要由砂岩砾石组成，20～40 mm 粒径，来源于附近的岬地。圆度和扁平度都很低，只有 0.45 和 0.3(图 8)，泥沙在低潮位处由分选差的粗砂组成。剖面 7 可见显著的季节性变化，在冬季风暴中砾石累积，但在夏季被从中部蚀去，堆积在海滩的低部。

(3) 单一前滨——以剖面 10 在鼓丘崖基部的单斜坡为例(Zencovich，1962)(剖面 10，图 7)。向岸冲流到达海崖崖麓，水流击撞，形成回流，在上部海滩主要堆积砾石，砾砂及中砂堆积在低海滩附近。由于鼓丘崖后退，而海滩剖面加积。因为在沿岸泥沙粒径中无系统性变化，所以很可能是没有明显的泥沙沿岸移动存在。

由于海滩沙主要由细砂组成且大部分砾石皆被磨圆，说明绝大部分沉积物不可能直接来自鼓丘侵蚀。经历长时期的泥沙改造在近滨处得到证实，该处泥沙源于前期侵蚀的鼓丘之冰碛物，只有部分海滩沙是由现代侵蚀的鼓丘崖或岩岬所提供。

在 Fourchu 湾的这一段鼓丘崖和砾石坝都表明了有限的侵蚀后退。因而这段海岸提供的泥沙有限。但由于来波被蚀余鼓丘岛和沙洲反射而波能大大降低。因而，岸线相对稳定。

在 Belfry 南部，沿 Marcoche 湖的外滨，岸线从北向南延伸，海岸面向东边外海，由于有比较长的吹程，东北向波浪很强，结果发生了广泛的越流与移沙，导致形成了相对坦缓的沙坝剖面(图 7 剖面 6)。波浪从近滨向岸边运移砂砾，并由于泥沙供给丰富，因此沙坝不断增长，为当地居民提供了建筑所需的泥沙。沙坝沉积物是分选好的细砂，夹少量砂砾，由于人工挖掘影响，海滩剖面非自然状态。

从 Marcoche 湾向南有两个小鼓丘崖，各有 100 m 长、5 m 高。崖脚几乎是裸露的泥质海滩，部分覆盖一薄层粗砂。泥滩滩坡 7°，而沙滩部分则 4°。再向南有一小河口，把这两个鼓丘与第三列鼓丘分开，后者走向也是 60°。该河口或潮流汊道是新近形成的。该处沙坝被强劲的海流冲开，很可能是底流作用。因为，沙坝两端的尾部均向陆湾伸。废弃的潮流汊道曾在第三个鼓丘的南端，现却已是一个新形成的新月状砾石坝所在。砾坝的两坡不对称，高出海面约 3 m，坝年龄很新尚无植被覆盖。在新月形坝下，堆积了巨砾，由强力迅移的海流塑造成了平坦的海底，显然，形成原潮流通道被废弃。

4. Framboise 湾

Framboise 湾沙坝有 3.5 km 长，从该海湾的鼓丘东岬向西延伸，海湾能分成三部分：一是沙坝末端；二是鼓丘崖基部；三是常态海滩。通向 Bagnell 湖的潮流通道分开了从 Framboise 湾的西鼓丘岬角延伸出沙坝，Framboise 河从内陆 10 km 处流入的 Bagnell 湖，因此在沙坝处无明显的河口形态。

剖面 1(图 7)位于 Framboise 湾沙坝的西端，是具双坡的完整剖面型。沉积物为中砂混杂着一些 20～30 mm 粒径的砾石(图 8)。沙坝低平，海滩剖面测量没显示出季节性变化，也许是因为人工开挖泥沙的缘故。

剖面 2(图 7)是在 No. 2 鼓丘崖处单一倾斜的海滩。海滩物质主要是砂质砾石，正偏度而分选性差。由于混入大量悬崖后退的崩塌物质，剖面测量表明泥沙在此积累，尤其在冬季，推测是因为冬天风暴蚀去海崖，并向海滩提供物质。在夏天中等波高的波浪把海滩砂侵蚀运移入海，因此海滩面微微下降。

剖面 4 和 5(图 7)位于 Framboise 湾坝的中段，这里沙坝面朝外海，因该处远离鼓丘崖，导致泥沙蚀积过程仅在坝上有所表现，沙坝由具双坡的完整剖面组成，邻近潟湖的后滨岸坡较缓，有植被覆盖；前滨岸坡较短，顶部有一 0.5 m 陡坎，是由波浪侵蚀造成。

沿这段沙坝低于高潮位的泥沙由细砂组成；在高潮线附近的剖面 5 中有砾石，但剖面 4 中只有中砂；粗砂和细砂在坝顶可见。这种在坝上部以粗粒泥沙粒级为主而随坡向下逐渐变细的分布标识着泥沙是由波浪运移。极细砂沉积在海湾底部。沿岸泥沙分布，尤其是细砾级粒径分布，表明从两边岬角向海湾中心有泥沙沿岸输送。泥沙平均粒径从坝的两端向中心减小，因此，剖面 4 的泥沙较其他任一剖面的都细些，剖面 4 的细砾的磨圆度和扁平度较高，并混以分选较好的砂(图 8、表 7)。虽然坝顶经浪蚀成陡坎，海岸剖面测量表明仅有轻度的季节性变化，沙坝似乎较稳定或有轻微的加积。

5. 结语

Fourchu 湾和 Framboise 湾均具有双重岸线，弯曲的鼓丘内海岸线被砂砾坝封闭形成外部的平直岸线。在这两条岸线之间是有着小型鼓丘岛屿的大型潟湖群，无明显的河沙输入，海岸侵蚀也没提供许多泥沙物质。海岸泥沙最初来自于由蚀余鼓丘老沉积物组成的现代水下岸坡。这种沉积物源有限，该海岸虽面朝外海，但几列残余鼓丘岩礁影响到达岸线的波浪强度。在这种情况下，可认为 Fourchu 湾和 Framboise 湾在一段时期内动力状况是稳定的，可忽略海平面的变化。

(四) Fourchu 岬角到红岬

从 Fourchu 岬到红岬，这类海岸特征突出，原是一基岩平台或一系列小山地，后来发育为小型的岩核鼓丘。这些表面具薄层冰碛物的鼓丘都坐落在基岩平台上，高出海面 5～10 m。鼓丘走向 055°～070°(北东)。海湾几乎与鼓丘列长轴平行。该处海岸面向东南外海，由于长吹程和波浪在突出于海的岬岸折射，因而波能高。高能波浪切蚀砂岩和页岩形成不规则的基岩海岸。崎岖不平的岩滩与经常有的岩块在岸线掩蔽岸段屡见不鲜。一些地区的浪蚀海滩超过 400 m 宽，例如，从 Fourchu 岬到 Framboise 东岬的水下浪蚀滩宽是 700 m。沿该海岸强烈的波浪蚀去鼓丘上的冰碛物，从而阻碍了海岸植被的生长。

近滨区域地质

(一) Gabarus 湾

在 Gabarus 湾 3.5kHz 的地震剖面图显示出三种海底类型(图 9)。

(1) 小范围的基岩(有一粗糙，极不规则的反射面)；

(2) 相对平坦的冰碛区，局部具有薄层砂、砾分布；

(3) 厚层沉积物区(可达 30 m/s)覆盖在略有起伏的硬地面上，推测为冰碛物，附图 9。

在 Cape Gabarus 东部的基岩露头和东北向的 Gabarus 湾是一连续突出的孤立岩岛或岩礁基岩岸线。岛、礁系列走向和分布仍保持原基岩鼓丘的风格，表面冰碛物被海侵所剥掉。平坦的冰碛面很可能是冰碛悬崖海岸后退的结果，夷平形成浪蚀平台。冰碛物表面上的等深线再现了鼓丘崖岸在全新世海进时的连续位置(图 3)。最深的平坦冰碛物表面位于现代海面以下 50 m 处。

相对而言，保存在沉积物之下的微伏不平冰碛物表面并没经历浪蚀，仍保存原形，它们出现在水深 50 m 以深，反映出最大的早全新世低海面约在 −50 m，也存在于西南 Gabarus 湾的许多浅海区域。在这些区域鼓丘可能沉溺于潟湖中(就像现在在 Harris 湖所见)，被岸外沙滩坝保护，与外海隔绝。因此，这些海滩可能与鼓丘岬连接，因此它们原始位置也能被判定(图 3)。

自早全新世以来，Gabarus 岬的冰碛崖已向西后退 5 km，后退的等深线表明可能有一半的泥沙被沿岸漂流移向 Gabarus 湾，另一半沿西南移向 Fourchu 湾。假定冰碛物崖曾为 5 m 高，大约有 10^8 m^3 的冰碛物从该处被侵蚀掉，另有 0.2×10^8 m^3 泥沙从 Gabarus 湾被蚀掉。

（二）Fourchu 湾和 Framboise 湾

要解译 Fourchu 湾和 Framboise 湾外的海岸的地质历史并不容易。基岩质浅滩，或许是常见于本区鼓丘的基岩核，这些浅滩间，海底沉积物包括砂与砾石以及厚度的声学穿透未见其沉积序列。

如果最初广布于大陆之上的鼓丘向海延伸至少 4 km，那么在全新世海进时，大量冰碛物则被蚀去，而供应海岸区域。

海岸发育

多数鼓丘海岸位于中高纬地区的沉降地区。鼓丘海岸的成因和冰川作用过程紧密相关。鼓丘原野在冰川作用时形成，而鼓丘海岸发育在随之而来的全新世的高海面期，典型的鼓丘海岸是由小海湾和无数小岛组成。冰碛物形成的泥沙粒径从巨砾到细泥不等，沙坝主要由砂和砾石组成。不同的海岸作用过程结果，鼓丘海岸最终变成为平直的。

鼓丘海岸的动力过程取决于鼓丘的原始形态，鼓丘所在的基岩表面地貌以及下沉的水下岸坡的地形特征。这些因素影响海岸沉积物供给和波浪的强度与波浪特性。

泥沙供给源自河流的很少，河流作用过程在高纬地区，在冰川作用过的高地，是不像在中低纬地区显得那么重要。因为许多冰川作用过的湖泊捕捉了流域区的泥沙。由于潮湿的气候，地面有着葱茏的草地与灌丛，因此减少了海岸侵蚀后退，在新斯科舍陆地和 Cape Breton 岛，河流提供的泥沙很少(Piper 和 Keen 1976)。

鼓丘海岸的泥沙由鼓丘冰碛泥直接或间接供给，或来自冰碛物崖或海底的老的海岸泥沙，可能最初源于鼓丘上的冰碛。鼓丘冰碛物分选差，鼓丘海岸沉积亦包括各种粒径的泥沙。沿着 Breton 岛的东南岸泥沙组成主要是卵石、砾石、粗砂和粗粉砂。

在研究区内，有两种相反的冰碛物发育过程：在崎岖的山地，冰碛物细，且多发育为岩核鼓丘；在谷地区基岩位置低，形成厚层冰碛物层鼓丘。这两种冰碛地貌对海岸地貌影响效应不同。我们能通过研究鼓丘海岸的发育阶段来证明这点(图 10)。在此分析中，假设海进过程是很缓慢的。

（一）具有岩核鼓丘的山地海岸

(1) 初始阶段——早全新世海面上升开始淹没与侵蚀鼓丘岸线阶段。起初有一系列鼓丘岬地和小海湾。随着时间推移，基岩岬和蚀余冰碛崖在海湾的两边形成，砾石坝逐渐封闭了海湾成小潟湖。在这阶段海洋侵蚀是主导作用过程，泥沙来自岬角地段，Gabarus 岬东岸反映着早全新世海岸的发育过程。而 Gabarus 岬角南部到 Sugar Loaf 的现代岸线是该阶段的典型(图 10)。

(2) 成熟过程——海蚀过程的继续发展，当鼓丘排列走向与岸斜交或一致于岸线方向，两个或两个以上的小海湾可连接成为一个大海湾，单一的沙坝变得更长，而且仍在两个基岩岬角的保护之下。泥沙来自两方面，或直接来自岬角地段，或再经过近滨地区，例如 Harris 湾情况。

岸线平行于鼓丘的大体方向，海洋侵蚀和海面上升导致岸线后退，形成一系列小岛屿或残留的鼓丘，泥沙堆积在岛或礁后的波影区，形成发育很完好的陆连岛或连岛坝。原始海岸也因此为连岛坝或陆连岛体系所包围，强烈的浪蚀不再是主导作用过程，因为波浪为蚀余的岛礁所折射，如海面不变化，海岸将达相对动态平衡。Winging 角是该过程的一个范例。

（3）晚期——鼓丘的这一阶段仅发育在面朝开阔的外海或该地海岸坡度非常陡峭处，由于海蚀，鼓丘近乎消失，而形成不规则弯曲的蚀余基岩低地海岸，例如，Fourchu 岬西海岸。

（二）厚层冰碛物鼓丘的谷地地区

（1）早期阶段——谷地区，因为地势低，在海进时，形成深海湾或汊道，因而相对隐蔽，受外海浪击作用少，海进淹没的鼓丘形成穹形的森林茂密的小岛。

在海蚀的初期，松树被蚀去露出冰碛崖。从崖上侵蚀下砾石和粗颗粒物质沿海岸运移，主要堆积在鼓丘岛的向陆侧，形成连坝或小沙嘴，而细物质被移入海中，鼓丘海岸的这一阶段持续时间很短，由于海蚀在未固结的沉积物海岸上作用迅速，这类海岸可在潟湖的内侧部分或有狭长的海湾地区看到，例如在 Fourchu 湾潟湖和 Framboise 湾。

（2）成熟阶段——由于海进，因为鼓丘被侵蚀，海岸线不规则地后退和较多的泥沙堆积，沙坝广泛发育而围封了一系列的潟湖和鼓丘冰碛崖。外滨发育一些小岛和鼓丘浅滩，泥沙大多在近滨带掀起，很少来自当时的海蚀作用，因而，海岸演变的这一过程，如在 Fourchu 和 Framboise 湾，也是一种动态平衡过程。

在此阶段，蚀积过程都在改造海岸地貌，每一过程都在一定时段中起主导作用，蚀、积两过程可能在不同的季节或长时段动力变迁包括海面变化影响下，互相更替，周而复始。

（3）老年期——这一发育阶段只在面向外海的地区才能看见。

鼓丘崖被侵蚀并被冲刷掉，留下的仅是毗近水面的鼓丘的基岩基底。在极端情况下，所有的鼓丘岛与基岩岩礁皆被蚀掉而消失，留下连坝的片段与岸线直交，当海浪侵蚀作用继续进行时，残余沙坝也会消失掉。只有巨砾随机零星地留在海岸边。就在这阶段，可能有新月形冰碛物崖或发育完好的双连坝或连岛坝在隐蔽良好的海湾内侧。Breton 岛的东南岸，还未达此阶段。相应的参照物形成在 Yarmouth 地区，在新斯科舍省陆地沿岸 Shag 港和三浔湾也可见到。

结　论

可以区分出鼓丘海岸的两种基本类型：山地海岸具有岩核鼓丘，谷地海岸有厚层冰碛物鼓丘。它们的发育可阐述为三个阶段。这些阶段并不指示不同的海岸年代，因为所有三阶段可在同一区域看到，而且这些鼓丘海岸的作用过程起源于同时，开始于全新世海侵时。这些阶段反映出海岸发育速度的不同，主要取决于波浪作用强度和可能的泥沙供应量的多少。然而，上述因素却被原始地形和海岸结构所制约。

致　谢

作者对 A. J. Bowen，H. B. S. Cooke，D. L. Forbes 和 R. B. Taylor 诸位博士，阅读本文并给予有益的意见表达衷心的感谢。同时，对 R. B. Taylor，D. Frobel，M. Gorveatt 和 Tom Duffet 诸位所给予调研工作的支持帮助表达深切之谢意。

Surface Textures of Turbidite Sand Grains, Laurentian Fan and Sohm Abyssal Plain*

INTRODUCTION

The Laurentian Fan is a major Pleistocene sediment accumulation on the south-eastern Canadian continental margin. Near-surface sediments consist principally of sands and red muds (Stow, 1977) deposited from turbidity currents (Stow, 1979). Because the outer continental shelf at the head of the Laurentian Fan is over 400 m deep, current- or wave-sorted sands do not appear to reach the Laurentian Fan (Piper, 1975). Turbidite sands on the Fan are probably derived directly from slumped glacial till (Stow, 1978).

The Sohm Abyssal Plain (Horn *et al.*, 1971) is fed by three turbidity current channel systems: the northwest Atlantic mid-ocean channel (Chough & Hesse, 1982), the Laurentian Fan system, and a channel leading from North-eastern Channel off the Gulf of Maine. A series of cores from the extreme southern Sohm Abyssal Plain (Fig. 1) contain red and greenish or olive-grey turbidite muds and rare sands. The foraminifera, clay mineralogy, heavy mineralogy, and detrital coal petrology (Buckley, 1981; Hacquebard, Buckley & Vilks, 1982) all suggest derivation via the Laurentian Fan. These sediments must have travelled over 1 000 km by turbidity current.

The sand samples from the southern Sohm Abyssal Plain thus provide an excellent opportunity to study the rate and style of surface texture development during turbidity current transport, since the source grains are probably derived from marine tills and thus lack surface textures due to current transport, but have pronounced glacially induced markings. Samples have also been examined from turbidite sand beds from the Scotian Rise, Laurentian Fan, and proximal Sohm Abyssal Plain representing various distances of turbidity current transport (Fig. 1 and Table 1).

The stratigraphy of the late Pleistocene turbidites of the Laurentian Fan and Sohm Abyssal Plain is not well defined, because of the lack of biostratigraphical markers, and the rapid southward transition from sub-Arctic to tropical planktonic assemblages. On the Laurentian Fan and Scotian Rise, Stow (1977) distinguished Holocene hemipelagic sediment, resting on late Wisconsinan (30 000 BP) turbidite mud that overlies a

* Ying Wang, David J. W. Piper, Gustavs Vilks: *Sedimentology*, 1982, Vol. 29, No. 5, pp. 727－736.

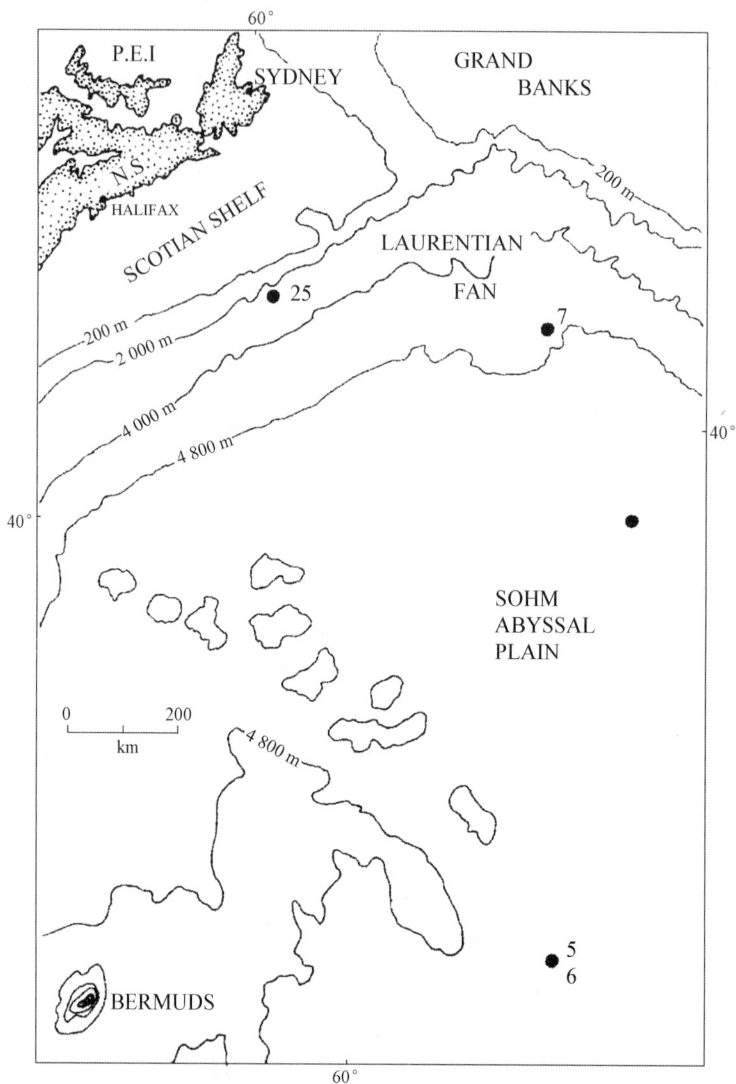

Fig. 1　Map showing location of samples from Laurentian Fan and Sohm Abyssal Plain (note irregular contour interval, in metres)

Table 1　Location and age of samples

Sample no.	Core no.	Depth in core	Location	Water depth (m)	Modal sand size (ϕ)	Quartz grain size (ϕ)	Probable age
5	80－016－45	1 248～1 259	32°29. 27′N 55°56. 2′W	5 560	5. 0	1cu－1	Mid-Wisconsinan
6	80－016－45	1 279	32°29. 27′N 55°56. 2′W	5 560	3. 5	1cu－1	Mid-Wisconsinan
7	74－021－34	130	42°20. 3′N 54°44. 3′W	4 429	3. 0	1. 5 cu 0	Basal late Wisconsinan
8	DSDP 381－1－2	230～237	39°14. 88′N 53°21. 18′W	5 280	2. 0	1cu－1	Late Pleistocene

(Continued)

Sample no.	Core no.	Depth in core	Location	Water depth (m)	Modal sand size (ϕ)	Quartz grain size (ϕ)	Probable age
25	73-011-1	400～401	43°12.56′N 60°24.63′W	1814	4.0	1cu-1	Basal late Wisconsinan
18	M15	Beach surface	High tide level of Martinique Beach, Nova Scotia		2.5	1cu-1	Modern

mid-Wisconsinan composite turbidite-ice rafted succession. Samples 7 and 25 are from near the base of the late Wisconsinan turbidite sequence, on the Laurentian Fan and lower Scotian Slope, respectively. Sample 8 is from a DSDP hole that penetrated several metres of sand of presumed late Pleistocene or Holocene age on the northern Sohm Abyssal Plain (Tucholke & Vogt, 1979).

Three stratigraphic divisions are recognized in cores from the southern Sohm Abyssal Plain (Buckley, 1981). The area lies about 1 500 m below the present carbonate compensation depth, so that the indigenous benthic foraminifera are the agglutinated siliceous type. The calcareous foraminifera found in cores must have been buried rapidly to be preserved and, at least in the case of benthic forms, must have been resedimented from shallower depths.

(1) 0～40 cm: this division consists of dark yellowish to pale yellowish-brown soft clay. Foraminifera are abundant except for the surface layer. The assemblage comprises subtropical planktonic foraminifera and both calcareous and agglutinated siliceous benthic types.

(2) 40～800 cm: the sediment is generally similar to division 1. Much of it occurs in sharp based sedimentation units 10～30 cm thick with a basal, well sorted, fine silt overlain by clay, which are similar in sedimentary structures and textures to type 3 turbidite silt of Piper (1970). The sediment is barren of foraminifera except for occasional layers of calcerous remains in varying states of preservation. The foraminifera include tropical and subtropical planktonics and bathyal calcareous and agglutinated benthics.

(3) 800～1 200 cm: the sediment consists of silty clay with sandy lenses and beds. All cores stopped in a layer of fine sand that was not completely penetrated. Sands in this setting are presumably turbidites. In the sand, foraminifera are well preserved and include subtropical, subArctic, and Arctic planktonic species. *Neogloboquadrina pachydema* (60% of which 39% are sinistral) predominates, along with *Globigerina bulloides* (17%), and *G. quinqueloba* (9%). The excellent preservation is illustrated by the occurrence of the fragile *Globorotalia scitula*. Benthic foraminifera are dominated

by species common on the continental shelf off Eastern Canada, but not found living in deeper water. The major species are *Elphidium excatavum* f. *clavata*, *Cassidulina reniforme*, *Islandiella helenae*, *Cibicides lobalatus*, *I. norcrossi*, *I. islandica*, *Bulmina marginata*, and *Buccella frigida*. The assemblage is similar to that in late Wisconsinan sediments on the Scotian Shelf (Vilks& Rashid, 1976; Mudie, 1980), except for the presence of up to 4% *Cassidulina* cf. *teretis*. In Eastern Canada this species has been found only in pre-late Wisconsinan sediment from Hudson Strait (60 000 BP, R. H. Fillon, 1981, personal communication) and Baffin Island (50 000 BP, Feyling-Hansen, 1980) but it is absent in Scotian Shelf cores (which are no older than 30 000 BP).

Division 1 is tentatively interpreted as Holocene. Many of the clay beds in division 2 are similar to those in the late Wisconsinan of the Laurentian Fan, and are interpreted as distal turbidites. The foraminifera data suggest that division 3 is mid-Wisconsinan or older, and that the turbidites are derived from the eastern Canadian shelf. A Scotian Shelf-Laurentian Fan source is suggested by heavy mineral assemblages (unpublished data) and the presence of coal similar to that in the Sydney coalfield (Hacquebard *et al.*, 1982).

METHODS

As pointed out by Krinsley& McCoy (1977), only very small numbers of sand grains are needed to obtain representative surface textures. In this study, 15~18 quartz grains ranging in size from -1ϕ to 1.5ϕ (2~0.4 mm) were taken from each sample studied.

The grains were boiled in concentrated hydrochloric acid for 10 min, washed with distilled water, boiled in stannous chloride for 20 min, then rewashed with distilled water, boiled in 30% hydrogen peroxide for 10 min, and finally washed with distilled water again and thoroughly dried (Krinsley & Doornkamp, 1973). The grains were coated with a 200° Å thick layer of an alloy of gold-palladium and were examined with a Cambridge 180 stereoscan.

SURFACE TEXTURES OF QUARTZ GRAINS

The outline shape of each grain was examined. The shape was recorded only if the grain was particularly irregular or smooth, or if there was clear crystal control of overall shape. The roundness of edges was classified according to the Powers (1973) scale. The presence or absence of 13 distinctive types of surface texture was noted (Table 2). Selected grains were photomicrographed (Figs 2 and 3). The distinctive features of each sample are summarized below.

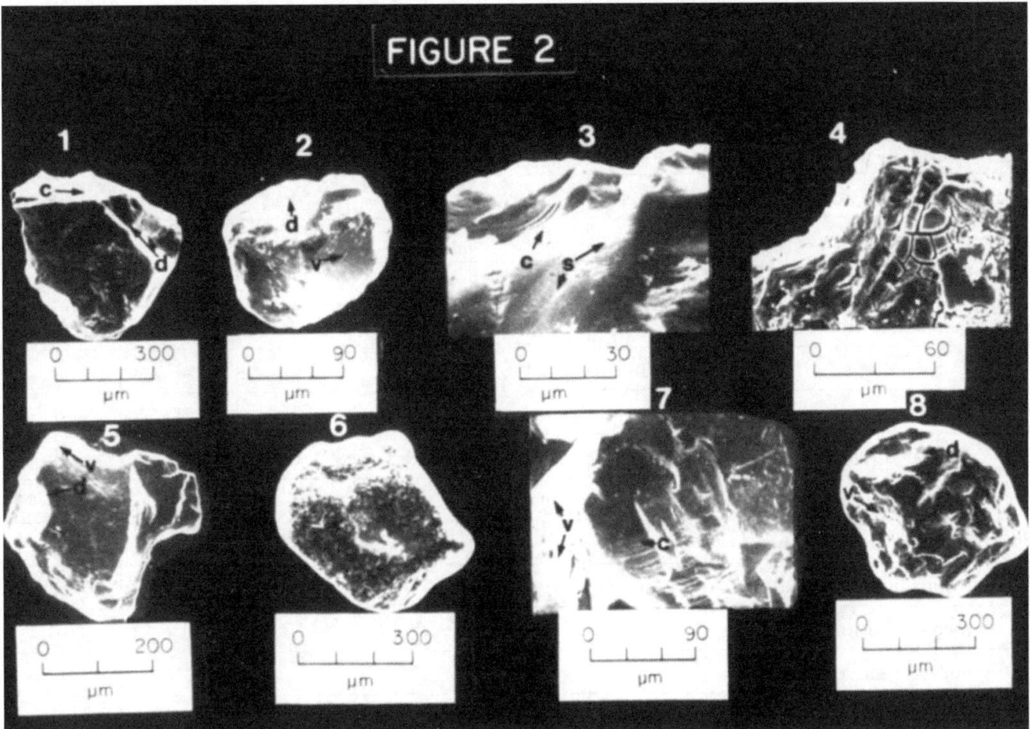

Fig. 2　Photomicrographs of representative quartz grain surface textures from samples 5 and 6.

On all pictures: c=conchoidal fracture, d=dish-shaped concavity, h=solution hole, s=striae, v=mechanical V-marks, p=collision pits. (1) Subangular to subrounded grain with dish-shaped concavities and traces of conchoidal fractures superimposed by V-marks (sample 5). (2) Cubic-shaped grain with subrounded edges, V-marks, and small amounts of surface silica deposits. Conchoidal fractures and depressions are seen on the top of the grain (sample 5). (3) Detail of 2(2), showing conchoidal fractures (c) in the depressions and striae (s) on elevated surfaces. (4) Silica coating with shrinkage cracks (sample 5). (5) Irregular shaped grain with subangular to subrounded edges. V-marks are superimposed on the surface of the grain (sample 6). (6) Subrounded to rounded grain with traces of concavities. Surface extensively covered with V-marks (sample 6). (7) Conchoidal fractures in the depression on the grain surface superimposed by V-marks (sample 6). (8) Rounded grain with many V-marks and concavities on the surface (sample 6).

Table 2　Number of grains in each sample showing particular surface textures

Sample	5	6	7	8	25	18
number of grains	18	16	15	15	15	15
Shape						
irregular	8	5	2	6	5	0
smooth	0	0	0	0	0	0
crystal control	5	3	3	5	3	3
Edges						
angular	1	0	0	0	2	0
angular-subangular	0	1	1	1	0	0
subangular	7	2	2	4	6	1

(Continued)

Edges						
subangular-subrounded	3	2	3	5	1	1
subrounded	3	4	7	2	5	6
subrounded-rounded	1	0	1	1	0	0
rounded	3	7	1	2	1	7
Glacial textures						
conchoidal fractures	11	7	5	6	7	3
parallel steps	3	1	0	0	1	0
striae	3	0	0	1	1	0
upturned plates	0	0	0	0	5	3
dish-shaped concavities	6	5	5	7	6	3
Current transport textures						
mechanical V-marks	17	15	11	15	15	12
mechanical impact pits	17	15	11	15	15	10
curved grooves						6
Diagenetic textures						
cracks	9	6	0	0	5	0
surface solution features	8	2	7	6	4	7
solution holes	?	?	?	?	11	6
silica coatings	13	3	3	2	5	0
silica fragments	8	11	9	14	15	0
tube-shaped silica deposits	0	0	1	0	0	0
weathering features	0	0	0	0	0	4

Sample 5 is from division 3 of the south Sohm Abyssal Plain core, from a 10 cm thick muddy sand bed with a modal size of 5ϕ. Most of the grains show either irregular shape or cubic shape as a result of crystal control. Most grains contain relict conchoidal fractures, dish-shaped concavities, and, less frequently, parallel steps or striae (Fig. 2(1), (2), (3)). Most of these markings are characteristic of glacial processes. All the grains, including those with irregular outlines, have their edges subrounded or even rounded (Fig. 4), suggesting substantial reworking following their primary glacial origin. Mechanical V-forms (or V-marks in short) are the main features of all the grains in sample 5 (Fig. 2(1), (2)). They are of variable size, but are frequently deep, and are accompanied by lesser numbers of dish-shaped concavities. These markings are characteristically produced by current transport. The V-marks are superimposed on other features, indicating that this mechanical abrasion was the latest erosional event in shaping the surface features of the grains. Most grains are partially coated by silica precipitates (Fig. 2(4)) making the grains smoother. Small pits or holes produced by surface solution are also common.

Sample 6 is a fine sand bed from the base of the same core as 5. It shows similar features as sample 5, except that edges of grains are much more rounded. Only about one-third of the grains have an irregular shape, and have dish-shaped concavities. More than half of the grains contain relict conchoidal marks (Fig. 2(5)). Deep and abundant mechanical V-marks and rounded edges are dominant features of all the grains and these features are superimposed on other markings (Fig. 2(6), (7)). Weathering cracks appear along conchoidal fractures. There are surface solution features and silica fragments appearing on the surface of some grains (Fig. 2(8)).

Sample 7 is from a 1 cm thick, very fine sand bed in the late Wisconsinan overbank turbidite sequence from the middle part of the Laurentian Fan. The quartz grains have subrounded or rounded edges; one-third of the grains retain traces of conchoidal fractures, which are preserved in depressions or dish-shaped concavities (Fig. 3(1)).

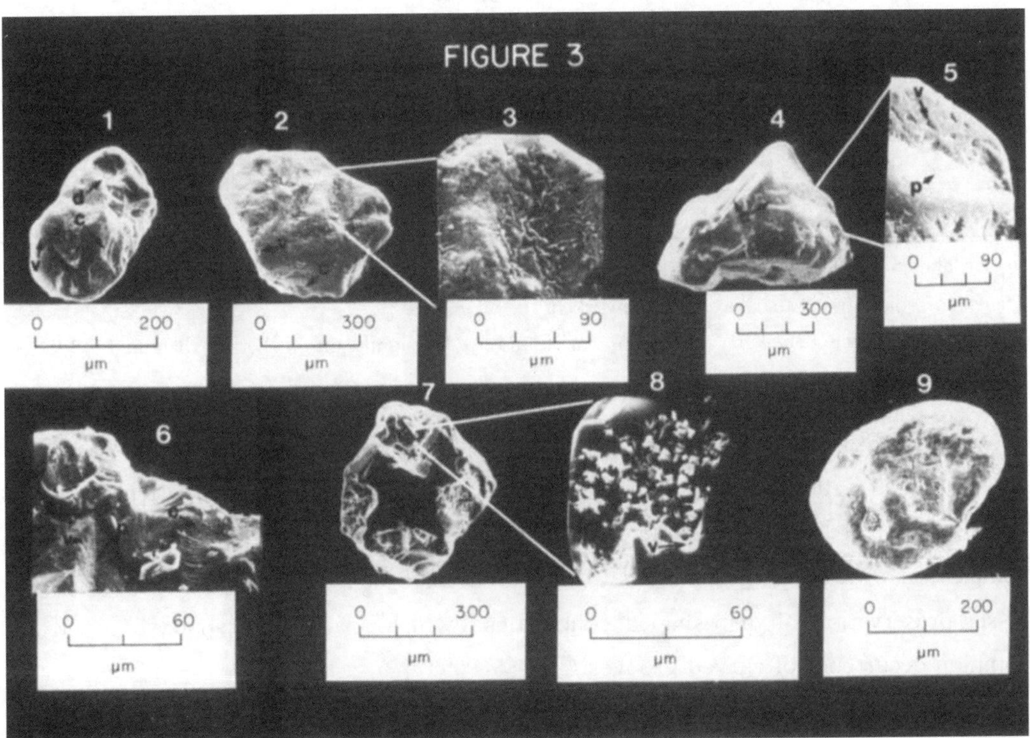

Fig. 3 Photomicrographs of representative grain surface textures from samples 7, 8, 25, and 18.

(1) Irregular grain with subrounded edges. In concavities there are traces of conchoidal fractures superimposed by V-marks (sample 7). (2) Grain with silica deposits in the deeper parts of depressions (sample 7). (3) Detail of 3(2) showing tube-shaped silica deposits. (4) Triangular shaped grain (controlled by crystal form). Traces of conchoidal fractures on grain surface with deep V-marks (sample 8). (5) Detail of 3(4) to show mechanical V-marks and collision pits (p). (6) Grain with conchoidal fractures preserved in concavities, superimposed by V-marks and deposited silica fragments (sample 8). (7) Cubic-shaped grain with subangular edges. Grain has dish-shaped concavities and conchoidal fractures with few V-marks (sample 25). (8) Detail of 3(7) showing silica deposits in the concavities. (9) Rounded grain of beach sand with V-marks and collision pits (sample 18).

V-marks are the commonest surface texture. Over one-third of the grains have surface solution features and fragmentary silica deposits. Only this sample shows tube-shaped silica deposits, which probably were precipitated by silica flow along the lowest part of depressions on the surface of the grains (Fig. 3(2), (3)).

Sample 8 is from a thick sand sequence near the transition from the distal Laurentian Fan to the Sohm Abyssal Plain. Subrounded grains retain the dish-shaped concavities and traces of conchoidal fractures (Fig. 3(3)). Most grains have deep V-marks, which are superimposed on the other markings (Fig. 3(4), (5)). Over one-third of the grains have solution features and most have silica deposits (Fig. 3(6)).

Sample 25 is from an overbank proximal turbidite sand bed on the lower Scotian Slope that is glacially derived. It shows surface textures similar to the other samples, except that edges tend to be more angular and diagenetic features are prominent (Fig. 3 (7), (8)).

Table 3 Number of V-marks 10^{-6} μm^2 of surface area of quartz grains

Sample	5	6	7	8	25
Number of grains*	15	10	10	11	15
Mean†	43.0	118.9	49.0	42.5	21.9
Standard deviation	17.3	27.3	28.9	19.2	12.7

* The 15% of grains with the most V-marks have been omitted in calculations as a precaution against including grains that have been reworked.

†The difference between each mean was found to be significant at better than 0.98 level of probability, using a *t*-test.

Sample 18 is a beach sample from Martinique Beach on the Atlantic coast of Nova Scotia, which is derived from glacial drumlins. The sand has thus experienced only a small amount of beach reworking that is superimposed on primary glacial features. In comparison with the turbidite sands, this immature beach sand has greater roundness, lesser preservation of dish-shaped concavities (which are always shallow), and the common occurrence of curved grooves (Fig. 3(9)).

Several surface textural characteristics appear to result from the current transport of glacial material, so that these current features are superimposed on the original glacial grain morphology; two of these have been quantified. Grain edges can be classified as angular, subangular, subrounded, or rounded; the abundance of each class is displayed in a histogram in Fig. 4. Angular edges are most common in the proximal turbidite sample 25; sample 6 has the highest proportion of rounded edges, while samples 8 and 5 are intermediate and are indistinguishable from one another. Sample 7, in which finer grains were examined, has a much smaller proportion of irregular grains and hence anomalously well rounded grains.

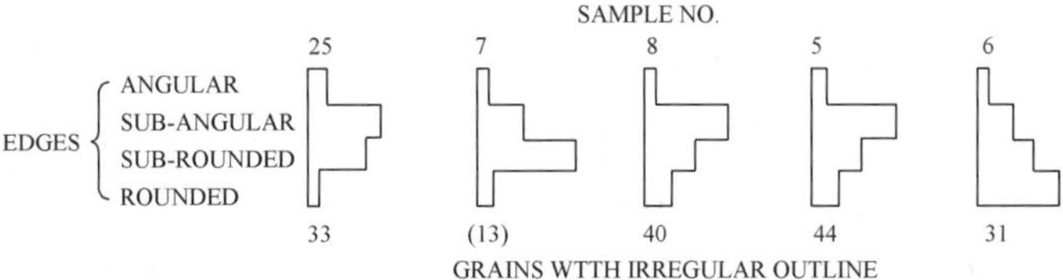

Fig. 4　Variation between samples in grain outline and roundness of edges. (Note that grain size used in sample 7 was finer, accounting for lower percentage of irregular grains; cf. Table 2.)

The number of mechanical V-marks on representative grain faces has been counted within a 1 100 × 800 μm area (Fig. 5). The results show that V-marks are most abundant on sample 6 and at least abundant on sample 25. This trend is in agreement with that shown by the roundness of grain edges, suggesting that the two processes are interrelated. V-marks are most common on rounded edges of grains.

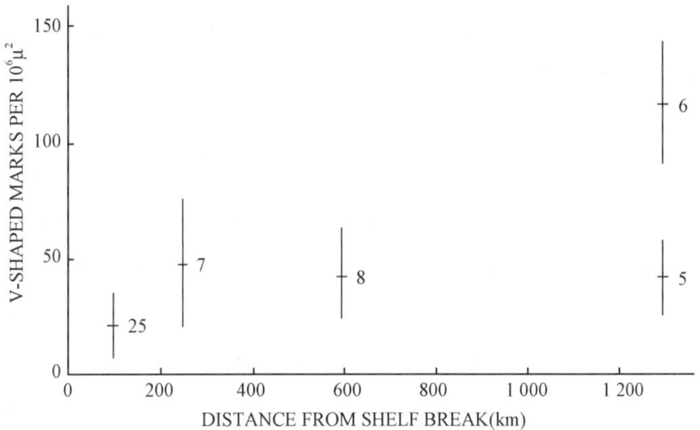

Fig. 5　Mean and standard deviation of number of mechanical V-marks plotted against distance from shelf break (Table 3).

DISCUSSION

Almost all the sand grains studied have a rather irregular outline and show primary glacial features, such as angular edges, conchoidal fractures, dish-shaped concavities, parallel steps, and striae (Krinsley & Doornkamp, 1973; Vilks & Wang, 1981). These glacial features have modified in two main ways. First, many exposed grain surfaces have current markings, particularly mechanical V-marks, super-imposed on the glacial features. Second, angular edges have been rounded. In addition, many grains have diagenetic features superimposed on both the glacial and the current markings; these include surface solution features and silica coatings.

The current markings might be produced by turbidity currents, or might result from river, beach, or shallow marine transport prior to turbidity current initiation. However, in related studies of fluvial sands from the Yangtze River (for example), V-marks are larger but less abundant as compared with the turbidite sands examined in this study. Furthermore, the grains are much more rounded and smooth after having been moved a long distance by the rivers. Although V-marks are common in beach and near-shore sands, such sands are subjected to intense abrasion, rapidly rounding edges, and obliterating inherited surface textures. Beach and nearshore sands that we have examined were derived from till and show fewer mechanical V-forms and poorer preservation of primary glacial features than the turbidite sands described here. In particular, dish-shaped concavities with traces of conchoidal fractures are found almost exclusively in turbidite sands. In contrast, the number of grains with irregular outline in our samples is independent of the degree of rounding of edges (Fig. 4).

We therefore suggest that the rounding of edges and the presence of mechanical marks combined with the dish-shaped concavities result from turbidity current transport. There is no simple relationship between the intensity of current-produced markings and the distance of sedimentary transport, although the least modified sample is the most proximal, and the most modified is the most distal (Fig. 5). Neither is there a clear relationship with parameters such as grain size or bed thickness, although of the two distal Sohm Abyssal Plain samples, more current-produced markings occur in the coarser sand. The intensity of current-produced markings does not vary systematically with indicators of current velocity (such as modal size or bed thickness) at the site of deposition. The characteristic that may correlate most closely with the intensity of current markings is the inferred total size or power of the turbidity current that deposited the sand bed (Fig. 6).

Three size classes of turbidity currents can be distinguished on the Laurentian Fan-Sohm Abyssal Plain system (Fig. 6). Small currents do not spill out of the Fan valleys, and deposit sand on only the proximal part of the Sohm Abyssal Plain (A in Fig. 6). The 1929 'Grand Banks slump' current was of this type (Fruth, 1965; Stow, 1977). Intermediate sized currents spill out of Fanvalleys giving overbank sands, but again transport sand only to the proximal part of the Sohm Abyssal Plain (B in Fig. 6). The occasional occurrence of late Wisconsinan overbank sands on the Laurentian Fan (Stow, 1977) suggests that there were rare intermediate sized currents in the Late Wisconsinan, but the much greater abundance of sand beds on the proximal Sohm Abyssal Plain (Horn *et al.*, 1971) shows that most currents were of the small class. Large sized currents transport sand as far as the distal Sohm Abyssal Plain (C in Fig. 6), and last occurred in the mid-Wisconsinan. Their extent on the Laurentian Fan is unknown, but presumably they also spilled out of the channels.

The distal Sohm Abyssal Plain samples must come from large sized currents. The

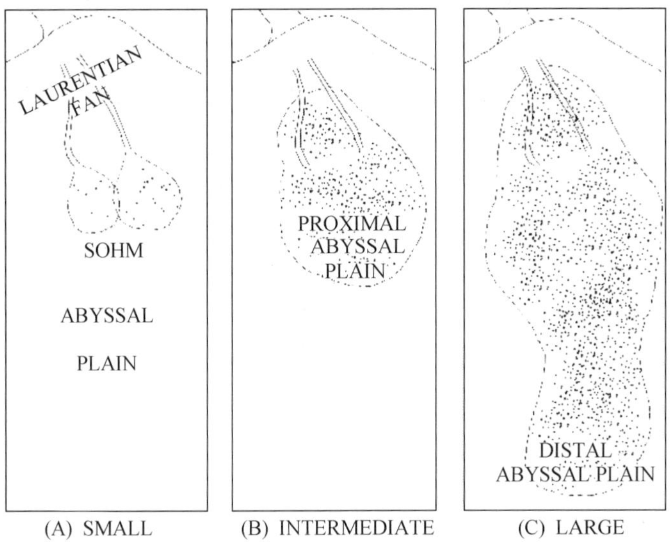

Fig. 6　Schematic model showing extent of sand deposits from small, intermediate, and large turbidity currents on the Laurentian Fan-Sohm Abyssal Plain system (based on data in Stow, 1977, Fruth, 1965 and Horn *et al.*, 1971).

coarser sand bed (sample 6), suggesting the more powerful turbidity current, has more current-produced markings than sample 5.

The grain size of sample 8 on the proximal Sohm Abyssal Plain, which is close to the average size found by Horn *et al.* (1971), suggests that it was deposited from a small sized turbidity current. The late Wisconsinan overbank sands of samples 7 and 25 were deposited from intermediate size currents.

The observed intensity of current-produced markings, except in sample 25 which is extremely proximal, appears to be related directly to the estimated total size (and hence power) of the depositing turbidity current. Since it is difficult to further quantify this concept of power from the available data, it is not possible to evaluate this hypothesis statistically.

Thus we conclude that only very large and powerful turbidity currents were capable of carrying sands to the southern Sohm Abyssal Plain during the mid-Wisconsinan or earlier. These very large events also produced many impact marks on the transported sand grains. Sample 6 is representative of such sands, having nearly three times as many V-marks as compared to sands deposited on the proximal Abyssal Plain or on the Laurentian Fan. The number of V-marks in these more proximal sands also appears related to the estimated total size or power of the current that deposited them.

Mechanical V-marks result from grain collision (Krinsley & Doornkamp, 1973). Collision may take place during three main processes in turbidity current transport: (1) true suspension (including intermittent suspension); (2) grain flow; and (3) bed load tractional processes.

Bed load traction is an unimportant process during turbidity current transport of sediment (Middleton & Hampton, 1976). Grain flow, in which grains remain dispersed through intergranular collision and move inertially, is clearly much more likely to produce grain collision marks than is suspension in turbulent liquid, in which all but the rare coarsest grains move viscously so that collisions are cushioned by fluid. Theoretical analysis (Bagnold, 1956) and studies of ancient turbidite sequences (Stanley & Unrug, 1972) suggest that grain flow occurs only in proximal turbidity currents, probably on the continental slope.

If grain flow is responsible for the distinctive surface textures of these turbidites, then the duration of grain-flow conditions (a function of turbidity current velocity and hence size) will determine the abundance of current markings on grains. The occurrence of these grains in beds clearly deposited from suspension and not from grain flows implies mixing of grain-flow sediment with the main body of the turbidity current, probably through turbulent processes in the head of the turbidity current (Fig. 7).

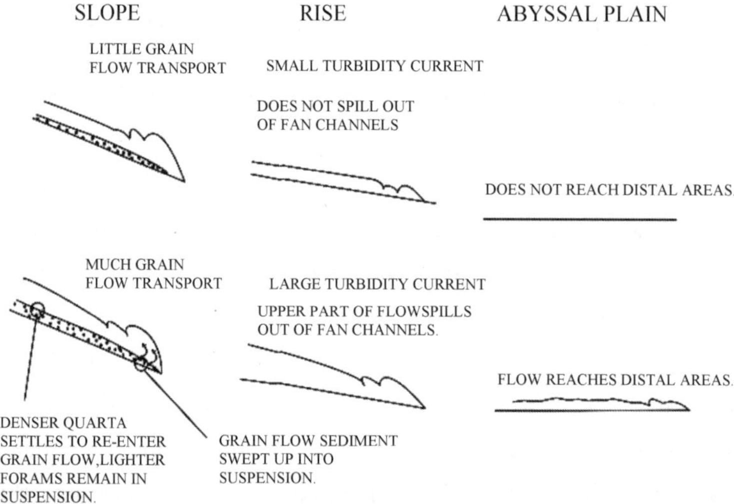

Fig. 7 Schematic model contrasting the development of grain flows and the distribution of turbidite sediment in small and large turbidity currents, based on data presented in this paper.

The excellent preservation of resedimented shelf foraminifera on the distal Sohm Abyssal Plain contrasts with the abraded quartz grains, suggesting that the foraminifera are transported without significant abrasion, presumably in suspension. Any foraminifera transported by grain flow would presumably be rapidly smashed. The lack of foraminifera in sands on the proximal Laurentian Fan suggests the action of powerful grain flow or other near-bed conditions immediately prior to deposition in that region.

Superimposition or turbidity current markings can only be recognized where the source sand lacks current markings, likely to occur only when the source material is glacial till. Krinsley & McCoy (1977) suggest that mechanical breakage of large quartz grains is characteristic of turbidite current transport; however, the earlier glacial features will comminute all but the most resistant quartz grains.

CONCLUSIONS

(1) Distal turbidites occur on the southern Sohm Abyssal Plain more than 1 000 km from their source on the Laurentian Fan.

(2) Turbidite sands on the Laurentian Fan-Sohm Abyssal Plain system are derived principally from glacial till.

(3) Coarse quartz grains have glacially induced surface textures on which are superimposed subrounded edges and current-induced (collision) markings, particularly mechanical V-forms. However, they maintain their irregular shape, thus preserving dish-shaped depressions containing characteristically glacial markings.

(4) The intensity of current modification of surface textures depends primarily on estimated total power and size of the turbidity current and mode of turbidite flow and distance of transport.

(5) The current modification takes place during grain flow. Mixing of grain-flow transported sands takes place with the overlying suspended sediment, probably through turbulence at the head.

(6) Resedimented shelf foraminifera are transported exclusively in suspension.

ACKNOWLEDGMENTS

Mr B. Deonarine is gratefully acknowledged for the micrographs. We wish to express our gratitude especially to Drs D. E. Buckley and J. P. M. Syvitski for their critical comments and encouragement in the surface textures studies.

REFERENCES

[1] BAGNOLD, R. A. 1956. The flow of cohesionless grains in fluids. *Phil. Trans. R. Soc. A*, 249: 335 - 397.

[2] BUCKLEY, D. E. 1981. Geological investigation of a selected area of the Sohm Abyssal Plain, western Atlantic: C. S. S. HUDSON cruise 80016. *AECL Tech. Rep. series.*

[3] CHOUGH, S. K. & HESSE. R. 1982. The Northwest Atlantic mid-ocean channel of the Labrador Sea. I. Morphology and sedimentary facies. *Can. J. Earth Sci.* (in press).

[4] FEYLING-HANSEN, R. W. 1980. Microbiostratigraphy of young Cenozoic marine deposits of the Qivitug Peninsula, Baffin Island. *Mar. Micropal.* 5: 153 - 184.

[5] FRUTH, L. S. 1965. *The* 1929 *Grand Banks turbidite and the sediments of the Sohm Abyssal Plain.* M. Sc. Thesis. Columbia University, New York. 258.

[6] HACQUEBARD, P. A. , BUCKLEY, D. & VILKS, G. 1982. The importance of detrital particles of coal in tracing the provenance of sedimentary rocks. *Bull. Cent. Rech. Expl.-Prod. Elf-Aquitaine*. 5 (in press).

[7] HORN, D. R. , EWING, M. , HORN, B. M. & DELACH, M. N. 1971. Turbidites of the Hatteras and Sohm Abyssal Plains, western North Atlantic Ocean. *Mar. Geol.* 11: 287 - 323.

[8] KRINSLEY, D. H. & DOORNKAMP, J. 1973. *Atlas of Quartz Sand Surface Textures*. Cambridge University Press. 91.

[9] KRINSLEY, D. H. & McCOY, F. W. 1977. Significance and origin of surface textures on broken sand grains in deep-sea sediments. *Sedimentology*, 24: 857 - 862.

[10] MIDDLETON, G. V. & HAMPTON, M. A. 1973. Sediment gravity flows: mechanics of flow and deposition. In: *Turbidites and Deep-Water Sedimeniation* (Ed. by G. V. Middleton & A. H. Bouma). Society of Economic Paleontologists and Mineralogists, Pacific Section Short Course. 1 - 38.

[11] MUDIE. P. J. 1980. *Palynology of later Quaternary marine sediments, Eastern Canada*. Ph. D. Thesis. Dalhousie University, Halifax, Nova Scotia. 638.

[12] PIPER, D. J. W. 1970. Transport and deposition of Holocene sediment in La Jolla deep sea fan, California. *Mar. Geol.* 8: 211 - 227.

[13] PIPER, D. J. W. 1975. Late Quaternary deep-water sedimentation off Nova Scotia and western Grand Banks. *Mem. Can. Soc. Petrol. Geol.* 4: 195 - 204.

[14] POWERS, M. C. 1973. A new roundness scale for sedimentary particles. *J. sedim. Petrol.* 23: 117 - 119.

[15] STANLEY, D. J. & UNRUG, R. 1972. Submarine channel deposits, fluxoturbidites, and other indicators of slope and base of slope environments in modern and ancient marine basins. In: *Recognition of Ancient Sedimentary Environments* (Ed. by J. K. Rigby and W. K. Hamblin). *Spec. Publs Soc. econ. Paleont. Miner.* 16: 287 - 340.

[16] STOW, D. A. V. 1977. *Late Quaternary stratigraphy and sedimentation on the Nova Scotian outer continental margin*. Ph. D. Thesis. Dalhousie University, Halifax, Nova Scotia. 360.

[17] STOW, D. A. V. 1978. Regional review of the Nova Scotian outer margin. *Mar. Sedim.* 14: 17 - 32.

[18] STOW, D. A. V. 1979. Distinguishing between fine-grained turbidites and contourites on the deep-water margin off Nova Scotia. *Sedimentology*, 26: 371 - 387.

[19] TUCHOLKE, B. & VOGT, P. 1979. *Init. Rep. Deep Sea drill. Proj.* 43: 827 - 845.

[20] VILKS, G. & RASHID, M. A. 1976. Postglacial paleoceanography of Emerald Basin, Scotian Shelf. *Can. J. Earth Sci.* 13: 1256 - 1267.

[21] VILKS, G. & WANG. Y. 1981. Surface textures of quartz grains and sedimentary processes on the southeastern Labrador shelf. *Pep. geol. Surv. Can.* 81 - IB: 55 - 61.

加拿大海洋地质学研究现状*

一、加拿大海洋地质学组织机构特点

加拿大有 14 万 km 长的海岸线，广大的国土与世界三大洋——太平洋、大西洋、北冰洋相接，由于渔业、交通及军事活动方面的需要，加拿大很早就开始从事海洋方面的调查研究，在 1883 年建立了水深测量部，1893 年建立了潮汐局。但是，系统的现代海洋学研究事业是从第二次世界大战后、于 1946 年建立加拿大海洋学委员会之后发展起来的。起步迟这点与我国相似，但他们坚持不断，持续前进。加拿大海洋地质事业是 20 世纪 50 年代末和 60 年代初开始发展的。虽然历史短，但保守思想少，借各国之长，故发展速度较快。根据是：1959 年开始了“极地大陆架研究项目”，而于 1962 年建立了贝德福德海洋研究所（简称 BIO）。这个研究所在 70 年代初期便成为西半球的三大海洋研究中心之一（Scripps、Woods Hole、Bedford），在世界海洋研究中居领先地位。

贝德福德海洋研究所是综合性的海洋研究所，分属加拿大政府的三个部：能源矿产资源部、渔业与海洋部、环境部。属于环境部的是个很小的单位——环境保护服务机构，主要是检测地方工业对环境的污染。属于渔业与海洋部的是个大研究单位，由两部分组成：其一为渔业管理局下属的海洋渔业研究部门；另一个是海洋科学与调查分部，它又细分为六个部门：人事、管理服务、研究设备、加拿大水深测量服务、海洋生态实验室及大西洋海洋研究所。大西洋海洋研究所是其中最大的部门，其下又分：气象、化学海洋学、物理海洋学、海岸海洋学等分支。海洋地质学的研究是由“大西洋地球科学中心”进行的，后者属于能源矿产资源部下的加拿大地质测量部。三个部各自的研究机构共同使用 BIO 的房屋设备、船只仪器设备与资料情报网，从而成为一个协调很好的综合研究所。有一个总的所长——A·R·郎豪尔斯特博士，但他实际上只代表渔业与海洋部，对生物、化学和物理各部门进行领导。海洋地质是由大西洋地学中心的 M. 金博士直接向加拿大地质测量部、进而向能源矿产资源部负责、报告工作。BIO 位于大西洋畔的达特茅斯市。依海湾建造了既漂亮又实用的办公室、实验室、车间、码头及服务设施。全所约 800 人，研究人员与技术员（相当于我国的辅助人员，但均具大学毕业水平）约 320 人，其中高级研究员 125 人（博士学位以上并经过实际工作训练的）；130 人属于海图测量系统；350 人为中心支持部门（包括船舰人员、计算机、仪器研制等方面的工程师、技师、摄影、打字、图书资料及行政人员）。据 1981 年统计，BIO 全部不动产约 2 亿加元，每年开支约 3 000 万加元。该所有一研究船队，其中 3 艘为具有全天候航行能力的大船，船只属于政府，由 BIO 代管。因此，除该研究所以外，附近有关大学与研究单位亦可使用这些船，但需每年 9 月前交出用船计划，由 BIO 统一安排。加拿大海洋机构之间，通过协调可以使用彼此之间的设备，研

* 王颖：《海洋科技动态》，1983 年第 9 期（总第 196 期），第 2 - 8 页。

究人员可据工作需要自由地使用实验室。这种办法值得效法。

“大西洋地学中心”(简称 AGC)是加拿大政府所属的主要海洋地质研究单位,共有97人,其中高级研究员与研究员32人,自然科学家(一般是年轻具有博士学位而无实际经验者,也有些是未取得博士学位,但有一定实际经验者)16人,技术员30人及人事、行政管理人员21人。“大西洋地学中心”分为五个组。

(1) 行政管理组。

(2) 环境海洋地质组。由海岸地貌与沉积学家、海洋地球化学家与第四纪微体古生物学家组成,主要研究第四纪与冰后期以来海滨与海底的地质作用过程。

(3) 区域普查组。由地质与地球物理学家组成,主要研究东部加拿大大陆边缘现代与古代的发展史,其工作不仅完成大陆架表面地质调查与制图,并且从事北美与极地深部地壳构造研究。

(4) 东部石油地质组。由生物地层学家、岩石学家、地球物理学家与石油地质学家组成,是作为加拿大政府对地方或石油公司进行监督的研究部门,根据在东部沿海的石油钻孔资料,从事大陆架深部构造研究与判断油气资源。

(5) 项目支援组。由技术员与工程师组成,主要支援野外工作并照顾海上仪器与设计实验新仪器。

各研究组虽有分工,但在区域与课题上是互有交错的,实际上主要包括表面地质(地貌、沉积、地化与生物过程、生物地层)与深部地质(岩石地层、构造、微古)两大部分。包括了各方面的科学家,主要工作区域为大西洋及北方诸海。对极地亚极地海洋地质作用过程、特点及扩张的大西洋海底有相当深入的研究。不足之处,对内、外动力作用本身的研究注意不够。

贝德福德研究所是加拿大从事海洋学研究(包括海洋地质研究)的主要单位,素为加政府引为骄傲。它的研究工作与研究水平可以代表加拿大的海洋学水平。同时,为西方国家公认为西半球三大海洋研究中心之一。1982年世界海洋学大会即在加拿大哈里法克斯市召开。

属于国家研究机构的还有“太平洋地球科学中心”,设于西海岸的温哥华岛的派垂莎湾。该所于20世纪70年代中期建立,形似BIO,但规模小,可能是由于大西洋为加拿大的主要研究中心之故。该所的建筑精致美观,采光条件好,环境优美,是研究工作的极好场所。其研究工作以地质构造、地震与石油地质见长。该所有一个很好的地震观测台站,人员以地球物理学家、微古学家为主,亦有少数沉积学家。主要研究太平洋沿岸及北方岛屿海域。

设立于安大略湖畔的国立水科学研究所,有从事海滨与湖岸的地貌与沉积学家。

从事海洋地质研究的第二个大部门是石油公司的研究机构。如加拿大石油公司、埃索石油公司等,各有一支庞大的海洋地质研究力量,主要基地在西部的卡里哥瑞市,研究工作涉及到构造、沉积、微古、工程地质与土力学等。研究区域主要集中于北方的蒲福海、东部的拉布拉多海、纽芬兰与新斯科舍海域。石油公司的研究工作特点是:经费充足,工作量大,但工作不够深入细致。近年来这些研究单位已注意到需要有较多的沉积学家参加原有的传统性地质工作。

从事海洋地质及海洋学研究的第三个大部门是大学，它们的研究项目受政府资助。在哈里法克斯市的达尔豪谢大学于20世纪50年代后期建立海洋研究所，嗣后即专门培养海洋地质、物理海洋学、化学海洋学与生物海洋学人才。达尔豪谢大学与贝德福德研究所密切合作，共同使用船只与技术设备，故而形成加拿大东部的一个重要的海洋学中心。达尔豪谢大学以物理海洋学、海洋沉积学、微古以及应用重磁法研究海洋深部构造而见长，近年来增加了海洋法与环境研究。魁北克的拉瓦尔大学是从1937年起从事研究圣劳伦斯湾。在蒙特利尔的麦基尔大学在50年代即开始极地渔业与海洋学的研究。1964年该校又建立了海洋科学中心，致力于包括海洋地质在内的各项海洋学研究。纽芬兰的纪念大学主要专长于寒冷海域的海洋工程学，为此开展了有关海底地貌与沉积过程研究。位于西海岸温哥华的不列颠哥伦比亚大学于1949年建立了海洋学研究所，初始，该校着重于峡江式河口湾的研究，因为这是西海岸的主要类型。他们结合动力过程从事地貌研究，嗣后又开设了化学海洋学、生物海洋学与海洋地质学各项课程，但以物理海洋学为特长。除培养专门人才外，各大学中的海洋地质学教授与研究生进行的科学研究亦组成加拿大海洋地质工作中的一支各具特色的研究力量，如位于安大略省的麦克马斯特大学地质系即以沉积学见长。

第四支力量是属于地方上的研究组织或私人小型公司，如新斯科舍的海洋基金组织为了该省的海底煤田与油气开发，专门研究仪器设备与进行基本调查搜集资料。西海岸有些私人公司为开发海滨砂石、海岸防护及工程建设的海岸制图而发展动态研究。

加拿大海洋地质学与海洋学人员的来源和培养途径经历两个阶段：早期是从世界各地聘请有经验的专家来从事研究工作。在国外，人员流动性大，且英语系国家无语言上障碍，容易集中各方面人才，能迅速地开展工作。但是，这毕竟为数太少，不足以使用。嗣后又派遣留学生与访问学者去美国学习，仍感收效不大。因此，1946年建立了加拿大海洋学联合委员会(1959年改组为加拿大海洋学委员会)，决定在国内各有关大学设立海洋研究机构，专门培养海洋人才。因此，从20世纪60年代起，加拿大的海洋学人才(包括海洋地质人才)主要由本国培养，其中尤以不列颠哥伦比亚大学与达尔豪谢大学为两个主要培养基地。到60年代中期，大量人才源源不断地输送到加拿大、美国、北欧及澳大利亚等有关单位。如美国第一个航天女宇宙地质学家，即是达大地质系60年代的毕业生。在加拿大从事海洋地质的人员，一般是取得博士学位的。各大学中是不设立海洋学的学士学位的。从物理系、生物系、地质系、化学系或心理系的毕业生，再进一步深造学习海洋学某一方面专业，取得硕士学位后，再进行海洋地球物理、海洋沉积、海洋地球化学或海洋微体古生物学，取得博士学位(或进行过一定实践的相当于博士学位水平)的人才能成为海洋地质的研究人员。大学毕业生主要从事技术员工作。技术员安心于本职工作，按科学家的工作计划从事技术操作、实验或搜集整理资料。因为只有科学家提出的课题通过并得以完成，才给予他们工作的机会，而且工作多年后，可提升为自然科学家或再上学深造，但大多数人只愿意从事技术员工作，工作后即可回家享受家庭生活，不必再像科学家那样不分早晚地苦干。科学家亦重视技术员的工作，互相尊重，两相彰益。

二、研究工作方式与特点

研究课题范围涉及到海岸、大陆架、大陆坡、深海平原、洋中脊与深海沟。可以说是概括整个海洋环境，内容涉及到表面地质(地貌与沉积)与现代地质过程；深部构造与发展历史。但偏重于大陆架表面地质及大陆边缘的沉积与构造。

研究课题的选择，在初期多为自选题，可选任一海域工作，范围涉及太平洋、大西洋以及加勒比海。可能由于联合用船的缘故，工作一开始即具有多学科协同合作的特点。逐渐地，研究课题多半为解决实际任务而拟定。如：

(1) 北冰洋与极地诸岛考察。1959 年，苏联的军用飞机在极地岛屿着陆，这些岛屿最早为挪威人发现，美国在极地岛屿有基地。但北极诸岛紧连加拿大海域，为维护该群岛主权，制定了“极地大陆架研究计划”，这是多学科的合作项目，大部分加拿大海洋地质学家参加了此项工作，已完成极地群岛海岸与蒲福海陆架的调查。现今，为开辟北方航道与石油开发等问题，仍在进行极地、罗蒙诺索夫山脊、兰卡斯特海峡、巴芬湾以及戴维斯海峡等处的调查工作。

(2) 石油资源开发。由于中东石油价格上涨引起国内消费价格和薪给上涨，促使加拿大近年来加强了本国石油资源的开发利用，除阿尔贝塔州的陆地石油资源外，已加强了对蒲福海、纽芬兰大浅滩以及新斯科舍陆架的塞伯岛地区油气资源的勘探与部分开发工作。此项任务带出一系列研究问题：如石油开发中人工岛选址、浮冰流动与永冻冰影响管道铺设、海底冰刻痕(Ice Scour)与冰核丘(Pingo)形成与分布规律的研究；新斯科舍与拉布拉多大陆架表面地质研究，由于查明了冰蚀形成的陆架缺少巨厚沉积层的良好储油条件，而带动了对大陆边缘——陆坡与深水扇的研究等。还有些与石油有关的研究课题是为政府服务的，如某石油公司在一海区勘探或打井时，必须将每孔岩芯一分为三：1/3 交省石油部门；1/3 交联邦政府——大西洋地球科学中心的东部石油组；1/3 留给公司自己。交给 AGC 的 1/3 样品，即做检验分析以掌握各石油公司实际工作进展情况。

(3) 美国和加拿大海域边界问题研究。

(4) 深海海床原子能废物埋藏问题研究。这是一项国际合作研究，参加国有美、英、法、荷、加(大西洋)与日本(太平洋)。问题涉及到可否将高位的原子能废物埋于深海床。英、日等岛屿国家已在倾倒废物，而加拿大持反对态度，为此专门选择了大西洋的梭木、尼尔斯深海平原进行沉积过程研究，同时研究环境地质等。

在进行实际问题研究时，也进行了理论研究，有专门性理论课题研究，如大西洋洋中脊、拉布拉多海底扩张、深海浑浊流沉积以及第四纪生物地层学研究等。

到目前为止，各项工作进展情况如下：

(一) 深海研究

(1) 大西洋为主及太平洋区(秘鲁海外，参加美国的工作)的原子能倾废问题。已完成百慕大附近的梭木深海平原调查，该区 5 600 m 海底的 12 m 深处有更新世中期的浑浊流砂层。根据石英砂特征，煤炭岩屑与有孔虫分析皆说明此砂层来自北方加拿大新斯科舍地区，系低海面时的浑浊流沉积。1982 年又进行了尼尔斯深海平原调查。海洋地球化学与热流测验等资料表明两处深海平原有一深海活动层，即该处海底的沉积与其上水层之间有着热力、水分、地球化学元素及生物的活动交换层。因而，即使是深海平原亦不适宜埋藏原子能废物，此项研究仍在继续进行。

（2）深海钻探。参加北大西洋梭木深海平原北部及北非大西洋沿岸，应用钻孔样品进行沉积相与岩矿分析，并结合地震剖面，分析该区中生代至新生代的沉积作用过程。

（3）试验 BIO 设计的电钻。在大西洋洋中脊的“FAMOUS”区域（法美洋中脊研究项目）进行地震重磁测量与试钻，1981—1982 年皆取得生物灰岩下的玄武岩岩芯，工作成果与冰岛附近的洋中脊调查相互比较。

（二）大陆边缘研究

研究集中于新斯科舍—纽芬兰—拉布拉多海盆一线。已总结了该处大陆边缘发展阶段与特点、沉积盆地结构与油气藏等问题。该项研究获得 1982 年加拿大地学重要成就奖。另一项是研究劳伦庭深水扇与大浅滩浑浊流沉积特点，亦取得较显著进展。

（三）大陆架工作

已完成新斯科舍、拉布拉多与蒲福海大陆架表面地质调查与制图，并分析了区域第四纪冰期、冰川活动范围、各地冰退的时代以及更新世与全新世地层分界问题。冰刻痕研究使判断出 1 000 个/年冰山与浮冰活动范围以及对海底的影响，结合极地和大陆架工作对部分罗蒙诺索夫山脊之构造、沉积与发展过程取得大量资料与重要进展。

（四）海岸工作

加拿大岸线长度约为我国岸线的 8 倍。但人口仅及我国的 2%，海岸研究工作开展得极不平衡，工作主要集中于人口密集的河口、港湾与大湖地区。

芬地湾是狭长的由中生代红砂岩组成的港湾岸，潮差大（最大达 16 m），潮流急（15 kn），风浪冲刷红砂岩海崖产生大量红色黏土物质，加上强烈的潮流作用，发育了淤泥质潮间浅滩。当初不了解海岸特性，在温莎尔镇的一个海湾支叉上修建了一条数百米宽的公路坝，结果，仅三年的时间招致了坝前形成方圆数百米、高约 4 m 的淤泥浅滩。这件事教育了工程师们，在计划利用芬地湾潮能发电时，开展了淤泥岸地貌与沉积过程以及生态环境的研究，并预测建大坝后海岸之变化。目前在芬地湾仍进行岸滩剖面定位观测与定期的遥感照片判读分析。芬地湾是一个详细研究的地区。在加拿大西海岸，对福瑞斯河三角洲与附近的峡湾岸进行了动力地貌与沉积作用过程的研究。最近，为研究冰川退缩对泥沙运动的关系，在冰川覆盖的巴芬岛这一天然实验室进行了峡江海岸的调查研究。

海岸工作大致有两类：一种是踏勘性的，如对极地群岛海岸调查，即采用直升飞机踏勘摄影，然后在图上填绘出岩岸、沙岸等简单类型；另一种是小范围的，如一个小湾或一段海滩，作动力泥沙观测，分析其季节变化或估算其输砂量。

海洋地质的研究课题通常是由科学家根据任务和自己的专长提出的。每年 9 月有一个项目汇报会议，常持续 2 天，由各科学家或一个组的代表报告研究课题的目的、意义、进展计划、所需经费及报告上一年成果。当计划批准后，一个项目的负责人常用 1～2 个助手或合作者。在夏季进行野外工作时，为共同使用船只与统一筹划，在一组科学家中常常推举主要项目的负责人为多学科研究组的首席科学家，任期 2～4 周不等，主要在出海工作时起学术上组织领导作用。

三、海上工作方法与装备

(一) 海岸带工作

加拿大地处寒带，为了在短暂的夏季及寒冷气候条件下工作能取得必要的资料，除延长每日工作时间外，主要靠改善装备提高工作效率。现介绍主要的工作手段。

在海岸踏勘中普遍应用直升飞机进行海岸摄影，对照航片进行记录，因此需要准备数只多镜头的摄影机与小型录音机，这样可在有限的飞行时间内做好记录。

陆上调查普遍使用轻型卡车，一个工作组 5 个人(两名科学家，两名技术助手，一名船员)，用两辆越野工具车，装满各种仪器用具，包括橡皮艇与海上装置，车后拖带一卧车，居住、做饭、办公全可在内。这种装备可在沿岸任何地点停留，节约时间，便于调查工作。除一般调查方法外，在详查阶段，在海岸各个地带采用不同方法。

(1) 海滩　用两根 1.5 m 高具有间隔标记的埃默里杆测量海滩剖面，测杆可求出两点间相对高差，两杆间联以 3 m 长测绳，测水平距离，用手水准可测出整个海滩的剖面，一直可测到低潮滩。至水边线处需记下时间以为改正潮位用，低潮线以下需穿潜水防护服进行。

(2) 近滨　用一只 3.6 m 长的橡皮艇上载一小型测深仪测水下剖面，配一小型抓泥器采样。岸上以水准仪观测基线与船位夹角来定位，橡皮艇可以通过激浪带浅水域，将海滩剖面与水下岸坡剖面联结起来。

(3) 外滨　用机动小艇工作，船上有雷达导航系统，一个船员驾驶，其余四人进行海岸工作。当船只沿岸测量时，用一微型导航仪装置、三个无线电发射器置于岸上，船上有一接收器把接收到的信号直接表示为各点与船间距离，从而可定出船位。船上设有回声测深仪、采样器，适于海岸浅水域的旁侧声呐及 3.5 kHz 的地震地层剖面仪，这样通过调查可取得断面地形、底质、海底地貌以及浅地层资料。

贝德福德研究所设计了一种小型仪器“RALF”，投放海底可自计水深、海流、波浪以及拍摄底沙运动的影片。该仪器多半用在经过详查后的海底，以取得泥沙运动的实测资料。

(二) 大陆架与深海工作

加拿大著名的海洋考察船有“巴芬”号、“哈德森”号、“道森”号及“潘德拉Ⅱ”号等，皆为全球性的航船，具破冰能力。“潘德拉Ⅱ”号为下潜器“PISCES Ⅳ”号的工作母船。各船性能装备类似，但以“哈德森”号最好，该船造于 1963 年，全长 90.4 m，4 870 t，可载 62 名船员与 25 名科学家，有 5 个固定实验室及 2 个活动实验室，船甲板有直升飞机降落场，前甲板有长 7 000 m、直径 20 mm 的钢缆及能起吊 20 t 重物的装置，右舷有绞车房，后甲板门式吊杆可投放各种观测仪器。

(1) 调查船采用BIO定位系统,这是一种综合卫星导航、劳兰C、雷达、罗经以及船速改正的资料,每分钟通过电视屏显示出电子计算机提供的经度、纬度、航向与船速等资料。同时,这些定位资料均记录在磁带上。在5 000～6 000 m深海,为了定出海底取样孔的准确位置,又加用了BIO海底声学导航系统,在一个小范围的海底投放若干只辅助性脉冲转发器,当取样器徐徐地向海底下降时,根据取样器上脉冲转放器的信号与海底各个辅助转放器发出信号的转播时间不同,可以确定取样器位置。电子计算机每隔3分钟就给取样器及船只定位。这方法也可用于小间距断面线测深定位。

(2) 在加拿大海洋工作中已不采用方格取样法。工作开始于断面或航线测深、测地层并伴之旁侧声纳海底摄影。深海工作主要是先测深、测地层,然后用地层剖面确定取样点及钻孔位置,这样可减少时间与精力,且效果较好。在大陆架多采用高分辨率的3.5 kHz的Huntec地震系统。此仪器发声在水中传播快,可穿透100 m厚的沉积层,分辨出0.5 m厚的地层。

在水深200～5 000 m时常用空气枪,误差约1%～5%,超过5 000 m水深为加大动力,则加放气泡,可测出良好的深海平原浅地层剖面。在洋中脊工作时,水深变化急剧,需注意迅速转换测深挡。近来,已开始采用“洋底地震仪”探测洋底表面至60 km深处地层结构。

(3) 取样方法。底质多用抓泥器取表层样,然后用箱式采样器与活塞取样管取保持原结构的泥质柱状样。活塞取样管可获得12 m的柱状样品。在砂砾质海底用震动取样管,所有样品皆按设计好的流程处理。柱状样的1/2按原状保存于冷藏库内,另1/2进行摄影与分析,即进行土力学、热流测试、沉积与地化分析以及微古鉴定。

深海中取箱式样品常遇到困难,但活塞取样管或加长活塞取样管仍可取得超过10 m的沉积柱。

BIO设计了一个轻型电钻,在洋中脊上一次进钻能取出6 m长的岩芯。

(4) 下潜工作。在加拿大进行大陆架、大陆边缘工作中已广泛采用水下摄影及下潜器的直接观测,由加拿大设计的“派塞斯Ⅳ”号下潜器(已生产11艘)已下潜2 000次,其最大下潜深度2 000 m。下潜器配有摄影、录像、采样装置,可乘一名操艇员与两名科学家。用下潜器对特定海域(一般属详查区)进行工程、海底地质与生物学方面的研究,可取得良好结果。

此外,观测冰山漂移及结构,以及在北冰洋工作时,大多使用直升飞机调查。同时用雪橇式的活动实验室,在冰上建立观测台站,通过实验室的底板钻冰成孔,然后将特别设计的小型取样器通过冰孔采取水样与底质,并通过冰孔观测冰下海水的物理特性。

在加拿大,调查船是属于国家的。因船只养护、出海费用大、设备多样,因此,政府要求各单位尽可能充分利用船只及设备,相同地区的项目尽可能排在同一航次中进行。用船时间是由有关项目的科学家与行政单位商定的。船长执行用船计划,并尽力配合首席科学家的工作。各科学家一旦取得用船计划后,即全力以赴地做好准备。出海时充分利用给予的用船时间。若是遇到风暴不能作业,也不能顺延工作日期,而必须按期转给下一个项目工作,故十分珍惜海上的每一分钟。用船计划一旦制定,即不更改。因此,每只调查船一年都工作10个多月。船上设备尽量为科学家服务,并且尽量在生活起居上给予方便，创造一种安静与愉快的环境。

船上实验室设备水平是很高的，达到准确与计算机化。如进行海水温、盐、密（CTD）观测时，伴随着仪器通过海水水层，室内计算机即打出并用图像表示出深度、温度、盐度等数据与曲线。陆上实验室设备也是很先进的，研究人员可以根据需要使用研究所或大学的有关实验室，没有人为的或管理上的障碍。

加拿大的大学或研究所的图书、刊物、资料均较齐全，与各国互有交流，使用方便。一般图书馆均全年昼夜开放，提供充分的利用。各大图书馆均有计算机终端与西方各大研究单位图书资料中心相联，可很快地为读者提供有关资料的目录、索引或复制品。对涉及范围较广的专题研究，只要能提出研究范围与内容的关键字句，那么，经过一周或十天，图书馆可为读者查出该专题的现有文献（10 年或 20 年）的目录、作者及摘要。这就为科学研究提供了有力的支持。

总之，加拿大的海洋学研究属后起之秀。几十年来，从渔业生产开展生物海洋学研究开始，而后为国防、交通服务的物理海洋学，至今是为矿产资源服务的海洋地质学，发展壮大，已建立起一个以中壮年学者为主力的研究体系。其海洋地质学是以结合本国的实际问题为主，同时也重视一些重大的理论问题的研究，如海底扩张、浑浊流沉积机制以及第四纪冰川作用与海面变化等。经过 20 多年的迅速发展，加拿大已在世界海洋学研究领域中居先进地位。中加两国在海洋学发展经历中，有很多相似之处。有许多研究课题具有相似的环境条件，可相互对比研究。例如，蒲福海与麦肯齐河三角洲可与我国的渤海湾及黄河三角洲相对比；芬地湾的泥滩岸与浙江沿岸的泥滩岸；扩张的拉布拉多海盆及沉溺的大陆架与南海及冲绳海槽；圣劳伦斯湾水下峡谷及深海扇与长江、杭州湾以及大陆架沉溺河谷等，分析它们的异同，探讨其成因、发展及其资源的开发等。显然，这类课题的分析研究，对我国海洋地质学的研究亦将是个促进。

图 1　"哈德森"号上的一般项目实验室

图 2　下潜器"派塞斯Ⅳ"号的工作母船"潘德拉Ⅱ"号

核废物安置与海床研究*

二十世纪七十年代以来，随着核能的广泛应用，人们面临着对核燃料废物的处理问题。一些发达的大陆国家采取了将核废物埋藏于结晶岩地盾区的深井中或者藏于岩盐层的洞穴内的办法。而岛屿国家以及西欧某些国家则直接把核废物抛掷于海洋中，这引起了世界各国的关注。海洋是联通的，与生态环境关系密切。因此提出了深度超过 4 000 m 的洋底能否作为置放强放射性废物的场所，投放后对自然环境以及生物圈可能产生的影响，投放的适宜地点与方式等一系列问题。为此，联合国海床委员会专门组织了美、英、法、荷、日等有关国家对这个问题进行调查研究。加拿大于七十年代后期参加了这一研究活动，并于 1980 年起连续地开展了工作。作者曾与加拿大的海洋地球化学家、微体古生物学家、地热学家以及土木工程师一起参与该项工作。西德与瑞士以观察员的身份参加有关会议，注视着这些工作的进展。1980 年以来，每年举行会议进行学术与工作交流[1~3]。

一、各国的观点

(1) 英国　持可将核废物投掷在海床之上的观点。英国使用的是经过再加工的核燃料，因此，核废物放射性以及毒害程度较失效的核燃料低(美国与加拿大的高强度放射性废物是来自失效的核燃料)。同时，由于是短龄裂变物质，残留的放射性与毒害性衰减很快。因此，可将核废物暂时储放于贮存所内约 100 年，核废物冷却后再投放于海床上。所以英国人主张，对于加工后稀释 10%的核废料，经过相当时期的储存，冷却后置于可保持 1 000 年的容器内，可以投放于海床上。1 000 年后，这些含核废物的玻璃最终与海水直接接触时，它所具有的毒性仅仅与自然界中铀的放射性相当。这样，目前有足够的时间进行各种处理办法的可行性调查，其中包括把高强度核废物投掷于海床上的调查。英国认为，到目前为止还没有任何技术上的理由可以反对把核废物置于海床上。

(2) 美国　认为应把核废物埋藏于海床内，因为海水水体可起稀释作用，但远不足以构成屏障使放射性核废物免与生物圈接触。由于对生物圈循环过程的了解很不够，人类已面临着放射性短周期循环所造成的威胁。为了避免放射性核废物与生物圈接触，必须将它储存于自然的屏障内。关于洋底有效自然屏障的选择，美国的科学家认为，玄武岩裂隙过多，而具吸附性的远洋黏土最为适宜。在储存罐周围大约 15 m 以内，洋底氧化黏土层可以吸附所有的阳离子，而阴离子却能很快地穿过黏土层。美国在技术上已可能将阳离子分离出来，储存于非氧化环境的深海盆地或陡深的峡江海湾内。

(3) 法国　与美国相似，但寻求多元化的自然屏障。法国不相信利用工程技术办法

* 王颖：《海洋通报》，1984 年第 3 卷第 6 期，第 84－88 页。

把核废物放在海床上部比自然屏障更有效。强调需从地质学角度来研究多元化的自然屏障，以最终解决核废物的埋藏问题。

(4) 荷兰　认为洋底沉积物应作为防止核废物放射性扩散的第一屏障。

(5) 日本　作为向海底投掷核废物的国家之一，按照他们对于屏障的理解，认为目前尚不存在严重的问题足以进行讨论。

(6) 西德　强调不要向公众兜售把核废物放于海床上的观点。

(7) 加拿大　主张将核废物埋于陆上结晶地盾区的深井内，反对将核废物向深海投放，认为投掷于大洋后难于有效控制。核废物放射性会污染海床底土，再经过物理海洋过程、地球化学元素转移过程以及生物作用等各种途径，影响到人类及生活环境，后果是严重的。但是，加拿大仍进行了海床环境地质调查，藉以了解发生于海底沉积物与上部水层间的自然过程以作出对核废物埋藏于海底利弊的正确判断。

国际舆论似乎认为，放在海床上的办法不好，会造成海底环境变化。即使经技术处理可减少毒害，堆放物也可形成海床上的障碍，会阻碍海底水流，造成对鱼群的吸引等，以致不能求出核废物堆放与海床环境关系间的正确模式。大部分国家倾向于寻求自然环境的蔽障，认为将核废物埋藏于海底是经济有效的办法。因此，寻求作为自然屏障的地质体，推进了深海平原的沉积物以及海床环境的调查研究工作。

二、研究工作概况

由于各国观点不同，因而研究工作各有侧重。

英国较少考虑自然屏障体系而强调研究人为蔽障，如核废料的型式与容器的完整性。英国认为硼硅玻璃是使核废物固体化的最适型式。因此，从事于研究超过 100 年以上还能保持完整性的储存容器。为了防止过分的热熔影响，因此在永远埋藏前，需要一个长达百年的预存期。

英国亦进行投掷场所研究，把可能范围扩大到大西洋北部的较浅的冰碛物沉积区以及距其本土遥远的深海平原。

结合北大西洋动力研究项目，英国进行了物理海洋学调查，连续观测海流以建立北大西洋水流扩散与东大西洋中等规模的海水循环模式。

生物学研究关系到食物链问题，尤其是受放射性影响的生物群对上部营养生物层间的放射性传递问题。同时在爱尔兰海研究了温德斯开核电站流出物的影响，在海洋生物中积累的放射性^{287}Np 与^{99}Tc 的浓度问题，这项研究是颇有成效的。英国从事于核废物储放于海洋环境的可行性研究，其 1979—1980 年度的经费为 0.4 亿镑，至 1980—1981 年度增至 1 亿镑。

美国海底放置核废物项目是由国家能源部的核能源组领导进行的，这个单位负责全国核废物的最终储放工作项目(NWTS)，以及负责调查稳定的地质建造，包括岩盐、玄武岩、页岩、火山凝灰岩以及海洋沉积物。海床沉积物的研究在美国的多元屏障项目中具有重要地位。与英国相反，美国认为容器材料的研究是次要的。

美国认为，把核废物放置海床内的同时，亦应放置一个透度计以测量伦琴射线的穿透力，或者放一个用绞车控制的投放器。这样，在置放核废物后，海面监测船可以控制投放位置以及了解核废物容器被安置的状况。美国的研究计划中没考虑容器回收的可能性，因为如果有必要取回一些容器，他们在技术上是不成问题的。

为了建立核废物投放后对环境影响的模式，美国加强了对核废物容器埋藏点附近两个地段环境体系的研究，一个是近容器地段，另一个是远区段。近容器地段仅包括容器周围 100 ℃等温线内的数米范围。在近容器地段内，具有较大的热力和地球化学元素的变化梯度，需要研究放射性热能对海床沉积物特性的影响。由于放射性热力散失与冷却的时间较长，所以对近地段和远地段海底沉积物都要进行长达 15 年周期的热力实验研究。现已知道，埋藏于海床内 50 m 深处的核废物容器所产生的热脉冲，大约经过 10 年即可穿透沉积层到达与海水接触的海床表面。

美国花在这项海床研究上的经费是充足的，并逐年增加。1979 年度的开支是 3.6 亿美元，1980 年为 5.9 亿美元，到 1981 年增加到 9.2 亿美元。虽然美国不一定将高放射性核废物安置于海床内，但其研究计划是按照在 2010 年将高能核废物安放于海床内的预期目的在进行工作。

法国认为海床的老沉积物保存核废物的性能更好，因此，他们集中研究数百米厚的沉积层。法国在东大西洋开普维德海盆进行了海底地形探测，并应用微型富来啸(microflexichoe)进行地震反射测量，已取得长达 1 200 km 的地震剖面，弄清楚了厚度为 600～2 000 m 的沉积层位置，并通过钻孔资料分析沉积结构特征。

海洋生物学家研究了该处洋底深部的底栖生物群体，生物对沉积物的扰动作用以及放射性的扩散途径。

此外，法国还对沉积物中的^{99}Tc、^{293}Pu、^{241}Am 诸元素的行径和影响进行了追踪调查。

法国的研究经费是迅速增加的，1979 年为 2.5 亿法郎，1980 年则增至 7 亿法郎。

荷兰计划利用盐穹作为储放核废物之场所，但因公众极力反对而转向海床。荷兰政府支持其科学家参加国际海床研究会议及海底地质调查活动，荷兰对京斯海槽和开普维德海盆进行了选址研究。为搜集海床地质和地球物理资料，1980—1983 年研究经费为 350 万荷兰盾。

日本 1978 年开始从事高放射性核废物投放于海床的研究，主要在西北太平洋搜集有关地质、物理海洋与海洋生物资料。因洋底地形变化大，在 36°～26°N，147°～160°W 范围内，选取四个地点进行详细调查。日美曾合作利用美国“维玛”号调查船进行了活塞取样与地层测量。

加拿大从 1980 年到 1983 年用“哈德森”号考察船在北大西洋的梭木(Sohm)深海平原和尼尔斯(Nils)深海平原各进行了两年的调查。同时，参加了美国“维玛”号在太平洋的调查。梭木深海平原位于百慕大群岛东侧，尼尔斯深海平原位于安德列斯群岛的北部，工作目的是寻找一个适宜于埋藏核废物的理想投放区，了解该区的沉积物特性以及海底作用过程，以判断核废物投放后可能产生的影响与治理措施。

三、海床研究的实例分析

下面介绍加拿大对海床研究的内容、步骤与结果来说明核废物埋藏问题。

(1) 寻找并选定一个水深超过 5 000 m 的平坦海底，该海底由未胶结的松散沉积物组成，沉积层厚度至少为 200 m，并具有未经扰动的细粒黏土层。

(2) 在面积约 10 km^2 的选定区，用 12 kHZ 的回声测深仪及 40～50 in_3 的空气枪式剖面仪进行海底地貌与沉积层特性的详细测量。

(3) 应用箱式取样器采集未经扰动的沉积层样品，进行沉积结构、上浮水、氧化还原度以及硫、铁等离子活动分析。应用活塞取样管及重力取样管采集海底柱状样品，进行沉积层结构、X 射线摄影、沉积物粒径、矿物、微体古生物及石英砂表面结构分析。

(4) 分层取样，在船上实验室进行土工实验，并于每平方呎加 20～50 磅的冷氮气压力将沉积物中的孔隙水析出，进行溶解氧化硅、碱度、酸度、铬及其他痕量元素分析。

(5) 对整个水层进行温度、盐度和密度测量，采取水样，特别是测量近底层水样中的悬浮质泥沙含量和化学特性，以便比较海水与底土之间的交互作用。

(6) 测定洋底表层 3～5 m 沉积层内的热流量及整个沉积孔芯各层的导热度与热量的水平分布状况。

(7) 进行海底摄影以发现一些由于海底水流或远洋生物活动所造成的微型地貌，这些地貌反映出海床底土与海水间的交换作用。

(8) 测量距洋底 150 m 及 50 m 处的水流流速与流向。

通过调查，沿 56°W，33°30′N 和 32°30′N 间，找到一块略微向南倾斜的平坦海底（斜坡率约 1.7×10^{-4}）。海床由发育良好的水平沉积层组成，沉积层厚度超过 370 m，但在 32°40′N 以北，有几个相对高度在 200～300 m 的小海山分布在海床上；32°40′N 以南无海山分布，但在海床下 12 m 深处却普遍发育着一层细砂层。

海床表层沉积物为充分氧化的红棕色黏土（Eh 值为＋470～＋180 mV），含水量为 42%～65%，有机炭含量为 0.3%～0.6%，黏土颜色向下变浅，并过渡为具水平层理的粉砂层；黏土—粉砂组合重复出现三次，但粉砂层下面的黏土层呈橄榄灰色。沉积柱中层的黏土含水量为 34%～56%；有机碳含量为 0.22%～0.7%，钙质炭含量为 26.5%～2.6%；氧化还原度 Eh＝＋120 mV。至 11 m 以下粉砂层出现增多，并含细砂；12 m 深处普遍出现砂层。

砂层中有保存完好的亚热带、亚极地种的浮游有孔虫。同时出现了在加拿大东部大陆架广泛分布的底栖有孔虫（*Elphidium excatavum f. Clavta*，*Cassidulina reniforme*，*Islandica helenae*，*Cibicides lobalatus*，*I. norcrossi*，*I. islandica*，*Bulmina marginata*，*Buccella frigida*）。这些种大部分是晚威斯康星冰期的底栖有孔虫，而绝非该深海平原的产物。由于上述有孔虫中出现了 4%的 *Cassidulina Cf. teretis*，因此，该沉积层的时代大致是 50 000～60 000 年。

砂层中还夹有 0.05%的煤屑，鉴定表明该煤屑来源于加拿大东部海底的石炭纪煤层。

砂粒表面有残留的红色，砂层的重矿物组合与加拿大东部的红色砂岩层相同。

种种迹象表明，砂层来源于北方的加拿大海岸与大陆架。用扫描电镜观察石英砂表面结构，确定该砂层系深海浑浊流堆积。在威斯康辛冰期低海面时，强大的浊流将堆积于加拿大浅海区的冰碛物向南搬运 1 000 多 km，而在梭木深海平原形成广泛的堆积。1982

年调查表明，尼尔斯深海平原亦分布着来自北方的浑浊流砂层。看来，即使是大洋底部亦能分布着渗透性良好的砂层，而且洋底并不是想象中的那种宁静稳定的沉积环境。砂层上部沉积物的剪切力，在表层为 0～30 g/cm^2，至 5m 深处为 20 g/cm^2，到 10 m 处为 80 g/cm^2，该海区沉积物的热传导速度约为每年 80～200 cm。

不仅在 12 m 深处有混浊流砂，并且地球化学与悬移质泥沙分析表明，在大洋深处存在着一个“深海边界层”，即在海床表层底土与水层间，水流、生物和热力活动使得泥沙、微量元素以及水、汽、生物之间存在着对流交换以及相互渗透、相互影响的关系。

因此，将海床作为核废物最终埋藏点时，需慎重考虑并采取措施，以防止高放射性核废物通过深海活动层使放射性污染传播开来。

我国历来主张和平利用原子能，自六十年代以来已经应用核动力，现又着眼于兴建核电站，同样面临核废物的安置处理问题。另外，由于岛屿国家已经将核废物投掷于海洋，其影响也是全球性的。我国地处太平洋西海岸，受海洋环境直接影响。因此，介绍世界有关国家对核废物安置与海床研究，对我国具有一定的意义。

参考文献

[1] Buckley, D. E. 1981. Geological investigation of a selected area of the Sohm Abyssal Plain, western Atlantic: C. S. S. Hudson Cruise 80 - 016. 6th Annual Meeting, NEA-Seabed Working Group, Paris, France.

[2] Vilks, G. 1980. Postglacial basin sedimentation on the Labrador Shelf. The fifth NEA-Seabed Working Group Workshop Bristol, U. K.

[3] Wang, Y., Piper, D., Vilks, G. 1982. Surface textures of turbidite sand grains, Laurentian Fan and Sohm Abyssal Plain. Sedimentology, 29(5): 727 - 736.

深海浊流沉积特征与浊流砂表面结构*

一、概 念

浊流(Turbidity Current)是饱含泥沙的密度流,在重力作用下通过或越过一种流体或流经另一种流体之下沿着海底斜坡向低处流动[1]。浊流运动取决于悬浮物的密度、粒径分配、海底坡度与粗糙度、扰动量与上覆层的混合量等。当密度差促进了作用在斜坡上的推进重力,而此重力又超过了固定的底界与运动的上层之间界面上的剪切应力时,流体就发生流动。悬浮体载荷量大则有效密度大,流动快;坡度增大速度亦快。流动越快越湍,就有更多的泥、粉砂保持悬浮状态。但是,悬浮物的密度和黏度过大会妨碍扰动。所以,对于最大的流动能力和容量有一最佳雷诺数。因此,顺坡流动的浊流处在速度、湍流及上下界面的混合之间的平衡中。密度流的下界对于保持大量的粗物质处于悬浮状态是太低了,其上界则因与上覆水的广泛混合而破坏密度流。只有中段适合形成密度流。

二、浊流的类型

据洋底浊流的研究,按其活动特点可细分为四种流[2],各具其沉积结构特点(图1)。

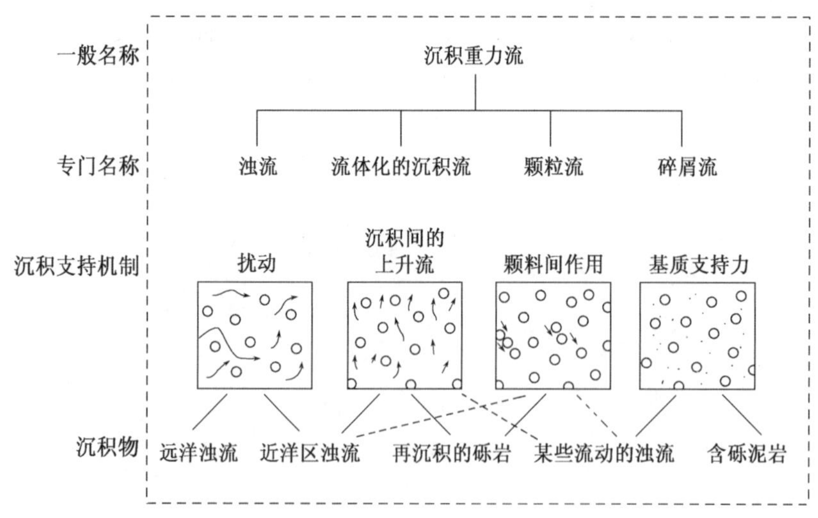

图1 水下沉积重力流分类(G. V. Middleton and M. A. Hampton)

* 王颖:《南京大学学报(地理)》,1987年总第8期,第1-7页。

(一) 浊流(Turbidity Current)

浊流主要是水流的扰动作用。浊流头部具有强烈的扰动与侵蚀作用,体部是快速沉积与尾部的稍慢沉积(图 2)。非常快的沉积成结构不明显的组(A);快速沉积的扩散形成纹层(B),适宜的粒度会形成爬动或旋卷波痕(C)。如沉积过程相当缓慢,则会出现大量拖曳质及上部流所造成的平面波纹层或波痕状交错纹层(D)。Bouma(布马)层序反映着流体的强度减小,以及从下到上的层次反映着悬移质沉积速率减小,而拖移质份量逐渐加大的过程(图 3)。Bouma 层序是理想化地反映着当浊流头部水流破碎时的初始沉积到浊流体以及尾部这一连续过程所形成的沉积。

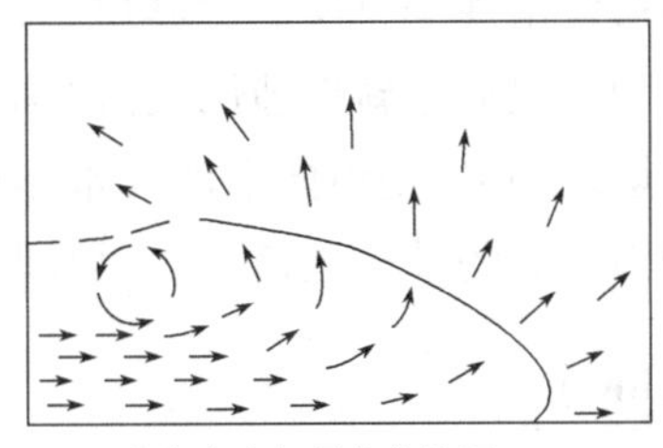

浊流头内与周缘之流况

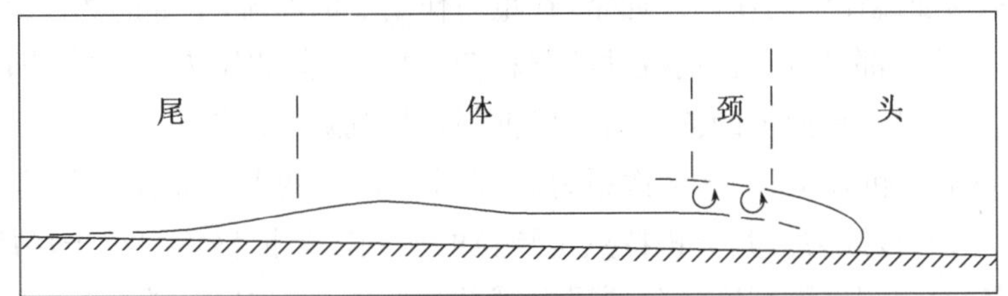

图 2　浊流组成图解(G. V. Middleton and M. A. Hampton, 1976)[2]

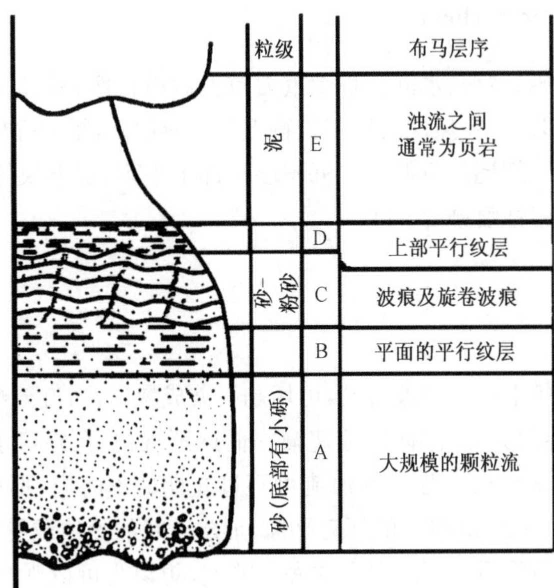

图 3　理想化的布马(Bouma)层序

(二) 流体化的沉积流(Fluidized Sediment Flow)

或称悬浮流。水下沙层结构不稳定,受震动会失去支持发生液化,继之受孔隙水压力及黏滞力之支持而移动,或沿海底缓坡向下流动,或随水流上升,使砂砾质成悬浮扩散分布。这种流与浊流、颗粒流伴生、多发生于水下峡谷的源头[3]。当孔隙水流逸出,泥沙会突出"凝结"而沉积。液化过程形成快速的砂质沉积如沙坎、沙墙或投射状沉积。流体化沉积流多发生于粗中砂层中。沉积层上部多为细砂具有水流向上逸出所形成的孔道与顶部的砂锥结构,状似雏形火山锥与管道,或成平顶的卷曲波痕;中部具有碟形构造(dish structures)[4][5],碟形 4~50 cm 宽,1~2 cm 厚,弯曲形;近底层之碟形构造较平坦,中底处为交织形小碟。碟形构造常出现于厚层的中砂层中,相邻层中出现平行纹层与凸透镜纹层。这种构造出现于加利福尼亚始新世地层及白垩纪到古新世的地层中[6]。碟形构造是一种连续过程中的微弱层理构造,是由向上的失水作用以及同时伴生的上部粘滞力小的砂层对下部的水与黏滞力较大的砂层之负载作用所形成的构造。

(三) 颗粒流(grain flow)

颗粒流是与浊流伴生的作用。通常,在重力作用下的浊流,上部的泥沙是由扰动作用所支持的,而浊流下部具有一快速流动的颗粒流,是由分散的压力所支持,即剪切应力在颗粒间的转移,这是一种向上的支应力可以抵制颗粒沉降的趋势。

F. P. Shepard 和 R. F. Dill 多次报道[3]:在水下峡谷的上游部分常出现颗粒流—"沙河",并发现它具有侵蚀力。其沉积特点是颗粒的放列。当推移重力小于场应力时,泥沙即于当地停滞,而沉积层无粒序或分选层次等其他标志,偶尔保存着印痕、重载或滑落的痕迹,沉积层顶部多平坦,颗粒分布与浊流方向一致。

(四) 碎屑流(Debries flow)

碎屑流是洋底的砾、砂泥、水混合物受重力拉曳后沿自然斜坡的滞缓移动,当推力小于碎屑间的应力时,即发生突然性的停积。特点是物质混杂,粗的碎屑浮于细粒的基质之上。粒径分配曲线有一细尾的负偏态。沉积层顶部不平坦,水下块体运动形式,分选差,顶部具有剪切作用与侵蚀痕迹(图 4)。

三、浊流

浊流沿大陆架坡折下的大陆坡流动,可切割斜坡成峡谷,当流至峡谷末端平坦的洋底时,因内部阻力与边界阻力,悬浮体随流速减小而逐渐沉积,在谷口水深 2 000~3 000 m 处发育深海扇(deep sea fan)。强大的浊流可穿过先期的深海扇,形成扇形串珠体,有的甚至可泛流堆积到达 4 000 m 以深的深海平原,形成席状的沙体或粉砂体沉积。我以为,可以将长时期积累碎屑物质而短间爆发的海底浊流,想象为可以观察到的干燥与半干燥区的山麓洪积扇,浊积扇上也有漫流水道及天然堤,而沉积颗粒比洪积扇细。由于深海环境相对稳定,沿大陆架边缘或斜坡上堆积的泥沙含水量大,其休止角很小(1/1 000 左右)。

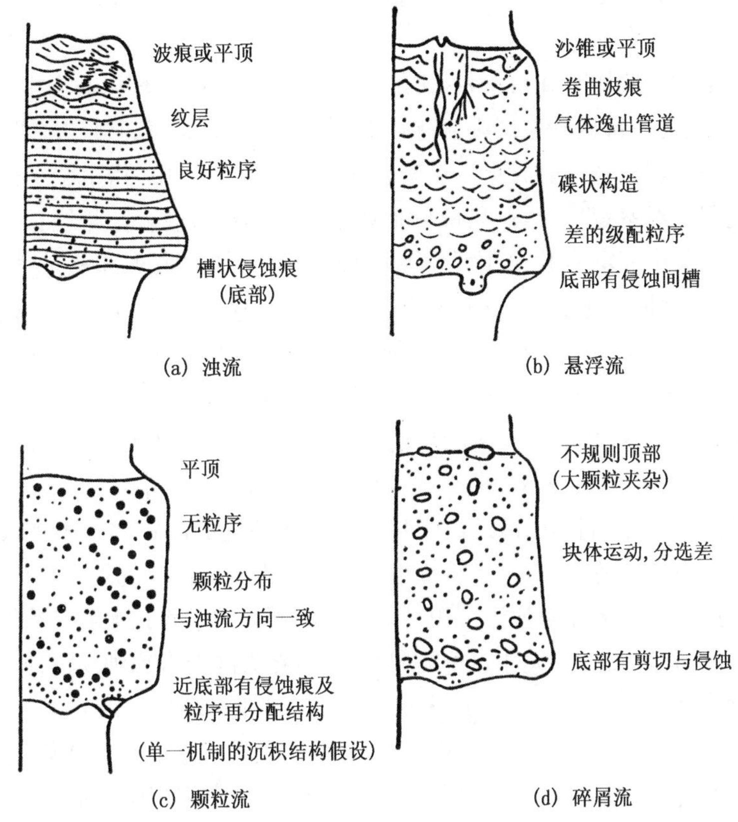

(a) 浊流　(b) 悬浮流　(c) 颗粒流　(d) 碎屑流

图 4　各种流的沉积结构(G. V. Middleton and M. A. Hampton，1976)

在外界扰动(如地震)使沉积物尤其是粉砂层发生液化，甚至在自重情况下沉积物亦可沿斜坡(3°～5°)向下移动。深海扇表面坡度小于 1°，其上发育漫流沟槽及天然堤，有的堤高度可达 100 m 以上。大规模的浊流出峡谷后，向深海平原扩散可超过 1 000 km 以上[7]。

1980 年我在大西洋梭木深海平原(Sohm Abyssal Plain)考察时，从洋底沉积孔芯中发现了砂层。但同船科学家们不以为然，认为该处不可能有砂。但的确是砂，奇怪的是有些海洋地质学家对这种变异都感到不能理解。为什么人们不认为大洋底会有砂层呢？因梭木深海平原水深大于 4 000 m，距大陆已有 1 500 km，用 12 kHz 的回声测深仪与 40～50 in^3 的空气枪剖面仪，几乎测不出红黏土层有粒度的变化。但活塞取样管孔孔均打到此砂层，在 10 km 水深 5 600 m 测区洋底的 12 m 深处普遍有一砂层，是个“席状砂体”。

梭木深海平原，在 56°W，33°30′N～32°30′N 间，洋底平坦微向南倾斜(坡度 1.7×10^{-4})[8]。海床底发育良好的水平沉积层，厚度超过 370 m，在 32°40′N 以北，有一些相对高度 200～300 m 的海山，32°40′N 以南无海山，但海床 12 m 深处却普遍有一细砂层。该处洋底表面沉积物为红棕色黏土，含水量 42%～65%，有机炭含量 0.3%～0.6%，表层氧化还原值+460 mV，黏土颜色向下变浅，并过渡为具有水平层理的粉砂层。黏土-粉砂组合重复出现 3 次，但粉砂层下面的黏土为橄榄灰色。沉积柱中层(深 4 m)的黏土含水量 34%～56%，有机炭含量 0.22%～0.7%，钙质碳含量 26.5%～2.6%，氧化还原值 Eh=+120 mV。橄榄灰色黏土层底部有粉砂层，这与表层红棕色黏土不同。红棕色黏土有黑色碳质斑，但绝无石英质砂。至 11 m 深粉砂薄层增多，并含细砂；12 m 深处普遍出现砂层。

在粉砂层中发现底栖有孔虫[9]，这又是一个异常，在大陆架环境我们多次在淤泥层中见到有孔虫，而此处是每遇砂质层则含有孔虫。在 5 600 m 深海超过碳酸盐补偿深度，该深度以下不应该出现完整的有孔虫。这些有孔虫保存在砂层中，指示了它是来自与砂层一致的动力过程。经鉴定后，是亚热带与亚极地区的有孔虫混合。是经长途搬运过程来自不同气候带。其中底栖有孔虫不是深水品种，而是与加拿大东部斯科舍(Scotia)大陆架环境的有孔虫相同。大部分是晚威斯康星(Late Wisconsinan)的沉积：*Elphidium excatavnm f. Clavta*, *Cassidulina reniforme*, *Islandica helenae*, *Cibicides Lobalatus*, *I. norcrossi*, *I. islandica*, *Bulmina marginata*, *Buccella frigida*。其中有 4%的 *Cassidulina Cf. eretis*，这是晚威斯康星前的沉积，在哈德森海峡这个品种时代为距今 60 000 年，巴芬岛(Baffin Island)是距今 50 000 年前，但在斯科舍是不老于 30 000 年的[9]，总之是最后一次冰期的中后期沉积物。在深海环境的沉积物中出现大陆架浅海环境的有孔虫。与当地环境不同，这个特点很重要。要重视沉积物的综合分析，尤其要注意寒冷相的有孔虫。有许多深海浊流沉积是来源于陆源碎屑堆积。尤其是冰期低海面大陆架边缘沉积。大西洋西北部，大冰盖着陆域及海域，在大陆架有冰蚀地形使内陆架崎岖不平，中部有深槽 300～700 m，是冰盖边缘水流沿冰蚀洼地的侵蚀，或沿地层交界线或构造交界处侵蚀而成。外陆架是冰水扇沉积，大量松散物质堆积于外缘与坡麓，这与中国陆架不同。当冰后期海面上升时，这些沉积物被冲刷，沿陆坡下泻堆积成深海扇。这就是在这一特定的历史环境背景下，于稳定大陆边缘发育了大规模的深海扇。浊流作用在现代以及其他地质时期也有发生，但大规模的洋底浊流沉积与深海扇，是形成在更新纪末与全新世初期，在有大量碎屑物质堆积及海面上升的条件下。

在梭木深海平原的砂质沉积层中，尚含有 0.05%的煤屑，经鉴定系加拿大东部悉尼(Sydney)煤田石炭系煤层中的煤屑。

浊流砂表面锈红色，与石炭纪的红砂岩层物质一致。砂层的重矿物组合与新斯科舍-劳伦庭深水扇(Laurentian Fan)的重矿物一致。

综合物质组成表明，该砂层来源于东部加拿大大陆架，经深水扇搬运至 1 500 m 以远的深海平原，这是海水底层浑浊密度流沉积，而非深海静水沉积。时代是晚冰期中后期，沉积层结构是浊流沙层与远洋红黏土呈互层，即：多发性的、来自大陆架浅水环境的浊流砂夹着浊流间断期沉积的深海红黏土，这就是远洋浊流的特征。

四、深海浊流砂表面结构

在确定了梭木深海平原的砂层为浊流沉积后，对该砂层石英砂表面结构作了研究，企图求得浊流砂表面的特征结构。

坚硬的石英砂表面保存着其经历的风化、侵蚀、搬运与沉积作用的标志。微小的砂粒也像地球表面一样，具有高处侵蚀低洼处沉积的特征。用扫描电子显微镜观察与摄影，分析这些表面结构特征，可以帮助恢复沉积物来源、搬运过程与沉积环境。

梭木深海平原呈T形，北接斯科舍大陆架与纽芬兰浅滩(Grand Bank)大陆架，向南伸到百慕大群岛西部，由三个浊流体系所补给：西北部缅因湾外水道体系、北部圣劳伦斯湾体系、东北部大西洋洋中峡谷浊流体系。样品取自水深5 600 m，洋底下12 m。

砂样经筛选取出细砂(2ϕ)，用双目镜选取石英砂，经过去污、镀膜处理，并经电子探针鉴定，用作扫描电镜的观察与摄影。

梭木深海浊流砂的特征[9,10]，主要是：

(1) 具有不规则的形态，受晶形控制。虽在海洋环境，从大陆架、大陆坡至深海平原搬运距离很长，但在陆上及海岸环境的搬运距离较近，而保存着原始晶体形态。

(2) 圆度，次棱到次圆，这与搬运距离及搬运特性有关。浊流砂为基质，水流及其他颗粒所支托，悬浮于物质流中，加之颗粒小，所以与流水搬运的底砂不同，虽经1 000 km以上长距离搬运，磨圆程度仍差，但颗粒间撞击痕多。

(3) 特征标志。① 砂粒具有贝状断口与碟形洼坑、平行阶、擦痕，这些是冰川作用的标志，反映浊流砂的起源，但这类标志仅保存于洼坑之中，系残存形态。砂粒的棱脊已经磨蚀，而与冰川沉积有区别。② 具有大量v痕与机械撞击点，叠加于原来的冰川营力标志上。v痕密度大。海滨砂多v痕，但却无洼坑，因全被磨蚀掉，河流砂v痕密度小，形态不同。

(4) 具成岩风化标志。系水下沉积。均有溶蚀孔与次生沉淀物，叠加在各种标志形态上。

梭木深海平原的浊流，主要是颗粒流，颗粒流搬运细砂至深海平原，细砂互相撞击而具v痕且密集。悬浮流在小的浊流爆发时，并不溢出谷道。只是在大的浊流爆发时才越过天然堤，并沿谷系至远方，北方种属的有孔虫即随悬浮流带到南部洋底沉积的。小型浊流，仅在劳伦庭峡谷末端出口处沉积深水扇；中等程度浊流可达到近端深海平原；高浓度强力的浊流可到达远洋深海平原如梭木深海平原的南部。深海平原的浊流砂，主要来源于冰期低海面时的冰川沉积或冰川海洋沉积。砂粒具有经寒冻风化作用解析出来时所具有的原始晶形或冰川挤压所形成之贝状断口。冰后期海面上升过程中，海底突发性的浊流作用频繁，物质浓度大，使颗粒互相撞击而形成密度大的v痕，而使冰川擦痕、贝状断口等标志仅在蝶形洼坑中得以保存，棱脊则经磨蚀成次棱或次棱到次圆级。由于堆积于水下环境并经初步成岩作用变化，叠加了溶蚀孔与次生硅堆积。深海浊流砂表面结构记述了这一沉积动力过程。

参考文献

[1] F. J. Pettijohn, P. E. Potter, R. Siever. 1977. 砂和砂岩. 李汉瑜译. 北京：科学出版社，279 - 324.

[2] Middleton, G. V. and Hampton, M. A. 1976. Marine Sediment Transport and Envionmental Managent. John Wiley, New York. 197 - 218.

[3] Shepard, F. P. and Dill, R. F. 1966. Submarine Canyons and Other Sea Valleys. Rand McNally Co. Chicago. 381.

[4] Wentworth, Carl M., Dish Structure. 1967. A Primary Sedimentary Structure in Coarse Turbidites (Abstract). *American Association of Petroleum Geology Bulletin*, 51: 485.

[5] Stauffer, P. H. 1967. Grain-flow Deposits and Their Implications, Santa Ynes Mountains. *California*: *Jour. Sedimentary Petrology*, 37: 485 - 508.

[6] Chipping, D. H. 1972. Sedimentary Structure and Environment of Some Thick Sandstone Beds of Turbidite Type. *Journal of Sedimentary Petrology*, 42: 587 - 597.

[7] 任明达,王乃樑编. 1981. 现代沉积环境概论. 北京:科学出版社,157.

[8] 王颖. 1984. 核废物安置与海床研究. 海洋通报,3(6):84 - 88.

[9] Wang, Y., Piper, D. J. W. and Vilks, G. 1982. Surface Textures of Turbidite Sand Grains, Laurentian Fan and Sohm Abyssal Plain. *Sedimentology*, 209(5): 727 - 736.

[10] 王颖,B. 迪纳瑞尔. 1985. 石英砂表面结构模式图集. 北京:科学出版社,8.

大西洋:过去、现在和未来*

国际地理学联合会(IGU)98年区域会议于1998年8月28日—9月2日,在葡萄牙里斯本召开。会议的主题是:The Atlantic:Past,Present and Future,与1998年为国际海洋年相适应。同时,在里斯本新建的展览中心举办了20世纪末规模最大的国际海洋博览会(Expro'98),参展的国家有121个,国际组织17个,到9月初,参观的人数超过600万。

海洋博览会由Independent World Commission on the Oceans(IWCO)的43个成员国支持,以"海洋与蓝水星球的遗产"为主题,强调人类与占地球2/3面积的"水球"的关系。因1998年是葡萄牙Vasco do Game(达迦马)首航印度的500周年纪念,IGU98区域会议突出蓝色海洋,会标反映出此特色。

出席会议的代表约400多人,来自44个国家,中国有5名代表(南京大学2名,台湾师范大学3名,陈国彦教授、徐胜一教授及张瑞津教授)。有20名各方面的代表人物作特邀报告讲座,其中法国代表4名,美国3名,英国2名,意大利2名,中国1名,即,我本人,其他为国际组织的代表。

会议的第1个专题为大西洋:航海之路与不同文化交流。有两个紧扣主题的特邀大会发言,继以两个分组报告会,各有8人发言。分会结束后,又继以一个小时的专题报告大会。

第2个专题是大西洋:海洋与海岸的特点与管理。有3个特邀的大会发言。一个小组报告会,一个含6项内容的展贴会和一个大会报告。

第3个专题是大西洋地区区域间矛盾和合作。有4位在大会作特邀报告,2个分会报告。

第4个专题为:全球与区域间的管理。我是应邀参加此会的。

大会特邀报告共有5个:第1位是葡萄牙女律师Maria Eduarda Conealaes,报告有关海洋法问题。第2位报告精彩,是美国Delaware大学的教授,著名的海岸管理、海洋海岸管理杂志主编,Biliana Cicin-Sain,她系统地阐述了国际海岸管理工作的发展阶段、特点,强调在区域、部门与专业方面的结合或一体化过程。第3位发言的是国际海滨组织东亚区域项目负责人Ehua Thia-Eng教授,他系统地阐述了东亚地区在环境挑战方面的政策与促进一体化的工作,和每年区域工作实际,讲演很精彩。

王颖为第4位发言,综述东亚海岸一体化管理的问题,以中、日、韩三国为例,阐述黄、东、南海的环境、资源、人口、经济增长的特点以及所带来管理的问题,指出在东海大陆架与南海岛礁之纷争,是大陆架窄狭的国家希冀以岛礁为据掠取海域资源。而历史与国际法原则表明这些纷争的岛礁与海域是中国的主权范围。目前应采取中国政府所提出的"搁置争议,共同开发"的缓解矛盾而促进环境持续发展的有益途径。在东亚区域项目选

* 王颖:《国际学术动态》,1998年第11卷,第74-77页。

厦门为例，用一体化管理办法减轻海洋环境污染、优化海岸环境的实例，证述区域海岸管理的有效办法。报告中扼要阐明东亚区域海岸管理的实质症结以及应采取的科学态度和有效办法。报告会后，会议主持人 Jorge Gaspar 教授表达了他诚挚的合作愿望。Jorge Gaspar 是葡萄牙地理学会理事长，本次大会科学组主席，是"葡萄牙区域地理"专著撰稿人，近 60 岁。他曾多次赴中国澳门工作，来过中国大陆，我表示欢迎他到南京大学访问的愿望。

第 5 位报告人是 IGU 副主席、意大利 Genoa 大学的 Adalberto Vallega 教授。他的发言很扼要，后在闭幕式上又做了系统发言，他强调 21 世纪海洋对人类生存、持续发展之重要性，地理学应和海洋学结合建立赋有多学科特性的 Ocean Geography(不是原提的 Marine Geography)。海岸含海岸陆地与海岸海洋两部分，其余为深洋，应摒弃陆地地理与海洋地理截然分隔的老概念。他提出 Ocean Geography 包含一体化的海岸地理、深海地理与海洋区域地理，建立跨部门的物理、生态、经济、地缘政治、海洋地理信息系统体系等。在闭幕式上他向 UNESCO 提议正式建立加强跨部门的 Ocean Geography 体系的工作，获全体代表起立鼓掌通过。会后，他邀请我参与 Ocean Geography 方面编委工作。

第 4 专题为大会的核心部分，下设 3 个分会报告课题：① 人类对海岸边界之压力，含 6 个报告；② 海洋资源——开发与管理，有 5 个报告；③ 人文地理，有 8 个报告。

第 5 专题为全球变化、海面上升对海岸带的影响。

有 4 个特邀大会报告，分别为荷兰 P. Augustinus 教授的"海面变化对欧洲海岸的潜在影响"，他是国际科联下属的海洋学研究委员会，在我担任主席的第 106 组的成员，专门研究海平面变化与淤泥海岸。他以荷兰为实例讲明海平面上升使岸陆沉降所采取的措施，科学总结性强，但未判明海平面变化的未来趋势；巴西 Dieta Muehe 教授，讲述海平面的变化对大西洋、南美海岸之潜在影响，为一般常识性讲演，附以巴西海岸类型图片；比利时 Jacques Charlier 讲海平面变化对港口的影响，科普性强，手绘草图，讲演很一般；德国 Hannover 大学的 Hanns Buchholy 讲述海平面上升对海岸的影响及教育，他概括海岸体系具有海陆交互作用，自然与人类构造过程的双重影响，从各学科交叉进行综合海岸教育，从认识过程进入解决问题与加强管理，来对待海平面上升的影响，此报告有力，接触到核心问题，并提出了针对性的办法。

第 5 专题下设 5 个分课题：① 海岸体系与利用间之影响，7 个报告；② 环境与应用地理，7 个报告；③ 相对海平面上升对海岸地带的影响，4 个报告；④ 有关持续的海岸与海洋管理的教育，3 个报告；⑤ 小岛屿，3 个报告。

此外还有 1 个展贴分组，包含 4 项研究成果：巴西 2 项，南非 1 项，中国台湾师范大学张瑞津教授 1 项(关于台湾海岸平原地貌变化)。在分会后，由美国新泽西大学、海洋与海岸科学中心主任 Norbert P. Psuty 教授做一般性大会讲演：海岸地貌中的时、空研究，内容及图表均完臻优秀。

整个会议的报告已进入到将海岸研究成果付诸策略与管理应用阶段，尤其是第四专题中 Biliana 与 Ehua Thia-Eng 的讲演中的图表展示，是科学性与表达效果恰到好处的优秀范例，表明当前海洋管理研究工作有了显著的进展。我本人的报告，在内容上下了功夫，达到了简明扼要，使听众有新知的目的。

大会闭幕式的出席人数增至近 400 人。大会主席与 IGU 主席秘书长均肯定此次为一成功的会议，并对欧洲参加地理奥林匹克的大学生优胜者：波兰、葡萄牙、法国等队成员颁奖，大会秘书长提出希望 2000 年在韩国举行的地理学大会上能有亚洲国家的学生参加地理学奥林匹克比赛。我不了解此比赛的规程，但深望能将此信息通过地理学会带给中国青年大学生，争取及早准备参赛。IGU 副主席 A. Vallega 作了“海洋学与地理学结合”的学术讲演倡议，获全体与会者的支持与赞同。在闭幕会上的学术讲演，将会议又推向一个新的高潮，这是与过去不同的。

在 IGU 会议期间，举行了新一届国际地貌学(IAG)会理事会，我和其他理事共 13 人出席了会议，缺席会议的 4 人：Barsch(德国)、Netto(巴西)、JeJe(尼日利亚)、Benazzonz(阿尔及利亚)。会议由主席 Olive Slaymaker 教授(加拿大)主持，同时，还有副主席 M. Panizza 教授(意大利)及秘书长 Piotr MiGon 博士(波兰)协同会议事项。会议讨论的内容要点如下：

(1) IAG 2001 年大会。IAG2001 大会定于 8 月 23—28 日在东京私立中央大学举行，中央大学距日本成田机场 60 km。经费方面日本代表强调“低花费，高质量”的办会原则，会议前安排了 7 条考察路线。

(2) 区域会议。① 通过 1999 年 7 月 17—21 日于巴西里约热内卢(Riode Janeiro)举行 IAG 区域大会，届时有为期两天的执委理事会议。会议有 8 项专题：流水地貌、构造活动与景观发育、海岸地貌、土壤与地貌、坡地地貌含谷地侵蚀讨论会、地貌学在环境与管理中的应用、遥感地貌过程与地貌学研究、喀斯特。会后考察安排 5 条路线。② 2000 区域会议，中国南京。会议主题，建议为：季风过程与人类活动的影响，后者在中国尤为特色，包括：水动力过程及地貌，主要是河流(长江、黄河)，海岸以及河海交互作用，冰川作用；黄土及土壤侵蚀作用与地貌；喀斯特作用过程与地貌；地貌学在工程与管理中的应用与信息系统等。野外考察集中于几条路线：A. 黄山—杭州湾；B. 长江三角洲，宁、镇、太湖、苏、沪；C. 九寨沟、海螺沟、冰川及路南石林等；D. 黄土高原、陕、甘、新；E. 长江、三峡。

(3) 奥地利的 Christine EMBLETON-Hamann 女士为 IAG 的出版负责人，她负责 IAG Newsletter，每年出 4 期 Newsletter，3 月号(截止收稿日期为当年 1 月底)、6 月号(4 月底截止收稿)、9 月号(7 月底截止收稿)和 12 月号(10 月底截止收稿)，寄给 IAG 主席与各国家代表，原各寄 3～4 份，因印刷及邮费开支大，现决定每个国家各寄一份，由各国家代表再分寄。希望中国的同行能投稿，内容可包括会议进展、研究进展和其他与地貌学有关的活动，扩大影响。中科院南京地理与湖泊研究所的陈志明研究员曾投稿两次，报道中国地貌图与亚洲地貌图事，Christine 在会上多次赞许此事。本次理事会她代表陈志明研究员提出申请资助出版亚洲图，IAG 副主席意大利的 M. Panizza 教授代表俄罗斯提出资助出版世界地貌图。该事例说明撰写通讯报道的有益影响。

(4) IAG 制定了“宪法”章程草案，将分发，供讨论。

国际海洋博览会盛况空前，近 140 个国家与国际组织用不同手法展示了海洋环境、资源与人类发展的关系。我两次前往，看了美国馆：展示阿拉斯加的天然冰、Scripps 与 Woods Hole 海洋考察、Titanic 号之发现等，突出重点而并非样样俱展，由美海军负责展项。中国馆大、气魄宏伟之中华门内展示海上油气开发、西康卫星发射站、郑和下西洋与昆明国际花卉世界博览会等，展示内容好，但缺乏英文说明，多为中文说明，虽有英文讲解录音，但效果差。中国澳门地区馆两部分：现代建设与文化传统，后者为庭院建设与工艺品，吸引人排长龙。亚洲各国中以日本馆、南美以阿根廷馆引人入胜，每日长龙，日本立体电影讲海洋生物链故事，非常有趣。

里斯本地区海岸类型丰富。北部欧洲之角为欧洲最西端—CABO DA ROCA，大理石水平岩层的海岸悬崖陡峭，大西洋涌浪翻滚，至岸边溅落，抛起砂砾，景象壮观。CASCAIS 与 ESTORIL 仍为水平砂岩与大理岩基岩岸，但高度由北向南降低。海蚀型海岸岬湾交错，海湾深度小，沿岬礁人工筑堤挡浪，海湾内人工填沙喂养海滩，成为海滨浴场，游人如潮。游泳与帆船活动盛行，由于受自然条件限制，陡坡、基岩岸、激浪不发育，故无滑板或 windsurf 活动。

里斯本市沿海（西部）傍河而建，该河受构造控制，在基岩夹峙处具河道形状，在平原开阔为湖泊或海湾而形成串珠状河系。里斯本港扼海峡出口建设，里斯本市南部均沿海湾发展，街市清洁，古堡建筑保存完好，配以新建筑的文化广场，新桥与旧桥相映，形成新的旅游区。海湾南部有盐田、渔港及农业区，新桥建立已使原海湾渡船事业渐向旅游业转变。Expo′98 展馆建筑费用高达 20 亿美元，为欧洲几个大型项目之一，由欧共体支持，以发展欧洲经济，拟接待 400 万人参观，每票约 30～40 美元，形成新的旅游业点。

里斯本市中心为老市，系经济中心，街道狭窄，块石铺路，以适应年降水达 800 mm 与地面起伏不平之自然特点。块石路与 20 世纪早期的经济实力相适应，旧建筑维持好。里斯本大学在新老城之间，由理学院、文学院几个建筑组成，学生 600 人左右。葡萄牙人饮食文化、习俗有古老传统，中国人易于适应。他们食大米与面粉，相对于北欧或西欧人士，位居欧洲南部的葡萄牙人个子多低矮。

红树林海岸*

一、红树林的特性与分布

(一) 概述

红树林海岸是指热带地区由红树丛林与沼泽相伴而组合成的一种海岸，是一种生物作用的海岸。因此，研究这类海岸时必须注意到有关植物的问题。

红树林是由若干种并不相近的常绿乔木及灌木所构成的。但它们有类似的生境要求与相似的外貌。这些植物共有二百余种，其中有代表性的是：红树属(*Rhizophora*)、海榄雌属(*Avicennia*，亦称黑红树)、野薮木属(*Sonnertia*)、拉贡木属(*Laguncululaia*，亦称白红树)和锥果木属(*Conocarpus*)。这些植物有某些特殊的生态，能适合于经常浸泡盐水的环境。因此，它们常分布在低缓的冲积海岸的潮间带内，形成一种特殊的植物海岸。

红树林是一种植物群落的复合体，它可以有高达 30 m、直径 1 m 多的乔木(*Rhizophora* 等)，也有低矮的高约 1～2 m 的灌木丛(*Avicennia* 等)。同时红树植物在生态上表现出某些适应环境的共同特性。

红树植物的枝叶肥厚，叶子暗绿而光亮，它能抵抗热带地区强烈的阳光。同时，为适应长期在海水中浸泡与在淤泥等缺乏空气的环境中生活，红树又具有极为发达的气根系统。一种是“支柱根”，它将树干高高举出地面，加强了树干在淤泥中的稳定性。另一种是“下落根”，由枝干向下生长而插入泥内，它们在低潮时出露。一些高的下落根长 3～5 m，交叉分布，在泥滩上构成难以通行的“植物迷宫”。另外，海榄雌红树具有芦笋状的尖芽，它露出地面不高，但却阻挡人们的通行。

红树植物是“芽生”，种子果实大似鸡蛋，当种子还未脱离母体时就在果实内发芽，芽长 20～70 cm，粗如手指。当树木受海水推动或风吹摇动时，果实落下，幼芽插入淤泥中；假使没有被波浪、潮流扰动，那么它便可以在那里生长巩固起来，成为新的红树。种实也可以随潮水漂流传播。

(二) 红树植物的生境条件

据 R. C. 威斯特等人的研究，红树林在下述的环境条件下繁殖得最好。

(1) 热带的温度：发育良好的红树林需要热带温度，最冷月的平均温度需超过 20 ℃，季节温度差不能超过 5 ℃，海水年平均温度为 25～28 ℃。

(2) 细粒冲积物的底质：红树林在三角洲沿岸或靠近河口区发育最好，因这里具有丰

---

* 王颖：《地理》，1963 年第 3 期，第 110 - 112，138 页。

本文曾经韩慕康同志审阅，提供宝贵意见，特此致谢。

富的细粒粉砂与黏土所组成的软泥，并有丰富的有机质，有利于幼苗生长。也有些红树生长在其他底质上的，如澳洲东岸与我国海南岛红树林生长在珊瑚礁上；印度孟买湾象岛附近，红树生长在玄武岩上。但在缺乏丰富淤泥处，红树往往生长得比较矮小稀疏。

(3) 需要不受波浪强烈作用的海岸带：红树植物不能忍受强烈的激浪作用，尤其是红树的胎萌幼苗在海面漂浮及刚扎根的时期，一旦红树成长为成熟的乔木红树林之后，由于根系所固结的土壤大大加厚，乔木的支柱长得非常稠密和牢固，只有极猛烈的风暴才能把它们吹倒。一般说来，海岸沙坝的后侧和三角港等隐蔽处红树林繁殖得最好。

红树植物在具有丰富均匀降水的热带海岸上发育得很好。因大量降水的地区就有很多含泥沙丰富的河流，它们使海岸、河口地带有大量泥沙沉积。

盐分并不是红树植物生长的特殊要求，有一些红树（特别是 *Rhizophora*）在淡水中生长得比在盐水中还要好一些。因此，红树能耐盐，并不是喜盐。

（三）红树林海岸的分布

红树林海岸大致分布在南北回归线范围内。局部地区如墨西哥湾沿岸、莫桑比克海峡等处，由于暖流的影响，分布的纬度偏高一些，最高的可达 32°N。

在赤道带的海岸带分布着高大茂密的红树林所成的海岸。例如美洲的哥伦比亚和厄瓜多尔太平洋海岸，西非几内亚湾的喀麦隆海岸与东尼日利亚海岸，东非的坦喀尼喀海岸，亚洲的马来亚南部、巽他群岛等沿岸地区。

低矮的灌木红树林海岸分布范围比较广泛。在美洲有佛罗里达海岸，古巴的西南部与北部，墨西哥湾沿岸，中美洲的东西岸，南美洲自委内瑞拉至南回归线以内的巴西大西洋海岸。非洲自东北部红海海岸至莫桑比克海岸及马达加斯加岛西岸。亚洲是印度、中印半岛、伊利安岛。此外，还有澳洲的北部、东部海岸。在这些地方，沿着低平的冲积海滩上皆镶嵌着低矮的红树灌木丛林。

我国主要是低矮的红树林，大致从福建省的福鼎开始一直向南，沿着大陆海岸的低平地段，河口、海湾，以及海南岛沿岸及台湾东部海岸都有分布。在海南岛文昌、儋县、榆林红树生长良好，形成高达 6～10 m 的丛林。随着纬度增高、气温降低，组成红树林植物的种类也逐渐减少，红树林变得低矮稀疏。到 24°N 以北，主要是灌木丛了。近几年来，浙江南部海岸引种了红树，在永嘉到平阳海岸已获成功，形成一片 1 m 多高的红树丛。据研究大致在温州以南均适宜红树植物生长。

总观世界红树植物，可大致为两个群系。

(1) 西方群系：树种比较简单，分布于美洲西印度群岛及西非海岸。

(2) 东方群系：树种丰富，包含有西方群系的树种，分布于印度与西太平洋海岸。

澳洲的红树在某些方面与东方的红树群系不同，成为一个亚群系。

二、地貌特点与海岸演变

（一）海岸特点

红树林分布在低平的堆积海岸的海滩上，主要分布在潮间带内，并且常常沿着河口或潮水沟向内陆深入数千米。典型的红树林分布在背风浪而正在向海扩展的淤泥海岸上，由于它要求避风与淤泥质滩地，所以与潮汐作用关系密切。

红树林岸最主要的特征是，海岸可以划分为一系列与岸平行的地带。每个带里各有特定的植物群落与地貌发展过程，而且各带的地貌演变与植物的更替交织在一起互相影响。这些地带按从海向陆顺序如下：

(1) 浅水泥滩带：位于低潮水位线以下，即相当于海岸水下岸坡之上部，是淤泥质的浅滩，海水很浅。

(2) 不连续的沙滩带：位于低潮水位线附近，是一系列被潮水沟、三角港或泥滩所分隔开的沙滩。

(3) 红树林海滩带区：宽度各地不一。例如在哥伦比亚太平洋沿岸是典型的红树林海岸，它的最大宽度达 30 km，而在最窄的海蚀岸的背迭海滩上，只有几米宽。通常红树林区海滩带的宽度是 1～5 km。

(4) 淡水沼泽带：位于红树林海滩带区后侧，而经常受到潮汐的影响。它的内缘，通常是被热带雨林所覆盖的较高的陆地。

这种带状排列的现象在哥伦比亚、佛罗里达、喀麦隆及马来亚等海岸都较典型。但由于各地的盐分、底质、波浪和潮汐的作用强度等的差异，也有些地区分带现象不明显。

从整体来看，红树林海岸具有阶梯状的剖面，各个带不同高度的植物层阶，构成了植物海岸阶梯状的外貌，而与海面相接处是植物“陡崖”。

(二) 海岸地貌

红树林海岸带整个潮间浅滩为红树丛林所笼罩，形成了潮湿的分割破碎的泥沼滩面。地貌十分单调，主要是一些长短不一、曲折迂迴的潮水沟系(esteros)。低潮时潮沟变得狭小，宽度仅数米，成沿滩坡上溯数百米的小水沟。高潮时沟道内充满水流，水流可漫溢到沟旁滩地。潮沟伸入陆地很深，在红树林带前缘的岸坡与不连续沙滩之间，潮沟与岸平行，与潟湖或三角港相接，成为沟通沿岸的水道。高潮时可利用小船航行。但必须掌握住潮时，否则会搁浅在泥滩上。

潮水沟的沟坡是圆缓的。低潮时成为数米宽的泥滩，在沟坡上及沟间沉积了厚层蓝黑色的淤泥(粒径小于 0.02 mm)，淤泥中富含有机质残余。由于嫌气细菌活跃的结果，形成了多量硫化氢，使红树林沼泽具有恶臭的气味。低潮时穿越红树林泥滩是很危险的，步行于淤泥中可以没膝，而后将越陷越深，甚至没顶而遭难。

潮水沟的沟底平坦，淤积着厚层的粉砂与淤泥，有时会出现细砂。因此，颗粒比沟坡上还稍粗一些。

平缓的泥滩为潮沟分割。在林带的后方，潮沟之间的岸滩上也出现一些坚实地区，成斑状散布于沼泽内，通称为“小岛屿”(术语 firmes)。这些斑状坚地可能是过去陆地地貌的残迹，如滨河岸沙堤、沙丘或过去海滩的残余。它们大多是砂质构成，也有的是坚硬的腐殖土的地块。在坚地的地面以下 1 m 多深处常含有淡水层。所以，这些“小岛屿”常常成为进入林区内人们的临时居所。在南美有一些村镇就分布在这些“小岛屿”上。

红树林内动物贫乏，主要是些水禽和贝类。有两种动物对红树的发展有好处。蟹的穴居助长了土壤的通气，并使地面形成一些小凹坑，促进了局部的潮水堆积作用。牡蛎附着在红树的主干上，它的壳是红树生长所需钙和碳酸的来源。在我国福建沿海红树林中有一种有害的贝壳——蛇，它附着在红树幼苗上，吸收红树的养料以繁殖自己的后代，致使林木枯萎死去。为保护红树丛林，必须剥去蛇壳。

红树林带后侧的淡水沼泽带，宽约 1～3 km，平时地面干燥，只有在特大潮水时才会被浸没。夏秋季通常是河水泛滥区，繁殖着茂密的棕榈和草丛。这一带常被开辟为牧场，亦有人在此进行垦殖，种植稻米、高粱等。

（三）海岸演变

沿岸海滩上繁殖红树林后，原始海岸逐渐被围封，而于红树林带外侧形成新的岸线。红树形成后阻滞了波浪、潮流，起着消能作用，保护了海岸免受冲刷，并促进堆积作用。所以，红树林形成后通常造成一个良好的海积环境，使岸滩不断地向海伸展。

红树林海岸的演变过程与植物群落的更替有着不可分割的关系。随着植物群落向陆生植物过渡，海滩就不断地向海增长。其总的过程可概述如下：

先是在岸线外面的浅水泥滩上繁殖着少量水生植物。随着红树林海岸的发育，浅水泥滩逐渐淤浅，为红树林的侵入准备了条件。

浅水泥滩后面是沙滩与泥泽，那里经常被潮水浸没。沙滩间是潮沟、淤泥滩地，具备着红树生长的条件。一般认为红树属(*Rhizophora*)是促使海岸向海增长的主要角色。红树种子从后面的红树林中落下，随水流飘到低潮线附近的海滩上。只要波浪与水流没有过多的扰动而影响其扎根，幼小植物就在这经常为潮水淹没的沙滩、泥滩上生长，幼树生出层层的支柱根，固定了浅滩。成长的红树抵挡了风浪，改变了该处动力条件，并且逐渐“网罗”了被潮流带来的泥沙，泥沙及植物残体的堆积，使滩面加高。因此，红树林前面幼林的成长不仅扩大了原来红树林的宽度，并且使地面淤高，海滨线向前扩展。所以，红树属被称为“造陆者”。这时的红树幼林在植物学上称为“先锋红树的单种集团”。

有些地方(如我国华南)先拓者不是红树属，而是海榄雌属(*Avicennia*)。因为前者要求有机质丰富的淤泥，而后者能在纯砂质的海滩、浅滩上生长，并且比前者更耐盐。

当外围生长了红树林以后，加宽了潮流作用消能带，潮流大多沿途消耗掉，使后侧原来的红树林带逐渐变干、变淡，而向陆地转化。

由先锋红树林演化到成熟红树林后，乔木的根系便已非常稠密，成高约 9 m 的树林。成熟的红树“网罗”沉积物的规模更大，促进了海积与生物堆积。在原来的沿岸浅水泥滩又产生新的先锋红树阶段，也有些海岸在远离岸线处无先锋阶段，而是乔木向水中生长根系，沉积物堆积在根上，沼泽就慢慢向海扩展。

壮年红树林又向前推进，该处就演变为海榄雌的盐沼。它有规律地被淹水或偶尔淹水，水中盐分重，积水面积小，林间已发展了一些草地。

最后演变为锥果木，拉贡木林，称为半红树。此时有红树植物，有硬木，也有林下植物及草类。这里通常已是潮汐不能到达，有坚实的土壤与泥炭层。海岸已演化到平均高潮线以上的陆地了。这时典型的红树林已告终结。其后面接着热带雨林。从地貌上看，已是滨海平原或海滨沙丘地区了。

综上所述，植物在此类海岸发展演化中具有重要的作用。植物造成了海积环境，改变了海岸带的动力因素，使陆地不断向海伸长。在美洲、亚洲与非洲，红树植物的种类组成稍有不同，但其总的演化过程是类似的。因此可将这过程用下列图示(图 1)概括之。

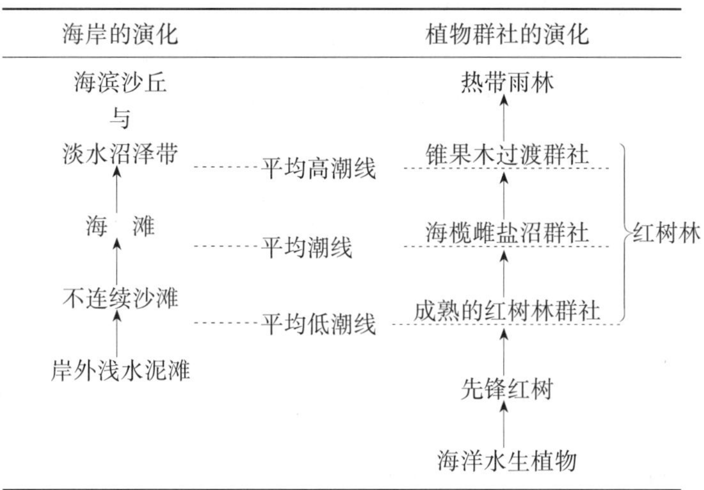

图 1　红树林海岸演化图示

(据 J. H. Davis、R. C. West 资料综合)

红树林海岸的演化趋势也有逆向发展的，并且通常是一处海岸前进，而另一处海岸在后退。海岸后退主要是由于局部波浪作用加强所引起的，波浪直接的侵蚀以及搬运来大量沙粒堆积在海滩上，扼杀了红树植物的生长。

三、研究意义

研究红树林海岸在实践上和理论上都很有意义。

首先，红树林有很大的经济价值。红树林的主要树种红树属等，木材色红，质地坚密，可作细木工用，因能长期保存在地底和水下，可作建筑和桥梁用材。枝杈可烧成优质的薪炭。红树还是单宁植物，树皮一般含单宁 15%～30%，是提炼栲胶的重要原料。果实可以酿酒，枝叶可作绿肥。同时，红树林是很好的护岸林，可以用它来巩固海堤，不被台风潮水冲破；可以保护农田村落，免受风暴海水入侵。林后的坚地可以垦殖，发展为农田或牧场。

其次红树林海岸地区的通行条件，对外交通的方式，林中淡水、食物的供给等，是研究该类海岸所需解决的实际问题。

在理论上，研究红树林海岸对丰富该类海岸特征、规律以及低平海岸的海陆演变过程是有意义的。研究该类海岸地貌必须注意树种、植物群社更替的分析，将它们与地貌组合、发育联系起来。

目前国外学者多偏重自然地理的观点来研究该类海岸，即注重植物、土壤、水文甚至动物等方面，而对该类海岸的地貌研究分析是不够的。在不多的地貌文献中，缺乏对海岸动力与发育过程的研究。对红树林海岸中很重要的问题，比如红树林海滩的成因，是先有少量海滨淤泥，然后由于发育了红树林海岸而扩展为大片泥滩，还是先有红树林生长，然后通过它的根系网罗了淤泥以形成大片泥滩，这类问题还未很好地解决。所以，直至目前，在各海岸类型中红树林海岸的研究还处于比较年轻的阶段。

我国华南、海南岛及台湾沿海，广泛分布着红树林海岸。那里的劳动人民积累了营造红树林以防止台风风浪危害的经验，并且还利用营造红树林以围垦海滨浅滩，增加耕地面积。因此，对我国红树林海岸的调查研究与总结，无疑地将会在理论上与实践上起着很大的作用。

参考文献

[1] Robert C. West. 1956. Mangrove swamps of the Pacific coast of Colombia. *Annals of the Association of American Geographers*, 46(1): 98 - 121.

[2] P. W. 理查斯. 1959 年. 热带雨林. 张宏达，何绍颐，王铸豪，刘健良译. 北京：科学出版社，334 - 349.

[3] 胡善美. 1960. 我国华南沿海的红树林. 地理知识，(7)：303.

[4] 丘思森. 1962. 我国的“海底森林”——红树林. 人民日报，1962 年 6 月 4 日，第二版.

[5] 侯宽昭，何椿年. 1953. 中国的红树林. 生物学通报，(10)：15 - 19.

珊瑚与珊瑚礁*

当海轮乘风破浪航行在西南太平洋及印度洋上时，你会看到：在波涛汹涌的海洋中，时时出现一个个为椰树笼罩的珊瑚礁小岛。它们出露在海面上，高度不大，银白色的沙滩在热带的阳光照射下，光亮耀眼。如果你在珊瑚礁的岸边稍停一下，你就会观察到在深蓝色的海水中，出露着形态不一的珊瑚丛，有的像树丛花枝，有的分枝似鹿角，有的成为扇状、芦笋状，还有的像卷心菜、人脑以及美丽的百合花，颜色也五彩缤纷，有淡黄、淡红、淡褐和浅绿色，其间夹有红色，给深蓝色的海水所映衬，更显得瑰丽动人。在我们祖国的南海亦分布着大量风光绮丽的珊瑚礁小岛。

生活在热带海洋中的珊瑚虫、红藻、绿藻、水螅和具有石灰质介壳的棘皮类、腹足类、瓣鳃类、有孔虫类等，它们的骨骼，特别是珊瑚虫的，积聚愈合，在海滩下形成一片石质平原似的礁块。潮流、波浪遇到它都被削弱，这就有利于砂砾的堆积，同时，水面下的珊瑚也不断地往上生长，这样就使礁块不断增大，由生物作用所造成的这类礁块，通常就叫做珊瑚礁。

一、珊瑚虫生长的条件

珊瑚虫是一种原始的动物，由于它与海藻共生，长期以来被误认是植物，实际上它是水母的亲族，是腔肠动物。它的形状像只袋子，食物的消化是在袋子中进行的。袋子边上有很多触手，当潮水上涨和夜间，它伸出花瓣状的触手并呈现最美丽的颜色来抓取小鱼、小虾和小虫等浮游生物，作为自己的食料。珊瑚虫可以生活在大部分海洋中，甚至在北方高纬度的挪威峡江中亦有。但是，造礁的六百多种珊瑚虫，一般只能繁殖在热带的浅海中，而且分布在三大洋的西部比东部来得多，换句话讲，造礁珊瑚虫蔓延生长要有一定的条件：

(1) 要求生长在暖水中。最适宜的温度是 25°～30 ℃，温度最高限度是 36 ℃，最低限度是 13 ℃。如我国台湾省的澎湖列岛的活珊瑚，在 1946 年二月一次寒潮中，就有大量死亡。因此，它只能在热带海洋中生长。但在某些温带海洋中，如有暖流经过，照样也可生长，如北大西洋的百慕大群岛，已达 32°N，仍有珊瑚礁出现，它是与墨西哥湾暖流有关。

(2) 要求足够的光线。珊瑚虫与一种单细胞藻类共生在一起，这种藻类需要充足的阳光进行光合作用来维持生活，以供给珊瑚虫所需要的氧气。因此，活珊瑚主要分布在低潮水位与一定水深之内。如果深度超过 40～60 m，由于光线不足，珊瑚便不能生活了。仅有极少的几种珊瑚虫可以生活在 100 m 上下。

---

* 肃波，曾昭璇：《地理知识》，1974 年第 1 期，第 27－30 页。

肃波，王颖原名王肃波，此处作笔名用。

(3) 要求海水具有正常的或较高的盐分，介于 27‰～40‰之间最为适宜。

(4) 珊瑚虫喜欢不断扰动的水。这样的水可以含有较多的氧气与营养料，通常向海侧的珊瑚构成体较背侧旺盛，因为向海侧有波浪的拍激，海水得以不断地渗含空气，水中溶解氧供应充足。但是，过于强烈的激浪作用也会妨碍珊瑚的生长。

(5) 要有合适的基础。如雅加达湾内某些直径达数百米的圆形珊瑚礁从泥质的海底生长起来，顶上还带有一些小岛(如埃担姆、恩库依曾等)，在周围坡面的下部找不到活珊瑚，就是因为石灰质细泥底质和缺乏阳光透入使珊瑚无法生存。只因为那里有着一些凸起的地形，才使珊瑚能够在比较接近海面的合适基础上立足。

基于上述条件，在混浊的、淡水的河口区，在停滞的海湾内，以及泥多、盐分不足和氧气缺乏的红树林海岸，都是不利于珊瑚生长的。此外，珊瑚虫一般只能生活在海水中，不能长期地出露水面，只有几种石珊瑚适应能力特强，能够长高到正常低潮面以上。

由于造礁珊瑚生长条件的限制，珊瑚礁的分布，以 32°S～32°N 以内的热带海洋为主。在我国以台湾岛以北各小岛为限，也就是大致分布到 26°N 附近。在南海海盆中散布着的东沙、西沙、中沙和南沙各群岛都是由珊瑚礁构成的岛礁区，丰富了我国岛屿的类型。

二、珊瑚礁的类型

珊瑚礁的分布以太平洋为最多，其次是印度洋和大西洋。但是，在太平洋东部却无珊瑚礁分布，因为那儿受秘鲁寒流的影响，海水太凉。同时，珊瑚礁也避开了印度支那、中国及热带大西洋等地由大河所形成的三角洲沿岸与沿岸潟湖地带。西南太平洋具有安静、清洁、深蓝色的海水，珊瑚得到了最好的繁殖条件。

根据珊瑚礁的结构和形态可以将它分为三类：

(一) 岸礁

岸礁是紧靠着陆地，围绕着岩石海岸的水下岸坡发育而成的。它常常分布在港湾海岸的岬角内侧与湾内海滩的下部，像一个短裙一样分布在海岸的低潮水边线以下。如东非赤道附近的海岸、南美的巴西海岸、西印度群岛、太平洋中的岛屿及我国华南局部地区的海岸多为岸礁形式的珊瑚礁。

要辨别清楚的是，珊瑚岛四周的珊瑚礁不能叫做岸礁，因为珊瑚岛本身就是一块巨大珊瑚礁上的沙帽，岛四周的珊瑚礁叫做"礁盘"。

岸礁具有一平缓的、微微向海倾斜的礁平台，并具有急陡的礁缘。礁平台上生长着角质的活珊瑚与海藻，也散布着一些巨大的礁块，突起其间，还有不少沟坑，所以礁平台上并非平坦，而呈不规则的起伏。当低潮时，礁平台上浅水荡漾，人们可在其上行走，采集珊瑚与贝壳。礁平台外缘是一陡坡，直临 5 m 以上的深水区。陡坡的坡度若大于 45°，并有网状沟谷的存在，表示这里的珊瑚礁块间还未长大到相互愈合的程度。礁平台内缘是砂质的海滩，这些沙滩多是由珊瑚粉碎后形成的，雪白细软，常成为海水浴场。与岛交接处若是较深水的凹沟，称之为"船沟"，表示这里常有泥沙及淡水流入，不利于珊瑚生长所致。

岸礁不是连续的，其间有许多缺口，尤其是在海湾湾顶，由于泥沙由海湾两侧向此处集中堆积，因此，珊瑚礁往往不发育。岸礁的这些缺口常成为渔船往返的通道。

(二) 堡礁(或称离岸礁)

在大陆外缘，距海岸较远的浅海中，成为岛链状分布的大珊瑚礁体。有的礁体已出露水面成为岛屿，有的仍为水下暗礁或浅滩。在堡礁与岸之间隔着一道浅海或潟湖，其水域宽度由几千米到几十千米不等，甚至超过一百千米。而深度一般不超过 100 m。

堡礁的宽度一般为几百米，很少超过一千米，个别的能达百千米以上。堡礁的长度由数百米达几千千米里，中间也有间断，成为浅海或潟湖与大洋中的水沟通的水道。

堡礁分布在西南太平洋、加勒比海及红海等处。我国海南岛四周也有堡礁的发育。举世闻名的澳大利亚大堡礁像城垒一样(图 1)，从托雷斯海峡南部(10°S 附近)到格拉德斯通(南回归线之南不远)，绵延伸展 2 400 多 km，与澳大利亚海岸相隔 13～180 km 的潟湖海面。它最宽处为 240 km，最窄处跨度仅 19.2 km。像海上冰山一样，它的大部分是位于海水下的，仅顶部出露成岛屿或浅滩。在这大堡礁内有 500 多个岛屿，它们间断地分布在南回归线以北的 900 多 km 长的范围内。这些岛屿的中央是茂密的热带丛林，在绿色丛林的边缘是白沙耀眼的海滩，它坡度较陡，再向外缘即为裸露的礁平台了。堡礁的基底是插入数千米深渊的巨大海底山脉。

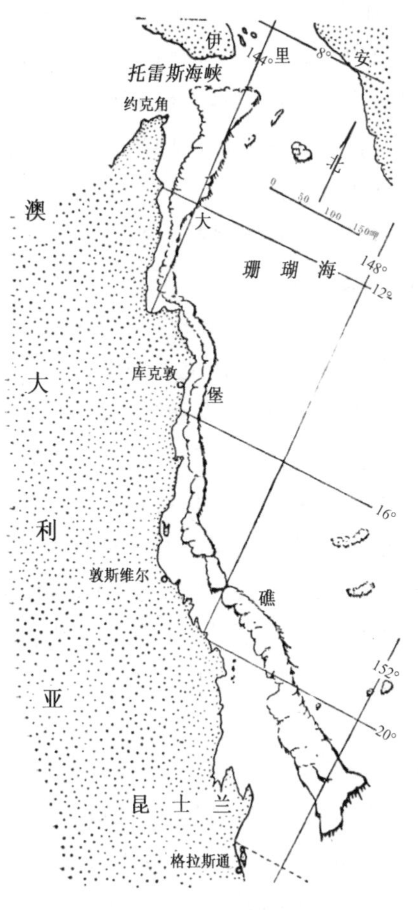

图 1　澳大利亚大堡礁

堡礁的主礁体——礁平台，广阔平缓，其上散布着许多低凹的积水洼地，叫做“礁塘”。因为每一个珊瑚生长体都向上生长，当它达到低潮水位时，它就向水平方向发展。活珊瑚的脊所围成的碟形构造，在落潮的时候，珊瑚碟内的水不能全部流出，因而形成了这类礁塘。礁平台的边缘，有时有礁堤，在高潮的时候可以露出水面，它是由风浪作用将珊瑚礁块堆积所成。堡礁四周的陡坡，以向海一侧为大，向陆一侧较缓，因为通常向海一面才是珊瑚繁生区。当大潮退落的时候，清澈深蓝的海水从堡礁外缘向大海流去，形成水珠四溅的白色瀑布，蔚为奇观。

(三) 环礁

环礁是环状的珊瑚，四周呈围墙状，封闭了其中的潟湖。环礁在太平洋发育极盛，如果你从飞机上俯视时，可以见到在碧波万顷的洋面上散布着一个个白色的花环，花环中央荡漾着一个平静的湖面。

珊瑚礁环有的完整，有的局部有断缺，形成水道与大海相通。环礁的外形与大小是形形色色的，有的呈圆形、椭圆形，或为不正规的三角形、多边形等。潟湖的直径变化很大，有的可大于 60 km。深度大部分在 60 m 左右，很少超过 100 m。由于环礁起到屏障作用，潟湖内风平浪静，又有水道与大洋相通，因而它是航船天然的优良避风港。

环礁的向海坡很陡，坡度常达 45°，有的甚至成为数米高的倒悬崖。在向海坡上沟谷地形发育。在崖顶上有彩色的由钙质藻组成的礁脊，礁脊高出海面 1 m 左右。受波浪冲蚀，使礁脊断断续续。礁脊的内侧是礁平台。它宽约数百米，台面崎岖不平，低水位时，礁平台可以出露一部分，礁平台主要是由死珊瑚所组成，掺杂有藻类、海绵以及为钙质藻所胶结起来的岩屑。礁平台上水源充分的地方亦有活珊瑚在繁殖着，它们又可形成小环礁。平台上还有由珊瑚沙所堆积的沙洲与沙堤，这些堆积物可以被钙质胶结为海滩砂岩，它又可受海水溶蚀成为崎岖不平的石芽地形。一些较高的礁体，遭受风、浪作用后会形成一个个石蘑菇。当沙洲堆积日高，到有草木生长，淡水贮存较多时，沙洲即为林木固定了，形成“绿岛”或“林岛”了。若海岸沙滩上堆积起较高的沙堤，把“林岛”包围起来，这时就呈碟形的低岛。这些珊瑚岛每呈圈状成串分布于浅海四周，形成大环礁。如我国的南海团沙群岛(即郑和群礁)就是个大环礁。

环礁的内坡，坡度和缓，覆着有珊瑚沙，并生长着活珊瑚。

在大陆边缘海(如南海、红海等处)以及在大堡礁的潟湖内，还有一些小环礁，它们外形呈圆环状或长圆环状，但规模小，如我国东沙岛是一环礁，礁环的长、宽不超过 5 km。小环礁中央的潟湖深度不过是数米或一二十米深。有趣的是这些小环礁的排列延长方向与当地常风向一致，甚至随季节风向的变化而改变着它们的伸长方向。

在海洋上并不是所有的珊瑚礁都是环礁，还有许多巨大的呈块状的珊瑚礁。礁的特点也是有一个广阔起伏和缓的礁盘和四周陡峭的礁坡，在礁盘上也有珊瑚岛的发育。这些岛屿大的可以建筑机场，例如南海的南威岛就是一例。这些块状礁上的珊瑚岛，其形成原因和地貌上的碟形特点与环礁上的珊瑚岛是一致的。但是，块状礁四周发育比中部要好，因此，礁体四边高起，中间成一凹地，并积水成一很浅的潟湖，有的还有红树林发育，在较高而阔的边缘，也有沙洲发育。

关于环礁的成因，有很多假说，其中最重要的是达尔文的假说。他在环球的航海中发现了珊瑚礁的秘密。达尔文认为：最初珊瑚以岸礁的形式环绕着岛屿生长，随着岛屿下沉，珊瑚礁向外扩展，而形成堡礁。随后岛屿继续下沉，这些礁体围绕着岛屿像一顶顶珊瑚冠冕一样，而当下沉的岛屿终于消失在海面以下后，那些在岛屿周围不断地向上生长的珊瑚礁就成为环礁的形态，原来的岛屿顶部的一部分被海水淹没而成为中央潟湖(见示意图 2)。当然，岛屿的下沉和珊瑚礁的生长，是经历了漫长的地质时期演变而成的，并非一朝一夕所能形成。达尔文的理论虽然不是尽善尽美。但近期在马绍

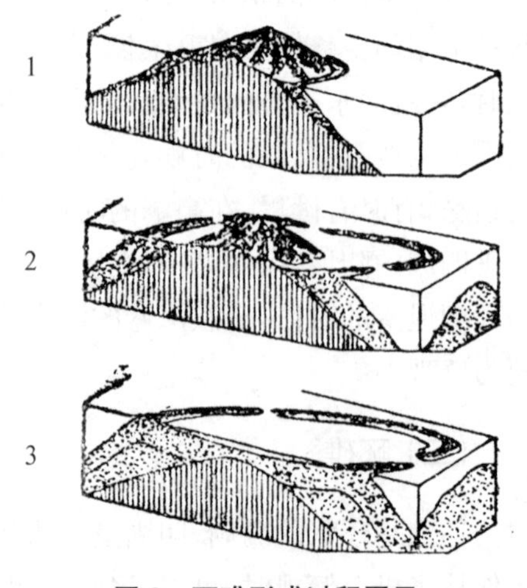

图 2　环礁形成过程图示

1. 岸礁　2. 堡礁　3. 环礁

尔群岛的安尼维托克岛打了深达 1 402 m 的钻孔，穿透了珊瑚礁，打到了火山岩——即达尔文所说的下沉岛屿，同时在环礁发育的海底，也发现了原来陆地上的河流、海湾和陆地上的沉积，这些都进一步证实了达尔文的沉降论假说。

珊瑚礁是快速生长的生物礁，对海岸下的地形改变很快。如印尼不少港口在十九世纪通航无阻，现因珊瑚礁发育而淤塞，使港湾变浅、港口障塞，以至码头不断伸展越过礁区，航道发生障碍，对渔场拖网作业也有影响。而散布在大洋中的珊瑚岛是航行的中途站，对远洋渔业、航行的补给，海洋高空、海面气象情报的取得，航空和军事基地的设立都有重要价值。因此，进一步研究珊瑚礁的形成、发展的规律对航运、捕鱼、军事等各方面都有意义。另一方面，可根据珊瑚礁目前的厚度与位置，了解这个地区自然地理环境的演变过程，对弄清这个地区的地壳构造运动、海面升降、海盆的形成和第四纪的历史都有很大的帮助。例如，前述的马绍尔群岛的比基尼岛、安托维克岛，通过打钻，得知珊瑚礁的厚度达 1 100 m 以上，这是海岸下沉的有力证据，因为活珊瑚生长的范围很少超过 100 m 深，一般是 60 m 深处。在巽他群岛，高出现代海面 1 500 m 处有珊瑚礁存在，我国海南岛南部在高潮面以上有礁平台分布，这些都是海岸上升的标志。

在珊瑚岛周围，繁殖各种珍奇的海洋生物；海龟、五光十色的贝类以及无数的海鸟。它们提供了珍贵的药材、稀有的珠宝和丰富的肥料。而珊瑚礁本身就是取之不尽的建筑材料。我国海南各地多用珊瑚礁烧制石灰和制作水泥。因此，研究珊瑚礁，利用珊瑚礁在经济上亦具有很大意义。我国南海有无数的珊瑚礁等待着我们去进一步研究它、开发它。

海洋工程环境条件调查研究[*]

一

海洋是流动联通的，各自然因素间是互相影响的。海洋环境包括：河口海岸、浅海与大洋。影响因子涉及海洋水圈、大气圈、岩石圈以及生物活动。海洋工程涉及海洋水域与岸底，因此，它不是一般的工程地质问题，而是要调查研究：① 海洋动力作用，它涉及水文与气象条件以及长时期的海面变化资料。② 海岸与海底地貌及沉积作用过程，即海洋圈所及的表面地质环境与现代地质过程。它涉及地貌结构与稳定性，沉积组成、泥沙运动、沉积速率与基底地层结构关系，沉积物与海水的地球化学特性，生物活动与生态特征以及海底热流活动等各方面。③ 工程地质特征，包括对海底沉积物的土力学研究，基底岩层分布特点与稳定性、现代火山与地震活动等方面的调查研究。在进行海洋工程环境条件调查研究中，要注意把海洋动力与海底过程结合起来，要把海底过程与底部基础的结构条件结合起来进行调查研究与综合分析。因而，海洋工程环境条件项目是一个多学科的综合调查研究项目。它应该根据不同海域部分的不同工程项目的要求，进行专项调查并辅以区域普查以及海洋台站的资料分析，而不是笼统的大海区调查。

15 年后，在我国应具有以下成果与力量：

(1) 积累专项工程环境条件调查与研究成果以及若干年后的效应检验成果。

(2) 建立专门的储存各海区不同海域部分的系统资料与单项因素的数据库。

(3) 具有为各海区从事环境条件调查的基本队伍、装备以及关于环境条件预测、监督、效果检测以及应用措施的管理办法与管理机构。

二

能源、交通与通讯、资源开发与环境污染等方面皆涉及对海洋环境与海洋自然过程的研究。大量的灾害性事故是由于海底条件的不稳定所造成，如：电缆断裂、平台翻倾等，一些重大的工程项目建立后，改变了自然条件的动态均衡，招致了灾害性事故。加拿大于 1970 年在芬地湾温莎尔附近的一个潮汐河汊上(Avon River)建立了一条公路坝，由于阻挡了潮流通过，水流动能骤减，造成了坝下淤积，至 1973 年，堤坝前已形成直径为 0.5 km、厚度近 5 m 的舌形的淤泥滩，大大改变了该处的海湾状况，影响了排洩洪水。加拿大政府在开发利用芬地湾潮汐能时，重视对芬地湾建立大坝的调查研究，既进行建坝前的自然环境调查，包括海洋水文、地质、地球化学与海洋生物方面的工作，同时又进行数字模拟以预

* 王颖：(南京大学海洋地貌与沉积研究室)《海洋开发战略课题 77：海洋工程环境条件调查研究》，1－5 页，1984 年 7 月。

测建坝后的效应。现阶段查明，如果在芬地湾湾顶的迈纳斯盆地(Minas Basin)湾湾口建立大坝，则会使芬地湾水面增高 15 cm，而芬地湾水面的抬高，在风暴天气时，建立于潮滩上的波斯顿机场则会遭受海水浸淹。因此，加拿大决定继续对芬地湾湾顶进行系统的调查研究，同时，在安那波利斯(Annapolis)建立小型的堤坝与实验性的潮汐发电站，既开发海洋能源，又进行实践性的建坝效应观测。由于海洋工程不断发展的需要，西方发达国家重视环境海洋地质的调查研究工作。如深度超过 4 000 m 的洋底能否作为置放或埋藏高强度的放射性废物的场所？投放后对自然环境以及生物圈可能产生怎样的影响等。为此，美、加、英、荷、法、日等国对此开展了专门系统的环境海洋地质学研究。我国自二十世纪七十年代以来，铺设中日海底电缆，特别是近期对海洋石油资源的勘探开发活动促进了对海洋环境条件的调查工作，但仍属于初期的专项调查阶段，尚需积累各有关海区的系统资料与准确数据，尚需把初期调查与建设中与建成后的检验性观测结合起来，并期建立系统的工作规范与熟练的技术力量。当前我国主要的海洋工程项目系引进外国资金与技术力量项目，一开始就重视对海洋环境条件的调查研究工作，这会减少盲目性与危害，是个良好的开端。

三

有关的海洋工程与涉及的海洋环境大致概括如下：

1. 海岸与河口环境

海港选址与码头航道建设中的海岸动力与泥沙运动研究，建坝围海工程对自然环境动态平衡以及生态环境的变化影响研究；电站、化工厂、炼油厂的排废污与环境保护问题；油轮事故与油田开发所造成的原油污染海岸事件与急救工程。

2. 浅海环境

钻井与油井平台；海底电缆与管道铺设；人工岛；倾废场地的选择；打捞以及其他水下工程等。涉及到海况、海底稳定性以及海底沉积特性等问题。

3. 深海或大洋环境

航天与潜海设备的试验场地；海底矿产的采掘条件；高强度的核废物安置问题以及深海钻探与打捞等。

这些工作属于多学科的专门研究项目，除完成一定海区的自然特性调查外，还需要回答或提出具体任务所要求的具体数据与系统资料。以核废物安置与海床研究问题为例：由于各国所持观点不同，所进行的项目不同。但是各国的研究项目都是专题性的海洋学研究，而不仅是泛泛的区域普查。其中，英国着重于容器的研究，但对投掷场所进行了物理海洋学调查，多年地连续观测海流以建立北大西洋水流扩散与东大西洋中等规模的海水循环模式；英国还在研究核电站流出物在海洋生物中所积累的放射性 ^{237}NP 与 ^{99}Tc 的浓度方面取得了成效。美国在核废物安置中重视对海床地质建造作为多元屏障的研究，他们广泛地调查岩盐、玄武岩，页岩、火山凝灰岩以及海洋沉积物对核废物的屏障效应。美国在核废物投放于海床内的同时，用透度计测量伦琴射线的穿透力，研究近容器段（100 ℃等温线以内的数米范围）与远地段的热力与地球化学元素的变化梯度，研究放射热能对海床沉积物特性的影响。由于放射性热力散失与冷却的时间长，因此对放射性投掷物的近、远段海床沉积物要进行长达 15 年的热力学实验研究。已查明埋藏于海床内 50 m 深处核废物容器所产生的热脉冲，经过 10 年即可穿透沉积层而达到与海水接触的海床表面。加拿大大西洋地质中心的环境地质部，自 1980 年至今，在大西洋地区进行多年的调查以探求核废物于 5 600 深处的海床埋藏可能性问题，其工作包括系统的海洋地质调查（在 10 km^2 小范围内进行测深；测地层；采表层样、箱式柱状取保持原结构的沉积层与上浮水；CTD 测量与取样；热流量探测等）。对所有样品进行沉积结构，微古，地球化学元素、氧化还原、孔隙水、土工学及热力学等方面的实验，是一项综合性的海洋地质学调查研究，其结果除取得大量专题研究成果外，还查明在大洋深处有一“深海边界层”，即深海海床的表层底土与水层之间，由于水流、生物与热力活动，使得泥沙、微量元素以及水、气、生物皆有着对流交换，互相渗透并互有影响。我想，这一成果不仅对深海埋葬废弃物问题做了回答，并且也是一项重大发现。这个例子说明了海洋工程环境条件调查研究的性质与工作特点。了解已进行的工作实例可能对我们进行这项工作的规划有所彰益。

海岸带资源开发研究*

海岸是海洋与大陆相互作用的地带，范围包括沿岸陆地、潮间带与水下岸坡，是在海岸动力与海岸岩石圈相互作用下形成和发展的。海岸动力主要是风、浪、潮汐、海流和入海河流等极为活跃的因素，它们不断地改变着海岸的地貌。岩石圈是组成海岸的物质基础，因受岩性、构造及原始地貌的不同而具不同性质，从而改变了海岸动力作用效果，使海岸发育具不同特性。因此，海岸是受海陆两方面多种营力作用，具有独特自然环境的"两栖"地带，在这个地带内有着独特的自然资源可供开发利用。从海陆关系分析海岸资源主要是：动能资源（风和海洋能）、砂矿资源、海涂与土地资源、港湾资源、食物资源以及旅游资源。它们具有用之不竭、成本低、少污染以及可以补偿的特性，开发利用这些资源既需进行充分的调查研究与全面规划，同时要给以技术指导与制定管理法规，避免在开发利用资源过程中破坏自然环境的生态平衡。

一、动能资源

（一）风能

由于气流下垫面位于海陆交界带温压条件的差异，海陆相互作用的效应之一是沿海岸形成一条狭长的风速变化增强带。以江苏省为例，据 20 年风速资料统计分析，风速（$\bar{U}$）与海边距离（X）呈指数幂的关系[1]，经验式为：

$$Y=\bar{u}_1\sqrt{u_0}=0.70+0.30e^{-0.0415X}$$

式中：$\bar{u}_1$ 为距海边 X_i km 处的累年平均年最大风速；$\bar{u}_0$ 为海边的累年平均年最大风速。

在 X 为 40 km 至 0 km（海边）的狭长带内，u_1 急剧增加；当 $X\geqslant60$ km 以后，Y 值近一常数为 0.7，即比海边常定小 3 成，Y 值计算平均误差为±0.029，沿海边累年平均最大风速为 18 m/s，目前全国使用的风力机有效风速范围是 3.0～20.0 m/s。江苏沿海滩地平均有效功率密度为 110～140 W/m^2，有效风速时数为 4 000～5 000 h；海上 10 km 范围内，平均有效功率密度为 200～300 W/m^2，有效风速时数达 5 000～7 500 h，占年总时数（8 760 时）57%～86%。这在缺柴少煤的长达 1 000 km 的苏北海岸带是一重要自然能源。如兴化县应用风力机提水灌溉，在 4 级风的推动下，每小时可提水 100 m^3，自动迎风和调速，8 级以上大风会自行停车。目前全县有 1 000 部风力机为灌溉的辅助功能，全年可节约柴油 10 000 t。目前欧美沿海地区亦重新发展使用风力能源。风力是取之不尽，使

---

* 王颖：中国科学院地学部编，《中国科学院地学部第二次学部委员大会文集》，219－222 页，北京：科学出版社，1988 年。

用方便，适合当前经济技术条件而无污染危害的能源。海岸是风能资源最丰富的区域，如以风力为能源淡化海水，解决海滨城镇与岛屿淡水匮乏问题，将会对海岸带发展带来新前景。

(二) 海洋能

海洋能指海水所具有的动能、势能和热能，包括海浪能、潮汐能、海水温差与盐度差能。据估计，全球海洋潮汐能的蕴藏量约 10 亿～27 亿 kW。我国大陆岸线潮汐能蕴藏量[2]为1 亿 kW。可开发利用的装机容量为 2 000 万 kW，年发电量为 1.5 亿 kW，而且集中于能量消耗量大而最缺乏能源的华东沿海。钱塘江潮汐能装机容量为 396 万 kW，可发电 100 亿度，超过葛洲坝水电站的能力。潮汐发电，规律性强，不受枯水季节影响，不淹没农田或搬迁村镇，更无战争或地震毁坝而造成灾害之虞。法国朗斯河口潮汐电站是目前最大的，该处潮差 13.5 m，双向进水堤坝长 750 m，24 台单机容量为 1 万 kW，年生产 5.44 亿度电。我国已在浙(温岭、象山、玉环、江夏)、鲁(乳山)、苏(太仓)、闽(长乐、平潭)、桂(龙门港)建成 9 个小型潮汐电站，为沿海及岛屿提供电力并积累了丰富经验。存在的问题是需加强建站前的调查研究与规划，解决防淤、排淤、防海水腐蚀、防生物附着以及与邻近火力电站并网等项问题，以保证长期有效运转。

我国波浪能蕴藏量达 1.5 万 kW，目前沪、穗等地波浪发电供港口航标灯能源。有关院校正进行研究期以解决海岛用电。

南海水深平均 1 000 多 m，全年海面水温为 25 ℃～28 ℃，与深层水温相差 20 ℃。受地形影响，在南海诸岛周围海域形成海底上升流，它不仅把海底丰富的营养盐带到上部形成渔场，而且可利用上升流冷水温差发电，供给南海诸岛电力，尤其供淡化海水之用，这对开发祖国南海疆域意义重大。目前，在台风频繁之海域环境研究与工程技术方面尚存在一系列待解决的问题。

利用河口区海水盐度的渗透压力差发电始于 20 世纪 70 年代，目前日、美、以色列、瑞典正在研究，尚属早期阶段。

二、砂矿和砂土资源

砂矿与砂土资源，这是海洋动力与沿岸岩石圈交互作用之产物。分散于原岩中之贵重金属与稀有元素矿物经风化离析，再经风浪与水淘洗堆积而成，如石岛附近沙坝中锆石砂矿品位高达 2 kg/m^3，海南岛东南岸钛铁砂矿已在开采。砂矿床的分布范围与原岩有关，山东半岛最富集的锆石砂矿与中生代的正长斑岩相伴分布，冲沟沉积砂矿颗粒大，但是，真正成为矿床而具工业开采价值的砂矿是由波浪作用簸选而被浪流堆积于高潮海滩上。重矿砂质纯、量大而颗粒细圆，表明为海浪堆积。风暴天气时，波浪扰动深度增加而砂矿富集量大，反映出海滨砂矿主要来源于水下古海岸堆积。砂金矿堆积与锆石矿不同，以冲积物居多。山东半岛尚有制玻璃的纯净石英砂矿，砂粒浑圆，分布于濒海的高大砂丘中，砂粒已相当程度地分选搬运。大砂丘形成时海平面应低于现代(即冰期低海面时期)，当时有较宽坦的海滩为风力吹扬海沙提供足够的场地，才能在海滩上部沿滩脊发育大型沙丘，故海滨砂与砂矿是海陆交互作用所产生的资源。值得注意的是，沿海民众挖砂土出售为建材，由于获利快，此现象发展迅速，严重地破坏了海滩物质供应的动态平衡，造成海滩冲刷与海岸后退。这是一项世界性的“砂权”问题，在美国是通过环境、工程、管理与立法四方面人员结合制定法规来制止挖砂。

三、环境资源

(一) 海涂与土地资源

我国有近 3 000 km 的潮滩海岸，或分布于大河平原沿海，或分布于狭长港湾之中，由于岸坡平缓或吹程短，波浪作用达不到岸边而以潮流作用为主，加之有大量泥沙供应，渐堆积发育了潮滩海岸[3]。若以理论深度基面以上计算，这类潮滩海涂的总面积约 2 500 万亩(1 亩＝667 m^2)，是东部沿海一项重要的“新生”土地资源。潮滩发育经历过低潮粉砂滩，中潮混合滩和高潮泥滩以至成为草滩等几个阶段，开发海涂需据自然地理条件不同而异。潮滩不同部分有不同用途：宜在平均高潮位建堤，可取泥质沉积为建堤材料，取土处挖成水渠排灌，泥质边坡稳定，围堤高程低。低潮位土层薄不利使用，而且堤坝易被冲垮[4]。围堤后的上部潮滩可以辟为农田种植稻棉(如江苏降雨与淡水资源丰富处)，可建立牧场(黄河三角洲某些缺淡水灌溉的地区)或开辟盐田(日照充分、无淡水影响、滩平土质黏重处，如渤海湾塘沽盐场)；下部潮滩以发展贝类养殖为宜，泥滩适宜养蚶，粉砂滩适宜养蛤类，而细砂质滩地则可养蛏。滩涂养殖投资少、收效快、单位面积产值高，但需加强技术指导。目前全国海洋水产养殖占海洋水产(350 万 t)12.5%，主要在海涂养殖。海涂还可作为食物网中的一个环节开辟为自然保护地(如苏北射阳丹顶鹤、大丰麋鹿自然保护区)或辟为野营场地。潮下带浅滩适宜于鳗鱼和对虾育苗以提供养殖。

(二) 砂质海滩

滩坡较缓而砂质松软，激浪作用活跃而空气清新，是良好的运动与娱乐场所，加上阳光与海珍食品，构成重要的“三 S”旅游资源。我国已重视开发沿岸与海岛旅游区。砂质海滩在风暴天气与泥沙供应不足段落常遭受海浪冲蚀后退，采用修建岸外顺坝与沿岸突堤组合的工程可以防冲促淤，但石材与工程花费大且影响观瞻。处于世界性海平面上升的环境背景，海岸侵蚀防治已成为当前研究的中心课题之一。美国已在珍贵的旅游点及某些私人海滩用人工“喂养”海滩的办法(Beach nourishment)在一定时期内保持天然沙滩的优越环境，而花费比建堤低。

(三) 港湾资源

港湾是联系海陆的天然桥梁，是重要的港口资源。我国海岸线长，但港口密度小，平均 500 km 有一个港口，大港口主要集中于北方，远不敷需要。浙、闽、粤沿海港湾多，有近百个港湾可辟为大中港口。建港需要妥善解决“水深与浪静”这一对矛盾，需进行海岸冲淤变化、泥沙运动与发展趋势研究。选取适宜的天然港湾深水岸段，加以工程措施，开辟为不同等级的港口。原则上深水岸段应供建港使用,并注意保护自然环境与海岸动态平衡。浅水的小海湾可辟为海藻、鱼、虾等养殖场,据海岸特点不同而有不同项目。在华南不正规全日潮岸段有众多的潮汐汊道型港湾,具有落潮延时短而流速强的天然冲刷力以维持航道水深,以及开辟为多功能港口的水域面积[5]。但这类港口的稳定性及动态均受潮量的大小和变化所决定[6],任何整治工程(如围垦、开辟盐田等)如果改变了其口门狭窄,内侧纳潮水域宽广的特点,即 P－A 关系,减少纳潮量,则会造成港口与航道回淤的严重后果。

总之,海岸是一独特的环境,在这里,气圈、水圈、生物圈与岩石圈“相遇”与相互作用并提供了丰富的动力、食物、矿产、土地、港湾与旅游资源。由于海面变化与地壳运动,在长久的地质过程中,昔日海岸带的河口、三角洲、潮滩、潟湖、海湾与一系列海岸沙体又成为良好的生储油场所。开展多学科的海岸带综合研究与规划,具有重要的理论意义与应用价值。在开发海岸资源与利用海岸环境时,要重视立法与综合管理。中国海岸资源丰富,就近工农业经济发达区,并且是通往海外的门户,应支持优先发展,以促进经济的发展速度。

参考文献

[1] 任美锷主编. 1986. 江苏海岸和海涂资源综合调查报告. 北京:海洋出版社,226.

[2] 许启望,李桂香. 1985. 2000 年我国海洋能源开发. 我国海洋开发战略研究论文集,256.

[3] 王颖. 1983. The Mudflat Coast of China. *Canadian Journal of Fisheries and Aquatic Sciences*, 46 (1): 160－171.

[4] 朱大奎. 1986. 中国海涂资源的开发利用问题. 地理科学,6(1):34－40.

[5] Wang Ying, Schafer C. T., Smith J. N. 1987. Characteristics of Tidal Inlets, Designed for Deep Water Harbour Development, Hainan Island. Proceedings of Coastal & Port Engineering in Developing Countries, China Ocean Press, 1: 363－369.

[6] 任美锷,张忍顺. 1984. 潮汐汊道的若干问题. 海洋学报,6(3):352－360.

海岸带与近海*

一、海岸带与近海科学的发展趋势与作为优先领域的理由

海岸带与近海是地球表层岩石圈、水圈、生物圈与大气圈相互作用、物质与能量交换、各种因素作用影响与变化最为活跃的地带，是人类生存环境重要组成部分和生产活跃的重要场所，是沿海国家划分海洋国土与经济权益纷争的地带。

海岸带与近海的范围各国划分不一致。海岸带范围包括沿岸陆地、潮间带与水下岸坡，是在风、浪、潮汐、海流、入海河流等活跃的海岸动力因素与海岸带岩石圈相互作用下形成、变化发展的，是具有海陆交互作用影响、变异的"两栖"性地带，既有陆上部分，亦有海域部分。但是海岸带环境与资源又具有既不同于陆地，亦与海洋差异的特性，人类活动间接地或直接地影响到海岸演化。因此，确切的海岸范围，由其独特的环境因素与环境特征所决定。近海的范围，从科学定义分析应包括近滨(NearShore)与外滨(Offshore)两部分，皆属海陆交互作用地带，与人类活动影响密切相关。1994 年 5 月联合国教科文组织在比利时列日大学召开的海洋工作会议明确地提出海岸海洋(Coastal Ocean)，范围包括海岸带、大陆架、大陆坡及坡麓的大陆隆，系海陆相互作用关联的体系，与人类生存发展关系密切。因此，这里所列的海岸带与近海研究的范围和领域与当代国际海洋科学发展的重点方向相适应，即"海岸海洋"科学(Coastal Ocean Science)，而与深海或大洋科学相区别。

海岸带与近海是一个特殊的地区。例如，由于气流下垫面温、压条件之差异，海陆相互作用的效应之一是沿海岸形成一条狭长的风速变化增强带，自海岸带向陆向海风速皆会递减。风速变化这一特点，促使波浪向岸增强、风沙运移堆积，风力亦可提供风能利用与发电。波浪向海岸传递过程中受地形影响而发生变形，波能增强，形成了与深水不同的浅水波及海岸激浪，波浪对海岸作用效应巨大。潮汐作用与潮流亦在海岸带形成其独特的形式与动能。风、浪、潮汐作用的结合与变化，在海岸带形成与深海大洋不同的动力体系。

海岸动力体系中另一特殊的动力是河流作用。世界河流每年自陆向海输送的固体与悬浮体达 200 亿 t，中国河流每年向海输送的泥沙达 20 亿 t。首先输至海岸带，再形成泥沙流沿岸运移或向外海扩散，参与海岸带与近海的沉积动力过程。另外，陆地的有机质、重金属及其他溶解质污染物，以中国为例，约有 1/3 是被河流输送至海岸带海洋中。陆地的有机物与营养盐促进了沿海的鱼类繁殖，而石油、汞、镉、铅、锌、砷、铬、铜等将污染近海环境，且污染物的数量与种类在日趋增加，扼杀海洋生物，构成海岸带与近海急需解决的

* 王颖：《走向 21 世纪的中国地球科学》调研组，《走向 21 世纪的中国地球科学》，147－156 页，第三章第 7 节，郑州：河南科学技术出版社，1995 年。

环境问题。

海岸类型与环境，受地质构造、陆地地貌发展阶段以及海岸动力过程的差异而具有区域性特点，如：河流三角洲、海岸平原、潮滩与湿地、海滩与沙丘、沙坝与潟湖、珊瑚礁、红树林等不同特征。所以，海岸带与近海的环境因素、动力过程与海岸类型、海岸自然环境各具特点。其特征可归结为：① 海岸带与近海是一个具有不断变化的生物、化学、物理及地质特性的动态区域；② 它是为多种海洋物种提供繁衍生境、高生产力和生物多样化的生态系统；③ 海岸地貌体系，诸如：珊瑚礁、红树林、潮滩与湿地、海滩与沙丘、沙坝与潟湖以及岛、礁等是抵御风暴、潮浪灾害以及侵蚀作用的重要天然屏障；④ 海岸生态系可以通过吸附、沉淀、固封等作用减轻陆源污染的影响；⑤ 海岸带与近海沉积保存着海陆作用过程，古海岸与大陆架发展变化的信息。研究古海岸海洋的演化过程，有助于分析海岸海洋未来变化趋势，对人类持续开发利用海岸带与近海环境具有重要意义。

海岸带与近海有着丰富的自然资源：风与海洋能电力资源；砂矿与油气矿产资源；海涂与土地资源；港湾与深水航道资源；食物资源以及旅游资源（阳光与空气、海水、沙滩与海鲜食品）等。它们具有成本低、少污染以及绝大部分可再生补偿的特性。由于这些丰富的生物和非生物资源，吸引着大量人口到海岸带居住，并从事多项、频繁的开发事业，如：海运、商贸、房地产、食品以及旅游业等。估计在沿海国家中，约有一半的人口居住在海岸带。

中国沿海地带经历过封建王朝时代的禁闭，20 世纪初期与中叶之开拓，70 年代与 80 年代大规模调查研究与规划发展，90 年代已成为我国的“黄金地带”，集中了 41%的人口与 60%以上的工农业产值，地位重要。由于人口密集，经济发展迅速，出现了对沿海资源近期消费（利用需求）与资源的长期供给之间的矛盾。在许多国家，这种矛盾已发展到危机程度，大批海岸带被当地或来自内陆的污染物所污染、渔业资源严重退化或被破坏、湿地干涸、红树林砍伐、珊瑚礁被炸毁，破坏了海岸生态系与供人类享受的海滨，出现了急需解决的新问题。以长江三角洲为例，平均人口密度愈 900 人/km^2，为全国平均值的 8.4 倍，人均耕地不足 1 亩，为全国平均值的 60%，是我国土地承载量最高的地区之一。近年区内人口约以 1%的年率增加，耕地平均以 0.5%的年率递减，农业产值占工农总产值的比重已降到 10%，人地关系协调发展决策迫在眉睫。长江三角洲环境容量的压力亦大，在占国土 1%的土地上承受着全国工业废水的 9%、废气的 5%和废渣的 4%，约 3 500 万以上的农村人口饮用水不符合标准，抵抗灾害的能力脆弱，1991 年涝灾和 1994 年旱灾损失达 500 亿～1 000 亿元以上。

人类活动直接或间接地向海洋输送的污物，60%集中于海岸带近海，深海亦已遭受影响，危及空气与降水的更新，进一步又反馈于人类。由于海岸带与近海急需采取有效的行动，保护与恢复海岸环境资源，因而产生一体化的海岸带管理计划（Integrated Coastal Zone Management，ICZM），并在 1992 年 6 月的联合国环境与发展大会所制定的“21 世纪议程”（Agenda 21）的 17 章中予以肯定。人口、资源与生存环境持续发展是当代面临的巨大挑战，海岸海洋科学是承担解决这项历史性任务的重要环节。无论从海岸带与近海环境的独特性、资源丰富性以及在人类生存环境持续发展中的重要地位等方面，海岸海洋科学（海岸带与近海科学）皆具有优先发展的战略地位。

1982 年第三次联合国海洋法大会通过的《联合国海洋法公约》，已于 1994 年 11 月贯彻执行，使沿岸国家向海延伸范围发生重大变化。

(1) 大陆架是毗邻国领土向海自然延伸至大陆边缘的外沿，在窄狭陆架区则自领海基线至 200 海里的距离。大陆架主权限于毗邻国家，对其他国家的开发需要有限制。

(2) 200 海里专属经济区是一多功能的法定区，海岸国可以根据本国人口的需要，也可据其他国家以及世界共同体的需要开发管理该水域的资源。

(3) 群岛水域、领海、海峡可供国际航行，但是领海权属海岸国家。

根据此公约，原属于公海的 1.3 亿 km^2 海域将划归沿海国家管辖，其面积略小于地球全部陆地面积(1.49 亿 km^2)，以此计，我国管辖的海域面积约为 300 万 km^2，相当我国陆地总面积的三分之一。管辖范围的划分带来权益之争，例如：200 海里专属经济区确定后，在开阔海域中丧失一个具备人类生存条件的岛屿，就会失去 43 万 km^2 原管辖海域，因而出现小岛与巨大海洋区的特殊组合，如南太平洋诸岛国——斐济岛面积为 18 272 km^2，而海域面积为 1 290 000 km^2。因而出现了"海洋国土"新概念，21 世纪是"海洋世纪"或"太平洋世纪"已为学者们所共识，"海岸海洋"成为海洋科学的重点领域，发达国家将研究热点放在海岸带与大陆架浅海不是偶然的。管辖海域划分中涉及领海基线，通常为海滨外界，但外界如何确定？管辖区域的资源环境特点如何？皆需开展研究。美国近十年来致力于大规模海底调查与制图，基本上完成大西洋、墨西哥湾与太平洋岛域周边海区，其目的是为疆域与资源权益服务。我国需加紧这方面的工作步伐，海疆权益急需海岸带与近海海洋的研究亦表明了海岸海洋科学的重要性。

二、海岸海洋科学特点、我国的特色和研究基础

海岸带与近海科学的发展是与其环境、资源的特点、经济开发与国家权益密切相关的。海岸海洋科学具有复合型科学的特点，表现在三个方面：

(1) 多学科交叉，具有海洋科学、大气科学、地质科学、生物科学交叉渗透之特点。原因在于海岸海洋环境因子的多方面性，涉及气、水、生物、岩石圈，需要多学科的交叉、渗透和融合。

(2) 自然科学与社会科学、技术科学的结合与相互渗透。海岸海洋科学研究内容涉及海陆环境、海岸资源、海岸经济开发、海岸与近海疆界、立法与管理等方面。海陆交界的两栖地带，日变化、月、季、年变化大，庞大的体系必须采用处理、分析、建立模式或规划决策等系列性工作，这又决定了它的复合科学体系特性。

(3) 基础理论研究与经济开发应用方面相结合。在解决经济开发、疆域政治与立法等方面项目的同时发展了基础理论，原因在于海岸海洋科学调查研究的装备与船只花费高昂，进行经济开发应用工作有可能获取丰富的海域信息。这将促进对自然环境的了解以及对海岸海洋发展规律的认识。

中国濒临亚洲东部与太平洋及其边缘海(黄海、渤海、东海与南海)之交，大陆岸线长达 18 000 km，包括 6 000 多个岛屿，其岸线长达 32 000 km。海岸带跨越寒、温、热三个气候带，以季风波浪作用为特征；地质构造具有隆起与凹陷相间、构造运动活跃、区域变异大

等特征；中国海岸带与近海具有显著的潮汐作用并以黄海与东海北部海域潮流作用强烈；发源于青藏高原的大河给海岸海洋供给了巨量泥沙，构成中国与东亚海岸带的重要特色；同时，人类活动直接地或通过河流作用而间接地对海岸带与近海环境的发展变化予以重大影响。这些都是中国海岸带与近海环境区别于世界其他海岸海洋的特色。

现代的中国海岸海洋科学始于20世纪50年代后期，一开始即重视多学科交叉（海岸水文学、地质地貌学、沉积学与港口工程学），结合海港泥沙来源、回淤速率与港址选择来研究海岸演变与泥沙运动，重视海岸“两栖性”的特点，将海岸的陆上部分与水下岸坡结合一起进行研究。由于方向正确与手段更新，海岸研究在海港选址工程中取得较大成就，至今仍为海岸海洋学研究中的重要应用方面。

20世纪80年代以来，中国先后进行了“海岸带与滩涂资源综合调查”“海岛资源环境综合调查”以及多种海岸发展的工程项目，中国已获得了海岸海洋全方位、多学科的系统资料、图件与研究成果，为开展海岸带与近海功能区划分、发展规划和立法管理打下了坚实的基础。我国在海岸演变与泥沙运动，大河河口与三角洲发育，平原海岸与淤泥质潮滩发展规律，潮汐汊道港湾类型与发育，海平面变化与低地海岸效应，人类活动对海岸发育影响等方面取得了重要的科学进展。其中大河与海岸发育、潮流作用的淤泥质海岸研究等方面具国际领先地位。与此同时，中国海岸海洋科学形成了具学科综合特点的老、中、青研究队伍，他们活跃于国内海岸海洋科学领域，具有一定的影响。海岸海洋科学工作方法与设备在高新技术装备方面初具规模，如中国科学院、国家海洋局、海洋石油总公司、地质矿产部等有关单位拥有海洋考察船及先进设备，南京大学、青岛海洋大学等高校也有相关的重点实验室及先进仪器。

上述表明，我国海岸带与近海科学已具备相当良好的基础与一定的发展规模，可以依靠本国力量在跨世纪发展的科学工作中做出重要贡献。

三、主要科学问题

作为优先发展领域的主要科学问题，既要反映我国海岸海洋科学的基础特点和在国际上领先的方面，亦需考虑国民经济发展的需要和经费支持的可能性。在跨世纪研究课题中，应注意与国际及中国21世纪议程的主要课题衔接。

（1）沿海经济发达地区海岸带及其邻近海区资源环境承载力与可持续发展研究

包括：研究资源、环境现状、发展演变规律；着重于水、土、生物、海洋资源环境承载力研究，建立数据库；环境质量监测系统研究；脆弱性分析、开发利用模式研究；分期建立海岸带及邻近海区资源可持续利用与保护示范区：辽河口、黄河口、胶州湾、苏北潮滩湿地、长江口、浙—闽港湾、珠江口及海南岛等地区。

（2）海底地貌和海底矿产资源

重点开展河口湾、大陆架、海岛周边海域、专属经济区的海底地貌分布（地形、物质组成）、运动演变规律与底床稳定性分析、非生物资源调查与分布规律研究、研制1∶20万海底地貌、矿产或非生物资源制图信息系统软件；开展以油气开发、海洋工程需要为目的的环境地质、工程地质及灾害地质调查与预测；建立重点河口湾、大陆架、海岛周边及专属经济区环境数据库，为海岸带及近海海疆或资源开发利用、保护与管理以及管辖海域界限划定服务。

(3) 海岸带海-陆相互作用

海-陆界面存在着复杂的物理、化学、生物和地质过程及其相互作用，对它的研究是为气候预测、海面变化和人类活动对海岸带的生态系统、资源、环境影响提供依据。其主要内容是：外力变化或边界条件变化对海岸通量的影响；海岸带生态系统与海平面变化之间的关系；碳通量和痕量气体排放的研究；全球变化对海岸系统社会经济活动的影响。

(4) 河口动力学、动力沉积与古海洋环境

从海-陆相互作用与河海体系相关科学研究不同类型的河口(包括河口湾、三角港、三角洲)动力体系与作用过程、生物化学过程、河口与海岸碳循环、泥沙交换、搬运与通量、沉积与地貌等在海岸与陆架发展演变过程的作用、影响及古河道体系分布与变化规律。

(5) 海平面变化、海岸效应与反馈

完善海平面监测网、建立海平面变化数据系统，开展地区、国际海平面变化对比研究；分析晚更新世、全新世、世纪性海平面变化过程，预测 21 世纪前期、中期，世纪末海平面变化速率与变化趋势；人类活动对海平面变化影响、中国不同类型与地区相对海平面变化、海岸效应与反馈影响研究；海平面变化对中国沿海经济发展影响及对策研究与战略方案制定。

(6) 海底介质的力学特性与海洋工程

海洋工程建设与海底环境：动力状况、浑浊流活动、水团的物化特性等；底床的稳定性；海底沙波的移动变化规律、礁、滩分布与局部动力影响、地形坡度变化，地震、泥火山、盐丘、局部天然气等爆发、滑塌、底床沉积物力学特性(如黏土质膨胀性、粉砂液化作用等)密切有关。包括：海洋工程海区动力场、海水特性、灾害天气影响与频率分析；不同的海底地貌单元结构及物质组成特点(基岩、黏土、砂、砾、粉砂等)；不同底质的物理、化学与抗剪、抗压、黏滞性、可塑性与流变分析；海底工程类型与布局：管道、平台、水下建筑群、倾废场所等。

(7) 海岸带、近海环境与海洋灾害

包括：环境要素监测系统高新技术应用：建成环境立体监测网接收卫星遥感信息，灾害海况监测等，内含机载激光测深技术，激光雷达海面风场探测技术，温度、盐度、密度测试技术，污染监测系统，海啸、风暴潮、水位监测系统等；灾害预报与减灾防灾技术系统研究：风暴潮预报与应急系统，海啸预报与应急疏散系统，赤潮爆发影响范围预报防灾应急系统，海冰预报，大河口、低海岸防涝与防潮侵系统工程，海岸侵蚀机制与防护以及人工海滩、人工岛礁系统研究设计；重点河口与海岸段减灾防灾信息系统软件研制。

(8) 中国领海基线、岛屿、海峡、水道、大陆架与 200 海里专属经济区管辖权限划定专题研究

组织外交、海洋法、历史、地质、测绘、经济以及东南亚、日本、朝鲜等区域研究方面的专业人员共同研究，阐明历史沿革、证据，阐明上述地区自然环境，资源与经济发展基本情况，划界原则与方案，矛盾纷争方面与解决途径等，这些将服务于海洋国土权益，以及子孙后代的持续发展利益。

(9) 中国一体化的海岸带与近海管理系统(ICZM)

一体化的海岸管理是多目标与多方位的,其内容包括:分析发展带来的各种相关影响;分析互相冲突使用导致的相关影响;分析自然过程与人类活动间相互关系及其相关影响;促进区段海岸与海洋活动各部门间的和谐联系,社区的群众必须参加确定研究和发展的议程。

一体化海岸管理要素包括:资源用户的参与;交叉学科的合作;减轻对生物多样性的威胁;海岸环境承载力;水产养殖与天然捕捞的联系。

总之,海岸带与近海是陆地、大洋和大气之间物质、能量交换以及各种因素变化最活跃的区带,是人类生存环境的重要组成部分和生产活动的重要场所,又是沿海国家维护海洋权益和专属经济区纷争的对象。海岸带有大量的溶解和悬浮物质被搬运、储存,并受陆地、大洋和大气的影响。各种物理、化学、生物和地质过程的变化和相互作用,对陆地、大气和水下环境与生态系统产生重要的影响。所以说,对海岸带与近海的研究,涉及到沿海国家的海洋权益以及经济专属区的划定,对全球生物地球化学循环和气候系统变化具有重大影响。

Sediment Transport and Terrigenous Fluxes*

1. Introduction

Rivers are the major transport agent of terrigenous sediment to the ocean, including solid and dissolved material, and greatly influence the sediment dynamics in coastal waters. The widely quoted value for fluvial suspended sediment flux to the world's ocean is 13.5 billion metric tons per year (Milliman and Meade, 1983). That estimate has been revised to about 20 billion metric tons per year (Milliman and Syvitski, 1992). The estimates are based on discharge loads and denudation yields and are partly influenced by elevation and area of river basin (Mulder and Syvitski, 1996). The Milliman and Syvitski estimate involves 280 rivers throughout the world (roughly 8.0 billion tons per year), and takes into account the contribution of small rivers from mountain areas (9.5 billion tons per year), and from upland and lowland areas (1～2 billion tons annually), and adding another 1～2 billion tons for undocumented larger rivers. This 20-billion ton estimate still involves uncertainties, as world observations on rivers are still not widely available. The developing trend may actually be toward decreasing fluvial sediment fluxes as the result of artificial controls (dams) and the continuing rise of sea level, which often acts to decrease river bed slopes eventually decreasing the transportation capacity (Wang and Wu, 1995). We review recent publications, including Wang et al. (1986), Meade et al. (1990), Milliman and Syvitski (1992) and Mulder and Syvitski (1995), on global and regional river fluxes and influences. We survey world river loads with a description of factors affecting sediment delivery by rivers with emphasis on American and Chinese rivers, and conclude with a discussion of the impact of river input to the ocean.

2. River Loads and Feedbacks

2.1 Sediment Delivery

Rivers are the chief mechanism for delivery of terrestrial sediment to the ocean. Sediment is delivered as bedload (sediment moved along the river bed by rolling,

---

* Ying Wang, Mei-e Ren, James Syvitski: Kenneth H. Brink, Allan R. Robinson eds. *The Sea*, Chapter 10, pp. 253－292, John Wiley & Sons, Inc. 1998.

skipping or sliding) and suspended load (sediment that is fully supported by fluid flow and maintained by the upward component of fluid turbulence). Bedload is flow dependent and thus hard to measure but easier to predict; suspended load is source dependent and easy to measure but hard to predict. The proportion of suspended load to bedload increases with the size of the drainage basin, from >50% bedload (Qb) in some small proglacial streams (Church, 1972), to <5% Qb for the large Mississippi River (Meade et al., 1990). In general, bedload accounts for <10% of the present-day transfer of sediment from the continental uplands to the continental margins (Meade et al., 1990).

For the more ephemeral desert rivers, periodic surges appear to play a significant role in the discontinuous process of sediment mobilization (Heggen, 1986). Surged flow increases suspended sediment and bed material transport. Mobilized particle diameter is increased manyfold in ephemeral watersheds. Unlike a flood peak, which rises and subsides within hours, and occurs only once per event; surges impact almost instantaneously, diminish almost as quickly, and may occur many times per event. Whereas the force necessary to turn nonsurge flow is developed by water surface superelevation and distributed by hydrostatic pressure around the outside of a bend, the reaction needed to turn a surge is localized at the point of impact.

Rating curves (Cs or $Qs = aQ^b$) provide an empirical method to convert discharge hygrographs into sediment yield estimates, and provide information necessary to predict sedimentation within coastal basins (Syvitski, 1989c). Rating patterns can vary from simple to complex. Of the rivers that flow directly into the ocean, only those that drain high-yield mountainous areas demonstrate a simple first-order relationship between concentration Cs, or suspended load Qs, and discharge Q (Fig. 1a).

The power function relation between Qs and Q implies that most sediment is transported during only a few days of the year. Most North American rivers, for example, transport 50% of their load in 1% of the year and 80% to 90% of their load during 10% of the year (Meade et al. 1990). At the most extreme, the Eel River, which drains the steep-sloped coast mountains of California, delivered more sediment in 3 days than it had carried in the previous 8 years combined (Brown and Ritter, 1971). Exceptions include glacial streams, where the suspended sediment delivery is spread over the melt water season (Fig. 1a), and coastal plain rivers, where Qs is not greatly affected by sharp increases in discharge (Fig. 1e).

The rating coefficient (b of $Cs=aQ^b$) may vary between 0.5 and 1.5 but can be as high as 2.0 (Nash, 1994; Mulder and Syvitski, 1995). Seasonal differences in the coefficient b reflect variable sources of water and sediment: for instance, ice melt versus snowmelt or rainfall (Fig. 1b), or snowmelt and winter rain versus intense convective rainfalls of summer (Fig. 1c). For upland maritime rivers, rating predictions are

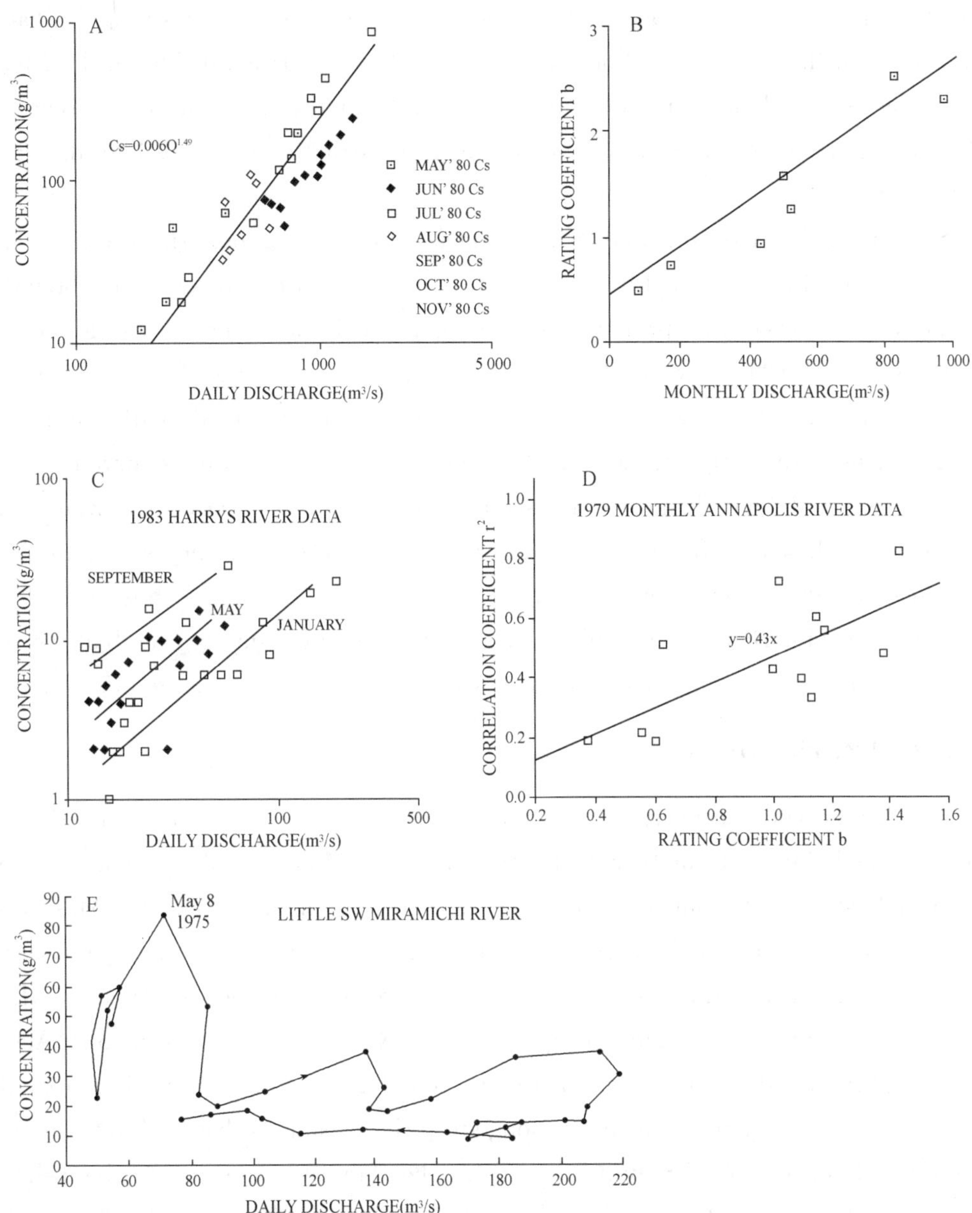

Fig. 1　Some sediment rating patterns for North American rivers of concentrations of suspended sediment. *Cs*, versus river discharge. *Q*, given in metric units, and the rating coefficients *a* and *b* (of $Cs=aQ^b$, based on daily average values over a 1-month period). (a) A simple first-order relationship of daily concentration versus daily discharge for the Stikine River, a very high-sediment-yield river draining the mainland of British Columbia. (b) The rating coefficient *b* increases linearly with increasing discharge of the Stikine River, B. C. (c) Harrys River, Newfoundland, demonstrates highly seasonal racing relationships, with the most turbid waters found in September. (d) For the Annapolis River, Nova Scotia, the statistical reliability of the rating predications increases as the rating coefficient *b* increases (r^2 is the linear correlation coefficient of log *Cs* versus log *Q*). (e) The time history curve of the severely sediment-limited Little Southwest Miramichi River (natural flow, drainage area of 1.340 km²) demonstrates initial sediment flushing of the river in the early spring, followed by less turbid water flowing at higher discharge levels. (Data from Water Survey of Canada, 1975.)

reliable only when the rating coefficient b is large (Fig. 1d). Rarely, there will be no predictable pattern between concentration and discharge; more rarely a strongly negative rating relationship can exist. For example, the Little northwest Miramichi River, Canada, is severely limited in sediment sources and after the snow introduced sediment is flushed out in the early spring, little sediment is carried even during high discharge levels (Fig. 1e).

Seasonality is important when examining sediment rating patterns (Syvitski and Alcott, 1995). If a river is fed by ice fields that melt in mid-to late summer, the initial high discharge levels of the spring when snow rapidly melts will tend to carry less turbid water (Hickin, 1989). The result is a negative rating loop (Fig. 2a). If a river is dominated by snow discharge, a positive rating loop will be observed, with the highest concentrations in the early spring (Fig. 2b). Double looping in a rating curve is typical of river basins with limited and seasonal sediment source, such as snowmelt, where once the seasonal sediment is flushed out, the river remains clean for the rest of the year (Fig. 2c). Random looping is typical for large, clean rivers (i. e. , those with large lakes in their hinterland), in which secondary tributaries supply the bulk of the sediment, and at times that are out of sync with the main river discharge (Fig. 2d).

2.2 Tectonic Activity

The height and steepness of mountains often reflect their youth in terms of tectonics, and Milliman and Syvitski (1992) and Mulder and Syvitski (1996) convincingly show that sediment yield is very dependent on mountain height. Most large river with annual sediment discharges of more than 100 million tons originate in active tectonic settings (e. g. , the Himalayas, the Alps, the Rockies and the Andes), where the snow, glaciers and abundant precipitation supply large volumes of runoff. Active tectonic uplift provides continuous energy for rapid river erosion, which causes the transport of huge amounts of sediment along steep river channels that lead to the ocean. Rivers that drain the tectonically influenced Qinghai-Tibet Plateau include the Yellow, Changjiang (Yangtze), Brahmaputra, Ganges, Irrawaddy, Mekong and Indus (Table I). These seven rivers originate from the highest parts of the earth's surface and may account for a third of the world sediment load.

Other tectonically influenced examples involve smaller rivers, such as those which drain south from the Alps. These rivers have much higher yields than rivers draining northern Europe (Milliman and Meade, 1983). More than 20 tectonically influenced alpine rivers with drainage basins of less than 0.22×10^6 km^2 collectively discharge more than 140 million tons of sediment annually (Milliman and Syvitski, 1992). Small drainage basins have little ability to store sediment, and thus the associated rivers have larger yields than those with larger drainage basins (Milliman and Syvitski, 1992).

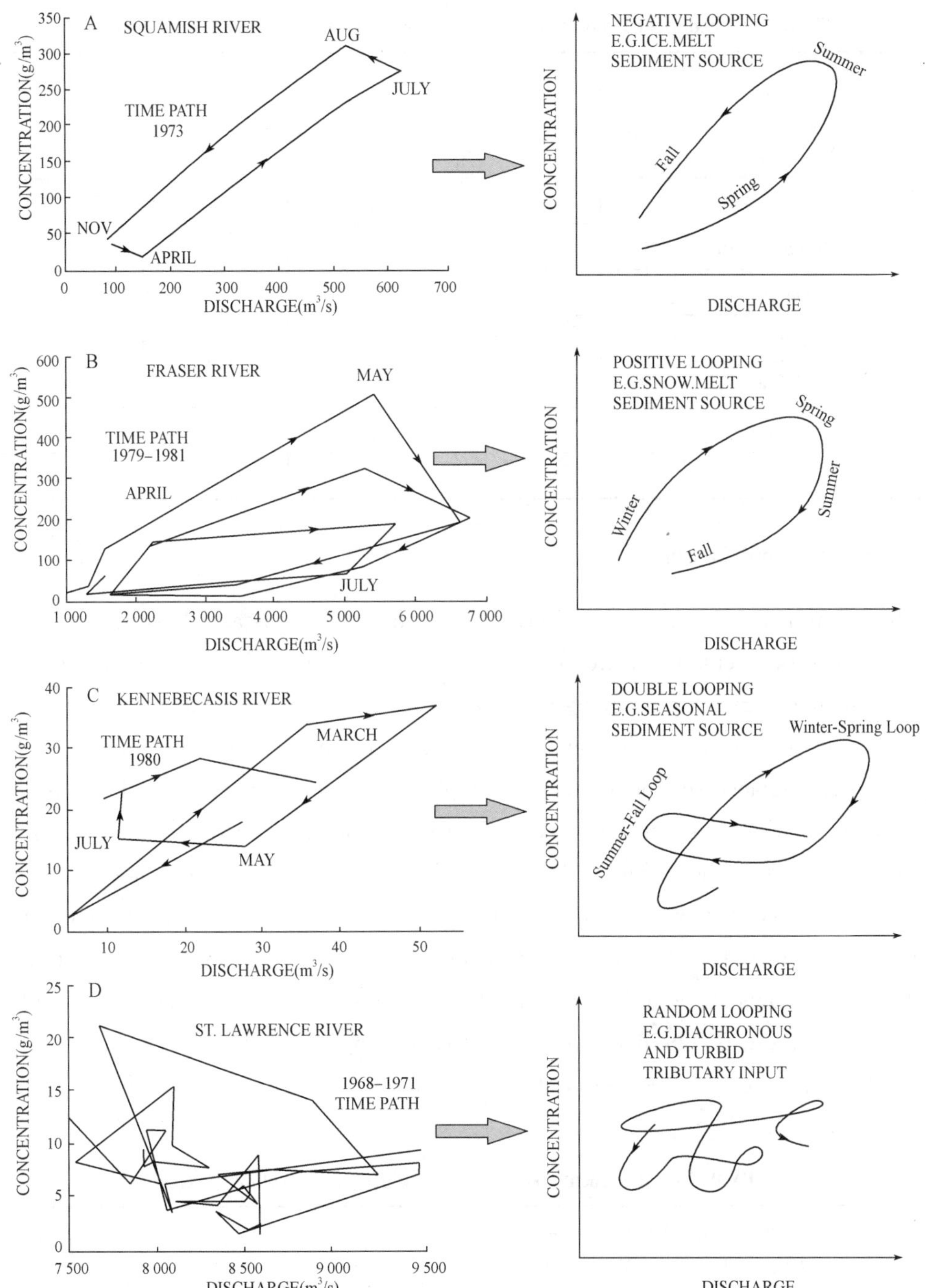

Fig. 2 Time-history plots of sediment rating patterns of four rivers in North American and their idealized rating loops (hysteresis). Mean monthly averages of sediment concentration and river discharge are used to accent the seasonality of these rating curves. (a) The Squamish River drains the southern Coast Mountains of British Columbia and is fed by snow and ice melt and seasonal rains. As an example of negative looping, the highest sediment concentrations are in the summer, due to delayed sediment source associated with glacial ice melt. (b) The Fraser River drains the Coast Mountains and the dry interior of British Columbia, and is an example of positive looping, with the highest concentrations in the early spring associated with snowmelt. (c) The Kennebacasis River drains the southern uplands of New Brunswick and provides an example of double looping: a result of a seasonal sediment source, such as snowmelt, that once flushed from the river is removed until the next year. (d) The St. Lawrence River drains the Great Lakes and the uplands of the St. Lawrence Estuary and provides an example of random looping, typical of relatively clean rivers (i. e., large lakes in the hinterland) wherein secondary tributaries seaward of the lakes supply the bulk of the sediment and at times not in synchroneity with the main river discharge. (Data from Water Survey of Canada files.)

TABLE I　Sediment Load of Rivers Originated from the Qinghai-Tibet Plateau

River	Area ($\times 10^6$ tons yr^{-1})	Load ($\times 10^6$ tons yr^{-1})
Yellow River	0.75	1 115
Changjiang (Yangtze)	1.81	461
Brahmaputra	0.61	540
Ganges	0.98	520
Irrawaddy	0.43	260
Mekong	0.79	160
Indus	0.97	59
	6.34	3 115

2.3 Regional Features

Regional features or geologic structure can exert substantial influence on the sediment load of rivers. Geologic factors that affect yield include the bedrock type being eroded (sedimentary material being more easily eroded than crystalline material) and the development of extensive soils, and the sedimentary material left behind by the last glaciation (paraglacial sediments). For example, the runoff of the Yellow River is about 1/15 of that of the Mississippi River, 1/183 that of the Amazon River, and 1/2 that of the Nile River, but the sediment load of the Yellow River is three times that of the Mississippi, twice that of the Amazon River and nine times more than the Nile River. The difference in sediment concentration is even higher. The reason for this is that the Yellow River flows through the region of the Loess Plateau, and the unconsolidated loess is easily eroded. The major load of silt carried by the Yellow River is eroded from its middle and lower reaches [i. e., the Loess Plateau and ancient fluvial fans (Wang et al., 1986); Table II].

TABLE II　Sediments-Source Relationships of the Yellow River

Station	Drainage Area (10^4 km^2)	Annual Average Runoff (10^8 m^3)	Annual Average Load (10^8 tons)	Average Concentration (kg/m^3)	Period
Lanzhou	22.3	343	0.96	2.80	1956—1989
Shanxian	68.8	436	16.10	36.90	1956—1979
Lijin	75.2	379	9.47	22.35	1950—1992

Source: Chai (1993).

The Mississippi River sediment load reflects the regional features of a glacial-fluvial fan provenance, which occurs in the river's middle and lower reaches. Badlands cover only 2% of the drainage basin of Red Deer River in Alberta, yet contribute 80% of the sediment that reaches its mouth (Meade et al., 1990). The Columbia River increased its suspended load by a factor of 4 following the ash contribution to the drainage basin from the Mount St. Helens volcanic eruption (Meade et al., 1990).

Although Canada and Alaska have significant areas of land under agricultural production, much of the sediment that reaches the Atlantic, Pacific, and Arctic Oceans results from the natural erosion of alpine regions, rock flour generated from the ablation of glaciers and ice fields (particularly in the eastern Arctic Archipelago), and erosion of Late Quaternary (glacial, preglacial and glacial-marine) deposits that cover much of the Canadian and Alaskan landmass (Meade et al., 1990). An example of the effect of glaciation is seen by comparing two Alaskan rivers that are similar in drainage basin size and runoff: the Cooper (glacial-fed) River has a yield over 10 times larger (at 1 200 tons km^{-2} yr^{-1}) than the Kuskokwim (nonglacial) River.

2.4 Climatic Factors

Mountain ranges that face humid oceanic climate settings receive plentiful precipitation. Other ranges may have more stable water resources from high mountain glaciers. Under either of these circumstances, abundant runoff is generated. Regional climate and seasonality can be important factors that influence annual river discharge. For example, the average runoff to the Changjiang River is a linear function of annual average rainfall (Fig. 3). However, between the monthly runoff and sediment discharge for the Changjiang River there is a disproportionately strong direct relationship with high sediment transport during summer and early autumn months (Fig. 4).

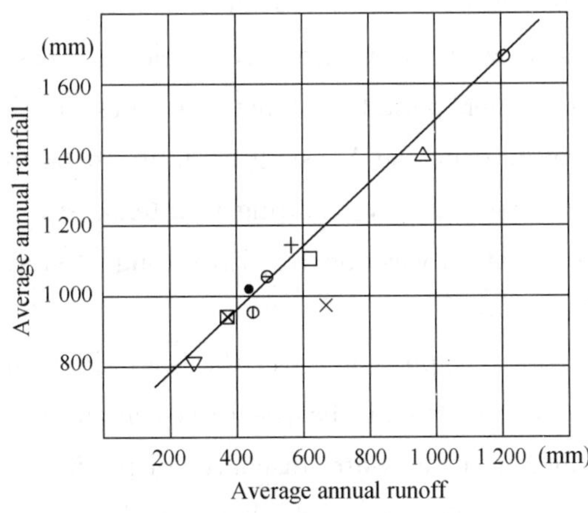

Fig. 3　Relation between average annual rainfall and average annual runoff of the Changjiang River and its major tributaries (from Geographic Institute of Chinese Academy of Sciences et al., 1985).

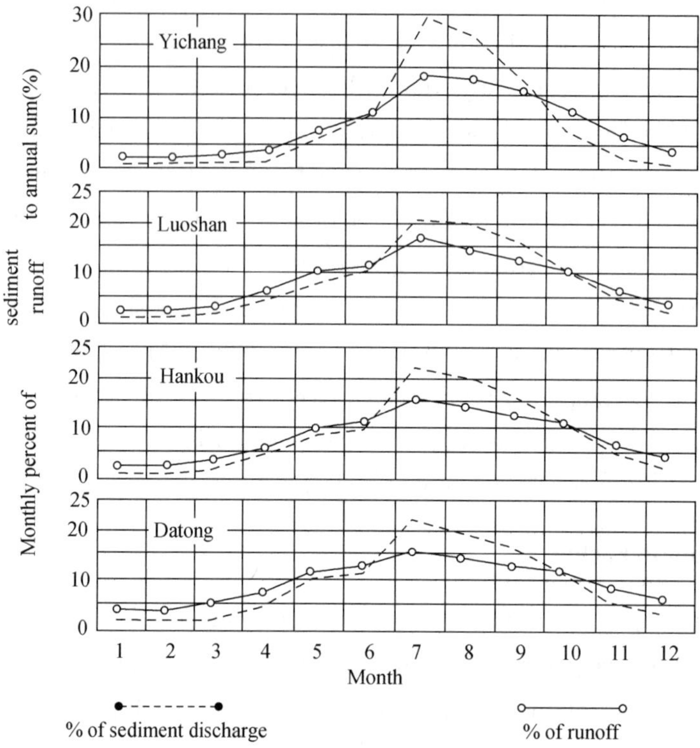

Fig. 4 Percent of monthly sediment discharge and runoff in relation to total annual sediment and water discharge at major stations in the lower reaches of the Changjiang River.

In North America, sediment yield in midcontinental areas reaches a maximum when the effective discharge is about 300 mm yr^{-1}. Below this value, sediment yield decreases due to low stream velocity, and above it also decreases, due to increased vegetative cover. On the coastal mountains of northwestern California, yield increases exponentially with increasing runoff. Over longer periods, climate changes in the atmospheric circulation can occur. For example, over the Mississippi drainage basin, before 1895 and after 1950, the circulation was meridional, resulting in intense storms in midlatitude; between 1895 and 1950 the circulation was zonal, with strong westerly flow and fewer storms (Meade et al., 1990).

Annual precipitation, especially rainy season precipitation, over the Loess Plateau, where the Yellow River collects much of its sediment load, has decreased about 10% since 1970 (Table III). There has been a reduction in the water discharge of the Yellow River since 1969. The Yellow River system's annual sediment discharge is determined largely by the number of days experiencing over 40 mm of precipitation. The decrement ratio of sediment discharges of the Yellow River during the 1980s is considered 63% affected by changes in precipitation and 37% by human activity(i. e., water and soil reservation) (Wang, 1994).

TABLE III　Comparison of Annual Average Observed Water and Sediment Discharge Between 1950—1970 and 1971—1985 in the Region of Loess, China

Sation	Item	Annual Average, 1950—1970 (10^8)	Annual Average, 1971—1985 (10^8)	Decrease	
				Value(10^8)	%
Hekou town	Water discharge (m^3)	255	246	9.1	3.6
	Sediment discharge (t)	1.6	1.2	0.5	29.0
Longmen Hua county	Water discharge (m^3)	444	378	66	14.9
Hejin Futao	Sediment discharge (tons)	17.4	11.5	5.9	33.8

Source: Shi (1989).

2.5　Anthropogenic Effects

Modern sediment loads in rivers seldom represent their natural load, increased diversion and damming of many rivers has decreased sediment discharge dramatically. The Nile and the Colorado deliver little sediment to the ocean. The Mississippi, Zambesi and Indus have experienced marked decreases in sediment discharges in recent years (Milliman and Syvitski, 1992). Dams on the Ganges have similarly decreased sediment discharge, although increased erosion in the mountains of Nepal as a result of deforestation has increased the load of the confluent Brahmaputra (Hossian, 1991). In the continental United States, yields have increased locally by a factor of 10 over preagricultural time. This effect is most dramatic in the cultivated loess soil regions of northern Mississippi, where sediment yields from cornfields can be 100 times those from woodlands within the same drainage basin. Logging in the western coastal mountains has also created extraordinary increases in sediment yields.

Other human-induced changes in sediment yields are caused by mining operations, urbanization, hydroelectric projects and irrigation. Over 1 km^3 of sediment was added to the Sacramento River system as a consequence of hydraulic mining in the Sierra Nevada foothills between 1849 and 1909 (Gilbert, 1917). Urbanization effects in the Washington-Baltimore area, for instance, doubled the sediment delivery of the Potomac River to its estuary. Other anthropogenic influences decrease yields, particularly through the proliferation of reservoirs along river systems in the continental United States and southern Canada over the last 100 years (Ashmore, 1993). For instance, the Colorado presently discharges three orders of magnitude less sediment (as measured at Yuma, Arizona) since the building of the Hoover Dam in the mid-1930s (from $>10^8$ tons yr^{-1} to 10^5 yr^{-1}). Similarly, the Mississippi presently discharges 50% to 60% less sediment to the ocean, due to dam construction, even though agriculture usage has greatly increased. Fig. 5 shows the effects on sediment delivery of reservoir construction on major rivers of Georgia, South Carolina and North Carolina.

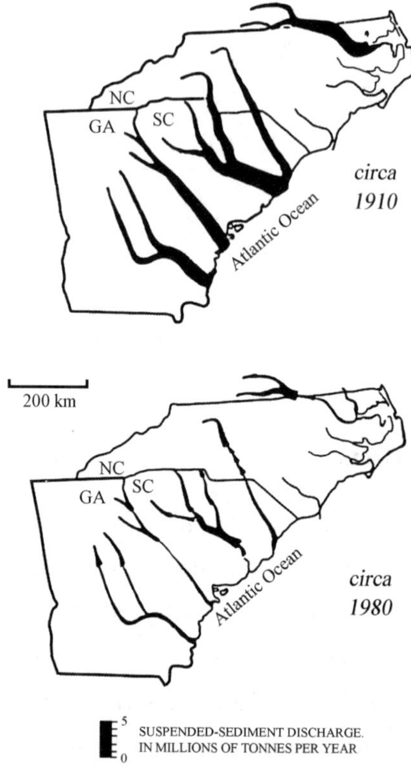

Fig. 5　Effect of construction of reservoirs on the major rivers of Georgia. South Carolina and North Carolina, shown by comparison of sediment discharge levels pre-dam (circa 1910) and after dam (circa 1980) construction. Width of the river represents the mean annual suspended sediment discharge. (From Meade et al., 1990)

As a result of human intervention in the Mediterranean Basin, sediment discharge of many rivers to the sea has been reduced. For example, as a consequence of the reduction of sediment flux of the Ebro River (Spain) by 96%, the delta of the Ebro is seriously threatened by erosion (Jeftic et al., 1996). In northern China since the 1980s, construction of reservoirs and diversion of river water for city water supplies has nearly eliminated the sediment lux from the Luan River to the Bohai Sea. Similarly, lock construction at the river mouth in 1958 has blocked nearly all the water and sediment delivery of the Hai River to the Bohai Sea. There are about 10 000 large water reservoirs in the world, storing 10 000 km^3 of water as of 1991, and concomitantly storing large volumes of fluvial sediment (Chao, 1991). In China, more than 80 000 different-sized reservoirs presently store 475 km^3 of water, equivalent to 26% of the total discharge to the sea of all Chinese rivers.

To emphasize the effect of humans on rivers, we examine anthropogenic influences on China's main river, the Yellow. During historical time, the increase in Yellow River sediment load started primarily after the Tong Dynasty (1 400 years ago). As a result of increased soil erosion caused by farming (including deforestation) of the pastureland on the Loess Plateau, the sediment load of the Yellow River has been increasing. The current plan for the economic development of the Upper and Middle Yellow River Valley will impose a huge demand in water resources. It is estimated that in the early twenty-first century, an additional 23 km^3 of water per year will be needed for the development of coal mining (projected production 6×10^8 metric tons per year), for a thermal power plant (projected capacity 30～40 million kilowatts; it is estimated that a 1-million kilowatt thermal power plant consumes about 44×10^6 m^3 of water a year) and for other water-consuming industries and irrigation. The Yellow River Commission projects that the irrigation rate will be doubled. This volume is approximately 50% of the mean annual flow of the Yellow River at Huayuankou (average, 1949—1979). Therefore, a significant decrease in the flow of the Yellow River seems inevitable in the future (Ren, 1994).

Lijin station is located at the apex of the Yellow River Delta, 113 km from the sea. Temporal data (Fig. 6) show a reduction in both sediment and water discharge since 1969 (Shi, 1989). There are three explanations for this trend. (1) A water and soil conservation project started in 1955 has operated effectively since 1977. Remedial land reached 10^5 km^2 by the end of 1985. This represents 23.3% of the total area of the eroded Loess Region (Table IV). (2) There has been decreased precipitation in the region since 1970, as noted above (Table IV). (3) Since the 1970s, water diversion projects in the Yellow River Basin have increased rapidly because of the rising demand

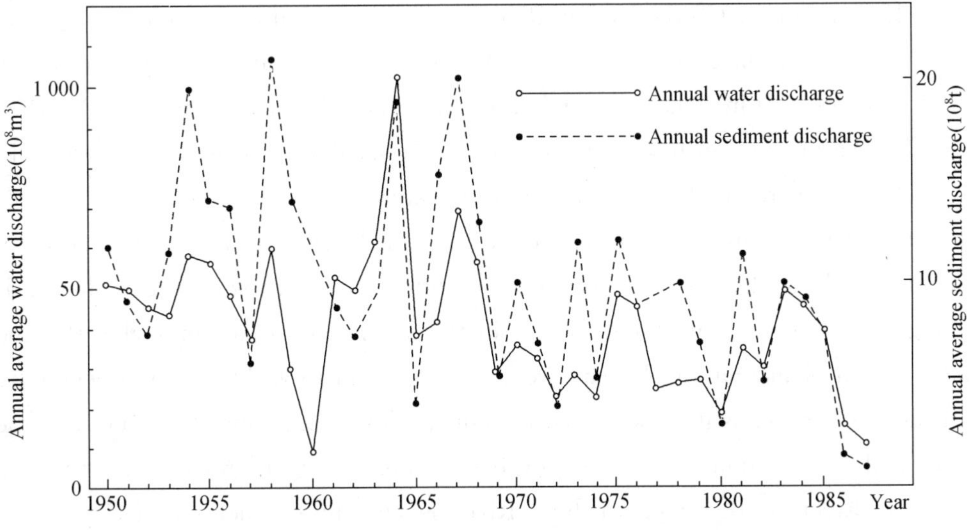

Fig. 6 Variation of annual water and sediment discharge at Lijin Station on the Yellow River between 1950 and 1987 (from Shi, 1989)

for water resources for industry and agriculture: 274×10^8 m^3 of water is being used each year, a 2.2-fold increase over the average 1950s value. This volume is about 47% of the annual average water discharge of the entire Yellow River. About 1.8×10^8 metric tons of sediment is also taken away from the river. As a result, Yellow River discharge of sediment to the sea (Lijin station) was reduced dramatically, to 4.1×10^8 tons yr^{-1} between 1986 and 1993.

TABLE IV Comparison of Precipitation Before and After 1970 at Three Locations in the Loess Region

Station	Annual Average Precipitation (mm)		Decrease		Annual Average Precipitation (mm)		Decrease	
	1954—1970	1971—1983	mm	%	1954—1970 July—Sept.	1971—1983 July—Sept.	mm	%
Taiyuan	475.1	441.7	33.4	7.0	301.9	269.6	32.2	11.0
Yanan	578.2	531.8	46.4	8.0	352.1	322.1	30.0	9.0
Yuling	444.8	367.3	77.5	17.4	295.7	244.0	51.7	17.0

Source: Shi (1989).

Rapidly growing water demand for agriculture and urban water supply has withdrawn an increasing amount of water from the Yellow River. The proportion of natural flow (water discharge recorded at a station plus amount of water diverted for agriculture, industry and other uses in the basin upstream from the station) discharging to the sea has gradually decreased. The proportion was 79.9% in the 1950s and 73.3% in the 1960s, decreasing to 54.7% in the 1970s. Forthese reasons, and also because of the drier weather, the mean annual flow in 1992 at Lijin was only 18 km^3, and the mean annual sediment load between 1986 and 1992 was about 40% of the long-term average (1949—1985). Since 1972 the Yellow River at Lijin has dried up for tens of days every year. The river remained dry for 103 days in 1980, for 128 days in 1992, for 133 days (March 4 to July 14) in 1995, and for 142 days in 1996 (February 24 to July 14) when it practically became intermittent. With further economic development of northern China, it is likely that increasingly large amounts of water will be diverted from the Yellow River. For example, by 2000, an additional 2 km^3 of the river's water will be needed for the modem Yellow River Delta, coastal cities of Shandong Province (Qingdao, Zibo, Weifang, and Yantai), and for irrigation in the southeastern part of Hebei Province. Other areas in Shandong and Henan Provinces also plan to withdraw more water from the Yellow River. Thus it is likely that the annual flow at Lijin may be reduced further.

To add to the complexity, the interbasin water transfer project from the upper Changjiang River to the upper Yellow River is currently under consideration. The easiest route, from the Yalong Jiang, a tributary of the upper Changjiang, will probably add 4.5 km^3 water per year to the upper Yellow River. The total completion of the project would divert 20 km^3 yr^{-1} of water from the upper Changjiang. This volume will roughly balance the additional water demand in the upper and middle Yellow River Valley over the next 20～30 years. Moreover, since more than 50% of the water diverted from the Yellow River is for agriculture, improvement in irrigation techniques and other advancements in water-saving agricultural methods may cut per water consumption considerably.

The increasing diversion of Yellow River water for irrigation and other purposes over the next decade will continue to withdraw sediment from the Yellow River. The construction of the large Xiaolangde Reservoir, scheduled to be completed by A. D. 2000, could trap 3.3×10^8 tons of sediment per year between A. D. 2001 and 2300. Recently, sediment has been pumped from the Yellow River channel to widen and strengthen the main dikes along its lower reaches (total length about 1 400 km). In Shandong Province, by the end of 1991, 3×10^8 km^3 of sediment had been pumped from the river. A similar amount of river sediment was also dredged in Henan Province. Perhaps 0.5×10^8 m^3 of river sediment will be pumped each year for this purpose over the next decade. Progress in soil conservation in the Loess Plateau will further reduce the river's sediment load. It estimated that owing to economic development and the construction of large reservoirs (25 reservoirs) and other water conservation works, discharge of Yellow River sediment to the sea may be reduced by 4×10^8～5×10^8 tons yr^{-1} in the next 20～30 years (Ren and Zhu, 1994).

Besides natural sediment, land-based contaminants are diffused by river flow. Along the coast of China, there are about 281 point sources of pollution, including industrial, agricultural, urban/domestic, harbour, vessel, marine-petroleum exploited, etc. Nearly one-third of the pollution reaches the coastal ocean through fluvial transport processes (Table V).

TABLE V　Amounts of Inland-Source Contaminants Reaching the China Seas Through River Transport Mechanisms, 1980s

Sea	Pollutant (tons yr^{-1})										
	Oils	Hg	Cd	Pb	Zn	As	Cr	Volatile	Cyanide	COD	Cu
Bohai Sea (river-induced)	5 038	8.9	17	44.4	120	63	10	430	40	304 811	
Yellow Sea (river-induced)	3 160	3.3	0.7	103.0	1 156	205	154	645	200	170 598	
East China Sea	58 300	70	139	5 111	5 900	3 154	3 125	3 654	6 248	3 355 600	1 530 000
(Yangtze River)	42 000	60	90	3 000	4 300	300	3 000	3 600	6 000	3 000 000	13 800
South China Sea (river-induced)	25 750	11.4	253	10 529	16 280	1 139	309	493	114	1 033 840	

Source: Fu and Li (1995).

3. World River Sediment Delivery

We have corrected some of the Milliman and Syvitski (1992) tabulations of sediment loads for various world rivers. The Milliman and Syvitski (1992) data set is large and so is not reproduced here. Corrections include (1) maximum altitude of the Yellow River: 4 368 m, thus classifying the Yellow River as a high-mountain river, (2) maximum altitude of the Mekong is higher than 4 000 m, thus a high-mountain river; (3) the Daling River is shifted from the 1 000 to 3 000 m mountain rank to the rank of upland river (500～1 000 m); and (4) Luan, Yalu, Qiantang and Min Rivers are now placed in the 1 000～3 000 m mountain rank. The sediment discharges by major rivers from different continents (Table VI), the impact of hinterland elevation (Table VII) and the delivery to each of the oceans (Table VIII) are summarized. The primary terrigenous sediment input is from the highest mountain areas of Asia, and the secondary is from the American continents. The Pacific Ocean appears to have accumulated most of the sediments.

TABLE VI Sediment Discharge by Major Rivers of Different Continents

River Inputs of Each Comment	Load($\times 10^6$ tons yr^{-1})
Asia	4 507
North America	595
South America	1 731
Oceanic Islands	380
Europe	388
Africa	295
	7 897

TABLE VII Estimated River Sediment Sources Based on Elevation Rank of River Headwaters

River Originates from:	Load($\times 10^6$ tons yr^{-1})
High mountain (<3 000 m)	5 370
Mountain (1 000～3 000 m)	2 139
Upland (500～1 000 m)	331
Downland (100～500 m)	55
Coastal plain (<100 m)	1.3
	7 897

TABLE VIII　Estimated Fluvial Sediment Inputs to Major Oceans and All Inland Seas

River Discharges into:	Load ($\times 10^6$ tons yr^{-1})
Pacific Ocean	3 229
Atlantic Ocean	2 507
Indian Ocean	1 940
Arctic Ocean	202
Inland seas	19
	7 897

Estimated annual sediment input by major North American rivers is 595×10^6 tons yr^{-1}, much less than sediment input from either Asia or South America. Our reason is the luxuriant vegetation cover. Much of the sediment leaving the contiguous United States and Canada comes from three large rivers-the Mississippi, Mackenzie and Colorado (now dammed)-and smaller west coast rivers (e. g. , Eel, Columbia, Fraser), most of which drain mountains. Large discharges of sediment also come from rivers draining parts of western Canada and Alaska. The Susitna, Cooper and Stikine Rivers, for example, collectively drain an area <4% that of the Mississippi but discharge nearly a third as much sediment. Many other rivers along the west coast must also contribute large amounts of sediment; the average thickness of Holocene sediment on the southeast Alaskan shelf is 55 m (Molinia et al. , 1978), and fjords into which many of these rivers discharge have Quaternary sediment thicknesses greater than 500 m (Syvitski et al. , 1987). Most rivers draining eastern North America have relatively low sediment loads.

Eastern South America is drained by four major rivers (Magdalena, Orinoco, Amazon, Parana) all having their headwaters in the Andes. Collectively, they drain more than half the continent (17 million km^2). The total sediment discharge by major rivers from South American rivers is $1\,730 \times 10^6$ tons yr^{-1}. The Amazon River is the main contributor, since it is the world's largest river, with an average discharge of 200 000 $m^3\ s^{-1}$ (Nordin and Meade, 1986). The Amazon affects a large region in the tropical Atlantic. Sediment delivery by the Amazon is between 11×10^8 and 13×10^8 metric tons of suspended matter (Meade et al. , 1985) and between 0.3×10^8 and 0.8×10^8 metric tons of bedload (Eisma et al. , 1991), the suspended load is dispersed from river mouth by the strong tidal current along the shelf toward the northeast. Annually about $6.3 \times 10^8 \pm 2.0 \times 10^8$ metric tons of mud is deposited on the continental shelf off the river mouth (Kuehl et al. , 1986). Approximately, 1.5×10^8 metric tons annually is transported along the coast of the Guianas to the northwest in the suspension. About 10^8 metric tons is transported in the form of large migrating mud banks, 20～35 km long, and move along the coast at the speed of between 1 and 5 km yr^{-1} with a spacing of 15 to 25 km (Eisma et al. , 1991). Dispersal of Amazon mud reaches from the river mouth to the Orinoco Delta and eastern Caribbean, distributed over 1 500 km (Eisma et al., 1991). Their studies indicate that the mud dispersal has been alternated from erosion to deposition along the coast of Surinam almost periodically during historical time. The river-sea dynamic systems of Amazon demonstrate a pattern among world rivers.

European rivers have the lowest sediment discharge (e. g., Holeman, 1968; Milliman and Meade, 1983). Their annual sediment discharge to the ocean is about 390×10^6 tons yr^{-1}. The Alps are a major sediment source, and short rivers draining south into the Mediterranean have high to very high yields (>1 000 tons $km^{-2}yr^{-1}$): Table VI (note that this figure does not include sediment discharge from small rivers). For example, the Semani River (Albania) has more than twice the annual discharge (22×10^6 tonnes) of the collective sediment discharges of the north-flowing rivers such as Garonne, Loire, Seine, Rhine, Weser, Elbe, Oder and Vistula, most of which drain upland or lowland terrain (Milliman and Syvitski, 1992). The Rhine is the only large alpine river that drains north to the sea, and most of its sediment load is trapped in Lake Constance. Euro-Asia has several large rivers (Ob, Lena and Yenesi) that drain north to the Arctic Sea. They originate from hilly and lowland areas, with low rates of soil erosion and long periods of frozen ground. Their sediment discharges to the ocean are anomalously low. The rivers draining the Caucasus Mountains and the Anatolian and Taurus Mountains in Turkey drain a tectonically active area and have high sediment yields. Before dam construction in the 1950s, the three largest Turkish rivers emptying into the Black Sea discharged an estimated 50×10^6 tonnes of sediment annually (Hay, 1989).

Rivers draining Africa discharge relatively little sediment to the sea, about 290×10^6 tons yr^{-1}. Africa is one of the highest-standing continents in terms of average elevation, but not in terms of maximum elevation. Most of the land is under a dry climate environment (Walling, 1985) and is drained by many large nonmountainous rivers with low loads (e. g., Senegal, Niger) (Milliman and Syvitski, 1992). The major sediment discharge comes from rivers draining the rift mountains in eastern Africa (Nile, Zambesi, Limpopo, Rufiji) or from those draining the mountains in Morocco, Algeria and Tunisia (e. g., Mouloura, Sebou, Cheliff).

Sediment discharge to the ocean by major Asian rivers is the highest in the world, at $4\,500\times10^6$ tons yr^{-1} (Table VI): the world's highest mountains, active tectonism, copious monsoon precipitation, serious soil erosion caused by deforestation, very high rock weathering and ubiquitous unconsolidated and thick sediment cover exist in the Loess Plateau. The combination of these factors produces a huge supply of sediment to the Asian rivers, especially the rivers from the Qinghai-Tibet Plateau (Table I). Seven rivers (Yellow, Changjiang, Brahmaputra, Ganges, Irrawaddy, Mekong and Indus) deliver $3\,120\times10^6$ tons yr^{-1}, larger than the total sediment flux from the large rivers of

North America, South America, Europe and Africa combined (Table VI). Note that this does not include sediment flux from small rivers. Rivers from eastern Asia have lower sediment loads relative to the mountainous rivers. In Japan, the mountains are forested to protect the hill slopes, and the construction of dams has reduced sediment discharge to 8×10^6 tons yr^{-1} compared with Korea, where the sediment input by rivers is about 19×10^6 tonnes annually.

The continent of Australia is generally low standing and arid and only those rivers draining mountain areas in the north (e. g. , the Ord) and east (e. g. , the Murray) have high loads. Millimanand Syvitski (1992) estimate the sediment load of New Guinea, the Philippines, Java and New Zealand as unusually high: Oceanian rivers drain high relief regions receiving heavy rainfall, arid thus some sediment yields reach 20 000 tons km^{-2} yr^{-1}. The rivers of Oceania contribute at least 10% of the global sediment discharge (Milliman and Syvitski, 1992). This figure comes from their analysis of larger rivers and does not include the contribution from smaller rivers.

4. River Input to the Ocean: Case Studies on Sedimentary Dynamics Processes

River discharge is the major contributor to delta formation and growth of continental margins. Longshore current transport, or cross-shelf transport through the submarine channels or canyons via turbidity currents, subsequently reworks the river-borne sediment. Not far from the river mouth of the Ganges, for example, a submarine canyon channels riverine sediment onto the huge Bengal Submarine Fan. Therefore, although the Brahmaputra-Ganges has one of the largest sediment loads in the world, the effect of the river on the continental shelf sedimentary patterns is rather insignificant.

Insight to sedimentary dynamic processes involving terrigenous fluxes to the coastal ocean are reviewed briefly through an examination of examples from China and North America. The authors are more familiar with these regions and they present a wide variety of terrigenous flux processes.

4.1 Land-Sea Interaction: Examples from China

The modern continental shelf of the China Seas was a coastal plain during the late Pleistocene. River processes were the major factors in both forming the plain and distributing sediment. Understanding the sedimentary processes of different types of rivers and the characteristics of the sediment they carry is very important for the understanding continental shelf sedimentation.

A total of 2×10^9 tons yr^{-1} of fluvial sediment is discharged into the China Seas

(Table IX). Of this, approximately 1. 21×10^9 tons yr^{-1} is deposited in the Bohai Sea, 15×10^6 in the Yellow Sea, 0. 6×10^9 in the East China Sea and 0. 16×10^9 in the South China Sea and the Pacific. Each year, 1 km^3 of sediment is carried by Chinese rivers to the continental shelf. The highest rate of deposition is in the Bohai Sea, with an average annual sedimentation rate of 8 mm. At this rate, and given an average water depth of 18 m, the Bohai Sea basin would be completely filled in 2 250 years if basin subsidence processes were not active (Wang et al., 1986.) The ancient Yellow River and the Changjiang (Yangtze) River once extended across the full width of the continental shelf. Detritus on the continental shelf was supplied by rivers during geological time (Ren and Zhang, 1980). In this chapter we present five different types of rivers showing different characteristics of sediment movement and distribution in the estuaries and their influence on the adjacent continental shelf (Fig. 7).

TABLE IX Average Water Discharge of Major Rivers of China

Sea Area	River	Area of Drainage		Average Water Discharge		Average Sediment Discharge		Yield (t/km²/a)
		km²	%	10^9 m	%	10^4 t	%	
Bohai Sea	Liao River	164 104	12. 3	87	10. 9	1 849	1. 5	113
	Luan River	44 945	3. 4	49	6. 1	2 267	1. 9	505
	Yellow River	752 443	56. 3	431	53. 7	111 490	92. 2	1 482
	Sum of rivers above	961 492	72. 0	567	70. 7	115 607	95. 6	1 202
	Sum of all rivers entered	1 335 910	100. 0	802	100. 0	120 881	100. 0	905
Yellow Sea	Yalu River	63 788	19. 1	251	44. 8	195	13. 3	31
	Sum of all rivers entered	334 132	100. 0	561	100. 0	1 467	100. 0	44
East China Sea	Changjiang (Yangtze)	1 807 199	88. 4	9 323	79. 7	46 144	73. 1	255
	Qiantang River	41 461	2. 0	342	2. 9	437	0. 7	105
	Min River	60 992	3. 0	616	5. 3	768	1. 2	126
	Sum of rivers above	1 909 652	93. 4	10 281	87. 9	47 349	75. 0	248
	Sum of all rivers entered	2 044 093	100. 0	11 699	100. 0	63 060	100. 0	308
South China Sea	Hanjiang River	30 112	5. 1	259	5. 4	719	7. 5	239
	Pearl River	452 616	77. 3	3 550	73. 6	8 053	84. 2	178
	Sum of rivers above	482 728	82. 4	3 809	79. 0	8 772	91. 7	182
	Sum of all rivers entered	585 637	100. 0	4 822	100. 0	9 592	100. 0	164
Pacific Ocean		11 760	100. 0	268		6 375		
Total	Sum of all rivers in Chinese territory	4 311 532		18 152		201 374		467

Source: Cheng and Zhao (1985).

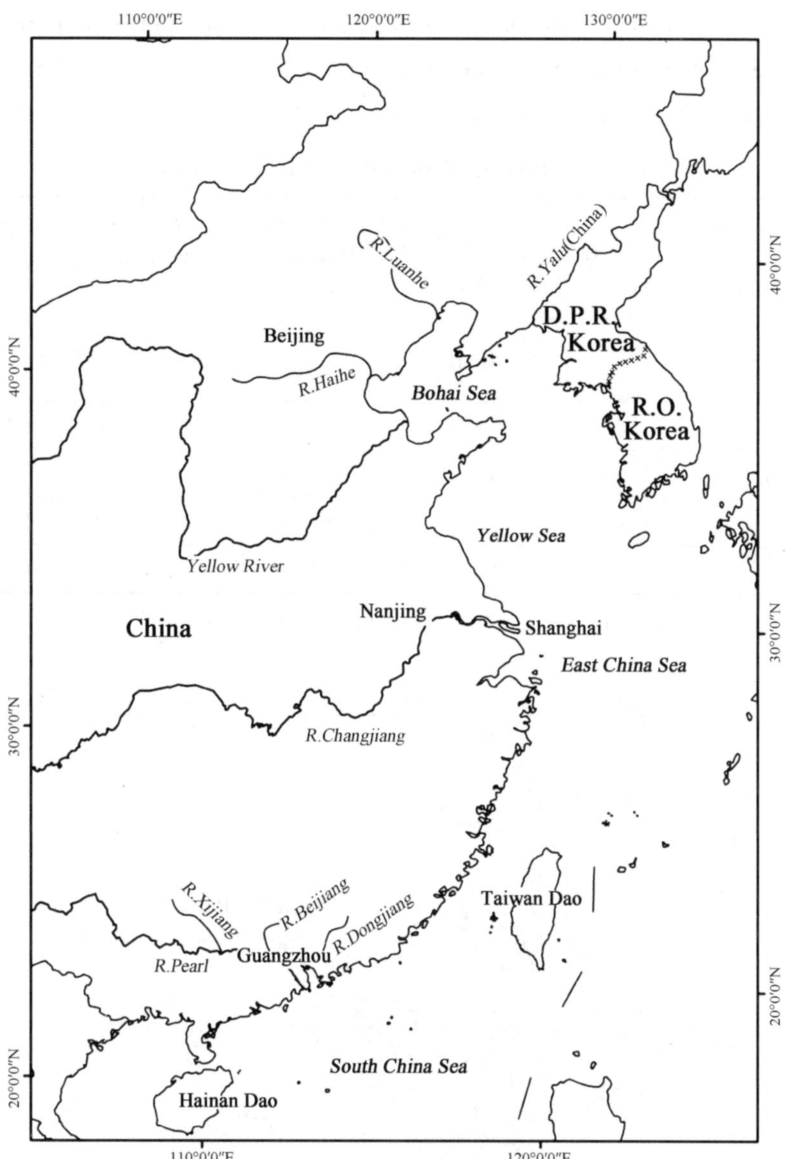

Fig. 7 Location map of the major rivers entering the China Seas (After Wang et al., 1986)

Yalu River Mouth: Pattern of Macrotide Control

The Yalu River flows through the Changbai Mountains in northeast China and discharges into the north end of the Yellow Sea (Table X). Of its total water and sediment discharge, 80% occurs during summer flood, from June to September. The river channel has a relatively steep gradient along its lower reaches, and only a small amount of sandy sediment is deposited in the river channel. Macrotidal processes at the river mouth have formed a wide, funnel-shaped estuary (Fig. 8). The powerful tides of the estuary result from the 6.9 m maximum tidal angle. Tidal currents run perpendicular to the coast, with an average velocity of 1.25~1.50 m • s^{-1}. The flood tide period is longer at the river mouth, decreasing upstream; the ebb tide period is correspondingly shorter and increasing upstream. The difference between the period of flood and ebb ranges from 15 to 18 minutes at the river mouth to 2 hours 28 minutes at Dandong.

TABLE X　Drainage Basin Data for Five Chinese Rivers

River	Drainage Area (km^2)	River Length (km)	Annual Water Discharge (m^3)	Annual Sediment Discharge (metric tons)	Average Sediment Concentration (kg/m^3)	Average Tidal Range (m)
Yalu	64 000	859	27.8×10^9	4.75×10^6 (ebb-flood)	0.33～0.42	4.48
LuanHe	44 900	870	38.9×10^8	24.08×10^6	3.94	1.50
Yellow	752 443	5 464	48.5×10^9	11.9×10^8	37.7	0.80
Changjiang	1 807 199	6 380	9.25×10^{11}	4.86×10^8	0.544	2.77
Pearl	452 616	2 197	3.70×10^{11}	$0.85\sim1.0\times10^8$	0.12～0.334	0.86～1.63

Source: Data from Tong and Cheng (1981) and Wang et al. (1986).

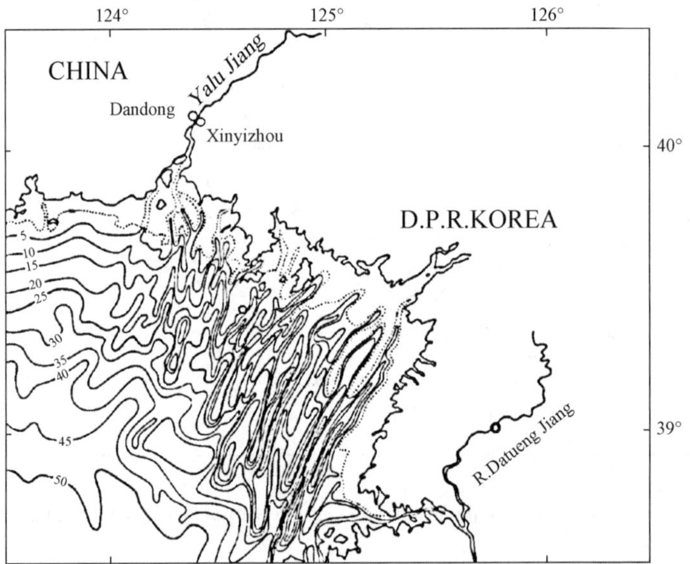

Fig. 8　Sketch map of the Yalu Estuary in relation to the inner continental shelf. Water depths are in meters.

The normal wave height at the mouth of the estuary is 1 m, with a maximum wave height of 3.3 m at Dalu Island outside the estuary. Estuarine sediment consists of sandy and coarser material. Pebbles and shingles are found in the upper estuarine channel, and medium to fine sands covered with a thin layer (less than 1 m thick) of silty mud in areas of sandy shoals. Outside the estuary, fine to medium sand is distributed on the inner shelf to water depths of 10 m, some 20～30 km away from the river mouth. Strong tidal currents outside the estuary have prevented the development of a submarine delta. Currents have created a series of linear sand ridges and troughs parallel to the direction of the tidal currents (Fig. 8). The convergence and divergence of the tidal prism during each tidal cycle aids in developing the sandy ridge fields. During the period of a tidal cycle, the mean sediment concentration is 0.42 kg m^{-3} during the flood tide and 0.33 kg m^{-3} during the ebb. The net input of sediment is 2 100 tons per cycle. Several sandy shoals on the inner shelf, in water depths of 1 m, have their fine sand winnowed by wave action (Fig. 9) and moved toward the west by long-shore drift.

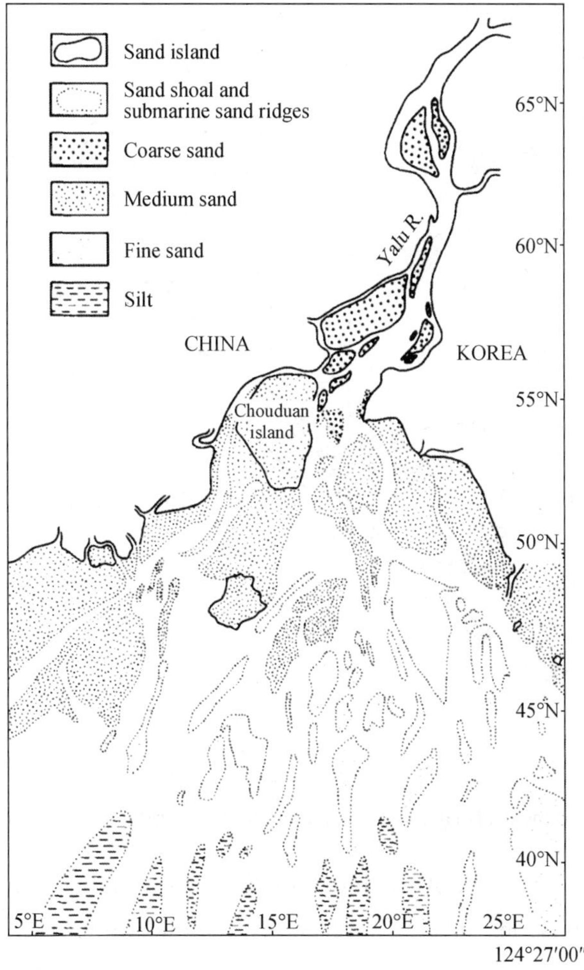

Fig. 9　Map of submarine topography and sediment distribution in the Yalu Estuary

Luanhe River Mouth: Sedimentation Pattern of Wave Dominant

The delta of the Luanhe River is located in the transition zone between the Ianshan Mountain Uplift Belt and the subsiding basin of the Bohai Sea. Subsidence is therefore the major controlling factor on coastal architecture. The mean sediment concentration at 3.9 kg m^{-3} is very high and contributes to the prograding delta. The Holocene delta consists of four lobes (Fig. 10): (I) the east lobe, formed during the early to middle

Holocene; (II) the west lobe, deposited during the middle and late Holocene, (III) the middle lobe, formed during the historical periods; and (IV) the newer lobe, formed since 1915 (Wang, 1963; Zhu, 1980).

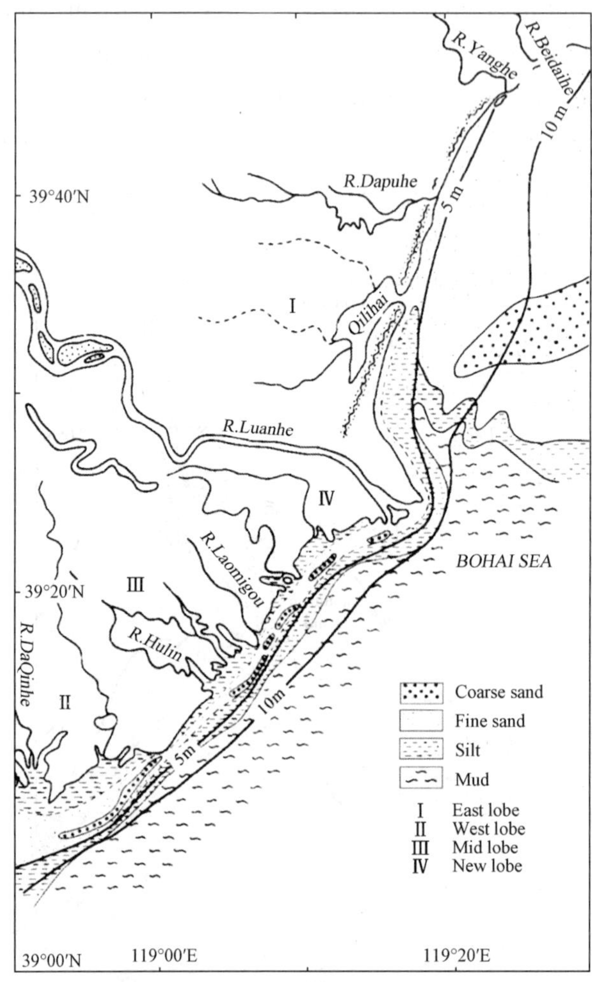

Fig. 10　Map of sediment distribution in the Luanhe River Delta and adjacent continental shelf

Wave patterns in the area of the inner shelf off the mouth of the Luanhe River are dominated by the monsoon. In the winter, north and northeast waves prevail, and east and southeast waves prevail in the summer. Wave dynamics are the major force in the transport of sediment. The significant wave height ($H^{1/3}$) is 3.8 m, with a period of 7.8 m. Breaker depth is 5.5 m. The ebb current is larger than that of the flood; the tidal range of the area is about 1.5 m.

To the south of the Luanhe River, there are narrow sediment belts parallel to the shoreline. The belt of fine sand is well sorted (S_0=1.5～1.7) and meets the coarse silt belt at the breaker zone, where the water depth is 5 m. The belt of dark gray muddy clay is located in waters more than 9 m deep. Small authigenic grains of pyrite in the muddy clay form in concentrations of 0.05%～2% by sediment weight. This indicates a quiet, reducing environment on the bottom. The present shelf deposit is only 0.4～0.8 m thick. North of the Luanhe River sandy deposits extend 20 km from the shoreline to a water depth of 1.3 m. Fine and medium sands are well sorted, with a coefficient of 1.3 to 1.8ϕ. Medium sands extend to a water depth of 13 m. Based on size frequency characteristics (Fig. 11) and surface features of the rounded quartz grains, these deposits are interpreted as having formed in a beach dune environment that was later submerged during the rise in global Holocene sea levels.

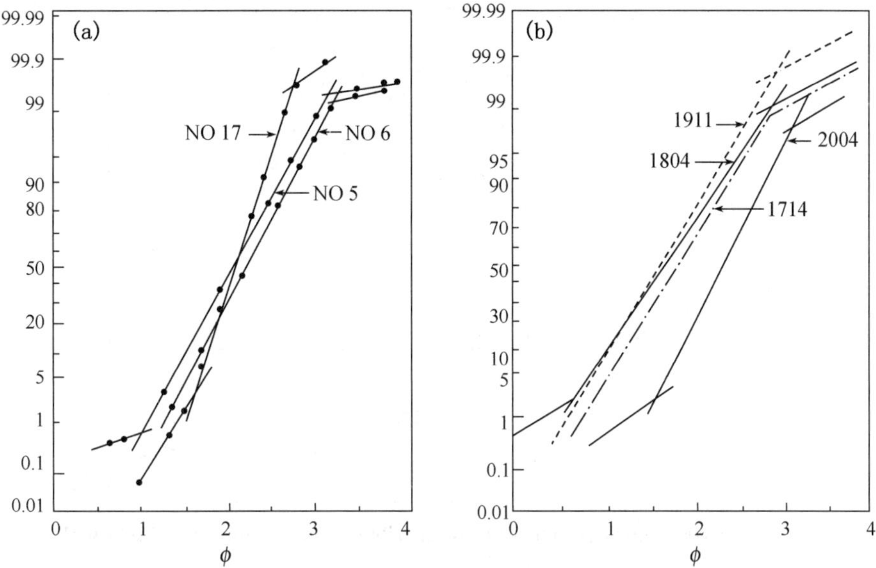

Fig. 11　Grain size distribution of (a) beach dune sands and (b) inner shelf sands of Luanhe River Delta

Some of the inlets of the Luanhe River were ancient river mouths formed as the river channel migrated from south to north during historic time. Sandy bodies such as point bars, river mouth bars and sand shoals were left in the area of ancient river mouths. When the river moved away, the sediment supply was cut off and erosion of these sandy bodies began. Sediment distribution in the area has shown a dynamic balance with wave action. The abundant modern river sediments enterthe sea near shore and then are pushed offshore by summer east and southeast waves that advance toward the coast transversely. Offshore sand barriers enclose the river mouth and coastal zone. River sediments are deposited mainly in the shallow-water environment protected by the sand barriers. Well-laminated and interstratified silt and fine sand are developed leeward of the sand barriers. The fine sandy materials come from the sea, either by overwash of sand barriers or by tidal currents through the inlets between the barriers. Along the coast, in a zone of decreased wave action behind offshore sand barriers, tidal flats develop and overlap the delta plain (Fig. 12). The river effluent reaches the foot of the delta at a water depth of 7 m, between 2.5 and 5.0 km from the river mouth.

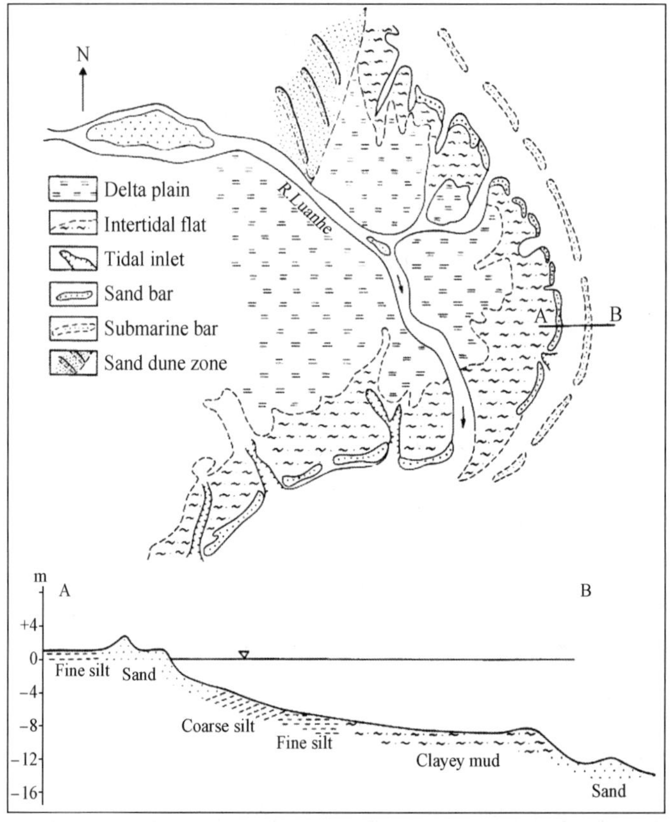

Fig. 12 Geomorphology and sediment distribution in Luanhe River Delta with a cross-section profile. Distance A—B=1000 m.

Yellow River Mouth: Microtides and Runoff Pattern

The Yellow River sediment is deposited mainly as a delta, and the remainder is caried into the Bohai Bay by tidal and residual currents. During the past 4 000 years, there have been eight major shifts in the course of the Yellow River (Pang and Si, 1979; Wang, 1983) (Fig. 13). The northernmost course flowed north of Tianjin and passed through the Haihe River into northwestern Bohai Bay, the southernmost course passed through the Huai River into the Yellow Sea. In 1855, the Yellow River migrated from the Yellow Sea back to the Bohai Sea, where it remains today. The annual sediment discharge of the river is 12×10^8 tonnes (Tong and Cheng, 1981). Sixty-four percent of the sediment is deposited on the delta and mud-flat and the remainder is passed farther out to sea (Pang and Si, 1980; Wang, 1983) (Fig. 14). Since 1976, the Yellow River has changed its course to the south, carried into Laizhou Bay and deposited in 15 m of water, where the foreslope of the submarine delta is building. Hydrological data at Lijin station shows that the average water discharge is 379×10^8 m^3, and the average annual sediment discharge is 9.5×10^8 tons (1959—1992). The annual sediment discharge range is between 4.1×10^8 tons (1987) and 21×10^8 tons (1958), During the period

1976—1988, 65×10^8 tons of sediment has accumulated in the area of new river mouth, and the maximum accumulated thickness is 12.4 m at river mouth bars.

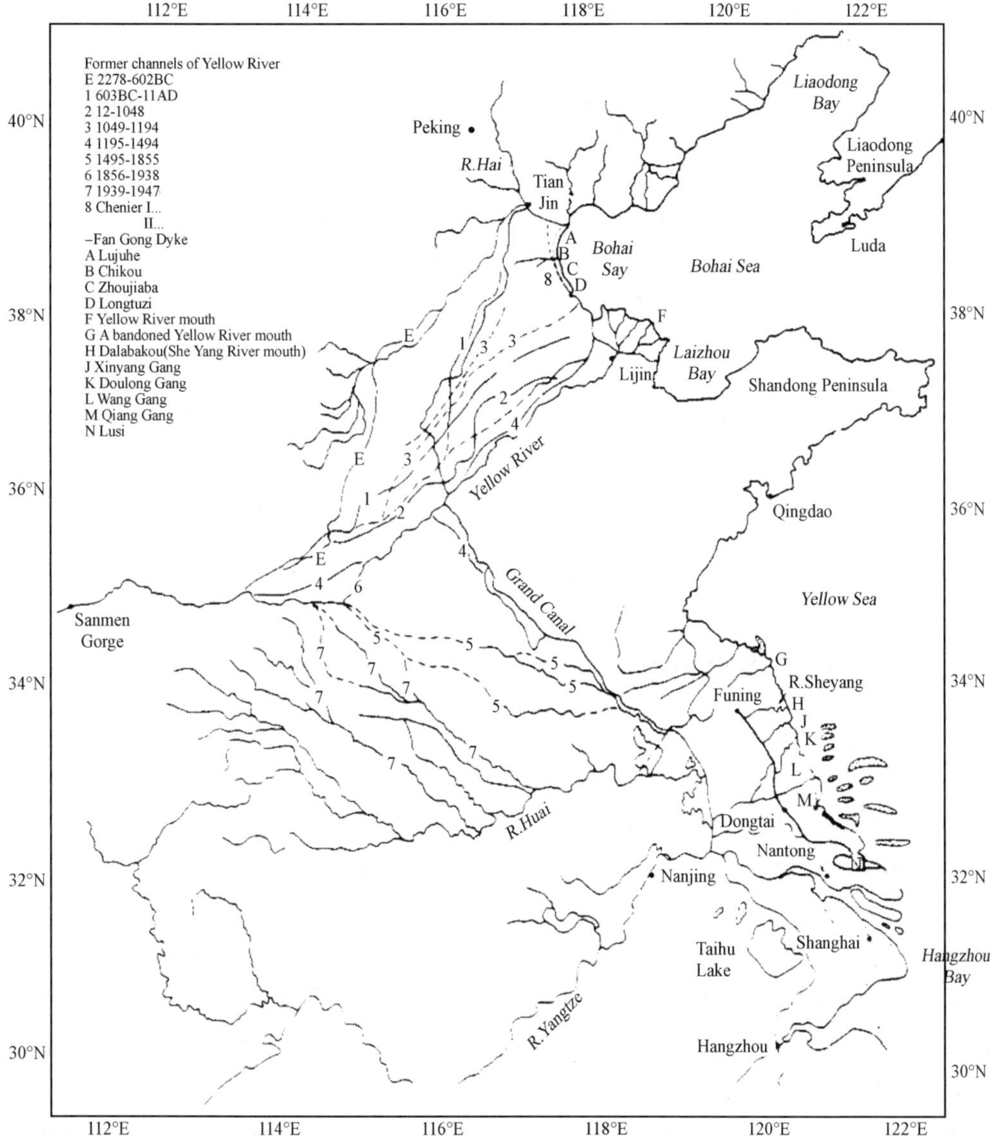

Fig. 13 Map of the present-day Yellow River and locations of older abandoned channels.

Wave action is minimal at the river mouth, resuspending the sediments only during heavy storms. Tidal currents and residual currents are the major agents in transporting the sediment. The tidal range is 0.5～0.8 m at the river mouth, increasing gradually to 2.0 m on both sides of the delta. The maximum velocity of the tidal current is 1.4 $m \cdot s^{-1}$ at the river mouth, gradually decreasing to both sides of the river mouth. Sediment distribution follows the dynamic pattern: very fine sand, up to 68%, and silt, 22%～30%, are deposited in the river channels; coarser silt and fine silt are up to 50% and 20% deposited on the floodplain; coarser silt and fine sand, up to 80%, are deposited as

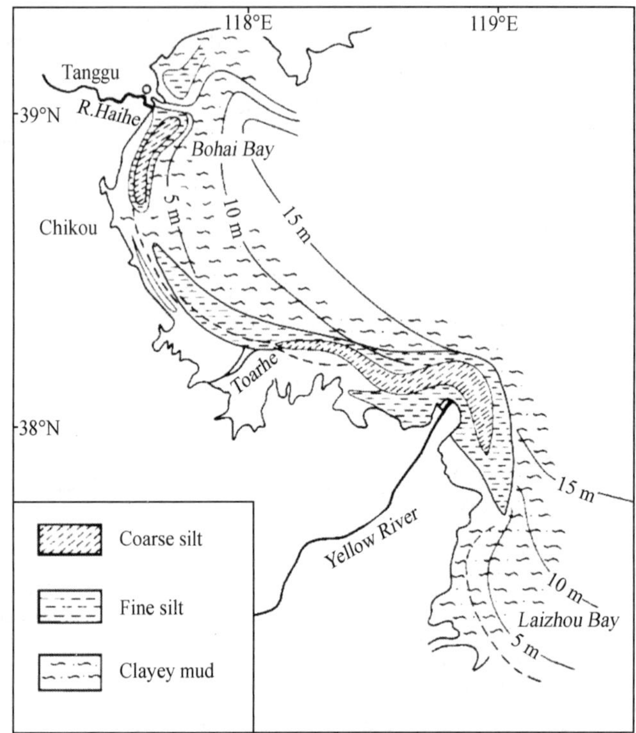

Fig. 14　Sediment distribution of Bohai Bay and the Yellow River Delta.

levees on both sides above the river channel. Coarse silt is deposited along both sides of the river mouth in the form of finger bars. The growth of the finger bars is rapid, especially in the early stages of a new channel; 10 km of progradation in a single year occurred in 1964 (Wang, 1980). On either side of the river mouth, bars protect bays in which fine silt (40%～60%) and silty clay (30%) collect as fluid mud. The lower Yellow River changes its course every 6～8 years, forming a new set of finger bars. The abandoned channel finger bars are eroded quickly, up to 17.5 km in a year by waves. Thus coarse silt covers the fine sediment in the muddy bays, while muddy deposits appear on the eroded finger bars.

In cross section, there are alternate beds of subaerial silt and marine mud in the lower layer and marine mud and shell-fragment sands in the upper layer. During a period of more than 100 years a fan-shaped delta prograded seaward from Lijin, with its distal margin in 15 m of water. The total coastline of the delta is 160km long from the Taoerhe River in the north to the Xiaoqing River in the south (Fig. 15). The total progradation of coast has been 20.5～27.9 km since 1855, with an average progradation of 0.2～0.3 km yr^{-1}. Between 1855 and 1988, the total volume of deposition is 33×10^9 tons, providing an annual deposition rate of 2.6×10^8 tons; A total of 2,100 km^2 of land has been accumulated since 1855 (Chai, 1993).

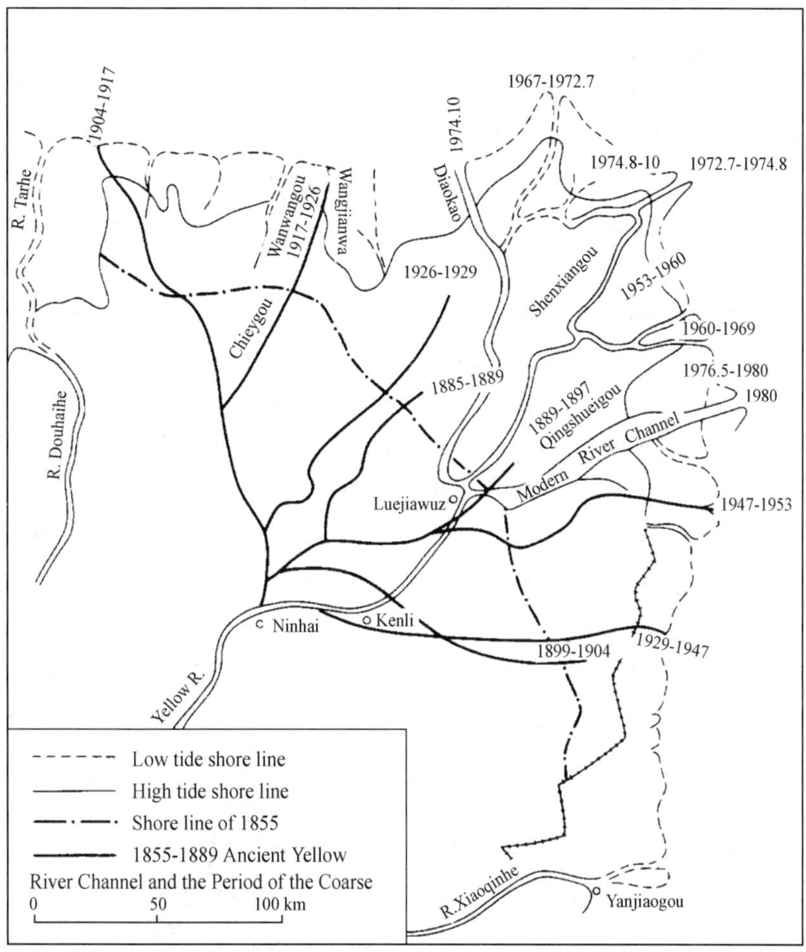

Fig. 15 Map of the modern delta of the Yellow River.

Changjiang River: Sedimentation Pattern from Runoff and Longshore Current Control

The sediment distribution off the Changjiang River Estuary is influenced by the Changjiang runoff and the longshore current running from north to south. The Changjiang, the largest river in China, is the main source of sediments to the continental shelf of the East China Sea (Table X). The semidiurnal tidal range varies from 0.17 to 4.6 m. The average velocity of flood tidal currents is 0.95 $m \cdot s^{-1}$ (maximum 2.08 $m \cdot s^{-1}$), and of ebb currents is 1.11 $m \cdot s^{-1}$ (maximum 2.5 $m \cdot s^{-1}$). The observed maximum wave is H=2.3 m, L=40 m, T=5.6 s. Waves are caused mainly by southeast winds during the summer and autumn and by northeast winds in the spring (Huang, 1981). Nearly 50% of total of 486 metric tons of fluvial sediment is trapped in the river mouth (estuarine reach) to form a shoal and submarine delta (Shen et al., 1992).

Shen et al. (1995) indicate that: (1) The dominant mechanisms of net transport of water and suspended sediment is that of nontidal steady advection transport, Stokes drift, tidal trapping and net vertical circulation. (2) The net transport of water and suspended sediment induced by mass Stokes drift is of considerable importance and helps restrain the seaward transport of suspended sediments and induces the increase of suspended sediment concentration within the estuary. (3) The magnitude of suspended sediment transport induced by tidal trapping is greater than that induced by vertical shearing diffusion. Tidal trapping with the resuspension of bedload plays an important role in the formation of turbidity maximum in the area where the action of tidal trapping is distinct. The resuspension of bedload and the asymmetric sediment transport are the main factors that induce the distinct action of tidal trapping. (4) In different tributaries, the contribution of each suspended sediment transport term to net sediment transport varies with the relative importance of runoff and tidal current. Mass Stokes drift and tidal trapping play important roles in the formation of turbidity maximum in the South Passage, where tidal current is stronger. In the North Channel, where runoff is relatively strong, the nontidal steady advection transport and net vertical circulation occupy dominant positions, and the latter is the main factor in the formation of turbidity maximum (Shen et al., 1995).

There is a system of two currents off the Changjiang Estuary. One is a low-temperature, high-salinity longshore current from the north, which passes between 122° and 123° E toward the south; the other is a high-temperature, high-salinity branched tongue, mainly from the surface water of the Kuroshio (Yu et al., 1983). The sediment output of the Changjiang River is carried mainly south by the longshore current. However, part of the sediment is carried by low-salinity water between the two saline tongues to the east (Fig. 16). Sediment concentration is 1 kg m^{-3} in the area between 121° 50′ and 122°20′ E; maximum concentration in the bottom layer is 8 kg m^{-3}. This is twice the concentration of the Changjiang Estuary. A submarine delta system has been developed in

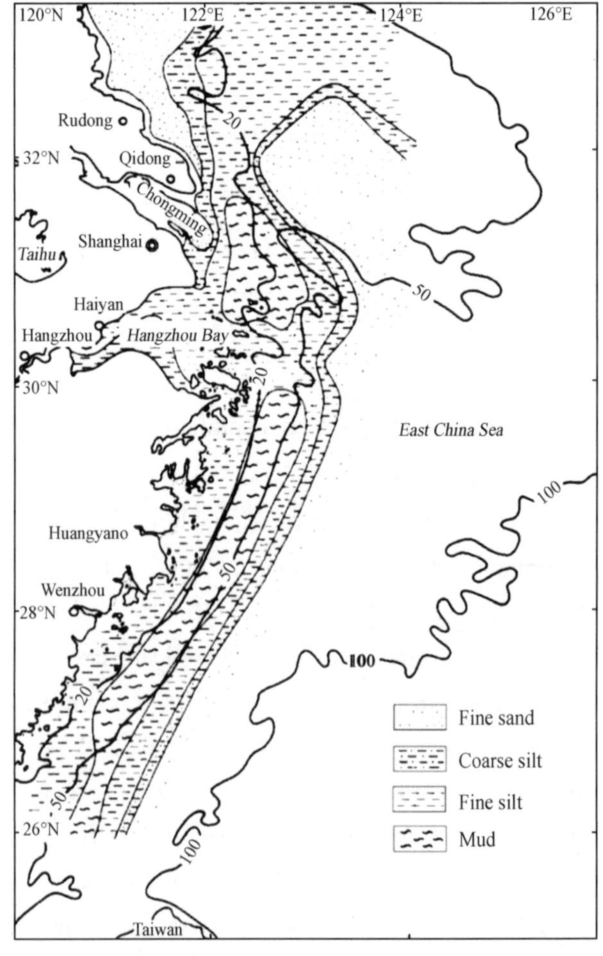

Fig. 16　Sediment distribution in the Changjiang Estuary and adjacent inner shelf. Water depth in meters.

water between 5 and 40 m deep, which consists of clayey silt, with a 20-to 90-mm-thick layer of fluid mud on the top. The sedimentation rate is 5.3 m per 1 000 years (You et al., 1983): 9.0 mm yr^{-1} as determined by ^{210}Pb. A large submarine delta, 40 000 km^2 in area, formed about 36 000 yr B.P. according to ^{14}C dating. It extends to a water depth of 60 m at its outer boundary. A smaller modern delta extends to a water depth of 25～30 m; formed since 7 000 yr B.P., and is superimposed on the older delta (Ren and Zhang, 1980).

Pearl River: Sedimentation Pattern from Embayment Trapping

The Pearl River is located in tropical South China, which is associated with abundant precipitation and vegetation. Sediment discharge into the estuary is relatively small compared to that of the Chinese rivers discussed previously. The discharge rate is 11 000 m • s^{-1} (Table X). At the Pearl River mouth, tides are irregularly semidiurnal with an average tidal range of 0.86～1.63 m. The Pearl River sediments are deposited in a sheltered environment and fill the embayment; a transition between estuarine and delta coastline is produced. River sediments have accumulated as point bars, river mouth banks or sand shoals, overwash fans, and tidal deltas around each river mouth. Suspended materials, both from the sea and from the tributaries, are deposited as a mud cap (Ren, 1964; Li, 1983). The sediments on the continental shelf adjacent to the Pearl River mouth can be divided into two belts (Fig. 17). The inner shelf deposit of river-

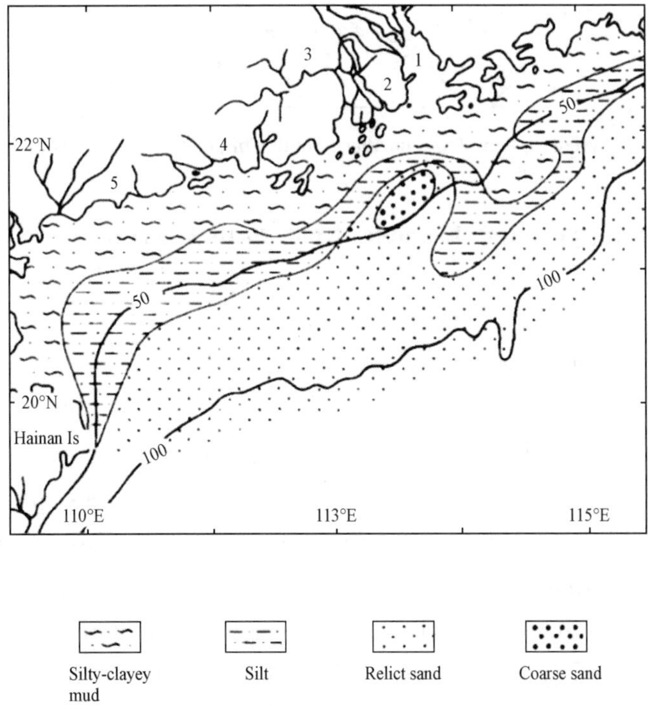

Fig. 17　Sediment distribution on the Pearl River delta and on the adjacent inner shelf. Water depth is in meters. 1, Pearl River; 2, Zhongshan; 3, Gao Yau; 4, Yang Jiang; 5, Dian Bai.

produced sediments is 35 km wide, with its outer boundary at a water depth of 20～30 m. From the coast seaward, sedimentation occurs in the order of sand, silt and clayey mud, showing a modern sedimentary series of wave dynamics. Sediments in this belt (Li, 1980) contain little carbonate (5%～10%), a few heavy minerals (<0.55%) and a large amount of organic minerals (1%～2%). The outer shelf deposit, which is separated from the inner by a narrow transition zone, is characterized by relict well-sorted fine and medium sands containing more than 2% heavy minerals. Shell fragments comprise 25%～50% of the material in this belt, with many foraminifera which reach 90% concentration at the continental shelf break. The organic matter is lower than 1%, and there is no organic matter in the fine sand belt (Li, 1980). There is a longshore current from northeast to southwest during all seasons of the year which carries the muddy clay westward (Fig. 18) (Zhao, 1983). This is a major source of mud for the tidal inlets along the coast west of the Pearl River Estuary. During summer, some muddy materials are also diffused to the outer shelf by northeasterly drift.

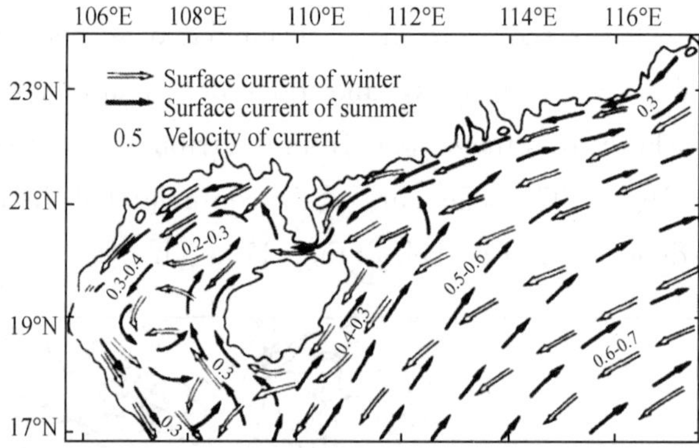

Fig. 18　Current system of the northern South China Sea. Velocity is in m · s^{-1}.

4.2　Land-Sea Dynamics: Examples from North America

Floods

In Canada, sediment yield ranges from 5 to 750 tons km^{-2} yr^{-1} and can locally exceed 1 000 tons km^{-2} yr^{-1} (Binda et al., 1986). In the continental United States, sediment yields can reach 1 700 tons km^{-2} yr^{-1}, particularly for steep mountain rivers such as the Eel River. Flood conditions for North American rivers can range from a few days to a month (Fig. 19) (Mulder and Syvitski, 1995). The nival freshet of the Fraser River is associated with a rapid dropoff in sediment concentration immediately after peak flow conditions are attained (Fig. 19a). Summer flood conditions of the Stikine River are a result of rapid melting of hinterland ice fields. Due to the large quantity of wash load (glacial flour), there is a more gradual decrease in water turbidity after peak discharge is attained (Fig. 19b). Flood conditions on the Mackenzie River, for example, often show as a slug of highly turbid water that suddenly appears, although river discharge has changed little (Fig. 19c). The slug of dirty water is from an upstream glacifluvial tributary, the Liard River. The tributary wash load is large enough to influence the ambient Mackenzie suspended sediment concentration, but its associated discharge is too small to affect ambient discharge of the main river. On smaller rivers, such as the Kennebecasis River, the annual nival freshet conditions the sediment availability within the river for the remainder of the year: a large freshet causes a large sediment supply throughout the year (Fig. 19d). Details on other flood conditions of North American rivers can be found in Matthai (1990).

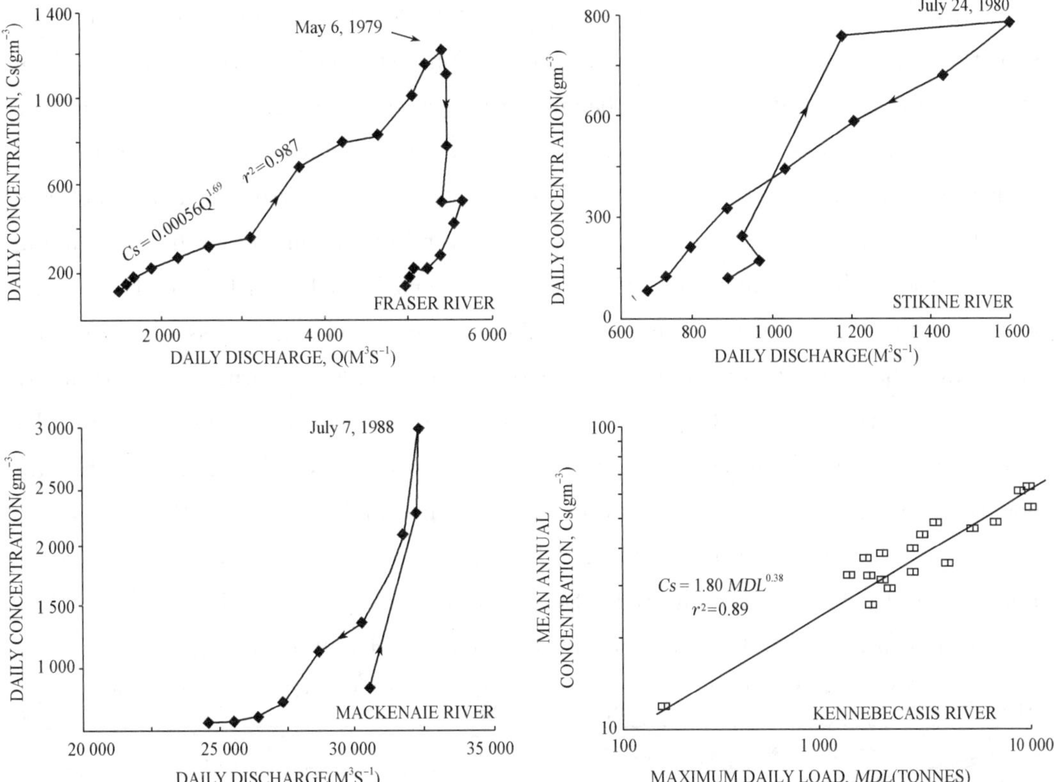

Fig. 19. (a)～(c) Three examples of rating curves (daily concentration versus daily discharge) for flood conditions of North American rivers. (a) The spring freshet of the Fraser River, B. C., showing the rapid drop off of concentration immediately after attainment of peak flow conditions. (b) Summer flood of the Stikine River. B. C., a result of rapid melting of hinterland ice fields. Note the more gradual decrease in water turbidity after peak discharge is attained. (c) Flood condition on the Mackenzie river, N. W. T. The apparent lack of relationship between flow conditions and suspended concentration relates to the Liard river as the major source of sediment, but only a small contribution to discharge. (d) On smaller rivers, such as the Kennebecasis River, N. B., the annual spring flood (as here indicated by the maximum daily load) conditions the sediment availability within the river for the remainder of the year (as here indicated by the mean annual concentration). (Data from Water Survey of Canada files.)

River-Sea Dynamics

Fluvial sedimentation is most pronounced at the mouths of rivers, dominated by four processes: (1) deposition of bed material load at the river mouth, the main point for hydraulic transition, (2) sedimentation under the seaward-flowing river plume that carries the wash load, (3) sediment bypassing processes, such as turbidity currents, that may result from delta-front failure; and (4) diffusive processes that work to smear sediment downslope and include tidal and wave action, creep and small slides. Together these processes are further conditioned by geologic-scale fluctuations in sea level, tides, waves, discharge, hurricanes and earthquakes (Coleman and Roberts, 1989).

Fluvial sedimentation affects three coastal North American environments: estuaries, fjords and deltas. Estuarine circulation depends in part on the level of stratification of the water column: a balance of buoyancy forces set up by river discharge and processes such as those associated with tidal action that work to mix the fresh water with the denser saltwater layers. See Fig. 20. Types of North American estuaries include (1) salt wedge (Fig. 20a), where the river flow dominates with little mixing between the fresh and salt water layers (e. g. , Fraser River, B. C. : Geyer, 1990), (2) two-layer flow with entrainment (Fig. 20b), where mixing across the interface (halocline) becomes important (e. g. , most Canadian fjords: Farmer and Freeland, 1983), (3) partially mixed (Fig. 20c), where river-flow buoyancy is balanced by tidal mixing (e. g. , Miramichi, New Brunswick: Vilks and Krauel, 1982) and (4) vertically homogeneous (Fig. 20d), where mixing processes predominate (e. g. , Bay of Fundy, Nova Scotia: Amos and Long, 1980). These ideal estuaries (e. g. , Jay and Smith, 1990) may be further complicated (Syvitski, 1986) by the Coriolis effect, flow accelerations around bends or constrictions of inlets and bathometric features, lateral

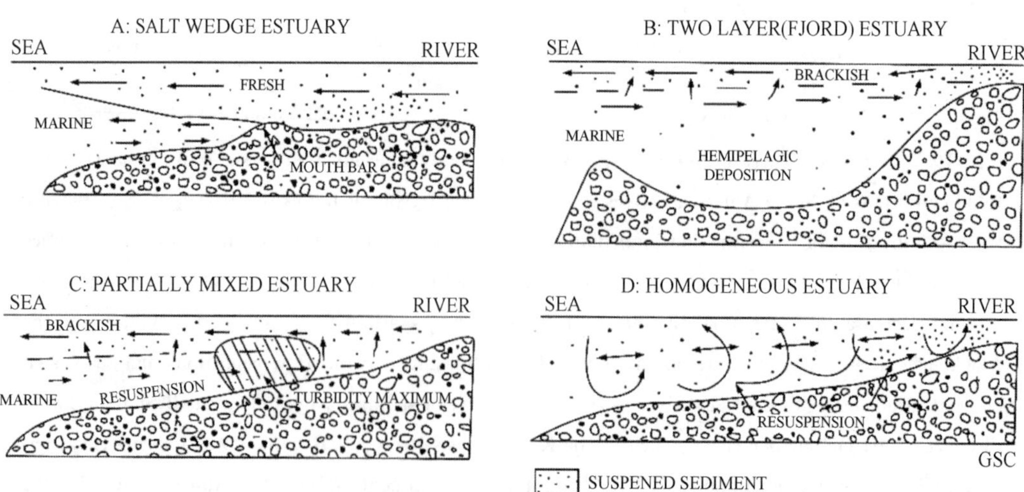

Fig. 20 Simplified diagrams of the pattern of circulation and turbidity within the four principal types of estuaries found along the coast of North America

separation of currents along multiple channels, barometric pressure gradients, surface mixing from strong winds, the breaking of internal waves and isohaline instabilities developed during the process of salt rejection associated with sea ice formation. Furthermore, the type of estuarine circulation may change with the season (Vilks and Krauel, 1982) and as affected by river regulation (Cataliotti-Valdina and Long, 1984) and coastal dredging.

A fjord is a deep high-latitude estuary which has been, or is presently being, excavated or modified by land-based ice (Syvitski et al., 1987). North American fjord examples range from those with temperate drainage basins free of snow and ice to frigid permanent sea ice-covered fjords, from tide-or wave-dominated fjords to river-dominated fjords, from permanently anoxic fjords to continuously oxygenated systems, and from low-to high-sedimentation fjords. Circulation within the upper waters (30～50 m) is usually of the two-layer estuarine type and related to river runoff. Stream discharge is typically in the form of a two-dimensional buoyant jet, although some streams may enter at depth through the tidewater fronts of glaciers in the form of three-dimensional free buoyant jets (Syvitski, 1989b). Deepwater circulation (some North American fjords are deeper than 1 000 m) depends on a number of critical factors, including sill depth and tidal mixing over the sill and deepwater renewal through periodic exchanges of the denser shelf waters with waters of the fjord basin. Fjords are sites of net sediment accumulation, and globally have retained one-fourth of the sediment removed from the land over the last 100 000 years (Syvitski et al., 1987). Fluvial deposits accumulating in fjords provide a highly resolvable record of terrestrial and marine events that have occurred since the last ice retreat. Typically, fluvial deposits are spread over a great distance from the river mouth, due to the fjord margins confining the fluvial plume emanating from the river and the propensity of fjords to experience long-traveling sediment gravity flows.

Deltas have nearly flat surface expressions of alluvial deposits formed around the mouth of a river (Coleman and Roberts, 1989). They result from a radical change in the fluvial hydraulics as the river enters a standing body of water, causing (1) deposition of bedload along the subaerial and intertidal topsets and foresets and (2) subtidal sedimentation of the suspended load onto the subaqueous portion of the delta, the prodelta environment. A marine delta survives only if sediment supply and accumulation are greater than sediment removal by waves, tidal action, submarine slides and longshore transport. In Canada, deltas are common to the heads of fjords and inlets (e.g., Syvitski and Farrow, 1983), along the paleofjord coastline of the north shore of the Gulf of St. Lawrence (e.g., Sala and Long, 1990) and in the large and turbid rivers that drain the western mountain ranges, such as the Fraser (Milliman, 1980; Kostaschuk et al., 1989) and the Mackenzie (Hill et al., 1991). Typically, there is a direct relationship between drainage basin area and the size of the delta plain, and as a result, the Mississippi is the largest delta in North America (Coleman and Roberts, 1989).

The complexity and size of the North American coastline offers a wide spectrum of delta/estuary development, including the long-term effects of rising sea level (Mackenzie), rapid delta progradation (Fraser), complex sea level fluctuations (Chaleur Bay) and changes in tidal amplitude (Bay of Fundy). There is also a wide spectrum of stream hygrographs (continental, maritime and island; pluvial, nival, temperate, arctic and icefield influenced). Tidal amplitudes vary from less than 1 m in the Arctic Archipelago to greater than 10 m in the Bay of Fundy, Hudson Straight and Ungava Bay region.

River Plume Sedimentation

Normally, a river plume will enter the sea as a hypopycnal flow: fresh water flowing into an ambient basin of saline water. The type of hypopycnal flow will strongly influence the behavior and character of the suspended particulate matter (SPM). In the case of inertial and friction-dominated flows, mixing between the fresh water and seawater can be very strong (Kostaschuk, 1985). The plume salinity will therefore be high close to the river mouth and allow the early flocculation of mineral grains. Frictionally influenced plumes can also be associated with resuspension and thus with the input of mud chips off the floor of the estuary. In deep coastal fjords, gravity flows associated with slide-generated turbidity currents can similarly resuspend mud chips from the seafloor.

A hyperpycnal plume is a particular kind of turbidity current that occurs at a river mouth when the sediment concentration in river water is heavy enough to create a plume with density higher than the local ambient seawater density. The plume can then plunge and flow along the seafloor, and possibly erode the seafloor if the inflowing velocity is great enough. Hyperpycnal plumes often are generated during flood conditions of many middle-sized North American rivers that drain alpine hinterlands (Mulder and Syvitski, 1995). Large rivers such as the Mississippi, Fraser. St. Lawrence and Mackenzie do not have such underflows at their mouth because of sediment retention within their expansive coastal flood plains.

River sediment that enters as buoyant hypopycnal flows are carried as finely divided individual sediment particles, which often contain attached hydrous oxides, organic coatings, freshwater microflora, and mud clumps eroded from raised terraces (Syvitski and Murray, 1981; Schafer et al., 1983). As the surface layer mixes with the marine water, the salinity of the turbid surface plume increases. Flocculation of silts and clays often begins within the river waters and accelerates with increasing salinities (3～5 ppt). In the flocculated state, particles settle faster than if they were to settle individually, and velocity enhancement can exceed four orders of magnitude for clay-size particles (Syvitski et al., 1985). Grains suspended in the ocean settle in four types of units of individual particles: (1) aogregates: inorganic particles strongly bonded by intermolecular or intramolecular, or atomic cohesive forces; (2) agglomerates: organic and inorganic matter weakly held together by surface tension, organic cohesion and adhesion: (3) planktonic fecal pellets, where individual particles are mechanically compacted and packaged, sometimes within a peritrophic membrane; and (4) floccules: inorganic particles held together by van der Waals forces, after the electrostatic repelling forces associated with natural mineral grains are neutralized within a saline solution greater than 3 ppt.

Except during the initial conditions of flocculation, the influence of organic matter dominates particle dynamics in the ocean. River water is carbon-rich and is mixed with nitrogen-rich seawater in estuaries. Below the halocline, organic cohesive/adhesive forces allow the attachment of organic detritus and biogenic debris (mostly phytoplankton material in various stages of mechanical or chemical destruction) onto the flocs. Thus floccules composed wholly of mineral grains are transformed into agglomerates of both organic and inorganic matter. Agglomerates may reach 10～15 mm (Syvitski et al., 1985, 1995). The composition of agglomerates may range from those dominated by mineral matter, or biogenic detritus, or mucoid matter. Interestingly, even in the cold arctic fjords of Baffin Island, where land vegetation is almost nonexistent, fluvially transported sediment settles to the seafloor in agglomerates composed in part of 10%～30% organic carbon (Syvitski et al., 1990).

Zooplankton graze on flocs and agglomerates, subsequently producing mineral bearing fecal pellets. Fecal pellets are compact and hydrodynamically shaped, and they settle rapidly. A pellet's settling rate depends on its volume, shape, composition and compaction. The rate at which fecal pellets are produced depends on both the distribution of zooplankton in the water column and the zooplankton egestion rate. The spatial distribution and density of zooplankton are highly seasonal, reaching maximum populations in the late spring and early fall, coincident with floods and maximum sedimentation. Directly under the halocline, zooplankton populations are very large, and although they can be found at all depths, they reach a penultimate population maximum near the seafloor. The egestion rate of zooplankton may vary by one order of magnitude depending on the food made available. Proximal to the river mouth, but in deep water, the pellets contain mostly mineral grains, a reflection of the composition of the ambient suspended matter. In more distal locations, the fecal pellets contain little mineral matter and are dominated more by biogenic (phytoplankton) detritus.

The seasonality in hydrologic and biologic events can affect the agglomeration of marine particulate matter. For instance, in temperate regions of southern Canada, the peak runoff of a lowland river is synchronous with winter rainstorms, while the peak runoff of a glacier-or snow-fed river would occur during the spring-summer period. Since the maximum biological productivity occurs during the latter period, organisms would mostly affect sedimentation of particles discharged from glacier-and snow-fed rivers (Lewis and Syvitski, 1983).

Scavenging Rates

Buoyant hypopycnal (hemipelagic) sedimentation is a function of grain size, initial load and particle travel time as a subsequence of both setting and advection into the coastal basin. The velocity distribution within highly stratified marine basins is typically modeled by a buoyancy-dominated, free, two-dimensional jet. The dynamic zones of such a river plume (Syvitski et al., 1988) are (1) proximal to the river mouth where the plume is not yet established and the core of the plume continues to behave as a plug flow, (2) further seaward where the established flow decreases as the plume spreads, and sometimes (3) where the spreading plume is affected or constrained by basin boundaries. Each particle size is characterized by a unique removal constant or scavenging rate.

Deposition of Coarse Sediment in Deep Coastal Basins

There are a number of basin properties that affect the quality of basin fill, including (1) the initial basin geometry; (2) the type and amount of bedload, Qb, and suspended load, Qs, brought down by rivers; (3) the oceanographic environment of the basin (fluvial discharge, tidal and wave environment, sea level fluctuations, climate), and (4) the tectonic regime. If a shelf basin receives mostly sediment input from suspended load, offshore sedimentation is dominated by a flux of hemipelagic particles. North American examples include the Saguenay Fjord and the St. Lawrence Estuary. Both are deep basins that receive little bedload, due to the filtering of a large hinter-land lake (Syvitski and Praeg, 1989). Sand is deposited relatively close to the river mouth (within the first few kilometers).

If the bed material load of a river is more significant, mouth bars form and prograde onto steep foresets and contribute to semicontinuous failures along the delta lip. Moderately sorted sand is eventually transported seaward along submarine channels by turbidity currents. For example, Knight Inlet, British Columbia, has sand deposited between 10 and 50 km offshore in the form of turbidites (Syvitski et al., 1988). Sand is not deposited over the first 10 km, as the turbidity current flows over steep slopes and sediment remains suspended within the turbidity current until a critical angle is reached.

For those rare marine basins that receive sediment from rivers transporting mostly bedload, such as Baffin Island sandurs (Church, 1972), the basin fill will be controlled by failure-generated diffusion of the proximal prodelta slopes with some form of bypassing as an invariable consequence. Itirbilung Fjord is a type of basin in which liquefaction-induced failures may occury anywhere near the river mouth, and cohesionless debris flows and turbidity currents smear sand downslope along the entire prodelta environment (Syvitski and Farrow, 1989). During a 40-day period in the summer of 1985, nine significant gravity-flow events were detected within one of the submarine channels that dissect the seafloor off the Itirbilung River Delta (Syvitski and Farrow, 1989). Each event lasted for 1～5 h and the maximum current detected was 35 cm s^{-1}. Sediment collected in nearby moored traps contained nine sand layers interlayered with hemipelagic mud (Syvitski and Hein, 1991).

Much of the fluvial bedload is contained within the lower reaches of the river, which form part of the distributary channel system of deltas. For example, virtually all of the gravel fraction transported by the Fraser River is contained between the Agassiz-Mission reach, and little reaches the river mouth (Mclean and Church, 1986). In salt-wedge estuaries, the river flow will decouple from the bed during periods of flood tide. In this situation, the distance depends on the stage of the river discharge. For example, the salt wedge migrates into the Fraser Channel by more than 30 km during periods of low discharge and by less than 7 km at high discharge. Inland mirgation of the salt wedge is associated with rapid deposition of the suspended bed-material load (Kostaschuk and Luternauer, 1989). On falling tide, the resuspension of bed-material load increases with the passing of the salt wedge due to enhanced turbulence. Channel shoaling at the river mouth is thus related to the low-tide position of the salt wedge (Kostaschuk and Atwood, 1990).

Deposition of River Sediment onto Open Shelf-Slope Environments

In the shallow marine and shelf setting, tide-, storm-, and wave-driven currents operate to produce an equilibrium shelf gradient (Swift, 1970; Ross et al., 1994) and change the configuration of the coastline (Coleman and Roberts, 1989). In the upper slope region, sediments are delivered from the shelf via traction and hemipelagic sedimentation and are then reworked downslope by a variety of processes characterized as diffusive (i. e., submarine landslides, mudflows and creep). When diffusive processes build a slope above a critical angle, a portion of the available sediment is transported offshore as turbidity or debris flows into a base-of-slope position (Syvitski et al., 1988). Sediment gravity flows comprise the major processes operating in the mid-to lower-slope region of the profile. Upper-slope diffusive processes operate in concert with sediment gravity flow processes in the lower slope and basinal environments to control the depositional grade (i. e., the profile of equilibrium) in sub-wave-base environments (Brunsden and Prior, 1984; Kenyon and Turcotte, 1985).

5. Conclusions

Rivers are the major natural agent of terrigenous sediment transport to the ocean and are part of the sedimentary dynamics system in the coastal ocean. Complete worldwide measurements of river data are not available, and the widely quoted values for global fluvial sediment flux may actually be decreasing as a result of artificial controls and rising sea levels. It is assumed that the present annual flux of global sediment is between 10 and 20 billion tonnes. The predominant terrigenous sediment input is from rivers draining tectonically active and mountainous regions of Asia and the Americas. More than one-third of the total sediment discharge is supplied by seven rivers originating from the Qinghai-Tibet Plateau. The Pacific Ocean appears to be accumulating the majority of this sediment.

Fluvial sedimentation effects coastal environments of estuaries, fjords, delta sand continental shelves. Terrigenous sediment delivered by rivers has been greatly affected by human activities, thereby affecting large areas of the coastal zone. Studies on five Chinese rivers have been highlighted to demonstrate the systematic mechanism of the river-sea interaction of terrigenous fluxes. Three dynamic patterns have been demonstrated. (1) High-energy wave regimes are recognized as showing mainly transverse sediment movement, with sediment distributed as narrow belts oriented parallel to the coast (e. g. , the Luanhe River Delta). (2) Tidal current regimes are recognized as macrotidal estuaries, such as the Yellow River, where sediment distribution follows the direction of tidal currents. Off the Yellow River mouth, sediments have been deposited as finger bars pointing to the sea or as sediment tongues oriented parallel to the coast. (3) River runoff and longshore current regimes are recognized, where sediment diffuses to the sea following the direction of the joint forces between the two types of currents (e. g. , the Changjiang and Pearl River Estuaries).

Examples of land-sea interaction in North America are often related to the terrigenous fluxes from high-latitude glacier-and snow-fed rivers. These fluxes have produced in the geological past (Pleistocene), and in many cases continue to produce a wide assortment of types of estuaries, fjords and deltas. Influencing factors include underlying topography, seasonal changes and human-induced changes. The influence on sediment delivery by river plume sedimentation and coastal turbidity currents greatly influences the development of continental shelves.

Bibliography

[1] Amos, C. L. and B. F. N. Long. 1980. The sedimentary character of the Minas Basin, Bay of

Fundy. In: *The Coastline of Canada: Littoral Processes and Shore Morphology*, S. B. McCann, ed. Paper 80～10. Geological Survey of Canada, Ottawa. Ontario. 123 - 152.

[2] Ashmore, P. 1993. Contemporary erosion of the Canadian landscape. *Prog. Phys. Geogr.*, 17: 190 - 204.

[3] Binda, G. G., T. J. Day, and J. P. M. Syvitski. 1986. Terrestrial sediment transport into the marine environment of Canada: annotated bibliography and data. In: *Sediment Survey Section Report IWD-HQ-WRB-SS* - 86 - 1. Environment Canada, Ottawa, Ontario.

[4] Brown, W. M., III and J. R. Ritter. 1971. Sediment transport and turbidity in the Eel River basin, Califomia. In: *Water-Supply Paper* 1986. U. S. Geological Survey, Washington, D. C.

[5] Brunsden, D. and D. B. Prior. 1984. *Slope Instability*. Wiley. Chichester, West Sussex, England.

[6] Cataliotti-Valdina. D. and B. F. N. Long. 1984. Évolution estuarienne d'une riviére régularisé en climat subboréal; la rivière aux Outardes (côte nord du golfe du St. Laurent, Québec). *Can. J. Earth Sci.*, 21: 25 - 34.

[7] Chai, Mingli. 1993. The evolution of the Yellow River Delta development and its influence on the Bohai-Yellow Sea. Ph. D. thesis, Nanjing University, Nanjing, China.

[8] Chao, B. F. 1991. Man, water and global sea level. *Trans. Am. Geophys. Union*, 72: 492.

[9] Cheng, Tian-wen and Chu-nian Zhao. 1985. Water and sediment discharges to the ocean and the affected to the coastal zone by che major rivers of China. *Acta Oceanol. Sin.*, 7(4): 460 - 471.

[10] Church, M. 1972. Baffin Island sandurs: a study of arctic fluvial processes. *Bulletin* 216. Geological Survey of Canada. Ottawa. Ontario.

[11] Church, M. and J. M. Ryder. 1972. Paraglacial sedimentation: a consideration of fluvial processes conditioned by glaciation. *Geol. Soc. Am. Bull.*, 83: 3059 - 3071.

[12] Coleman, J. M. and H. H. Roberts. 1989. Deltaic coastal wetlands. *Geol. Mijnbouw*, 68: 1 - 24.

[13] Eisma, D., P. G. E. F. Augustinus and C. Alexander. 1991. Recent and subrecent changes in the dispersal of Amazon mud. *Neth. J. Sea Res.*, 28(3): 181 - 192.

[14] Farmer, D. M. and H. J. Freeland. 1983. The physical oceanography of fjords. *Prog. Oceanogr.* 12: 1 - 73.

[15] Fu, Winxia and Guangtien, Li. 1995. Marine environment projection. In: *Marine Geography of China*, Ying Wang, ed. Science Press, Beijing.

[16] Geographic Institute of Chinese Academy of Sciences, Water Conservancy and Hydroelectric Science Academy of Changjiang. Institute of Changjiang Channel Planning and Designing. 1985. *The Characteristic and Evolution of Middle and Lower Reaches of Changjiang*. Science Press, Beijing.

[17] Geyer, W. R. 1990. The advance of a salr wedge front: observations and dynamical model. In: *Physical Processes in Esruaries*. J. Dronkers and W. van Leussen, eds. Springer-Verlag, New York. 181 - 195.

[18] Gilbert, G. K. 1917. Hydraulic-mining debris in the Sierra Nevada. *Professional Paper* 105. U. S. Geological Survey. Washington, D. C.

[19] Hay, B. J. 1989. Particle flux in the western Black Sea in the present and over the last 500 years:

temporal variability, source, and transport mechanisms. Ph. D. thesis, WHOI-MIT, Cambridge, Mass.

[20] Heggen, R. J. 1986. Periodic surges and sediment mobilization. In: *Drainage Basin Sediment Delivery*, R. F. Hadley, ed. Publication 159. IAHS, London. 323 - 333.

[21] Hein, F. J., J. P. M. Syvitski, L. A. Dredge, and B. F. Long. 1993. Quaternary sedimentation and marine placers along the north shore, Gulf of St. Lawrence. *Can. J. Earth Sci.*, 30: 553 - 574.

[22] Hickin, E. J. 1989. Contemporary Squamish River sediment flux to Howe Sound, British Columbia. *Can. J. Earth Sci.*, 26: 1953 - 1963.

[23] Hill, P. R., S. M. Blasco, J. R. Harper and D. B. Fissel. 1991. Sedimentation on the Canadian Beaufort Shelf. *Cont. Shelf Rcs.*, 11: 821 - 842.

[24] Holeman, J. N. 1968. Sediment yield of major rivers of the world. *Water Resources Res.*, 4: 737 - 747.

[25] Hossian, J. N. 1991. *Total Sediment Load in the Lower Ganges and Jumuma*. Bangladesh University of Engineering and Technology, Dacca.

[26] Huang, Shen. 1981. General description and deformation of the river bed of the Yangtze Estuary [in Chinese]. *Res. Navigat. Yangtze Estuary*, 1: 2 - 30.

[27] Jay, D. A. and J. D. Smith. 1990. Residual circulation in and classification of shallow. stratified esruaries. In: *physical Processes in Estuaries*, J. Dronkers and W. van Leussen, eds. Springer-Verlag, New York. 21 - 41.

[28] Jeftic, L., S. Keckes and J, C. Pernrtta, eds. 1996. *Climate Change and the Mediterranean*. Edward Arnold, New York. 8.

[29] Kenyon, P. M. and D. L. Turcotte. 1985. Morphology of a delta prograding by bulk sediment transport. *Geological Society of America Bull.*, 96: 1457 - 1465.

[30] Kostaschuk, R. A. 1985. River mouth processes in a fjord-delta, British Columbia. *Mar. Geol.*, 69: 1 - 23.

[31] Kostaschuk, R. A. and L. A. Atwood. 1990. River discharge and tidal controls on salt-wedge positionand implications for channel shoaling: Fraser River, British Columbia. Can. *J. Civil Eng.*, 17: 452 - 459.

[32] Kostaschuk, R. A. and J. L. Luternauer. 1989. The role of the salt-wedge in sediment resuspension and deposition: Fraser River Estuary, Canada. *J. Coastal Res.*, 5: 93 - 101.

[33] Kostaschuk, R. A., M. A. Church and J. L. Luternauer. 1989. Bedforms, bed material and bedload transport in a salt-wedge estuary: Fraser River, British Columbia. Can. *J. Earth Sci.*, 26: 1440 - 1452.

[34] Kuehl. S. A., D. J. Demaster and C. A. Nittrover. 1986. Nature of sediment accumulation on the Amazon continental shelf. *Cont. Shelf Res.*, 6: 209 - 225.

[35] Lewis, A. G. and J. P. M. Syvitski. 1983. Interaction of plankton and suspended sediment in fjords. *Sediment. Geol.*, 36: 81 - 92.

[36] Li, Chueizhung. 1980. The characteristic of the surface sediments of the northern shelf of the South China Sea [in Chinese]. *Nanhai Stud. Mar. Sin.*, 1: 35 - 50.

[37] Li, Chunchu. 1983. Dynamics and sedimentations of Madao-Men Estuary. *Trop. Geogr.*, 1: 27 -

34.

[38] Matthai, H. F. 1990. Floods. In: *Surface Water Hydrology*, M. G. Wolman and H. C. Riggs, eds. Geological Society of America, Boulder, Colo., *Geol. N. Am.* 0-1, 97-120.

[39] Mclean, D. G. and M. A. Church. 1986. A re-examination of sediment transport observations in the lower Fraser River. *Sediment Survey Section Report IWD-HQ-WRB-SS*-86-6. Environment Canada, Ottawa, Ontario.

[40] Meade, R. H., T. Dunne, J. E. Richey, V. DeM. Santons and E. Salati. 1985. Storage and remobilization of suspended sediment in the lower Amazon River of Brazil. *Science*, 228: 488-490.

[41] Meade, R. H., T. R. Yuzyk and T. J. Day. 1990. Movement and storage of sediment in rivers of the United States and Canada. In: *Surface Water Hydrology*. M. G. Wolman and H. C. Riggs, eds. Geological Society of America, Boulder, Colo. *Geol. N. Am.*, 0-1, 255-280.

[42] Milliman, J. D. 1980. Sediment in the Fraser River Delta and its estuary. *Estuarine Coastal Mar. Sci.*, 10: 609-633.

[43] Milliman, J. D. and R. H. Meade. 1983. World-wide delivery of river sediment to the oceans. *J. Geol.*, 91: 1-21.

[44] Milliman, J. D. and J. P. M. Syvitski. 1992. Geomorphic/tectonic control of sediment discharge to the ocean: the importance of small mountainous rivers. *J. Geol.*, 100: 525-544.

[45] Molinia, B. F., P. R. Carlson and W. P. Levy. 1978. Holocene sediment volume and modern sediment yield, northeast Gulf of Alaska (abs.). *Am. Assoc. Petrol. Geol. Bull.*, 62: 545.

[46] Mulder, T. and J. P. M. Syvitski. 1995. Turbidity currents generated at river mouths during exceptional discharges to the world's oceans. *J. Geol.*, 103: 285-299.

[47] Mulder, T. and J. P. M. Syvitski. 1996. Climatic and morphologic relationships of rivers: implications of sea level fluctuations on river loads. *J. Geol.*, 104: 509-524.

[48] Nash, D. B. 1994. Effective sediment transporting discharge from magnitude frequency analysis. *J. Geol.*, 102: 79-95.

[49] Nordin, C. F., Jr. and R. H. Meade. 1986. The Amazon and the Orinoco. In: *McGraw-Hill Year Book of Science and Technology*. McGraw-Hill, New York. 90-385.

[50] Pang, J. and Shuheng, Si. 1979. The estuary changes of Huang He River. *Oceanol. Limnol. Sin.*, 10: 136-141.

[51] Pang, J. and Shuheng, Si. 1980. Eluvial processes of the Huang He River Estuary. *Oceanol. Limnol. Sin.*, 11: 293-305.

[52] Ren, Mei-e. 1964. Geomorphological characteristics of the Pearl River Estuary and vicinity [in Russian]. *Acta Sci. Nat. Univ. Nankinensis*. 8: 136-146.

[53] Ren, Mei-e. 1974. The subaqueous delta of Hai He River and silting problem of Tiantsin new port. *Acta. Sci. Nat. Univ. Nankinensis*, 1: 80-90.

[54] Ren, Mei-e. 1981. Sedimentation on tidal mudflat in Wanggang area, Jiangsu Province, China. *Mar. Sci. Bull.*, 3: 40-50.

[55] Ren, Mei-e. 1994. Impact of climate change and human activity on flow and sediment of the rivers: the Yellow River (China) example. *Geo. J.*, 33(4): 443-447.

[56] Ren, Mei-e and Chenkai Zang. 1980. Quaternary continental shelf of East China Sea. *Acta Oceanol*. Sin., 2: 94-111.

[57] Ren, Mei-e and Xianmo Zhu. 1994. Anthropogenic influences on changes of sediment discharge of the Yellow River, China during the Holocene. *The Holocene*, 4(3): 321 - 327.

[58] Ross, W. C., B. A. Halliwell, J. A. May, D. E. Watts and J. P. M. Syvitski. 1994. The slope readjustment model: a new model for the development of submarine fan/apron deposits. *Geology*, 22: 511 - 514.

[59] Sala, M. and B. Long. 1990. Evolution des structures deltaïques du delta de la Riviére Natashquan, Québec. *Geogr. Phys. Quaternaire*, 43: 311 - 323.

[60] Schafer, C. T., J. N. Smith and G. Seibert. 1983. Significance of natural and anthropogenic sediment inputs to the Saguenay Fjord, Quebec. *Sediment. Geol.*, 36: 177 - 195.

[61] SCOR Working Group 89. 1991. The responses of beaches to sea level changes, a review of predictive models. *J. Coastal Res.*, 7(3): 117 - 130.

[62] Shen, Huanting, Songling He Pingan Pan and Jiufa Li. 1992. A study of turbidity maximum in the Changjiang, Estuary. *Acta Geogr. Sin.* , 47(5): 472 - 479.

[63] Shen, Jian, Huanting Shenm, Dingan Pan and Chenyou Xiao. 1995. Analysis of transport mechanism of water and suspended sediment in the turbidity maximum of the Changjiang Estuary. *Acta Geogr*. Sin., 50(5): 411 - 420.

[64] Shi, Yunliang. 1989. The changes of water and sediment discharges of the Yellow River and the river mouth renovating project. *J. Nanjing Univ. (Geogr)*, 10: 24 - 33.

[65] Swift, D. J. P. 1970. Quaternary shelves and the return to grade. *Mar. Geol.*, 8: 5 - 30.

[66] Syvitski, J. P. M. 1986. Estuaries, deltas and fjords of eastern Canada. *Geosci. Can.*, 13: 91 - 100.

[67] Syvitski, J. P. M. 1989a. The process-response model in quantitative dynamic stratigraphy. In: *Quantitative Dynamic Stratigraphy*, T. A. Cross, ed. Prentice Hall, Upper Saddle River, NJ. 309 - 335.

[68] Syvitski, J. P. M. 1989b. On the deposition of sediment within glacier-influenced fjords: oceanographic controls. *Mar. Geol.*, 85: 301 - 329.

[69] Syvitski, J. P. M. 1989c. Modelling the sedimentary fill of basins. In: *Statistical Applications in the Earth Sciences*. F. P. Agterberg and G. F. Bonham-Carter, eds. Paper 89 - 9. Geological Survey of Canada, Ottawa, Ontario. 505 - 515.

[70] Syvitski, J. P. M. and J M. Alcott. 1995. RIVER3: Simulation of river discharge and sediment transport. *Comput. Geosci.*, 21: 89 - 151.

[71] Syvitski, J. P. M. and G. E. Farrow. 1983. Structures and processes in bayhead deltas: Knight and Bute Inlets, British Columbia. *Sediment. Geol.*, 36: 217 - 244.

[72] Syvitski, J. P. M. and G. M. Farrow. 1989. Fjord sedimentation as an analogue for small hydrocarbon-bearing fan deltas. In: *Deltas: Sites and Traps for Fossil Fuels*, M. K. G. Whateley and K. T. Pickering, eds. Special Publication 41. Geological Society, London. 21 - 43.

[73] Syvitski, J. P. M. and F. J. Hein. 1991. Sedimentology of an arctic basin: Itirbilung Fiord, Baffin Island, Canada. Professional *Paper* 91 - 11. Geological Survey of Canada, Ottawa, Ontario.

[74] Syvitski, J. P. M. and J. M. Murray. 1981. Particle interaction in fjord suspended sediment. *Mar. Geol.*, 39: 215 - 242.

[75] Syvitski, J. P. M. and D. B. Praeg. 1989. Quaternary sedimentation in the St. Lawrence Estuary

and adjoining areas: an overview based on high-resolution seismo-stratigraphy. *Geogr. Phys. Quanteranaire*, 43(3): 291 - 310.

[76] Syvitski, J. P. M., K. W. Asprey, D. A. Clattenburg and G. D. Hodge. 1985. The prodelta environment of a fjord: suspended particle dynamics. *Sedimentology*, 32: 83 - 107.

[77] Syvitski, J. P. M., D. C. Burrell and J. M. Skei. 1987. *Fjords: Processes and Products*. Springer-Verlag, New York.

[78] Syvitski, J. P. M., J. N. Smith, E. A. Calabrese and B. P. Boudreau. 1988. Basin sedimentation and the growth of prograding deltas. *J. Geophys. Res.*, 93: 6895 - 6908.

[79] Syvitski, J. P. M., K. M. G. LeBlanc and R. E. Cranston. 1990. The flux and preservation of organic carbon in Baffin Island fjords. In: *Glaciomarine Environments: Processes and Sediments*, in from p. Special Publication 53. Geological Society, London. 217 - 239.

[80] Syvitski, J. P. M., K. W. Asprey and K. W. G. LeBlanc. 1995. In situ characteristics of particles setting within a deep-water estuary. *Deep-Sea Res. II*, 42(1): 223 - 56.

[81] Tong, Qicheng and Tian-wen Cheng. 1981. Runoff. In: *Physical Geography of China* [in Chinese] Bingwei Huang, ed. Science Press, Beijing. 6 - 121.

[82] Van Leussen, W. 1989. Aggregation of particles, setting velocity of mud flocs: a review. In: *Physical Processes in Estuaries*, J. Dronkers and W. van Leussen, eds. Springer-Verlag, New York. 347 - 403.

[83] Vilks, G and D. P. Krauel. 1982. Environmental geology of the Miramichi Estuary: physical oceanography. *Paper* 81 - 24. Geological Survey of Canada, Ottawa, Ontario. 1 - 53.

[84] Walling, D. E. 1985. The sediment yields of African rivers. *Int. Assoc. Hydrol. Sci. Publ.*, 144: 276 - 316.

[85] Wang, Ying. 1963. The coastal dynamic geomorphology of the northern Bohai Bay [in Chinese]. *Collect. Ocean. Works*, 3: 25 - 35.

[86] Wang, Ying. 1980. The coast of China. *Geosci. Can.*, 7, 109 - 113.

[87] Wang, Ying. 1983. The mudflat coast of China. *Can. J. Fish. Aquat. Sci.*, 40 (Suppl. 1): 160 - 171.

[88] Wang, Yuenzhang. 1994. The characteristics of 1980's precipitation and its influences to the water and sediment discharges of Yellow River between the Hekou town and Longmen area (middle reaches). *People's Yellow River*, 12: 5 - 9.

[89] Wang, Ying and Xiaogen Wu. 1995. Sea level rise and beach response. *ACTA Geogr. Sin.*, 50 (2): 118 - 127.

[90] Wang, Kangshan, Jilan Su and Lixian Dong. 1983. Hydrographic features of the Changjiang Estuary [in Chinese]. *In: Proc. International Symposium on Sedimentation on the Continental Shelf, with Special Reference to the East China Sea*. 1. 137 - 147.

[91] Wang, Ying, Mei-e Ren and Dakui Zhu. 1986. Sediment supply to the continental shelf by the major rivers of China. *J. Geol. Soc. London*, 143: 935 - 944.

[92] Water Survey of Canada. 1977. Sediment Data: Atlantic Canada and British Columbia, 1975. Inland Water Directorate, Water Resources Branch, Water Survey of Canada, Ottawa, Canada, published by authority of the Minister of Environment by the Minister of Supply and Services, Canada.

[93] You, Kunyuan, Liangren Su and Jiang-chu Qian. 1983. Modern sedimentation rate in the vicinity of the Changjiang Estuary and adjacent continental shelf [in Chinese]. In: *Proc. International Symposium on Sedimentation on the Continental Shelf, with Special Reference to the East China Sea*. 1. 590 - 605.

[94] Yu, Honghua, Dacheng Zheng and Jingzheng Jian. 1983. Basin hydrographic characteristics of the studied area [in Chinese]. In: *Proc. International Symposium on Sedimentation on the Continental Shelf, with Special Reference to the East China Sea*, 1. 295 - 305.

[95] Zhao, Huanting. 1983. Hydrological characteristics of the Zhujiang (Pearl River) Delta. *Trop. Oreanol.*, 2: 108 - 117.

[96] Zhou, Fugen. 1983. Automorphous calcite crystal in seawater of the northeastern East China Sea [in Chinese]. In: *Proc. International Symposium on Sedimentation on the Continental shelf, with Special Reference to the East China Sea*, 1. 447 - 461.

[97] Zhu, Dakui. 1980. The coastal evolution of the east Hebei Province [in Chinese]. In: *Proc. Symposium of China's Coast and Estuary*, Shanghai.

Sea Shore[①]

1. Definition and Progress of Epistemology

The term of sea shore, beach and coast are loosely used with varying significance by terrene and marine scientists, and mostly mixed use by ordinary speech.

In the early stage of 20 century, Douglas Wilson Johnson summarized (Johnson, 1919) the variety of usage and naming shore features, by using the terminology as following:

"Sea Shore" is the zone over which the waterline, the line of contact between land and sea, migrates. The position of waterline at high tide marks the high tide shoreline. The low tide shoreline marks the seaward limit of the intermittently exposed shore, and the term of shoreline is used as the line of low tide level.

The shore is sub-divided into two minor zones by D. W. Johnson.

"ForeShore", the zone lies between the ordinary high and low water marks, where is daily traversed by the oscillating waterline as the tides rise and fall.

"BackShore", back of foreshore there is covered by water during exceptional storm period.

While the "Coast", defined by Johnson, is a broad zone of indeterminate width located just landward from the shore.

Seaward from shore, two units have been defined by Johnson as:

"Shoreface" is the zone between the low tide shoreline and the beginning of more nearly horizontal surface.

"Offshore" is a comparatively flat zone of variable width extending from the outer margin of the rather steeply sloping shoreface to the edge of continental shelf (Fig. 1).

The above concepts have been adopted widely, such as the book of "Beaches and Coasts" written by Cuchlaine A. M. King (1959, 1972), and the book of "Submarine Geology" by Francis Shepard (1948, 1963, 1973). King has defined more beach features in detail, and Shepard defined the offshore as the seaward of foreshore (Fig. 2). These terms have been used up to date.

---

① Ying Wang: Yang Wencai, Paul Robinson, Fu Rongshan, Wang Ying, eds. *Geodynamic Processes and our Living Environment*, chapter7, pp. 191－200. Beijing: Geological Publishing House, 2001.

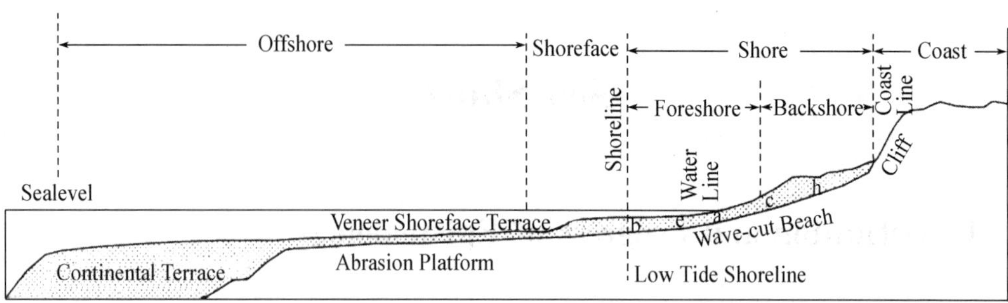

Fig. 1　Elements of the shore zones in an advanced stage of development (by D. W. Johnson. 1919)

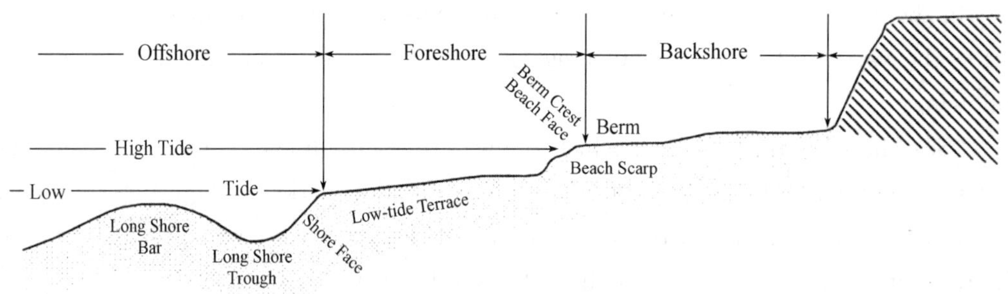

Fig. 2　The principal subdivisions of beaches and of adjacent shallow-water area (by F. P. Shepard, 1973)

As a professional term, "sea shore" indicates the ground between ordinary high and low tide levels, which is similar to the meaning of foreshore. However, as an ordinary speech, sea shore is sometime used as the same meaning as "coast" named by Johnson. For example, in the Glossary of Geology, 2nd edition 1980, "sea shore" is a narrow strip of land adjacent to or boardering a sea or ocean. Thus, it needs to clear up.

Following the progress of marine science since 2nd World War of 20 century, the nature of coast has been more and more further understood. Present scientific conception about coast, is a zone where the land meets the sea, a transitional zone between land and ocean, and an active dynamic zone of land-ocean interaction.

As an "Amphibious" zone, the coastal zone includes coastal land, upper limited by the break waves (swash) or tidal current acted zone, sea shore or inter-tidal zone, and the submarine coastal slope offshore, ended at the outer boundary area where wave acted on the sea bottom of coastal slope, where the water depth is normally equal to 1/2 or 1/3 of average wave length of the region (Fig. 3, Y. Wang, 1994).

It was first time that the coastal zone with clearly boundary defined by the nature of coastal dynamic progress (V. P. Zenkovich, 1962). Based on the practice in the coastal engineering projects it is successfully since 1960s. China adopted the definition of the coastal zone especially for its whole nation wide environment and resources survey along the mainland coasts, and larger islands during 1980s. The definition indicates the

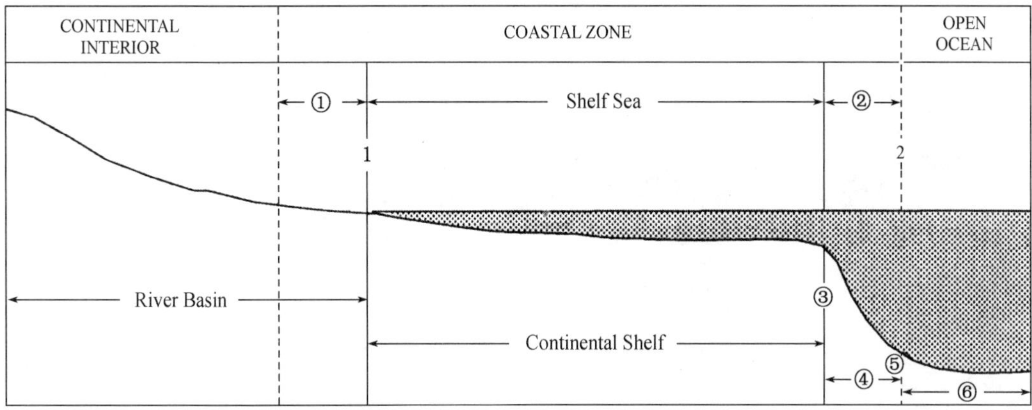

Fig. 3　Coastal Zone—Land and Ocean Boundary

1—Land/Sea Interface; 2—Shelf Sea/Ocean Interface; ① Serf Zone; ② Shelf Edge; ③ Shelf Break; ④ Continental Slope; ⑤ Continental Rise; ⑥ Ocean Floor

progress on the coastal science. People realize that the coastal zone is a dynamic system, the change of upper coastal feature is the result of the change by submarine coastal slope, and the inter-tidal zone or sea shore, is the one of most dynamic parts in the coastal zone.

The 21th century, human kind faces a challenge that the ocean has a core role in providing living and non-living resources for the world populations, but it is much less understood than the terrestrial part of our planet especially.

Following the Global Change studies and "the Law of Sea" of United Nations is in effect since 1994, the conception of "coastal ocean" has been employed and is gradually adopted in a broad sense and as the legal criterion, such as:

LOICZ Core Project (IGBP 1993), defined the coastal zone as: "extending from the coastal plains to the outer edge of continental shelves, approximately matching the region that has been alternately flooded and exposed during the sea level fluctuation of the late Quaternary period". However, the continental slope and continental rise should be included in the land-sea interaction zone or dynamic zone of coastal water either during the later Quaternary period or present time. The book of "The Global Coastal Ocean Processes and Methods", as the volume 10 of "The Sea" published with the financial assistance of UNESCO/IOC (K. H. Brink and A. Robinson, 1998) is a mile stone to indicate the science advancing and responding to the need for scientific knowledge arising from the UNCED Agenda 21. "Coastal Zone" is concerned with the coastal systems, extending seawards up to outer edge of the continental margin by physical criterion or the outer limit of the national maritime jurisdictional zone by Legal Criterion (Fig. 4 Coastal Zone), (Adalberto Vallega, 1999).

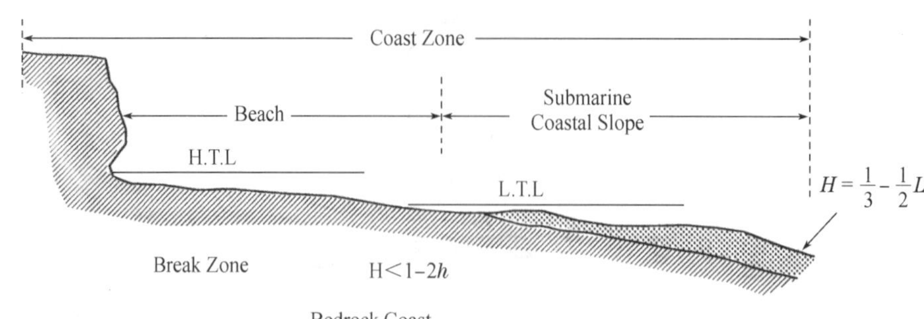

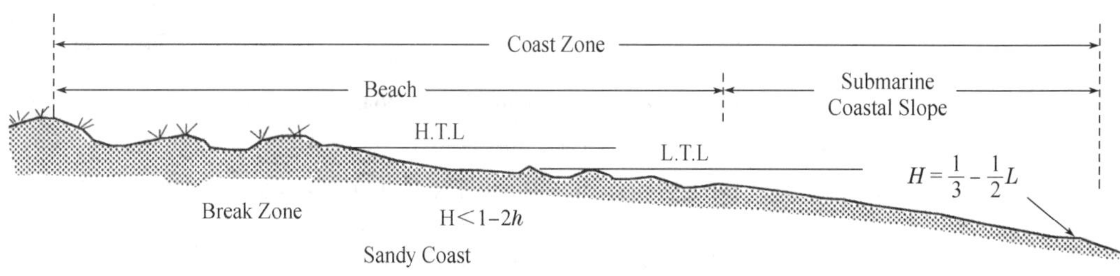

Fig. 4　Coastal Zone (by Wang Ying, 1994)

The epistemological setting of Coastal Ocean will be implement in such a way as to optimize the contribution of geosciences to Ocean Science as an effective inter-disciplinary approach to coastal ocean or coastal sea system—an independent and closed inter-related system characterized by air-sea, land-sea interaction and human beings impacts.

As a summary review on the definition of sea shore, coast, coastal zone and coastal ocean, it can be seen that sea shore has its specific content as inter-tidal ground or with a broader sense can be used as an "Amphibious" coastal zone by general speech. The coastal ocean indicates that the whole area of land-sea interaction can not be instead of "Sea shore".

2. The Characteristic Nature of Sea Shore

Sea shore or coastal zone has an independent environment, which differs from neither oceanic environment nor territorial, because of the land-ocean interactions produce the coastal dynamic with their own natures.

Wind speeds up in velocity and changes their direction in the coastal zone. According to statistics data of 20 years from Jiangsu coasts of China, that coastal wind speed is an index power relation with the distance to the sea shore (x). The empirical equation is:

$$Y=\bar{U}_1/\bar{U}_0=0.70+0.3e^{-0.0415x}$$

$\bar{U}_1$ is a maximum wind speed of several years' average in the x_1 distance to the sea shore.

$\bar{U}_0$ is a maximum wind speed of several years along the sea shore.

While the x is 0 (shoreline) to 40 km, $\bar{U}_1$ increases rapidly;

While the x is more than 60 km, the value of Y is almost a constant of 0.7, i.e., the wind speed is 30% less in the outside of the sea shore or coastal zone.

- Waves produced by the wind action are the most important type of sea waves. While waves approach to sea shore or coastal shallow waters, they also change their nature of wave rays, speed up the velocity and shorten the period. Wave refract, reflect and form standing waves, diffract, converge or diverge their energy, produce break waves and wave current, storm surge etc., these processes are only happened in the coastal zone.
- Tidal dynamics are even with special features, accumulate enormous energy and act as an active dynamic factor in the coastal zone.
- River-sea system bears the most unique processes in the coastal ocean. Rivers transport terrigenous sediment to the ocean, including solid and dissolved material, i.e., 20 billion metric tons per year, and greatly influence the sediment dynamics and ecological environment in the coastal water (Wang et al. 1998, Milliman and Syvitski, 1992). Human impacts are direct or indirect through rivers onto coastal environment. In China, the contaminated material was carried to sea through rivers, and more than 70 % of them concentrated in the coastal zone (Fu and Li, 1996). For example, the area of Changjiang River delta has taken 9% of waste water, 5% of waste gas, and 4% of other residual from industry, and the delta land area is only 0.6% of the nation.

Under marine dynamic processes, the nature of lithosphere of coastal land—geological structures, lithological and geomorphological natures etc., has formed different coastal type and evolutional processes. For example: Global Plate Tectonics control the largest scale of coastal types as Collinsion coasts, trailing-edge coasts, and marginal sea coasts; secondary scale coast features controlled by geological structures, such as bedrock embayed coast, plain coast, volcanic coast…… and drumlin coast, sand dune coast are typical examples of geomorphological controlled coasts. The debris from coastal erosion enter coastal zone consisting the sedimentary dynamics. With the different kind of Bedrocks, the sediment size, mineral component and quantity of sediment supply are completely different.

Thus, marine dynamics and nature of lithosphere and tectonic movement of coastal land are two groups of factors to control the coastal evolution. However, under global warming trend and sea level rising processes, sea shore area is sensitive response for the changing. Beach and shoreline erosion, sand barrier and coastal retreating, storm surge strengthen, low land drowned, salt-water intrusion etc. are all kind of problems that sea shore area is facing!

Coastal zone has important natural resources. There are mostly with low cost, less polluted and can be nourished. These resources are:

- Energy resources: wind wave, tidal, potential energy of temperature and salinity of coastal seawater;
- Sandy material and placer resources;
- Space resources, including tidal flats and wild land, sandy beaches, harbour and estuarines etc;
- Food and medical resources, coastal zone has 1/4 of global primary productivity;
- Recreation and tourism resources.

As a whole, sea shore is an important area both for study processes and history of land-sea interaction, and to sustainable development of human beings in 21 century.

Bibliography

[1] Adalbelto Vallega. 1999. *International Charter on Ocean Geography*. In Press.

[2] Cuchlaine A. M. King. 1972. *Beaches and Coasts*. Second edition. Edward Arnald Lt. d.

[3] Douglas Wilson Johnson. 1919. *Shore Process and Shoreline Development*. John Wiley&Sons, ING, New York. 159 - 163.

[4] Francis P. Shepard. 1973. *Marine Geology*. Third edition. Harper & Row. New York.

[5] Fu, Wenxia and Li, Guang Tian. 1996. *Ocean Pollution and Environment Protection*. In: *Marine Geography of China*, edited by Wang Ying. Science Press. China. 474 - 517.

[6] Kenneth H. Brink and Allan R. Robinson. 1998. *The Global Coastal Ocean Processes and Methods*. John Wiley & Sons. Inc.

[7] Milliman, J. D. and J. P. M. Syvitski. 1992. Geomorphic/tectonic control of sediment discharge to the ocean, the importance of small mountain rivers. J. Geol. , 100: 525 - 544.

[8] Robert L. Bates and Julia A. Jackson, editors. 1980. *Glossary of Geology*. Second edition. American Geological Institute.

[9] Report on the comprehensive survey of coastal zone and tidal flat resources in Jiangsu. 1986. China Ocean Press. 226 - 229.

[10] Wang Ying and Zhu Dakui. 1994. *Coastal Geomorphology*. China Higher Education Press, Beijing.

[11] Wang Ying, Mei-e Ren and James Syvitski. 1998. *Sediment Transport and Terrigenous Fluxes*. In: The Sea Volume 10 (Chapter 10). The Global Coastal Ocean. Processes and Methods. 253 - 292.

全球火山地震活动与海岸海洋环境效应*

海岸海洋是陆地与深海大洋相互过渡地带，并与之共同构成地表三大基本组成单元。它是既区别于陆地又区别于深海大洋的独立环境体系，受人类活动影响深刻，其资源环境与人类生存发展关系密切，是研究大气、水、岩石、生物各圈层相互作用的最佳切入点。海岸海洋的范围包括海岸带、大陆架、大陆坡及坡麓的大陆隆（Continental rise），涵盖了整个的海陆过渡带（图 1）[1-4]。1994 年联合国海洋法正式生效，沿海国 12 海里领海、24 海里毗连区，200 海里专属经济区以及大陆架归属沿岸国管辖，兴起了各国对“海洋领土”的关注，推动了对海岸海洋研究的进展与海岸海洋科学（Coastal Ocean Science）的迅速成长。它是一门具有自然科学、人文社会科学与技术科学相互交叉渗透的复合型科学，专门研究陆海过渡带的表层系统作用过程、环境资源特性变化发展规律，以及人类生存活动与之和谐发展的科学。它的应用基础型科学特性，吸引着公众的注意。

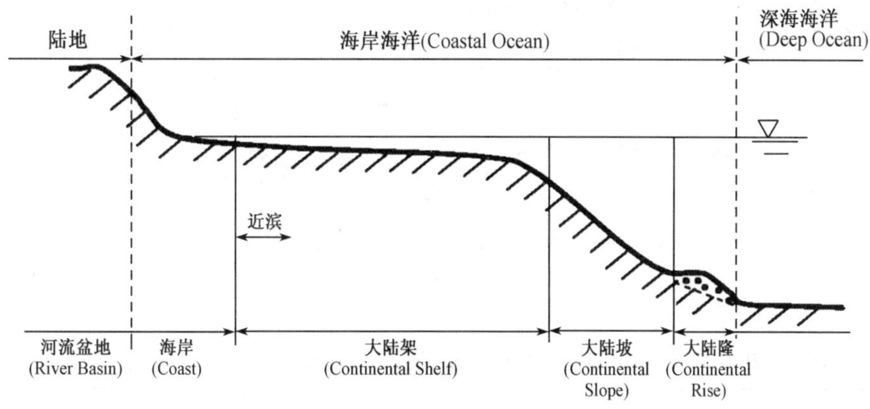

图 1 海岸海洋图示

全球变化是反映在气、水、岩石、生物圈的事件性变动形成全球性的频发与持续效应，对人类生存环境影响深刻。气候变化、海平面变化与人类活动是与海岸海洋密切相关的，已形成全球变化研究的热点课题，大气与海平面环境监测及趋势性分析等方面已取得重要进展。40°N 以北春季增温快。北京自 2000 年 6 月以来，形成近 50 年来的持续高温纪录。北美多伦多市亦是。高山冰川退化，南极冰融期增加到三周，随着中高纬度地区气温增高，海平面持续上升，大河三角洲区面临着土地淹没与风暴潮频袭城市的境界。

海陆交互作用影响的另一例证是气温增升，内陆沙漠干热，沙尘暴活动频频侵袭东部沿海城市，甚至长江以南的南京今春多昏霾的白昼。但是东亚大陆沙尘暴却使得北美沿岸的东太平洋海域鲑鱼丰收，加拿大太平洋渔业研究机构已有较多的研究成果。应追索

* 王颖，张永战，牛战胜：卢演俦，高维明，陈国显，陈杰主编，《新构造与环境》，349－355 页．北京：地震出版社，2001 年．

求源，进行对比研究，以此探北半球西风盛行带的迁移规律，追索地表圈层系统的相互影响。

当前全球变化与海陆交互作用影响研究的一个闪光点是：地震火山活动与海岸海洋效应。全球有活火山500多座，主要分布于板块边缘、岛弧海沟系、洋中脊和板块内部四处构造活动带，其中沿太平洋沿岸密集分布着370多座活火山，形成著名的“火圈”；其次沿东西向“关闭的地中海”带分布。20世纪90年代以来，火山爆发频频，逐年递增，尤以海岸海洋地区为突出(图2、表1和表2)。至2001年太平洋“火圈”火山爆发异常频繁，南太平洋、大西洋、墨西哥湾及大陆地区亦相继频频爆发，形成全球性构造活动，使地震与火山喷发频繁活动期的表现，已对地表系统相互作用过程环境特性形成重大影响。

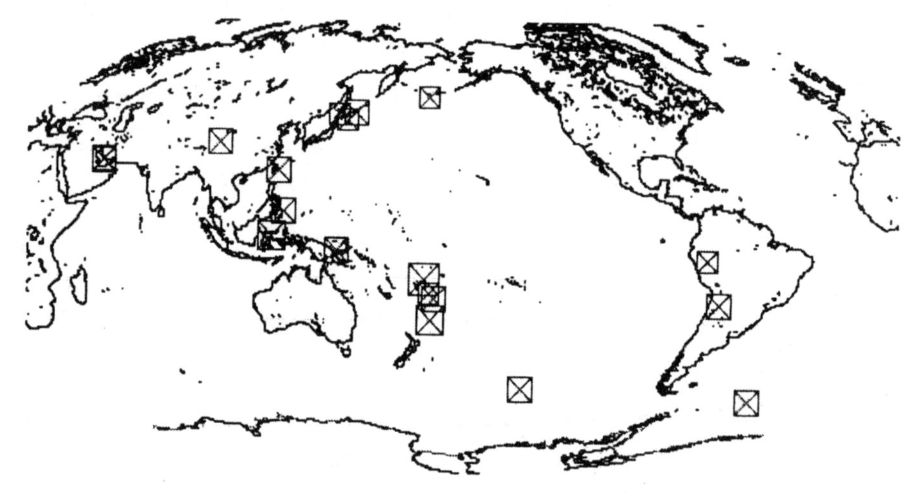

图2 全球火山爆发分布

(据美国国家航空和宇宙航行局，2000年10月；☒表示火山)

表1 1995—2001年5月29日全球火山活动情况(据美国国家航空和宇宙航行局数据统计)

火山爆发		
年代	次数	位置
1995	5	4次位于太平洋地区(2次位于南太平洋)，1次位于印度洋地区
1996	10	9次位于太平洋地区(1次位于南太平洋)，1次位于大西洋地区
1997	14	13次位于太平洋地区(2次位于南太平洋)，1次位于大西洋地区
1998	15	11次位于太平洋地区(4次位于南太平洋)，2次位于大西洋地区，1次地中海，2次陆地(美洲、非洲)
1999	13	12次位于太平洋地区(2次位于南太平洋)，1次位于大西洋地区
2000	50	33次位于太平洋地区(3次位于南太平洋)，6次位于墨西哥湾，3次非洲(刚果，喀麦隆)，1次地中海，3次印度洋，1次大西洋，3次南美洲
2001(截至5月29日)	26	19次位于太平洋地区(8次位于南太平洋)，3次位于墨西哥湾，1次印度洋，1次大西洋，大陆1次

表 2　2001 年全球火山爆发统计表

（据美国国家航空和宇宙航行局，2001 年 6 月 4 日）

火山名称	最初爆发日期	持续爆发日期	位置
1. Sheveluch, Kamchatka, Russia	May 21, 2001	May 29, 2001	56.65N, 161.36E
2. Mount Oyama, Miyakejima, Japan	May 24, 2001	May 24, 2001	34.08N, 139.53E
3. Tungurahua, Ecuador	May 15, 2001	May 24, 2001	1.467S, 78.44W
4. Lokon, Sulawesi, Indonesia	May 20, 2001	May 24, 2001	1. 36S, 124.79W
5. Mayon, Philippines	May 16, 2001	May 18, 2001	13.3N, 123.7E
6. San Cristobal, Nicaragua	May 10, 2000	May 16, 2000	12.7N, 87.0W
7. South Sister, Oregon	May 8, 2001	May 11, 2001	44.1N, 121.75W
8. Ulawun, New Britain, Papua New Guinea	April 30, 2001	May 2, 2001	5.1S, 151.3E
9. Masaya, Nicaragua	April 23, 2001	April 24, 2001	12.0N, 86.2N
10. Popocatepetl, Mexico	April 16, 2001	April 23, 2001	19.0N, 98.6W
11. Krakatau, Indonesia	March 27, 2001	April 12, 2001	6.10S 105.42E
12. Maroa, New Zealand	March 30, 2001	April 11, 2001	38.4S 176.1E
13. Jackson Segment, Northern Gorda Ridge	April 3, 2001	April 9, 2001	42.66N 126.78W
14. Canlaon, Philippines	January 2001	April 5, 2001	10.4N 123.1E
15. Piton de la Fournaise, Island of Reunion	March 27, 2001	April 4, 2001	21.23S, 55.71E
16. Arenal, Costa Rica	April 4, 2001	March 24, 2001	10.46N, 84.70W
17. Merapi, Java, Indonesia	March 10, 2001	March 21, 2001	7.54S, 110.44E
18. Cleveland, Chuginadak Island, Alaska	March 20, 2001	March 19, 2001	52.49N, 169.57W
19. Kliuchevskoi, Kamchatka, Russia	March 4, 2001	March 9, 2001	56.06N, 160.64E
20. Pacaya, Guatemala	February 21, 2001	March 2, 2001	14.4N, 90.6W
21. Colima, Mexico	February 22, 2001	February 28, 2001	19.51N, 103.62W
22. Ijen, Java, Indonesia	February 5, 2001	February 15, 2001	8.1S, 114.2E
23. Karangetang, Siau Island, Indonesia	January 25, 2001	February 9, 2001	2.78N, 125.48E
24. Nyamuragira, Congo, Africa	February 6, 2001	February 7, 2001	1.4S, 29.2E
25. Kelut, Java, Indonesia	January 29, 2001	February 2, 2001	7.9S, 112.3E
26. Rotorua, New Zealand	January 26, 2001	February 1, 2001	38.1S, 176.3E

火山爆发的同时，地震活动亦极为频繁。1990 年以来几乎是年年有大地震，2000 年在全球范围内，几乎每日有地震发生，强地震亦引起火山爆发。现代地震与火山活动分布

带基本相同，强地震发生在环太平洋构造带与东西向的“地中海”构造带。不同的是，地震活动还出现沿南大洋地区呈东西向的分布(图 3 和图 4)。

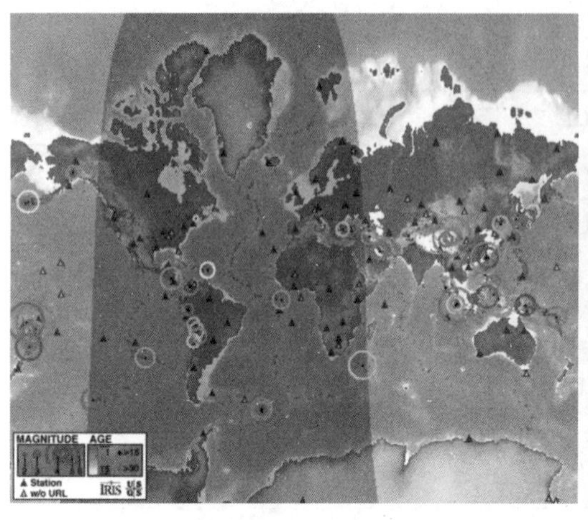

图 3　全球地震分布图

(据 NEIS of USGS IRIS 监测，2000. 9. 18)

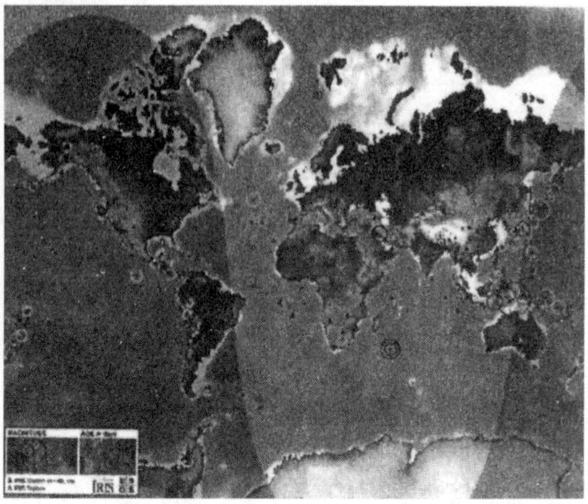

图 4　全球地震分布图

(据 NEIS of USGS IRIS 监测，2001. 3. 29)

对比美国地震信息中心发布的全球地震分布图(图 3、图 4、图 5、图 6)，其中黑圈表示发生于 24 小时之内的地震；白圈代表发生于 15 日内之地震；三角代表超过 15 天以上及 5 年以内的地震情况；圆圈大小表明震级。对比表明震级强度的转移，一次强地震爆发，能量释放以后震级变小，但同一构造带的另一处震级加强，如苏门答腊、青藏高原东部。这一现象为地震预报提供启示。

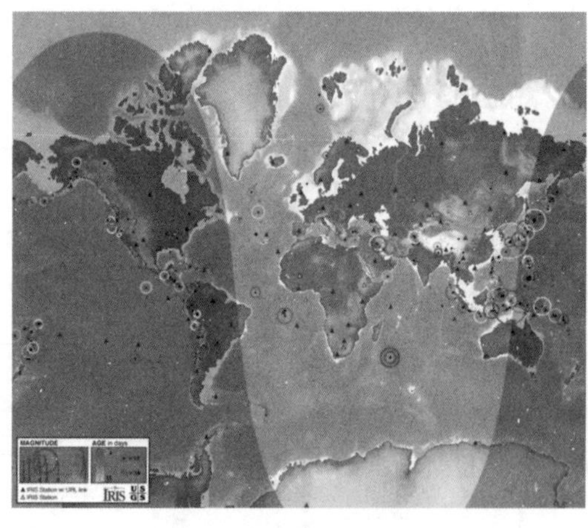

图 5　全球地震分布图

(据 NEIS of USGS IRIS 监测，2001. 4. 5)

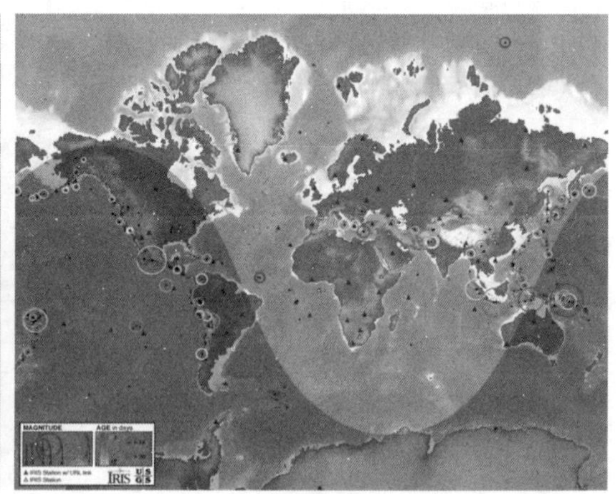

图 6　全球地震分布图

(据 NEIS of USGS IRIS 监测，2001. 5. 31)

地震与火山活动释放着地球内部的能量，太平洋中巨厚水层的压力、密度及温度变化聚集着较大的能量，与地球内部能量的结合，可能是大洋中能量较陆地更大的原因。其能量往往释放于板块结合部、构造裂隙断裂带与压力减轻处。如集中于太平洋两侧，尤以西侧的边缘海岛弧带最密集；关闭的古地中海地带，亦与俯冲挤压带有关。强地震多发生在海洋中，或发生于子夜，或发生于月半大潮期间，是否反映了月球引潮力对地震的激发或加助了地球内部能量的向外释放？或者说，白天太阳引力或陆地人类各种活动，削弱了月球引力与分散了部分地壳中所聚集的能量？归纳分析地震发生的时间，以及地震带强度转移变化的规律，有助于找出防震减灾措施——人为地释放、控制或分散能量、削弱地震发生的强度。

以发生在海岛的两次强地震为例：2000 年 6 月 4 日 23 时 28 分印尼苏门答腊岛南部明古鲁省发生里氏 7.9 级地震，震中位于雅加达西部 640 km 处（4.6°S，102.1°E 的印度洋底），震源深度 32 km，为浅源地震，是印尼近年最强烈的地震，死亡人数超过 100 人。一次强烈地震爆发后，地震活动仍继续，虽强度减小，但构造断裂活动带的潜在危机，使苏门答腊海岛地震仍频频发生。

1999 年 9 月 21 日 1 时 47 分 12.6 秒，台湾日月潭西南 6.5 km 处集集镇发生里氏 7.3 级地震，震中位于 23.85°N，120.78°E，震源深度 7～10 km[5]。整个台湾岛均可感到地震活动*。系台湾西部山麓地带的系列逆断层，长期受吕宋岛弧推挤，积蓄大量能量，从车笼埔断层发生错动，因能量释放而造成地震（图 7）。

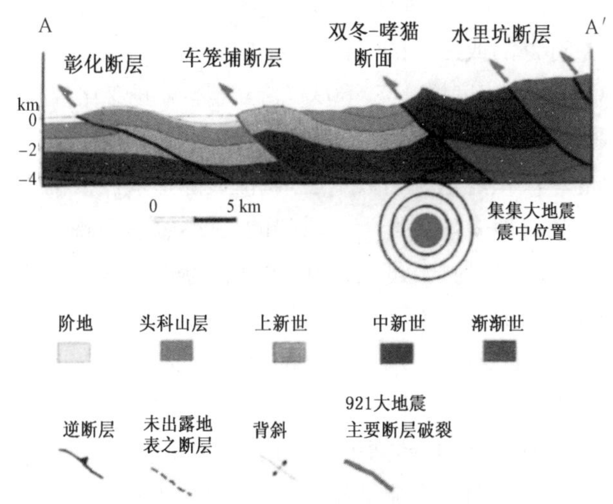

图 7　集集地震震中与车笼埔断层位置关系与地质剖面构造示意图（据[5,6]）（附彩图）

车笼埔断层东侧上盘上升 2.2～4.5 m，向 NW 及 NNW 方向水平滑移 7.1 m；下盘最大沉陷 0.3 m，向相反方向平移 1.1 m，最大水平地表加速度达 1 g，强震延时 25 s，波及相距 150 km 的台北市。损坏倒塌房屋约 2 万幢，死亡人数超过 2 300 人，伤者约 8 000 人。地震造成的环境效应与自然灾害包括：地表断裂、抬升形成断崖；河流中断，形成瀑布；山崩地塌，造成堰塞湖与滑塌；土壤液化，发生喷沙、喷泥与地层陷落等，伴随之造成对道路、桥梁、水利与交通设施、港湾设施、动力与工业设施以及建筑等的严重损害[5,6]。

作者近日考察了海南岛北部铺前—东寨海湾。该处位于琼州海峡断陷带，于 1605 年（明万历三十三年）发生了可能超过 7 级的大地震。地震沿河谷两侧断陷，使得 72 个村庄陷落成海，海湾口狭内宽，呈长方形，并且湾套湾，系沿支流陷落所致。湾内在古河道处水

* 台湾气象局地震预报中心第 043 号有感地震报告，1999；王鑫，林俊全，2000，九二一台湾地震纪念选址及地震登录之研究。

深超过 4 m，大部为浅水区，交通不便。2001 年 5 月 21 日，趁小潮落时，登上了露出的沉陆，它位于林市村东北海湾中。位于海水中经过 400 多年，沉陆上已淤积了 20～50 cm 厚的淤泥，但仍见碎瓦残缸，遍地皆是，但均有牡蛎附着生长。经仔细观察，见到一残石臼约 40 cm 的直径，已残破，一只小磨盘，均有牡蛎附着。在条石杂横处，出露一石棺，宽约 70 cm，仅一半出露，另一半在淤泥中。估计横陈的条石均为墓地遗迹。当地因地少，至今仍有墓葬于家居附近之习俗。据东寨红树林保护区站长讲，其祖先于大地震前一年因人口增多，而迁离该村，地震时，几乎无人得以逃出，他们一支因外迁而得以延续至今。瞬时发生之地震灾害不可低估。

火山喷发与强地震活动一样，危害性巨大，长时间影响大范围的生存环境[7]。熔岩流温度高达 500 ℃～1 400 ℃，速度每小时可达数千米，所至之处燃树焚屋；炽热的火山灰沙沉降，掩埋田野、道路和建筑物；火山泥流灾害是由炽热的火山灰融化山顶积雪或促成局部暴雨，结果沿坡而下的泥流袭夺堵塞河谷而又促使洪水暴发成灾，这类灾害更易发生于水气充沛、交换活跃的海岸海洋地区；炽热的火山云，是炽热气体和细火山灰的混合物，温度高达 1 000 ℃。它沿火山向下运动时，速度可达 100 km/h，因此，人被烧伤与窒息的危害大；火山喷发出大量有毒的或无毒的气体，主要成分是水气和二氧化碳，虽是无毒的，但是，若在局部水域中积聚，使浓度增高（＞50％），当气体温度变化或振动使这些无色无臭的气体再被释放时，使人窒息死亡[8]；大量海水渗透到火山岛岩层内，与下伏岩浆相接触而形成水气，当炽热的蒸汽积聚至相当数量后，可沿裂隙形成巨大的爆炸。

此外，火山喷发活动对地表环境产生一系列的长期效应，表现于地貌、气象、生态等多方面[7]。如火山活动形成新的岛屿，迅速造陆，并由于板块移动，形成著名的夏威夷火山岛链。由于熔岩流凝结，形成崎岖不平的熔岩原野、台地和陡崖等地貌组合；火山喷发使周边范围内的原始地貌发生变化，海南岛西北部火山喷发自陆向北部湾年代渐新，全新世火山口形成火山海岸，有火山颈、海蚀柱、盆状海湾、断流河口，抬升的海底形成高出海平面 10～20 m 的阶地，涠洲岛是由北部湾中的火山口形成的海岛。火山喷发的造貌活动，亦是短时间内的突出效应，火山熔岩流堆积于河谷洼地，形成新的高地，使地形发生倒置现象。海岸火山喷发的同时，邻近多有断陷海湾伴生。

夏威夷的火山口形成马蹄形海湾，湾顶发育珊瑚礁。岛下的热点使夏威夷岛利用 360 ℃的地热蒸汽发电，提供现在岛上电力需求量 5 倍的能源。火山活动所引起的气候变化效应，历时长、范围广。1815 年印度尼西亚坦博拉火山喷发，产生了更加明显的气候变冷效应，使 1816 年成为全球闻名的“无夏之年”。火山喷出的富硫气体进入大气层中则可形成酸性水滴云团，或遮蔽日光或降酸雨，亦造成气候变化。

火山活动形成玄武岩堆积，使局部海域增温，为珊瑚的生长繁殖创造了稳定的礁基和合适的水温。火山海岸与珊瑚礁的伴生关系，是生态效应的体现。热带海洋性气候条件下的火山岩风化形成富含矿物质的肥沃土壤原野，为林木花丛和多种飞禽走兽提供了良好的生活环境。同时，丰富的营养物质和大量珊瑚虫则维持了众多的龟、鱼等热带生物群落的生存条件，形成了极富特色的热带海岸生态系，不仅为人类生存发展提供了优越便利的自然条件，而且成为海岸带资源的重要组成部分。大量的热带植物，亦成为潜在的新型能源——生物能源。

重视研究现代地震与火山活动在海岸海洋的表象及影响，对海陆环境变迁、减灾防害、保护人类生存健康发展有重要意义。研究与对比古今火山与地震构造活动，有助于判断构造活动的效应与分析判断构造活动的发展趋势，做出对火山、地震活动发生与灾害程度的预报。

参考文献

[1] 王颖，张永战，邹欣庆. 1998. 面向 21 世纪的海岸海洋科学. 见：97′海岸海洋资源与环境学术研讨会论文集. 香港：香港科技大学理学院及海岸与大气研究中心出版. 68－73.

[2] 王颖，张永战. 1995. 海岸海洋科学研究新发展. 见：中国海洋学会第四次代表大会文集. 5－13.

[3] Pemetta, J. C. and Milliman, J. D. 1995. Global Change-IGBP Report No. 33: Land-Ocean Interactions in the Coastal Zone, Implementation Plan. Stockholm: IGBP Secretariat, The Royal Swedish Academy of Sciences. 1－21.

[4] 王颖. 1999. 海洋地理国际宪章. 地理学报，54(3)：284－286.

[5] 台湾大学地理系. 1999. 集集大地震灾害分布图. 台北：台湾营建署.

[6] 台湾大学地理系. 1999. 台湾地区天然灾害分布图. 台北：台湾农业委员会.

[7] 王颖，张永战. 1997. 火山海岸与环境反馈——以海岛火山海岸为例. 第四纪研究，(4)：333－343.

[8] Montgomery, C. W. 1992. Environmental geology. Third Edition. Dubuque: Wm. C. Brown Publishers. 88－113.

全球变化与海岸海洋科学发展*

1. 海岸海洋科学的发展

海岸海洋是陆地与大洋相互过渡的地带，它是既区别于陆地，又有别于深海大洋的独立环境体系，受人类活动影响密切，是研究水、岩、气、生圈层交互作用的最佳切入点。研究陆海过渡带的表层系统作用过程、环境资源特性及发展变化规律，以获求人类生存活动与之和谐相关等，构成了海岸海洋科学的研究对象与任务，是基于地理学、地质学与海洋学相互交叉渗透形成的新学科，具有自然、人文与技术科学相互交叉渗透的复合型科学特点[1,2]。

联合国《21 世纪议程》指出："海洋是生命支持系统的基本组成部分，也是一种有助于实现可持续发展的宝贵财富。"海岸海洋仅占地球表面积的 18%，其水体部分占全球海洋面积的 8%，占整个海洋水体的 0.5%，却拥有全球初级生产量的 1/4，提供 90%的世界渔获量，为 60%的世界人口的栖息地，目前全世界人口超过 160 万的大城市中约有 2/3 分布于这一地区[3]，海岸海洋与人类生存关系密切。

海岸海洋是既有别于陆地，也有别于深海大洋的独立环境体系。水、岩石、大气、生物圈层在此"界面"相互作用影响，成为物理、化学、生物与地质过程极为活跃的动态区，保存了丰富的海陆作用过程信息。全球河流悬移质及其所吸附的元素与污染物的 75%～90%被带至这一地区沉积，拥有全球 50%以上的碳酸盐和 80%的有机残体沉积以及 90%的沉积矿产。同时，作为人类最主要的居住区，人口密度极高，大城市林立，通过河流输移的人类活动产物的沉积速率相当或超过自然产物的输送率，90%的陆源污染被排入海岸海洋地区[3]。海岸海洋通过生态系统沉淀、吸附、固封等作用，减轻或消除了陆源污染对大洋的直接影响与危害。同时，活跃的动力、上升流与陆源物质的汇入为海洋生物提供了繁衍生境，使之成为一个高生产力和生物多样性的体系，为人类提供了食物、药物、动能、空间、旅游等多种具有再生性与可补偿性的资源[1]。

海岸海洋科学的兴盛与 1994 年正式生效的"联合国海洋法公约"密切相关，公约对沿海国主权的 12 海里领海、24 海里毗连区、200 海里专属经济区以及大陆架是沿海国陆地领土自然延伸原则等规定，使海洋权益及管辖范围发生巨大变化，推动了沿海国对"海洋领土"的关注，全球涉及海洋划界的有 370 处。基于主权与资源开发的需要，推动海岸与大陆架浅海成为海洋科学领域的新热点，明确地认识到，海洋是由两个主要环境组成，即海岸海洋与深洋[4]。

---

* 王颖，牛战胜：《海洋地质与第四纪地质》，2004 年第 24 卷第 1 期，第 1－6 页.

2006 年载于：路甬祥主编，《科学与中国——院士专家巡讲团报告集・第三辑》. 北京：北京大学出版社，207－217 页.

海岸海洋(Coastal Ocean)定义是 1994 年 UNESCO 政府间海洋委员会(IOC)在比利时列日大学召开的国际海岸海洋科技会议(1st COASTS of IOC)上正式提出，明确了海岸海洋的范围包括海岸带、大陆架、大陆坡与大陆隆，含整个海陆过渡带(图 1)。会议正式出版的“The Sea”系列第十卷“Global Coastal Ocean”[5]，成为国际海洋学界正式确定海岸海洋的里程碑。国际地理学家联合会(IGU)1996 年发表“海洋地理宪章”正式将全球海洋区分为 Coastal Ocean(海岸海洋)与 Deep Ocean(深海海洋)两部分[4]。

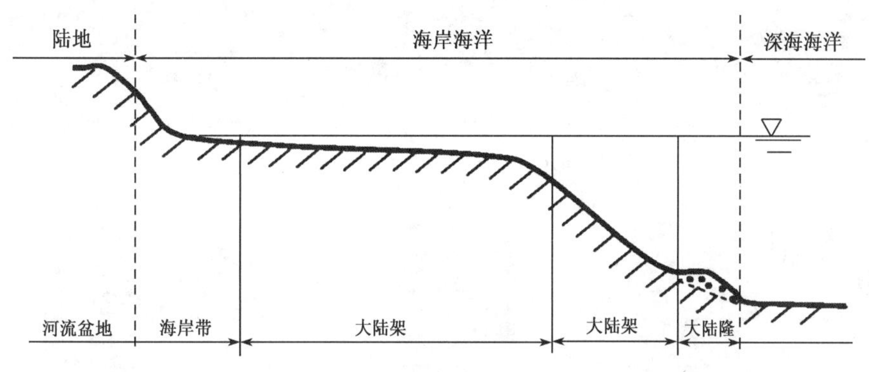

图 1 海岸海洋图示

20 世纪初，经典文献将海岸定义为沿海滨分布的狭窄陆地[6]；20 世纪中期，海岸工程实践明确了现代海岸带是包括沿海陆地及水下岸坡的“两栖地带”：上界止于风暴潮、激浪作用于沿海陆地的上限，下界始于水深相当于 1/3～1/2 当地波长处；至 90 年代，形成了包括海岸带、大陆架、大陆坡及大陆隆，涵盖了整个海陆过渡带的海岸海洋(Coastal Ocean)。经历了 20 世纪两次科学认识上的飞跃，加深了对海岸海洋环境特点的认识与资源环境的利用，发展形成具有交叉学科特点的应用基础型新学科。

海岸海洋研究方法具有外业工作的多学科综合性，包括陆上调查，浅海水、岩、气、生方面的观测，空中的同步监测；实验室多项分析及计算模拟。加上因时、因季节与因地观测，投入的人力与经费大，但其科学成果严密，能直接服务于生产建设与国家权益。

2. 全球变化与海岸海洋科学研究

(1) 全球变化是反映在气、水、岩石、生物圈的事件性变动，形成全球性的频发与持续效应，对人类生存环境影响深刻。

气候变化、海平面变化与人类活动是与海岸海洋密切相关的全球变化研究课题。大气与海平面环境监测及趋势性分析已取得重要进展。近 200 年的验潮资料反映，海平面上升趋势与大气温度及海水温度增高趋势呈良好相关。长时期的地质记录反映出海平面、大气温度和海水温度三者间存在正相关[7](图 2)。百年来水动型的海平面上升值为 1～2 mm/a，随气温升高、海平面持续上升，至 2100 年，海平面可能上升 1 m(IPCC-WG1，1990)。2000 年以来，中高纬度地区气温增高明显，南极洲 2001 年融冰期增长 3 个星期，为近 20 年之最。随着海平面持续上升，平原海岸与大河三角洲区面临着土地淹没与风暴潮频袭城市的境界。

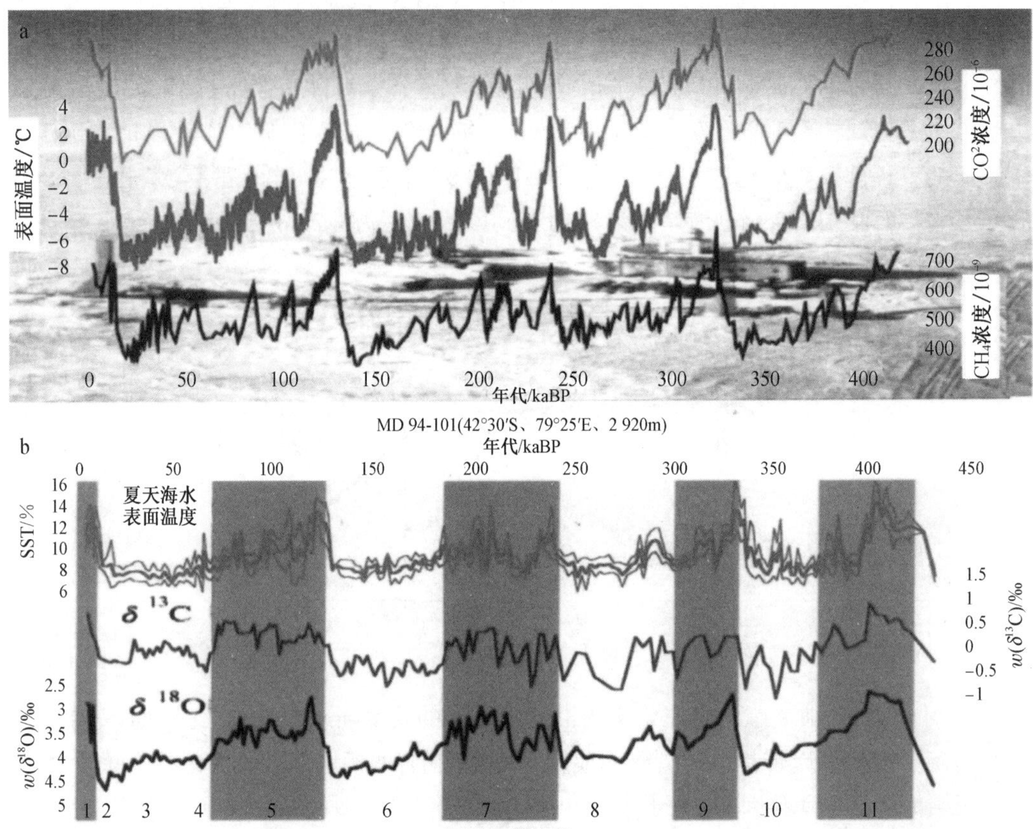

图 2　海平面、大气温度和海水温度三者间正相关示意图(据 PAGES Newsletter, 1999)(附彩图)

a. Vostok 冰心所揭示的 400 kaBP 以来大气 CO_2、CH_4 浓度和气温变化曲线;

b. MD94－101 大洋钻孔所揭示的 450 kaBP 以来氧、碳同位素与夏季 SST 变化曲线

海陆交互作用影响的另一例证是气温增升,内陆沙漠干热,沙尘暴活动频频侵袭东部沿海城市,甚至在长江以南的南京春季形成昏霾的白昼。海水增温使北美太平洋沿岸鲑鱼减产,但是,中国大陆沙尘被漂移的西风带输送至北太平洋东岸降落,又为鲑鱼带来富含 Fe 质的营养盐,使鲑鱼丰收。如追索求源,进行中、加两岸海陆对比研究,不仅加深对北半球西风盛行带迁移规律与效应之认识,而且是研究地球表层系统岩、气、水、生相互作用的最好切入点。

(2) 应引起重视的全球变化与海陆交互作用影响的另一方面,是发生在岩石圈的地震火山构造活动。

现代全球 500 多座活火山中,370 多座活火山沿太平洋岸分布,其余沿东西向"关闭的地中海"带分布。20 世纪 90 年代以来,火山爆发频频,并逐年递增,尤以海岸海洋地区最为突出(表 1)。自 2000 年来太平洋"火圈"火山爆发异常频繁,南太平洋、大西洋、墨西哥湾及大陆地区亦相继频频爆发,形成全球性构造活动,对地表系统造成巨大影响。

表 1　1995—2002 年 9 月 30 日全球火山活动情况(据美国国家航空和宇宙航行局数据统计)

年代	次数	位　　置
1995	5	4 次位于太平洋地区(2 次位于南太平洋),1 次位于印度洋地区
1996	10	9 次位于太平洋地区(1 次位于南太平洋),1 次位于大西洋地区
1997	14	13 次位于太平洋地区(2 次位于南太平洋),1 次位于大西洋地区
1998	15	11 次位于太平洋地区(4 次位于南太平洋),2 次位于大西洋地区,1 次地中海,2 次陆地(美洲、非洲)
1999	13	12 次位于太平洋地区(2 次位于南太平洋),1 次位于大西洋地区
2000	50	33 次位于太平洋地区(3 次位于南太平洋),6 次位于墨西哥湾,3 次非洲(刚果、喀麦隆),1 次地中海,3 次印度洋,1 次大西洋,3 次南美洲
2001—2002	57	40 次位于太平洋地区(8 次位于南太平洋),4 次位于墨西哥湾,2 次印度洋,1 次大西洋,大陆 1 次,1 次位于墨西哥湾,3 次非洲,2 次地中海,3 次南美洲

火山爆发的同时,地震活动亦极为频繁。1990 年以来在全球范围内年年有大地震,2000 年几乎每日发生地震,强地震亦引起火山爆发。地震频发在环太平洋构造带与东西向的“地中海”构造带,同时,地震活动还出现于南大洋地区,呈东西向的分布(图 3 中 a,b,c,d)。我国西部与台湾岛东部地震频频发生,正是全球构造活动变化的反映。

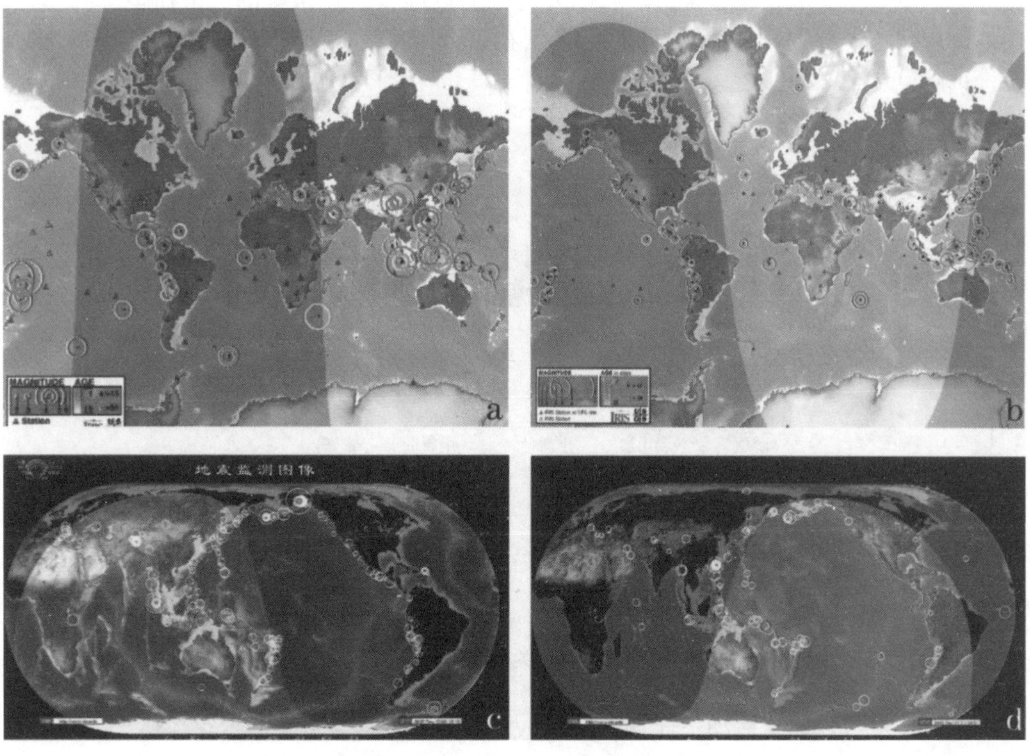

图 3　全球地震分布(附彩图)

a. 据 NEIS of USGS IRIS 监测,2000-09-18;b. 据 NEIS of USGS IRIS 监测,2001-04-05;c. 据 NEIS of USGS IRIS 监测,2002 年 11 月;d. 据 NEIS of USGS IRIS 监测,2002 年 11 月,其中红圈表示发生于 24 h 之内的地震;黄圈代表发生于 15 d 内之地震;紫三角代表超过 15 d 以上及 5a 以内的地震情况;圆圈大小表明震级。

对比图 3a 与 3b 可看出震级强度的转移。一次强地震爆发，能量释放以后震级变小，但同一构造带的另一处震级加强。这一现象为地震预报提供启示。

地震与火山活动释放着地球内部的能量，太平洋巨厚水层的压力、密度及温度变化聚集着很大的能量，与地球内部能量的结合，可能是大洋中能量较陆地更大的原因。强地震多发生在海岸海洋地带，或发生于子夜，或发生于月半大潮期间，可能反映了月球引潮力对地震的激发或加助了地球内部能量的向外释放，或者白天的太阳引力与陆地人类各种活动，削弱了月球引力并分散了部分地壳中所聚集的能量。归纳分析地震发生的时间及地震带强度转移变化的规律，有助于找出防震减灾措施，可尝试人为地释放、控制或分散能量、削弱地震发生的强度。

强地震造成的环境效应灾害以海岛地震为例，1999 年 9 月 21 日 1 时 47 分 12.6 秒，台湾岛日月潭西南 6.5 km 处集集镇发生里氏 7.3 级地震，震中位于 23.85°N、120.78°E，震源深度 7～10 km[7]，系台湾西部山麓地带的系列逆断层长期受吕宋岛弧推挤，积蓄大量能量，从车笼埔断层发生错动，因能量释放而造成地震(图 4)。车笼埔断层东侧上盘上升 2.2～4.5 m，向 NW 及 NNW 方向水平滑移 7.1 m；下盘最大沉陷 0.3 m，向相反方向平移 1.1 m，最大水平地表加速度达 1 g，强震延时 25 s，波及相距 150 km 的台北市，整个台湾岛均可感到地震活动①②。地震造成地表断裂、抬升形成断崖；河流中断，形成瀑布；山崩地塌，造成堰塞湖与滑塌；土壤液化，发生喷沙、喷泥与地层陷落等，伴随之造成对道路、桥梁、水利与交通设施、港湾设施、动力与工业设施以及建筑等的严重损害[8,9]。居民房屋倒塌约 2 万幢，死亡人数超过 2 300 人，伤者约 8 000 人。自 2002 年 3 月至今，台湾地震仍频繁活动。

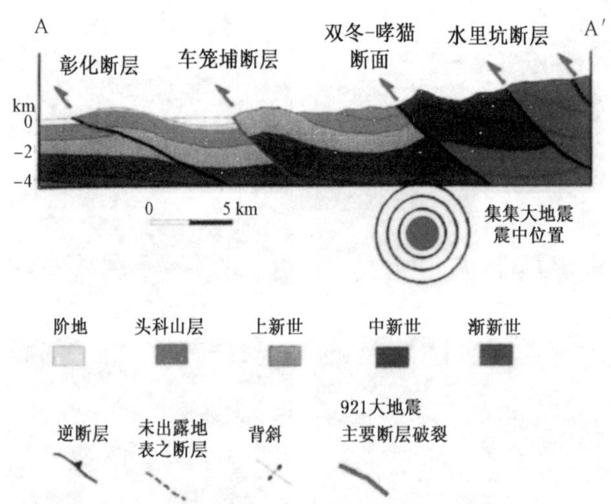

图 4　台湾岛集集地震震中与车笼埔断层位置关系及地质剖面构造示意图(据[5],[6])(附彩图)

火山喷发与强地震活动一样，危害性巨大，长时间影响大范围的生存环境[9]。熔岩流

① 台湾气象局地震测报中心第 043 号有感地震报告，1999.

② 王鑫，林俊全. 九二一台湾地震纪念地选址及地景登录之研究，2000.

温度高达 500 ℃～1 400 ℃，每小时可达数千米，所至之处燃树焚屋；炽热的火山灰沙沉降，掩埋田野、道路和建筑物；火山泥流灾害是由炽热的火山灰融化山顶积雪或促成局部暴雨，结果沿坡而下的泥流袭夺堵塞河谷而又促使洪水暴发成灾，这类灾害更易发生于水气充沛、交换活跃的海岸海洋地区。炽热的火山云是炽热气体和细火山灰的混合物，温度高达 1 000 ℃。它沿火山向下运动时，速度可达 100 km/h，因此，烧伤与窒息的危害大；火山喷发出大量有毒的或无毒的气体，主要成分中水气和二氧化碳是无毒的，但是若在局部水域中积聚，使浓度增高（>50%），当气体温度变化或振动使这些无色无臭的气体再被释放时，会使人窒息死亡[10]。大量海水渗透到火山岛岩层内，与下伏岩浆相接触而形成水气，当炽热的蒸汽积聚至相当数量后，可沿裂隙形成巨大的爆炸。

火山喷发活动对地表环境产生一系列的地貌、气象与生态效应[9]。如火山活动形成新的夏威夷火山岛链，形成崎岖不平的熔岩原野、台地和陡崖，夏威夷岛火山口形成马蹄形海湾，湾顶发育珊瑚礁，岛下的热点，可利用 360 ℃的地热蒸汽发电，提供岛上需求的电力能源。火山喷出的富硫气体进入大气层中则可形成酸性水滴云团，或遮蔽日光或降酸雨，亦造成历时长、范围广的气候变化效应。

热带海洋性气候条件下的火山岩风化形成富含矿物质的肥沃土壤原野，又为林木花草和飞禽走兽提供了良好的生活环境，形成了极富特色的热带海岸生态系与潜在的新型生物能源。

综上所述，无论全球气温与海平面变化或岩石圈构造活动变化，均对海陆交互作用的海岸海洋造成显著影响，影响到海陆环境变迁、灾害与人类生存健康发展构成海岸海洋科学关注的重要内容。

3. 沿海城市化发展与海岸海洋科学

伴随着我国经济的持续发展，东部沿海地区城镇发展迅速，社会经济发展与人民生活水平提高的需求在增长。城市化发展过程，以珠江三角洲城镇发展为先声，继深圳、珠海新市区建设之后，广州市自 2000 年实施城市发展战略总体规划，在珠江滨水环境优化，新机场与大学城建设，现代化的会展中心屹立于珠江之滨，轻轨地铁的发展及老城区疏导等方面成果卓著，广州已展现出现代大都市新风貌。2003 年，广州又在总结实践前三年成就的基础上，向海洋扩展，与珠江三角洲城市加强合作；同一时期，浙江省完成了新杭州湾城市的发展规划，建设长江三角洲南部经济与生态环境和谐发展的城市群；山东省也在进行半岛城市群建设。

江河湖海是水陆交互作用带，地理与生物种群结构复杂，物种变化活跃，生产力较高，适于人类聚居与城镇发展。河流与海洋在城市发展的早期既有舟楫之利，又为防御的天然屏障。当代全球化经济体系的发展，更突出了滨海城市的重要性。

城市的发展规划应体现以人为本，人居环境应具有清新的空气、明亮的阳光、清洁的淡水、安全的宅舍与宜人的风貌。同时需要有繁荣的商贸街市、便利的交通运输、咨询网络、文化历史内涵、科教医卫与综合设施安全的保障、服务系统及有效的管理体系等，这些均涉及自然环境的特点与滨水水体的发展变化规律，需要充分考虑海岸海洋生态环境与城市发展规划的有机结合，人地关系和谐发展的思想应渗透贯穿于设计体系中。宋代张择端所绘的《清明上河图》展示了北宋临水都城之布局，突出了河流在城市建设中的功能。具有复合型特点的海岸海洋科学已在并继续在 21 世纪中国城市化发展中发挥一定的作用。

4. 结　语

海岸海洋科学是在地理学、地质学与海洋科学相互交叉渗透的基础上发展形成，赋有科学活力，与人类生存活动所需的清洁淡水、新鲜空气、空间、动能、食物与医药资源密切相关，是 21 世纪地学发展的重要支柱，宜从人才教育、科学研究及建设应用项目等方面给予重视与支持。

参考文献

[1] 王颖，张永战，邹欣庆. 1997. 面向 21 世纪的海岸海洋科学[A]. 见：97′海岸海洋资源与环境学术研讨会论文集[C]. 68 - 72.

[2] 王颖，张永战. 1995. 海岸海洋科学研究新发展[A]. 见：中国海洋学会第四次代表大会文集[C]. 5 - 13.

[3] Pernetta J C, Milliman J D. 1995. Global Change — IGBP Report No. 33: Land-Ocean Interactions in the Coastal Zone, Implementation Plan [R]. Stockholm: IGBP Secretariat, The Royal Swedish Academy of Sciences. 1 - 21.

[4] 王颖. 1999. 海洋地理国际宪章[J]. 地理学报，54(3)：284 - 286.

[5] Johnson D W. 1919. Shore Processes and Shoreline Development [R]. John Wiley & Sons Inc.

[6] 台湾大学地理系. 1999. 集集大地震灾害分布图[G]. 台北：台湾营建署.

[7] PAGES IPO. 1999. PAGES Newsletter [J]. Bern, Switzerland, 7(3): 1 - 2, 9.

[8] 台湾大学地理系. 1999. 台湾地区天然灾害分布图[G]. 台北：台湾农业委员会.

[9] 王颖，张永战. 1997. 火山海岸与环境反馈——以海岛火山海岸为例[J]. 第四纪研究，(4)：333 - 343.

[10] Montgomery C W. 1992. Environmental geology [M]. Third Edition. Dubuque: Wm. C. Brown Publishers. 88 - 113.

海岸海洋海陆过渡带环境变化及其资源环境特征*

一、海陆过渡带研究的发展与海岸海洋学科之兴起

海洋占地球表面积的71%，早期人类逐水、食而居，入海河流的淡水，海滨的鱼、虾、贝、蟹易于获取为食，因此，沿海的岩穴、阶邱多有先民遗迹。人类居住繁殖在陆地上，随着畜牧、农耕、工业、商贸的发展，开发利用与研究陆地环境日益深入，陆地环境科学得到发展，并转注人类尚未居住的环境。发达国家的社会经济是沿江、湖向沿海发展，进步飞跃。海洋科学是伴随资本主义的扩张与地理大发现而兴起，在20世纪第二次世界大战中引起重视：德国潜艇频袭同盟国船舰，封锁海域，严酷的战争促使英、美重视海洋领域的斗争；二战后，海军舰艇转为民用，推动了海洋科学研究及后起的极地探索。随着人口增长，对资源环境需求呈迫胁性增强，人们又注视新一轮的海洋开发。

但是，人们忽略了在陆地与海洋交界处，存在一个海陆过渡带，它既不同于陆地，也区别于深海大洋，是一个独立的地域环境体系。地球表层的大气圈、海水圈、岩石圈与生物圈层在此带交融、相互作用影响，自然过程复杂多变，生物种群丰富、生产量高，与人类生存关系密切，受人类活动影响大。海陆交互作用过程亦赋予海陆过渡带的气、水、岩、生圈层体系具有与陆地、与深海大洋不同的自然属性，也是重大海洋灾害的源发区。

20世纪后期，人们虽认识到浅海环境与深海之区别，却未将海陆过渡带作为统一体系加以研究：虽然认识到海岸带包括陆地与海下，具有水陆两栖性，但仅至波浪扰动海底泥沙的数十米水深处，未包含大陆架及大陆坡；将陆架浅海以"近海海洋"概括，却未包括大陆坡及坡麓的海底扇(陆隆)，近海海洋调研不包括海岸带陆地，均未涵盖整个海陆过渡带。

海陆过渡带作为一个体系正式提出是在20世纪90年代，缘于《联合国海洋法》(United Nations Convention on the Law of the Sea)于1994年11月16日正式实施。海洋法形成两个全新的国家管辖海域：200 n mile① 专属经济区(Exclusive Economic Zone，简称EEZ)与群岛国家水域(Archipelago Nation's Water)。同时明确："大陆架是沿海陆地领土的自然延伸，沿海国对大陆架内自然资源拥有主权权利"；"海峡可无害通过"；"公海资源是全人类共同继承的财产"。因此，海洋法赋予沿海国拥有12 nmile领海、24 nmile毗连区，以及对大陆架或200 nmile海域的管辖权利。实质上，涉及整个伸向海下之陆地。据此，我国的管辖海域达300×10^4 km^2，太平洋岛国(斐济、关岛)形成百倍于岛域面积的小岛大海洋的现状。联合国海洋法促动沿海国重视"海洋国土"权益，推动了对

* 王颖，胡敦欣：中国科学院地学部地球科学发展战略研究组，《21世纪中国地球科学发展战略报告》(第六章 第三节)，149－154页. 北京：科学出版社，2009年.

① 1 n mile＝1.852 km.

海陆过渡带——海岸海洋之研究，也掀起全球 370 多处海洋划界之纷争。我国面临着与周边诸国油气资源与海洋疆域之纷争。

英国自然环境研究会(NERC)于 1992—1998 年组织 11 个研究机构和 27 所大学的 360 位科学家参加"陆海相互作用研究项目"——(Land Ocean Interactional Study, LOIS)，包含河流过程、河口过程、海岸过程、大气过程与大陆架过程 5 个课题，其优点是陆海一体，量化和模拟受人类活动影响，河流汇入海岸带及海洋的泥沙、营养物与污染物之通量与变化过程。

1995 年后，国际科学联合会(ICSU)将海岸带海陆相互作用(LOICZ)作为国际岩石圈—生物圈计划项目(IGBP)的重要组成，开展研究河流—河口—海岸—大陆架物质输移过程与通量，列为前沿的研究课题。其范围未包括大陆坡和海底扇，未将整个海陆过渡带作为一体的环境，加以研究其动力作用过程、环境发展变化趋势与和谐开发利用途径。

海陆过渡带研究具有里程碑意义的科学进步。1994 年 5 月由联合国教科文组织的政府间海洋委员会(IOC)在比利时列日大学(Liege)召开了"海岸海洋先进科学与技术研究第一次国际会议"(COASTS, Coastal Ocean Advanced Science and Technology Study)，获得 IOC 欧洲委员会、海洋研究科学委员会(SCOR, Scientific Committee on Ocean Research)、美国科学基金会(NSF, National Science Foundation)及海军研究办公室(ONR, Office of Naval Research)的支持。参加会议的代表 100 多名，中国正式代表 4 位：国家海洋局的苏纪兰、李海清，中科院海洋研究所的胡敦欣，南京大学的王颖。王颖以中国为例的"河海交互作用和陆源通量"被列入大会报告。这次会议正式提出 Global Coastal Ocean(全球海岸海洋)，会议论文"The Sea"(Ideas and Observations on Progress in the Study of the Sea)的第 10、11 卷以 The Global Coastal Ocean 专著形式出版(John Wiley & Sons, Inc, 1998)。两部巨著内容包括：① 海岸海洋的作用过程与研究方法(Volume 10)；② 综合性的海岸海洋区域：西部的、东部的、极地的、半封闭的海洋与群岛(Volume 11)。对"Global Coastal Ocean"明确提出："它是居于地球的陆地与深海之间的客体……"包括："陆架海与陆坡海以及海岸带"，即整个的海陆过渡带(Robinson, Brink, 1998)。

此后，国际地理学联合会(IGU)发表"海洋地理国际宪章"，将全球海洋明确地区分出海岸海洋(Coastal Ocean)与深海海洋(Deep Ocean)两部分(图 1，王颖，1999)。

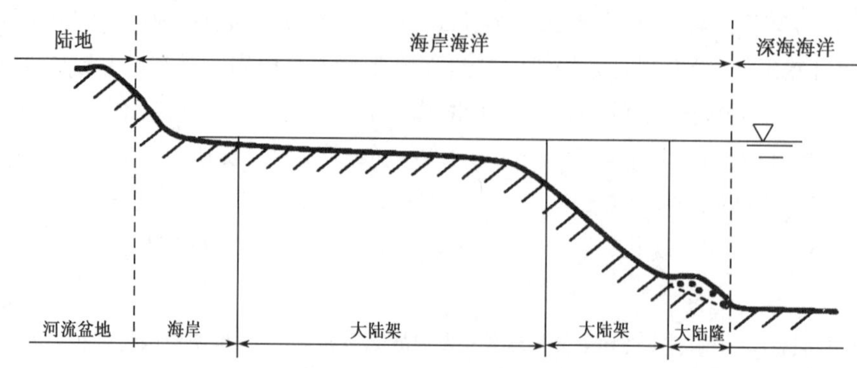

图 1　海岸海洋图示

LOICZ曾统计：海岸海洋面积相当于陆地面积的18%，占全部海洋面积的8%，仅为整个海洋水体的0.5%；但是，海岸海洋拥有全球1/4的初级生产力，90%世界捕鱼量；占全球生产力值的14%，占全球反硝化作用的50%，容纳全球有机残体的80%，沉积物矿体的9%以及全球悬移质、相关元素与污染物的75%～90%。

上述表明海岸海洋与人类生存关系密切。但是，“海岸海洋”的理念尚未在我国被广泛应用，正如海洋科学尚未如陆地科学一样地获得重视。

二、海岸海洋的特点与我国发展的需求

包括海岸带、大陆架、大陆坡及坡麓海底扇（大陆隆）在内的海陆过渡带具有“海陆两栖性”，海陆交互作用变化复杂，受人类活动影响巨大，形成明显的动力环境特性。其调查研究工作较陆地或大洋更具有多元性与复杂性。“中国海”是亚洲与太平洋之间的边缘海，除台湾以东濒临太平洋及南海中部为深海盆外，全属于海岸海洋，并以海岸带与大陆架浅海为主体，在南海与东海具有不同类型的拉张沉降的大陆坡与深海盆（槽），构成中国海陆过渡带环境特色与科学内容。

（1）由于海陆交界，气温、气压的不同，昼夜之差异，海岸海洋的风力较陆地与海洋均为增强约30%①。因海、陆风之昼夜变换，向岸风风速与频率均大于来自陆域的向海风。与潮流相伴的风吹流是大陆架海域最具能量的动力，导致渔业的高生物产量（Brink，1998）。其他如三维陷波、风暴潮、风成射流等，均反映出大陆架与海岸带一体相关互为影响。

（2）潮汐作用的海面振动从深海向海岸海洋逐渐加强并形成潮流作用；浅海海底地形与海岸轮廓对潮汐作用影响大，在一些突然收缩的河口湾形成涌潮（如杭州湾）；潮流主导了浅海的动能收支平衡（Simpson，1998）；潮动能量产生扰动效应，控制影响水体的密度结构，成为泥沙循环性悬浮与沉淀的动能；反复的涨潮与落潮，由于潮流的剪切力扩散效应及非线性过程的卷流效应而形成溶解质与悬浮质的传输。上述作用主导着陆架海的环境特性。潮汐扰动作用影响到水体的稳定性、营养盐的再循环以及光照性，进而影响到海域的初级生产力（Tett，et al.，1998）。如果没有潮汐作用，则陆架海环境会完全不同，水体的通换作用差，缺乏对日益加多的海岸排泄物的降解能力。

（3）海洋体系内能量、动量、化学的与生物量的预算取决于海洋与大气、海洋生物的动力和生物化学的交换过程。在海岸海洋，大气与海洋、海水与海洋生物交换的多边界作用过程相互依赖：浅水区海洋生物的物理特性影响到波场，进而影响到其糙度和气-海交换的动量、热量与质量（Geernaert，Larsen，1998）。

（4）海岸海洋的波浪作用显著增强，以风浪为主，受风速、风时与风区的影响。波浪向海岸传播时，受海深的影响发生变形，形成激浪与激浪流。受海岸轮廓与岸坡坡度影响会形成波浪折射、反射或绕射以及裂流，进而影响到泥沙输运与海岸蚀、积效应。

（5）伸向海洋的陆地岩性、结构与构造活动，对海岸海洋作用过程影响大，形成不同

① 任美锷．1986．江苏省海岸带和海涂资源综合调查（报告）．北京：海洋出版社，226.

的原始海岸类型与海岸发育效应;海平面变化在海岸带与大陆架浅海效应显著;现代海底火山与地震活动是在大陆边缘岛弧海沟系频频发生,效应突出。

(6) 入海河流是海陆作用过程中的活跃因素,河流向海输入的水体与泥沙,抵消着波浪、潮汐和海流的动能;建造着海岸与海底地貌;河流与潮、浪两类性质不同的水体相互作用而产生了复杂的结果。全球源自青藏高原的 8 条大河,有 5 条流入中国边缘海。中国河流入海年径流量约为 $18\ 200\times10^{8}\ m^{3}$,占全国径流量的 69.8%,为世界径流量的 3.9%,输沙量占全球的 10%,在未兴建规模化的水利工程前,输入海中的泥沙约 2.01×10^{9} t/a。其中约有 1.21×10^{9} t/a 堆积在渤海,14.67×10^{6} t/a 堆积于黄海,0.63×10^{9} t/a 堆积在东海,0.16×10^{9} t/a 堆积在南海及台湾以东太平洋,因而,每年约 20.1 亿 t 泥沙由中国海流输入到边缘海大陆架(王颖等,2007)。

上述表明海陆过渡带的环境特性,与气、水、岩、生圈层复杂的交互作用过程,既是独立的环境体系,又互为依存,交叉影响:大陆边缘构造活动与岛弧-海盆体系,被河流携运泥沙填充发育了堆积型大陆架;地质构造与岩性对海岸类型之控制;海岸轮廓与坡度差异对浪、流之变化影响;河流泥沙既为渔场提供营养盐饵料,亦带来人类活动——经济与城镇化发展所形成的污染物。海岸海洋科学具有多学科交叉融合的特性,同时具有自然科学、社会经济科学与技术科学相结合的复合型科学特点。

25 年来,大规模经济建设的快速发展,亦出现资源环境利用与海岸海洋的安全性警示:长江三峡建坝、上游水系梯级开发、中部与东部南水北调,沿江大规模开发等,未从全流域及海陆结合一体化的系统研究规划。长江年平均径流量原为 $9\ 051\times10^{8}\ m^{3}$,洪水流量 $8\times10^{8}\sim9\times10^{8}\ m^{3}/s$,宜昌站最大洪水流量达 $11\times10^{8}\ m^{3}/s$,长江原年均入海泥沙量 4.7×10^{8} t。大坝建设拦水,中线与东线引水 2 000 m^{3}/s 及三峡水库拦截 1/2 泥沙,2006 年长江大通站径流量为 $6\ 886\times10^{8}\ m^{3}$,年输沙量 $8\ 480\times10^{4}$ t①,水沙之锐减必对河口与三角洲水土环境影响,招致盐水入侵距离延长与海岸侵蚀,对长江口河海动力均衡、南部海岸和东海大陆架环境与生态系亦会产生重大影响。阿斯旺高坝建设与尼罗河三角洲之变化已有先例。黄河因沿河引水与水坝、水库建设,入海径流及泥沙剧减,2006 年入海径流量为 $191.7\times10^{8}\ m^{3}$①,输沙量仅为 $14\ 900\times10^{4}$ t,招致下游断流、三角洲海岸受蚀后退、湿地退化与土地盐化。珠江三角洲城镇发展和农田大量用水及陆源污染,影响到淡水供应、盐潮入侵与赤潮频发。近 30 年,中国沿海海平面上升 39 cm,预计 2010 年长江三角洲吴淞口海平面较 1990 年上升 15~20 cm,2030 年上升 25~35 cm,2050 年上升 40~70 cm,风暴潮频繁,将加速海岸侵蚀后退,津、苏、沪、杭低地洪涝灾害或淹没,沿河道海水入侵与水质恶化等系列环境问题与资源危机。三大河流域是我国经济发展的杠杆,须重视从全流域着眼,进行水量平衡分析,陆海结合地研究水利工程效应与综合规划,从"源"及"汇"的系统调控解决。

海岸海洋环境的自然独立属性、海洋权益、海洋纳污日益污染严峻以及人类生存发展潜在的环境资源危机,唤醒人们将海陆过渡带的海岸海洋纳入学科体系,开展当务之急的研究。

---

① 中华人民共和国水利部. 2007. 中国河流泥沙公报.

三、建议研究的项目目标、内容与技术路线

(一) 目标

以 2020 年为阶段性期限，以三大河流域和关键岛礁为杠杆，以我国海岸带与大陆架浅海为重点研究区，向外延伸至我国所属的岛弧与深海，陆海结合、内外动力与人类活动因素相结合地研究陆海交互作用、海陆过渡带资源环境特性、变化机制与海洋安全，探索人类生存活动与海岸海洋和谐发展的途径与安全保障，建设海岸海洋新学科。

为经济发达的沿海城乡安全用水，洪、涝、风暴潮灾害防治，环境安全，海洋资源持续利用，疆界纷争与国防安全等，提供科学决策保障。

建议对此涉及国家长远利益、具有科学前瞻性与重大应用价值的课题立项。

(二) 研究途径与内容

1. 研究途径

(1) 以长江、珠江、黄河三大流域，渤海海峡，西沙群岛五处为主干与代表性节点，“源”“汇”一体、陆海结合的开展气、水、岩、生多学科结合的调查研究。

(2) 从国家利益统筹安排调查研究，异地同步地开展 5～10 年期的季节与年度调查研究，以大型计算机集存多站点的资料，建立数字海岸海洋信息系统，分析演绎，逐步做出流域与关键地区陆海交互作用系统模型，三大流域水资源环境利用与工程效应的趋势性分析，胁迫性灾害的预警与防、减灾模拟等。

(3) 建设五条干线的监测网络与“管理”系统，持续积累资料，扩大应用研究，为国家宏观战略与决策提供支撑。

2. 科学选题

(1) 全球变化与中国海岸海洋动力作用、变化趋势及对海岸、海底的影响。

(2) 中国主要河流水、沙、营养与污染物入海通量的特性变异及海岸海洋生态效应。

(3) 长江、珠江、黄河流域开发及大型水利工程，对下游、河口及海岸海洋环境变化的效应与和谐发展的利用研究。

(4) 世纪性海平面变化趋势与中国海岸海洋效应与对策。

(5) 三角洲与平原海岸侵蚀动力过程与海岸演变动态预测，海水入侵的基本特征与变化规律。

(6) 沿海经济高速发展与海岸海洋环境承载力分析。

(7) 中国及东亚沙漠扬尘、沙尘传输途径对中国沿海环境与鱼类生物效应。

(8) 南海大陆坡西沙与南沙群岛岛礁地质与海洋环境特点、资源开发前景、划界依据、疆域纷争与安全保障研究。

(9) 建设三大流域—河口—近海一体化的水、土、气、生资源环境监测体系与信息库，以及灾害预警系统设计研究。

以上选题建议涵盖面较大，但已有研究基础，对青藏高原与冰雪源之研究，三大河流

局部开发研究，海岸与海岛调查及进行中的近海海域调查等。关键在于立项，全国统一领导部署，部门联合地开展同步研究工作。

参考文献

[1] 王颖译. 1999. 海洋地理国际宪章. 地理学报，54(3)：284－286.

[2] 王颖，傅光翮，张永战. 2007. 河海交互作用沉积与平原地貌发育. 第四纪研究，5(27)：674－689.

[3] Allan R. Robison，Kenneth H. Brink. 1998. Preface. The Sea，Volume 10：the Global Coastal Ocean. John Wiley & Sons. Inc.

[4] Gerald L. Geernaert，S ϕ ren E. Larsen. 1998. Air-sea Interaction in the Coastal Zone. The Sea，Volume 10：the Global Coastal Ocean. Chapter 4，89－109.

[5] John H. Simpson. 1998. Tidal Processes in Shelf Sea. The Sea，Volume 10：the Global Coastal Ocean. Chapter 5，113－150.

[6] Kenneth H. Brink. 1998. Wind-driven Currents Over the Continental Shelf. The Sea，Volume 10：the Global Coastal Ocean. Chapter 1，3－20.

[7] Tett. P. B.，I. R. Joint，D. A. Purdie，et al. 1993. Biological Consequences of Tidal Stiring Gradients in the North Sea. Philos. Trans，R. Soc，London A 343. 494－508.

全球变化与海岸海洋科学进展*
——中国科学院院士、海洋科学家王颖在澳门科技大学讲座
（2006 年 10 月 18 日）

尊敬的许校长、各位老师和同学，大家下午好！很高兴来到澳门这座美丽的海滨城市，与大家分享我对全球变化与海岸海洋科学进展的一些认识。鉴于澳门、珠海位于珠江三角洲，具有海陆过渡带的特点，故而选择海岸海洋科学讲座，内容涉及以下方面，一是海岸海洋的范围，国际重视之热点；二是海岸海洋科学（Coastal Ocean Science）的研究对象，指出海陆交互作用带的资源、环境特点与重要性；三是介绍全球变化中的海岸海洋地带，包括气温升高与海平面上升的影响，地震火山活动对海岸海洋的影响实例，以及中国海岸带城市发展的关键问题。具体分为以下五个内容：一是海洋意识与海洋的重要性；二是海岸海洋科学的内涵和特征；三是全球变化与海岸海洋研究；四是全球构造活动与海岸海洋效应；五是海岸海洋科学与城市化发展的若干建言。

一、海洋意识与海洋的重要性

联合国“21 世纪议程”（Agenda 21）指出：“海洋是生命支援系统的一个基本组成部分，也是一种有助于实现持续发展的宝贵财富。”因为 21 世纪是生命的世纪，也是海洋世纪与高新技术开发应用的世纪。而地球科学之所以重要，在于它为人类提供生存发展的环境、资源与必需的财富。如果从太空看，地球是个蓝色的星球，地球表面积的 71% 是海洋（图 1）。

图 1　太空看地球（附彩图）

海洋的重要性，可用一些资料来说明。海洋是生命的摇篮，是风雨的故乡。生命的起源、人类的生存与海洋密切有关。海洋吸收 4/5 的太阳能，向大地释放热能。海洋植物通过光合作用产生氧 360 亿 t/a，大气中 70% 的氧来自海洋。海洋是 CO_2 储存器，吸收 60 倍于大气中

* 王颖：许敖敖，唐泽圣主编，《聆听大师　走近科学——澳门科技大学科技大师讲座院士讲演录》，25－40 页，2013 年 6 月澳门科技大学出版。

的 CO_2 含量。海洋蒸发 44 亿 km^3/a 淡水，以降水形式返回陆地。海洋、大气中的水分，每 10～15 天完成一次更新。没有新鲜的空气与清洁的淡水，人类是难以生存的。海洋具有反硝化作用的净化能力。如果海洋受到污染，人类会面临生存的危机(图 2)。

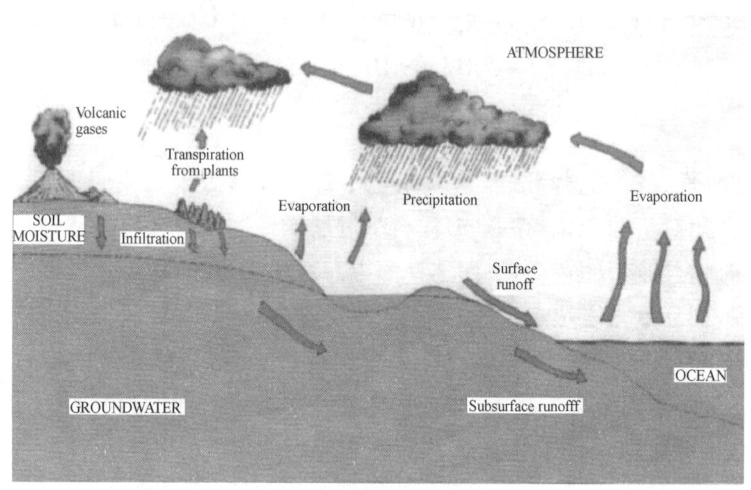

图 2　海洋是生命的摇篮

海洋更是食物、医药与矿产资源的巨大宝库。海洋生物资源是人类重要的蛋白质来源，海洋中有 22 万多种生物，为全球提供 22%的动物蛋白质。世界海洋渔业生产 8 000 万 t～9 000 万 t/a，95%来自国家管辖的海域。联合国粮农组织(FAO)统计，世界人均鱼消费量为 13 kg(1992)，主要来自海洋产品。中国海洋年捕捞量 760 万 t(除 7%为远洋渔产，绝大多数来自海岸海洋)，人均水产品占有量是 20 kg(1995)。大洋底锰结核矿产含有锰 2 000 亿 t，镍 164 亿 t，含铜 88 亿 t，含钴 58 亿 t，锰结核是重金属结核，其重金属含量较陆地的大 40～100 倍。海底磷、硫以及稀有金属与贵重金属矿产藏量十分巨大。海底石油天然气储量，石油 1 350 亿 t，天然气 140 万亿 m^3。21 世纪初，世界海洋石油产量 15 亿 t/a，为石油产量的 50%(主要集中于海岸海洋：北海、墨西哥湾、马拉凯博湖、中东与中国海)。海洋再生产能源(潮汐能、波浪能、海流能、海水热能、温差能、盐度差能及可燃冰)，多集中于海岸海洋。这些能源的理论储量约 1 500 亿 kW，相当于全世界发电量的十多倍。当代陆地资源日益短缺，必然转向海洋，探求所需的资源。以我国为例：中国人均占有的陆地面积为 0.008 km^2，世界人均量为 0.3 km^2；中国人均淡水量约为世界平均水准的 1/4；矿产总量居世界 18 位，为世界人均量的 1/2。占消耗量 90%以上的 45 种矿产，近年内将有 1/4 不能满足需求，估计我国矿产总短缺量为所需量的 1/2。

因此，当代面临着人口、资源与环境的挑战，出路之一在于开发利用海洋资源。不仅是地学专业人员，而且全民都应具有海洋意识。

二、海岸海洋科学的内涵和特征

说到海岸海洋科学，得先简单谈它的定义。所谓海岸海洋科学，是研究海陆过渡带表层系统作用过程、环境与资源特性、自然发展规律，以及人类生存活动与之相和谐的科学。

它应用于环境资源开发与保护，涵盖海洋经济、疆域政治、立法管理与新技术等方面，探求人类生存活动与海岸海洋环境资源和谐相关可持续发展的科学。

海岸海洋是介于大陆（硅铝层）与深海海洋（硅镁层）之间过渡的一个独立的环境体系，范围包括海岸带、大陆架、大陆坡及坡麓的大陆隆，它是海陆、海气、水、生相互作用活跃，相互交换的过渡地带[1]。它的特点是：人与自然的相互联系紧密，互相影响突出；发生变化迅速、生物量巨大、生态系统丰富、缓冲作用突出等环境特性，既不同于陆地，也有别于深海海洋（图 3）。

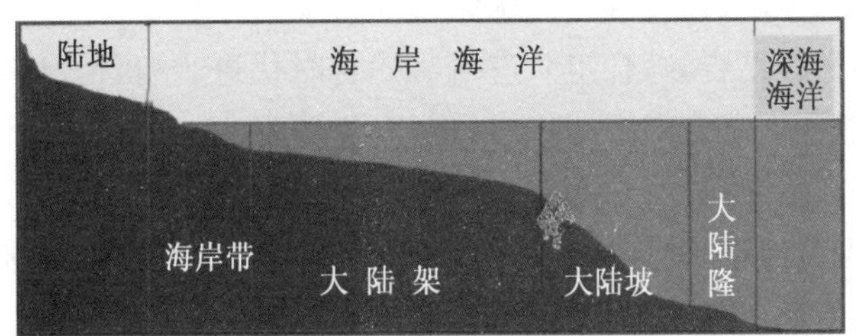

图 3　海岸海洋（Coastal Ocean）

据 LOICZ 统计：海岸海洋面积相当于陆地面积的 18%，占全部海洋面积的 8%。为整个海洋水体的 0.5%。但是，海岸海洋拥有全球 1/4 的初级生产力，提供 90%的世界捕鱼量。全世界 60%的人口聚居于沿海地带，人口超过 160 万的大城市之中 2/3 位于海岸带。海岸海洋占全球生产力值的 14%，占有全球反硝化作用的 50%，容纳全球有机残体的 80%，沉积物矿体的 9%，全球硫酸盐沉积的 50%，以及全球悬移质、相关元素与污染物的 75%～90%。

海岸海洋科学的性质，是基于地理学、地质学与海洋学相互交叉渗透形成的新学科，具有自然、人文社会与技术科学相互交叉渗透的复合性特点。海洋科学研究具有的特点是：研究对象的范围广，花费高，研究方法难度大，且需多学科综合与采用新技术的手段。尽管难度大，但我们必须展开研究，因为与人类生存发展及国家权益关系实在密切。

这门科学涉及的领域包括：① 自然科学方面，例如：海洋物理：季风、气温、雾、水的形成变化，海浪、潮汐流、水团等海水的物理特性。海洋化学：海水的化学性质与作用变化，泥沙与生物的化学成分、变化。海洋生物学：浮游生物，鱼，虾贝，蟹，藻类种，生态系，分布演化，资源等。海洋地质学：海岸、海底的结构，形成，地壳运动，沉积组成与矿藏。海洋地理学：疆域、区域、自然环境特性、人文、经济特色与文化历史传统等。② 技术科学方面，例如：声学与电子仪器的应用，如全球定位系统、导航、遥感遥测、海洋观测；港口航道工程，人工台站；电脑资讯系统等。③ 社会与人文科学方面，例如：海洋经济、海洋规划与管理、海洋法等。

海岸海洋科学研究的方法：由于海岸海洋范围大，人难以直接接触，其调查研究必须利用船只，应用仪器进行调查，实验分析，以及运算、归纳、存储、演绎、决策、规划并应用于建设。因此，调查涵盖海、陆、空，从三方面进行。例如，空中利用航测及监测卫星，如红外与雷达探测及资源环境卫星应用，可观测同地异时的变化情况，以及观测异地同时海况（如浪、流、混浊度、水色、鱼群分布等）。

至于海洋观测，依靠船只与电子仪器进行。一是定位：从观星、六分仪发展到无线电定位，雷达导航至全球定位系统（GPS）的应用，使精度得以提高。二是测量：按测点定站位，沿断面测量，以断面组合控制海域。三是测深：回声测深仪，测断面沿线海底的起伏。应用双频或多频道测深仪，可以测海底表层淤泥厚度：① 测海底表层地形起伏及物质组成，可应用旁侧声纳扫描测海底沙波。② 应用电火花、电脉冲、声学测量等探测技术，探测海底以下地层剖面，厚度达数十米、数百米等。四是观测海水物理化学特性与动力状况：按全潮 25 小时或半潮 13 小时为测量周期，每时观测水位（定实际深度、测波浪、波压仪等），测海水（温度、盐度、密度等），测水流（方向、流速），等等。

海上工作应多学科结合，综合组队进行。随着时代进步，目前已装备有多波速、多项组合的同步测验装置，但是定点观测仍需与断面测量相结合，才可控制地段或海域的动力状况。另一方面是在沿海陆地，需关注地质地貌沉积调查，剖面探测，河、湖水文泥沙观测，村落海岸变迁考古遗证，渔业生产力等。其次是室内实验分析：包括水质分析、酸碱度分析、浊度、化学元素、有机质、重金属分析，需经过过滤水样，定量定性地研究；水文测验的计算，制表分析；生物样品鉴定，微量元素与有毒质分析，生物量测定；沉积物分析，如细微性、矿物、黏土矿、微体古生物、磁化率、有机质的鉴定。

应用海洋地理资讯系统，采集自然环境与资源咨询，采集社会、经济、海域疆界与历史沿革等资料，存储资料，计算归纳，或图表演绎、规划，以供决策方案并发展推广。总之，海岸海洋科学必须进行多学科、多领域的调查研究与综合分析，才能获取真知。

时至今日，海岸海洋已成为国际海洋科学发展的新重点。这里简单回溯一下。1994 年以来，由于海洋权益的日趋重要，兴起了全球性的海岸海洋科学研究。1998 年，国际权威著作《The Sea》第 10 卷正式以“全球海岸海洋”（Global Coastal Ocean）命名。1999 年，国际地理学家联合会（IGU）颁布《海洋地理国际宪章》，将全球海洋正式划分为“海岸海洋”与“深海海洋”两部分[2]。

至于“海岸海洋”概念的兴起，与由 150 多个国家签署、1994 年 11 月 16 日正式生效的《联合国海洋法公约》（The United Nations Law of Sea Convention）有关。该公约中规定：沿海国有权划定 12 海里领海，24 海里毗连区和 200 海里专属经济区；确定大陆架是沿海国陆地领土自然延伸的原则；规定沿海国对专属经济区和大陆架内的自然资源拥有主权权利，并对防止污染和科研活动等享有管辖权；国际海底及其资源是人类共同继承的财产原则。

《联合国海洋法公约》提出的下列术语，各有特定的内涵。譬如，“内水”是指大陆岸线以外，领海基线以内水域，为领土部分，拥有完全排他的主权。“领海”（又名领水），是指邻接陆地领土及其内水的一带海域，宽度从领海基线向外 12 海里。领海是国家领土在海中的延续，属于国家领土的一部分。国家对领海内的一切人和物享有专属管辖权，主权及于领海上空，海床及底土。“毗连区”（又名连接区、特别区）是指领海基线外邻接领海的一带海域。宽度从领海基线向外 24 海里。该区是保护沿海国权益的重要海域之一，在该区域内，沿海国为了保护渔业、管理海关和财政税收、查禁走私、保障国民健康、管理移民，以及为了安全的需要制定相应的法律和规章制度，行使某种特定管辖权。“专属经济区”，是指从领海基线向外 200 海里的海域。该区在地理位置或法律性质上介于领海与公海之间。

沿海国家享有以勘探和开发、养护和管理自然资源为目的的主权权利，以及对于人工岛屿、设施和结构的建造和使用，从事海洋科学研究、海洋环境保护和保全的管辖权。其他国家则享有航行、飞越、铺设海底电缆和管道等自由。

《联合国海洋法公约》的实施，引起全球关注海洋权益。事实上，岛屿的纷争源于小岛大海洋。沿海国管辖海域的范围，均在海岸海洋地带。1966 年 5 月 15 日，我国第八届全国人民代表大会常务委员会第 19 次会议批准实施此公约。根据调查，我国“海洋国土”面积约为 300 万 km^2(包括海岸线、海岸与海底类型)。在 2004 年 3 月 10 日的中央人口资源环境工作座谈会上，胡锦涛指出：“开发海洋是推动我国经济社会发展的一项战略任务。”

三、全球变化与海岸海洋研究

全球变化是反映在气、水、岩石、生物圈的事件性变动，形成全球性的频发与持续效应，对人类生存环境影响深刻[1]。现有的工作侧重对全球变化的古资讯的研究，探根索源，讨论发展。今后应注重现代过程的研究，利用高新技术手段进行海洋监测，分析判断当前海洋变化的特点，探索变化规律，建立预测其趋势性变化的模型。

首先，我们应对“全球变化”的科学内容作出新的认识。

现代科学告诉我们，气温变化与海平面变化相关。如果对比 20 世纪末与 20 世纪初，可见全球气温升高约 0.5 ℃。科学家预计在未来 30 年，全球将继续增温，每 10 年将平均升高 0.3 ℃。到 2025 年，全球气温将升高 1 ℃。长时期的地质记录，反映出海平面、大气温度和海水温度三者间存在正相关[3]。近 200 年的验潮资料反映，海平面上升趋势与大气温度及海水温度增高趋势呈良好相关。有资料表明，百年来水动型的海平面上升值为 1～2 mm/a，2050 年海平面将上升 30～50 cm(3～5 mm/a)，到 2100 年海平面可能上升 1 m(图 4)。

气候变化、海平面变化与人类活动，是与海岸海洋密切相关、在研的全球变化热点课题。目前的严峻形势是，中高纬度地区气温增高显著，海平面持续上升，大河三角洲区面临着土地淹没与风暴潮频袭城市的境界。至于南极，由于冰盖减薄，融冰期延长，南极企鹅栖息地减少(图 5)。

由于中高纬度升温，暴雨侵袭延长。例如，2000 年秋季以来，暴雨侵袭英格兰和威尔士长达 3 月，低地被淹。2004 年 8 月下旬，英格兰西部山溪洪水成灾，汽车房屋损失惨重(图 6)。随着气温增加，海面上升，使三角洲与海岸低地更严重地遭受土地浸淹、海岸侵蚀、风暴潮频袭大城市等一系列威胁。根据《时代》(Times)2001 年 3 月的一份资料，可以看到密西西比河、尼罗河与恒河三角洲平原的受淹图示。下面这幅图，是 2005 年 8 月 28 日“科特琳娜”飓风登陆美国新奥尔良，造成 100 多万人流离失所(图 7)。

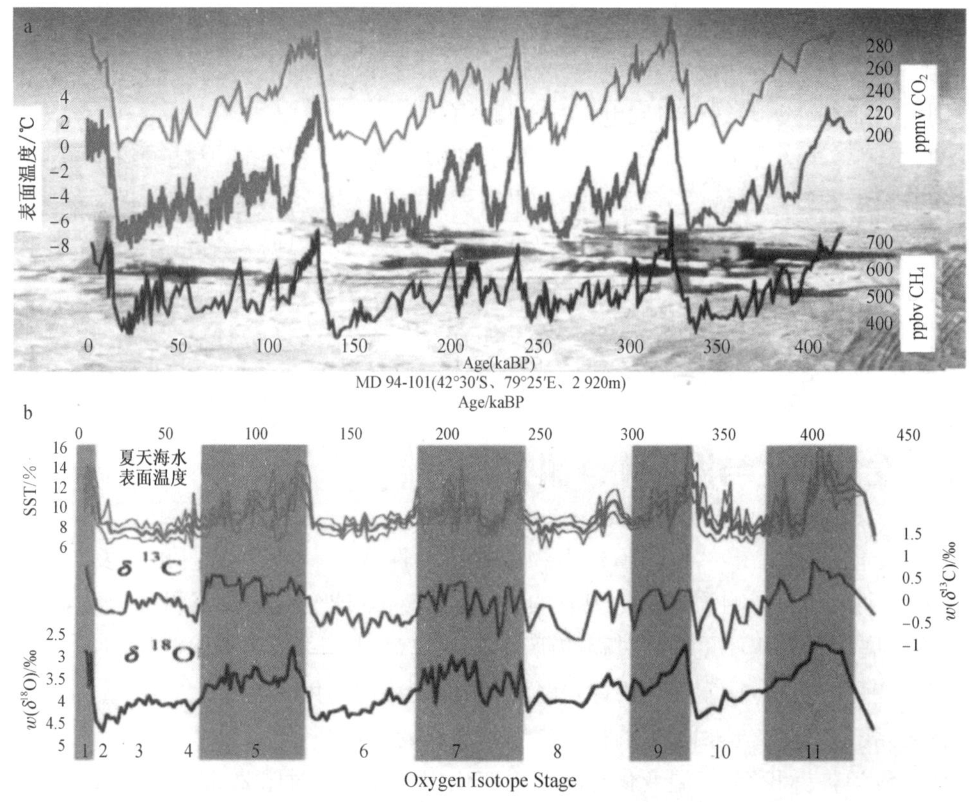

Vostok 冰芯所揭示的 400 ka BP 以来大气 CO_2、CH_4 浓度和气温变化曲线（上图）。MD94－101 大洋钻孔所揭示的 450 ka BP 以来氧、碳同位素与夏季 SST 变化曲线（中图），及其位置（右图）

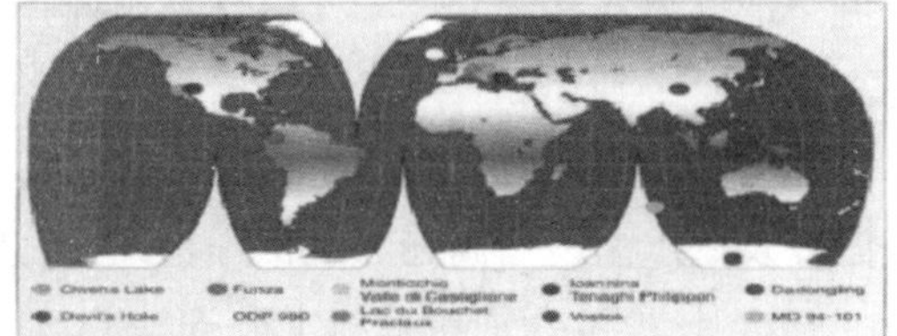

图 4　气候变化、海面上升及与人类活动的相互影响，与海岸海洋密切相关（附彩图）

(a) 南极冰盖减薄，融冰期延长，南极企鹅栖息地减少

(b) 冰山融溶

图 5　海冰减薄

图 6　中高纬度升温，暴雨侵袭延长

图 7　飓风影响

海陆交互作用影响的另一个例证是气温增升，内陆沙漠干热，沙尘暴活动频频侵袭东部沿海城市。但沙漠尘暴经西风气流搬运，在东太平洋沉降，却促使加拿大海域的鲑鱼生产丰度增加(图 8)。另一方面，当海水温度升高，又造成鲑鱼丰度降低。沙尘沿西风带飘移，促使东太平洋海域鲑鱼丰收，是地球表层系统地、气、水相互作用的典型例证。

图 8　沙漠尘暴经西风气流搬运，在东太平洋沉降，却促使加拿大海域的鲑鱼生产丰度增加

四、全球构造活动与海岸海洋效应

首先，我们看看火山活动与海岸海洋效应。全球有活火山 500 多座，主要分布于板块边缘、岛弧海沟系、洋中脊和板块内部。太平洋沿岸“火圈”分布着 370 多座活火山，其次沿东西向“关闭的地中海”带分布[4]（图 9）。

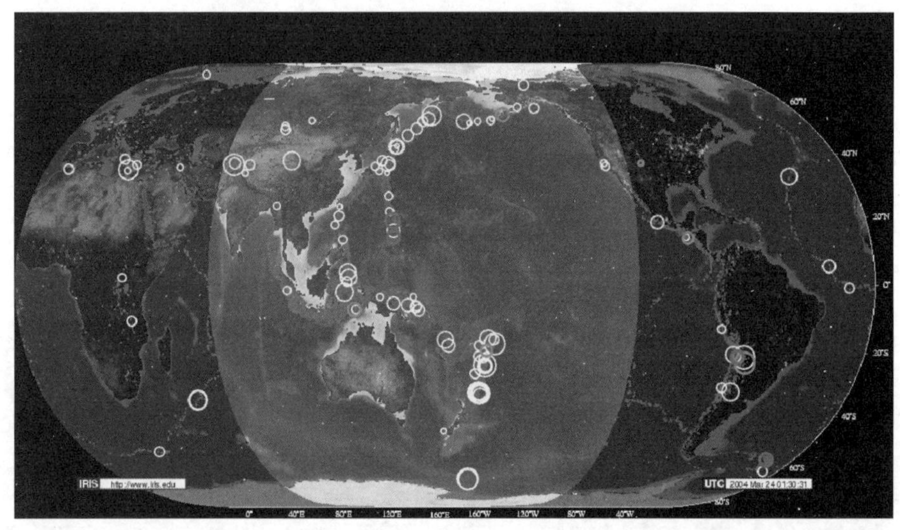

图 9　太平洋圈与全球火山地震分布图（据美国国家航空和宇航局，2004 年 3 月 24 日）（附彩图）

20 世纪 90 年代以来，火山爆发频频，逐年递增，以海岸海洋地区为突出，形成全球性构造活动，对地表系统造成巨大影响[4]（图 10）。火山活动在海岸海洋地区的其他环境效应，包括以下方面：一是地貌效应，火山活动是控制海陆分布格局的重要因素；二是气候效应，火山活动所引起的气候变化效应，历时长、范围广。冷期、富硫气体与酸雨；三是生态效应，如火山海岸与珊瑚礁的伴生关系，富有特色的热带海岸海洋生态系（图 11），等等。

（a）火山喷发的炽热的火山碎屑

（b）火山喷发的浓烟碎屑

图 10　火山喷发

（a）富饶的热带火山熔岩海岸

（b）美丽的珊瑚海

图 11　火山熔岩海岸

这里谈谈地震后活动情况。1990 年以来，几乎年年有大地震。2000 年以来，全球范围内几乎每日有地震发生，强地震亦引起火山爆发。强地震发生在环太平洋构造带、东西向的"地中海"构造带，并出现沿南大洋呈东西向分布。对比表明震级强度的转移，一次强地震爆发，能量释放以后震级变小，但同一构造带的另一处震级加强。这一现象，可为地震预报提供启示（图 12）。

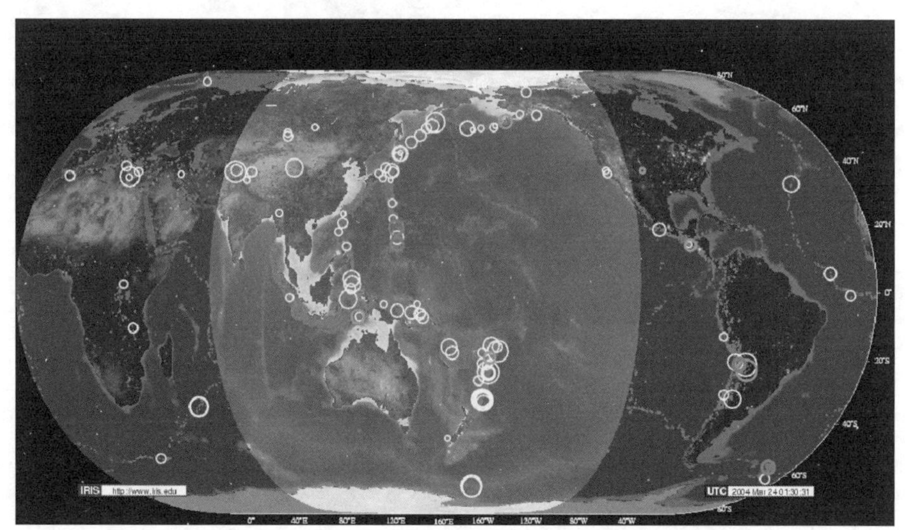

○：发生于 24 小时之内的地震；○：发生于 15 日内之地震；

▲：超过 15 天以上及 5 年以内的地震；圆圈大小：震级

图 12　全球地震活动分布

所谓强地震，多发生在海洋中，或发生于子夜，或发生于月半大潮期间。这是否与月球引潮力、太阳引力及陆地人类活动有关？地震火山活动释放地球内部能量，太平洋巨厚水层的压力、密度及温度变化聚集着较大的能量，与地球内部能量的结合，是否是大洋中能量较陆地更大的原因？这些问题，仍然有待科学家们探索。

我们现在已经知道，地震能量往往释放于板块结合部、构造裂隙断裂带与压力减轻处。这方面的典型事例很多。

例如，2000 年 6 月 4 日晚 23:28，印尼苏门答腊岛南部明古鲁省发生里氏 7.9 级地震，震中位于雅加达西部 640 km 处(4.6°S，102.1°E 印度洋底)，震源深度 32 km，为浅源地震，死亡人数超过 100 人。强震过后，强度减小，余震仍频频发生。

又如，1999 年 9 月 21 日 1:47，台湾岛日月潭西南 6.5 km 处集集镇发生里氏 7.3 级地震，震中位于 23.85°N，120.78°E，震源深度 7～10 km。整个台湾岛均可感到地震活动，系台湾西部山麓地带的系列逆断层，长期受吕宋岛弧推挤，积蓄大量能量，从车笼埔断层发生错动，能量释放而至。据考察，车笼埔断层东侧上盘上升 2.2～4.5 m，向 NW 及 NNW 方向水平滑移 7.1 m；下盘最大沉陷 0.3 m，向相反方向平移 1.1 m，最大水准地表加速度达 1 g，强震延时 25 秒，波及相距 150 km 的台北市。此次地震，损坏倒塌房屋约 2 万幢，死亡人数超过 2 300 人，伤者约 8 000 人，破坏力是非常惊人的[5](图 13)。

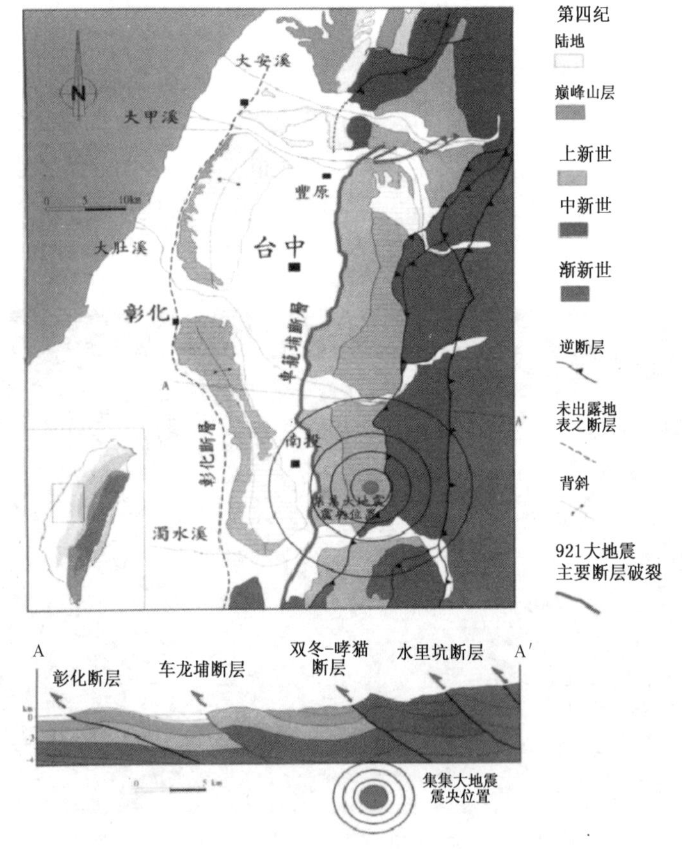

图 13　台湾岛集集镇地震(附彩图)

地震对海岸海洋的效应，主要表现为以下方面：形成断崖、裂点与瀑布；山崩地塌，造成堰塞湖与滑塌；土壤液化，发生喷沙、喷泥与地层陷落等。这种破坏效应，对城市水利与交通设施、港湾设施、动力与工业设施及建筑等，都会造成严重的损害。例如，大甲溪上因断层而造成了河床上的瀑布即为典型[5,6](图 14)。

下面的这组图片更为直观。丰原丰势路因地震挤压而形成阶地与阶地崖，推挤出的土甚至掩盖了路边的轿车[5,6](图 15)。丰原丰势路上的稻田，受到断层带的影响，向上隆起约 5 m 左右。雄县燕巢乡的泥火山，系地下水夹带泥浆与天然气沿地下裂隙上涌出地

图 14　大甲溪上因断层而造成的河床上的瀑布

表所致，隆起的地块造成铁路近 4 m 的抬升。丰原市大甲溪河段，由于车笼埔断层通过，使河床抬升 8～9 m 形成瀑布，并破坏了桥梁(图 16)。丰原中等教师研习会，受到挤压扭曲变形，上方的建筑物都扭曲破坏而倒塌[5,6]。

图 15　丰原丰势路因地震挤压而形成阶地与阶地崖，推挤出的土掩盖了路边的轿车

图 16　丰原市大甲溪河段，由于车笼埔断层通过，使河床抬升 8～9 m 形成瀑布，并破坏了桥梁

再举一个例子。海南岛北部铺前—东寨海湾，位于琼州海峡断陷带，1605 年(明万历三十三年)发生了可能达 7 级以上的大地震，造成沿河谷两侧断陷，使得 72 个村庄陷落成海(图 17)。海湾口狭内宽，呈长方形，并且湾套湾，系沿支流陷落所致。人们利用沉陷湾老河谷为通航水道，落潮时沉陆出露，尚残留石棺、石臼、石磨盘及大量砖、瓦、缸碎片，多被淤泥掩埋。东水港海湾与西部的洋浦湾，亦为受火山地震影响，沿河谷陷落的海湾[4]。因此，地震活动急剧地改变了自然地貌。

图 17　村庄陷落成海

五、海岸海洋科学与城市化发展

江河湖海是水陆交互作用带，地理与生物种群结构复杂，物种变化活跃，生产力较高，是人类聚居与城镇发展、分布的集中地带。环顾全球，可知亚洲大河流域古文明，美加环五大湖、沿大西洋与太平洋城市群，西欧等国沿海与沿河城市群，以及长江三角洲与珠江三角洲城市群，都与水路交互作用息息相关。

我国的城市建设在 20 世纪 90 年代发展迅速，城市面貌显著改变，发达的沿海城市，如深圳、广州开始注意城市建设发展中发挥滨海环境的特点与优势。滨水环境与城市发展，成为海岸海洋科学的又一个新的研究点。但是，沿海城市需根据海岸海洋环境特点，才能做出科学的城市发展总体规划。

中国有众多傍水而建的城镇。沿长江的大城市有重庆、武汉、南京、上海均为大都市；水乡城市有苏州、无锡、镇江、扬州等名冠古今的滨水城市。在中国古代，水体的主要功能是运输及赏玩，在四方汇集的滨水地区，兴港建镇，逐渐发展为城市甚至市镇群。

大家都知道，宋代著名画家张择端有一幅传世名画《清明上河图》。它展示了北宋临水都城之布局，生动地描绘出河流在古代城镇发展中发挥了景观、水体联系、运输灌溉、积聚人气、商贾与集市扩展的功能。从画中可以看到，汴河流贯城乡，沿河曲缓流处，船只停泊，桅杆林集，近河有修造船厂；拱桥为中心，商肆林立，形成新的集镇；桥上行人熙熙攘攘，行人驻足观景，桥下舟楫川流，篙师缆夫牵舟，逆河而上。清明时节的桥外城郊，田亩阡陌纵横，疏林薄雾村舍酒家，村外大道有骑而宾士，有踏青扫墓而归，田野情趣意浓。桥内城郭老街，商店林立，行人南来北往，酒店、茶肆、旅店、雅货百物，酒招牌匾，名师字画，鳞次栉比，市景文华繁荣；深宅大院、官宦宅邸则现于市侧(图 18)。

目前，我国城市的滨水带，以建设公园绿地居多，或辅以钓鱼舟艇及水上活动设施，增强了人居环境的健康因素。但仅有日间活动，入夜则冷清，使亲水临街缺少人流；也有的建立了高档宾馆，增加了河上景点，但价格不菲，有如阳春白雪未能聚集大量客流；有些岸段建立了大量酒肆茶楼，残渣油污却损毁了滨岸与沙滩的洁净。尚不少大城市的临江、滨海段，却空旷荒凉或游人寥寥，需重视聚集人气。

图 18　桥内城郭老街，商店林立，行人南来北往

以南京为例。江阔水深、浩浩荡荡、气势磅礴的长江，带给南京市无限生机，可陶冶胸怀、焕发活力、令人鼓舞竞进。但是，由于战乱、血吸虫病害与过去无序的开发，留下了荒芜、杂乱与破旧的景象(图 19)。南京大都市的建设，必须重视滨江两岸的规划，改变城不见江、江不见城的现状，建设自然生态环境与历史人文要素和谐发展的沿江城市风光带。南京近几年相继实施的中山陵环境综合整治、秦淮河水环境综合整治、明城墙风光带环境综合整治、老城墙增绿等工程，已见成效。以此领衔，南京又陆续启动北极阁风貌区、小桃园、玄武湖环湖风光带、东西干长巷、汉中门广场等点上工程，使南京老城逐步“显山、露水、见城、滨江”(图 20 和图 21)。但是，滨江带建设发展仍待提上重要的议事日程。

图 19　南京滨江考察(2003－10－30)
(王颖　摄)

图 20　南京滨江考察(2003－10－30)
(王颖　摄)

在滨岸规划中，需要考虑预防的，即如何处理集聚客流与维持环境洁美。这方面，丽江古城处理得很好，保持了小桥流水与坊、店、酒肆，溪水流畅、空气清新与民俗文化浓郁之佳，具有个性、特色。

我国城市建设自 20 世纪 90 年代以来发展迅速，城市面貌显著改变。在发达的沿海城市发展中，逐渐开始重视发挥滨海环境的特点与优势。

随着城市化进程在我国方兴未艾，滨水环境与城市发展也成为海岸海洋科学的又一个新的研究点。例如，广州市 2000 年实施城市发展总体规划以来，在珠江滨水环境优化，新机场、大学城、会展中心屹立于珠江之滨，轻轨地铁及老城区疏导等成果卓著，展现出现

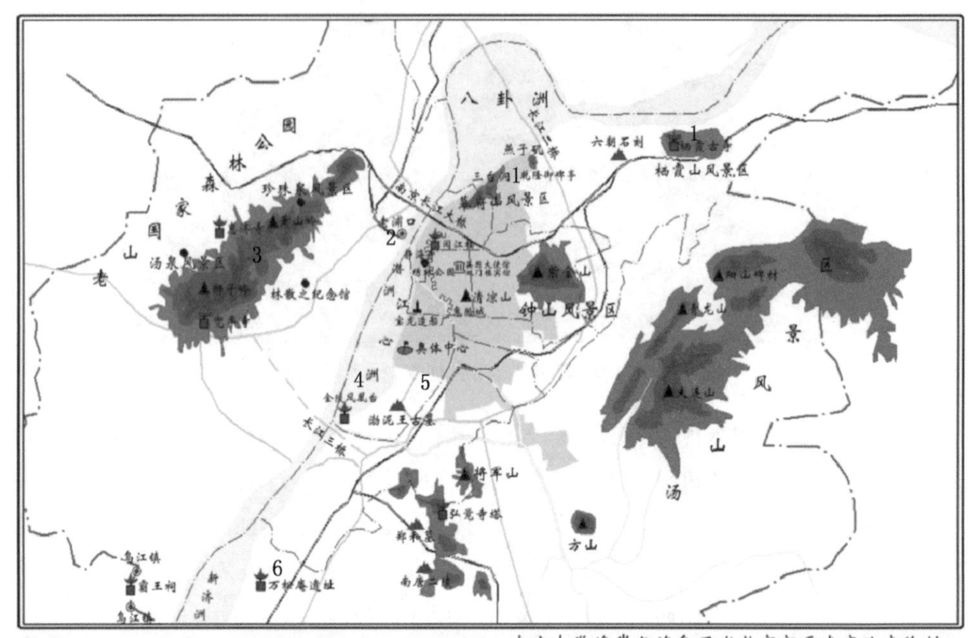

1. 幕府山、燕子矶基岩岸板块；2. 下关—浦口南京外滩板块；3. 江北一山三泉森林地质景观带；4. 江滩沙洲生态绿岛带；5. 河西新民居—宝船厂遗址—奥体中心板块；6. 板桥、梅山、新济州风光带

图 21　南京沿江宜由绿色通道连接六个板块组成串珠状风光带

代化大都市新风貌(图 22)。2003 年，又向海洋发展；2006 年又进行新订正。浙江省完成了新杭州湾城市的发展规划，建设跨湾大桥，建设长江三角洲南部经济与生态环境和谐发

图 22　今日珠江——广州

展的城市群。山东省也在进行半岛城市群建设。

城市的发展规划,应以人为本。它应当包括这样一些指标:一是清新的空气、明亮的阳光、清洁的淡水、安全的宅舍与宜人的风貌。二是繁荣的商贸街市、便利的交通运输、咨询网络、文化历史内涵、科教医术与宗教设施安全的保障、服务系统及有效的管理体系等。

总之,城市发展规划需充分考虑与海岸海洋生态环境的有机结合,人地关系和谐发展的思想应渗透贯穿于设计体系中。在这方面,南海之滨的湛江、三亚亚龙湾国家旅游度假区,均在构思新一轮的发展规划中,重视海岸海洋的生态环境特点与功能的最宜利用,是值得我们关注的。古今例证反映出具有复合型特点的海岸海洋科学,将在21世纪中国城市化进程中发挥一定的作用。

参考文献

[1] 王颖,牛战胜.2004.全球变化与海岸海洋科学发展[J].海洋地质与第四纪地质,24(1):1-6.

[2] 王颖译,任美锷,吴传钧校.1999.海洋地理国际宪章[J].地理学报,54(3):284-286.

[3] PAGES IPO. 1999. PAGES Newsletter[J]. Bern, Switzerland, 7(3):1-2,9.

[4] 王颖,张永战.火山海岸与环境反馈——以海岛火山海岸为例[J].第四纪研究,1997(4):333-343.

[5] 台湾大学地理系.集集大地震灾害分布图[G].台北:台湾营建署,1999.

[6] 台湾大学地理系.台湾地区天然灾害分布图[G].台北:台湾农业委员会,1999.

【附注:图5、6、7、8据Google Earth.】

海岸海洋——海陆过渡带之环境资源特征与变化研究*

一、海陆过渡带——海岸海洋

海陆交互作用带是位于以陆壳(SiAl 层)组成的大陆与以洋壳(SiMa 层)组成的大洋之间的过渡地带，兼有陆壳与洋壳结构，成为一独立的客体单元。海陆过渡带被传统的研究所忽略，自 1994 年《联合国海洋法公约》实施以来，却因涉及沿岸国家的海洋权益而被重视。联合国教育、科学及文化组织(United Nations Educational, Scientific and Cultural Organization, UNESCO)有关会议已确定这一地带为"coastal ocean"，即海岸海洋(Brink and Robinson, 1998)。

海岸海洋涵盖整个海陆过渡带，即陆地延伸至海洋全部(图 1)。据 1994 年 UNESCO 在比利时列日会议定义：其范围包括海岸带、大陆架、大陆坡及坡麓的堆积体——大陆隆(continental rise)。

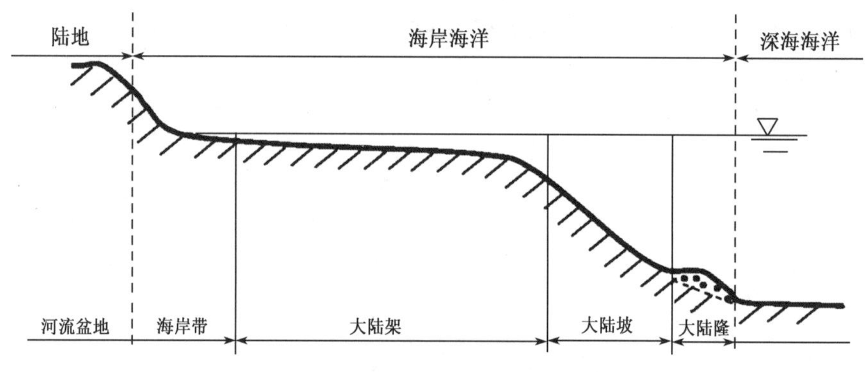

图 1　陆海过渡带——海岸海洋

(资料来源：王颖和季小梅，2011)

海岸海洋是独立的环境体系，因有海水覆盖而区别于陆地，又因有陆地延伸为海底而区别于深海大洋。地球表层的气、水、岩，在此地带交互作用变化频繁，风力、波浪、潮流均在海陆交互带加强，生物量巨大，生态系统丰富，缓冲作用突出，与人类生存活动关系密切。

1994 年《联合国海洋法公约》实施，将大陆架与 200 海里①专属经济区划归沿海国管辖，这些均在海陆过渡带的范围之内，全球有 370 多处海域发生权益争执。我国的管辖海域有 300 万 km^2，在东海大陆架岛屿归属与油气资源开发，以及在南海岛礁与海域疆界归

* 王颖：中国科学院(王颖主编)《中国学科发展战略·海岸海洋科学》，1－6 页，2016 年北京：科学出版社。本文内容为 2009 年 10—12 月向中国科学院地学部的建议稿的综合。

① 1 海里＝1.852 千米。

属方面与周边国均存在着海疆权益纷争。海洋国土权益推动了对海岸海洋的关注，国际上兴起了对海岸海洋科学研究的热潮。

海岸海洋科学是研究海陆过渡带表层系统作用过程、环境与资源特性、自然发展规律，以及与人类生存活动相关的科学，具有地理学、地质学与海洋学交叉，自然科学、人文与社会科学和技术科学相互复合的特点。其研究具有对象范围广（从海岸带、滩涂到浅海）、调查方法难度大与花费高的特点，需从陆海空三方面观测，调查工作量大并受天气与海况之限，船只与探测仪器费用高，滩涂具有从陆、从海均难深入的困扰。调研工作要求采用多学科结合的手段，进行调查与实验分析，并以海洋地理信息系统进行运算、归纳、存储、演绎及决策、规划与实践。服务于环境资源开发与保护、海洋经济、疆域政治、立法管理和军事方面。我国对海岸海洋领域尚未进行系统性或体系性研究。

中国海陆过渡带，南北纵向跨越 4°N～41°N，气候与生物资源环境多样、差异大。横向范围，除台湾以东濒临太平洋，其余均位于由岛弧列岛环绕的边缘海内（continental marginal seas），越洋联系受制于海峡通道。我国海区均具潮汐作用，季风波浪动力活跃，河-海交互作用效应为一突出的特征。海岸与浅海陆架地貌反映着成因动力特色：长江以北的平原海岸、淤涨的潮滩、宽广的大陆架与内陆架巨大的堆积沙脊群与潮流通道组合；南方港湾海岸，华南的生物成因海岸，众多的岛礁、海峡，拉张沉降的边缘海与洋壳隆升，均孕育着重要的科学机理与应用价值。我国地球科学领域对陆地研究十分深入，20 世纪 60 年代关注并重视对海洋的研究，但忽略了对陆海过渡带这一独立环境体系的整体关注。海陆是沟通的，地质时期渤、黄、东、南诸海的堆积型大陆架均为来自江河入海的陆源物质，现代的海岸海洋又成为江河等陆源污染之“汇”。2004 年，我国海域污染面积就已达 16.9×10^4 km^2，水环境恶化，生态系统脆弱，赤潮爆发 96 次，海洋灾害损失 54 亿元，也造成了生命伤亡。海岸海洋被污染后既减弱其沉淀过滤作用，又成为深海大洋污染之源，影响到新鲜空气与淡水的形成机制。因此，我国在关注海洋开发研究之际，必须重视对海岸海洋——陆海过渡带，作为一个体系而加以研究，尤其我国人口众多，沿海与通海大河流流域开发频繁，这一研究意义尤为重要。

二、对海岸海洋认识的重要进展

1994 年 5 月由联合国教科文组织的政府间海洋学委员会（Intergovernmental Oceanographic Commission, IOC），在比利时列日大学举行了“海岸海洋先进科学技术研究”第一次国际学术大会（Coastal Ocean Advanced Science and Technology Study），是具有里程碑意义的重要科学进步。该会获得 IOC 欧洲委员会、海洋研究科学委员会（Scientific Committee on Ocean Research, SCOR）、美国国家科学基金会（National Science Foundation, NSF）及美国海军研究办公室（Office Of Naval Research, ONR）支持。会后出版了海岸海洋论文专辑（上、下册）。

1999 年国际地理学家联合会以英国、法国、中国等 5 国文字出版的“国际海洋地理宪章”明确地将全球海洋区分为“海岸海洋”（coastal ocean）和“深海大洋”（deep ocean）（国际地理学联合会海洋地理专业委员会，1999）。上述内容表明国际海洋科学与地理科学学界对海岸海洋的正式确定。但是，当时尚未被我国海洋界完全采纳。专家们认可UNESCO的划分，但却仍沿袭旧法将coastal ocean译为近海。但是，近海的研究工作主要是在海岸带与大陆架海域，未明确地将沿岸陆地与大陆坡海域纳入统一的研究体系。

2006年以来，国家海洋局统一组织领导开展了“我国近海海洋综合调查与评价”项目，将海岸带、大陆架、陆坡海结合为一体进行调查。调查内容与方法基本一致，但因同一海域参加单位较多，又在深水、浅水带，在专业项目上存在分工等，故尚难以获得不同海区海岸海洋的系统化概念。因此，如能设立海岸海洋科学研究专项，整合科学成果，将会促进我国对海陆交互带的研究，为填补学科空白做出贡献。

三、海陆过渡带环境特征与变化研究

在对亚洲—太平洋边缘海海陆过渡带作为一个系统进行研究时，首先应针对中国海的环境特点，以河海交互作用过程为关键。因为发源于世界屋脊青藏高原的8条大河，有5条汇入中国海，输入了巨量泥沙，成为陆、海沟通的干道，控制着海岸与海底地貌发育。一个显著的例证是尚保存于东海与南黄海的古扬子大三角洲，其范围界于26°N～32°N，122°E～127°30′E，外缘至水深200 m处（王颖等，2012）。因此，研究过程中，有重点地向长江、黄河、珠江等大河河口及流域衔接延伸，研究陆源泥沙、人类活动产物向海洋的输送，在海岸海洋动力环境下，物理、化学与生物过程的变化与发展以及综合效应至关重要。以河海交互作用为切入点，研究陆海过渡带的边缘海环境特点，资源蕴藏及其发展变化趋势，具有重要的科学价值。事实上，人类沿大河的开发活动也间接地向海洋施加影响：长江流量3×10^8～4×10^8 m^3/s，洪水期流量8×10^8～9×10^8 m^3/s，宜昌站最大洪水量为11×10^8 m^3/s，三峡大坝建立与中、东线引水约占径流量的22%；原入海泥沙量4.7×10^8 t/a，三峡水库拦储泥沙占总输沙量的1/3～1/2，加上沿江数百个水库拦截泥沙，使入海泥沙量锐减达总量的2/3，必然影响到河口与口外三角洲的变化，形成外冲内淤（海底三角洲冲蚀，岸滩淤涨）之势；黄河沿途引水，入海径流量及沙量剧减，引起三角洲海岸侵蚀与湿地退化；珠江三角洲城镇发展及农田大量用水，影响到淡水供应及盐水入侵等。当代城乡与工业化发展，江、河汇集了大量陆源污染物入海，形成近岸海水质劣与富营养化灾害，频发且势猛。因此，陆海交互过渡带的河海交互作用与人类活动影响相互关联密切，是研究的关键。

其次，需重视研究海平面变化、动力机制与作用效应。

全球气候持续变暖使海水膨胀，极地冰盖、冰帽、冰川加速减薄与融化，形成全球海平面持续性上升。地面升降、季风和海流等局部因素引起区域性海平面变化。沿海与河口三角洲大城市，由于超采地下水及大型建筑物重压，使平原松散沉积层压实而下沉，加强了海平面上升效应。海陆交互作用带是海平面上升直接影响的易招灾区。

据全球海平面观测系统（global sea level observing system, GLOSS）第10次会议的约定，我国自2008年起，将1975—1993年的平均海平面定为常年平均海平面，该期间的月平均海平面定为常年月平均海平面。根据海平面监测与分析结果，1975年以来，我国海平面呈波动上升趋势，平均上升速率为2.6mm/a，高于全球平均水平。据上述常年平均海平面,相对于2008年海平面,预测未来30年海平面将持续上升(表1),比2008年升高80～130 mm,长江三角洲、珠江三角洲、黄河三角洲和天津沿海是海平面上升影响的主要脆弱区。

表1 未来30年中国海域海平面上升预测

上升预测	渤海	黄海	东海	南海	中国全海域
常年平均海平面上升速率/(mm/a)	2.3	2.6	2.9	2.6	2.6
与2008年相比,未来30年上升值/mm	68～120	89～130	87～140	73～130	80～130

资料来源:国家海洋局2009年发布的《2008年中国海平面公报》。

国际有关科学家研究认为:到2100年全球海平面将升高18～88 cm,会使许多沿海城市处于海平面以下,将严重地威胁人类生存。海平面持续上升,对沿海地区、海陆交互作用带及大河流域带来一系列影响:风暴潮频繁,海岸侵蚀加剧;海水入侵,潮汐影响范围扩展;大河潮流界与潮区界受海水顶托而沿河上溯,长江潮流界内限,已沿江阴西移,潮区界上限也自大通西迁,影响到水质与水环境变化。在洪水与风暴潮交会期间,沿海与河流三角洲平原受洪、潮灾害频繁,尤其是东部沿海为经济发达区,其损失严重。而人类的一些活动如围填海,减少纳潮洼地与减堵潮流通道,上中游兴建堤坝水库,蓄水拦沙,改变了河道行水、行沙方式与季节等,在海平面上升的大背景下,必然改变了河海水沙交换的自然状态,加强了海面上升与海水入侵的效应。在倡导低碳经济,限定温度上升的目标前提下,温家宝提出"中国在2050年使温度升值限于2 ℃"。未来海平面变化的趋势、速率与效应如何?需要进一步研究以期找出相适应的对策。

四、建议关注研究的课题

以下为海陆交互带体系新兴海岸海洋学科的基础性研究,为国际与我国的关注点。

(1) 亚洲—太平洋边缘海陆交互带动力体系特征、变化与效应研究。

① 海陆交界气温与气压差异、昼夜差异与风力差异及效应,季风作用效应以及风能利用,不同海域陆架风吹流分布活动特点与生物量效应。

② 海陆过渡带波浪传播变化,季风波浪变化特点及效应。

③ 潮汐作用与海平面震动,潮波变形辐聚、辐散与潮流效应,海陆交互带浅海动能收支、物质输送与陆架海环境特征研究。

④ 海陆交互带海洋体系能量、动量、化学与生物量交换过程与效应研究。

(2) 伸向海洋的陆地的岩性、结构,及亚太大陆边缘岛孤海沟系构造活动效应研究。

(3) 晚第四纪与全新世以来,海平面变化与海陆过渡带的变化效应。

(4) 世纪性海平面上升、发展变化趋势与不同海区相关效应与应对措施研究。

① 比较研究近50年、未来30年以及2050年限制升温2 ℃后的海平面变化趋势,以及对我国海陆过渡带的效应。

② 海平面上升，海水入侵，潮汐与潮流活动影响范围分析，江、河水质水环境的变化，河海水沙交换活动规律与效应。

③ 气温与海平面变化所导致的风暴潮频繁程度与影响范围，海浪侵蚀强度与海岸侵蚀效应、速度与对策。

（5）人类开发活动——沿大河中上游兴建水库、拦沙蓄水，中下游引水，河道改变，围填海工程——与陆源水沙通量变化规律，以及在海平面上升交互作用下对海岸带大陆架及大陆坡环境的影响效应。

（6）黄、东、南海环境资源远景，开发历史，海洋权益与国家安全研究。

参考文献

[1] Brink K H, Robinson A R (ed.). 1998. The Global Coastal Ocean[M]. John Willey & Sons Inc.

[2] 王颖，季小梅. 2011. 中国海陆过渡带——海岸海洋环境特征与变化研究. 地理科学，31(2)：129-135.

[3] 王颖译，任美锷 吴传钧校. 1999. 国际海洋地理宪章(中文版)[J]. 地理学报，54(3)：384-386.

[4] 王颖，邹欣庆，殷勇，张永战，刘绍文. 2012. 河海交互作用与黄东海域古扬子大三角洲体系研究. 第四纪研究，32(6)：1055-1064.

[5] 国家海洋局. 2009. 2008 年中国海平面公报.

The Surface Texture of Quartz Sand Grains from the Continental Shelf Environment: Examples from Canada and China*

INTRODUCTION

The surfaces of quartz grains often preserve the effects of dynamic processes that have acted on them. By recognizing the sequence of surface markings, the sequence of environments experienced by the grain and hence the pathways of grain transport can be established. Thus it should be possible to use the surface textures of quartz sands to help determine sediment provenance in addition to sedimentary environment. This paper attempts to relate characteristic surface markings of quartz grains, as seen on scanning electron micrographs, to the continental shelf environments, and to the sequence of processes that have occurred through the history of sedimentation.

The report is based on the study of 16 samples (Table 1). One sample came from the outer shelf of the East China Sea as an example of the relatively warm water environments of a Pacific marginal sea. One sample came from the Beaufort Shelf of the Arctic Ocean, where a large amount of sediment is being supplied by the Mackenzie River. Four samples were taken from a piston core containing a sequence of continental shelf sediment deposited since late glacial times on the inner Labrador Shelf off eastern Canada. One sample was taken from the outer Scotian Shelf off Nova Scotia, on the eastern Canadian coast of the Atlantic Ocean (Fig. 1). The grain surface characteristics of these sands were compared with the sands from possible source environments, such as glacial till, fluvial, littoral, beach and coastal-dune. In this paper, we refer in particular to seven samples from the coast of Nova Scotia and Sable Island and samples from the Changjiang and Mackenzie Rivers (Fig. 2 and Table 2).

* Ying Wang, G. Vilks, D. J. W. Piper: *Proceedings of International Symposium on Sedimentation on the Continental Shelf with Special Reference to the East China Sea*, pp. 920 - 931. Beijing: China Ocean Press, 1983.

Table 1 Sample Locations and Environment

Sample No	Latitude	Longitude	Environment
Core 95	54°06.9′N	56°41.0′W	Labrador Shelf, 135m depth
1. (0～5 cm)			
2. (200～205 cm)			
3. (450～455 cm)			
4. (775～780 cm)			
1023	45°57.2′N	62°15.1′W	Scotian Shelf, 200m depth
1076	70°08.5′N	130°03.0′W	Beaufort Shelf, 10m depth
1092	29°15.0′N	127°09.0′E	East China Sea Shelf, 130.7m depth
F_3—PF_3	45°42′N	60°19′W	Framboise, Nova Scotia, Canada (glacial till)
PF_8	45°42′N	60°19′W	Framboise, Nova Scotia, Canada (beach)
1018	44°41′N	63°07.5′W	Martinique Beach, Nova Scotia (at high tide)
1019	44°41′N	63°07.5′W	Martinique Beach, Nova Scotia (beach dune)
1077	43°55′N	59°50′W	Sable Island, Nova Scotia (beach berm)
1078	43°55′N	59°50′W	Sable Island, Nova Scotia (old beach dune)
1079	43°55′N	59°50′W	Sable Island, Nova Scotia (windblown sand)
1039	32°05′N	119°18′E	Changjiang River at Zhejiang, China
1074	69°11.50′N	134°00′W	Mackenzie River, lower reach

Fig. 1 Location map for sample sites in Labrador, Scotian, and Beaufort Shelves

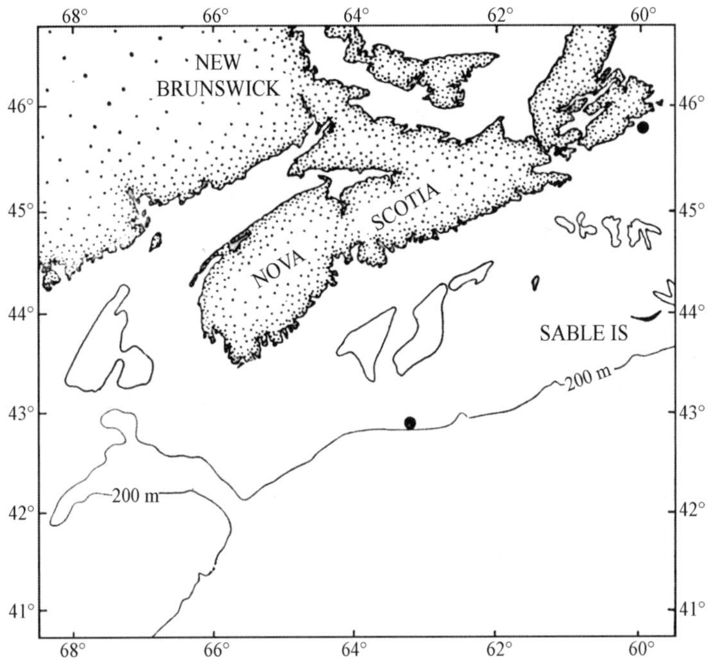

Fig. 2 General physiography of the Scotian shelf showing sample locations

Methods

The grains were hand picked and boiled for 10 minutes in concentrated hydrochloric acid, 20 minutes in stannous chloride, and 10 minutes in 30% hydrogen peroxide, as described in detail by Krinsley and Doornkamp (1973). After each boiling, the grains were washed thoroughly in distilled water and dried. This procedure removed the organic matter and iron oxide. About 15 to 20 clean grains were selected at random from each sample (Krinsley and McCoy, 1977). Care was taken to ensure that the full range of grain shapes present in the sample had been selected for analysis. The selected grains were glued on a metal specimen holder with double-sided sticky tape, coated with a 200 Å layer of an alloy of gold-palladium, and examined with a Cambridge 180 stereoscan. Each grain was examined for overall shape, angularity of edges, characteristic markings, diagenetic features, and the sequence of superposition of markings. Angularity of edges uses the scale of powers (1953). The nomenclature of surface markings is largely that of Krinsley and Doornkamp (1973); examples of these surface markings are shown in plates 1～10. Observations on the samples are detailed in Table 2. The occurrence of the commonest markings is summarised in Table 3, where only those markings found in at least 50% of the grains in a sample are listed.

Table 2　Number of Grains With Particular Shapes, Edges and Surface Markings Per Sample

Sample No. Surface Markings and Grain Shapes	Illustration Plate	Labrador Shelf				1023 Scotian Shelf	1076 Beaufort Sea	1093 China Shelf	Till F3～pF3	pF8	1018	Beach		1078	1079	River	
		1	2	3	4							1019	1077			1039	1074
Grain shape:																	
irregular		8	6	7	10		2	5	7	2		1	1	1		4	1
crystal control									1							2	1
smooth						9	1	1		1		6	3	3	7		5
other		7	7	7	4	6	7	8	7	5	*	8	10	9	3	9	
Grain edge:																	
angular		3		1	2			1	4								
angular-subangular			6	1	2				6	1							
subangular		7		3	4	1		2	4	2	1						2
subangular-subrounded			3	4	2	2	5	6	1	3	1					3	1
subrounded		4	4	4	4		2	4		2	4	3	1	3		8	7
subrounded-rounded						2	1	1			2	1	5	1		3	1
rounded		1		1		10	2				7	11	9	7	10	1	4
Surface markings:																	
conchoidal fractures	3	8	5	7	9	2	1	7	13	7	3	5		2		2	7
parallel steps	8					2			6								1
striae	7					2		1	1								
upturned plates									8							8	2

(**Continued**)

Sample No. Surface Markings and Grain Shapes	Illustration Plate	Labrador Shelf				1023 Scotian Shelf	1076 Beaufort Sea	1093 China Shelf	Till F3～pF3	pF8	1018	Beach		1078	1079	River	
		1	2	3	4							1019	1077			1039	1074
dish-shaped concavities	6					3	9	8	7		3	1		1			
mechanical V depressions	10					5		1		7	3	2	14	9	4	2	5
mechanical V marks	11	9	13	10	5	15	9	12	2	8	15	15	15	11	9	15	14
mechanical impact pits	1	2	9	10	5	15	5	4		7	15	15	15	11	10	15	10
curved grooves	4					4	1				7	4	9	5	3		1
scratched lines	9					1	1				1	1	5	7	3		
cracks or fissures		3	5	2	1	3		2	3		1	1					3
surface solution features							2	2					1	1		1	1
solution holes	2	12	8	11	13	1	10	9	6	2	6	9	5		1		3
silica coating								1									
silica deposits	5	8	2	5	123	1	7	2	7	1		2					6
weathering features											11						
Total grains		15	13	14	14	15	10	14	15	8	15	15	15	11	10	15	15

* undulated surface

Table 3 Surface Markings Recorded in More than 50% of Grains in Each Sample

Possible Source				Continental Shelf Samples			
Environment Surface textures	Glacial Till	Beach, Dune	River	Scotian Shelf	Labrador Shelf	Beaufort Shelf	East China Sea Shelf
conchoidal fractures	×	×			×		
upturned plates	×		×				
dish-shaped concavities						×	×
mechanical V depressions		×					
mechanical V marks		×	×	×	×	×	×
mechanical impact pits		×	×	×	×		
curved grooves		×					
scratched lines		×					
solution holes		×	×		×	×	×
silica fragments					×		
weathering features		×					

SOURCE ENVIRONMENTS

Sand from till. A sample (F_3—PF_3) from a till cliff of a glacial drumlin in Framboise Cove, Nova Scotia, Canada, shows distinct features of glaciation: angular to subangular grains with irregular shape, conchoidal fractures, parallel steps, striae, upturned plate, and dish-shaped concavities. Evidence of weathering, such as solution holes and silica deposits, probably indicates that sediment had remained undisturbed for a long time since deposition.

Beach sands derived from nearby glacial till were examined from two localities. A sample (PF_3), 2 km from the till cliff described above, contained mostly subangular to subrounded grains, and was thus more rounded than the till sand. Conchoidal fractures are still major features but are superimposed by V marks and small pits. Dish-shaped concavities and other smaller glacial features are absent and the grain surface is significantly smoother. Two further samples were taken from Martinique Beach, Nova Scotia (samples 1018, 1019), one from the beach and one from coastal dunes. Most grains have rounded or subrounded edges, but coastal-dunes sands (sample 1019) seem to be more rounded than grains from the high tide level of the beach. Mechanical V marks, collision pits, and curved grooves are dominant features on all of the grains (Fig. 3). Many grains contain conchoidal fractures, reflecting the local glacial till source.

Coastal sands from Sable Island. Mechanical V depressions (Fig. 4(3)), V marks, impact pits, and scratched lines are the major features on the beach sand (sample 1077). Coastal-dune sands (samples 1088) and windblown sands from the

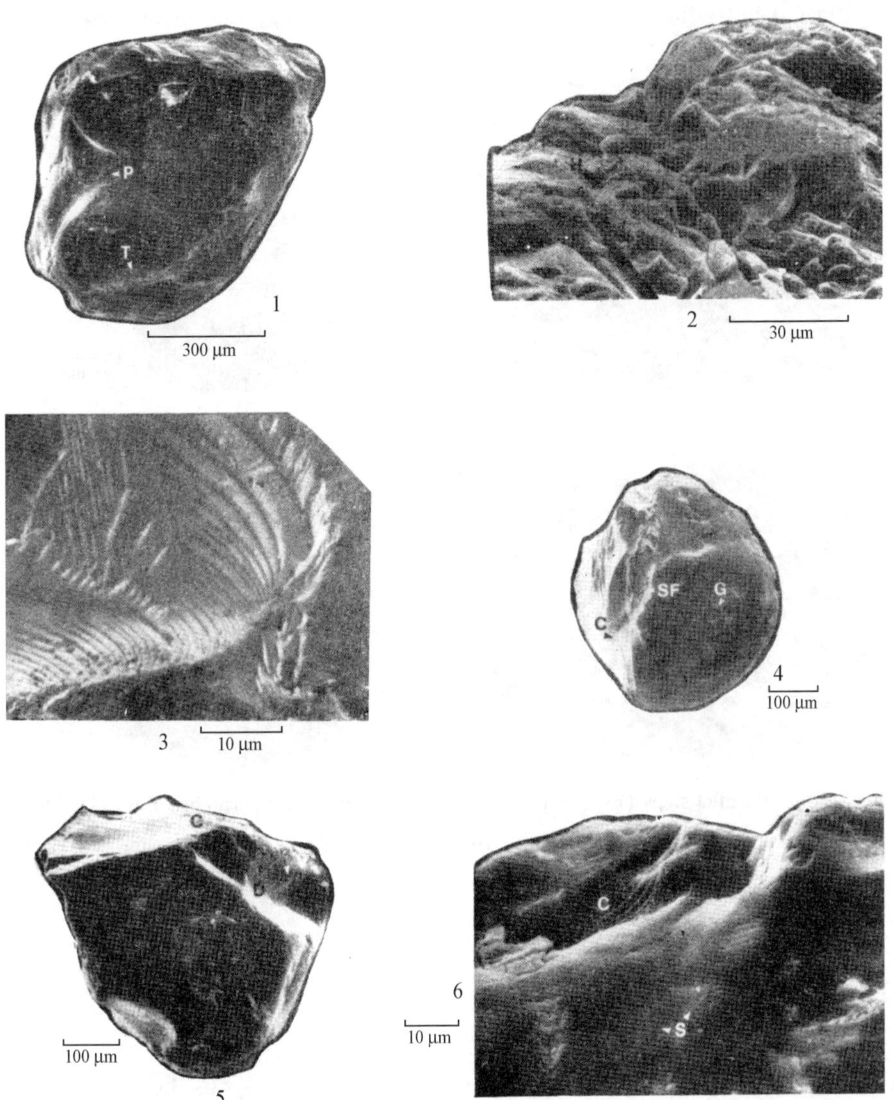

Fig. 3 1 - Mechanical holes (P), Labrador Shelf 135 m depth (core 95); 2 - Solution holes (H); 3 - Conchoidal fracture (C); 4 - Curved groove (G) and silica fragments (SF), Mackenzie River (sample 1074); 5 - Dish-shaped Concavity (D), Sohm Abyssal Plain 32°29.2′N, 55°56.2′W; 6 - Striae (S).

backshore (sample 1089) are similar to the beach sands, except that they are smoother with more small pits and scratched lines (Fig. 4(2)).

River Sands

Both the Mackenzie (1074) and Changjiang River (1039) samples contain mostly subrounded grains which show a similar range of mechanically induced surface markings dominated by the V-marks and impact pits. The Changjiang River Channel sample (1039) is characterised by the lack of diagenetic features. The Mackenzie River sample contains significantly larger number of grains with conchoidal fractures and diagenetic feature (Fig. 3(4)) which are almost lacking in the Changjiang River sample.

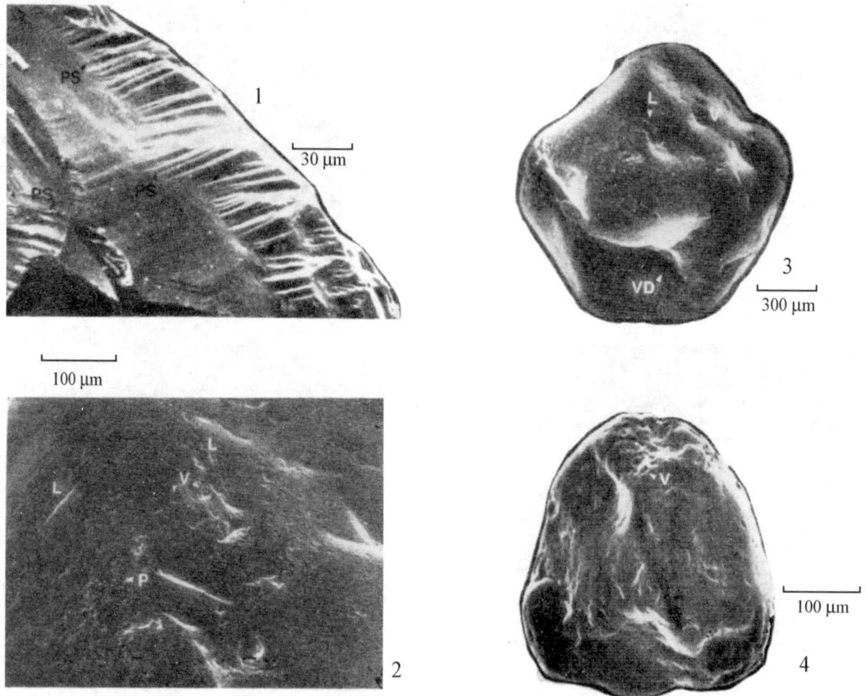

Fig. 4 1 – Parallel steps (ps), A-ja glacier, Tibet, China; 2 – Scratched lines (L), Sable Island, Nova Scotia. (sample 1077); 3 – Mechanical V-depressions (V); 4 – Mechanical V-marks (V), Beaufort Sea continental shelf (sample 1076).

DISCUSSION

To establish a possible set of diagnostic surface markings between sand grains collected from till, beach or river deposits, only the commonly occurring markings (Table 3) are considered. Many markings are common to all the three environments. However, the variety of common markings increases along the probable pathway of grain transport. Only conchoidal fractures and upturned plates were common in glacial till. River samples commonly contained four markings: upturned plates, mechanical V marks, mechanical impact pits and solution holes. In beach sands, as many as eight types of surface marking are common. Thus, the diversity of grain markings of the continental shelf samples could be used to establish the transport and environmental history of the grains.

SCOTIAN SHELF

The Scotian Shelf is about 115 to 140 km wide and on the geomorphologic features it can be subdivided into an inner, middle and outer shelf (King 1970). The inner shelf has a highly irregular surface of glacially sculpted bedrock, gravelly sand or till. The middle shelf consists of a series of depressions, generally in the depth range of 140～180 m, but a few major basins reach a depth of 275 m. These contain Quaternary glacial tills, silt and mud. The outer shelf is dominated by a chain of banks defined by the 100 m isobath and a minimum depth of about 60 m. At the extreme shelf edge the sediment consists mainly of medium to fine-grained and moderate to well-sorted sand. The quartz content of the sand is generally 65% to 75%.

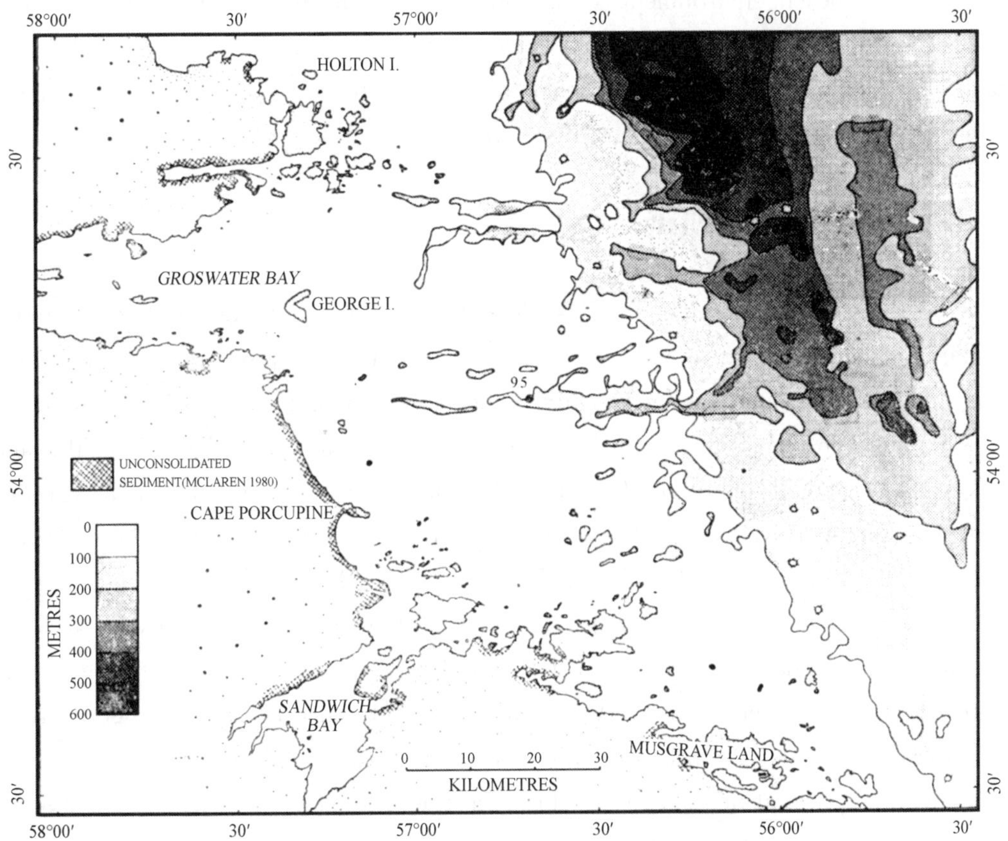

Fig. 5　Bathymetry of southeastern Labrador Shelf showing location of Core 95

Sample 1023 was taken from a saddle between Emerald Bank and Western Bank on the outer continental shelf at a water depth of 200 m (Fig. 2). This is an area of generally strong currents (Hill and Bowen, 1983). The size range from which the grains were taken is 0.5 to 1.0 phi. The surface textures of the sands are shown in Table 2.

Most grain surface of sample 1023 are smooth with rounded edges. Large mechanical V-marks and abundant mechanical impact pits are the dominating features. Most grains have fresh surfaces, suggesting little solution and silica deposition. Only a few grains contain dish-shaped concavities, upturned plates, parallel steps and conchoidal fractures that suggest a previous phase of glacial transport.

DISCUSSION

Sample 1023 from the outer Scotian Shelf includes some grains with surface textures suggesting glacial processes, so that the sample most likely originated from a glacial till or a glacio-fluvial deposit. There are many mechanical features similar to those developed on beaches. Such intensity of V-marks and impact pit formation does not occur under simple current transport (Wang et al. , 1982) suggesting the sands were modified in a beach environment. The sample is taken from 85 m below the late Wisconsinan coastline, suggesting either that the textures are relict from a much lower stand of sea level, or more probably, that the sand has been transported from shallower water, perhaps by the powerful currents in the area.

LABRADOR SHELF

The Labrador continental shelf is close to 200 km wide considering the outer boundary to be the 300-m isobath. It can be divided into three physiographic units: the inner, middle and outer shelf. The inner shelf, extending to the 100 m isobath, is extremely uneven due to minor shoals, depressions, and transverse channels (Vilks and Wang, 1981). The distribution of the bottom sediment on the inner shelf is as variable as its topography (Reinson, 1979). Sand is present in the flat areas between the bedrock highs, cobbles and boulders are present along the flanks of the highs and in saddles between banks. In many depressions, sand is found as a thin veneer overlying late-glacial clay. The middle shelf consists of a series of basins forming a longitudinal depression known as the Labrador Marginal Channel (Grant, 1972). The depth of the depression is 400 to 700 m, containing clay and silt. Quaternary deposits of the Labrador Shelf are at least 200 m thick in the basins (H. Josenhans, personal communication, 1983) but vary from a few metres to 100 m on the banks. The outer shelf is flatter and consists of several banks, between 100 and 200 m deep. Sand is the major sediment type on the banks, but gravel or gravelly sand most commonly covers the saddles between the banks.

The dominant oceanographic feature on the Labrador Shelf is the Labrabor Current, which is responsible for the transport of ice along the coast of the western North Atlantic as far south as 45°N latitude (Dinsmore, 1972). In March close pack ice covers the whole shelf as far south as the northern Grand Banks. The Labrador Shelf is covered with sea ice during the period from December to July when it is protected from the effect of large winter storm waves.

In addition to the extensive annual sea ice, the environment of the Labrabor Shelf is modified by the presence of icebergs. Extreme concentrations of icebergs (20 icebergs per 12 000 km^2) occur in spring (February-April) along the marginal channel of the southeastern Labrador Shelf (Gustajtis, 1979). Bottom sediment and the sea floor are also modified by the grounding of icebergs (van der Linden et al., 1976) and by ice-rafted sediment being deposited from the ice.

In summary, the surface physiography of the Labrador Shelf is typical of high latitude glacial shelves with the marginal channel as the most characteristic feature. Most of the sediment was deposited as glacial drift during major episodes of glaciation. During post-glacial time sediment reworking has been the major process: finer sediments have been redeposited in the marginal channel, leaving lag deposits on banks and on the inner shelf. New sediment is primarily delivered via ice rafting.

The study of grain surface textures at the various levels of core 95 reflect the changing environment towards higher energy levels in the recent times as deduced from microfossil and textural analysis of the sediments (Vilks, 1980). Although all grains contain a considerable number of glacially induced features (Fig. 3(3)), there is a slight difference in degree of subsequent modification of grain surfaces in samples downcore. Fewer mechanical V-marks and impact pits (Fig. 3 (1)) are found in sample 4 at the bottom of the core, where diagenetic features (Fig 3(2)) predominate. The sediments at the bottom of the core are silty clays (Vilks and Wang, 1981) containing very few, but well preserved, microfossils. This preservation is most likely due to rapid sedimentation rates, and the low numbers of mechanically-induced markings may likewise be due to rapid burial removing grains from environments in which they can be reworked.

BEAUFORT SEA

The sedimentary environment on both the Beaufort Sea and East China Sea continental shelves is influenced by the effluent of large rivers. The Beaufort Sea is a southward extension of the Arctic Ocean and is under the influence of the polar ice pack. The sea ice normally retreats from the nearshore zone in August and returns in October. During the summer the margin of the polar ice may retreat as far as 300 km off the Canadian coast, but persistent northwesterly winds may drive the polar pack shoreward (Vilks et al, 1979).

Three major physiographic features dominate the area (Fig. 6):

(1) The moderately wide and gently sloping continental shelf, which extends approximately 100 km off the Yukon coast to a depth of 100 m;

(2) The continental slope, which falls steeply at the 100-m isobath to a depth of 2 000 m in the Canadian Basin approximately 500 km offshore; and

(3) The Mackenzie Canyon, which transects the continental shelf and upper slope

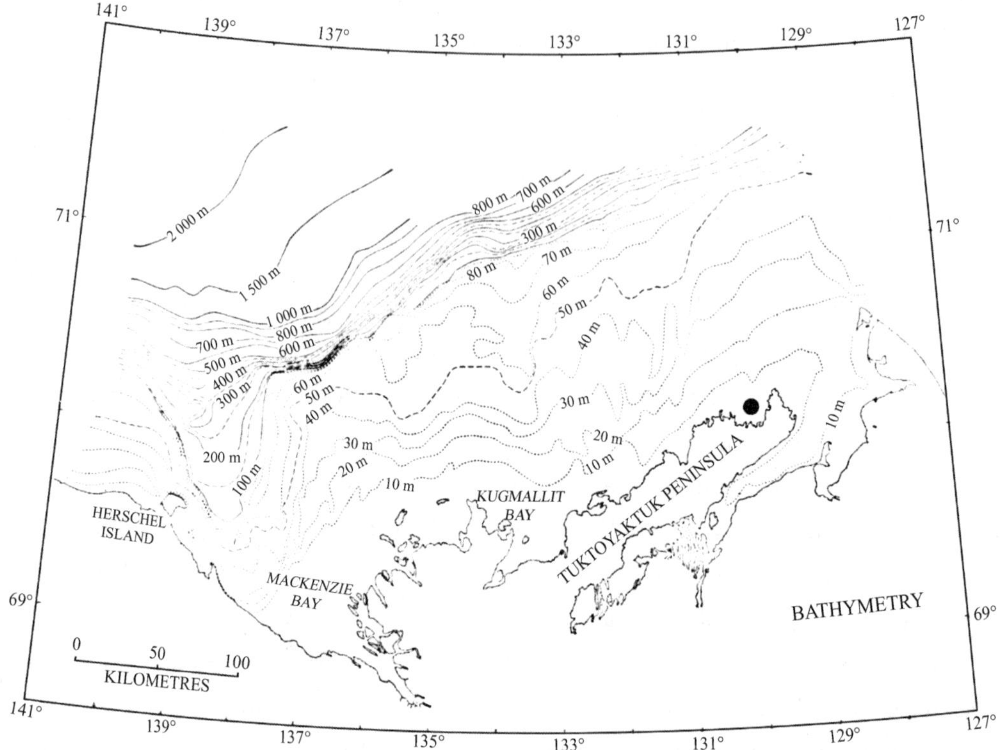

Fig. 6 Bathymetry of beaufort sea showing location of samples.

in a pronounced V-shaped pattern, with the headward portion lying immediately adjacent to the Mackenzie delta. From the mouth of the delta, the Mackenzie Canyon extends about 120 km along a northwest axis to a depth of some 500 m, and thence to the upper slope of the Canada Basin.

The surface sediment of the continental shelf consists mainly of silt and clay. Sand and gravel occur west of Herschel Island, along the coast, and in isolated areas across the eastern section of the Beaufort shelf. Silt dominates the inshore zone from the Mackenzie delta out to about the 10 m isobaths.

Fine sediment from the river is carried from the Mackenzie delta along the coast towards the east and along the outer shelf back towards the west in an anticlockwise transport path. Thus the main region of deposition is the Mackenzie Canyon and its vicinity. Here the Holocene rate of sedimentation is more than 100 cm/1 000 years, whereas on the continental shelf to the east the rate is only 3 to 30 cm/1 000 years.

The SEM analysis of surface textures was carried out on grains taken from the Mackenzie River (1074) and the nearshore continental shelf (1076). The pattern of surface textures reflects the multiple source of sand on the Beaufort shelf. Subangular to subrounded grains with dish-shaped concavities, mechanical impact pits, and V marks (Fig. 4 (4)) dominate the surface texture of many grains: these are believed to have recently arrived from the river. A few rounded grains show abundant V marks and curved grooves: these are the characteristics of beach sand and represent the fraction of grains that have been derived from beaches. The surfaces of some grains are dominated by diagenetic features, such as solution holes and silica fragments superimposed on the mechanically imposed markings: these have probably resided on the continental shelf or in the delta for a considerable time. The abundance of sand grains on the Beaufort Shelf with surface markings similar to those on sands from the Mackenzie River suggests that the Mackenzie River is the major source of the sandy material on the Beaufort shelf.

EAST CHINA SEA

Like the Beaufort shelf, the continental shelf of the East China Sea also receives large amounts of fluvial deposits, but it is not exposed to a polar climate and annual sea ice. The continental shelf can be divided into two parts:

(1) The inner shelf which is a great Pleistocene delta-alluvial plain with several old deltas that can be traced from the coastal area out to a water depth of about 60 m.

(2) The outer shelf, which was once a late Pleistocene coastal plain where several old strandlines can be recognized at a depth of around 100 m, 120 m, and 150 and 160 m.

During the eustatic lowering of sea level in glacial times the shelf was a broad alluvial plain, accumulating sediment. During the post-glacial transgression the coastal plain submerged to become a prograded continental shelf. Subaerial landforms, such as river channels, barrier bars, lagoon and beach ridges, on the continental shelf provide evidence for subaerial exposure. At the present modern river processes, monsoon waves, and currents affect the inner part of the shelf, and the Kuroshio current influences the outer part as it passes by the edge of the shelf from south to north.

The East China Sea sample was taken from the outer shelf at a water depth of 131 m. It contains a significant number of grains with conchoidal fractures, dish-shaped concavities and mechanical V-marks superimposed by solution holes. The range of surface markings of the East China Sea shelf grains is not significantly different from the Canadian continental shelves (Table 2) reflecting both river and beach sources. Apparently, the higher latitude climate and the relatively recent glaciation in Canada is not reflected by a distinct grain abrasion or diagenetic processes.

CONCLUSIONS

The analysis of surface textures of quartz sand is complicated because of the large number of markings that can occur on different grains. The overlap of markings on grains collected from widely different environments is most likely due to two reasons

(1) different environments may impose similar markings, and (2) previous markings from earlier sedimentary environments have not been completely obliterated. By considering only the major markings, those that occur on more than 50% of the grains in a sample, the predominant processes can be determined and minor processes may be eliminated (Table 3).

According to the major surface markings, the Beaufort Sea and East China Sea shelf samples are similar in terms of grain source (beach and river). Both shelves have dish-shaped concavities as major markings, suggesting similar grain reworking/transport mechanisms on the sea floor of the shelf.

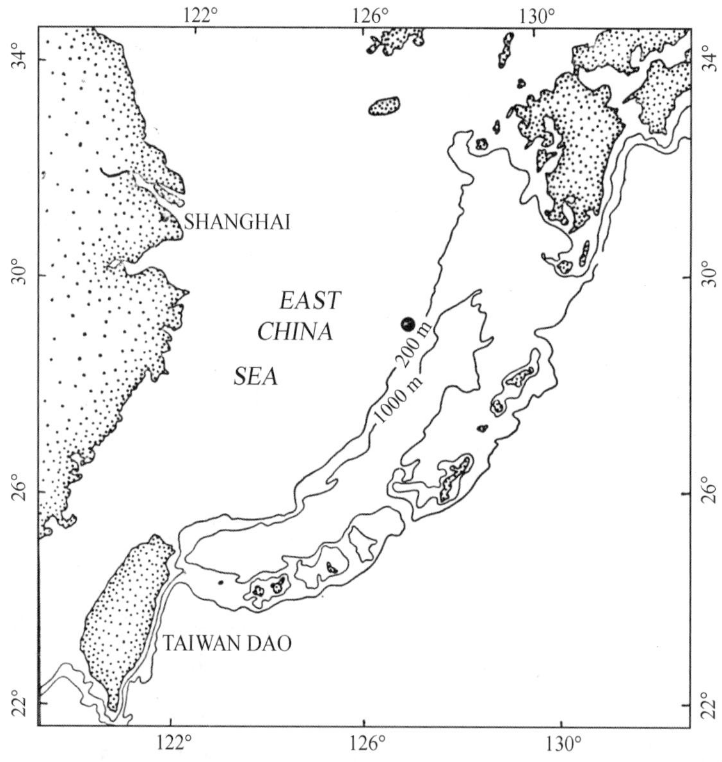

Fig. 7 Bathymetry of the East China Sea showing location of samples

The sediment-starved Scotian and Labrador shelves are different in terms of major surface markings of the sand grains. The major markings on the Labrador shelf grains are solution holes and silica fragments of diagenetic origin suggesting a relatively long residence time in the marine environment. They may reflect the fact that the Labrador shelf grains were collected from downcore subsamples. The Labrador shelf grains also contain conchoidal fractures as major markings, demonstrating a closer relationship with glacially-derived sediments in comparison with the other shelves studied here.

The sediment sample from the outer Scotian shelf shows a strong beach influence (Table 3). The sample contains a surprisingly low variety of other types of surface markings. The absence of diagenetically-induced markings is unusual for continental shelf and may suggest a constant abrasion in a highly dynamic environment along the outer continental shelf.

The increase in the range of markings on the continental shelf grains reflects the complicated pathways of grain transport through the littoral zone and through cycles of reworking on the continental shelf before ending up on the outer shelf.

REFERENCES

[1] Dinsmore, R. P. 1972. Ice and its drift into the North Atlantic Ocean. *ICNAF, Special Publication*, (8): 89 - 128.

[2] Grant, A. G. 1972. The continental margin off Labrador and Eastern Newfoundland-morphology and geology. *Canadian Journal of Earth Sciences*, 9: 1394 - 1430.

[3] Gustajtis, A. K. 1979. Iceberg population distribution study in the Labrador Sea. Data Report, C-CORE (Centre for Cold Ocean Resources Engineering), Publication, 79 - 8, 41.

[4] Hill, P. R. & Bowen, A. J. 1983. Modern sediment dynamics at the Shelf-Slope boundary off Nova Scotia, In: The Shelf-Slope Boundary: critical interface in continental margins, editors D. J. Stanley and G. T. Moore, *Society of Economic Paleontologists and Mineralogists, Special Publication*.

[5] Krinsley, D. & Doornkamp, J. 1973. *Atlas of quartz sand surface textures*, Cambridge University Press, New York, N. Y. , 91.

[6] Krinsley, D. H. & McCoy, F. W. 1977. Significance and origin of surface textures on broken sand grain in deep-sea sediment. Sedimentology, 24: 857 - 862.

[7] King, Lewis H. 1970. Surficial geology of the Halifax-Sable Island map area. Department of Energy, Mines and Resources, Ottawa. *Marine Sciences Paper*, 1.

[8] Powers, M. C. 1953. A new roundness scale for sedimentary particles. *J. Sedim. Petrol.* 23: 117 - 119.

[9] Reinson, G. 1979. *Inner Shelf Sediment Dynamics*. Cruise Report, 69 - 108, Bedford Institute of Oceanography, Dartmouth, N. S. , 4 - 7.

[10] Van der Linden, W. J. , Fillon, R. H. & Monahan, D. 1976. Hamilton Bank, Labrador margin, Origin and evolution of a glacial shelf. *Geological Survey of Canada*, Paper 75 - 40, 31.

[11] Vilks, G. 1980. Postglaciation basin sedimentation on Labrador shelf. *Geological Survey of Canada*, Paper 78 - 28, 28.

[12] Vilks, G. & Wang, Y. 1981. Surface textures of quartz sands and sedimentary processes on the southeastern Labrador Shelf. *Geological Survey of Canada*, Paper 81 - 1B, 55 - 61.

[13] Vilks, G. , Wagner, F. J. E. & Pelletier, B. R. 1979. The Holocene marine environment of the Beaufort shelf. *Geological survey of Canada Bulletin*, 303: 43.

[14] Wang, Y. , Piper, D. J. W. & Vilks, G. 1982. Surface textures of turbidite sand grains, Laurentian Fan and Sohm Abyssal Plain. *sedimentology*, 29: 727 - 736.

海岸沙丘成因的讨论*

海岸带主要受海洋气象与水文要素的影响。大陆气象要素及海洋水文要素到达海岸带发生突变，构成了区别于陆地亦区别于海洋的一个气象与水文要素特殊的地带[1][2]。海岸带是海陆交互作用最剧烈的地带，而沙丘海岸是这种交互作用的突出产物。

沙丘海岸的发育条件是：① 海岸带有丰富的沙质沉积物。通常是陆源的，来自河流或来自海底古海岸沉积物。② 干寒或半干燥的气候，有强劲的并方向恒定的海风作用。③ 平坦的海岸坡度，平均坡度 1∶1 000，中等潮差，可形成宽广的海滩，落潮时宽广的海滩上风力不断吹拂海沙向岸堆积成沙丘。

本文的中国和西非两个典型的沙丘海岸实例，说明海滨沙丘是海陆相互作用下形成的，特别是在冰期极地气候条件下，低海面时形成广大的岸外砂质浅滩，极地气旋风暴作用，促进海岸沙丘的形成发育。

一、中国滦河三角洲北部沙丘海岸

河北省东部滦河口北岸到洋河口（39°47′N～39°20′N），在 45 km 长的岸线上，分布着中国规模最大的沙丘海岸。沙丘岸宽度 1～4 km，沙丘系列的主要延伸方向为 NE20°～30°，与海岸线平行。沙丘高度自南向北，由 40 m 逐渐降低到 10 m，沙丘的系列与分布宽度亦由南向北减小（图 1）。沙丘海岸受河流分割，可分为三段。

（1）滦河北岸至七里海，该处沙丘规模最大，沙丘带的宽度在滦河以北的河口区为 4 km，向北平均宽 2 km，至新开河潮流通道口，宽度减小为 1 km。靠海第一例沙丘 NE20°延伸，规模大。沙丘高度在 30 m 以上。最高的沙丘峰高达 42 m。向陆侧，沙丘系列转为 NNW 与 NW 向，为新月形沙丘链，沙丘的向风凸坡朝向 NE24°，背风坡向 SW29°，沙丘高度向陆降低 20 m 以下，新月形沙丘向陆地高度降低，沙丘链之间散布着小型的洼地沼泽（图 2）。沙丘组成物质为石英长石质砂，丘顶为分选良好的细砂，向海坡为中细砂，至坡麓则粗颗粒成分增高（表 1），沙丘中重矿物主要为角闪石、绿簾石、石榴子石、磁铁矿、钛铁矿、锆英石与独居石等，属变质岩火山岩类矿物组合，成分与滦河物质一致，但磨圆度较高。沙丘的形态反映：自沙丘形成至今，主要受 NE 及 ENE 风的影响。目前海岸受蚀形成冲刷坎，潮间带海滩宽 80 m，自岸线至 5 m 深处的水下岸坡坡度为 1∶180～1∶300，平均约 1∶250，陡峻的岸坡与大沙丘相邻，表现为沙丘临海的不对称现象。

* 王颖，朱大奎：《中国沙漠》，1987 年第 7 卷第 3 期，第 29 - 40 页。

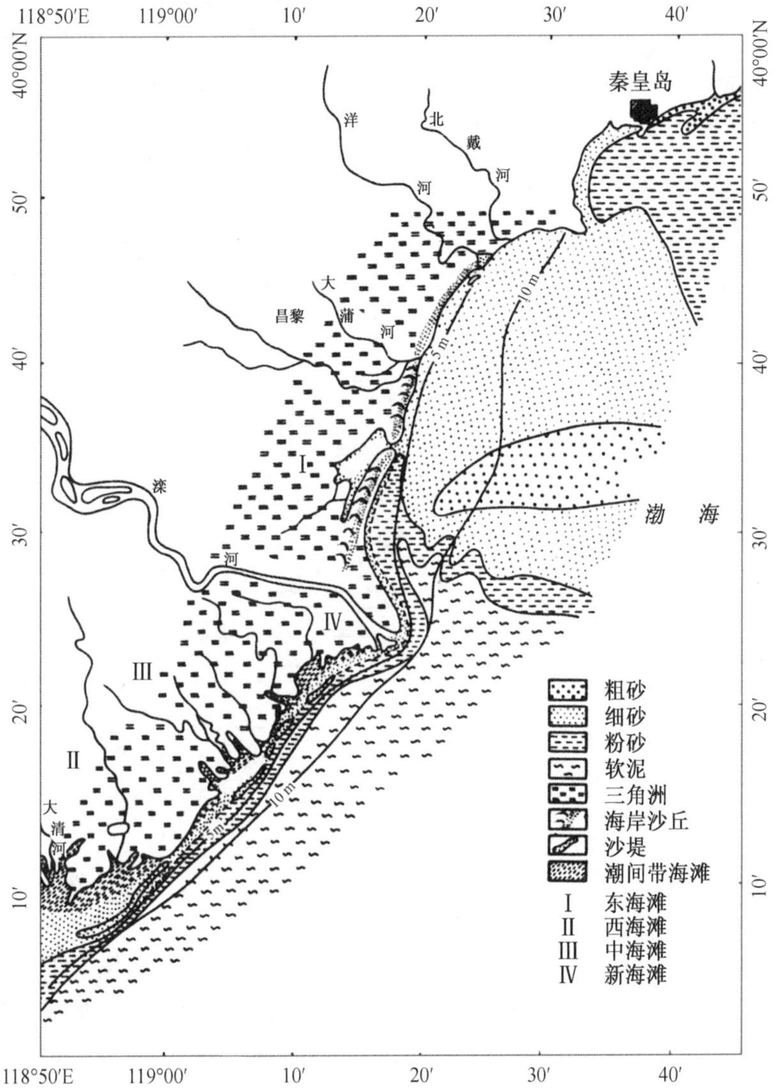

图 1 滦河三角洲海岸图

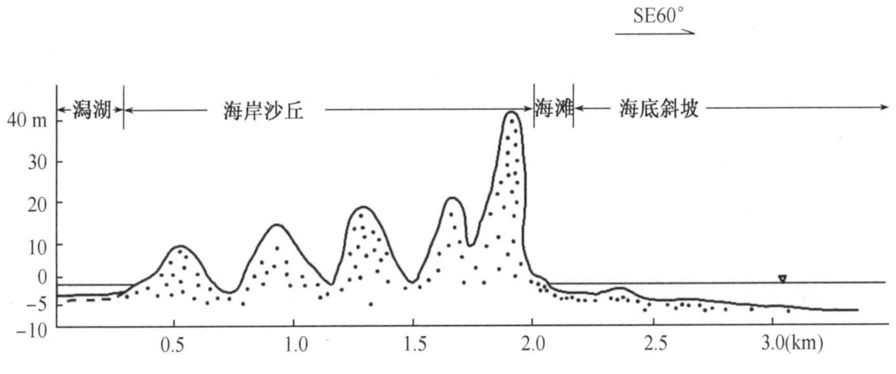

图 2 七里海海岸沙丘地形剖面

表 1　滦河北岸—七里海海岸沙丘颗粒组成

样号	位　置	0.5～1 mm(%) 粗　砂	0.25～0.5 mm(%) 中　砂	0.1～0.25 mm(%) 细　砂	0.1～0.01 mm(%) 粗粉砂	So	Sk	d_{50}
21	沙丘顶		9.9	88.8	1.3	1.680	1.025	0.16
22	向海坡	0.6	52.4	46.4	0.6	2.24	0.903	0.26
23	向海坡麓	0.5	62.0	37.3	0.2	2.00	0.389	0.30
S1	第一列沙丘		14.8	84.8	0.4	1.76	1.011	0.165
S2	第二列沙丘	0.1	50.0	49.8	0.1	2.118	0.979	0.25
C1	北　岸	0.5	32.2	66.0	1.3	2.074	1.05	0.19
C2	北　岸	2	51.4	47.3	0.25	2.059	0.95	0.25
C6	南海向海坡		37.7	62.1	0.20	2.22	1.033	0.205
C5	沙丘顶部	0.1	31.4	68.2	0.30	2.074	1.047	0.19
C4	沙丘背侧	0.1	1.4	49.8	47.9	2.88	0.896	0.25
C8	洋河南岸	2.4	59.5	37.7	0.4	1.947	0.836	0.29
C9	洋河南岸	3.5	58.1	37.3	1.1	1.846	0.78	0.30
C11	洋河北岸沙丘坡麓	4.0	52.4	42.0	1.6	1.84	0.93	0.26
C12	洋河北岸沙丘坡麓	6.1	19.5	73.2	1.2	2.0	1.021	0.175
C13	洋河北岸沙丘顶	6.1	4.3	50.6	0.3	2.332	0.992	0.35

(2) 大蒲河口为三角形河口湾，湾口向东而河道略向 SE20°偏转，河口两岸为沙丘夹峙，河口外为水下沙坝两列，与岸平行分布于水深 1～3 m 处。现代海岸水下斜坡坡度为 1∶300，潮间带宽 150 m，北岸为中细砂质组成，具有三列滩脊。南岸为细砂与粗粉砂质的单斜坡，坡度 6°～8°。海滩砂分选好(S_O=2.2)，高潮线处海滩受蚀形成陡坎(25 cm)。高潮以上为高海滩，中细砂质分选好(S_O=2)，亦有三列滩脊，高海滩已长草固定。高海滩与现代海滩系海岸沙丘前坡改造而成。高海滩背侧之沙丘带宽 2 km。第一列沙丘沿岸平行连绵分布，高度在 20 m 以上，最高 24 m。中砂与细砂组成，圆度高，内侧沙丘高度减低，走向为 NW30°，与岸呈羽状斜交，形成纵横相交的沙丘链(图 3)。

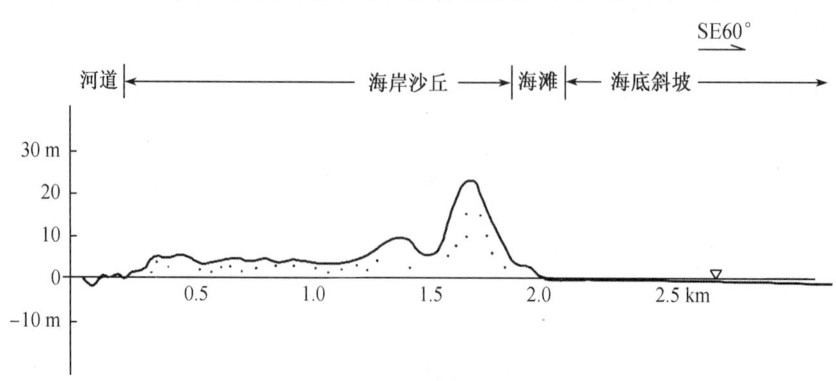

图 3　大蒲河海岸沙丘地形剖面

(3) 洋河口附近岸线，走向 NE30°，河口流向 SE20°，口外为沙坝掩护。南岸水下岸坡为 1∶360，北岸为 1∶440。现代海滩为中细砂组成，磨圆度与分选性好($S_O=1.88$)，潮间带宽度 100 m，高海滩宽 60 m。大蒲河口至此，海岸沙丘高度明显降低，宽度减小。至洋河南岸沙丘带宽 1 500 m，高度 10 m，已零星植树，与前二段海岸的裸露沙丘有明显不同。洋河以北的沙丘已造林绿化固定，沙丘带宽 9.8 km，高度在 10 m 以下，河口南部受蚀，岸外发育雏形堡岛。

因此，整个沙丘海岸是连续分布的，外侧因海滩冲蚀，使沙丘如同山岗状紧贴现代岸线，内侧新月形沙丘链向内陆高度减低。在七里海与大蒲河之间，沙丘阻挡小河入海，在沙丘间形成沼泽湿地。海岸沙丘组成物质细砂与中细砂，成分与滦河泥沙一致，但分选度与磨圆度皆高，石英砂表面具有 V 痕及撞击点，并有硅藻附着[3]，这些反映了砂来源于近岸海底与海滩。

研究区属滦河三角洲平原，海岸带泥沙主要来自滦河。滦河发源于变质岩、花岗岩与火山岩组成的燕山山地，系多沙河流，河长 870 km，年径流量 36.7×10^8 m^3，年平均输沙量 24.08×10^6 t。山地为砾石粗砂河床，比降 1%～2%，平原为细砂质河床，比降 0.25%，善冲善淤，摆移不定，其北支达洋河，南到大清河。因此这个范围内，海岸泥沙主要来自滦河。其他河流影响小，如洋河，年输沙量最大 4.79 万 t，最少仅 0.28 万 t。

现代滦河入海泥沙，以细颗粒为主，主要向口外即东南向扩散，细砂沉积于河口及口外水深 5.5 m 范围以内。粗粉砂分布在水深 5.5～9.7 m，粉砂黏土可分布到水深 13 m 及以远(图 1)，其次，有部分滦河入海泥沙沿岸向西南运移约 30 km，达到湖林口一带[4]。滦河泥沙向北漂移的数量少，最远到沙丘带南竭距河口 4.8 km 处。滦河沿岸输沙在 SE 向(河口南)及 NE 向(河口北)波浪作用下，破浪带沉积为砂，其上发育了海岸沙坝、坝后潟湖与潮滩，破浪带外侧为粉砂质淤泥，−5 m 水深以外，至海岸带下界沉积为黏土淤泥(图 4)。现代海岸水下岸坡平均比降为 1∶400，至 10 m 深坡度为 1∶500。在有现代滦河泥沙影响的南部海岸，沙丘不发育，滦河三角洲平原受潮侵而发育了淤泥质浅滩，滦河口南北海岸形态迥异。

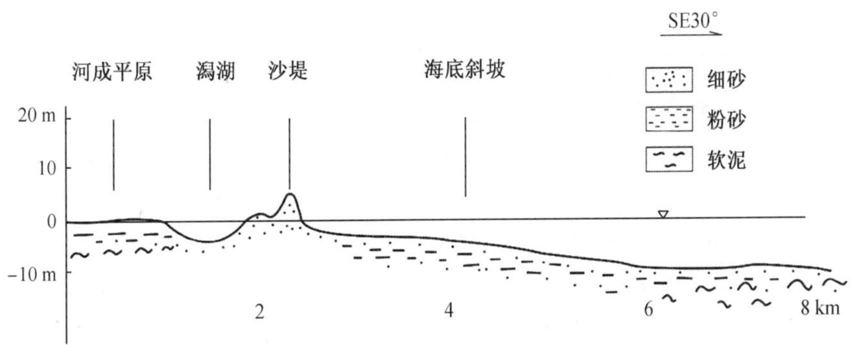

图 4　滦河三角洲岸外沉积物分布

北部沙丘岸物质成分与滦河一致，但泥沙粒径较粗，为中细砂。由于现代滦河泥沙主要向南部，达不到北部沙丘岸，该处海岸因供沙不足而发育着冲刷陡坎。但是在沙丘岸外广阔平坦的海底上却分布着大面积的砂质沉积。其范围自岸边至 14 m 水深处。在 20 km 宽的海底上，是细砂与中砂(图 1)，细砂黄褐色 0.1～0.25 mm 粒级占 60%～90%，

0.25～0.5 mm 粒级占 10%～30%粉砂少量，一般很少含黏粒，分选好，S_O－1.3～1.8，少数 S_O－2.4。中砂为黄褐色 0.25～0.5 mm 粒级占 60%～70%，0.1～0.25 mm 级占 10%～38%，没有黏土颗粒，分选性 1.5～1.8，个别达 2.2，主要分布 10 m 以深。

本区海岸，水深大于 10 m 处，已在波浪基面以深，通常为泥质沉积，而此处大面积出露纯净砂质沉积，是与目前动力条件不适应的，属残留砂，由于目前滦河泥沙主要向南扩散，未在该砂质分布区沉积，故砂质出露海底，按其粒度组成，海底残留砂与海岸沙丘相似(图 5)，是低海面时的宽广的海岸浅滩与海滩带，是海岸沙丘形成时的供沙源地，如按 10 m 等深线为界(它包括了大片的纯净砂质堆积区)，沙丘海岸各段的平均坡降是 1∶1 000。这个坡度既有充分发育的激浪带，亦可在潮差不大的情况下形成宽广的落潮浅滩，使沙粒在海风作用下向岸输送，积聚为海岸沙丘，因此，岸外平坦的砂质海底，可以满足形成海岸大沙丘的物质与地形条件。

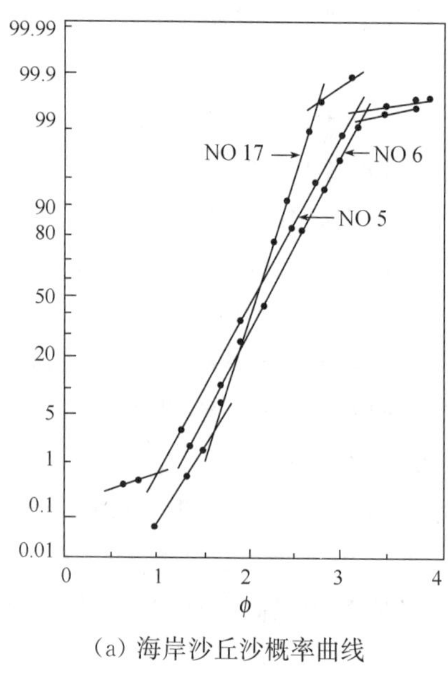

(a) 海岸沙丘沙概率曲线

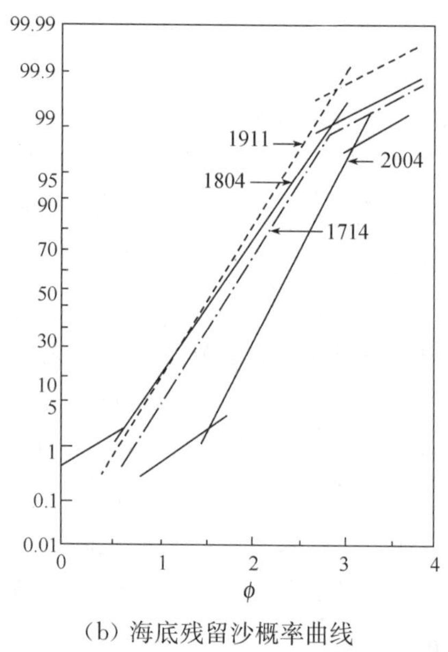

(b) 海底残留沙概率曲线

图 5

滦河北岸与南岸，风和风浪的条件没有明显的变化，北岸的沙丘，南岸的沙坝表明，海岸带泥沙均受 SE 向风浪作用，以横向运动为主。滦河以北受 NE 向强风的影响较大。现代海岸地貌特征表明：① 现代滦河入海泥沙颗粒细；② 目前动力作用下主要形成沙坝而未发育高大的沙丘。这反映着气候条件变化，招致在温度、湿度以及风力强度或者强风频率的变化，研究区位于渤海北侧，面向 SE 方开阔海面。现代是温带海洋性气候，季风明显，冬季盛行东北风及北风，平均风速 7.0 m/s，大于 6 级的风(11 m/s)主要为 NE 及 NEN 向的，其频率达 2.3%，最大风速可达 20 m/s，亦为 NEN 向。冬季低温(一月平均气温－10 ℃)，平均气温湿度小(平均相对湿度 65%～70%)。夏季，盛行偏南风，对研究区有影响的 SE 向及 S 向风风速达 6 m/s 的频率分别为 8.9%与 5.7%，平均风速 4～5 m/s，夏季温暖(8 月平均气温 26 ℃～28 ℃)，湿润(相对湿度大于 85%)，春、秋季风向转变，春季以南风出现次数增多，秋季以北风出现次数增多，平均风速 5～6 m/s。

波浪与风况一致，NE 与 ENE 方向强浪影响下泥沙运动最活跃的范围平均在 4.0 m

水深，最大可能范围 5.5m。SE 向波浪影响下泥沙最活跃的范围平均在水深 3.0 m，最大达 4.1 m①。其他方向影响小，滦河以北潮为不正规日潮，涨潮流 NE 向，落潮流 SW。滦河口以南为不正规半日潮，涨落潮流方向与北部相反。本区潮差约 1.5 m。

按现代的风与波浪，北部海岸的下界为 5～6 m，南部为 4 m，南部海岸坡度 1∶400～1∶500，发育了小型海岸沙坝，北部坡降为 1∶250～1∶350，现代海岸坡度陡，遭受轻微冲刷。狭窄的海滩不宜发育高大的海岸沙丘(图 6)，高大的海岸沙丘紧临岸边，是与现代海岸过程不相称的(图 2 和图 3)。

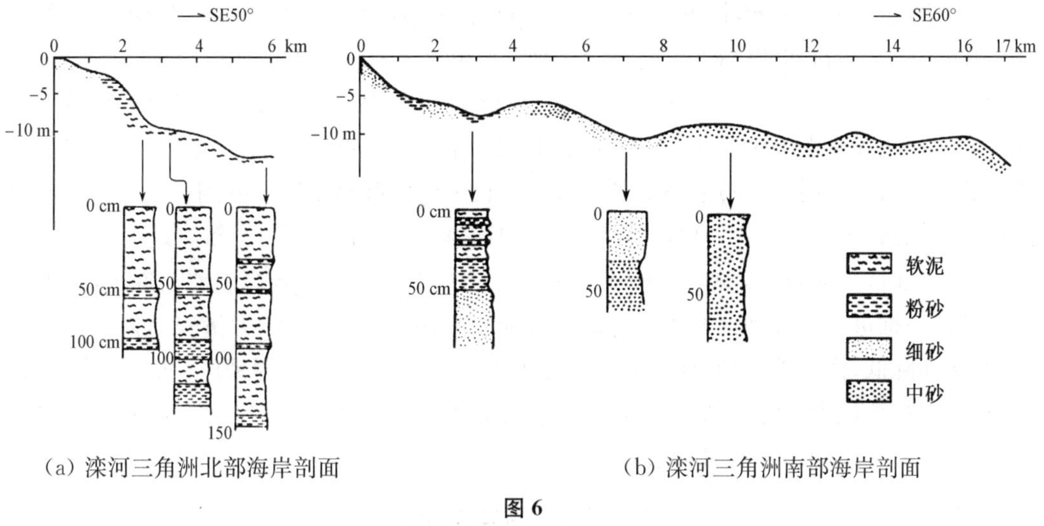

(a) 滦河三角洲北部海岸剖面　　(b) 滦河三角洲南部海岸剖面

图 6

滦河三角洲平原是由四个三角洲瓣所组成(图 1)[3][5]，滦河北岸属于三角亚工瓣，其沉积层系海陆过渡相(图 7)[3]。

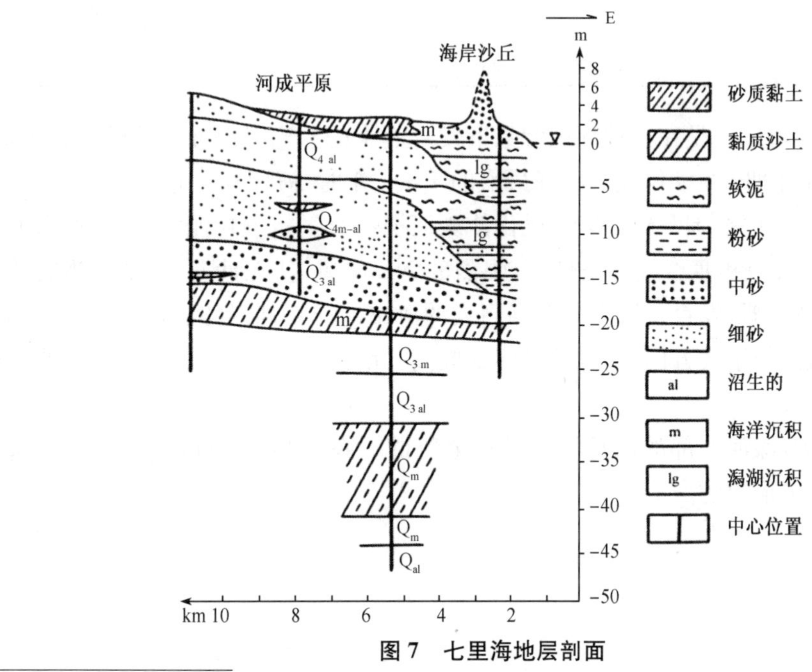

图 7　七里海地层剖面

① 南京大学海洋地貌研究室、交通部一航院. 1978 冀东海岸与深水港选址调查报告。

（1）基底埋深 28～32 m，中砂夹砾（10%），砾径 3～5 cm，深灰色系滦河河床沉积。

（2）海相层，灰色、灰绿色砂黏土和粉砂，含有机质、钙结核及少量有孔虫。

（3）滦河冲积层，顶面埋深 16～18.5 m，厚 3～5 m，灰白色中粗砂夹圆砾石，分选好，含陆相介形虫化石，属晚更新世。

（4）海相层，灰色细砂，含有机质及有孔虫（Ammonia beccarii var，Miliammina fasca (Brady)，Trochammina inflata (Moutagu)，Pseudononiou minutum S. Y. zharg Elphldium magellaulcum Heron Alleu D Earlaud）。

（5）埋深 2.6～7.7 m，棕黄色中砂夹小砾石，古河道洼地深 4 m 泥炭。^{14}C 测定 8 025±105 aB. P.，为全新世初期沉积，也就是沙丘海岸基底为 8 000a，当时海面在今－13 m 处。

据冰后期的气候波动与海面升降[6][7]，全新世时，中国东部有四个寒冷时期（表 2），几次寒冷期之间有温度上升 1～3 ℃温暖期，尤以距今 7.5 ka～7.0 ka 及 4.0 ka～3.5 ka 为显著。在新冰期时，北京地区以云杉、冷杉为代表的针叶林都曾下降到海拔 500 m 的丘陵地带，气温较今低 5 ℃～6 ℃，寒冷气候带向南移 5～7 个纬度[8]。新冰期一、二、三期的冷锋出现时，海面曾下降幅度为 2～4 m[6]。这时，中国西部地区高山冰川推进，冻土发育以及沙漠推进。东部地区则由于寒冻天气而机械风化作用加强，大量碎屑物质入海或入湖，海面降低，而湖泊充填，泥炭发育。北方冷空气爆发，直入平原地区，加之气旋活跃，冷空气南移，故风力增强。新冰期时高空急流将黄土搬到 28°～30°N[8]。亚洲大陆北方冷空气频繁，干燥度亦随低温且增加，使泥炭停止发育。

表 2　北半球全新世气候波动[6]

寒冷期名称	新冰期第一期	新冰期第二期	新冰期第三期	新冰期第四期
持续时间(距今)	8 200～7 000	5 800～4 900	3 300～2 400	(小冰期) 450～30
冷峰时间(距今)	7 800	5 300	2 800	200

滦河北部沙丘海岸正是在这种环境背景下发育起来的。当时海面普遍较今日为低，均在－10 m 处，新冰期 8 000 年或当冷峰最盛的 7 800 年时，海面在－13 m 处（图 8），这与沙丘海岸沙源分布的范围一致。寒冻风化使滦河携带大量粗颗粒泥沙入海，低海面时平缓的岸坡（1∶1 000）和寒冷气候条件下，强烈的 NE 向风，使泥沙向岸，堆在当时的海岸——滦河三角洲工瓣，发育了沿岸分布的高大的海岸沙丘系列。随后，由于海面上升，虽风力减弱，但风向与岸线方向未变，几经演变成高大的海岸沙丘，今日的是残留改造的沙丘岸。

因此可以认为，滦河北部的沙丘海岸是在全新世新寒冷期低海面时，干冷气候条件下，海陆相互作用下形成的。

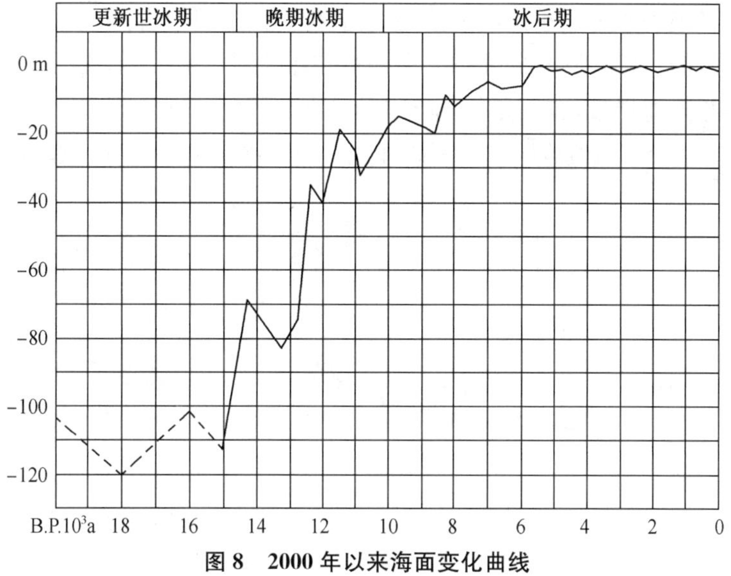

图 8　2000 年以来海面变化曲线

二、西非毛里塔尼亚海岸沙丘[9,10]

毛里塔尼亚处撒哈拉沙漠的西南端(图 9)，盛行东北信风。海岸带的强风在北北东—东之间，海域来的强风为西—北西风。

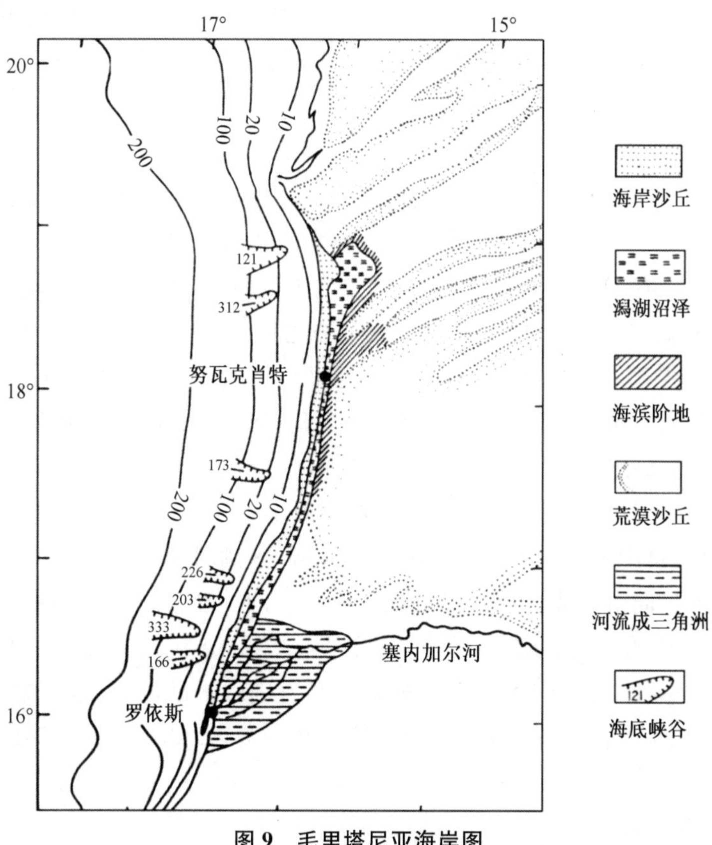

图 9　毛里塔尼亚海岸图

从撒哈拉沙漠中心有北东走向大沙岗，一直伸到海边，长数百千米，这些古沙丘岗地，大多已固定，表面有钙质硬壳，硬壳被人类开拓破坏处则有流沙活动。

沿毛里塔尼亚海岸带，有绵延长达 315 km 的海岸沙丘(16°N～19°N)。该处的海岸剖面(图 10)，自陆向海为海成阶梯—潟湖洼地—海滨沙丘带—海滩—水下岸坡。

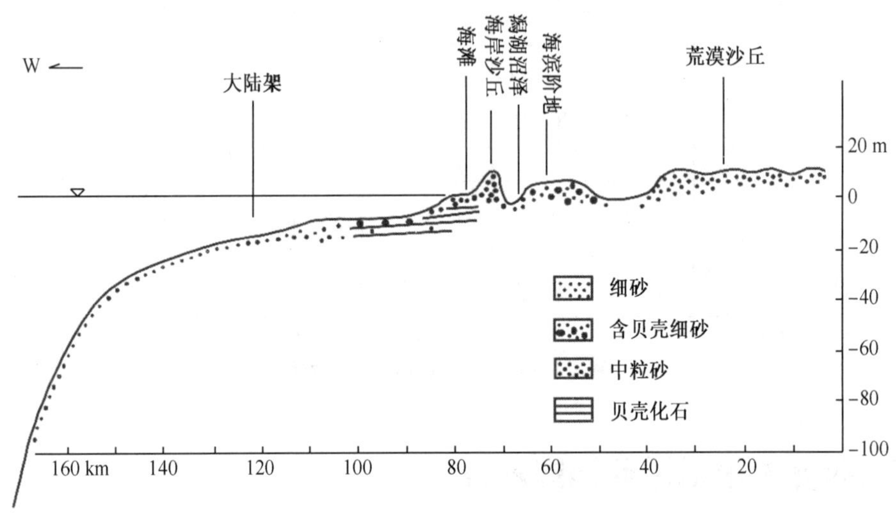

图 10　毛里塔尼亚海岸带剖面

海成堆积阶地，高程 3～4 m，是贝壳砂层；贝壳种属多，是砂质浅水岸的生物。在激浪作用下贝壳及砂堆积的高海滩、堆积阶地最大宽度达 80 km，^{14}C 测定年代为 5 755±120 aB. P. *，它把海岸带与撒哈拉沙漠区的大沙岗地带分隔开，在地貌与沉积组成上均为一明显的界线。

潟湖洼地带宽 2～3 km，表面高程±0 m，一般已干涸，无积水，或短期积水(风暴潮时)。表面有盐壳，沉积为砂、黏土，或白色石膏与粉砂互层，潟湖洼地表层沉积^{14}C 定年为 5 910±115 aB. P. 。北部潟湖沼泽(Timirst 盆沼)^{14}C 测定年代为 5 745±100 aB. P. *。

整个海岸带均有海岸沙丘分布，沙丘链走向平行海岸(南北向)，平均高度 30 m。努瓦克肖特附近最宽(400 m)，向北至阿斯玛，蒂尤利特为 200 m 宽。努瓦克肖特以南 30 km，海岸沙丘带受侵蚀宽仅 20 m，高度不足 2 m，大浪时海水可越过沙堆进入潟湖凹地。至塞内加尔河三角洲，海岸沙丘又增宽至 3.5 km，高 10 m，沙丘上灌木丛茂密。

海岸沙丘由中砂组成(0.35 mm)，砂质纯净，分选性好，它们是向岸风将海滩砂向陆吹刮堆积而成。该处虽位于撒哈拉沙漠外围，但海岸沙丘不是内陆沙漠沙丘的继续，而是浅水区波浪作用，将岸外海底泥沙向岸堆积为海滩，落潮时海滩砂再被海风吹扬，形成海岸沙丘。沙丘的走向由形成沙丘的主要风向决定。内陆沙丘形成于 NE 向风，海岸沙丘是由 NW 向风形成的。其形成时代按砂中贝壳的^{14}C 测定为 5 270±170 aB. P. 。

毛里塔尼亚岸外海底至水深 20 m，普遍分布着细砂中砂，该处水下岸坡的平均坡度 1∶1 000，砂粒表面纯净，无杂质污染，分选性较好(S_O≈1.63～1.72)，磨圆度好(49.2～

* 南京大学大地海洋科学系同位素实验室测定。

52.92)。根据石英砂表面结构分析,海底砂粒中 59.2%～89.9%具风成砂的特征。沙漠沙丘的石英砂表面特征具风成砂特征的占 69.36%,海底砂起源与撒哈拉沙漠有关,但它堆积于海底后,却成为海岸带新的沙源。

毛里塔尼亚海岸位于撒哈拉沙漠边缘,处东北信风带,从大范围的区域特点看,内陆沙漠沙受信风作用向西,似乎可到达海滨,但通过野外工作,可用下列两点说明内陆风吹沙不能构成海岸沙丘。

(1) 砂的粒度组成。内陆沙丘为细砂粒级,而海岸沙丘为中砂(表 3),据野外测验,砂粒在风作用下主要在地表 10 cm 高度内,作跃移或蠕动,占 91.1%～93.3%,而随气流悬浮运动的仅占 0.4%,这些是粒级小于 0.074 mm 的粉尘。海滨区内陆沙丘表面已结有薄壳,大多有植被,沙岗均已固定,只城市四周及公路两侧,经人工破坏处才起风蚀流沙,一般无直接由地表作跃移蠕动的流沙进入海岸带。悬浮空中的粉尘可由内陆向海中输送,因粒径较小,入海后将随海流向外海扩散消失。沿岸 315 km 长的距离内均有潟湖凹地。潟湖中沉积黏土、粉砂、石膏,这也说明没有内陆流沙层越过潟湖进入海岸带。

表 3　海岸沙丘与内陆沙丘组成

类　型	粒径(mm)	颜　色	植　被	沙丘走向
海岸沙丘	中砂 0.3～0.4	灰　白	覆盖好	N—S
内陆沙丘	细砂 0.11～0.16	浅　棕	差	NE—SW

(2) 海岸带主要受海洋气象要素的影响。据努瓦克肖特栈桥码头 1977—1979 年测风资料分析,大于起动流沙的风速为 5.5 m/s,其向岸风频率为 35.35%,离岸风频率为 24.87%。因此,海岸带主要受海风作用,海滩砂向陆运移堆移的。沙粒组成与海滩砂相联系,而与内陆无关。

与中国滦河三角洲海岸沙丘相似,毛里塔尼亚沿岸目前主要亦受海蚀,海滩狭窄,沿岸高大的海岸沙丘与侵蚀海岸极不适应。沿岸沙丘亦为残留形态,形成于海岸堆积阶地之后(5 755±120)aB.P.,即沙丘年龄为(5 270±170)aB.P.,当时为新冰期第二期的冷峰年代——5 300a,海面降低岸滩展宽,沿岸具备了风沙运动发育沙丘的条件。

上述两个沙丘海岸的实例说明,高大的海岸沙丘形成在岸外有宽广的浅滩、丰富沙源的区域。目前一些海岸,海滩狭窄,无显著的泥沙供应,岸线甚至在受侵蚀,但却分布着高大的海岸沙丘,这是与环境不相适应的。这些海岸沙丘是新冰期低海面时期,岸外有大片浅滩暴露沙源充足,干冷强劲的海风作用下形成的残留地形。

参考文献

[1] 张正元. 1982. 江苏海岸带气候特征分析. 见:江苏海岸带与海涂资源学术讨论会文集.

[2] C. A. M. King. 1972. Beaches and Coasts. Edward Arnold, 165 - 190, 314 - 364.

[3] 高善明,李元芳,安凤桐,李凤新. 1980. 滦河三角洲滨岸沙体的形成和海岸线变迁. 海洋学报,2(4):102 - 114.

[4] 王颖. 1964. 渤海湾北部海岸动力地貌. 海洋文集,(3):25 - 35.

[5] Wang Ying, Ren Mei-e, Zhu Da-Kuei. 1986. Sediment Supply to the Continental Shelf by the Major Rivers of China. *Journal of Geological Society, London*, 143: 935 - 944.

[6] 杨怀仁,谢志仁. 1984. 中国东部近 20 000 年来的气候波动与海面升降运动. 海洋与湖沼,15(1): 1 - 13.

[7] G. H. Denton and W. Karlen. 1973. Holocene Climatic Variation—their pattern and possible Cause. *Quarternary Research*, 3: 155 - 205.

[8] 于革. 1985. 中国东部全新世气候变化研究. 南京大学硕士学位论文.

[9] 朱大奎,王玉定. 1981. 西非毛里塔尼亚滨海区地貌与第四纪地质. 世界地理,91 - 96.

[10] 朱大奎,王玉定. 1983. 西非毛里塔尼亚海岸动力地貌. 南京大学学报(地理学版),(S1):11 - 25.

A Comparative Study of Open Coast Tidal Flats: The Wash (U. K.), Bohai Bay and West Yellow Sea (Mainland China)*

REGIONAL SETTING

Coastal tidal flats are developing in a wide range of climatic environments, ranging from the Canadian Arctic (Frobisher Bay) to the tropical coastlines of Guyana and the Amazon Estuary. In this contribution, three areas have been selected as examples for a comparative analysis; these are the wash on the eastern coastline (located at 53°00′～53°30′N, 0°～0°30′E) of the British Isles, the western Bohai Bay coastline (at about 38°00′～39°00′N, 117°E) of the Bohai Sea and the north Jiangsu coastline (in the area of 32°～35°N, 121°E) along the western part of the Yellow Sea, mainland China.

The main factors which control the development of tidal flats are tidal dynamics, the slope of the tidal flat and the abundance of fine-grained sedimentary material. The width of any particular tidal flat is governed mainly by tidal range; similarly, morphological characteristics and their associated sedimentary structures are controlled by the tidal energy (current velocities). Large tidal ranges are accompanied always by well-developed flats, such as in the Bay of Fundy and the Hangzhou Bay and several estuaries.

Typical tidal flats in China are developed where the tidal range is about 3 m and the velocity of tidal currents is not as large as in the estuaries referred to previously. Such tidal flats are between 6 to 18 km in width and are much more extensive than in the estuarine areas. These major tidal flats of China are developed along the fringing zone of the North China Great Plain, the slope of the coastal zone is very gentle and the gradient is less than 1/1 000. As a result of this setting, wave action is restricted to offshore areas and tidal currents are the dominant dynamic processes on the flats. For comparison, some of the tidal flats in South China along the East China and South China Seas, and in the northern and western parts of the British Isles are not developed along

---

* Ying Wang, M. B. Collins, Dakui Zhu: *Proceedings of International Symposium on the Coastal Zone 1988*, pp. 120－134. Beijing: China Ocean Press, 1990.

plain areas of the coastline; rather they are associated with narrow embayments or other types of sheltered coasts. Fetch within such environments limits the wind/wave interaction; hence, because of this limitation, tidal currents are, once again, the major agents controlling sedimentary processes.

The sediments deposited over the largest part or areas of tidal flats are, in general, fine-grained. Silt is the major constituent of the deposits, ranging from fine to coarse silt. Very fine sands and clays are also present on most tidal flats. Along the Atlantic Ocean and the ice-covered Arctic Ocean, coarser materials (such as coarse sand, gravel and even boulders) are deposited on the tidal flats, because of the influence of waves and floating ice. All such sediments deposited on tidal flats are carried mainly into the flats by flood tidal currents, bringing material from the adjacent seabed; this infers transport from seaward to landward. The original source of such sediments may be terrigenous, however, with sediments on the Chinese tidal flats being fluvial in origin, having been supplied to the coastline at lowered sea level by the large Yellow River and Changjiang Rivers.

Sediments on tidal flats in the British Isles were also originally terrigenous, but were laid down offshore as glacial deposits. Such sediments were from land originally, being deposited during lower sea level stands during the last glaciation. This material is now being, and has been, resuspended by wave action and transported landwards by tidal currents during the Holocene sea level rise. Even with a supplementary local sediment supply from contemporary coastal erosion and small river systems, the main source of sediments is to seaward.

Tidal currents carry the sediment in traction, in saltation and in suspension towards the land, with gradually decreasing velocities, caused by friction from the intertidal zones. Thus, the sediment decreases in grain size on intertidal flats, from low water level both towards high water and in the adjacent deeper waters. The sediment morphology and benthic zonation features of intertidal flats are also a reflection of changes in the tidal dynamics. The flood tides generally attain higher velocities and shorter duration than the corresponding ebb; further, the extensive accretion on such flats suggests that flood-tide deposition has been dominant over ebb-tide erosion. Even though the flood tide may be of longer duration, but as with additional energy by sea, waves, thus the flood tide is still dominant. Micro-morphological features, such as ripples and small creeks, exposed on intertidal flats during the passing of the ebb tide, reflect the last phase of the ebb currents.

Intertidal flats develop as part of the newly formed coastal landscape. There are some exceptions to this generalisation, however, as the accreted land retreats due to wave erosion. In the area of the abandoned mouth of the Yellow River, for example, the land has retreated by an average rate of 20 m per year since 1855. This erosion has taken place because the Yellow River has reverted back to discharging into the Bohai Sea, to the north, and sediment supply has been terminated. Similarly, in the wash, part of the flats have suffered from wave erosion, to expose the old polygon features over the middle part of the intertidal flat; this is considered to be the result of setting up an artificial embankment to reclaim land, which has steepened the slope of the coastal zone.

Hence, these plain coasts, which form the basis of most tidal flat development reflect a response to the dynamic balance between sediment supply and erosion due to wave action.

SEDIMENTARY CHARACTERISTICS

The tidal flat coastline of the wash, in Britain, and the two flats in China are located in temperate climatic zones which were developed along coastal plains deposited during the Holocene. The plains in China were formed by the larger Changjiang and Yellow rivers, supplying enormous quantities of well-sorted fine-grained material, within the clay, silt and fine sand size range.

The Bohai Sea receiving discharge from the Yellow River, is a shallow inland sea and the Yellow Sea, into which the Changjiang drains is a marginal sea of the Pacific. These regions are not as open to wave approach as the North Sea. The mean tidal range is 2～4 m, which is less than the tidal range of 6 m in the wash. The wash coastline is developed along an outwash apron (glacio-fluvial), with the sediments being coarser grained (sand size) and less well-sorted than in China.

The Wash is a sheltered embayment, but with water depths greater than in the other two areas. The embayment is bordered by the North Sea and faces the open sea towards the northeast; thus, there is a strong influence of wind/wave action and relatively strong tidal current action. There are several rivers discharging into the wash, but the fluvial sediment supply, which is mainly silt and clay, is only of minor importance and accounts for only about 20% of the total volume of annual sediment deposition on the tidal flats.

Being coastal types of intertidal flats, both British and Chinese regions have the same basic sedimentological characteristics, as described above. Different coastal evolution and environmental dynamics between the two settings lead, however, to there being different features within each of the three areas (Tables 1 and 2).

Morphological and sedimentary zonation is typically representative of intertidal flats everywhere in the world. (Table 3). Basically, there are three zones on an intertidal flat. A mudflat forms the upper part of the flat, around high tide level, where there are mainly deposits of fine-grained material, from suspension, as silty-clay or clayey-silt thin laminae. A silt flat, with ripples as its dominant feature, is located in the lower part of an intertidal flat, immediately above low tide levels. Such areas form the widest part of the intertidal flat and are deposited mainly from sediment moving in traction, within the coarse silt or very fine sand size range. It is an homogeneous silt deposits, with thicker laminae structures. A silt or silty-mud flat is located over the middle part of the intertidal zone, with transitional deposits derived mainly from fine siltsized material moving in saltation, with clay material being deposited from suspension, alternating thin laminae are formed. Erosional or microerosional features are quite common over this zone, which may reflect the fact that this part of the flat has a steeper slope; this tends to concentrate the stronger ebb current activity.

Table 1 Environmental Elements of Tidal Flats

Location		Bohai Bay, Bohai Sea, China	North Jiangsu Coast of Yellow Sea, China			The Wash, North Sea, U. K.	
Tidal Pattern		Semidiurnal	Semidiurnal			Semidiurnal	
Average Tidal Range(m)		3.0	2.0～4.0			3.6 Neap	6.5 Spring
Tidal Period	Flood(hr)	4	2～4			6～1/4 4	
	Ebb(hr)	7	10～8			6～1/4 8 1/2	
Flood Current	Velocity(cm/s)	25～60	35～100			20～60	
	Direction	210°～310°	Rotary(Clock-wise)			360°～300°	
Ebb Current	Velocity(cm/s)	20～40	45～90			20～40	
	Direction	60°～90°	Rotary			360°～180°	
Wave Condition	Hgt. (m)	0.1 0.4	0.6～1.2				
	Period(s)	1～2.2 3～4	4～6				
Width of Intertidal Zone (km)		3～6 (N) (S)	3.3 (erosion coast, N)	7.7 (depositional coast, S)	14～18 (with offshore submarine sand ridges)		
Slope of Intertidal zone ‰		0.69	1.26	0.3	0.2		
Slope of Submarine Coastal Slope ‰		0.21～0.51	1.66	1.66～2.27	4.39		

Table 2　Comparison of Tidal Flat Zonations

Location (Author)	The Wash, North Sea, U. K. (after G. Evans)	North Jiangsu Coast, Yellow Sea, China (D. Zhu and T. Xu)	Bohai Bay, Bohai Sea, China (Y. Wang et al.)
Zonation	A. *Salt Marsh*. Floods only during Spring High Tide. Flat with muddy depressions. Well-laminated silt-clays, clays and clayey silt. Very few sands. B. *Higher Mud Flats*. Located between Spring High Tidal level with High Tidal level with forms of elongate depressions, shallow channel systems, ripples and mud cracks. Laminated silty sands and sandy silt.	Ⅰ. *Supratidal Zone* (grass flat) Located above Spring High Tidal level. Only flooded during Storm High Tide. Very thin laminae clayey silt with organic layers. Ⅱ. *Suaeda salsa mud flat*. Located between Spring to Mean High Tidal levels. Laminae of clayey silt with thin layers of silt. Warm holes and Polygons (mud cracks) on the tidal flat surface.	Salt Marsh Plain and lagoon depressions. Ⅰ. *Polygon Zone*. Clay or silty clay deposits with mud cracks. Only flooded by Spring High tide. Ⅱ. *Inner Depositional Zone*. Muddy. Located below Mean High Tidal level.
	c. *Inner Sandy Flats*. Less well laminated sands and silty sands, with muddy layers. Ripples and scour and fill structures. d. *Arenicola Sand Flats*. Poorly stratified sands and silty sand, with minor fine-grained sediments mixed by the worm *Arenicola marina*. e. *Lower Mud Flat*. Poorly sorted laminated sands and silty sands intermixed with muddy layer.	Ⅲ. *Mud-Silt Flat*. Located between Mean High Tidal level to Neap High Tidal level. Alternating laminae clayey silt and silt, medium laminae with cross-bedding.	Ⅲ. *Erosional Zone*. Around middle tide level, alternating laminae of fine silt and silty clay. Overlapping fish scale structure or tidal creeks.
	f. *Lower Mud Flats*. Fine, medium sands and silty sands with thicker laminae, transverse larger ripple marks with steep lee slopes facing the direction of ebb.	Ⅳ. *Sandy Silt Flat*. Fine sands and silt flat with wave and current ripples.	Ⅳ. *Outer Depositional Zone*. Very fine sands and coarse silt with ripple forms.
Conclusion	Sand Flat. Tidal dominant flat with strong influence by wave action especially during storm periods to deposit sandy materials.	Silt Flat. Tidal current dominated flat with slightly or occasional wave influences. Silt deposits, mainly from offshore submarine delta of old Changjiang River.	Mud Flat. Typical tidal currents control intertidal zone flat with very fine sediment originally from the Yellow River. Very gentle coastal slope to form very classic intertidal zonation of four along 400 km long coastline.

Table 3　Morphological & Sedimentary Zonation Features of Tidal Flats in the World

Author	Van Straten 1954	Wang et al. 1963	Evans 1965	Reineck 1967	Thompson 1968	McCave 1972	Zhu & Xu 1980	Wang 1981	Dale 1985
Location	The Netherlands	Bohai Sea, China	The Wash	North Sea	Colorado River Delta		North Jiangsu Coast, Yellow Sea, China	Bay of Fundy	Frobisher Bay, Arctic Ocean
Zonation	Salt Marsh	Salt Marsh Plain with Lagoon Depression	Salt Marsh	Marsh	Supratidal Flat	Supratidal Zone	Supratidal Flat (or Grass Flat)	Beach Marsh	Beach
	Higher Tidal Flat	Polygon Zone	Higher Mud Flat	Mud Flat	Intertidal Zone/Tidal Zone	Intertidal zone	Suaeda Salsa Mudflat	Mud flat	Finer Flat Boulder Flat
		Inner Depositional Zone							
		Erosional Zone	Inner Sand Flat *Arenicola* Sand Flat Lower Mud Flat Lower Sand Flat	Sandy Mud Flat			Mud-Silt Flat	Tidal Creek Zone	Boulder Ridge
	Lower Tidal Flat	Outer Depositional Zone		Sand Flat			Sandy Silt Flat	Silty Ripple Flat	Very Bouldery Flat Graded Flat
		Submarine Coastal Slope			Subtidal Zone/Tidal Flat	Subtidal Zone	Subtidal Zone or Submarine Coastal Slope	Subtidal Coastal Slope	

Zonation of the north Jiangsu coastline may be a typical example of such an intertidal zone; it is an accretional intertidal flat, with silt as its main component over the complete range of the flat, which reflects completely tidal current processes. Even though the flat suffers sometimes from storm wave erosion, it soon returns to a normal profile. Such a flat may be referred to as a "silt flat". A sedimentation model for this area is shown as Figure 1. In the figure (A) are vertical deposits of mud suspension, carried by the flood tide. The thickness of the layer is limited by the tidal range between mid-and spring tidal levels. (B) is a transverse deposit, from material moving as a traction load, of silt or fine sands. The maximum thickness of this layer may be the same as the vertical distance between Low Tide Level and the coastal wave action base.

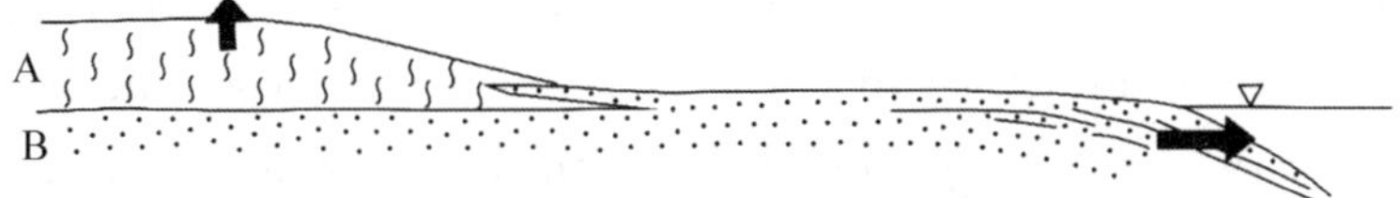

Fig 1 Sedimentation model of tidal flat

A. Vertical deposits of S. P. M., maximum thickness is 2～3 m.

B. Transverse deposits of traction load with the thickness of 15～20 m.

The intertidal flats along the western Bohai Bay are depositional in character, with the tidal dynamics being the dominant process. The main components of the flats are muds or silty clays; these are special fine-grained sediments which have originated from the loess plateau and have been carried into the Bohai Sea by the Yellow River. The coastal slope around the Yellow River delta and the western Bohai Bay is very gentle, with a gradient of the order of 1/3 000. There are 4 zonations in the Bohai Bay which are clearly distinguishable along the 400 km long coastline; a polygon zone consisting of clay cracks is distributed on the upper part of the intertidal zone which is only flooded during spring high tides. The classical feature of the mud flat is a 0. 5 to 1. 0 km wide belt of semi-fluid mud located below mean high tidal level as an inner depositional zone and it is very difficult to walk through this mud pool zone. Only in the Bohai Sea, where there is a large quantity of sediment supplied by the Yellow River plume to support the coast, is the mud flat well developed with a wide zone of semi-fluid mud. Outside this zone, there appear to be clear erosional features of overlapping fish scale structures and elongated depressions or creeks which follow the trend of the ebb current direction. A silty ripple zone is known as the outer depositional zone located on the lower part of the intertidal zone.

This type of intertidal flat described above may be termed a mud flat, with the Bohai Sea region being a classical example of such an area. Such deposits may exist also in the vicinity of large muddy river mouths, such as the northern side of the Amazon River in South America. A sedimentation model for this kind of intertidal zone shows there are fine-grained clayey sediments in the upper flat and subtidal zones, with an intermediate coarser-grained area of alternate muddy and silty layers subjected to wave action.

The intertidal flat of the Wash, even though it is still a tidally-dominated accumulating flat, containing sediment which is coarser than that present in the other areas being considered. Sand is the main component of the sediment and is distributed over the whole of the intertidal zone. The coarser sediments here not only reflect the submarine moraine source, but also the larger tidal range (of up to 6 m) and the strong dynamic processes. Sandy veneers or laminations appear in the upper part of the intertidal flat; these are formed normally by storm wave action in other areas, appearing as a sandy layer deposited on top of the mudflat. In the Wash, natural processes of sediment transport are influenced by the presence of an artificial embankment, which has changed the coastal slope and caused readjustment to commence on the flat. Local erosion and deposition of coarser-grained materials may take place because of changing wave/bed interaction. Such an intertidal flat may be referred to as a sand flat, providing an example of an area wherein there is strong tidal and storm wave influence.

OBSERVATIONS OF SEDIMENTATION PROCESSES

Sedimentation on intertidal flats is caused mainly by tidal processes. Due to differences of tidal inundation, the flood tide covers the flats, for different parts of the tidal cycle in relation to location across the flats. Water depths, current velocities and the wave action change both in time and location across the flats. Thus, sedimentation rates and patterns over intertidal flats are variable and have particular characteristics. Hydrographic data of 25 hours' duration, including measurements of current and sediment concentration across the whole of the intertidal flats along different parts of the Bohai Sea and Yellow Sea coastline of China, have indicated typically (Wang and Zhu):

1. Coastal suspended sediment concentrations, below the low tide level and further offshore, generally do not show any appreciable changes in pattern or level with changes in water depth and current speeds. Approaching high tide level, however, the maximum suspended sediment concentration occurs during the first hour of the flood tide on the intertidal zone when the water is shallow (20～40 cm only) and turbulent over the flats. The concentration then decreases rapidly with time, being only 50% of the initial concentration during the second hour of the flood and 5% in the third hour. The concentrations decrease to a minimum at high tide, as the water becomes less turbulent (Figure 2). During the ebb, the concentrations are low, because of the extended ebb phase of the tide with lower current speeds. As a result, the principal transport direction for material in suspension is created by the flood tide, in a landward direction.

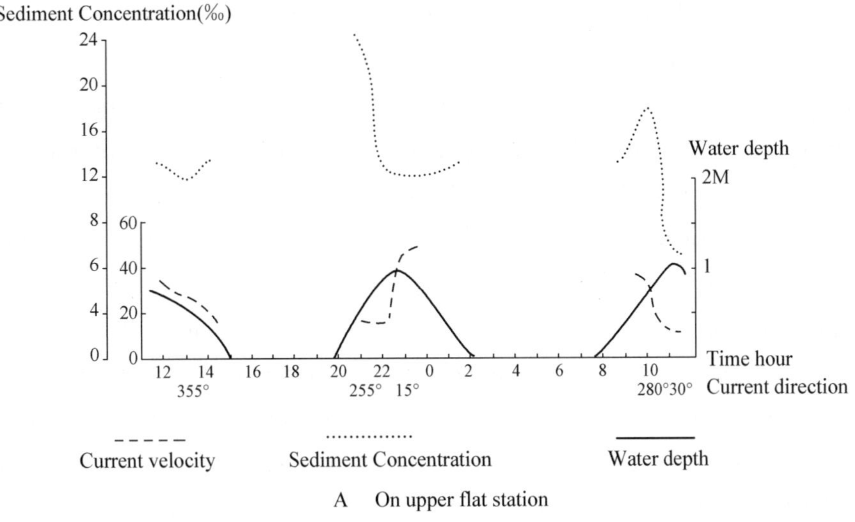

A　On upper flat station

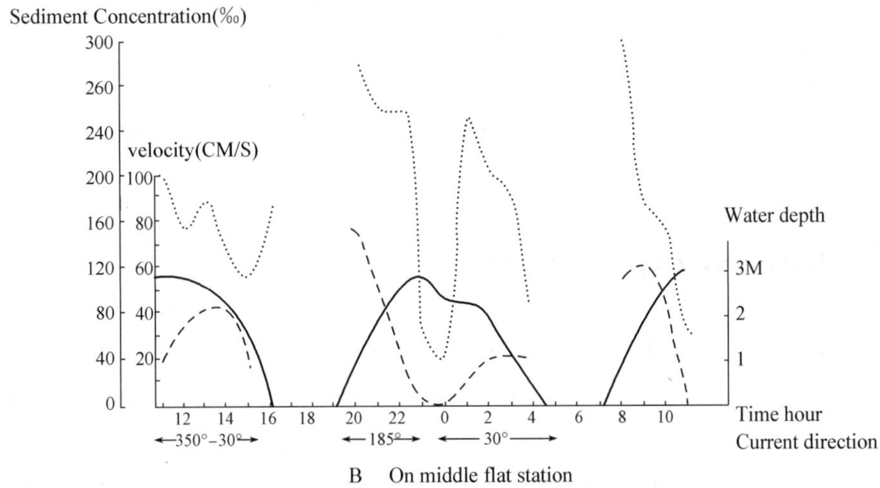

B　On middle flat station

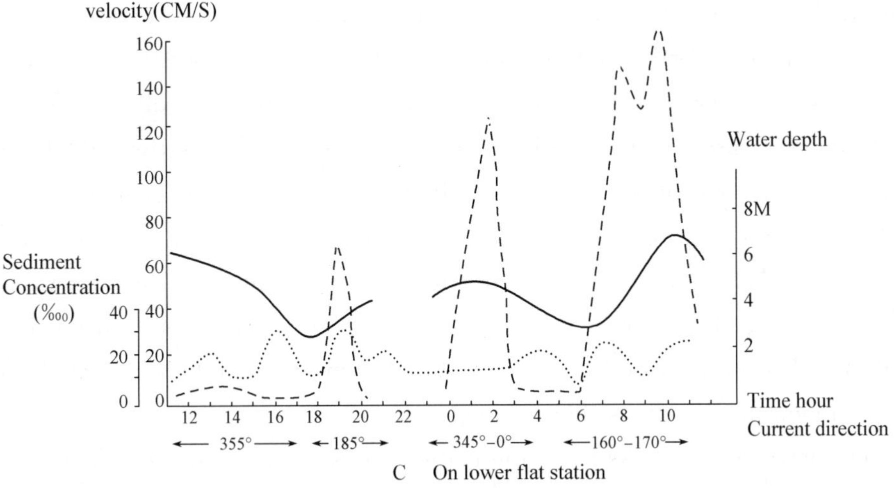

C　On lower flat station

Fig 2　Section records of tidal currents sedimentary processes on tidal flat during spring tide, Oct., 1980 in Xing Yang Gang area, Jiangsu Province, China

For most of the silt and mud flat, very fine sand and coarse silt are transported throughout the tidal cycle, mainly as a traction load. Fine silt is transported in suspension in the bottom 20 cm of the water column, with clay being restricted to the upper layers of the water column. Even in areas where the speed of the flood tide is equal to that of the ebb, because of tidal flow varying with time as a sinusoidal function, the suspended particulate matter is still removed to landward to cause the intertidal flat to accrete. This process conforms to the principle of "settling lag" and "scour lag" identified in European areas (postma, 1954, 1961; Van Straaten 1957).

2. Sometimes, even if the sediment concentration is low at low tide level along the Yellow Sea coastline, it can reach 10-times higher at high tidal levels of suspended sediment concentration during the same period. It indicates that part of materials are resuspended from local flat itself.

3. The maximum sediment concentrations and depositional rates are found over the middle part of the muddy-silt zone. Even the flat surfaces contain erosional features. It might be that the accumulated sediment has increased the flat surface, which caused the erosion by ebb current, and also may be a residual flow coming into tidal creeks even during flood when it causes the erosional process by creeks.

TIDAL CREEK SYSTEMS

Tidal creeks extend across the whole tidal flat zone, as a series of meandering creeks which, in some parts, are bordered by levees. Tidal creeks play an important role in the morphological evolution and sedimentary formation of tidal flats; they are developed widely on Chinese intertidal flats, occupying 10% of the total area of the flat of the north Jiangsu coastline. Here is a summary of a series of observational data of tidal creeks from the north Jiangsu coastline, China, as an example to discuss the natural processes in tidal creek systems. Ebb currents are the controlling agent in the creek development, as a seaward residual flow exists in the creeks through the whole tidal cycle. The water volume of the flow is 10% of the tidal prism, coming into creeks from intercreek flats during the flood. The residual flow is along the creek slope towards the sea. Thus, the creeks are ebb-dominated tidal controls. The tidal creeks are not only dissecting the intertidal flat, but also acting as a pass transporting the sediment. Creek deposits form a mosaic formation within the intertidal flat deposits, providing an indicator for this sedimentary environment.

Current dynamics in creeks show there are a series of asymmetry phenomena (Ding and Zhu, 1985) as 10% of the complete tidal prism which comes into the creek on the flood results in a residual down-channel flow on the flood. Within the creeks, the ebb discharge is greater than the flood discharge throughout a complete tidal cycle.

Maximum ebb currents are higher than on the flood and the ebb duration is longer. Both the flood and ebb velocities will increase considerably and get asymmetrical gradually as tidal level rises above the adjacent flat. The highest velocities occur in creeks whilst the flat is being exposed. Tidal currents within the creeks are much stronger than that on the adjacent intercreek flats.

Sediments found in tidal creeks are coarser than those deposited on the intertidal flats and there is a general decrease in grain size from the low to the high water mark. In this direction, the sediments change from sands through to silty to clayey sediments.

Within the high tidal mudflat zone, most of the creeks are smaller than elsewhere on the flats and have only a minimal number of distributaries in comparison with the lower areas of the intertidal flat. The average current speed is also here, being$<$40 cm/sec in creeks of the western Yellow Sea coastline. Slow deposition of fine-grained material takes place over this part of the system.

Tidal creeks are well developed in the middle tide level of the muddy sand zone. The main creeks are incised into the flat and meander with many distributaries. The creeks dissecting the middle part of the intertidal flats form a badland feature; the main creeks erode strongly the intertidal flats such as those along the northern Jiangsu coastline. The ebb current speed here sometimes reaches 80 cm/sec, with erosion at the head of the tidal creek taking place at 5 m/hr. The meandering creeks intensify erosion of the concave outer bank, which is accompanied by deposition of inclined strata with dips of up to 20 on the convex bank of the creek. In this way, the cross-bedded tidal creek deposits create a mosaic within the laminae bedding of the intertidal flat. Thus, the coarser-grained silt and sand deposits of the creeks, with side slopes of fine-grained clayey silt and silty clay of less coarse levee deposits, and intercreek flat deposits of laimnae interbedding of mud and silt are the whole sedimentary facies, and will show an in intersected profile while the creeks meandered.

In the low tidal sandy or silt zone, the creeks are shallow and wide such as at Xingang in northern Jiangsu. At its mouth, the tidal creek is wide and shallow, ranging from 300～500 m to 1～2 km in width and with a depth of only 1～2 m. As stronger currents are always present in this zone, ripple or even megaripple marks build up to form bidirectional cross-bedding, with creek migration, sandy inclined strata will be formed. Basically, the tidal creeks dissect the intertidal flat and act as sediment transport passes. Field data have demonstrated that both the upper and lower parts of tidal creeks are gradually being infilled, whilst the middle parts are undergoing erosion.

As a summary, the sedimentation structures of the tidal flats of the Chinese coastline can be shown in the accompanying vertical profile and description, Table 4.

Table 4 Sediment Structures of Silt Flat (After Zhu & Xu)

Profile	Sediment Structure	Sedimentary Environment
I.	Brownish Clay Silt. Surface layer: laminae contain organic matter layers and abound with plants and worm holes. Lower layers: inter-beddings of laminae and cross strata.	Supratidal zone, grass flat; high tidal level, mud flat. Suspended material deposits and underwater silt ripples formed during the high tidal period.
Ⅱ.	Brownish Clay Silt. Upper part: unclear laminae as biological turbulence, with various worm holes. Lower part: alternating laminae and cross beddings of clay and silt. Boundary with silt incline beddings	Middle tidal level, mud-silt flat with larger quantity of benthos. Stronger current action, and with many tidal creeks intercrossing.
Ⅲ.	Greyish Yellow Silt 1. Valley-shaped cross beddings. 2. Fine silt laminae-cross beddings; and 3. Mini-cross bedding of silt. 4. Coarse silt cross bedding. 5. Coarse silt laminae. 6. Greyish-black fine sand and coarse silt cross bedding. 7. Eroded discontinuity surface. 8. Inter laminae of fine and coarse silt. 9. Fine sand inland beddings.	Low tidal level silt flat. Tidal creek depression with current ripples. Silt flat ripples. Bed layer of current ripples. Megaripples, wave marks. Tidal creek margin.

As a conclusion of the above comparative studies, three types of tidal flats can be recognized, as follows:

SANDFLAT of the Wash is an example of intertidal flats with high wave energy influences. Original glacial deposits were submerged offshore during post-glacial sea level rising. Now they are winnowed by wave action, especially during storm periods and carried by flood tidal currents to a relatively open embayment of the Wash to form a coarse sandy material flat.

SILTFLAT well developed along the northern Jiangsu plain coastline to face the open Yellow Sea is a typical type of tidal flat with three zonations: there are mudflats on the upper flat, mud-silt flats on the middle part and sandy silt flats on the lower part of the intertidal zone. A medium level of wave energy stirs up the silt material from submarine sandy ridges of the old Changjiang River delta as a main sediment source and then transports it onto shore by the flood tides.

MUDFLAT has developed along the plain coastline of Bohai Bay. Tidal currents are only predominant factors there with minimal wave influences but with large quantities of clay and silt material supplied by the Yellow River.

The differences between these three types of tidal flats have shown the control of environmental factors such as dynamic processes, sediment sources and the factor of coastal slopes. Tidal flats are developed where the tidal currents are the predominant agents of dynamic processes, but as with the different levels of wave energy input they will change the features and sediment components of the tidal flats.

UTILISATION

Tidal flats are of widely-distributed coastal type which require basic investigation; they are also areas where new land is developing, i. e. they provide a potential land resource. The utilisation of the tidal flats can have successful or unsuccessful outcomes, in relation to the extent of the knowledge of coastal characteristics. To set up an embankment for reclamation, for example, the best location is at the average high tidal level where more land can be reclaimed and the bank can be kept standing (Zhu, 1986). Above the embankment, in some areas such as the Wash and the Yellow River delta, the land is used normally as pasture during the first stage of reclamation. Land use is then for agricultural purposes as mature land, when and where there are freshwater resources for irrigation. Over the major part of the western Bohai Bay coastline, however, the climate is dry (500～800 mm of precipitation/year) and the mudflats consist of clayey sediments. The lowered land within an embankment is an ideal place for the development of salt fields, except for areas near the Haihe River, where the land can be reclaimed as rice fields even at an early stage.

Most parts of the north Jiangsu tidal flats can be reclaimed as cotton fields, as there is an abundant rainfall of 1 000～1 500 mm annually. The presence of a sandy/silt soil and enough warm days in a year make this possible. The level of precipitation and the sandy land are not suitable for creating salt pans. There were quite unsuccessful examples for building salt pans in north Jiangsu coast (with an annual precipitation of 1 500 mm/yr), even though it is also a wide tidal flat coast.

The lower parts of tidal flats are ideal sites for the aquaculture of shellfish and other aquatic products, but are not so good for navigation purposes and harbour construction (except along river estuaries and within deep tidal channels). Tianjin New Port, for example, is set upon the tidal flat of the Bohai Bay. As a result, the port suffers from problems of siltation and 6×10^{6} m^{3} of silt has to be dredged annually. This example has triggered a multidisciplinary study of the tidal flat coastline in China.

References

[1] Amos, C. L. 1974. Intertidal Flat Sedimentation of the Wash, E. England. Unpub. Ph. D. Thesis, University of London.

[2] Bayliss-Smith, T. P. 1979. Tidal flows in salt marsh creeks. *Eatuar. Coast. Mar. Sci.*, 9: 235 - 255.

[3] Carling, P. A. 1978. Intertidal Sedimentation. Unpub. Ph. D. Thesis, University of Wales.

[4] Carling, P. A. 1982. Temporal and spatial variation in intertidal sedimentation rates. *Sedimentology*, 29: 17 - 23.

[5] Collins, M. B., Amos, C. L. and Evans, G. 1981. Observation of some sediment transport processes over intertidal flats, the wash. U. K. *Spec. Publ. Int. Assoc Sediment*, 5: 81 - 98.

[6] Collins, M. B. 1981. Sediment yield studies of headwater catchments in Sussex, S. E. England. *Earth Surf. Proc. and Landforms*, 6: 517 - 539.

[7] Ding, X. R. and Zhu, D. K. 1985. Characteristics of tidal flow and development of geomorphololgy in tidal creeks, southern coast of Jiangsu Provice. In: Proceeding of China-West Germany Symposium on Coast Engineering.

[8] Evans. G. 1965. Intertidal flat sediment and their environments of deposition in the wash. *Quat. Jnl. Geol. Soc. London.*, 121: 209 - 245.

[9] Hantzschel, W. 1955. Tidal flat deposits (Wattenschlick) Recent Marine Sediments. Dover Publications, New York. 196 - 206.

[10] Reineck, H. E. 1967. Layered sediments of tidal flats, beach and shelf bottoms of the North Sea. *Esturies Am. Assoc. Advan. Soc.*, 83: 191 - 206.

[11] Thompson, R. W. 1986. Tidal flat sedimentation on the Colorado River delta, Northwestern Gulf of California. Geol. Soc. Amer., Mem. 107, Boulder.

[12] Van Straaten, L. M. J. U. 1976. Sedimentation in the tidal flat areas. In: Holocene Tidal Sedmemation. 25 - 46.

[13] Wang, Y. 1983. The mudflat system of China. *Can. Jnl. Fish. Aquat. Sci.*, 40(Suppl. 1): 160 - 171.

[14] Zhu, D. K. 1986. The utilisation of coastal mudflats of China. *Scientia Geographica Sinica*, 6(1): 34～40 (in Chinese).

[15] Zhu, D. K. and Xu, T. G. 1982. The coast development and exploitation of middle Jiangsu. *Acta Scientiarum Naturalium Universitata Nankinesis*, 3: 799 - 818 (in Chinese).

Sand Dune Coast—An Effect of Land-Sea Interaction Under the New Glacial Arctic Climate*

Coastal zone is a special area where both continental and oceanic meteorologic and hydrologic environments change sharply. Thus, air and water circulations are different to those of terrestrial and the ocean's. Coastal zone is also a belt along which very strong interaction between land and sea takes place and sand dune coast may be a dominant expression of this interaction[1]. Sand dune coast is developed under the following conditions:

1. The coast abounds with sandy sediment which normally is of terrigenous origin and is carried to the sea from another source by rivers, or retransported from a submarine old coastal deposits.

2. Dry cold or semi-arid climate with strong prevailing sea winds.

3. A gentle subaquatic coastal slope, with optimum average gradient of 1/1 000 and/or a medium tide range with sufficiently wide exposed beach zone during ebb tide for wind to pick up sand.

This paper discusses two examples of typical sand dune coast (one is in North China and the other in West Africa) to explain how the dune coast is formed by land-sea interaction. Under Arctic climate influence during the New glacial lower stand of sea level, large areas of sandy coast were exposed to frequent cyclonic storms. The strengthened winds shifted large volumes of sand gradually to accumulate in huge sand dunes in large scale of coast area.

Ⅰ. Sand Dune in the Northern Luanhe River Delta of North China

The 45-kilometre-long sand dune coast in China is located along the west of Bohai Sea from the northern part of the Luanhe river (滦河) delta to the Yanghe river(洋河) mouth (39°47′N～39°25′N, Fig. 1)[2-4]. The coast dune belt is 1～4 km wide, and the trend of the dune ridges is mainly in the northeast direction of 20°～30° which is close to the local coastline. The height of sand dunes decreases from 40 m in the south to 10 m in the north. The width of the dune belt also decreases from south to north. Several small

* Ying Wang, Dakui Zhu: *The Journal of Chinese Geography*, 1992, Vol. 3, No. 1, pp. 37 – 54.

rivers cut through the coastal dune belt and divide the coast into three parts.

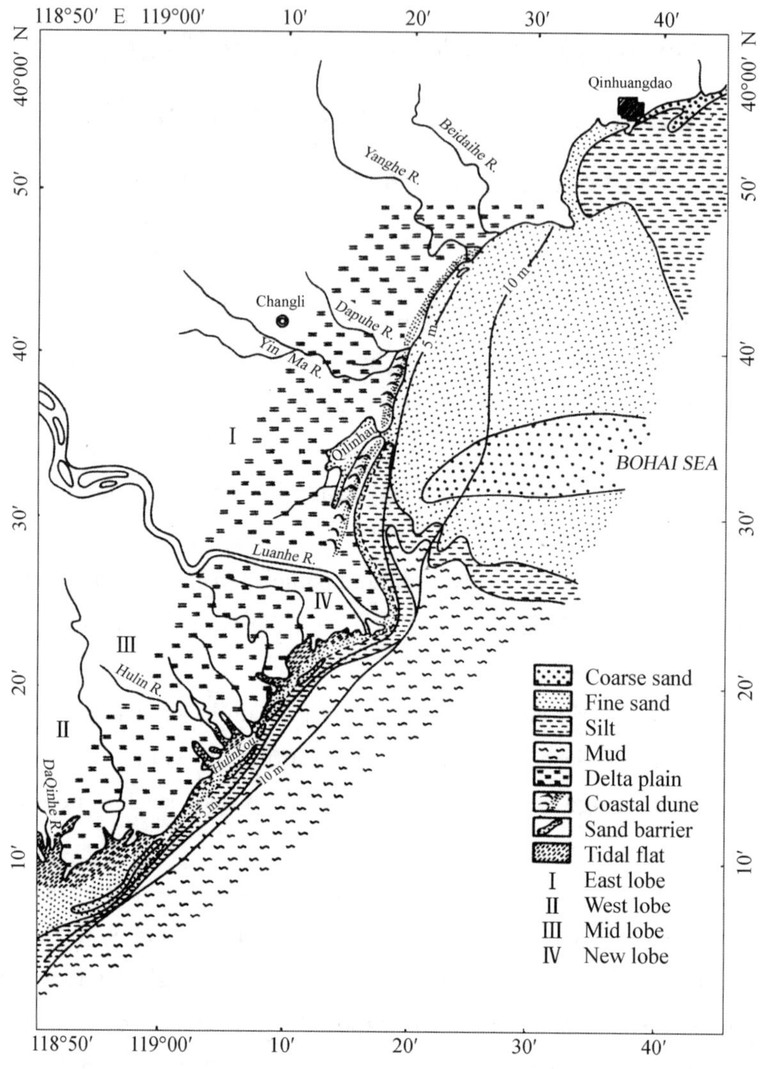

Fig. 1 Coastal Map of Luanhe River Delta

1. First part: From Luanhe River northward to Qilihai Lagoon

Sand dunes are larger in this area; the width of the zone is approximately 4 km near the Luanhe River in the south, but averages 2 km for most part, and only 1 km near the inlet of the Qilihai Lagoon(七里海潟湖). The zone consists of transverse dune ridges of the first rank of the coastal dune series and is in parallel to the sea shore that trends NE 20°～30°. The first rank is also the largest one with average height of over 30 m. The highest peak of sand dune is 42 m located near the Qilihai Lagoon. The inner dunes are chains of barkhans with the average distributed trend of NW and NNW and are intersected by the coastline. The convex wind slopes of the barkhans face NE and ENE, which is the direction of the dominant sea winds. The steep slip slopes face southwest to the leeward. The heights of the barkhans decrease landward down to less than 30 m. There are several swales and swamps dotted in the chain of dunes. The coastal dunes consist mainly of well sorted quartz and feldspar sands. The fine sands are on the top part of dune ridges and medium-fine sands are distributed on the seaward slope of the coastal dunes (Table 1). The heavy mineral assemblage is characterized by hornblende, epidote, garnet, magnetite, limenite, zircon and monazite. The heavy mineral content is similar to the Luanhe River sediment, and comes from the metamorphic and pyrogenic rocks of the Yanshan Mountains located to the west of the delta plain on the upper part of the river in this region. However, the sand grains of coastal dunes are more rounded than the grains from the river.

Table 1　Analytical Sediment Composition of Sand Dune Coast Around the North Luanhe River Delta

Sample Code	Size % / Location	0.5～1.0 mm Coarse sand	0.25～0.50 mm Medium sand	0.10～0.25 mm Fine sand	0.10～0.01 mm Coarse silt	S_o	S_k	d_{50}
21	Top of sand dune, Qilihai Lagoon		9.9	88.8	1.3	1.680	1.025	0.16
22	Seaward slope of dune, Qilihai Lagoon	0.6	52.4	46.4	0.6	2.24	0.903	0.26
23	Seaward foot slope of dune, Qilihai Lagoon	0.5	62.0	37.3	0.2	2.00	0.389	0.30
S1	First rank of sand dune, Qilihai Lagoon		14.8	84.8	0.4	1.76	1.011	0.16
S2	Second rank of sand dune, Qilihai Lagoon	0.1	50.0	49.8	0.1	2.118	0.979	0.25
C1	Northern Coast to Dapuhe River	0.5	32.2	66.0	1.3	2.074	1.05	0.19
C2		2	51.4	47.3	0.25	2.059	0.95	0.25
C6	Seaward slope dune Southern coast to Dapuhe River		37.7	62.1	0.20	2.22	1.033	0.20
C5	Top of sand dune, Southern coast to Dapuhe River	0.1	31.4	68.2	0.30	2.074	1.047	0.19
C4	Landward slope of dune, Southern coast to Dapuhe River	0.1	1.4	49.8	47.9	2.88	0.896	0.25
C8	Southern Coast to Yanghe River	2.4	59.5	37.7	0.4	1.947	0.836	0.29

(Continued)

Sample Code	Location \ Size %	0.5～1.0 mm Coarse sand	0.25～0.50 mm Medium sand	0.10～0.25 mm Fine sand	0.10～0.01 mm Coarse silt	S_o	S_k	d_{50}
C9		3.5	58.1	37.3	1.1	1.846	0.78	0.30
C11	Northern Coast to Yanghe River	4.0	52.4	42.0	1.6	1.84	0.93	0.26
C12		6.1	19.5	73.2	1.2	2.0	1.021	0.17
C13		6.1	4.3	50.6	0.3	2.332	0.992	0.35

Dune shapes show that the formation of the sand dunes are influenced mainly by northeasterly and east-northeasterly winds. The sand dune belt is closer to the modern coast where the shore beaches are less than 100 m wide. Scarplets appear along the coastline formed by wave erosion. The subaquatical coastal slope is about 1/180 to 1/300 from the coastal line to where the water depth is 5 m, and the average slope is 1/250. Thus, the steeper coastal slope is associated with large asymmetric sand dunes.

2. Second part: Around Dapuhe River mouth

Here the estuary faces eastward towards the open sea, with the river channel facing the southeast (110°). Sand dunes are present on both sides of the estuary. Parallel to the coastline a series of submarine sand bars are located in water depth of 1～3 m outside the river mouth. The gradient of modern coast slope is 1/300, with the shore beach in the wide range of 150 m. The Northern coast to Dapuhe River(大蒲河) consists of medium-fine sand with three series of beach ridges. The Southern Coast to Dapuhe River consists of fine sand and coarse silt to form a single slope with a dip angle of 6°～8°. Beach sands in the Southern Coast are well sorted (S_o=2.92). There is an eroded scarplet about 25 cm high along high tide level, and an old beach terrace is located above the high tide level. The old beach consists of medium-fine sand, with three series of beach ridges. However, the old beach has been anchored by the growth of vegetation. Both the old beach and the modern beach are reshaped from the seaward slope of coastal dunes. The coastal dune belt behind the old beach is 2 km wide. The first rank of the coastal dunes is oriented transversely to the coast line with an average height over 20 m. The highest peak of the dune hill is 24 m. The dune sands are well rounded in medium and fine size range. The height of the inner sand dunes decreases landward and the feather patterned dune chains are oriented at an angle to the modern coastline.

3. Third part: Coastline on both sides of the Luanhe River mouth

The river flows into the sea along the direction of SE (20°), forming submarine sand bars around the river mouth. The coastline trends to the SW (20°), with the submarine coastal slope 1/360 in the southern part and 1/440 in the northern. Modern beach consists of medium and fine sands. They are well sorted (S_o=1. 88) and the grain shape is well rounded. The intertidal zone is about 100 m wide, and above the high tide level there is a 60 m wide old beach. The height and width of the coastal dune belt decrease compared with that of Dapuhe River. The dune belt is about 10 m high and 1 500 m wide to the south of the Yanghe River mouth, and plants dotted over the sand dunes create a different view from the bare sand dunes surrounding Dapuhe River estuary. The sand dune belt in the northern part of the area has been fixed as a result of afforestation. The dune belt is 0. 8 km wide, and the height of dunes is less than 10 m. Marine erosion takes place along the south coast and there are several initial sand barriers developed in the near shore zone. As a whole, the sand dune coast extends uninterruped from the northern bank of Luanhe River to the northern coast of the Yanghe River mouth. The sand dune hills are very close to the modern shore line, and the inner barkhan dune chains are gradually becoming lower landwards, because the coastal beach suffers from wave erosion.

There are several swamps and marshes in the coastal area between Qilihai Lagoon and Dapuhe River as the dunes here block the flow of several small rivers into the sea. Sediments of the coastal dunes consist mainly of fine sands or medium-fine sands. They are similar to the Luanhe River sediment but with higher roundness and better sorting. There are collision V marks and small pocks on the surface of the quartz sands. Attached to several grains are diatoms, showing that the dune sands came from the beach and near shore sea bottom.

The research area is within part of the Luanhe River delta plain. Coastal sediment is mainly supplied by the Luanhe River. The total length of Luanhe River is 870 km, and the annual water discharge before the water diversion of 1984 was 36.7×10^8 m^3, and the annual sediment discharge was 24.08×10^6 tonnes. In the mountain, the river bed of upstream is gravel-sand, with a gradient of 0. 1/1 000～0. 2/1 000. Fine sand river bed is present in the lower reaches of the plain area where the gradient is only 0. 25/1 000, and where the shifting river channel transects the large delta plain. To the north the shifting course reached the Yanghe River and to the south it reached the Daqinghe River(大清河) channel. Thus, the whole range of coastline along the delta consists of sediment supplied mainly by the Luanhe River. Sediment supply by other rivers in this area is negligible, such as the maximum sediment discharge by the Yanghe River is 48 000 tonnes per year, and the minimum is only 2 830 tonnes. Fine grained materials are mainly supplied by the modern Luanhe River. The sediments that enter the sea are diverged mainly towards the southwest reaching the area of Hulinkou[2]. There is only a little longshore drifting to the north of the Luanhe River mouth, at the maximum distance of only 4. 8 km from the Luanhe River mouth. However, these longshore sediments are removed by prevailing south-easterly and northeasterly waves acting transversely towards the coast. Sands are deposited in the breaker zone above which sand barriers are developed with a lagoon behind, and mudflats behind the lagoon.

Silty clay is deposited outside the breaker zone, and the clayish mud beyond 5 metres of water depth marks the lower boundary of modern coastal zone. The average gradient of subaquatic coastal slope at the present is 1/4 000, but it changes to 1/5 000 at 10 m water depth. Thus, the environment is not suitable for the development of sand dune coast in the southern part of the Luanhe River delta. So, despite the sediment supply, the modern Luanhe River's sediment is transported mainly towards the south.

The sediment composition of sand dune coast is similar to the Luanhe River sediment, but the size range is larger than the modern river sediment. Although the sands of the coastal dunes have the same source, they have been released and deposited under different climatic environment. It is believed that frost-shattered debris of cold zone environment normally form large size ranges than debris due to chemical weathering in the tropical or temperate climate environment. Therefore, the coastal dune sands which were originally supplied mainly by the Luanhe River may have been formed by mechanical weathering not in the modern time.

Because the modern Luanhe River sediment is transported mainly to the south coast, the sand dune coast in the north suffers from wave erosion due to shortage of sediment supply. Thus, scarps or sandy cliffs appear along the dune coastline. What is more important is that, there is a large area of sandy sediment on the flat sea bottom about 20 km from the shoreline down to the water depth of 14 m (see Fig. 1). These are yellow-brownish fine to medium sands. Fine sand of 0. 1～0. 25 mm in size makes up 60～90 percent, whereas that of 0. 25～0. 50 mm account for 10～30 percent. There is very little silt and almost no clay. The fine sand is well sorted (S_o=1. 3～1. 8, few S_o= 2. 4). The medium sand in this area is also well sorted, with average S_o=1. 5～1. 8 and few of S_o = 2. 2. The size distribution of medium sand is 60～67 percent of 0. 25～0. 5 mm, and 10～38 percent of 0. 1～0. 25 mm with no grains of clay size. This is the characteristics of wind blown and wave acted sands in a vast flat area during lower sea level. However this fine-medium sands are now exposed to the sea floor where water depth is over 10 m. This figure is beyond the modern wave base of this coast, therefore, sediment of this depth is normally mud. However, pure sands exposed to the larger area of the sea bottom indicate that the presence of relict sand is not in equilibrium with the present dynamic setting. Because the modern Luanhe River sediment is transported to the south, the relict sands are not covered. Size and mineral composition of the relict sands are similar to the coastal dune sand (Fig. 2a and 2b) and it was the source of dune sands when the area was a beach or flat coast during the time of lower sea level. If in the present time the area of the relict sand is approximately within the 10 m isobath and the average slope gradient is close to 1/1 000 within this slope gradient the breaker zone is well developed. Even if the tidal range was relatively small it is still possible to expose a vast sand flat during ebb tide. When the dry sands are blown by strong winds coming from the sea towards the coast it accumulated in coastal sand dunes. Thus, the flat sandy seabottom off the dune coast is sufficient to satisfy the material and topographic requirements for sand dune coast formation. There is no apparent change in wind and wave climate between the north and south coasts during present time. Both the northern dunes and southern sand barriers indicate that coastal sediment movement is mainly transverse up to coast under the control of southeasterly wind and wave action. The northern coast has also been under the influence of strong northeasterly winds but the protruding delta of Luanhe River has protected the leeward south coast. The characteristics of the modern coastal geomorphology reflect that: (1) present river sand is finer than in the past, (2) under modern coastal dynamic processes it is a suitable environment for the development of sand barrier system but not huge coastal dune. The existence of sand dune coast in the north indicates a climate change, such as temperature, moisture, wind speed or frequency of strong winds during the post glacial time. The research area is located at the north side of Bohai Sea and faces the open sea to the southeast. The present climate is oceanic temperate and monsoonal. Northeast and north winds dominate during winter, average wind speed is 7. 0 m/sec. , but speeds over 11 m/sec. occur at 2. 3 percent of the time, maximum wind speed is 20 m/sec. and is also NEN ward. The average January temprature is −10 ℃ and average relative humidity is 65～70 percent. During summer season southerly winds prevail, southeast and south winds have influence over the research area, average speed of these winds is about 4～5 m/sec. The frequency of speeds over 6 m/sec. is 8. 9 percent from southeast and 5. 7 percent from the south. Summer is warm and humid, the average temperature is 26 ℃～28 ℃ in August and relative humidity over 85 percent. Transitional winds during spring are normal, from the south, and during autumn from the north. The average wind speed is 5～6 m/sec.

Wave climate reflects the climate of the winds, under NE and ENE wave action, the depth range of active sediment movement is about 4. 0 m, the deepest can reach 5. 50 m. Under the action of southeasterly waves, the sediment movement is at a water depth of about 3 m, but may reach 4. 1 m. The influence of waves from other directions is minimal. Along the northern coast of Luanhe River, flood current flows to the northeast, and ebb current flows to the southwest. Along the southern coast the semidiurnal tidal currents flow in opposite direction to the north. The tidal range is about 1. 5 m in this area (Table 2).

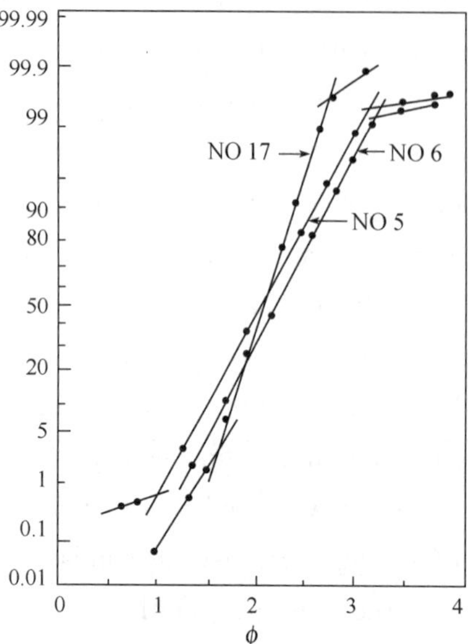

Fig. 2a Cumulative Frequency Curve of Coastal Sand Dune Having a Probability Ordinate Scale

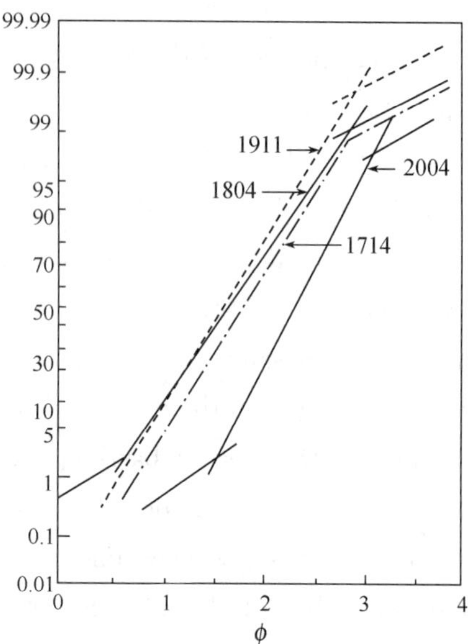

Fig. 2b Cumulative Frequency Curve of Submarine Relic Sand Having a Probability Ordinate Scale

Table 2 Sediment Discharge and Wave Climate Data of Luanhe River Delta

Items \ Location		Northern Coast of Luanhe River Mouth	Southern Coast of Luanhe River Mouth
Subaquatic slope	Modern Past (to 10 m contour)	1/250 to 1/350 1/1000	1/400 to 1/500 1/1500
Modern coast wave base		5～6 m	4 m
Prevailing wind	Direction Ave. speed Max. speed	NE. ENE. SE. 7.0 m/sec 20 m/sec	SE. NE. ENE. 4～5 m/sec 20 m/sec
Prevailing wave	Winter Summer	NE. ENE. SE	NE. ENE SE
Tidal form	Range Pattern	1.50 m Irregular diurnal	1.50 m Irregular sedimental
Direction of	Flood current Ebb current	NE SW	SW NE

According to modern dynamics of wind and waves, the depth boundary of the coastal zone is 5～6 m in the north, and 4 m in the south. The gradient of submarine coastal slope is 1/400～1/500 in the south, where small sand barriers have been formed (see Fig. 3b). In the north the gradient of coastal slope is 1/250～1/350. The steeper coastal slope has been undermined by wave attack to form a narrow shore beach, which is not suitable for the development of coastal dunes (Fig. 3a and 3b).

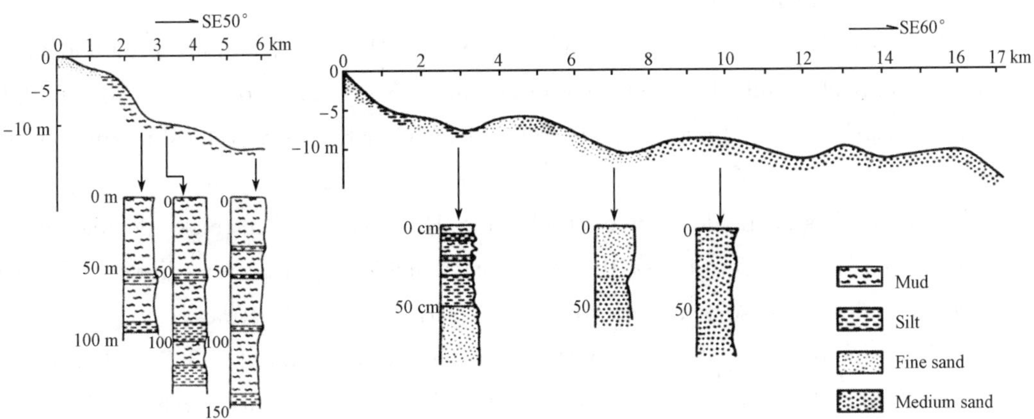

Fig. 3a　Northern Coastal Profile of the Luanhe River

Fig. 3b　Southern Coastal Profile of the Luanhe River

The delta plain of Luanhe River consists of four delta lobes (*. Fig. 1), and the sand dune coast is developed along one of the lobes. The sediment strata represent transitional facies of fluvial and marine deposits (Fig. 4) [2].

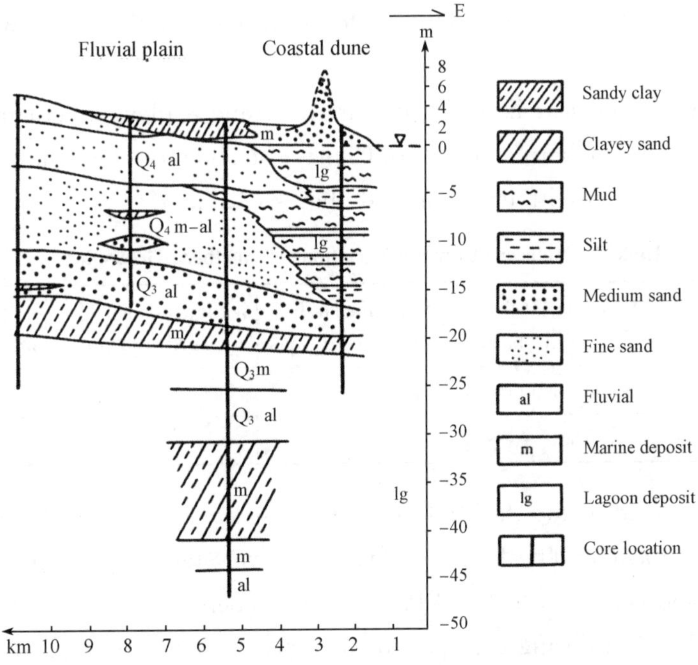

Fig. 4　Stratigraphic Profile of Qilihai [2]

(1) The base of the delta sediments is located 28～32 m under the river bed deposits of the old Luanhe River. Dark grey medium sand and pebbles (3～5 cm in diameter)

* Report on deepwater harbour site selection of East Hebei Province, Nanjing University, 1978

account for 10 percent of the total sediment.

(2) Marine deposits at the subsurface depth of 21.2～23.4 m are grey and grey-greenish sandy clay and silt. They contain organic matter, calcareous concretion and few foraminifera of *Ammonia beccarii*. The ratio of Ca/Ca+Fe=0.87 typical of shallow sea deposits.

(3) Late pleistocene deposits of the Luanhe River are 3～5 m thick and the surface of this layer is 16～18.5 m below the sea floor. They are light grey medium sands with rounded pebbles, well sorted, and contain terrestrial fossils of ostraca.

(4) Marine deposits of grey sand layer are at the intervals of 3.3～14.4 m below the sea floor. They contain organic matter and foraminifera *Ammonia beccarii*, *Elphidium simplex cuehman*, *Elphidium hughesi foraminosum cuehman*, *Cribronomion* sp. *Ammonia annectens* (Parker and Jones), *Elphidium magellanicam* (Heron-Allen & Earland), *Protelphidium compressum* (S. Y. Zheng), and *Elphidume* sp.

(5) Fluvial deposits buried between 2.6～7.7 m are brown-yellowish medium sands and small pebbles. The ^{14}C date of a peat layer at the 4 metre interval of this sediment is 8 025±105 yr BP. The delta lobe I, ie. the base of sand dune coast was formed 8 000 years ago in the early Holocene. The sea level during that time was 13 m lower than today.

According to the research on climate variations and sea level fluctuation during post glacial period[5,6] in northern and eastern China, the climate, since the beginning of the Holocene, has experienced four major cold phases (Table 3)[6].

Table 3 Holocene Climatic Variations of Northern Hemisphere[6]

Cold phases	New Glacial Ⅰ or first cold phase	New Glacial Ⅱ or second cold phase	New Glacial Ⅲ or third cold phase	New Glacial Ⅳ or fourth cold phase
Period (yr BP)	8 200～7 000	5 800～4 900	3 300～2 400	450～30
Time of cold peak	7 800	5 300	2 800	200

During each cold period the sea level receded by 2～4 m, during the warm intervals the sea level rose again. The Holocene high sea level stands occurred 10 000～8 300, 8 000～7 000 and 6 000～5 500 yr BP. The air temperature during warm interval was 1 ℃～3 ℃ higher and during cold periods 5 ℃～6 ℃ lower than present. During the cold periods in eastern China cold climate advanced the mechanical weathering processes, which produced abundant debris that entered lakes or seas. Abundant peat also developed during the beginning of this period. Waves of cold air entered from north over the China plains, in addition to cyclonic storms that also came from the north more frequently. High level rapid air currents transported loess to the 28°～30°N in Southern China during the new glacials[7]. Under strong gale and violent storms, a dry, cold environment was formed in the north-eastern China, which stopped the peat deposit but formed coastal sand dunes. The sea level stand was 10 m lower during the early Holocene; in the new glacial of 8 000 yr BP or cold peak time of 7 800 yr BP, the sea level was 13 m below the present level (Fig. 5, after Yang and Xie 1985). This low level coincides with the water depth of relict sand offshore of the sand dune coast. During the cold periods river sediment discharge was high and with coarser debris due to the increased frost weathering processes under the cold arctic climate. Along the flat coastal slope of 1/1 000 gradient during low sea level stand, strong northeasterly winds carried sands landward forming huge sand dune series along the coastline of the existing delta lobe I of the Luanhe River. After the sea level had risen and wind speeds were reduced, the northeasterly winds still prevailed. During this time coastal sand dunes may have migrated landward. It is a relict or reformed sand dune coast.

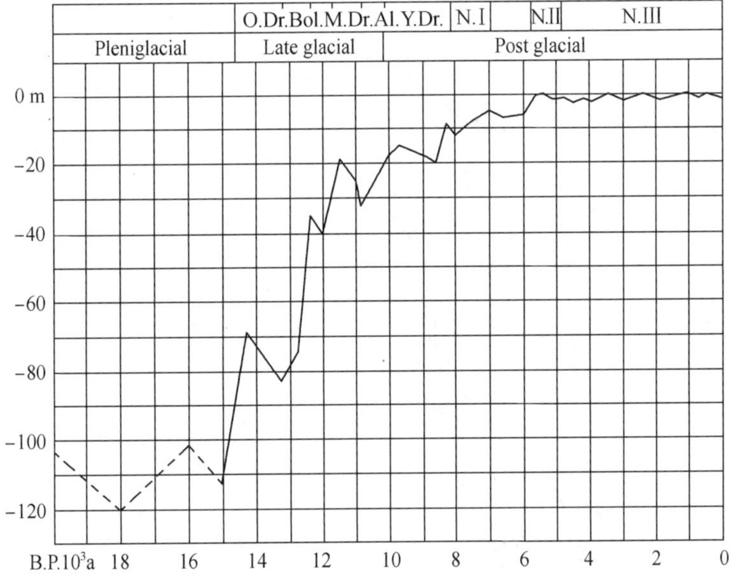

Fig. 5 Curve of Sea-level Change Since 20 000 yr BP

(After Yang and Xie, 1985)

In conclusion, the sand dune coast in the northern Luanhe River delta was formed during the lower sea level of the first cold phase of the Holocene as a result of Land-Sea interaction under Arctic climate influence.

Ⅱ. Sand Dune Coast along West Africa of Mauritania Atlantic Ocean

Mauritania is located at the southwest end of Sahara Desert, where northeasterly trade winds are prevailing with directions from north-northeast to east. Strong sea winds are from west and northwest. Several sand dunes trend northeasterly to southwesterly from the centre of the desert to the coast but the dune ridges are with hard cover of calcium on the crest. Loose sand is present where the calcium cover is broken by human activities.

Along Atlantic Mauritania Coast, sand dunes extend 315 km from 16°～19°N[9]. Coastal profile across land to sea follows the sequence of desert-marine terrace-lagoon depressions to coastal dune belt-beach-subaquatic coastal slope (Fig. 6 and 7). The marine terrace 3～4 m above sea level and 0.5～80 km in width, consists of shell sands and fragments. The various mollusc species represents mainly shallow waters of sandy coast. Raised breaker zone deposits form a terrace that gave a ^{14}C age of 5 755±120 yr BP. The terrace separates the coastal zone from inland desert sand dune ridges as a distinct geomorphologic and sedimentary structure.

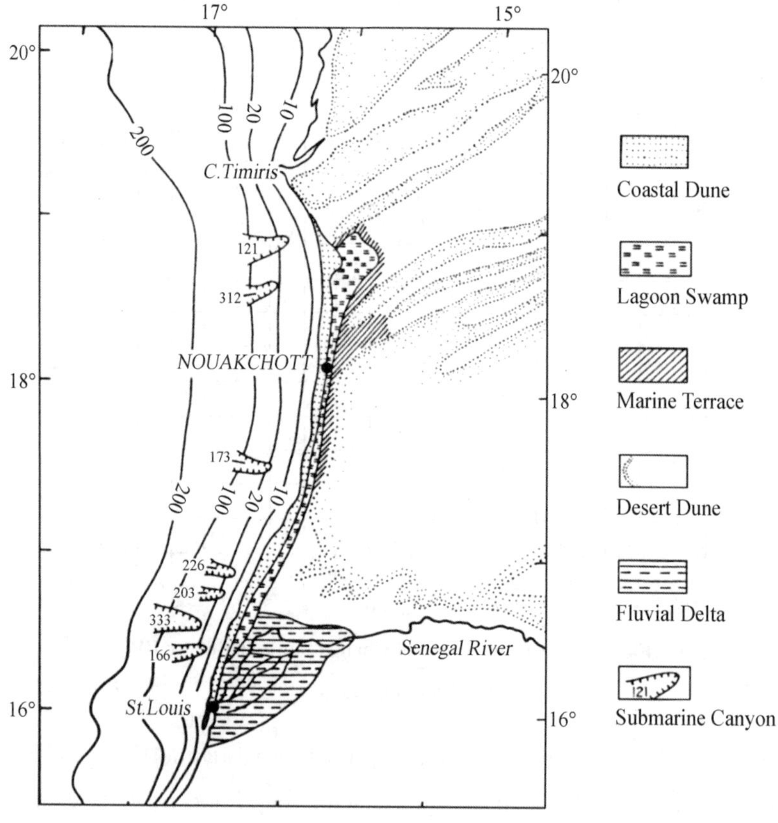

Fig. 6 Coastal Map of Mauritania

The lagoons are 2～3 km wide, and located between the marine terrace and coastal dunes. The bottoms of the lagoons are about at sea level, and most are dry depressions with a salt crust on the surface or temporary water in the depressions. Lagoon deposits consist of sandy clay interbedded occasionally with gypsum and silt. ^{14}C dates of surface deposits of the Trimirst lagoon in the north are 5 745±100 yr BP, 5 910±115 yr BP of middle lagoon.

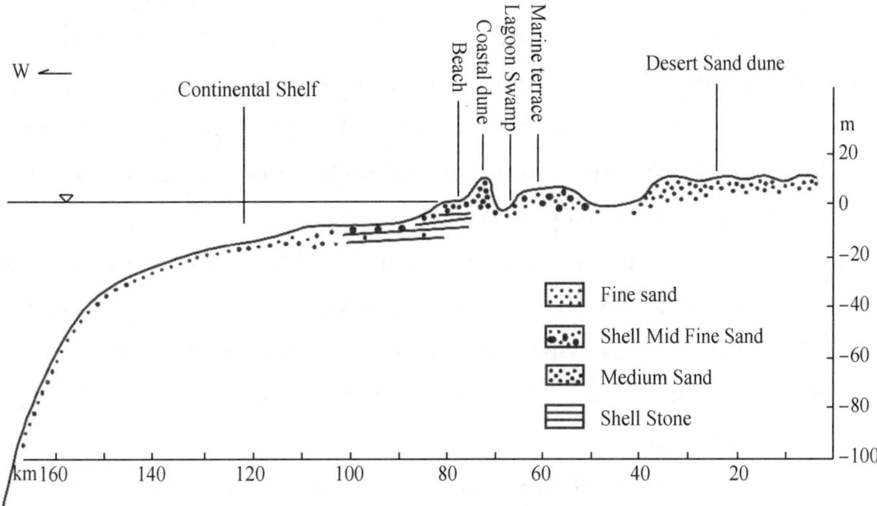

Fig. 7 Coastal Zone Profile of Mauritania

Sand dune chains extend along the whole coast zone of Mauritania, with the trend parallel to the coastline from north to south. These are 30 m high on the average, the width of sand dune belt is 400 m around the area of Novakchott, 200m wide in northern area. In the southern part 30 km from Novakchott, the width of the dune belt is only 20 m and the heights of dunes are less than 2 m as a result of wave erosion especially during storms. Sand dune belt is 3. 5 km wide in the Senegal River delta, the dune ridges are 10 m asl and covered with luxuriant bushes.

The coastal dunes consists of well sorted pure medium sands (0. 35 mm), they were blown from beaches by landward winds. The coastal dunes distributed around the outer fringes of the Sahara, are not the continuation of the inland desert dunes. Coastal dune sands are coarser than desert sands and were sea bottom sands. Shallow water waves transport sea bottom sands up to shore to form sandy beaches. The dominant northwesterly sea winds, blow beach sands onshore during ebb tide that form coastal dunes. The time for the dune formation is 5 270±170 yr BP, according to ^{14}C data of shells in the dune sands (Table 4).

Table 4 ^{14}C Data of Dune Coasts in China and Mauritania

Location		^{14}C data (yr BP)
China	Delta lobe Ⅰ of the Luanhe River	8 025±105
Mauritania	Marine terrace	5 755±120
	Trimirst Lagoon Middle Lagoon	5 725±100 5 910±115
	Coast sand dune	5 270±170

Fine to medium sands are widely distributed along Mauritania coast to the water depth of 20m, where the gradient of subaquatic slope is 1/1 000. The sands are clean, well sorted (S_0=1. 63～1. 72) and with high roundness (49. 4～52. 92). Surface texture of quartz sands indicates that 59. 2～89. 9 percent of sea bottom sands are marked with the characteristic eolian features. It is possible that these sandy sea bottom sands originated from the desert, but, after a period of Holocene transgression, the sands became a new source of the coastal dune.

The Mauritania coast is located at the outer boundary of the Sahara Desert and under the northeast trade wind zone. This would support the view that inland desert sands might be blown by these winds towards west reaching Modern Seashore to supply sands for the coastal dunes. However, the results from field work do not support this theory by following two reasons:

1. This size range of sand grains is different in that the desert dunes consist of fine sands and the coastal dunes of medium sands (Table 5).

Table 5　Comparative Data of Coastal Dune and Desert Dune Sand

Location	Grain Size	Colour	Vegetation	Trend of dune ridgs
Coastal dune	Medium 0. 3～0. 4 mm	White/Grey	good	N-S
Desert dune	Fine Sand 0. 11～0. 16 mm	Light Brown	poor	NE-SW

Observations show that the wind transports 91. 1～93. 3 percent of sand in saltation and only 0. 4 percent of dust in traction about 10 cm above the ground. Grains smaller than 0. 074 mm are transported in suspension by air currents. Most inland dunes around sea shore are fixed by calcic cover and vegetation on the crest; in only few locations near high way city flowing sands move towards the coast. Suspension dust may be transported from inland to sea, but it is located inside the coastal dunes along 315 km of the coastline. Sediments in the lagoon depressions are sandy clay. Clayey-silt and gypsum, which also indicate that these are not inland sands passed over to the coast zone.

2. Coastal zone is mainly influenced by oceanic meteorologic factors according to wind records from 1977 to 1979 at the Novakchott harbour station. Sand start to be moved while wind speed reaches 5. 5 m/sec. , the frequency of this speed for landward winds is 35. 35 percent, and is 24. 87 percent for seaward winds. Thus, seawinds are important in the processes of moving beach sand to form coastal dune.

Mauritania sand dune coast is another example of processes that show: 1) Sand dunes, that trend parallels to the coast line, consist of sands mainly from the sea bottom and have been formed by sea winds. 2) Mauritania Sand Dune Coast formed during 5 270 ±170 yr BP. This corresponds with the second phase of New Glacial, i. e. 5 300 yr BP in northern hemisphere. This was an environment of lower sea level stand and cold windy climate.

The above two examples demonstrate that the formation of sand dune coasts is a process closely related to land-sea interaction under Arctic Climate.

References

[1] C. A. M. King. 1972. Beaches and Coasts (2nd Ed.). 165 - 190 and 314 - 364.

[2] Gao Shanming, Li Yuanfang, An Fengtong and Li Fengxin. 1980. The Formation of Sand Bars on the Luanhe River Delta and the Change of the Coastline. *Acta Oceanologica Sinica*, 2(4): 102 - 114 (In Chinese).

[3] Denton, G. H. and Karlen, W. 1973. Holocene Climatic Variation—Their Pattern and Possible Cause. *Quaternary Research*, 3(2): 155 - 174.

[4] Wang Ying. 1964. The Coastal Geomorphology of the Northern Part of Bohai Bay. Oceanographic Theses, (3): 25 - 35 (In Chinese).

[5] Wang Ying, Ren Mei-e and Zhu Dakui. 1986. Sediment Supply to the Continental Shelf by the Major Rivers of China. Journal of Geographical Society, London, 143: 935 - 944.

[6] Yang Huairen and Xie Zhiren. 1984. Sea-level Changes Along the East Coast of China Over the Last 20 000 Years. *Oceanologica et Limnologia Sinica*, 15(1): (In Chinese).

[7] Yu Ge. 1985. Holocene Climatic Change in the Eastern China. Master Theses of Nanjing University.

[8] Zhang Zhenyan. 1982. The Analysis of the Characteristics of Coast Zone Climate of Jiangsu Province. In: Proceedings of National Symposium on Coast and Beach Resources of Jiangsu Province, Nanjing (In Chinese).

[9] Zhu Dakui and Wang Yuding. 1981. Coastal Geomorphology of Quaternary of Mauritania, West Africa. World Geography, 91 - 96.

[10] Zhu Dakui and Wang Yuding. 1983. Dynamic Geomorphology of Mauritania Coast, West Africa. *Journal of Nanjing University (Geography)*, (S1): (In Chinese).

The Sedimentation Processes of Tidal Embayments and Their Relationship to Deepwater Harbour Development Along the Coast of Hainan Island, China —A Summary on the Canadian-Chinese Cooperative Research Project*

The joint project was carried out by the Marine Geomorphology and Sedimentology Laboratory (MGSL) Nanjing University (China) and the Atlantic Geoscience Center. Bedford Institute of Oceanography (Canada) during a four-year period of December 1988 to December 1992. The project had been approved by the State Commission of Science and Technology of China (file No. 1988 010), and supported by the International Development Research Centre of Canada (IDRC, file No. 3-p - 87 - 1003 - 02. China, Harbour Siltation Project), which has contributed 361 230 Canadian dollars for the acquisition of equipment to support MGSL's ocean survey, and 130 790 Canadian dollars to MGSL for the expenses of scientific exchange, conference, data publication, etc. The joint cooperative team had also received strong support from the State Education Commission of China and the Bureau of the Hainan Harbour Administration. Professor Wang Ying was appointed as chief scientist of the joint studies. The cooperation in Hainan Island was going smoothly and harmonically through four stages, and concluded with great success.

1. The first stage of the project was to set up two complete new labs with advanced equipment in Nanjing University: one is Coastal-Ocean Survey Laboratory with the equipment as: Transponder navigation system, Dual frequency echo sounder, Geopulse seismic profiler, Lehigh core system, Scuba diving equipment, under water video camera system, Rubber boat and Personal computers; the other is the isotope Pb210 Analysis Laboratory with chemical analyses lab for sample treatment and the alpha spectrometer, and also the improved size analyses equipment with computer system; then, followed by a training program for younger scientists and technicians for a total

* Ying Wang: *Annual Science Report—Supplement of Journal of Nanjing University*, 1996, Vol. 32(English Series 4): pp. 18 - 19.

number of ten persons.

2. The second stage of the project involved a survey along the southern and western coasts of Hainan Island, including the embayment of Sanya, to Yangpu, Haikou, and Dongshui by a joint Chinese-Canadian Scientific team during November to December of 1988. 14 stations were organized for a full tidal period hydrologic survey; 27 stations for the survey of temperature, salinity and density of sea water, suspended sediment and the data analysis of wind, wave and current. It was the first time in history to survey sea bottom sediment strata by seismic profiler for a total of 130 km in that coastal area. 19 sediment cores were collected from coastal sea bottom, and a total of 1 100 samples were analyzed for size, heavy mineral, clay mineral, pollenspore, foraminifer and sedimentary rate. Through the survey 6 younger scientists and 6 post graduate students were also trained to operate the ocean survey equipment.

3. The third stage of the project involves the analysis of data collected by the joint team and the summarizing of 58 papers written through the joint study, and 8 reports for applied items. The Pb210 data got excellent remarks from the world wide comparative studies carried out by the International Atomic Energy Agency in Monaco.

4. The fourth stage was the holding of the International Symposium on Exploitation and Management of Island Coast and Embayment Resources (ICER) during 2～10th November of 1990 in Haikou city, Hainan Province, sponsored by the joint project and chaired by both Professor Wang Ying (China) and Dr. C. T. Schafer from Canada. 81 participants from 10 countries (Australia, Bangladesh, Canada, China, Denmark, Germany, India, Japan, Korea and U. S. A.) attended the Conference and contributed 81 abstracts. As a result, a book of "Island Environment and Coast Development" was published in 1992, which included 13 papers from the joint studies.

The joint study was examined and evaluated by IDRC in 1993 as their end-of-project review, and forwarded a summary to both MGSL of Nanjing University and AGC of B. I. O, during March 1994. That the evaluation has recognized that the China harbour siltation project is "the one of ten successful projects in IDRC during last two decades" and "the research carried out by the two partners is at the leading edge of sedimentation and erosion studies today, the quality of the data and the validity of the conclusions. The research partners have succeeded not only in generation good sciences, as evidenced by the number or scientific publications produced, but also in disseminating their result to user and in making decision-makers aware of the consequences of different development scenarios". "Capacity-building through the procurement of equipment and through the practical training of scientists and engineers was an important part of the project". In December 1994, the cooperative study was evaluated by a joint evaluation committee consisting of Chinese, Canadian and American marine authoritative scientists, which was organized by State Education Commission of China. The conclusion is "the research carried out by the project partners is at the leading edge of modern sedimentation and erosion studies. The results represent a significant contribution to: (1) the theory of coastal erosion and deposition (2) embayment tidal dynamics and sediment transport (3) the coastal evolution process and (4) application in harbour construction. The partners have reached the international leading level in coastal scientific studies".

The scientific results of the cooperative studies can be summarized as follows:

1. Tidal inlet embayment in Hainan may be divided into three types according to genetic factors, dynamic environment, coast evolute geomorphology and sedimentary characteristics. (1) The sand barrier and lagoon type of Sanya inlet can be expanded as a larger harbour, but need to retain the tidal presume; the critical value is $P \geqslant 1\,510\,000\ m^3$. (2) River mouth inlet type can be used as harbours of medium or smaller size. (3) Tectonic-submerged inlets as Yangpu Bay can be set up for a larger harbour, and two of 20 000 ton docks had been set up since then without siltation problems.

2. To indicate the sedimentary dynamics and sedimentary rate of the above embayments.

3. To conducted the Holocene volcanic coast evolution and related coral reef development as the geothermal condition there is favourable for their growth. There might be a crustal extension happening under the sea bottom of Beibu Bay.

4. Through the Canadian-Chinese cooperation a firm foundation has been set up for developing the advanced State Pilot Laboratory of Coast & Island Exploitation in Nanjing University today.

火山海岸与环境反馈
——以海岛火山海岸为例*

火山海岸是指沿火山锥、火山口和裂隙喷溢带所发育的海岸，与单纯由火山岩构成的海岸有区别。在这一特定环境下的海陆交互作用，具有表面地质过程与深部构造活动相结合而互有影响的特点，其结果不仅仅影响到局部海岸变化，而且会对区域环境产生一系列影响。现今地球上的活火山约有 500 多座，死火山则在 2 000 座以上[1]，主要分布于地表 3 个不同地区，即板块边缘、洋中脊和板块内部[2]。其中，最著名的火山密集分布带为环太平洋火山带。环太平洋火山带沿南、北美洲西岸、阿拉斯加半岛、阿留申群岛、堪察加半岛、日本、我国台湾岛、菲律宾群岛至新西兰，共有活火山 370 多座，占全球活火山数量的 75%以上，因而被称为“火环”[2]。这些火山以中酸性岩浆型居多，其中多处发育为火山海岸。本文以作者调查过的 3 处海岛火山海岸为例，探讨火山海岸特征和环境影响：以中国海南岛西北部的母鸡神—龙门火山海岸为主，与处于相同地质构造背景——板块内部的美国夏威夷火山，以及环太平洋列岛的日本樱岛火山等做对比，后两处火山海岸仍处于频繁的火山喷发环境中。

1　现代火山海岸类型

1.1　中国海南岛西北部火山海岸

该火山海岸沿北部湾东部分布，从陆向海经历过更新世晚期及全新世中期的 5 次火山喷发活动，现代火山海岸分布于洋浦以北的莲花峰和笔架山沿岸，是沿 NE 向断裂带的全新世火山活动带发育的[3]。火山活动系沿密集成串的小火山口发生的裂隙式喷发，火山喷发岩墙依附于早期火山的岩壁上，未形成明显的火山锥。火山喷发物在堆积过程中经构造变动出现断层，再经海浪冲蚀成崎岖的海蚀型火山港湾岸。该火山海岸以母鸡神段和龙门段最为典型。

1.1.1　母鸡神火山海岸

在长度约 1.8 km 的母鸡神火山海岸岸线上，沿断裂带发育着 15 个串珠状的小火山口，经 3～4 次喷发的玄武岩层呈间断性接触。喷发口多为椭圆形，直径 7～14 m，外围是气孔玄武岩，具垂直节理；内壁为致密玄武岩，表层成为烘烤层；口内堆积着熔渣状玄武岩和集块岩。这些小火山口喷发物不整合地覆盖在莲花山火山堆积物(第 4 次喷发)之上，向海侧，火山口喷发物与海相沉积物呈互层状，层理清晰。沿这一系列小火山口、火山颈和裂隙喷发岩发育了海蚀型火山海岸。火山颈呈现为岬角或海蚀柱，后者仍保留着高出

---

* 王颖，张永战：《第四纪研究》，1997 年 4 期，第 333－343 页。

现代海平面 1.7 m、4.0 m 及 6.0 m 的三层海蚀穴。火山喷发口或形成凹入式海蚀崖，或形成环状岩墙和堆积着角砾的磨蚀台地。

母鸡神系沿火山颈发育的海蚀柱，因有数层海蚀穴相间，使此"颈柱"构成抱窝母鸡形态，故名之。自母鸡神向南至神尖岬之间为一海湾，湾顶为棕黄色石英长石砂质海滩，与火山口基岩性质迥异，海滩砂并非现代海岸海蚀产物，而是火山喷发所阻滞的河流自上游花岗岩山地带来的泥沙。海湾围崖是 10 m 高的海积阶地，由海滩砂与海湾粉砂相物质组成，其上覆盖着晚更新世喷发的玄武岩层，抬升海滩砂与现代海滩砂质地相同，显然，该处原为一河口湾，后来，由于德义岭和莲花山火山喷发，熔岩流迫使河道南迁并堵塞了河道，从而使之发育成小海湾。全新世母鸡神和神尖一带火山喷发，使海湾的玄武岩及下伏之老海滩和海湾沉积抬升 10 m，成为海积阶地和现代小海湾背侧之围崖。当年的海岸沙坝被抬升成高达 15～20 m 的沙丘状堆积，但丘顶组成物质却为粗砂并含海滩岩块和珊瑚块砾。因此，母鸡神和神尖段火山海岸以小火山口形成海蚀柱、海蚀崖，以及全新世火山爆发使局部海湾抬升成海岸阶地为特征。小范围内的局部抬高突起，反映出火山构造抬升活动并非均一性的缓慢运动。

1.1.2 龙门火山海岸

龙门火山海岸位于母鸡神海岸段的东北方，由密集的火山口群与熔岩流和岩块构成，构造-海蚀地貌显著。呈 NW 向的岸线沿笔架山的火山喷发物（玄武岩与凝灰岩互层）构成岬角和海湾，并被全新世火山喷发活动抬升为 20 m 高的海岸阶地。海湾则成为高5 m、具有海相砂砾层覆盖的基座阶地，海湾形态被保存，形似巨型扶手椅。龙门兵马角盾形火山叠加在 20 m 高的海岸阶地上，两者共同组成现代海蚀崖。崖前方由一系列火山口组成数十个小海湾。火山喷发口直径 15～20 m，底部保留着熔岩沸腾时的形态，岩浆溅起高达 40 cm，犹如沸腾的粥锅骤然冷却而凝固。此熔岩经热释光定年为 4 000a. B. P.[3]，系全新世中期火山喷发形成。沿火山口冲蚀的海蚀崖呈圆盆形或弧形斜坡、弧形悬崖、箱形巷道及石柱兀立的岩滩。由于海浪沿玄武岩节理（NE10°或 NW30°走向）冲蚀，发育了海穹和海蚀穴，龙门即是沿节理形成的高 8 m、宽 10 m 的海穹，目前海穹底部已高出现代岩滩 4 m，反映了近 4 000 年来的海岸抬升量。

1.1.3 火山活动特点

海南岛西北部火山海岸外缘，沿玄武岩岸基发育了珊瑚岸礁，仅有的离岸礁——大铲礁和小铲礁亦发育于 NE 向的玄武岩基础上。海南岛珊瑚礁发育始自全新世，北部湾岸礁分布超过 20°N，与玄武岩喷发后形成礁基并使海域增温密切相关。北部湾中的涠洲岛，是由火山口发育而成的基岩岛和港湾，沿涠洲岛基岩上亦发育着珊瑚礁。火山海岸与珊瑚礁伴生的关系反映出火山喷发活动对海域环境之影响。

海南岛西北部火山活动具有继承性、多发性的特点，自东至西火山喷发的时代渐新（从晚更新世至全新世），喷发规模减小，喷发形式由盾形火山、马鞍岭火山、多孔火山锥发展为系列小火山口裂隙喷发。火山活动在短期内改变了区域环境特征，尤以河流断流、改道及地形负正倒换为显著。整个海岸带的地势、坡度、动力条件、沉积特征以及海岸演变趋势均受到火山构造活动控制。火山海岸发育受到构造内动力、波浪、潮流、风化作用等外动力以及生物作用的多重影响。

海南岛西北部晚更新世前3次喷发的玄武岩为橄榄拉斑玄武岩和橄榄玄武岩，第4次喷发的莲花山火山为拉斑玄武岩；全新世喷发的龙门和母鸡神火山为拉斑玄武岩。据$MnO\text{-}TiO_2\text{-}P_2O_5$判别图式，该区橄榄拉斑玄武岩介于洋岛玄武岩与洋中脊玄武岩区间界限上；橄榄玄武岩介于洋岛玄武岩区内，而拉斑玄武岩则属于洋中脊喷发型。此分析结果极为重要。北部湾火山分布反映出自海湾向陆地时代由新至老的扩展活动。该处地壳厚度约为30 km，上地幔拉斑玄武岩上涌出露，反映湾底的张裂构造活动[3]。该处可能为发展中的裂谷。

1.2　美国夏威夷火山海岸

夏威夷群岛位于太平洋中部，是近7 000万年来热点持久活动所导致的火山喷发、造陆，以及太平洋板块缓慢移动而逐渐形成的一系列岛链(图1)。当代的活动火山主要位于夏威夷岛、Kilauea及Mauna Loa火山，仍呈周期性喷发，使岛屿面积不断地扩展。该岛最高的火山峰冒纳凯阿海拔4 200 m，Kilauea火山口海拔高度为1 247 m，Mauna Loa为4 169 m，当地太平洋深度达4 900 m，该火山基底的直径达100 km，考虑到压实和风化侵蚀作用，则Mauna Loa火山堆积厚度至少达10 000 m，火山堆积物体积可达26 000 km^3。所以，火山造陆和板块移动形成岛链是大洋火山活动的两大重要环境效应。

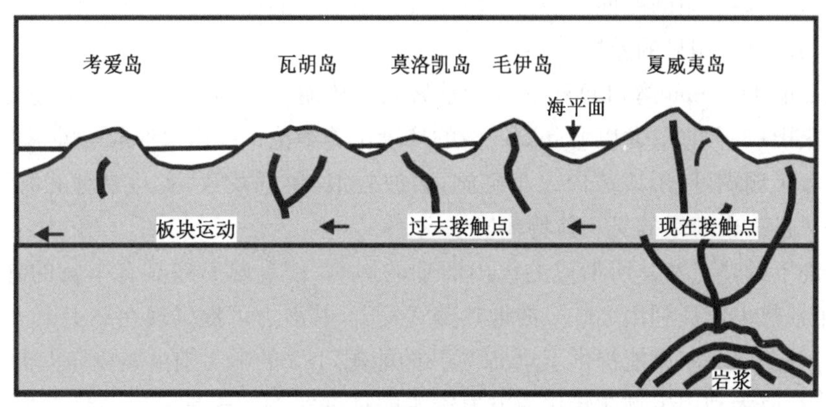

图1　夏威夷火山岛链形成图式

夏威夷火山岩浆为基性，SiO_2含量小，介于45%～52%间，铁、镁氧化物含量少于20%，其温度介于1 000 ℃～2 000 ℃间或更高，黏滞性小，气体可自岩浆中逸出形成众多气孔，喷发时不形成强烈爆炸，多以涌流状外溢，故其堆积成盾形火山。海底熔岩流则凝结为绳状或枕状堆积，目前遍布于Kilauea火山海岸带。岩石主要为拉斑玄武岩，种类多样性高于典型的海洋玄武岩，可能含Si，Al较多，而含Ca，Fe，Mg等较少，包括粗面岩、响岩、流纹岩等[2]。Kilauea火山呈间歇性喷发，熔岩浆自3 300 km深处的地幔上涌，沿主通道或裂隙中喷溢而出，以东裂隙喷出为主，其次为南裂隙及西南裂隙。位于其西北侧的Mauna Loa火山，熔岩流沿山地的NE裂隙和SW裂隙外溢。黑色的玄武岩流原野崎岖不平，遗留下1940年、1942年、1949年、1950年、1954年、1959年、1969年、1971年、1974年、1975年、1982年、1983年、1984年、1990年和1991年等不同时期喷发的岩层。

Mauna Loa-Kilauea 复合火山锥体沿太平洋形成长达 100 多 km 的火山海岸：岬湾曲折的海湾由不同时期的岩流舌构成，海湾顶已发育了松软、耀眼的黑砂海滩。沿着熔岩层内的空洞和通道可窥见熔岩滚动、气体逸出及冷却时的情景。海浪冲刷形成的海穹、海蚀通道和海蚀天窗，其景象蔚为壮丽。热带海洋气候下的火山岩风化形成了新的沃野，繁殖着茂密的椰林、野菠萝、羊齿类植物和灌木花丛以及禽、兽、龟、鱼等，形成新的火山海岛-海岸生态系，为人类生存发展提供了优越、便利的条件。夏威夷火山海岛、海岸的这一环境效应亦是十分突出的。

1991 年 5—6 月，作者考察夏威夷岛南部火山海岸，当时，尚处于 1990 年 Kilauea 火山喷发的后期阶段，炽热的岩流沿山坡低洼带缓缓地向海推流，吞没田野，推燃树木和房屋，所至之处一片火海。更为壮观的是，熔岩流自海岸悬崖倾流入海时，形成了耀眼的熔岩流瀑布。初势强，海浪熄不灭岩火，进而形成沸腾的热气柱直冲云霄，烟雾腾腾，数千米外仍散发着浓烈的硫磺气味。

1992 年夏，作者再次考察夏威夷岛南部火山海岸时，Kilauea 火山岩浆喷发活动已经停息。新凝结的海岸地面呈圆环状的垄岗，崎岖难行。被推斜的残留椰树已形成披挂着岩浆的熔岩树，新凝固的玄武岩表面呈针状和放射状的晶形结构，清晰地保留着岩流推挤、移动、卷蚀、覆盖等各种形态。我们急步行走意欲赶赴熔岩流瀑布处观看其凝固形态，但在相距约 2 km 处即被制止了，因为浓烈的熔岩水蒸气弥漫于空中，会使人窒息。返程的汽车，每每为熔岩阻断路面而绕行。海岸带火山活动的环境效应、持续的时空范围等与陆地火山活动又有明显的差异。

瓦胡岛的 Hanama 湾和 Diamond 岬是著名的旅游火山海岸，均位于该岛的东南端，是由两个火山口所组成，尤以 Hanama 湾的环境优美绝伦。火山口形成马蹄形海湾，湾内沿岸又发育了珊瑚礁，形成湾内之小礁湖，碧波轻浪，鱼群戏泳，实乃罕有的海岸憩游海湾。这是热带火山海岸的又一独特环境效应。

由于岛下的热点及其所形成的火山活动的影响，使夏威夷岛具有丰富的地热资源。目前，一个地热发电厂利用 360 ℃的地热蒸气发电，其潜力可提供现在岛上电力需求量 5 倍的能源。在瓦胡岛，风能提供了全岛 10%的能源，1/3 的电力则由燃烧岛上大量的甘蔗渣所提供[4]，成为利用生物能转化为电力资源的良好实例。这是夏威夷热点型火山海岸的另一环境效应。

1.3 日本樱岛火山海岸

位于大隅海峡北岸、鹿儿岛海湾东侧的樱岛火山锥高耸入云，黑色山顶烟云弥漫，给人以阴森神秘之感。

樱岛火山系于 3 000 万年前开始爆发的周期性喷发的火山锥。1914 年喷发的“大正熔岩”使樱岛与大隅半岛相联，其上植物生长茂密。1940 年喷发的“昭和熔岩”地面植物稀少，两期火山喷发物界限明显。鹿儿岛海湾系沿几个沉陷的火山口所构成，与樱岛火山构成现代火山海岸。作者于 1983 年 10 月 21—22 日在樱岛火山考察时，火山频频爆发，吼声隆隆，山体颤动，黑烟滚滚，呈蘑菇云状上升，继之，黑烟弥漫笼罩了山顶两个火山口，渐渐地固体喷发物——火山灰、火山砂、火山渣随气体散逸，飘落于山体周围和山下。爆发时作者于半山观测台处摄影、录音，衣、帽、脸、颈皆为火山灰所玷污。火山灰掩埋了山坡下数米高的神社，仅露出其顶部。山麓和海湾沿岸有多种热带植物繁殖，并有多处被辟为农田，仅在湾岸岬角处出露玄武岩，海湾湾顶处则为玄武质黑色沙滩。

由于火山喷发频繁，沿山坡堆积着大量的火山灰和碎屑物质，因此，一旦降暴雨(在10分钟内形成30～40 mm的降水时)，樱岛火山区未胶结的火山碎屑堆积物即可形成爆发的泥石流，短瞬间沿坡冲下而形成灾害，当地人称为“黑神川”。鹿儿岛大学工学部采用沟道引排的方法，以减少泥石流灾害。

三种不同类型的现代火山海岸，虽在地貌和沉积形式、形成演变方式等多有区别，但对于邻近地区的环境影响却都是显著的，它涉及到大气、降水、地表、生物群落等各个方面，而且造成多种灾害，危及环境和人类生存。因此，有必要研究火山海岸及其环境效应。

2　海岸火山活动的环境效应

火山活动对环境演变的影响是显著的，与海陆交互作用带的其他众多作用因素相结合，火山海岸环境变异更具有特殊的内涵。其环境影响可概括为地貌效应、海岸效应、气候效应、生态效应和灾害效应等几个主要方面。

2.1　地貌效应

火山活动可能是控制地球形状和决定地表海陆分布格局的重要因素之一。有一种全球构造理论认为，早期地球在星云收缩过程中，由于陨石撞击、放射性同位素蜕变和重力收缩等原因所产生的巨大能量，使地球内部温度不断增高，含水系统中的硅铝物质首先被熔化。因受到当时地球南方宇宙空间里存在着的天体的强大引力作用而向早期地球的北极地区移动并不断集聚，最终冲破原始地表从北极地区大规模涌出，成为现今世界大陆最初的物质基础，随后南移，并逐渐形成现在的海陆分布格局。因此，今日的北冰洋是当时地球这一座“超级火山”的“火山口”，是地表陆地的源头。正由于这种构造活动控制，使地球表现为北极凹陷为洋，而南极凸起为陆的形状。同时，北极地区是一个近乎圆形的广阔海洋——北冰洋(真正的“地中海”)，南极地区则为接近圆形的巨大陆地——南极洲(真正的“洋中陆”)，它们分别为地表陆地由北向南运动的起点和终点，地表其他陆地(除南极洲外)则围绕着北冰洋呈放射状向南展布[1]。根据这一理论，火山活动可能是控制地表海陆分布格局的重要因素。

从地表最基本的大地貌而言，火山活动改变了地表海陆分布格局。海底扩张、大陆漂移、板块构造等都与火山活动密切相关。例如，红海在30 Ma B. P.原是一个稳定的大陆区，经29 Ma～24 Ma B. P.红海南部地区的柱状玄武岩熔岩喷发和21.5 Ma B. P.的再次火山活动，古老的大陆地壳开裂扩张，至20 Ma B. P.红海洋壳开始发育，并在持续的扩张过程中逐渐形成这一地区现代的海陆分布格局。红海由于现代海底火山活动，目前仍处在不断扩张的过程中[5]。

从小范围而言，火山活动形成新的岛屿，迅速造陆，并由于板块移动，在夏威夷形成著名的火山岛链地貌景观；在海岸带动力作用下，形成海蚀柱、海穹，造成岬湾曲折的海岸线和火山砂砾质海滩等火山海岸地貌组合；由于熔岩流的作用和凝结，形成崎岖不平的熔岩流原野、台地和陡崖等地貌组合。火山活动一方面有抬升作用，抬升海岸为阶地，抬高海滩、沙坝为丘垄，抬升河口形成深切河口湾，并使河流断流和改道，不仅改变了局部海岸动力过程，而且造成地形正负倒置；另一方面也有回弹性的沉降作用，从而使火山喷发口周边范围内的原始地形发生变化。海南岛火山海岸剖面中海岸沉积层厚达 16 m，反映出该段海岸曾经历过沉降，至第 5 次火山喷发，因紧邻喷发口又被抬升为海岸阶地。由此，也形成了丰富多彩、引人入胜的独特景观资源。

2.2　海岸效应

火山活动改变了海岸坡度，由此使海岸海洋动力因素和作用过程发生变化，从而改变了海岸类型及其发育演变史。海南岛母鸡神火山海岸剖面[3]反映出由于火山活动致使该段海岸发育经历了三次潮滩环境和两次海滩环境。海岸坡度增大时，波浪作用为主，堆积砂砾质海滩；海岸坡度减小时，平缓宽阔之潮间带潮流作用堆积粉砂、黏土质而形成潮滩。

2.3　气候效应

火山活动所引起的气候变化效应，历时长，范围广。长期以来，有一个未能完全证实的假设，即大冰期与火山活动增长期之间可能存在着某种因果关系。一方面，全球性火山活动时期数量惊人的火山灰尘把巨量的太阳辐射遮掩，使得全球气候变冷，地表冰川发育；另一方面，在大冰期中，水流被固定于大陆冰盖内，海面大大降低，结果使得大洋地区和大陆地区的正、负重力异常值明显增大，因而在海陆交界带上由此所产生的不平衡力之间的相互作用也会随之加剧，导致火山活动的相应增强[1]。显然，在冰期与火山活动增长期之间存在着一定的正、负反馈机制，已有一些实例从不同侧面证实了这种关系。1883 年喀拉喀托火山喷发，火山尘埃沉降缓慢，由于大量的火山尘埃笼罩地表，长期遮天蔽日，使全球温度下降约 0.5 ℃，这种变冷效应持续约 10 年。1815 年印度尼西亚坦博拉火山喷发，产生了更加明显的气候变冷效应，使 1816 年成为全球闻名的“无夏之年”[4]。

火山活动对我国的气候影响十分显著。由于喷发到平流层的火山灰尘幕在此长期滞留，使得我国东部长江以北地区夏季短波辐射加热显著减少[6]。据 1951—1975 年 6 次强火山爆发后一年与火山爆发当年 6—8 月雨量距平百分率差值分布情况的统计资料，火山爆发后一年我国东部 6—8 月的雨量有两个增加区：一是以长江中游为中心的长江干流到江南地区，另一个为华北到东北大部地区[7]。据洞庭湖洪涝灾害的统计资料，1951—1988 年间的 8 次强火山爆发后洞庭湖地区发生过 5 次重大洪涝事件(而同一时期，这一地区共发生过 6 次重大洪涝事件)，其中有 4 次发生于强火山爆发的次年，1 次发生在强火山爆发的当年[8]。因此，由于火山活动明显的气候效应，使得强火山爆发事件成为预测洞庭湖地区重大洪涝事件发生的先兆指标。

此外，火山喷出的富硫气体进入大气层中可形成酸性水滴云团，或遮蔽日光或降酸雨，亦造成气候变化。

2.4　生态效应

火山活动不仅直接形成玄武岩台地，而且使局部海域增温，为珊瑚的生长繁殖创造了适宜的条件(稳定的礁基和合适的水温)。因此,火山海岸与珊瑚礁往往表现为伴生关系,现在的海南岛火山海岸、夏威夷火山海岸均如此。地质历史中的红海火山海岸也体现出这一特点。

热带海洋性气候条件下的火山岩风化形成富含矿物质的肥沃土壤原野,为椰林、菠萝、大量鲜艳的灌木花丛和多种飞禽走兽提供了良好的生活环境,同时,丰富的营养物质和大量珊瑚虫则维持了众多色彩斑斓的龟、鱼等热带海洋生物群落的生存条件。两者结合,形成了极富特色的热带海岸生态系,不仅为人类生存发展提供了优越便利的自然条件,而且成为海岸带旅游资源的重要组成部分。大量的热带植物,亦成为潜在的新型能源——生物能源。

2.5　灾害效应

火山活动形成美丽的火山海岸环境,对人类生存环境有益。同时,它所造成的灾害效应亦是多种多样,其危害往往非常巨大。

熔岩流温度高达 500 ℃～1 400 ℃,其速度每小时可达数千米,人畜尚可躲避,但其所至之处燃树焚屋,破坏性强。

火山灰砂沉降,掩埋田野、道路和建筑。因其喷出具突发性和爆炸性,并且散开得更快更远,因而比熔岩流更具有危害性。樱岛火山海岸所接受的火山降尘每日达数厘米。美国圣海伦火山于 1980 年 5 月 18 日喷发后,抛掷出的碎屑物质达 1 km^3,使得 150 km 以外中午的天空变黑,约 600 000 t 的火山灰降落在 100 km 外的华盛顿州亚基马市,埋葬了房屋、道路。当火山灰降落仅几毫米厚时,炽热的路面已造成汽车熄火停驶。火山灰降落是大量的、短时间的,居民往往来不及躲避,其危害极大。意大利庞培城即是在公元 79 年时被维苏威火山爆发喷出的炽热火山灰所埋葬的。

火山泥流灾害是由炽热的火山灰融化山顶积雪或促成局部暴雨,使沿坡而下的泥流堵塞河谷继而又促使洪水暴发成灾。这类灾害更易发生于水气充沛、交换活跃的海岸带。1985 年哥伦比亚 Nevado del Ruiz 火山喷发时,灼热的火山灰融化雪盖,形成突发性的火山泥流,顺陡坡倾泻而下,结果造成火山脚下的城镇中两万多人死亡[4]。

炽热的火山云,系火山喷发所形成的、比空气密度大的炽热气体与细火山灰的完全混合物,可发生于火山喷发的任何阶段,其温度高达 1 000 ℃,它沿火山向下运动时,速度可达 100 km/h,因此,造成烧伤和窒息的危害大。1902 年,马提尼克的加勒比岛上的培雷火山持续喷发,5 月 8 日早上,突然喷出大量的炽热火山云横扫圣皮尔城及其港口,在 3 分钟内就造成 2.5 万～4.0 万人死亡、烧伤或窒息[4]。

有毒气体危害。火山喷发出大量有毒的或无毒的气体。其中主要成分是水气和 CO_2,虽是无毒的,但是,若在局部水域中积聚,使浓度增高(＞50%),当气体温度发生变化或受到振动使这些无色无嗅的气体再被释放时,会使人窒息死亡。1986 年 8 月 21 日,在喀麦隆位于裂谷带的 Nyos 湖附近,释放出的 CO_2“气团”飘移至附近城镇使 1 700 人死亡[4],这些 CO_2 是从火山岩浆中释放出来的。同时,大量的含硫及 CO、氢氯酸等成分的有毒气体,不仅直接、迅速地危及生物,而且影响到气候,形成延滞性的影响和危害。

蒸气爆炸灾害。大量海水渗透到火山岛地区岩层内,与下伏岩浆相接触而形成水气,当炽热的蒸气积聚至相当数量后，可沿裂隙发生巨大的爆炸。例如，印度尼西亚的喀拉喀托火山岛于1883年发生的蒸气爆炸[4]，其威力相当于1亿t TNT炸药，3 000 km外的澳大利亚仍可以听到其爆炸声。爆炸形成的火山尘冲入80 km的高空，使多年后仍红尘蔽日，火山灰污染的面积达750 000 km^2。该岛虽无人居住，但由于爆炸震动所形成的高达40 m的海啸，却造成被淹没地域的36 000人死亡。

3　海岸火山活动的预报与减灾

太平洋岛屿火山海岸分布是广泛的（环太平洋火山活动带和太平洋板块内部火山链），但其现代火山活动则是局部的。研究火山海岸特征，可启示人们了解和掌握火山海岸环境发展变化的规律，找出相适应的途径和方法来预报火山活动，从而有效地减灾和防灾，达到海岸资源环境合理利用和持续发展的目的。

环太平洋火山活动带是板块边缘型火山活动带，对其火山活动的预报，要从研究和监测板块边缘区的活动历史和现代活动状况着手。而太平洋板块内部的热点型火山活动，则要监测板块运动速度和研究热点活动的规律，力求达到对火山活动的长期监测和预报。例如，通过对Emperoe——夏威夷火山链的研究发现，火山链最西北部的火山活动于75Ma B. P. 开始，向SE方向火山活动渐新，并在40Ma B. P. 太平洋板块由向WNW方向运动转向NW方向运动[4]。计算表明，夏威夷火山链的火山移动速度达10 cm/a，而太平洋板块离开东太平洋脊的运动速度为6～7 cm/a，因此，热点移向东太平洋脊或东太平洋脊移向夏威夷火山链的速度约为3～4 cm/a[9]。这为预测新的火山活动提供了必要的依据。

火山活动造成的灾害是可以通过短期监测预报而防治或减轻的。其原因在于火山爆发有前兆：

（1）从深部微震发展到地震次数增多而震源渐浅。1959年8月在Kilauea火山下的55 km深处测到地震，此深度相当于该处岩石圈的基底。嗣后，由于岩浆上涌，地震活动变浅，而次数增多，至11月底，日记录小地震达1 000次，结果，熔岩流沿火山侧翼裂隙溢出。这表明，地震活动也可触发火山爆发。

（2）地面膨胀，倾斜上升。

（3）地温增高或火山气体逸出等。

火山活动预报目前还不可能达到短期无误，但当地居民必须警惕并及早疏散到安全区，且应做好不止一次的疏散准备，这是在目前状况下最大程度减灾的最有效措施。同时，以往的一些经验表明，合适的工程措施在一定程度上对于减少灾害损失有益。例如，泥石流活动宜采取樱岛火山区的办法，即开挖预留泥石流通道、人工控制引流等。

位于大西洋洋中脊上的冰岛西南岸外的赫马岛提供了人工控制熔岩流的良好例证[10]。该岛以优良的渔港而著称，它提供了冰岛主要出口产品——鱼类产品产量的20%。1973年，由于裂谷的拉张活动，赫马岛被分裂开而且形成了数月的火山喷发。当地居民在火山活动前数小时内安全撤离，而房屋、商店、农场被炽热的火山碎屑点燃、焚烧或被熔岩流埋葬。但也有很多建筑物因为从屋顶移去炽热火山碎屑负载而得以保存。当熔岩流流至港口威胁到居民的主要生活来源时，政府采取了不寻常的措施：动用所有船只的抽水机，汲取海水来冷却熔岩流，使它变冷、变稠、增厚而阻滞其流动，使熔岩流前缘形成固定的堤坝以阻滞后来之岩流，并浇水使后至的岩流冷却。结果，港口仅部分被熔岩流填充，而大部分港口被保留下来，并因新的岩坝蔽护而使港口的泊稳条件受益。

当火山喷发或涌溢物减少或熔岩流遇到天然的或人工的屏障阻滞时，在火山喷发期间熔岩流的运动被暂时地减缓或停止。熔岩流固体外壳之中的岩浆可以在几天、几周甚至几个月的时间内保持熔融态。这时，将它从一个可能造成大量损失的地区转移到处于威胁之中的财产价值相对较小的地区是有可能的。1983 年初，意大利埃特纳火山出现了一系列断断续续的喷发活动。在此期间，政府采取措施在老火山岩坝上炸开了一个洞，希望将新的熔岩流导引到一个宽浅而无人居住的地区，这将不仅使人口密集区免遭其害，而且可以使熔岩流更大地展布，并迅速冷却而凝固。这种努力仅仅获得了暂时性的功效，部分熔岩流被改向。4 天后，新喷出的熔岩流离开了人工通道，又恢复到其原来的流路，并进而危及到人口居住区[4]。

樱岛火山海岸和赫马火山海岸成功地减灾的实例与意大利不完全成功的举措，启迪着人们进一步采用更有效的措施来防治和减轻火山灾害。

参考文献

[1] 陈年. 1985. 大陆扩散与全球构造. 福州：福建科学技术出版社，133 - 142.

[2] Kennett J. 1982. Marine Geology. Englewood Cliffs: Prentice-Hall, Inc.

[3] 王颖，周旅复. 1990. 海南岛西北部火山海岸的研究. 地理学报，45(3)：321 - 330.

[4] Montgomery C W. 1992. Environmental Geology. Third Edition. Dubuque: Wm. C Brown Publishers. 88 - 113, 319.

[5] 英国开放大学教材研究室编. 1983. 海洋变迁——英国开放大学海洋学教程第十三、十四单元. 北京：海洋出版社，33～35.

[6] 徐群. 1986. 1980 年夏季我国天气气候反常和 St. Helens 火山爆发的影响. 气象学报，44(4)：426 - 432.

[7] 张先恭，张国富. 1985. 火山活动与我国旱涝、冷暖的关系. 气象学报，43(2)：196 - 207.

[8] 叶愈源. 1991. 太阳活动、火山爆发、ENSO 事件与洞庭湖重大洪涝事件. 南京大学学报(自然灾害成因与对策专辑)，129 - 135.

[9] McDougall I. 1979. Age of shield-building volcanism of Kauai and linear migration of volcanism in the Hawaiian Island Chain. *Earth and Planetary Science Letters*, 46(1): 31 - 42.

[10] Decker R and Decker B. 1981. Volcanoes. San Francisco: W. H Freeman. 91 - 93.

潮滩沉积动力过程与沉积相*

潮滩(Tidal Flat 潮坪)为平坦海岸和潮间带滩涂。其范围宽广,坡度近于水平,由未固结的细粒泥沙所组成;涨潮时为海水淹没,落潮时出露;或为裸露的光滩,或者上部长草发育为盐沼湿地;潮滩滩面地貌与沉积具有明显的分带现象,冲淤变化迅速[1]。潮滩广泛地发育于各种气候带,从热带的红树林海岸、温带的平原海岸到极地的永冻土热力海蚀岸皆有分布,温带平原海岸与三角洲顶部发育的潮滩规模宏大。

形成潮滩的自然因素有三方面[2]:

(1) 潮流为控制动力。潮差大(>4 m)的海岸,由于潮量大、潮流急,潮滩发育完好。中等潮差(2～4 m)或小型潮差(0～2 m)则波浪控制作用显著。中小潮差区,由于有适宜的地形配合,也可发育潮滩。如在口门狭窄而内湾水域宽阔的潮汐汊道型港湾或河口湾内,由于纳潮量大,潮流速度具有一定强度携运泥沙并逐渐堆积为潮滩,美国东部的切萨比克湾即为一例[3],其潮差为 1 m,口门很小而湾内水域很大。

(2) 平缓的海岸坡度,通常在 1/1 000 以下,即使在开阔岸段,波浪近岸变形,波能衰减迅速而达不到岸边,因而潮流成为海岸形成的主要动力,在隐蔽岸段(岛屿下风方向的波影区)或狭长的海湾内(吹程小),波浪作用受限制,因而潮流是海岸的主要动力。

(3) 海岸带有丰富的细颗粒泥沙供给,主要是粉砂、黏土、极细砂与细砂,粒径范围自 3Φ 至 9Φ,以 4Φ 至 8Φ 的粉砂物质为主。潮滩的泥沙是来自潮下带或水下岸坡的古海岸带,被潮流横向搬运向岸堆积。中国沿海潮滩的泥沙主要起源于黄河与长江等大河源远流长的入海泥沙;欧美沿岸砂源来自于低海面时期的冰碛物;加拿大芬地湾的细粒泥沙来自沿岸红色砂页岩的风化海蚀产物。全新世海面上升过程中,已成为浅海的泥沙被风浪簸扬再悬浮,继而被潮流搬运向岸,在潮间带逐渐沉积形成潮滩。

典型的潮滩海岸如:中国黄河下游的渤海湾沿岸、废黄河三角洲地区与长江口北部的苏北平原海岸;南美亚马逊河口北部的圭亚那沿岸;北美芬地湾沿岸、美国东南部马里兰沿岸、密西西比河口两岸、华盛顿州沿岸;欧洲的北海沿岸、英格兰东部的沃什湾岸;澳洲东北部的淤泥质海岸等。潮滩宽度自数百 m 至数十 km 不等。

潮流是潮滩形成演变的主要动力,潮汐是周期性海面涨落与水位变动现象,在涨落潮间的干出带(潮间带)形成复合型的动力环境——潮流、波浪、日晒、气流、降水以及生物作用等。潮流是水质点沿海水等压面节奏性振动的坡度上,从高向低的周期性运动。浅水区的潮波,在波峰与波谷时,潮流速度为零,只是在高潮与低潮的中间,流速最大。潮流速度与潮高并不是同步变化的,近岸潮波变形产生的时间不对称导致了流速不对称,涨潮流速大于落潮流速。当潮波向岸运动到达坡度非常平缓的潮滩时,低潮水边线始涨潮,流速从零开始;随着水位增高,潮水淹没了低潮滩部分,流速逐渐增大;至潮位达到高潮与低潮

* 王颖:苏纪兰、秦蕴珊主编,《当代海洋科学学科前沿》,177－182 页,北京:学苑出版社,2000 年。

位之间的半潮位，相当于潮滩中部被淹没时，潮流流速最大，实测表明最大涨潮流速出现于涨潮后一小时（涨潮初期），此时低潮水边线处流速亦达到最大值；过了潮间带中点向岸，最大流速又逐渐降低。这样的流速分布符合滞后效应导致悬移质堆积的基本条件[4]，但对潮间带中下部的砂质物质（推移质）影响甚微[5]。潮间带落潮流速分布规律类似于涨潮流速。低潮位处最大落潮流速小于涨潮最大流速。在潮间带中上部，涨潮流速随时间的变化规律是涨潮后流速很快增至极大，然后逐渐减小；而落潮时流速逐渐增至极大，然后快速减小。潮滩水体最大含沙量出现在最大涨潮流时段。涨潮流含沙量大于落潮流含沙量。外海含沙量稳定，进入潮滩水边线后含沙量增高，并向高潮位增大，高潮位含沙随潮时迅速变化，最大含沙量出现于涨潮开始 1 小时以内，以后含沙量迅速减为 1/2，第 3 个小时约为最大含沙量的 1/19。这种现象反映出：部分泥沙来自潮滩中下部，涨潮时被掀起并随潮流向岸搬运，使悬浮物质逐渐富集。最大含沙量及最大沉积量出现于泥-沙混合沉积带。落潮最大含沙量则在落潮流出现后 1 小时。图 1 为潮滩各部分潮位流速与含沙过程曲线，以苏北新洋港潮滩剖面实测资料为例。

潮流进入潮滩时大体与岸线以锐角相交（20°～30°）。由于水深变浅受滩底摩擦而潮流与岸线夹角变大（中潮位时约 40°），流速递减，至高潮位潮流方向与海岸近似直角（70°～

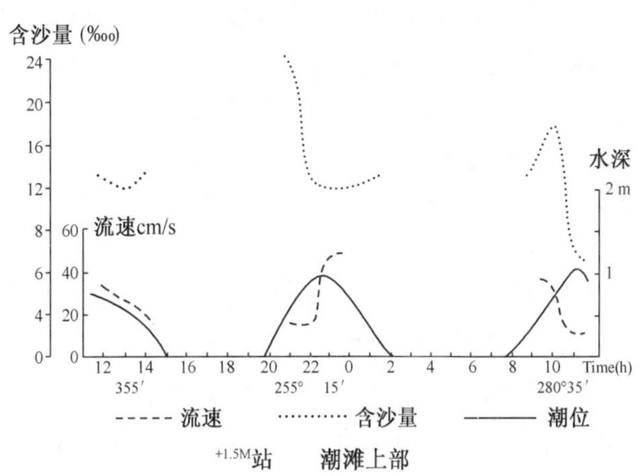

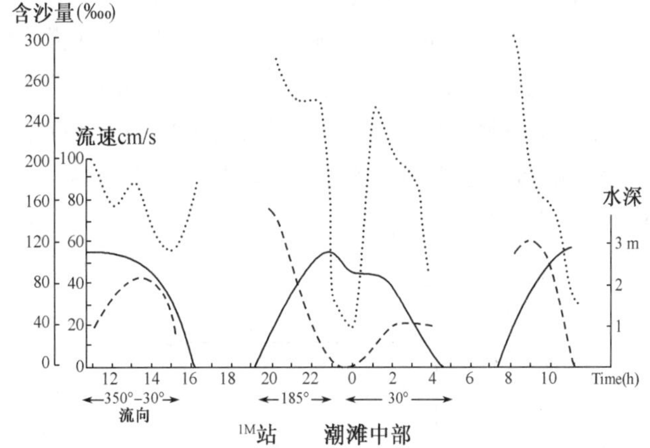

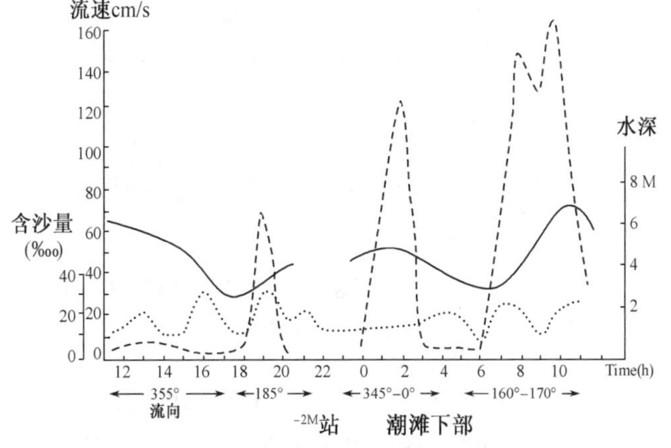

图 1　江苏潮滩水文泥沙过程曲线

（新洋港断面；1980. 9 大潮）

80°)，故呈现为向岸与离岸的往复流。这样，在一个潮周期中泥沙做向岸的净运动。据动力特点，潮间带主要具有三个不同的动力沉积带：潮间带上部为滞后效应起主要作用的泥质堆积带；潮间带中下部为受流速-时间不对称效应控制的砂质堆积带；在两大堆积带之间的潮间带中上部应存在同时受滞后效应与不对称效应影响的过渡地带，发育泥-沙混合带。潮滩沉积物的搬运过程是：潮滩刚上水时水层薄，流速较快，近底层紊动强，使滩上沉积扰动。大风天时滩上呈泥浆状，风浪起着扰动掀沙作用，潮流则将悬移质搬运。滩上沉积物不仅从外海带来，还有很大一部分是中下部滩面上沉积物被风浪水流作用再悬浮，而随涨潮流向高潮位搬运。当潮位增高，水层增厚，底层水流扰动减弱，含沙浓度也逐渐减低，至高潮时为最低值。落潮时因流速小于涨潮流，落潮历时长，含沙浓度较涨潮时小。这种情况促使高潮滩泥质增多。波浪作用使沉积物再悬浮而使潮滩水层含沙浓度增加，泥沙沉积形成各种波痕和斜层理，风暴天气时波浪对潮滩侵蚀和沉积物分选作用明显，形成各种侵蚀形态与风暴沙层。上述为一个潮周期中沉积过程的剖析，而潮差大小的变化受大小潮的半月周期变化影响最为显著。以中国的平原潮滩海岸为例，大潮潮差可达小潮潮差的3倍以上，对潮滩沉积作用造成显著影响。其他不同周期的潮差也类似。图2[5,6]中 A 点为大潮时砂质沉积的上界，B 点为小潮时砂质沉积的上界，在一次小潮—大潮—小潮的周期中，将在潮滩中留下一个向岸尖灭的砂质薄层。大小潮周期的多次重复，在 A、B 之间潮滩剖面上形成泥-砂混合沉积层。在渤海湾大潮与小潮的水位差为1.91 m。这将使 A 与 B 的水平距离达1 500 m，整个潮滩相带发生位移。江苏潮滩大小潮水位差，将使沉积相带水平移动1 900 m，导致潮滩沉积剖面复杂化。大潮的流速是小潮的2倍，在江苏潮滩大潮汛沉积物粒径 $Md=5.4\Phi$(中粉砂)，小潮汛时沉积物粒径 $Md=6.3\Phi$(细粉砂)，这使泥滩带出现粉砂与黏土质纹层。

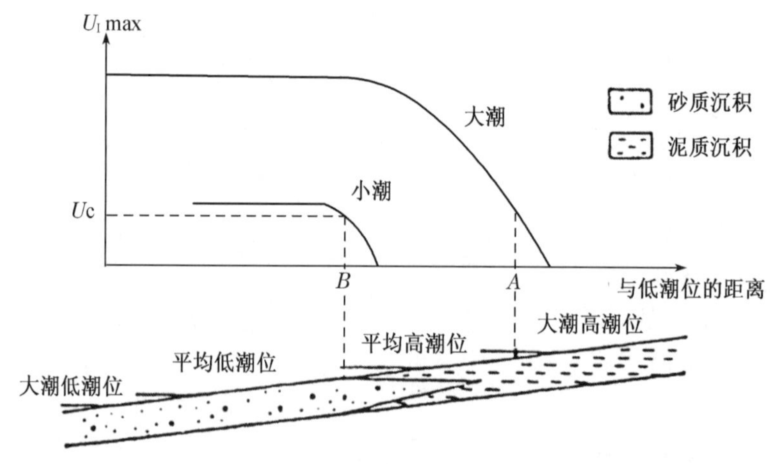

图2　大小潮对潮滩沉积的影响

$U_{l\max}$—大潮涨潮流速　U_c—小潮涨潮流速

潮滩海岸沉积物分布的规律与波浪作用的海滩沉积物不同，海滩组成以砂、砾质为主，由激浪流搬运，沉积颗粒自海岸向陆变粗，岸上多大浪堆积物。潮滩沉积以粉砂为主，沉积物分布是低潮水边线处最“粗”，主要是极细砂和粗粉砂，向岸沉积物颗粒逐渐变细为粗粉砂与细粉砂、细粉砂与黏土，以及黏土质。由于动力作用与沉积物的递变，使得潮滩地貌具有分带性：

（1）低潮位附近为滩面广阔的细砂与粗粉砂组成的波痕带。反映出推移质泥沙与浪、流作用结合的特征。

（2）中潮位附近为粗、细粉砂叠层，或细粉砂与黏土质粉砂叠层。中潮滩由于落潮流的集中水流作用与潮流流速加大而发育成明显的滩面冲刷地貌：滩鳞或冲刷体、冲刷洼地与潮水沟。砂泥混合滩的冲刷现象亦由于不同潮型作用于混合滩的差别侵蚀效应[5]。该处在大潮时为强动力带，小潮时为弱动力带。在大潮向小潮过渡的时期，逐渐堆积的细粒物质有较长的固结时期，需较大的起动流速才能再悬浮，这部分物质即使大潮时的高流速亦不能被全部冲走，而粗粉砂颗粒则不然。这样使得中潮位附近的砂泥混合滩或泥砂叠层带冲蚀现象明显。潮水沟集中发育于潮滩中部，自此向陆与向海延伸。向陆成溯源侵蚀型；向海初为集中冲刷，后成为涨潮流的漫浸通道与落潮流的集水通道而逐渐展宽与加深。潮水沟沟底起伏大，堆积物颗粒较粗，沟壁因粉砂液化与塌陷作用而扩展。平原海岸的潮水沟众多，规模不大，但在某些强潮区，如芬地湾、苏北中部潮滩，潮水沟规模巨大，沟内涨落流急，是潮滩上的危险地带。潮滩中部冲刷带的存在，使分带现象突出。

（3）高潮位附近是由粉砂质黏土或黏土落淤而成的泥沼带。沉积物颗粒细，经常为海水淹没，极难通行，成为泥沼带的特征表象。

（4）特大高潮位附近的黏土质泥裂带、虫穴、湿地草滩等。仅特大高潮时偶被潮水淹没的黏土质落淤带，曝露时被日晒形成龟裂，降水或有淡水自岸堤渗出而发育盐沼湿地。

潮滩分带现象是普遍性的规律，是潮流动力在滩涂上的递变反映。分带的名称各不相同，但是，潮滩低部的砂质-粉砂滩、上部潮滩的泥质滩与中部潮滩的泥、砂混合滩却是共同的。如图 3 潮滩分带沉积相，据 Darlymple. R. W。

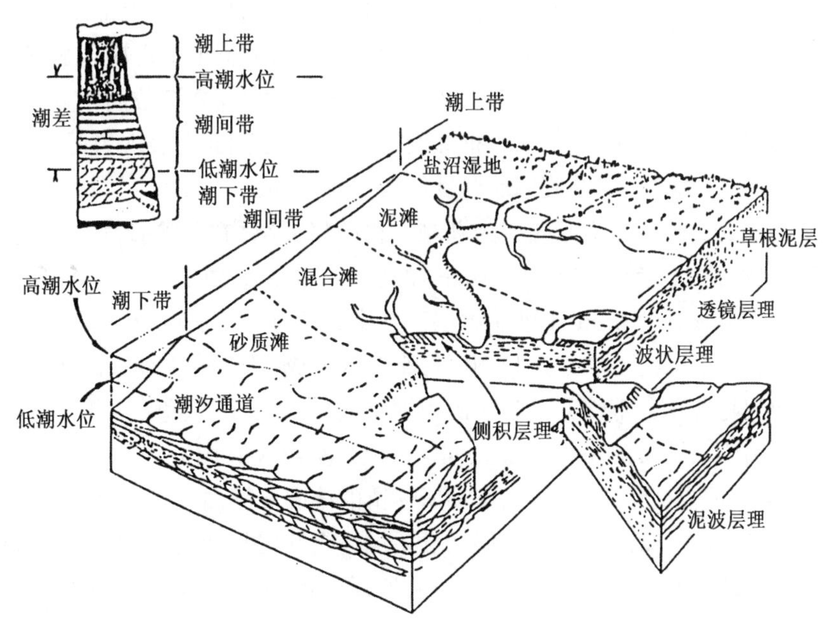

图 3　潮滩分带图式(据 Dalrymple. R. W)

根据潮滩主要物质组成的差异，反映出波浪作用在潮滩沉积过程中的影响程度。将潮滩划分为三种类型[2]：

（1）粉砂质潮滩　主要由粉砂物质组成，粉砂出现于潮滩各部分，具有明显的泥滩带、砂-泥混合滩带与砂质粉砂滩带三部分。这类潮滩以潮流作用为主要动力，仅具有中等程度的波浪影响。以苏北潮滩、加拿大芬地湾潮滩为代表。

（2）淤泥质潮滩　潮流作用控制潮滩发育，具有微弱的波浪作用影响，但有大量淤泥供应，潮滩上发育较宽的泥沼带，以渤海湾潮滩为代表。

（3）砂质潮滩　潮流动力叠加有较强或频繁的波浪作用，这类潮滩的上部常有较窄的海滩分布，系风暴潮作用的遗迹。如英格兰的沃什潮滩，中国闽、粤沿海的港湾内潮滩等。

红树林潮滩发育于热带与亚热带沿海的港湾、沙坝后侧潟湖内或河口湾湾顶，系植物根系滞留潮流与径流泥沙所成。

加拿大的极地海岸有砾石与淤泥质潮滩，砾石系浮冰筏运物。

潮滩是由潮流堆积形成的。如果泥沙供应中断或潮滩坡度突变（如围海造田），则潮滩的动态平衡遭到破坏，潮滩会因动力加强而受到冲刷，冲刷的潮滩常会在激浪作用下形成贝壳滩脊，同时，老的龟裂带会重现于滩上，形成与现时动力不适应的表现。

潮滩的分带性表现为相应的沉积相带：① 草滩或盐沼湿地沉积带；② 泥滩沉积带；③ 泥-粉砂滩（叠层）沉积带；④ 粉砂-细砂滩（波痕）沉积带。

随着潮滩的淤长发展，这四个带向海推进，原来向陆一侧的沉积会依次地覆盖在向海一侧的沉积带上。当潮滩发育趋于“成熟”时，就形成一个完整的潮滩沉积相序，下部是砂质沉积，上部为泥质沉积。下部的粉砂、细砂是低潮位高能流态下的推移质沉积，界于低潮位与波浪作用下界（海岸斜坡的下限）间，沉积层厚度约 15～20 m，这是潮滩沉积的基础。砂质沉积以推移质运动，从海向岸推进，使潮滩向海增加宽度。上部泥质沉积是潮流搬运的悬移质，它使滩面增高。当滩面达到大潮高潮位以上就很少再被海水淹没，一般不再堆积加高，所以泥质沉积层的最大厚度相当于中潮位至大潮高潮位间距，约为 2～3 m。这两者之间既有悬浮的泥质沉积，又有推移的砂质沉积，在剖面上表现为泥层与砂层的交替沉积。所以，一个潮滩环境的旋回层厚度约为 20～30 m。

水平纹层或页状的砂泥交互叠层为潮滩沉积的特征结构，反映着潮汐往复作用的特性。大潮时流强，微层（带）的厚度大，小潮时的层（带）薄。砂质层带色浅，泥质层带色深。同一组叠层中，泥质微层厚度小于砂质微层。在半日潮地区，每 14.77 天为一大、小潮循回，因此可形成 28 个潮流薄层。全日潮区每半个月形成 14 个或更少些的薄层[3]。

潮滩微地貌与沉积分带性亦反映于沉积层结构中，如：泥裂、龟裂纹以及其内的次生填充物（砂质、钙质或铁质沉积物），虫穴、虫塔、根系空隙及其内充填物，波痕、流痕、豆荚状或鲕状体、斜层理或交错层理，潮沟摆动所遗留的镶嵌堆积砂层、透镜体，风暴潮侵蚀痕迹与丘状砂层等。这些微结构与沉积层（泥层、泥沙交互层与砂质层）相结合，就形成潮滩沉积环境所赋予的沉积相特点。潮滩沉积层所反映的特点与天然剖面类似，可概括为几点：

（1）潮滩上部沉积物中黏土含量高，质地均匀，具良好的水平纹层，粉砂层厚 1～2 mm，黏土层更薄，一般仅 0.5 mm。此层内具龟裂构造，剖面中有上翘碟形层与粉砂楔辟。有虫穴扰动结构，草根有无取决于堆积速率，堆积速率快的层中无草根，堆积速率慢的有草根，多半为钙质充填根穴之孔隙。风暴沉积出现于不同层位中，其结构为：质地较粗，有冲刷痕与扰动结构。

（2）潮滩中部沉积结构表现为粉砂与泥质呈交互叠置的带状结构，每个薄层可自数 mm 至数 cm 不等，泥质层厚于粉砂夹层；夹极细砂透镜体，具有显著的潮水沟镶嵌层-砂质斜层理，透镜体与交错层理。

（3）潮滩下部多为粗细粉砂水平纹层或纹层状交错层理，具细砂与粉砂交错层、潮水沟沉积的细砂斜层理以及侵蚀面。

水边线沉积层具有侵蚀间断面、细砂透镜层以及泥饼或泥丸等潮滩侵蚀残余物。

潮滩各相带可有多次反复交替，但一个大层序的厚度多在 20 m 左右，潮滩沉积层中亦会镶嵌有河口堆积层，后者具二元相结构以及咸淡水交汇相的贝类残遗物。

潮滩是增长中的土地资源，中国的潮滩面积约 1 270 000 km^2，在海域疆界、水产养殖、农、林、牧以及浅海石油等工农业开发的土地利用中皆具有重要的价值。了解潮滩沉积过程有助于合理开发利用潮滩资源环境，认识潮滩沉积相有助于分析古沉积层所反映之环境特点，为分析石油、天然气形成与储存环境提供判别依据。

参考文献

[1] Wang, Ying. 1983. The Mud flat Coast of China. *Canadian Journal of Fisheries and Aquatic Sciences*, 40(Supplement 1): 160 - 171.

[2] Wang, Ying, Collins, M. B., Zhu Dakui. 1990. A Comparison Study of Open Coast Tidal Flat: The Wash (U. K), Bohai Bay and West Yellow Sea (China). In: Proceedings of International Symposium on the Coastal Zone, 1988. China Ocean Press. 120 - 134.

[3] Dalrymple, R. W. 1992. Tidal Deposition System, Facies Models Response to Sea Level Change. Geological Association of Canada. 195 - 218.

[4] Postman, H. 1967. 瓦登海水文和泥砂运动状况. 1983. 4. 21 在南京大学的讲座. 地理科技资料 1983(28): 5 - 19. 原文载 Estuaries, 158 - 179.

[5] 朱大奎，高抒. 1985. 潮滩地貌与沉积的数字模型. 海洋通报，4(5)：15 - 21.

[6] 王颖，朱大奎. 1990. 中国的潮滩. 第四纪研究，10(4)：291 - 300.

江河湖海与城市发展相关剖析*

一

当今世界超过100万人口的90个城市，60%分布于沿海地带，尤其是河口海岸平原区。这些地区自然环境适宜，易于生息繁殖：土质肥沃，淡水采用与灌溉便利，易于农耕，方便于交通运输，工商业发达，服务迅捷。城市发展的早期，河流尚为城镇防守的天然屏障。逐渐地沿河、湖、海的村镇聚落发展成大城市，这样的例子遍及全球。著名的如：美国大西洋沿岸的波士顿、纽约、菲拉德菲亚、巴尔的摩与华盛顿，组成大城市带；美国与加拿大大湖区的多伦多、芝加哥、底特律等环湖城市圈；太平洋沿岸的温哥华、西雅图（海岸山脉间断）、旧金山、洛杉矶、圣地亚哥城市链。澳洲的城市均分布在沿海地带：坎斯、汤士弗、布里斯班、悉尼、墨尔本和佩斯，这种分布受气候因素控制，内陆沙漠干旱，降水主要分布在海陆交互带。我国的主要大城市沿海分布，具有外资、商贸与文化侵入的后遗影响。20世纪80年代以来，北始大连，经天津、青岛、上海、厦门达广州、深圳与香港，已发展成为我国人口密集、经济最发达的沿海城市带。其他还有环太平洋西岸的东京、神户、大阪、名古屋等。欧洲城市的分布与发展与河流关系密切，特点突出。如泰晤士河畔的伦敦、塞纳河畔的巴黎、易必河畔的柏林、沿提瓦尔（TEVIER）河分布的罗马、涅瓦河口的圣彼得堡、沿河发展的莫斯科等，阿姆斯特丹具有河海交汇之优势，发展成为世界级的港口城市，而威尼斯更以水城而著称。这些城市，随时代的发展而不断地扩展与更新，但始终保持着临水、亲水分布的特色。

二

城市的发展规划应以人为本，人居环境应具有清新的空气、明亮的阳光、清洁的淡水、安全的宅舍与宜人的风貌，同时，又需要有繁荣的商贸街市，便利的交通与运输、资讯网络，文化历史内涵，科学、教育，医疗卫生与宗教设施，安全的保障，多种多样的服务网络以及有效的管理体系。总之，需通过科学系统的城乡发展规划与设计，达到人居环境的合理建设，促进人地关系的和谐发展。

河流是地球表层系统的动脉，它焕发城市的青春活力，是食物与淡水的源泉、农业与交通运输的动脉，是发展商贸、旅游、运动的一项重要载体。反之，河流干涸或湖、海消亡，丧失了水源与湿润空气，导致城市或王朝废弃：楼兰古国的废墟周围仍保留着死亡胡杨林带沿干涸河谷分布的遗证；西夏王朝的灭绝；安西风沙侵袭，从唐代至今的1200年间，安

* 王颖，盛静芬：广州市城市规划局主编，《滨水地区城市设计》，85－92页，北京：中国建筑工业出版社，2001年。后附英文版。

西境内又有37座城池被风沙掩埋成废墟。剖析中外城市布局，可归纳水体在城市发展中的功能为三类。

（1）城市的兴起以河流为依托，河流横贯发展着的大都市，成为城市的主轴线。如巴黎、伦敦与汉城，以塞纳河、泰晤士河、汉江为城市建设的中心，道路呈放射状沟通环城的公路与铁路干线（图1）。巴黎的宫殿古建筑与现代建筑融合，层次错落地沿河分布，若干大景点与广场形成团簇，辐射出主要的街市，巴黎铁塔、凡尔赛宫与巴黎圣母院构成中心的景观轴，集古今于一体，融合了天然与人工之美。宜人行、宜车舟、宜赏析、宜探索、宜休憩、宜购物……，是城市规划建设为人着想的典范。流动不息的塞纳河孕育着世代相继的文化精华。伦敦河沿岸建筑雄伟辉煌，利用废弃河道洼地建设成若干公园绿地，成为园林与城市的"绿肺"，为伦敦增加了活力。但是中心带的下游，旧日的码头街市予人以没落感

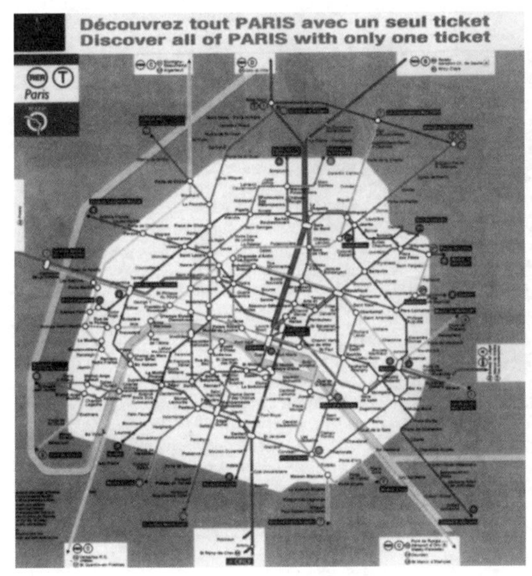

图1　巴黎城市轮廓

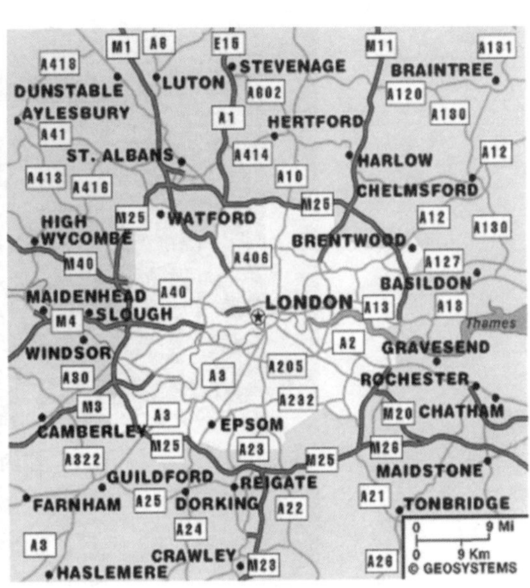

图2　伦敦城市轮廓

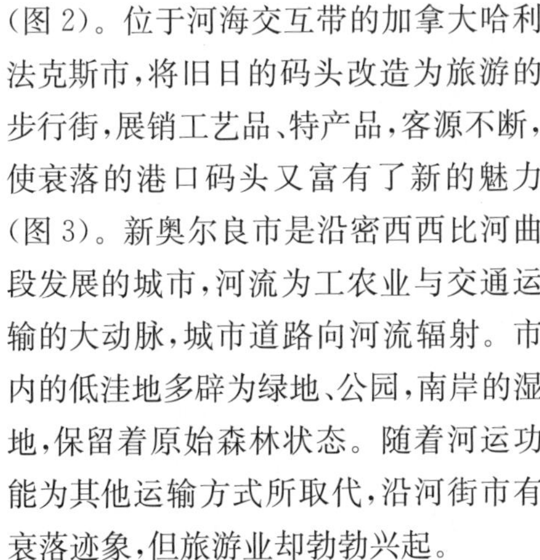

（图2）。位于河海交互带的加拿大哈利法克斯市，将旧日的码头改造为旅游的步行街，展销工艺品、特产品，客源不断，使衰落的港口码头又富有了新的魅力（图3）。新奥尔良市是沿密西西比河曲段发展的城市，河流为工农业与交通运输的大动脉，城市道路向河流辐射。市内的低洼地多辟为绿地、公园，南岸的湿地，保留着原始森林状态。随着河运功能为其他运输方式所取代，沿河街市有衰落迹象，但旅游业却勃勃兴起。

（2）沿海湾或河海交汇的河口湾发展的城市，这类城市很多，如纽约、波士

图3　哈利法克斯市海岸城市

顿、旧金山、圣地亚哥、温哥华以及澳大利亚的大城市。具有早期移民与多民族文化汇合的特点，已发展成金融、贸易、科学教育的中心。纽约沿 Hudson 河、Manhattan 是世界级的金融中心，相应地，文艺与第三产业高度发达。滨水的沙滩、沙岛多辟为绿地公园与高级住宅区（图 4）。圣地亚哥市建立海军基地于海岸岛、滩区，沿河为新发展区，旅游业与商业点在兴起。波士顿与温哥华利用潮滩建设了国际机场。澳大利亚大堡礁水碧清澈、阳光充足，2 000 多 km 的岛礁海岸带是海洋生物与生态环境保护区，集国家公园旅游业与体育活动于一体，是著名的黄金海岸，吸引世各地的游人，充分享受 3S 健康环境，增长知识，陶冶情趣。东京是海港性的大都市，市内的皇宫保留了大片绿野园林，四周围以人工河道，这是利用水体发挥阻隔的功能，使皇宫园林成为禁苑（图 5）。

图 4　纽约城市景象

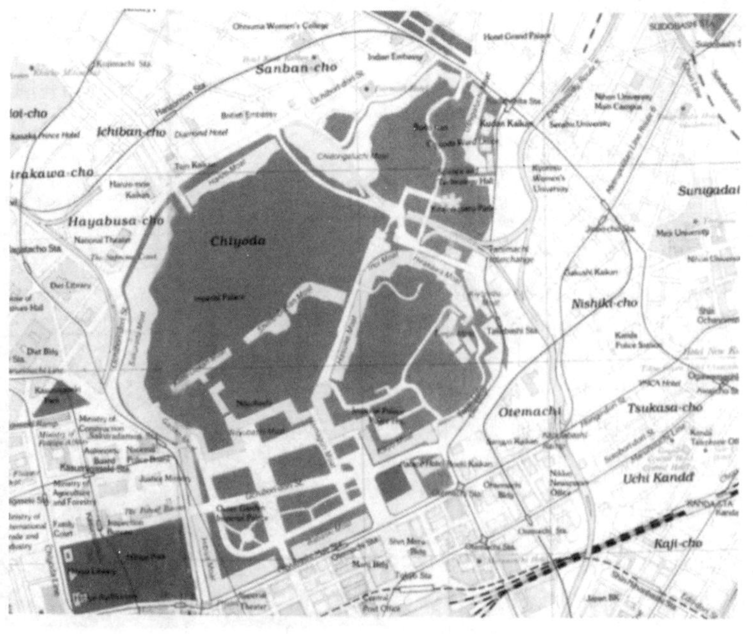

图 5　东京皇宫水体的防御功能

(3) 湖滨城市。以多伦多市及瑞士的城市为代表，沿湖是贯通四方交通的节点，湖滨、沙滩及沙岛，多辟为公园与绿地、体育活动(游水、划船与溜冰)、科技与文艺博览场馆。相应地，旅馆、餐饮与服务业发达。持续不断地吸引着城市居民与外来游客。

中国城镇傍水而建的更多，沿长江的大城市重庆、武汉三镇、南京、上海均为大都市。三角洲水乡中苏州、镇江、扬州滨水建城，名冠古今。即使当年的北京宫苑亦引水贯通，临水建亭、台、楼、阁。在中国古代，水体的主要功能是运输及赏玩，在四方汇集的滨水地区，兴港建镇，逐渐发展为城市。宋代翰林画家张择端所作的《清明上河图》写景逼真，充分反映出汴河在北宋都城的功能以及世俗人事。汴河流贯城乡，沿河曲缓流处，船只停泊、桅杆林集，近河有修造船厂。以拱桥为中心，商肆林立，形成新的集镇。桥上行人熙熙攘攘，或驻足观景，拥挤不堪。桥下舟楫川流，篙师缆夫牵舟寸进，逆河而上。桥外城郊，清明时节，田亩阡陌纵横，疏林薄雾村舍酒家，村外大道有骑而奔驰，有踏青扫墓而归，田野情趣意浓。桥内城郭老街，商店林立，行人南来北往，市、农、工、商、医、卜、僧、道，或乘骑赶集，或牛马拖车载货。妇孺乘轿，货郎以物易货。酒店、茶肆、旅店，雜货百物，酒招牌匾，名师字画，鳞次栉比，市景文化繁荣。深宅大院，官宦宅邸则现于市侧。清明上河图生动地记录了宋代临水都城之布局，河流在城市发展中的功能：景观、水体联系、运输灌溉、积聚人气、商贾与集市扩展。

当代我国城市的滨水带，以建设公园绿地居多，或辅以鱼钓舟艇及水上活动设施，增强了人居环境的健康因素。但仅有日间活动，入夜则冷清。无夜间与假日繁华的关键是缺少吸引人流的亲水街市，以供人们于工余之后赏游。虽也建立了高档宾馆，增加了河上景点，但价格不菲，有如阳春白雪未能集聚大量客流：有些岸段建立了大量酒肆茶楼，残渣油污却毁损了滨岸与沙滩的洁净。集聚客流与维持环境洁美的矛盾是在滨岸规划中需要考虑预防的。这方面丽江古城处理得很好，保持了小桥流水与坊、店、酒肆，具溪水流畅、空气清新、与民俗文化浓郁之佳。但是，不少大城市的临江、滨海段却空旷荒凉或游人寥寥，需重视人气的聚集。

沿海城市的建设中，厦门市曾经过围海成湖，湖水污染，环境劣化：沟通活水，清污、治理与环境建设；实施海岸带一体化管理，达到建设发展与环境健康至臻的初步境界，值得推广效法。

通过三类国外城市及古今中国城市的实例，归纳出水体在城市发展中具有：景观功能，水源与水体活力的功能，联系与运输的功能，游乐与运动的功能，生态环境陶冶与更新功能以及文化历史渊源及延续的功能等。

滨水地段是城市的珍贵资源，需很好地从“人地和谐相关”与“生存环境持续发展”的观点指导进行城市规划与建设。

三

广州文化历史渊源深厚，人文荟萃，商贸发达，经济繁荣，是我国东南沿海的华南重镇。城市依山临海，位于珠江三角洲的顶点，珠江蜿蜒贯流市区，具有水土源泉与运输的功能，她与广州市的过去、现在和将来息息相关，共存一体(图 6)。珠江三角洲是从三角港因珠江泥沙淤积而发展成的充填式三角洲。其环境特点：来自南海的涨潮流沿珠江河口湾东岸伸入，而落潮流与径流汇合却沿西岸流出；河流与潮流交汇与交互作用，促进泥沙落淤；三角洲淤涨又导致了河流弯曲与分叉；后者又促进沙洲、沙岛之形成，新生土地与潜在的土地资源条件良好。位于热带亚热带季风气候带，广州地区的降水水源充沛，而夏秋季台风与风暴潮影响频繁，低洼的平原土地，易受潮涝积水之害。

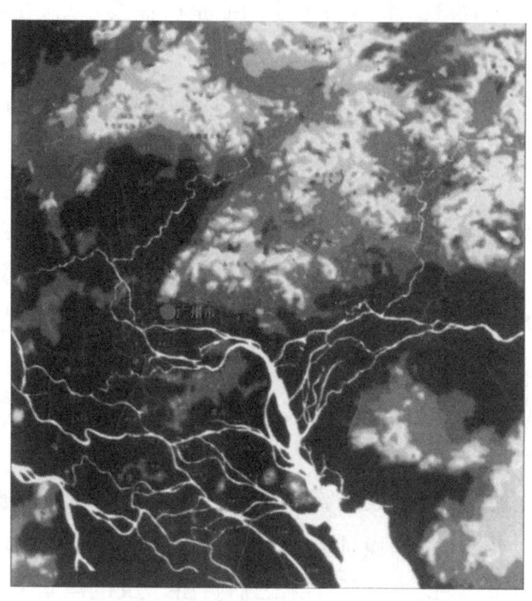

图 6　广州在珠江三角洲

广州市的自然与经济地理位置，使它成为珠江三角洲都市群的政治与经济的中心都市。除陆路四通八达外，广州通过河流水系联系周边与大陆腹地，水运条件优越，为其他大都市所不及。她通过东江联惠州，北江达韶关，西江通梧州；她位居珠江口湾岸中心，东翼有深圳、香港，西翼为珠海、澳门，南面通过南海与东南亚，澳、非、欧、美相通。河港、海港的都市，滨江、滨海岸段长，是广州的自然优势特点。城市发展的新规划中，结合广州市规划及充分发挥珠江在城市中的水域水源、生态、游乐、运动休闲、交通运输的功能，体现出滨水城市的水体功能，利用水体增强广州的中心枢纽地位，扩大影响范围。基于此设想，提出几点思路性建议：

（1）浚深与疏通内河航运，使珠江干流成为多功能的水道：农、工、加工产品的运输与转运；客运与旅游线路；洪水与泥沙的疏通干道。分流陆路交通，减少空气污染，增强河流活力，发展城乡间旅游事业，并且由此带动一系列经济活动。

（2）利用深水潮流通道，扩大建设“虎门”海港。虎门古要塞扼珠江湾口狭道，束水流急，进一步浚深，可建设 5 万～10 万 t 级的港口。沿珠江口与广州湾衔接，内接河道航运，发挥其水路转运的优势。这是邻近其他港口所不及的。新海港可集散客、货，尤其便利海外游客巨轮停泊，使旅客沿河上溯深入珠江三角洲腹地。河海港口周边可建设国际性的高科技合作基地以及国际性的休闲游乐购物场所。在黄埔港下游地区，建设先进的集装箱拼箱中心，提高联运能力。

（3）珠江滨水带景观规划必须具备层次性与地带性。在滨水带生态景观规划原则基础上，根据腹地功能确定滨水带的具体规划项目。层次性上保证从珠江观赏岸段的视线宽阔，建筑景物设计有层次感、整体感，沿江不宜有突兀的摩天建筑隔离、断开视线。地带性体现在滨水地带规划的区域特点，沿岸规划既不能纷繁随意，各取所需，也不能统一规划成绿地或公园，失去必要的土地利用经济效益。宜以生态、人居要求为首，兼顾经济效益，开发使用滨岸土地和水域。

(4) 保证沿岸留有充分的公共空间、娱乐场地和绿地林带。目前有些岸段大量占据水面空间，建设高架桥，完全破坏了城市的水空间的完整与优美。根据广州市城市规划，琶州岛至龙穴围岛岸段及其腹地的东涌至南沙地区将成为未来大广州市的行政、科研、教育、高新产业、旅游休闲中心以至珠江三角洲的 CBD 核心。从山水城市的广州市城市形象出发，应该致力形成美丽的沿岸风景线，绿树成荫的滨江大道，隔离喧嚣与宁静的景观隔离带，自东向西形成一条从自然到人文景观、从现代到历史的具有丰富内涵和节奏感的珠江景观通道。

目前，延伸的滨江岸段，堤岸建设需更新。马路扩建，公园绿地在规划，但是河水不清澈流畅，两岸缺少林木，太空旷，有些大学的校园，围墙临岸，不能透视，借景美化环境，予人以偏僻、单调感，不热闹，无夜景。因此不能吸引客流，不能予人以多种享受。

类似的感觉如汉城，城市建立于汉江两岸阶地上。河谷宽阔，以坝拦水增加河床水深，但水流迟缓，堤岸上无树林，堤下混凝土河滩，空旷、少设施，游人可数。

珠江沿岸宜增设古民居、历史文物与历史景点，分层次地建设，平行沿堤的树木草地带，宜有店、铺、堂、馆、“雅货百物”吸引人们观赏、购买，品茗，餐饮与小憩，聚人客、有气氛，才能经营有发展。无论是当地居民或外地游客，均喜傍晚漫步，浏览，珠江两岸夜晚要亮起来，舟艇亮起来、动起来，成为“上河”风光带。提高人居环境质量，增加居民健康与文化生活，沿江旅游休闲等设施要发展起来。

(5) 沿海、沿江低洼区易受洪水风暴潮威胁的地区，提高堤围的设防标准至 50～100 年一遇的防洪标准。保证沿岸生产生活建设设施的安全。

(6) 沿海淤滩、沙洲、岛屿，宜在原养殖与农业开放的基础上，开展田园风光与度假旅游。菜蔬、海鲜既供应城市，亦可让游人垂钓、采摘。海滨还可以建设嬉水、游泳、快艇运动等多种活动，增加趣味性：建设农家小住型宅舍，供市内与港口居民假日合家小住，或供新婚夫妇度蜜月。平日可供会议或体训应用，丰富生活内容，提高生活文化质量，集聚人气，繁荣城郊经济。

(7) 沿岸水污染控制和治理。珠江在广州接纳了整个市区工业废水和生活污水，珠江沿岸的居民企业基本上将生活、生产废水排入珠江，河流成为纳污的通道。在新规划中应该严格控制沿岸污水的排放，保证河段空气清新，河水洁净，为沿岸居民提供良好的环境氛围。尤其黄埔岸段受附近黄埔工业群的影响较大，空气污染和水污染严重，应增加治理和控制力度。

广州市滨水环境具有与伦敦(潮汐河流)、巴黎(历史文化名城)相类似之处，学习其城市规划建设的经验，扬长避短，可使广州市沿水建设达到更为完臻的境地，要防止广州港的发展走英国伦敦 Dockland 港的发展老路。

参考文献

[1] (英)奥雷尔,斯坦因.路经楼兰.肖小勇,巫新华译.2000.南宁:广西师范大学出版社.
[2] (宋)张择端. 清明上河图.
[3] 阎水玉,王祥荣.1999.泰晤士河在伦敦城市规划中的功能定位、保证措施及其特征的分析.国外城市规划,(1):34-35.
[4] 中国城市设计研究院.广州市总体发展概念规划.

附(英文版):

The Function of Waters Related to Urban Development

Part Ⅰ

Nowadays there are 90 cities with a population of more than one million in the world, 60% of which are located in coastal areas, especially on the plains near river mouths or seashores, where the natural environment is good for human reproduction, and where the soil is fertile and suitable for farming. The usage of fresh water in these places is no problem and it is true with irrigation too. There are convenient transportation, prosperous industry and commerce, and accessible services in coastal areas. In the early stage of city development, rivers served merely as natural barriers for defense purposes. Gradually, villages, towns and tribes along rivers, lakes and seas developed into big cities. Cases in point are seen everywhere among which the following are famous: The Megalopolis Belt of Bosnywash consisting of Boston, New York. Fairfield, Baltimore and Washington in U. S. A; the city circles around the Great Lakes between the United States and Canada made up of Toronto, Chicago and Detroit; the city chain along the Pacific Ocean consisting of Vancouver, Seattle (a disconnection of the coastal mountain ranges), San Francisco, Los Angeles and San Diego. Australian cities such as Cairns, Townsville, Brisbane, Sydney, Melbourne and Perth are scattered around the coastal areas. The location of the cities finds its reason in the climate, which is dry in the mainland desert, and precipitation mainly occurs at the intersections of land and sea. In China, major big cities are located coastally with leftover influences from previous foreign investment, trade and business, and cultural invasion. Since 1980s coastal cities ranging from Dalian in the north through Tianjin, Qingdao, Shanghai and Xiamen to Guangzhou, Shenzhen and Hong Kong, have become a densely populated and mostly flourishing economic belt. Other examples include Tokyo Kobe Osaka and Nagoya, the coastal Japanese cities along the west side of the Pacific. The distribution and development of European cities is conspicuous and has much to do with rivers. For instance, London on Thames, Paris on Seine, Berlin on Elbe R. , Rome on Tevere R, St. Petersburg at the mouth of Neva R. , and Moscow along the river etc. Amsterdam has become an international port city because of the advantage of being at the cross of the river and the sea, and Venice is well known as a water city. These cities mentioned have kept expanding and renovating in pace with the changing eras, maintaining their waterside characteristics.

Part Ⅱ

Plans for city development should take people's needs into first consideration. Living environment demands for fresh air, bright sunshine, clean and fresh water, safe housing and pleasant surroundings. Meanwhile, it needs flourishing business downtowns; convenient transportation, and information network; cultural and historic sites; scientific, educational, medical and religious facilities; security; and a variety of service networks and effective managerial systems. In all, it calls for scientific, systematic urban and rural development plans and designs to realize the building of reasonable living environment and to promote the harmonious human-environment development.

Rivers are arteries of the surface layer of the earth. They radiate youth vigor of the cities. They are sources of foods and fresh water. Also as arteries of agriculture and transportation, rivers are important carriers of the development of commerce, tourism and sports. Otherwise, the drying-up or disappearing of rivers, lakes or seas would result in the abolition of cities or dynasties. Evidence of the distribution of dead diversiform-leaved poplars along the dried-up river valley can still be found around the relics of Loulan, an ancient city. And for the same reason Xi Xia Dynasty came to an end. Anxi has been attacked by sand wind for 1 200 years since Tang Dynasty. Furthermore, 37 cities have been buried and ruined. So by analyzing the distribution of cities both home and abroad, we come to the following three functions of waters related to urban development.

(1) Cities depend on rivers to spring up. Rivers traverse the developing metropolis as principal axis. For example, Paris, London and Seoul take Seine, Thames and Han River as the center of city building respectively. Roads radiate to reach the main highways and railways around the city (picture 1). In Paris, ancient palaces and modern buildings are interspersed along the river with several major tourist spots surrounding the square reaching out in all directions to the main streets. In the center are the Eiffel Tower, Versailles Palace and the Cathedral of Notre-Dame, which combine in themselves both the past and the present, integrating natural and artificial beauty. These places set good examples of urban planning that concern with people. They are ideal for people to take a walk, to take a ride by bus or by boat, to go sightseeing and exploring, to relax and to shop. The everflowing Seine carries the cultural essence generation after generation. There are magnificent buildings on both sides of London River. Some greens are made by using the abandoned watercourse and low-lying land. Thus they make the city look more lively and become the "green lungs" of the gardens

and the city. Anyhow, at the down stream of the downtown, old docks and streets appear deserted (picture 2). Halifax, Canada, which is located in the intersection of the river and the sea, was transformed into a pedestrian commercial street for tourism purposes, displaying handicrafts and special local products. As a result, it is now crowded with tourists and is no long a port on the wane (picture 3). New Orleans is a city that developed along the bank of Mississippi River, which acts as the main artery of industry, agriculture and transportation. All city roads lead to the river. Lowlying lands in the city have been turned into Greens and parks, while marshes in the south bank keep the original way as it was. Since other transportation means gradually take the place of river transportation, trade in cities and streets along the river begin to decline. Instead, tourism springs up.

(2) There are many cities that developed along the bays or estuaries such as New York, Boston, San Francisco, San Diego, Vancouver and the big cities of Australia. These cities are characterized by early immigration and multi-cultures. And they are financial, trade, scientific and educational centers now. Along the sides of Hudson River in New York, Manhattan boasts its most prosperous arts and tertiary industry as an international financial center. Most of the sand beaches and islands have been developed into green parks and highclass residence blocks (picture 4). In San Diego, a naval base is established at the coastal islands and shores while new developed regions are along the river. Tourism and commercial spots are being set up. In Boston and Vancouver international airports have been built on the tidal flat. The Great Barrier of Australia is a place where the sun shines brightly and the water is crystal. Over - 2000 - kilometer-long coast made of isles and reefs is set aside as a preserve of marine organisms and environment. It is called golden coast for its dual roles of national parks and sports grounds. Tourists from all corners of the world come to the Great Barrier to enjoy healthy environment, gain knowledge, and mould temperament. Tokyo is a harbor metropolis within which much greens and groves remain in the palace with man-made river way surroundings. The reason why it is designed like this is that waters can separate. In this way the royal gardens become an exclusive one (picture 5).

(3) Lakeside cities, taking Toronto and Swiss cities as examples, have, around the lake, joints of transportation to all directions. The majority of lakesides, sands and isles have been developed into parks, greens, sports places (for swimming, boating and skating), and exhibitions for scientific and artistic uses. Accordingly, restaurants, hotels and service sectors are flourishing. Urban residents together with foreigners are attracted in ceaselessly.

In China, there are more cities and towns close to the waters. Along the Changjiang River are big cities such as Chongqing, Wuhan, Nanjing and Shanghai. Among the delta cities, Suzhou, Zhenjiang and Yangzhou are the most famous ever since the ancient ages. Even when Beijing Palace was built, water was introduced to run across it so that pavilions, platforms, and towers could be built on it. In ancient China the function of waters lied mainly in transportation and entertainment. In regions bordering waters from all round, ports and towns were built and gradually they developed into cities. Zhang-zheduan, a master painter of Song Dynasty, drew a picture titled "A River Scene on Pure Brightness Day", which vividly and fully reflects how Bianhe River functioned in Bei Song Dynasty as well as the life style at that time. Bianhe River runs across the city and the town. Boats, with a forest of masts, are berthed on the bends or where the water flows slowly. Nearby there are shipyards and repair shops. Centered on the arch bridge, there are numerous shops, forming new fairs. On the bridge, people are bustling about. Some are stopping to appreciate the scenery. It's very crowded indeed. Under the bridge, boats are busy shuttling back and forth. The boatmen and the boat trackers are struggling to pull their boats forward against the current. In the suburbs away from the bridge, since it is Pure Brightness Day, such a picture is unfolding before us: crisscross footpaths in the fields; hazed cottages and pubs with sparse woods; passing horsemen on the outside road; folks returning from spring outing or after paying respects to the dead by sweeping the tomb. That's idyllic life full of countryside charms. Within, there are shops on old lanes. Passersby go up and down. They are townspeople, farmers, workers, merchants, doctors and monks, riding on a horse or donkey to the fair. Some of them have horse-drawn carriages loaded with goods. Women and children are taking sedan chairs. Peddlers exchange goods for goods. Pubs, teahouses, hotels and shops are closely placed side by side with a variety of brand names of goods, calligraphic works and paintings by celebrities. Town culture is really quite prosperous. On the other hand, some imposing dwellings and spacious courtyards in which officials and lords live are shown at the side of the marketplace. "A River Scene on Pure Brightness Day" records vividly the layout of a Qing Dynasty city close to the waters. And it also indicates the function of waters related to urban development: landscape, waters relationship, transportation and irrigation, assembling of people, commerce and market expansion.

Most regions in China close to waters have been turned into parks and greens. Some of these regions are supplemented with fishing, boating or water sports equipment in order to help improve the living conditions, especially in health aspect. But with night activities lacking, day recreations alone is far from enough. The fact that there is no sign of prosperity at night and on holidays is caused by the lack of waterside markets and streets. No such places to attract people when they are leisure. Top grade hotels do add to the scenic spots on river, but, with high prices, which reminds people of highbrow art and literature that appeals to only a few social groups. On some sections of the riverbank there are a lot of pubs and teahouses from which leftovers contaminated the beach and the sands. When one is making a coast planning, one should take the above-mentioned contradiction into account: the contradiction between attracting more customers and maintaining clean environment. Lijiang city has done very well in this field. It keeps in it the bridges and streams, mills, shops and pubs. Flowing river, fresh air and strong cultural atmosphere all are perfectly interwoven. However, there are still many other open, desolated waterside regions in big cities with few visitors. Something must be done to attract people.

In the construction of coastal cities, Xiamen once reclaimed lake from sea. Consequently, the lake was polluted and the environment was deteriorated. Then people linked up waters, engaged in environmental cleaning, managing and building. An integrated coastal administration was implemented and as a result, the initial level set for the constructive development and environmental health has been reached. This method is worth promoting and learning.

By analyzing cases of Chinese cities in the post and at present as well as three categories of foreign cities, we conclude the following functions of waters related to urban development: the functions of presenting landscape; of being water sources and providing flowing water; of linking and transporting; of supporting recreations and sports; of offering environmental renovation and education; and of remaining and continuing cultural history, etc.

Waterside areas are precious resource to a city. Urban planning and construction should be properly instructed from the view of "Human-Environment Harmony" and "the Sustainable Development of Living Environment".

Part Ⅲ

Guangzhou has a long history and profound culture with refined humanity, flourishing business and prosperous economy, with which it has been one of the most important South-China cities along the southeast coast in our country. The city is at the foot of the mountain and with the sea nearby. It lies at the head of the Pearl River Delta with the Pearl River flowing through out the urban area, which functions as the water sources and transportation passages. She is closely bound up to the past, present and the future of Guangzhou City (Picture 6). The Pearl River Delta is an estuary stuffed with deposited silt from the Pearl River. The rising tide trend from the South Sea extends into the Pearl River alongside the estuary's eastern bank, while the falling tide trend, which joins into the runoff, flowing out of the estuary alongside the western bank. The river flow and the tide flow join together and interact with each other, accelerating the deposit of silt. On the other hand, the rising deposit of the delta results in river curves and branches. The latter helps to bring about shoals and sand islands.

The neo-land and the potential land both are good-conditioned land sources. Lying in the tropical and sub-tropical monsoon climate zone, Guangzhou area has plentiful rainfall. There are frequent attacked by typhoon, windstorm and tide in spring and autumn. So the low-lying land in Guangzhou is susceptible to moist, water logging and overstock of water.

The geographical location and the economic importance enable Guangzhou to become the central political and economical city among the urban groups in the Pearl River Delta. Besides the convenient transportation in all directions, Guangzhou has advantageous waterways conditions, which makes it possible to exchange with the periphery and the hinterland and to be superior to the other big cities. It links to Huizhou through the Eastern River, to Shaoguan through the Northern River and to Wuzhou through the Western River. It lies in the central point of the estuary of the Pearl River, with Shenzhen and Hongkong to the east, Zhuhai and Macao to the west. And to the south, the Southern Sea is the passageway between Guangzhou and the Southeast Asia, Australia, Africa, Europe and America. Urban areas along the river or the coast generally have long banks and coasts. Guangzhou has such geographical advantages. According to the new plan of the city development, we should fully exert the functions of Pearl River in aspects of the water source, ecosystem, entertainment, sports and the transportation. That is to say, we should emphasize the functions of waters, enhance Guangzhou's central position as a pivot, and extend its influence. Thus, we come up with the following proposals:

(1) Being dredged for the transportation, the trunk of the Pearl River is multifunctional, such as the transportation and transshipping of the agricultural, industrial and processed products, the passenger transportation and tourist routes, and dredging the flood and the silt. Thus the river can share the responsibilities in transportation with the land routes, help to alleviate air pollution, activate the river's vitality and develop tourism of the urban and rural areas. A series of economic activities is brought about.

(2) Expand the construction of the Humen Harbor. The ancient fort of Humen guards the narrow passage of the river estuary with torrents from the east. If dredged deeper, it can be constructed into a harbor of 50 000～100 000 tons berths. This harbor links to Guangzhou Bay alongside the estuary, connects to the river way transportation toward the inland, and has transshipping advantages. All of these make it superior to the nearby harbors. This new harbor can transship passengers and cargoes. It is especially convenient to anchor large ocean-going ships. Passengers are made possible to go into the hinterland of the Pearl River Delta. In the periphery areas, international hi-tech cooperative bases and international entertainment centers and shopping malls can be built. In the area of the down-stream in Huangpu Port, advanced container-reallocation

centers can be built; thus the capability for cooperative transport is improved.

(3) It is necessary to have gradational layout and regional identity in the riverside landscape plan. Based on the ecological planning principles and according to hinterland functions, the detailed riverside programming is to be determined. On gradational layout, extensive vision from the Pearl River should be guaranteed. It is better for architectures to display proper profiles and entirety. Abrupt skyscrapers are not good for they sever the vision. Regional identities lie in expressing the characteristics of the waterside regions. The riverside programming can be neither numerous or disorderly, nor merely a grassland or a park. The ecological and residential demands should be firstly taken into account. Meanwhile the economic benefits should also be considered when exploiting riverside lands and waters.

(4) Secure ample public space, recreational places and grasslands and trees alongside the river. At present, some bank segments occupy large water space, which, along with elevated roads, destroy the entirety and beauty of the urban water space. According to the urban planning of Guangzhou, the bank segment from Pazhou island to Longxue island and its hinterland area from Dongchong to Nansha will be the center of future Guangzhou for its administration, science and research, education, hi-tech industry, tourism and recreation. And this area will even be the CBD core of the Pearl Rive Delta. To establish the urban image of Guangzhou as a city with mountains and rivers, efforts should be put into creating beautiful riverside landscape, green shading riverside roads and a scenic segregation between noise and tranquility. Thus from east to west forms a scenic passageway with natural and artificial landscape, and modern and historical contents with a sense of rhythm.

Now, extended riverside banks need rebuilding. Roads are being extended; parks and grasslands are being programmed. But the water is neither clear nor smoothly flowing. It looks too spacious with few scattered trees along both banks. Some universities' campuses alongside the river are enveloped by wall, which blocks the vision. As a result, riverside looks isolated and dull. There is no night scenery. Put together, the banks fail to attract the visitors and bear no enjoyments.

A similar case can be found with Seoul, which was built on the terrace on both sides of Han River. It has wide and spacious river valley, with embanked water to increase the water depth. But the water flows slowly and there are no woods on the banks. The flood land is built with concrete, which is spacious, and has few and scattered facilities and visitors.

It is appropriate to built ancient style residential houses, to supplement historical relics and tourist spots and to construct green zones parallel to the banks alongside the river. There should be shops, stores, traditional houses, hotels and goods in good taste to attract people to admire, purchase, sip tea, eat, drink and have a short break. To have more and more customers is the way to develop. Whoever they are, the residents or travelers all like to take a walk and glance over at nightfall. So the riverbanks and ships should be lit up at night, the ships should set movingas well to be the scenery in the river. To improve the residential environment and enrich the cultural life, it is important to develop tourism and recreations alongside the river.

(5) The low-lying lands of coasts and riversides are susceptible to attacks by flood, typhoon, windstorm and tide. Flood control standards should be up to encountering once in 50 to 100 years to safeguard the production, residential life and construction facilities.

(6) Take the existing breeding and farming, agricultural tourism is suggested to develop along the silted seashores, shoals and islands. The city does not lose its vegetables and seafood supplies, and at the same time people can go fishing and plucking fruits here. On the seashore, facilities for people to swim, waterski and drive speedboat can be built to enhance the delightfulness. And small cottages can be offered to city people on holiday, both for a family and for the newly weds on honeymoon. At other time, these cottages can serve as meeting halls or physical training centers. In this way, life is enriched and improved, more and more people are assembled, and the rural economy is becoming prosperous.

(7) Control and harness water pollution alongside the river. The Pearl River receives all the industrial and residential waste water of Guangzhou. The residents and industries drain off their waste water into the Pearl River that becomes the passage way of wastes. In the new plan, the drainage of the waste water from the riverside should be strictly controlled to keep the air and the stream water clean and clear, which in turn gives the residents living along the river a good environment. The HuangPu segment should be especially on alert since it is badly polluted by the nearby industrial groups. Along this river segment, the air and the water is seriously polluted and appropriate measures should be put forth to control and harness the pollution.

The riverside environment of Guangzhou City has some in common with London (Tide River) and Paris (the famous historical and cultural city). We can learn a lot from their experience of the urban programs and construction. To exert the advantages and refrain from the weakness can better the construction of the urban riverside in Guangzhou. We should learn a lesson from London Dockland Port and avoid making the same mistakes.

References

[1] (UK) O'neil, Stein. Passing by Loulan Translated by Xiao Xiaoyong and Wu Xinhua. 2000. Publishing House of Guangxi Normal University.

[2] (Song Dynasty) Zhang Zeduan. A River Scene on Pure Brightness Day.
[3] Yan Shuiyu, Wang Xiangrong. 1999. Thames in London's City Planning, Analysis of Its Function Orientation, Secure measures and the Characteristics. Overseas Urban Programming, (1): 34-35.
[4] Guangzhou Comprehensive Development. China Academy of Urban Planning & Design Concept Plan.

Research Issues of Muddy Coasts*

1 INTRODUCTION

SCOR Working Group 106 was established with terms of reference which focused on the dynamics of muddy coasts. The original terms of reference included, *inter alia*:

- review of the oceanographic geomorphic and sedimentary dynamics of muddy coasts;
- to assess the impact of sea level rise on the sedimentary evolution of muddy coasts;
- to assess the impact of sea level rise on the mangrove forest and salt marshes of muddy coasts; and
- to recommend necessary research, data collection, and monitoring for the future management of muddy coastal zones.

The Working Group met in Nanjing, China, in October, 1995 and again in Wilhelmshaven, Germany, in September, 1997. It became clear at the first meeting that the consensus view of the Working Group was that many fundamental processes of muddy coast formation, function, ecosystems, and evolution were but poorly understood, and that to assess the impacts of relative sea level change on muddy coasts would require better understanding of the fundamental physical and biological processes leading to the formation and modification of muddy coast sedimentary, ecological, and physical oceanographic systems. Accordingly, the Working Group decided to address the terms of reference by focusing on a review of the fundamental processes leading to formation and modification of muddy coast, the functioning of the mangrove ecosystems, and to present examples of specific types of muddy coast in a global context.

In addition the Working Group addressed four specific questions relating to muddy coasts in a global context, *viz.*,

i) What are the perceived essential research issues of muddy coasts?

---

* Y. WANG, T. HEALY and Members of SCOR Working Group 106, being P. AUGUSTINUS, M. BABA, C. BAO, B. FLEMMING, M. FORTES, M. HAN, E. MARONE, A. MEHTA, X. KE, R. KIRBY, B. KJERFVE, Y. SCHAEFFER-NOVELLI, and E. WOLANSKI.: Terry Healy, Ying Wang and Juddy-Ann Healy, eds., *Muddy Coasts of the World*: *Processes*, *Deposits and Function*, chapter one, pp. 1～8. ELSEVIER, 2002.

ii) What are the urgent research problems of muddy coasts?

iii) What are the major gaps in knowledge pertaining to muddy coasts? and

iv) What recommendations can be made for furthering the environmental health and utilization of muddy coasts?

2 THE ESSENTIAL RESEARCH ISSUES OF MUDDY COASTS

The essential, urgent and perplexing research problems of muddy coasts identified by Working Group 106 may be categorized into two areas, namely (i) understanding the fundamental natural processes causing their formation and evolution, and (ii) research issues pertaining to their exploitation and management.

2.1 Fundamental processes of muddy coast formation and evolution

Sedimentary processes

The Working Group accepted that there were many unknown and unquantified aspects of the sediment transport and depositional processes, in part due to the typically heterogeneous composition of mud sediment. For example, fully predictive formulations for the critical stress for mud entrainment are not available because of the variety of physical and bio-geochemical factors that influence the bed shear strength. Similarly, the settling velocity of flocculated sediments is influenced by the chemical composition and temperature of water in addition to particulate makeup. A critical need is to understand the manner in which fluid mud, a high density suspension, is formed, entrained, and transported. There are other specific questions: at what applied shear stress does suspended sediment no longer deposit? Is there a critical suspended sediment concentration above which deposition tends to occur? Does this vary for different clay mineral particles? Is there a micro-organism influence? —or a freshwater flocculation influence?

The major issues identified by the Working Group were:

- sediment transport processes including mechanisms for fine particle erosion and deposition;
- the role, function and mechanism of bio-geochemical processes in the sedimentation dynamics of muddy coasts;
- accurate predictive modeling of sediment transport at muddy coasts;
- the role of flocculation in the deposition process of mud particles, and
- the role of fluid mud in configuring bottom profiles under tides and waves.

Morphology

While the Working Group recognized that the distinctive geomorphology of muddy coast is dominated by essentially flat broad inter-tidal surfaces, there are nevertheless additional distinctive elements of muddy coast geomorphology associated, for example, with mangrove stands, salt marsh, sandy-gravelly-shelly beaches surmounting inter-tidal mud flats, chenier deposits, and low sea cliffs. The essential research problems are seen as:

- quantifying coastal erosion and muddy coast land loss consequent solely upon sea-level rise;
- measurement of muddy coast morphological changes through time; and
- clarifying the fundamental process of morphodynamic change of muddy coasts.

Ecology of muddy coasts

The Working Group recognized an existing emphasis on the importance of mangrove ecology, as manifest for example in the global and regional programs sponsored by IOC and UNEP. However the Working Group felt that there should be wider recognition of the importance of muddy coast ecosystems, and additional research should be focused on:

- understanding the role of mangroves and salt marsh vegetation in the evolution of muddy coast.

Oceanographic processes

The Working Group recognized that oceanographic processes and hydrodynamic forcing clearly play a special role in the formation and morphodynamics of muddy coast and tidal flat deposits. In particular, research needs to address:

- investigation, monitoring, and evaluation of the role of flooding from abnormal tides and storm surges in muddy coast morphodynamics;
- monitoring effects of, and forecasting impacts of, storm-surge on muddy coasts;
- the process of saline water intrusion into farmland on lowland muddy coasts; and
- monitoring the response of muddy tidal flats, marsh, and mangrove systems to long term, small scale changes in regional relative mean sea-level.

2.2　Exploitation and management issues

Apart from the fundamental "process" research issues identified above, the Working Group recognized that there is considerable international interest, especially from Developing Countries, in undertaking research related to exploitation, development and management of muddy coasts. The Working Group identified the issues broadly as relating to (i) coastal management, and (ii) human impacts:

Coastal management

In this area we identify the main research issues as:

- designing strategies for conservation, planning and integrated management of muddy coasts; and

- identifying impacts of significance from global climatic change on the muddy coast morphodynamic and ecological systems.

Human impacts

Monitoring of human impacts opens a potentially wide and important area for research, as follows. Research should address:

- identifying and quantifying (including statistical methods) impacts from disturbance of vulnerable ecosystems, e. g. effects of clearing mangroves along muddy coasts;
- effect on muddy coast morphodynamics from mangrove degradation and the change from one ecosystem type to another;
- impacts on muddy coast from land clearing and accelerated soil erosion, and associated increased catchment flooding and sediment loads delivered to the muddy coast;
- quantifying effects of continuing developments on pollution levels of muddy coasts; and
- effects of coastal development on maintaining navigation channels, ports and sediment disposal sites.

3 KNOWLEDGE GAPS IDENTIFIED FOR MUDDY COASTS

The Working Group perceived several "gaps in knowledge" about muddy coasts. These again may be categorized as issues of fundamental science (sedimentation processes, geomorphology, oceanography), as well as global change effects, and coastal management effects.

Sedimentation processes

- The role of biological products in the flocculation/deposition process and in enhancing resistance of mud particles to erosion;
- the understanding of floc behaviour with respect to cohesive characteristics and non-Newtonian flow behaviour;
- the fundamental mechanism of mud particle entrainment and deposition in a macro-tidal setting; and
- modelling of muddy deposit erosion from the sea floor, its transport, and redeposition with respect to the forcing hydrodynamics.

Morphology

- Short and long term macroscale evolution of muddy coast topography due to episodic events against a background of longer term environmental forcing and

human influences;
- modelling of muddy coast morphodynamics; and
- understanding the roles of sediment sources and sinks in governing the erosion/accretion cycles on muddy coasts.

Oceanography

- Details of tidal dynamic processes influencing the different types of muddy coasts;and
- analyses of relative sea-level changes in the major muddy coastal regions.

Global change effects

- Effects of global climate change and human usage on muddy coastal systems;
- the mechanisms linking muddy coast and sediment transport dynamics with sealevel changes; and
- the morphological response of muddy coasts to sea-level rise.

Management

- Scientific and engineering techniques for restoration and management of mud dominated shores and shore habitats; and
- the modelling of muddy coast dynamical response to land management, especially in the tropics where changes have been drastic. Such modelling should include economic impact modelling.

4 HUMAN IMPACTS ON MUDDY COASTS

The Working Group considered human impacts on muddy coasts, and identified the following impacts and issues as likely to require ongoing monitoring and research on a global, regional, and local scale:

- Channel dredging and dredged sediment disposal at ports and harbours and navigation approach channels on muddy coasts;
- river channel migration at major river mouths and deltaic coastline erosion;
- muddy sediment accretion and progradation on muddy coasts from increased catchment soil erosion caused by deforestation and degradation of pasture land;
- damming of rivers affecting delivery of sediment to the coast thereby affecting coastal sediment budgets and changing the coastal ecological environment;
- overpumping of ground water along extensive coastal plains of muddy coasts;
- ground subsidence from overloading due to construction, and increased relative sea-level rise;
- reclamation of lagoons and inlets accelerating harbour and tidal channel siltation as a result of decreased tidal prism;

- harmful algal blooms and 'red tides' in extensive areas of shallow shelves fronting muddy coasts exacerbated by nutrient enrichment due to discharges from terrestrial catchments;
- pollution of waters and sediments arising from urban, port, industrial and agricultural activities;
- real estate development encroachment onto muddy coastal flats;
- reclamation of muddy coast for agricultural and aquacultural developments;
- removal of mangrove forests for fuel;
- pollution of mud sediment by heavy metals;
- destruction and trapping processes, such as mangroves and salt-marshes liberating the mud and destabilizing ecological community abundance and diversity in the coastal zone;
- delta building (e. g. Mississippi) and sediment starvation of coastal wetlands;
- reclamation of muddy coastal areas rich in pyrite resulting in acid sulphuric soils; and
- hydrocarbon (oil and gas) and groundwater extraction in adjacent to muddy coastal areas inducing accelerated relative sea-level rise.

5 RECOMMENDATIONS FOR ENHANCING ENVIRONMENTAL HEALTH AND SUSTAINABLE UTILISATION OF MUDDY COASTS

The Working Group considered the methodological approach for certain problems facing muddy coasts, and especially the focus of research and monitoring that would be suitable and necessary for furthering both the sustainable utilization and the enhancement of environmental 'health' of muddy coasts. These recommendations are:

- *Basic knowledge of muddy coastal systems needs to be increased.*

Comment: This can be applied by undertaking investigation and monitoring of special or representative sites either via long term fundamental research programs, or through dedicated projects, such as major port, airport, tourist facilities or acquaculture constructions.

- *National governments and international funding agencies (such as SCOR, UNEP, IOC) are urged to allocate sufficient funds for seeding basic research and conservation of representative muddy coastal habitats.*

Comment: The need to undertake basic background research and baseline studies on the nature of muddy coast ecosystems is seen by the Working Group as a matter of priority which is in the national interests of sovereign states.

- *Ongoing development of conceptual, analytical and numerical models which can be used for simulating sedimentary, ecological oceanographic and morphodynamics of muddy coasts, needs to be supported at regional national and international levels.*

Comment: Although over the past decades initial research has been undertaken on muddy coasts many of the processes are poorly understood. This requires renewed fundamental research effort linking field measurements and numerical modelling of both the physical (hydrodynamic forcing, sediments) and biological realm. There are complicating effects of indirect processes such as fresh-saline water interaction, flocculation of clay particles, the role of fluid mud, de-watering, benthic organism effects, organic matter complexing, micro-organisms, and chemical reactions within the sediment, the effects of which may not even be recognized. It is not simply a physical sedimentation problem. The basic understanding of these processes requires ongoing fundamental research and modelling, especially linked numerical simulation of biological, hydrodynamic, sedimentation, and morphodynamic processes at various time scales.

- *Development of monitoring strategies, techniques and programs which allow international comparison for muddy coasts should be supported.*

Comment: In many parts of the globe muddy coasts are affected markedly by human impact and thus standardized strategies and techniques for monitoring are necessary. In the modern context, monitoring of the large expanses of muddy coast is likely to involve both direct field measurement as well as remote sensing.

- *Muddy coast management policies should integrate models based on the function of mangroves and salt marshes in stabilizing shorelines and trapping muddy material for the further physiographic evolution of muddy coastal landforms and ecosystems.*

Comment: The function of mangroves and salt marsh vegetation in stabilizing the muddy coast shorelines has in the past been overlooked in coastal management policies, and needs to be considered as an integral component of coastal management practices.

- *Developments within and over muddy coasts should be based upon EIA and Integrated Coastal Zone Management strategies.*

Comment: It is axiomatic that coastal developments should only proceed after consideration of environmental impacts, and that developments should fall within planning strategies of coastal management which link the development with environmental effects which do not contravene the concept of sustainable development. Moreover, where possible, developments should also include components of "environmental enhancement".

- *Development of muddy coast should proceed within the context of "integrated coastal zone management".*

Comment: Development of a new port facility, for example, would ideally follow a multidisciplinary approach with interaction of engineers, physical oceanographers, benthic and mangrove ecologists, geo-scientists, social scientists and planners.

Development needs to take cognizance of the impact on the sedimentary and ecological systems, as well as human impacts, and where possible mitigate such impacts.

• *Development of innovative techniques is required for restoring or maintaining degraded muddy coast shores and nearshore habitats based upon a sound knowledge of scientific principles of muddy coast behaviour and functions.*

Comment: Human impacts are widespread over many muddy coastal zones and innovative strategies are required to repair, restore or enhance utilization of muddy coast following a philosophy of "sustainable management".

6 CONCLUSION

The above ideas summarize the findings of SCOR Working Group 106 on the issues of, and urgent research problems for, muddy coasts, including identified gaps in knowledge of muddy coasts, human impacts on muddy coasts, and recommendations for furthering the health and "sustainable utilisation" of muddy coasts.

The following chapters of this volume comprise contributions from the Working Group on systematic aspects of muddy coasts within its general terms of reference. This general theme is broadened with additional contributions on specific aspects, types, properties, and phenomena affecting the various muddy coasts of the world.

Definition, Properties, and Classification of Muddy Coasts*

1 WHAT DO WE MEAN BY "MUDDY COAST"?

From review of a wide range of coastal environments, SCOR Working Group 106 defines "muddy coast" as:

a sedimentary-morphodynamic type characterized primarily by fine-grained sedimentary deposits-predominantly silts and clays-within a coastal sedimentary environment. Such deposits tend to form rather flat surfaces, and are often, but not exclusively, associated with broad tidal flats.

A coastal sedimentary environment is understood as one in which processes of erosion, transportation, and deposition of sediments are due to coastal and estuarine forcing mechanisms. It extends more or less from 'wave base' -typically of order 6～10 m water depth on the open coast, but much less in sheltered embayments and lagoons-inland within estuaries or embayments, to the limit of direct marine influence. Muddy depositional environments typically occur in the lower energy part of the coastal environmental hydrodynamic energy ranges.

Predominantly, but not exclusively, muddy coast sediments are most obviously observed as deposits surfacing a broad intertidal zone. However, the adjacent shallow sub-tidal sea floor is likewise very muddy and typically contains fluid mud (for example, see Kirby 1988; Healy et al. 1999). Often this acts as an adjacent source which supplies the mud deposited in the intertidal and supratidal zones under suitable hydrodynamic forcing conditions. Thus from the perspective of muddy coasts, the muddy depositional environment exists within coastal fringe waters, and extending as far inland, or within an estuary, as the limit of marine influence (waves, tides, littoral sediment transport, and estuarine circulation processes) which result in the deposition of muddy sediments.

---

* Y. WANG, T. HEALY and Members of SCOR Working Group 106, being P. AUGUSTINUS, M. BABA, C. BAO, B. FLEMMING, M. FORTES, M. HAN, E. MARONE, A. MEHTA, X. KE, R. KIRBY, B. KJERFVE, Y. SCHAEFFER-NOVELLI, and E. WOLANSKI.: Terry Healy, Ying Wang and Juddy-Ann Healy, eds.,《Muddy Coasts of the World: Processes, Deposits and Function》, chapter two, pp. 9－18. ELSEVIER, 2002.

2 PROPERTIES OF MUDDY COASTAL DEPOSITS

In sedimentological terms "mud" is easily, if somewhat arbitrarily, defined. Texturally "mud" is a detrital deposit composed primarily of particle sizes smaller than 62.5 microns (i.e. 0.062 5 mm or 4ϕ), and includes the textural classes of silt, 0.062 5~0.039 mm or 4ϕ~8ϕ in the sedimentological (phi) classification (Folk 1968; Friedman and Sanders 1978), and clay, which is generally accepted as being composed of particle sizes smaller than 0.0039 mm or 3.9 microns (8ϕ).

Muddy coastal deposits almost always contain a proportion of organic matter-typically 3%~5%-originating from the importation of fine suspended particulate organic material to the zone of deposition of the muddy sediments. The organic matter originates from both terrigenous (from fine vegetation detritus) and marine biogenic sources, as well as from in situ biogenic processes, such as mucus and faeces from worms and other benthic organisms inhabiting the surficial muds. In many locations organic matter contributions include fine plant detritus originating from such diverse sources as mangrove and salt marsh stands, or adjacent 'pastures' of seagrass, marsh grasses, and wetlands. Biogenic contribution to muddy sedimentary deposits may also include small shell fragments, minute sea urchin spines, and fragments of living and dead diatoms, foraminifera, ostracods and coccoliths.

Additional mineral matter occurs in the mud deposits, including fine sands, terrigenous matrix fragments, and shelly gravel fragments. Composition of the mineral matter is influenced by local supply and may consist of quartz, feldspar and mica, and small proportions of heavy minerals.

An important property of mud is its rheology, i.e. its flow and deformation characteristics (Berlamont et al. 1993). Typically mud deposited in the intertidal zone is soft, pliable, of high plasticity, unctuous, thixothrophic and contains a large volume of water in its physical structure. In the field, a person may sink up to ones thighs in wading through intertidal muddy deposits. This thixotrophic property suggests that in geotechnical terms mud behaves variously as either a visco-plastic or a fluid substance, and behaves rather differently to non-cohesive clasts in the sedimentary environment.

In mud-rich shallow waters of muddy coastal areas, the sea floor may be characterized by "fluid mud" -a state in which the particles occur partially as concentrated colloidal state, and the viscous substance possesses extremely high water content (Kirby 1988; Mehta and Hayter 1989; Mathew and Baba 1995). Fluid mud may be defined as a highly concentrated sediment suspension with concentrations exceeding 10 000 mg · L^{-1}, a concentration at which hindered particle settling effects are first observed (Kirby 1988).

Muddy coastal sediments also possess interesting geochemical characteristics, related to the chemical reactions associated with micro-organism activity within the mud deposits. Below the oxic surface layers, usually only a few cm, the mud often looks blue-black and exhumes a pungent sulphurous smell. Both colour and smell are due to the formation of FeS in a reducing environment. Under these conditions sulphides (S^{2-}) are released from sulphates by sulphate reducing bacteria, the required energy being drawn from the oxidation of organic matter (Moorman and Pons 1975).

3 MORPHO-SEDIMENTARY CHARACTERISTICS OF MUDDY COASTS

3.1 Physiographic types of muddy coast

For expansive open ocean tidal flats, for example those of the Dutch-German North Sea coast or of China, muddy deposits are an integral component. The muddy deposits may range from several hundred meters to more than ten km in width (see Wang, Zhu and Wu, chapter 13). But muddy coastal deposits may also exist in as limited an extent as a few tens of meters wide.

• *Tidal fiats* are the typical form of expansive muddy coasts and are characterized by inter-tidal surfaces with low slopes of order (0.5～1.0) : 1 000. Distinctive morphological and sedimentary zonation features may be identified (see for example, Evans 1965; Wang 1983; Wang et al. 1990). Generally one can identify a low relief salt marsh zone developed within the supra-tidal zone, which may be surmounted by sandy or shelly chenier beaches or ridges, or artificial dikes on populated extensively flat lowland coasts (Augustinus 1989). Below high tide level, true 'mudflats' are located in the upper part of the inter-tidal flat. Clay and silt layers often become deposited in the central parts of the tidal flats, and the lower-to-sub-tidal sectors become more sandy-silty or silty-sand, with wave ripple bedform features developed on the surface. Dendritic meandering tidal creeks are common on extensive mud tidal flats, formed by the scouring of ebb-tidal waters as they drain off the large areas of inter-tidal flat upon the falling tide. Erosion scour features, both tidal current and wave induced, are often observed on the central sectors (Pethick 1996). For broad muddy tidal flats, spatially varying morpho-sedimentary zonation is evidently associated with different hydrodynamic forcing processes acting on the flat.

• *Enclosed sheltered bay deposits*. Muddy deposits of varying types may occur in sheltered embayments. Geomorphically such embayments may be tectonically induced by block faulting, or evolve from active muddy sedimentation within drowned river valleys consequent upon the Holocene sea level rise. Muddy sediments normally originate from the surrounding catchments and deposit particularly in the tributary valley heads (Pethick 1996; see also chapter 14 by Healy).

• *Estuarine drowned river valley deposits*. Drowned river valleys, frequently manifest as funnel-shaped estuaries, typically have significant river input and exhibit strong mixing phenomena of the fresh and marine saline waters. The resulting "estuarine circulation" causes mud to be transported landwards at the bottom of the estuary, so that deposits become markedly more muddy toward the head of the estuary or bay (Nichols and Biggs 1985; Pethick 1996). The fresh waters also carry suspended sediment loads of silt and clay, especially in river floods, and these tend to undergo flocculation and deposition upon mixing with the saline waters in the upper part of the estuary. The result is a concentration of muddy deposition at the heads of estuaries.

• *Inner deposits of barrier-enclosed lagoons*. Within the landward side of barrier enclosed lagoons, along the southeast coast of the USA, for example, a veneer of deposited mud is typically found. Such muddy deposits are features of barrier enclosed lagoons, and may be associated variously with wetlands, oyster beds, and estuarine re-entrant valley head deposits. Source of the muddy deposits is mainly from input from the surrounding land catchment, and the deposits are usually thin (0. 5～2 m) but relatively extensive in the available area, and overlie sands of the barrier lagoonal system.

• *Supra-tidal (storm surge) deposits*. Lowland coasts with a broad shallow shell especially where they exist in the form of a large funnel, such as the German Friesland Bight, the Bohai Bay in China, the Gulf of Thailand, or the Gulf of Bangladesh, may be subject to strong storm surges (see chapter 10 by Bao and Healy). Where there are abundant nearshore submarine muddy deposits, during the storm surge event waters of high suspended load sediment concentration can be swept onshore, and the sediment load become deposited in the supra-tidal zone. Such supra-tidal deposits are found in China, and in Bangladesh as a result of storm surges and tropical cyclones. The material deposited tends to be mainly silt.

• *Swamp marsh and wetland deposits*. This type of muddy coast deposit occurs particularly in temperate coastal environments beyond the limits of mangroves. Good examples are found along the low terrain coasts which occur on a broad scale around the Dutch and German North Sea coasts, and in The Wash of England. Other well known examples are found on the landward side of the extensive barrier enclosed lagoon system that is found around the southeast USA and the Gulf coast (Davis 1983).

• *Mangrove forest and swamp deposits*. Mangrove forests occur widely in the tropics, and their root mat assists in the trapping of fine suspended sediments, so that typically they are surrounded by mud deposits. Mangrove stands also occur on the landward shores of coral enclosed lagoons, and likewise may induce a veneer of muddy deposits to occur over the coral (see Wolanski and Duke, chapter 21).

• *Chenier plains*. Shelly chenier ridges can form plains as described by Schofield (1960) and Woodroffe et al. (1983) for the Firth of Thames in northern New Zealand. The cheniers sit upon thick deposits of structureless mud, and the seaward-most chenier comprises the active beach face surmounting a broad intertidal mud flat. Augustinus (1989) has described mainly sand cheniers on the shore side of the elongated mud banks along the Suriname coast.

• *Mud deposits veneering eroded shore platforms*. In this type of muddy coast, an eroded shore platform has subsequently been covered with a veneer of muddy sediment. These types of muddy coast occur on a relatively small scale and in sheltered harbor environments where there is a high source of muddy material, either from adjacent catchments or from sub-littoral deposits in the nearshore. Volumes of mud involved are relatively small, and examples occur in the harbours of northern New Zealand.

• *Ice deposited mud veneer*. On Arctic shorelines, a muddy flat may be formed by ice rafts transporting mud and boulders, grounding and melting out in the thaw so that mud and boulders are jointly deposited on the tidal flats (see chapter 19 by Dionne).

• *Sub-littoral mud deposits*. Along many coasts in sheltered and semi-enclosed bays where there is an abundant source of fine sediment supply and not particularly strong tidal currents, mud deposits occur in the sublittoral zone. These deposits may be of the fluid mud type (Kirby 1988; Healy et al. 1999) or occur as offshore mud banks, as off the coast of Kerala, India (Mathew and Baba 1995) and the coast of Suriname (Augustinus 1989). On occasions of strong wave agitation and shorewards wind-induced surface current, the wave agitated mud sediments may be eroded, and suspended by the wave turbulence in the estuarine "turbid fringe" and swept onshore and deposited as a mud layer, either in the high tidal zone or the intertidal zone.

The above physiographic classification of muddy coast types is not necessarily comprehensive, and there may be some overlap in classification. One could recognize deltaic muddy coast as a specific type, for example, and for specific sites there could be sheltered embayments with muddy deposits as well as estuarine drowned valleys. But not all estuarine types are sheltered embayments exhibiting muddy deposits, and vice versa.

3.2 Sedimentary structures and stratigraphy of muddy coastal deposits

SCOR Working Group 106 identify that tidal flat muddy deposits are often stratified in fine laminae, the delicate bedding being caused by the inter-lamination of thin layers of very fine sandy material in the more argillaceous basic substrate. The lamination of thin bedding is the result of varying hydrodynamic conditions, including the obvious flood and ebb oscillating tidal flows, but also reflecting neap and spring, storm and seasonal changes. This type of thinly laminated bedding was identified as tidal bedding as early as 1841 (Hantzschel 1955). The bedding is seldom strictly parallel, but instead is generally streaky and lenticular. The alternating fine sandy and clay layers wedge out rapidly and their thickness is not uniform. Such alternating bedding indicates frequent reworking of sediments under the influence of tidal currents of varying strength and direction, and the influence of meteorological factors, especially storm winds and associated waves.

Stratigraphically, modern coastal mud deposits tend to be thin relative to true deep sea marine muds. This reflects the limited time-only some 6 500 years-since the Holocene sea level reached its approximate present level, as well as the shallow depositional environment, which is typically energetic on a day to day time scale. In some areas the deposits, especially clays, appear to be structureless and massive, but in other deposits there are often rare lenses of sand and shell-the result of episodic storm processes (Rine and Ginsburg 1985; and see chapter 16 by Park and Choi).

In stratigraphic section, across broad tidal flats, such as occur in China, shelly chenier deposits overlie the supratidal clay-rich muds. Typically these clay muds, laid down by storm surge processes, are massive, lensoid shaped, and also contain relatively high organic matter, especially plant fragments. The supra-tidal clays overlie the silty muds of the upper tidal flats, which overlie the alternating fine sand and clay lenses of the mid tidal area. These in turn overlie the very fine sand and coarse silt of the low tidal zone flat.

Sedimentary structures found in muddy coastal deposits include ripple bedforms on the sandier sediments. Within the larger tidal creeks with high current speeds and transporting sandy bedload, bedforms of megaripples, and current-cross-bedding is typical. Upon the mud flats, bioturbation structures produced by organisms inhabiting the tidal flats are also evident, while polygon cracks are to be found on the supra-tidal deposits. These sedimentary structures reflect a range of dynamic processes on the muddy coast tidal flats.

4 GLOBAL DISTRIBUTION OF MUDDY COASTS

Muddy coasts are developed in a wide range of coastal climatic and oceanographic environments in the world, from Arctic oceanic coast to the tropics. Their detailed geographic location and nature are presented in chapter 6 by Flemming. Extensive well formed, broad muddy coasts are present around the Yellow River delta in the Bohai Sea of China, as well as the abandoned Yellow River delta in North Jiangsu Province along the Yellow Sea, and the southern embayments of the Changjiang River. Extensive sectors of muddy coast also occur along the east coast of the Americas, with the Amazon River mouth and the northern coasts of Guiana being of immediate note. But extensive muddy coast also occurs within Canada's Bay of Fundy, around the seasonal ice covered St. Lawrence estuary, along the Atlantic east and southeast coasts of the United States, the Mississippi delta and Vicinal coasts of the Gulf of Mexico, and inside several of the larger bays along the west coast of North America from California to Washington. The North Sea and Baltic coasts of Europe, such as Dollart, Lay and Jade Bays in Germany, the funnel-shaped mouths of large rivers (Elbe, Weser), in the east and north Friesian Bight and The Wash and other estuaries of east England, likewise all exhibit different examples and types of muddy coast. Muddy coast is widely evident in the tropical and monsoonal countries of Asia, including much of the coasts of India (see chapter 15 by Baba), Bangladesh, Thailand, Malaysia, Indonesia, the Gulf of Papua, and Vietnam. The northern coasts of Australia (Wolanski and Chappell 1996), as well as the islands of the Pacific including New Zealand, New Guinea, and even the coral reef islands, contain areas of muddy coast. For the African continent muddy coast occurs mainly within the tropical belt of both the east and west coasts.

It is evident that muddy coasts are characteristic of all continents (perhaps with the exception of Antarctica which typically exhibits a steep rocky profile for its minor sectors of ice free coastline). They occur in a wide variety of the global coastal environments, but can vary widely in size and morphogenetic environment. However, muddy coasts tend to be particularly conspicuous in tropical zones, especially in the Asia region, and the muddy coastal deposits of largest extent are associated with major continental river discharges.

5 FACTORS FACILITATING FORMATION OF MUDDY COASTS

The above summary review of the types and occurrence of muddy coasts indicates that they develop in a wide variety of morphogenetic environments. However the Working Group recognizes there are some common factors which facilitate and influence their evolution.

5.1 Abundant fine-grained sediment supply

This is the most important factor in facilitating the formation of muddy coastal deposits of various types. Fundamentally, if a significant source continuously provides fine sediments to the coastal zone at a rate faster than the hydrodynamic conditions can remove them, then muddy deposits will form. But where there is not an abundant and continuing supply of silts and clays to the coast, then muddy coastal deposits do not form.

Sources of sediment for concentration by sedimentary processes and subsequent deposition to form muddy coast are many and varied, but include:

Adjacent rivers. Rivers with high concentrations of fine suspended sediment loads are a prime source of sediment to form muddy deposits in the coastal zone. The classic example is the Yellow River of China, which drains a huge catchment including highly erodible loess regolith, which provides the bulk of the fine sediments to the Bohai Bay. Other examples include the Amazon, draining a huge and partly steep catchment of highly weathered tropical soils, the Fly River of New Guinea, the Mekong River of Vietnam, likewise draining large and partly steep catchments of highly weathered tropical soils, and the Waipaoa River of New Zealand, draining a steepland catchment of highly erodible montmorillonitic clay-rich Tertiary rocks. In the latter cases, the mountainous steepland catchments are also subject to episodic, intense, orographically induced rainfalls, factors which enhance the sediment load delivered to the coastal zone. Thus secondary factors which contribute to high concentrations of suspended sediment loads debouching into coastal waters include:

- large fluvial catchment size;
- abundant erodible soils and regolith;
- steepland terrain, and an active seismic regime, which promote slope instability and soil erosion; and
- episodic or monsoonal storm events to exacerbate catchment erosion and provide high suspended loads to the coast.

Supply of sediment from offshore. Often sediment that forms or contributes to coastal muddy coast may be of "secondary origin" in the sense that it is re-worked from offshore deposits and brought onshore to become deposited as muddy coastal deposits. In this sense the reworked mud deposits are "palimpsest". Examples include the onshore transfer of fine sediments from the ancient river deposits along the northern Jiangsu coast (see chapter 13 by Wang, Zhu and Wu), and the onshore transfer of eroded glacial till deposits from around the North Sea and Baltic.

Erosion of coastal sedimentary deposits and cliffs. In many enclosed and semi-enclosed seas, active erosion of the cliffs provides a limited source of fine material for muddy coastal deposits. In the Kiel Bay of the Western Baltic the surrounding coast is formed of glacial till containing a high proportion of fine sediment (Healy and Werner 1987). Rapid erosion of the boulder clay cliffs (−0.3~1.0 m/a) provides considerable material for local muddy coastal deposits. Where there has been rapid cliff erosion and continuing supply of muddy sediment the shore platform may subsequently become veneered with mud. Likewise in sheltered steepland embayments containing an estuarine lagoon fronted by barrier dunes erosion of the cliffs produces mud which deposits to overlie the early Holocene marine sands.

5. 2 Tidal range

It is sometimes claimed that a macro tidal range, defined as >4 m by Komar (1998), is a necessary factor in explaining extensive broad muddy coast deposits, but that is not always the case. A general dearth of sediment supply or high wave energy may restrict tidal flat development (Davis 1983). Certainly the broad expansive inter-tidal flats and associated muddy deposits of the Jiangsu coast of China are very wide and have associated a macro tidal range. Other extensive muddy coastal deposits associated with large tidal ranges include the Bay of Fundy and The Wash of England (Amos 1995).

But in other locations of muddy coasts the tidal ranges may be meso (2～4 m) or micro (<2 m), and still extensive muddy deposits of various types occur. Such occurs around the coast of Indonesia, India, the southeast coast of the USA, off the mouth of the Yellow River of China, and around northern New Zealand. However even in northern New Zealand, the areas of widest intertidal muddy coast are associated with the larger tidal ranges of ～4 m, for example within the Firth of Thames or within the Manukau and Kaipara Harbours (see chapter 14 by Healy).

A fundamental question is: why should a large tidal range be associated with broad intertidal muddy coastal deposits? Evidently the answer lies in a fundamentally different morpho-depositional mechanism from sandy and mixed sand gravel beaches, as well as the thixotrophic nature of the muddy deposits. Firstly sandy beaches are characterized by non-cohesive grains, and thus undergo regular morpho-dynamic change in response to wave properties but typically assume some modal morphodynamic beach state (Wright and Short 1984). However beaches on muddy coasts tend not so obviously to change their morphology, although the beach is active over time. This is evidently because:

(i) the individual muddy sediment clasts are cohesive and thus the surface layer of sediments is not moved around by the waves (as are non-cohesive sandy sediments). Rather, in energetic wave conditions the surface layers of mud particles are re-agitated and taken into suspension within the water column;

(ii) the sediments in purely muddy beaches tend to be deposited in episodic storm surge events with a net onshore movement of suspended sediment which becomes deposited on top of the existing muddy flat sediments; and

(iii) the muddy bulk sediment possesses little strength, is subject to fluid and plastic flow, and thus results in a low angle of natural repose within the intertidal zone.

Thus, strictly muddy coast, (i. e. without a landward surmounting narrow beach face of sand or shelly gravel or chenier beach) typically has a muddy upper inter-tidal and supra-tidal subdued relief upper beach which exhibits relatively little morphodynamic change compared to a sandy beach.

The essential point seems to be that the mode of formation of the inter-tidal muddy deposits is by net onshore sediment movement and initially fluid mud deposition, which because it has little strength, causes an essentially flat topographic surface to result. Then providing there is abundant and continuing fine sediment supply the extent or width of a muddy inter-tidal flat becomes a function of the tidal range.

6 CONCLUSION

Muddy coast geomorphology and sedimentary depositional systems occur widely around the earths' coastlines, but compared to sandy coasts have been relatively little studied. Texturally, mud is defined based upon particle size, and comprises mineral material of silt and clay sizes. Muddy coasts are defined as possessing distinctive muddy deposits, typically but not exclusively manifest as broad intertidal muddy flats. Additional geomorphic features found on muddy coasts include chenier ridges and beaches, veneer muddy deposits and shallow offshore muddy deposits, while biologic features may occur such as mangrove stands and salt marshes. Muddy coasts occur widely in geographic distribution from the tropics to the Arctic. Their fundamental *raison d'existence* is abundant mud sized sediment particles in the coastal environment, but the existence of broad inter-tidal mud flats is often associated with a large tidal range.

REFERENCES

[1] Amos, C. 1995. Siliclastic tidal flats. In: G. Perillo (editor), *Geomorphology and Sedimentology of Estuaries*. *Advances in Sedimentology* 53. Elsevier, Amsterdam. 273 - 306.

[2] Augustinus, P. G. E. F. 1989. Cheniers and chenier plains: a general introduction. *Marine Geology*, 90: 219 - 229.

[3] Baba M. and S. R. Naya. 2002. Chapter Fifteen: Muddy coasts of India. In: Terry Healy, Ying Wang and Juddy-Ann Healy, eds., Muddy Coasts of the World: Processes, Deposits and Function. ELSEVIER. 375 - 390.

[4] Bao C. and T. Healy. 2002. Chapter Ten: Typhoon storm surge and some effects on muddy coasts. In: Terry Healy, Ying Wang and Juddy-Ann Healy, eds., Muddy Coasts of the World: Processes, Deposits and Function. ELSEVIER. 263 - 278.

[5] Berlamont, J., M. Ockenden, E. Toorman, and J. Winterwerp. 1993. The characterization of cohesive sediment properties. *Coastal Engineering*, 21: 105 - 128.

[6] Davis, R. A. 1983. Depositional Systems, A Genetic Approach to Sedimentary Geology. Prentice-Hall, New Jersey, 699 p.

[7] Dionne J. -C. 2002. Chapter Nineteen: Sediment content of the ice-cover in muddy tidal areas of the

turbidity zone of the St. Lawrence estuary and the problem of the sediment budget. In: Terry Healy, Ying Wang and Juddy-Ann Healy, eds., Muddy Coasts of the World: Processes, Deposits and Function. ELSEVIER. 463 - 478.

[8] Evans, G. 1965. Intertidal flat sediments and their environments of deposition in The Wash. *Quarterly Journal of the Geological Society of London*, 121: 209 - 245.

[9] Flemming B. W. 2002. Chapter Six: Geographic distribution of muddy coasts. In: Terry Healy, Ying Wang and Juddy-Ann Healy, eds., Muddy Coasts of the World: Processes, Deposits and Function. ELSEVIER. 99 - 202.

[10] Folk, R. L. 1968. *Petrology of Sedimentary Rocks*. Hemphills, Austin, 170 p.

[11] Friedman, G. M. and J. E. Sanders. 1978. *Principles of Sedimentology*. John Wiley and Sons, New York, 792 p.

[12] Hantzshel, W. 1955. Tidal Flat Deposits (Wattenschlick). In: P. D. Trask (editor), *Recent Marine Sediments*. Dover Publications, New York. 195 - 205.

[13] Healy, T. 2002. Chapter Fourteen: Muddy coasts of mid-latitude oceanic islands on an active plate margin—New Zealand. In: Terry Healy, Ying Wang and Juddy-Ann Healy, eds., Muddy Coasts of the World: Processes, Deposits and Function. ELSEVIER. 347 - 374.

[14] Healy, T., A. Mehta, H. Rodriguez and F. Tian. 1999. Bypassing of dredged muddy sediments using a thin layer dispersal technique. *Journal of Coastal Research*, 15(4): 1119 - 1131.

[15] Healy, T. and F. Werner. 1987. Sediment budget for a semi-enclosed sea in a near homogeneous lithology. The example of Kieler Bucht, Western Baltic. *Senkenbergiana maritima*, 19 (3): 32 - 70.

[16] Kirby, R. 1988. High concentration suspension (fluid mud) layers in estuaries. In: J. Dronkers and W. van Leussen (editors), *Physical Processes in Estuaries*. Springer-Verlag, Berlin. 463 - 487.

[17] Komar, P. D. 1998. *Beach Processes and Sedimentation* (2 nd Edition). Prentice-Hall, New Jersey, 544 p.

[18] Mathew, J. and M. Baba. 1995. Mudbanks of the southwest coast of India. II: Wavemud interactions. *Journal of Coastal Research*, 11(1): 179 - 187.

[19] Mehta, A. J. and E. J. Hayter (editors). 1989. High Concentration Cohesive Sediment Transport. *Journal of Coastal Research*, *Special Issue* 5, 230 p.

[20] Moorman, F. R. and L. J. Pons. 1975. Characteristics of mangrove soils in relation to their agricultural land use and potential. *In: Proceedings of the International Symposium on Biology and Management of Mangroves*, Honolulu. 529 - 547.

[21] Nichols, M. M. and R. B. Biggs. 1985. Estuaries. In: R. A. Davis (editor), *Coastal Sedimentary Environments*. Springer-Verlag, New York. 77 - 186.

[22] Park Y. A. and K. S. Choi. 2002. Chapter Sixteen: Late Quaternary stratigraphy of the muddy tidal deposits, west coast of Korea. In: Terry Healy, Ying Wang and Juddy-Ann Healy, eds., Muddy Coasts of the World: Processes, Deposits and Function. ELSEVIER. 391 - 410.

[23] Pethick, J. S. 1996. The geomorphology of mudflats. In: K. F. Nordstrom and C. T. Roman (editors), *Estuarine Shores*, *Evolution*, *Environments and Human Alterations*. John Wiley and Sons, New York. 185 - 211.

[24] Rine J. M. and R. N. Ginsburg. 1985. Depositional facies of a mud shoreface in Suriname, South America. A mud analogue to sandy shallow-marine deposits. *Journal of Sedimentary Petrology*, 55: 633 - 652.

[25] Schofield, J. C. 1960. Sea level fluctuations during the last 4000 years as recorded by a chenier plain, Firth of Thames, New Zealand. *New Zealand Journal of Geology and Geophysics*, 3: 467 - 485.

[26] Wang, Y. 1983. The mudflat coast of China. *Canadian Journal of Fisheries and Aquatic Science*, 40 (supplement 1): 160 - 171.

[27] Wang, Y., M. B. Collins and D. Zhu. 1990. A comparative study of open tidal flats: The Wash (U. K.), Bohai Bay and west Yellow Sea (Mainland China). In: *Proceedings of the International Symposium on the Coast Zone*. China Ocean Press. 120 - 130.

[28] Wang Y., D. Zhu and X. Wu. 2002. Chapter Thirteen: Tidal flats and associated muddy coast of China. In: Terry Healy, Ying Wang and Juddy-Ann Healy, eds., Muddy Coasts of the World: Processes, Deposits and Function. ELSEVIER. 319 - 346.

[29] Wolanski, E. and J. Chappell. 1996. The response of tropical Australian estuaries to a sea level rise. *Journal of Marine Systems*, 7: 267 - 279.

[30] Wolanski E. and N. Duke. 2002. Chapter Twenty-One: Mud threat to the Great Barrier Reef of Australia. In: Terry Healy, Ying Wang and Juddy-Ann Healy, eds., Muddy Coasts of the World: Processes, Deposits and Function. ELSEVIER. 533 - 542.

[31] Woodroffe, C. D., R. J. Curtis and R. F. McLean. 1983. Development of a chenier plain, Firth of Thames, New Zealand. *Marine Geology*, 53: 1 - 22.

[32] Wright, L. D. and A. Short. 1984. Morphodynamic variability of beaches and surf zones, a synthesis. *Marine Geology*, 56: 92 - 118.

Integrated Coastal Zone Management for Sustainable Development—With Comment on ICZM Applicability to Muddy Coasts①

INTRODUCTION

Coastal environments globally, including particularly the environmental aspects of coastal water quality, coastal ecosystems, scenic coastal landscapes of nature and natural resources, are under threat. Coastal ecosystems are highly productive and are valuable both ecologically and economically, but extremely sensitive to development. Healthy coastal wetlands provide a buffer zone against the impacts of climate change and rising sea levels. Coastal waters are the nurseries for most commercial fish and shellfish stocks. Tourism focussed on the coast is a major component of the global tourist industry (Miller, 1993), which since the new millennium is the largest global industry—and an increasingly important source of revenue. As the global population continues to migrate from inland to the coastal zones (Hinrichsen, 1999; Fletcher et al, 2000) and from rural to urban areas, some irreversible impacts on coastlines themselves, and the coastal zones will inevitably occur.

If an overarching environmental, sociological and economic integrated management is not implemented, there is considerable danger of irreversible environmental despoilment and economic blighting (Yu, 1994; Murray and Cook, 2002). While coastal development is an important economic driver of employment and increase in living standards, it is axiomatic that uncontrolled *laissez faire* development of the coast without consideration of ongoing sustainability can within a short time lead to "economic blighting", and in the interim the coastal environment resource has been severely degraded.

Integrated Coastal Zone Management (ICZM) evolved to avoid such problems and as a forward thinking tool for long term environmental, economic and social sustainability of coastal resources. The concept of ICZM has been widely advocated in the early 1990s (Clark, 1996; 1997) and since the 1992 Rio "Earth Summit" conference. But the concept has been rather difficult to apply and gain widespread acceptance

① Terry Healy, Ying Wang: *Journal of Coastal Research*, 2004, Special Issue, No. 43, pp. 229 – 242.

(Ngoile and Linden, 1997; Basiron, 1998; El-Sabh et al, 1998; Anon., 2002; Munoz et al, 2003), especially for muddy coasts. And in some regions implementation of the concept has stalled (Westmacott, 2002). Initially it was applied to sandy beach coasts, rather rarely to muddy coasts.

Accordingly, this paper outlines the essential issues of IZCM and some internationally recognized protocols for implementing sustainable development and management, particularly as applied to muddy coastal zones. The underpinning ideas draw heavily on the protocols established for the European Union countries, which were themselves catalysed by concept developments of UNEP and the 1992 Rio Earth Summit Conference, and in particular Chapter 17 of the Agenda 21 document.

DEFINITIONS OF THE COASTAL ZONE

Over the past few decades there have been numerous definitions of "coastal zone". Ketchum (1972) defined the coastal zone as "*the band of dry land and adjacent ocean space in which land ecology and land use directly affect ocean ecology*".

The US Commission on Marine Science, Engineering and Resources in 1986 defined the coastal zone "*as the area where land and sea interact, with its landward boundary defined by the limits of ocean influence on the land and the seaward limit being the limit of influence of land and freshwater on the coastal ocean ...*" (Healy et al., 2001). The coastal ocean in this context was defined by UNESCO in 1994 as the land-ocean transfer zone.

Put another way by the International Conservation Union (IUCN), the coastal zone is "*that part of the land affected by its proximity to the sea and that part of the ocean affected by its proximity to the land. The inland and ocean boundaries are not however spatially fixed ...*"

The World Bank (1993) defines the coastal zone as: "*the interface where the land meets the ocean, encompassing shoreline environments as well as adjacent coastal waters. The limits of the coastal zone are often arbitrarily defined, differing widely among nations, and are often based on jurisdictional limits or demarcated by reasons of administrative ease. For practical planning purposes, the coastal zone is a special area, endowed with special characteristics, of which the boundaries are often determined by the specific problems to be tackled.*"

And the European Commission more or less follows the definition by Ketchum (1972) envisaging the coastal zone as: "... *a strip of land and sea territory of varying width depending on the nature of the environment and management needs. It seldom corresponds to existing administrative or planning units. With regard to fisheries, it is common to limit the coastal zone to territorial waters as defined in the Convention on the Law of the Sea, although this limit does not correspond to any distinct biological or management unit. The natural coastal systems and the areas in which human activities involve the use of coastal resources may therefore extend well beyond the limit of territorial waters, and several kilometres inland.*"

For the case of oceanic muddy coasts (of China for example) with broad intertidal and sub-tidal flats, the coastal zone is taken to include the morphodynamic sedimentary systems extending from the limit of land processes affecting the primary ecosystems (mangroves or salt marshes), out to beyond "wave base" which is taken as a depth of about half the average wave length typical for the area. This is around 30～40 m depth along the ocean margins. The coastal zone includes estuaries and lagoons, as well as estuarine margins and wetlands.

INTEGRATED MANAGEMENT OF COASTAL ZONES

The Philosophy of Integrated Coastal Zone Management

Integrated Coastal Zone Management (ICZM) is increasingly regarded as a tool to sustainably manage development in coastal regions. There are various terms used to describe this process such as Integrated Coastal Management (ICM), or Integrated Coastal Area Management (ICAM) each of which is defined or approached somewhat differently (Sorensen 1997; Vallega, 1999). For the purposes of this paper we use the term Integrated Coastal Zone Management (ICZM) generically to refer to the full range of approaches pertaining to both planning and management, as applied to both land and sea components of the estuarine and muddy coastal zone, and to "*manage development so as not to harm environmental resources*" (Clark, 1996). As perceptively recognised by Ketchum (1972, p. 347) two decades before the name ICZM as such evolved, "*effective management of the coastal zone implies viable and economic integration of complex life systems and varied human activities within a confined area which is itself subject to the influence of major exterior forces both from the land and from the sea*".

Various definitions of ICZM abound. The European Commission concept of ICZM is a continuous process of administration the general aim of which is to put into practice sustainable development and conservation in coastal zones and to maintain their biodiversity. To this end, ICZM seeks, through more efficient management, to establish and maintain the best use and sustainable levels of development and activity (use) in the coastal zone, and, over time, to improve the physical status of the coastal environment in accordance with certain commonly held and agreed norms (Healy et al., 2001).

The World Bank (1993) emphasises public participation in the decision making process:

"*Integrated Coastal Zone Management (ICZM) is a governmental process and consists of the legal and institutional framework necessary to ensure that development and management plans for coastal zones are integrated with environmental (including social) goals and are made with the participation of those affected.*"

UNEP emphasises particularly the importance of sustainability of the environment as the key issue for ICZM:

"*Integrated Coastal Area Management (ICAM) is defined as an adaptive process of resource management for sustainable development in coastal areas. Sustainable development requires that the quantity and quality of coastal resources are safeguarded in order that they not only satisfy the present needs but provide a sustained yield of economic and environmental services for future generations.*"

According to the social scientists,

"*Integrated means incorporating several disciplinary concepts, i. e. integrated management refers to management with recognition that human behaviour, not natural resources land or water is the focus of management*"(Kay and Alder, 1999),

and

"*Coastal Zone Management involves an ecosystem approach—any human activity or development along the coast must be undertaken only with due consideration of the environmental needs of erosion accretion, habitat creation and stability, i. e. protecting the coastline from damage and change from any activity*"(French, 1997, p. 191).

In practical applications for development of muddy coasts we envisage ICZM ideally as a multidisciplinary approach to problem solving, which involves several iterations in the planning cycle. For example, for a tidal flat and mangrove development management plan we might expect the following procedures:

- development identification and design limitations;
- potential options selection;
- environmental impact assessments *e. g.* oceanography, sedimentology, ecology impacts;
- interdisciplinary input on development: social impacts including cultural sensitivities, traffic, and infrastructure, noise, air pollution, run-off and effluents, navigation and piloting, design engineering;
- stakeholder consultation;
- consideration of stakeholder inputs and objections;
- reassessment of Best Practical Option (BPO);
- revised initial design;
- initial costings; and
- iterations of steps above to final development.

The ICZM approach is meant to enhance development and planning models which treat single issues separately, or are implemented by individual administrative units. It is a continuous and iterative process, active before, during and after coastal planning (Chua, et al, 1997; Olsen, 2002). The concept of integration therefore encompasses a wide variety of factors (Cicin-San, 1993):

- integration of planning and development by the full range of socio-economic sectors;
- integration of approaches between different levels of government at the international, national, regional and local levels, and/or administrative units;
- integration of economic, environmental and social issues;
- integration of planning management across geographic components of the coastal zone, encompassing land and the coastal oceans, as well as inland areas which have a significant influence on coastal processes, and taking account of different coastal landscapes and habitats;
- integration of planning and approaches across various time scales from long (50 years and more) to short-term (e. g. annual), and
- integration of the knowledge, understanding and views of different scientific disciplines, non-government organisations, and the public.

In other words the process is meant to combine physical, biological and human elements into a single management framework encompassing both land and marine coastal area, and ensure that the most important issues receive the highest priority of attention (UNEP, 1995). For the case of muddy coast tidal flats, ideally ICZM would take account the economic and social impacts of development along with the impacts on the muddy coastal morphodynamic and ecosystems as a whole from the edges of the marine influence to the sub-tidal depths, regardless of differing administration or jurisdictional units.

The Benefits of Integrated Coastal Management

The benefits of the ICZM approach are not always simple to define, primarily because relatively few such initiatives have progressed from planning to implementation (OECD, 1996; Clark,1997; Sorensen, 1997; Anon. , 2002). Problems that have arisen in the absence of ICZM, include:

- unnecessarily reactive management (responding after the fact to problems which should have been anticipated and avoided);
- cumulative impacts (where the many small decisions made by different levels of government add up to major problems for the coastal environment);
- transfer of problems from one sector to another;
- predominance of short-term economic interests (often at the expense of nature and the environment, and in many cases having a negative long term economic or

social impact); and

- fragmented geographical planning (lack of co-ordination between managers of land and marine areas, managers of different economic activities, or neighbouring communities bordering a single coastal ecosystem).

In short, it can be argued that a lack of integrated planning and management will almost surely result in the degradation of the coastal environment and in negative economic trends in the longer term (Goldberg, 1994). The converse is not necessarily the case, however: ICZM will only promote sustainable coastal management if this is an express goal of the planning process, which, however, is the central concept of Chapter 17 of Agenda 21, discussed below.

CONTRIBUTION OF NEW ZEALAND'S RESOURCE MANAGEMENT ACT 1991

This is an all-embracing statute relating to environmental management, which controls all resource developments within New Zealand (Healy et al., 2001). It is especially applicable to the coastal zone. Much of the philosophy behind the act has subsequently been espoused in the Agenda 21 arising from the Rio Earth Summit conference of 1992. The purpose of the act is "to promote the sustainable management of natural and physical resources". Sustainable management in the context of the act "means managing the use, development, and protection of natural and physical resources in a way, or at a rate, which enables people and communities to provide for their social, economic, and cultural well being ... "

Matters of National Importance

Section 6 of the RMA 1991 outlines matters of national importance to be recognised and considered by the environmental managers and Environment Court in administration of the act. These include the:

- preservation of the natural character of the coastal environment (including the coastal marine area) ... and the protection of them from inappropriate subdivision, use, and development;
- protection of outstanding natural features and landscapes from inappropriate use and development;
- protection of significant indigenous vegetation and significant habitats; and the
- maintenance and enhancement of public access to and along the coastal marine area.

Section 7 outlines other matters of high importance, including the:

- efficient use and development of natural and physical resources;

- maintenance and enhancement of amenity values;
- intrinsic values of ecosystems; and the
- maintenance and enhancement of the quality of the environment.

Many of these points have subsequently been in corporated within the Agenda 21 (1992), and encapsulated within the concepts of ICZM adopted by the European Commission (1996), and the Coastal Code of Conduct adopted by the European Parliament in September 2000.

INFLUENCE OF THE 1992 UNITED NATIONS RIO "EARTH SUMMIT" CONFERENCE ON ENVIRONMENT AND DEVELOPMENT

Agenda 21

Arising from the United Nations Conference on Environment and Development, was a far reaching document, "*Agenda* 21". Chapter 17 of Agenda 21 addresses the protection of the oceans and coastal areas and the protection, national use, and development of the coastal and marine resources (Cicin-Sain, 1993; Cicin-Sain and Knecht, 1995).

Chapter 17 sets forth the obligations of states to pursue the protection of sustainable development of the marine and coastal environment and its resources. This required new approaches to marine and coastal area management and development at both the national and global levels, and envisaged approaches that are "integrated and precautionary in ambit."

Identified Issues and Objectives

Among the identified programme areas include the issues of (Hong and Lee, 1995):

- integrated management and sustainable development of coastal areas ...
- addressing critical uncertainties for the management of the marine environment and climatic change.
- integrated management and sustainable developments of coastal and marine areas.

The objectives of Chapter 17 include that coastal states commit themselves to integrated management and sustainable development of coastal areas and the marine environment, and to this end it is necessary, *inter alia*, to "apply preventive and precautionary approaches when project planning and implementation".

Coastal development activities should "include implementation of integrated coastal and marine management leading to improvement of coastal human settlements" and promote "environmentally sound technology and sustainable practices".

Under item 17. 19, degradation of the marine environment was recognised as resulting from a wide range of activities on land including human settlements, construction of coastal infrastructure, and urban development. Coastal erosion and siltation were noted as of particular concern.

Chapter 17 emphases that a "precautionary" and anticipatory, rather than a reactive approach, is necessary to prevent ongoing degradation of the marine environment (paragraph 17. 21). This requires, *inter alia*, the adoption of precautionary measures. "Any management framework must include the improvement of coastal human settlements in the integrated management and development of coastal areas".

Unfortunately, in some countries the bureaucrats at all levels will speak of 'sustainable development' but essentially they will mean 'continued economic growth' when they use the phrase. Agenda 21 policies may be developed, but priority will continue to be given to economic expansion (Murray and Cook, 2002).

ENVIRONMENTAL IMPACT ASSESSMENT (EIA)

The Importance of Assessing Environmental Impacts

A central tenet of ICZM is the process of environmental impact assessment (EIA)—perhaps the most important steps in the coastal planning and management process. It is a procedure designed to identify the potential consequences for the scenic, natural landscape and the coastal environment arising from the proposed development (Harropand Nixon, 1999). This information is then used by decision-makers to assess whether, or in what form, proposed activities should go forward.

EIA is now practised in many countries around the world. Specific EIA protocols vary between countries but there are certain core procedures. These include:

- *Screening*: the procedure for determining whether a particular proposed activity (project) will require a full EIA or a less rigorous environmental assessment procedure.
- *Scoping*: the procedure for determining which issues are likely to be important and should be examined in an EIA.
- *Production of an Environmental Impact Assessment or Statement*: the document which describes the potential environmental impacts of a proposed activity.

It should also contain a discussion of possible alternative developments, including a non-development option, along with an analysis of their expected environmental impacts. In addition, the EIA should describe how eventual impacts will be monitored, and any mitigation techniques that will be applied.

- *Baseline Studies*: a detailed description and compilation of critical data of present environmental and socio-economic conditions against which subsequent changes can be assessed.

- *Review*: a peer review of the EIA/EIS is undertaken and its acceptability assessed.

Free access to full information, consultation, and stakeholder participation, and transparency of decision making are integral to the process of environmental impact assessment. When vigorously pursued, and begun at the earliest stages of a project, the benefits of public information dissemination, consultation and participation of affected stakeholders can be significant. For example, if controversies are discussed early in the process while there is still time to alter plans and mitigate possible environmental despoilment, there is a greater likelihood of eventual stakeholder commitment to, and acceptance of, decisions. This may help to reduce costly delays later in the process.

Public Participation in the EIA Process

Under the recommendations of Agenda 21, wide stakeholder and public participation should be vigorously pursued at all stages of project planning (Clark, 1996). When specific new projects are planned, public hearings should be held to solicit views at the earliest possible stage (i. e. scoping) before vested interests take hold, and certainly prior to taking any decision about whether to proceed. Public opinion should be incorporated into the plans, and a mechanism for appeal should be available where this has not occurred. All consultants' reports, feasibility studies, safety studies, should be publicly and conveniently available. In modern electronic communications that would typically be via a web site. And stakeholder advisory groups should be established as an integral component of the project development to allow continued involvement while the project is being carried out.

Applied to muddy coasts of various types, the EIA process is a difficult assignment because of the wide variety of physiographic types of muddy coast (Healy, Wang and Healy, 2002) and because typically we do not fully understand the sedimentary morphodynamics and ecosystem processes of muddy coasts. In particular we do not have good information on the resilience of the vegetative ecosystems, although the deleterious impact of muddy waters on the coral reefs is well explicated (Wolanski and Duke, 2002). Detailed research on a wide front is an ongoing requirement for muddy coasts, and that is not aided by the physically difficult nature of the muddy environment.

DEVELOPING COASTAL MANAGEMENT PLANS

Procedural Sequence

There is a progression of steps generally considered to be essential in developing

ICZM plans, all of which require extensive consultation and co-operation amongst the various stakeholders, who typically include government agencies, local authorities, sectoral planners, non-government organisations, and others (OECD, 1993; World Bank, 1993; UNEP, 1995):

- preparation of detailed and appropriate baseline information about the physical environment(Department of the Environment, 1995; 1996), coastal processes and ecosystems, cultural features, and establishing the geographical scope of the proposed development;
- establishing a mechanism to ensure stakeholder participation in the process;
- assessment of existing management and legal structures and establishing the necessary institutional, legal and administrative framework for integrated management;
- undertaking an audit of good and bad elements within the natural/human matrix and identifying priority issues;
- setting clear objectives and priorities for planning and management as well as for all sectoral activities;
- drawing up the initial plan and proposed projects, including proposed regulatory measures(including an enforcement system) and economic incentives to ensure wise use of resources;
- environmental and strategic impact assessment of the proposed plan and projects;
- stakeholder comment on the proposals, based on information which is made freely available throughout the process;
- revision of proposals;
- implementation of the plan;
- monitoring and evaluation of the outcome;
- in-built mechanisms for response during emergencies arising between the various phases; and
- review and revision of plans as research becomes available, or as new circumstances arise which require changes in the plan.

Social and Cultural Goals to Ensure Durable Sustainability

Promoting socially and economically sustainable livelihoods for the local population is necessary for the long-term maintenance of coastal areas. In developing coastal management plans, there is a need to encourage innovative, low-impact economic activity. It must also be recognised that new sources of financing are often needed to cover the costs of switching to lower impact activities or to compensate local communities.

Economic Instruments and Incentives

Application of economic instruments and incentives can be an effective and economically efficient means to promote environmentally sustainable development in the coastal zone. They encourage rather than coerce changes in behaviour and they exert continuous pressure over time. However, there are many theoretical and practical limitations to this approach that policy makers should consider before deciding upon this approach or upon which instruments to apply (Tol et al., 1996). Difficulties include:

- how to value nature, the ecology and landscape and other non-monetary benefits derived from coastal areas;
- how to avoid a disproportionate impact on lower income groups;
- how to avoid undesirable market distortions and impacts on competitiveness; and
- how to incorporate the value of coastal resources for future generations.

Nevertheless, the use of economic instruments and incentives can help to internalize external costs such as damage to the environment, and induce companies or individuals to achieve environmental goals in a cost-effective manner. They are of special interest where regulatory instruments may not be applicable or deemed to be particularly harsh in certain cases. They may also spur innovative approaches to environmental problems. Finally, the potential for such instruments to raise revenues for re-investment into further measures to reduce environmental impacts and the loss of biodiversity should not be overlooked.

A range of economic tools are now in use, including:

- *Eco-taxes*: polluters are required to pay a tax on each unit of pollution emitted in order to raise the cost of polluting to the level of the social costs incurred as a result of these emissions.
- *User charges*: users of services and products (or nature areas) are charged a fee that covers the full cost of using that service or product.
- *Subsidies*: companies or individuals are given cash rewards for producing or using products or services which are beneficial (or less harmful) to the environment.
- *Rights based instruments* (e.g. emissions trading): rights to use or pollute environmental resources are provided up to a pre-determined limited. Excess rights can then be traded or sold.
- *Tax incentives/Green investments*: green investments are directed at raising investment funds for projects that are defined by the government as being environmentally friendly. Governments can encourage such investments by making approved investments tax-free.

In summary, one of the most important aspects of integrated coastal zone management is that it is forward looking. Many economic sectors focus far too heavily on short-or medium-term economic profit in place of the longer-term perspective required for the sustainable management of coastal resources. A good ICZM plan will examine the potential consequences of development over the long term. Secondly, the importance of stakeholder participation in coastal planning cannot be overstated. All those with a legitimate interest in the management of the coastal area should have the opportunity to be involved in the identification of key issues and the development of policies designed to resolve conflict.

GUIDELINES FOR INTEGRATED COASTAL MANAGEMENT

Identification of the Issues

Before any plan is produced it is important to identify the major issues to be addressed, and at what level of priority, through a process of discussion between the relevant stakeholders. One method of achieving this is to use the data gathering process as a means of bringing the vested interest sectors together in a neutral forum, which may help overcome the vested interest barriers to meaningful dialogue.

National, regional and local authorities should ensure that all development occurs within the context of an integrated coastal management plan, in which areas are designated for certain kinds of development or as areas to be left free from development altogether (although even development free areas require a different type of managing). A zoning system designed to accommodate a diversity of specified and permitted uses or activities could be helpful in this regard.

ICZM plans should specifically recognise the need to conserve nature as a precondition for all development, as this is the only way to ensure that development is truly sustainable. ICZM plans need to obtain data and research to establish the carrying capacity of the coastal and marine environment, taking into account the vulnerability of coastal environment types and habitats, and ensure that development is not allowed to exceed this capacity (Beatley et al, 2002).

For the case of muddy coast, the geographical coverage of ICZM plans should be extensive enough to encompass the primary ecosystems, be they mangroves, saltmarsh, supra-tidal flats, intertidal flats, and sub-tidal flats, as a whole. Involvement and co-operation between neighbouring communities, and local and regional authorities, should be encouraged, recognising the transboundary nature of most muddy coastal environmental issues.

Stakeholder Participation in Coastal Management

Sustainable development and management of coastal regions generally requires a combination of top-down and bottom-up approaches, and public participation in the process is essential. In order to ensure adequate public participation in coastal planning, decision-making and management authorities should (Healy et al., 2001):

- make sure that the decision-making process is consultative and open to all parties who want to or should be involved, and encourage such parties to do so;
- establish along the coast coastal forums for ongoing discussions;
- support education and mobilisation programmes in schools and universities and other community programmes;
- hold stakeholder consultation workshops as public meetings;
- involve the public in monitoring human activities along the coast, impacts on the coastline, and implementation of laws, agreements, or other decisions; and
- involve local businesses in programmes, and work with them to advertise the issues in their outlets.

Techniques that can be used to assess public opinion include public hearings and inquiries, questionnaires and surveys, and stakeholder interest groups.

IMPORTANCE OF PROTECTED AREAS IN ICZM

Establishing coastal and marine protected areas is an integral component of coastal management programmes. While the primary purpose of a protected area is to conserve operating natural ecosystems and coastal landscapes, it does not necessarily require the cessation of all human activities within the area, and indeed, eco-tourism based upon coastal-marine protected areas is a continuing growth industry (Agardy, 1993; Stewart, 1993). A variety of uses may be permissible within a protected area, provided that sufficient controls exist to ensure sustainable use of resources.

The success of a protected area designation depends upon a variety of factors, including: definition of the area so that it can be managed as a unit; acceptance by local inhabitants; and the existence of appropriate legal, administrative and enforcement frameworks (Alder et al., 1994). Benefits of marine reserves include:

- *Biodiversity and habitat protection.*—Marine reserves provide long term and secure protection from a wide range of potential threats, such as over-harvesting or impacts from other human activities. Populations and communities within them have more natural behaviour, interactions and food webs and therefore assist to retain the full diversity of marine species and ecosystems for the future. Secure habitat may also be required to adequately protect populations of marine threatened species.
- *Non-extractive activities*—Many activities that depend upon undisturbed habitat, such as education, snorkelling, diving, eco-tourism, and underwater

photography, benefit from the undisturbed marine life and more natural marine communities in marine reserves.

- *Research*—The more natural communities in marine reserves provide opportunities for increased knowledge, and can also act as benchmark reference sites.
- *Nurseries*—Marine reserves act as nurseries where the young of a species are allowed to regenerate.
- *Genetic variation*—Marine reserves help to protect the marine gene pool and maintain genetic variation within a species, by protecting populations of species.
- *Spillover*—Marine reserves contribute to marine life in nearby areas via eggs and larvae being carried out of the reserve by currents. Larger individuals and denser populations within marine reserves tend to produce more eggs. Adults of mobile species may also migrate out of a marine reserve.
- *Impact on Fisheries*—There is a growing body of international research that shows marine reserves have a ppositive effect on the size and abundance of local marine species and fish stocks.

Benefits of Coastal Reserves or Non-development zones

Apart from marine reserves, there is increasing need to preserve coastal land areas of significant conservation value. These typically occur as coastal pockets of ecological value for vegetation and wildlife, and rare disappearing natural ecosystems. But also important is the need to conserve distinctive coastal landscapes (Brouwer, 1997; Healy, 2003), principally because of the damage to future tourism earnings if the essential scenic values of the coast are lost.

The general points made above are all applicable to muddy coasts.

FISHERIES AND AQUACULTURE ISSUES FOR ICZM

Muddy coastal waters serve as important nursery, feeding and spawning areas for fisheries. Serious depletion of fish stocks in muddy coastal areas (from overfishing, improved fishing methods, climatic change and pollution) has often led to the forced reduction in catches and decreased employment opportunities (OECD, 1993; Newkirk, 1996). There is an increasing trend towards fish farming, however, with particular interest in shellfish. In the European Union, for example, farmed fish and shellfish account for around 7% of aquaculture output worldwide, providing employment for more than 80 000 people in coastal areas.

Environmental impacts of fishing in the muddy coastal zone include overfishing of commercial fish stocks (Newkirk, 1996), damage to the marine ecosystem due to the incidental take of non-target species (including marine mammals), oil pollution, and litter from fishing vessels, and pollution from fish processing plants. These impacts, combined with problems of coastal and marine habitat destruction, land and marine-based pollution, and the introduction of alien species, have serious consequences for aquatic diversity, and ultimately for the health of fish stocks themselves.

Impacts from aquaculture along muddy coasts are also a serious concern, particularly if a larger industry develops. The discharge of wastes (which may include alien species, anti-fouling pesticides, antibiotics and other pharmaceuticals, organic matter and nutrients) to the surrounding shallow environment can affect local fish populations, contribute to eutrophication, and upset the ecological balance, particularly in semi-enclosed coastal areas (FAO, 1995; UNEP, 1995).

Fisheries management should adopt a *precautionary approach* in which the fundamental health of coastal and marine ecosystems is maintained. A lack of adequate data should not be considered grounds for postponing effective conservation measures along muddy coasts. The optimum sustainable yield should be considered as a replacement for the maximum sustainable yield of any given fishery. For stocks which are currently over-exploited, particularly where spawning stocks are depleted or where the ecosystem has been seriously damaged, fishing efforts should be reduced or ceased until stocks have recovered.

Aquaculture along estuaries and muddy coasts can be developed in ways which do not degrade coastal and marine biodiversity. Integrated systems will be more sustainable than monocultures, for example by the combined culturing of seaweeds, mussels and fin fish. Due to their extensive space and water requirements, hatcheries will have fewer impacts on the surrounding environment if they are located within developed areas of the coast and where road access already exists. If possible, they should be contained within existing buildings. Small-scale production units should be located onshore, set back from the shoreline, in areas with good road access. Fish tanks should be set in the ground, with pipelines and services buried underground.

Fish farms located within muddy coasts can minimise nutrient discharges and losses by developing and utilising appropriate feeding methods and fish feed (predominantly dry). Special treatment plants can be an effective way to eliminate solid wastes, chemical and pharmaceutical additives, and nutrients from effluents or other discharges to the sea. In some cases, treated effluent water from fish farms can be used for agricultural irrigation.

INDUSTRIAL DEVELOPMENT IN THE COASTAL ZONE

Excavation, dredging and reclamation for a variety of purposes in shallow muddy coasts may be particularly damaging for coastal ecosystems in near-shore areas, causing alterations in muddy sediment transport mechanisms and possibly erosion, but often undesirable deposition. Dumping of dredge spoils and reclamations for all kinds of purposes, in muddy embayments can smother benthic flora and fauna, alter local hydrodynamic regimes, and release toxic contaminants to the marine environment. On the other hand replacement back into the littoral system of muddy sediment captured in artificially dredged sediment traps, such as navigation channels, may cause no particular damage to the shallow near shore environment. (Healy et al, 1996; Healy et al, 2000). Indeed, under certain circumstances the deposition of dredged material (e. g. for beach nourishment or habitat re-creation) may be beneficial for coastal ecosystems. For example Healy et al. (2003) report on the use of non-polluted dredged material from a nearby navigation channel, for adjacent beach renourishment.

Industries, including the future tourist industries, operating within the muddy coastal zone have a special obligation to ensure their activities have minimal impact on the marine and coastal environment. The potential for adopting cleaner operational processes, with an emphasis on prevention of pollution and eutrophication should be fully integrated into the coastal management plan for muddy coastal zones.

ICZM ISSUES FOR TOURISM

Tourism is the world's fastest growing industry, and has been an important part of the economic development of many developing nations. Trends indicate that tourists are becoming increasingly interested in higher quality tourism experiences with particular interest in cultural, historic, and natural sites. Eco-tourism is likewise a growing trend.

The sheer speed and scale of global tourism development has had a major impact on the coastal environment (Clark, 1996). In addition although many local people have benefited from the increase in prosperity which tourism brings, the social and cultural effects are significant. Unfortunately, over-development and environmental degradation have led to many areas losing their appeal. As tourist numbers drop off the ability to maintain the infrastructure becomes more difficult. Where beach erosion has become a problem, as has happened in many areas where development has occurred in vulnerable zones, the cost of maintenance can be particularly high and often has to be borne by the local tax-payer. Development which is in harmony with the coastal environment will be more sustainable in the long term and less costly to maintain (Wong, 1998).

Major impacts of tourism in coastal areas arise from the construction of infrastructure (e. g. hotels, marinas, transport, waste treatment facilities, groynes) and from recreation (golf courses, water sports, thematic parks, beach access and parking etc.). The problems in tourism industries differ from other economic sectors in that the degradation of the environment results in the degradation of the industry itself with knock-on effects in other industries. It is becoming self-evident that if the coastal areas damage their appeal, for example, by water quality degradation, their main source of income dies, and this also destroys the opportunities for attracting other activities besides tourism. The coastal areas lose their strong attraction and with it their value to important networks.

Infrastructure

Massive tourist facilities, particularly hotels and apartment complexes, often cause significant harm to coastal systems due to their preferred siting in areas protected from the open sea, e. g. on shallow muddy coasts or lagoons which may require dredging. They can also have a devastating impact on coastal processes on the down drift side of the construction leading in many cases to severe erosion and loss of beaches, and threats to hotels and other facilities (Wong, 1998).

Tourism also has impacts on environmental quality, especially in shallow muddy coastal waters—the treatment and disposal of solid and/or liquid wastes, particularly during peak tourist seasons, may be inadequate or at worst non-existent. A large quantity of water is consumed not only for drinking but also for showers, laundry, and swimming pools, which can be a major problem in regions where freshwater resources are limited. Air pollution from cars and buses transporting tourists to and from their destinations is a problem which extends far beyond the local tourism site. Given that coastal regions are primary tourist destinations, sensitive marine and coastal environments and coastal communities potentially may suffer dramatically.

Guidelines for Tourism Development in Muddy Coastal Areas

Tourism in the coastal zone, and notably eco-tourism, is a continuing growth industry. For coastal zones in general, tourist developments should be carried out in such a way as to ensure that the environmental, cultural, and social diversity of the coastal zone is protected and enhanced (Wong, 1998). Primarily, it needs to meet the needs of the local host community without compromising the natural or cultural values which are attractive to tourists in the first instance, or the economic viability of existing sustainable commercial activities. Local communities can be supported, for example, by the use and promotion of locally produced food and souvenirs.

The attitudes of local communities, and stakeholders in general, should be incorporated into development plans at the earliest stages, well before planners become wedded to any particular decision. In addition, it would be astute to determine whether the carrying capacity of the local muddy coastal environment can sustain the impacts of a new tourism development, adhering to the *Precautionary Principle*, before further planning is allowed to proceed.

Zoning of muddy coastal lands for specific recreational uses, or for nature and wildlife conservation, should be encouraged, and allow for the possibility of establishing disturbance-free zones generally, but especially in the habitats of threatened or endangered species. This will enhance the eco-tourism as a sustainable development, both environmentally and economically. Where new facilities are considered necessary, they should be compatible with the architecture and environment of the surrounding area (Lyle, 1994). Large buildings which impair quality of the coastal scenic landscapes should be avoided. Innovative designs, technologies and construction techniques should be encouraged and supported. Facilities should be designed to avoid changes in near-shore sediment transport patterns, the geomorphology of the coastline, and/or water quality. Before any coastal installation is built in or near to a muddy coastal location, a thorough study of the geomorphological sedimentation regime is essential (Healy et al., 2002). This will reveal the way in which sedimentary patterns affect existing habitat development and provide a basis for assessing the likely changes to sedimentary transport systems consequent upon erecting any structures.

Natural vegetation along muddy coasts should be left intact as much as possible. Where this is not possible, indigenous species should be used for landscaping and environmental remediation.

ICZM ISSUES FOR WATER MANAGEMENT

All phases of water management from reducing water demand and ensuring a stable supply of fresh drinking water to the appropriate disposal of wastewater—are central to sustainable development, including in muddy coastal zones. Water—both riverine and marine—is an essential resource for agriculture, industry, energy production, tourism, urban life, nature, and wildlife. The transboundary nature of water resources makes it at the same time one of the most difficult resources to manage sustainably in the muddy coastal zone.

The volume and quality of water reaching estuaries, muddy tidal flats, and ecosystems has important ramifications for biodiversity. Poor water quality, as indicated by the presence of viruses, bacteria and toxic chemicals, is harmful to both humans and wildlife. This is particularly so in high suspended load muddy coast rich in organic matter. The concentration of harmful substances, and hence their potential to cause damage, is partly affected by the volume of water flowing into the system. The flow of water is also an important factor in sedimentation patterns and hence the existence and resilience of muddy sedimentary coastal systems (e. g. mangroves, mudflats and deltas).

Water quality and volume of the muddy coasts is affected by a range of human activities: excessive demand, over-pumping, pollution from sewage outfalls or direct discharges (including by industry); thermal pollution from power stations; runoff from agricultural lands and roadways; construction of dams and reservoirs; irrigation; and operational discharges from vessels at sea, to name just a few. The depletion of groundwater is particularly detrimental to muddy coastal systems in a variety of ways (Han, 2002; Wang et al, this volume). It can cause saltwater intrusion into aquifers, soil erosion and land subsidence and salinisation, to name a few (UNEP 1995; Wang et al, 2002).

Beaches throughout Asia are littered with materials which have been flushed into sewage systems, many of which still provide primary treatment only. Damage to muddy coastal ecosystems occurs not only from the discharge of sewage, but also from rivers and streams carrying a host of other waste materials which find their way into sewage systems: industrial and other hazardous wastes (e. g. hospital wastes), plastics, fishing wastes, sanitary products, and so forth. Overflows are probably the single greatest cause of sewage related pollution. Litter is a major problem for coastal areas, because of its potential impact on wildlife and human health and because of the high costs to local communities. And once these contaminants get deposited, they are unlikely to be removed due to the low hydrodynamic energies typical of muddy coastal systems.

Although the average person may not be as 'caring' about the environmental health of muddy coast as for open ocean sandy beaches, actually a careful and stringent waste management regime is more necessary because of the muddy sedimentation regime which tends to accumulate material on the tidal flats. Once accumulated, waste material is difficult and expensive to remove. Thus, in reality, it is much more difficult to restrain environmental degradation in muddy coastal environments than in high energy open coastal areas, where the hydrodynamics play an active cleansing role.

ICZM ISSUES FOR MANGROVE MUDDY COAST

In general mangrove ecosystems are reasonably resistant to many kinds of environmental perturbations. However they are sensitive to excessive sedimentation, hypersalinities, surface water impoundment, herbicides in runoff, and major oil spills (Clark, 1996). On the other hand mangroves help maintain water quality by extracting chemical pollutants.

Today many mangrove forests on muddy coast are being converted to aquaculture ponds an activity responsible for the major loss of mangroves (Clark, 1996). Moreover, many of the drivers which detrimentally alter mangroves originate from outside the mangrove ecosystem. Thereby illustrates the need for integrated planning and management, for to develop effective management plans for mangrove stands it is necessary to relate them to management of the adjoining tidal flats and estuarine waters. Mangroves should be viewed as part of a complex system of interrelated habitat and dependent biota which is maintained by natural drainage patterns from the land in conjunction with tidal and salinity regimes.

CONCLUSIONS

As Ketchum (1972, p. 347) perceptively explicated more than three decades ago: "*Effective management of the coastal zone implies viable and economic integration of complex life systems and varied human activities*" within a confined coastal area which is itself subject to the influence of major exterior forcing, both from the land and from the sea. Obviously coastal areas must be managed from a perspective of an admixture of natural processes (biological, physical, chemical), socio-economic developments, and long term external environmental changes such as accelerated sea level rise and climatic changes. The management of coastal areas thus needs to be focused on these processes and developments in an integrated manner.

Integrated Coastal Zone Management strategies and procedural recommendations as outlined in this paper provide an internationally accepted protocol for future developments, particularly in developing countries without a strong ethic of coastal planning based upon Agenda 21 (Beatley et al., 2002). For the case of muddy coasts, known protocols may now be aimed at sustainable developments in fishing, aquaculture, tourism, dredging, and sediment extraction, while at the same time enhancing the functioning of the coastal estuarine environment. But given the increasing population and development pressures on the large tracts of muddy coast, and concomitant requirement for long term sustainable management, a strong conservation ethic and an equally strong consideration of social impact and community participation, is paramount. This ultimately provides a win-win situation for the community, the economy, and the coastal environment.

The process of integrated coastal management requires an effective legal and administrative framework (Chiau, 1998). In states where these are not yet established, their development should now be considered an urgent priority (Gerrard et al, 2001).

LITERATURE CITED

[1] Agardy, M. T. 1993. Accommodating Ecotourism in Multiple Use Planning of Coastal and Marine Protected Areas. *Ocean and Coastal Management*, 20: 219－239.

[2] Alder, J., Sloan, N. A., and Uktolseya, H. 1994. A Comparison of Management Planning and

Implementation in Three Indonesian Marine Protected Areas. *Ocean and Coastal Management*, 24 (3): 179 - 198.

[3] Alder, J., Zeller, D., Pitcher, T., and Sumailer, R. 2002. A method for evaluating marine protected area management. *Coastal Management*, 30: 121 - 131.

[4] Anonymous. 2002. Towards an Integrated Coastal Zone Policy. Policy Agenda for the Coast. The Hague: National Institute for Coastal and Marine Management, 47p.

[5] Basiron, M. N. 1998. The implementation of Chapter 17 of Agenda 21 in Malysia. Challenges and opportunities. *Ocean and Coastal Management*, 41(1): 1 - 17.

[6] Beatley, T., Brower, D. J., and Schwab, A. K. 2002. An Introduction to Coastal Zone Management 2nd Edition. Washington: Island Press, 329p.

[7] Brouwer, C. 1997. Managing our coastal paradise. Managing the Visual and Cultural Values of Regional Coastal landscapes. Pacific Coasts and Ports '97. 313 - 317.

[8] Chiau, W-Y. 1998. Coastal zone management in Taiwan: a review. *Ocean and Coastal Management*, 38: 119 - 132.

[9] Chua, T.-E., Yu, H., and Guoqiang, C. 1997. From sectoral to integrated coastal management: a case in Xiamen, China. *Ocean and Coastal Management*, 37(2): 233 - 251.

[10] Cicin-Sain, B. 1993. Sustainable Development and Integrated Coastal Management. *Ocean and Coastal Management*, 21(1 - 3): 11 - 43.

[11] Cicin-Sain, B. and Knecht, R. W. 1995. Measuring progress on UNCED implementation. *Ocean and Coastal Management*, 29(1 - 3): 1 - 11.

[12] Clark, J. R. 1996. Coastal Zone Management Handbook, Boca Raton: CRC Press, 694 p.

[13] Clark, J. R. 1997. Coastal zone management for the new century. *Ocean and Coastal Management*, 37(2): 191 - 216.

[14] Department of the Environment (UK). 1995. Coastal Planning and Management: A Review of Earth Science Information Needs. London: HMSO, 187 p.

[15] Department of the Environment (UK). 1996. Coastal Zone Management, Towards Best Practice. London: HMSO.

[16] El-Sabh, M., Demers, S., and Lafontaine, D. 1998. Coastal management and sustainable development: From Stockholm to Rimouski. *Ocean and Coastal Management*, 39: 1 - 24.

[17] European Commission. 1996. Ensuring a Common Understanding of ICZM Concepts within the Terms of the European Demonstration Programme on Integrated Management of Coastal Zones, CZ Demo 96 - 2, November 1996.

[18] FAO (Food and Agriculture Organization of the United Nations). 1995. Code of Conduct for Responsible Fisheries, Rome.

[19] Fletcher, C, Anderson, J., Crook, K., Kaminsky, G., Larcombe, P., Murray-Wallace, C, Sansone, F., Scott, D., Riggs, S., Sallenger, A, Shennan, I., Thieler, E., and Wehmiller, J. 2000. Coastal sedimentary research examines critical issues of national and global priority, *EOS*, 81 (17): 181 - 186.

[20] French, P. 1997. Coastal and Estuarine Management. London: Routledge, 251 p.

[21] Gerrard, S., Turner, R. K., and Bateman, I. J. (eds) 2001. Environmental Risk Planning and Management. Cheltenham: Edward Elgar, 615 p.

[22] Goldberg, E. D. 1994. Coastal Zone Space. Prelude to Conflict? Paris: UNESCO, 138 p.

[23] Hammond, A, Adrlaanse, A, Rodenburg, E., Bryant, D., and Woodward, R. 1995. Environmental Indicators: A Systematic Approach to Measuring and Reporting on Environmental Policy Performance in the Context of Sustainable Development. Washington: World Resources Institute, 43 p.

[24] Han, M. 2002. Human influences on muddy coasts. In: Healy, T., Wang, Y., and Healy, J. (eds.) Muddy Coasts of the World: Processes, Deposits and Functions. Amsterdam: Elsevier Science. 293 - 317.

[25] Harrop, D. O. and Nixon, J. A. 1999. Environmental Assessment in Practice. London: Routledge, 219 p.

[26] Healy, T., Hull, J., and Hume, T. 2001. Dispersal of muddy dredged material on broad intertidal flats in a fetch-limited environment. *Shore and Beach*, 69: 15 - 22.

[27] Healy, T., Mathew, J., and Black, K. 2001. Integrated Coastal Management for Sustainable Development— Concepts Applicable to the Ashtamudi Estuary, Kerala, India. In: Black, K. and Baba, M. (eds) Ashtamudi Management Plan, Hamilton: ASR Ltd and Centre for Earth Science Studies. 160 - 209.

[28] Healy, T., Mehta, A, Rodriguez, H., and Tian, F. 1999. By passing of dredged muddy sediments using a thin layer dispersal technique. *Journal of Coastal Research*, 15(4): 1119 - 1131.

[29] Healy, T., Stephens, S., Black, K, Cole, R., and Beamsley, B. 2002. Port redesign and planned beach renourishment in a high wave energy sandy-muddy coastal environment, Port Gisborne, New Zealand. *Geomorphology*, 48: 163 - 177.

[30] Healy, T., Wang, Y., and Healy, J-A. (eds.) 2002. Muddy Coasts of the World: Processes, Deposits and Functions. Amsterdam: Elsevier Science, 542 p.

[31] HELCOM PITF MLW. 1995. "Technical Guidelines on Elaboration of Integrated Coastal Zone Management Plans for HELCOM MLW Task Areas", January 1995. Hinrichsen, D., 1999. Coastal Waters and the World: Trends, Threats and Strategies. Washington: Island Press, 275 p.

[32] Hong, S-E. and Lee, J. 1995. National level implementation of Chapter 17: the Korean example. In: CicinSain, B. and Knecht, R. W. (eds.), Ocean and Coastal Management Special Issue Earth Summit Implementation: ogress Achieved on Oceans and Coasts, 29(13): 231 - 249.

[33] Kay, R. and Alder, J. 1999. Coastal Planning and Management. London: Routledge, 375 p.

[34] Ketchum, B. (ed.) 1972. The Water's Edge. Critical Problems of the Coastal Zone. Cambridge MA: MIT Press, 393 p.

[35] Lyle, J. T. 1994. Regenerative Design for Sustainable Development. New York: John Wiley & Sons, 338 p.

[36] Miller, M. L. 1993. The Rise of Coastal and Marine Tourism. *Ocean and Coastal Management*, 20(3): 181 - 199.

[37] Munoz, J. M., Dadon, J. R., Matteucci, S. D., Morello, J. H., Baxendale, C, and Rodriguez, A. 2003. Preliminary Basis for an Integrated Management Program for the Coastal Zone of Argentina. *Coastal Management*, 31: 55 - 77.

[38] Murray, G. and Cook, I. 2002. Green China: Seeking Ecological Alternatives. New York: Routledge Curzon, 254p.

[39] Newkirk, G. 1996. Sustainable coastal production systems: a model for integrating aquaculture and fisheries under community management. *Ocean and Coastal Management*, 32(2): 69 - 83.

[40] Ngoile, M. A. K. and Linden, O. 1997. Lessons learned from Eastern Africa: the development of policy on ICZM at national and regional levels. *Ocean and Coastal Management*, 37(3): 295 - 318.

[41] OECD. 1993. Coastal Zone Management, Integrated Policies, OECD.

[42] OECD. 1996. Review of Progress Towards Integrated Coastal Zone Management in Selected OECD Countries, ENV/EPOC/CZM(96).

[43] Olsen, S. B. 2002. Assessing Progress toward the Goals of Coastal Management. *Coastal Management*, 30: 325 - 345.

[44] Sorensen, J. 1997. National and International Efforts at Integrated Coastal Management: Definitions, Achievements, and Lessons. *Coastal Management*, 25: 3 - 41.

[45] Stewart, M. M. 1993. Sustainable Tourism Development and Marine Conservation Regimes. *Ocean and Coastal Management*, 20(3): 201 - 217.

[46] Tol, R. S. J., Klein, R. J. T., Jansen, H. M. A., and Verbruggen, H. 1996. Some economic considerations on the importance of proactive integrated coastal zone management. *Ocean and Coastal Management*, 32(1): 39 - 55.

[47] UNEP. 1995. Guidelines for Integrated Management of Coastal and Marine Areas—With Special Reference to the Mediterranean Basin. UNEP Regional Seas Reports and Studies No. 161. Split, Croatia, PAP/RAC (MAP-UNEP).

[48] Vallega, A. 1999. Fundamentals of Integrated Coastal Management. Dordrecht: Kluwer, 264 p.

[49] Wang, Y., Zou, X., and Zhu, D. 2000. The utilization of coastal tidal flats: a case study on integrated coastal area management. In: Flemming, B. W., Delafontaine, M. T., and Lieberzeit, G. (eds) Muddy Coast Dynamics and Resource Management. Amsterdam: Elsevier. 287 - 294.

[50] Westmacott, S. 2002. Where should the Focus be in Tropical Integrated Coastal Management? *Coastal Management*, 30: 67 - 84.

[51] Wolanski, E. and Duke, N. 2002. Mud threat to the Great Barrier Reef of Australia. In: Healy, T., Wang, Y., and Healy, J-A. (eds.) Muddy Coasts of the World: Processes, Deposits and Functions. Amsterdam: Elsevier Science. 533 - 542.

[52] Wong, P. P. 1998. Coastal tourism development in Southeast Asia: relevance and lessons for coastal zone management. *Ocean and Coastal Management*, 38: 89 - 109.

[53] World Bank. 1993. The Noordwijk Guidelines for Integrated Coastal Zone Management, The Netherlands: National Institute for Coastal and Marine Management/RIKZ—Ⅰ11. Coastal Zone Management Centre Publication no. 4.

[54] Yu, H. 1994. China's Coastal Ocean Uses: Conflicts and Impacts. *Ocean and Coastal Management*, 25(3): 161 - 178.

地理信息系统技术在海洋测绘中的应用*

一、21世纪是海洋世纪

海洋占地球表面积的71%，生命起源于海洋；是人类赖以生存的环境——空气、淡水的调节与供应地。因为大气中70%的氧来源于海洋，并具有大于大气60倍的能力吸收储存CO_2。海洋每年蒸发44亿km^3的淡水，它以降水形式（雨、雪、冰）返回陆地与海洋，大气中的水分每10～15天完成一次更新；海洋生物资源是人类的食物及蛋白质重要来源，它有20多万种生物，为全球提供22%的动物蛋白，世界渔产8 000～9 000万t/a，85%来自管辖海域。

海洋中有锰（2 000亿t）、镍（164亿t）、钴（58亿t）多金属含量相当于陆地的40～1 000倍，稀有金属与贵重金属矿藏大；海底石油（1 350亿t）、天然气（140万亿m^3）储量大，产油已占50%；海洋能源，可再生的（潮汐能、波浪能、海流能、海水热能）理论储量1 500亿kW，相当于世界总发电量的十倍。

海洋中药物、化学元素及空间资源丰富。

当代面临着人口、资源与环境的巨大挑战，利用海洋是当务之急，尤其是与人类生存密切相关的海岸海洋（Coastal Ocean）[1]，《联合国海洋法公约》于1994年11月16日正式生效，中国于1996年5月15日，经第八届人大第15次会议批准正式实施，全球已兴起“海洋国土”热潮：

沿海国有权规划12海里领海，24海里毗连区和200海里专属经济区；

确定大陆架是沿海领土自然延伸的原则；

规定沿海国对专属经济区和大陆架的自然资源有主权之利，并对防止污染和科研活动等享有管辖权；

国际海底及资源是人类共同继承的财产原则。

《公约》形成海洋权益再分配，全球有1.3亿km^2的近海（占海洋面积的35.8%）脱离公海，370多处国家间海洋划界，并制定国内法律政策。

因此，21世纪是全球规模开发利用海洋的新世纪，已形成国际共识。但是，海洋的厚层海水覆盖，使我们不能透视，直接接触，海洋国土的环境特点，疆界划分的根据主要靠海洋测绘：

（1）定位：从望星空、六分仪进步到无线电定位、雷达导航，后又发展到GPS卫星定位系统。

（2）应用点、线、测深方法反映海底表面起伏。

（3）测海底地形起伏，船舷摄影仪旁侧声呐扫描。

---

* 王颖，马劲松，李海宇：（中华人民共和国海事局主办）《中国海事》，2003年3月，第28－31页。

(4) 采取底质，点、孔采样，取岩芯。

(5) 地震剖面测海底地层。

(6) 数据存储、计算，归纳与制图。

这是个复杂的系列过程。

二、应用4S技术(GIS，RS，GPS，Geopulse)系统分析，对黄海辐射沙脊群进行研究，揭示了该海上迷宫真面目[①]

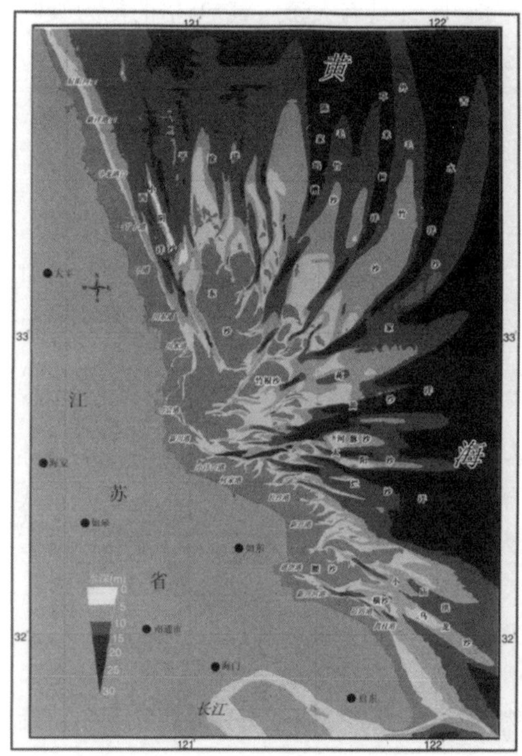

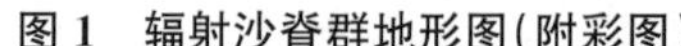
图1　辐射沙脊群地形图(附彩图)

图2　辐射沙脊群海底地形三维效果(附彩图)

应用GPS在黄海辐射沙脊群海域进行导航、勘测和采样，获取现场实地资料作为控制样本。应用GIS技术将该区域20世纪60年代、70年代1∶20万海图及90年代局部航道测量资料，进行了数据订正、转换于统一的高斯投影坐标下，建立不同时代的数字高程模型(DEM)及可视化的海底沙脊群地形图像，给人以直观效果(图1和图2)；再将不同时代的DEM叠置分析，获取近40年来的地形冲淤演变情况及演变模式。应用RS获取沙脊群最近10年的遥感图像，使用PCI及其他图像处理软件对图像进行配准、剪切、重采样、线性增强及主元素分析等多项处理，进行多时相遥感复合分析与分类，获得近10年沙脊群各段冲淤变化(枢纽部淤长，南部冲淤动态平衡，北部冲刷)(表1)，再结合Geopulse(地脉冲)地震地层剖面分析获得近期地质时期沙脊群演变。4S在对辐射沙脊

① 李海宇，2000，“4s”技术在大陆架辐射沙脊群冲淤变化研究中的应用趋势性分析与技术方法建设。

群的分析当中各有侧重又相互支持、相互结合，实现了从空中到地面以至海底以下岩层的全方位立体观测与研究，传统地学论理方法与现代海洋技术手段有机结合，长时间尺度的定性研究与短周期定量分析结果相互比对，相得益彰。

表 1

部位	潮流通道名称	潮流通道长度(km)	水深(m)
枢纽部	烂沙洋 黄沙洋	9 7	10～25(局部 29) 11～20
南部	小庙泓	5.5	8～14
北部	西洋王港	5	9～14(局部 28.5)

4S 技术在监测大陆架浅海沙脊群应用中，揭开了海底沙脊群真面目，进行了其形成演变及趋势性分析及为该海域资源环境的有效与可持续性地开发利用提供第一性科学依据。

(1) 辐射沙脊群是由古长江泥沙入海物质经辐聚状潮流冲刷而成，是新生的土地资源。地理信息系统分析方法，揭示出其面积为 22 470 km^2，其中露出海面以上的沙洲与滩地面积为 3 782 km^2，水下部分：0～5 m 水深的沙脊群面积为 2 611 km^2，5～10 m 水深的沙脊群面积为 4 004 km^2，10～15 m 水深的沙脊群面积为 6 825 km^2，15 m 水深以下的沙脊群面积为 5 045 km^2。不同水深的沙脊为土地、滩涂的开发利用提供了多种途径。其中，位于毛竹沙沙脊上的外磕脚，水深 0.5 m，每年出露一次，被选为我国大陆沿岸 49 个领海基线点之一，使该处领海基线向东推出 4 km，海洋测绘增进了疆域划定准确性与加宽内海范围。

(2) 地震剖面与钻孔揭示出辐射沙脊群枢纽部分的烂沙洋-黄沙洋潮流通道是沿着 30 000 年前古长江从苏北入海的主通道发育的(图 3 和图 4)[2]：

① 40 m 沉积层下有一宽 10.42 km，深度 22～33 m 的河谷；

② 37 m 沉积层下有一宽 2.2 km 的 V 型谷；

③ 20 m 厚全新世沉积层下有宽谷。

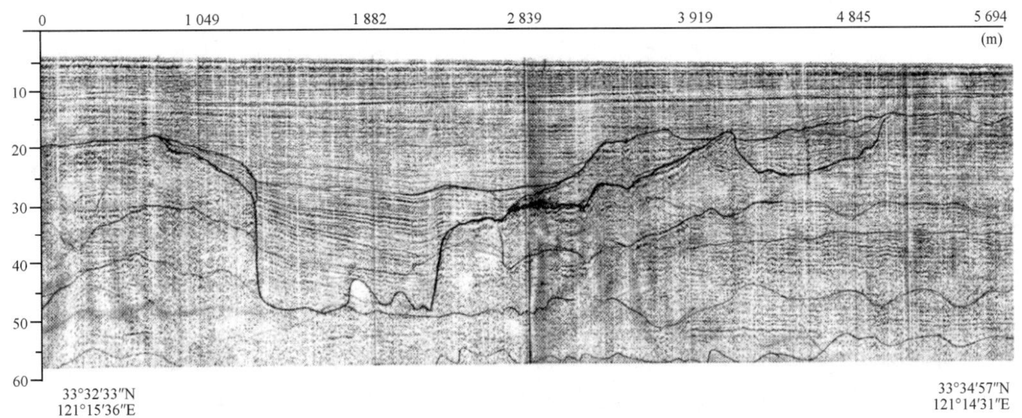

图 3　大北槽地震地层剖面图

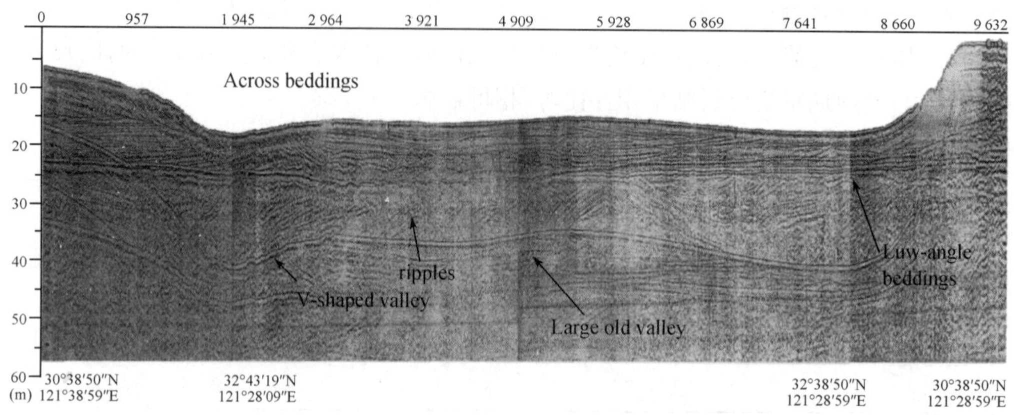

图 4　黄沙洋中段地震地层剖面图

卫星照片(1996.2.6)(图 5)仍隐现此埋藏谷。

小庙泓水道是近 10K 年长江李堡或遥望港入海时形成。

目前,烂沙洋与黄沙洋通道潮流作用强,最大潮差 9.28 m,平均潮差 4.18 m,涨潮流流速 1～1.5 m/sec,每潮纳潮量达 19.5×10^8 m^3,落潮流流速 1.2～2.0 m/sec,每潮纳潮量为 21×10^8 m^3,强大的潮流使通道维持水深 17～20 m,使苏北 883 km 长的淤泥平原海岸上最优的深水航道,可辟建 10 万 t 级大型港口,其意义重大。

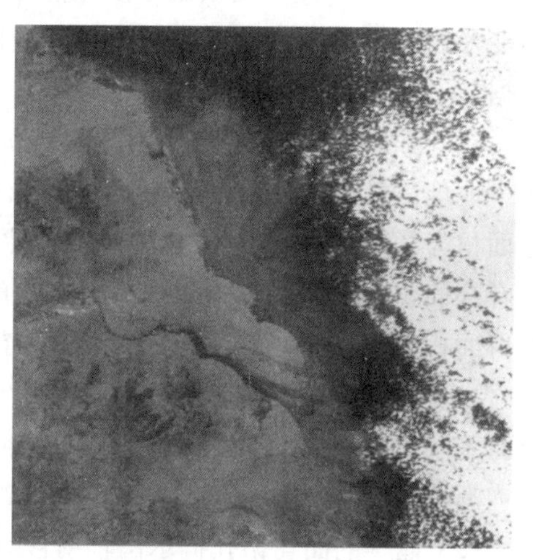

图 5　辐射沙脊群卫星遥感相片(附彩图)

(3) 完成高程数字模型图像给人以直视,并通过多时相遥感图像与海图的复合分析,提出位于黄沙洋-烂沙洋潮流通道内侧南岸的西太阳沙脊,其沙基层厚度大,冲淤变化具动态稳定的特性。它与深水航道相伴分布,提供了天然的陆地基础,供建设码头与临港工业使用。

三、地理信息系统技术在南海海域疆界中的贡献[3]

(1) 20 世纪 70 年代后期,随着对南海丰富的自然资源的认识,越南、菲律宾、马来西亚和印度尼西亚等国纷纷提出对南海主权的无理要求,抛出多种南海划分界线,造成南海国界问题的矛盾日益突出。

(2) 海域疆界划分主要以陆地为依据。海洋法的贯彻,在群岛海域形成"小岛大海洋"的新概念。一个人居的小岛,它可据有内水、12 海里领海、24 海里毗连区及 200 海里专属经济区。因此,南海的岛、礁、滩、沙十分重要,即使暗沙,如果一年中可出露一次,也可为划界依据。但曾母暗沙等位于巽他陆架上(图 6),西沙群岛也不位于我国现代的大陆架上,其归属必须借助历史疆域加以认定。

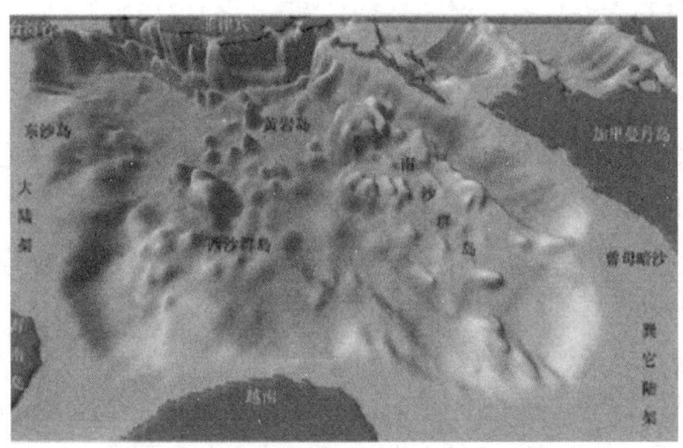

图 6　南海巽他陆架立体地形(附彩图)

从历史上看,断续线是我国在南海的传统疆域线,国际上对其承认已达半个多世纪之久。

1833 年德国曾在南沙、西沙群岛进行考察和测量,广东省政府提出抗议,德国停止了作业。

1911 年辛亥革命后,国际组织多次要求中国政府在南海诸岛上建立气象台、灯塔等航海设施,以保证航行安全,表明国际公认南沙、西沙等岛礁是中国领土。第二次世界大战期间,南海诸岛虽被日本侵占,而波茨坦国际公告规定,日本侵占中国领土必须归还中国。

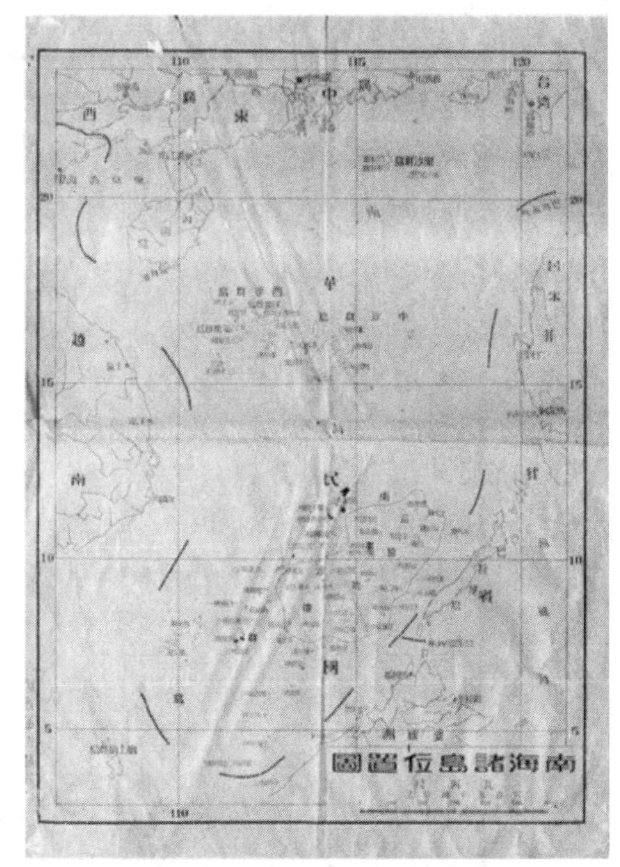

图 7　民国内政部《南海诸岛位置图》

抗战胜利后的 1946 年,中国政府接收西沙、南沙,在岛上升国旗、立主权碑,昭告世界,并派兵驻守南沙群岛的最大岛屿—太平岛,国际上对此皆予承认。

1947 年,中国政府正式公布了《南海诸岛位置图》(图 7),图上画有断续线形式的国界线 11 段①,以后出版的地图均遵照政府规定画有这些断续线。中国对南海诸岛的主权得到国际公认。

① 原件存于南京,中国第二历史档案馆。

1953 年，经政府批准去掉北部湾 2 段断续线，改成 9 段，沿用至今，已超过半个世纪。

为了从历史疆界的角度确定我国对南海的领土主权，因此，有必要对我国南海国界断续线给予确定性的认识。

由于对 9 条断续线的位置需进一步确定其经纬度，因此，我们根据 1947 年原内政部《南海诸岛位置图》的原图复印件和 1983 年我国政府正式出版的《南海诸岛》图，采用数字化投影变换测绘技术，利用地理信息系统分别量测出新旧各 9 条断续线的经纬度坐标，以度为单位，舍入到小数点后 3 位，具有较高的精确度。

然后，将各新旧断续线按经纬度制成南海新旧国家断续线示意图（图 8）。通过对比发现，新旧各 9 条断续线的位置并不完全重合，表现出多种位置关系：新旧线是相互衔接，如西 1（旧）与西 1（新），东 3（旧）与东 2（新）；新旧线是相互平行，有外展，如东 2（旧）与东 1（新），南 2（新）与南 2（旧）；有的新线在旧线的延伸线的位置上，如南 3（旧）与南 3（新）；一般而言，旧断续线长度较长，通常在 130～210 km 之间，且长度变化较大，不规则；而新断续线长度较短，一般在 120～130 km 之间，长度变化不大，较为规则。对应编号的新旧断续线之间的距离从最接近的 18 km 左右（南 2 新旧之间），到相距最远的 333 km 左右（东 1 新旧之间）。

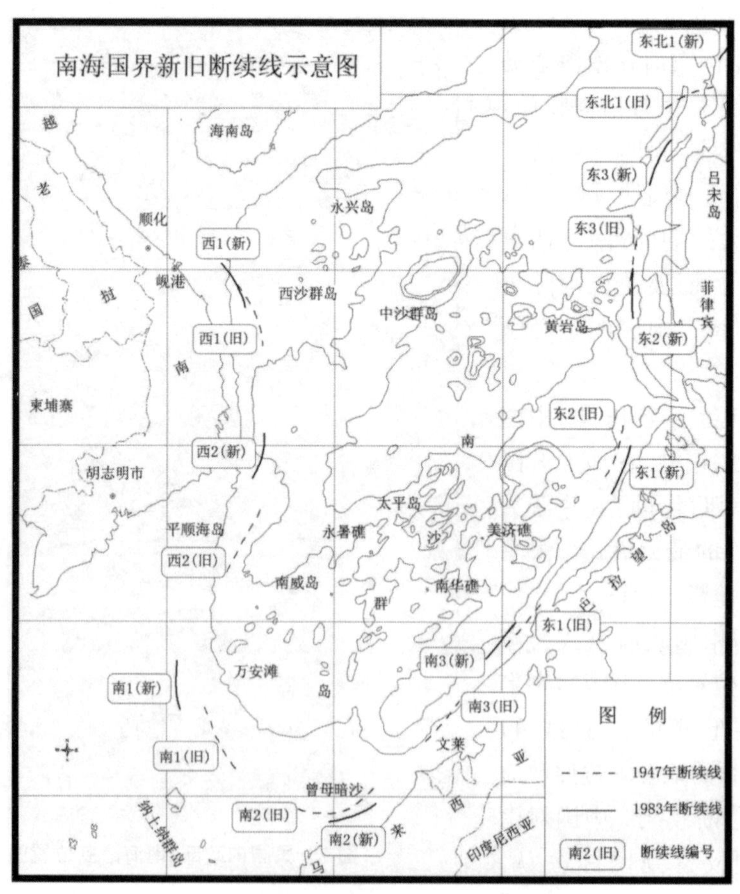

图 8 南海国界新旧断续线对比示意图

虽然各条新旧断续线有着空间位置上的差异，但新旧断续线共同界定的领土范围是基本一致的，经地理信息系统计算得出，旧断续线以内的南海海域面积约为 192.7 万 km^2，新断续线以内的南海海域面积约为 202.5 万 km^2，误差小于 5%，显示出中国南海领土主权范围的历史延续性。我们认为作为疆界依据，应明确采用一致的断续线画法，且应该是以 1947 年旧图为准，更有历史依据。通过地理信息技术量计断续线位置，为我国划分海域疆界获得准确的地理位置。

四、结束语

应用海域多项目实测资料与新旧图对比分析，通过地理信息系统技术的集成、归类与分析，并建立“4S”技术，为现代海洋国土勘测、对海洋环境进行全方位立体监测与分析，开辟了新技术领域。在我国南海疆界位置的确定，维护我国海洋权益，开发利用海洋环境与资源，为国民经济建设、可持续发展做出重要的贡献。

参考文献

[1] 张永战，王颖. 2000. 面向 21 世纪的海岸海洋科学. 南京大学学报（自然科学），36(6)：702－711.

[2] 王颖，朱大奎，周旅复，王雪瑜，蒋松柳，李海宇，施丙文，张永战. 1998. 南黄海辐射沙脊群沉积特点及其演变. 中国科学（D 辑），28(5)：385－393.

[3] 王颖，马劲松，朱大奎. 2002. 南海海底特征、资源区位与断续线. 见：海南南海研究中心：南海问题研究论文集.

论江河湖海与城市发展
——兼议南京市滨江段建设*

江河湖海是水陆交互作用地带，地理环境与生物种群结构多样，物种变化活跃，生产力较高。丰富的天然产出，平坦的土地与易于采用的淡水，易于农耕，易于生息繁殖。古人类遗迹多沿江沿海分布，亚洲大河流域的古文明等，是历史的遗证。

城镇发展的早期，江海既是防守的屏障，亦是交通运输、工商贸易与服务的通道，逐渐地沿江、河、湖、海的市镇发展成大城市。当今，人口超过100万的90个城市，60%分布于沿海与滨江、滨湖区。例如：亚洲环太平洋西岸的东京、神户、大阪、名古屋等。美国大西洋沿岸大城市带：波士顿、纽约、菲拉德菲亚、巴尔的莫与华盛顿；加、美五大湖区的多伦多、芝加哥、底特律等环湖城市圈；太平洋沿岸的温哥华、西雅图、旧金山、洛杉矶、圣地亚哥城市链。澳洲的大城市均分布在沿海地带：坎斯、汤士弗、布里斯班、悉尼、墨尔本和佩斯，主要受气候环境影响，内陆沙漠干旱，降水主要分布于海陆交互带，也具有海外移民来澳的遗留效应。旧时我国的大城市多沿海分布，具有外资、商贸与文化融入的特色。20世纪80年代以来，深圳、广州、香港、上海、天津、青岛、大连与厦门，已发展为人口密集、经济最发达的沿海城市带。

欧洲城市的分布与发展仍保持城市早期发展的特色——与河流关系密切。如：泰晤士河畔的伦敦、塞纳河畔的巴黎、易北河畔的柏林，沿着提瓦尔河(R. TEVIER)分布的罗马、涅瓦河口的圣彼得堡、沿河发展的莫斯科等。阿姆斯特丹具有河海交汇之优势，发展成为世界级的港口城市，而威尼斯更以水城而著称。这些城市，随时代发展而不断地扩展与更新，但始终保持着临水、亲水分布之优势。

一、水体在城市发展中之功能

城市的发展应具有以人为本的理念与承前继后的总体规划。既考虑人居环境应具有清新的空气、明亮的阳光、清洁的淡水、安全的宅舍与宜人的风貌，同时，又需要有繁荣的商贸街市、便利的交通与运输，资讯网络、文化历史内涵，科学、教育、医疗卫生与宗教设施，安全的保障，多种多样的服务网点以及有效的管理体系。总之，需通过科学系统的城乡发展规划与设计，达到人居环境的合理建设，促进人地关系的和谐发展。

从太空俯视，人类繁衍的地球是一个蓝色的水球，其中海洋占据了71%的面积。人类生存发展与海洋密切相关：海洋吸收4/5的太阳能，并向大地释放热能；海洋植物通过吸收太阳能与光合作用产生氧，每年达360亿t；大气中70%的氧来自海洋。海洋又吸大气过剩的CO_2，60倍于大气中的CO_2含量；海洋蒸发44亿km^3/a的淡水，以降水形式返

* 王颖：《江苏科技信息》，2005年10月，院士话题，4－10页。

回陆地(王颖、牛战胜,2004)。

应意识到人类赖以生存的新鲜空气与清洁淡水来源于海洋。如果无视严重性地向海洋排污倾费,其结果是自断生路,不堪设想。更何况,海洋还提供人们以丰富的渔产、珍惜的医药、能源与矿产。

河流是地球表层的动脉,它焕发城市的青春活力,是食物与淡水的源泉、农业与交通运输的动脉,是发展商贸、旅游、运动的一项重要载体。反之,河流干涸或湖、海消亡,丧失了水源和湿润空气,导致城市或王朝废弃:楼兰古国的废墟周围仍有死亡的胡杨树带沿干涸河谷分布;西夏王朝的灭绝;安西受风沙侵袭,从唐代至今的 1 200 年间,境内又有 37 座城池被风沙掩埋成废墟。

剖析中外城市布局,可归纳水体在城市发展中的功能为三类(王颖、盛静芬,2002)。

(一) 城市的兴起以河流为依托,河流横贯发展着的大都市,成为城市的主轴线

巴黎、伦敦与汉城,以塞纳河、泰晤士河、汉江都为城市建设的中心,道路呈放射状沟通环绕的公路与铁路(图 1)。巴黎的宫殿古建筑与现代建筑融合,层次错落地沿河分布,若干大景点与广场形成团簇,辐射出主要的街市,巴黎铁塔、凡尔赛宫与巴黎圣母院构成中心的景观轴,集古今于一体,融合了天然与人工之美。宜人行、宜车舟、宜赏析、宜探索、宜休憩、宜购物……,流动不息的塞纳河孕育着世代相继的文化精华,是城市规划建设为人着想的典范。伦敦沿河岸建筑雄伟辉煌,废弃河道洼地被建设成若干公园绿地,成为园林与城市的“绿肺”,为伦敦增加了活力。但是中心地带外围,泰晤士河旧日的码头街市予人以没落感(图 2)。位于河海交互带的加拿大哈利法克斯市,将旧日的码头改造成为旅游的步行街,展销工艺品、特产品,客源不断,使衰落的港口码头又富有了新的魅力(图 3)。新奥尔良市是沿密西西比河曲段发展的城市,河流成为工农业与交通运输的动脉,城市道路向河流辐射;市内的低洼地多辟为绿地、公园;南岸的湿地,保留着原始森林状态。随着河运功能为其他运输方式所取代,沿河街市有衰落迹象,但旅游业却勃勃兴起。

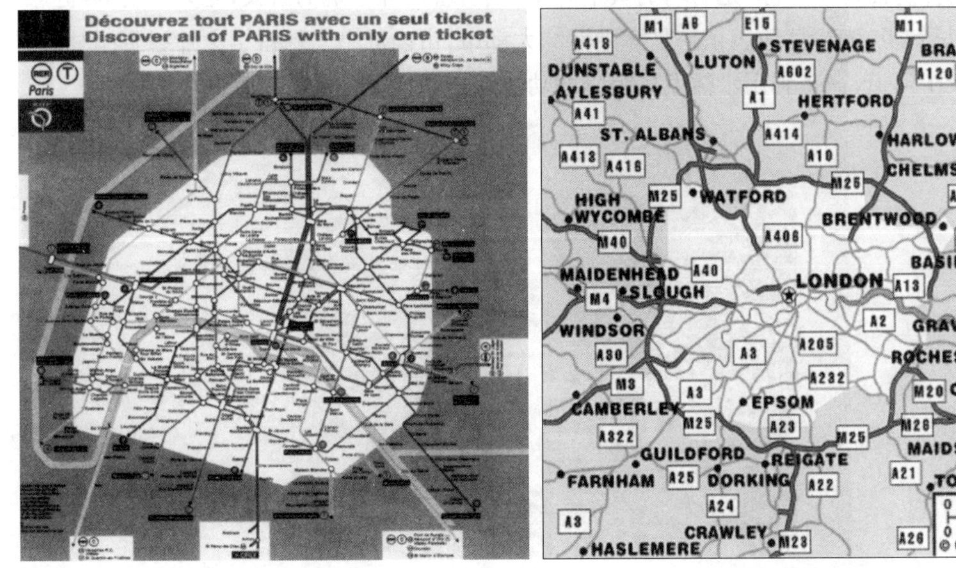

图 1　巴黎城市轮廓　　　　图 2　伦敦城市形态

图 3　哈利法克斯市(Halifax)海岸城市

南京濒临长江，随着城市的扩展，具有了长江横贯市区的特点。相类似的城市之发展与布局，值得总结与效法。

(二) 沿海湾或河海交汇的河口湾发展的城市

这类城市很多，如纽约、波士顿、旧金山、圣地亚哥、温哥华，及澳大利亚的大城市，都具有早期移民与多民族文化汇合的特点，已发展成金融、贸易、科学教育的中心。纽约市沿哈德逊、曼哈顿是世界级的金融中心，相应地，文艺与第三产业高度发达。滨水的沙滩、沙岛多辟为绿地公园与高级住宅区(图 4)。圣地亚哥市建立海军基地于海岸岛、滩区，沿河为新发展区，旅游业与商业点在兴起。波士顿与温哥华利用潮滩建设了国际机场，温哥华住宅区建筑在几级海湾海岸阶地上，高住观海却又安全避风浪袭击。澳大利亚大堡礁水碧清澈、阳光充足，2 000 km 的岛礁海岸带是海洋生物与生态环境保护区，集国家公园旅游业与体育活动于一体，是著名的黄金海岸，吸引着世界各地的游人，充分享受 3s 健康环境，增长知识，陶冶情趣。东京是海港性的大都市，市内的皇宫保留了大片绿野园林，四周围以人工河道，这是利用水体发挥阻隔的功能，是皇宫园林成为禁苑(图 5)。

图 4　纽约城市景象

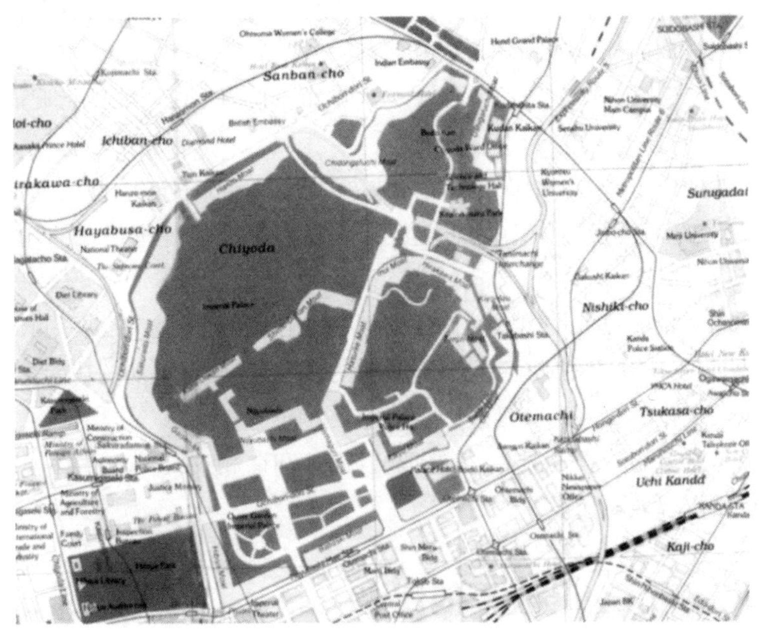

图 5　东京皇宫水体防御功能

（三）湖滨城市

湖滨城市以多伦多市及瑞士的城市为代表，沿湖是贯通四方交通的节点。湖滨、沙滩及沙岛，多辟为公园与绿地、体育活动（游水、划船与溜冰）、科技与文艺博览场馆。相应地，旅馆、餐饮与服务业发达。持续不断地吸引着城市居民与外来游客。

中国城镇傍水而建的更多，沿长江的重庆、武汉、南京、上海均为大都市三角洲水乡中苏州、镇江、扬州滨水建城，名冠古今。即使当年的北京宫苑亦引水贯通，临水建亭、台、楼、阁。在中国古代，水体的主要功能是运输及赏玩，在四方汇集的滨水地区，兴港建镇，逐渐发展为城市。宋代翰林画家张择端所作约《清明上河图》写景逼真，充分反映出汴河在北宋都城的功能以及世俗人事。汴河流贯城乡，沿河曲缓流处，船只停泊、桅杆林集，近河有修造船厂。以拱桥为中心，商肆林立，形成新的集镇。桥上行人熙熙攘攘，或驻足观景，拥挤不堪。桥下舟楫川流，篙师缆夫牵舟寸进，逆河而上。桥外城郊，清明时节，田亩阡陌纵横，疏林薄雾村舍酒家，村外大道有骑而奔驰，有踏青扫墓而归，田野情趣意浓。桥内城郭老街，商店林立，行人南来北往，市、农、工、商、医、卜、僧、道，或乘骑赶集，或牛马拖车载货。妇孺乘轿，货郎以物易货。酒店、茶肆、旅店，雅货百物，酒招牌匾，名师字画，鳞次栉比，市景文化繁荣。深宅大院，官宦宅邸则现于市侧。清明上河图生动地记录了宋代临水都城之布局，河流在城市发展中的功能：景观、水体联系、运输灌溉、积聚人气、商贾与市集扩展。

当代我国城市的滨水带，以建设公园绿地居多，或辅以鱼钓舟艇及水上活动设施，增强了人居环境的健康因素。但仅有日间活动，入夜则冷清。无夜间与假日繁华的关键是缺少吸引人流的亲水街市，以供人们于工余之后赏游。虽也建立了高档宾馆，增加了河上景点，但价格不菲，有如阳春白雪未能集聚大量客流；有些岸段建立了大量酒肆茶楼，残渣油污却毁损了滨岸与沙滩的洁净。集聚客流与维持环境洁美的矛盾是在滨岸规划中需要考虑预防的。这方面丽江古城处理得很好，保持了小桥流水与坊、店、酒肆，具溪水流畅、空气清新、与民俗文化浓郁之佳。但是，不少大城市的临江、滨海段却空旷荒凉或游人寥寥，需重视人气的聚集。

沿海城市的建设中，厦门市曾经过围海成湖，湖水污染，环境劣化；沟通活水，清污、治理与环境建设；实施海岸带一体化管理，达到建设发展与环境健康俱臻的初步境界，值得推广效法。

通过三类国外城市及古今中国城市的实例，归纳出水体在城市发展中具有以下功能：景观功能，水源与水体活力的功能，联系与运输的功能，游乐与运动的功能，生态环境陶冶与更新功能以及文化历史渊源及延续的功能等。

滨水地段是城市的珍贵资源，需很好地从"人地和谐相关"与"生存环境持续发展"的观点指导进行城市规划与建设。

二、对南京市滨江段发展的建议

南京市位居长江三角洲的顶点，据内陆与江海联运之枢纽。受潮水顶托、江潮相互作用，长江在南京地区形成一系列沙岛与浅滩：新生洲、新济洲、江心洲、八卦洲及潜州等。沙岛与浅滩使江水分为夹江与主河道。陆地上河、渠、湖、塘成星网状分布，反映出三角洲地貌属性。沿江两岸除南部有幕府、栖霞低山丘陵与北部老山、龙王、灵岩山地遥相对峙外，其余均为坦荡的平原河段长江流贯东西，孕育了两岸肥田沃野，赋予南京市生存活力。六朝古都历史悠久、民国首府、华东重镇人文荟萃，跨江大都市生机勃勃，南京市发展宜突出地体现其地理环境与历史发展特色，建设长三角山水生态园林历史文化名城。

江阔水深、浩浩荡荡、气势磅礴的长江，带给南京无限生机，是横贯东西的动脉，衔接海内外的通途，组合了大江南北的重镇，地位十分重要。但是，由于战乱、血吸虫病害等遗留的影响，无序的开发与采砂，使滨江段留下荒芜、杂乱与破旧的景象。即使是地质遗迹三台洞的沿江段也不例外。很可能由于传统的陆地经营观念，视开阔的大江为天堑，难以将长江水域与城市建设联为一体。南京市现况仍明显的具有城不见江、江不见城，江、河、湖、溪水系未全沟通之不足。

滨江地段是南京市最珍贵的自然环境资源，长江对南京具有水源与沙岛土地资源功能、景观与生态环境功能、交通运输功能、游憩与运动娱乐等功能，以及具有深厚文化、历史渊源的最宜人居的功能。如何将宽阔江面中隔的浦口与南京主城联为和谐的一体，并扩展建设城市滨江带，颇具难度。但是，可以参考比较国际相类似的大城市：纽约、蒙特利尔、温哥华、巴黎等处，发挥南京市科技发达与人文荟萃之优势，能够做到，而且会做出新模式。关键是以"人地和谐相关"的理念，以滨水、亲水、爱水、惜水的情态，发展山水生态历史文化名城的目标等为指导，统筹考虑沿江风光带建设、经济发展与生态保护等方面，作好沿江发展规划，融入南京市城市发展的总体规划中。

拟从地理环境角度提供几点城市滨江段建设的意见（王德滋等，2003）。

(一) 发挥黄金水道功能，建设沿江深水段港口，增强城市发展的经济动力

南京至吴淞口的长江约长 366 km。南京段长江江宽约 1.5 km，最宽处在新济洲尾东北侧，约 3.5 km；最窄处为长江大桥一带，江宽约 1.4～1.5 km；江心洲夹江河道仅宽 250 m。水深一般为－15～－30 m，最深处约 70 m，是一条黄金水道，天然水道可通行 5 万 t 级海轮，具有 6 000 标箱装载的集装箱船直达南京，节约中转费用与船时，经济效益大。

南京港地居江海港的顶端，地位重要，但水运功能受限于：

(1) 跨越下关与浦口间的长江大桥净空 24 m，阻挡了 5 万 t 级海轮向上游行驶。长江大桥以东建设须重视设计桥下的净空不再碍航。

(2) 长江口拦门沙碍航，天然水上深－6.5 m，经一期浚深至－8.5 m，趁潮(平均潮差 2.6 m)可通行 5 万 t 级海轮。长江口二期整治工程于近期完成后，长江口航道水深－12.5 m，趁潮可通行 7 万～10 万 t 级海轮。但是，南京以下的三角洲段河道在江阴福姜沙及焦山丹徒水道，需整治——裁湾顺直，拓宽浚深，使维持 15 m 水深，则 7 万 t 海轮可通行至南京，10 万 t 级船可乘潮满载至南京(据南京港务局长江航道整治报告，2002.9.4)。

目前，新生圩石油与机械箱码头，龙潭深水集装箱码头，及北岸仪征石化码头，前沿水深为－11～－13 m，均为深水深用的合理安排。修造船厂码头安排亦适当。下关老码头及火车轮渡码头可结合旅游业建设航运历史博物馆，开辟游艇码头及专门的水文勘测码头。建设水陆交通一条街，繁荣大江南北往来。

沿江的小码头与采砂码头(图 6)，须加以规范整治和强化管理，严格制止江岸采砂，防止江岸坍塌，堵绝陆地与人身损害。

图 6　沿江采砂码头照片

(二) 建设南京市沿江风光带，可组合 6 个板块由绿树廊道联通，呈串珠状沿江分布

(1)燕子矶、幕府山、栖霞山基岩江岸风光带

① 以南京长江二桥为中心点，西联燕子矶、幕府山，东联九乡河与栖霞山。

这段江岸由古老的下奥陶统仑山组石灰岩构成，悬崖临水，岩洞成群。

② 西侧燕子矶位于长江南岸直渎山，海拔 36 m，如乳燕展翅临水，矶头远眺，二桥历历在目，江亭和风，御碑辞赋，浑然一气。

幕府山宜修复植被为绿色丘陵，山下头台洞、二台洞、三台洞为沿江的大型喀斯特岩洞，岩洞成层发育，印证着山体的隆升与地下水下降归流入长江之过程。三台洞既有吴道子的碑画，亦有着多次长江洪水位遗迹，其中标高为 8.5 m、9.4 m、10.5 m 及 12.9 m 的四次洪水位，高出于现代长江江面（标高 3.498 m）之上，断续保留于头台洞至燕子矶之间，为珍贵的古洪水“记录”，需注意保护（何华春等，2004）。岩洞前沿江岸零乱，宜整顿清理与绿化。

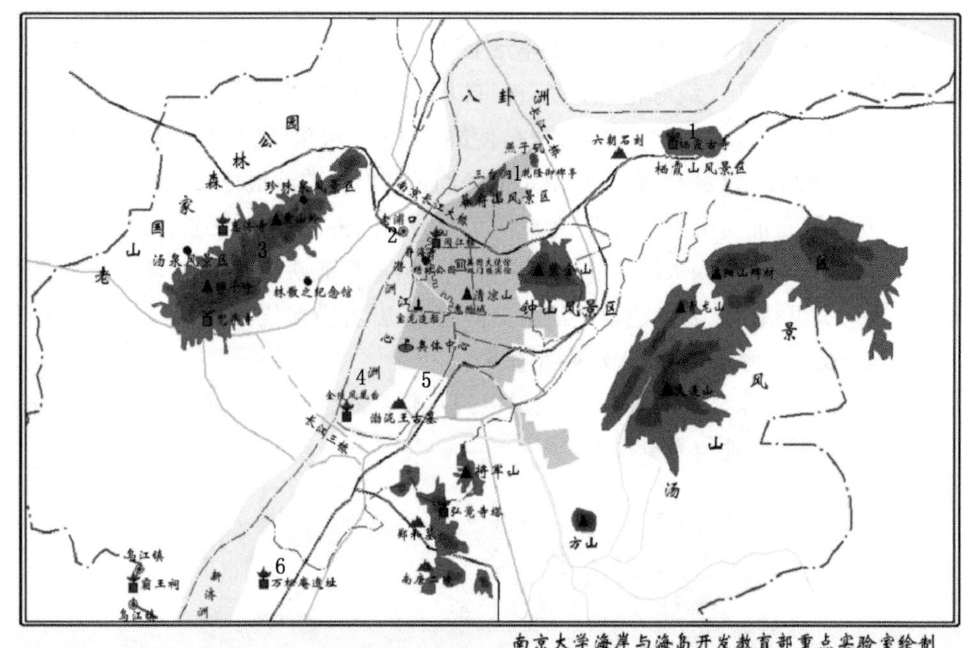

图 7　沿江六大板块分布图

③ 南京长江二桥东侧，栖霞山麓江水蜿蜒，空气清新，极目远眺，丹枫秋色古寺，飞天石刻遗迹与石雕经文交相辉映，古城与龙潭新区山野，已成为老少皆宜的远足胜地。

(2)下关—浦口老轮渡码头建设南京滨江中心文教商旅与风光带

两岸码头与火车轮渡曾为大江南北往来功勋卓著。宜保留原有码头与轮渡设施、开设旅游轮渡与江上夜游，组建动态式的轮渡历史博物点。加上下关地区人文历史丰富景点：狮子山、阅江楼、挹江门、明清古城墙、静海寺、江南水师学堂、英国使领馆、渡口战役纪念碑以及小桃园、绣球公园景区，构成南京市滨江带明、清、民国、近代历史文化中心，辅以传统的街市餐饮、手工艺品作坊与商肆，繁荣沿江中心带经济。自下关—浦口滨江中心带向南北延伸，均与国内重点大学相联，增添现代文化气氛。

(3) 江北“一山三泉”——森林地质生态景观带

浦口老山林区面积为紫金山山地森林之两倍，两处森林夹峙长江南北，是南京市的两个“绿肺”（图 8），对南京市空气更新，水源保护至关重要。中科院王德滋院士倡议：将长江北岸的老山森林，珍珠泉、琥珀泉、汤泉的自然风光与沿江民居、古文化遗迹融为一体，列为新市区建设的重点之一；老山森林公园，又是地质公园，其岩层是6亿年前震旦纪的浅海沉积形成，是南京与华东地区最古老的岩层。老山与三泉可为中小学生建造学习自然科学知识和外野生活的夏令营地。

对江北兜率寺与惠济寺进行保护性修建，尽量保留周围的田野风光和自然村落本色，给南京市民营造有一个田野风光区。

(4) 江滩与沙洲生态绿岛建设

潜洲绿林草地自然景观，可供泛舟憩游；江心洲景色宜人，沿主流江滩，或柳林成荫，或芦花翻白燕子飞，令人心旷神怡，近年已渐形成季节性周末旅游场所。仍需关注污水处理厂的环境管理。

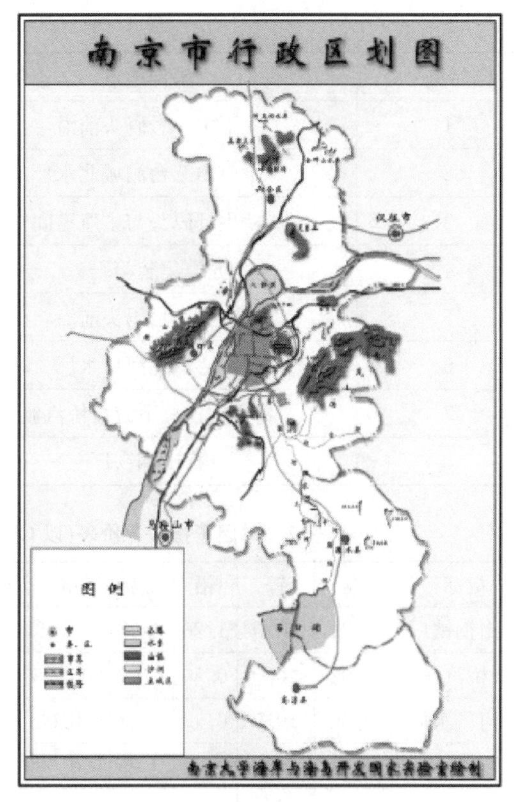

图 8 南京绿肺图(附彩图〈上册〉)

八卦洲既为现代化的路、桥中途岛与巍峨二桥的展览中心，亦宜发展为四季风光旅游绿洲："春游芳草地与芦蒿尝新，夏绕碧荷池与沙滩戏水，秋饮黄花酒与肥蟹品尝，冬吟白雪诗与环岛车骑"融景观与体育活动于一体。

(5) 河西新民居——宝船厂遗址——奥体中心，建设营造中的南京新亮点。石头城—秦淮河滨水公园初具规模，已成为市民喜爱的胜地：春日风和日丽、风筝轻扬云端，夏夜纳凉话星空，秋花明月幼童嬉戏池畔，冬阳温暖漫步河岸健身。进一步需沟通秦淮河水系，形成江河自然互补；需大力整治秦淮入江河口段。

(6) 板桥—梅山工业区与新济洲两岸

保持天然平原江岸，建设绿荫大道绿廊与中心带相联，为工厂区及南郊山村居民提供滨江风光带。加强沿江地带的管理，保护水源区，规范民用小码头，严格制止开采江沙，以防江岸坍塌。

(三) 加强长江水质的监测管理，防止江水进一步污染

2003年10月下旬，在宁院士对南京市发展咨询工作时，在南京市委组织部与南京港务局支持下，本人负责组织了市域长江水质调查。沿江采样点见表1和图9。南京大学环科院刘华梁同志负责及许叶华同志参加，采用国家标准GB3838-2002地表Ⅱ类水标准，对水质进行单因子评价。计算所得水质单因子评价指数及水质类别如表2。

表 1　长江江水采样点位置

样品编号	采样地点	时间
1	下关 4 号码头前沿	2003 - 10 - 26　9:20
2	幕府山三台洞城北水厂	2003 - 10 - 26　10:50
3	新生圩码头与主航道间	2003 - 10 - 26　11:15
4	龙潭集装箱码头	2003 - 10 - 26　12:30
5	下关 1 号码头前沿	2003 - 10 - 30　13:30
6	江心洲北河口水厂	2003 - 10 - 30　13:56
7	梅山钢铁厂下游板桥汽渡	2003 - 10 - 30　15:01
8	梅山钢铁厂	2003 - 10 - 30　15:20

表 2　单因子指数评价表(以 GB 3838 - 2002 Ⅱ 类水为标准)

指标	铜	锌	铅	镉	铬	氨氮	总磷	COD	石油类	溶解氧	pH	水质类别
梅山钢铁厂	0.04	<0.04	1.20	0.05	<0.08	0.25	0.79	1.40	2.40	0.73	0.33	Ⅲ
板桥汽渡	0.02	<0.04	0.50	0.02	0.24	0.74	1.29	0.75	2.20	0.62	0.37	Ⅲ
北河口水厂	0.00	<0.04	0.40	0.13	<0.08	0.44	1.73	<0.5	2.00	0.68	0.41	Ⅲ
下关 1 号码头	0.00	<0.04	0.30	0.05	<0.08	0.43	1.18	<0.5	2.80	0.51	0.38	Ⅲ
下关码头	<0.00	<0.04	0.30	0.05	<0.08	0.49	1.15	0.50	4.00	0.51	0.31	Ⅲ
三台洞	<0.00	<0.04	0.40	0.05	<0.08	0.45	0.67	<0.5	1.00	0.51	0.41	Ⅱ
新生圩	<0.00	<0.04	0.10	0.05	<0.08	1.00	0.79	<0.5	4.00	0.70	0.37	Ⅲ
龙潭码头	<0.00	<0.04	0.60	0.05	<0.08	0.78	1.21	<0.5	2.20	0.48	0.26	Ⅲ

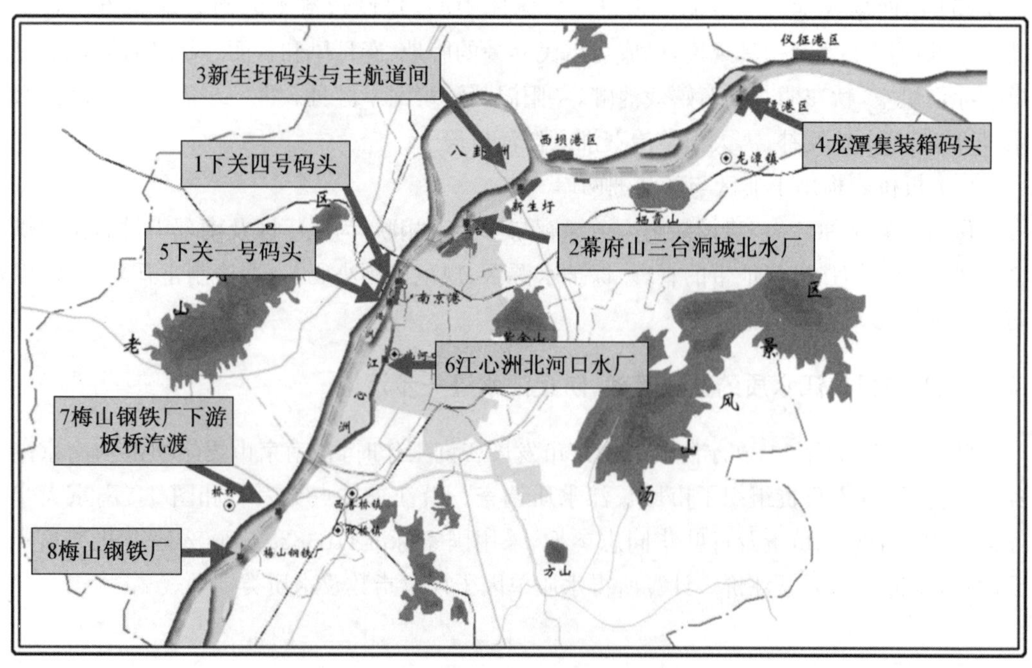

图 9　长江江水采样点与分布图

（1）重金属污染

铅超标，出现于梅钢段水域，南京北河口取水口位于其下游。

铜：实测值范围为 0.002～0.035 mg/L，断面超标值为 0，最高值出现在梅钢段水域。

锌：实测值全部低于检测限，即小于 004 mg/L，断面超标率为 0。

铅：实测值范围为 0.000 1～0.021 mg/L，断面超标率为 12.5%，最高值出现在梅钢段水域，超标 1.2 倍。

镉：实测值范围为 0.000 1～0.000 7 mg/L，断面超标率为 0，最高值出现在北河口水厂段。

六价铬：实测值范围为 0.004～0.012 mg/L，断面超标率为 0，最高值出现在板桥汽渡段。

（2）营养盐污染

氨氮：实测值范围为 0.124～0.502 mg/L，断面超标率为 0，最高值出现在新生圩段，单因子指数为 1.00，处在超标的边缘。

总磷：实测值范围为 0.067～0.173 mg/L，板桥汽渡段、北河口水厂段、下关一号码头段、下关码头段、龙潭码头段共五个断面超标，断面超标率达到 62.5%。最高值出现在板桥汽渡段，五个断面平均超标倍数为 1.31 倍。各断面浓度比较如图 10 所示。

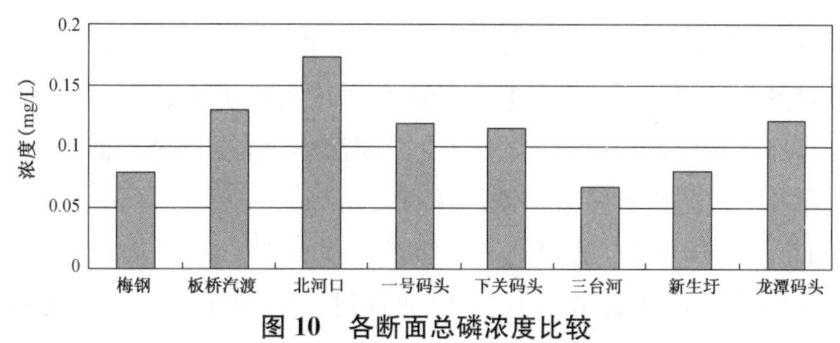

图 10　各断面总磷浓度比较

（3）有机物污染

COD：实测值范围为 2～5.6 mg/L，断面超标率为 12.5%，最高值出现在梅钢段，超标 1.4 倍。

石油类：实测值范围为 0.05～0.20 mg/L，除三台洞段外其余 7 个断面都超标，断面超标率达到 87.5%。最高值出现在下关码头段，7 个断面平均超标倍数为 2.8 倍（图 11）。反映出南京长江段大部分遭受石油污染。

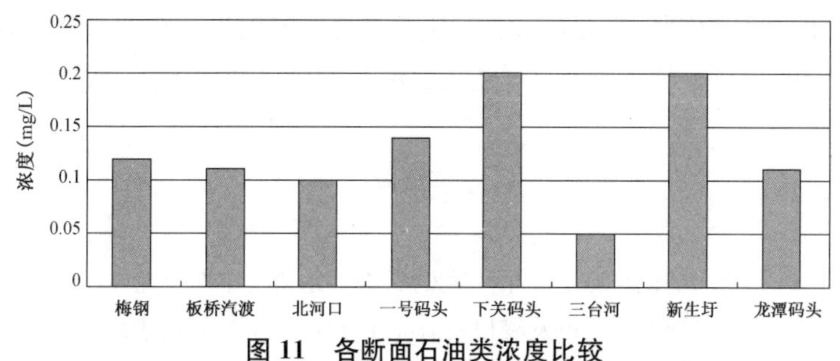

图 11　各断面石油类浓度比较

（4）其他

pH、溶解氧：未超标。

水温：检测温度介于 19±0.3 ℃之间，经调查，基本不存在人为造成水温的变化，该指标全部达标。

总上述，南京长江段水质检测可归结为以下五点：

① 所测试的 12 个指标中，8 个达到 GB3838－2002 地表水Ⅱ类水标准，分别是水温、PH、溶解氧、氨氮、铜、锌、镉、六价铬。出现超标的因子有：石油类、总磷、COD、铅(图 12)；

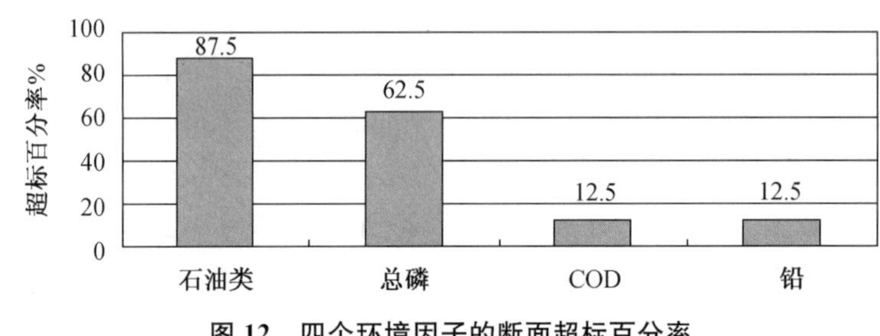

图 12　四个环境因子的断面超标百分率

② 石油类、总磷化学耗氧量、铅超标断面百分率分别是 87.5%、62.5%、12.5%、12.5%；

③ 石油类、总磷化学耗氧量、铅平均超标倍数分别是：2.8、1.3、1.4、1.2 倍；

④ 除三台洞段为Ⅱ类水外，其余七个断面水质均为Ⅲ类水；

⑤ 检测表明南京长江段江水大部分遭受石油污染，可能与船舶泄露及洗舱有关；铜、铅、磷与有机质污染，主要源于梅山钢铁厂水域。

加强长江南京段水体保护，已到刻不容缓的关键时刻。南京市在城市建设发展与沿江开发中，从关注南京段长江的水质出发，应加强对已建的化工厂所排废水、废气之处处理与监测，即使达标排放，亦会有长年积累的危害；尽可能地限制沿江兴建化工、造纸等大量排放废弃水、气的工业与产业。切实保护南京市民的饮水源与水环境安全是生存环境可持续发展之关键。

结束语

本文是由中科院在宁院士办事处组稿、《江苏科技信息》约稿，促使我对学习、工作与生活长达 45 年的南京市抒发自己的感想，对其建设发展提出一些真挚的意见。深深的期望，南京市在 21 世纪新一轮的发展中，更加展示其深厚文化底蕴的新风采。

参考文献

[1] 王颖，牛战胜. 2004. 全球变化与海岸海洋科学发展. 海洋地质与第四纪地质，24(1)：1－6.

[2] 王颖，盛静芬. 2001. 江河湖海与城市发展相关剖析. 广州市城市规划局主编. 滨水地区城市设计. 北京：中国建筑工业出版社，85－92.

[3] 王德滋，王颖，朱大奎. 2003. “南京市沿江发展专题”—对南京市港口与沿江风光带发展建设建议. 南京大学.

[4] 何华春，王颖，李书恒. 2004. 长江南京段历史洪水追溯. 地理学报，59(6)：938－947.

火山地貌例述*

自 20 世纪 90 年代中晚期以来，火山与地震构造活动时有发生，在全球范围内，几乎是年年有大地震。2000 年以来，地震活动日益频繁，几乎是每月有地震报道，强地震又引发火山爆发，火山活动显现于当代，已处于全球性的火山地震构造活动期。火山与地震活动是短期间的突发事件，但对自然环境与人类生存活动却有着持续性影响。归纳六十年来，作者考察与经历的火山活动，进行总结对比，可供教学、研究火山活动效应与灾害防治的参考。

现今地球上约有 500 多座“活”火山、2 000 多座“死”火山。活动火山主要分布在：板块的边缘，大洋中脊以及板块内部构造交接带，即：地质结构脆弱、构造断裂与岩浆喷溢区。地震活动与火山分布区相伴生，环太平洋火山带最为突出——沿着南、北美洲西海岸，阿拉斯加半岛，阿留申群岛，堪察加半岛，千岛群岛，日本列岛，琉球群岛，中国台湾岛，菲律宾群岛，印度尼西亚群岛至新西兰——共有 370 多座活火山，占全球活火山数量的 75%以上，是地球表面的火环地带[1]（图 1）。在环太平洋火山带中，尤以北美、堪察加半岛、日本、菲律宾及印尼一带的岛弧-海沟系构造带的活火山活动频繁。2000 年全球火山

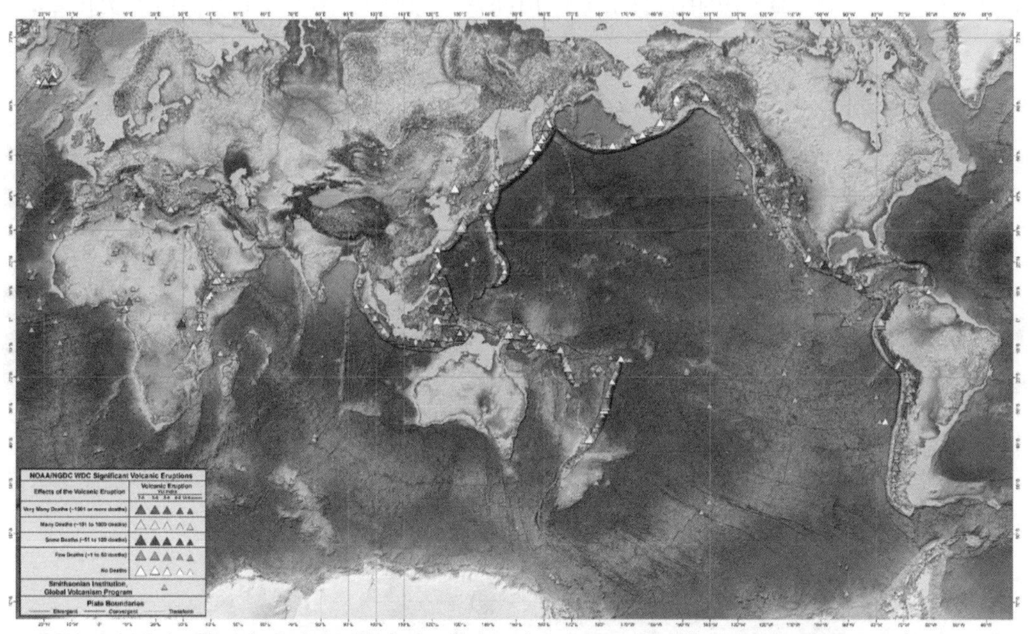

*注：图中总结了最近 6000 多年间 630 次的火山喷发，其中 39%在中、南太平洋，17%在欧洲，17%在东亚，7%在中美与加勒比区，7%在北美与夏威夷，6%在南美，3%在非洲，2%在堪察加与库页岛，1%在中东，1%在南美。

图 1　公元前 4360 年至公元 2013 年间主要火山喷发的分布图(附彩图)

（据 NOAA 地球物理资料中心，世界地球物理与海洋地质资料中心及海啸信息中心资料）

* 王颖，王敏京：《海洋地质与第四纪地质》，2018 年第 38 卷第 4 期，第 1－20 页.

爆发达 50 次，其中 33 次发生在太平洋周围[2]。环太平洋带火山喷发以中心式喷溢为主，强度大，多为中性的钙碱系列岩浆，形成自海沟向两侧分布的安山岩、拉斑玄武岩、钙碱性岩石与碱性系列岩石。

火山爆发，山崩地裂，炙热的硫气喷出，熔岩流倾溢，推房倒林，掩埋田野生物，作者曾亲历夏威夷火山岩流缓缓流动，却形成推林、岩流巨瀑注海、水气柱冲天的磅礴景象。6 000 多年来全球 630 次火山喷发活动，造成约 30 多万人死亡，损失金额超过 20 亿美元[3]。另一方面，火山喷发却形成新生的土地、巨大的能源、突发地改变着自然环境特性。经风化的熔岩层却是肥沃土壤与生物繁殖的矿物营养源。作者在国内外多处海岸与海岛考察中经历过多处火山活动地貌环境，主要是新生代中后期、历史时期，以至当代的火山喷发活动，对人类生存环境有显著的影响，启发并与王敏京合作总结成文，以供研究火山活动为实录参考。

1　火山各论

据火山地貌形成时代与火山活动动态，将考察过的火山分为三类加以例述。

1.1　死火山(55.9 Ma～3.82 Ma～0.23 Ma)

(1) 最早接触到的火山地貌，是 1957 年在山西大同盆地阳高县，分布于桑干河北岸的火山群。一系列低缓的火山锥体分布于山间盆地原野上(图 2)，山顶保留着火山口，山体基岩裸露无植被(图 3)。盆地旷野人稀，南山与北山两山夹峙，时逢雷雨阵阵，景象恐怖。

图 2　大同火山群低缓的火山锥成列分布

图 3　大同火山(附彩图)

大同火山群有 32 座低缓火山丘，相对高度约 100 m，最高的黑山相对高度 211 m，小的火山相对高度 5～30 m，寄生火山约 5 m(表 1)[4]。火山群呈现为 NW－SE 与 NE－SW 两个方向的系列交叉分布。火山分别为玄武岩锥体、火山渣锥体及玄武岩与火山碎屑物混合层锥体三种类型。岩层主要为碱性橄榄玄武岩、碱性玄武岩及少量拉斑玄武岩，呈现为自火山口向四周外倾的层状堆积，多气孔，含浮石，表现为岩流喷溢与喷发岩屑两种堆积。火山喷发始于第四纪早期(约 98.5×10^4～84.3×10^4 aB. P.)[5]，剧烈的喷发在第四纪中期(54×10^4～36.9×10^4 aB. P. ，24×10^4～15×10^4 aB. P. ，22×10^4～12×10^4 aB. P.)[5]，约在第四纪晚期、距今 9 万年前停止喷发，现为死火山。火山喷发有多期差异与间歇性，取样层次与地点不同，所获的火山岩年龄不同：大同册田水库以南的拉斑玄武岩上、中、下层年龄分别为(0.54±0.05)Ma，(0.59±0.10)Ma，(0.74±0.22)Ma[5,6]；桑干河南龙堡村与秋林村拉斑玄武岩分别为(0.23±0.14)Ma，(0.23±0.23)Ma[5,6]；大同黑山脚下碱性玄武岩为(0.38±0.13)Ma，(0.49±0.06)Ma；金山顶碱性玄武岩火山弹年龄为(1.27±0.08)Ma[5,6]；大同肖家头村碱性橄榄玄武岩年龄为(0.458±0.025)Ma；余家寨橄榄玄武岩年龄为(0.455±0.036)Ma[5,7]等。定年数据是不同地点的非系统采样，反映了华北地区更新世中、晚期喷发的火山活动，以分布于内陆平原断裂带的火山活动为特征。

表 1　大同主要火山高度特征[4]

火山名称		山顶海拔/m	相对高度/m	火山口直径/m
黑山		1 431.00	211.00	320
艾家凹		1 367.90	177.50	200
狼窝山		1 335.10	140.10	530
阁老山		1 276.10	106.10	350
马蹄山		1 233.45	105.15	250
东山		1 199.00	119.00	150
昊天寺		1 145.80	105.80	100
老虎山		1 210.40	50.40	
双山		1 179.00	20.00	
马蹄山东北小火山	(1)	1 195.30	25.30	
	(2)	1 206.10	36.10	
	(3)	1 178.00	7.50	
	(4)	1 165.30	10.00	
	(5)	1 174.10	12.50	
	(6)	1 169.10	5.00	
狼窝山东南坡寄生火山——小山		1 215.00	30.00	
狼窝山东北坡寄生火山		1 260.00	15.00	
黑山南坡寄生火山		1 305.00	15.00	
南山东坡寄生火山		1 305.00	10.00	

（2）南京江宁、六合与仪征地区，在秦淮河与滁河平原上，散布着二十多座独立的方山与桌状熔岩山丘，高度不大，江宁方山海拔高度 208 m，六合方山为 188 m，桂子山高 52.6 m。岩性主要为碱性橄榄玄武岩和碧玄岩，含有尖晶石二辉橄榄岩包裹体。多数火山保存着火山口洼地与层状火山堆积。六合桂子山玄武岩厚度约 30 m，剖面显示出完整的六边形柱状节理，形成柱状石林地貌（图 4）。经南京地矿研究所以 K-Ar 法测定，时代＞5.9 Ma[8]。南京地区的火山活动划分为三个时期：第一期自古新世末至始新世中期约 55.9 Ma～43.8 Ma，以南京长山林场（55.9 Ma）、安徽当涂釜山（53.18 Ma）、江苏溧阳（43.8 Ma 及 48.0 Ma）为代表；第二期火山活动发生在中新世早、中期（20.7 Ma 和 16.27 Ma～13.9 Ma 两个时段），包括溧阳竹箦（20.7 Ma）和上兴（15.0 Ma）、六合塔山（16.27 Ma）、句容赤山（13.9 Ma）；第三期火山喷发位于中新世晚期（11 Ma～7 Ma），火山活动达到高潮，江宁方山（10.89 Ma～10.32 Ma）、六合方山（9.41 Ma～8.60 Ma）、瓜埠山（9.35 Ma）、东海安峰山（7.58 Ma）与平明山（6.10 Ma）等[5]。如：江宁方山，基底（位于海拔 100 m 以下）为上白垩统赤山组砂岩和中新统洞玄观组砂岩及砂砾岩，上部为方山玄武岩构成火山体，出露约 1.5 km^2（图 5）。

图 4　六合桂子山石柱林（张永战　摄）（附彩图）

（3）海南岛北部海口与西北部洋浦地区分布着典型的新生代火山群[10]。

沿琼州海峡断陷及北部湾构造张裂活动区，多次火山喷发堆积层，叠加贴覆于海南岛中部岩浆岩山地的北部与西北侧，扩大了岛域面积。

琼北是我国新生代火山构造活动最为显著、持续时间最长的地区之一，火山分布具有自陆地向海、自东向西喷发时代渐新的特征。海口市石山火山群分布面积约 108km^2，40 座火山[11]，具有火山熔岩锥、火山碎屑锥、熔岩与碎屑混合锥，保存着低平的火山口洼地或破火山口。石山火山锥完整，峰顶开敞为火山口，口底尚存熔岩流通道（图 6）。

海口地区火山曾有多次喷发：金牛岭为石英拉斑玄武岩，具有（16.77±0.47）Ma 与（11.68±0.75）Ma 两个时代堆积[5,12]；秀英村中埋深于 104 m 处的石英拉斑玄武岩定年为（6.27±1.2）Ma[5,13]；金牛岭地表石英拉斑玄武岩测定为 3.82Ma[5,10]，该处玄武岩为（3.82±0.11）Ma[5,14]，与石山荣堂村地表岩洞内粒玄岩的定年（3.81±0.99）Ma

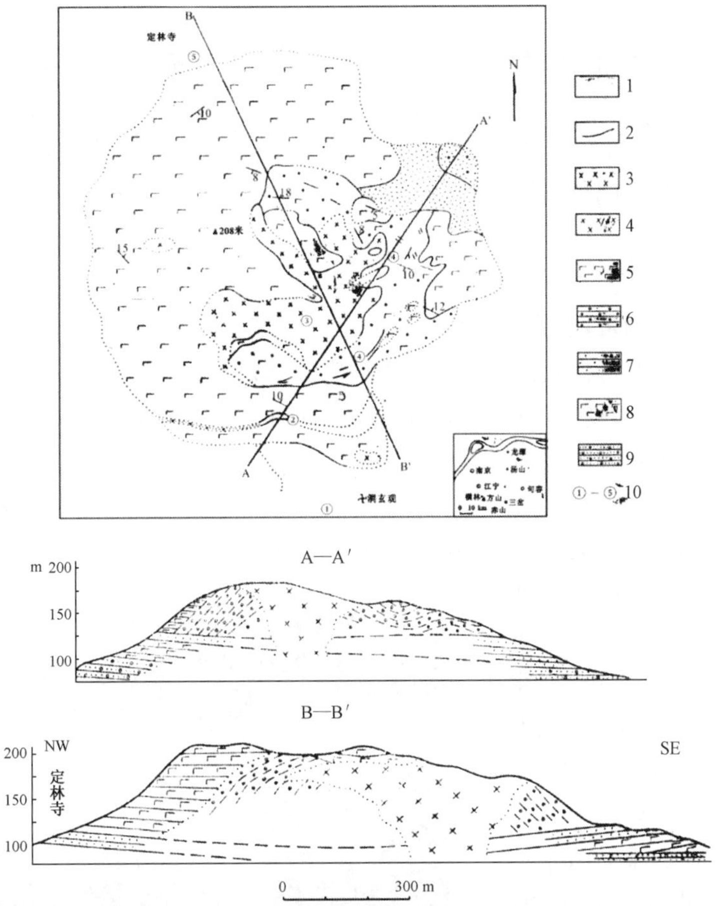

1. 第四系；2. 上方山玄武岩-碱性橄榄玄武岩墙；3. 上方山玄武岩-暗色碱性橄榄辉绿岩；4. 上方山玄武岩-碱性橄榄辉绿岩；5. 上方山玄武岩-碱性橄榄玄武岩；6. 上方山玄武岩-玄武质火山碎屑岩；7. 火山砂岩；8. 上方山玄武岩-碱性橄榄玄武岩；9. 洞玄观组，砂岩、砂砾岩；10. 观察点

图 5　江宁方山地质简图[9]

图 6　海口市郊石山火山群

（马蹄形火山锥与 2 座低平火山口）

相当[5,15]。

马鞍岭在琼北火山中喷发年代最新，约在 13 000 年前喷发[11]。由两座火山组合：南面风炉岭火山，海拔 222.2 m，是海口市最高点，具有 130 m 直径的火山口，建有步道下达 69 m 深处的火山口底部；北峰包子岭，海拔 186.75 m。在风炉岭南侧还有两个寄生的小火山，称为眼镜岭，共同组成马鞍岭火山族系。

1.2 休眠火山

休眠火山很难确定，实为过渡类型。因为有些火山的喷发间歇达数万年，甚至数十万年，分类具有相对性，可据火山所处的地质结构综合判断。休眠火山属活火山，或者是趋向死亡发展的活火山，现时没爆发喷出熔岩流，却时有地震、山体内轰鸣声与火山气体逸出。这些特点也与死火山区别。典型的例子如：

（1）乞力马扎罗火山

非洲的乞力马扎罗火山（Mt. Kilimanjaro）是坦桑尼亚东北部与肯尼亚交界的分水岭，位于东非大裂谷以东相距约 160 km 的边缘地带，属于东非大裂谷的中心式喷发火山，是多火山口的多火山锥体组合，东西向延伸 80 km。最近的一次大爆发在（15～20）×10^4 年前[16]，大部分火山已死亡，而主体火山基博峰（Kibo）仍有水蒸气与含硫气体逸出，并有温泉发育。远眺乞力马扎罗屹立于非洲高原上的雄伟山体，平顶有白雪环覆与汽云飘逸。估计是东非大裂谷张裂过程中残留的火山（图 7）。

图 7　乞力马扎罗山峰

（2）富士山火山

富士山（Mt. Fuji）海拔 3 775.63 m，位于东京西南相距 80 km，横跨静冈县和山梨县，居于富士火山带上，是日本的第一高峰。

富士火山带起自马里亚纳群岛，经伊豆群岛、伊豆半岛到达本州北部。富士山是锥状成层堆积的新生代火山，基底为第三纪地层，第四纪初火山熔岩冲破第三纪地层，喷发堆积形成山体。大约从 8 万年前开始至 1.5 万年前多次火山喷发堆积为古富士山，是火山渣砾与火山灰堆成层状火山锥，高度接近 3 000 m[17]。距今 1.1 万年前，古富士山的山顶西侧开始喷发出大量熔岩，形成富士山主体的新富士。约在 2 800～2 500 年前，古富士山顶因风化作用与大规模崩塌，仅剩新富士山顶。距今 11 000 年～8 000 年前时，新富士山顶曾不断地喷溢熔岩流，嗣后，山顶部分没有新的喷发[17]。而侧火山——长尾山（Mt. Nagao）和宝永山（Mt. Hōei），仍断续地喷溢熔岩，自公元 781 年以来，富士火山喷发了 18 次，最后一次喷发是在 1707 年，嗣后成休眠火山[17]。1707 年喷发形成一个巨大的火山口，喷发的浓烟可上达平流层，在相距 100 km 的东京堆积了 4cm 厚火山灰，火山周围的平原地区，常有地震活动。富士火山山体呈圆锥状，平顶，火山口直径约 800 m，深达 200m[17]（图 8）。处于构造断裂带上的富士山在积累相当能量后，可能会形成新爆发。

图 8　富士山火山口(附彩图)

（3）海南岛西北岸火山

琼西北海岸火山发育于儋县一带，有晚更新世及全新世中期两期火山活动的 5 次喷发（表 2），火山活动具有继承性、多发性以及自东向西即从岛陆向北部湾海域转移而规模减小的特点。地处雷琼拗陷南部的邻昌凹陷内，沿两条主要的断裂：NW 向的兵马角-木棠-儋县断裂与 NE 向的松林-木棠-干冲断裂成 X 相交，沿 X 相交断裂处发生火山活动，控制了海岸动力地貌与环境特征[18]。

表 2　海南岛西北岸火山活动与地质地貌特征[18]

喷发期次		时　　代	岩石特征	分布与地貌特征
第一期	第一次	晚更新世玄武岩烘烤层热释光定年为 52 000 ± 400 a.B. P.	气孔状玄武岩层厚 5～10 m，经鉴定为橄榄玄武岩及橄榄拉斑玄武岩	洋浦，沿 NEE 向断裂多中心溢流喷发，形成火山区的玄武岩台地和丘陵，向海盖在海相沉积层上，火山弹与火山碎屑堆积易侵蚀成低地，岩流形成 8 m 高海岸阶地。有多个喷发孔，使地面起伏为火山丘陵
	第二次		气孔状橄榄玄武岩 橄榄玻基玄武岩 岩屑玄武岩 层凝灰岩	峨蔓为中心，沿 NW 向断裂成多火口喷发，喷发剧烈，形成 20～40 m 崎岖不平熔岩低丘，沿断裂带成一系列火山锥，笔架山三峰高度为 208 m、121 m、171 m，春历岭 100～167 m，火山锥兀立于第一期玄武岩台地上
	第三次		橄榄玄武岩	干冲断裂北侧及德义岭，成盾形火山及熔岩台地，穿插、覆盖在第二玄武岩上，德义岭火山高 97.4 m，火山口完整、直径 200 m
	第四次	火山岩覆盖在海岸潮滩沉积层上，其中的扇贝 ^{14}C 定年为 26 100 ±960 a B. P.，喷发时代在更新世末	多孔的熔岩、拉斑玄武岩、玻屑凝灰岩、玻璃火山碎屑岩及海水具微层理与波痕结构的凝灰质砂砾岩	莲花山及德义岭以北，龙门以南，喷发强烈，黏滞性大。火山锥兀立，坡度大，穿插于熔岩台地（第三次）间，莲花山有一中心火山喷发口，高 65.0 m。四周有几个火山喷发孔
第二期	第五次	经热释光定年为 4 000 ± 300 a B. P.	炉渣状熔岩 气孔状玄武岩 拉斑玄武岩	母鸡神、龙门、沿海岸断裂带，成串珠状小火山喷发，孔裂隙式喷发，全新世沉积受火山活动而抬升构成现代火山海岸

火山岩属洋岛型橄榄玄武岩及大洋中脊型拉斑玄武岩，表明在海南岛-北部湾海底的扩张、地壳减薄处，地幔物质沿断裂喷发溢出[18]。

沿着北东向断裂带在全新世的第 4 次与第 5 次间的火山喷发带发育了现代火山海岸，分布在海南岛洋浦港以北的神尖与龙门之间，沿裂隙带呈密集成串的小火山口喷发，呈岩墙或成层贴附于早期火山的岩壁上。堆积过程中经历构造变动，多断层，再经海蚀成崎岖的海蚀型火山港湾岸，外缘发育着珊瑚礁坪。火山海岸受构造抬升，使母鸡神火山颈抬升为海蚀柱(图 9)，龙门岸海湾与海底抬升，岩滩与海穹形成高阶地[18]。全新世火山喷发提高了北部湾海水温度，促进了珊瑚礁发育。

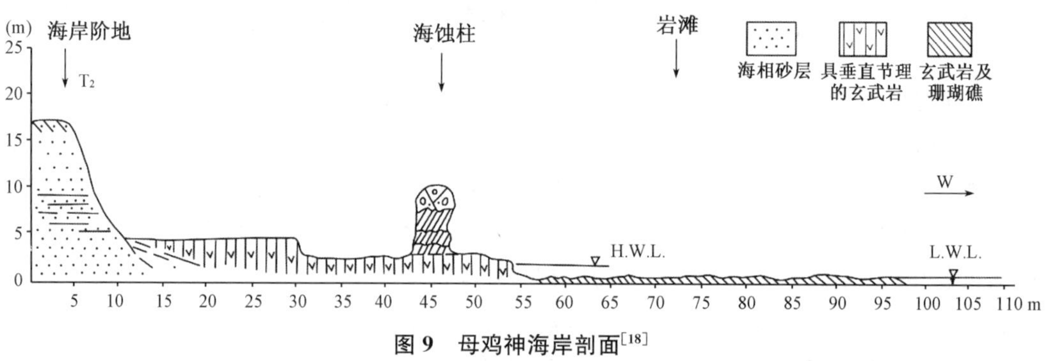

图 9 母鸡神海岸剖面[18]

(4) 涠洲岛

位于北部湾东北部，北海半岛以南、相距 38 km 的海域中。涠洲岛南端位于 21°N 线上，北端至 21°05′N，东端在 109°09′E，西侧距 109°E 线约 8.5 km。

涠洲岛是沿沉溺的火山口发育的玄武岩基岩岛，面积 24.98 km^2，岛屿南侧的海湾是沉溺的两个火山口(图 10)。

图 10 涠洲岛火山口海湾

地属雷琼火山带，在板块边缘裂谷，于更新世火山喷发形成，具有多中心、多小火山口、多期次喷发的特点(表 3)。早更新世喷发的火山岩夹层于第三纪湛江系地层中；中更新世喷发玄武岩规模大，中心层厚约 120 m，四周减薄为 20～25 m；晚更新世早期堆积火山碎屑岩与凝灰岩层状沉积层，晚期堆积为玄武岩[19]；全新世为碧玄岩及火山碎屑与海浪交互沉积层。

表 3 涠洲岛火山岩年代测定

地点	岩性	测年方法	测得年龄	数据来源
石螺村潮间带	球状风化玄武岩	K-Ar	(1.42±0.12)Ma	樊祺诚等,2006[20]
涠洲岛东岸	碱性橄榄玄武岩	K-Ar	(1.26±0.26)Ma	冯国荣,1992[13]
鳄鱼嘴	熔岩角砾	K-Ar	(0.86±0.07)Ma	樊祺诚等,2006
横路山锅盖岭	球状风化玄武岩	K-Ar	(0.79±0.05)Ma	樊祺诚等,2006
北港	球状风化玄武岩	K-Ar	(0.75±0.04)Ma	樊祺诚等,2006
横路山锅盖岭	火山碎屑岩	K-Ar	(0.60±0.02)Ma	樊祺诚等,2006
航标塔	早期玄武岩捕虏体	K-Ar	(0.59±0.05)Ma	樊祺诚等,2006
西角村	球状风化玄武岩	K-Ar	(0.58±0.03)Ma	樊祺诚等,2006
斜阳村	球状风化玄武岩	K-Ar	(0.57±0.05)Ma	樊祺诚等,2006
涠洲岛横岭北	球状风化玄武岩	K-Ar	(0.49±0.03)Ma	樊祺诚等,2006
斜阳岛东码头	贝壳	AMS^{14}C	36135 a±185 a(距 1950 年)	樊祺诚等,2006

地貌组合特点反映出三个阶段的火山活动效应:早-中更新世(1.42 Ma~0.49 MaB.P.)火山喷发岩浆溢流,形成玄武岩台地与盾形火山丘;晚更新世(36 ka~33 kaB.P.)[20],气流爆发式喷射,形成低缓火山口与激浪堆积,呈现出已受海水浸淹效应。全新世多期火山喷发形成小火山口。约在 7 000 aB.P.,多处小火山口沉溺水下,部分位于海平面附近而形成小海湾,少量仍出露于陆地为熔岩流岩滩。类似地貌亦见于西部洋浦-峨蔓及龙门一带海岸。

北部湾火山活动与涠洲岛的形成,反映出构造张裂活动具有自海南岛西部火山海岸,又向西至北部湾中部的迁移,在北部湾张裂过程中内部岩浆冲出形成火山喷发。从小火山口与裂隙喷发相组合,火山活动似趋向于停息。尚难以定断涠洲岛火山是死火山。

1.3 活火山

活火山是当代仍具爆发活动的火山。本文例举在 20 世纪中期后,作者考察的活火山,居环太平洋火山带及新生代东西向构造带的火山。

(1) 樱岛火山(Sakurajima)

位于日本列岛南端,以火山、温泉与茂密的森林为特色的鹿儿岛县。县域均覆盖着火山灰,近半地区已成为灰砂台地。

樱岛地理位置在 31°35′N、130°40′E,与阿苏山(Mt. Aso)、三原山(Mt. Mihara)、浅间山(Mt. Asama)组成日本的四大活火山[21]。1914 年前樱岛是一个海岛,3 000 年前开始火山爆发、周期性喷发堆积为火山锥;1940 年喷发的"昭和熔岩"地面少植物[1];1914 年 1 月 12 日,樱岛火山喷溢出数十亿吨岩浆("大正熔岩"),填充了与大隅半岛之间宽 400 m,深约 72 m 的海峡[21],使樱岛与陆地相连,并生长茂密植物,两期熔岩界限明显。樱岛呈现为削平顶尖的圆锥体,底座直径 5 km,面积 77 km^2,山顶有三个喷火口——北岳(Alt. 1 117 m)、中岳(Alt. 1 060 m)、南岳(Alt. 1 040 m)[22],鹿儿岛海湾系沿几个沉陷的火山口所构成，与樱岛火山周缘组成岬湾相间的现代火山海岸。

樱岛火山自 1955 年以来，每年喷发超过 400 次，1978 年 6 月—12 月间火山碎屑与灰尘堆积达 865×10^4 m^3[23]。作者于 1983 年 10 月 21 日—22 日，在樱岛火山考察时，火山频频喷发，相间数分钟，吼声隆隆，山体颤动，黑烟滚滚呈蘑菇云上升弥漫笼罩山顶，火山口喷出火山灰、砂及火山渣，随气体飘落于山体周围和山下(图 11)。在半山观测台摄影时，帽顶、衣领、面、颈皆积下亮黑的火山砂，抖掉，又积下。经长期连续堆积，火山灰掩埋了山坡下的神社牌坊，仅顶部露出。樱岛火山区有 10 000 多居民，似习以为常，或在对岸鹿儿岛工作，或在樱岛从事农耕与观光业。樱岛山麓和海湾沿岸有繁茂的热带植物与多处农田，仅海岸岬角出露玄武岩，海湾顶部发育了玄武岩质黑色沙滩。

图 11　2017 年 4 月 28 日上午 11 时 01 分左右，位于日本鹿儿岛市樱岛的昭和火口喷发

火山喷发频繁，沿山坡与山麓堆积了厚层的火山碎屑，每当暴雨时(30～40 mm/10 min)，火山周围会爆发泥石流，沿山坡冲下，形成灾害，当地人称其为“黑神川”。鹿儿岛大学工学部采用沟道引排疏导，减轻瞬间泥石流突涌灾害。

(2) 长白山天池火山

长白山天池火山地跨中朝边境，是大型的层状堆积的复式火山锥，由中-晚更新世喷发的碱性粗面岩质熔岩与碎屑岩，及少量碱流岩质熔岩和碎屑岩所组成[24]。天池是沿多次喷发的破火山口积水而成火口湖，海拔高度为 2 189 m，湖水面积 9.82 km^2，平均水深 204 m，最深 373 m(图 12)。

由于火口湖位置高，水汽与云雾弥漫，作者两次登顶未见其貌，故选清晰的火山口湖照片以示之。据刘若新、李继泰等人研究，天池火山在历史时期有四次主要喷发[24]：

① 玄武岩，喷发覆盖在晚更新世的粗面岩的不整合面上，厚度约 100 m，为暗紫与暗红色碎屑熔渣状玄武岩、气孔状玄武岩和少量深灰色块状玄武岩，推测其喷发年代为 5 000～1 000 aB. P.(老虎洞期)。

② 火山碎屑岩流，喷发规模大，碎屑岩流分布在天池破火山口四周的沟谷中，堆积层最大厚度 60 m，为粗面岩质或碱流岩的混合碎屑堆积。它覆盖在上述的玄武岩上，时代晚于 1 000～5 000 aB. P.(八卦庙期)。

图 12 长白山天池火口湖(附彩图)

③ 浅色碱性流纹岩类浮岩，是一次大规模喷发，堆积在破火山口周缘、锥体斜坡及火山周围地区，覆盖面积达 8 000 km^2，最大厚度达 70 m，喷发出的火山灰漂浮跨海至日本。据浮岩堆积中夹杂的炭化木定年，时代约为 990±95 aB. P.(天文峰期)。

在浅层浮岩层顶部有薄层黑色-灰黑色浮岩堆积，为后期小规模喷发堆积层。

④ 黑曜岩与凝灰岩互层，沿山坡呈舌状分布，覆盖于浮岩层上。火山喷发时代晚于 990±95 aB. P.(气象站期)。

嗣后，刘若新等[25]根据对一棵约 270 年树龄的炭化木，自边缘至中心进行系统的^{14}C 年代测定值，与高精度树轮校正曲线匹配拟合，获得了长白山天池火山的一次大喷发年代为 AD1215±15a。

金东淳与崔钟燮(1999 年)[26]，根据查阅文献与大量中、韩历史古籍资料中记录有关天池喷发火山粉尘(雨土)、浮岩、火山灰(雨白毛)、森林焚毁之炭灰、炭渣(雨炭、雨灰)等，认为长白山天池火山于 1 000 年前曾发生过大规模喷发，影响达 500 km(开城与汉城)，最远达 1 000 km；天池火山于公元 1018 年、1124 年、1200 年、1265 年、1375 年、1401 年、1573 年、1668 年、1702 年和 1903 年多次喷发，喷发规模较小，是气体喷发并伴随着灰渣与水汽。喷发年代的间隔时段规律不明显，却反映出一次喷发释放后，伴有相当时间的岩浆动力积聚，再爆发。现代，天池火山口仍时有气体沿裂隙喷溢，1991 年 8 月 29 日 12 时 29 分，朝鲜人员观察到天池白云峰有爆炸声，中方在当月观察到地震与喷气活动。迹象表明：历史时期至今，天池火山仍有间断的喷发活动。在树线高度以下山地有茂林繁殖，而火山口周边仍为裸岩，时有气体沿裂隙逸出(图 13)，经常发生有感地震，表明长白山天池

图 13 长白山火口洼地气体逸出

火山仍在活动,或为趋向休眠发展的活火山。需关注火山动向与预防灾害措施。

纵观我国大陆活动火山,具有沿 NE - SW 向断裂发育,并自西向东迁移,基本稳定或趋向死火山发展。

(3) 台湾大屯火山

我国沿海与海岛多具有火山活动地貌。大屯火山群位于台湾岛东北端,20 座火山分布于呈 NE - SW 向延伸的金山与崁脚断层带间的谷地中,在南北长 22 km、东西宽 20 km 的地带内分布着观音山、大屯山、竹子山、七星山、烧焿寮山、内寮山、磺嘴山、南势山与丁火杇山,共九个火山亚群(图 14),分别具有数个层状火山体,或火山锥体,高度从近海平面,陡增至 1 120 m(七星山)。据台湾学者研究[27],大屯火山喷发始于 280 万年前蓬莱运动后期——北吕宋岛弧与亚欧大陆之碰撞;距今 80 万年前,弧陆碰撞缓停,北部地质构造由挤压转为张裂,岩浆沿裂隙上升形成密集的火山活动;喷发约在 20 万年前结束,形成安山岩为主的火山锥与马蹄形火山口。现今,山体内的岩浆活动与气体溢出,形成喷汽、硫气孔与温泉,从北投经硫磺谷、小油坑、马槽、大油坑至金山,沿途可见。以小油坑为例,喷出的气体中 92%为水蒸气,其余为二氧化碳、硫化氢与二氧化硫。大屯火山群地底似有岩浆库,可能再次爆发。

图 14　大屯火山群[28,29](附彩图)

泥火山　在高雄县燕巢乡,地下水夹带火山灰泥浆随天然气沿裂隙上涌喷出,泥浆沿坡流动(图 15a),泥浆干缩后形成的泥火山体(图 15b)。

a

b

图 15　高雄县燕巢乡的泥火山和和泥火山体[29]

台湾岛外围分布着澎湖群岛及一系列小岛；彭佳屿、棉花屿、花瓶屿、基隆屿、龟山岛、绿岛、兰屿等，均为火山喷发形成的基岩岛，目前无火山喷发，但地热资源丰富。火山、地震、温泉、硫磺气喷射等现象，反映出位居大陆与大洋交接带的台湾岛，仍为地质构造活动带，与大陆东部的死火山系列，有质的区别。

(4) 维苏威火山(Mt. Vesuvius)

维苏威火山位于意大利半岛西南部的那不勒斯湾东北方的海湾平原上(图 16)，西部山基却延伸于海湾内，是“欧洲最危险的火山”，海拔 1 281 m，距那不勒斯市东南侧约 11 km。

图 16　耸立于那不勒斯海湾平原的意大利维苏威火山(附彩图)

公元 79 年维苏威火山喷发猛烈，摧毁了拥有 2 万居民的庞贝古城。18 世纪中期，考古学家从数米厚的火山灰中发掘出庞贝古城，古老建筑和姿态各异的尸体都保存完好，反映出火山喷发掩埋古城的迅猛灾难。

界于欧亚板块、印度洋板块和非洲板块边缘，在各板块的漂移和相互撞击挤压下，维苏威火山在更新世晚期(2.5 万年前)爆发形成。曾长期处于沉寂的休眠，自罗马时期至今又爆发过 50 多次。

1980 年火山锥高 1 280 m，而每次大喷发后高度都有很大变化。在维苏威火山的 540 m 高处，耸立着高大半圆形山脊的索马山(Mt. Somma，海拔 1 132 m)，从北面围住火山锥。索马山与火山锥间为巨人谷(Valle del Gigante)。火山锥顶的火山口深约 300 m，直径

600 m，是 1944 年喷发后形成[17]。

公元 79 年大喷发，庞贝和斯塔比伊两城被火山灰和火山砾埋没，赫库兰尼姆城也被泥流掩埋。嗣后，火山多次喷发于公元 203 年、472 年、512 年、787 年、968 年、991 年、999 年、1007 年和 1036 年。其中 512 年的喷发严重，国王 Theodoric the Goth 免除了维苏威火山山麓居民的赋税。几个世纪静止后，于 1631 年发生持续 6 个月且强度逐渐增加的地震，至 12 月 16 日火山大喷发，山坡上很多村庄被毁，约 3 000 人死亡，熔岩流直抵海边，天空昏暗达数日之久。1631 年后火山活动持续不断，火山活动特征变化，具有静止期与喷发期：静止期火山口封闭，喷发期火山口几乎持续张开。1660—1944 年间火山喷发约有 19 个周期，猛烈爆发之后，即由喷发期转入静止期。其间大喷发的年份是 1660 年、1682 年、1694 年、1698 年、1707 年、1737 年、1760 年、1767 年、1779 年、1794 年、1822 年、1834 年、1839 年、1850 年、1855 年、1861 年、1868 年、1872 年、1906 年和 1944 年。每一喷发期为时从 6 个月至 30.75 年不等，静止期从 18 个月至 7.5 年不等[17]。

1845 年在海拔 678m 处建立观测站，20 世纪在不同高度分设观测站，并建一大型实验室和一深隧道从事地震重力测量等项科学研究[17]。

(5) 美国西海岸圣海伦斯火山(Mt. St. Helens)

在所考察的活火山活动中，以日本的樱岛火山频频不断的地震摇晃，山体喷气轰鸣，火山沙尘降雨般地纷落；夏威夷基拉韦厄火山(Kilauea)奔腾升空的热气与火花，黏稠的熔岩流沿坡谷缓缓蠕动，以及海岸崖麓堆积的岩流扇等，深感天威地力之撼。但是最为惊异的是，美国西海岸华盛顿州的圣海伦斯火山的当代喷发，却是火焰地狱再现。

圣海伦斯火山位于美国太平洋沿岸，地理位置于 42°12′N，122°11′W，原海拔高度 2 975 m，属于环太平洋火山带中的玄武岩层状火山。曾经反复多次激烈喷发，最激烈的是，经过 123 年休眠后，于 1980 年 3 月 27 日复醒，历经半年至 9 月，共有 5 次大喷发。5 月 18 日持续 9 小时的火山爆发最为激烈[30]：该次喷发使圣海伦斯山年内降低 400 m(图 17)；喷发期间地震活动频繁；山体以 1.5 m/d 速度膨胀是喷发的前兆；喷发时发生大规模山崩，岩石崩落形成岩屑流，温度 200 ℃～300 ℃，沿坡下流速度约 160 km/h；在山麓为泥

图 17 1980 年圣海伦斯火山大爆发[31]

石流伴以快速强风；喷发在山腰地带形成巨大的火山口（宽度 1 500～3 000 m，深达 3 400 m）；喷出物高达平流层，巨量火山灰漂浮广，3 日间涉及整个美洲，并逐渐波及欧洲及东亚；火山口底部不断形成熔岩圆顶丘，其规模增长，从 190 m×40 m、高 60 m，发展至直径 800 m、高 100 m，以后却遭破坏。该次喷发导致 57 人吸入火山灰窒息而死，导致野生动物死亡。2004 年 10 月，圣海伦斯火山再次轰鸣爆发，烟柱升空达数百米，火山口突出新的熔岩丘顶。

圣海伦斯火山喷发，造成相关范围内海陆环境、城乡人民生命的灾难性变化，同时，也展示了火山地质实况，使人类对自然环境客观的时空演变有了深入认识：

① 圣海伦斯火山休眠了 123 年后，于 1980 年 3 月 27 日突然复活，5 月 18 日剧烈喷发，烟云冲向 20 000 m 高空，火山灰随气流扩散至 4 000 km 以远，在距火山 800 km 处的洒落堆积达 1. 8 cm 厚；火山附近河流被堵改道，道路摧毁；熔岩流引起森林大火，摧毁 595. 7 km^2 的森林；火山热力使山地冰雪融化，形成急流；上升气流在高空凝结为雨降落而暴雨成灾，冲刷火山灰为泥浆洪流，从山上倾泻，毁坏农田、村落；9 个小时火山爆发引发大规模山崩，山峰的海拔高度从 2 950 m 降为 2 550 m，山顶形成一个长 3 km、宽 1. 5 km、深达 125 m 的马蹄形火山口；喷发出的火山灰与碎屑物积达 2. 3 km^2。圣海伦斯火山一次爆发造成 57 人死亡，390 km^2 变为废墟，是 20 世纪地球上最大的火山活动！被认为相当于 100 万年的环境自然蜕变①。

② 1980 年 6 月 12 日 21 点到 24 点三小时内，火山喷出的火山沙尘至山顶 14. 48 km 高处，嗣后形成炽热的、汹涌流动的火山碎屑泥浆沿北坡流下，在山谷中形成超过 100 层的沉积地层。即在几秒钟或几分钟时间内，可形成约 1 m 厚的沉积层①。更新了人们对沉积层堆积速率的认识。

③ 1982 年 3 月 19 日，火山爆发熔化了火山口的积雪，混合了山坡松散物质，形成大量泥流，经历 9 小时后，人们发现泥流在山谷中形成一个排水系统，将泥水汇入太平洋①。灾难性事件形成侵蚀地表的流水系统。

④ 火山爆发后 5 个月，泥浆和火山尘暴在火山口外坡形成两个峡谷型碎屑物排出口，是持续性改变地貌的效应。

⑤ 被火山熔岩流推入湖中的百万棵树木，经过数年浸泡下沉到湖底，却保持着直立形态，被误解为经历百年以上浸淹的木石化的森林，实为一次灾难性火山爆发于几年中形成的①。

圣海伦斯火山爆发过程与效应展示给人类深刻的"教育"，改变着人们对自然环境具常态与灾变不同变化之认识，充实人们对环境变化趋势与突变迅猛之应对。

（6）夏威夷火山

夏威夷群岛位于北太平洋中部、接近北回归线南缘的海域。群岛由 124 个小岛与环礁组成，是近 7 000 万年以来，地壳内热点持久活动导致的火山喷发与造岛，伴随着太平洋板块缓慢地向西北方向移动，热点位置改变，逐渐形成了一列岛链[1]（图 18 和图 19）。现代火山主要集中于夏威夷岛，珊瑚环礁是依附于火山岛的基础上发育，伴随着火山岛的回弹沉降，礁体不断增厚。

---

① 据：http://www.creationism.org/sthelens/MSH1b_7wonders.htm

图 18　夏威夷火山岛链是伴随着板块 NW 向移动，远离热点火山活动渐息而形成

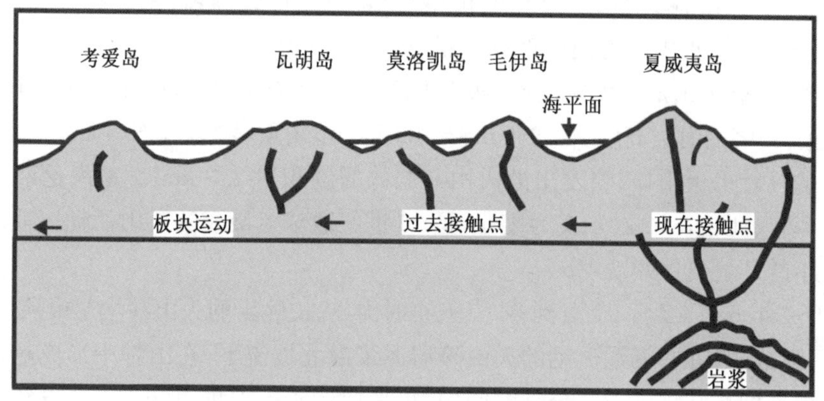

图 19　夏威夷火山岛链形成图式[1]

夏威夷岛上有 5 座火山，其中以冒纳罗亚火山（Mauna Loa）及基拉韦厄火山（Kilauea）最为活跃，具周期性地持续喷发。

① 冒纳罗亚火山（19.475°N，155.608°W）是该岛第一大火山，海拔 4 170 m，若从海平面以下 5 182 m 深处起算，则为 9 352 m，比珠穆朗玛峰还高。该火山的堆积厚度超过 10 000 m，体积大，可知其造陆效应。据 200 年的记录，冒纳罗亚火山喷发过 35 次，山顶保留着一个宽约 2 700 m 的大型火山口莫卡维奥维奥（Mokuaweo-weo），其内有数个小型锅状火山口[22]（图 20），大火山口底为凝结的岩浆与火山渣堆积，并有一些深坑，在山体侧

图 20　冒纳罗亚火山口（附彩图）

面的裂隙中亦有岩浆喷出。

夏威夷国家火山公园自冒纳罗亚山顶的火山口，一直延伸到海边。可以见到火山喷发时形成的硫磺质堆积平原、熔岩隧道、自地面裂隙中喷发含硫的热水蒸气。在活火山的几个亚喷火口中，还有沸腾的熔岩岩浆在翻滚，尚可见到断落的岩层掉进熔浆里，溅起几十米高的火炬。在火山喷发活动强烈时，会从火山口溢出熔融状态的岩浆，沿着山坡向下流，一直流淌到远在几十千米外的太平洋里，发出咆哮的声响，并堆积了熔岩流扇。熔岩流过的地方，房屋树木全被熔岩吞没。岩浆冷却后，便形成山坡上坚硬的熔岩覆盖层，寸草不生。

冒纳罗亚火山喷发约在 40 万年前露出海平面，目前，在当地已探知的古老岩石年龄不超过 20 万年。随着太平洋板块持续地缓慢漂移，冒纳罗亚火山最终会被带离热点，并将在 50 万～100 万年后停止喷发[17]。

② 基拉韦厄火山位于夏威夷岛的东南 19.43°N，155.29°W 处，海拔 1 222 m，山顶有一巨大的破火山口，直径约 4 027 m，深 130 多 m，其中有多个火山喷发口(图 21)。在破火山口西南角还有一个翻腾着炽热熔岩的哈雷茂茂(Halemaumau)火山口(直径 1 000 m，深约 400 m[17]，图 22)。

图 21　夏威夷基拉韦厄火山口之一

图 22　哈雷茂茂火山口

夏威夷火山岩浆为基性，SiO_2 含量小，介于 45%～52%间，铁、镁氧化物含量少于 20%，其温度介于 1 000 ℃～2 000 ℃间或更高，黏滞性小，气体可自岩浆中逸出形成众多气孔，喷发时不形成强烈爆炸，多以涌流状外溢，故其堆积成盾形火山。海底熔岩流则凝结为绳状或枕状结构堆积，遍布于 Kilauea 火山海岸带。岩性主要为拉斑玄武岩，种类多样性区别于典型的海洋玄武岩，可能含 Si、Al 较多，而含 Ca、Fe、Mg 等较少，包括粗面岩、响岩、流纹岩等[32]。

基拉韦厄火山呈间歇性喷发。熔岩浆自 3 300 km 深处的地幔上涌，沿主通道或裂隙中喷溢而出，以东裂隙喷出为主，其次为南裂隙及西南裂隙。位于其西北侧的冒纳罗亚火山，熔岩流沿山地的 NE 裂隙和 SW 裂隙外溢。黑色玄武岩原野崎岖不平，遗留下 1940 年、1942 年、1949 年、1950 年、1954 年、1959 年、1960 年、1969 年、1971 年、1974 年、1975 年、1982 年、1983 年、1984 年、1986 年、1990 年和 1991 年等不同时期喷发的岩层[1]。当火山熔岩从山顶迅速下流过程中，岩流顶部与两侧冷却形成硬壳，而中部岩流仍流动向海，结果形成一中空的穹形隧道，瑟斯顿熔岩隧道(Thurston Lava Tube)是熔岩流塑造的奇迹地貌(图 23)。沿着熔岩层内的空洞和通道可窥见熔岩滚动、气体逸出及冷却时的情景。

图 23　瑟斯顿熔岩隧道

Mauna Loa-Kilauea 复合火山锥体沿太平洋形成长达 100 多 km 的火山海岸：岬湾曲折的海湾由不同时期的岩流舌构成，湾顶已堆积了松软、耀眼的黑砂海滩。海浪冲刷玄武岩海岸形成海穹、海蚀通道和海蚀天窗，其景象蔚为壮丽。热带海洋气候下的火山岩风化形成了新的沃野，繁殖着茂密的椰林、野菠萝、羊齿类植物和灌木花丛以及禽、兽、龟、鱼等，形成新的火山海岛-海岸生态系。夏威夷火山海岛的环境效应是十分突出的。

1991 年 5—6 月，作者考察夏威夷岛南部火山海岸时，处于 1990 年基拉韦厄火山喷发的后期阶段，炽热的岩流沿山坡低洼带缓慢地向海推流，吞没田野，推燃树木和房屋，所至之处一片火海。当熔岩流自海岸悬崖倾泻入海时，形成了耀眼的熔岩流瀑布，初势强，海浪熄不灭岩火，进而形成沸腾的热气柱直冲云霄，烟雾腾腾，数千米外仍散发着浓烈的硫磺气味。

1992 年夏，作者再次考察夏威夷岛南部火山海岸时，基拉韦厄火山岩浆喷发活动已经停息。新凝结的海岸地面呈圆环状的垄岗，崎岖难行。被推斜的残留椰树已形成披挂着岩浆的熔岩树，新凝固的玄武岩表面呈针状和放射状的晶形结构，清晰地保留着岩流推挤、蠕动、卷蚀、覆盖等各种形态结构。曾急步行走意欲赶赴熔岩流瀑布处观看其凝固形态，但在相距海崖约 2 km 处即被制止了，因为熔岩入海激发的浓烈的熔岩水蒸气弥漫于空中，会致人窒息。返程的汽车，每每为熔岩阻断路面而绕行。海岸带火山活动的环境效应、持续的时空范围与陆地火山活动又有明显的差异。

基拉韦厄堪称当代最活跃的海岛火山，火口喷溢与岩流活动延续至今，是可以直接观察火山喷发与岩浆流活动过程的火山，虽岩浆溢流过程中有对地表的破坏，但缓流的速度可以随时、随地跟踪观察，增进人们对火山岩流的地表效应之认识，实独为可贵。其次，基拉韦厄火山曾存在着一个大岩浆湖，面积广达 10 万 m^2，通红炽热的岩浆深达十几米。岩浆湖边缘部分，因冷却常形成暗红色的结壳，结壳如破裂可再次沉入炙热的岩浆中。岩浆湖上还不时出现高几米的岩浆喷泉，喷溅着五彩缤纷的火花，当地称它为“哈里茂茂”，意为“永恒火焰之家”。

1960 年基拉韦厄火山大爆发时，熔岩流从高处奔腾下泻，涌入大海，在海边填造了一块约 2 km^2 的扇形岩流扇。1986 年又一次喷发，为大岛增加了 6.8 km^2 的新陆地。多年来，基拉韦厄火山持续不断涌出的大量岩浆已经在夏威夷岛东南形成几个新的黑沙滩并使岛的面积不断扩大。2002 年 7—8 月间，滚滚岩浆从基拉韦厄火山喷涌而出，流入大海，水火交融，形成壮观的景象。

2014 年 10 月 28 日，夏威夷岛的小镇帕霍瓦受到火山岩浆威胁，居民被迫撤离，缓慢流动的岩浆已经烧毁了该镇的第一栋房屋；11 月 10 日，基拉韦厄火山喷发出的岩浆流吞噬了帕霍瓦镇上的民居与农田。2015 年 5 月 5 日美国地质调查局照片显示，夏威夷基拉韦厄火山的哈里茂茂火山口 5 月 3 日崩塌，激起火山灰和熔岩爆炸，状如烈焰。熔岩湖中的岩浆多次溢出。2016 年 7 月 9 日，基拉韦厄火山喷发后，熔岩流向夏威夷火山国家公园，吞噬森林；9 月，基拉韦厄火山喷发，熔岩湖翻滚喷涌。2017 年 5 月以来，美国夏威夷基拉韦厄火山持续喷发，岩浆从火山口流出，绵延数英里，流入太平洋。冷却凝固的岩浆继续地形成新的陆地。

夏威夷瓦胡岛的 Hanama 湾和 Dimond 岬位于该岛的东南端，是由两个火山口组成的马蹄形海湾，沿湾岸发育了珊瑚礁，形成湾内之小礁湖，碧波轻浪，鱼群戏泳，实乃罕有的海岸憩游海湾。这是热带火山海岸的又一独特环境效应。

岛下的热点及其所形成的火山活动效应，使夏威夷岛具有丰富的地热资源，已建一个地热发电厂利用 360 ℃的地热蒸气发电，其发展潜力可提供现今岛上电力需求量 5 倍的能源。在瓦胡岛，风能提供了全岛 10%的能源。1/3 的电力则由燃烧岛上大量的甘蔗渣所提供[33]，成为利用生物能转化为电力资源的良好实例。这是夏威夷热点型火山海岸的另一环境效应。

(7) 冰岛火山

冰岛紧邻北极圈南侧，界于 63°30′N～66°30′N，13°W～25°W，是大西洋洋中脊上的海洋岛，海拔最高处 2 009 m，但岛屿的总体地势缓平。冰岛地处大西洋裂谷北部中的一个热点之上，有 30 个火山系统贯穿全岛，约有 130 座活火山与死火山，冰岛由大洋火山喷发形成，体现着大西洋裂谷中，火山从喷发活动、休眠及死亡的几个过程。自第四纪时火山喷发延续至当代，反映着大西洋裂谷与洋中脊发育的长期性、持续性。外观总体上，冰岛是北大西洋中的黑色玄武岩质岛，实质上，岩层组成有火山熔岩、火山碎屑岩、凝灰岩、玄武岩与安山岩。火山岩年龄从 300 万年前至今，分别为 0～1 100 a～1 1000 a～0. 8 Ma～3. 3 Ma[34]。大西洋裂谷的地质基础，使冰岛上有多条裂谷与断陷盆地，最主要的裂谷带为北东向延伸，贯穿全岛长达 400 km。冰岛首府具黑色玻璃质的岛陆、众多溢气的裂隙与丰富的地热资源环境奇观(图 24)。部分岛屿覆盖着冰川，冰川或冰融湖之下却有熔岩流动，冰火两重天，是冰岛火山的又一重要特点。

图 24　冰岛火山

冰岛有九大火山(图 25)①。

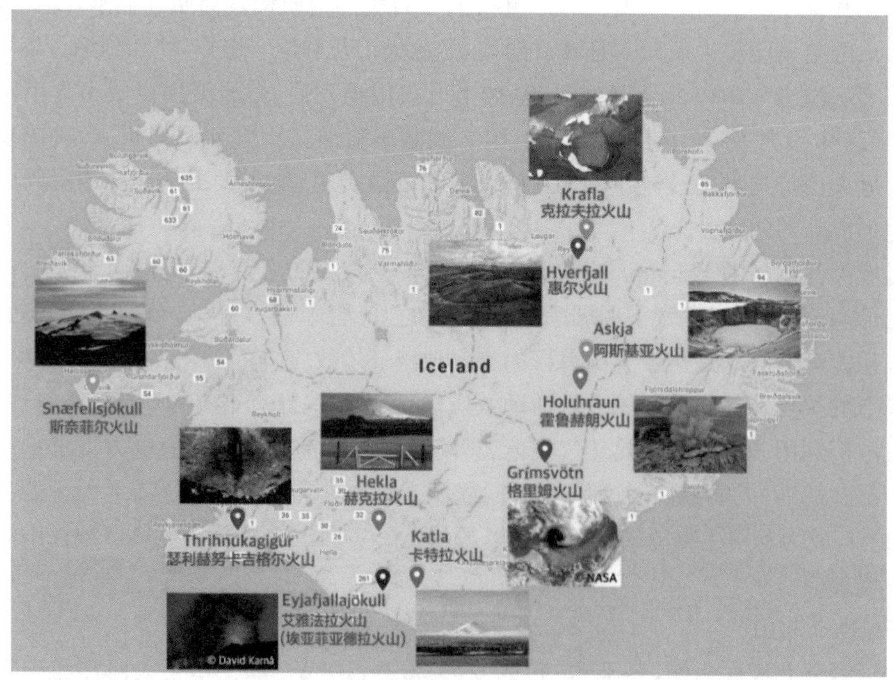

图 25　冰岛最著名的火山(附彩图)

① 据：GUIDE TO ICELAND。http://cn. guidetoiceland. is/nature-info/the-deadliest-volcanoes-in-iceland

① 最著名的艾雅法拉火山(Eyjafjallajökull)埋藏在同名的冰盖之下,曾在1821—1823年间有规模不大的喷发,在2010年喷发规模增大(图26),当火山喷发时,人们群起前往观看,这是冰岛火山区别于其他火山区的特点,从一个侧面反映出火山爆发的裂度与局限性,没有地势高差而形成的快速岩流。

图26 艾雅法拉火山喷发(附彩图)

② 斯里赫尼卡居尔火山(Thrihnukagigur)是座沉睡4 000年的小火山。具有一个4 m×4 m的火山入口与联通的巷道,可乘升降机进入火山内部120～200 m,了解火山内部结构。开辟火山体内旅游观察点,是冰岛的特色(图27)。

图27 深入火山内部的通道

③ 拉基火山(Laki)属格里姆火山(Grímsvötn)位于格里姆湖群的下面,而湖群位于冰川之下。1783—1784年拉基火山爆发时,迅速融化为冰层,以巨大爆破力,喷发大量火山灰,有毒的气体造成9 350人与大量牲畜死亡。

④ 赫克拉火山(Hekla)为冰岛火山活动最频繁的,喷发尚无规律可预测,人类于公元

874 年在冰岛定居以后，每 9～121 年就有喷发。1104 年喷发射出百万吨火山灰，波及全岛。1300—1301 年喷发使 Skagafjörður 峡湾和 Fljót 镇造成 500 人死亡。1341 年爆发使牲畜大量死亡。1693 年爆发破坏最强，火山灰形成火山泥流与地震海啸毁坏了许多农场。1845 年突发使火山灰覆盖了整个冰岛，有毒的灰尘引起牲畜大量死亡。最近在 2000 年 2 月 26 日爆发，破坏较小。赫克拉火山以频繁喷发，带毒火山灰破坏性强。

⑤ 卡特拉火山(Katla)位于 Mýrdalsjökull 冰川之下，公元 934 年爆发时引起冰川河流洪灾，破坏农场与房屋。1918 年、1974 年间，有多次爆发引起冰川洪灾。

⑥ 斯奈菲尔火山(Snæfellsjökull)，埋藏于同名的冰川之下，为复式层状火山，喷发形成熔岩台地。

⑦ 阿斯基亚火山(Askja)与赫克拉火山、卡特拉火山组成冰岛三大火山，喷发气势强，火山灰有毒害性，1875 年大爆发，火山灰传播至挪威与瑞典。Askja 意为火山口，已积水为火口湖，面积 12 km^2，湖深 220 m，(图 28)，临近有温泉湖。

图 28　阿斯基亚火山口湖

⑧ 克拉夫拉火山(Krafla)最高处海拔 818 m，火山口深 2 000 m，火口湖直径 10 km，为冷水湖，1975—1984 年间曾喷发 29 次。

⑨ 惠尔火山(Hverfjall，Hverfell)为一死火山口(图 29)，已经历 4 500 年没爆发。

图 29　惠尔死火山口(附彩图)

冰岛是大西洋洋中脊上最活跃的活火山岛，并以近半数的火山位于现代冰川或湖泊之下为特点。具有裂隙喷发与火口喷发的多种形式，火山爆发既形成泥流与毒害气体喷溢之灾，也提供了丰富的地热资源，为能源利用于发电、取暖、医疗及暖房种植提供了优异条件，冰川覆盖下之火山更为海岛提供了淡水之源。环境结合的特点使高纬度大西洋的玄武岩海岛成为人居的"天堂"。火山作用过程与环境资源的利弊效应，均值得加以重视进行深入的研究。

2　结语

汇集分析所经历的火山地貌，形成几点认识：

(1) 火山爆发于地壳减薄、地质构造脆弱的断裂地带。火山爆发释放了聚集于地壳下之地幔能量，形成地表一定范围内环境改变与爆发时所形成的突发灾害。能量释放却避免了因能量持续集聚所形成大范围地壳破坏的潜在危害。

(2) 火山爆发具有地带性特点。

① 当代火山频频爆发于大洋洋中脊裂谷，该处是地幔物质溢出地壳的主要通道。火山岩浆喷溢以玄武岩类为主，爆发强度与岩流漫溢速度相对缓和，形成平顶的层状火山堆积与大洋中的火山锥岛。冰岛是这类火山活动的典型，提供研究火山地质环境的天然实验基地。

② 现代火山亦爆发于板块边缘与板块交界地带。

太平洋两侧：太平洋与欧亚大陆交界带（以 NE－SW 向构造为主体，日本列岛火山、中国台湾与澎湖列岛火山活动，菲律宾与东南亚诸岛火山活动等）；太平洋与美洲间板块边界（北美圣海伦斯火山、墨西哥、南美秘鲁、智利等，NW－SE 向，S-N 向构造带火山等）；太平洋中部群岛的多火山活动等，表明太平洋火山活跃是特点之一。

极地与大陆板块间，极地与大洋板块间，欧亚板块与印度板块之间，南北美板块之间，以 E－W 向的构造带为特征，喜马拉雅山脉、古巴、巴拿马群岛等具代表性。

这类爆发的间歇期长，是集聚了相当能量后的爆发，冲击力强；喷发物既有广泛散布的火山气体与大量飞溅的火山碎屑，更以炽热的熔岩岩浆流为主体，其活动速度快，改变火山区地貌突出，伴以频繁的地震活动；火山爆发造成地表山崩地裂，岩层垮塌，危害巨大。日本樱岛火山、中国台湾大屯火山及北美圣海伦斯火山是这类火山活动与环境效益的天然课堂。

③ 火山爆发于板块内岩浆聚集的"热点"——估计为老的裂谷带之残留？伴随着板块漂移，而形成沿板块移动方向的系列火山。

(3) 火山爆发形成新的陆地，是造陆的主要过程，不同时期火山爆发，形成不同的山脉、高地与岛陆，伴随着陆地形成、稳定，而活动火山，趋向于休眠与死亡。

(4) 火山活动带具有迁移性。自洋中脊断裂谷，或自板块内巨大断裂带向两侧迁移，即从爆发的活火山向两侧推移为间歇爆发火山→休眠火山→死亡火山迁移。另一种转移是随板块移动而渐离地壳下热点，火山爆发渐停，后续移至原热点之上的板块部分又形成新的火山爆发，嗣后，随板块移动，再离开"热点"后，火山爆发逐渐停息。如此因板块移动，从热点位置所在、逐渐移开而远离热点，因形成一系列沿板块移动轨迹的火山群岛。现代火山是位居热点上部的。

(5) 火山爆发宣泄了地球内部炽热岩浆所聚集的热能，消力、减压、避免持续积累的广泛性灾害。地热可为能源发电，地热改变了局部区域环境，繁殖生物、改变生态系与人类生存活动方式，其利亦巨大。

(6) 关注与监测活动火山，发出灾害预警与实施减灾措施。

火山爆发是地内热量之释放，虽爆发时间间距不同，但迟早爆发不可避免。必须实施火山分布带的监测与发布预警信息。建议重点关注：

① 大洋洋中脊地带；

② 大陆边缘、洋陆板块交界带；

③ 板块内部脆弱地带，尤其是东西向的喜马拉雅山与地中海式构造带；

④ 极地海洋地带，南极洲周缘宜重点观测，重力、冰流融化速度改变等。

参考文献

[1] 王颖，张永战. 火山海岸与环境反馈——以海岛火山海岸为例[J]. 第四纪研究，1997，17(4)：333－343.

[2] 王颖，张永战，牛战胜. 全球火山地震活动与海岸海洋环境效应[C]. //卢演俦，高维明，陈国星，陈杰主编. 新构造与环境. 北京：地震出版社，2001：349－355.

[3] National Geophysical Data Center / World Data Service（NGDC/WDS）. Significant Volcanic Eruptions Database. National Geophysical Data Center，NOAA. doi：10.7289/V5JW8BSH.

[4] 安卫平，苏宗正. 山西大同火山地貌[J]. 山西地震，2008，(1)：1－5，9.

[5] 刘嘉麒. 中国火山[M]. 北京：科学出版社，1999：13－52.

[6] 陈文寄，李大明，戴橦谟，等. 大同第四纪玄武岩的 K-Ar 年龄及过剩氩[C]. //刘若新主编. 中国新生代火山岩年代学与地球化学. 北京：地震出版社，1992：81－92.

[7] 王慧芬，杨学昌，朱炳泉，等. 中国东部新生代火山岩 K-Ar 年代学及其演化[J]. 地球化学，1988，(1)：1－12.

[8] 钱迈平. 江苏六合桂子山石柱林[J]. 江苏地质(现更名为：地质学刊)，1996，(3)：192.

[9] 夏邦栋主编. 宁苏杭地区地质认识实习指南[M]. 南京：南京大学出版社，1986：135.

[10] 黄镇国，蔡福祥，韩中元，等. 雷琼第四纪火山[M]. 北京：科学出版社，1993：1－17.

[11] 王海雪，蒙传雄. 琼北火山，用一夕生命抒写万年传奇. 中国国家地理，2013，(2)：74－83.

[12] 朱炳泉，王慧芬. 雷琼地区 MORB-OIB 过渡型地幔源火山作用的 Nd-Sr-Pb 同位素证据[J]. 地球化学，1989.(3)：193－201.

[13] 冯国荣. 华南沿海新生代玄武岩基本特征及其与构造环境的关系[J]. 中山大学学报论丛(自然科学)，1992，(1)：93－103.

[14] 葛同明，陈文寄，徐行，等. 雷琼地区第四纪地磁极性年表——火山岩钾-氩年龄及古地磁学证据[J]. 地球物理学报，1989，32(5)：550－558.

[15] 孙嘉诗. 南海北部及广东沿海新生代火山活动[J]. 海洋地质与第四纪地质，1991，11(3)：45－65.

[16] Nonnotte P，Hervé G，Bernard L，etc. New K-Ar age determinations of Kilimanjaro volcano in the North Tanzanian diverging rift，East Africa[J]. *Journal of Volcanology and Geothermal*

Research, 2008, 173 (1-2): 99-112.

[17] 齐浩然编著. 经典科学系列高温下的火山[M]. 北京:金盾出版社,2015:17-29.

[18] 王颖,周旅复. 海南岛西北部火山海岸的研究[J]. 地理学报,1990,45(3):321-330,图版Ⅰ、Ⅱ.

[19] 广西壮族自治区地矿局. 广西北海市 1:5×10^4 区域综合地质调查报告[R]. 1990.

[20] 樊祺诚,孙谦,龙安明,等. 北部湾涠洲岛及斜阳岛火山地质与喷发历史研究[J]. 岩石学报,2006,22(6):1529-1537.

[21] 朱宝天. 活火山胜地樱岛[J]. 航海,1987,(3):41.

[22] 探索发现丛书编委会编. 闻名世界的壮观火山[M]. 成都:四川科学技术出版社,2013,95-114.

[23] 王中刚. 访日归来话见闻[J]. 地质地球化学(现更名为:地球与环境),1981,(6):57-60.

[24] 刘若新,李继泰,魏海泉,等. 长白山天池——一座具潜在喷发危险的近代火山[J]. 地球物理学报,1992,35(5):661-665.

[25] 刘若新,仇士华,蔡莲珍,等. 长白山天池火山最近一次大喷发年代研究及其意义[J]. 中国科学(D辑),1997,27(5):437-441.

[26] 金东淳,崔钟燮. 长白山天池火山喷发历史文献记载的考究[J]. 地质论评,1999,45(增刊):304-307.

[27] 王鑫. 台湾的特殊地景——北台湾[M]. 台北县新店市:远足文化,2004:102-127.

[28] 王执明等撰. 台湾土地故事[M]. 台北市:大地地理出版事业股份有限公司,2000:16-27.

[29] 林俊全,齐士峥,刘莹三,等. 台湾的地景百选 1[M]. 台北市:行政院农业委员会林务局,台湾大学地理环境资源学系,2010:108-109.

[30] 泽田可洋. 圣海伦斯火山的大喷发[J]. 李福田译,韩贞镐校. 地质地球化学(更名为地球与环境),1982,(8):25-30.

[31] 李念华. 圣海伦斯火山爆发[J]. 大自然探索,2005,(1):8-19.

[32] Kennett J. Marine Geology[M]. Englewood Cliffs: Prentice-Hall, Inc., 1982.

[33] Montgomery C W. Environmental Geology. Third Edition [M]. Dubuque: Wm. C Brown Publishers, 1992: 88-113, 319.

[34] 易荣龙. 冰岛火山地质[J]. 石油与天然气地质,2010,31(4):131.

Island Volcanic Activity and Coastal Environment Responses
—Example From Comparitive Studies*

The volcanic eruption during Holocene of Hainan Island, China has elevated the coastal bays to be terraces, blocked the river mouths, forced river channels migration and elevated the abundant river mouths to be deep erode-meandered bays, and volcanic rocky coast offered the basements for coral reef development. The volcanic activities have changed coastal landforms, slops, dynamic conditions, sedimentary characteristics and the trends of coastal evolution. The development of volcanic coast has shown a multi-affection of both inner and outer dynamic processes of tectonic, marine hydro-dynamics and biological factors. Comparatively, Hawaii volcanic coast is characterized by a volcanic-island chain formed by the lateral movement of the oceanic plate moving over a stationary hot spot. The ages of the islands increase toward the left, and new islands will continue to form over the hot spot. Recently, lava flows of Kilauea eruption continue the island-building processes, produce a barren volcanic landscape of creator bay, lagoon etc. , that served as a foundation for coral reefs and lives: plants, animals and humans. Also, the eruption has changed the local atmospheric environment. Plenty precipitation of the tropic islands develops fertilized soil, which produces favorable condition for luxury forests and bio-diversity. The Pacific volcanic islands are the precious location of tropic biological resources and ecological environment. Volcanic debris daily erupted by active volcano of Sakura Island, Japan have buried the parts of island, highways, houses, temples, and have caused muddy flow disasters often, especially during thunder storm periods, as a result, they have changed the dynamic processes. What is more important, besides the change of coastal geomorphology, dynamic processes and the trend of coastal evolution, is that the volcanic activities might lead to a series of severe disasters.

* Ying Wang, Yongzhan Zhang: The International Association of Geomorphologists (IAG), The Brazilian Geomorphological Union (UGB), *Regional Conference on Geomorphology*, pp. 49 – 50, Hotel Glória, Rio de Janeiro—Brazil, July 17 – 22, 1999.

Some Global Issues of Mud Sedimentation on Tidal Flats*

Tidal flats are belatedly receiving scientific attention on a global scale. SCOR Working Group 106 was established to review the oceanographic, geomorphic and sedimentary dynamics of muddy coasts, assess the effects of sea level change on muddy coasts including on salt marsh and mangrove forest, and to recommend necessary research and data collection for monitoring and future management of muddy coast.

Muddy coast typically, but not exclusively, occurs as extensive intertidal flat deposits. SCOR Working Group 106 highlighted that many fundamental processes of muddy coast formation, ecosystems and evolution require better understanding of the fundamental physical and biological processes—especially if one attempts to assess the effects of relative sea level change.

Some essential research issues relating muddy tidal flat sedimentation identified by SCOR WG106 included the role of biogeochemical processes in the sediment dynamics of muddy coasts, the accurate predictive modeling of sediment transport on tidal flats, and the role of turbidity maxima, fluid mud, and flocculation, in configuring bottom profiles under hydrodynamic forcing of tides and waves. A related issue is the morpho-sedimentary dynamic link between sub-tidal sedimentary processes and the intertidal flats in different environments.

Particular forcing processes identified as requiring further research included the role of abnormal tides, impacts of storm surge, and saline water intrusion in adjacent low lying lands. To this should be added the effects of tsunamis on tidal flats. These processes, as well as small scale sea level changes, should be monitored to determine the time frames required for morphodynamic adjustment of tidal flats.

There are several issues in the exploitation development and management of muddy tidal flats. Theses include designing strategies for the conservation, planning, and integrated coastal management of muddy coast tidal flats, and identifying impacts of significance from global change processes. Monitoring of human impacts opens a wide and important area for research, including quantifying impacts from disturbing vulnerable benthic ecosystems; mangrove forest degradation and removal; marsh grass

* Terry Healy, Ying Wang: *INTERNATIONAL CONFERENCE ON TIDAL DYNAMICS AND ENVIRONMENT—ABSTRACT OF CONFERENCE PAPERS*, pp. B - 21, August 8 - 13, 2002, Hangzhou, China.

degradation and removal; adjacent catchment land clearing and accelerated sediment transport onto the tidal flats; impacts from transmission of industrial pollutants to muddy tidal flats; effects of coastal developments; maintaining navigation channels; and dredging and sediment disposal sites.

北美 Sedona 红岩地貌
——与中国丹霞地貌之相关比较*

Sedona—红砂岩、灰岩地貌之乡位于 Colorado 高原西南部界于 Flagstaff 及 Phoenix 之间，属于 Colorado 高原与 Phoenix 盆地—山脉之间的过渡带[1]（图 1①）。

Sedona 砂岩呈红色，是由于氧化铁之锈色涂染砂中之石英颗粒而致，红色是铁质氧化后所成。铁质氧化亦可呈现为棕色、黄色、黑色，甚至绿色。Sedona 砂岩之源为古老 Rocky Mountain 的花岗岩，花岗岩中通常含铁质，当铁质受氧化与水的作用，形成红色氧化铁，成为很稀薄的粉红色层涂在砂粒中的石英表面，经过数百万年的作用，松散的砂粒被紧压胶结成砂岩，Sedona 区的石灰岩通常是浅灰色的，但是在 Sedona 上层的 Apache 灰岩与 Sedona 下层的 Redwall 灰岩是红色的，是由于红砂岩受淋溶后，氧化铁从砂岩中渗出，而浸染了其下层[2]。

Sedona 红砂岩是在 280 Ma～275 Ma 前（相当于二叠纪及三叠纪时），始沉积于 3 600 m～4 000 m 深海底的海相沉积。随着地质时代之发展，形成了一系列砂岩与灰岩的系列约 4 000 m 厚（图 2）。由于在 8 000 万～6 000 万年前，北美板块与南太平洋板块碰撞形成 Rocky Mountain，抬升约 5 000 m。Sedona 等沉积被抬升为 Colorado 高原的一部分，但仍保持着沉积岩层的水平层理（图 3），嗣后，受风化剥蚀与相伴的河流侵蚀，形成今日奇特的红砂岩地貌：悬崖陡壁的山地，塔状地貌、各种山峰、峡谷、溪流、草地与生物群落（图 4～图 13），以及早期人类于红砂岩乡的史迹（图 14）。这个例证，藉以说明红岩地貌—Sedona 与丹霞，在岩层形成时代、地貌特征以及生物组合方面均具有全球的相关性，与气候以及构造活动大背景有关。中国丹霞地貌以其范围特征，所代表的沉积、成岩、气候环境与地貌发育均有特色，为红岩地貌的东亚地区之代表，可与北美洲及非洲红岩地貌对比，以推动区域、洲际及全球环境变化过程与效应之研究，对人与自然和谐相处发展，具有重要意义。中国丹霞红岩地貌应列入世界自然遗产。

参考文献

[1] Kathleen Bryant. 2002. Sedona Treasure of the Southwest. Northland Publishing.

[2] Mike Ward. 2007. So, why are the Rocks Red? A Guide to the Magnificent Geology Surrounding Sedona. Nueva Science Press.

* 王颖，朱大奎，傅光翮：《第一届丹霞地貌国际学术讨论会 世界的丹霞（第二卷）》，第 46 - 47 页，广东丹霞山，2009. 5. 26 - 28.

① 图片见后文英译文，后同。

附(英文版):

Sedona-Red Rock Country of the Southwest Colorado Plateau in U. S. A. , for Comparison

The red sandstone-limestone geomorphological Country of Sedona is located in the Southwest part of Colorado Plateau, as a transition zone between Flagstaff hill land and the basin and range area of Phoenix. Red Rock Country, approximately 500 square miles of canyon, colorful buttes, speres, cliffs and sparkling springs and creeks[1]. (Fig. 1)

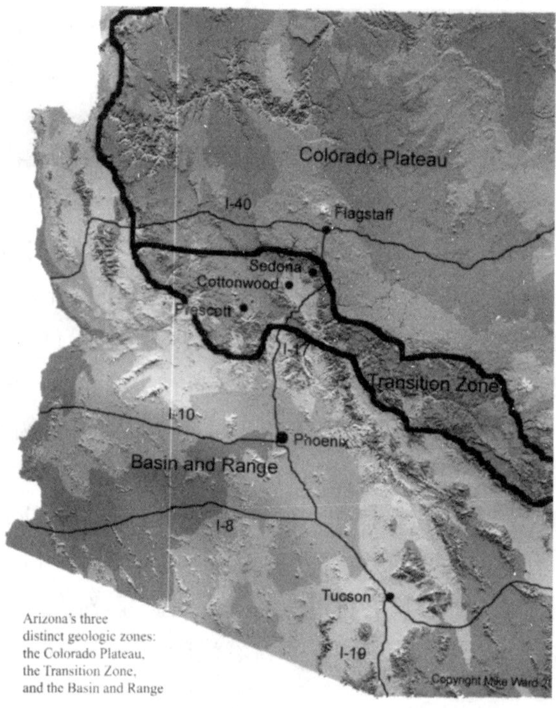

Fig. 1

The red colour in Sedona's sandstone is due to rust, a very thin layer of oxidized iron coats each grain of quartz sand in the sandstone formation around Sedona[2]. This iron oxide is a powerful coloring pigment, just a very small amount can produce vivid colors in soils and rocks. Sedona's sandstone originates from the breakdown of the granite rocks of the ancestral Rocky Mountains. Granite is comprised of minerals that contain some iron, while the iron—bearing minerals react with oxygen and water, red iron oxide is created and forms a very thin, paint-like coating on the quartz sand grains. Over millions of years, the loose sand grains are compressed and cemented into the sandstone.

The area's limestones are normally a light gray colored rock. The pink to red coloration of the Fort Apache Limestone above Sedona and the RedWall Limestone beneath Sedona is a result of the red iron oxide in the sandstone layers above it. Water leaches the iron oxide out from the sandstone formations. As the iron oxide rich water passes over the exposed limestone, the dissolved iron oxide stains the rock red[2].

Sedona red sandstone and limestone was formed between 250～270 million years ago (the time period of Permian and Triassic). It was deposited in 3 600～4 000 m depth of deep ocean. Originated a marine deposits, following the geological development to form a series of sandstone and limestone formations, about 4 000 m thick (Fig. 2). While the North American Plate collided with the South Pacific Plate during 80～60 million years ago, elevated 5 000 m to form the Rocky Mountain. Sedona's sediments were elevated to be part of the Colorado Plateau, but still retains the level beddings of sedimentary rocks (Fig. 3). During hundred of millions of years, it took a succession of ancient seas, deserts, rivers, lava flows, and shifts in the earth surface, under all of processes of weathering, denudation, erosion, biological and human activities, as a result to form the splendid landforms of cliff surrounded hills, buttes, spires peaks, canyon creeks and colorful grassland (Fig. 4～13), biological settlement and ancient human living remnants of alcoves, paintings in the sandstone beddings (Fig. 14).

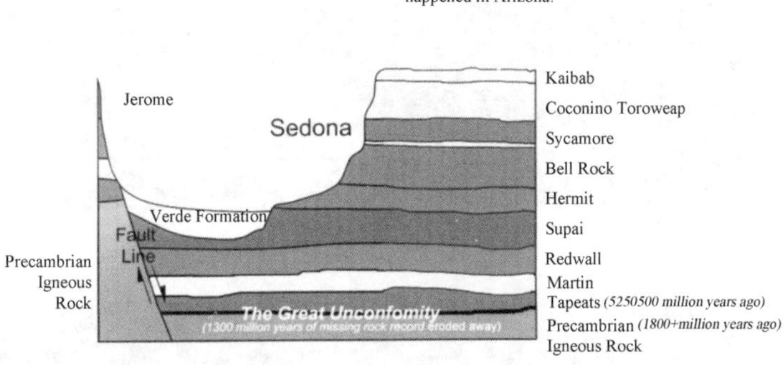

Fig. 2

Fig. 3

Fig. 4

Fig. 5

Fig. 6

Fig. 7

Fig. 8

Fig. 9

Fig. 10

Fig. 11

Fig. 12

Fig. 13

Fig. 14

Here, I would like to emphasize, that the landform of Sedona can be compared with Danxia landform in China, as the similarities of bedrock formation, geological time period, geomorphological characteristics, dynamic processes, and biological settlement, all indicates a global environmental relations of climate change and tectonic activities. The geomorphological characteristics, distribution zones of Danxia has its specific meaning from Asia can be compared with North American, Africa and others, for study the regional, continental and global environment change history and its impact result. It will improve the understanding and to find way out for settling up a harmonic relations between natural and human beings. Thus, the Danxia land forms should be ranged as one of the World Natural Heritages.

第六篇
学科发展与教学科普

总结集成海岸海洋调查研究方法，以实践探索、积累真知，建设海洋地球科学系列："海岸动力地貌学"；海洋地质学当代发展；海洋地理学发展；"海岸海洋科学"。

关于加强地球系统科学教育、发展海洋科学与海洋科学教育的论述与建议。

谈谈海岸动力地貌学*

海岸动力地貌学是地貌科学中一门新兴的学科，这门学科的迅速发展是由于它与工程建设和开采矿产等生产实践有着密切的联系。在工程方面，选择和改建港口，开凿与维护航道，保护海岸免受冲刷以及建立潮汐电站等工程中都需要研究该海岸地段的动力地貌。在矿产方面，主要是寻找海滨砂矿及各种化学沉淀矿。因为海滨地带在波浪、海流等动力作用下，可使一些稀有元素富集成极有价值的砂矿床，这些砂矿或沉淀矿床是在海岸地貌形成过程中堆积的，因此必须按地貌分布与发育规律去寻找和开采它们。这两方面的任务都与当前国家经济建设与国防建设有密切的联系。此外，在滨海地带进行围垦，改良滨海盐碱地，排洩滨海低地的积水，防止潮水、海浪侵入滨海农田等农业经济活动中，都需要很好地研究该地段的海岸动态与发展过程。

海岸动力地貌学研究各种类型海岸的特点、成因、发展规律及其开发利用。因此，它的研究对象是海岸。但是，海岸的科学涵义及其范围还不是很明确的。常有人把海岸仅仅理解为水边线以上的陆地部分。因此，明确地了解海岸的涵义是必要的。

海岸是指地球表面岩石圈与大洋水圈接触交界处，即大陆和海洋相互作用的地带。它的范围包括沿海的陆地，同时也包括陆地延伸到海面以下的斜坡(通称水下岸坡)。这个地带内是受着统一的动力作用(主要是海洋边缘的波浪、潮汐和海流)的影响。这些因素在大陆和海洋交界地带内有着和深海不同的作用特性，它们的动能消耗范围很广，从岸边向海可达数千米至数十千米，即深度相当于 1/3～1/2 波长范围之内，因而在这个地带内地貌变化的过程是相互关联、相互影响的。而水边线以上的陆地部分只是波浪等动力因素动能消耗的边界。水上部分的地貌密切地受着比它范围远为广大的水下岸坡变化的控制。因此，不研究海岸的水下部分就不可能得到关于海岸发展的正确概念。正确地理解海岸是包括沿岸陆地和水下岸坡的统一体。它的上界相当于激浪与涨潮水流经常活跃的地方，在冲蚀性海岸地段以海蚀崖的上缘为起点；在堆积海岸地段则起于海滩的内侧边缘。海岸的下限是止于海洋的浅水区外缘，即相当于 1/3～1/2 波长的深度处，在这个深度以内，沿岸海底都受到波浪的扰动作用。

由于地壳运动及大气环流变化的影响，海面会发生升高和降低，使海岸位置变迁。海面的上升使海岸被淹没在水面以下，而海面下降将使海岸被抬高到海面上，凡经过这样变动的海岸称为古海岸线(图 1)① 。通常将海岸线以上有古今海岸形态的地区称之为海滨地带。

---

* 王颖：中国地理学会，中国科学院地理研究所编，《地理》，1962 年第 3 期，第 99－101 页，科学出版社。

① 不是所有的海岸带皆同时保存着上升或下沉的海岸。

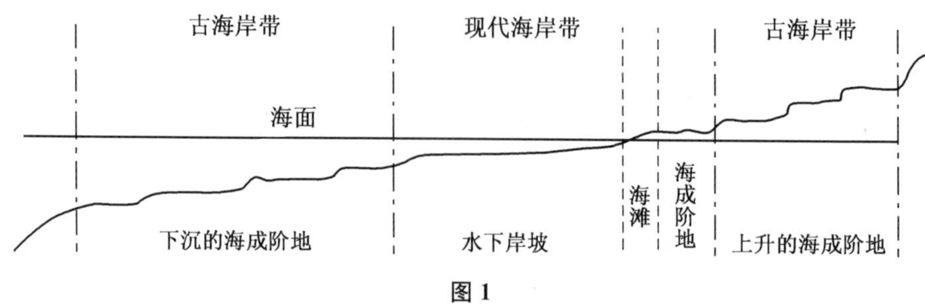

图 1

海岸正位于海陆交互作用之处，因此形成海岸的因素是很复杂的。这些因素可以归纳为两类：① 构成海岸带的物质基础，即分布在海洋边缘的岩石圈所组成的沿岸带陆地，它的特性随着沿岸带地质与地貌结构的差异而变化。② 作用于沿岸带的内外动力因素，它们主要是波浪、潮汐、海流、河流、生物、风等外力以及地壳运动等内力作用。

这两类因素涉及地球表层大气、水、生物及岩石四个圈，在不同地区不同因素的作用下，将形成各种类型的海岸。例如，在热带、亚热带的气候条件下，生物作用活跃，由于一些造礁生物的活动(珊瑚、钙质藻类及一些海绵)，可以形成各种珊瑚礁，而珊瑚礁受到波浪不断地冲刷与溶蚀，会形成具有各种冲蚀和堆积地貌的珊瑚礁海岸。其次，由于某些灌木植物能够长期生长在潮间带上，因而形成一种特殊的灌木林海岸(红树林海岸、棕榈林海岸)，这些植物的生长保护了海岸免遭波浪冲刷，并促使该地成为潮水淤积的地区。

在寒带地方，由于常年低温形成由冰及冻土所组成的海岸，这种海岸地带波浪作用退居次要的地位，而以大气和海水热力融解作用为主，称为热力喀斯特海岸。其他如河流、风力等都会形成独特的海岸地貌。但是上述因素除一些特殊条件下成为海岸发育的主导因素以外，一般地讲它们是次要的因素。

影响海岸发育的主要因素是波浪、潮流等海洋动力作用与海岸带岩石圈。波浪、潮流是最积极最活跃的因素，它们日夜不停地冲击海岸，形成各种海蚀、海积地貌形态，不断地改变着海岸带的面貌。由于波浪、潮流的强度、方向及作用性质的不同，会形成各种特性的海岸，因而海水动力因素在塑造海岸的过程中居于主导地位。显而易见，没有海水的作用，就不会形成海岸地貌。但是沿岸带陆地也积极地影响和改变着波浪作用的效果，濒临不同的陆地地貌会形成不同类型的原始海岸轮廓。例如冰川作用地区会造成峡江式海岸、羊额石式海岸，河流作用地区会形成各种河口式海岸，而且它们进一步发育的途径也是不同的。此外，由于组成海岸的岩石不同，坚硬致密的岩石抗蚀性强，海蚀作用表现得微弱，海岸破坏后退的速度极慢，能长期保持原始海岸状态。而黏土岩组成的海岸耐蚀性差，在波浪、潮流作用下善冲善淤，变化很快。

因此，海水动力与岩石圈形成海岸的作用力和物质基础。它们构成了海岸动力地貌的主要矛盾。研究这对矛盾两方面的斗争和统一，才能深入掌握海岸发育的实质。

基于海岸动力地貌学研究对象的特性，决定了它的研究方法应从三方面着手。

1. 动力分析法

采用水文气象观测分析的方法，研究海岸带的波浪、潮汐、海流、河口水流及风等动力因素。浅海区的水动态与深海区的不一样，一方面它影响改变海岸与海底的地貌形态，另外它也受到海底与海岸地貌的影响，改变了它的作用特性。所以必须观测研究海岸带水动力的特性，才能得到海岸地貌形成变化的原因。如同研究侵蚀地貌时，必须了解河流水文特性是一样的。有些学者分析海岸地貌只是从地质、地形方面着手，而不研究海水活动特性，因而不能找到海岸变化的原因。

进行动力分析时，野外观测工作是很重要的。通常是在海上一定距离内设立垂直海岸的定位观测断面，断面的范围从水边线开始到浅水区的外缘。在断面上的各定点上，利用船只来进行对于该区波浪、潮汐、海流、风以及泥沙活动的观测，来了解该区各种动力因素的状况和特性。苏联采用了一些新的自动记录仪器，如电动波压仪、遥控测流器等，以及用架空吊索观察暴风天气下的海况，来取得长期的、各种天气情况下的、准确的水文资料。这些都是进行这项工作的新方法。

在室内要总结海上的观测资料，进行计算分析与推理。同时，可按小比例尺制作海岸地段的模型，在模型上进行波浪、潮汐与泥沙运动的人工实验，借此分析各种水动力状况下海岸地貌变化的状况，以及更精确地掌握水动力活动特性。

2. 地貌形态分析

海岸的各种形态是自然条件的产物。因此，研究海岸地貌的形态及其表层结构，就可以了解海岸发展、成因及其结构特点。

形态分析的方法尤其在堆积地貌类型中更为重要。例如，沿岸堤是激浪在海滩上的堆积体。它总是与海岸线平行分布的，并且与现代海面相适应的沿岸堤不论它的数量多少，它们的高度大体上相近。因此，可以根据沙堤的分布来了解该处海岸受到波浪的作用，并可确定其最大的影响范围。根据新、老沙堤的位置，判断古海岸线的位置以及岸线发展的状况；根据新、老沙堤的高低变化，可以推论海面升降的状况；根据沙嘴的形状、规模和方向，来分析海岸泥沙的来源、数量、移动方向与泥沙饱和度等问题。还可以根据每种堆积形态的位置，来推论该处波浪作用的主要方向。

形态分析法是研究海岸动力地貌学的重要方法，但并不是唯一方法。为了很好地分析地貌形态，必须多做室内外的观察，并且要应用新技术。在陆地上可以进行直接观察与测量。

对水下地貌的形态分析，依据一般的测深剖面是不够的，这方面的研究，目前以苏联最有成就。在苏联，海岸工作人员直接潜入水下进行观察，用蜡板制图和进行水下摄影，同时应用水下电视机，直接在船上进行观察研究。

为了全面、清楚地掌握海岸形态与表层结构特征，还需要进行航空目测，利用飞机的低空飞行，可以一目了然地观察到全部海岸的概况。为了取得长期的详细资料，可以选取典型的海蚀或海积地段来进行定位观察与多年重复测量。这可帮助我们掌握海岸发展的动态，了解海蚀与海积作用的速度。此外，在室内进行大比例尺的航摄照片判读，也可以深入地分析地貌演变过程。

3. 岩性动态分析法

它主要是研究海岸带泥沙、砾石的运动和堆积规律。海岸带泥沙是很活跃的动力因素，是波浪、潮流冲刷岩石的产物，以及由河流带来的冲积物。在波浪作用下，泥沙不断地迁移运动，在泥沙大量地迁走处形成海蚀地貌（海蚀穴、洞道、海蚀崖、海蚀阶地等）；在泥沙移动过程中，由于动力减弱就形成各种堆积地貌（沙嘴、沙坝、连岛沙洲等）。因此，海岸地貌实质上是波浪与泥沙相互作用的结果。为了掌握海岸的发生与演变过程，必须研究海岸带泥沙的运动和堆积规律。同时，分布在表层的物质是最活跃的，只有它才能反映当前海岸的动态，因而在解决有关海岸的工程问题时，更须加强对表层泥沙的研究。

进行这项工作时，首先是在海岸的水上、水下部分进行采样，然后分析这些泥沙样品的颗粒形态与矿物成分，根据颗粒的磨耗状况分析搬运的远近与搬运的营力，根据其主要矿物成分来追溯物质的来源。例如，在我国渤海湾进行过泥沙样品的岩性分析。根据大量泥沙样品的粒级及其矿物成分与黄土特性相似可以推论它的来源与黄河冲积物有关。此外，在砾石质海岸地段可以用标记砾石来分析海岸带砾石的动态。将砾石染色或用彩色水泥制造人工砾石，制成特殊的标记，投入海中，然后沿岸取样，根据大批砾石中标志砾石的数量及重量来分析砾石运动的路线与磨耗状况。

在研究砂或更细的物质时，采用荧光染胶涂在砂上制成染色砂投入海中，然后用荧光灯来照射沿岸取得的天然砂样，根据荧光砂的含量来分析泥沙的运动路线与距离。

最近，苏、美、英、法、日各国相继使用放射性同位素（铁、钪、磷等）制作出示踪砂来研究泥沙运动的规律。这个方法速度快、结果准确，但应用时有一些技术问题，并且施放同位素示踪砂时应考虑到海水污染的影响。

研究海岸带的沉积地层是研究海岸沉积物的另一方面，从它们的产状结构、层次韵律及包含化石的状况来分析沉积动态与沉积环境。目前，在苏联创造了“震动活塞取样器”，利用它可以在陆上与水下采取厚约 6 m 能保持原状的沉积层，因而提高了分析的精确性。

上述的三种方法是研究海岸动力地貌时最基本的方法，这三方面必须同样的重视，而其他有关的方法也需兼顾。例如：① 历史地理的方法：搜集各时期海岸地带的地图、岸外水下深度的变迁图、历史时期海岸变迁动态、河口位置分布等资料，对这些资料按历史时期排列对比，可为历史时期海岸变化情况及今后变化趋向的推断提供有益的资料；② 考古的方法：比较古代文物埋藏位置与现代岸线的距离，与现代海水面的高差，可以了解海岸与海面的变迁；③ 生物地理的方法：研究古代岸边生物残体分布地点及埋藏的位置（深度），如芦苇、贝壳等埋藏层，可以表示出古岸线的分布。

不论采用哪种研究方法时，都应重视向群众进行调查访问。沿海的渔民、船工等劳动人民最熟悉海边的情况，访问他们可以获得最真实具体的动态资料，用这些资料与其他观测材料相互参证后，有时可以得出很有意义的结论。

为了更好地开展海岸的研究，必须同时进行陆上与水上的工作。在陆上进行海岸地貌与第四纪地质的调查研究，而在水上开展水文、气象观测，水下地貌观测与采样工作。水上工作是比较繁重的，在技术与装备上也比陆上工作复杂一些。可是由于海岸的特殊性，必须进行这种两栖性的工作。

由于海岸动力地貌学具有自己的研究对象、独特的研究方法以及为生产实践服务的部门，因而它已经形成为一门独立的学科，正如 1952 年 4 月全苏海岸委员会会议的决议中所指出：“海岸动力地貌学是一门独立的学科，它是介于地貌学、海洋学和工程技术（港工和工程地质）间的一门边界科学。”它有着为其他科学所不能代替的重大作用。我国海岸动力地貌的研究工作还是在1960年以后开始的。我国地域广大，自然条件复杂，在绵延一万三千多千米的海岸线上，有着丰富多彩的海岸地貌类型，在工作中必须抓住中国海岸的特点，即研究东亚大陆具有潮汐作用、季风作用以及大河作用影响的海岸类型与海岸发育，特别是对基岩质港湾式海岸、粉砂淤泥质平原海岸、中国浅海珊瑚礁海岸与红树林海岸应加强研究。这些研究在解决实际问题与丰富世界海岸理论宝库方面，都将起着巨大的作用。

海岸地貌学现状的初步分析*

一、历史概况

海岸地貌学是研究各种海岸的特点、成因、发展规律与开发利用的科学。很早的时候人们就对海岸产生广泛的兴趣，专门性的研究是从18世纪末开始的。首先进行这方面工作的是港工人员，为了防止海岸冲刷与保护海港建筑，他们注意到海岸带波浪作用与泥沙运动问题，代表性的人物是法国、英国和意大利等国的港工技术人员，如英国的J.斯密顿(J. Smeaton, 1769)提出波浪作用是海岸冲积物移动的主要因素，并首先提出建筑人工海滩保护海岸；法国的J.E.兰白拉德(J.E. Lamblardie, 1789)第一个详细地描述大范围内海岸的动态与地貌(法国诺曼底海岸)，并第一个提出海岸冲积物的移动速度决定于波浪射线与海岸线之间的夹角；19世纪初，F.J.格瑞斯特涅(F.J. Gerstner, 1804)研究了深水波浪的特性，提出了"转迹波"的理论；法国工程师A.R.埃梅(A.R. Emg, 1831)研究了波浪水分子的轨迹及对底质运动的影响；在海岸学说中最具有重要意义和影响的是意大利人P.柯尔纳利亚(P. Cornaglia, 1881)的研究工作，他指出了海岸波浪速度的不对称性，提出了"中立线"这一重要的科学概念，为以后海岸均衡剖面的研究打下了基础。其次，研究海岸的是地质与地理学家，他们注意到：

(1) 地表很多形态是海蚀作用的结果。

(2) 应用现代滨海沉积相分析来进行古海滨相的研究。

这方面的代表性学者有：法国地质学家E.拜蒙特(E. Beonmont)在他的《实用地质学教程》(*Lecons de Geolgie Pratgue*, 1845)一书中，叙述到海岸带的现象，提出了水下沙坝的形成是冲积物横向运动的结果，这个论点至今仍是有意义的。H.达尔文提出了古海岸线的概念，拟定了珊瑚礁的形成过程(1831—1836)，一直到现在，这些理论仍具有科学价值。F.李希霍芬(F. Richthofen)对海岸新构造运动、海岸分类及海蚀作用的学说(1886)。美国地质学家G.K.吉尔伯特(G.K. Gilbert, 1885)进行了美国东部海岸与五大湖湖岸的研究，他提出：必须在研究水动力及海岸堆积地貌中进行详细的形态结构分析，他指出"沿岸流"的影响，但是过分夸大了它对海岸地貌的塑造作用。俄国学者，H.A.苏科洛夫(H.A. Соклов, 1884)在海岸地貌理论上有重要的贡献，他提出波浪移动的特性，在海蚀与海积过程中坡度的意义，海岸带冲积物的来源及沙坝形成过程，并着重提出了海滩和海滨沙丘形成中的互相影响。日后，这些问题曾被苏联学者所证实(В.Г.乌尔斯特[В.Г. Ульст, 1959]，О.К.列昂杰夫[О.К. Леонтьев, 1960])。德国的H.凯勒

* 王颖，朱大奎：《南京大学学报(地理学)》，1963年第1期，第74－82页，照片16张。本文蒙任美锷、杨怀仁两位教授审阅、修改，谨致深切谢意。

(H. Koller, 1881)及地质学家 A. 彭克(A. Penck)对海蚀作用及海岸发育理论方面也有重要的贡献。

20 世纪初,美国的地貌学家对海岸研究有着突出的作用,N. H. 芬涅曼(N. M. Fenneman, 1902)对海岸均衡剖面的发育进行了分析。1909 年戴维斯(W. Davis)根据构造、营力、时间的原理进行了关于海蚀循环的概括,提出了海岸发育阶段的基本图式。后来,他的学生 D. W. 约翰逊(D. W. Johnson)继承了他的工作,专门从事于海岸地貌的研究,并在海岸方面发表了两本重要的著作:

《海滨作用与海滨线发育》——*Shore Processes and Shoreline Development* (1919).

《新英格兰—阿凯第海滨线》——*The New England—Acadian Shoreline* (1926).

约翰逊的《海滨作用与海滨发育》一书是出现于 20 世纪初期的第一本海岸地貌专著,他总结了前人和他自己观察研究的资料,进行了系统的理论概括,奠定了海岸地貌学的基础,标志着海岸地貌学已开始形成为一个独立的学科。但是,由于当时科学水平的限制,从目前看来,其内容尚存在一些重大的缺点:

(1) 偏重于研究海岸带的陆上部分,而对变化最活跃的水下岸坡部分研究不足。前者仅是海岸地带的边缘部分,它的特征取决于水下岸坡地貌变化的结果,若仅研究陆上部分就抓不住海岸发育的内在规律。

(2) 偏重于观察研究地质构造、岩性对海岸地貌的影响,而未从形成海岸的动力(波浪、潮汐、海流与泥沙流)的作用特性来分析海岸的形成和演变,因此不能掌握海岸的成因规律,仅是机械地划分海岸发育阶段,影响了它的理论价值与实际应用。

但是在其后若干年中,几乎没出现过这方面的系统著作,虽然在一些地貌、地质著作中列出海岸部分章节,但基本原理仍是采用 D. W. 约翰逊的,在理论上没有重大的新发展。

二、现代海岸地貌学

现代海岸地貌学是近 30 年来随着军事和经济发展的需要而迅速成长起来的,这个成就主要归于苏联科学家的工作。由于苏联开展大规模的社会主义建设,在扩建旧海港、建设新海港,开辟航道,进行海滨建筑,开辟疗养地、水库护岸与开采海滨矿产中向海岸地貌工作者提出很多新的任务与要求,当时,作为苏联采用的理论基础只有 W. M. 戴维斯和 D. W. 约翰逊的海蚀循环学说,在应用时发现它不能解决实际问题,因为它缺乏对海水动力与泥沙运动的研究,而这些正是海岸地貌变化的关键所在,是对生产部门最有帮助的。这样,生产建设要求产生新的科学体系。由于有辩证唯物主义世界观为指导,针对海岸地貌学的主要矛盾,在苏联研究海岸[①]是从动力、形态及沉积物分析三方面着手的,海岸地貌、地质人员与海洋水文、港工技术人员密切结合在一起,既研究海岸形态,也研究造成这些形态的动力因素,以求得海岸演变的内在规律。在研究地质时期的海岸变迁时,更重视海岸的现代过程及其发展动态,因此,海岸地貌学获得了新的生命力,在解决大量实际问

① 苏联研究水库库岸的工作,是包括在海岸工作计划之内的。

题中蓬蓬勃勃地发展壮大起来，并形成以 В. П. 曾柯维奇（В. П. Зенкович）为首的海岸动力地貌学派。他们拟定了新的工作方法，建立了新的理论。在 1952 年 4 月苏联海岸委员会决议中指出：海岸动力地貌是一门独立的科学，它是介于海洋水文、地貌与港工之间的一门边界科学。苏联的海岸理论成就主要有两点：

（1）海岸是海洋与大陆交界地带，是包括沿岸陆地与水下岸坡这两部分的统一体。它们共同在沿岸浅水波浪，特别是在激浪作用下发展变化的。陆上沿岸只是海岸水动力消耗的边界，沿岸地貌变化是水下岸坡变化的结果，因而要全面了解海岸地貌，必须要重视研究水下岸坡，而这在过去恰恰是个空白地区。

（2）海岸带岩石不仅是波浪冲刷的对象，并且冲刷下来的物质也是海岸堆积的来源，它也消耗着波浪的能力，这些泥沙沿着海岸移动，改变着海岸带的面貌。研究海岸发育也就是研究海蚀与海积作用过程。因此，研究海岸带物质的来源、移动和堆积状况是研究海岸的核心问题。

其次，他们建立了一系列的研究方法：

（1）形态分析法：进行航空目测、航空照片判读；开展大比尺的地貌测量；选择典型的冲刷海岸与堆积海岸段落进行定位观测与多年重复测量；对水下地貌进行潜水观察与水下摄影，并充分应用各种测深资料进行分析等。可以说是进行了海陆空的全面研究，取得对海岸地貌形态的“立体”资料。在这方面工作有卓越成就的，如 О. К. 列昂杰夫在里海海岸的工作，А. С. 约宁（А. С. Ионин）在新地岛，白令海海岸的工作，П. А. 卡普林（П. А. Каплин）对楚克奇沿海的研究。

（2）动力分析法：观测研究海岸带波浪、潮汐及风等水文气象因素，结合动力要素对地貌形成演变进行精确定量的分析。在工作中设置海岸剖面进行系统的水上观测，应用 ВДК 波压计、遥控测流器等自记仪器进行连续测记，架设跨到海上的栈桥以进行在风暴天气下海况的研究。在室内进行相应的计算分析与模拟试验，以掌握动力因素与海岸变化过程间的相互关系，在这方面工作最有成效的是 В. В. 朗金诺夫（В. В. Лонгинов），К. 克纳普斯（К. Канапс），А. М. 日丹诺夫（А. М. Жданов），М. А. 波波夫（М. А. Попов），Е. Н. 耶哥罗夫（Е. Н. Егоров）与 Н. Е. 康德拉采夫（Н. Е. Кондратьев）。

（3）沉积物分析法：研究海岸带泥沙运动与堆积规律，分析泥沙的形态、矿物成分，采用染色砂和标志砾石以及制作放射性示踪砂（Fe^{59}）来研究沉积物来源、搬运途径与搬运方式。利用震动活塞取样器取得保持原层理结构的沉积层，分析其形成环境与相对年龄。这方面工作有成效的主要是，Н. Е. 涅维斯基（Н. Е. Невесский），其次如 А. М. 日丹诺夫在黑海应用标志砾石研究泥沙运动以及 Н. А. 阿伊布拉多夫等（Н. А. Аибулатов）。建筑在上述的各种专门性工作①基础上的综合研究分析是最有成效的方法。必须提出：В. П. 曾柯维奇是全面应用上述方法进行研究、总结理论最具成就的。

通过长期的系统工作，苏联在海岸研究方面取得了巨大的成就，特别是关于海岸带波浪特性与作用过程方面；关于阐明砂质与砾石质的冲积物运动规律方面；关于海岸均衡剖

---

① 在苏联还有专门从事海岸仪器研究的工作人员，研究设计各种仪器及装备，为上面三种工作不断地提供新式的轻便的、有效的工具。

面的塑造、演化与水下岸坡问题；关于水下沙坝的成因与潟湖海岸的发展过程；关于海滩、沙咀、沙堤的成因与动态分析；港湾式海岸演变及堆积型河口等方面。同时，制定了新的海岸分类，完成了世界海岸类型图的编制。总之，苏联在以波浪作用为主及砂砾冲积物活跃的港湾式海岸研究方面有独到的成就，对一些特殊的海岸类型——热力海蚀岸的研究也有所专长，并且海岸的研究与港工结合得最为密切。对已工作过的黑海、亚速海、里海编出了"海岸水册"，正编制白令海与巴伦支海的水册。苏联对于海岸带潮汐作用特性、淤泥质冲积物以及某些热带亚热带海岸类型缺乏系统的研究。目前，开始注意加强远东潮汐海岸的研究。

进行海岸工作的机构有苏联科学院海洋研究所、水文气象仪器研究所、莫斯科大学、全苏交通运输、建筑科学研究所、渔业和军事部门。它们拥有众多的专业研究人员、优良的船只与实验室设备及新技术方法，整个工作由全苏海洋委员会海岸组领导，统筹计划，统筹工作方法与出版论文集，经常刊载海岸文章的杂志有：

(1) 海洋研究所著作集——*Трудш Института Океаногин*.

(2) 海洋委员会著作集——*Трудш Океанографической Коммнсий*.

不定期的出版海岸问题研究专集、专著很多，著名的有：

(1) В. П. 曾科维奇：海岸的动态与地貌——*Динамика и морфология морских берегов* (1946).

(2) В. П. 曾科维奇：海岸发展学说原理——*Основы учение о развитии морских берегов* (1962).

(3) О. К. 列昂杰夫：海岸与海底地貌学——*Геоморфология морских бергов и дна* (1955).

(4) О. К. 列昂杰夫：海岸地貌学原理——*Основых геоморфология морских берегов* (1961).

苏联对海岸地貌研究的成就标志着目前世界的水平，从动力着手研究海岸的特点、成因与演变是目前海岸地貌学发展的主要趋势。

在美国对海岸地貌的研究也是极其重视的，其特点是海洋地貌的研究为军事活动服务、为商业资本服务。在第二次世界大战期间，这方面研究得到迅速的发展，其中，海底地貌的研究(从研究水平、所取得的成就及人力装备来看)是居世界首位，而海岸地貌比较逊色，自 D. W. 约翰逊以后，工作进展不是很突出的。近年来的研究项目主要是：

(1) 研究暴风浪、沿岸激浪流与保护海滩问题。

(2) 港湾海岸的特点与泥沙运动规律。

(3) 各国海岸地貌的特点与登陆(通行与给养)的条件。

(4) 各海盆中海水透明程度、海水层的活动特性和水下地形的情况与潜艇活动及水下发射导弹的可能性。

进行这方面工作的机构主要是：美国海军研究处(简称 O. N. R.)，海滩侵蚀测量局、哥伦比亚大学拉蒙特(Lamont)地质观察台、路易斯安那大学、耶鲁大学、乔治亚大学及其海洋研究所、斯科利普(Scripps)海洋研究所、渥斯候尔(Woods Hole)海洋研究所及渔业交通部门等。

美国海洋方面的实验室有 60 个，其中 6 个是规模较大的大学实验室，19 个是规模较小的大学实验室，6 个较大的和 24 个较小的属于渔业部门，5 个属于海军部门，海岸地貌是他们研究课题之一。1957 年美国科学院和科学研究联合会组成了海洋委员会咨询和奖励有关海洋学方面的研究工作。

海滩侵蚀测量局主要是组织各单位进行东部沿海各州海滩疗养地及五大湖湖岸滩地的保护工作。

O. N. R. 领导与组织了大规模的海岸研究工作，特别是对一些具有战略意义的问题进行研究，如拉丁美洲及非洲的红树林沼泽海岸的通行与供给条件的研究，太平洋珊瑚礁的研究及对各国海岸进行航空摄影研究等。

各个研究机构的分工情况是：

Woods Hole 海洋研究所主要工作地区是大西洋海岸，近年来对美国东南部各州岛屿海岸的成因有所著述(J. M. Zeigler，1959)。

加利福尼亚的 Scripps 海洋研究所，主要研究太平洋沿岸，F. 施巴尔特(F. Shepard)综合地研究了太平洋沿海的大陆棚和大陆坡，根据海岸作用过程进行了海岸分类。

路易斯安那大学主要在密西西比河河口及墨西哥湾进行工作，从 1950 年以来参加了 O. N. R. 计划，在非洲及拉丁美洲热带地区进行海岸研究。R. J. 罗塞尔(Ruosell)对密西西比河三角洲的研究，甚为著名。近年发表的 C. W. 罗伯特(C. W. Robert，1956)关于哥伦比亚太平洋地区的红树林海岸所进行的研究是目前关于这类海岸较为全面的著述。美国第二次海岸地理学会议(1959)即在该校举行。

哥伦比亚 Lamont 地质观察台主要是在大西洋中北部海岸工作，近年又开展太平洋北部海岸与地中海沿岸的工作。

耶鲁大学参加了 O. N. R. 关于太平洋地区的珊瑚礁研究，1959 年发表了其中关于环礁的研究论文。

美国拥有 40 多艘考察船，作为综合考察用，船上设备齐全，具有新式的电磁、无线电仪器，并配有专门的考察飞机。在海岸工作中，从 1955 年以来应用放射性同位素(P^{32}、C^{14})研究沿岸流、泥沙运动及底质年龄。应用新式的海岸摄影机与潜水装备研究水下地貌，与苏联比较，在海岸定位观测及水下岸坡的研究等方面是较差的。

由于海岸研究与军事目的结合，所以重要的文章发表不多，一般文章散见于地质学杂志、地理学杂志及专门的海滨与海滩杂志(*Shore and Beach*)中，在理论著作方面，当推 F. 施巴尔特的《海底地质学》(*Submarine Geology*，1958)及《海洋地质学》(*Marine Geology*，1954)。

英国是一个传统的航海国家，海洋科学研究素具较高的水平，但是，由于老大帝国的腐朽衰落，所以在研究耗费、物资及人力方面都感到贫乏，因而限制了科学的发展。

英国对本土的地质研究是很详尽的，因此，在海岸研究中，海岸地貌与岩性构造的关系是很清楚地得到阐明的。由于英格兰南部海岸由中生代白垩软岩层组成，海蚀破坏使岸线后退很迅速，因此，对冲蚀海岸地貌类型，海岸变化动态与护岸方面研究较好。同时，由于淤泥海岸的河口及三角港内受潮流影响，淤积严重，如泰晤士河口的淤积影响了伦敦港的使用。因此，英国重视对潮汐作用、对淤泥湿地海岸的研究，在这些方面有独到之处。

这些方面的成就在J. A. 司蒂尔斯(J. A. Steers)的《海岸》(*The Sea Coast*，1954)一书中有所反映。

目前,英国重视海岸动力与泥沙运动的研究,即研究海岸动态的方向占优势。近几年,英国采用示踪砂(Sc^{46}，1954)与示踪砾石(Ba^{140}，1957)研究沙咀、沙坝与海滩的动态,同时,开展了对于水下岸坡的研究。诺廷汉姆(Nottingham)大学地理系金氏(C. A. M. King)综合大量实验观察工作,证实了冲积物在波潮作用下移动状况的推论,并且在她的著作《海滩与海岸》(*Beach and Coast*，1959)一书中全面系统地论述了对海岸研究的工作方法,拟定和总结了对陆上与水下进行形态、岩性及动力分析的方法和室内实验、计算分析。这本书可作为英国目前研究水平的代表。

在英国进行海岸地貌研究的机构有国立海洋研究所、利物浦观测台、潮汐研究所、利物浦大学和剑桥大学等单位。

在英联邦国家中,澳大利亚开展了对东北部珊瑚礁海岸及东南部海岸在波浪、潮汐与强烈风力影响下的砂质海滩、沙咀等堆积地貌以及海滨砂矿的开采与富集规律的研究。

新西兰的地理学家C. A. 考顿(C. A. Cotton)对新西兰的原始海岸类型火山岩流岸及冰川作用海岸及海岸分类方面进行了研究,但其水平仍未超过D. W. 约翰逊。

法国对海岸地貌研究开展得比较早,其特点是海岸地貌研究与港口工程密切结合。对海岸动力与泥沙运动(特别是砾石质与砂质冲积物的运动)以及沉积相的分析有着深入的研究,从1955年起有效地使用示踪砂(Cr^{51})研究泥沙运动。海港工程方面的著作丰富,并且有很高的水平。此外,法国很早就注意开发使用潮汐能以及防止海岸沙丘(西北部海岸)内移危害的问题。他们对海岸的研究始终注意到动力与地貌的结合,并且注意开展水下岸坡工作。

法国海洋研究设备不多,但比较精巧实用,使用效果好。例如,著名的加里普斯号(Calypso)考察船。船上有21名专职潜水员,他们都有最新的水下摄影和传真设备,在船首的下部(水面以下2～5 m)有水下观测舱,船的中部有舱口,通过这个舱把潜水员放入水中,此外,船上备有深水沉箱,它的最大沉没深度是300 m,沉箱中可容两个人(观察者及驾驶员),氧气设备可供24 h,电能储备可供6～7 h,沉箱的发动机是喷水式的。加利普斯号考察船是进行综合考察的,它有效地取得了海岸与海底地貌资料。

法国的海岸类型是很丰富的,有港湾式海岸,海侵冰川岸、岛屿岸、沙丘海岸,泥质平原岸、湿地型海岸,各种河口岸及构造海岸等。著名的地理学家Emm. de马东进行了海岸分类,但是缺乏统一分类标准与统一命名方法。

在著作方面,1954年出版了A. 格里舍(A. Guilcher)的《海岸与海底地貌》(*Coast and Submarine Morphology*)一书,他较全面地总结了西方国家在海岸方面的成就(如对珊瑚礁岸、三角港岸、石灰岩海岸及海蚀崖类型等问题的研究),缺点是缺乏动力分析。

德国在地貌学和海洋学的研究上原有比较深厚的基础。第二次世界大战以后,海岸地貌的研究主要集中在波罗的海及北海沿岸的冲刷问题,与工程相结合。如K. 布罗(K. Bulow)的著作《南波罗的海的海岸动力与海岸保护》(Allgemaine kustendinamik und küstenschutz an der Südlichen ostsee zwischen Trave und Swine，1954)结合护岸工程,研究了海岸冲积物的运动,在动力方面比较重视风成增水减水及其对海岸冲刷的影响,并提出了保护海岸的措施。此外，也有综合性的著作，如 H. 瓦伦丁(H. Valantin)的世界海岸一文(Die küsten der Erde，1954)内容论及整个世界的海岸，但因讨论的范围较广，深入分析不够。但其中比较深入地研究了世界大洋水准面的变动及海岸地带性问题。同时，他根据海面变动、地壳升降运动与海蚀、海积过程等因素制定了新的海岸分类。在这分类中强调了海岸的现代化过程，这是一个进步。但是，他混淆了成因因素与演化过程；抓住了海岸变化动态的差异，却未明确其变化的主要原因；列出了各种影响海岸过程的因素，但是缺乏进行系统分类的中心思想。所以，分类仍是形式的与折中的，在理论上与实用上都感到不足。

北欧诸国如挪威、瑞典，很早就从事航海探险活动，著名的探险家南森亦注意对海岸地貌方面的研究。

斯堪地半岛各国地处高纬，第四纪时累经大冰盖的活动，冰川后退迟，冰川遗迹清楚，海水入侵了冰川谷及羊额石地区，形成典型的峡江式海岸与羊额石岛屿港湾岸。此外，沿海发育了很多冰冻磨蚀的海岸平台(Strand flat)。因此，对这些高纬地区的海岸类型有着详细地记述与研究。同时，由于冰后期以来，海面上升，斯堪地半岛亦有地壳隆起，岛屿沿岸保存了较完整的抬升古海岸(古环滨线)，在这些国家海岸地貌的研究是与第四纪地质学的研究密切结合，它们有记录最久的验潮资料，对上新世以来海面变化，海岸新构造运动与古海岸变化研究得很深入。

丹麦从 1930 年起，即计划疏干日德兰半岛西部和北弗里安群岛之间的丹麦瓦特托夫海地段。因而，对该地区进行了 1∶10 000 的地形测量，研究浅海海底地形，海滨湿地、沙丘、湿地上的植物生态与潮流、波浪流动力与搬运悬浮质的关系等方面的问题。从 1953 年在研究对该区疏干的方法时，又进行了详细的地貌研究，特别是对潮流动态、湿地植物与湿地海岸动态方面做了大量的观察、实验与分析工作。

荷兰也是个传统的航海国家。由于第四纪以来，海面上升，沿海平原下沉，因而修建了很多海堤，利用风车抽水来淤造土地，在这些工作中对湿地海岸特点，海岸平原的微地貌及海面上升，海岸相对下沉的动态方面有深入的研究。此外，对太平洋中南部的珊瑚礁、南洋群岛的火山海岸及海底水下谷地等地形进行了系统研究。代表性的著作是 Ph. H. 奎恩(Ph. H. Kuenen)的《海洋地质学》(*Marine Geology*，1950)，以及他近几年来在地质杂志所发表的文章。Ph. H. 奎恩对珊瑚礁特别是环礁的研究是有很大成就的。

在日本，为了解决港口淤积以及由于海底地震或风暴侵袭引起的风暴浪对海岸冲刷方面，展开了一系列的研究工作，海岸地貌与海岸工程密切结合，形成"海岸工学"并成立相应的学术组织。近年来，日本积极从事河口及海岸水文与漂砂问题的研究，广泛应用了新技术如制作放射性示踪砂、示踪砾石(Zn^{65}，Co^{60}，1941、1957、1960、1961 年)来研究海岸泥沙运动。目前总的趋势是重视海岸动力与泥沙运动的研究，但是在海岸发育、分类等理论问题上，基本上仍袭用 D. 约翰逊的观点。

日本从事海岸工作的机构有海岸工学会、北海道大学、德岛大学、中央气象台、水文、农业及土木建筑学会等研究机构。

我国专门性大规模的海岸地貌工作开始得较晚①，主要是1958年后，在苏联专家B. П. 曾科维奇、O. K. 列昂杰夫与H. E. 涅维斯基等人帮助下开展起来的，进行的时间虽然还不长，但是仍具有自己的特色。

我国海岸线很长，地质构造复杂，海岸带动力因素较为特殊（受季风、潮汐、河流的影响很大），海岸类型丰富。有基岩港湾式海岸、粉砂淤泥质平原海岸、珊瑚礁海岸、红树林沼泽海岸、各种河口岸、大规模的断层海岸及各种岛屿。这些海岸的形成发育是不同的，其开发利用也有区别。各种基岩港湾式海岸是与沿岸带地质构造特性及地形特点有关的，它们多发育在长期上升的陆台区与复杂的褶皱构造区所形成的山地或丘陵的边缘，如胶辽两半岛海岸、浙闽沿岸即属此类型。这类海岸地区岸线弯曲，临岸坡陡水深，波浪作用强烈，海湾内水面宽阔、风平浪静，是天然的良港。由于波浪的冲蚀，一些岩浆岩变质岩破坏崩解，有用矿物离解出来由波浪携带堆积逐渐富集成各种稀有元素及贵重金属的砂矿。我国海滨砂矿蕴藏丰富，它们是与基岩海岸的堆积地貌相伴生的，基岩海岸堆积地貌的发育与分布，是普查海滨砂矿的重要前提。

另一类海岸——北方大规模的粉砂淤泥平原海岸（渤海湾沿岸与苏北海岸），它们的形成是与黄河冲积物——大量粉砂淤泥物质——的供给，潮流的搬运堆积有关系的。这类海岸地势低、岸坡缓，沿岸水浅，由于是松散物质组成海岸，抗蚀性低，岸线及水下堆积浅滩善冲善淤，变化极快。这类海岸不会有砂矿，对港工建筑也不利。沿岸低平，潮水内侵范围大。在河道中潮水内侵，不利灌溉，修建挡潮闸后，闸下淤积严重，河道逐年淤积，过水断面减小，在雨季时不能通畅地排涝，促使闸上农田内涝，解决这一系列问题要求深入地研究粉砂淤泥质海岸的特性与动态，掌握它的演变规律。

此外，我国的珊瑚礁海岸、红树林海岸及各类河口岸，都是有特殊的风格，对这些海岸开展调查研究，进行理论上的总结，对丰富世界海岸理论具有重大意义。

三、现代海岸地貌学基本问题

在对海岸地貌学研究现代状况讨论后，可以看出，目前世界各国对海岸地貌学的研究主要集中在以下几方面：

（1）研究海岸带主要的动力因素（波浪、潮汐、海流与河流）及其对海岸的影响；研究在波浪、潮汐、河流作用下泥沙运动状况。

（2）研究海岸带岩性、构造、新构造运动特性及其对海岸发育过程的影响。

（3）研究各种典型海岸的特点、演变规律与开发利用的可能性。

（4）从动力与沉积的角度分析海蚀地貌与海积地貌的成长变化过程。

（5）研究第四纪及现代海面变化与海岸升降问题。

（6）对海岸进行科学分类。

（7）海岸研究方法及新技术的应用问题。

---

①　新中国成立前后，曾有一些地质、地理学家在沿岸陆地部分进行一些研究，但大多未涉及到水下岸坡。

主要参考文献

[1] В. В. Логинов. 1959. Истории изучения динамики береговой зонш. Тр. Океанограф. Комиссии，Т. 4.

[2] О. К. Леонтъев. 1955. Геоморфология морских бергов и дна.

[3] О. К. Леонтъев. 1962. Основы геоморфологии морских берегов.

[4] В. П. Зеикович. 1958. Морфология и динамика советских берегов чвервого моря，Т. 1.

[5] В. П. Зенкович. 1962. Основы учение о разнитии морских берегов.

[6] В. П. 曾科维奇. 1959. 海岸地貌. 商务印书馆.

[7] 江美球编译. 现代海洋学的成就. 北京大学地质地理系(未刊).

[8] П. Джеймс(P. E. James). 1962. Направление географиических исследований в сша после 1954 Г.，Изв. Веес. Геогр. Т. 94，Вып. 1.

[9] H. J. Wiens. 1959. Atoll development and morphology. *Annals of the Association of American Geographers*，49(1)：31 - 54.

[10] J. M. Zeigler. 1959. Origin of the Sea Islands of the Southeastern United States. *Geographical Review*，49(2)：222 - 237.

[11] C. W. Robert. 1956. Mangrove swamps of the Pacific coast of Colombia. *Annals of the Association of American Geographers*，46(1)：98 - 121.

[12] A. Guilcher. 1954. Coastal and submarine morphology .

[13] J. A. Steers. 1954. The sea coast.

[14] D. W. Johnson. 1919. Shore processes and shoreline development. New York：John Wiley & Sons.

海洋地质学的发展现状与趋势*

新技术革命提出了海洋开发问题，海洋开发实质上包括开发海洋资源与利用海洋环境这两个方面的问题。海洋地质学恰在这两方面都可起到重要作用。作为江苏省地质学会的新会员，我愿就此机会谈谈海洋地质学的发展现状与趋势问题。

海洋地质学是属于海洋科学的，海洋科学的特点是具有技术复杂性、综合性和全球性。

首先是技术复杂性，海洋地质研究的是海岸与海底，即被海水覆盖的岩石圈。这与陆上不同，看不见，摸不着，至今人类可登月，拍摄月面照片，采取月岩，但却不能揭开海底帷幕的一角，要通过回声测深与船舷摄影装置来了解海底地形起伏，通过地震剖面测量了解地层结构，通过采样、取沉积孔、深海钻探以取回沉积物与岩芯，通过重、磁测量了解深部构造，所有这些工作必须要用船只，甚至要用深潜器进行，涉及高度的技术与重型装备。

综合性，海洋是个综合复杂体，任何过程都是互相影响的，海底山或起伏处，常会形成上升流，上升流带来丰富的营养盐往往会在该海山、海岛处形成渔场。可以说，在海洋中水圈、岩石圈、生物圈、大气圈都是在这里交汇、相互作用的。综合性强，要求多学科综合，从事海洋地质学工作不仅要具有地质学基础，还要具备有关海洋动力、泥沙运动、生物与化学等方面的知识。加之技术复杂，进行海洋调查花费大，所以，每出海一次，在同一航次中尽可能安排多学科多项目调查以求得最大效益。

全球性，大洋是互相沟通的，如海洋环流、南纬和西风怒吼区，形成波浪海流会影响很远。夏威夷、智利发生海底地震，经 4 个半小时，地震波会越过太平洋形成海啸，对日本造成灾害性的影响。厄尔尼诺现象是海、气交互，海流与生物相关，太平洋东西岸互为影响的典型例子。因此，对海洋现象不能仅仅从一个地区孤立地研究，需要加强国际合作与交流。最后，海洋科学的发展和应用分不开。如海洋水产，各国的海洋学都是从渔捞开始而以海洋生物学发展较早；进一步由于交通航海，需要知道风浪、海洋水文与天气预报，所以海洋物理学自 20 世纪 40 年代后兴起了。而 60 年代，由于对海底矿产的注意，海洋地质学以蓬勃的气势登上海洋科学舞台。

生产活动的需要促使海洋地质学发展，最主要有以下几项。

一、海底矿产资源的开发利用

石油、天然气。据法国石油研究所估计：全球石油储量为 3 000 亿 t，海洋石油为 1 350 亿 t，占总储量的 45%；日本学者估计占 27.5%。今年海洋石油开采量超过 7 亿 t，

* 王颖：(江苏省地质学会编)《江苏省地质学会会讯(江苏省地质学会第四次会员代表大会专辑)》，1985 年第 1 期，第 28－31 页。

占石油总产量的20%。我国的油气资源丰富，目前已知六大沉积盆地（渤海、南黄海、东海、南海珠江口、北部湾、莺歌海盆地）面积达80万km^2，油气丰富。美国总统能源顾问于1982年估计中国近海石油可采储量500亿～700亿桶（70亿～105亿t）。日本估计我近海石油储量约42亿t，天然气100兆m^3。我们自己估计：石油资源的地质储量可能为100亿～300亿t，可采储量42亿～100亿t。世界舆论普遍认为"中国大陆架的海上油田是世界上屈指可数的原油宝库"。根据世界油田的分布特点来分析，油气田主要分布于生储油条件良好的海岸沙坝潟湖地带、海湾地带、三角洲，世界三大油极都分布于三角洲地区，以及离岸珊瑚礁岛、大陆架沉积盆地及大陆坡下的深海扇浊流层中。如：委内瑞拉马拉开波油田为海湾潟湖，美墨西哥湾是三角洲与古珊瑚礁；英国北海大陆架油田与我国南黄海等处为沉积盆地，珠江口是陆架上的古三角洲沉积盆地；加拿大的冰蚀大陆架、基岩裸露，沉积物于大陆架坡麓深水扇中如Grand Bank（大浅滩），为油气田所在，当然该处有复杂的盐楔构造。

南黄海沉积盆地位于江苏岸外，面积达10万km^2，沉积层厚达2 000～3 000 m，下第三系是生油层，层厚100～1 000 m。此外尚有上古生界与三叠系海相生储油层，据估计，南黄海沉积盆地石油储量为2.9亿t。美国估计，黄海的可采石油储量大，低值为2.2亿t，中值为5.7亿t，高值为21.6亿t，仅次于东海沉积盆地而居第二位。这是一项与苏北油田衔接的巨大油气资源。南黄海沉积盆地已由英国石油公司等外商向中国海洋石油总公司承包。自1983年以来，合作勘探达10 595 km^2，外商投资已达11 175万美元。在外商投资的两口油井中，已有一口具有良好的油气显示。南黄海油气资源开发利用会根本改变江苏省能源短缺之现状，改变苏北经济结构，省地质学会有义务对此加以注意。海岸与大陆架主要是石油天然气资源。其次是海滨砂矿。分散于原岩中的稀有元素（锆英石、金红石、独居石）及贵重元素（金、金刚石），经过风化剥蚀，水流特别是波浪分选富积成矿，如山东半岛锆英石砂矿，在有些大沙坝中品位达2 kg/m^3。山东半岛北部的砂金及海南岛的金、红石钛砂矿等。美国在阿拉斯加大陆架开采砂金，对该区地貌、沉积与动力作用进行了详细的研究；印度、澳大利亚开采金红石、独居石砂矿；南非金刚石砂矿促进了河口、海滨及水下峡谷的研究。总之勘探砂矿促进了对海岸与大陆架形成发展过程的研究。

深海矿藏主要是多金属矿瘤，即所谓的锰结核。主要分布于2 000～6 000 m深的海底，如在太平洋6°～20°N，110°～180°W之间有600万km^2。水深3 200～5 900 m的海底，锰结核的富集程度每平方米在10 kg以上。现在世界上有8个跨国财团约100多家公司从事勘探与试采。到1990年锰矿产将进入商业性开采；至2000年，将成为稳定的金属供给源。

二、港口交通

人们交往是从海岸带开始，即从港口与海上交通开始。海岸过去认为是条线，实际上是个地带——两栖地带，除沿岸陆地外，向海延伸到波浪开始扰动泥沙之处开始。即1/2～1/3当地波长处，通常为水深10～20 m。从建港条件看，不是所有的海岸地段都可建港，大连位于天然港湾，水深条件好，同时建港条件优越，而淤泥平原海岸不宜建港。港口要求满足“水深、浪静”两个条件，水深才能保证船只不搁浅，港口是船舶停泊之处，浪要小。在自然界恰是水深处浪大，所以要解决这一对矛盾，要对海岸进行波浪、海流、河口、海岸冲刷、泥沙运动与海岸堆积特点及海岸发展变化趋势等一系列研究，所以港口发展促使了对海岸的研究。同时，海岸带区由于建坝、围垦等一系列工程所造成的关于生态系平衡与生态环境研究。

三、通讯

如铺设海底管道与海底电缆问题。要求在海底选择一条地形适宜，波浪、水流扰动较小及泥沙活动强度较小的地方，这需要了解海底地貌与海底表面地质过程。在北海铺设海底电缆管道，需防止沙浪移动及浊流活动切断海底电缆。加拿大在蒲福海要防止浮冰切割海底管道，需研究海底冰刻痕、冰核丘等问题。这些都是陆架现代地质过程即动力、地貌与沉积。现代地质过程是海洋地质学工作中的一个重要方面。

四、海洋环境

石油资源开发，海上钻井平台，涉及环境海洋地质问题，如沙浪移动规律影响海上平台稳定性；海底沉积层如粉砂层、砂层、黏土层等不同的土层与海底工程地基的稳定性问题。这些工作与陆地工程地质不大相同，国外称为环境海洋地质问题，就是研究水流特点，泥沙运动与沉积层稳定性，及海底地貌特点与变化规律。开采海洋石油需进行环境地质特点研究，还有核废物安置涉及对深海床的研究。核电站的废物具高放射性，加拿大与美国将它储藏于古老地盾深井或盐穹内，但主张研究在深海平原具有 200 m 厚的松散沉积中能否埋藏核废物、沉积物特征、活动或稳定程度及水热传导问题。法国亦是研究厚层沉积(600～1 200 m)结构及生物对沉积物的扰动与放射性扩散的途径问题。英国则研究容器的可靠性而把核废物投放于海床之上。日本亦是如此，岛国缺乏土地而投核废物于洋底上。荷兰主张置核废物于盐穹中，竭力反对投放于海床上。因此联合国海床委员会专门组织有关国家进行专门研究，我参加了加拿大 80 年和 81 年度工作。这项工作主要是深海平原的海洋地质学工作，参加的人员有沉积学家、微古、海洋地球化学家，要求进行测深、地貌摄影、地层测量、采样与取沉积柱，还需进行微古、地化(上浮水、孔隙水、氧化还原度，痕量元素)、土力学以及热流量等项测量。研究结果，发现在 5 600 深的梭木深海平原，12 m 松散沉积下有一层分布广泛的砂层。砂来源于加拿大东部(红色砂岩之剥蚀产物，砂的颜色、重矿物组合及含东部煤层碎屑，均为佐证并且有孔虫为寒冷性)，是由于浑浊流所搬运(据石英砂表面结构)，在晚更新世被浊流搬运了 1 000 km。砂层上部的热传导速度约 80～200 cm/a，剪切力自沉积表层向下部递增，悬移质与泥沙分析表明，在海床上与上部海水之间存在着一“深海边界层”。水流、生物和热力活动使得泥沙、微量元素、水、汽、生物之间存在对流交换，因此，必须采取措施以防止核废物在海底扩散，防止从生物传播到人类。上述研究主要是研究表面地质 Surface Geology 过程，涉及地貌与沉积两大部分，这是现代海洋地质学的重要组成部分。

五、领土问题

促进了对大陆架、对极地海洋的研究。

一方面，应用问题提出了要求，促进了海洋地质学的发展；另一方面，不属应用问题，而是通过实践所建立的理论有力地促进海洋地质学的发展。这主要是海底构造即深部地质研究。环太平洋的火山带与地震带、喜马拉雅与地中海的地震分布带涉及海洋的起源与发展演变问题。板块构造理论的兴起，使人们逐渐意识到占地球表面70%的洋底是研究地球变动的关键地区，尤其是洋中脊的大裂谷与俯冲带的深海沟，是沟通地球内外之通道，是研究洋壳的形成、发展、消亡以及向陆地过渡、演化的关键地带。洋壳演变与板块运动的动力机制是当代地球科学研究的焦点之一。由于研究手段困难(多频道测量、深海钻探与深潜器观测)，搜集到的资料有限。因此，人们的活动着眼于大陆边缘。海沟、岛弧与弧后盆地带是我国具有的独特的构造地带，同时，我国还具有洋壳向陆壳进一步演变的喜马拉雅构造带及欧亚板块与太平洋板块，与印度板块两个构造相交体系的过渡地带——南北向的横断山脉带。从海洋地质学观点看，后两者属古海洋学范畴。这三个枢纽地带投入力量从事研究会取得具全球性重大意义的理论成果。

海洋地质学是研究海岸与海底，研究被海水覆盖作用的岩石圈的形态、成因和变化。研究内容涉及三个方面：

(1) 海岸与海底地貌，海底山脉、海沟、高原，主要是构造地貌，它反映着成因，是海底地质工作的着手点，所以受到当代重视。

(2) 研究海洋沉积与地层，包括微体古生物与洋壳的岩石结构特征(层3、层4等)，主要属现代地质过程与表面地质问题。

(3) 海底构造，包括内部动力机制与深部构造。

研究目的是：① 开发利用海岸海底资源与环境；② 了解全球性大地质现象与地壳变动的动力机制。海洋地质学在世界上有三个研究中心：① 北美中心(美、加)先进，应用深海钻探、潜海地质从事板块运动基础理论、新发现及海洋矿产资源开发等项研究工作。② 欧洲中心(英、法、荷、西德、苏联)在海岸三角洲工程实验、大陆架油气开发以及潜海地质方面有特色，苏联从事全球海洋调查。③ 亚太中心(澳、新、日、中)的工作发展不平衡，在大陆边缘与极地海洋工作方面具有特色。

我国海洋地质工作始于二十世纪五十年代末期，初期以海岸、河口工作为主，研究较深入。七十年代，由于海洋石油资源需要促进了对浅海大陆架的研究。八十年代又开展了太平洋与南大洋调查，工作范围已涉及海底的主要领域。由于我们有独特的地理环境，具有丰富的海底矿产资源以及已形成的研究主调查力量，可以相信，我国会在海洋地质学方面做出重大贡献。

Coastal Geomorphology*

The coast of China lies between the eastern continent of Asia and the marginal seas of the West Pacific. The coastline of the mainland extends 18 400 km. The total length of the coastline including that of the 6 500 islands, is approximately 32 000 km.

COASTAL RESEARCH

The systematic study of sea coast of China started in the late 1950's. A group of coast experts, including the famous coast authority V. P. Zenkovick, who joined the Chinese siltation study of Tianjin New Port in 1958, held a series of lectures in Peking University and conducted reconnaissance work along the Chinese coastline from Liaodong, Shandong Peninsulas to the Pearl River delta and Hainan Island. We adopt the definition of sea coast as a coastal zone, which includes the land along the sea shore, the beaches or intertidal flat and the submarine coastal slope. Thus we investigate the sea coast from back beach to shore and off shore, out to the water depth where waves start to move sediment. The work is therefore carried out on land over the intertidal zone and at the sea, and to study the coast from the point of view of geomorphology, dynamics and sedimentology. This includes coastal dynamic geomorphology, and the work is carried out by a team of coastal geomorphologists, sedimentologists, physical oceanographers and coastal engineers. The aim of the research is to define the source and movement of sediments, the coastal characteristics of erosion and deposition, the genesis of the coast, and to predict the future evolution of the coastline.

Much of the work is directed towards the location of sea harbours, quays and navigation channels. Which means that coast research in China is connected closely with applied projects. Sea port construction is an important step for Chinese coastal zone development, as a limited number of sea harbours can hardly meet the increasing volume of transportation. The average is one harbour for each 500 km coastline, with very little number of sea ports scattered along the quite longer coastline. Up to 1985, we have a total of 400 berths, of which 199 are deepwater berths with a depth of 10 m or more. The total volume of traffic has increased from some ten million tons in 1985. However, the majority of larger sea ports are located in the north of China. In the 1 000 km long

* WANG Ying：曾呈奎，周海鸥，李本川主编《中国海洋科学研究及开发》，157－161 页（中文），539－545 页（英文）。青岛：青岛出版社，1993 年。后文为中文译文。

coastline of north Jiangsu, there is only Lianyungang Harbour at the north end of the province. Bedrock embayments are distributed widely in the south of China, nearly one hundred coastal bays can be developed into middle or large sea harbours. For harbour construction, there are the contradictory factors of "deepwater and calm sea" in the natural coastal environment especially along China's coast zone with the great sediment supplied by rivers, even mud flats have developed inside longer bays in the areas of Zhejiang and the northern part of Fujian Province owing to long shore drifting from the mouth of Changjiang River. Thus, the siltation of harbours has become a serious problem. A typical example is the siltation of Tianjin New Port. The port was first constructed by the Japanese from 1939 to 1945. It is situated on the mudflat of the Bohai Sea with the Haihe River at the south and the Jiyunhe River at the north, emptying into the vicinity of the bay. The conditions of the harbour grew rapidly worse and the water depth in the approach channel was reduced to only 3 m during 1949. Extensive dredging has been carried out since 1949, and by the end of 1952, the water depth was increased to 9 m. But the main trouble remained the siltation of the harbour and the water depth could be maintained only by powerful dredging. Each time the harbour area was extended to meet the demands of increasing traffic, the amount of siltation was increased accordingly. A total of 600×10^4 m^3 of silt was removed annually. This example triggered extensive coastal research programs. As a result, it was ascertained that silt from the Yellow River was the main source of material forming the entire mudflat of the Bohai Sea, but the Haihe River was also responsible. To reduce siltation, in 1958, sluice gates was built at the mouth of Haihe River. This has a prominent effect in reducing the harbour siltation.

At present, the 24 km approach channel has been widened from 60 m to 120 m and 150 m with a water depth of 11 m. Subsequently, many harbour sites were chosen on the basis of thorough studies of coastal dynamic geomorphology. Thus practical requirements have advanced coastal science. For this type of research, there is a team of geoscientists and engineers working on the same project. The work of the coastal geomorphologists, who normally come from universities and institutes, forms the basis of most of these studies. The Ministry of Transportation supports this work with technicians, money, ships, vehicles and drilling equipment. When the project is finished, a report is submitted to them. Scientific research was promoted by these projects, for example, mudflat coast and Yellow River Delta Research in the Tianjin New Port Project; Investigation of the sand barrier coast with the Qinhuangdao Oil Port Project; Study of the estuaries with Changjiang River Mouth Project, Study of the Tidal Inlets and Coral reefs with the South China Sea Port Project. This research accelerated the advancement of knowledge about the coast of China. For example, we were able to establish the mudflat zonation, the proper significance of shell beach ridges, and the sedimentary dynamics of tidal inlets.

Another type of applied study is the exploitation of the valuable placer deposits. These are zircon, rutile, titanic, iron, monazite and gold, which are mostly deposited in the fine sediments of sand beaches, sand spits, bay mouth bars, sand barriers and tomboloes, but are limited in sand dunes and river mouth deltas. The placers were formed during late Pleistocene and Holocene eras in areas such as Liaodong, Shandong peninsula, Fujian, Guangdong, and Hainan Provinces there are mesozoic granites. The coast there belongs to the marine erosional-depositional type of bedrock embayment coast. This kind of project has also improved the study on coasts, sand-body formations and submarine coastal slope sediments.

Specific coasts of coral reef and mangrove swamps have been studied in detail from the southern coast of the East China Sea to the coasts of Guangdong and Hainan Provinces and the South China Sea Islands. The study has achieved greater understanding of coral reef evolution during the Holocene period and has provided the evidence for analysis of the tectonic movement and sea level changes in different coast areas. Six natural reserves of mangrove forest have been set up in Fujian, Guangdong and Hainan Provinces to preserve the natural beauty and to benefit the ecosystem.

From 1980 to 1985, a comprehensive survey was carried along the whole coast zone of the mainland to understand natural resources of the coast and tidal flat as well as its quantity and quality; to get the basic data of coast environment and social economics; to make comprehensive review on the environment and resources of China's coast; and to work out the plans for full scale exploitation of coastal resources. Large amount of scientific data include: Coastal lithology and structure, geomorphology, sedimentology, oceanography, climate, soil, fishery and aquatic production, botany and zoology, agriculture, coastal engineering etc. have been collected. It encompassed a great deal of work and was the first time in ten coastal provinces and cities of China that systematic coast data has been ascertained. It is also a great help to the development of China's coastal zones.

MAJOR PROGRESS OF CHINA COAST RESEARCH

The speed of coastal research in China has increased since the 1980's. Because of the limited space I can only summarize a few of the major topics here.

(1) The Plain Coasts are still hot spots of research both theoretically and practically. The plain coast is formed along the depressed belts, usually in the lower reaches of a large river. The Liaohe River Plain, the Yellow River Delta Plain and the North Jiangsu Plain are the major types of plain coast discussed here.

1) The intertidal flat is a predominant feature of this extensive plain coast. The

main factors which determine the development of tidal flats are the very gentle slopes of plain coasts, tidal dynamics, and the abundance of fine-grained sedimentary material. The width of any particular tidal flat is governed mainly by the tidal range; similarly, morphological characteristics and their associated sedimentary structures are controlled by the tidal energy (current velocities), large tidal ranges are always accompanied by well developed flats, such as Hangzhou Bay in China, Bay of Fundy, and the Severn Estuary abroad. Typical tidal flats in China are developed where the tidal range is about 3 m and the velocity of tidal currents is not as large as in the estuaries referred to previously. Such tidal flats are between 6 to 18 km in width and are much more extensive than in the estuarine areas. The major types of tidal flats in China are developed along the fringe zone of the North China Plain, where the slope of the coast zone is less than 1 : 1 000. As a result of this, the wave action is restricted to offshore areas and tidal currents are the dominant dynamic processes on the flats. In comparison, some of the tidal flats along the East China and South China Seas are associated with long narrow embayments of other types of sheltered coast. Fetch within such environments limits the wind/wave interaction, hence tidal currents are, once again, the major agents controlling sedimentary processes. Silt is the major sediment, ranging from fine to coarse silt, deposited over the largest part of tidal flats. Very fine sands and clays are also present on most tidal flats. All such sediments deposited on tidal flats are carried mainly onto the flats by flood tidal currents, bringing material from the adjacent seabed; this infers to transport from seaward to landward. The original source of such sediments may be terrigenous, however, the sediments on the Chinese tidal flats are fluvial in origin and have been supplied to the coastal zone at lowered sea level by the Yellow and Changjiang Rivers.

The sediment, morphology and benthic zonation features of intertidal flats are distinct phenomenon in a wide range of climate environments, and are also a reflection of changes in the tidal dynamics. Three types of coastal tidal flats have been recognized in China and the U. K. by comparison.

A silt flat well developed along the North Jiangsu coast of the Yellow Sea is typical of tidal flat with three zones, namely, mud zone on the upper flat, muddy silt zone on the middle part and sandy silt zone on the lower part of the intertidal zone. A medium level of wave energy stirs up material from the submarine sandy ridges of the old Changjiang River delta and transports it onto the shore by the flood tidal currents.

A mud flat has developed along the shore of Bohai Bay with only minimal wave influences but with large quantities of fine grained material of clay and silt which is supplied mainly by the Yellow River. There are four distinguishable zonations along the 400 km long coastline: a polygon zone consisting of clay cracks is distributed on the upper part of the intertidal zone which is only flooded during the spring high tides. The classic feature of the mudflat is a 0.5～1.0 km wide belt of semi-fluid mud located below the mean high tidal level as an inner depositional zone. Only in the Bohai Sea, where there is a large quantity of sediment supplied by the Yellow River plume to support the coast, is the mudflat well developed with a wide zone of semi-fluid mud. Outside of this zone, there appears to be an erosional zone with interbedded silt, clay and silt deposits, and a silt ripple zone, known as the outer depositional zone, is located on the lower part of the intertidal zone.

In comparison, a sandy flat developed in the wash along the coastline of eastern England, U.K. is an example of a tidal flat with high wave energy influences. Coarse sand and shingles are present offshore as glacial deposits during a period of lower sea level. They are now reworked by wave and deposited by tidal currents as a sand flat.

Differences between the three types of tidal flat are considered to be in response to the varying levels of wave energy input. A sedimentation model of the typical tidal flat is shown in two-layers:

A. The top layer consists of vertical deposits of mud suspension carried by the flood tide. The thickness of the layer is limited by the tidal range between mid and spring tidal levels.

B. The lower layer is a transverse deposit of silt of fine sand moving as a traction load. The maximum thickness of the layer may be the same as the vertical distance between the low tide level and the coastal wave action base.

However, during the typhoon season, a strong tropical cyclone wind (Beaufort scale 10～11) causes high waves which are 5 times greater than the average wave height. Strong waves coupled with swift tidal current cause widespread surface erosion of the tidal flats, scouring especially in the lower intertidal zone. But in other parts of the tidal flat, the early phase scouring is followed by aggradation and accretion occurs only in the upper part of the supratidal zone. The sediments deposited during the typhoon season are distinctly different from the sediments deposited during ordinary weather. In the upper intertidal zone, typhoon deposits are coarse silt ($Md=4.930$) with good sorting i.e. with only very little fine portion, while the underlying sediments are mainly medium and fine silt ($Md=6$) with poor sorting. Typhoon deposits are also characterized by its sediment structures: thin horizontal beddings in their middle and upper parts.

Tidal creeks meandering across the whole tidal flat zone, are bordered in some parts by levees. Tidal current dynamics in creeks show that there is a series of asymmetrical phenomena, as 10 % of the complete tidal prism which comes into the creek on the flood results in a residual down-channel flow on the flood. Within the creeks, the ebb discharge is greater than the flood discharge throughout a complete tidal cycle. Maximum ebb currents are higher than on the flood and the ebb duration is longer.

Thus, the creeks are ebb-dominated tidal controls. The tidal creeks are not only dissecting the intertidal flat but also acting as a pass transporting the sediment and creek deposits, resulting in a mosaic formation within the intertidal flat deposits. Sediments in the creeks are coarser than those deposited on the flats, and sometimes appeared as silty cross-beddings or sandy lens in the sediment profiles.

In summary, tidal flat sediment have the following structures:

a. Thin bedding layers, especially the alternating laminate of fine silt and silty clay, are the main component and structure of tidal flats, indicating the systematic sedimentary processes of tidal cycles.

b. Coarse silt or sandy silt is the lower deposit and has pod, ripple beddings or horizontal beddings, pod features are common in the coarse silt bedding as micro-ripples shown in the section.

c. Sandy silt or very fine sand beddings with cross or herringbone shaped beddings, are normally a mosaic formation in the laminate groups.

d. Silty clay laminate as upper part of tidal flat deposits have clay cracks, worm burrows and root mottlings, sometimes the burrows are filled with lots of small pills or iron pyrite materials, the root fissures are filled with calcaeous materials.

e. Storm deposits of thinly interlayered silt, with cross beddings.

2) Shell beach ridges (cheniers) are distributed in the upper part of silty tidal flats, and are distinct land-form associated with the extended plain coast. The cheniers consist mainly of shells and shell fragments with a little proportion of silicate materials of silts and fine sands, which is different to normal beach ridges. Due to fresh water precipitation remaining in the shell deposits, the ridges are the green island on the salty lowland. Cheniers are interesting topics to archaeologists, coastal geomorphologists and geologists. Since the early 1960's scientists have carried out investigations on a large scale and dated the shell ages by using ^{14}C method. According to recent publications four series of cheniers are distrubuted in the coastal plains of the west part of Bohai Sea and the West Yellow Sea, more than five cheniers are in the vicinity of Shanghai, and also four cheniers have been dated on the outer continental shelf of the East China Sea.

Most of the ^{14}C data of molluse shells are reliable to identify the old coastline environment. However, it is normal for beach ridges to be formed with shell fragments, but not all the ridges are cheniers, some deposited forms might be old sand barriers as shown in the North Jiangsu coastal plain, some might be shallow water deposits with shell fragments as seen in the West Bohai Coast Plain. Attention should be paid to the under water features such as ridge forms and sediment structures to identify the chenier series. Cheniers present a special coast environment when tidal flat of the plain coast is retreating due to insufficient sediment supply. The molluse shells can be dug out by break wave action, and carried landwards by swash to accumulate shell beach ridges on the upper part of the inter tidal flat. The evolution of both cheniers I and II (counted from sea to landward) at the west side of the Bohai Bay, as well as along the Jiangsu coast, appears to be related to the history of the shifting lower course of the Yellow River. Coastal progradation occurred in the vicinity of the active delta of the Yellow River. On the contrary, erosion took place along the abandoned deltaic coast and it's in this erosional and wave dominated tidal flat environment that coarse shell cheniers are formed. According to the history of the shifting Yellow River Mouth, this occurred alternatively on the coastal plain of Bohai Sea and Jiangsu. As an illustration, the behaviour of the lower course of the Yellow River and the related coastal development in the last 2 000 a can be briefly summarized.

3) The plain coast represents a balance between wave erosion and deposition of sediment supplied, principally by rivers. If the amount of sediment supply is larger than the amount eroded, the coastline progrades; if the sediment supply is reduced or stopped, wave erosion causes the coastline, to retreat. Relative changes in sea level do not affect to any great extent the development of the plain coastline. For example, at the present location of the Yellow River, in Southern Bohai Sea, major shoreline accretion has occurred during the past 130 a, the delta forming seaward at 20～28 km, despite a steady regional subsidence since the late Miocene period. Along the abandant Yellow River mouth in the North Jiangsu coast, where the Yellow River emptied from 1128 to 1855, erosion has been extensive. The coast has retreated about 17 km since 1855. Thus, efforts have been made to construct detached breakwater parallel to the coast, and groins at the end of the longitudinal breakwaters in a combined project to protect the tidal flat coast erosion.

(2) Tidal inlet embayments are distributed widely along the coast zone of the East China and South China Seas, eg. Xiangshan Bay, Sanmen Bay, Leqing Bay (Zhejiang Province), Quanzhou Bay, Xiamen Bay (Fujian Province), Shantou Bay, Zhan Jiang Bay (Guangdong Province), Yulin Bay and Yangpu Bay (Hainan Province) are all large tidal inlets.

Tidal inlets should be distinguished from tidal channels which connect open sea at both ends, such as the Jintang channel near the Xiangshan Bay where the deep water port of Bailun is situated. Three types of tidal inlets have been summarized in China as: 1) the embayments—lagoon type on sandy or rocky coasts such as at Beihai port and Shantou Bay; 2) estuarine inlets on mouths of small or medium rivers which may be on mud plain coasts, such as the west channel of Yalujiang Estuary and the Huangpujiang Estuary in Shanghai; 3) Artificial inlets enclosed by breakwaters such as the Tianjin New Port. Improvement of navigation channels of these inlets follows the principle of the O'Brian *P-A* formula. Where accurate oceanographical and littoral drift data are not available, a careful analysis of coastal morphology and sedimentology may provide a useful clue for the evaluation of the inlets in navigation.

Recently, the mean tidal prism (P) of 32 tidal inlets along the South China Sea coasts and mouth cross-sections area below mean sea level (A) have been calculated. The P-A optimum regression equation with a minimum standard deviation (solved through using the regression analysis) is in the power function form with the power approaching unity, however the coefficient C doubles the sandy coasts in the United States, and is only 1/10 as the rocky bay coasts in Japan. The P-A relation here shows either a very high correlation, or a very obvious unstability.

Not only is the coefficient exchangable in P-A relation formula, P and A points deviate apparently from the regression line, frequency distribution of the P/A ratio is discrete, but also the evolutionary directions of the P/A ration are not consistent during the developmental process of the same inlet, when tidal prism reduces dramatically under the impact of reclamation. P-A relation is a quite complex problem which should be further studied and improved. Tidal prism is an important factor, and so are the geologico-geomorphologic conditions and hydrodynamics of the different inlets.

Tidal inlets count most among 64 embayments around Hainan Island; and the inlets are of three genetic patterns: 1) Tidal inlets developed along structural fault zone, where there is also a boundary weak zone of volcanic rock and sandy deposits of old terraces, as the Yangpu inlet, the largest inlet located in the northwest part of the island. 2) Sandy barriers enclosed embayment, such as Sanya, Xincun, and Qinglan Inlets. 3) The Holocene transgression flooded mountains or river valleys. Yulin and Yalong Bays belong to this type. All three types of tidal inlets are suitable for building sea harbours and can serve multipurposely. There are two factors in the maintenance of the inlet bay and the water depth of navigation route. A. The factor of the tidal prism is decided by tidal range and the duration ratio of flood and ebb tides. The ebb current is the natural dredge washing out sandy sediments to benefit the navigation channel. It is important to keep or extend the larger area of the inlet bay for retaining larger quantities of flood sea water. B. Determining the sediment sources along coastline and inlet, and decreasing the coast sediment supply are both important steps producing natural flushing out of the sediment or sediment by passing by long shore drifting. Sedimentation of the larger tidal inlets of Hainan Island consists of three dynamic phases according to our recent Lehigh core data.

Inner bay phase: River influences at the bay head, sediments are coarse material with sand or gravel on the form of sand banks or shoals. Deposits in the outer part of inner bay are mainly found on flood tidal deltas, and are well sorted fine sediments.

Deep channel phase: Coarse sand or gravel deposited in the narrow pass, and sometimes mixed with fluid mud, fine materials of silt deposited in the deep channels extended to both sides of the narrow pass.

Outer bay phase: Mainly ebb tidal deposits of mixed finer sedimentation, but with storm deposits of sandy lens as its typical features. In the ebb tidal delta at the end of the deep channel in the outer bay, sediments are coarser and may be mixed with local shells or coral fragments.

Outside of the outer bay: Seabed sediments are more homogenous but with storm lenses and less animal trace marks than in the bays.

(3) Many other interesting topics can not be introduced in detail because of the limited space. The publication of "Modern Sedimentation in Coastal and Nearshore Zones" of China edited by Prof. REN Mei-e., has summarized the major results of Chinese coast research, which is valuable for detailed references. Given here are only a few supplements for further studies. 1) Sand barriers and coastal dunes are developed along alluvial plain coasts such as the Luanhe River delta and the Hanjiang River delta. The sand deposits are controlled by monsoon-produced waves, E and SE waves advance towards the coast transversely with limited longshore drifting from the river mouths. They form off shore sand barriers enclosing the river mouth and the coast. Some of the sand barriers are the top parts of sand shoals, which rest uncomfortably upon the bottom sediments. The foreslopes of the barriers are steeper than the lee slopes, and the river sediments are mainly deposited in the shallow water environment protected by the sand barriers. Well-laminated and inter-stratified silt and fine sand are developed leeward of the sand barriers. The fine sandy materials come from the sea, either by the overwash or by tidal currents through the inlets between the barriers. Along the coast in a zone of decreased wave action behind offshore sand barriers, tidal flats develop over the delta plain of the original coastline. With delta progradation, several series of sand barriers can be formed parallel to the coastline. These features reflect the constant directions of monsoon wave dynamics.

Several larger coast dunes, with a beach zone of less than 100 m wide and 25～42 m high, are very close to the present seashore. They are presently being undermined by wave attack, such as the dune coast along the northern part of Luanhe River delta plain. This refers to different circumstances of air, sea, and coastland interactions. The larger sand dune coast is developed along the seashore with abundant sands; dry, cold or semi-arid climates with a strong prevailing wind on land, and a gentle submarine coastal slope, i. e. 1 : 1 000 or a medium tide range which allows wide enough exposure of beach zone during ebb tides for wind blowing sands to form the huge sand dunes. A comparative study on the sediments among sand dunes, on the nearshore and offshore seabed and the ^{14}C data of the foundation of coast dunes, have proved that the huge sand dunes were formed during 8 015±105 a B. P., the first cold phase and the lower sea level of the Holocene period. This result is also proved by the dune coast along the Mauritanian coast of the Atlantic Ocean.

2) The evidence of coastal tectonic regimes.

There are a number of terraces along the bedrock embayed coasts of China's seas, especially along the coasts of small islands.

Due to lack of sedimentary evidence it is often hard to decide whether they are elevated wave cut platforms or surfaces of denudation.

By using a combination of coastal features, such as sea notches, sea benches, sea stacks and sea cliffs it is possible to identify an elevated coast terrace. For example, a series of elevated coast terraces with heights of 5 m, 15～20 m, 40 m, 60～80 m, 120 m, 200 m, 320 m, 450 m, and 600 m, around the ancient island of Yuntai mountain in Jiangsu Province can be distinguished via geomorphologic evidence.

The 120 m high coast terraces at Zhenshan mountain, Xiamen of Fujian Province can be identified by the combined features of trough-shaped sea notches, rocky bench and sea stacks.

Sixty sea notches single or trough-shaped with a few pebbles have been found at heights of 120 m, 200 m and 250 m in the vicinity of Dalian area of the Liaodong Peninsula.

Elevated sand barriers and lagoon systems have been found in the southern part of Hainan Island at heights of 10 m, 15 m and 30 m. Three periods of volcanism are documented in the North West Coast of Hainan Province. The first eruptions occurred during late Pliocene to early Pleistocene, forming a basaltic hinterland with some lava covering the lower sand barriers and a veneer of contact metamorphic alteration. The second eruption was in the Pleistocene, depositing volcaniclastics with well-preserved ripple beddings. The latest eruptions were in the Holocene with many small breached craters forming volcano coasts and islands in the area of Beibu Bay and the Yinggehai Sea. The last eruption elevated adjacent Shingle beaches by 15 m and raised the tidal flat and shallow bay deposits to form a 20 m high terrace. The volcanic belts, with tholeiite basalt lithologies, migrate through time from the east to the west. These tectonic events might have developed the taphrogenic fault zone in the Beibu Bay and Yinggehai Sea. Present barrier coral reefs have developed only on the northwest side of Hainan Province. This perhaps reflects recent seabed volcanic activity.

(4) "Man is a geological agent". Human impact on morphology and sedimentation of the coastal zone of north China gives sufficient evidence that river and sea can form an intergrated system. Both conclusions are drawn from a specific research on the larger river influences on the natural environment, and the long history of human activity directly or indirectly affected the major coastal landforms and sedimentation.

① The coastal plain of North Jiangsu and the associated subaqueous delta of the abandoned Yellow River were the direct result of human diversion of the Yellow River southward between 1124—1854 A. D.. During that time the coastal plain prograded seaward by 30～50 km.

② The present Yellow River delta is river-dominated and micro-tidal, but has an accurate form due indirectly to man's cultivation of the loess plateau resulting from accelerated erosion. The huge sediment load of 1. 1×10^9 t per year.

③ The accumulation rate of the Bohai Sea is also greatly affected by human activity. Detailed analysis of historical data shows that during the last 5 000 a, the Yellow River did not flow into the Bohai Sea for 758 a, and annual sediment input into the Bohai Sea reached 11×10^9 t for only 840 a. This data together with coring and seismic profiling data seems to indicate that the sediment accumulation rate of the Bohai Sea averages about 0. 6～0. 7 m/ka in the Holocene period.

Literature

[1] Chen Jiyu, Yu Zhi-Ying and Yun Cai-Xing. 1959. Geomorphologic development of The Changjiang delta. *Acta Geographica Simica*, 25(3): 201－220 (in Chinese with Russian abstract).

[2] Dale Janis E. 1985. Physical and biological zonation of intertidal flats at Frobisher Bay. N. W. T. In: Abstracts 14th Arctic Workshop, Dartmouth, Canada.

[3] Ding Xianrong and Zhu Dakui. 1985. Characteristics of tidal flow and development of Geomorphology in tidal creeks, Southern Coast of Jiangsu Province. In: Proceeding of China-West Germany Symposium on Engineering.

[4] Evans G. 1965. Intertidal flat sediments and their environment of deposition in The Wash. *Quat. Jun. Geol. Soc. London*, 121(1～4): 209－240.

[5] Gao Shanming, Li Yuanfong, An Feng-Tueng and Li Fengzin. 1980. The Formation of Sand Bars on the Luanhe River Delta and the Change of the Coastline. *Acta Oceanologica Sinica*, 2(4): 102－114 (in Chinese).

[6] Han Mukang. 1986. The discovery of High Sea level remnants in the Dalian area. *Acta Oceanologica Sinica* 3(6): 793－796.

[7] Li Shih-Yu. 1962. A Preliminary Study on the Remains and Underground Cultural Relics of the West Coast of the Ancient Bohai Bay. *Archeology* (12): 652－657 (in Chinese).

[8] O'Brian M. P. 1967. Proceedings of the 10th Coastal Engineering Conferences I. 676－686.

[9] O'Brian M. P. 1980. Proceedings of the 17th Coastal Engineering Conference III. 2. 504－2016.

[10] Ren Mei-E and Zeng Chenkai. 1980. Late Quaternary continental Shelf of East China. *Acta Oceanologica Sinica*, 2(2): 106－111 (in Chinese).

[11] Ren Mei-E *et al.* 1985. Modern Sedimentation in Coastal and Nearshore Zone of China. China Ocean Press, Springer-Verlag.

[12] Ren Mei-E. 1989. Human impact on the coastal morphology and sedimentation of North China. *Scientia Geographica Sinica* 9(1): 1－7 (in Chinese).

[13] Ren Mei-E and Zhang Renshun. 1985. On tidal inlets of China. *Acta Oceanologia Sinica*, 4(3): 423－432 (in Chinese).

[14] Ren Mei-E, Zhang Renshun and Yang Juhai. 1983. Proceedings of International Symposium on

Sedimentation on the Continental Shelf, with Special Reference to the East China Sea 1. 1 - 19.

[15] Ruan Ting. 1988. Types of the Marine Placer Deposits in China. *Journal on Ningbo University*, 1 (2): 49 - 62 (in Chinese).

[16] Wang Ying, Ren Mei-E and Zhu Dakui. 1986. Sediment supply to the continental shelf by the major rivers of China. *Journal of the Geological Society*, *London*, 143(3): 935 - 944.

[17] Wang Ying. 1980. The Coast of China. *Geoscience Canada*, 7(3): 109 - 113.

[18] Wang Ying and David Aubrey. 1987. The characteristics of the China Coastline. *Continental Shelf Research*, 7(4): 329 - 394.

[19] Wang Ying, Schafer C. T., Smith J. N. 1987. Characteristic of Tidal Inlets Designated for Deep Water Harbour Development, Hainan Island, China. Proceedings of Coastal and Port Engineering in Developing countries. China Ocean Press, 1. 363 - 369.

[20] Wang Ying. 1983. The Mudflat System of China. *Canadian Journal of Fishery and Aquatic Science*, 40(1): 160 - 171.

[21] Wang Ying, Collins M. B. and Zhu Dakui. 1992. A Comparative Study of Open Coast Tidal Flat: The Wash (U. K), Bohai Bay and West Yellow Sea (Mainland China). In: Proceedings of International Symposium on the Coastal Zone, (in press).

[22] Wang Ying. 1964. The Shell Coast ridges and the old coastlines of the West Coast of the Bohai Bay. *Journal of Nanjing University*, 8: 420 - 440 (in Chinese with English abstract).

[23] Wang Ying and Zhu Dakui. 1987. An approach on the formation causes of coastal sand dunes. *Journal of Desert Research*, 7(3): 29 - 40 (in Chinese with English abstract).

[24] Wang Ying. 1983. Some problems on the coast morphologic analysis of the sea level changing. *Journal of Nanjing University* (*Natural Sciences*), 4: 745 - 752 (in Chinese with English abstract).

附(中文译文):

海岸地貌学研究*

中国海岸位于亚洲大陆东部与西太平洋陆缘海之间。大陆海岸线长 18 400 km。包括 6 500 个岛屿在内,岸线总长约为 32 000 km。

中国海岸综合研究始于 20 世纪 50 年代后期。一些海岸专家,包括著名的海岸研究权威 V. P. Zenkovich(曾科维奇)于 1958 年参加了对天津新港的回淤研究,在北京大学举行了一系列讲座,并对辽东半岛、山东半岛至珠江三角洲和海南岛的广阔海岸线进行了勘察。

我们所用海岸定义是海岸带,它包括沿海陆地、海滩或潮滩及水下斜坡。因此,我们的海岸调查是从后滩到海岸和滨海,向外至波浪开始搬动沉积物的深度。研究涉及整个潮间带的陆地部分和海洋部分,并从地貌学、动力学和沉积学的观点来研究海岸,包括海岸动力地貌学。这一研究由一些海岸地貌学家、沉积学家、物理海洋学家和海岸工程师共同完成,目的是要确定沉积物的来源和搬运、侵蚀和堆积海岸的特征、海岸发育和岸线演变的预测等。

海港位置、码头、航道是研究的重点。不论哪一方面,中国的海岸研究都紧密地与实际应用相结合。海港建设是中国岸带开发的重要步骤,有限数量的海港很难满足运输吞吐量的增长,平均 500 km 才有一个海港,这些为数不多的海港,分布在漫长的海岸线上。到 1985 年,总计有 400 个泊位。其中 199 个是水深 10 m 以上的深水泊位。1985 年船舶的总吨位已经增加到几千万吨,但是大多数较大的海港位于中国北方。江苏北部 1 000 km 长的海岸线上,只有连云港这唯一的港口。基岩海湾广泛分布在中国南部,近 100 个海湾能够发展成中等或大型海港。对于港口建设,尤其是在有河流补给大量泥沙的中国海岸带的自然环境中,存在着"深水和平静海域"的相互矛盾的因素。由于长江口岸的较大迁移(大摆动),在浙江范围和福建北部海湾发育着大片泥滩。因此,港口淤积成为一个严重问题。典型的例子是天津新港的淤积。天津新港建于 1939—1945 年,至 1949 年进港水深减少到仅 3 m,1949 年后加强了疏浚,至 1952 年底水深增至 9 m,但仍存在着严重的淤积问题,每年要挖泥 $6\times10^6\ m^3$。这个事例促进了对加强海岸研究的认识,通过调查,查清了泥沙来自泥滩,原来是黄河物质所形成,其次为海河。为减少泥沙淤积,在海河口建筑了泄水闸,对减少港口淤积起了重要作用。现在,一条长 24 km 的进港航道已经从 60 m 拓宽至 120 m 和 150 m,水深为 11 m。此后许多港口选址都是通过海岸动力地貌学的研究来确定的。这种建设需要与海岸动力地貌研究的结合,促进了海岸科学的

---

* 原刊的本文系英文原稿之代译稿,原刊登于"中国海洋科学研究及开发",主编:曾呈奎、周海鸥、李本川,青岛出版社,1992 年。存在缺句与术语翻译的不正确:将"贝壳堤"或"贝壳沙堤"译为"沼泽沙丘";文句中年以 a 表示。现经作者校正,并补充缺句。以前文之英文稿为准。

发展。如淤泥质海岸与黄河三角洲的研究，天津新港的设计，秦皇岛油港的沙坝调查，长江口的河口研究，中国南海港口的潮汐汊道和珊瑚礁的研究。这些研究都增强了对中国海岸的认识，以使我们现在有可能建立淤泥质海岸分带、海滩贝壳堤特征以及潮汐汊道的沉积动力学等。

其他类型的应用研究是那些有经济价值的砂矿床的开发，包括锆石、金红石、钛、铁、独居石和金等，它们通常堆积在砂质海滩与沙嘴、湾口坝、沙坝和连岛沙洲的细粒沉积物中。在辽东和山东半岛、福建、广东和海南省等中生代花岗岩地区，在晚更新世和全新世时期形成了这些砂矿，这里海岸属基岩港湾岸的侵蚀堆积型。对此类项目的研究同样促进了海岸、沙体形成和水下岸坡沉积的研究。

独特的珊瑚礁和红树林沼泽海岸，从东海南部海岸至广东和海南省海岸以及南海岛屿，都已进行了研究。此类研究为认识全新世时期珊瑚礁的演化并为不同海岸地区构造运动和海面变化提供了证据。目前，在福建、广东和海南省已建立了 6 个自然红树林保护区，以保护这一自然景观和有利于生态系统。

从 1980—1985 年对大陆整个海岸带进行了综合考察，定性定量地了解了海岸和潮滩的自然资源，获得了海岸环境和社会经济等一些基本资料，对中国海岸环境与资源做出综合评价，制订了海岸资源完全开发的工作计划。已知获得的大量科学资料包括：海岸岩石学和构造学、地貌学、沉积学、海洋学、气候学、土壤学、渔业和水产捕捞、植物学、动物学、农学和海岸工程等，对开发中国海岸带有明显的积极作用。

一、海岸研究

20 世纪 80 年代以来，中国对海岸的研究速度日益加快，平原海岸仍然是理论上、实践上的研究热门。

平原海岸主要形成在大河下游河口地段，沿着低平的地带形成。辽河平原、黄河三角洲平原和苏北平原是平原海岸的主要类型。

潮间浅滩是开阔平原海岸的主要特点，决定其发育的主要因子是平原海岸很平缓的坡度、潮汐动力和丰富的细颗粒沉积物。某一特定的潮间浅滩的宽度主要受潮差控制；同样，地貌特征及其相应的沉积构造主要由潮能（流速）控制。大潮差往往发育宽阔的潮间浅滩，如中国的杭州湾、加拿大的芬地（Fundy）湾和若干开阔的河口湾。在中国典型的潮间浅滩其潮差约 3 m，潮流速度不像过去参考文献所说的那么大。这种潮滩宽度为 6～18 km，比河口区更加宽大。中国北方平原边缘地带发育的潮滩是中国潮滩的主要类型，其海岸带坡度小于 1∶1 000。有调查结果认为，波浪的作用仅限制在近岸地区，潮流控制潮滩的动力过程。比较起来，东海和南海有些潮滩是与有掩护海岸狭长的海湾联系在一起，这样一种内部环境限制了风浪的相互作用，因而潮流再次成为控制沉积过程的主要因素。粉砂是主要沉积物，潮滩的绝大部分地区沉积着细粉砂和粗粉砂，同时还存有极细砂与黏土。在潮滩上堆积的这些沉积物主要是通过涨潮流从附近的海床携带进来，这意味着泥沙运送是由海向陆输送的。中国潮滩的沉积物来源于河流，是在低海面时长江和黄河输送给海岸地区的。

潮间浅滩的沉积物、地貌和底部分区的特征，是在一种大范围气候环境中的独特现象，也是潮汐动力变化的反映。通过对中国和英国潮滩海岸的比较研究，确定三个海岸潮滩类型。

苏北的黄海沿岸发育得十分良好的粉砂潮滩，是典型的具有三个带的潮间浅滩，即潮滩上部为泥带，中部为泥质粉砂带，下部为砂质粉砂带。中等水准的波能可掀起古长江三角洲水下沙脊的沉积物，由涨潮流向岸搬运。

渤海沿岸主要由黄河供给大量的黏土和粉砂细粒物质，发育成为淤泥滩。沿着400 km长的海岸线有成带现象可以识别：由黏土裂缝组成的龟裂带分布在潮间带的上部，它仅在大潮高潮期间被淹没；这类泥滩的特点是有一条0.5～1.0 km宽的半流动淤泥带，它位于平均高潮面以下，是为内淤积带；该带的外侧，呈现一条具有夹层粉砂、黏土和粉砂的沉积带，因沉积物结构不同的差别侵蚀成潮水沟或滩面冲刷体，是为滩面冲刷带；其外侧为宽阔的由粉砂或含淤泥的粉砂组成的沙波带，是为外淤积带，它位于潮间带的下部。

比较而言，英国的英格兰东部的侵蚀海岸线内发育的砂质滩是一个受高波能影响的潮滩例子。外滨存在的粗砂和砾石，是低海面时期的冰川沉积，现在被波浪破碎，受潮流堆积成为沙滩。三种潮滩类型之间的差别被认为是不同程度的波能输入的结果。典型潮滩的沉积作用模式有以下两层：① 顶层由涨潮搬运的悬浮淤泥的垂直沉积构成，该层厚度受中潮面与大潮面之间的潮差制约。② 下层是粉砂和细砂呈推移质横向搬运的沉积，该层最大厚度等于低潮面与近岸浪作用基面两者间的垂直距离。

但是，在台风季节，强热带风暴(蒲氏风级为10～11级)产生的波高，比平均波高大5倍，强浪与急速潮流互相耦合造成潮滩大面冲刷，特别是在潮间带下部。但在潮滩的其他部分，仅在潮上带的上部早先侵蚀的被填淤并加积。在台风季节沉积物的堆积明显不同于正常天气期间沉积物的堆积。在潮间带上部，台风堆积的是粗粉砂($Md=4.930$)且分选好，只有很少一点细砂，而底层沉积物主要是中粉砂和细粉砂($Md=6$)，且分选很差。台风沉积的特点也在于它的沉积结构：在其上部和中部有一薄层水平层理。

横穿整个潮滩带的弯曲的潮沟的一些部分与堤岸相毗邻。潮沟处的潮流动力反映出一系列不对称现象：当涨潮的时候，全部进潮量的10%进入潮沟，形成涨潮时剩余的顺流水道；潮沟内一个完整的潮汐周期落潮流量大于涨潮流量；最大落潮流比涨潮流大，且落潮延时长，潮沟被落潮流所控制，它不仅切割潮间带，而且还起到输运沉积物和形成潮沟内沉积的作用，结果在潮间带沉积内导致镶嵌结构的形成；潮沟内的沉积物比在潮滩上的沉积物更粗，在沉积剖面中有时呈现出粗砂质交错层理或砂质透镜状。

总之，潮滩沉积有如下一些结构：

(1) 薄层层理，特别是细粉砂和粉砂质交错层是潮滩的主要成分和构造，表明潮汐周期的系统沉积过程。

(2) 粗粉砂或砂质粉砂沉积在下层，并且有扁豆层、流纹层理或水平成层，粗粉砂层中扁豆层呈细微流纹，出现在断面下部。

(3) 砂质粉砂或极细砂层具有的交错层理或鱼骨形层理，在该层状中通常呈一种镶嵌结构形态。

(4) 粉砂黏土层处于潮滩沉积的上部,有黏土裂缝、蠕虫穴和根状斑痕,有时洞穴被许多小丸或硫化铁矿物充填,裂缝基部被石灰质物质所填塞。

(5) 风暴潮沉积含有交错层理的薄层泥砂。贝壳滩脊分布在粉砂质潮滩上部,其特殊地貌伴随平原海岸而扩展。沼泽沙丘主要由贝壳和贝壳碎屑组成,粉砂和细砂中含有少量硅酸盐物质,与一般的沙脊不同。由于在贝壳沉积中保留有降雨淡水,滩脊成为盐碱低洼地上的绿岛。

自20世纪60年代以来,对海岸带贝壳堤进行了大规模调查,并用^{14}C方法测定贝壳年代。据目前发表的资料,在渤海西部和黄海西部沿岸平原上分布有4条贝壳堤,在上海附近有5条以上的贝壳沙堤,还有4条分布在东海外陆架上。

大部分软体动物贝壳的^{14}C资料能可靠地确定古海岸线环境。通常滩脊由贝壳碎屑构成,但并不是所有的滩脊都是贝壳沙堤。有些堆积能形成古沙坝,正如在苏北海岸平原中见到的;有些可以是带有贝壳碎屑的浅水沉积,可在渤海海岸平原西部见到。通过对水下的一些特征,如滩脊的形态和沉积结构,可鉴别出贝壳堤的类型。当平原海岸的潮滩由于缺乏沉积物的补给而退缩时,软体动物贝壳受破碎浪作用被掏出,并向陆地方向携带,在潮间浅滩的上部堆积成贝壳滩脊。渤海湾西部以及江苏海岸沙堤Ⅰ和沙堤Ⅱ的演变(从海向陆)看来是与黄河的历史演变有关。海岸推进发生在目前黄河活动的三角洲附近;相反,冲蚀发生在废三角洲沿岸。在这种冲蚀中,波浪控制着潮滩环境,形成粗颗粒的贝壳沙堤。黄河口演变的历史,是在渤海和江苏沿岸平原交替发生。

平坦海岸体现出波浪冲蚀与主要由江河提供沉积物的堆积两者之间的平衡。如果沉积物供给量大于冲蚀量,海岸线就向前推进;如果沉积物供应数量减少或停止,海岸线即被波浪冲蚀并后退。海平面相对变化不会对平原海岸发育以重大影响。例如,目前黄河位于渤海南部,海岸线加积主要发生在过去的130年间,三角洲向海推进20～28 km,虽然自晚中新世以来,区域稳定的下沉。苏北海岸的废黄河口是1128—1855年黄河流入这一地区形成的,现正大范围地受到冲蚀。1855年以后此段海岸后退约17 km,于是人们分段建筑顺岸防波堤和J型坝,以联合预防海岸潮滩的侵蚀。潮汐汊道广泛分布在中国东海和中国南海的海岸带上,例如象山湾、三门湾、乐清湾(浙江省),泉州湾、厦门湾(福建省),汕头湾、湛江湾(广东省),榆林湾和洋浦湾(海南省)都有较大的潮汐汊道。

潮汐汊道不同于潮汐水道。潮汐水道的两端与公海相连,如象山湾附近的金塘水道(此地有北仑深水港)。中国三种类型的潮汐汊道已作过阐述,即:① 砂质岸或基岩海岸上的海湾-潟湖类型,如北海港或汕头湾;② 淤泥质平原岸上小的或中等河流河口的河口汊道,如鸭绿江河口水道和上海的黄浦江河口;③ 由防波堤围成的人为汊道,如天津新港。这些汊道航道的改造是遵循了O'Brian $P-A$ 公式的原理。在此不可能获得准确的海洋和沿岸输沙资料,从海岸形态学和沉积学进行仔细分析,可对汊道通航的评价提供有效的线索。

目前,已对南海海岸32个潮汐汊道的平均纳潮量(P)和平均海平面以下口门横断面积(A)进行了计算。具有最小标准误差(利用回归分析取得的解)的 $P-A$ 最佳回归方程是具有指数趋于1的指数函数方程,其系数 C 是美国砂质岸的两倍,却仅为日本基岩港湾岸的 1/10。这里的 $P-A$ 关系表明，或是有很高的相关性，或是有很明显的不稳定性。关系式不仅其系数可交换，P 和 A 点明显偏离回归线，P/A 比率的频率分布是离散的，而且在同一汊道的发育过程，P/A 比率的发展方向也是不一致的。围垦使纳潮量迅速减少。$P-A$ 关系是一个十分复杂的问题，应进一步研究与改进。纳潮量是一个很重要的因子，也是不同汊道地质-地貌和水文动力学的重要因子。

在海南岛 64 个海湾中，大多数都有潮汐汊道，并有三种成因模式：① 潮汐汊道沿着构造断层带发育，这里存在火山岩软弱带边界，并有古海底台地的砂质堆积，如洋浦汊道，最大的汊道位于海南岛西北部。② 沙坝封闭的海湾，如三亚、清澜汊道。③ 被全新世海侵淹没的山地或河谷，如榆林湾等均属此类。这三种潮汐汊道适合于建筑海港并可作其他用途。两种因子可利用维护航道：① 纳潮量因子由潮差和涨、落潮的历时比率决定，落潮流是天然的挖掘者，它向外冲刷砂质沉积物而有利于航道，维持并扩大内湾使有较大的面积，对保持较大的涨潮量是十分重要的。② 确定海岸和汊道的沉积物来源和减少沿岸沉积物的供应，以使沉积物产生自然冲刷或者使沉积物被沿岸飘砂带走都是重要的办法。

根据我们最新的 Lehigh 岩芯资料，海南岛较大的潮汐汊道沉积由三个动力区组成。

(1) 内湾区　湾顶受河流影响，沉积物是砂和砾石粗物质，形成砂质海滩或浅滩；其外部主要是涨潮三角洲的沉积区，并具分选很好的细粒沉积物。

(2) 外湾区　主要是混合较细沉积物的落潮沉积，且具有砂质透镜体的风暴沉积。在外湾潮汐末端的落潮三角洲内，沉积物是较粗的，也可能是局部带有贝壳或珊瑚碎屑的混合体。

(3) 外湾的外侧　海底沉积更均匀并带有风暴透镜体，而且动物踪迹比湾内少得多。

沙坝和海岸沙丘是沿着冲积平原海岸发育起来的，如滦河三角洲和韩江三角洲，砂质沉积受季风引起的波浪所控制。E 向和 SE 向浪垂直于海岸推进而限制了河口的沿岸输砂，形成外滨沙坝封闭河口和海岸。有些沙坝是在砂质浅滩的顶部，不整合地建立在底部沉积物上。沙坝的前坡比后坡陡，河流沉积物主要沉积在受沙坝保护的浅水环境中。层理分明的层间粉砂和细砂发育在沙坝的向陆面。细砂物质来自于海洋，由越浪或者由潮流通过沙坝间的汊道携带来。位于近岸沙坝的后面——波浪作用减弱地带的海岸，潮滩发育在原来海岸的三角洲平原上。随着三角洲的推进，平行于海岸线形成几个系列的沙坝。这些反映出季风波浪动力的方向不变的特点。

在宽度小于 100 m，高 25～42 m 的海滩地带上分布一些较大的海岸沙丘，它们非常接近现在的海滨，其基部正在被海浪冲刷，滦河三角洲平原北部的海岸沙丘即属此例。

海岸沙丘的形成取决于大气、海洋、海岸等因子的相互作用。海岸带砂源丰富，陆地盛行风强劲，干、冷或半干旱的气候，平缓的潮间带岸坡(1∶1 000)或中等潮差海滩，在退潮期间形成开阔的海滨地带，受风吹作用形成大沙丘和大规模的沙丘海岸。在近岸和离岸海底对沙丘沉积物进行比较研究，并根据海岸沙丘基底的 ^{14}C 资料，已证明大沙丘在距今 8 015±105a 期间形成，即全新世第一次寒冷期和低海面期。这种结论也可从大西洋毛里塔尼亚沿岸沙丘岸得到证明。

二、海岸构造变动证据

中国海域沿着基岩港湾海岸，特别是沿着一些小岛的海岸有大量的阶地出现。由于缺少沉积资料，常常很难确定它们是上升浪蚀台或是剥蚀面。

综合利用一些海岸特点，如浪蚀龛、浪蚀台、海蚀柱、海蚀崖，有可能识别上升岸海岸阶地。如江苏省云台山的古老岛屿周围分布的一系列高度为 5 m，15～20 m，40 m，60～80 m，120 m，200 m，320 m，450 m 和 600 m 的阶地即可通过其地貌形迹来识别。

福建厦门镇山的 120 m 高的海岸阶地可通过槽形浪蚀龛、海蚀台和海蚀柱等综合特征来识别。

辽东半岛大连地区附近，在高度为 120 m，200 m 和 250 m 处发现 60 处单个或槽形成群的浪蚀龛，其上有一些卵石。

在海南岛南部高度为 10 m，15 m 和 30 m 的地方发现有上升沙坝和潟湖系统。该岛西北海岸有 3 次火山活动的记录：第一次喷发在晚上新世至早更新世期间形成玄武岩的腹地，其下部沙坝有部分熔岩覆盖并有薄层接触蚀变沉积层。第二次喷发是在更新世，火山碎屑沉积中有保存良好的波纹层理。最后一次喷发是在全新世，许多小型的缺火山口在莺歌海和北部湾地区形成火山海岸和岛屿。此次喷发抬高邻近的 Shingle beaches 15 m，并使潮坪、浅海湾沉积形成 20 m 高的阶地。具有拉斑玄武岩的火山带随时间自东向西迁移。这些构造变动在北部湾和莺歌海可能已经发展成为断裂带。目前，珊瑚堡礁仅在海南岛西北部发育，这可能反映当前海底的火山活动。

“人是一种地质营力”。人类活动对中国北方海岸带的地貌和沉积作用给予了很大影响。这一结论可从大河流影响自然环境的一项专门研究，及人类历史上对主要海岸地形及沉积作用的直接与间接影响上得出。苏北平原海岸和有关的废弃黄河水下三角洲是 1124—1854 年人为改道的，海岸平原向海推进 30～50 km；现在的黄河三角洲尽管是以河流为主潮汐作用较小的三角洲，但由于人类开垦黄土高原加速水土流失而影响了三角洲的地形，每年有 11×10^{8} t 的巨量沉积物输入；渤海沉积速率也受到人类活动的巨大影响，历史资料分析表明，在过去 5 000 年期间，有 758 年黄河并不流入渤海，仅有 840 年沉积物输入渤海，输入量为 11×10^{8} t/a。这些资料与岩芯、地震剖面等资料结合起来可看出，渤海沉积速率在全新世期间平均为 0.6～0.7 m/ka。

海洋地理学的当代发展*

海洋地理学是地理科学的新分支，具有自然科学、社会科学、技术科学相互交叉渗透的特点。从学科发展分析，它属于地理学与海洋学之间新的结合点学科。学科术语中地理学(Geography)与海洋学(Oceanography)是平行的地学学科。海洋学通常包括四大学科：海洋物理学、海洋化学、海洋生物学与海洋地质学，而未明确的划分出海洋地理学。海洋地理学研究的客体是海洋，包括海岸与海底，范围涉及到气、水、生物与岩石圈。研究内容包括海洋环境、海洋资源、海洋开发利用与保护、海洋经济，疆域(海岸、岛屿、领海、大陆架、专属经济区、公海等)政治、立法与管理，海洋新技术发展、应用及影响等。概言之，海洋地理学是从宏观调控与区域入手，从立法、政策、区域经济、管理着手进行海洋资源开发、环境利用与保护。当代面临着人口、资源与环境巨大挑战，开发利用海洋已是当务之急。1982 年通过的《联合国海洋法公约》，使人们的海洋观念发生了深刻的变化，原属于公海的大约 1.3×10^8 km² 海域将划归沿海国管辖，其面积与地球整个陆地面积(1.49×10^8 km²)相当[1]。公约对沿海国所管辖的范围、所占有的海洋资源与沿海国权益均发生重大变化。有的国家已把管辖海域作为国土的海洋部分，出现了海洋国土这一概念。按这一规定，估计可能划归我国管辖的海域面积约占我国陆地总面积的三分之一。管辖范围的划分带来权益之争，例如，200 海里专属经济区确定后，在开阔海域中丧失一个具备人类生存条件的岛屿，就会失去 43×10^4 km² 的管辖海域。21 世纪为“海洋世纪”或“太平洋世纪”已为学者们所共识。因此，海洋地理学是适应时代发展而兴起，具有重要意义的新学科。

中国地理学会第四届理事会于 1984 年即提出要发展海洋地理。1987 年 4 月，中国地理学会五届二次理事会决定成立“海洋地理专业委员会”，经过充分酝酿和协商，海洋地理专业委员会于 1988 年 4 月正式成立。

1 国际海洋地理专业委员会及其活动

1986 年国际地理联合会(International Geographical Union)正式成立海洋地理研究组(Study Group On Marine Geography)，中国是发起国之一。研究组的核心课题集中于“人类活动对海洋管理之相互关系”。它引起各方面的广泛关注，通过研讨提出三方面的课题：① 海洋法所涉及的地理课题；② 海洋的利用与管理；③ 海洋地理学的国际合作。

1988 年 8 月，在澳大利亚悉尼举行的第 26 届世界地理大会期间，国际地理联合会投

---

* 王颖：《地理学报》，1994 年第 49 卷(增刊)，第 669－676 页。

1995 年载于：《南京大学海岸与海岛开发国家试点实验室，海岸与海岛开发国家试点实验室年报(1991—1994)，海平面变化与海岸侵蚀专辑》，261－269 页，南京大学出版社。

票批准成立“国际海洋地理专业委员会”(Commission on Marine Geography)。当时共有三个研究组升格为专业委员会,而国际地理联合会下属的42个专业委员会中有20个专业委员会被取消。新升格的一般是地理科学中新生长点,交叉、综合、横向联合的新分支学科。这种重视海洋领域的研究,同样也反映在1986年澳大利亚召开的国际沉积学家大会(IAS)上,其主题报告中:“Something's Old, Something's New, Something's Blue”,蓝色指的就是海洋领域研究。国际海洋地理专业委员会有成员国40多个,首届主席是意大利热那亚大学地理学教授Adaberto Vallega,由意、法、英、中、美、澳、德、印、坦桑尼亚、象牙海岸、智利等11国组成常务委员会,中国海洋地理专业委员会副主任委员杨作升教授为首届常务理事。1992年第27届世界地理大会后,由英国威尔士大学的Hance Smith教授任第二届海洋地理专业委员会主席,兼任《海洋政策》杂志(*Marine Policy*)副主编。中国海洋地理专业委员主任委员王颖教授任常务理事并兼任“海洋政策”编委。该专业委员会主要对全球性问题进行协调、讨论以及从事专题研究,以推动海洋地理学发展,它每年发布一次新闻通讯,两年举行一次学术会议,四年举行一次大会,出版有关著作与刊物,经费自筹。1994年6月于加拿大召开学术讨论会,主题是海岸带管理。

国际海洋地理专业委员会具有较强的社会科学背景,着重于海洋法与疆界划分问题,主要涉及政治地理领域。1990年以来,则明显地转向海洋管理,增强了自然科学(海洋环境与资源研究)与技术科学(遥感与地理信息系统应用)的内容与研究力量,加强对区域海洋研究的关注。这一转变,反映了对海洋特性认识的深化。海洋是大气、海水、生物与岩石圈相互搭界共同作用的场所,既是生命的摇篮、风雨的源地,也是人类发展所依赖的食物、土地、矿产、动能以至气、水、资源所在。海洋环境的发展变化与人类社会经济活动息息相关,必须采用多学科交叉、渗透观点与方法调查、研究海洋的区域特性与变化规律,才能达到综合开发与合理利用的目的。1992年6月在意大利热那亚召开的“海洋地理学国际学术讨论会”是以“全球变化中的海洋管理”为主题,亦为纪念哥伦布发现美洲500周年、纪念联合国人文环境大会20周年以及纪念联合国海洋法10周年。这一内容也反映了海洋地理学广泛的领域及重要意义与发展进程。这次有1 000名学者出席的盛会,标志海洋地理学的成熟。会议期间,还举办了“人与海洋”大型博览会,反映出人类开发海洋之发展进程、海洋物产与海洋高科技。

2 海洋地理学的主要内容及发展趋势

当代海洋地理学研究的主要内容与学科发展趋势,可以归纳为以下五个方面。

2.1 海洋管理

海洋管理的历史及长期发展形成的海洋利用管理的概念是:海洋自由权限、开发资源与分区管理[2]。世界其他海域人与海洋环境关系的例证,更正了以欧洲为中心的观点,海洋资源分散的特点加强了开发过程中新技术的应用。海洋管理政策仍局限于领海以外的自由捕鱼权与航行问题。由于出现了较多的区间问题,促使多学科交叉概念的新发展,联合国的全球政策制约了现存的国家地区管理方法,而最新的发展表明,已开始了“海洋一体化管理”新时机。

海洋管理的理论：自17世纪以来新发展的一系列关键思路，主要限于海洋某方面的应用，而不是一体化的理论主体。由于社会科学与自然科学的不断进步，并参照传统的社会与区间概念，以及环境与人相关联的全球变化机制，全球变化的海洋管理理论是主导的。其应用管理强调人与海洋之相互作用为第一位。相互作用的基础是人类对海洋的技术管理与一般管理。由于区域文化与环境的特点形成不同的管理系统，要从全球观点来建立海洋管理的理论，要按新的海滨地带的概念与应用管理方法，来建立地区间的管理体系[3]。

海洋管理的实践：海洋管理的历史与理论皆推进了“一体化的海洋管理”，地中海1975行动证实了这种一体化海洋管理是可行的。北海诸国，加、美新大陆国家都通过一系实践而努力推动一体化的海洋管理。这是当前海洋管理的新趋向[4]。

2.2　海平面上升与人类活动关系问题

2.2.1　海平面上升对海岸带之影响

环球海岸带由于人口的急剧增长与相应的发展而遭受越来越大的压力。海岸带既以其丰富的资源与栖息地为主，亦因其活跃的动力环境而易遭受自然或人为的灾害，灾害发生的缘由是海岸带开发、旅游地与倾废物增长，以及长周期自然海岸过程的间接影响。海岸侵蚀与沼泽栖息地丧失是世界诸国面临的主要问题，源于复杂的自然因素与人为影响之综合。海平面上升是陆地丧失的主要原因，其变化在过去15万年间，曾自－130 m至＋7 m。世界潮位记录表明，近百年来，水动型的海平面上升12～15 cm，并且继续在上升。三角洲陆地的下沉，如密西西比河路易新安那地区，海面上升速度十倍于世界海面的平均速度。威尼斯受海面上升威胁，一年数次为潮水淹没。已预报因全球气候变暖造成的海面持续上升，在21世纪将更严重，对世界人口和海岸发展的影响将更大。为减缓这种影响而进行的海岸规划与管理是建立于海岸地质过程的当代知识基础之上的，如建设护岸工程、用海滩喂养重建海滨的方法稳定岸线、调节和控制人类在海岸的发展等[5]。

2.2.2　海洋循环对地区海域与全球之影响

风与地球自转科氏力决定了洋流型式，而洋流的影响是世界性的，如El-Nino现象发生是气流、水流与气候、生物之相关的最好例证。南大洋缺少陆地阻隔而形成极地环流系统，是最强有力的海流。为此，开展了广泛的国际合作，应用最先进的技术进行观测研究。海平面变化的研究也需要广泛的国际合作。大洋是互相沟通而互为影响，海洋地理学研究需多学科交叉渗透与国际合作，亦说明海洋管理是跨越地区的，需重视自然科学与新技术的应用[6]。

2.2.3　海平面上升对珊瑚礁与礁岛之影响

由于温室效应导致海平面上升而预报珊瑚会在潮间带礁滩上复苏生长。根据晚更新世从18 000年至6 000年海侵中珊瑚礁向上繁殖的证据，根据测量珊瑚礁生长速率、生物化学量收支估算以及珊瑚礁对构造下沉的反应等情况表明：珊瑚礁的向上繁殖是伴随着较慢的海面上升，其速率为＞10 mm/a，若海面上升速率快，则珊瑚礁会沉没。珊瑚繁殖对海平面上升的效应亦取决于生态情况，特别是人类在珊瑚礁区的活动情况。面临海面上升，珊瑚礁岛均遭受侵蚀，除非有大量珊瑚在周边繁殖，并供给足够的珊瑚碎屑使岛屿处于加积状态[7]。

珊瑚礁的变化是世界性的海洋地理课题，涉及领土、资源与岛屿人民的生存环境发展变化。这是海平面变化引起之社会效应之一。

2.3　海洋科学、技术与海洋管理

(1) 全球温度持续的微量上升达数世纪，使气候、海平面、植被生物等都发生不同的变化，它必然改变了陆地与海洋生活条件。为了解全球变化及所引起海洋过程的变化，作为管理与决策的基础，需要进行海、气界面分析及研究海洋特性由于失去平衡所引起的变化，为海洋渔业、航运、海洋工程服务，并提供新的科学资料[8]。

(2) 海洋管理中的遥感应用，近三十年来为取得对海洋过程系统的概貌测量，曾发展了一些多学科的科学方法，空中台站与轨道飞行器获得海面的总体现象，探测资料的数字化处理程序获得直观资料以了解海面与海底、大陆过渡带前沿与大气之间物理、生物与化学等相互作用。空间遥感利用光学、紫外线、红外线与微波感应器连续观察与监测海洋的动力过程，有助于判断维护、开发利用海洋环境资源之管理政策[9]。

(3) 地理信息系统是一体化的海洋管理工作中的关键技术。海洋管理信息系统涉及海洋环境各要素的资料收集与处理；空间资源利用、生物资源、非生物资源、倾废与再生资源等信息搜集、处理、储存；渔业、航海、港口等技术管理；地方、国家、海区、全球海洋；加之，资源环境的长期、短期变化趋势分析；以及进行海洋管理的决策与实施。这个复杂的大系统需依赖于一体化的海洋管理系统与研究所之间的广泛合作。首先需要决策者果断地建立一体化的海洋管理系统以有利于决策的推行[10]。

2.4　海洋政策

2.4.1　目标与行动

20 世纪 90 年代三大因素导致一体化的海洋政策：① 1992 年联合国环境与发展大会的召开，形成新的政治势头，政府的焦点着眼于全球尺度的环境问题，而海洋是此课题的关键。在环境与发展中，海洋/海岸组体将是关键之关键。② 海洋法的加强驱使沿岸各国限量地、保护地与保守地利用海洋资源与环境。③ 海岸与海洋活动的强度日益增加，包括海洋交通、海洋油气资源开发、渔业、旅游、海洋文化等，并具有巨大发展潜力。当前的海洋政策已超出 20 世纪 60 年代和 70 年代海洋法的政策以及 80 年代初期建立国家为主体的海洋政策，再进入中长期目标的一体化海洋政策，在双边与海区之间发展更为密切的合作关系，加强已有的研究机构合作网是不困难的，关键仍是提供进行多功能管理的工具，加强为开发海洋而进行的长期、综合的高科技力量，提高政策制定的水平与规划管理的技能，为决策者提供一个坚实的信息与资料基础。海洋政策正处于发展的初始阶段，尚无成型的模式，一体化的海洋政策必将在 90 年代建立[11]。

2.4.2　海洋管理中的国家管辖区作用

传统的海洋法划定的沿岸国家向海延伸的管辖范围是很窄的。第三次联合国海洋法大会后,发生“革命性”的变化,形成两个全新的国家管辖海域(专属经济区与群岛国家水域),大陆架与海峡通道内容亦有了重要的改变。200海里专属经济区是一多功能的法定区,海岸国可以据本国人口需要,也可据其他国家以及世界共同体的需要去开发与管理此水域。大陆架是毗邻国领土向海的自然延伸至大陆边缘的外沿,在窄狭陆架区则自领海基线至200海里的距离范围。大陆架限于毗邻国家而对其他国家开发需要有限制,这与专属经济区不同。群岛水域、领海、海峡可供国际航行,但是领海管辖权属海岸国家[12]。

2.4.3 疆界与海洋管理

疆界的概念:① 人为划定的界限。内界为领海基线,它划出海滨的外界;外界所划定的区间为国家管辖界限,同时港口的界限亦考虑为疆界。② 划分出海洋单位与以陆地为主的单位,大单位如大堡礁(Great Barrier Reef),小的如一个河口或红树林滩。这些单位实际上是一个带,此带可能与人为活动有关,亦可能为自然过程所形成[13]。疆界的划定需依据1982年海洋法,疆界确定后,即进行海洋管理——建立组织与对人类在一定海洋空间活动的指导,出现划界的纷争需通过协商加以解决。

2.4.4 海洋渔业政策

世界渔捞生产已接近其资源最高限1亿t,加强渔业管理包括政策、计划、资料收集、研究、法规实施及区间合作。建立于1945年的联合国粮食与农业组织(FAO)下的渔业部门与项目,一直在进行这方面的管理工作,卓有成效。例如欧洲共同体在其海域的地理范围内,各国可享受同等的活动权力,建立了捕鱼、保护渔业资源、生产、价格体系与贸易制度等,为当代海洋渔业政策与管理提供了实践的范例。

2.4.5 保护海洋环境是开发海洋政策的关键组成

2.4.6 海港与航海管理政策

2.5 国家、地区或自然体的海洋管理

(1) 东南亚诸国由于开发资源(食物、能源)、海上交通以及历史传统等方面的因素,在向海洋发展时,各自划定的领海界限成重复交叉,引起纷争,尤其是第三次海洋法会议以来,对专属区的划定也产生多边矛盾[14],需妥善地、通过协商解决,并处理好历史传统与当前划界原则之间的关系。

(2) 北海渔业的矛盾始于11世纪,通过多边会议、协议以及非政府的科学机构(ICFS)向诸国提供建议等而逐步解决。当代北海开发由于工业排污倾废形成新的矛盾,通过建立新的会议机构、利用与该区有关的其他机构,如欧洲共同体,以及少量的海事法庭仲裁而逐渐解决,趋向于建立法律机构作为解决矛盾的关键措施[15]。

(3) 人类对海洋多项开发的关注产生两类矛盾:① 对特定海域过分利用者与未利用者之间的矛盾;② 政府内与海洋法、政策有牵连的各机构间之矛盾。这些需要应用“反、正”反应的方式及比较研究的办法解决矛盾[16]。

(4) 厄瓜多尔的海岸管理体现出针对海岸经济的特点:养虾与捕鱼的食物生产、出口与外汇收入。管理工作着眼于虾养殖、河口捕鱼、海岸土地利用、环境卫生、生活服务,发展机制使海岸管理法规执行、研究机构主持管理以加强技术和运行能力等。组织资源开

发、当地居民由美国罗德岛大学与相关研究所支持建立特殊管理区，以实现海岸带的持续发展。该项目获得厄瓜多尔政府支持[17]。

(5) 中国海岸管理自 20 世纪 80 年代以来经历四个阶段：海岸带资源环境综合调查、开发规划、健全管理组织系统与海岸管理立法。海岸带经济发展迅速使其成为中国的黄金地带，虽人口密度大，但资源潜力巨大，需要综合的、系统的、现代化的海岸带管理以促进经济与生态环境和谐地持续发展[18]。

(6) 南太平洋中的岛国，由于第三次海法会议确立的新海洋法，形成了小岛与巨大海洋区的特殊组合。如南太平洋诸岛国：法属玻利尼西尼，陆地面积为 3 265 km^2，但所属的海域达 5 030 000 km^2；斐济岛陆面积为 18 272 km^2，而海域面积为 1 290 000 km^2；关岛陆地面积为 541 km^2，但海域面积达 218 000 km^2。专属经济区管理是增长中的重大任务，同时，近岸带亦需成为综合的区域规划体系的组成部分，因此这些面积很小的岛国面临着艰巨的任务。这是当前世界海洋地理学中的新课题[19]。

(7) 都市海滨区尤其是老海港城区面临着更新发展的新问题，如：热内亚、利物浦、伦敦、上海等大的港口城市。应考虑海滨带发展的起源与特点，建立特定的组织机构，制定长期城市发展规划，需注意到将老城区海滨与相邻区结合在一起，考虑结合的发展规划以取代单一的"更新"计划[20]。加拿大新斯科舍省会哈利法克斯市的海滨改造发展了新的经济区与旅游点，提供了大量就业，是一好的例证。

(8) 河口区既接受陆地径流汇入，也受海洋影响，它既具有自然过程的特点，也是人类环境的产物。河口区受未来海平面上升、管理机制以及持续发展周边土地利用趋势等的影响，河口区管理反映出海洋地理学人与海洋交叉及多学科综合的特性。类似的自然体管理问题，尚有潟湖——受生物过程及社会和经济情况影响的生态过程的影响，封闭与半封闭水体等[21][22]。

(9) 北冰洋与南大洋资源环境的管理，先进技术的应用对北冰洋的开发使政策与管理问题已提到议事日程；南大洋获得较多的关注：捕获鲸鱼、海豹、磷虾海底资源，全球环境影响以及南极条约等，环境演化观测及极地探险在继续进行[23][24]。

以上概述了海洋地理学的主要方面与任务。海洋是地球表面的组成部分，而海洋科学又是与地理学相当的大科学体系。海洋是海洋地理学研究的主要客体，海洋具有的多功能性，同一空间存在多种资源，适合多种产业发展，因此，开发同一海域资源产业层次结构繁多，互相影响制约，协调管理困难。海洋是流动贯通的，鱼类洄游、污染散播及灾害波及等，易引起国际间矛盾斗争，国际合作协调比陆上资源开发与管理更为迫切。海洋与岛屿划界任务重大而艰巨，争夺资源与战略地位之岛屿涉及国家、民族的长期利益。海洋环境的利用与保护、海洋资源开发具有极大潜力，是人类生存发展的重要依赖，但海洋开发难度大、技术性强、花费大，必须有雄厚的产业群支持(船只、机械、电子通信、遥测遥控、航海、工程等)，才能做到有效开发。同样，海洋开发必带来重大效益与形成新的科技与产业群。当今，仍处于海洋大规模开发的前奏，21 世纪将是海洋经济开发的时代，需唤起人们的海洋意识，赢得更大投入以加速开发海洋的步伐。

3 我国海洋地理学的研究现状

中国海疆辽阔，濒临太平洋与边缘海，所辖海域面积约 300×10^4 km^2，蕴藏着丰富的渔业、矿产、动能、港湾、土地、化工与旅游资源。有海洋生物 2 000 多种，其中鱼类 1 500 种，经济价值较大的鱼类 150 种，最大持续渔获量为 470×10^4 t。已发现 16 个中新生代沉积盆地，总面积约 130×10^4 km^2，石油资源 308×10^8 t，天然气资源量约 7.2×10^{12} m^3，海洋能源约 6.3×10^8 kW，浅海滩涂面积达 2×10^4 ha，港湾岸线 5 000 km，海水资源取之不竭。

历史上，中国航海、对外贸易与文化交往具有光辉的成就，至今仍为海外传扬。而后，由于封建锁国政策与科技落后，抑制了中国向海洋开拓，国民的海洋意识薄弱。直至 20 世纪早期，沿各大河口兴起了津、沪、穗等大的港口城市，才有稀少的海岸开发与移民点。新中国成立后，重视发展海上交通、海洋调查、制图、油气资源勘探、海底通信工程、沿海渔、盐与种植业等，以及参加讨论有关海洋法的国际会议。60 年代提出“向海洋进军”的号召，当时的口号是：“摸清中国海、进军三大洋、登上南极洲”，组织了队伍、建立了海洋局与教、研机构，开展了实质性的海洋调查与开发事业。直至 80 年代，科技事业繁荣、国际交流加强以及市场经济的发展，我国海洋事业有了显著的稳步发展。完成了全国海岸带与海涂资源综合调查、全国海岛资源环境综合调查研究，获取了系统的科学资料，开展了规划与实施开发，海岸带已由昔日的禁区成为今日经济发展的黄金地带。海洋油气资源勘探、开发已进入国际招标、开发与正式生产。太平洋锰矿瘤调查已取得开发权，西太平洋海气交换、黑潮等项国际合作研究取得突出的科学成果。实现了南大洋考察与南极大陆建立科学考察站连续观测的夙愿，60 年代的口号基本实现。海洋环境监测与保护工作在实施，管理机构逐步建立，沿海各省市相继制定了海岸带管理条例，经各级人民代表大会通过，并拟定与试行有关的管理法规。高新技术在海洋中的应用已列入日程。海洋科学、地理科学在生存环境持续发展项目中占有举足轻重的地位，两学科交叉结合点的海洋地理学有着广阔的发展前景。当前的国力尚未形成对海洋开发管理的大量投入，海洋法规尚待建立与健全，对一些重要的国家海洋权益急需维护，海洋开发过程中出现的新矛盾问题有待解决等。海洋地理科学宜从海陆结合的观点与方法进行资源、环境与新技术开发应用的实践与总结，促进一体化的海洋管理，为迎接 21 世纪海洋开发新时代的到来做出贡献。

参考文献

[1] 艾万铸，倪轩，隋绍生. 1991. 我国海洋资源管理工作的建议. 见：我国海洋资源管理研究报告.

[2] Alastair D. Couper. 1992. History of Ocean management. In: Ocean management in the global changes. Elsevier Applied Science. 1 - 18.

[3] Hance D. Smith. 1992. Theory of Ocean Management. In: Ocean management in the global changes. Elsevier Applied Science. 19 - 38.

[4] Gerard Peet. 1992. Ocean Management in Practice. In: Ocean management in the global changes.

Elsevier Applied Science. 39－56.

[5] Dallas L. Peck and S. Jeffress Williams. Sea Level Rise and Its Implication in Coastal Planning and Management.

[6] Andre Guilcher. 1992. Impact of Ocean Circulation on Regional and Global Change. In: Ocean management in the global changes. Elsevier Applied Science. 74－89.

[7] Eric C. F. Bird. 1992. The Impacts of Sea Level Rise of Coral Reefs and Reef Islands. In: Ocean management in the global changes. Elsevier Applied Science. 90－107.

[8] Andre Vigarie. 1992. Ocean Science and Management. In: Ocean management in the global changes. Elsevier Applied Science. 108－123.

[9] Renato Herz. 1992. Remote Sensing in Ocean Management. In: Ocean management in the global changes. Elsevier Applied Science. 124－133.

[10] Adam Cole-King and Cjandra S. Lalwani. 1992. Information and Data Processing for Ocean Management. In: Ocean management in the global changes. Elsevier Applied Science. 134－152.

[11] Stella Maris A. Vallejo. 1992 Integrated Marine Policies: Deals and Constraints. In: Ocean management in the global changes. Elsevier Applied Science. 153－168.

[12] Moritaka Hayashi. 1992. The Role of National Jurisdictional Zones in Ocean Management. In: Ocean management in the global changes. Elsevier Applied Science. 209－226.

[13] Victor Preslott. 1992. Borndaries and Ocean Management. In: Ocean management in the global changes. Elsevier Applied Science. 227－231.

[14] Phiphat Tangsubkul. 1992. A Review of Disputed Maritime Areas in Southeast Asia. In: Ocean management in the global changes. Elsevier Applied Science. 255－279.

[15] Patricia Birnie. 1992. Compartive Evalution in Managing Conflicts: Lessons from the North Sea Experience. In: Ocean management in the global changes. Elsevier Applied Science. 308－324.

[16] Biliana Cicin-Sain. 1992. Multiple Use Conflicts and Their Resolution: Toward a Comparative Research Agenda. In: Ocean management in the global changes. Elsevier Applied Science. 280－307.

[17] Luis Arriaga M. 1992. Coastal management in Eaidor. In: Ocean management in the global changes. Elsevier Applied Science. 440－459.

[18] Wang Ying. 1992. Coastal management in China. In: Ocean management in the global changes. Elsevier Applied Science. 460－469.

[19] Hanns Buehholz. 1992. Small Island states and Huge maritime Zones: Management tasks in the South Pacific. In: Ocean management in the global changes. Elsevier Applied Science. 470－480.

[20] D. A. Pinder and B. S. Hoyle. 1992. Urban Waterfront Management: Historical patterns and Prospects. In: Ocean management in the global changes. Elsevier Applied Science. 482－501.

[21] Norebrt P. Psuty. 1992. Estuarines: Challenges for Coastal management. In: Ocean management in the global changes. Elsevier Applied Science. 502－520.

[22] S. G. F. Zabi. 1992. Comlaxity of coastal lagoons Management. In: Ocean management in the global changes. Elsevier Applied Science. 521－538.

[23] Walker H. Jesse. 1992. The Arctic Ocean Management in global changes. In: Ocean management in the global changes. Elsevier Applied Science. 550－525.

[24] Juan Carlos, M. Beltraimno. 1992. Management of the Sorthern Ocean Resources and Environment. In: Ocean management in the global changes. Elsevier Applied Science. 576－594.

海岸海洋科学研究新发展*

（会议文集）

一、海岸科学进展

海岸带是海陆过渡与相互作用活跃的地带，其环境与动力因素：风、浪、潮汐与河流等具有其与陆域或海洋各不相同的特性；自然资源包括港湾、滩涂、砂土、矿产、食物、动能等，具有更新与再生之特点。海岸是人类通向海洋的桥梁，人口密集、经济发达。中国沿海地带经历过封建王朝时代的禁闭，二十世纪初、中叶之开拓，七八十年代大规模调查研究与规划发展，九十年代已成为我国的“黄金地带”，集中了41%的人口与60%以上的工农业产值，地位重要。但是由于人口密集，经济发展迅速，也出现了一些亟待解决的新问题。以长江三角洲为例①，平均人口密度逾900人/km^2，为全国平均值的8.4倍，人均耕地不足1亩，为全国平均值的60%，是我国土地承载量最高的地区之一。近年区内人口约以1%的年率增加，耕地平均以0.5%的年率递减，农业产值占工农业总产值的比重已降至10%。人地关系协调发展决策迫在眉睫。长江三角洲也是环境容量压力最大的地区之一，在占国土1%的土地上承受着全国工业废水的9%、废气的5%、废渣的4%……约3 500万以上的农村人口饮用水不符标准，抵抗灾害的能力脆弱。加之三角洲地面低，水网密集，洪涝灾害频繁，严重制约区域经济发展与人民生活安定。如1994年旱灾，江苏省损失超过1 000亿元，1991年江淮特大洪涝灾害使4 500多万人受灾，经济损失500亿元。人类活动直接地或通过河流向海洋输送污染物，其中60%集中于海岸带，深海已遭受影响，危及空气与降水的更新，进一步又反馈于人类。人口、资源与生存环境持续发展是当代面临的巨大挑战，海岸科学是承担这项历史性重任的重要环节。1982年《联合国海洋法公约》将原属于公海的1.3亿km^{2}海域划归沿海国管辖，兴起了海洋国土新概念。1992年联合国“环境与发展”大会集中了占全球98%以上人口的国家政府首脑与有关专家，共同研定了“21世纪议程”（21 Agenda），其中心内容是善待地球，使人类生息发展得以制约于地球环境、资源所能容忍的范围内。1994年联合国教科文组织与发展总署组织召开了海岸海洋工作会议，制定计划着重研究与人类生存密切相关的海、陆过渡带。“Coastal Ocean”是一新概念，其范围包括海岸Surf Zone，也包括大陆架、大陆坡以及坡麓的大陆隆。大会报告内容涉及：

1. 海岸海洋物理与动力过程

风效应；海陆交互作用、波浪及表面边界层；潮汐效应；浮力效应；深海强制力与交换

* 王颖，张永战：中国海洋学会秘书处，《中国海洋学会第四次代表大会文集》，5－13页，1995年。

① 据南京大学谢志仁：长江三角洲研究，1994。

过程；地形效应、海峡与底面边界层。

2. 全球与多项交互作用过程

水体构成、水动力和气候效应；全球海平面变化、海岸影响与反馈；生物生产力与生态系动力；生物化学过程、海岸海洋与碳循环；沉积物搬运与陆源物流量。

3. 模式与资料

遥感测量——物理的、生物的、化学的；水文、生物和化学测量；流及海平面测量；边界通量；统计方法与资料分析；海岸海洋循环数值模式；结合的物理、生物的和化学的海岸模式；资料吸收与模式核实。

4. 区域综合与总结等

总之，多学科结合地研究海岸海洋环境、资源、权益，以期获得持续有效地合理利用，已成为当今的重点课题。国际海洋科学与地球科学研究中心与热点具有向海岸海洋转变的趋势。

二、世纪性海平面上升与海岸效应

（一）世纪性的海平面上升

验潮站水位记录反映出世界范围的海平面上升，源于冰川的进一步融化与海洋水体热膨胀。近 200 年的验潮资料（瑞典 Brest 站位记录始自 1704 年，荷兰 Amsterdam 站位始自 1682 年，以 Brest 站自 1807 年的记录最为标准）反映出海平面上升趋势与大气温度、海水表层温度变化趋势呈良好的相关，并且自 1930 年以后，海平面上升速率增加。例如，构造活动与人类影响使陆地水准发生变化，使海平面上升值形成明显的地区差异。纽约验潮站代表美国东海岸状况，近百年海平面上升速率 3 mm/a，是海平面上升与相当数量陆地下沉的综合效应；南部的得克萨斯站位资料表明海平面上升速率的平均值达 6 mm/a，原因在于抽取地下水与原油而引起的地面沉降；美国西海岸俄勒冈站几乎未表示出相对的海平面上升，因为水动型的海平面上升与陆地抬升量相当。D. Aubrey 与 K. O. Emery 试图将新构造运动上升值与全球性的水动型上升区别开来。虽然验潮站在南半球分布稀少，但从全球范围的验潮站记录进行相近比较与趋性分析，表明在过去 50 年到 100 年间，水动型的海平面上升值变化为 1～2 mm/a。尽管测算方法不同，但结果相近（表 1）。

表 1　据验潮资料所确定的全球水动型海平面变化

（据国际海洋研究科学委员会第 89 工作组）

研究者	上升速率(mm/a)
Gutenberg (1941)	1.1±0.8
Kuenen (1950)	1.2～1.4
Lisitzin (1958)	1.1±0.4

续 表

研究者	上升速率(mm/a)
Wexier (1961)	1.2
Fairbridge and Krebe (1962)	1.2
Hicks (1978)	1.5±0.3
Emery (1980)	3.0
Gornitz et al. (1982)	1.2
Barnett (1982)	1.51±0.15
Barnett (1984)	2.3±0.2
Gornitz and Lebedeff (1987)	1.2±0.3
Braate and Aubrey (1987)	1.1±0.1
Peltier and Tushingham (1989)	2.4±0.9
Douglas (1991)	1.8±0.1
谢志仁(1992)	0.7～1.2

近数十年来海面加速上升与地球的温室效应有关。预测由于全球变暖使冰川融溶与海水热膨胀,可使海面上升的数值如下:政府间气候变化专门委员会(IPCC - WG1,1990)估计至2050年海平面上升30～50 cm,至2100年海平面可能上升1 m。美国环境保护局预测到2100年海平面将上升50～340 cm,相当于5～30 mm/a上升速率[1]。全美研究委员会的二氧化碳评估组(The committee of the National Research Council on Carbon Dioxide Assessment)提出,至2100年,海平面上升速率为7 mm/a[2];Van Der Veen(1988)估计,至2085年海平面上升率为2.8～6.6 mm/a[3]。这些数据比过去100年来海平面1～2 mm/a的上升速率速高出2～4倍。

国家海洋局1990年公布了据44个站位的验潮资料分析结果:到1989年为止的近30年来,中国沿岸海平面平均上升速率为0.14 cm/a①。国家测绘局于1992年7月发布根据9个观测站的资料分析结果②:在过去100年中,中国东海与南海沿岸海平面分别上升19 cm与20 cm;中国海平面的年上升率为2～3 mm;未来海平面仍呈上升趋势。同时,发表了对世界的102年验潮站海平面记录的计算分析结果:在过去100年中,全球海平面平均上升15 cm,太平洋海平面上升10 cm,大西洋海平面上升29 cm,印度洋海平面上升39.6 cm。上述资料表明,海平面变化存在着海区差异与时段的差异,但过去100年的海平面上升数值是相近的,未来海平面上升趋势与速率增加是为大多数学者所肯定的。

短周期海平面变化,是由于大气与海洋作用过程的变化,如海水温度的地区性变化、海岸水流强度的改变、气压与风作用力与方向的改变等所造成的海平面年度变化、季节变化或日变化等。最突出的短周期海平面变化与太平洋的厄尔尼诺(EL Nino)的发生有关。在太平洋东岸赤道附近的岛屿验潮站重复记录到,在不到一年的时间内,海平面变化达到40～50 cm。在美国西岸,由于EL Nino形成的海平面高达10～20 cm。1982—1983

① 国家海洋局:1989年中国海洋环境年报,1990年3月。

② 国家测绘局:中国海平面每年上升2～3 mm,人民日报海外版,1992年7月8日第3版。

年间，俄勒冈州海岸由于 EL Nino 与海平面季节变化造成海面在 12 个月内抬升达 60 cm。风暴潮所形成的增减水在孟加拉湾形成年海平面差异达 100 cm 的记录[4]。人类活动的影响，如过度抽取地下水或建筑物重载，使河口三角洲地区大面积沉降，加大海平面上升值，如天津新港码头自 1966—1985 年下沉达 0.5 m（国家测绘总局，1992 年）。从某种意义上讲，这类变化可归为短周期变化，通过人工措施可控制。短周期海平面变化对海岸带会形成灾害性破坏。而对海岸潜在效应的推析，研究工作应致力于世纪性的全球范围的水动型海平面上升。这种世纪性的、全球性范围的变化促进了风暴潮与 EL Nino 现象发生频率的增加。

（二）海岸侵蚀效应

全球海平面上升在海岸带的主要反应是海滩侵蚀和海岸沙坝向岸位移。组成海滩与沙坝的沉积物主要是砂级的，属波场中的沉积物，是由波浪自水下岸坡海底掀带，并被浪、流进一步搬运（以横向运动为主）至岸坡上部堆积的。泥沙或来源于河流供给，或来自海蚀岸段以及由近岸海底供沙。后者主要是古海岸堆积（如中国沿岸），或为冰期低海面时的冰川作用沉积（如欧、美沿岸）。由于海平面持续上升，加大的水深，使波浪对古海岸带的扰动作用逐渐减小而形成海底的横向供沙减少，却加强了激浪对上部海滩的冲刷。同时，逐渐升高的海平面，降低了河流的坡降，减小了河流向海的输沙量。因此，世界上大部分海滩普遍出现沙量补给匮乏。海平面上升伴随厄尔尼诺现象与风暴潮频率的增加，使水动力作用加强，加上泥沙量匮乏的综合效应，使海滩普遍遭受冲蚀，而沙坝向海坡受冲刷与越流扇（overwash fan）的形成过程，综合表现为沙坝的向陆迁移。

海平面上升与海滩侵蚀是全球性现象，为此，海洋研究科学委员会（SCOR）已成立专门工作组（WG89）进行研究。中国科学院地球科学部要求：预估当 21 世纪海平面上升 50 cm 时，中国主要海滨旅游海滩的变化数据。本文选择了大连、秦皇岛、青岛、北海、三亚等五处著名海滨旅游区内的若干海滩进行了分析计算。依据多年考察的实测剖面数据，海岸段近滨带外界水深，除北海为 −2 m 外，其余均采用 −5 m，再结合大比例尺地形图与海图，可以获得有关参数。作者经过对多处海岸剖面重复测量结果对比研究后，认为 SCOR89 工作组研究成果及 Bruun 公式①基本反映海滩变化的自然规律，并结合我国海滩研究的实际，对 Bruun 图示加以修正（图 1）[5,6]。图中滨线采用低潮海滨线；Δy 为预定的海平面上升幅度；Δx_1 表示因海平面上升使部分海滩受淹没而产生的后退量；Δx_2 为海平面上升而产生的海滩侵蚀后退量；由于海平面上升而形成的海滩总后退量为 Δx_1 与 Δx_2 之和，计算的基本依据是海滩趋向于在海平面变动情况下形成新的均衡剖面。表 2 总结了各海滨沙滩在 21 世纪海平面上升 0.5 m 后的淹没与冲蚀后退数值[7,8]。各海滨沙滩面积损失的最小值为 12.7%（亚龙湾），因为该处海滩位于湾顶，背叠沙岸高。最大值达 66%（北戴河海滨面向开阔外海），上述海滩面积总损失量可达 266 万 m^2。实际损失

---

① Bruun 公式即：$R=\frac{L}{B+h}S$，其中：R 是后退速率；S 是海平面上升速率；h 是近滨沉积物堆积水深；L 是海滩至水深 h 间的横向距离；B 是滩肩的高度。

值可能要大于上述预算数值，因为激浪与风暴潮作用将更加频繁，其影响的范围更大，大部分海滩均会遭受海水淹侵冲蚀。

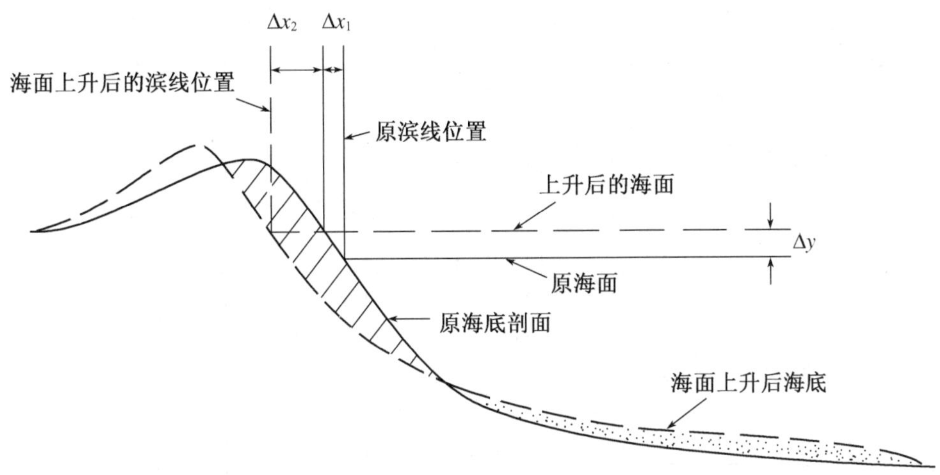

图1 海面上升使海滩遭受淹没与侵蚀

（Δy 为海面上升幅度，Δx_1、Δx_2 分别为海滨线因海滩遭受淹没和侵蚀而产生的后退量；假定海面上升前后的海滩剖面均已达到平衡）

海平面上升在淤泥质低地海岸的效应更为复杂，因为除水动型与地动型的自然因素变化外，还受人类活动的影响。为此 SCOR 于 1994 年专门成立 106 工作组，专题研究世界淤泥海岸的海平面变化效应，参加者有中、英、美、荷、印与巴西等国，由作者担任研究组主席。工作组将会进一步促进海平面变化研究与淤泥海岸研究的交流。

此外，由于温室效应招致海平面上升而预报珊瑚会在潮间带礁滩上复苏生长。根据晚更新世从 18 000～6 000 年前海侵中珊瑚礁向上繁殖的证据、实测珊瑚礁生长速率、生物化学量收支估算以及珊瑚礁对构造下沉的反应等情况来看，珊瑚礁的向上繁殖是伴随着较慢的海面上升，其速率为＞10 mm/a，若海面上升速率快，则珊瑚礁会沉没。珊瑚繁殖对海平面上升的效应亦取决于生态情况，特别是人类在珊瑚礁区的活动情况。面临海面上升，珊瑚礁岛均遭受侵蚀，除非有大量珊瑚在周边繁殖而供给足够的珊瑚碎屑使岛屿处于加积状态。

珊瑚礁的变化是世界性的海洋课题，涉及到领土、资源与岛屿人民的生存环境之发展变化。这是海平面变化引起之社会效应例证。

海平面变化、海岸侵蚀、海岸带森林破坏、海岸带污染及海岸带建筑物发展等一系列自然的与人为的因素，引起了海岸发生变化。海岸是人类居聚、工作、食物、矿产与能源、旅行、休假、防卫及排污之场所，与人类关系密切，在美国有近 50%的人口居聚于海岸带。海岸带变化招致了资源环境水平下降，引起当前海洋科学与地球科学界的高度重视，纷纷探求解决这一系列的矛盾，诸如：发展与保持、掠夺与保存、污染与清除、海岸后退与防护、地方与全球、私人与公众、低税与增税等，以期获得海岸带的持续性发展。出路是关注海岸带变化的研究，推行海岸带的一体化管理。

表 2　海平面上升 0.5 m 对我国重要海滨旅游区海滩的影响

海滨位置		现代海平面之海滩				海平面上升 0.5 m 之海滩响应预测								
						海滩淹没			海滩侵蚀			综合效应		
		长（m）	平均宽（m）	相对高（m）	面积（m^2）	滨线后退（m）	损失面积（m^2）	损失率（%）	滨线后退（m）	损失面积（m^2）	损失率（%）	滨线后退（m）	损失面积（m^2）	损失率（%）
大连	星沙公园	2 125	68.5	6.1	145 613	6.8	14 450	9.9	26.5	56 314	38.7	33.3	70 764	48.6
	东山宾馆	510	42.2	3.8	21 645	7.4	3 774	17.4	15.8	8 078	37.3	23.2	11 852	54.7
	大沙滩	756	56.3	3.8	42 560	7.9	5 972	14.0	24.7	18 674	43.9	32.6	24 646	57.9
	小　计	3 391	61.9		209 818	6.8～7.9	24 196	11.5	15.8～26.5	83 066	39.6	23.2～33.3	107 262	51.1
秦皇岛	北戴河	7 850	87.1	5.9	683 456	8.7	68 295	10.1	48.8	383 080	56.1	57.5	451 375	66.1
	西向河塞	3 124	223.6	6.4	698 466	6.7	20 930	3.0	41.5	129 650	18.6	48.2	150 580	21.6
	山东堡	756	88.2	3.5	66 672	7.5	5 670	8.5	25.4	19 202	28.8	32.9	24 872	37.3
	小　计	11 730	123.5		1 448 594	6.7～8.7	94 895	6.6	25.4～48.8	531 932	36.7	32.9～57.5	626 827	43.3
青岛	青岛湾	1 356	72.8	6.0	98 650	8.5	11 526	11.7	37.9	51 455	52.2	46.4	62 981	63.9
	汇泉湾	1 124	70.6	6.0	79 356	7.0	7 868	9.9	38.6	43 386	54.7	45.6	51 254	64.6
	浮山所口	1 625	193.1	5.4	313 857	8.9	14 462	4.6	26.4	42 932	13.7	35.3	57 394	18.3
	小　计	4 105	119.8		491 863	7.0～8.9	33 856	6.9	26.4～38.6	137 773	28.0	35.3～46.4	171 625	34.9
北海	外　沙	2 530	60.8	6.2	153 750	5.8～9.5	17 254	11.2	27.9	70 587	45.9	33.7～37.4	87 841	57.1
	大墩海至电白寮	5 516	258.4	5.0～9.2	1 425 588	5.4～9.8	41 926	2.9	48.1	265 335	18.6	53.5～57.9	307 261	21.5
	电白寮至白虎头	5 165	183.2	5.0～7.2	946 363	5.4～8.7	36 457	3.9	45.2	233 458	24.7	50.6～53.9	269 915	28.6
	小　计	13 211	191.2		2 525 701	5.4～9.8	95 637	3.8	27.9～48.1	569 380	22.5	33.7～57.9	665 017	26.3
三亚	大东海	2 650	81.5	5.9	215 905	7.9	20 935	9.7	12.2	32 330	15.0	20.1	53 265	24.7
	亚龙湾	8 880	166.1	5.4～13.4	1 475 184	6.8～9.8	74 592	5.1	12.7	112 776	7.6	19.5～22.5	187 368	12.7
	三亚湾	16 360	296.2	3.3～11.6	484 6024	5.6～10.2	137 654	2.8	43.6	713 296	14.7	49.2～53.8	850 950	17.5
	小　计	27 890	234.4		6 537 113	5.6～10.2	233 181	3.6	12.2～43.6	858 402	13.1	20.1～53.8	1 091 583	16.7
总　计		60 327	185.9	3.3～13.4	11 213 085	5.4～10.2	481 765	4.3	12.2～48.8	2 180 553	19.4	20.1～57.9	2 662 314	23.7

注：表中海滩面积指低潮为海滨线以上包括沿岸沙坝在内的海滨沙滩面积。

三、一体化的海岸带管理

一体化的海岸带管理是当前世界海岸科学与海岸带发展中的关键课题，尚处于研究探索的早期阶段。一体化涉及到各海岸段落间的相关问题，如海陆作用相关、不同层次的政府之间的相关、国家之间的相关以及自然科学、社会科学与工程技术科学间的相关等。

一体化海岸带管理的总目标是减少海岸带遭受自然灾害侵袭的脆弱性，使其维持基本的生态过程、生命支持系统与生物多样性，实现海岸带地区的持续发展。

一体化的海岸带管理是多目标、多方位的，它包括：分析发展带来的各种相关影响，包括互相冲突的使用招致的相关影响，自然过程与人类活动间相互关系及其相关影响，以及促进区段的海岸与海洋活动各部门间的和谐联系，强调社区群众必须参加确定研究和发展议程。

一体化的海岸带管理要素与内容涉及：资源用户的参与，交叉学科的合作，减轻对生物多样性的威胁，研究与确定海岸环境承载力，以及水产养殖与天然捕捞的联系等。

海岸生态系具有部门的完善固定的支配系统，但在制定一体化的规划、政策或贯彻执行政策方面，这一支配系统对生态却难以发挥作用，这是“一体化的海岸带管理”提出的一个挑战。

成功的一体化海岸带管理项目具有的特点与标志是：有明确的有限的主题，成功地参与制定与贯彻执行政策，提供一个合理、有效的决策制定过程，使项目被增值地执行。一体化地综合运用生态系功能领域可利用最佳的知识，并促进所有参与人员的互相学习。

举例领航项目应考虑的各方面：

——地方/区域科学能力与水平的提高

——环境与发展之间的相互依赖性

——建立区域一体化海岸带管理数据库

——惯例管理策略的进步与发展，如：紧急反应措施

⊙ 识别相邻海岸各系统的相互联系的管理策略

目标：保持海岸资源系的功能一体化

——减少资源使用者之间的冲突

——便利了多部门的发展进步

——社团基础知识的一体化

——更新的公共教育事业

——重建政府与个人、区段与环境与经济的关系

⊙ 制定海岸带应急计划

基础上结合的大学与社区训练

目标：初期合作管理联系的建立

——公众有效地参与决策

——年度活动计划的进展

——确定可度量的项目目标正效应

——方案的战略
——与过去经验结合、并建立在中国本国的能力上
——符合当地人民的需要
——为双方创造经济效益的机会
——中央政府的支持是项目的战略性保障
——一体化海岸带管理的边界有哪些方面涉及到尊重本地居民的特殊权益
——环境及其利用特征的研究
——未来应用的规划
——项目的策略，项目与政治体系类型的适应
向其他地区（国内与国际）传输知识的潜在性
目标：评定当前在地区和国家水平上海岸带管理的成熟性
——对国家的、省级的海岸带管理基础数据（数据库）做出贡献
——有助于建立紧急反应计划的资料
——评价鱼类等水产资源及其栖息地的减少与水质恶化状况
——目标：长期的一体化海岸带管理哲学的能力建设
——公众教育以提高海岸带管理意识

主要参考文献

[1] Hoffman. J., Keyes, D. and Titus, J. G. 1983. Projecting Future Sea Level Rise: Methodology, Estimates to the Year 2100, and Research Needs. U. S. Environment Protection Agency. Washington D. C.
[2] Revelle, R. R. 1983. Probable Future Changes in Sea Level Resulting From Increased Atmospheric Carbon Dioxide, Changing Climate. National Research Council Report. National Academy Press, Washington, D. C. 433 - 448.
[3] Van Der Veen, C. J. 1988. Projecting Future Sea Level Surveys in Geophysics, 9, 389 - 418.
[4] SCOR Working Group 89. 1991. The Response of Beaches to Sea Level Changes A Review of Predictive Models. *Journal of Coastal Research*, 7(3): 895 - 921.
[5] Bruun, P. 1962. Sea-level rise as a cause of shore erosion Journal Waterways and Harbours Division. *American Society Civil Engineers*, 88 (WWI): 117 - 130.
[6] Bruun, P. 1988. The Bruun Rule of erosion by sea-level rise: A discussion of large-scale two-and-three-dimensional usages. *Journal of Coastal Research*, 4: 627 - 648.
[7] 南京大学海洋科学研究中心. 1988. 秦皇岛海岸研究. 南京：南京大学出版社.
[8] 庄振业，陈卫栋，许卫东. 1989. 山东半岛若干平直砂岸近期强烈蚀退及其后果. 青岛海洋大学学报，19(1)：90 - 98.

中国的海岸科学与海洋地理学的新进展*

一、中国海岸研究简况

在中国，系统的现代海岸科学研究始于20世纪50年代中后期。主要由于海港建设的需要，并受到苏联海岸科学发展的影响。1958年，天津就新港回淤研究邀请了苏联专家工作组，海岸学权威B. П. 曾柯维奇在北京大学做了海岸科学讲座，对山东半岛、辽东半岛、珠江三角洲及海南岛作了踏勘性考察。他以海岸演变的动力学观点，对中国一些海岸区域作了概括论述，发表在他的专著《海岸发展学说》中，这为当时中国海岸研究开拓了新的领域。

20世纪50年代后期以来，中国海岸研究逐步形成了一些新概念：海岸的范围包括沿海陆地、海滩或潮间带浅滩，以及水下岸坡三部分。研究海岸需进行沿岸的及水下的两栖性工作。研究工作具有多学科的性质，它包括地貌学、海岸动力学、沉积学，并发展为海岸动力地貌学。海岸研究的队伍包括海岸地貌学家、沉积学家、物理海洋学家以及海岸工程师。主要研究目的是确定海岸沉积物来源、海岸带泥沙运动、海岸冲蚀与堆积特性、海岸成因以及推论海岸发展的趋势。调查方法包括广泛使用观测仪器从事浅海与沿岸陆地调查与定位重复测量；实验室多项分析与计算；应用计算机进行系统总结、制图与做出工程或决策方案。

中国海岸研究的主流是重视现代过程与密切联系生产实践，一开始即重视定性分析与定量分析相结合，并从事模型实验。大量研究工作是为选择海港、确定码头位置以及选挖深水航道服务的。当时，我国漫长的海岸线上分布着有限的港口，平均每500 km才有一个港口，远不能满足日益增长的货运需要。至1958年，中国已有400个泊位，其中199个是水深大于10 m的深水泊位。1989年，深水泊位达214个，而中级水深泊位为306个，年吞吐量为4.08亿t，预计“八五”期间还要增加100个深水泊位及80个中级泊位，总吞吐量可达5.8亿t。然而，已有的港口布局是大港主要分布于北方，江苏省长达1 000 km的岸线上，只有一个连云港位于省域的北部。基岩港湾海岸于华南广泛发育，约有100多个海湾可利用发展为大中型港口，这为海岸科学工作者提供了广阔的工作场所。港口建设需妥善处理海岸自然环境中“水深浪静”这一对矛盾。在中国这一矛盾尤其突出，因为河流供给海岸带大量泥沙，细颗粒泥沙易于携带、搬运而难于沉淀堆积稳定。潮流作用与细粒泥沙运动是中国海岸环境的特点，港口回淤是中国海港建设中的突出问题。典型的例子是天津新港。1949年至1952年对该港进行了大量疏浚，使港口水深达

---

* 王颖：中国地理学会主编《面向21世纪的中国地理科学》，168－182页，上海教育出版社，1997年。

到 9 m。由于港口回淤严重，只有依靠挖泥才能维持使用水深，而回淤量伴随着疏浚量增加，平均每年的挖泥量达 600 万 m^3，挖泥费用大，港口不能充分使用。回淤研究引起大规模的海岸调查工作，研究表明：淤泥物质主要来源于黄河汇入的泥沙，它是形成渤海湾潮间带淤泥滩的物质基础；海河泥沙对新港所在的潮间浅滩提供了补给；新港泥沙主要来自港口所在的浅滩。海岸研究为减轻新港回淤提供了科学依据，其工程措施是：延长南北防波堤并堵住缺口，疏浚航道与港池，吹填多余的港域等。多年的实践证明当时提出的工程措施是正确的。

目前，天津新港的入港航道长达 24 km，宽为 60 m、120 m 及 150 m，水深 11 m。港口泊位达 70 个，年吞吐量约 2 000 万 t。继天津新港回淤研究以后，很多港口的选建是在进行了充分的海岸动力地貌研究后才规划施工的。

生产实践推动了海岸科学研究，例如，天津新港项目促进了对淤泥潮滩海岸以及黄河三角洲的研究；秦皇岛油码头选址项目促进了对砂砾堤与沙坝海岸研究；长江口项目促进了河口学；华南建港项目促进了对潮汐汊道的研究等。这些研究推进了对中国海岸的认识与理解，建立了具有中国特色的海岸学说，如潮滩分带性、贝壳堤的形成与古海岸指标、潮汐汊道的沉积动力等。

另一类实践研究是勘探海滨砂矿，如锆英石、金红石、钛铁矿、独居石和砂金等矿。砂矿富聚于山溪性河床粗砂中，海滩、沙嘴、沙坝等的细砂沉积中，以高潮线附近堆积为主，海滨沙丘或大河三角洲中则无富聚的砂矿。现代海岸砂矿源主要为古海滨堆积，矿物来源于中生代的花岗岩类岩层，在辽东与山东半岛，福建、广东与海南等地的海蚀-海积型港湾海岸带更为发育。砂矿项目促进对海岸砂体结构以及水下岸坡沉积物的研究。

特殊类型的海岸如珊瑚礁与红树林曾在浙闽南部海岸、广东海岸、海南岛沿岸及南海诸岛进行了细致的研究。该项研究认识到全新世的珊瑚礁发育，为分析不同海岸段落海平面变化与构造运动的影响提供了判别依据。在福建、广东、海南、广西分别建立了红树林自然保护区作为美化环境与生态效益之示范。

自 1980—1985 年，中国进行了大规模的大陆海岸带综合调查，主要目的是摸清海岸、滩涂的自然资源(数量与质量)，获得海岸自然环境和社会经济状况的基本资料，对中国海岸环境与资源作出综合评价，制定全面开发海岸带的规划。通过海岸带综合调查，汇集了大量科学资料，包括海岸带岩层、构造、地貌、沉积、海洋、气候、土壤、渔业与水产、植物和动物、农业、海岸工程等。作为海岸调查工作的继续，90 年代中国又进行了海岛调查，着重于中等规模或乡以上的海岛。上述调查为海岸海洋开发、管理与研究奠定了基础。

二、中国海岸研究的主要进展

中国海岸研究 1980 年以来得到迅速发展，以下对几个主要方面进行评述。

1. *平原海岸*

平原海岸仍为热点课题，在理论与实践两方面均有显著进展。平原海岸形成于构造沉降带，通常为大河的下游，如黄河三角洲平原及苏北平原。

潮滩(潮间带滩)是广阔平原海岸的特征地貌。非常平缓的岸坡、潮流动力与大量的细粒沉积物供应是控制潮滩发育的三个主要因素。潮差大小影响着潮滩发育的宽度，潮滩地貌特性与其沉积结构受潮能(流速)所控制。中国典型的潮滩海岸潮差约 3 m，但是潮滩的宽度界于 6～18 km。中国主要潮滩海岸发育于华北大平原边缘，该处海岸坡度小于 1/1 000，波浪作用被限于外滨而达不到岸边，因此，潮流是潮滩上的主要动力。作为对比，在南方，东海和南海沿岸，潮滩发育在狭长海湾内以及岛后隐蔽地段。这类环境水域面积有限，吹程小而限制了风浪作用。粉砂是潮滩的主要沉积组成，包括细粉砂及粗粉砂，亦含黏土及极细砂成分。这些物质是由波浪从外滨海底扰动悬浮、再被涨潮流向陆携运而堆积于潮滩之上。泥沙系陆源物质，在中国沿海主要是冲积物，是低海面时黄河与长江等大河的堆积。

潮滩具有地貌、沉积、底栖生物的分带性，在各种气候环境——从寒带、温带到热带皆有发育。分带特征反映出潮流动力沿滩地的变化规律。通过中国与英国潮滩对比研究，总结出三种类型的海岸潮滩。

(1) 粉砂质潮滩，以中国黄海的苏北沿岸潮滩为代表。从岸向海具有三个带：潮间带的上部为泥滩，中部为淤泥粉沙滩，下部为砂质粉沙滩。中等强度的波浪扰动岸外辐射沙脊群——古长江三角洲堆积，泥沙被流速较强的潮流搬运向岸。

(2) 淤泥质潮滩，发育于中国内海，尤以渤海湾为著。黄河汇入大量细颗粒的泥沙—粉砂与黏土。潮间带浅滩上明显出现四个带：潮间带的上部为龟裂带，由黏土组成，平时出露，仅特大潮时被海水淹没。特征地貌是在平均高潮位以下有一个宽约 0.5～1.0 km 的泥沼带，称为内淤积带，只有在渤海湾因有大量的黄河泥沙的补给，才会发育这类泥沼带。中潮位附近为滩面冲刷带，鱼鳞般的冲刷体或沟槽剥露出该处沉积为黏土质与粉砂质的交叠薄层。低潮水边线附近为大范围的粉砂质波痕微地貌，是为外堆积带。

(3) 砂质潮滩，发育于英国东海岸，仍以粉砂成分为主，但在潮滩上部、下部及潮沟中均出现较多的砂质堆积。这类潮滩代表着经常遭受风浪作用的海岸环境。砂砾物质来源于外滨海底。

典型的潮滩沉积具有页状薄层理结构，潮滩堆积具有双层结构：顶层是由悬浮淤泥沉淀所形成的垂直堆积，主要由涨潮流带入，层厚由大潮与中潮潮差决定；底层是横向堆积，由粉砂与细砂等底移质组成，最大的层厚相当于低潮位与海岸基底之间垂直距离。

台风季节，强风浪与急潮流形成大范围的潮滩冲刷。高潮滩及潮上带遭受冲刷后却伴以加积作用。台风期间在高潮滩沉积了粗粉砂而分选良好，只含有很少部分的更细物质，而其下层的沉积物代表平常天气状况的沉积，为细粉砂，分选较差。

潮水沟蜿蜒曲流纵贯潮滩而发育，部分潮水沟侧尚有小型天然堤。潮流动力在潮水沟中呈现系列性的不对称现象。在同一潮周期中，潮沟中的落潮流量大于涨潮流量，因为潮沟中 10%潮流量系剩余流，是涨潮时进入潮沟中的。由于滩面水归入潮沟，故最大的落潮流速较涨潮流强而且落潮流延时长。因此，潮沟动力是以落潮流居主导。潮水沟不仅分割了潮滩，同时亦是传送沉积物的通道，并形成潮滩沉积中的镶嵌沉积，其沉积颗粒较潮滩沉积粗，或形成粉砂质交错层，或成为砂质凸镜体。

贝壳堤分布于粉砂潮滩上部，成为平原与潮滩间的明显海岸界限。贝壳堤主要由贝壳和壳屑组成，夹有少量的硅质成分，如粉砂或细砂。近河口的贝壳堤亦能含少量黏土物质。贝壳堤与海滩上部的滩脊或沿岸沙堤在成因及组成物质上皆有不同。贝壳堤高于潮滩可防潮水淹浸，贝壳又为暂时淡水(降水)储水层，成为"苦海盐边"的绿岛。因此，引起考古学家、海岸地貌学家与地质学家的兴趣。据近期的著作，在渤海西岸与黄海西岸有 4 条贝壳堤，在上海附近有 5 条贝壳堤。同时，也曾对东海大陆架的古贝壳沉积作了定年研究。

大部分软体动物贝壳的^{14}C定年资料是可信的，并且是判定老的海岸环境的依据。但是，海岸沙堤、砂质堆积体皆含有贝壳，不是所有含贝壳的堤脊皆为贝壳堤。例如，苏北海岸平原的含贝壳堆积体实为老的海岸沙坝，渤海湾西岸的"老贝壳堤"，实为海岸带浅水堆积。确定贝壳堤要有足够的沉积与堤脊形态的证据，同时需注意水下岸坡的结构特点。贝壳堤形成标志着一种特定的海岸环境：平原海岸的潮滩由于泥沙供应骤减而遭受波浪冲刷，海岸后退，在蚀退的平原海岸带激浪活跃，贝壳从潮滩沉积中被激浪刷出并被进流带至潮滩上部堆积成岸堤。据黄河下游河口变迁史，淤泥质潮滩与贝壳堤交替地发生于渤海湾与苏北平原海岸带。

平原海岸体现着波浪冲蚀与泥沙堆积的动态平衡。如果泥沙供应量大于侵蚀量，则岸线淤进；如果泥沙供应量骤减或停止，则平原海岸会遭受波浪冲刷而后退；两者相当，则海岸保持稳定。大河泥沙供应对平原海岸发育有关键性的影响，其影响超过相对海平面变化之效应。例如，现代黄河汇入渤海南部，近 130 年来三角洲海岸向海推进达 20～28 km。虽然渤海海区近期自中新世以来一直处于下沉趋势，近百年的海平面亦是上升的；而苏北废黄河口海岸，自 1855 年以来海岸后退达 17 km。为此曾建立一系列丁坝、顺岸防波堤以及采取块石坡护堤等办法防止海岸蚀退。处于相同沉降构造与海平面上升背景的苏北中段海岸，由于有岸外沙脊群的掩护与泥沙补给，潮滩海岸是不断堆积加宽的。

2. *潮汐汊道港湾海岸*

潮汐汊道港湾海岸广泛地分布于中国东海与南海沿岸。

潮汐汊道是海洋深入陆地的分汊，它是由一个入口水道与纳潮水域组成。纳潮水域的面积、形状和潮汐特征决定了口门段过水断面的大小及内外两侧浅滩与水道的特征。应用潮汐汊道为航道，常遵循 O'Brian 的 P-A 定律(即 $A=CP^n$)而进行航道治理。

潮汐汊道是海南岛 64 个港湾中的重要组成类型，根据其成因可区分为三种潮汐汊道。

(1) 沿构造断裂带或火山岩与古海岸砂层交界带发育的潮汐汊道。如位于海南岛西北岸的洋浦湾，是海南岛最大的潮汐汊道港湾。

(2) 沙坝封闭的潟湖湾，如三亚湾、亚龙湾、清澜港等。

(3) 全新世海侵沉溺的山谷或河谷。此为大型汊道湾，水深条件优越，榆林港属此类型。潮汐汊道内湾的维护与航道水深的维持受两个因素的控制：① 纳潮量。纳潮量与潮差及涨、落潮流的延时有关，强力的落潮流可冲刷航道的淤积，是天然的疏浚力，落潮流的流速强弱与纳潮量的大小有关。因此，需注意保持汊道内湾的面积与水深，以蓄纳大量潮水，加大纳潮量。② 泥沙来源。确定海岸与潮汐汊道的泥沙来源，尽量减少沿岸的泥沙供应，保证沉积物天然冲刷或泥沙沿岸越过，而不在航道落淤。

据南京大学海岸与海岛开发国家试点实验室在海南岛的 Lehigh 钻孔分析结果，反映出海南岛大型潮汐汊道(如洋浦港、三亚湾等)的沉积层具有三种沉积动力相。

(1) 内湾相:由于有小河汇入的影响,沉积物以粗颗粒为特征,砂或砂砾堆积成浅滩或暗滩;在内湾外侧的堆积系涨潮流三角洲,主要为分选好的细粒泥沙,具虫孔有机斑等扰动结构。

(2) 深水道相:在内湾出口的狭窄通道内主要堆积粗砂与砾石,并有浮泥于表层;自狭窄通道延伸出的深槽中主要为细颗粒的粉砂质沉积。

(3) 外湾相:主要为落潮流三角洲沉积,表现为混合的细粒沉积物,但是具有风暴堆积的砂质透镜体或夹层,这是该环境的特征标志,反映着经常遭受风暴浪潮之扰动。位于深槽终端的落潮流三角洲上,沉积物较粗并混杂贝壳与邻近珊瑚礁的破碎产物。

潮汐汊道港湾以外的浅海沉积物质地均一,具较多的风暴砂夹层,但底栖生物扰动较内湾减少。

3. 中国海岸研究的重要成果

中国海岸研究有较多重要文献,反映了近年中国海岸研究的主要成果,如《中国海岸带与近滨沉积》(英文版任美锷主编)等,本文略加综合阐述。

(1) 海岸沙坝与海岸沙丘是冲积平原海岸的主要地貌组成。对于滦河三角洲、韩江三角洲海岸发育的沙坝与海岸沙丘系列的物质来源、沉积机制、演化顺序等方面的研究,结合全新世海面变化,得出的若干结论,可与大西洋东海岸的情况相互佐证。

(2) 海岸带构造抬升遗证。海岸阶地的研究为地质、地理学家关注的传统课题。重视现代过程,注意地貌标志的综合分析与沉积证据,致力于区别地壳升降标志与海平面变化效应,重视年代学数据等是当前研究工作中的主要趋向。对江苏云台山(古海岛)、辽东半岛大连地区、海南岛三亚地区、厦门市郊的曾山等处的各级海蚀阶地的研究,都已取得了相应的成果。对于海南岛西北部沿岸更新世晚期及全新世中期两期火山活动的研究,结合年代学数据等的分析,认为众多的地质地貌现象综合反映出北部湾海底具有扩张趋势。

(3) 全球气候变化引起海平面上升成为海岸侵蚀的动力,温室效应将加速海平面上升,这是当前的热点课题。

海平面上升引起风暴潮频率加剧,加大对上部海岸冲蚀;降低河流坡降与减少向海输沙量;加大了古海岸沉积的水深,从而减少了对海岸的横向供沙。因此,砂质海滩普遍遭受侵蚀,海岸后退,影响到海岸带丧失土地、建筑与旅游资源;三角洲低地海岸与湿地受淹没或移位,影响到食物链与生态系的变化;平原海岸与河口地区的海水入侵,降低水质;改变海岸带潮差,改变泥沙沉积类型等。这一系列自然环境的变化必然形成对社会经济的影响。

当前,对世纪性海平面变化、全新世海平面变化、晚更新世海平面变化等均有专题研究,并具有多学科结合的特点,预计不久将会在海平面变化历史、变化趋势、变化影响与对策方面皆会出现系列的显著成果。

(4) "人类是一种重要的地质营力"。人类活动通过对河流的影响,直接或间接地影响海岸地貌的塑造。如人为挖堤引起黄河夺淮入黄海,造成废黄河水下三角洲影响苏北平原海岸地貌。人为垦耕引起黄土流失加剧,造成现代黄河三角洲的弱潮型河口堆积。

人类活动影响入渤海的泥沙沉积速率。人类过量开采地下水及建筑重载，造成上海、天津等地的地面沉降加剧，引起风暴潮及洪涝灾害。苏北沿海各河口建闸造成闸下河道淤积和河口过程的变化。冀东平原河流因建水库及分水工程，减少了入海径流泥沙，影响河口动态平衡，造成河口沿岸发生海蚀后退。这些都是人类活动所造成的海岸变化。

三、海洋地理学的发展

海洋地理学是地理科学的新分支，具有自然科学、社会科学、技术科学相互交叉渗透的特点。从学科发展分析，它属于地理学与海洋学之间新的结合点学科。海洋地理学研究的客体是海洋，包括海岸与海底，范围涉及到气、水、生物与岩石圈，研究内容包括海洋环境、海洋资源及其开发利用与保护，海洋经济，疆域（海岸、岛屿、领海、大陆架、专属经济区、公海等）政治、立法与管理，海洋新技术发展、应用及影响等。概言之，海洋地理学研究是从宏观调控与区域入手，从立法、政策、区域经济、管理着手进行海洋资源开发、环境利用与保护。当代面临着人口、资源与环境的巨大挑战，开发利用海洋是当务之急。1982年通过并于1994年11月16日正式实施的《联合国海洋法公约》，使人们的海洋观念发生了深刻的变化，原属于公海的大约1.3亿 km^2 海域将划归沿海国管辖，其面积约与地球整个陆地面积（1.49亿 km^2）相当。公约又引起沿海国所管辖的范围、所占有的海洋资源与沿海国权益均发生重大变化。有的国家已把管辖海域作为国土的海洋部分，出现了海洋国土这一概念。按这一规定，估计可能划归我国管辖的海域面积约占我国陆地总面积的1/3。管辖范围的划分带来权益之争，例如，200海里专属经济区确定后，在开阔海域中丧失一个具备人类生存条件的岛屿，就会失去43万 km^2 的管辖海域。21世纪为"海洋世纪"或"太平洋世纪"已为学者们所共识。因此，海洋地理学是适应时代发展而兴起、具有重要意义的新学科。

1986年国际地理联合会（International Geographical Union）正式成立海洋地理研究组（Study Group On Marine Geography），中国是发起国之一。研究组的核心课题集中于"人类活动与海洋管理之相互关系"。它引起各方面的广泛关注，通过研讨提出：① 海洋法所涉及的地理课题；② 海洋的利用与管理；③ 海洋地理学的国际合作等三个方面的课题。

1988年8月，在澳大利亚悉尼举行的第26届世界地理大会期间，国际地理联合会投票批准成立"国际海洋地理专业委员会"（Commission on Marine Geography）。当时共有3个研究组升格为专业委员会，而国际地理联合会下属的42个专业委员会中有20个专业委员会被取消。国际海洋地理专业委员会有成员国40多个，首届主席是意大利热那亚大学地理系教授 Adalberto Vallega，中国地理学会海洋地理专业委员会副主任委员杨作升教授为首届常务理事。1992年第27届世界地理大会后，由英国威尔士大学的 Hance Smith 教授任第二届海洋地理专业委员会主席，中国地理学会海洋地理专业委员会主任委员王颖教授任常务理事。该专业委员会主要对全球性问题进行协调、讨论以及从事专题研究，以推动海洋地理学发展，它每年发布一次新闻通讯，两年举行一次学术会议，四年举行一次大会，出版有关著作与刊物，经费自筹。1994年6月于加拿大召开学术讨论会，主题是海岸管理。

国际海洋地理专业委员会具有较强的社会科学背景，着重于海洋法与疆界划分问题，主要涉及政治地理领域。20世纪90年代以来，则明显地转向海洋管理，增强了自然科学（海洋环境与资源研究）与技术科学（遥感与地理信息系统应用）的内容与研究力量，加强对区域海洋研究的关注。这一转变，反映了对海洋特性认识的深化。海洋是大气、海水、生物与岩石圈相互搭界共同作用的场所，既是生命的摇篮、风雨的源地，也是人类发展所依赖的食物、土地、矿产、动能以至气、水资源所在，海洋环境的发展变化与人类社会经济活动息息相关，必须采用多学科交叉渗透的观点与方法调查研究海洋的区域特性与变化规律，才能达到综合开发与合理利用的目的。

海洋地理学的主要内容及发展趋势，可归纳为五个方面：① 海洋管理。我国已加强海洋管理的立法、执行及有关的研究。但当前国际海洋管理已赋予全新的概念，即以环境与人相关联的全球变化机制，建立全球变化的海洋管理理论。② 海平面上升与人类活动关系问题。这在我国已在学术界引起广泛重视，大体与国际上同步进行有关科研，今后任务是地理学者要有更多切实可用的成果，使政府领导、决策部门及广大工程技术界重视与应用。③ 科学技术与海洋管理。其中地理信息系统是海洋管理的关键技术。④ 海洋政策。主要是各海洋政策的制定，国家对海洋管辖区的作用，以及海域疆界问题。⑤ 国家、地区的自然体的海洋管理。

中国海疆辽阔，濒临太平洋与边缘海，所辖海域面积约300万km^2，蕴藏着丰富的渔业、矿产、动能、港湾、土地、化工与旅游资源。有海洋生物2 000多种，其中鱼类1 500种，经济价值较大的鱼类150种，最大持续渔获量为470万t。已发现16个中新生代沉积盆地，总面积约130万km^2，石油资源308亿t，天然气资源量约7.2万亿m^3，海洋能源约6.3亿kW，浅海滩涂面积达2万公顷（1公顷=10 000 m^2），港湾岸线5 000 km，海水资源取之不竭。

历史上，中国航海、对外贸易与文化交往具有光辉的成就，至今仍为海外传扬。以后由于封建锁国政策与科技落后，抑制了中国向海洋开拓，国民的海洋意识薄弱。直至20世纪早期，沿各大河口兴起了津、沪、穗等大的港口城市，才有稀少的海岸开发与移民点。新中国成立后，重视发展海上交通、海洋调查、制图、油气资源勘探、海底通信工程、沿海渔盐与种植业等，以及参加讨论有关海洋法的国际会议。20世纪60年代提出“向海洋进军”的号召，当时的口号是：“摸清中国海，进军三大洋，登上南极洲”，组织了队伍，建立了海洋局与教研机构，开展了实质性的海洋调查与开发事业。直至80年代，科技事业繁荣、国际交流加强以及市场经济的发展，我国海洋事业有了显著的稳步发展。完成了全国海岸带与海涂资源综合调查、全国海岛资源环境综合调查研究，获取了系统的科学资料，开展了规划与实施开发，海岸带已由昔日的禁区成为今日经济发展的黄金地带。海洋油气资源勘探、开发已进入国际招标、开发与正式生产。太平洋锰矿瘤调查已取得开发权，西太平洋海气交换、黑潮等项国际合作研究取得突出的科学成果，实现了南大洋考察与南极大陆建立科学考察站进行连续观测的夙愿，60年代的口号基本实现。海洋环境监测与保护工作在实施，管理机构逐步建立，沿海各省市相继制定了海岸带管理条例，经各级人民代表大会通过，并拟定与试行有关的管理法规。高新技术在海洋中的应用已列入日程。

海洋科学、地理科学在生存环境持续发展项目中占有举足轻重的地位，两学科交叉结合点的海洋地理学有着广阔的发展前景。当前的国力尚未形成对海洋开发管理的大量投入，海洋法规尚待建立与健全，对一些重要的国家海洋权益亟须维护，海洋开发过程中出现的新矛盾问题有待解决等。海洋地理科学宜从海陆结合的观点与方法进行资源、环境与新技术开发应用的实践与总结，促进一体化的海洋管理，为迎接 21 世纪海洋开发新时代的到来做出贡献。

主要参考文献

[1] Wang Ying. 1980. The Coast of China. *Geoscience Canada*, 7(3): 109 - 113.

[2] Wang Ying and David G. Aubrey. 1987. The characteristics of the China coastline. *Continental Shelf Res.*, 7(4):329 - 349.

[3] 朱大奎，许廷官. 1982. 江苏中部海岸发育和开发利用问题. 南京大学学报(自然科学版)，(3): 217 - 236.

[4] Ren Mei'e, Zhang Renshun, Yang Juhai. 1983. Sedimentation on Tidal Flat of China—with Special Reference to Wanggang Area, Jiangsu Province. In: Proceedings of International Symposium on Sedimentation on the Continental Shelf, with Special Reference to the East China Sea.

[5] Ding Xianrong and Zhu Dakui. 1985. Characteristics of Tidal Flow and Development of Geomorphology in Tidal Creeks, Southern Coast of Jiangsu Province. In: Proceeding of China-Germany Symposium on Engineering.

[6] 陈吉余，虞志英，恽才兴. 1959. 长江三角洲的地貌发育. 地理学报，25(3):201 - 220.

[7] 李世喻. 1962. 古代渤海湾西部海岸遗迹及地下文物的初步调查研究. 考古，76(12):652 - 657.

[8] 王颖. 1964. 渤海湾西部贝壳堤与古海岸线问题. 南京大学学报(自然科学版)，8(3):424 - 440，3 页照片.

[9] Zhao Xitao. 1989. Cheniers in China, an Overview. *Marine Geology*, 90(4): 311 - 320.

[10] Wang Ying and Ke Xiankun. 1989. Cheniers on the Coastal Plain of China. *Marine Geology*, 90(4): 321 - 335.

[11] Wang Ying. 1983. The Mudflat Coast of China. *Canadian Journal of Fishery and Aquatic Science*, 40(Supplement 1):160 - 171.

[12] 王颖. 1994. 海洋地理学的当代发展. 地理学报，49(增刊):669 - 676.

海洋地质学的发展与展望*

一、海洋地质学定义、研究领域、特点与意义

海洋地质学以海水作用与覆盖的岩石圈为研究对象，其研究范围涉及海、陆交互作用的边界与洋底，包括海岸与滩涂、大陆架、大陆坡、大陆隆以及深海洋底。研究内容包括海岸与海洋底部地貌，相关的浪、潮、流、生物、化学与泥沙作用过程，沉积与岩层（尤其是洋壳的结构）特点、构造变异成因与演变，以及由此而产生的对海水、海洋作用与海洋生态环境的影响。研究对象与内容具有范围广、地球表层与深部作用相互关联的特点。

海洋地质学具有综合性、全球性和技术复杂性。厚层海水"帷幕"覆盖与作用的海底，不能直视与直接测量，必须通过技术装备（船只与仪器）探测（深度、地层、重力与磁学等）；海底过程互相影响并演变，大气圈、水圈、岩石圈与生物圈在海岸与海底是相互交汇、互相作用影响的，综合性突出，所以研究工作需要多学科协同合作。海洋地质学涉及的学科群与结合点领域，可分为表层地质学、深部地质学与应用海洋地质学。

（一）表层地质学部分

研究海底表层岩石圈及沉积层的形态、结构及发生发展变化。如：

（1）海岸海底地貌学。海岸与海底地貌是构造活动与外力作用的直接表现与结果。研究海底首先接触的是海底地貌——洋中脊、裂谷、海山、海盆、海沟、峡谷与堆积扇等，是研究海底现代作用过程的依据。洋底地貌图的完成，推动了板块学说的发展。海洋军事活动与维护海洋权益，十分重视海洋地貌的变化。

（2）海底与大洋沉积学。海洋沉积中包含着环境变化过程的信息，是研究海洋形成发展与全球变化的重要依据，由此发展而形成古海洋学。

（3）海洋环境地质学。研究水圈、气圈、生物圈、岩石圈相互作用的综合效应，分析判断气候与海平面变化，以及对人类生存环境的影响，探求相对稳定的地域与时期，以适应海洋工程建设的需要。

（二）深部地质学部分

研究地壳与洋壳的岩性、构造特点和分布变异；研究上地幔热液活动效应，以及通过洋中脊裂谷的活动过程；研究洋底动力机制、海底扩张与板块运动对深部地质过程的反映等。海底深部地质的研究可促进对地球的结构与发展变化历史的认识。同时，深部动力机制可阐明火山、地震、热流点等活动规律。深部地质研究需借助于地球物理探测方法与

---

* 王颖：中国海洋学会编《21 世纪中国海洋科学与技术展望》，49－59 页，海洋出版社，1998 年。

潜海调查手段，同时是动力地质、岩石、矿物、矿产地质、构造地质与地球物理学科的新结合点。

（三）应用海洋地质学部分

海洋地质学的重要性在于其研究对象是人类生存发展所依赖的环境资源巨大潜在源泉。自20世纪后半期以来，海洋环境资源的开发利用日渐广泛，发展了海洋地质调查技术，也促进了海洋地质学的发展。目前，这部分含下列三方面：

（1）海洋环境工程地质学。海洋石油资源的开发、海上平台、钻井的设置、大洋矿产的开挖、海底倾废等，涉及到工程所在的环境特性、工程地质条件、海岸与海底稳定性等，需进行最适宜场点的选择、不良环境地质的工程措施以及建设后工程地质环境变化的预测等。

（2）海底矿产地质学。研究海底矿产资源的类型、丰度、成矿机制、分布规律、再生速率，以及开采方法等。海底矿产涉及到砂石建材、稀有与贵重矿物砂矿、石油与天然气矿产、多金属软泥、矿瘤（结核）与结壳、磷（硫）块状矿产、海底火山与洋中脊热液矿产等。

（3）海洋地质调查研究技术。包括空间与海面的遥感遥测技术，海面与海底的定位技术，海洋场动力环境、海水层理、化、生物要素系统剖面测试与记录技术，海底起伏测量、摄像与制图技术，海底沉积与岩层结构的探测技术，底质采样与采取沉积层与岩芯技术，地震、重力、磁力与热液等地球物理与地球化学特性测试与记录技术等调查技术，藉以获取第一性资料。上述工作在不同深度与不同气候条件的海域要求有不同灵敏度的仪器与不同适应性船只。海洋地质调查设备昂贵、消耗性大，经济有效的办法是根据任务性质组装有相关学科实验设备的综合性调查船，以便在一个航次中获得海域多项相关资料。

总之，海洋地质学的研究目的与意义在于：认识了解占地球表面70%以上的海洋盆地的地貌起伏变化特点，它所反映的地球内部与表层的作用过程；研究海洋沉积层与岩层的结构与分布特征，以阐明它所记录与反映的地球发展与变化过程；认识了解地球内部（地幔与地核）动力活动机制，阐明地震、火山与海底矿产的形成与发展变化规律；开发利用海洋、海岸与海底资源，以及由此而涉及的管辖权益与新技术开发利用等方面。海洋地质学的科学与应用意义重大，它是了解地球发展变化——过去、现在与将来的关键学科，是人类生命、生产与生活赖以存在与发展的重要领域——海洋的三大支柱学科之一。

二、海洋地质学的发展与现状

（一）历史回顾

早期的海洋地质学工作是在航海的发现中积累了有关资料。第二次世界大战以后，海洋地质学与地球物理学发展很快，对海底展开了系统测量，测深精度达到1/5 000。B.黑曾（Bruce Heezen）致力于调查研究贯穿大西洋南北，高达3 000 m的洋中脊体系，阐明了地球是被洋中脊所环绕，洋中脊的顶部分布着一条裂谷，在20世纪50年代与60年代期间，在各大洋进行了重力、磁力、地震与海底热流量测量。至60年代中期明确了地磁场周期性的倒转是地球历史发展过程的记录。这一系列研究成果发展形成洋底扩张与板块学说。

(二) 海洋地质学的当代发展

20世纪60年代后期开始了现代海洋地质科学的新阶段,以深海钻探与沉积研究为特点,执行钻探莫霍面计划。已进行的深海或大洋钻探深度达2 km,虽未能穿透在洋底下5～10 km厚的地壳,未能达到莫霍面(地壳与上地幔间的不连续界面),但有重大的发现与科学进展。在洋中脊裂谷、海底隆起高原上的海脊,以及弧后盆地中均发现上地幔物质上涌的热液矿床堆积、大洋锰结核与多金属矿瘤堆积,对深海扇浊流沉积层特点的确定,以及对地震、火山活动机制的进展等,推动了海底构造与地层研究,发展了沉积学理论与技术,推动了新的学科分支的兴起。浅海油气藏与大洋多金属矿瘤的开发研究,形成了全球对海洋矿产资源的关注。海底石油和天然气蕴藏丰富,分布于海湾、海岸潟湖沉积、三角洲沉积、大陆架沉积盆地、大陆坡下的深海沉积扇或沉积盆地中,如美国墨西哥湾油田,英国、挪威北海油田,阿拉伯湾油田,中国渤海与南海油田。深海锰结核(多金属矿瘤,以锰、镍、钴、铜为主)分布在2 000～6 000 m深的海底,在太平洋6°～20°N、110°～180°E之间的600 km^2 海域,锰结核富积达10 kg/m^2 以上。在2 000 m水深的海底山基岩上形成厚度可达15cm的锰壳系热液堆积。大洋矿产的开发活动促进了对深海的研究。此外,含有稀有元素与贵重金属原岩,在浪流冲蚀、搬运下于海岸带富集形成海滨砂矿,如:海南岛东部钛铁矿砂矿、山东半岛南部的锆英石砂矿、澳大利亚的金红石和独居石砂矿、南非的金刚石砂矿、美国阿拉斯加大陆架的金砂矿等。

《联合国海洋法公约》于1994年11月16日正式生效,为全球海洋资源与环境的持续发展奠定了国际海洋法律的基础,海洋权再分配引起了全球的关注:12海里领海制度、200海里专属经济区制度、大陆架制度、公海和国际海底区域与资源是全人类共同继承的财产及国际共同管理公海的制度。海洋权益、海洋管理、海洋环境资源与人类生存发展成为新的研究热点,并促进兴起海岸海洋科学(Coastal Ocean Science)新领域。海岸海洋范围包括海岸、大陆架、大陆坡与大陆隆,涵盖整个海陆过渡带,与人类生存密切相关。

在我国,海洋地质学规模性的发展始于20世纪50年代后期,对渤海、黄海、东海海底沉积与地貌进行综合性普查,进而结合石油、天然气等矿产资源勘探,进行了海域地球物理探测与钻探。伴随着海洋石油、天然气的勘探与开发,又推进了海洋环境工程地质学的发展,研究"活动"的表层海底过程:海底沙波的移动规律、风暴潮活动与海底稳定性等。

海岸带研究是我国海洋地质工作的先锋,为解决港口回淤与泥沙来源,发展了海岸动力、地貌、沉积结合的海岸动力地貌学,逐渐形成具有中国特色的河口科学。20世纪80年代完成了全国海岸带与滩涂资源环境调查,获得了系统的多学科成果。在太平洋赤道水域、中太平洋海盆和东太平洋海盆进行了十几个航次调查,在太平洋7°～13°N、138°～157°W范围内圈出具有商业开采价值的采矿作业区30万 km^2,多金属结核储量约20亿t。我国政府向联合国提出登记为国际海底开发先驱投资者的申请于1991年获得批准,使我国拥有了15万 km^2 海底矿产的勘探权,最终将拥有7.5万 km^2 海底矿产开采权。开展了西北太平洋、南沙群岛海域及环大洋海洋地质及地球物理调查。与美、法、德、加、俄、荷等国进行了多项海底与海岸地质合作调查研究项目。

三、21 世纪的研究领域

21 世纪是人类全面认识、开发利用和保护海洋的新世纪，海洋地质研究工作将深入了解海洋作用岩石圈的自然过程与发展规律；按照持续发展的观点，有计划、有步骤地开发利用海洋资源环境，保护人类所赖以生存发展的海洋。结合我国的环境特点与社会经济发展，并考虑国际海洋地质学的进展，21 世纪我国海洋地质学应着重于海岸海洋、深海与大洋以及全球变化三个领域。

（一）海岸海洋领域

（1）边缘海的地质过程及动力学。包括边缘海的地质构造、岩石圈结构与动力机制、边缘海形成演化研究、边缘海的古海洋学研究，以及边缘海成矿作用与矿产资源等。

（2）河海交互作用体系与大陆架沉积作用。河流是海陆交互作用的结果与输送陆源物质的载体，与边缘海相似，河海体系是具有中国特色的重要课题。内容包括：不同类型的河口动力体系与作用过程、生物与化学作用过程与效应、泥沙（包含附着的污染物）交换、搬运通量、分布特点、沉积过程与沉积相以及堆积型陆架发展演变等。

（3）海岸类型、环境特点与功能。潮流控制、受大河作用与人类活动影响的淤泥质平原海岸发育演变以及开发利用模式。

（4）中国海海底地貌、沉积、非生物资源信息系统与 1∶200 000 万海底立体图系，以及应用 3S 方法对海疆监测系统。本课题为海洋权益、划界、资源开发以及海底环境与工程的基本依据。

（5）中国海岸环境资源（水、土、生物、海洋）承载力与最宜开发模式研究。

（6）海岸海洋环境工程与地质灾害预防研究。

（7）领海基线、岛屿、海峡、水道、大陆架与 200 海里专属经济区权限划定综合研究。包括历史沿革、海洋地理与地质依据，以及海洋法等。

（二）深海与大洋研究

这方面课题是参与世界公海海域研究项目，分享人类共有的资源与财富；通过国际合作交流促进科学家间的了解与科学水平共同提高，而且有些海域必须通过国际合作才可开展工作。

（1）大洋锰结核、多金属矿瘤与富钴结核成因机制研究。

（2）洋中脊与俯冲带热液活动、地球化学、成矿过程研究，动力效应与地震、火山活动相关机制研究。

（3）参与国际大洋钻探项目。目前以我国为主体的研究项目如：南海古季风记录、南海海盆地质过程与发展演变、印度洋深海扇与青藏高原隆起的相关沉积研究。

（4）东亚、太平洋与印度板块活动趋势与相关影响效应研究。

（5）台湾东部板块俯冲带活动机制与效应。

（三）全球变化

涉及海岸海洋与深海大洋两部分，着重于国际合作与全球对比课题。

（1）全球海平面变化与中国近海相对海平面变化及海岸效应对比研究。

（2）两极海洋地质过程的对比研究。

（3）环太平洋地震、火山带与地中海型地震火山带对比研究。

（4）海洋地质环境演变在全球气候变化和全球生态系统变化中的地位和作用。

上述三部分课题是从学科发展趋势及我国国情所做出的预测性分析与建议。进一步工作需根据国力与需要的紧迫程度，有选择性地、有步骤地进行。当代海洋地质学工作，需重视多学科，如地质科学与海洋科学、大气科学、地理科学，以及生物科学的交叉渗透；同时，重视自然科学与社会科学的交叉渗透、资源与环境开发、疆界和海洋权益与国际海洋法、21世纪议程以及管理科学。因为研究的目的在于应用，重视应用新技术可加速研究工作开展与提高科学水平，所以在工作中宜重视与相关部门的合作，相关项目应与有相关能力的部门结合，可以采取强强合作或国际合作的办法加强研究力量。这样才可达到事半功倍，有利于海洋科学的发展。

主要参考文献

[1] 蒋平. 1983. 王颖谈海洋地质学——海洋学中的一门新兴学科. 世界科学，10：1-6.

[2] 王颖. 1995. 海岸带与近海. 见：走向21世纪的中国地球科学. 郑州：河南科学技术出版社. 147-156.

[3] 王颖. 1996. 面向21世纪的海岸海洋科学. 见：中国海洋学会四届二次理事会论文集.

[4] Bates L. and J. A. Jackson. 1980. Marine Geology and Geological Oceanography. In：Glossary of Geology，2nd edition. American Geological Institute，258，381.

[5] Shepard P. 1973. Submarine Geology，3rd edition. Harper& Row，Publishers. 1-8.

[6] 国家海洋局. 1996. 中国海洋21世纪议程. 北京：海洋出版社.

[7] 喻普之，许东禹. 1995. 边缘海的地质过程及动力学. 见：走向21世纪的中国地球科学. 郑州：河南科学技术出版社. 199-204.

面向 21 世纪的海岸海洋科学*

（研讨会论文）

海岸带是海陆相互作用、过渡、互为关联的体系，与人类生存发展密切相关。海岸带环境资源特点与海洋、与陆地都有不同，是一个特殊区域。例如，由于气流在海、陆下垫面的温、压条件差异，海陆相互作用的一个突出效应是在海岸带形成一条狭长的风速变化增强带，而向陆、向海风力皆减弱。风力强度变化、海岸坡度的变化，又引起一系列的变化[1]。概言之，海岸带的风力、波浪、潮汐与泥沙运动均有增强，河流作用是海岸带的重要环境因素。中国河流向海输送的泥沙年最大量达 20 亿 t，占全球河流泥沙入海量的 10%。陆地的有机质、重金属及其他溶解质污染物总量的 1/3 是被河流带入海洋的[2]。人类活动通过河流对海岸海洋发生影响，例如，历史上黄河泥沙含量增大是由于唐时移民边寨，屯垦黄土高原，变牧为农而促使水土流失加强造成的。黄河 8 次大改道，其中有两次是人为的。其中如公元 1128 年冬开封府尹掘黄河堤以黄河水阻金兵，迫使黄河夺泗水入淮河从江苏入海，肇始黄河夺淮泛滥之端，大量黄河泥沙积聚使苏北沿海形成淤泥质海岸。1885 年黄河北归至渤海，苏北海岸遭受侵蚀，而渤海黄河三角洲淤积迅速[3,4]。20 世纪 80 年代以来，人工分水活动使现代黄河断流日增长，这些是人类活动改变了海岸发育方向的明证。

海岸环境特点[1]：

(1) 它是大气圈、水圈、岩石圈、生物圈共同作用的界面，是物理、化学、生物与地质作用活跃的动态区域。

(2) 活跃的海岸动力、上升流与陆源物质汇入为多种海洋物种提供繁衍生境，是一个具极高生产力和生物多样化的生态系。

(3) 海岸地貌体系是抵御风暴、潮浪灾害及侵蚀作用的天然屏障。

(4) 海岸生态系可以通过吸附、沉淀、固封等作用发挥减轻陆源污染之影响，是自然与人为过程排向海洋各种物质的缓冲地带。

(5) 保存着海陆作用过程信息，研究古海岸有助于分析海岸变化之趋势，有利于人类持续开发海洋。

1994 年联合国教科文组织于比利时列日召开的海洋会议，明确地提出“海岸海洋”(Coastal Ocean)，范围包括海岸、大陆架、大陆坡及坡麓大陆隆，反映出海陆相互作用带，尤其在第四纪末期以来曾出露与淹没的海岸地带[5,6]（图 1）。这一个新的定义可与我国通常应用的海岸带与近海相适应，它标志着海岸科学进入一个新的阶段：从 20 世纪 30 年代视海岸为沿海的狭窄陆地[7]（图 2）；50～80 年代定义海岸为一个两栖性地带，水下部分

---

* 王颖，张永战，邹欣庆：《’97 海岸海洋资源与环境学术研讨会论文集》，68－73 页，1998 年，香港科技大学理学院及海岸与大气研究中心出版。

至－20 m 左右[8,9]（图 3）；及 90 年代中期提出的海岸海洋。“海岸海洋”概念的提出，表明人们对海岸认识的进步，从全球变化与海陆交互作用入手，明确海岸海洋范围。同时，反映着国际海洋科学重视海岸海洋的新趋势，因为海岸与近海同人类生存发展密切相关。据 IGBP 的 LOICZ 项目（Land-Ocean Interactions in the Coastal Zone）概括的海岸诸因素①。

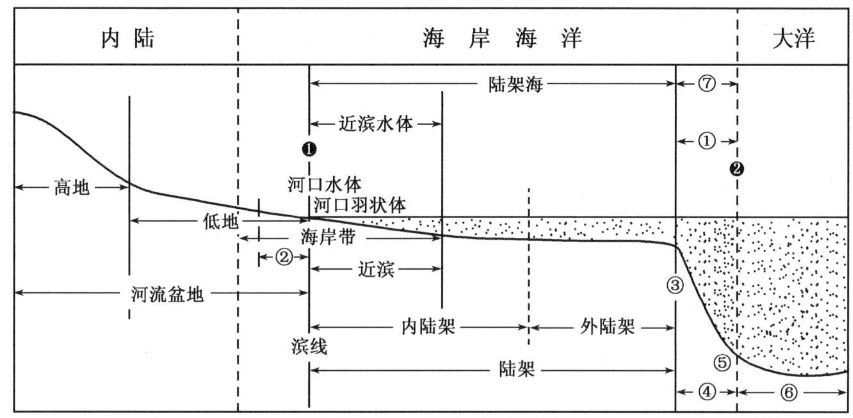

图 1　海岸带——海陆相互作用区

❶ 陆/海界面；❷ 陆架边缘海/大洋界面

① 陆架边缘带；② 盐沼、湿地、沙丘；③ 陆架坡折；④ 陆坡；⑤ 陆隆；⑥ 洋底；⑦ 陆架边缘海

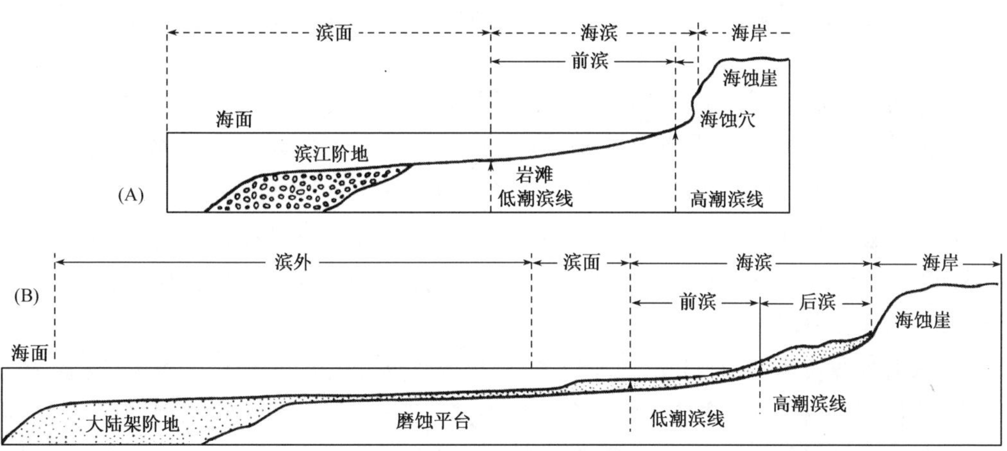

图 2　约翰逊的海岸概念

（约翰逊的海岸与海滨分带及述语）

（1）海岸带的领域（现代海平面±200 m 的范围）内：占地球表面积的 18%；约拥有全球初级生产量的 1/4；全球约 60%的人口的居住地；世界大城市（人口超过 160 万）中的 2/3 的坐落地；占有世界捕鱼量的近 90%。

（2）现代海岸海洋的几个数据：占海洋表面积的 8%；占海洋水体不到 0.5%；占有全球海洋生产力的约 14%；占有全球海洋反硝化作用的近 50%；占有全球有机质残体的约

① IGBP-LOICZ. News Letters. 1996，10(1).

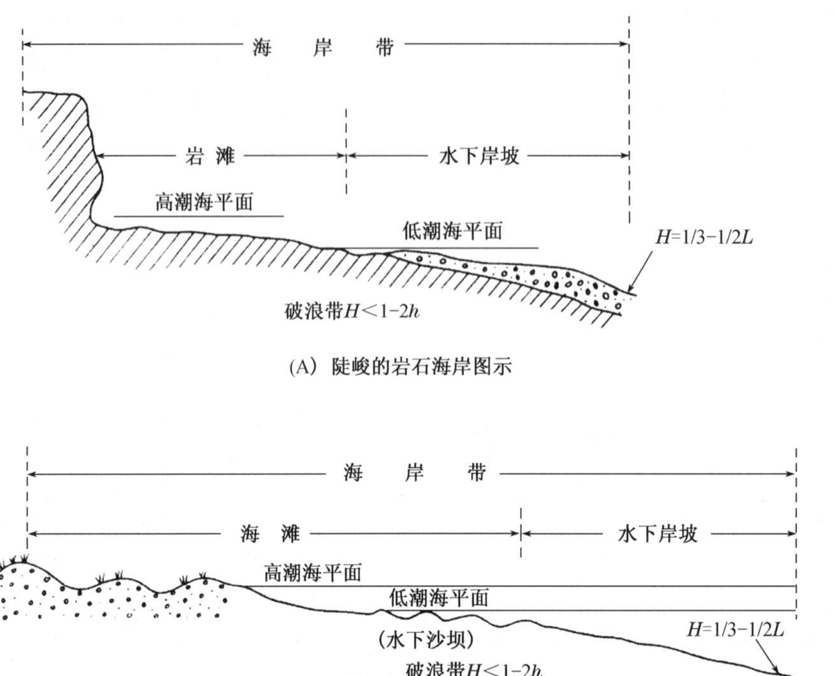

(A) 陡峻的岩石海岸图示

(B) 平缓的砂质海岸图示

图 3 海岸——海陆过渡的两栖地带

80%;占有全球沉积矿体的90%;占有全球河流悬移质及与之相关的元素/污染物的75%~90%;占有50%以上的全球碳酸盐沉积。

海岸带与近海蕴藏着丰富的自然资源:风与海洋动能资源;砂矿与油气矿产资源;滩涂与荒地等空间资源;港湾与深水航道资源;食物、医药等生物资源,以及旅游资源(阳光与空气,海水、沙滩、岛礁岸岬以及海鲜产品)等。它们具有成本低、尚少污染并且绝大部分可再生补偿的特性。由于这些丰富的生物和非生物资源,吸引着大量人口聚居于海岸带,从事着海运、商贸、房地产、食品及旅游等多项开发事业。估计,在沿海国家中约有一半的人口居住在海岸带。

中国海岸线长达32 000 km,地跨温带、亚热带及热带地区,海岸带受季风、波浪、潮汐与大河等作用,具有平原、港湾岛屿、河口、珊瑚礁与红树林等多种海岸类型,海岸环境、海岸类型与作用过程具有特色[8],并且绝大部分海岸尚保持着优美的自然环境。中国沿海地带经历过封建王朝时代的禁闭,20世纪初期与中叶之开拓,70年代与80年代大规模调查研究、规划与发展,90年代已成为我国的"黄金地带",集中了41%的人口与60%以上的工农业产值,地位重要[1]。由于人口密集,经济发展迅速,出现了对沿海资源近期消费(利用与需求)与资源的长期供给之间的矛盾。以长江三角洲为例①,其人口占全国人口的3.9%,但生产总值达13.5%,但是平均人口密度逾900人/km²,为全国平均值的8.4倍,人均耕地不足1/15 hm²,为全国平均值的60%,是我国土地承载量最高的地区之一。近年内人口以1%的年率增加,耕地平均以0.5%的年率递减,农业产值占工农业总产值

① 谢志仁:长江三角洲研究,1994.

比重已降到10%。环境容量的压力亦大，在占国土1%的土地上承受着全国工业废水的9%，废气的5%和废渣的4%，农村饮水质量降低，抗灾能力脆弱，人地关系协调发展决策迫在眉睫。人类活动直接或间接地向海洋输送的污物，60%集中于海岸带与近海，深海已遭受影响，危及空气与降水的更新，进一步又反馈于人类。因此急需采取有效的行动，保护与恢复海岸环境资源，因而产生一体化的海岸管理计划（Integrated Coastal Zone Management）并在联合国“21世纪议程”的第17章中予以肯定。人口、资源与生存环境持续发展是当代面临的巨大挑战，海岸海洋科学是承担解决这项历史任务的重要环节。海岸海洋科学具有显著的综合性或复合性特点，表现在：

（1）地球表层各圈层交互作用，海岸海洋的形成发展与气、水、地、生及人类活动密切相关，这一特性确定了海岸海洋科学的主干体系。

（2）自然科学、社会科学与技术科学，尤其是高新技术应用，互相结合、渗透。科学领域涉及到资源与环境的合理开发利用、减灾防灾举措与海洋权益问题。

（3）基础研究与开发应用以及规划管理相结合。

这些特点比单一科学更适应21世纪社会、经济与生存环境持续发展之需求。

“联合国海洋法公约”是于1994年11月16日生效，我国人大已于1996年5月批准，公约中有关领海、大陆架、200海里专属经济区、海峡与群岛水域等规定，均与沿海国的海岸与近海关系密切，资源、环境与海洋权益要求加强海岸带与近海的研究与管理。因此，海岸海洋科学应作为我国优先发展科学领域的主要项目，与国际接轨以适应21世纪人类社会发展之需要。根据我国海岸海洋科学的基础特点和在国际科学中的领先方面，并考虑国民经济发展的需要和经费支持的可能性。建议重视下列几方面的科学工作：

（1）沿海经济发达地区海岸与邻近海域资源环境承载力与可持续发展研究。

（2）海岸与海底立体成像与制图。

（3）海陆相互作用过程及海岸演变。

（4）河口、海岸沉积动力与沉积相。

（5）海平面变化、海岸效应与发展预测。

（6）海岸带、近海环境与海洋灾害。

（7）海底地貌和海底矿产资源；海底介质的力学特性与海洋工程。

（8）中国领海基线、岛屿、海峡、水道、大陆架与200海里专属经济区管辖权益专题研究。

（9）一体化的海岸带与近海管理系统。

在国际海洋科学领域，中国海洋研究工作在某些方面刚刚崭露头角，在一些方面已形成相当的影响，少数具国际公认的领先地位。总体发展不平衡，似乎有生物、物理与地质这样的顺序，尚未形成在海洋科学领域整体性的重大影响，可能是由于中国缺乏一体化的海洋学术领导所致。需分析海洋领域的形势、我们的优点与不足，打破部门界限，有目的、有步骤地组织力量，加强支持使得优势点形成长足性进展，从而在某一方面形成影响、发挥重大作用。另一方面，需重视当前国外海洋科学发展的新趋势，组织力量迎头赶上。我认为海岸海洋资源环境与持续发展研究以及海岸海洋一体化管理这两方面的研究十分重要。基于这一认识，深刻感到需促进多学科的交叉渗透与高科技的应用。海洋科学人才，尤其是研究生，需打破过细的专业分工界限，加强多学科交叉综合训练，培养、发展具有多功能适应的素质，在嗣后的科学实践中形成复合型的新专业力量。海洋科学研究项目若能被广泛应用到社会生活与经济发展的各方面，则获益多，科学价值大。我国海洋科学是能够逐步从实验性科学阶段发展到理论模式的建立、从趋势分析到决策实施，并适应21世纪人类生存、环境发展变化与科学进步的需要。作为结语，作者提出海岸海洋一体化管理模式(图4)，作为探索海岸海洋这一重要生存环境持续发展的初步设想。

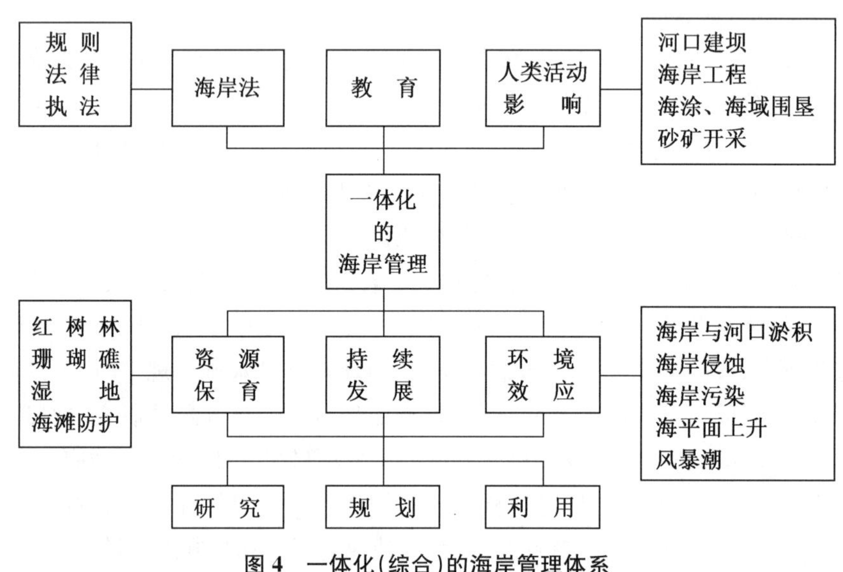

图4　一体化(综合)的海岸管理体系

主要参考文献

[1] 王颖. 1995. 海岸带与近海. 见：走向21世纪的中国地球科学. 郑州：河南科学技术出版社，147－157.

[2] Wang Y, Ren M E, Zhu D. 1986. Sediment supply to the continental shelf by the major rivers of China. *The Journal of Geological Society*, 143(6): 935－944.

[3] 王颖. 1964. 渤海湾西部贝壳堤与古海岸线问题. 南京大学学报(自然科学版)，8(3)：424－442.

[4] 王颖. 1994. 海洋地理学的当代发展. 地理学报，49(增刊)：669－676.

[5] Wang Y. 1994. Effect of military activities on environment in eastern and southeastern China. *Annual Science Report-Supplement of Journal of Nanjing University*, 30 (English Series 2): 43－46.

[6] Pernetta J C and Milliman J D. 1995. Global Change-IGBP Report No. 33: Land-Ocean Interactions in the Coastal Zone, Implementation Plan. Stockholm.

[7] Johnson D W. 1919. Shore Processes and Shoreline Development. John Wiley&Sons Inc.

[8] Wang Y and Aubrey D. 1987. The characteristics of the China coastline. *Continental Shelf Research*, 1(4): 329－349.

[9] 王颖，朱大奎. 1994. 海岸地貌学. 北京：高等教育出版社.

海洋地理国际宪章*

一、变化中的海洋作用

在迈向21世纪之际，人们面临着一个矛盾：海洋覆盖着大于2/3的地球表面，在提供着世界人口所需的生物与非生物资源方面起着核心作用，但是它却远不如我们地球陆地部分那样被人们所了解。

必须解决三方面的重要需求：第一，由于缺乏了解海洋在不断变化的地球生态系中的作用，必须加强研究海洋全球变化；第二，缺乏有关人类社会与海洋生态系相互作用之知识，要求一方面加强物理海洋学与生物海洋学间的合作，另一方面，加强海洋科学与社会科学间的合作；第三，显然需加强从持续发展观点研究深海、海岸与小岛的管理。

海洋生态系统正经受气候变化，尤其是随之而起的海平面变化以及相联系的生物化学过程的影响，因此要求科学研究不仅仅提供上述过程的全球性总结，并且需要集中关注海洋生物多样性、复原能力和生产力等方面剧烈的和深远的变化，因为这些会对人类生存有重要影响。

这些影响将在海岸人口中体现(在20世纪90年代末期，大约60%的世界人口居住在海岸60 km内)并将继续增长，造成加速城市化与海岸大城市扩张。伴随着都市增长与扩张，将会增加人类对海岸陆地与海洋资源的压力，但是各处的海岸生物资源已经被充分开发或者是过度开发。当地球的人口增加到100亿人这一临界值时，则对海洋生物资源的需求更要增加2～3倍。这种情况将招致全世界人类社会的营养出现前所未有的问题，因为人类食品中60%的蛋白质来源于鱼类食品。此外，非生物资源的开发将以迅猛增加开发海底石油与天然气为标志，并且可能开始开发洋底多金属矿核与其他矿产资源。

伴随着资源开发的发展、海岸与岛屿的人口的逐渐增加，也增加了不同的利用方式以及不同经济的和社会的利益间的冲突。这种过程导致着建立日益复杂的世界性的法律结构，多种的海岸司法管辖区，环绕着这些区的国家间的冲突将有所扩散。深洋将划分为两种体制：公海无产权制(资源对所有国家开放开采的体制)和深海床共有财产制(资源属于全人类的继承物，其开发权由国际权力机构行使)。

---

* 国际地理学联合会海洋地理专业委员会，王颖译，任美锷，吴传钧校：《地理学报》，1999年第54卷3期，第284-286页。

2004年载于：《"中国海洋资源环境与工程学术研讨会(联合学术年会)"论文摘要集》，6-8页．宁波，舟山，2004.4.23-29.

本文由国际地理联合会副主席 Adalberto Vallega 教授执笔，英译者：John R. Smith，Institute of Geographical Science，London，U.K.

海洋地理国际宪章可以翻译成英语以外的其他语言，但是，只有英文译本具有权威性。

二、地理学认识论的调整

与变化中的海洋管理方式相比，由一连串界定明确的研究部分所组成的常规的地理学认识论，应由整体的探讨所补足，即所有的地理学分支应从个别分支进展到为建立综合知识体系做贡献。整体性要求抛弃简化主义者和烦琐无遗的调查，所设计的知识体系应聚焦于组织空间体系上，空间体系的组成应是主导的概念。它将促使地理学加入其他科学的行列并致力于设计有效的学科间渗透。研究工作最基本的社会效应在于地理学研究与联合国环境发展大会的“21 世纪议程”及相继的政府间文件等所阐明的目标相一致。

因此，仅仅只由海岸与海洋地理学家关心设计与提出的海洋地理学（Ocean Geography）的观点应当被抛弃。被采纳的相反的观点是，能够在人类社会与海洋相互作用的评估方面做出贡献，所有的地理学分支均应被热烈邀请向海洋进军。应当抛弃将地理学研究领域划分为“陆地的”和“海洋的”之分界观点。

据《21 世纪议程》中全球变化的观点和政治趋势，海洋（ocean）的特点是由两个主要环境所组成，即深洋（Deep-Ocean）与海岸海洋（Coastal Ocean）。虽然这两种环境体具有一些共同的重要形态，每一个环境体还是不同的生态过程与社会过程的研究客体，这些过程招致不同的效果以及在不同时间与地理尺度上的相应的独特形态。结果在地理科学内，海洋地理将归结为三个领域：

海岸海洋地理：涉及海岸体系，向海延伸至大陆边缘外沿（自然标准），或者是国家海洋管辖区外限（法律标准）。

深洋地理：涉及的海洋自大陆边缘外沿向海延伸（自然标准），或者包含公海的深洋海床（法律标准）。

区域海洋地理：涉及海洋的各个区域，或由于多国间合作或其他自然的或社会的因素所引起的海洋区域化问题。

这种认识论的调整将提供一种途径，使地理学对海洋体系包括海岸与岛屿提供一个有效的学科间方向。沿着这一途径，地理学家将与其他科学家合作去设计、实验和贯彻逻辑上的和方法论上的背景以适应海洋世界整体化的趋势，并将进行经验研究使对海洋空间有更好的了解，以满足人类发展需求，以及在全球、区域和地方三种不同尺度上对海洋进行最佳的管理。

三、21 世纪的海洋地理

在上述三个领域内，地理学家将寻求由联合国环发大会“21 世纪议程”所需求的科学知识，并为此做出学科间的积极反应。据此，地理学家将集中注意于由于气候变化引起的自然过程与经济社会力量的相互作用所形成的空间表现形式。

关于海岸海洋所涉及的方面，地理学家将做出有效的贡献，为设计出以持续发展为目的的海岸管理项目去研究海岸体系。着重于分析因气候变化所形成的海岸生态系变化和人类对海岸地区的压力之间的相互作用。研究内容包括海平面上升、不断增加的人类压力及海岸城市化、海岸资源利用的增加、海岸生态系统的经济价值、文化遗产在海岸发展中的作用以及21世纪议程中其他重要问题所引起的空间表现。

关于深洋，地理学家将与其他海洋学科合作集中致力于海洋的两个客体领域的空间表象：气候变化与大洋生物和非生物过程间的相互作用；人类加注于大洋的压力和活动，包括实施对可再生与不可再生资源的利用及其对海洋环境的影响。

研究地理学在行政管理和政治边界中的作用，国家海洋管辖区和由于联合国大会海洋法(1982—1994)所建立起来的国际海洋制度将会被研究。在这方面，地理学将于海洋应用它在陆地从事法律与政治学科间合作的成功经验。

基于区域化的理论和有关的认识论，构成了地理学的核心组成部分，海岸与深洋管理的空间结果将要按照了解海洋是如何属于近期由提出和扩展空间系统而加快区域化的进程这一目的而进行调查。研究中应特别注意这些体系与持续发展概念的连贯性。

在海岸与深洋两领域研究工作的结构中，地理学研究将着重于小海岛，以尽量发挥在保护其生态系与文化传统方面的作用。

在海岸与深洋的区域和区域化过程的调查中，地理学家将与自然和社会科学的科学家一起去适应整体化的进程，以及应用落实资料管理及海洋信息系统。

四、国际地理学联合会的作用

面临着1998国际海洋年在海洋政策与海洋研究两方面意义深远而迅速的进展，国际地理学联合会应集中努力于三方面：

第一，研究方面：促动自然地理学家与人文地理学家去综合部门的观点，适应整体化的精神和适应持续发展为目的的研究需求，因此有效的学科间合作将会形成特别是国际研究项目的网络结构。

第二，教育方面：主要是设计学科交叉的课程以及相应的方式来传播海岸和深洋地区可持续发展的各方面知识及专家技术。

第三，公共机构的结构：由政府间的组织、主要是联合国的机构以及非政府组织的战略、项目和行动计划所组成，在这些部门地理学家将以满足地方团体的需要为目的，以优化政治和科学两方面交互作用的观点与之合作。

在这些领域，受益于观念的和认识论的专门知识，集中关注于自然环境和人类生存的交互作用以及在海洋的多种活动，地理学家将与其他科学家联合对付海洋的全球变化和追求海洋的持续发展。

面向21世纪的海岸海洋科学*

工业革命以来，科学技术加速发展，生产力水平迅速提高，促进了经济工业化和社会城市化，人类活动对自然界的生物地球化学循环的影响日益突出，地球系统中的人为-自然变化过程加剧，人口膨胀、资源枯竭、环境恶化、社会动荡等一系列重大的全球性社会环境问题以多种多样的形式表现出来，人与自然的矛盾日趋尖锐，人类生存发展面临时代性的巨大挑战，全球变化成为科学研究的热点，并引起政府首脑与世界人民的普遍关注。人类被迫重新审视自己追求和曾经标榜的社会经济行为和生产生活方式，开始致力于探求经济社会发展与资源环境相协调的可持续发展之路。

1　海岸海洋科学的兴起

20世纪50～70年代，随着地球科学与生物科学领域众多大型国际合作研究计划的制定与实施，获得了一系列新的发现与重大成果，导致人类的地球观发生变化，"人化自然"的哲学思想引起人们的普遍关注。20世纪80年代中期，提出并逐渐形成了地球系统科学(Earth System Science，ESS)思想。在其指导下，与全球变化研究结合，世界气象组织(World Meteorological Organization，WMO)与国际科联(The International Council of Scientific Unions，ICSU)共同组织进行了"世界气候研究计划"(World Climate Research Programme，WCRP)。随后，针对全球变化，ICSU于20世纪80年代末至90年代初期，进一步组织并实施"国际地圈-生物圈计划"(International Geosphere-Biosphere Programme，IGBP)。20世纪90年代，作为对IGBP的补充，ICSU又组织了"全球变化中的人类作用因素研究计划"(International Human Dimension of Global Environmental Change Programme，IHDP)，形成当代研究全球变化的三个主要的大型计划[1]，将自然环境与人类社会有机结合(图1)，在世界范围内全面推动全球变化研究，使之成为当代最活跃的研究领域。随着研究范围的日益扩大，研究内容的迅速增加，研究水平的不断提高，获得了越来越多的阶段性与地区性成果，推动人们的认识程度不断加深，海陆相互作用区——海岸海洋(Coastal Ocean)在全球变化研究中的优越性与重要性日渐突出。20世纪90年代中期，以IGBP新的核心项目——"海岸带海陆相互作用研究"(Land-Ocean Interactions in the Coastal Zone，LOICZ)的建立与实施为标志，海岸海洋已成为当代全球变化研究的一个关键地区[1](图2)。

---

* 张永战，王颖：《南京大学学报(自然科学)》，2000年第36卷第6期，第702－711页。
基金项目：国家自然科学基金(49236120)。

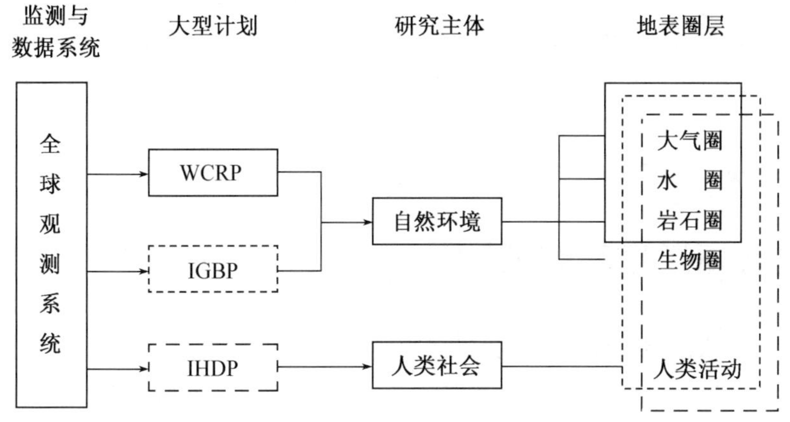

图 1　当代全球变化研究系统

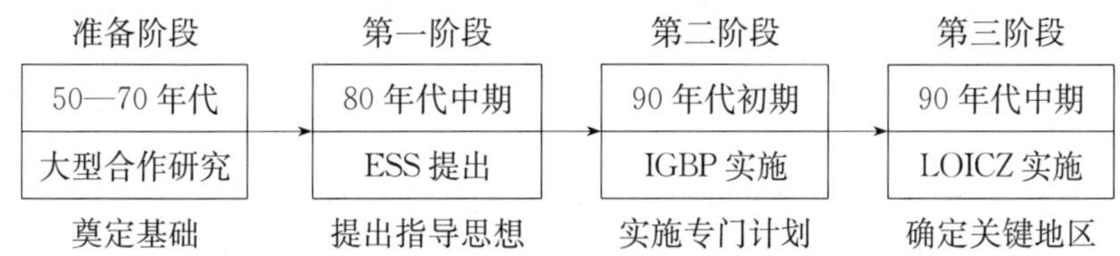

图 2　当代全球变化研究的进程

1992 年联合国"环境与发展"大会制定了"21 世纪议程(21 Agenda)",其中心内容是善待地球,协调人地关系,保证人类社会的可持续发展。人类最后的、目前尚未被完全破坏且有待开发的丰富资源宝库与仅存的"避难所"——海洋受到了前所未有的重视,提出了保护与恢复海岸带资源环境的一体化海岸带管理计划(Integrated Coastal Zone(Area) Management,ICZM 或 ICAM)[2]。

1994 年 UNESCO 政府间海洋学委员会(Intergovernmental Oceanographic Commission,IOC)在比利时列日大学(University of Liege)发起与组织国际海洋科技会议(First COASTS Workshop of IOC),制定计划着重研究与人类生存密切相关的海陆过渡带,正式提出了"海岸海洋"的概念①。LOICZ 所定义的"海岸带"(coastal zone)概念[3]与"海岸海洋"概念基本一致。

1994 年 11 月 16 日,"联合国海洋法公约"(The 1982 United Nations Law of Sea Convention)正式实施,将 1.3×10^8 km^2 的公海(全球陆地面积为 1.48×10^8 km^2)以领海、毗连区和 200 海里专属经济区等形式划归各沿海国家管辖,海洋权益及范围发生急剧变化,国际海洋秩序产生重大调整,"海洋国土"概念兴起,并立刻成为各国普遍关注的焦点。随之,许多国家相继提出了各自的"海洋 21 世纪议程",海岸海洋的资源环境开发利用与保护、海洋国土权益等是其主要内容。

① 王颖,张永战. 海岸海洋科学新发展. 中国海洋学会,中国海洋学会第四次代表大会文集,1995:5-13.

因此，海岸海洋受到了国际社会的普遍关注，国际海洋学研究的重点由深海大洋转向海岸海洋，相关领域的各方面专家亦开始重视并涉足这一地区。这样，在 20 世纪最后 10 年，推动了以海陆相互作用区——海岸海洋为研究对象的海岸海洋科学（Coastal Ocean Sciences，简称 COS）在世界范围内兴起。受全球变化研究的推动，并与之紧密相连，海岸的概念随之发生了重大变化。

2　海岸海洋的定义与特点

海岸是地球表层岩石圈、水圈、大气圈与生物圈相互交错、各种因素作用影响频繁、物质与能量交换活跃、变化极为敏感的地带。是海岸动力与沿岸陆地相互作用、具有海陆过渡特点的独立的环境体系，也是受人类活动影响极为突出的地区。回顾历史，人们对海岸这一概念的认识，大致经历了三个发展阶段。

20 世纪 20～30 年代，将海岸与海滨分开，视海岸为沿海的狭窄陆地。以 D. W. Johnson 为代表，提出海岸边缘有 3～4 个带，自陆向海分别为海岸、海滨、滨面和滨外[4]。

第二次世界大战，大大推动了世界范围内的海洋科学研究，使之进入了一个通过国际合作和利用现代立体观测技术进行全球规模的海洋考察时期，对海岸的概念也有了进一步的认识。20 世纪 50～80 年代，将海岸定义为“海陆两栖地带”。以 B. П. Зенкович 为代表，认为海岸是一个地带，即海陆交互作用带，由沿岸的陆地、潮水周期性作用的潮间带和水下岸坡三部分组成，范围从波浪作用的上界向海延伸至波浪开始扰动海底泥沙之处[4]。我国的海岸研究人员，将这一概念运用于生产实践，适应了生产建设与港口开发的需要，成功地解决了一系列的实际问题。20 世纪 80 年代，我国进行全国海岸带和海涂资源综合调查时，采用的海岸带范围为自高潮线向陆延伸 10km，向海扩展至水深 15～20 m 处[5]，也反映了海陆过渡、两栖地带的特点。

20 世纪 90 年代中期，随海岸海洋学的兴起，海岸海洋反映了现今人们对海岸概念的新认识，从全球变化的观点，赋予海岸以新的定义。“海岸海洋”是海陆相互作用的地带，范围从沿海平原延伸至陆架边缘，包括海岸平原、潮间带、水下岸坡乃至大陆架、大陆坡及大陆隆，主要分布于地表现代海平面上下 200m 的区域，大致相当于第四纪晚期以来海平面起伏波动交替性地被淹没或被暴露的地带，覆盖相对完整的整个海陆相互作用区[6]（图 3）。这样，地球表面被划分为 3 个大的地貌单元，即陆地、深海大洋与两者间的过渡地带-海岸海洋。

开阔的大洋和暴露的陆地高度分层（分带），环境相对稳定，变化缓慢；大气则迅速混合，风云突变。海岸海洋作为水、陆、气的相互作用区，相对于大洋和陆地，动力活跃，变化迅速；与大气相比，则相对稳定，成为一个在时间与空间上具有相当独特性的地表环境（表 1）。如，由于下垫面温、压条件的巨大差异，气流在海岸地区形成一条狭长的风速增强带，向陆、向海风力皆迅速减弱[5]。事实上，海岸地区的风力、波浪、潮汐和泥沙运动均有增强。同时，河流作用也是这一地区活跃的环境因素。河流向海岸海洋输水供沙，我国河流每年排泄入海的沙量最大达 20×10^{8} t，不但影响与控制了海岸线的进退演变，堆积形成

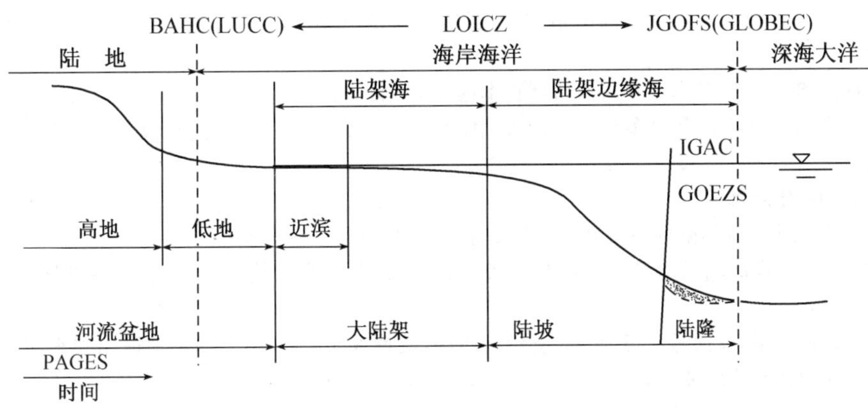

图3 海岸海洋——全球变化研究的关键地区

广阔的沿海平原、三角洲和大陆架，且由于大量有机质、重金属与其他污染物的排入(表2)，极大地影响了海岸海洋的自然环境。海岸海洋是人类由陆地通向海洋的桥梁，人口密集、经济发达、大城市众多，人与自然的相互联系更加紧密，相互影响更为突出。同时，由于河流的作用，使得自然环境变化与人类活动对河流流域内其他地区的影响，亦在海岸海洋地区表现出来，甚至强度更大。

表1 海岸海洋的特点

• 相对独立的自然环境　系陆地和大洋的相互作用与过渡区，陆地地层以 Sial 层为主，大洋地层以 Sima 层为主，而海岸海洋地带兼具两种地层结构。加之，大气圈和生物圈的作用，形成与陆地和大洋不同的自然环境，并与之共同构成地表 3 大环境体系。
• 动静共存的变化特点　系气、水、岩石和生物圈共同作用的“界面”，物理、化学、生物过程显著，物质转移活跃，能量输送持续，信息交换频繁。与陆地和大洋相比，动态特征突出，更新周期短，频率快；与瞬息万变的大气作用密切，但又相对稳定。因此，海岸海洋对地球表层系统中各子系统间的正负反馈过程反映敏感，且具可识别性。一方面，随气候变化和海、陆力量相对强弱的消长，海岸海洋处于不断发展、演变的过程中；另一方面，在演变过程中的每个阶段，海岸海洋又表现出巨大的缓冲性。因此，海岸海洋是一个的典型的开放系统。
• 丰富多彩的生态系统　由于活跃的动力与上升流作用，以及大量陆源物质的汇入，海岸海洋为多种生物提供了繁衍生境，是一个具有极高生产力和生物多样性的生态系统。其平均生物量约为 200 g/m^2，是大洋的 5 倍，它在海洋生境中所占的比例很小(其水体不到海洋水体的 0.5%)，但其生物总量约占海洋生物总量的 35.6%。
• 作用巨大的缓冲地带　作为由于自然和人为过程排向海洋的各种物质的缓冲带，海岸海洋是地表最大的沉积物捕获区(数量与面积)，不仅形成多种沉积地貌体系，抵御了风暴、潮、浪等灾害，有效减缓了海岸侵蚀，并通过吸附、沉淀、固封、降解等物理、化学、生物过程，减轻与消除了陆源污染的影响。同时，保存了完整而丰富的海陆相互作用过程的信息，为多种科学研究提供了良好的材料。

(据[1],[3],[7],[8]综合)

表 2 海岸海洋与人类活动的关系

• 海岸海洋是目前人类生存活动的最主要地区，人口密度极高，大城市林立，人类活动频繁，影响巨大。
• 由于移民和定居者，使这一地区人口增长率远高于内陆地区。
• 热带国家肥沃的适农土地多在海岸海洋，因此这里亦是土地利用和植物覆盖层变化最频繁的地区。
• 众多的国际旅游基地。
• 20 世纪初人类活动由陆向海带来的沉积物流，已经超过其自然输送率。
• 人类活动由陆向海输送溶解营养物质的速率，现在已与其自然输送率相当，某些地区甚至已经超过。
• 90%的陆源污染，包括污水、营养物、有毒物质等，被排入海岸海洋，三角洲地区出现水质型缺水，局部海岸海洋水体已被污染。
• 由于水坝的修建和大型灌溉项目(一个与自然趋势相反的作用)及人工调水工程的实施等，入海径流减少，甚至河流下游出现季节性断流，陆架上的沉积物输送率显著降低，使海陆相互作用发生变化，已深刻地影响了海岸海洋环境和人类生存发展。
• 生产生活耗水量大，过量抽取地下水，加之众多高大建筑物的重压，使地面沉降形成漏斗，盐水入侵，并使地下水与地表水一样也被污染，进而影响到土壤与动植物及食物等，并加重了缺水局面。同时，海平面上升导致风暴潮等灾害更加频繁，使得海岸海洋地区人类的生存环境更加脆弱。

(据[1],[3],[9],[10]综合)

海岸海洋虽然仅占有相对很小的地表面积(占地球表面积的 18%)与很少的海洋水体(其水体部分占海洋水体不到 0.5%)[3]，却蕴藏着丰富的自然资源：巨大的国土资源(包括海洋国土及潜在的与新生的土地)、丰富的空间资源(领空、水体、底床及港湾与深水航道等)、能源(波能、潮能与梯度能等)、砂矿与油气资源、食物药物资源及旅游资源等，其中绝大多数成本低、污染少并可更新。同时，由于人口密集，经济发达，活动频繁，膨胀的人口与有限的资源之间的矛盾、资源的近期消费与长期供给之间的矛盾更加突出。因此，海岸海洋对于人类社会和地球生物化学循环具有全球性的突出的重要性(表 3)。

表 3 海岸海洋的重要性

海岸海洋	• 有全球初级生产量的近 1/4 • 提供世界捕鱼量的约 90% • 系全球约 60%人口的居住地 • 世界上人口超过 160 万的大城市中有 2/3 在此带建立
其水体部分	• 全球海洋生产力的约 14% • 全球海洋反硝化作用的近 50% • 目前全球碳酸盐沉积的 50%以上 • 全球有机物质残体的 80% • 全球沉积矿产的 90% • 全球河流悬移质及与之相关的元素/污染物的 75%～90%

(据[3]补充)

海岸海洋对全球变化的反映敏感而多样。气候变暖将引起全球平均海平面上升，因此，海岸海洋受到气候与海平面变化的双重影响。加之，河流作用与人类活动，使得这一地区受全球变化的影响敏感而脆感，成为当今全球变化的信息库。另一方面，随全球定位技术(GPS)、遥感技术(RS)和地理信息系统技术(GIS)等高科技的发展与应用，对海岸海洋地区进行全方位的立体监测、综合调查与分析研究已成为可能。因此，20 世纪 90 年代中期，IGBP 专门建立并实施 LOICZ 核心项目，以海岸海洋为对象，研究其在地球生物化学过程中的地位和作用。至此，以 LOICZ 为核心，并同 IGBP 已经建立并实施的与海岸海洋有关的其他核心项目相互补充(表 4)，海岸海洋被置于一个时、空多维立体交叉网中(图 1)，围绕地表这一地带已基本形成了一个相对完整的研究体系，海岸海洋已成为当代全球变化研究的一个关键地区[1]。通过共同努力，有望从这一敏感地带打开突破口，推动全球变化研究进入新的阶段。

表 4　IGBP 与海岸海洋有关的核心项目

- 海岸带海陆相互作用研究(Land-Ocean Interactions in the Coastal Zone，LOICZ)，以海岸海洋为研究对象。
- 生物圈对水循环的影响(Biospheric Aspects of the Hydrological Cycle，BAHC)，研究汇水盆地模型，涉及海岸海洋的陆上部分。
- 土地利用与覆盖层变化(Land-Use/Cover Change，LUCC)，研究人类活动与地表变化的相互反馈关系，涉及海岸海洋的陆上部分。
- 共同的全球海洋通量实验(Joint Global Ocean Flux Study，JGOFS)，研究陆架边缘的物能相互交换，涉及海岸海洋水体部分的外界。
- 国际全球大气化学计划(International Global Atmospheric Chemistry，IGAC)，研究海气相互交换，涉及海岸海洋水体表层与大气的物能交换过程。
- 全球海洋真光层研究(GOEZS)，探索海洋真光层物理、化学、生物特征间基本关系，涉及海岸海洋水体的上部。
- 全球海洋生态系统动力学研究(Global Ocean Ecosystem Dynamics，GLOBEC)，研究海洋物理过程与生物资源变化间的关系，海岸海洋是其重要研究地区。
- 过去全球变化(Past Global Changes，PAGES)，研究 15 万年以来的全球变化，包括海岸海洋的变化过程，尤其是晚更新世以来的海岸海洋环境演变。

我国位于世界最大的大陆——欧亚大陆东部与最大的大洋——太平洋及其边缘海之交，大陆岸线长达 18 000 km，6 500 多个岛屿岸线长达 14 000 km，海洋国土面积达 300 多万 km^2。海岸类型丰富，作用过程独具特色，并且绝大部分至今还保持着良好的自然环境，海岸海洋特点鲜明(表 5)。我国沿海经历过封建王朝时代的禁闭，20 世纪前期的开拓，70 和 80 年代的大规模综合调查研究，90 年代的规划与发展，已成为“黄金地带”，集中了全国 41%的人口与 60%以上的工农业产值，地位非常重要。①

① 王颖，张永战. 海岸海洋科学新发展. 中国海洋学会，中国海洋学会第四次代表大会文集. 1995：5－13.

表 5　中国海岸海洋的特点

- 南北跨越 39 个纬度，具有温带、亚热带、热带 3 个气候带的不同环境特点，海岸类型多样。
- 位于欧亚板块与太平洋板块的俯冲作用带，地质构造具隆起与凹陷相间、构造运动活跃、区域变异大等特征。
- 东亚季风波浪作用突出，尤以东北风与东南风的影响明显，在很大程度上控制了沿岸流的作用。
- 潮流作用显著，并以黄海与东海北部海域最为强烈，发育典型的独具特色的淤泥质海岸和一系列潮流沙脊群和潮成的砂体。
- 发源于世界屋脊——青藏高原的大河作用显著，不仅输送了大量的淡水，更为海岸海洋供给了巨量的泥沙，移山填海，在太平洋西岸岛弧-海沟型板块边缘的弧后盆地堆积形成广阔的大陆架。
- 第四纪晚期，由于气候变化和海平面变化，海水在现代海岸海洋地区大范围地进退，河流长距离地伸缩，广阔的陆架区交替性地被淹没或暴露，环境发生重大的变化。
- 历史悠久而频繁的人类活动直接地或通过河流作用间接地对海岸海洋环境的发展变化予以重大影响。

（据[1]，[4]，[9]，[10]，[11]，[12]，[13]，[14]，[15]，[16]综合）

3　海岸海洋科学的展望

“21 世纪是海洋世纪”，面对人口、资源、环境与发展的巨大挑战，国际海洋科学研究的重点由深海转向海岸海洋。适应 21 世纪社会、经济与生存环境持续发展的需要，新兴的 COS 与全球变化研究密切相关，具有显著的复合性特点[6]。突出表现在：① 研究对象是地球表层各圈层相互作用的“界面”，气、水、土、生物与人类活动在这一地区密切相关，形成一个有机的整体；② 自然科学、社会科学与技术科学相互结合、渗透；③ 基础研究、规划管理与开发利用相结合。在研究资源环境发展规律的基础上，进行合理规划与科学管理，以组织有效的开发利用，同时，涉及减灾防灾举措与海洋权益维护和政策制定等众多领域。

以 LOICZ 为代表，IGBP 在海岸海洋地区进行研究的过程中，非常重视自然科学与社会科学的结合，重视人与自然的相互关系，目前已建立了自然环境变化与社会经济活动紧密相关的 P(环境压力)-S(环境变化)-I(不利影响)-R(防治措施)模型[17]，在此基础上，对越南红树林海岸、印尼局部海岸及日本东京湾等地区，从不同方面进行了实例研究。国际最高海洋研究机构——国际海洋研究委员会（Scientific Committee on Oceanic Research，SCOR）新近成立的各工作组亦与全球变化密切结合，重视气、水、岩石、生物圈之间的相互作用，研究重点由深海向海岸海洋转移，并同 IGBP 结合，积极参与 GLOBEC 和 JGOFS 核心项目的研究。国际地理学联合会（International Geographical Union，IGU）新近制定的“海洋地理国际宪章”中亦指出，海岸海洋着重于分析因气候变化所形成的海岸生态系统变化和人类对海岸地区的压力之间的相互作用[18]。这些均充分体现了交叉、渗透、综合的边缘科学特点。目前，国际 COS 研究与全球变化研究紧密结合，研究的重点集中在海岸海洋动力与物质通量；海-陆-气相互作用与碳、氮循环；海岸海洋生态系、生物地貌与全球变化；全球变化对海岸海洋生态、社会经济的影响及一体化的海岸海洋管理等方面。

20 世纪 50 年代，我国海岸研究人员重视海岸“海陆两栖性”的特点，适应生产建设的实际需要，将陆上部分与水下岸坡结合，研究泥沙运动与海岸演变，大大推动了我国海岸科学研究。80 年代以来，先后进行了全国海岸带与海涂资源综合调查、局部地区的大陆架考察、全国海岛资源综合调查等，获得了海岸地区多方位、多学科的基础资料，并在潮成海岸、潮滩发育、大河河口与三角洲演变、海平面变化与海岸效应、人类活动对海岸影响等诸多方面取得了一系列富有特色，并居国际前沿的研究成果。90 年代进一步开展了全国海岸带功能区划，为海岸地区的发展规划、立法管理打下了良好的基础。但这一系列工作所应用的海岸概念仍基本停留在 20 世纪中期的水平。近年来，研究范围逐步扩展至内陆架地区，开始了海陆结合、注重海陆相互作用进行研究的新阶段。然而地貌学界，至今仍有不少人对海岸的认识还停留在 20 世纪初的水平。

1996 年 5 月，我国人大正式批准“联合国海洋法公约”。同年，我国“海洋 21 世纪议程”正式提出，海洋高技术成为我国已批准的第七个高技术领域。COS 正逐渐成为我国优先发展科学领域的主要项目所在。1999 年，海洋学成为一级学科，必将更大地推动 COS 在我国的发展。同时，我国已积极地参加到众多的海岸海洋国际合作研究计划中，包括 PAGES、JGOFS、LOICZ 等，地区间的合作研究也已展开，如中韩合作进行黄海大陆架的对比研究等。与国际接轨，根据我国海岸海洋的基本特点和在国际海洋科学中的领先方面，并考虑国民经济现阶段发展的需要和经费支持的可能性，以及维护海洋权益的迫切需要，当前我国 COS 优先考虑并深入进行的研究内容应包括：

(1) 海岸海洋海底立体监测与制图技术。研制海底地貌、矿产、非生物资源制图软件，适应海岸海洋的动态特点，利用“3S”(GPS、RS、GIS)技术，建立快速制图系统，为划界与维护海洋国土和权益服务。

(2) 海岸海洋环境监测与海洋灾害预报、应急系统研究。建立数字海岸海洋，实时提供资源环境各要素的状况信息，为其开发利用与保护及减灾防灾等建立可靠的保障。

(3) 沿海经济发达地区，尤其是三大三角洲地区(长江、黄河、珠江)海岸海洋资源环境承载力与可持续发展研究。涉及海平面变化与海岸海洋效应、河海交互作用、人类活动与海岸海洋相互影响等众多方面，探索规律，建立人与自然相互协调的发展模式，为可持续发展服务。

(4) 海岸海洋物质通量与影响及生态系研究。揭示我国边缘海物质循环与生态系演变的关键过程。

(5) 一体化的管理系统研究。在海岸海洋资源环境监测、演变规律研究的基础上，行政、立法、司法等有机结合，打破地区界限和部门分割，建立高效的海岸海洋管理体系。

我国是人口大国、陆地大国，更是海洋大国。海洋国土面积广阔，资源丰富，环境多样，绝大部分均属于海岸海洋地带。目前，对其本底情况调查研究仍很初步，这里至今还是一片神秘的国土。21 世纪是海洋世纪，维护我国海洋国土权益，抓住时机，发挥优势，多学科协同，加强 COS 研究，必将进一步推动我国科学技术与海洋事业在 21 世纪的腾飞。

主要参考文献

[1] 张永战,朱大奎. 1997. 海岸带——全球变化研究的关键地区[J]. 海洋通报,16(3):69 - 80.

[2] 王颖. 1995. 海洋地理学的当代发展[C]. 见:南京大学海岸与海岛开发国家试点实验室年报 1991—1994 及海平面变化与海岸侵蚀专辑. 南京:南京大学出版社,261 - 269.

[3] Pernetta J C and Milliman J D. 1995. Global Change-IGBP Report No. 33: Land Ocean Interactions in the Coastal Zone, Implementation Plan[R]. Stockholm: IGBP Secretariat, The Royal Swdish Academy of Sciences. 1 - 12.

[4] 王颖,朱大奎. 1994. 海岸地貌学[M]. 北京:北京高等教育出版社.

[5] 任美锷. 1986. 江苏省海岸带和海涂资源综合调查报告[R]. 北京:海洋出版社.

[6] 王颖,张永战,邹欣庆. 1998. 面向 21 世纪的海岸海洋科学[C]. 见:中山大学近岸海洋科学与技术研究中心编辑. SCOR'97 海岸海洋资源与环境学术研讨会论文集. 香港:香港科技大学理学院及海岸与大气研究中心出版. 68 - 73.

[7] 王颖. 1995. 海岸带与近海. 见:《走向 21 世纪的中国地球科学》调研组. 走向二十一世纪的中国地球科学[C]. 郑州:河南科学技术出版社,147 - 156.

[8] 巴恩斯,休斯 R. N. 著,王珍如等译. 1990. 海洋生态学导论[M]. 北京:地质出版社. 1 - 12.

[9] 王颖,张永战. 1998. 人类活动与黄河断流及海岸环境影响[J]. 南京大学学报(自然科学),34(3):257 - 271.

[10] 任美锷. 1989. 人类活动对中国北部海岸带地貌和沉积作用的影响[J]. 地理科学,9(1):1 - 7.

[11] Milliman J C and Meade R H. 1983. World-wide delivery of river sediment to the oceans[J]. *Jour Geol*, 91: 1 - 21.

[12] Wang Ying, Ren Mei-e, Zhu Dakui. 1986. Sediment supply to the continental shelf by the major rivers of China [J]. *Jour Geol Soc, London*, 143: 935 - 944.

[13] Wang Ying and Aubrey D G. 1987. The characteristics of the China coastline [J]. *Contin Shelf Res*, 7(4): 329 - 349.

[14] Milliman J D, Qin Y, Ren Mei-e, et al. 1987. Man's influence on the erosion and transport of sediment by Asian rivers, the Yellow River (Huanghe) example[J]. *Jour Geol*, 95: 751 - 762.

[15] Wang Ying and Ge Chengdong. 1993. Several aspects of human impact on the coastal environment in China. In: Hopley D and WangYing. Proceedings of the 1993 PACON China Symposium-Estuarine and Coastal Processes. Published by the Sir George Fisher Centre for Tropical Marine Studies[C]. James Cook University of North Queensland, Townsville, Australia. 596 - 603.

[16] Wang Ying, Ren Mei-e, Syvitski J. 1998. Sediment transport and terrigenous fluxes [M]. In: Brink K H, Robinson A R. The Sea. Volume10. John Wiley&Sons, Inc. 253 - 292.

[17] Turner R K, Adger W N, Lorenzoni I. 1998. Towards Integrated Modeling and Analysis in Coastal Zones: Principles And Practices, LOICZ Reports & Studies NO. 11[R]. LOICZ Core Project Office, Texel, The Netherlands. 1 - 122.

[18] 王颖. 1999. 海洋地理国际宪章[J]. 地理学报,54(3):284 - 286.

《第四纪研究》海陆交互作用研究论文专辑后记*

【2006,26(3)】

承《第四纪研究》编辑部的大力支持与精心安排，本期文章主要选自中国第四纪科学研究会“海岸海洋专业委员会”与中国地理学会“海洋地理专业委员会”2006 年于北海召开的联合学术年会。两个专业委员会均具有学科交叉的特点，反映我国既是有 960×10^4 km^2的陆地领土，也有 300×10^4 km^2 的管辖海域这一事实。加强对海域环境、资源与权益的研究与交流，对我国构建全面小康社会意义重要。

“海洋地理专业委员会”是在 20 世纪 80 年代初，由黄秉维、周立三、吴传钧、施雅风 4 位院士及王乃樑教授等老一辈科学家支持下成立，挂靠在南京大学大地海洋科学系，旨在促进学科发展与人才培养。“海岸海洋专业委员会”是由刘东生院士大力支持，从 20 世纪 90 年代，在南京大学的“海岸线专业委员会”的基础上，适应科学的进步而发展建设的。专门研究海陆交互作用带的发展变化过程，及其环境资源开发利用。结合我国河海交互作用的特点，其研究领域涉及大河流域、海岸带、大陆架及大陆坡，即陆地伸向海洋的整个地带。海岸海洋(Coastal Ocean)既不同于陆地，也不同于深海。其环境特点在于地球的圈层系统：大气圈、水圈、岩石圈、生物圈在这个地带交汇，作用变化活跃，受人类活动影响频繁，与人类生存关系密切。

科学的进步与社会经济基础密切关联，海岸海洋科学的发展经历了一个世纪：20 世纪早期，人们关注到海滨线(shoreline)与海滨发展，以美国 D. Johnson 的巨著为代表；20 世纪中期，苏联海岸学泰斗 B. П. Зенкович(曾柯维奇)及他的学生 K. O. Леонтвев(列昂杰夫)建立了海岸带学说。海岸带范围包括沿岸陆地、潮间带浅滩及水下岸坡(达到当地波浪作用的下界，约相当于 1/3～1/2 波长处)。客观地反映了海岸带是水、陆“两栖”地带，波浪水流动力对水下岸坡的侵蚀、堆积变化，决定了海岸的自然发展趋势。海岸带学说在海港建设中获得成功应用。我国在 20 世纪 80 年代开展的全国海岸带与滩涂资源调查是以海岸带理论为科学依据，沿海省市的团结努力获得卓越成果。

1982 年通过的“联合国海洋法公约”(The 1982 United Nations Law of Sea Convention)于 1994 年 11 月 l6 日正式实施，我国于 1996 年 5 月 15 日经第 8 次全国人大常委会第 19 次会议决定，批准实施。继大陆架归沿海国管辖外，海洋法通过了两个全新的国家管辖海域：200 海里专属经济区(Exclusive Economic Zone)与群岛国家水域(Archipelago Nation' s Water)，以及大陆架与海峡通道内容有重要变化。全球将原属于公海的 1.09×10^8 km^2 的面积划给沿海国管辖，所占资源与权益也发生变化，兴起了“海洋国土”新概念。联合国教科文组织(UNESCO)政府间海洋学委员会(Intergovernmental Oceanographic Commission，简称 IOC)于 1994 年在比利时列日

* 王颖：《第四纪研究》，2006 年第 26 卷第 3 期，第 496 页。

(Liege)大学召开首次国际海岸海洋科学会议(COASTS),正式明确全球海岸海洋是“居于陆地与深海之间的客体”,包括“陆架海、陆坡海及海岸带”,即整个海陆过渡带。出席该次会议的代表有国家海洋局李海清、苏纪兰,中国科学院海洋研究所胡敦欣及我本人。会议论文于1998年在“The Sea”第10卷及11卷以“Global Coastal Ocean”命名出版。国际地理学家联合会(International Geographical Union,简称IGU)海洋地理专业委员会于1998年正式发表“海洋地理宪章”,将海洋地理研究明确划分为“海岸海洋”、“深海海洋”和“区域海洋”。国际学术界于1992年起即重视“陆海相互作用”研究,1995年后,国际科学联合会(ICSU)将海岸带陆海相互作用(LOICZ)列入国际岩石圈—生物圈计划项目(IGBP)。

第四纪科学是研究最近地质历史时期的环境变化,其中有关第四纪气候变化、海陆变迁是重要内容之一;人类活动直接地或间接地(通过河流入海)对海洋的影响,环境和资源变化,与我国沿海经济发达区的发展息息相关;海洋疆界的纷争是潜在的危机,这些值得我们关注与研究。本届第四纪科学研究会在刘东生、刘嘉麒、秦大河等院士大力支持下,将原“海岸线专业委员会”改为“海岸海洋专业委员会”,并报民政部于2005年11月批准。这是与国际科学发展俱进的高瞻远瞩之举,支持在北海召开专业委员会更名后的首次的学术会议,其中部分论文被选入《第四纪研究》2006年第3期刊出。这是个良好的开端,是半个世纪以来,以刘东生院士为代表的老一辈科学家在中国传播第四纪科学研究的又一茁壮新芽。在此,向刘东生院士、向《第四纪研究》编委会表示衷心的感谢。并愿以此集中海陆交互作用的论文,作为对南京大学新组建的、学科交叉型的“地理海洋学院”的一份献礼。

海岸海洋科学研究新进展①

海岸海洋科学的兴起源自“联合国海洋法公约”于1994年正式实施。依据公约的有关规定，$1.3\times10^8\ km^2$ 的公海(全球陆地面积为 $1.48\times10^8\ km^2$)将以领海、毗连区和专属经济区等形式划归各沿海国家管辖。海洋权益因此改变，国际海洋秩序随之调整，推动沿海国对“海洋领土”的关注。基于主权与资源开发的需要，推动海岸与大陆架浅海成为海洋科学领域的新热点。

海岸海洋(Coastal Ocean)概念于1994年UNESCO政府间海洋学委员会组织召开的国际海岸海洋工作会议上正式提出并使用。会后，出版了国际海洋学系列专著“The Sea”第十卷“Global Coastal Ocean”专著。随之，国际地理学家联合会颁布了“海洋地理国际宪章”，正式将全球海洋明确地划分为Coastal Ocean与Deep Ocean两部分。

国际科学联合会(ICSU)组织的IGBP计划自1992年以来，建立并实施了“海岸带陆海相互作用研究”(LOICZ)核心项目，其所定义的“海岸带”与“海岸海洋”概念一致，推动海岸海洋成为当代全球变化研究的一个关键地区。

从20世纪初，将海岸定义为沿海滨分布的狭窄陆地；到20世纪中期，明确了海岸带是包括沿海陆地及水下岸坡的“两栖地带”；再到20世纪末，海岸海洋科学的兴起。对海岸这一概念的认识，经历了两次飞跃。海岸海洋反映了现今对海岸概念的新认识，从全球变化的观点，赋予海岸以新的定义。

“海岸海洋”是海陆相互作用的地带，范围从沿海平原延伸至陆架边缘，覆盖整个海陆相互作用区。随海岸海洋概念的确立，地表被划分为三大地貌单元，即陆地、深海大洋与其间的过渡地带——海岸海洋。

海岸海洋是既有别于陆地，也不同于深海大洋的独立环境体系。它仅占地表面积的18%，其水体面积仅占全球海洋面积的8%，水体部分占整个海洋水体的0.5%，却沉积了全球河流悬移质及其所吸附的元素与污染物的75%～90%，拥有全球50%以上的碳酸盐沉积、80%的有机残体与90%的沉积矿产，占有全球初级生产量的1/4，提供90%的世界渔获量，系60%的世界人口的栖息地，是全世界2/3的大城市的分布地，其与人类生存关系密切。

海岸海洋科学是研究海陆过渡带表层系统作用过程、环境特性与自然发展规律以及人类生存活动与之和谐以可持续发展的科学，具有自然、人文社会与技术科学相互交叉渗透的复合性科学特点。突出表现在：① 研究对象是地球表层各圈层相互作用的“界面”，气、水、土、生物与人类活动在这一地区密切相关，形成一个有机的整体；② 自然科学、社

---

① 张永战，王颖：《地理学报》，2007年2006年第61卷第4期，第446页。

2001年曾以摘要刊登于《海峡两岸地理学术研讨会暨2001年学术年会论文摘要集》，30页，上海，2001年8月。

会科学与技术科学相互结合、渗透;③ 基础研究、规划管理与开发利用相结合,并涉及减灾防灾举措与海洋权益维护和政策制定等众多领域。

20 世纪中期,我国重视海岸"海陆两栖性"的特点,适应生产建设的实际需要,将陆上部分与水下岸坡结合,研究泥沙运动与海岸演变,推动了海岸科学研究。80 年代以来,先后进行了全国海岸带与海涂资源综合调查、局部地区的大陆架考察、全国海岛资源综合调查等,获得了大量基础资料。90 年代,开展了全国海岸带功能区划。近年来,研究范围逐步扩展至内陆架地区,开始了海陆结合、注重海陆相互作用进行研究的新阶段。

全国人民代表大会于 1996 年 5 月正式批准"联合国海洋法公约"。同年,我国"海洋 21 世纪议程"正式提出,海洋高技术成为我国已批准的第七个高技术领域。1999 年,海洋学成为一级学科。同时,我国已积极地参加到众多的相关国际合作研究计划中,地区间的合作研究也已展开。21 世纪初,"海域法"通过并开始实施。目前,海岸海洋地区的新一轮调查研究活动(908 项目),亦在组织实施中。

与国际接轨,根据我国海岸海洋的基本特点和在国际海洋科学中的领先方面,并考虑国民经济现阶段发展的需要和经费支持的可能性,以及维护海洋权益的迫切需要,当前我国海岸海洋科学优先考虑并深入进行的研究内容应包括:① 海岸海洋海底立体监测与制图技术;② 海岸海洋环境监测与海洋灾害预报、应急系统研究;③ 陆海相互作用与海岸海洋安全研究;④ 海岸海洋物质通量与影响及生态系研究;⑤ 一体化的管理系统研究等。

我国是人口大国、陆地大国,更是海洋大国。海洋国土面积广阔,资源丰富,环境多样,绝大部分均属于海岸海洋地带,这里至今还是一片神秘的国土。21 世纪是海洋世纪,维护我国海洋国土权益,多学科协同,加强海岸海洋科学研究,必将进一步推动我国科学技术与海洋事业的腾飞。

《第四纪研究》海陆交互作用研究论文专辑后记*

【2007,27(5)】

地处亚洲大陆与太平洋之交,海陆交互作用与悠久的人类活动影响,赋予我国在气候、地貌、生物,尤其是陆地延伸到海洋的"海陆过渡带"以独特的环境效应。长江、黄河贯通东西,是联系海陆与传输人类活动影响的主动脉,研究河海交互作用及其效应具有重要的科学意义。非常感谢刘东生院士的关怀指导,中国第四纪研究委员会主任刘嘉麒院士、韩家懋秘书长及编辑部杨美芳同志的鼎力支持,两次组织出版海陆(河海)交互作用论文专辑,有力地支持研究海陆交互作用带的新学科——海岸海洋科学之发展。

第四纪研究第27卷第5期共选登23篇论文,反映出我国第四纪科学工作的新进展:古今结合,阐明现代环境特征与分析未来发展趋势结合,基础研究与应用相结合。有关长江、黄河、珠江及淮河的论文共7篇,体现着本专辑的特色与研究成果之深入;几位青年学者分析洪泽湖底质的重金属含量与^{210}Pb定年相结合,获得洪泽湖近代变化的"自然记录",成果喜人。河海交互作用与滦河平原发育是继苏北平原发育文章后的第二篇,古为今用,滦河三角洲平原发育与曹妃甸深水港建设相结合;平原海岸、潮滩沉积与内陆架沙脊群研究反映海岸海洋体系与沉积动力研究的新发展。贝壳堤发育与平原海岸蚀积作用研究发展到阐明渤海湾平原发育与海陆相互作用的效应;对平原河口海岸牡蛎礁的研究,以及我国在海岸生态环境研究的重要进展:反映热带生物海岸珊瑚礁与红树林沼泽海岸与全球变化的区域响应等。沙漠绿洲则是生态环境与人类开发活动相关的实例,以及大运河一文继往开来,为申遗工作添砖加瓦……如此等等,不一一叙述。在本期出版之际,我愿表达对各位作者辛勤写作、热情投稿,对各位评委认真负责与真诚帮助,致以深深的谢意!

---

* 王颖:《第四纪研究》,2007年第27卷第5期,第900页。

海岸海洋科学与江苏海岸环境资源*
——与有色金属华东地质勘查局的学术交流

联合国“21 世纪议程”(Agenda21)指出:“海洋是生命支持系统的一个基本组成部分,是一种有助于实现可持续发展的宝贵财富。”21 世纪是生命的世纪,也是海洋世纪和高科技应用的世纪。海洋的重要性在于它是“与国家安全和权益维护、人类生存与可持续发展、全球气候变化、油气与金属矿产等战略性资源保障等方面休戚相关。人类对海洋的认识依赖于海洋科学的进一步发展,海洋事业的发展离不开现代海洋科技的强大支持”。

地球表面积的 71%是海洋。生命的起源、人类的生存与海洋密切有关。当代面临着人口、资源与环境的挑战,出路在于开发利用海洋。地球科学的重要性在于它有助于为人类提供生存发展的环境、资源与必需财富的途径。地学专业人员以至全民均应具有海洋意识。

江苏省海岸带经济建设提高到国家重点项目,反映出江苏海岸环境资源在我国东部经济持续地高速发展中,具有重要的战略地位。

一、全球变化与海岸海洋科学

全球变化是反映在气、水、岩石、生物圈的事件性变动,形成全球性的频发与持续性效应,对人类生存环境影响深刻①。全球变化不仅仅侧重对古信息的研究,探索根源,讨论发展,而且更应注重现代过程的研究,利用高新技术进行海洋监测,分析判断当前海洋的特点,总结、探索变化规律,建立预测趋势变化的模型,以期应对(适应)。对全球变化的科学内容宜有新认识:

(一) 全球变化反映在气温、海温、海平面与生物圈的效应

20 世纪末与 20 世纪初相比较,全球气温升高约 0.5 ℃,预计在未来 30 年时间,全球气温将继续增加,每 10 年将平均升高 0.3 ℃,2005 年全球气温将升高 1 ℃。

近 200 年的验潮资料反映,海平面上升趋势与大气温度及海水温度增高趋势呈良好相关。长时期的地质记录反映出海平面、大气温度和海水温度三者间为正相关(图 1)。百年来水动型的海平面上升值为 1～2 mm/a,随气温升高,海平面持续上升,2050 年海平面将上升 30～50 cm,至 2100 年海平面可能上升 1 m(IPCC－WG1,1990)。

---

* 王颖:《资源经济与管理研究》,2009 年第 2 期,第 6 - 13 页。

① 中国科学院地学部 2008“我国海洋领域若干战略性科技问题的建议”。

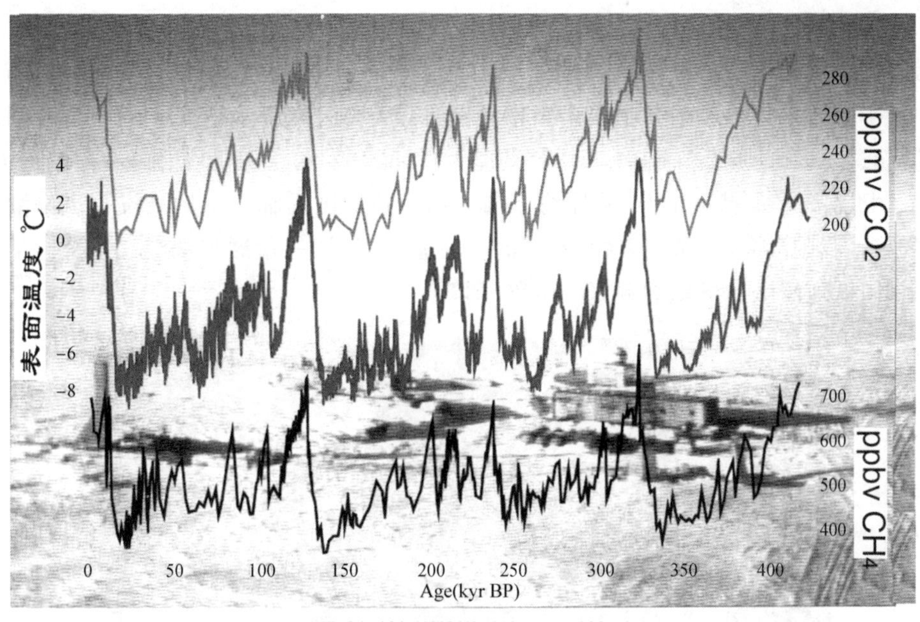

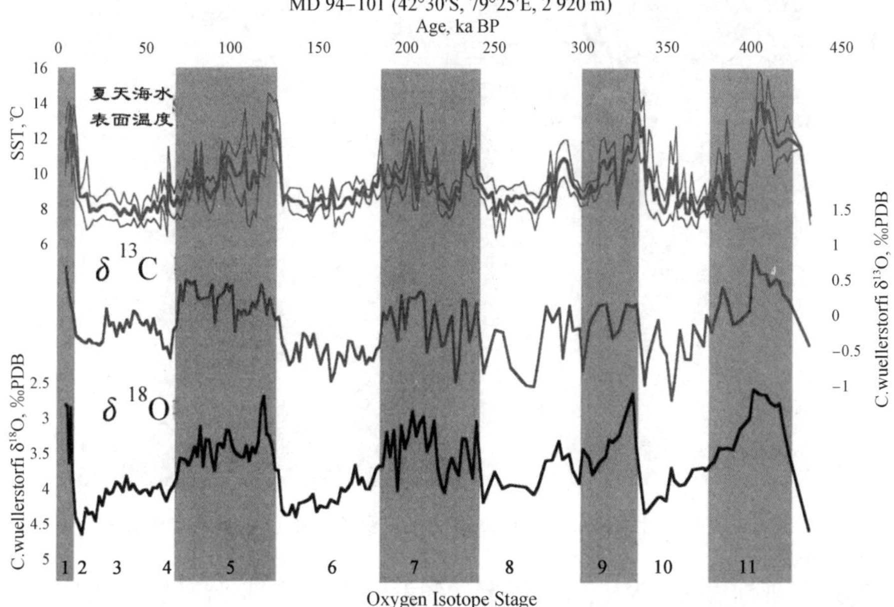

Vostok冰芯所揭示的 400 ka BP 以来大气 CO_2、CH_4 浓度和气温变化曲线（上图）。MD94－101 大洋钻孔所揭示的 450 ka BP 以来氧、碳同位素与夏季 SST 变化曲线（中图），及其位置（右图）

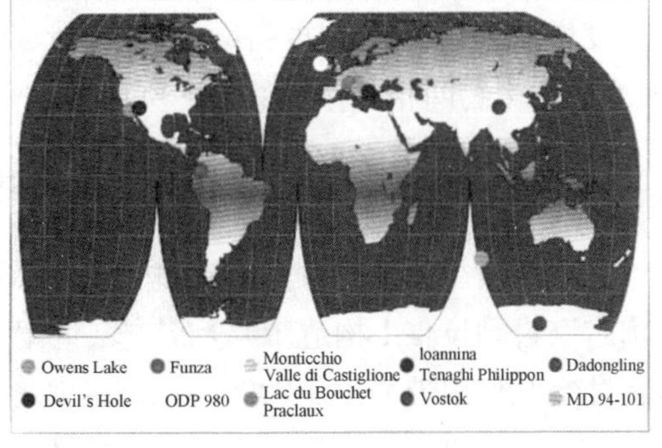

图 1 海平面、大气温度和海水温度三者间正相关示意图（据 PAGES News letter，1999）[2]（附彩图）

2000 年以来，中高纬度地区气温增高明显[1]，欧洲英格兰与威尔士地区于 2000 年秋季，暴雨侵袭绵延达 3 个月，低地被淹；2004 年 8 月下旬，英格兰西部山溪洪水成灾；2009 年 11 月 19 日，千年一遇的暴雨袭击英格兰北部，24 小时内降水达 314 mm，短时间内积水深度超过 1.5 m，城市被淹，坎布里郡两座大桥冲毁；爱尔兰 19 日暴雨 30 年来最大，积水 1 m。随着海平面持续上升，平原海岸与大河三角洲区面临着土地淹没与风暴潮频袭城市之灾。我国渤海湾、黄河三角洲、黄海平原海岸、长江三角洲与珠江三角洲会遭遇海平面上升，低地浸淹与风暴潮频繁的侵袭(图 2)。

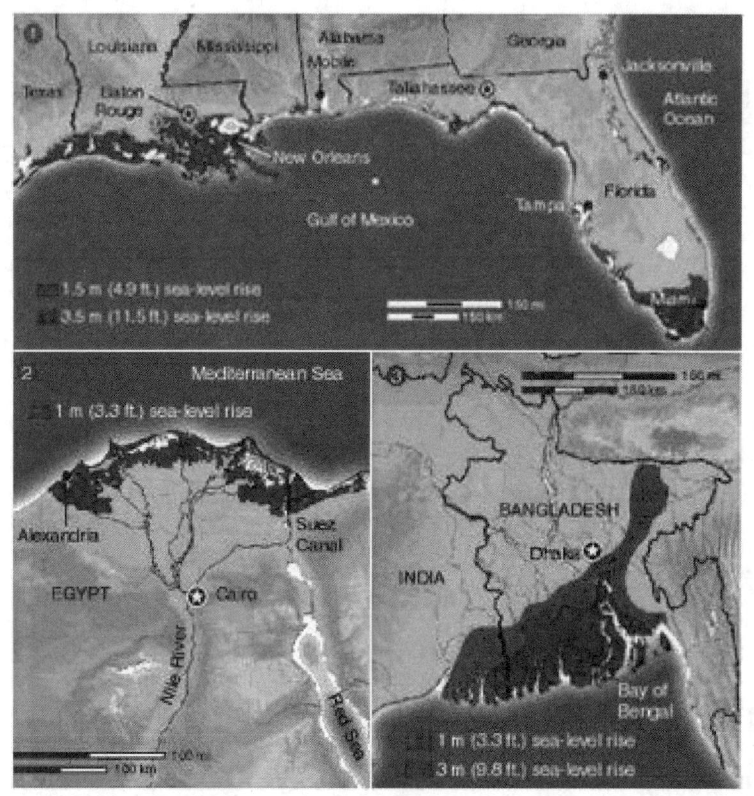

图 2　海平面升高 1 m，1.5 m，3 m 后海岸受淹范围(据 Times，2001.3)

气温增升，亚洲内陆沙漠干热，沙尘暴频频发生，袭击东部沿海城市；沙尘经高空西风气流搬运，跨越大洋在东太平洋沉降，带来富含 Fe 质的营养盐，促使加拿大海域鲑鱼生产丰度增加；但当海水温度持续升高，又造成鲑鱼丰度降低，这是地球表层系统岩石圈、气圈、水圈与生物圈相互作用的典型例证。

(二) 全球变化在海陆交互作用带岩石圈的火山地震效应

现代全球 500 多座活火山中，约 370 座沿太平洋沿岸分布，其余沿东西向“关闭的地中海带”(喜马拉雅山构造带)分布。20 世纪 90 年代以来火山爆发频频，逐年递增，尤其以海陆交互作用带突出。2000 年以来，太平洋大陆边缘“火圈”与大洋中脊火山爆发频繁，南大洋、墨西哥及大陆地区火山频爆，形成全球性构造活动(表 1 及图 3)，对地球表层系统造成巨大影响，对海陆过渡带效应更为显著。

表 1 1995 年—2005 年全球火山活动情况

时间	次数	火山爆发位置
1995	5	4 次位于太平洋地区(2 次位于南太平洋),1 次位于印度洋地区
1996	10	9 次位于太平洋地区(1 次位于南太平洋),1 次位于大西洋地区
1997	14	13 次位于太平洋地区(2 次位于南太平洋),1 次位于大西洋地区
1998	15	11 次位于太平洋地区(4 次位于南太平洋),2 次位于大西洋地区,1 次地中海,2 次陆地(美洲、非洲)
1999	13	12 次位于太平洋地区(2 次位于南太平洋),1 次位于大西洋地区
2000	50	33 次位于太平洋地区(3 次位于南太平洋),6 次位于墨西哥湾,3 次非洲(刚果、喀麦隆),1 次地中海,3 次印度洋,1 次大西洋,南美洲 3 次
2001	26	19 次位于太平洋地区(8 次位于南太平洋),3 次位于墨西哥湾,1 次印度洋,1 次大西洋,大陆 1 次
2002	31	21 次位于太平洋地区,1 次位于墨西哥湾,3 次非洲,2 次地中海,1 次印度洋,南美洲 3 次
2003	28	22 次位于太平洋地区,2 次在墨西哥湾,1 次在大西洋,亚洲 2 次,欧洲 1 次
2004	13	7 次位于太平洋地区,4 次在美洲,亚洲、欧洲各 1 次
2005	45	22 次位于太平洋地区,8 次在美洲,5 次在欧洲,4 次在墨西哥,1 次在非洲

资料来源:据美国国家航空和宇宙航行局数据统计。

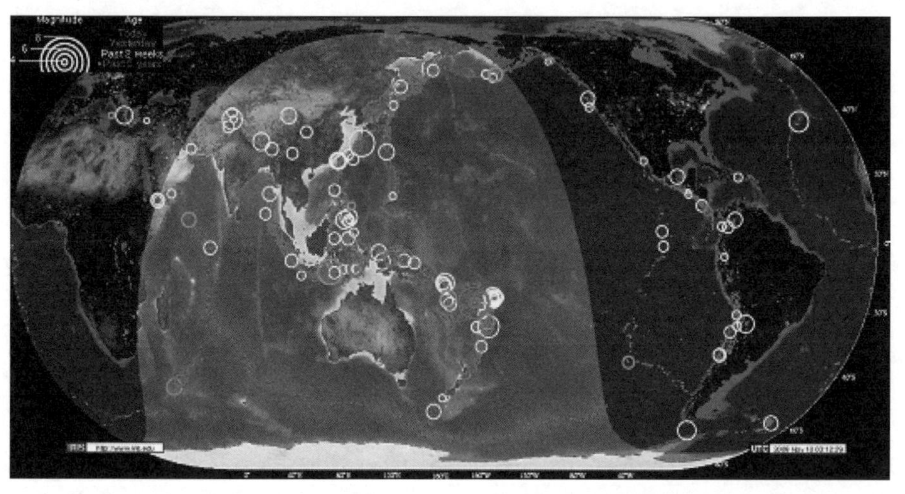

图 3 太平洋圈与全球火山地震分布图(附彩图)

资料来源:据美国国家航空和宇宙航行局,2009。

1990 年以来,地震活动亦极为频繁,在全球范围内年年有大地震。2000 年以来,几乎是每日发生地震。地震带与火山带分布是沿环太平洋、大洋中脊、南大洋、东西向的地中海构造带以及南北向构造带活动张裂,对比表明,震级强度具有转移的特性,一次强地震爆发,能量释放后震级减小,但却在同一构造带的另一处震级加强。这一现象可为地震预报提供启示。

火山喷发与强地震危害性巨大，长时间影响大范围生存环境。火山熔岩流温度高达500～1 400 ℃，每小时移动数公里，所至处燃树焚屋，炽热火山灰漂浮沉降，掩埋田野、道路与市镇；火山灰与山顶融雪形成泥流，沿坡而下堵塞河道，促使洪水暴发；炽热火山云，1 000 ℃高温快速运移，烧伤与窒息生物；喷发之气体若在水域积聚，会形成窒息呼吸之毒气，在空气中长期飘荡形成酸雨与低温层；海水渗透至炽热火山内，积聚蒸汽，会形成爆炸。但在另一方面，火山爆发亦会形成新的海岸、岛屿、陆地、沃野、火口湖与海湾，并促使珊瑚礁发育。

地震活动灾害对海岸与海岛更为显著，形成断崖，河床裂点，瀑布，山崩地塌造成滑塌与堰塞湖，地面陷落，土壤液化，喷沙、喷泥，破坏道路、港湾、市镇、水利与工程设施等。

(三) 海洋权益与全球对海岸海洋的关注

1994 年 11 月 16 日由 150 多个国家签署的“联合国海洋法公约”(The United Nations Convention on the Law of Sea)正式生效。我国是签约国，于 1996 年 5 月 15 日由全国人大八届常委 19 次会议批准实施。公约对沿海国主权的 12 海里领海，24 海里毗邻区，200 海里专属经济区以及大陆架是沿海国陆地领土自然延伸等原则规定，使海洋权益及管辖范围发生巨大变化，推动了沿海国对“海洋领土”的关注，全球涉及海洋划界的有 370 处。基于主权与资源开发的需要，推动海岸与大陆架浅海成为海洋科学领域的新热点。明确地认识到，海洋是由两个主要的环境组成：海岸海洋与深海海洋(图 4)[3,4]。沿海国海洋权益所涉及的范围主要在海岸海洋。

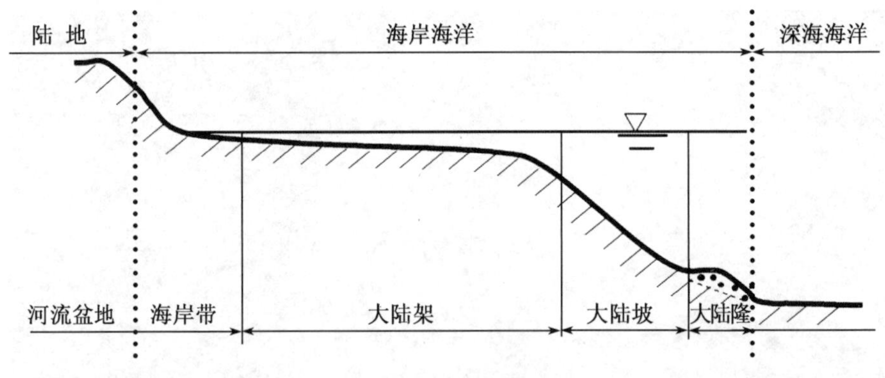

图 4　海岸海洋图示

海岸海洋是陆地延伸至大洋、海陆交互作用的过渡地带，既区别于陆地，又有别于深海大洋[5]。它是一个独立的环境体系：水、岩石、大气、生物圈层在此地带相互作用活跃，受人类活动影响密切，发展变化迅速，生物量巨大，生态系统丰富，缓冲作用突出。研究陆海过渡带的表层系统作用过程、环境特性及发展变化规律，以及人类生存活动与之相和谐的科学，构成“海岸海洋科学”(Coastal Ocean Science)的研究对象与任务，是基于地理学、地质学与海洋学相互交叉渗透的新学科，具有自然、人文与科学技术相互交叉渗透发展形成的复合型科学特点。

以上三点反映出全球变化对海岸海洋研究的关注及新学科点之发展。

二、江苏海岸海洋环境资源调查建议

(一) 江苏海岸海洋环境资源特点

濒临黄海，海岸北起苏鲁交界的绣针河口，南达苏沪交界的长江北口，地处 31°33′～35°N 之间。全省沿大陆岸线长 953.875 9 km，沿海 19 座岛屿岸线长 26.941 km。按 1985 全国海岸带调查时陆上 10 km 宽，水下至 −15 m 深处，江苏省海岸带面积 $3.5\times10^4\ km^2$，其中沿岸陆地约 $0.5\times10^4\ km^2$，潮间带滩涂 $0.5\times10^4\ km^2$，浅海海域 $2.5\times10^4\ km^2$。可以说江苏省海域全部属于海岸海洋的范畴(表 2)。

表 2　江苏省海域范围与面积

海域类型	内水	领海	毗邻区	专属经济区	江苏省
范围	海岸线—领海基线	基线向海 12 海里	基线向海 24 海里	基线向海 200 海里	海域总计
面积($10^4\ km^2$)	2.175 9	0.982 9	0.999 1	14.688 3	18.842 6

1. *海岸有三种类型*

(1) 砂质海岸。分布于海州湾龙王河以北，全长 30.062 5 km，沿剥蚀平原外缘发育，基底为古老的片麻岩与中生代花岗岩，上覆约 2 m 厚的冲积砂砾层，两层间为多钙质结核的晚更新世黏土层。砂质海岸激浪带宽，发育有三列沙坝：① 现代海岸高潮线上为一列高 6～7 m，宽 40～50 m 沙坝，由石英、长石质粗砂组成(向海坡 5°，向陆坡 25°～30°)，顶部为细砂质沙丘，已植林。② 现代海岸西侧 1 km 处在南北朱皋、海脐、梁东沙、匡口、里沙一带分布着第二列沙坝，石英质粗砂及砂砾组成，高出地面 1～3 m，宽 110 m，贝壳 ^{14}C 定年为 2 532±76 年。(3) 在第二列沙坝以西约 1 km，在董家庄、阎村东、海头、赵家沙一带还有一道沙坝，石英质粗砂，时代为 7 682±248 年。

三列沙坝反映海州湾北部是近 10 000 年来由海淤积成陆。沙滩曾是江苏最珍贵的旅游资源，由于当地在海底大量挖沙，破坏了海岸稳定性，招致海岸被侵蚀。嗣后，沿岸加铺水泥与钉防护，又造成沙滩与海岸景观之破坏。

(2) 基岩港湾海岸。分布于连云港市内山地临海处，自西墅至大板巃岸线长约 40 km，原是沿山麓分布的大海湾，由于黄河自 1128—1855 年夺淮入黄海，巨量泥沙汇入填海造陆，古海湾淤积成平原，海岛成为低山丘陵，海岛上古海岸遗迹是沧海桑田的证据。东西连岛北岸基岩小港湾中上部潮间带为砂质海滩，但低潮水边线以下又是淤泥沉积，系废黄河口海蚀泥沙向北部沿海扩散的影响，因此，形成一种基岩港湾砂质海滩与淤泥海岸之间的过渡类型。

(3) 粉砂淤泥质平原海岸。云台山以南至长江口，岸线长达 883.56 km。

海岸特点：江海交互作用的平原，宽阔平缓，岸坡<1/1 000，波浪作用达不到岸边，潮流是主要动力，发育了广阔的潮滩。滩涂地貌具分带性，系涨潮流沿滩坡上溯，受摩擦动力减弱，泥沙沿途卸下：在潮滩下部卸下细砂与粉砂质，中部卸下粉砂与黏土(互层)，最后

卸了黏土质。落潮时，潮流初始弱，渐而增强，在潮滩中部集中水流形成冲刷。因此，平原潮滩具明显的沉积与动力地貌分带性：

高潮线以上海岸为贝壳堤，向海渐次为→草滩（成熟的滩地）→淤积泥滩（平均高潮线以下）→砂泥混合滩之冲刷带（中潮滩）→具波痕的粉砂细砂滩（低潮滩）。不同的滩面带具有不同的开发利用途径。

平原海岸，泥沙松散，受冲刷与堆积变化快，取决于海岸带有无泥沙补给以及与海岸动力强弱之对比关系。

苏北海岸由长江泥沙与海洋作用堆积形成，同时受1128—1855年黄河夺淮入黄海之影响，使北宋时（11世纪）以范仲淹为首，利用6 000年前高海面时所形成的海岸贝壳堤为基础，加筑而成的防波堤，已远离海滨，即今日的盐城—东台以东为40 km宽的沿海平原与宽度超过10 km以上的潮滩（坡度仅为0.2‰）。实际上，当黄河夺淮入黄海时，河口曾向东淤积超过90 km。1855年黄河北归渤海后，废黄河口遭受侵蚀后退，1921年的海岸线已在距现代海岸线3 km的海底，1958年及1971年所建的岸上海堤均受侵蚀。废黄河口是冲刷海岸，泥沙向两侧扩散。近期研究表明，黄河泥沙曾在新近纪多次影响到江苏。

江苏海岸从射阳河口向南至吕四岸段，因外侧有水下沙脊群为掩护，使海岸免受冲刷；射阳河口以北至云台山，吕四港以南，因海岸带泥沙补给短缺而遭受冲刷。在辐射沙脊群掩护的地段，因沙脊群外缘受蚀，泥沙被涨潮流运向陆地方向的岸线，使潮滩淤长，形成新的土地资源。海岸开发需了解海岸发展动态。

淤泥质平原海岸因水浅泥滩宽，冲淤变化快，被视为苦海盐边，不适于农耕，封建王朝控制海盐生产，以及解放初期封滩防御及苏北平原河网化等因素，因此，苏北888.3 km长的海岸带曾为港口空白区，海运与海洋经济不发达，长期为内陆型农业经济。

2. 苏北浅海内陆架为古三角洲海底

（1）北部是废黄河三角洲，界于灌河口与新洋港之间，呈微凸的扇面向东延展至废黄河口以东72 km处，宽度约160 km，水深15 m，是细粉砂淤泥质海底，坡度倾斜约为1.5‰，系黄河南迁时形成，是江苏北部及中部海岸带黏土物质的主要来源。废黄河水下三角洲以东至40 m水深处是末次冰期与冰后期时的老三角洲，以粉砂为主，并含有河口环境繁殖的近江牡蛎、长牡蛎与褐牡蛎。

（2）南黄海辐射沙脊群，分布于江苏中部及南部，南北长度达199.6 km（32°00′N～33°48′N），东西宽140 km（120°40′E～122°10′E）。以弶港为中心，呈扇形展开，它由70多条沙脊与其间的潮流通道组成，是我国乃至世界上最大的浅海潮流沙脊群。它位于海陆交互作用带，水深为0～25 m，全部面积为22 470 km^2，其中约有3 782 km^2是出露于水面之上的沙洲（图5）。南黄海海域为正规半日潮，每日两涨两落，涨潮时，潮流自北、东北和东南三个方向涌向弶港海域，落潮时，以弶港为中心，呈150°扇面向外逸出，形成辐射状潮流场。潮波的辐聚与辐散，形成潮差大，平均潮差4.18 m，大潮潮差6.5 m，在核心区黄沙洋获得我国最大潮差记录9.28 m，潮差大、潮流强是江苏海岸动力的特色。涨潮时潮流辐聚，平均流速1.2～1.3 m/s，落潮时潮流辐散，流速急，平均流速1.4～1.8 m/s，落潮流大，每落潮通量可达21×10^8 m^3潮水，形成全球最具代表性的辐合-散射潮流系统之海域。辐射沙脊的基本轮廓与潮流场相符，是稀世的自然遗产。强大的潮流动力是维持潮流通道水深的强大动力，为江苏提供了天然深水航道的优越条件，江苏沿海潮汐能居全国第一。

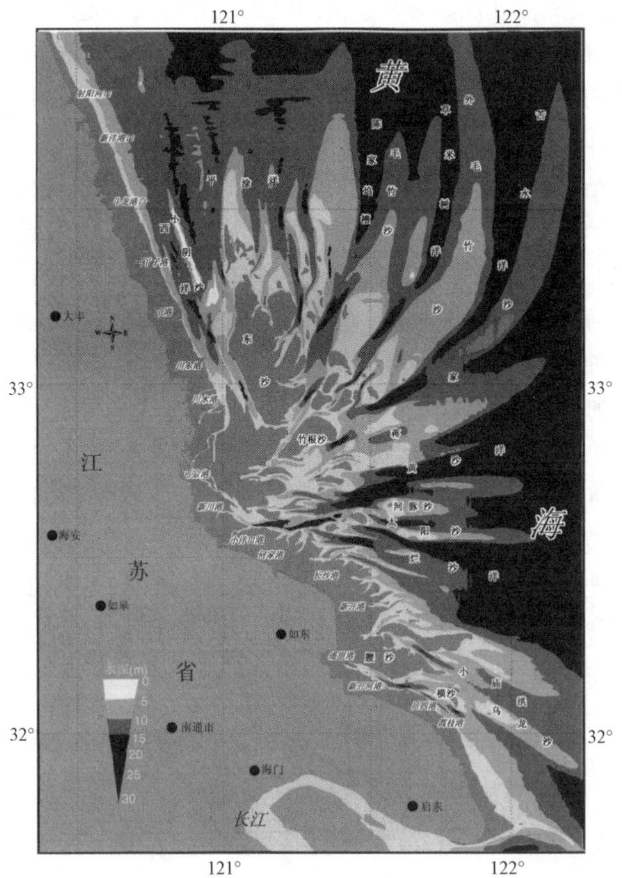

图 5　辐射沙脊群全貌(附彩图)

辐射沙脊群由细砂与粉砂组成，其泥沙粒度、矿物成分均与长江泥沙相同。“八五”期间，通过 600 km 地震剖面，探测出沙脊群枢纽部分的两条主潮流通道——烂沙洋与黄沙洋为晚更新世末(约 35 000 年前)古长江从苏北入海的主通道。沙脊群的泥沙是古长江从弶港一带入海时堆积的细砂物质，当时海平面位于现今－20 m 处。近 1 万年来，在全新世海面上升过程中，尤其在距今 6 000 年及 1 000 年两个明显的海侵冲刷期，由潮流冲刷改造古长江口泥沙形成与潮流方向一致的长条形沙脊。嗣后，历史时期中，黄河于 1128—1855 年夺淮入黄海又向辐射沙脊群供给了淤泥与黏土物质。因此，北部有黏土沉积。

辐射沙脊群是以晚更新世末期的古长江三角洲沉积为基础，而于全新世海平面上升海侵过程中成型。

(二) 开展江苏海岸海洋地质调查建议

(1) 发展海洋经济是当代江苏发展的必由之路，而且江苏海岸海洋具有优越的发展条件：平原海岸与具有发展潜力的广阔滩涂是宝贵土地资源，区域内充分的降水形成农业持续发展的优势条件；强大的潮流动力沿古河谷发育的深水潮流通道，是天然的建设港口群的宝贵资源；海岸带增强的风力、潮汐与日照组合形成新能源；发达的中、高等教育所形成的智力资源等，汇集成发展江苏海洋经济的巨大动力与强势基础。

（2）江苏在“八五”期间，是全国海岸带和滩涂调查的试点单位，调查研究实践所获得的工作方法与经验向全国推广，所取得的一系列科学成果，获得了国家科技进步特等奖。该项成果与科学著作应成为现今海岸带及浅海陆架研究的基础。当前，要与“908”国家海洋调查研究项目相结合，发挥华东有色地质勘查局的优势，加强江苏海岸海洋研究的基础地质工作；实现岸陆、滩涂与浅海的一体化调查研究；加强从空中遥感到地表地质及深部探测工作，尤其是加强海岸海洋地带的地球物理探测工作，除应用测深仪调查海底地形，地震剖面仪做海岸海洋连续地层剖面外，要加强海陆过渡带的重力、磁力勘测及钻探，这是地质勘查局的强项，而是江苏海岸海洋调查的弱项；要重视新近纪地质时期，并且将第四纪晚期、全新世的地质调查与历史时期的海陆变化结合起来研究。

（3）结合地质勘查局的特长与优势选定重点工作。油气资源，尤其是在辐射沙脊群巨大沙体中探索油气藏；江苏地面水已遭受污染，所以地下水资源调查相当重要：地下水的储量与分布规律，及地下水开采与海水入侵相关研究等；海岸与海洋工程地质工作，如：海港与临港工业基地及沿海城镇建设之地基稳定性，滩涂围垦与港口回淤影响，以及地震地质之效应等。

主要参考文献

[1] 王颖，牛战胜. 2004. 全球变化与海岸海洋科学发展[M]. 海洋地质与第四纪地质，24(1)：1－6.

[2] PAGES IPO. 1999. PAGES Newsletter[J]. Bern, Switzerland, 7(3).

[3] 王颖译. 1999. 海洋地理国际宪章(International Charter of Marine Geography)[N]. 地理学报，54(3)：284－286.

[4] Global Coastal Ocean. 1998. The sea, Vol. 10. John Wiley & Sons. Inc.

[5] 王颖，张永战，邹欣庆. 1997. 面向21世纪的海岸海洋科学[C]. 见：97’海岸海洋资源与环境学术研讨会论文集. 68－72.

海岸海洋科学的发展*

海岸海洋是陆地与大洋相互过渡的地带，它是既区别于陆地，又有别于深海大洋的独立环境体系，受人类活动影响密切，是研究水、岩、气、生圈层交互作用的最佳切入点。研究陆海过渡带的表层系统作用过程、环境资源特性及发展变化规律，以获求人类生存活动与之和谐相关等，构成了海岸海洋科学的研究对象与任务，是基于地理学、地质学与海洋学相互交叉渗透形成的新学科，具有自然、人文与技术科学相互交叉渗透的复合型科学特点。

联合国《21 世纪议程》指出：海洋是生命支持系统的基本组成部分，也是一种有助于实现可持续发展的宝贵财富。海岸海洋仅占地球表面积的 18%，其水体部分占全球海洋面积的 8%，占整个海洋水体的 0.5%，却拥有全球初级生产量的 1/4，提供 90%的世界渔获量，为 60%的世界人口的栖息地，目前全世界人口超过 160 万的大城市中约有 2/3 分布于这一地区，海岸海洋与人类生存关系密切。

海岸海洋科学的兴盛与 1994 年正式生效的"联合国海洋法公约"密切相关，公约对沿海国主权的 12 海里领海、24 海里毗连区、200 海里专属经济区以及大陆架是沿海国陆地领土自然延伸等原则规定，使海洋权益及管辖范围发生巨大变化，推动了沿海国对"海洋领土"的关注，全球涉及海洋划界的有 370 处。基于主权与资源开发的需要，推动海岸与大陆架浅海逐渐成为海洋科学领域的新热点，因此海洋是由两个主要环境组成，即海岸海洋与深洋。

海岸海洋(Coastal Ocean)定义是 1994 年 UNESCO 政府间海洋委员会(IOC)在比利时列日大学召开的国际海岸海洋科技会议(1st COASTS of IOC)上正式提出，明确了海岸海洋的范围包括海岸带、大陆架、大陆坡与大陆隆，含整个海陆过渡带。会议正式出版的"The Sea"系列第十卷"Global Coastal Ocean"，成为国际海洋学界正式确定海岸海洋的里程碑。国际地理学家联合会(IGU)1996 年发表"海洋地理宪章"正式将全球海洋区分为 Coastal Ocean(海岸海洋)与 Deep Ocean(深海海洋)两部分。

20 世纪初，经典文献将海岸定义为沿海滨分布的狭窄陆地；20 世纪中期，海岸工程实践明确了现代海岸带是包括沿海陆地及水下岸坡的"两栖地带"：上界止于风暴潮、激浪作用于沿海陆地的上限，下界始于水深相当于 1/3～1/2 当地波长处；至 90 年代，形成包括海岸带、大陆架、大陆坡及大陆隆，涵盖整个海陆过渡带的海岸海洋(Coastal Ocean)。经历了 20 世纪两次科学认识上的飞跃，加深了对海岸海洋环境特点的认识与资源环境的利用，发展形成具有交叉学科特点的应用基础型新学科。

海岸海洋研究方法具有外业工作的多学科综合性，包括陆上调查，浅海水、岩、气、生方面的观测，空中的同步监测，实验室多项分析及计算模拟。加上因时、因季节与因地观测，投入的人力与经费大，但其科学成果严密，能直接服务于生产建设与国家权益。

---

* 王颖：《地理教育》，2012 年第 10 期卷首语。

Coastal Geomorphology*

(Introduction)

The book was published by Higher Education Press of China in 1994. It was awarded a first rank prize as an excellent textbook from the State Education Commission of China in 1996.

Coastal zone is an independent environment system with the nature of transition from land to ocean. The formation and evolution of coastal zone are under the interaction of hydrosphere, atmosphere and lithosphere. The textbook integrates of coast into a multidisciplinary system. By using modern and classical concepts, theories, methods and examples from China and abroad, the book provides a way for recognizing different types of coasts, to study their natural processes and the evolutional history. The book has been widely used both in teaching and applied projects since the first drafting written in 1961, and gradually improved and accumulated in the last three decades. Even though the book was published recently the authors hope to add some new data and advanced principles for the revision of the book in the near future.

The present text is organized into 12 chapters as the following:

Chapter 1 Introduction

Chapter 2 Coastal dynamic

Waves, ties, currents, winds and rivers

Chapter 3 Geological background and geomorphological structures of coasts

Coast outline, tectonic movement, lithologic structures, coast sediments

Chapter 4 Sediment movements and geomorphological structures of coasts

Transverse movement, dynamic balance point, break wave base, accumulated equilibrium profile, marine erosional equilibrium profile, longshore sediment movement, sediment drifting and study methods

Chapter 5 Types of coastal morphology

Marine erosion features, marine depositional features

Chapter 6 Sea level changes and coast responses

Long term changes of sea level, short term changes, and coastal responses

---

* Ying Wang, Dakui Zhu: *Annual Science Report—Supplement of Journal of Nanjing University*, 1996, Vol. 32, English Series 4, pp. 35.

Chapter 7 Bedrock embayment coasts

Classification and its natures, marine erosional type, erosional & depositional type, marine depositional type, tidal inlet embayment

Chapter 8 Muddy plain coasts

Sediment movement and sedimentation of tidal flats, morphology of tidal flats, tidal sandy ridges, silty and clayey tidal flats-plain coast of China, muddy coasts of England and the Netherlands

Chapter 9 Estuaries and deltas

Current patterns of estuaries, evolutional factories of estuaries, major morphologies of estuaries, classification of delta, the evolution of delta, reform and construction in the area of estuary and delta

Chapter 10 Coral reef coasts

Chapter 11 Mangrove coasts

Chapter 12 Research methods and theoretic problems

面向21世纪的海岸海洋科学*

（摘要）

海岸带是海陆相互作用、过渡、互为关联的体系，与人类生存发展密切相关。海岸带具有与陆地、与海洋均为不同的环境资源特点。它具有不断变化的生物、化学、物理及地质的动态特性；它为多种海洋物种提供繁衍生境，是一个具有极高生产力和生物多样化的生态系；海岸地貌体系是抵御风暴、潮浪灾害及侵蚀作用的天然屏障；海岸生态环境可以通过吸附、沉淀、固封等作用减轻陆源污染的影响；保存着海陆相互作用过程的信息，有助于分析海岸变化趋势，有利于人类持续开发海洋。

20世纪，人们对海陆交互作用带——海岸带的认识是逐渐加深的：20～30年代，将海岸定义为沿海的狭窄陆地（D. Johnson，1919）；50～80年代定义海岸带为一两栖地带，包括沿岸陆地、潮间带与水下岸坡（В. П. Зенкович，1962），水下部分至－20 m左右（1980中国海岸带调查），这个定义适用于狭义的海岸带概念；1994年，联合国教科文组织于比利时列日大学召开的海洋学会议，明确提出“海岸海洋”（Coastal Ocean），范围涵盖整个海陆交互作用带海岸、大陆架、大陆坡与坡麓的堆积体——大陆隆。概念的变化反映着对海岸客观环境认识的深入，反映出国际海洋学界重视“海岸与近海”的研究，认识到几个部分是一个有机联系的客体，与人类生存发展至关密切。

海岸海洋占海洋表面的8%，约为海洋水体的0.5%，拥有全球1/4的海洋生产力，占有世界渔获量的90%，世界大城市（人口超过160万）中的2/3坐落在沿海地带，为全球60%的人口居住地。海岸海洋蕴藏着丰富的自然资源——能源、矿产、空间、食物与旅游资源，具有成本低、少污染及可再生补偿的特性。

海岸海洋科学研究的对象是整个海陆过渡带，涉及海岸体系，向海延伸至大陆边缘外沿（自然标准）或者是国家海洋管辖区外限（法律标准）。其研究内容包括海岸海洋体系的相互作用过程、变化与发展规律，环境资源容量、合理开发与可持续利用，以及减灾防灾举措与海洋权益问题。因此，海岸海洋科学具有自然科学、社会科学与高新技术科学交叉复合的特性，基础研究与开发应用、与规划管理相结合。这一特点使它更适应21世纪的社会、经济与生存环境可持续发展之需求。“联合国海洋法公约”实施所涉及的领海、毗连区、大陆架、200海里专属经济区、海峡与群岛水域等规定推动着沿海国对海岸海洋的关注与研究，海岸海洋科学成为新世纪初始的优先发展科学领域。

根据我国海岸海洋环境的特点，在科学的基础与发展趋势以及国民经济的发展需求

---

* 王颖，张永战：《迎接海洋新世纪暨纪念中国海洋学会成立20周年学术年会论文摘要汇编》，42－43页，1999年。

同年载于：《面向21世纪的科技进步与社会经济发展（上册）》，183－184页，1999年中国科协首届学术年会。

与经济实力，认为应重视下列科学项目：

(1) 海陆相互作用过程与边缘海发展变化以及资源环境效应预测。

(2) 河海体系沉积作用与海岸大陆架发展演变。

(3) 全球气候与海平面变化与海岸海洋效应及灾害发展机制。

(4) 21 世纪海岸海洋空间与能源利用。

(5) 沿海经济发达区、海岸与浅海资源环境承载力、开发管理与可持续发展研究。

(6) 领海基线、岛屿、海峡、水道、大陆架与 200 海里专属经济区管辖权益专题的多学科综合研究。

(7) 建立我国一体化的海岸海洋的科学与管理体系。

妇女与科技[*]
——评述与感受发生于自身周围的事例

（在第四届世界妇女大会发言，1995.北京）

记得二十世纪八十年代初期我在加拿大贝德福海洋研究所做客座研究员时，一位熟悉的科学家问我："颖，你为什么不做护士和秘书?"我愣了一下，因为我们将出航大西洋做海洋地质工作。我反问了一下"Why should I?"（为什么我一定要当护士与秘书?）他也愣了一下，然后微笑不语。继而我讲："中国有很多具聪明才智的妇女当护士及秘书，她们干得十分出色，赢得了人们的赞誉，如：中国的南丁格尔——李兰丁。但是，中国妇女有权力从事她所热爱与向往的事业，尤其在五十年代，青年时期的我，充分体会到中国是男女平等的。我喜欢地球科学，我喜欢森林，喜欢大海，所以我选择了海洋地质事业……"随着谈话的进行，我们都陷入了美好的沉思……这段对话，我时刻萦记心头：加拿大人对妇女职业是有传统看法的，尽管当时在我研修的大西洋地质中心，女海洋地质工作者有四位，而且在本行干的都不错。不论从理论或实际情况看，妇女都可以从事她所愿从事的工作，可上天、可下海，也可以深入洞穴，这些事就发生在我的身边。美国的第一位女宇航员，是我所在的加拿大达尔豪谢大学(Dalhousie University)毕业的。我的同学王飞燕副教授，从22岁就爬山钻洞，一直到62岁退休后仍不停息地从事喀斯特地貌研究与洞穴探察，她爬遍了贵州的洞穴，几十米的竖井只能靠吊绳上下，是什么力量在鼓舞她？几乎不可思议，她的双目因青光眼而视力减低。在南京大学地学院，从事地学工作的女科学工作者91名，其中副教授以上有38名，博导1名，广大的基础性研究与教学是由妇女承担的。

女科学工作者富于理想，对事业有一股执着、坚韧、不畏艰苦地追求，对待工作细心、认真、一丝不苟，一步一个脚印，不达目的不罢休，而追求的仅仅是理想的实现。我青年时代的一位女友，她是杨虎城将军的女儿，五十年代末，在大西北勘探石油，不幸于沙漠遇难，从牺牲的遗迹可辨识她是坚持到最后，多次爬越，希冀救出难友。"生命不息，努力不已"！每每，我以她的年轻形象鼓舞自己克服困难，我默默地注视着飞燕执着干劲，逐渐平息了内心的惶惑，科学实无坦途，天才出于勤奋。犹如马拉松赛跑，在起点时浩浩荡荡，只有迈步向前、坚持不懈才能达到终点。而科学攀登更是山外青山，登到一个高度后，会看到新的高度境界，没有坚韧不拔的毅力与不畏艰苦的工作是难以达到新高峰！蓝天白云、繁花似锦是令人流连，但是，研究任务促使我在水泥封闭的实验室工作，忘记了时间，进去时是日丽风和的假日清晨，出来时已是明月清空、树影斑斓的夜晚，但我内心是充实的，我完成了预定的分析。就这样一步一步，三年内完成了鼓丘海岸研究、石英砂表面结构与沉积环境、大陆架比较研究与深海埋葬核废物研究，一项项研究，一篇篇论文，表明成果丰硕，赢得了加拿大海洋地质学界的尊重，我以实际行动向外界展示了中国女地学工作者的

---

* 王颖：《南京大学报》，1995年6月10日。转载于：《南京大学报》，2016年1月8日。

智慧与毅力。

力量源于强大的祖国靠山：人民花费金钱供我外出学习，有生活习惯差异，但条件比当年下放好多了；思念家人，下放时也是三年，而地学工作外出习惯了；在国外日久渐渐体会出种族偏见，但我理直气壮；我无求于外人，祖国供我衣、食、住；我的研究成果却丰富中、加海岸成果，对加拿大亦有贡献。国内外海洋研究的经历、中华儿女的气质与道德观，使我毫不犹豫，学成返国，希冀增加一份力量推祖国地学向世界先进之林，这是内心实际感受，十年努力表明心迹。

返家后，人员倍亲，但条件依旧，彷徨徘徊。经受开放春风熏陶荡涤，认真思考，明确新起点再攀登。教学、研究、培养新一代，出海考察建设新港口，获取经费建设实验室，开展国际学术交流。辛苦耕耘，培育新苗圃。当年的成就、信誉促成新的中、加合作研究：海南港湾蚀积规律、沉积速率和深水港建设。五年合作完成了海南最大深水港——洋浦港建设研究项目，召开了国际会议，出版专著："Island Environment and Coast Development"。该合作项目加拿大无偿援助 40 万美元，在南京大学建设海洋勘测实验室与 Pb210 同位素实验室，培养了中、青年科技学者。加方合作者亦从深海研究转而掌握海岸工作方法，获得了中国海岸及热带海岸丰富的科学知识，双方合作富有成效，推动了海岸侵蚀演变规律研究，达国际领先水平。同时，中英、中日、中俄的合作亦连连开展。九十年代，在国家计委、科委与教委大力支持下，获得世行 42 万美元贷款，进一步扩充实验设备，并在国家自然科学基金会与教委亲切领导关怀下，建成海岸与海岛开发国家试点实验室(Pilot Laboratory of Coast and Island Exploitation)，由海岸海洋勘探、沉积物分析与定年实验以及海洋地理信息系统实验室三部分组成，以当代先进仪器装备的现代化实验室系列(其总价值达 130 万美元)，一支多学科人员组成的 20 人科技队伍，活跃在海岸海洋研究的科学领域。自八十年代末至今的七年时间内，负责组织召开了三次国际学术会议，皆正式出版了论文集。以新的实验设备完成国家度假村亚龙湾海底旅游规划、三亚港扩建、小东海人工海滩建设方案、石梅湾海滨旅游规划、江苏海岛调查中海域研究、黄海辐射沙脊群研究以及深圳湾人工岛沙源研究等重要项目，获得生产、研究经费、人才培养与国际交流的良性循环和丰硕成果，在国际海岸科学领域占有一定地位。

返国十年圆一梦，建成了先进的海岸实验室系列，推动了海岸科学与教育事业进步。理想追求、艰苦踏实工作，孕育着力量与发展潜力。中华儿女似海鸥般自由翱翔在辽阔的海空，拼搏撷取科学硕果。

试论大学本科地球学科教育*

一

地球科学是研究地球发展变化自然规律与人类生存发展活动相关的科学，发展地球科学至关重要，其作用可影响到地区、国家甚至全球发展的战略决策。

地球是由几个圈层组成的开放体系：气圈、水圈、生物圈、岩石圈、地幔、地核等。各圈层有各自的活动规律特征与表现，但亦互为动力、互为影响，是不能截然分开的相关体系。例如，地幔活动影响到海陆分布的变化，后者影响到气流运动、洋流的形成变化，气候的差异又影响到生物、人类生活等。EL-NINO 现象是气、水、生物及人类生活关联影响的明显例证。人类社会发展活动对气、水、生物及岩石圈也形成巨大影响，已为当今的共识。当代地球科学研究反映着开放体系及圈层的相互影响。

教育应当适应社会发展需求，反映科学技术进步，提高人类文化素质与创新能力。以此相律，当前的地球科学教育与社会发展需求、与科学发展水平之间存在着相当的差距。例如，地学教育按圈层区分为几个体系：大气科学、海洋科学、地质科学、生物科学等，地理科学尚具有地球科学体系的某些共性特点，但尚未形成共识。海洋科学出现学科交叉：物理海洋学、海洋化学、海洋生物学、海洋地质学、海洋地理学等，与海洋水体的联通性有关。但是，地学更多的趋向是区别分支学科的特性，深化专业的分工，忽视地球体系内在的联系和共性。结果大学毕业生专业口径狭窄、知识更新缓慢，与社会需求不适应，亦削弱了地球科学在人类社会的重要作用。探索地球科学体系的内在联系机理、过程与表现，必须重视多学科交叉、渗透与结合，必须重视人类生存活动的相关，从单纯的自然科学体系转变为以人与地球关系为核心，具自然科学、社会科学与技术科学交叉渗透特点的新的学科结构。

二

人类生存依赖自然界。基本需求是新鲜的空气供呼吸，洁净的淡水供饮用，利用土地获取衣、食、住与通行。但是人类生存活动带给自然界的消极影响，使地球面临着接近承载人类生活的生态极限之潜在危机：保护人类免遭太阳有害辐射的臭氧层正在变薄，在南极上空已形成空洞；严重的土壤侵蚀引起肥力下降，粮食生产的潜力减小；大面积毁坏森林与草地促进了沙漠化过程，加剧了生物种类与数量的消失，削弱了大气圈的自洁能力；海洋动物被滥捕而锐减，海洋遭受污染即将达到其所承受的极限；毒害废物排放加剧了对

* 王颖：《地理学的理论与实践——纪念中国地理学会成立九十周年学术会议文集》，32－36 页，北京：科学出版社，2000 年。

气、水和陆地的污染，引起了全球性的健康问题；环境恶化严重地破坏了全球生态平衡，20世纪后期人口激增，贫穷、饥饿、疾病和文盲同样威胁到人类的生存。资源、环境、生存、发展是当代面临的巨大挑战，也是地球科学以及地学教育改革与发展必须正视的重大课题。从全球与国家的宏观需求着眼，地学教育任务应当包括：增强人类依存地球的意识，培养促进人与自然环境和谐发展，有效地利用自然资源，使人类生存环境持续发展的科学技术和管理人才。这一任务不仅是本科，也包括研究生教育。

1992年在巴西召开的全球首脑会议通过的《21世纪议程》，是具有里程碑性质的历史文件。占全球人口98%以上的世界各国首脑聚会共商地球环境、资源与人类持续发展的大事，制定了全球性的发展纲领，动员全人类重视参与，以全面规划指导人类正确处理人与地球的关系，合理地利用资源和环境，以达到全人类生活水平的提高与持续发展。这与地球科学教育的使命是吻合的，全球重视人与地球的关系，是地学前所未有的发展机遇，因此，《21世纪议程》可接受为地学教育改革的指导性纲领。

根据《21世纪议程》的中心内容，我认为，有七个方面须在地学教育中分别增加、建组或加强。

1.《21世纪议程》的基本目标是使人类获得持续生存的方式

这与地理学的中心思想协调人地关系相一致。当前严重的问题在于世界人口急剧增长，每年新增近1亿人，给地球生态系统带来巨大的冲击与不堪负担的压力。出路在于协调人类生活和生产的整体水平和方式，使之与地球的承载力相适应。人口与地球承载力的关系历来是地理学研究的传统课题，今后需从新的认识水平在整个地球科学体系内建立人地关系的教育。

2. 自然资源

(1) 淡水资源保护、承载力分析、合理利用、防止污染与污水处理等。地学教育应打破以往的圈层分割，将海水水源、蒸发降水、地表水、地下水自然过程，水循环、水资源、淡水资源管理与海水水源的保护等作为地学教育的支柱体系课程，作为地学人才必须掌握的基础知识。

(2) 陆地资源。合理地、有限制地开发利用与保护陆地资源，是21世纪议程的要点之一，从山脉、沙漠等各种生态环境、地表土壤发育、地貌过程，到岩石圈结构、地球内动力、矿产形成分布等，是地学教育的另一个重要组成部分。

(3) 能源。减少传统的非再生能源——煤与石油燃烧所造成的环境污染，开发可再生的对环境无害的能源系统：太阳能、风能、水能、潮能、浪能以及海水温差能等，这些是地球自然环境要素的新技术应用，应该也必将成为地球科学教育的新生长点。

(4) 森林植被。涵养水土、调节气候、抑制沙漠化，植物、动物物种与数量，药用资源等生物地学是地球科学教育的传统项目，宜从人类生存发展中自然资源危机的新高度来加以研讨。

3. 大气与海洋

这是流动贯通全球的自然环境与共同资源，海洋巨大的水体是生命的摇篮、风雨的故乡与人类生存发展最大的潜在资源宝库——淡水、新鲜空气、食物、土地、能源、矿藏。以

1982 年通过的《海洋法》为依据，划归沿海国的海域管辖面积达 1.3 亿 km^2，与全球陆地面积相近。人类活动引起气候变暖，海平面上升，后者造成风暴潮、灾变天气频繁与低地海岸沉溺等一系列问题；世界海洋捕捞量已接近其最高限度——1 亿 t；公海倾废、油船倾溢、河流排污，空气携运污染物于海洋等，已很严重，将使海洋难以容纳与稀释，其后果将对人类带来不可挽救的灾难影响。海洋与大气科学必须成为地学基础教育必不可少的重要支柱内容。

4. 与人类居住环境有关的生存质量改善与管理

这主要是城镇、乡村的人口膨胀，供水、居住的条件与质量；废物、废水的处理；能源、交通条件以及医疗、教育等服务设施。中国经济发展快，城市亦发展快，现代化、国际化成了中国城市化的新一轮目标。我国的地球科学重视野外实践，但对城镇的调查与研究薄弱，城市地学的教育处于初期阶段，尚需重视加强。《21 世纪议程》提出了全球范围城镇地区的环境与发展战略，为地学在城镇与乡村的自然、社会、经济及信息等方面教学提供了纲领与体系。

5. 对化学品的利用、生活废弃物和工业废物的管理

人口增长、工业与生活废弃物积聚引起环境污染与疾病问题，发展中国家 80% 的疾病与 1/3 的死亡是由于污染食物和水、气引起。经济发展会因健康及环境质量下降而抵消。人口众多、经济发展的中国尤需注意减少化学品废弃物、处理与消灭有害废物，开发资源、发展经济必须注意到的环境效益，包含社会重视、技术处理、法规管理等项措施，地学教育应从地球开放体系角度重视并提出处理废物的正确途径，避免扩散与再污染。

6. 全球经济的持续发展

在各个层次的政治和经济决策中，必须考虑持续发展和良好的环境条件，经济核算需调整以体现对环境的影响及体现资源的价值。地学教育与有关的商贸、金融科学皆应适应这一调整的需要，培养具有持续发展观点的各层次的决策人与管理者。

7. 建立和谐的人地关系系统

每个国家与地区都需要建立现代化的地球数据信息系统，进行信息收集、规划、决策及传播服务，以及新知识的吸收与新技术的应用。这既是技术保证手段，也是国际交流合作的前提，地学教育需加强已有的遥感、遥测与地理信息系统教育训练，为国家、地区建立空间台站与地面接收体系，建立信息处理与数据库网络而提供相应的人才，逐步发展数字地球科学。

三

地球科学教育尤其是地理学教育应包括自然、社会经济与技术三方面的科学内容。这样的内容安排符合社会对人才的需求，也适应地球科学本身发展的特点：多学科交叉形成新结合点的复合科学。如海洋地理学，是地理学与海洋学两大学科群的结合点，其研究客体是海洋，包括海岸与海底，范围涉及到气、水、生物与岩石圈。研究内容包括海洋环境、海洋资源的开发利用与保护；海洋经济、疆域（海岸、岛屿、领海、大陆架、专属经济区、

公海等)政治、立法与管理;海洋新技术发展、应用及影响等。

当代科学的发展,新的项目或工程需要具有复合科学的特点。地学高等教育由于客体环境的多因素、交互作用与时空变化万千的特点,更应以学科群的主体支柱作为教学体系的主纲领,而以主结构间的交叉结合点形成专门分支。这种分支与专业有区别,是以不同学生所选课程的交叉结合不同而已,而不是成批的以专业为模式的培养。地学发展过程说明,有造诣的科学家多数是在原专业训练以后,不断吸收相邻科学的知识技能,而形成了新型的具复合科学特点的专家。凡具复合科学进步特点的科学,都能在获取项目、经费与研究成果方面反映出强大的竞争性与生命力。科学发展的客观实际如此,在经历半个世纪的发展后,地学教育工作者应意识到这种发展变化,责无旁贷地进行教学体系与教学内容的变革,不能漠然置之,仍坚持数十年一贯制的传统教学体系与专业建制。

实践结果亦反映出,当代地学科学人才不一定或不宜于长期地“成批”生产。因为自然环境的多样性,地区差异与需求也是多样的。地学人才培养宜遵循以下原则:基础教育(包括基础知识、基本理论与基本技能)以气、水、土三大支柱体系为纲,作为具有共性的主体课程;选修大量的临界课程或原有的基础课程,使得大多数学生在大学阶段即形成一些区别以适应不同部门的需求;少部分人(1/10)可进入研究院深造,在导师特长的基础上结合不同的相邻学科形成新的结合点专长。

所以,培养途径之一,是打破以圈层为界形成的人才培养界限,代之以主干体系课形成学生广博的地学知识基础,个人可据其选修课的组合不同而形成其特点,经过数年培养后,学生既具共性,又不划一,可适应不同服务部门的需要——大学、政府机构、工业、农业、旅游业等。从发展看,在当前更应重视从事宏观规划与决策的地方人才的培养,有意识地予以加强。

其次,可试行以学院为主体招生,地学各系仍保留,提供一定特色的基础课、实验与技术设备。大学生必须学习地学主体课程与数理化等基础课程,其比例可达总学时 1/2 左右;允许学生广泛选课,自选课比例为 1/4,余下的 1/4 学时用于野外实践、实验技术及计算机应用训练。

其三,必须发扬地学教育优良传统:理论联系实际,坚韧不拔、不怕艰苦、团结努力攀高峰的精神,培养一代又一代的中国地球科学人才。

最后,在大学生的教育中,应将地球科学作为一门重要的公共基础必修课,它是提高全民综合素质的重要内容。

人生在世,应正确理解与对待自然环境与资源,建立“人与自然”的和谐关系,对人类生存的可持续发展,至关重要。

主要参考文献

[1] 张知非,郭亚曦,应思淮译校.1990.2000 年地球科学进展的前沿——英国的研究重点.地球科学进展,(5):1-13.

[2] 谢宗强,马克平,丁洪美,贺金生编译.1990.警醒吧!人类行动起来拯救地球——《21 世纪议程》介绍.

对大学地球科学教育的几点思索*

一

地球科学是研究地球发展变化自然规律与人类生存发展活动相关的科学，发展地球科学至关重要，其作用可影响到地区、国家甚至全球发展的战略决策。

地球是由几个圈层组成的开放体系：气圈、水圈、生物圈、岩石圈、地幔、地核等。各圈层有各自的活动规律之特征与表现，但亦互为动力、互为影响，是不能截然分开的相关体系。例如，地幔活动影响到海陆分布的变化，后者影响到气流运动、洋流的形成变化，气候的差异又影响到生物、人类生活等。EL-NINO 现象是气、水、生物及人类生活关联影响的明显例证。人类社会发展活动对气、水、生物及岩石圈也形成巨大影响，已为当今的共识。当代地球科学研究反映着开放体系及圈层的相互影响。

教育应当适应社会发展需求，反映科学技术进步，提高人类文化素质与创新能力。以此相律，当前的地球科学教育与社会发展需求、科学发展水平之间存在着相当的差距。例如，地学教育按圈层区分为几个体系：大气科学、海洋科学、地质科学、生物科学等。地理科学尚具有地球科学体系的某些共性特点，但尚未形成共识。海洋科学出现学科交叉：物理海洋学、海洋化学、海洋生物学、海洋地质学、海洋地理学等，与海洋水体的联通性有关。但是，地学更多的趋向是区别分支学科的特性，深化专业的分工。忽视地球体系内在的联系和共性，结果大学毕业生专业口径狭窄、知识更新缓慢，与社会需求不适应，亦削弱了地球科学在人类社会的重要作用。探索地球科学体系的内在联系机理、过程与表现，必须重视多学科交叉、渗透与结合，必须重视人类生存活动的相关，从单纯的自然科学体系转变为以人与地球关系为核心，具自然科学、社会科学与技术科学交叉渗透特点的新的学科结构。

二

人类生存依赖自然界。基本需求是新鲜的空气供呼吸，洁净的淡水供饮用，利用土地获取食物。但是人类生存活动带给自然界的消极影响，使地球面临着接近承载人类生活

---

* 王颖：《中国地质教育》，2001 年第 2 期，第 12 - 15 页。

1995 年，以“对大学本科地球科学教育的几点思索”为题载于：《“面向 21 世纪地学人才培养研讨会”论文集，高教研究与探索，南京大学学报哲学社会科学版专辑》，1995 年 2 期，第 18 - 21 页。

1998 年转载于：杨承运主编，《地球科学发展趋势与人才培养》，33 - 39 页，河海大学出版社。本文稍作修改。

的生态极限之潜在危机:保护人类免遭太阳有害辐射的臭氧层正在变薄,在南极上空形成空洞;严重的土壤侵蚀引起肥力下降,粮食生产的潜力减小;大面积毁林促进了沙漠化过程,加剧了生物种类与数量的消失,削弱了大气圈的自洁能力;海洋动物被滥捕而锐减,海洋遭受污染即将达到其所承受的极限;毒害废物排放加剧了对气、水和陆地的污染,引起了全球性的健康问题;环境恶化严重地破坏了全球生态平衡,20 世纪后期人口激增,贫穷、饥饿、疾病和文盲同样威胁到人类的生存。资源、环境、生存、发展是当代面临的巨大挑战,也是地球科学以及地学教育改革与发展必须正视的重大课题。从全球与国家的宏观需求着眼,地学教育任务应当包括:增强人类依存地球的意识,培养促进人与自然环境和谐发展、有效地利用自然资源、使人类生存环境持续发展的科学技术和管理人才。这一任务不仅是本科,也包括研究生教育。

1992 年在巴西召开的全球首脑会议通过的《21 世纪议程》,是具有里程碑性质的历史文件。占全球人口 98%以上的世界各国首脑聚会共商地球环境、资源与人类持续发展的大事,制定了全球性的发展纲领,动员全人类重视参与,以全面规划指导人类正确处理人与地球的关系,合理地利用资源和环境,以达到全人类生活水平的提高与持续发展。这与地球科学教育的使命是吻合的,全球重视人与地球的关系,是地学前所未有的发展机遇,因此,《21 世纪议程》可以作为地学教育改革的指导性纲领。

根据《21 世纪议程》的中心内容,笔者认为,有七个方面须在地学教育中分别增加、重组或加强。

1.《21 世纪议程》的基本目标是使人类获得持续生存的方式

这与地理学的中心思想协调人地关系相一致。当前严重的问题在于世界人口急剧增长,每年新增近 1 亿人,给地球生态系统带来巨大的冲击与不堪负担的压力。出路在于协调人类生活和生产的整体水平和方式,使之与地球的承载力相适应。人口与地球承载力的关系既是地学研究的传统课题,也需从新的认识水平上加强这方面的教育内容。

2. 自然资源

与上题密切相关,地球的自然资源与承载力是有限的,要贯彻保护非再生资源与持续利用的战略。最主要的资源涉及:① 淡水资源保护、承载力分析、合理利用、防止污染与污水处理等。地学教育应打破以往的圈层分割,将降水、地表水、地下水至海水的自然过程,水循环、水资源、淡水资源管理等作为地学教育的支柱体系课程,作为地学人才必须掌握的基础理论与知识。21 世纪面临着水资源的严峻危机。② 陆地资源。合理地、有限制地开发利用陆地资源,是《21 世纪议程》的要点之一,从山脉、沙漠等各种生态环境、地表土壤发育、地貌过程,到岩石圈结构、地球内动力、矿产形成分布等,是地学教育的另一个重要组成部分。③ 能源。减少传统的非再生能源——煤与石油燃烧所造成的环境污染,开发可再生的对环境无害的能源系统:太阳能、风能、水能、潮能、浪能以及海水温差能等。这些是地球自然环境要素的新技术应用,应该也必将成为地球科学教育的新生长点。④ 森林。涵养水土、调节气候、抑制沙漠化,植物、动物物种与数量,药用资源等生物地学是地球科学教育的传统项目,应从人类生存发展中自然资源危机的新高度加以研讨。

3. 大气与海洋

大气与海洋是流动贯通全球的自然环境与共同资源,海洋巨大的水体是生命的摇篮、风雨的故乡与人类生存发展最大的潜在资源宝库——食物、土地、能源、矿藏。以1982年通过并于1995年实施的《海洋法》为依据，划归沿海国的海域管辖面积达1.3亿km^2，与全球陆地面积相近。人类活动引起气候变暖，海平面上升，后者造成风暴潮、灾变天气频繁与低地海岸沉溺等一系列问题；世界海洋捕捞量已接近其最高限度1亿t；公海倾废、油船倾溢、河流排污、空气携运污染物等于海洋，已很严重，将使海洋难以容纳与稀释，其后果将对人类带来不可挽救的灾难影响。海洋与大气科学必须成为地学基础教育的重要支柱内容。

4. 发展城市科学与人类居住环境及生存质量改善与管理密切相关

主要是城镇、乡村的人口膨胀，供水、居住的条件与质量；废物、废水的处理；能源、交通条件以及医疗、教育等服务设施。中国经济发展快，城市亦发展快，现代化、国际化成了中国城市化的新一轮目标。城市化的过程与加强城市管理对控制淡水，土地的消费，减少污染，提高人民生活质量提供了有效的途径与机制的可行性。地球科学重视野外实践，但对城镇环境、资源、文化、历史渊源与民俗方面的调查与研究薄弱，城市地学的教育处于初期阶段，尚需重视加强。《21世纪议程》提出了全球范围城镇地区的环境与发展战略，为地学在城镇与乡村的自然、社会、经济及信息等方面教学提供了纲领与体系。

5. 对化学品的利用、生活废弃物和工业废物的管理

人口增长、工业发展、废弃物积聚引起环境污染与疾病问题，发展中国家80%的疾病与1/3的死亡是由于污染食物和水、气引起。经济的发展会因健康及环境质量下降而抵消。人口众多、经济发展的中国尤需注意减少化学品废弃物、处理与消灭有害废物，开发资源、发展经济必须注意到的环境效益，包含社会重视、技术处理、法规管理等项措施，地学教育应从地球开放体系角度重视并提出处理废物的正确途径，避免扩散与再污染。

6. 全球经济的可持续发展

在各个层次的政治和经济决策中，必须考虑可持续发展和良好的环境条件，经济核算需调整以体现对环境的影响及体现资源的价值，这是必不可少的重要内容。地学教育与有关的商贸、金融教育皆应适应这一调整的需要，培养具有资源环境与可持续发展观点相结合的各层次的决策人与管理者，既是地学人才的重要出路，也是实现生存环境可持续发展系统管理的最佳人才来源。

7. 贯彻执行《21世纪议程》

每个国家都需建立现代化的数据信息系统，进行信息收集、评价与传播服务，以及新知识的吸收与新技术的应用。这既是技术保证手段，也是国际交流合作的前提，地学教育需加强已有的遥感、遥测与地理信息系统教育与应用训练，为国家、地区建立空间台站与地面接收体系，建立信息处理与数据库网络而提供相应的人才。

三

地球科学教育应包括自然、社会经济与技术三方面的科学内容。这样的内容安排符合社会对人才的需求，也适应地球科学本身发展的特点：多学科交叉形成新结合点的复合

科学。如海岸海洋科学，是一新兴的学科点，是在原海岸动力地貌学与海洋地理学基础上发展起来的，随着社会发展实践逐步认识到，包括海岸带、大陆架、大陆坡与陆麓在内的海岸海洋(Coastal Ocean)是一个独立的自然体系，既不同于陆地，也不同于大洋，它以海陆交互作用对人类生存关系密切为特征。第四纪以来，地球表层的交互作用与变化赋予此地带以深刻的烙印与特性，影响到现今的环境与资源。海岸海洋科学是一新学科点，在美、欧，在中国均获得科学界与管理部门的关注，原因在于它与《联合国海洋法》实施有关。海岸海洋科学本身具有自然科学(海洋学、地质学、地理学)与新技术结合，密切为工程、社会发展以及海洋权益服务的特性。

当代科学的发展，新的项目或工程需要具有复合特点的学科。地学高等教育由于客体环境的多因素、交互作用与时空变化万千的特点，更应以学科群的主体支柱作为教学体系的主纲领，而以主结构间的交叉结合点形成专门分支。这种分支与专业有区别，是以不同学生所选课程的交叉结合不同而已，而不是成批的以专业为模式的培养。地学发展过程说明，有造诣的科学家多数是在原专业训练以后，不断吸收相邻科学的知识技能，而形成了新的具复合科学特点的专家。如地图学与遥感技术与地理信息系统的结合；如地貌学与冰川学与技术科学(测量新技术，测试新技术——又可分地球化学，物理或声学的，生物的技术等)的结合；力学与海洋水文或物理海洋学的结合；经济地理与建筑科学、地理信息系统相结合形成具复合科学特点的城市规划学，但绝非原来的经济地理。凡具复合科学进步特点的科学，都能在获取项目、经费与研究成果方面反映出强大的竞争性与生命力。科学发展的客观实际如此，在经历半个世纪的发展后，地学教育工作者应意识到这种发展变化，责无旁贷地进行教学体系与教学内容的变革，不能仍坚持数十年一贯制的传统教学体系与专业建制。

实践结果亦反映出，当代地学科学人才不一定或不宜于长期地“成批”生产。因为自然环境的多样性，地区差异与需求也是多样的。地学人才培养宜遵循以下原则：基础教育以大的支柱体系为纲，譬如水、土、气(包括基础知识、基本理论与基本技能)为主要课程体系，而选修大量的临界课程或原有的基础课程，使得大多数学生在大学阶段即形成一些区别以适应不同部门的需求；少部分人(1/10)可进入研究生院深造，在导师特长的基础上结合不同的相邻学科形成新的结合点专长。

因此，建议培养途径之一，是打破以圈层为界形成的人才培养界限，代之以主干体系课形成学生广博的地学知识基础，个人可据其选修课的组合不同而形成其特点，经过数年培养后，学生既具共性，又不划一，可适应不同服务部门的需要——大学、政府机构、工业、农业、旅游业等。从发展看，在当前更应重视从事宏观规划与决策的地方人才的培养，有意识地予以加强。

其次，可试行以学院为主体招生，地学各系仍保留，提供一定特色的基础课、实验与技术设备。大学生必须学习地学主体课程与数理化等基础课程，其比例可达总学时 1/2 左右；允许学生广泛选课，自选课比例为 1/4，余下的 1/4 学时用于野外实践、实验技术及计算机应用训练。

其三，必须发扬地学教育优良传统：理论联系实际，坚韧不拔、不怕艰苦、团结努力攀高峰的精神，培养一代又一代的中国地球科学人才。

四

自 1994 年 11 月 8～10 日由国家教委在南京大学主办的“面向 21 世纪地学人才培养研讨会”后，南京大学地学教育在教育部、高等教育研究中心的支持下，有一些进步：

（1）校教务处在地学院教学中实施了三门公共基础课作为必修课：① 地球科学概论（地科系开设）；② 环境科学概论（环科系开设）；③ 地理信息系统概论（城资系开设）。同时，开设了一批适应当前社会发展需要的选修课，如可持续发展概论、环境地质学等；出版了教材如：地球科学（多媒体光盘，1999 年高教出版社）、地球科学现代测试技术（1999，南京大学出版社）、地下水动力学（1997，地质出版社）、天气学（1999，高教出版社）、地球物理流体力学（1996，气象出版社）、环境科学原理（1998，南京大学出版社）、地理信息科学导论（1999，中国科学技术出版社）及 21 世纪教程教材：区域分析与规划（1999）、环境地质学（2000，高等教育出版社）等 31 本，反映着当前地学进展的新教材。这些措施推动了地学教育进步。

（2）校研究生院支持在地学院成立“水资源与水循环研究中心”，促进学科交叉渗透。

（3）2000 年，校教务处联合地学院三系，共同申请了 21 世纪初高等教育教学改革项目：“地学类创新人才的培养方式与实践”，实际上是探索地球系统科学（大地学）人才培养途径与办法，已获教育部批准，提供 12 万元经费资助。

南京大学计划于 2001 年 7 月按地球系统科学招收本科生试点班（20～30 名），打破原系科界限，对现行课程进行改革和重组，建立三层次教学模式：第一层次：扎实的数、理、化、生、外语、计算机基础，由南京大学基础学科教育学院实施；第二层次：大地学学科群专业基础课和专业选修课，建成地质、地理、气象三个体系，学生可交叉选课，拓宽口径，加强大地学基础；第三层次：专业课，与研究生教育相衔接。

紧扣“创新”人才培养的要求，根据地球科学在 21 世纪发展趋势，改革课程内容和知识结构，加强高、新技术应用，建立现代化的教学方法、手段，总结和探索培养“创新”型人才的思路和措施，为 21 世纪国土资源合理开发利用、环境保护、灾害防治的迫切需要培养地学人才。

主要参考文献

[1] 张知非，郭亚曦，应思淮译校. 1990. 2000 年地球科学进展的前沿——英国的研究重点. 地球科学进展，(5)：1 - 13.

[2] 国家自然科学基金会. 世纪中国地球科学(讨论稿).

[3] 谢宗强，马克平，丁洪美，贺金生编译. 1990. 警醒吧！人类行动起来拯救地球——《21 世纪议程》介绍.

海洋在召唤*

20 世纪 50 年代，当时苏联专家 В · П · 曾柯维奇的科普读物《海岸》一书，对我有很深的影响与启迪。该书深入浅出，趣味盎然。它使我认识了海岸，引起我对海岸科学的浓厚兴趣。纷繁多姿的祖国海岸吸引着我，一直从事探索至今。我们是地球居民，人类只有一个地球，了解地球、善待地球，是我们每个人的职责，义不容辞。

海岸分布遍及世界各大洋（各海域），与陆地岛屿交接的地方，从极地到热带，海岸海洋面积占地球面积的 18%。

海岸介于大陆与海洋之间，它的自然环境特点不同于陆地，亦不同于海洋。它蕴藏着丰富的动能、矿产、水产、药物资源等。海岸带清新的空气与充足的阳光（Sunlight）、松软的沙滩（Sand Beach）与海鲜食品（Sea Food）构成了以"3S"著称的重要的旅游资源、运动与疗养胜地。

《联合国海洋法公约》已于 1994 年 11 月 16 日实施，它形成了全新的国家管辖海域：200 海里专属经济区与群岛国家水域。大陆架与海峡通道内容也有了重要改变。原属于公海的 1.3 亿 km^2 的海域将划归沿海国管理，其面积与地球上全部陆地面积 1.49 亿 km^2 相近。海洋国土的基线是以毗邻的海岸带为依据的。中国不仅有 960 万 km^2 的陆地面积，还有 300 多万 km^2 的管辖海域。

人类的发展活动自远古至今均与海岸密切相关，贝塚、绳纹陶器、夹砂红陶网坠、石质重锤（石砣）、沉船及其他器物在沿海与海域被广泛发现，反映出海岸带渔猎活动源远流长。但是，随着人类开发活动的进展，也给海岸带来了影响与变化。大约近 1/3 的陆地污染物直接从海岸进入海洋，使海水污染、赤潮灾害频频发生。因此，认识和了解海岸的环境特点与变化规律，分析确定海岸带自然承载力的限度至关重要，它将促进人们对开发利用海岸与保护海岸的持续发展，有利于海岸带为子孙万代的生存发展服务。

海岸海洋约占有全球海洋生产力 14%，占有全球反硝化作用的 50%，占有全球有机质残体的 80%，沉积物矿体的 9%，50%以上的全球硫酸盐沉积，全球河流悬移质及相关元素与污染物的 75%～90%。上述数字反映出海岸海洋与人类生存活动至为重要。

海岸海洋的环境特性赋予它蕴藏着丰富的、大部分可再生的自然资源：风能的动力资源、海洋动力资源（波浪能、潮汐能），可供海岸与岛屿发电而取之不竭；海滨稀有元素与贵金属砂矿如：锆英石、独居石、金红石（为航天与航海器材所需的特种金属材料）、沙锡、沙金及金刚石等；海底的石油与天然气等矿产资源；滩涂、岛礁、荒地等空间资源；旅游资源——美丽的海岸风光，清新的空气、明亮的阳光、清澈的海水、细软的沙滩、美丽的岛礁

* 王颖：《地理教育》，2003 年第 3 卷：序。

摘自：王颖，《海岸——通向海洋的虹桥》，广西教育出版社，1998 年。

与岸岬以及美味的海鲜产品等。这些丰富的生物与非生物资源，吸引着大量人口聚居于海岸带，并发展交通、商业、房地产、食品及旅游等事业，促使经济发达。在发展经济的同时，需重视保护海岸环境的质量与生态系统的良性发展，使子孙万代能享受这个美丽地带的恩泽。所以，认识海岸，了解海岸的特性与发展规律是十分重要的。

地球系统科学创新人才培养模式探索与实践*

一、地球科学高等教育教学改革背景分析

(一) 地球系统科学的特征

当代人类社会发展面临着严峻的人口、资源、环境等问题，作为一个研究人与地球关系的综合性学科，地球科学已突破传统学科界限，成为21世纪人们最关注的学科之一(王德滋等，2001)。当代地球科学发展具有显著的地球系统科学(大地学)特征：① 强调不同时空尺度基本地球过程及其相互作用，其时间尺度从几秒钟的地震到几十亿年的地球环境演化，空间尺度从微细矿物研究到全球环境变化；② 重视应用现代观测、探测、实验和信息技术对海量数据的系统采集、积累与分析；③ 强调地球系统的整体行为、各圈层的相互作用、自然系统中的物理、化学、生物过程和人文因素的耦合以及全球与区域、宏观与微观、地球环境与生命过程等紧密结合；④ 地球系统科学的复杂性和探测技术、实验技术以及信息技术的广泛应用，使前沿基础研究与高新技术发展融为一体。

地球系统科学的发展使人们对资源、环境、灾害的认识程度和研究方式发生了重大变化。对资源找寻的视野逐步从地球表层走向深部，从陆地走向海洋，走向近地空间，从单纯地注重矿产资源的找寻逐步转向以可持续发展为目标的资源合理利用与环境保护并重；对环境问题的关注已从局部走向区域，走向全球，从单一污染物的研究转向复合污染的形成机理与防治研究；从生态系统结构、功能的研究延伸到生态系统的维持与退化生态系统的修复，并从生态效应深入到人体健康，推动绿色生产；对自然灾害的研究也从定性走向定量监测、预警、预报以及灾情评估于一体的综合研究。地球科学又一场新的革命性突破正在来临，我国在世纪之交的地球科学战略部署中，已经十分明确地将地球系统科学定为基础研究与应用基础研究的总方向(赵连泽等，2006)。

(二) 地球科学高等教育的历史使命

长期以来，国内高校在地球科学人才培养中，都是按照地质、地理、大气、环境等各传统分支学科的各自模式独立培养的，这种培养模式限制了地球科学各分支学科的交叉与融合，培养的学生知识面较窄，难以适应21世纪地球系统科学发展的需要，也不利于学生创新能力的培养。

---

* 王颖，赵连泽，吴小根，蒋全荣，邵进，陈云棠：《首届大学地球科学课程报告论坛论文集(2004年4月7日)》，14－19页，北京：高等教育出版社，2007年。

同年载于：《中国大学教学》，2007年第6期，第13－16页。

近几年来，中国科学院地学部曾多次组织院士与教育部及有关高校负责人召开研讨会，探讨中国地学教育的未来发展问题，并完成咨询报告(中国科学院地学部地学教育研究组，2003)。咨询报告明确指出："社会对于地球科学的需求，和社会的发展程度密切相关，地学教育也是国民素质教育的重要组成部分，从我国地球科学教育的未来着眼，当代地学创新型人才的知识结构与综合素质培养也必须要与地球系统科学的发展趋势相适应。"2006 年 10 月 11—14 日在西北大学召开的全国高等学校地球科学教学指导委员会第一次全体会议上，主任委员张国伟院士指出："21 世纪人类社会发展向地球科学提出了新的重大需求，地球科学正处于新的重要发展时期，建立地球系统科学，并以地球系统科学的理念加强地学各分支学科的研究，从而发展、构建新的更高层次的地学理论与知识体系，已是当代地球科学发展的必然趋势和方向。"地球科学的发展要求地学本科人才培养的课程体系、课程内容与形式必须进行大幅度的改革与创新，建设创新人才培养模式，造就新一代高素质地学创新人才，以满足 21 世纪国家基础科学研究和国民经济发展对新一代地学人才的需求。

二、南京大学的探索与改革

地球复杂、开放的大系统，其表层的大气圈、水圈、岩石圈、生物圈及其相互作用和有机联系是地球系统科学的重要内容，也是 21 世纪地学教育的核心内容。针对地学教育按圈层设系，相互交叉融合不够的现状，南京大学地学院三系自 1999 年起酝酿如何联合培养地球系统科学(大地学)人才，提出了打破以圈层为界的人才培养界限，按"大地学"口径培养地学人才的设想(王颖，2001；王德滋等，2001)，得到了全国高等学校教学研究中心原常务副主任陈祖福的支持，并于 2000 年申报并获得教育部世行贷款 21 世纪初高等教育教学改革重点项目——"地学类创新人才的培养方式与实践"。在此教改项目带动下，南京大学地学院在国内率先开展了地球系统科学创新人才培养的探索与实践。

(一) 开办大地学基地试点班

南京大学自 2001 级起，每年从新生中挑选 60 名学生组成地球系统科学(大地学)基地班，其中地质学 24 名，地理学 15 名，大气科学 21 名。截至 2006 年已招收 6 届，累计招收本科学生 356 名，按上述新的培养模式实施培养。

为激发大地学基地班学生热爱地球科学的热情，培养他们学习地学专业的浓厚兴趣，地学院为地学基地班开设了地球系统科学系列专题讲座，主要安排在 1～2 年级，每学期由地球科学系、城市与资源学系(现地理与海洋科学学院)和大气科学系各安排 1～2 位著名学者(包括院士、资深教授和中青年骨干教师)为学生作相关专业基础知识和学科发展趋势的科学讲座。如王德滋院士、王颖院士、伍荣生院士分别为地学基地班学生作了题为"20 世纪地球系统科学重要进展"、"全球变化与海岸海洋科学进展"、"大气科学的现况、展望与问题"的科学讲座，开阔视野，启迪兴趣，深受学生的欢迎和好评。

（二）明确教学改革思路和目标

（1）更新教育观念。地学教育以培养基础研究和应用基础研究人才为宗旨，它的起点虽是本科，但后续教育应是硕士和博士；针对地球科学发展趋势和国家对创新人才培养的需求，在教学改革实践中，按照“专业培养与素质教育相结合、知识传授与能力培养相结合、教学与科研相结合”的教育思路；在加强基础、拓宽知识面的同时，又充分体现地球系统科学特点，培养学生个性发展，落实南京大学提出的“以学科群打基础，多次选择，逐步到位；学生早期进行科研训练，按一级学科交流、贯通；本科—硕士培养”的教育理念。

（2）创新培养模式。通过教学改革实践，探索创建具有大理科基础、大地学特色的地球系统科学创新人才培养模式：建设并完善三层次课程体系和教学内容，编写出版高水平的专业教材，建设现代化的实践教学基地；加强基础，拓宽知识面，注重培养学生个性发展；力争做到“两通”：即地学各学科横向打通，本科生、研究生教育纵向打通，培养基础宽、能力强、适应面广的新世纪地球科学人才；积累教学运行和教学管理的思想与方法，建设新型的教学队伍。

（三）改革教学内容重组课程体系

根据地质学、地理学和大气科学专业特点，以及地球系统科学发展趋势和国家对创新人才的需求，通过对现行的课程体系和教学内容进行改革和重组，创建了强化大理科基础、体现大地学特色与自主专业选修的三层次课程体系。课程体系总体结构包括通修大平台课程、大理科平台课程、地学模块核心课程、各专业核心课程、各专业主干课程、各专业选修课程、科研训练课程及毕业论文；相应的教学计划第一学年为公共基础课，设置通修大平台课程和大理科平台课程，第二学年设置地学模块核心课程和地质学、地理学、大气科学各专业核心课程（选修），第三、第四学年设置各专业选修课程、科研训练课程及毕业论文等。这样的课程体系和教学计划安排，旨在通过大平台课程加强通识教育，通过设置大量的选修课程可以为培养个性化人才提供足够的时间和空间。

较之改革前的传统课程设置，第一层次增加了数学、物理、化学、外语、计算机基础课程学分，由南京大学基础教育学院统一负责相关课程的教学安排，按南京大学理科要求强化基础，为后续课程铺路，启迪持久的学习能力和构建合理的知识结构；第二层次增设了地球科学学科群基础课，确定“地球科学概论”、“大气科学概论”、“环境科学导论”和“遥感与地理信息系统”4门课程为大地学基地班学生必修的学位课程，体现大地学学科交叉融合之特色；第三层次包括专业主干课程和专业选修课程，由地质学、地理学、大气科学三个学科推荐选出10门课程作为交叉互选课程，这些课程是：现代气候学原理、大气探测基础、天气学、天气学实验、海洋科学导论、自然地理学、人文地理学、第四纪地质学、普通地质学、现代测试技术与应用等，其余的专业主干课程和选修课程由三系自行确定，增加了专业选修课程特别是跨专业交叉选修课程的比例，有利于学生的自主专业选修。

在此基础上，我们还系统地组织教师修编了相关专业课程的教学指南和教学内容，在学校“985工程”一期教改经费的支持下开展了30多门专业课程的建设工作，在专业课程教学中注意培养学生个性发展并与研究生教育相衔接。

（四）加强实践教学环节

针对地球科学具有实践性、综合性强的特点，我们在安排教学工作过程中，坚持和完善“课堂教学—科学实验—社会实践”相结合的三元教学模式，采取相应措施加强基础课程和专业基础课程等课堂教学，不断完善实验课程体系和实验教学过程，同时特别注重抓好野外实习教学环节和地学野外实习教学基地建设工作，为培养学生的实践能力创造了良好的支撑条件。

在野外实习教学基地建设方面，“十五”以来我们进一步完善了南京汤山和安徽巢湖地质学教学实习基地、江西庐山和江苏连云港地理学教学实习基地、上海中心气象台、江苏省及安徽省气象台大气科学教学实习基地，为学生创造良好的野外工作条件和生活条件，使本科学生的个性化培养、创新能力训练有了理想的硬件条件和科研环境。近几年来，上述地学野外实习教学基地已实现校内教学资源共享，在我校地质学、地理学和大气科学等专业野外实习教学过程中发挥了重要作用；不仅如此，另有国内 20 余所高校也利用我校建立的野外实习教学基地开展了相应的野外实习教学，起了开放和辐射的作用，也扩大了影响。

为进一步加强学生实践能力的培养，我们在学校的大力支持下，在大地学基地班教学计划中增加了选修的野外调查实习教学内容：如为期 2 周的西部地区地学综合调查教学实习，内容包括黄河上游兰州段的河流地貌、黄土地貌与成因、水土流失与生态环境保护、西部干旱与沙漠化及防治，天山地质与地貌、生态特征、植被与土壤，西部内陆大陆性气候与东部沿海海洋性气候的差异等。

在探索实习教学的国际化方面，2006 年暑期，我们组织了 2003 级大地学基地班学生开展了两项国际合作教育交流实习教学：一是与俄罗斯伊尔库茨克国立技术大学合作开展了为期 19 天的贝加尔湖地区地质、地理、生物综合调查野外实习教学；二是与美国加州理工学院合作开展了为期 2 周的新疆天山地区地学综合调查野外实习教学。积极探索采用“研究性实习教学”方法，注重培养学生发现问题和分析、解决问题的能力。考察前，学生在教师的指导下确定研究课题，完成开题报告。考察结束后，学生以项目小组为单位进行研究课题的答辩。同学们通过上述实习直接观察到了大量地学景观和现象，与外国同行师生共同开展野外调查和面对面的交流，开阔了视野，促进了多学科知识的交叉互补，激发了他们对地学研究和地学事业的兴趣和热情，提高了他们开展地学野外调查和综合分析的能力以及锻炼了体力与意志。

（五）构建了一整套学生创新训练机制

在南京大学的统一规划下，地学院积极实施富有创造能力学生培养计划，大地学学生创新训练计划等，鼓励学生早期介入科学研究，形成了一整套学生创新训练机制。学院成立专门的学生创新训练专家指导委员会，设立学生科研创新项目，给学生配备研究导师，开放专业实验室，营造浓厚的学术环境和氛围，培养学生早期科研兴趣与意识。

南京大学地学院三系采取多种措施鼓励学生开展早期科研训练，除了实行导师制即在二年级末通过师生互选确定指导老师、实现师生一对一的科研指导外，自 2003 年以来，还针对三、四年级学生推行“大学生创新研究基金”，资助学生的早期创新性科研训练项目，每个项目资助额度为 1 000～4 000 元不等，由学生撰写申请书、自由申报，经各系组织专门评审后择优分级资助(资助率在 60%左右)，获得资助的项目，由学生自主管理研究费用，在老师指导下完成科研训练过程，编写结题报告(学年论文或毕业论文)，参加学校一年一度的基础学科论坛交流和论文评选，组织论文报告会，逐步形成了以导师制和“五种报告(论文)”为特色的早期科研训练机制。通过这一完整的科研训练，使学生的自由探索和参加科研的期望得以实现，并在实施过程中锻炼了学生的科研项目申请与管理、资料收集、报告编写、论文撰写、科研协作等的科研工作能力。

同时，我们还鼓励高年级学生积极参加由研究生主办的两周一次的地学沙龙，增加大学生和研究生的沟通与交流，模糊研究生和本科生的学制界限，让他们提前体验开展科学研究的学术氛围，激发他们探索科学的兴趣和钻研精神。此外，我们还积极资助学生参加各种专业夏令营、报告会等科普、科研活动，以锻炼学生的表达能力，拓宽专业知识面。

三、教学改革与人才培养的成果与展望

(一) 已经取得的阶段性成果

(1) 教学获奖情况。近年南京大学来地学院共获得省部级以上单项教学成果奖 7 项；18 门课程获得省部级以上精品课程称号或高校优秀课程奖，其中“地球科学概论”和“现代天气学原理”被评为国家精品课程；编写出版了各类教材 30 余部，其中出版普通高等教育“十五”国家及规划教材 6 部、入选普通高等教育“十一五”国家及规划教材 7 部，“地球科学”被列入国家精品教材建设计划；围绕地球科学人才培养以及相关课程和教材建设内容，撰写发表教学研究论文 30 多篇。

(2) 培养优秀学生情况。自 2001 年以来招收地球系统科学(大地学)基地班学生 6 届 356 名，实施“三三一”地球系统科学创新人才培养模式，大部分同学确立了稳定的专业思想，愿意为祖国地球科学事业贡献力量，至今已有 2 届 120 位本科毕业生，他们的研究生录取比例在 70%以上，为国内多所知名研究机构和高校输送了高质量的研究生生源，此外还有多名学生赴美国、加拿大、瑞典等国深造。

南京大学地学院按照新模式培养的大地学基地班学生普遍具有宽厚的理科和大地学基础、良好的科学素养和知识结构、较强的实践和科研能力、良好的深造和发展潜力的群体特征。

(二) 进一步推进教学改革的计划

(1) 完善通识教育平台建设。为了进一步深化教育教学改革，拓宽学生的知识基础，培养全面发展的高素质人才，提高学生的创新意识和创新能力，南京大学将积极探索由专业教育向通识教育人才培养模式的转变。为探索适合南京大学学科特点的通识教育模式和在通识教育基础上的个性化人才培养体系，探索并积累研究型教学的实践经验，南京大学已于 2006 年 3 月正式组建匡亚明学院。目前，匡亚明学院设置数理、化生、地学、人文

和社会5大模块13个一级专业分流方向。南京大学地学院大地学基地班被纳入匡亚明学院，实施“2+2”培养与分流模式：大地学基地班的教学计划和学籍在大一、大二两年由匡亚明学院统一管理和组织，第一学年末学生可以跨模块或专业流通，人数控制在各个基地班的20%～30%；从第五学期开始，各个专业基地的学生分流到其所在院系，并按各自院系的教学计划组织教学和进行学籍管理。

（2）将教学改革纳入“985工程”二期重点项目规划。2005年12月，南京大学地学院申报的“地球系统科学野外基础实习基地建设与创新研究及大地学实验教学示范中心建设”项目作为“985工程”二期教学改革重点项目获得正式立项，共获得投入近480万元。该项目旨在通过教学改革实践，促进地质学、地理学、大气科学等多学科的交叉融合，在相应的学科群基础课和专业选修课以及野外实习教学过程，促进各学科专业教师、实验室、野外实习基地等教学资源的共享，促进办学效益的提高。项目的侧重点在于建立和完善全新的野外实践教学体系，包括教学内容精选与拓展、野外研究性教学方法的创新和现代化教学手段的实现；在进一步完善野外基础教学实习基地建设的基础上，建设适应地球系统科学创新人才培养需要的现代化实验教学示范中心。

（3）建立一支结构合理、治学严谨、教学和科研水平高、掌握现代科技方法，能满足地学创新人才培养和知识创新需要的教师队伍。一是要吸引名教授尤其是科研水平高的中青年教授上本科生基础课，在教学教师队伍的建设方面真正做到科研和教学相融合，加强对名师教学经验与资源的总结、研究和推广，提高南京大学地学院教学研究的水平，努力构建一支“名师加教学创新团队”的优秀教学队伍；二是着力培训青年教师，通过规范管理，加强青年教师的教学工作培训，尽快提升青年教师的教学水平，确保教学质量。

主要参考文献

[1] 汪品先. 2003. 我国的地球系统科学研究向何处去[J]. 地球科学进展，18(6)：2-7.

[2] 刘东生. 2006. 走向“地球系统”的科学：地球系统科学的学科雏形及我们的机遇[M]. 中国科学基金，20(5)：266-271.

[3] 王德滋，赵连泽. 2001. 关于地球科学人才培养的实践与思考[J]. 江苏地质，25(3)：129-133.

[4] 赵连泽，吴小根，杨修群，陈云棠，王颖. 2006. 地球系统科学创新人才培养与实践. 见：地质与地球化学研究进展[M]. 南京：南京大学出版社，509-514.

[5] 陈云棠. 2004. 地球科学人才培养与教学改革的思考[J]. 中国大学教学，(9)：19-20.

[6] 中国科学院地学部地学教育研究组. 2003. 中国地学教育的未来[J]. 中国地质教育，18(2)：135-138.

[7] 王颖. 2001. 对大学地球科学教育的几点思考[J]. 中国地质教育，(2)：12-15.

[8] 王颖，赵连泽，吴小根，蒋全荣，陈云棠. 2005. 发展·创新·改革(第二集). 见：世行贷款21世纪初高等理工科教育教学改革项目结题成果汇编[Z]. 北京：高等教育出版社.

加强地球系统科学教育　培养一流地学人才*

现代地球科学已经突破传统各学科的界限，向地球系统科学方向发展，主要表现为：强调不同时空尺度的地球作用过程及其相互影响；重视应用现代观测、探测、实验和信息技术以采集、积累和分析数据；强调地球系统的整体行为，各圈层的相互作用，自然系统中的物理、化学、生物过程和人文因素的耦合，以及全球与区域、宏观与微观、地球环境与生命过程等紧密结合。从社会对地学的需求来看，地球系统科学的发展使人们对资源、环境、灾害的认识程度和研究方式发生了重大变化，强调高效、可持续、和谐发展等理念。这些新变化要求大学地学教育要与时俱进，适应时代发展的要求，培养新时期高水平专业人才。但是，长期以来，我们的高校在地学人才培养中，主要是按照各传统分支学科的模式培养，这种教学方法限制了学科的交叉与融合，难以适应 21 世纪地球系统科学发展的需要，也不利于学生坚实基础和创新能力的培养。在全国高等学校教学研究中心的支持下，南京大学从 20 世纪 90 年代末开始了地学教学改革实践，于 2001 年建立地球系统科学基地班，在一、二年级教学中加强地球科学、数理化和生物以及中、外文和计算机技术等基础教学，在地学教育中以地球系统科学的理论为指导，打破学科界限，强化基础、注重实践、鼓励创新。通过 9 年的教学实践，培养了一批具有宽基础和创新能力的大学生，为实现南京大学建设国际一流大学的目标，起到了积极推进作用。

一、发挥学科齐全优势，努力培养创新人才

南京大学具有学科多样、地学学科齐全的优势，我们整合多学科资源，构建了“三位一体”地学实践教学体系，成为南京大学地学教学的核心内容。所谓“三位”，包含两层内涵，一是地质、地理和大气科学三个学科相互贯通，二是课堂实验教学、野外实践教学和科研能力训练三者并重、层层递进。所谓“一体”，就是三学科融于地球系统科学一体之中，实验、野外实践和科研训练融于创新能力培养之中。具体目标是：创建三层次课程体系；建设成多学科、多类型、多层次野外实践教学基地；建设成注重基础和学科前沿、全方位开放的地球系统科学教学实验中心；构建高水平、多层次、多方向大学生科研训练平台。三层次课程体系包括：第一层次的数、理、化、生、外语、计算机基础（由南大基础教育学院负责，1.5～2 年）；第二层次的地球系统科学学科群基础课（由地学院负责）“地球科学概论”、“大气科学概论”、“环境科学导论”和“遥感与地理信息系统”；第三层次的专业主干课和专业选修课（由三院系分别负责），由三个学科推荐并选出 10 门课程作为交叉互选课程，包括“现代气候学原理”、“大气探测基础”、“天气学”、“天气学实验”、“海洋科学导论”、“自然

---

* 王颖，鹿化煜，胡文瑄，王元，邵进，王腊春：《中国大学教学》，2009 年第 8 期，第 11 - 12 页，85 页。2010 年载于：《大学地球科学课程报告论坛论文集(2009)》，3 - 6 页，北京：高等教育出版社，2010 年。

地理学”、“人文地理学”、“第四纪地质学”、“普通地质学”、“现代测试技术与应用”。

在教学理念上，坚持自然科学、人文科学和现代技术教育并行，重视地学学科交叉，注重地球圈层相互作用过程。在教学方法上，坚持把理论教学、认知实习和创新实践结合，多方位提高大学生的地学知识和科研素养。在课程设置上，打通地理、地质、海洋和大气科学的界限，开设互通式的基础课程。通过这些地学教学改革实践，使得大学地学教育突出了学科交叉、理论与实践教学并重的优势，提高了大学生实践创新能力。

二、加强实习和实验教学，提高大学生认知和动手能力

野外实习基地建设是地学人才培养的重点。为了在野外实践中贯穿学科交叉的理念，我们进行了反复论证，积极探索与实践，逐渐融合，最终形成了三个层次的实习基地：第一层次“基础实习基地”包括宁镇山脉普地实习和庐山自然地理实习基地，这些基地有50余年的建设历史；第二层次“综合实习基地”有东北长白山实习基地、台湾实习基地、阿尔卑斯实习基地；第三层次“创新实习基地”由贝加尔湖实习基地、兰州-天山-帕米尔实习基地、连云港实习基地组成。对于在不同阶段的实习，提出了不同的要求，包括认知、综合和创新等层次。从东海之滨的海岸带到世界屋脊的帕米尔高原，从烈日炎炎的海南岛到凉风习习的贝加尔湖，都有南京大学学生实习的身影。这些实习基地系列的建设，形成了在实习教学中有基础、有综合、有研究的不同形式的体系，为地学院本科生的野外实践提供了丰富多彩的场所，也为同学开展科学研究提供了广阔的空间。

先进的实验室建设是培养高水平地学人才的基础。在南京大学“985”二期经费的支持下，我们建设成以“地球动力学教学实验室”、“地理信息技术教学实验室”和“现代大气探测教学实验室”为主体的大地学实验教学体系。同时，基地班的学生还可以进入相关的国家重点实验室、省部级重点实验室和南京大学的各类实验室进行实验研究。在实验教学过程中，把实验内容(包括基本实验、提高型实验和研究创新型实验三个层次)与科学研究、工程实践和各类社会应用实践密切联系起来，既具有综合性、设计性、应用性，又具有系统性，大大提高了学生的实验和动手能力，增强了学生的科研素质。

三、注重海陆交互带的特色教学

南京大学最早涉及海洋科学是1959年，参加了为解决天津新港泥沙来源而开展的渤海海岸与浅海调查。1984年建立了南京大学海洋科学中心，包括地理、地质、大气、生物和法学。1990年，通过中加合作与世行贷款专项建立了海岸与海岛国家试点实验室，后为教育部重点实验室，研究领域从海岸带扩展至大陆架、大陆坡整个海陆交互带。2000年建立海洋地球化学研究中心和海洋大气相互作用研究中心。2006年，在南京大学地理与海洋科学学院成立了海岸海洋科学系，培养本科生和研究生。

南京大学重视海陆交互带的基础和应用研究工作，经过多年的发展和积累，形成了以海岸海洋(包括海岸带、大陆架、大陆坡及大陆隆在内的整个海陆过渡带)资源环境为重点，注重大学教育和基础研究，兼顾管理和开发的学科体系，具有一只老中青结合的精干

教师队伍。对海岸海洋科学系的学生，加强其基础理论教学，与南京大学国家基础科学人才培养基地班的学生享受同样的教学资源，并强化野外实习和实验室教学。学生在参加地理与海洋科学学院的普通地质学实习、庐山自然地理学实习之后，还要到具有海岸带特色实习基地——连云港和海南海岸海岛实习基地学习，使得学生得到了海岸海洋方面的实际锻炼，开阔了视野。同时，南京大学海岸与海岛教育部重点实验室的全部仪器设备向大学生开放，他们可以进行光释光和 Pb－210 的测年研究，也可进行粒度、有机质含量的沉积物分析；在高年级的时候，更多的是参与教授的课题，和教授共同参加实际工作，了解海岸海洋研究的全过程。

四、探索国际化交流协作，拓展实践教学内涵

在南京大学"985"教学项目的支持下，我们组织地质学、地理学、气象气候学和生物学的学生组成野外联合考察队，由相关学科的教师与俄罗斯伊尔库茨克大学地学教师共同带队，到贝加尔湖地区进行为期两周的地学和生命科学的野外考察。考察期间，有俄罗斯知名专家给学生讲解；同时，在出国考察之前，学生在老师的指导下作了相关的准备，带着科研问题观察、采样和分析。迄今，我们已经成功地组织了三次联合考察，使学生开阔了眼界，启迪了地球系统科学的思维习惯，取得了很好的效果。此外，地质学和地理学国家基础科学人才培养基地班的学生还有机会到欧洲阿尔卑斯地区进行野外综合实习。实习之前，我们精心组织了地理和地质等相关领域的知名教授和专家，聘请法国奥尔良大学知名教授（包括国际地科联副主席 Charvet 教授）连续 10 天给学生现场讲解地球系统科学的基础科学和前沿科学问题，使得学生的眼界大为开阔。在实习过程中，既可以看到教科书提到概念的经典原型，比如阿尔卑斯的冰川沉积和单面山，也可看到板块构造控制的太平洋火山和台风发育过程模拟等典型现象。这些都为他们理解地球系统科学问题和培养地学思维起到了积极作用。

五、强化大学生早期科研训练

大学生的科研能力训练和创新能力培养是地学教学的核心，我们一直重视本科生的科研创新能力的培养。目前，科研训练项目已经成为每一名大学生在校期间必须完成的学习项目。从 2002 年起，依托国家人才培养基金和教师的科研项目，每年设立 60 项左右科研训练课题，每个课题由 2～3 人的学生科研小组承担，配以相关指导教师，针对性地指导学生循序渐进地步入科学研究中。正在执行的学生训练项目包括"国家大学生创新训练计划"项目，南京大学"985 大学生创新训练计划"项目，以及国家基金委"地理学人才基金"资助项目。这些项目涵盖不同专业方向，供不同兴趣和特长的学生自主选题，开展科研训练研究，逐步提高学生的科研能力和创新能力。

同时，我们要求每一名毕业生的毕业论文与指导教师承担的科研项目和科研训练项目挂钩，积极利用国家、省部级和南京大学的各类实验室，进行实质性的研究，提高毕业论文的质量。指导教师亲自带学生出野外考察、采样；在专业人员的指导下，大学生到实验

室进行测试和实验分析;分析测试结果在小组上讨论,撰写的毕业论文在老师的指导下反复修改,直至达到专业水平。为了活跃学术气氛,强化学生所学到的基本知识,定期举办学术报告讲座和本科生学术研究报告会,并评选出优秀的学生论文和报告给予奖励。优秀的报告经导师指导撰写论文,在学术刊物发表。这些活动大大提高了学生的科研能力。

自 2001 年招收地球系统科学(大地学)基地班学生以来,按照新模式培养的学生普遍具有宽厚的理科和大地学基础、良好的科学素养和多学科交叉知识结构、较强的实践和科研能力、良好的发展潜力,为国内多所知名研究机构和高校输送了高质量的研究生生源,还有多名学生赴美国、加拿大、瑞典等国深造;工作之后受到用人单位的高度评价。

在新时期,为了实现南京大学创建国际一流大学的目标,我们将继续优化教学资源,提高办学质量和教学水平,加强地球系统科学基础理论的教学,深化国际交流和国际化办学水平,为把南京大学建设成国际知名的重要地学人才培养基地而奋斗。

[感谢全国高等学校教学研究中心和教育部高等学校地球科学教学指导委员会的支持与鼓励,感谢国家基金委员会基础科学人才培养基金以及南京大学“985”教学项目的资助,感谢中国台湾大学、俄罗斯伊尔库茨克大学、法国奥尔良大学及相关专家教授的大力支持]

关于地球科学教育发展的建议*

我国地球科学教育，自“八五”规划以来，经过多次会议的讨论研究与多年的教育实践，已有长足进步，反映在：

(1) 以地球系统科学思想为指导，多次召开地质、地理、大气、海洋、环境科学等各学科结合的研讨会议，一次次地深入探讨，专业设置与课程体系已有长足发展。

(2) 适应当前社会发展需求，在专业设置与课程体系中，将地球环境资源与城乡规划体系结合为一体，学以致用，密切为我国发展建设服务。资源环境与城乡规划管理专业是在第二届地理学教育委员会提出的。现在这个专业设置数目众多，遍及地学院系，反映出综合性学科之生命力；启示我们应进一步适应地球客体之特点，研讨地学教育的创新发展，培养具有全局观点、深入与综合分析能力的实干人才。

下面从适应人地和谐相关发展的需求与当前科学进步的可能性，提出几点建议。

一、现阶段地学教育与课程体系宜加强的三个方面

(1) 地球表层气、水、岩、生圈层系统相互作用，相互影响变化，与人类生存活动关系密切。因此，地学(系、科)教育，应将大气、海洋(与淡水)、地质、地理与生物五方面相关的基础理论列为必修课程。实例如：20 世纪 50 年代时南京大学地理学教育，除未涉及海洋科学外，其他四方面均强，学生毕业后，为国家建设作出重要贡献；1952 级的地貌班中，如今有 3 位院士。继承、认知、拓宽知识面与思路是培养人才的第一步，不打好基础是无法健步地前进。同样，局限于原有的专业教育，是不能探索到自然变化的真谛，何况人类活动影响使其变化过程更复杂。

例如，对全球变化的研究，不仅仅是气候变化。我理解是：“全球变化是反映在气、水、岩石、生物圈的事件性变动，形成全球性的频发与持续效应，对人类生存环境影响深刻。”[1]气候变化有人类活动的影响，但主要是大气的自然变化，气候带移动的纬度地带性变化。近 50 年来的个人体会是：当前南京的冷热气温变化，比 20 世纪 50 年代时“温和”多了，夏季不是那么酷热，而冬季不是那么寒冷，冰雪冻结减少了[2,3](图 1)；而北京夏季比 20 世纪 50 年代热，冬季不像 20 世纪 50 年代时那么寒冷[4,5,6](图 2 和图 3)。我理解全球变化不仅仅是气候变化，其自然变化还包括了火山、地震活动的影响效应。地震与火山活动自 20 世纪 90 年代至今在加强，尤其是 21 世纪以来，不仅在全球范围内日日有地震，而且强震发生频频[1]；火山爆发不仅在日本与夏威夷群岛，2010 年 3 月冰岛强烈的火

---

* 王颖：《中国大学教学》，2010 年第 12 期，第 17－18 页，70 页。

2011 年载于《大学地球科学课程报告论坛论文集(2010)》，3－7 页，高等教育出版社，2011 年。内容有增改。

山爆发，尘埃蔽日，飘移散布，影响航空；同年 10 月 26—29 日，印尼中爪哇省的默拉皮火山爆发，造成 33 人死亡，炙热烟尘影响范围巨大。这些岩石圈层的构造活动，影响到气候变化，确实是影响到地球表层的事件性变动。所以，地球科学系统教育，必须加强圈层系统的基础教育，不全面了解地球系统变化的基本情况，如何进行全球变化分析研究以探求其变化趋势并加以应对呢？

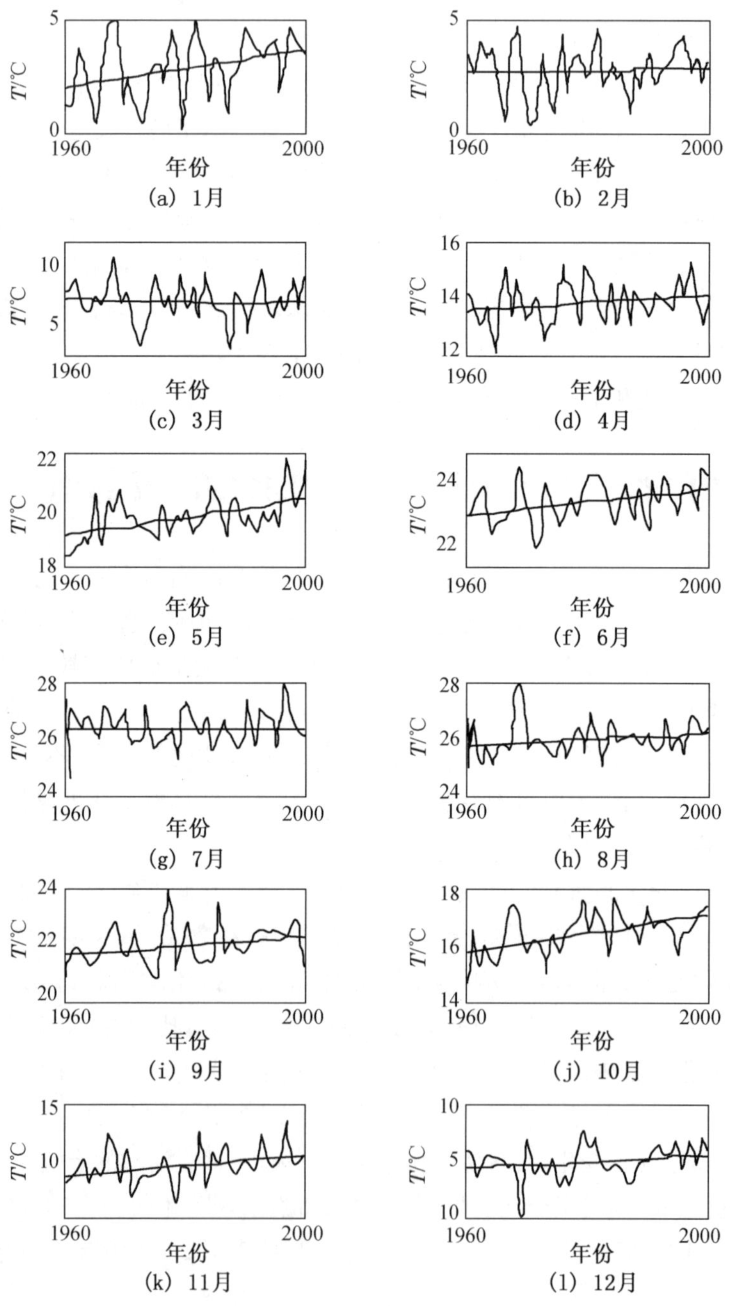

图 1　1960—2000 年南京地区地面平均气温变化

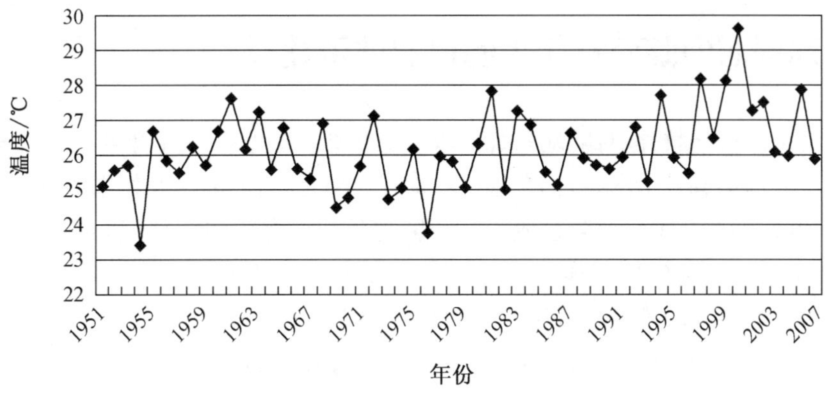

图 2 北京夏季年平均气温变化

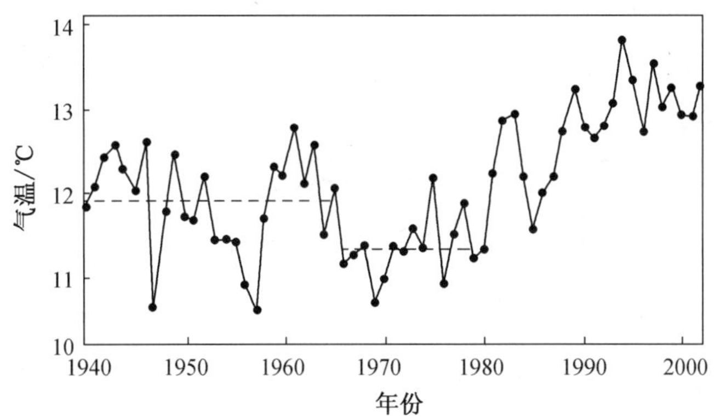

图 3 北京站年平均地面气温变化特征(虚线为年代间的平均)

(2) 地学各专业学生,既应学习地球发展与地球环境演变的历史,更要重视学习与研究现代作用过程与现世状况,以此为基础,探索未来发展变化趋势。与自然和谐共处是人类生存环境持续性发展的必要途径,地理科学的"天人合一"的核心思想,应体现在地球科学体系的教育之中。

(3)重视与强化实习与实践教学,圈层系统的基础理论教学,必须有实践教育配合。重视师生对自然环境的认知,重视综合性与区域性分析能力的提高,重视基础理论研究与应用发展。建议在实践教育中:① 加强完善原有的"宁镇山脉地质实习基地"与"庐山自然地理实习基地",这是基础教育基地,应对自然环境倍加保护并扩大开放利用。② 适应发展的需求,建议教育部领导组建"秦岭山脉(含南北两侧)地学教育基地"、"长白山(火山、森林、湖泊)教育基地"、"海南海岸与海岛环境资源教育基地"以及开始实施"吐鲁番盆地、戈壁沙漠及帕米尔高原大路线实习"与建站。这几处可作为本科高年级及研究生实习与研究基地。进一步与地方实习点结合,逐渐发展成为地学实践基地骨干网络,对地学人才培养,对公众教育,对促进我国资源、环境和谐发展利用均有推动。基地建设应由国家出资,教育部主持相关院校组办。实际上除过秦岭外,南京大学在地学教育改革中,已经进行了上述的基础地学教育实习,海岛海岸、长白山与西部大路线实习与考察,的确提高了学生的学习兴趣、观察与研究能力,增强了对伟大祖国的热爱。

二、加强对海陆交互作用的过渡地带教育

我国的地球科学教育，对陆地系统的教育历史久、力量强、成果累累。海洋科学教育以青岛为中心已形成骨干网络，仍在进一步加强。但是，对海陆交互作用的“过渡地带”教育很薄弱，并未形成共识。海陆过渡带是指陆地延伸至海洋的全部（图 4），包括海岸带、大陆架、大陆坡与坡麓的堆积体——大陆隆（Continental Rise），于 1994 年 UNESCO 在比利时列日大学召开的第一次海岸海洋大会定名为“Coastal Ocean”，即“海岸海洋”。相当于我国海洋学家所沿用的“海岸带”与“近海”。海陆过渡带因有海水覆盖而区别于陆地，又因有陆壳（硅铝层）延伸至海底而区别于深海大洋——海底主要是由硅镁层洋壳所组成。

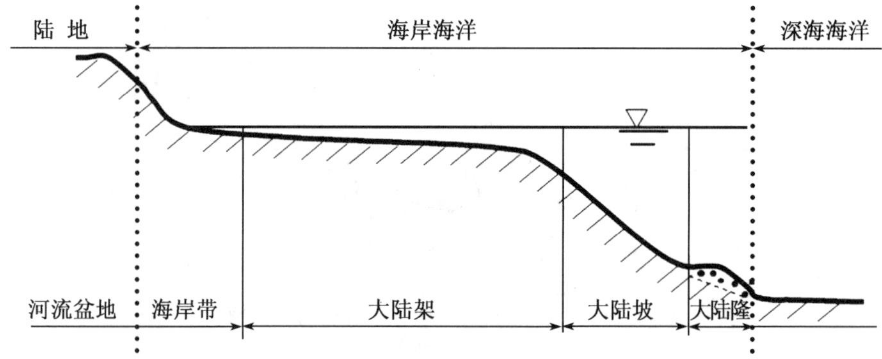

图 4　海岸海洋图示

海岸海洋是一个独立的环境体系，地球表层气、水、岩、生在此地带交互作用变化频繁，风力、波浪、潮流均在海陆过渡带加强，其面积占地表面积的 18%，水体占海洋面积的 8%，但生物量巨大，拥有全球 1/4 的初级生产力与 90%渔获量，生态系统丰富、缓冲作用突出，与人类生存活动关系密切[1]。必须认识到海陆过渡带的独立特性与重要性，在地学教育中宜尽早开设“海岸海洋科学”系列课程填补空缺。

1994 年《联合国海洋法》实施，将大陆架与 200 海里专属经济区划归沿海国管辖，这些均在海陆过渡带即海岸海洋范围之内。全球有 370 多处海域具有权益纷争，我国在管辖海域的海疆划界方面，与周边国亦有权益纷争。石油能源的“生命线”亦在海上。资源争夺与海洋国土权益推动了对海岸海洋的关注，国际上兴起了对海岸海洋科学研究热潮。自 1994 年 UNESCO 列日会议以来，我国关注了“海岸带与近海”的一体化调查研究，但是缺少对外陆架与陆坡的一体化关注，“海岸海洋”术语与内涵未被关注与采用，外陆架与陆坡实属陆地的一部分，有争议的岛屿主要在外陆架与陆坡上部。南京大学海洋研究与教育是以海陆结合为特色，以海陆过渡带为核心，设有海岸海洋科学系与博士学位点，南京大学“海洋地质学”博士点是江苏省建设的第一个海洋地质学重点学科，现正联合大气、地质、生物与法律等系科，努力建设海洋科学一级学科点。

海岸海洋科学是研究海陆过渡带表层系统作用过程、环境与资源特性、自然发展规律，以及与人类生存活动相关的科学。它应用于环境资源开发与保护，服务于海洋经济、

疆域政治、立法管理与军事活动等方面。

中国居于亚洲与太平洋之交，除台湾以东濒临太平洋外，渤、黄、东、南海均居于岛弧内侧，属边缘海(Continental Marginal Sea)，绝大部分海域属海岸海洋，独具特色：季风波浪，有潮海，大河入海径流及长江口北侧的平原海岸，使河海交互作用形成，河流泥沙汇入，促成发育了淤长的潮滩、宽广的大陆架、巨大的三角洲，以及内陆架叠置着巨大的沙脊群与潮流通道组合；南方的港湾海岸与三角洲平原相间分布，生物成因的海岸、岛礁，海峡，拉张沉降的大陆坡与深海盆中洋壳隆升……均蕴含重要的科学机理与应用价值。现代海岸海洋又深受人类活动影响，成为陆源污染物之汇：除外洋及苏北南黄海尚有部分Ⅱ级海水外，均污染严重，为劣Ⅲ级与Ⅳ级海水；赤潮爆发频频，水环境恶化，减弱海岸海洋沉淀过滤作用；海洋灾害损失严重，亦造成人员伤亡。内陆人类活动亦通过河流而带到海岸海洋：沿江河流域筑坝、引水，减少了泥沙汇入，入海泥沙从20世纪80年代前的 20.1×10^8 t[7]，减至目前的 $5\times10^8\sim6\times10^8$ t，加之海平面上升，风暴潮侵袭，水下三角洲与平原海岸普遍受蚀后退[8]，危及沿海大都市及经济带的安危。所以，我国在关注陆地与海洋开发之后，必须重视海陆过渡带(海岸海洋)作为一个环境体系的重要性，需加以研究，加强对海陆过渡带环境的关注，在地学教育与人才培养中贯彻，并应用于公众环境教育，意义重要。

三、组织优秀地学教师轮换、巡回讲学

应遴选基础课、稀缺专业课的优秀教师，组织轮换性教学或巡回讲学，促进沿海与内地，北京与边远地区的学识与科学教育成果相互交流。地球科学需积累，充分利用现有教师资源，使学生普遍受益，提高教学质量，充分共享优秀教学成果。

主要参考文献

[1] 王颖，牛战胜. 2004. 全球变化与海岸海洋科学发展[J]. 海洋地质与第四纪地质，24(1)：1-6.
[2] 周曾奎. 2000. 南京地区50年冬夏气温特征分析和演变趋势[J]. 气象科学，20(3)：309-316.
[3] 郑红莲，严军，张铭. 2002. 南京地区地面平均气温的变化[J]. 解放军理工大学学报，(5)：87-91.
[4] 何立富，武炳义，管成功. 2005. 印度夏季风的减弱及其对流层温度的关系[J]. 气象学报，63(3)：365-373.
[5] 马柱国. 2005. 我国北方干湿演变规律及其与区域增暖的可能联系[J]. 地球物理学报，48(5)：1011-1018.
[6] 刘春蓁，刘志雨，谢正辉. 2004. 近50年海河流域径流的变化趋势研究[J]. 应用气象学报，15(4)：385-393.
[7] 程天文，赵楚年. 1984. 我国沿岸入海河川径流量与输沙量的估算[J]. 地理学报，39(4)：418-427.
[8] 恽才兴. 2010. 中国河口三角洲的危机[M]. 北京：海洋出版社.

突出江海文化特色　优先发展海洋科学*

一、南通区位独特重要

扼守中国第一大河河口，万里长江连通中西部，物华天宝、腹地广大；邻接黄海与东海交互作用过渡带，海洋环境、动能与水产资源丰富；江海相连的坦阔平原海岸与淤涨的滩涂，为众多的人口提供新生的土地资源；暖温带风雨与充沛的阳光，稻、麦、棉、桑与瓜果生长丰硕；地居长江三角洲北翼，大潮差与强潮流动力，往返疏通沙脊群间的古长江干道河谷，构成深水潮流通道，提供了选建 30 万 t 级海港的优越条件；洋口港建设成功的例证，南通江港与造船业，与两港之间的吕四渔港和通州湾联合可建成大中小结合的江海港口群，促进海内外沟通与海洋经济发展，推进以上海为中心的长三角海洋经济、贸易、文化位居全球前位。但是，也应注意到，伴随着世纪性海平面上升与长江坝闸建设使河流入海泥沙量减少，今后，潮滩淤涨会减小，风暴潮侵蚀岸滩会加大。海岸带规划发展应注意到自然环境发展趋势的前景。

二、南通优秀人文传统历史悠久

青墩文化源远流长，西汉时改造海滨湿地，“九里十八墩”建墩田屯兵，创建潮滩海岸湿地利用之范例；北宋范仲淹沿贝壳滩自然标志，建海防长堤，固陆防灾；明清渔盐并举，开灶熬盐，建潮墩救民，抗击倭寇卫护家园；清末民初张謇围田屯垦，建设向阳新村居，耕植桑、棉、稻、麦，兴建蚕丝、棉纺与造船工业，发展刺绣与蓝染工艺，大力办学，提高南通人民教育与就业水平，文化与医药名家世代辈出；抗日与解放战争烽火凝练了南通人民柔韧结合勤劳勇敢的朴素品质；当代南通经济发达，中学教育水平与船舶制造居全国前列，大型海港与新能源兴起，现代工业蓬勃发展，建筑大军遍迹海外……进而也反哺南通农村，兴建成现代农居村落新风貌。

三、南通发展展望

作为海洋地质地理学者，基于参加长三角与江苏沿海调研与开发工作的经历，我认知南通的地域与文化特色，诚望在已有基础上，进步发展与升华，凝练成为 21 世纪的江海明珠灯塔，为建设屹立于东亚与太平洋之交的文明强国，光辉照耀海内外的现代化中国贡献巨大力量。

---

* 王颖：《南通日报》，2015 年 11 月 6 日 A6 版。

千里之行始于足下，关键是培养人才，巩固提升原有的中学教育力量，着力发展高等教育，重视教育、研究与应用建设相结合，突出海陆结合的江海文化特色，优先发展海洋科学，增强江苏在海洋经济领域的力量，以海洋科学为主干延伸发展相关的理、工、医、农、法、政学科，建设现代化的综合型大学，培养世代相继的新型建设人才。

发展战略目标宜充分发挥区位优势，重视南入沪、浙，北接苏、鲁，西沿大江通过川、青联陆桥，东跨岛弧越大洋。

发展步骤扎实前进，重视发挥海陆交互作用带的资源环境优势，以农、渔、养殖与轻工业为基础，以新能源，港航与建筑业为特色，工农商贸共进，在实践中持续前进发展，立足长三角，面向海内外，期待 21 世纪南通城乡崭新辉煌。

（2015 年 10 月 30 日于南京大学仙林校区）

《〈更路簿〉与海南渔民地名论稿》序*

刘南威教授是我国著名的自然地理学家，兢兢业业地执教于华南师范大学地理学系达 60 多年。我与南威相识于 1959 年调查“渤海湾海岸地貌”时，我们同在北队，从天津新港到滦河口，在岸陆、潮滩与浅水海域，进行海陆交互带海岸动力地貌与沉积调查，追溯天津新港泥沙来源，证明了：滦河泥沙对新港回淤没有影响。当时正逢国家困难时期，缺食、少穿，在北方调查时吃粗粮，在整个调查期间，刘南威同志始终如一地乐观、认真、积极负责地与队友合作，完成了调查任务。当年结下的科学友谊，延续至今。

也许源于在广州地区工作，刘南威教授从事的自然地理教学与研究工作，始终具有陆、海相关，自然与人文结合的特色。南海诸岛与海域，是中国最早从事开发经营与纳入海疆行政管理，《更路簿》是中国渔民世世代代在南海捕捞作业与落脚经营的真实记录，是将海上交通航路以岛礁、海空地物位置、罗经方位、航时与航行距离相结合而成的航海指南，是世世代代渔民海洋作业经历集成的导航体系。通过实践凝聚的南海航行指南，十分珍贵。

20 世纪 70 年代以来，刘教授从事西沙群岛考察与南海诸岛调研、南海诸岛地名普查和审定及《南海诸岛地名资料汇编》编著等，在科学考察、调查研究的实践活动基础上，发表《中国南海诸岛地名论稿》等学术著作，为我国南海诸岛地名的发掘、考证与文化传承等作出了重要贡献。在刘南威教授与张争胜教授合作领导的华南师范大学研究团队，积极搜集、整理并深入研究《更路簿》，以及收录的南海诸岛土名，运用数理统计等定量方法分析《更路簿》的版本传承与相关，利用 GIS 软件构建南海诸岛地名数据库并绘制海南渔民赴西沙、南沙群岛航线图，借鉴集体记忆、批判地名学、文化生态学等理论探究《更路簿》和南海诸岛地名的形成与演变机制，将汇集与精校的 21 个版本的《更路簿》与论著一并出版，展现《更路簿》和南海地图研究的最新成果。为维护我国南海海疆权益提供科学与历史的论证。

深深敬佩 87 岁老友刘南威教授勤恳笔耕所凝聚的南海科学成果，专此为序。

---

* 王颖：刘南威，张争胜编著，《〈更路簿〉与海南渔民地名论稿》，1－2 页，北京：海洋出版社，2018 年。

第七篇
科学交流活动与考察研究

海洋联通世界,学术交流与促进海洋科学发展。

拉布拉多海考察散记*
——在国外的海上生活与见闻

清晨，飞机穿越浓云，从加拿大东部的海滨城市哈利法克斯（Halifax）向东北方的纽芬兰（Newfoundland）飞去。我们一行十人，是前往加拿大东北面的拉布拉多海（Labrador Sea）进行大陆架研究工作的。当日，哈利法克斯大雾弥漫，但飞机航行在纽芬兰上空时，天却放晴了，自机上俯瞰，但见千百个湖泊闪闪发光，一些已干涸的湖泊的轮廓也清晰可见，可惜我没有带望远镜头的相机，不能把这珍贵的地貌景象一一拍摄下来。中午时分，飞机抵达纽芬兰岛上的甘德（Gander），然后换乘汽车到路易斯港（Luis），再登上海洋科学考察船“哈德森”号（C. S. S. Hudson），前往预定海区进行考察。

这次考察是由贝德福德海洋研究所（Bedford Institute of Oceanography）大西洋地学中心（Atlantic Geoscience Centre）环境海洋地质部组织的，负责人为著名海洋地质学家G. 维克斯博士（Vilks），他是这次考察的首席科学家。第二负责人为海岸及沉积学家G. 雷森（Reinson）。其他还有地质、地貌、地球化学、孢粉及技术物理方面的科学家共二十三人。调查区介于北纬五十四度至五十六度之间，主要是以哈密尔顿浅滩（Hamilton Shoal）为中心，沿测线进行测深、地震、采样、钻探、海底摄影，和进行海流流速、流向、温度、盐度、密度等特性的研究，以分析大陆架的沉积环境，确定第四纪末冰川作用的界限。

“哈德森”号是一艘现代化的海洋科学考察船，排水量四千八百吨，具有破冰能力和雷达定位、回声测深、长短距电子定位等航海装置。船上有六个实验室：海洋学实验室，综合目的室（设有定位、测深、地震与旁侧声纳等装置），地化和化学实验室，钻孔与样品室，重力实验室及工作甲板。采样有专门的绞车房，上层甲板有直升飞机起落场。钻探机具设于主甲板的前部，旁侧声纳、空气枪等则放在后甲板。此外，船上还设有海图绘制室、电子计算机房等。在生活设施方面也比较好，科研人员住在主甲板层，二人一室，第二科学家为单人室，首席科学家还另有一办公室。餐厅、咖啡室、图书室也在主甲板层，另外还有文娱活动室。值得称道的是伙食管理好。照顾到海上工作的特点，饭食清淡新鲜，营养丰富，餐餐变换口味。餐厅远离厨房，无油腻气味，以避免引起感觉不适而晕船。每日除供应三餐外，还专设一室，备有茶水、咖啡、牛奶、果汁、水果、面包、熟肉、鸡蛋、面条等各种食品，以及简单的炊具，全日开放，可供随意选用，以保证精力旺盛地进行繁忙的工作（船上工作是4班/24时）。

往常，我是晕船的，这次去拉布拉多海工作之前，我是既高兴又担心，高兴能有一次出海学习的机会，担心的是，北大西洋风大浪高，我一定会晕船，如果呕吐不止，不能坚持正常工作，那就糟糕了。出奇的是，这次出海却例外地没有晕船，而且充分享受了海上工作的愉快。分析原因，一是有加拿大朋友们的关心，鼓励我：“别紧张，船只很平稳，一两天后

* 王颖：《海洋》，1980年第4期，第1－2页。

适应了,你会感到很舒服。”另一方面是伙食好,清淡适口,餐厅服务员照顾非常周到,还特地给我煮了大米稀饭等我们中国人爱吃的食品。但最重要的原因是船只的稳定性能好,设有自动调节平衡的防晃油箱装置,再加上工作环境清洁卫生,没有油烟、鱼腥等怪味,虽然出海第一天就遇上比较大的风浪,但我的感觉却始终良好。

北大西洋天气多变,拉布拉多海经常云雾弥漫。在茫茫的海面上,只有海鸥伴随船只翱翔。当停船钻探的时候,它们就成群地在船旁水面上浮游嬉戏觅食,给我们的海上生活增添了无限情趣。

考察工作是紧张的。在整整十六天,科研人员夜以继日地三班干。值班时,思想要高度集中,排好工序,不停地在各种仪表间转动。工作室内极为安静,只听见机器有节奏的嚓嚓声,绝无人谈笑,即使交接班也是轻声细语。有些仪器和观测项目只有一两个人管,常常一干就是十二或二十四小时。工作这么紧张,却没听到有人埋怨。工作多,意味着他干得好,受到尊重与信任。当然,工作干得好,职业才稳定,才可得到较多的收入。记得在船上工作后期的一个下午,首席科学家对我讲:“颖,今晚你休息吧,不必值班了。”我很诧异,忙声明:“不,我不累。”但他坚持说:“工作几天了,你休息一夜吧,再说,有人要求值夜班多挣些钱。”我这才明白而同意了。

一次海上考察成果的大小,与主要科学家的工作能力有很大关系。维克斯博士精通业务,各方面的工作都能具体掌握,及时指导,并且与船长能很好地相互配合。他常常是一连两班地干,即使航行结束了,其他人员多登岸进城观光,而他仍在赶写航行报告,或是接待电视记者的采访。在这次考察过程中,加拿大科学家们充分利用航行时间与船上设备紧张地进行工作的精神,以及负责人精通业务,指导有力,和严谨的工作态度,给我留下了深刻的印象,很值得我们学习和效法。

海上考察成果的大小,与仪器设备有很大关系。这次拉布拉多海考察,使用了先进的仪器设备,比如,用电子计算机控制测定海水的物理性质,随着测量仪器进入水中,不同深度海水的温度、盐度、密度和混浊度的曲线立即显示在荧光屏上,各种数据与图表也随即出来了。

拉布拉多海离北极圈不远,气候寒冷,海面上漂浮着一座座形状各异的冰山。远处的洁白晶莹,近处的闪烁着绿光。这些冰山的规模不大,高出海面约 10～20 m。据一位研究冰山的学者介绍,冰山沉没在水下的部分要比水面部分大得多,一般可大到 4～5 倍,淡水冰山则为 9∶1。所以,冰山底部常可与大陆架海底接触,形成擦痕。研究冰山擦痕,可以了解冰山的移动路线,这对于航运以及石油钻探平台的设置都有实际意义。高纬度海区的另一奇景是北极光,当晴朗的夜晚,可见青白色亮光映射天际。另外,在航行中还可时常见到喷水的鲸和成群嬉游的海豚。这些美丽的景色,使我感到新奇,我时常凭栏眺望,久久不忍离去。

半个月的船上生活,与加拿大朋友建立了深厚的友谊。在行将结束航程的前夕,“哈德森”号举行了两次晚会,一次是海员晚会,另一次是科学家晚会。船上只有我是中国人,受到了多方面的关怀与照顾,很想借此机会与加拿大朋友们欢叙,略表谢意。由于船上海员大多来自纽芬兰,所以,海员晚会洋溢着纽芬兰地区的特有风情,男海员穿黑色橡皮裤,女海员着大褶裙,都戴着渔帽,吃着纽芬兰烤鱼,跳着一种欢快的纽芬兰民间舞。当我一进入会场，立即响起了热烈的掌声和“中国女朋友，你好！”的问候声。不少人邀我一起跳舞，这使我有些局促起来，不知怎么办才好。虽然我很想和他们一起跳舞，但不知怎么跳法。这时，年长的D. E. 贝克礼(Buckley)博士过来了，他说：“我也参加，你看，我也不会跳，就跟着音乐蹦就是了。”这倒是个好主意，于是，我也学着他的样子蹦了起来。这样一来，气氛更趋热烈。海员们一次一次地邀我参加他们的舞蹈，并且送来两杯不同的饮料让我喝。在边喝边谈中，有位老海员对我说：“我去过中国，那是个美丽的地方。”另一位海员说：“《白求恩大夫》这部中国影片真好，我们过去不知道白求恩，是通过电视台播放这部影片以后才知道的，他真了不起。”一位青年海员自我介绍他是诗人，当场向我朗诵他作的诗。在晚会上，我听说海员们几乎每天都打听：“那位中国女朋友在船上过得愉快吗？”当听说我适应船上的生活，过得很愉快时，都大为高兴。由于我参加跳了纽芬兰民间舞，喝了他们送的两种饮料，海员们授予我一张“荣誉纽芬兰人”证书，这对我来说是一个很为珍贵的纪念。

科学家的晚会比较文静，主要是饮酒谈天。但后来由于船上的大副、医生都来参加，也跳起舞来。目前，国外青年人多爱跳一种随音乐节奏而任意摇摆的舞蹈。但中年人和老年人多看不惯，认为是胡闹，中、老年人仍跳交谊舞。当晚，大副邀我跳“快三步”，我又犹豫了，我从来没有与外国人跳过这种舞，但人家客气相邀，不好拒绝。于是，我愉快地应邀了，科学家们都高兴地鼓起掌来。朋友们激动地说：“我们都喜欢你，你和我们打成一片，希望今后能有更多的中国朋友到这里来，同我们一起工作和生活，加强友谊交流。”这些友好的话语使我深受感动，我真没想到，由于无拘无束地置身于加拿大人民的工作、生活，甚至舞蹈之中，竟能得到如此深切的友情。这种真挚的友情，将促使我为发展海洋科学，为中、加两国人民友谊的交流，尽自己最大的努力。

（写于加拿大哈利法克斯）

圣劳伦斯湾潜水地质考察记*

1981 年 8 月中旬，我乘“潘德拉Ⅱ”号到圣劳伦斯湾，参加了贝德福德海洋研究所第 8151－52 航次用下潜器 Pisces Ⅳ号(派塞斯)进行“在自然水环境中悬移质泥沙的物理行迳”的专题调查。

圣劳伦斯湾位于加拿大东部，界于纽芬兰、魁北克、新不伦瑞克与新斯科舍省之间，系沿 NE 向的圣劳伦斯河支谷与 NW 向的主河谷，由张裂断陷而成的一个内陆海湾。NE 与 NW 向为加拿大东部两组主要的断裂走向，至今沿 NE 向的新构造活动仍很频繁。海湾东北侧纽芬兰岛分布的古生代地层、构造特征与海湾西南侧新不伦瑞克以及新斯科舍的地层构造遥相对应。海湾中的爱德华太子岛以及西北侧的安提科斯提岛是两边陆地张裂移开的残留见证。沿海湾内外，200 m 等深线仍可反映出张裂的圣劳伦斯河谷的轮廓，海湾南岸，绿树青山，为林牧场所，居民房舍建筑在海岸的 5～8 m 及 15～20 m 两级阶地上。东北侧纽芬兰岛，地处高寒，且为大冰盖最后退出之地，巉岩裸露，岩岸陡峭。

“潘德拉Ⅱ”号(照片 1)是加拿大第一艘科学潜艇的工作母船，1974 年在西海岸的北温哥华市建成，该船全长 58.22 m，载重量 1 200 t，最大航速 14 节，为平底方船，防浪性能差，但具有全球性持航能力。该船前甲板有一小型实验室可供小规模的地质测量与采样。中部为舰桥与住舱，仅能容纳 19 名船员与十几名科学家，除首席科学家外，其余两人同住一间舱房。后甲板的舱房被“开膛”，作为小潜艇(或称下潜器)Pisces Ⅳ号的“卧舱”。

照片 1　Pisces Ⅳ号之工作母船

51 航次是在原调查的基础上，利用 Pisces Ⅳ号下潜到海底选择理想的抛泥地，而我参加的 52 航次则是从圣劳伦斯海湾→圣劳伦斯河口湾→圣劳伦斯河口，沿三条横断面的 11 个站位，乘小潜艇直接目测与采取水样进行上述悬移质课题的研究工作。通过这 11 个站位工作拟了解：① 在不同沉积环境中，悬移粒径的级配范围；② 絮凝物、团块与小球粒等各种类型的悬浮物在水体各层中的相对负载；③ 这些负荷与粒径级配随环境而变化的情况；④ 各类悬移质的组成与结构；⑤ 各种悬移质的沉降速度。调查课题负责人是加

* 王颖：《东海海洋》，1983 年第 1 卷第 4 期，第 63－68 页。

拿大沉积学家捷姆·苏维茨基博士,他曾多次应用小潜艇进行悬移质泥沙运动研究。我参加这次考察主要想了解下潜器在海底地质调查中的功能与取得这类工作的感性体会。同组还有两位技术员与一位来自瑞木斯基的大学教师。

Pisces Ⅳ号小潜艇是1972年11月由加拿大北温哥华市的国际流体动力公司建造的。同年由渔业与海洋部以250万加元购进,1974年与潘德拉Ⅱ号母船结组,到1981年8月已下潜了1 250多次,最大下潜深度700多m,而设计下潜能力为2 000 m。加拿大前后共建造了11艘Pisces小潜艇,除Ⅰ号已退休外,其余均在工作。美、日两国曾经仿造,但均不如Pisces灵活和易于操纵。Pisces Ⅳ号(照片2)全长6.1 m,宽约3 m(具推冲器肘),高约2.5~3.4 m(具蓬状翼肘),重11 000 kg。应用铅酸电池为动力,最长下潜延续时间约8 h,一般多为2~4 h。艇前有三个观察窗,驾驶员蹲坐于中,而两名科学家平卧于两侧,艇顶开孔,人由孔进入舱内,艇外腹部有一滑行器(长2.0 m,宽19.7 cm,轨间距2.3 m),藉轨道下滑而入水。Pisces Ⅳ号具有32个传感器,可量计海水盐度、温度、电导率及流速等,并且直接记入磁带中以供研究与制海图用。潜艇的前臂装有电视摄影机与静物摄影机,下潜过程中可进行录音、录像以记录海底的真实状况。有一个机械手,可用于采样与取物,特别是采取具有放射性元素的或核污染物。观测孔下有一抽屉装置,由它来取送样品瓶。Pisces Ⅳ号自应用以来曾用于:① 水层混合过程研究:如冷、暖流混合;气体与化学元素混合;咸淡水混合以及清浊水混合等。调查研究成果用于渔业和天气预报。② 由于不少国家以海洋为天然垃圾倾废场,特别是大量的废钢铁与工业垃圾。Pisces Ⅳ号曾用来调查这些垃圾对海洋环境的影响与清理办法,以协助贯彻执行加拿大海洋环境法。③ 环境调查:与上题有关,但着重调查油船与油井对海区的污染状况与处理办法;油管道通过区的海底冰刻痕与冰核丘研究。后者受永冻冰与温压控制而变化无常,因而常使蒲福海石油开发及航行遇到困难。④ 地质研究:直接观察大陆架区地壳运动迹象与沉积作用过程。⑤ 曾被租用进行海洋生物与太平洋的虾类生产调查。⑥ 打捞失落于海洋的仪器与设备,而这些仪器常储有周或月的观测资料。总之,在直接观察海底并进行各种应用活动中,小潜水艇Pisccs Ⅳ号确实有其独到的作用。这次考察包括三个不同内容。海洋物理学家拟观察港口外抛泥地的水流和泥沙状况,看看是否由于水流扰动使泥沙趁潮回流入港;我们是进行地质调查;生物学家调查浮游与底栖生物。

照片2　Pisces Ⅳ号小潜艇被吊离母船

8月中旬地处亚寒带的圣劳伦斯湾,虽已进入仲夏,但其天气仍冷。当船只驶向第一站位时,即开始对Pisces Ⅳ进行安全检查,至达尔豪谢河口外海区,适逢落潮,表面水温17℃,盐度26‰,25m深处水温为1℃,盐度29‰,那日天气晴朗,Pisces下潜对抛泥地状况进行观测，但因流速和水下浑浊度大，视野不甚远，测知至65m深处，水流扰动仍强。次日阴云密布，但无风。Pisces Ⅳ再次下潜，但水下仍浑浊不清，水层中充满悬移质泥沙，稠似泥浆，致使未能得到良好的观测结果。第三日，飓风过境，狂风大作，小潜艇不能下水工作。“潘德拉Ⅱ”号成了名副其实的Pondora（把灾害带到地球上的希腊女神）；我曾两次随它出海，第一次是去新斯科舍大陆坡进行深水扇调查，遇上风暴，船摇晃颠簸不止，大部分乘客晕了。我几乎不能支持，怪不得“大西洋地学中心”没人愿跟这条船出海。此次，虽然在内海，颠簸仍甚，尤其仅我一个女科学工作者，被安排在通常为首席科学家使用的单人舱内，位置高，摇晃得厉害，同行的人皆悄悄地躺在床上不动了。我这次有了经验，到船底的酒吧间内（此时已无人光临），安稳平静，躺在沙发上看书避浪，效果很好。次日，是我的下潜班次。临我下潜的前夕，突然狂风大作。我心中和船只一样不平静，一再祈望第二日风平浪静能照常工作。下午风停了，人们都又重新活跃起来，船长与领航驾驶员来研究第二日的下潜。他们曾于1980年接待过来自上海和武汉造船厂的访加人员，并安排来访者参观了小潜艇和进行了尝试性下潜。因此对中国有一定的了解，甚至希望Pisces Ⅳ号能访华协助工作，他们说：“中方只需付船只的每日花费而不需付船员工资。”盛情感人，我表示了要转达他们的心意。次日，风半浪静，天气晴朗，船已驶进德芒特角附近海域。我与捷姆俩下潜，准备了数只相机、录音机等。捷姆建议我在450 m最深站位下潜时，将名字写在一牌子上，放入盒中，然后扔到海底作为纪念，并看它是否会压扁。我想，如果能实现450 m站位下潜，我一定做。当日小潜艇由最佳驾驶员阿伦驾驶，在下潜时，母船上有一组人为它服务。首先，船长于船桥值班，注视下潜的全过程。大副在后甲板现场指挥掌握全局，领航驾驶员掌握母船与小潜艇之间的通讯联系，小潜艇几乎每5分钟要向母船报告一次：水深、温度、能见度、海底底质与水流速度等。通讯必须保持绝对通畅，实际上领航驾驶员是下潜的全程指挥。此外尚有牵引组两人——身穿防水服与脚蹼，一个站在潜艇上，关闭小潜艇入口；另一个驾驶橡皮艇并将小艇牵引至预定的下潜点及拖返母船。我们下潜点位于河口湾外的深水部分，然后驶向德芒特角的陡峻基岩岸附近。

上午9时，一切准备就绪，驾驶员阿伦、“观察员”捷姆与我陆续进入潜艇。很小的空间，我们分别按指定的位置“就座”——驾驶员居中，我与捷姆分别半卧于左右两侧。

9时04分，小潜艇开始沿轨道滑行出舱，至船尾被吊起后再轻轻放入水中。继之，由潜水员与橡皮艇牵引至下潜点——观测台站No.7。我打开了相机与录音机，全神贯注地从面前的窗孔观察舱外。这时，舱内很安静，只听到通讯脉冲器发出有规律的“卡塔、卡塔”的信号以及电视录像时的轻微声响。

9时07分，潜艇开始进入水下，水很清，碧绿碧绿的犹如玻璃缸一般的明亮好看。驾驶员与领航员通过无线电联系交谈，母船上在呼唤：“Pisces！Pisces！不易听清你的讲话”。实际上无人发言。继之，驾驶员报告：“一切正常，科学家们已就位工作了。”

小艇在逐渐下潜，水中出现了很多细小悬浮质颗粒，一条小鱼安祥地向我们游来。渐渐水深加大，至16 m深处，光线减弱，水变成深绿色，此时的温度为12.4 ℃。下潜至18 m深处，水色变暗，水温为9.5 ℃。悬移质小颗粒减少，开始出现絮凝颗粒。至20～25 m深处，光线变暗，出现很多悬浮絮凝物质，絮凝团变大，直径为1 cm，此时潜艇下降减慢，有充分时间观测与摄影。至37 m深处，光线更暗，舱窗外已分不出海水颜色，而是一片昏暗。此时潜艇打开了前灯，但是水中小粒子与絮凝团很多，絮凝质的直径约 0.75～1 cm，就像下雪一样。至 56 m 深处，棉絮般的絮凝团与小颗粒似大雪般地下沉，无怪乎海洋沉积学家称此为海雪(Sea Snow)。摄像机记录下这海中奇景。下潜至 63m 深处，仍见絮凝团与小颗粒似大雪般地下降。仔细观察与摄影，发现此处有些团粒在蠕动，喔！悬浮游生物吧！再仔细观察，原来其中很多是很小很小的水母，透明的幼水母在蠕动飘移。至 70～80 m 深处，有很多透明的小鱼苗在游动，出现了较多的浮游生物及悬凝物质。

驾驶员不停地向母船报告下潜深度、水温、自然现象与工作状况。显然，他也为这些奇观而激动，他也全神贯注地在观察并不时地发出赞叹，而我则惊奇地发现他对生物与地质科学有广博的知识，因为向母船报告的内容都是符合科学家要求的。而领航驾驶员，大副以及在母船上工作的船员也都在认真地听取这些报告。

迅速下潜至 131 m，成群的小鱼穿梭游动于小潜艇周围。此时，我的感觉良好，但稍微或者偶尔有些恶心。到 144～150 m 深处，大的棉絮状的絮凝物质已减少，继之，很少了，只有小的悬浮颗粒飘荡于水中。此时，突然一个大水母向潜艇游来，正好进入摄影机镜头。此时水温为 4 ℃。在 184 m 水深处，测深仪示出距海底还有 40 多 m，此水层中有很多水母与浮游生物。204 m 深处，水温为 4 ℃，艇内感觉良好，我穿了一件毛线衣，与在水面时一样。鱼群增多，潜艇似接近底部了。

噢！到达海底了。很多扁平状的鱼在翩翩游过，此处似为软泥质海底。小潜艇轻轻贴近海底，引起了底部水层轻度浑浊。此处深度是 215 m。驾驶员报告："已至底部，将继续报告。"突然他转向我说："颖，感到流很急吗？""不，没感到。"因我没经验，所以不如驾驶员那样警觉和敏感。

216 m 深处为黄色的软泥海底，尚处于氧化环境，但在生物逸出孔附近却翻出灰色物质。扁平的鱼类(两眼长在顶部并有一细长的尾)以及一种灰色鳗状鱼不时游过我们面前，根本不在乎小潜艇在它们附近轻轻地沿海底滑行。喔，流确实很急，小潜艇已发生倾斜。虽说 Pisces Ⅳ操作灵活并适应于窄小的空间，但因为它太小太轻，以至于当底流流速大于 2 kn 时，小潜艇不宜活动了。此时驾驶员报告："平坦的海底，粉砂与淤泥沉积；常见鱼、海星与螃蟹；底部流速达 1 kn，能见度 7 m，小艇已接近陡崖，水温 4.8 ℃。"此处仍为黄色淤泥质海底，细长的海星如交叉的"十"字，很多瘦弱的螃蟹，背壳为红色，静静地伏在海底，无疑为死蟹，但是捷姆观察到蟹眼在翻动，后来它们果真缓缓地爬动了。海底还有生物滚动的泥球、单株的铁树、海螺以及数量不多的珊瑚状生物?!

接近陡崖的基部，出现了成 20°倾斜的岩层——它被覆盖在薄层的软泥之下，此处流急，于艇内即感到水流在冲击着小艇。突然，在岩石海底上出现了粉红色的海菊花般的生物，一株株的，个体不大，高约 4～5 cm。在一些尖锐的岩石碎块上长满了这种生物，这是"活珊瑚"。在亚极地气候的海底，出现了活珊瑚，虽亲眼见到，仍是出乎我的意料。据说西海岸温哥华一带也有类似生物。

小艇轻轻地飘移，仍在陡崖岸附近，细砂与粉砂质海底出现砂波痕，但此处波痕时隐时现，排列方向不规则，尚无明显波列，可能由于水流扰动力强所致。此处底栖生物为海星(五角星式的与角不明显而近似方形两种)。

海底坡度加大，潜艇轻微上升至水深 200 m。倾斜的海底，为细砂质，具较多闪光的贝壳类，出现明显波列的波痕，其排列方向成南北向或 N9°E，波高数厘米，峰谷明显，波峰顶呈黑色，为重砂。附近散布的小砾石上生长着 2～3 cm 高的粉红色、黄色的小海葵。

沿倾斜海底上升到 170 m 至 130 m 一段，基岩海底，岩层成 20°～30°倾斜，几乎全部为钙质藻类所覆盖。基岩海底景色最美，长满了粉红、米黄及白色高约 10 cm 的海葵以及很多白色的海绵，还有一些软体动物的贝壳及小海胆。此处出现了较多的鱼、蟹。其中有一只蝴蝶状的鱼翩然游过，就像美国地理杂志上所刊登的大堡礁中的蝴蝶状鱼一样。但在圣劳伦斯湾海底，我们只见到一条，"快把它拍摄下来"。亚极地的基岩海底也是这么五彩缤纷，真令人难以相信。

至陡崖岸附近的 115 m 处，仍为长满生物的基岩海底，岩层表面由于覆满钙质而呈白色与红色。当 Pisces Ⅳ触及底部，引起水流浑浊，使彩色缤纷的海葵扇动触手时，景色更为引人入胜。离开悬崖岸，上升至 94 m，粗砂海底，大沙波上有小沙波系列，其波痕高约 2～3 cm，峰顶有黑色重矿物，波谷约 10 cm。至 91 m 处，水温 1 ℃，很多石块和具棱角的砾石，与粗砂相伴沉积，砾石、岩块上长满海葵与海绵，并出现扇贝。升至 71 m 处仍为粗砂海底，具有带棱角的砾石，上面生长着海葵，并有很多扇贝。海葵高约 4～5 cm。60 m 处海底水温 1.2 ℃，粗砂海底出现大沙波，砾石上的海葵彩色缤纷极为茂盛，有许多海胆与长达 50～60 cm 的海蟹以及鱼群。升到 50 m 水深处，已远离陡崖，此处粗砂质海底，发育了排列不规则的巨型流痕。上升到 40 m、37 m、35 m、33 m 处，发现下沉的古海滩。砂砾质海底具有大量扁平状的海滨砾石，砾石长径约 4～6 cm，磨圆度高达 6～8，由于砾石长期沉积于海底，已被红色的钙质藻类所包裹，其上仍有海绵与海葵的单株，但已较底部稀少，粗砂底偶见小棵的海铁树与海葵。至 30 m 深处，仍有古海滨沉积，见到一把落入海底的八开小刀，"捞上来吧？""不，让它待在海底作为人类扰动层的证据吧！"海底砾石上有海葵，但个体很小，数量少。25 m 处，仍为砂砾质古海滩，向上逐渐过渡为砂质海底。至 24 m，底质变了，为粉砂淤泥质海底，伴随着大量细长的海星。远处有一个 2 磅的洗衣粉盒子，"A B C"的商标仍清晰可见。进入到现代海岸带下界，结束了这次观察。驾驶员报告母船："Pisces Ⅳ要求上升到水面"。返抵 10 m 水层，阳光耀眼，海水浑浊度大。

10 时 51 分上升至 4 m，潜水艇飘浮等候牵引。下潜近 2 个小时，突然停止工作，也许神经一下子松弛下来，在飘荡的海水中微感头晕。捷姆与驾驶员都很兴奋，认为是他们下潜观察中最精彩的一次。我仍继续看着舱孔外，悬移质小球如粒雪般地在飘荡沉降……我回想着，在这次下潜中观察到：悬移质泥沙与浮游生物沿水层的分布变化；从陡崖岸外最远端的深处到现代海岸带下界，由软泥、粉砂、细砂、粗砂、砾石到基岩陡崖麓的各种海底状况与底栖生物。明显地看出冰碛与古海滨砾石的不同分布与特点。因此，应用小潜艇进行特定目的海底地质调查，如：大陆架上的贝壳堤、水下沙波、泥火山、冰核丘与冰刻痕等调查，是一个相当有效的方法。当然也可用于进行海洋学其他分支学科以及海底工程等的调查。同样亦可用于内河。但需选择平静晴朗的天气与水底流速适中处……突然，橡皮艇曳引潜艇打断了我的遐想，海水亦被搅动引起了轻微的浑浊。

11 时整上升至水面，结束了这次海底考察。我怀着兴奋的心情记下了这次潜水地质调查的经历并准备下一个潜水航次。

访日结交的学者和友人*

在1982—1983年期间访问中结识了一些著名的日本学者和朋友，建立了友好科学交流关系。日本友人曾与中国有着密切的联系，至今仍然怀念中国。考虑到他们的学术水平、地位和影响，以及对我的态度，我认为有三位代表性人物宜进一步结交。

（1）奈须纪幸博士，现任东京大学海洋研究所所长。该所自建立以来已有十任所长，一般每任2～3年，奈须曾任第六任所长（自1968年12月至1972年10月），并为日本海洋科技中心评议员（相当于我国学术委员会成员）。其本人专长海底沉积学，近年来从事日本海沟-岛弧构造以及深海沉积的研究，是国际著名的海洋地质学家，是日本海洋界举足轻重的人物。

据奈须与我交谈，他父母是从闽南到日本的，祖先是中国人，他本人有中国血统，曾多次访华。其妻、女也访问过中国，对中国友好。目前任国家海洋局第一海洋研究所名誉教授。

（2）星野通平，理学博士，海洋地质学教授，现任日本东海大学海洋学部海洋资源学科（相当我国的系）主任。1923年生于日本，1949年毕业于东京大学，多年从事教学和海上考察，著作多。专著有《浅海地质学》、《深海地质学》和《海洋地质学》（部分章节）。他以传统地质学见长，对区域海底地质研究深入，多年教书育人，已形成一个海洋地质学流派。他的观点与苏联地质学家别洛乌索夫相近，并与苏联合作多年，共同开展了印度洋研究。他曾访问过中国，所作的学术讲演与交流，得到北京、青岛等地院校的好评。

（3）早坂祥三博士，海洋地质学教授，现任鹿儿岛大学理学部地学科主任、理学部就职委员会（评选教职）委员长。

早坂于1930年生于我国台北市，父亲是地质学教授，为台湾大学地质系的创始人。早坂祥三为老早坂的第三个儿子，图吉利而取名“祥三”。1946年后早坂祥三返回日本。至今仍讲些闽南话，自认为是半个中国人。

早坂留学和访问过欧美，英语水平高，知识见解新。与星野通平性格相反，早坂祥三稳重朴实，热诚耐心，而学术造诣深，对火山、地震与海洋沉积（特别是微体古生物方面）皆有深入研究。早坂领导下的大冢裕之在研究东海大陆架方面亦具相当造诣。由于地域接近，鹿儿岛大学有较多台湾省留学生，双方曾合作研究东海海域。早坂教授表示希望访问中国与中国海洋地质学家进一步合作。我认为，早坂教授不仅专长于海洋地质学，而且知识渊博，对火山、地震和第四纪地质学有相当深入的研究。

除上述三位外，还有一些日本学者热诚表示了访华愿望。如：

（1）西村实教授，东海大学海洋学部长，东海大学海洋研究所所长。

（2）掘川清司博士，东京大学海岸工学教授。

---

* 王颖：《国际学术动态》，1983年第3期，第116－117页。

(3) 田中则男，工学博士，运输省港湾技术研究所海洋水理部部长。

(4) 春山元寿，工学博士，鹿儿岛大学教授，对处理泥石流有一定水平。

(5) 镰田泰彦博士，长崎大学海洋地质教授。

(6) 永田丰博士，东京大学海洋物理学教授。

(7) 冈田安弘，东京水产大学"海鹰丸"通信长，随"海鹰丸"访问过青岛。对我代表团极热情友好，邀请我代表团有关人员家访和出游。

(8) 大槻一枝，生于青岛，与我国一干部结婚，因夫丧回国。现东京明华贸株式会社任课长(水产渔业方面的)，肯为中日贸易尽力。我代表团在东京期间，她担任日方正式翻译，淳朴热情。

(9) 榎田政子，1947 年参加八路军，曾在哈尔滨护理学校学习护士专业，抗美援朝时任外科护士赴朝三年，后去中国人民大学学习 4 年。其丈夫死后，于 1958 年返回日本，现在鹿儿岛市某医院任护士长。每年 8 月仍返回哈尔滨医学院一周。

(10) 伊田照子，日侵华军后裔，15 岁在中国成亲，在内蒙古生活多年，有六个儿女。其丈夫死后，于 1976 年返回日本。每年仍让其子回中国扫墓，表示死后还要与其丈夫合葬于中国。

关于加强地理科学国际交流的几点认识*

一

中国地处东亚大陆与太平洋边缘海之间，居亚洲、太平洋、印度洋三大板块之交，自然环境独具特色。有世界屋脊的青藏高原；独特的横断山脉；内陆盆地沙漠；黄土高原；横贯大陆入海的长江、黄河，后者又泥沙多，下游改道频繁而著称于世；东部的冲积平原三角洲保留着海陆变迁的遗迹，并有丰富的历史记载；岛弧边缘海是西太平洋活动大陆边缘所独有的，它处于火山地震带，是研究板块运动机制的关键地区之一。不断堆积的大陆架蕴藏着丰富的油气资源。所有这些均为地球科学上独特课题，每一项都有可能成为世界水平的，具国际意义的课题。因而，强烈地吸引着中外地学界与资源开发部门的注意。

中华民族有着悠久的历史，具有丰富的开发利用自然环境与资源的经验，保存大量人类活动遗迹以及人为因素塑造的地貌现象，如洞穴、丘陵及海岛、古人类活动遗迹，文化沉积层；古运河、海塘、堤防以及洪水标志等。据任美锷教授研究：黄河第5次改道(1495—1855年)，造成苏北沿海的海滨平原，即人为决口的，其性质与1939—1947年花园口决口相似。中国经济建设的发展，也引起世界各国及学者的关注。新中国成立后，重视对地理环境的调查研究与解决建设中的实际问题，加上丰富的历史记载。我国已积累了丰富的资料，在地理研究上已有许多新的发现与成就，可以说，我国地学的理论水平与实践能力都是高的，并非完全落后于世界先进水平。

基于上述原因，各国的地理学家、海洋学家、地质学家对中国均有巨大的兴趣，希望访问中国，与中国进行学术交流，对中国区域的各种地学课题，普遍地盼望进行合作研究。这远非数、理、化科学领域等方面能与比拟的。

地球科学本身需要世界性交流与区域对比研究。许多自然现象，如大气环流、海洋、地壳运动等是全球性的，地球上各部分是互有关联的，一些区域特征中国与外国是可以相互对比的，一些区域社会经济发展模式，也可在各国相互引证应用，因此，地理科学需要世界性的交流合作。我国地理学研究的技术装备、应用技术水平与科技情报资料方面，仍处于比较落后状态，远不能满足我国建设发展的需要。因此，我国地理科学需要加强国际交流与合作。这就像体育运动要请进来派出去，加强国际合作以提高我国体育运动水平一样。开展国际合作中的关键是，需要根据课题、地域、工作的不同阶段以及学科发展的需要，有计划、有步骤、有分工地开展。这项工作中，作为中国科协的一员——中国地理学会，应承担起组织与领导全国地理科学国际交流的重大任务。

---

* 王颖：《南京大学学报(地理版)》，1987年第8期，第238－241页。本文系在中国地理学会第五届第二次理事会上的发言。

二

近几年来，我个人从事于海洋地貌与沉积学方面的国际合作，访问交流活动，就此概要地谈谈国际学术动态，拟将各国分为几个中心来谈：

（一）西欧（英、法、德、荷）

1. 英国

英联邦的宗主，西方传统的地学中心，专长于航海、制图与区域研究。剑桥、牛津、伦敦大学是地学教育科研的主要阵地。1986年在澳大利亚召开的国际沉积学会议上，获奖的3人（英、美、加），全出身于伦敦大学帝国理工学院。英国本国领土小，每年有智力输出。或以学生实习，或作为联合国、英联邦或共同体组织项目，或以成员身份前往希腊、土耳其、西班牙或非洲从事调查研究，其研究成果或以专著（如 Pergamon press）出版，或刊登于伦敦出版的地学专刊杂志，活跃于国际学术界，影响很大。近年来，由于北海油田开发，经济复苏，世界油气勘探与开发能力向中国海集中。因此，英国地理学家、地质学家的注意力转向东方，由英联邦科学委员会负责的石油项目会议，邀请我方参加。中英建交，文化交流早于其他西方国家，但近年来美、日与中国的交往已超过英国。鉴于此，英国虽削减国内的教育经费，但英政府会同有关财界和私人财团，对中国的交流经费仍有增加。如女皇奖学金、包玉刚奖学金及石油集团支持的皇家学会对华交流。英国文化协会在上海专设机构，加强与华东院校交流项目，包括学者交往、合作研究与联合培养研究生。其中的自然地理是定向交流课题。牛津、剑桥的地理学家，如 M. Sweeting 与 D. Stoddart 等人已据近年访华所搜集的文章，撰写与出版《中国自然地理》。我个人认为，将出版的这本书，比我国的同类书籍，概括总结了更多近代研究，尤其是中青年科学家的研究成就。

目前南京大学大地海洋科学系，有三项中英合作研究：中英海岸研究对比；土地资源利用；西藏古喀斯特，由英国文化协会（British Council）支持，各项均为三年合作交流协议。

2. 德国

科学作风严谨，实力强，对中国尤其是上海地区有传统的交流史，是培养地理、海洋研究生的好场所。也许德语不如英语应用广泛，与德国的交往尚不多，但是德方经费较多，对德交流应引起重视。

3. 法国

重视动力地貌与深海调查。法国在太平洋西北部与日、美合作调查，仍居领先地位，在南中国海有石油基地，以助学金办法鼓励中国学者留法。目前为遥感技术应用及长江口海洋合作研究。语种受到限制，科学合作也不如与英、美、加那样广泛。可能不能与英国的智力输出政策相比拟。

4. 荷兰

比法、德与华交往还要进步些，得助于联合国资助，对华合作交流以及培养留学生皆有开展。特点是小规模但持续多年。合作项目同荷兰环境有关，涉及海岸、三角洲改造等。

（二）北美（美国、加拿大）

1. 美国

科学技术水平高、实力强，地学仍居高水平。从海洋科学看，中美合作项目多，但偏重于有实力的大单位进行，如美国海洋大气局与中国国家海洋局交流，或与石油部、中国科学院等，着重于对中国海域、中国环境方面的调查。这些尚属空白区，美欲捷足先登，已在长江口、杭州湾、黄河口、黄海与中国合作搞科学研究。另外是实业单位，如中国阿科(Arc China)公司，对南中国海台风、海洋环境地质的研究，部分应用中国台站资料，但大部分请外国机构，如英国的海洋、地学院校来华工作。环境的研究占经费开支的1/4，如果国内各单位能合作起来，返承包这部分工作，则能在技术、财政上双收入，还锻炼培养了人。美国国内期望对华了解，我校有教师前往讲基础课，如中国自然地理、自然资源、历史、数学等，这是其他国家所不曾具备的。美国基金项目多，合作潜力大。

2. 加拿大

与美国不同，多关注于本国利益、本国领土的研究，对外事务的兴趣不及美国。但是有"国际发展研究中心"(IDRC)，及"加拿大国际发展机构"(CIDA)两个机构支持第三世界国家发展经济文化科学，其中IDRC对我国提供500万加元(1985—1987年)，援助农业、海洋等方面。但项目落实渠道尚待通畅，据我知，已落实的有广州地理所的"水土流失"和我系的"海南岛港湾开发"两项目，其他如黄土、冰缘等项目尚在落实中。援助款项用于发展中国家的研究力量(技术、设备及人才培养)。这类合作中尚存在渠道不通畅，图件资料如何使用，成果的分享及有关馈赠设备，进口手续等简化问题。

财政经费仍是交流中的主要问题。美国政府机构着重于与实际开发任务有关的大课题，对于小课题或科学家有兴趣的合作单位，往往缺乏经费，如美国的People to people代表团有许多科学家自费来华，渴望交流，但缺乏经费及组织支持。

（三）亚太地区（日本、澳大利亚、新西兰、中国香港）

1. 日本

积极从事与本土及周围海域有关的国际合作研究，如日、法海沟与洋底山调查，中日美黑潮调查；大陆边缘地质研究，以及远洋资源考察，南极与印度洋调查(与苏联合作)。我个人感到：日本是近邻，其实验装备先进，若合作研究东亚大陆边缘、火山、地震或板块运动等重大地学课题，是有积极、重大意义的。日本地学家与中国台湾的大学交往密切。我们应重视与日本的交流，有些项目需要从专家之间的个人交往开始，有了共同的兴趣相互了解，才会进一步合作。

2. 澳大利亚与新西兰

地处南半球，国土大，与北半球的环境适成对比，气候不同，今古相对应，可能是北半球的过去。澳大利亚的悉尼大学与堪培拉国立大学地理系已多次访华建立联系，接受中国学者访澳，他们重视与亚洲，与中国交流，因为中文资料丰富。根据其工作与环境特点，在研究海岸、火山、沙漠、资源开发及环境管理方面可以开展合作。海洋方面已有合作。据我所知，澳大利亚接受中国留学生较多，一些学生一年即取得硕士学位，三年学博士，奖学金每年7 000多元，房租、家属另有津贴，条件较好，并且学成以后促使学生返回祖国工作，这些都是较实际且友好的。我认为应在南半球的澳大利亚建立一个地学合作科研的基地。

3. 中国香港

中国香港教育从属于英国，培养的大学生，优秀者赴英、澳、加等英联邦国家深造，双语制，但英文、中文都有困难。香港大学有不少外籍教师，地理系属社会科学院，但自然地理仍较强。香港中文大学地理系属文学院，教师多中国台湾学者，经济地理与自然地理皆有。香港地理学界的特点是教育经费充足(与近年经济繁荣有关)，但研究课题不多，大多附属于教学。地理出版物多，是个重要的对外联系的窗口。1997年香港由我国行使主权领导，目前香港学术界注意与内地交往。我系打算在城市规划管理、印刷出版及教学实习、区域研究等方面，与香港两所大学合作交流。

(四) 非区域性的

非区域性的，如联合国，有很多研究机构资助人才培养与合作交流。目前主要是英、美等国学者参与，另外像智利、秘鲁等国也有较多代表与学者，而我国学者在联合国科研机构中尚很缺乏。我想，中国地理学会需专门研究国际学术动态，组织力量，有计划地参加国际学术活动。由中国地理学会设置专门机构，组织一批专业外语与组织活动强的同志，定期召开专门会议，掌握情况、方向与重点，制定交流计划。

三

为加强地理科学的国际交流，首先应加强我国地理科学研究成果的对外介绍。学报及各种学术刊物应更活跃些，尽可能减少语言文字上的障碍，刊物周期要短，易于及时传播。

其次，地理人才的培养，仍应以国内为主。但亦可在国际学术交流中，合作培养研究生。出国研究生，以研究中国区域的课题为主，在国外学习基础与技术，回国考试答辩取得中外双学位。同时亦应鼓励国内大学地理系科招收外国留学生。中国教师有坚实的地学基础与丰富的实际工作经验，这些都是外国留学生的弱点，将会受到外国学生的欢迎。

合作交流中重要方面即双方互相讲学。出国讲学扩大我国地理科学在国际上的影响，邀请外国专家来华讲学，通过讲学研讨班，带动一大片，最为有利。短期讲学专家往往把数年心得于短期内系统加以介绍，我们受益一片。关键是邀请人选需有真才实学而为我所需的方面，翻译者要外文与专业皆强才能达到预期效果。南京大学地理系近几年来，举办各学科的外国专家讲学研讨班14次，对全系教学科研有很大的推动。同样，在国内举办国际学术会议，更是受益面广，是对人才培养与促进科研水平收效大的途径。

合作研究，共同出版成果，这是地学交流的主要方式。我们同英国一些大学订了三年合作协议，这可促使人才培养、提高学术水平。通过合作研究，可组织中外联合科研梯队，较快地取得科学工作进展，并可以支持第三世界其他国家。

国际海洋科技动态*

1991 年我几次出差国外，已有详细汇报，这里将与地海科学系有关的科技动态摘要介绍，以供参考。

一、太平洋海洋科学技术会议(Pacific Congress on Marine Science and Technology，简称"PACON International")

21 世纪是海洋时代，是太平洋世纪。太平洋的活动影响着气候、环境与灾害，太平洋的资源、环境与人力具有极大的发展潜力。因此，建立"太平洋圈"已成为沿岸国家领导与科学家的共识，而海洋科学是其核心。PACON 是由美国政府(美国海洋大气局)、日本政府以及一些国家的私人财团支持的海洋科技组织，总部在美国夏威夷大学。John Carey 任名誉主席。Co-chair man 为 Saxena Harmon(夏威夷大学)，以及日本的 Kenji Hotta (现任 UNESC 非生命矿产资源委员会副主席)、澳大利亚的 David Hopley(北昆兹兰 Jams Cook 大学热带海洋研究中心)。PACON 1984 年起每两年开会一次。1990 年 PACON 东京会议上侯国本、陈吉余、王颖联合提出在中国召开 PACON 会议。该建议得到国家教委、国家海洋局的支持，以及与 PACON 两次正式会谈，PACON 总部决定在中国召开 1993 年 PACON 区域会议。由于语言与通讯条件的方便，Saxena 建议由王颖代表中国方面与 PACON 联系。目前我系海洋室在协助筹办事宜。1991 年 2 月，我出席了在夏威夷召开的 PACON 理事会。其中，讨论落实了 1993 年在中国开会的事宜。主要是：1993 年 6 月 14 日—18 日在中国青岛召开 PACON 区域会议。会议内容：海岸过程与海岸带开发，包括河海体系、大陆架、大洋合作考察研究、全球变化、海洋遥感、地理信息系统、数学模型、海洋工程、海洋教育和科技合作。规模：中国代表 200 人，外国代表 100 人。以此活动将推动中国对 PACON 的了解与参与。

二、第十七次太平洋海洋科学会议

1991 年 5 月在夏威夷召开，到会代表超过 1 000 人。我应 UNESCO 东南亚办事处邀请作为海洋教育特邀代表，作了"多学科培养海洋科学研究生"的专题报告。报告后反映热烈，得到美国、中国台湾代表、泰国、印度等代表热情赞同，同意我阐述的观点。UNESCO 拟建立太平洋与印度洋地区海洋教育网。我们支持此建议并介绍了中国的海洋教育单位。第十八次会议于 1995 年在中国召开，主席是中国科学院周光召院长。

1991 年 8 月，我参加作为"国家试点实验室主任代表团"成员访问了美国、加拿大的

* 王颖：《南京大学学报(地理学专辑)》，1992 年总第 13 期，第 191 - 198 页。

相关大学与研究所。11 月参加江苏省科委组织的代表团访问了荷兰海洋所、荷兰地质局、荷兰 Delft 研究所、荷兰须德海与三角洲工程。

三、加拿大温哥华市英属哥伦比亚大学(University of British Columbia，简称 U. B. C)地理系

该校有地理系沉积实验室、遥感制图与地理信息系统实验室，这些实验室皆与我系海岸海岛开发实验室有相似性。系主任 Olav Slaymaker 教授邀请我们参观，并安排午餐与座谈。Olav Slaymaker 教授于 1991 年 9 月起任该校副校长，主管科学研究。

该地理系共有 21 名教授与副教授，三、四年级大学生 300 人，研究生 60 名。该系自然地理教师力量强，系主任研究地貌学(冰川与冻土)。该系对气候(现代、古气候与全球气候变迁研究)、森林遥感、灾害防治，以及上述各项的地理信息系统应用都有很好的研究成果。

接待我们参观教学与沉积室的是 Margaret E. A. North 教授，其工作为 Senior Instructor，专业为 Vegetation Mapping，属植物地理学方面(Plant Geography)。其气象气侯教学以现代的热敏、雷达设备观测，辅以传统的风速、风向仪、百叶箱、干湿温度计等，使学生了解基本技术之发展。沉积实验室有传统的粒度分析——以采砾石研究为主，与 B. C 省的河流多砾石沉积为主的特色有关。同时，有风洞、底质运动测定的管道装置。实验室以研究生工作为主，设备与研究课题内容有关。

遥感实验室负责人为 Grahan Thomas 博士。主要用遥感图像分析森林的覆被、成长、灾害以及所反映的气候变化，其中对森林的成长所反映出的滑坡灾害、年代与轮回很有特色。其遥感图像来自 MACDONALD DET TWILER 公司，他们收购了原属 U. B. C 的卫星接收站，再与 U. B. C. 人员合作，共同完成政府交予的项目，B. C 省以森林资源为特产。

地理信息实验室接待人为 Brian Klinkenbery 博士及其中国学生肖燕妮。该系共有 486 机约 20 台，工作站 9 台，所有计算机皆以服务器相连。486 微机主要供大学生实习使用，研究生与教师主要用工作站。其中有一个实验室有 12 台无硬盘 PC 机(486＋487)，与一个服务器相连接，并连接两个 300 MB 硬盘(在维修时可更换使用)。由于每一台微机本身无硬盘，因此故障率低，易于维护。此实验室是专供该系大学生作地图数字化、数字地图编辑、数据建库、制图、遥感图像处理、分析数据转换以及地理信息之应用等实习。功能很强，大大增加了学生的实践能力。多功能系统工作站用于研究与服务，承担各种应用项目，从森林监测至电话号码簿服务，从中取得经费。遥感图像分析判断与地理信息制图与信息采取相结合，是它的另一特点。该系所用的地理信息软件是 ARC INFO 及 SPANS(它是 Spatial Analysis System 之缩写)。前者是由美国加州的 ESRI 研究所开设，并被国际上广泛应用的地理信息系统；后者是由加拿大研制开发的地理信息系统，主要是用于空间数据的开发。SPANS 可在 PC(286/386)上运行，它基于操作系统 OS2 环境(而非 DOS 环境)。该系统在国际上也有很多用户。

该系还有一台挪威生产的 AP 190 像片量测仪，像点坐标量测精度高于 20 μm，通过

步进电机实时消除上下视差，保持主体观测。它实质是基于 PC 机的像片量测系统，价值 4 万加元。

在大学地理系三、四年级教育课程内容方面，为适应现代发展，开设了海气动力交换课、全球气候与气候变化、大气污染气象学、环境发展地理、资源管理地理、自然灾害地理、环境与资源、环境评估与分类，以及东亚与中国地理学等，内容较新，具有现实意义。这些课程取代了传统课程，值得我们效法。

四、加拿大安大略省滑铁卢大学(University Waterloo 简称 U. W.)

这是新型的理、工科大学，包括部分社会科学。我们主要参观该校环境科学中心的地理系与环境科学系的水质分析中心。地理系的 Renald A. Bullock 教授(该系系主任)和安大略省地质测量部(Ontario Geological Survey)Owenl White 博士接待了我们，该校国际项目组的 Pauline C. O'Neill 女士(International Programms Officer)陪同午餐，水质分析所所长 R. W. Gilliam 教授接待参观环科系水质分析室。

地理系为 1962 年建立，是加拿大的第一个地理系，该系文、理相结合，遥感与计算机应用强，并以与国外第三世界(东南亚与南美)广泛联系为特色，这同加拿大专注国内教育与研究事项的国策有所区别。但同时，该系与邻校 Wilfrid Laurier 大学地理系合作，集中了 45 位教师，150 个学生从事水质与水源研究，这是加拿大最大的地理学项目，由系际与校际联合向政府部门取得。该系有大学生 480 多人，研究生 150 人，教授与副教授 23 人，但感到人力困难，加之政府对华合作项目采取保留政策，故而该系与我们系虽已订有合作协议，却难以取得经费支持。

该系着重水环境、区域地理与国土规划研究，每个学生必须学习与使用计算机，这是由于 Waterloo 大学是以工程技术与计算机为主建校的。

150 个研究生共用 21 台 PC 计算机，480 名大学生拥有 40 台 PC，其他专门的工作站实验室拥有 16 台机器(DEC Station 5000/200)，遥感实验室有 14 台计算机，此外尚有 Vax11785。

地理信息系统用 16 台工作站联网进行，所用软件为 ARC INFO SPANS 及 IDRIS，其中 ARC INFO 开发使用有困难，各成系统，除大学生实习用机外，工作站与遥感实验室主要服务于专题项目，已在校内及校外与加拿大其他实验室联网，目前主要服务于政府的海水研究项目，北极海冰厚度、固结、流动直接影响到极地海洋石油开发，用遥感图像、雷达测试海水厚度与活动特性，用地理信息系统(GIS)建立模式。

以 U. B. C 与 U. W. 两实验设备之数量，型号与广泛应用程度，我们系在这个技术领域约相当他们 1984 年的初期水平。

环科系的水质分析室力量雄厚，不仅作水样分析，而且实验研究汽、水、土之间的相互作用过程，具体工程是采取柬埔寨(项目委托国)的地面土层，分析地表废污对水质影响的程度以及危害元素分析，将实测→实验分析→计算机模拟→推理结合起来，反映出高等学府的重点研究所应具备的特点与科学先导地位。

从加拿大两个大学的实验室观察感到，其实力雄厚，保持北美相当水平的先进与优势，实验室向学生开放，普及程度与更新发展都较强。

五、加拿大测绘与遥感局(Surveys, Mapping and Remote Sensing Sector)，渥太华

加拿大能源矿产和资源部(Energy, Mines and Resources Canada)的 Surveys, Mapping and Remote Sensing Sector，其基本任务是布设全国一等二等控制与 1∶50 000 的国家基本图。

我们参观了其中有关的分部：遥感中心、地图印刷分部(主要生产线划地图)、数字制图出版系统(主要生产影响图)、视频磁盘制图系统、GIS 技术中心、地形制图分部的数字测图系统以及卫星接收站(Canada Centre for Remote Sensing, Cartographic Reproduction Services, Digital Cartographic Publishing System, Video Disk Mapping System, GIS Technology Centre, Digital Mapping System in Topographic Mapping Division and Satellite Receiving Station)。其中加拿大遥感中心是于 1972 年成立。加拿大是世界上最早利用遥感技术的国家之一，它具有自己的地面接收站(ground receiving station at Prince Albert,Saskat,Chewan)。它不仅能接收美国的 LANDSAT 影像、法国的 SPOT 影像(所接收的 SPOT 影像还供美国用)，而且还能接收最近由欧空局(European Space Agency)发射的 ERS-I 卫星与日本的 J-ERS-I 卫星的合成孔径雷达影像(Synthetic aperture radar image)。我们参观地面接收站时，十分有幸见到刚刚接收到 ERS-I 的合成孔径影像。由于这种合成孔径雷达(Synthetic aperture radar image)具有全天候等一系列之特点，它的应用前景将是十分广阔并具有很大潜力的。为了大力推广其应用，加拿大遥感中心准备接收六帧 ERS-I 和 J-ERS-I 影像，为加拿大的有关部门免费提供少量的影像数据，其要求是他们应写出这种新的合成孔径雷达影像新的应用建议。

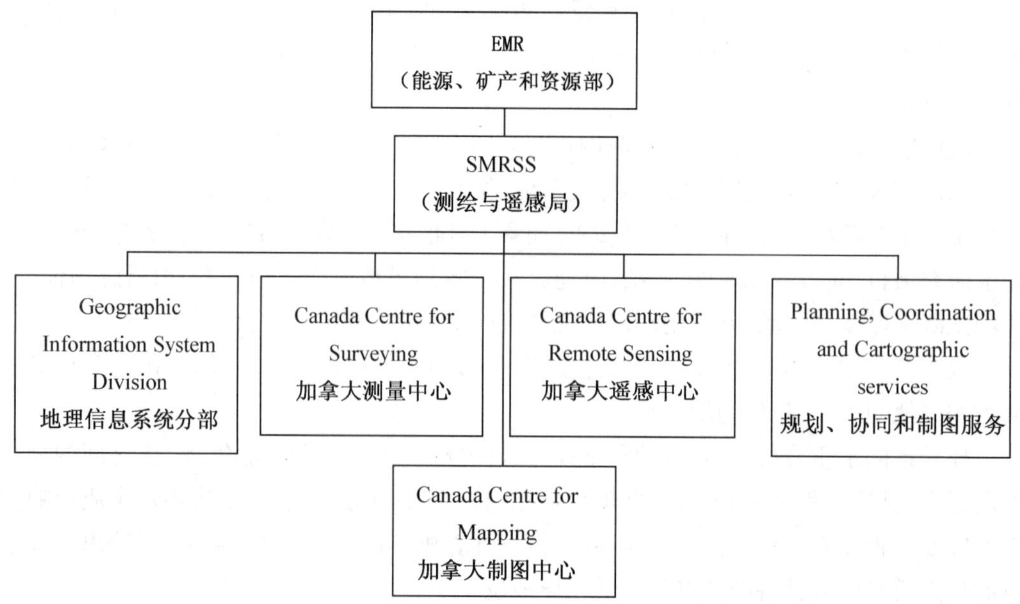

由于加拿大有不少地方常常受云层覆盖，有暴风雨气候，特别是北极地区冬天全部是黑夜，由于这些条件的限制，加拿大正在加紧研制自己的雷达卫星，标为 RADARSAT，预计其分辨率为 28 m，高度为 100 km，轨道通过极地。因此，它可以获取地球上任何一个地方的雷达影像。RADARSAT 卫星是由加拿大空间局设计制造发射，由 EMR 研制地面系统和它的应用，它预计在 1994 年发射。

加拿大遥感中心下设四个分部。

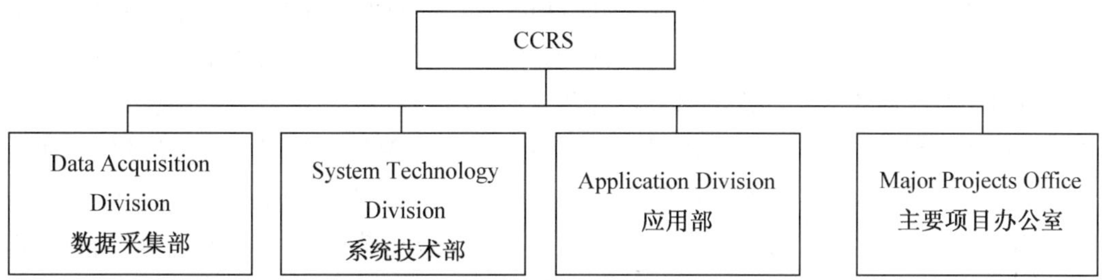

他们还成立了加拿大遥感咨询委员会，它由政府、各地方、大学、应用部门、工业部门等代表组成，它将为 EMR 部的遥感政策提出咨询，对加拿大遥感中心的政策、计划和活动提出意见，从而使 CCRS（加拿大遥感中心）更好地协调加拿大的遥感研究、发展和应用。

地理信息系统分部（Geographic Information System Division）是 1987 年成立。它的任务是帮助地理信息的用户，研究与制订地理信息系统的标准，协调所有政府级 GIS 的发展、支撑 GIS 工业的发展以及与用户一起改进和发展 GIS 技术的应用。地形制图分部（Topographical Mapping Division）利用光电扫描仪（electro-optical scanning equipment）用五年的时间完成了整个加拿大 917 幅 1∶25 万地形图。目前正在完成 1∶5 万地形图的数字化工作，这是它的全国性 GIS 之基础。

在该分部，具有很多种 GIS 软件系统，例如 ARC/INFO，SPANS，Geo Vision，CARIS，PAMAP，LASERSCAN 以及 MGE/TIGRIS 等。该分部对各个系统的性能、特点进行试验、分析和比较。然后，他们可以根据各个 GIS 用户之需要、任务、已有之设备以及今后发展提供咨询意见。该分部同样还有广泛的国际合作，去年曾在苏联举行过 GIS 研讨会，对各个 GIS 系统进行演示，讲解其性能与特点，以及 GIS 之应用与发展，并表示希望在中国进行类似的研讨会。

六、美国马萨诸塞州，Cape Cod，伍兹霍尔海洋研究所（Woods Hole Oceanographic Institution）

由该所海岸研究中心主任 David Aubrey 博士接待我们：① 观测了 1991 年 8 月 18 日 Bob 飓风对 Cape Cod 海岸破坏状况：冲岸、毁屋、滨海柏油路面平移 4～5 m、路面毁坏、拔树、折断电杆等；② 参观了海岸研究中心及地质、地球物理部的主要建筑与研究设施；③ 座谈有关研究、组织机构，人才培养与管理问题。

伍兹霍尔海洋研究所是西方著名的三大海洋研究所之一，建立于 1930 年，是私人非营利的研究所。联邦政府与省政府皆不给资助。该所每年经费是 8 000 万美元，主要靠项目申请，其中美国自然科学基金会的来源占总收入的 45%，海军提供的研究经费占 19%，美海洋与大气局提供的研究经费占 5%，其他机构为 11%，因此，人员素质好，思想新，开发研究适应国家需要的项目，写申请书方面严格要求等使其竞争性强，而获得经费后该所对工作的要求高，推动性强。竞争机制是推动实验研究发展的重要手段。

伍兹霍尔海洋研究所建于 Boston 城东南方 150 km 以外的 Falmouth 镇 Cape cod 的西南角，是世界海洋研究的中心，在 Woods Hole 区共有四个研究所：海洋生物研究所、伍兹霍尔海洋研究所及两个政府研究所：美国海洋渔业服务中心的东北渔业中心；美国地质测量局的大西洋地质分支。这些研究机构共有 50 个建筑群，13 艘研究船、艇，1 000 名雇员与 500 名研究生。伍兹霍尔海洋研究所下设 5 个专业部：海洋生物学、海洋化学、地质与地球物理、物理海洋及海洋工程学部。同时设有多学科结合的研究中心：海岸研究中心、海洋政策中心、海洋开发中心。这个研究所占地 200 公英亩，共有 800 名全年雇员，有一只由 5 艘船组成的船队——Oceanus 号，177 英尺长；Atlantis Ⅰ号 210 英尺长；Knorr 号 279 英尺长；Asterias 号为 46 英尺长以及 25 英尺长可乘 3 人，具有下潜至 6 000 m 水深的著名下潜器——Alvin 号。4 船可以续航 30～60 天，研究人员乘载从 12 名到 34 名不等，全天候工作，年工作率可达 11 个月。Alvin 号已下潜 2 500 次。各船只皆有相应实验室，具有全球定位、岸台与声学定位辅助系统，测试、采样、观测、分析、计算等设备，其中可于 4 000 m 深处拍照的水下相机，成功地找到冰海沉船——Titanic 号。该所重点研究世界大洋底部洋中脊扩张与热液矿床，地热，洋流，海底生物与环境反映，污染及全球气候与海平面变化，三个研究中心具有自然科学与社会科学交叉、研究与应用相结合之特点。该所也以技术研制见长，设计产品供商业部生产。目前研制潜水汽车，由船上操纵，不用缆，可在 6 000 m 海底行驶、拍照，已在地中海试用，亦将超级计算机与遥感相结合，应用于北极及大洋研究。

海岸研究中心项目：① 为本地区海岸防护、海滩喂养工程服务；② 与苏、罗、保、土合作研究黑海——富有机质、无氧环境下生物、沉积过程为石油形成、储藏机制研究服务；以及与加勒比海周边国合作研究珊瑚礁与全球气候变化问题，具有以当地的应用课题支持长远大项目研究之特征。

自 1968 年起麻省理工学院(MIT)授权伍兹霍尔所代培海洋学领域研究生，至今已培养出 300 名博士研究生。另外，为海军与海洋政策部门培养硕士生。接收来自中国大陆申请者不多，因为不了解研究生的实际水平，所以竞争后选中率不高。研究生费用每人每年约 1 万美元，由导师项目支付。博士生需学习 5 年，男、女学生各一半，取决于其才能来录取。

该所有一个大的样品库(600 m^2 面积，高达 3 m)，保持 4 ℃温度，储存 50 年调查大洋采集的海底样品，可供世界各地免费取样，而我们尚无此设备，所采的海底柱样保存也不当，该所现将柱状采样套管改为 T 形，可直接采到显示剖面之柱样，不需要剖析。

为适应科技发展，由美国自然科学基金会支持 500 万美元，于三年内建成大型同位素质谱仪，可以测海水(500 mL)及微量沉积样，直接测定 ^{14}C 数量，而不是测半衰期。该仪器年需 10 人工作，运行费为 100 万～150 万美元。测试样品价格是 250～500 美元/样品，共同项目人员的样品收费低。该所通过合作申请项目实行对外开放与交流。

关于研究人员职称系列与升迁等情况大致是：伍兹霍尔所共有130名科学家，其中Associate Scientists所约40名，需工作四年，至少6～10篇好文章，由外单位10名科学家评估同意，由本所学术委员会评定，然后可升至Associate Scientists，工作四年后，有12～20篇好文章，经过国际知名科学家评议，以及学术委员会评定可升为固定雇员Tenure，该处现有副研约40名，具有Tenure后，在Woods Hole仍有被辞退可能。具Tenure的副研再工作10～20年，或少于10年有突出贡献者，其国际知名度高，为学科带头人，经过国际著名科学家评定，再经由国际科学家组成的学术委员会评定可升为Senior Scientist，此在伍兹霍尔所有40～50人，所以该所实力强，竞争机制与淘汰办法始终保持了研究所的活力。D. Aubrey属Senior Scientist。

七、纽约市郊的哥伦比亚大学莱蒙特—道尔特地质观测台（Lamont-Doherty Geological Observatory of Columbia University）

该台是1949年由著名的地球物理、地质学家在哥伦比亚大学建立起来的，应用第二次世界大战后海军转业的船只，继续于海底的研究，制作出世界第一张大洋洋底图，展出洋底山脉、裂谷、平原的巨大地貌变化。并进一步根据磁测等确定了洋底运动机制以及陆地形成的原因。奠定了近代海洋地质学的基础，震撼了传统地质学基础。因此该所在全球地质研究中具领先地位。由于精密的地震观察仪器防震、防喧闹，而1948年Thomas Lamont的遗孀捐献了其在Hudson河西岸、Manhattan以北15哩之宅园，故而该所有了静谧、宽敞、美丽的研究场所，1969年Henary L. 和Grace Doherty又给予可观的资金捐助，使该研究所有了今日有面积为300英亩，掩隐于树林之中的建筑群。现任所长Gordon P. Eaton博士是1990年11月上任的，是地质学家，专长于西北美洲地质研究，在本任前曾任IOWA州立大学校长达4年半。现该所共有500名雇员，其中包括100多名研究人员与100多名研究生，其他200多名为行政、建筑维护、后勤与船员（60名）。研究所属哥伦比亚大学，地质系的教授为该所高级研究员，但行政管理与经费上独立。该所年经费需3 000万到4 000万美元。经费的70％来自美国自然科学基金会，30％来自美国海军研究办公室、海洋大气局等。与伍兹霍尔所的经费来源相当，但是，伍兹霍尔所以技术研制见长，而莱蒙特所以基础研究为特色，是全球范围的地学研究。该所共有三条研究船，最老的是著名船只Vema号，最新的船只为Maurice Euling号，尚有一只Conral号。安排我们访问的是George Kakla博士，是高级研究员，研究全球环境变化，与我国科学院黄土实验室有密切的关系。接待与介绍该所管理情况的是Loren C. Cox先生，是该所发展和特殊项目主任，该职位介于所长与副所长之间。向我们介绍了遥感与海底制图的是William Ryan博士。Constance Sancetta博士（女）向我们介绍了她应用硅藻分布研究海岸环境变化的成果。

莱蒙特所的研究属于学术前导地位。其研究课题为：

（1）大洋钻探与全球记录，设置钻孔研究所，从事全球研究，内容包括深海沉积、黄土与冰芯。

（2）地球的热流状况，以分析全球的环境系统。

（3）全球 CO_2 的动向，将研究资料汇集为海平面图以示 CO_2 的逸出与吸收带，做出环境与工业资料分布之分析，而不是就事论事。

（4）EL Nino 现象的模拟分析，已将 1982—1983 年、1988 年的厄尔尼诺现象与影响分析透彻，可望做出全球短期预报。

（5）生物海岸学研究浮游生物对大气海洋中 CO_2 的平衡作用，已可用卫星的测示了解浮游生物对大气中 CO_2 吸收情况，同时进一步研究，有多少碳化合物已被动物在吃草中吸收转换，以及有多少沉淀于深海中。最后亦研究深海的生态环境因素与食物网间的相互关系。这些皆是超前的基础研究，均广泛应用了新技术。

（6）多频道地震仪研究、地震活动以及石油地质，已获得三维的图像显示断裂与地震活动带分布。

（7）树年轮实验室，已据年轮获得地震、火山活动、气候变化、污染、水供应信息与成果等。

（8）海底制图，用遥感图像与声呐摄影拼接图，测知加利佛亚大陆坡各种规模地貌，甚至表示出大陆坡上的河流系统，可以用图像的不同色调分析海底泥沙来源与动向。下阶段以地理信息手段研究储存于大洋钻孔中的环境变化信息。

莱蒙特所的先导地位还在于广泛吸收年长的博士研究生参加研究，给予条件使其创造。在莱蒙特的中国留学生很多，而且以成绩优秀著称。如 Dr. Ryan 的 13 个学生中有 9 个是中国学生，而当我说到愿派学生前来学习时，他非常高兴，连连告诉 Kukla 博士：“我将有一个中国学生。”这说明莱蒙特所的学者深入了解情况，而且实事求是，我国地质学教育是严格而训练有素的，所以，一旦有先进设备即可做出好成绩，而且中国学生勤恳吃苦，尤以地学学生为著。

该所的研究组织是随时间的发展而改变，近 7 年来科学组织为：① 海洋地质与地球物理；② 海洋与气候；③ 地球化学；④ 地震、地质与构造地质。

这些组织既独立又联合承担课题，近年来地球环境的研究经费增加，另外石油公司也提供经费。每年项目发展负责人召开成果交流会向各界介绍新成果，同时扩大合同的建立。对外交流中的一个大项目是美—苏联合的南极浮冰观测站。

年经费中约 54.6%用作水、电、房屋维护与行政费用，以保证科研进行。伍兹霍尔所是 35%用于行政，但加上设备费用则达 55.2%，所以也是超过 1/2 的经费间接用于研究，不过所列的名目有区别。如果一个科研人员得到的项目经费达不到工资的 10%，则会被辞退。一个研究人员的聘期为 6 年，如果 6 年后达不到高级研究员的水平，则会被辞退。所以优胜劣汰，研究人员不断地得到更新替换，保持了研究所的活力。莱蒙特研究所的特点，安静的环境，不断更新的设备，前沿课题与众多的青年人员力量，值得我们重点实验室效仿。

八、荷兰海洋研究所（Netherlands Institute for Sea Research）

该所属荷兰政府教育科学部，位于 Texel 岛上。全所 180 人。有海洋物理、海洋生物、海洋地质与地球化学 4 个研究室。

海洋地质和地球化学研究室：主要研究现代沉积物(细颗粒)的运移和沉积过程；利用同位素示踪测定沉积速率；南大西洋的古海洋学；大西洋和南极硅(Si)循环；黏土矿物；有孔虫和硅藻。这个室是我们这次访问交流的重点，全室有30人。该室研究手段和仪器设备都很先进，有α、γ同位素极谱分析仪、沉积分析仪、钻孔扫描仪、现场定位悬浮体照相机、电子扫描镜等。其中现场定位悬沙照相机是较先进的设备，它能和现场由摄影法记录悬沙运移及絮凝情况，这比取样回室内分析测定更符合实际情况，对悬沙运移和絮凝规律的研究具有重大意义。为此南京大学海洋研究中心已邀请室主任Doeke Eisman教授于1992年5月携带该仪器(重700 kg)来江苏连云港海域，南京大学配备定位系统、海洋化学、取样及测年等技术，协同合作科研。该所有两艘调查船"Pelagi"号长66 m，用于北海调查；"Navicula"号，长22 m，用于须德海调查。还可使用国家的"Tyro"号、"Tydeman"号进行远洋调查。

荷兰地质局(The Geological Survey of the Netherland)

该局下设四个部门：深层地质部(包括地球物理、水库工程地质、构造地质、沉积岩地质、石油地质等领域)；浅层地质部(包括地质、沉积地质、海洋地质、浅层地球物理、地貌学、过程地质、环境地质、水文地质、工程地质等领域)；科学研究所(包括生物地层、有孔虫、介形虫、藻类、相分析、沉积结构、重矿、有机地质，无机地质、煤炭地质等实验室)；项目管理及应用地质部(包括人事部、管理部、设计部、成果部、计划部、后勤部等部门)，全局250人，其中：1/3为科研人员，1/3为技术人员，1/3为行政人员。这次考察，我们主要和浅层地质部接触。所谓浅层是指表面到200 m深的表层。该部从事的海洋地貌专业主要从事大陆架测量制图。装备有技术先进的设备，如多频道地震削面仪、震动取样管、水力震动取样管、汽车轮船两用车、10 000 m长轻型缆绳(用于深海勘测)、Pisces采水器等。另外还设有专门车间负责设备安装与实验组合等工作。该部的环境地质专业主要是编制荷兰第四纪地质时期以来环境变迁过程及各时期地质图，供工程与资源开发使用。

荷兰地质局虽是独立的政府机构，但经常和其他重要研究所开展各种联合研究项目，在荷兰地质科学界居于重要的地位，特别在以下领域具有技术专长：采用专门设计的设备和技术编制系统的浅部或深部地质图；石油、天然气和煤、地下水和盐等天然资源勘探；地质学咨询服务(包括城市、工程和环境方面的地质咨询、沿岸地区的地质状况、自然环境开发计划和土壤改良、管道系统、锚泊地和近海结构物的海底稳定性、天然资源评价等方面的咨询)。

有关海洋科学国际学术活动动态*

首先，我要检讨一下开会来迟了，因公出差去荷兰于十三日刚回来。这次在无锡开会，感到很高兴，也很惭愧。江苏省地居黄海之滨，有长达一千公里的海岸线，有面积占全国四分之一以上的沿海滩涂资源，有世界上独特的辐射状海底沙脊群，是全国首先完成海岸带调查的省份。但是，江苏省是全国沿海省市中唯一未建立海洋学会的省份。原因种种，主要是认识不统一，认为江苏已有省海洋湖沼学会，已有中国海洋学会的二级学会等等。作为江苏省科协副主席，我为此努力多次，征求老科学家严恺、任美锷等教授的众多支持，争取到广大海洋工作者支持，并向省科委、省科协有关领导同志反映，取得支持，但最终仍未办成。后来又逢社团整顿。这件事，我希望国家海洋局、中国海洋学会领导予以支持，以早日促成江苏省海洋学会的建立。

下面我想就国际学术交流委员会进行的一些活动，现向各位理事汇报。

一、海洋学会与海洋局领导对国际学术交流活动是支持的，例如：1990 年 11 月在海南岛召开的“国际海岛与港湾开发管理学术会议”葛副局长代表严恺理事长和杨文鹤副理事长出席了会议。陈吉余副理事长，苏纪兰常务理事以及老一辈科学家任美锷教授，王乃樑教授，仲崇信教授以及美、加、西德、丹麦等 11 个国家的代表都到会做了学术报告，会议收获大。

二、关于 PACON International：PACON 的全称是 Pacific Congress on Marine Science and Technology，即太平洋海洋科学与技术会议，简称 PACON 国际。总部设在夏威夷大学 Honolulu 本部中，每两年召开一次会议。它由私人的海洋工业公司及美、日、其他太平洋国家的政府机构所支持，如：美国是海洋大气局支持，John Carey 任名誉主席，其次是由二次大战后美海军转业后在夏威夷与加州发展起来的海洋技术工业财团的支持；日本主要由日本大学的海洋科技学部支持，如 Kenji Hotta 为日本大学副教授，毕业于夏威夷大学，现任联合国教科文组织非生命矿产资源委员会的副主席。日本大学的 Wataru Kato 博士既是该校教授，又是大通财团的实力人物之一，财力雄厚，支持了 1990 年在日本东京召开的 PACON 会议。日本方面偏重于海洋工程技术。澳大利亚的代表人物是 David Hopley 副教授，在 North Queensland James Cook 大学的 Sir George Fisher 热带海洋研究中心工作。PACON 以美、日、澳大利亚三国为主，其他尚有南朝鲜、新西兰及太平洋岛国的代表。PACON International 是于 1984 年开始的，每两年开一次会议，1986 年及 1988 年均在夏威夷召开，1990 年是在日本召开，中国代表是青岛海洋大学的侯国本教授，于 1984 年参加会议，现任 PACON 理事，台湾海洋大学前校长郑森雄博士亦为理事，其次尚有台湾海洋大学教务长、水产食品科学系的陈幸臣教授，他们的专业是海洋渔业与水产食品。作为教委代表团成员之一，我出席了 1990 年的东京 PACON 会议。侯

---

* 王颖：中国海洋学会秘书处编，《中国海洋学会三届二次理事会文集》，7－9 页，1991 年 12 月。

国本教授热情推荐让我担任中国理事。在东京 PACON 会议上，由侯国本教授、陈吉余教授和我本人积极争取，出国前先曾获得严宏谟理事长的支持，我们与 Saxena Harmon 以及 NOAA 的三位代表先后做了两次正式会谈，初步同意于 1993 年在中国召开一次 PACON 区域会议。由于通讯与语言条件的方便，Saxena 建议由王颖代表中国方面与 PACON 联系。1991 年初，我得到了 PACON International 的邀请，请我担任中国方面的理事(故而，目前中国大陆有两名理事，侯国本教授和我)并出席 1991 年 2 月中旬在夏威夷召开的 PACON 理事会，这次会议讨论确定了 PACON92 年会议的内容，分组主席人选，经费以及开始出版论文集。分会主席的名单是按报名先后而定，故而我们获得了海洋声学分组(关定华教授任主持人之一)与海洋遥感分组(马蔼乃教授任主持人之一)，几名主持同时确定：中国、东南亚太平洋岛国只要会员满 25 名即可建立一个分组(Chapter)。在这次会议上各国代表还支持 PACON International 在夏威夷取得州政府财政支持的倡议。这次会议讨论了在 1994 年澳大利亚会议前，于 1993 年在中国召开一次区域会议的倡议。我介绍了中国海洋研究的特点；在中国召开会议的主要议题；经费估算为 7 万美元。相当于 1992 年 PACON 会议经费的三分之一；会议的选择地点为北京、杭州或青岛，我把这三个地方放了幻灯介绍。最后会议表决，在 24 位理事中以 14 票的优势票通过。有几票弃权，如澳大利亚 Hopley，他担心 1993 年中国的会议会对 1994 年澳洲的 PACON 会议有影响。我做了解释，通过 93 年的中国会议，可能促进中国代表对 PACON 的了解，就可能有较多人参加澳大利亚(1994)会议。所以我们争取到在 1993 年召开的 PACON 区域国际会议，而不是一直等到 1998 年(PACON 每两次会议中必须有一次在夏威夷召开，所以按正常顺序最早是 1998 年)会后，Saxena 表示他可能由于参加美大洋底部制图(海军与 NOAA 项目)可能不能访华或出席会议，但他全力支持：

(1) 由 PACON 负责印发会议通知；

(2) 由澳大利亚出资与出版论文集；

(3) 日本和南朝鲜将给予一定的资助；

(4) 各代表包括 PACON 总部代表自付旅费和食住费；

(5) 中国会员只需要有 15 个代表即可建立一个 Chapter。

1991 年 7 月下旬至 8 月初，PACON International 副主席 Harmon 来华访问由中国海洋学会理事长严宏谟接见商谈了 PACON1993 年会议事宜，由国家海洋局李海清同志接待安排了他参观了北京国际会议中心，杭州、上海、青岛等地的会议和旅馆设施。最后确定于 1993 年 6 月 14 日—18 日在青岛召开 PACON 区域会议。出席会议代表初步定为中国代表 200 名，外国代表 100 名。会议内容着重于海岸过程与海岸带开发，同时包括了河海体系、大陆架、大洋合作考察研究“全球变化、海洋遥感、制图、地理信息系统、数学模式、海洋工程、海洋教育和科技合作”等等。这次国际会议将由中国海洋学会、国家海洋局、国家教委、国家自然科学基金会以及几所大学发起与组织，在此特向诸位理事汇报，希望通过我们的共同努力，团结一致地把这次会议办好。美国夏威夷方面对中国不了解甚至提出诸如：你们有译意风设备吗？有会议厅吗？有幻灯机吗？等等问题。我们要争口气，开一次成功的会议，显示我们的进步、实际力量与海洋科学水平。

1992 年 6 月 1 日至 5 日将在夏威夷的大岛召开 PACON92 年会议。希望代表们准备论文,1992 年秋季将遴选我国出席 1993 年区域 PACON 会议的论文。

今年五月,应联合国教科文组织东南亚区域办公室的邀请,我又出席了第 17 次太平洋科学会议,是在夏威夷檀香山举行的,出席该会议的代表超过 1 000 人,包括各国著名学者。该会议内容广泛,包括自然科学、社会科学与生命科学,由"东西方中心"支持,我是海洋教育分会特邀代表,做了专题发言:"多学科培养海洋科学研究生"报告后反映热烈,得到美国、中国台湾代表、泰国、印度及密克罗尼西亚的好评,同意我阐述的论点。教科文组织拟定建立太平洋与印度洋地区海洋教育网,互相交流与合作研究,我支持此倡议并介绍了中国的海洋教育单位,第十八次太平洋科学会议将于 1995 年在中国,主席是中国科学院周光召院长。

这两次会议使我感受到:

(1) 代表们普遍认为 21 世纪是太平洋世纪,是美、加、日、苏、中、澳等大国所在,太平洋的活动影响着气候、环境与灾害。太平洋的资源、环境与人力具极大的发展潜力。建立"太平洋圈"势在必行,而海洋科学是太平洋科学的核心。

(2) 科学与技术结合,科技与经济开发结合,科技发展得到财界的支持,这是 PACON 的特色。而太平洋科学的会议则广罗知名学者与活动家,得到了美国东西方中心的支持,该中心富有影响,实力强大。

三、访问美国、加拿大、荷兰及参加在北京召开的国际第四纪会议(INQUA)的体会。

今年 8 月我作为国家试点实验室主任代表团成员访问了美国渥斯侯 Woods Hole 海洋研究所,拉蒙特—道尔特地质观察台(Lamont-Doherty Geological Observatory)、麻省理工学院、哈佛大学、加拿大哥伦比亚大学与滑铁卢大学等等。11 月我又访问了荷兰海洋所,荷兰地质勘查局,荷兰 Delft 研究所及荷兰的须德海(Zuider Zee)与三角洲工程。主要的体会是:

(1) 海洋科学重视多学科交叉的综合研究,当前美、加重视环境研究,美国偏重于支持与苏联的合作课题。如荷兰的 delft 研究所,它是以海洋工程模型实验见长,但目前重视环境与全球变化的综合研究,并将之用于社会影响与决策咨询。这是很大的变化。Woods Hole 重视技术改进与应用研究,每年经费 8 000 万美元。Lamont 的经费与之相当或更多些。但其中的 2/3 用于环境研究,而不是洋底构造与石油矿产。这些反映出美国海洋学适应科学发展趋势的转变相当快。

(2) 技术进展快,荷兰的三大工程包括须德海围海工程,三角洲工程及莱茵河航道化。这些大工程的实现反映了先进的科技水平,强大的经济实力,以及荷兰政府与人民改造国土环境的决心。建起了海上长城。如风车发电已由古老的风车改为三叶片、二叶片的新型风车,每台风车日发电量 75 kW。麻省理工学院研究陶瓷的新材料,仪器设备由日、美大公司提供,条件是给予仪器性能的优缺点。其研究生来自电器电话等大公司,人工智能方面研究无人驾驶机的神经系统计算机软件以适应多变天气下的飞行。麻省理工学院(MIT)校园不大,但设备却不断更新,居于科技最前沿。

美、加各实验室普遍应用苹果 486 机,从事地理信息系统工作,全国与国际联网,工作站应用普遍,而小型机在减少。因为 PC 机易于操作,价格便宜,在海洋沉积研究中,已可反映出海底峡谷泥沙流动趋势与泥沙源。在应用计算机方面,我感到我的实验室相当其

80 年代中期的水平。

(3) 我国的海洋科技工作自改革开放以来取得了很大的进步,实现了 20 世纪 60 年代的口号:“查清中国海,进军三大洋,登上南极洲”,今年又取得了太平洋底多金属结核矿的开采权,这是造福中华民族的大事,不仅如此,在地球科学中有许多成果在国际上亦有较大的影响。如在国际第四纪大会上,中国的黄土研究、冰川研究,特别是青康藏冰川研究,中国的成果具有主导地位,为各国所侧目。全球环境研究重视海陆对比,如深海沉积的层序,黄土沉积的层序、结构,以及与冰岩芯对比,将这些沉积中所储存的环境信息结合比较可得到重要的全球环境变化信息。可是,我们的海洋工作者在深海沉积与冰岩芯研究方面的工作较黄土之工作逊色。海洋是第四纪研究的重要方面,如海平面变化是当前的重要课题,其研究内容着重于:① 30 万年来的海平面变化;② 冰后期海平面变化;③ 世纪性的海平面变化,其长期变化情况分析变化规律,原因及变化趋势。海洋科学最具有这方面的研究条件,但是,目前在我国长期从事海洋工作的人未能组织到海洋平面研究的重大课题中来。研究的分散、不突出。我国的海洋工作者未发挥出应有的影响,各个方面有些人才未能在科学工作中形成互补体系,未形成拳头课题,未形成特色,亦反映出我们的海洋科学工作不及地球科学及其他学科成熟。我们海洋工作者要团结起来,有意识、有领导地组织一些重大课题,发挥各家所长,做出有特色系统的重大成果,把海洋科学水平推向前进!

Opening Speech on the International Symposium on Exploitation and Management of Island Coast and Embayment Resources "ICER", 1990, Haikou, Hainan Island, China*

Mr. Chairman

Distinguished delegates and guests

Ladies and Gentlemen

Today, the international symposium on exploitation and management of island coast and embayment resources is now declared open. Please allow me on behalf of the Organizing Committee, which consists of Nanjing University, the Traffic and Transportation Department of Hainan Province, the Bureau of Hainan Harbour Administration, and the Bedford Institute of Oceanography Canada, to express our sincere greetings to all of the participants, and to extend a warm welcome to all of our guests who are attending the meeting. I also wish to express our hearty thanks to all of the sponsors who are supporting the symposium. I would like to specifically mention the International Development Research Centre of Canada, the People's Government of Hainan Province, the National Natural Science Foundation of China, State Education Commission of China, the Chinese Society of Oceanography, and the Chinese Society of Geography.

It is our honour to host the symposium in Hainan Island, as this island is second largest one in China. It is a beautiful tropical island with a total of 1 500 km of coastline and 60 embayments. The potential resources of its deep water harbours, aquiculture and winter farming, historical and tourist spots, its unique geological landscape that includes volcanic features, tidal inlets, small river delta plains, coral reef platforms, and mangrove swamps. All of these features are all attractive and are raising the interest and attention of both scientists and capitalists. We hope that through the symposium, we will improve our understanding of the island's natural resource potential, and

---

* Ying Wang(Chairman of the Organization Committee of the Symposium): Ying Wang, Charles T. Schafer eds., *Island Environment and Coast Development*, pp. 3 - 5, Nanjing University Press, 1992.

through scientific exchange, be able to raise the level of utilization and management of the island's resources. In general, we hope that through the last decade of this century, Hainan Island, which we refer to as a pearl of China, will show an increase in economic development but will retain its natural beauty and ecologically pure environment while contributing benefits to Chinese and international peoples.

Secondly, I should like to introduce to you all of the delegates of the symposium which is the third stage of four part joint research project by the Marine Geomorphology and Sedimentology Laboratory of Nanjing University and Bedford Institute of Oceanography, Canada. The title of the project in full is "The sedimentation processes of tidal embayments and their relationship to deep water harbour development along the coast of Hainan Island, China"; it started in 1988 and will finish in 1991. The project has been approved by the State Commission of Science & Technology of China, supported by the International Development Research Centre of Canada. We have also received strong support from the State Education Commission of China and the Bureau of the Hainan Harbour Administration. The first stage of the project was to set up a coastal survey laboratory and a ^{210}Pb Lab, followed by training program for younger scientists and technicians. The second stage of the project involved a survey along the southern and western coasts of Hainan Island including the embayments of Sanya, to Yangpu, Haikou, and Dongshui by a joint Chinese-Canadian scientific team. The third stage of the project involves the analysis of data collected by the joint team, and the convening an international symposium for promoting the international scientific exchange, and hopefully to encourage attention from capitalists to provide financial support in aid of promising development projects. The fourth stage will be to continue the analysis and data interpretation efforts and to publish the results of scientific research efforts.

I should like to say that Canadian-Chinese cooperation in Hainan Island is going smoothly and harmonically. We are enjoying our joint program and we all feel happy to be able to hold this international symposium here, on the beautiful Hainan Island. Everyone of the project participants really feel that we are a part of the Hainan people.

Personally, I have worked here on various coastal researches since 1960s, and continually studied Sanya, Yangpu and other harbours in 70's and 80's. I love the Island deeply. I have seen the island begin to develop quickly since the 1980's. At the present time, I think that the island's managers should be concerned more about its resources, the kinds of environment research, and for planning that will help to strengthen the management structure and legislation needed to develop this singular chinese resource. I hope that the symposium will be helpful for this purpose.

For the symposium, we have received 80 abstracts and today we have 81 participants from 10 countries, including Australia, Bangladesh, Canada, China,

Denmark, Germany, India, Japan, Korea and the USA. Among the participants, there are professors, postgraduate students, and also governors and managers. I should mention that I feel a great honour to have such a group of famous Chinese professors to attend the symposium, such as Prof. REN Mei-e from Nanjing University, who is the member of Chinese Academy of Sciences, and who has won the Victory Medal of Royal Society of the U. K.; Prof. CHEN Jiyu, Vice-Chairman of the Chinese Society of Oceanography, who is a leading professor in estuary & coast research in China; Prof. WANG Nailiang from Beijing University; Prof. SU Jilan, from the 2nd Institute of SOA; Prof. CHUNG Chonghsin who is famous for his research on *Spartina* plantation in tidal flat environment. The statistics indicate that the symposium is a middle sized, but with a high level of participants in multidisciplinary science.

The meeting has been arranged as: (1) two plenary sessions that will take place this afternoon and tomorrow morning; each paper will be allowed 30 minutes for presentation and 5 minutes for questions. (2) four concurrent sessions will be held on the mornings of Nov. 4th and 5th. Each paper will be 20 minutes long for presentation and questions. Excursions will start on the afternoon of Nov. 3rd, and will continue on the afternoon of the 4th and 5th with field trips to Haikou Harbour, mangrove preserves and volcanic features in the local area.

On Nov. 6th, we are going to leave Haikou and travel to Sanya. On the 7th and 8th, we will be in Sanya, where you will have an opportunity to enjoy the local coral sand beaches and the tropical forests. We will return here from Sanya on the evening of Nov. 9th. I think that this cross-island trip will cause you to fall in love with the beautiful scenery, rich resources and interesting local customs of Hainan Island.

We wish that the symposium will improve scientific exchange, friendship and the environmentally sound development of this island.

Thank you.

2nd, November, 1990

河口湾沉积学与地球化学*

河口湾可能是海洋学家进行系统研究的首批海洋环境之一。长期以来人们已经认识到河口环境是介于陆地与海洋环境之间的有机纽带。

现在它显得更加重要了，我们要确定污染物质是怎样通过河口湾而最终被传输送到海洋中去的，那就要了解发生在河口湾内各种过程中的复杂的相互作用。如果我们知道了这些过程是怎样地关联到自然循环，以及人类的活动如何加速河口变化，那么，我们就能够恰当地维护和管理这个活跃的地带。

虽然有着多种河口湾分类，但是所有的河口湾都具有从陆地流来的淡水与海洋输入的盐水相互交换的这一基本特征。河口湾内的淡水与盐水的比例是决定性的特点之一，很多分类是据此特性而作出的。虽然人们还会争论，但要了解和预报河口湾的行径，必须分别对每个河口湾进行详细的研究，不过，重要的是要认清一些为大多数河口湾所共同的作用过程。这些作用过程(如潮流流速和河流流量)既是沉积过程与地球化学交换的控制因素，同时也可决定生物群落的生态。

为了阐明河口湾的一些重要特点，我将要介绍一个在加拿大由我本人以及我的同事们共同进行的多学科研究的河口湾。这是一个小型的河口湾，它不能作为世界上众多河口湾的典型代表。但是，它具有与其他大多数河口湾一样的某些共同特性。

Miramichi 河口湾位于加拿大东岸的 St. Lawrence 湾(图 1)，流入此河口湾的河流其淡水流域面积仅有 13 680 km^2，但径流量的季节性变化大，从 50～2 000 m^3/s。这个河口湾是由两个地形单元组成的：① 海湾，通过一系列水道与 St. Lawrence 湾相联。② 内侧的沉溺河道，它是一个典型的海岸平原河口湾，全部河口湾长达 40 km。

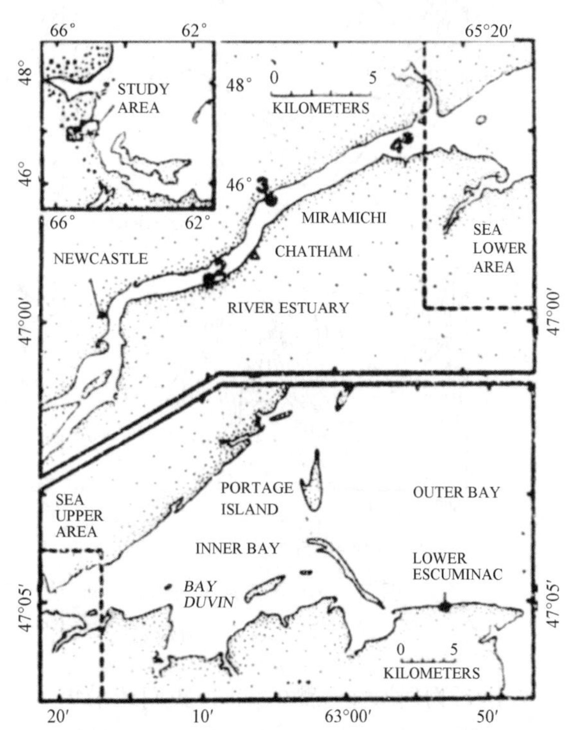

图 1　Miramichi 河口湾位于 St. Lawrence 湾的东南岸。图下部的内湾区在东部并毗邻图上部的河口湾沉溺河段

我们曾对这个河口湾进行了为期

* 王颖译：加拿大 Bedford 海洋研究所 Dale Buckley 研究员在国家海洋局第二、三海洋研究所的讲演，1982 年。

两年的多学科研究。这项研究包括物理海洋学方面的观测，采取水样；在悬浮质与底质的样品中分析多种金属；研究沉积物的分布与沉积动力学；研究底质沉积物中的生物与微体古生物。在本次演讲中，我主要介绍与讨论地球化学与悬浮质沉积物的研究结果。

由于我们预料到 Miramichi 河口湾具有较大的季节性变化特点，因此，我们进行了全年各个季节中一定时期的重复观测与采样。由于冬季该河口湾的水面冻结，所以需要某些特殊的野外观察方法。在有些情况下，观测站位就设在凿破冰层的孔洞内，水样是通过这些孔洞而取上来的。有时，还在冰孔穴上建立起临时性的实验室。

物理海洋观测资料主要包括测定盐度、温度以及流速和流向的资料（图 2）。是在河口湾的不同站位上进行了若干个潮周期间的观测。将全年不同季节的资料进行比较，并考虑到河流流量变化的影响，我们就得出了在河口湾上段的盐水入侵和混合程度。在冬季的和夏季的某些月份中，当径流量最小的时候，盐水的入侵程度最大。当从 Miramichi 河流来的淡水增加时，在盐水楔前缘的咸淡水混合度也增加。当河流的径流量接近于最大的情况下，盐水入侵仅仅发生于河口湾的底部，而使河口湾的上部也变淡了。

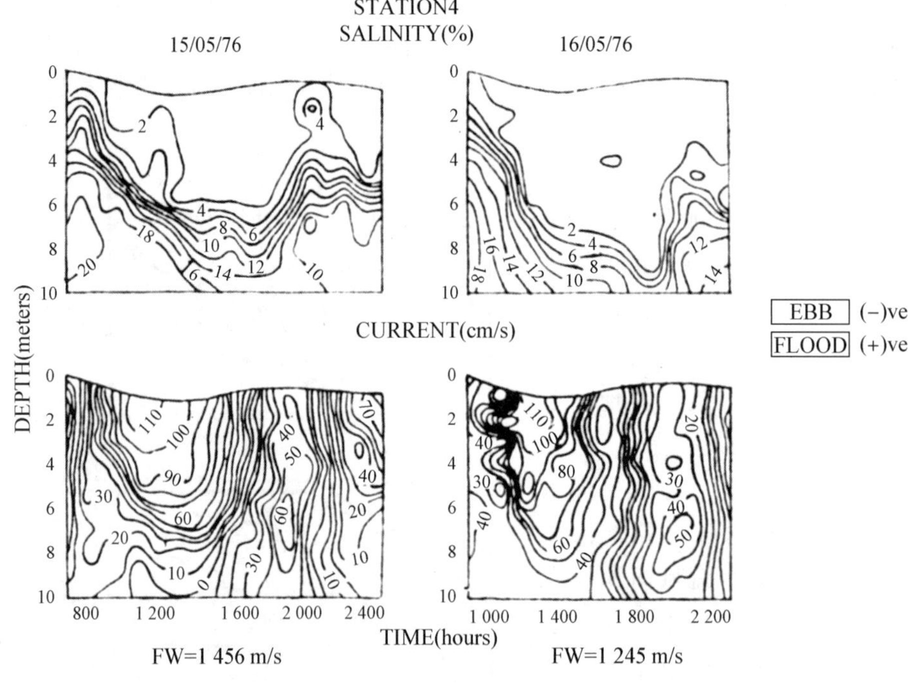

图 2　1976 年 5 月 15 日和 16 日，位于河口湾沉溺河段东部的 4 号观测站盐度和流速随时间变化图

在进行物理海洋观测的同时，我们还采集了不同深度水层的水样和悬浮质样品。而在采样周期之间，则应用光衰减仪测定了悬浮质的浓度。因此这些资料可以与连续观测的物理海洋资料进行对比（图 3～图 9）。结果表明，悬浮质浓度的变化与潮流流速的变化有关。在某些情况下，由于悬浮质从河水中扩散开来而导致河口湾的浑浊度增大。分析了河口湾不同断面水体输送的资料表明，潮流输送水体比在一年各个季节中变化较大的淡水余流的输送更为重要。在最大的春汛期间，由于落潮流量与径流量相当，这样大量的水体就被输送出河口湾。

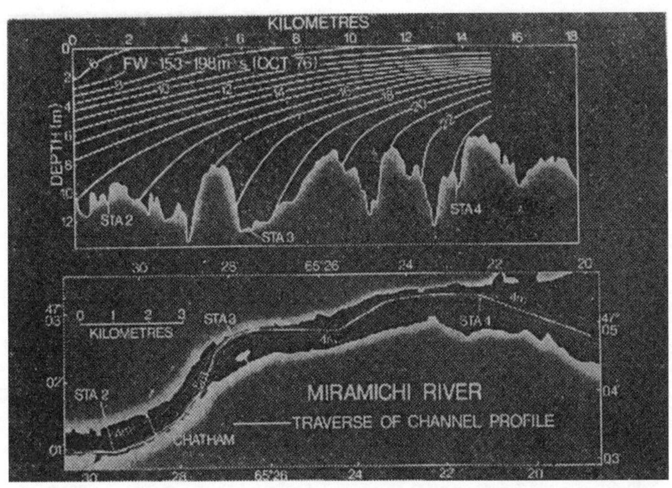

图 3　1977 年 2 月份，淡水流速为 70 m³/s 沿河流—河口湾各段咸水楔剖面图

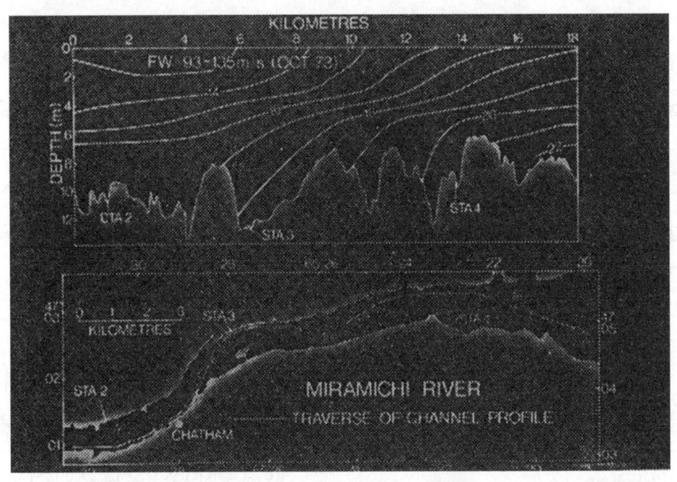

图 4　1973 年 10 月份，淡水流速为 93～135 m³/s，沿河流—河口湾各段咸水楔剖面图

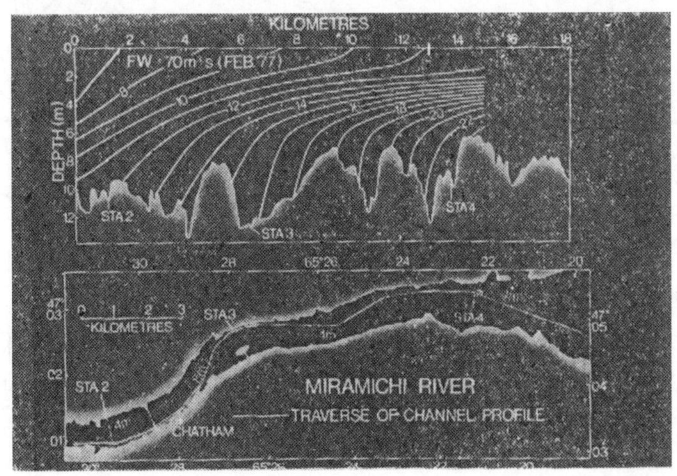

图 5　1976 年 10 月份，淡水流速为 153～198 m³/s，沿河流—河口湾各段咸水楔剖面图

KILOMETRES
FW=1250m³/s (MAY 76)
DEPTH (m)
STA 2
STA 3
STA 4
KILOMETRES
MIRAMICHI RIVER
TRAVERSE OF CHANNEL PROFILE
CHATHAM

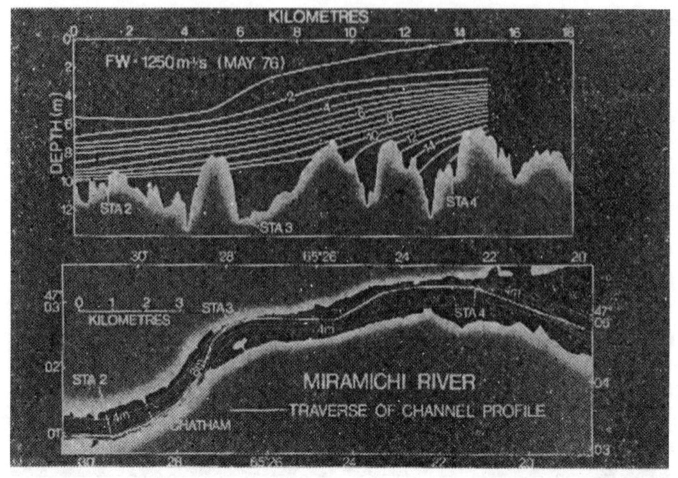

图 6　1976 年 5 月份，淡水流速为 1 250 m^3/s，沿河流—河口湾各段咸水楔剖面图

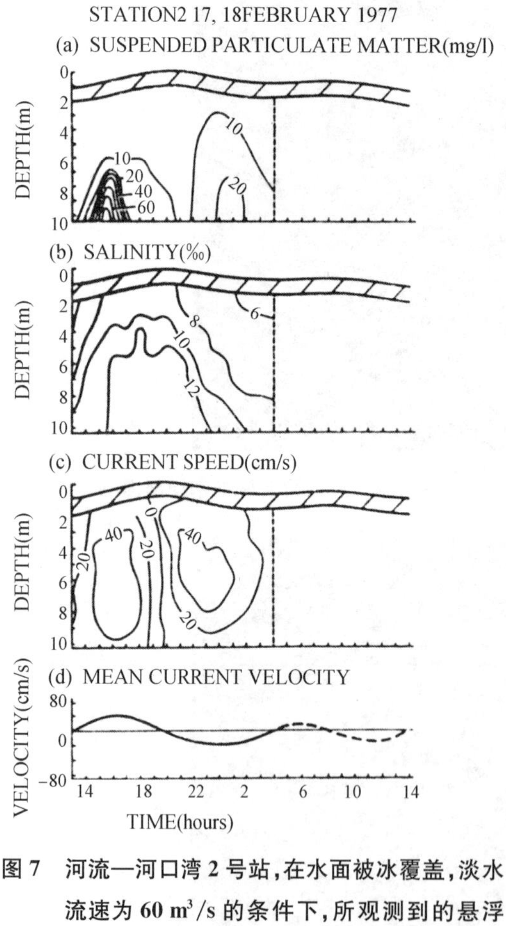

STATION2 17, 18FEBRUARY 1977
(a) SUSPENDED PARTICULATE MATTER(mg/l)
DEPTH(m)
(b) SALINITY(‰)
DEPTH(m)
(c) CURRENT SPEED(cm/s)
DEPTH(m)
(d) MEAN CURRENT VELOCITY
VELOCITY(cm/s)
TIME(hours)

图 7　河流—河口湾 2 号站，在水面被冰覆盖，淡水流速为 60 m^3/s 的条件下，所观测到的悬浮质浓度、盐度和流的变化

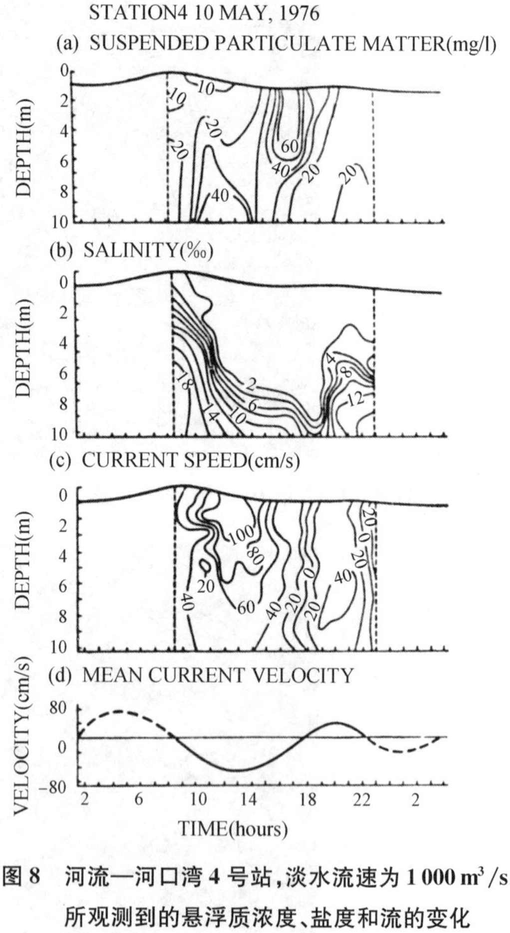

STATION4 10 MAY, 1976
(a) SUSPENDED PARTICULATE MATTER(mg/l)
DEPTH(m)
(b) SALINITY(‰)
DEPTH(m)
(c) CURRENT SPEED(cm/s)
DEPTH(m)
(d) MEAN CURRENT VELOCITY
VELOCITY(cm/s)
TIME(hours)

图 8　河流—河口湾 4 号站，淡水流速为 1 000 m^3/s 所观测到的悬浮质浓度、盐度和流的变化

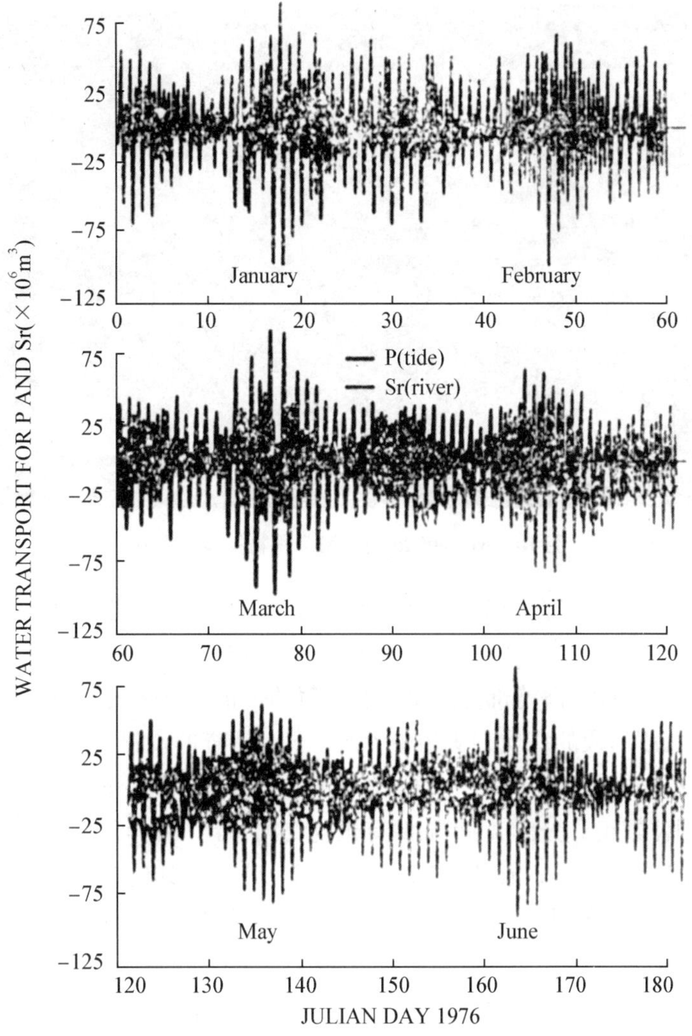

图 9 通过 Miramichi 河口湾的河流和潮流的水体传送。水体输送量由六个月内涨潮和落潮周期累计而得

根据底质采样和测量资料编制的沉积物分布图表明(图 10),底质结构上的变化与悬浮质的扩散以及从 St. Lawrence 湾来沙的再沉积作用有关。在湾口部分,砂和砾石被强烈的落潮和涨潮流重新扰动起并携带通过堡岛间的潮流通道。在堡岛的内侧发育了一个完好的具有沙波叠加的涨潮流三角洲。在湾中部,即在介于涨潮流三角洲与河口湾上部的沉溺河口之间的地方,堆积了大范围的软泥与砂质淤泥。淤泥同时也沉积在沉溺河段的河道内。细粒沉积物的这种分布表明,从河流以及河口湾上部来的悬浮质泥沙,沉积在河口湾内半稳定的河道与洼地中。

已知部分河口湾由于受到位于湾顶附近的木纤维、纸浆以及造纸厂所排出废物的影响而被污染。这些工厂排出了大量的木质素和有机纤维到河口湾上部的水中。这些有机物部分通过水体而沉积,部分则被河流或潮流携带而分散开来。这种分布的效果可以从底部沉积物的有机碳浓度中反映出来(图 11～图 14)。在接近湾顶处,有机碳的浓度占沉积物总量的 8%。虽然,湾底的沉积物结构在各处有较大的变化,但是有机碳的浓度是自

图 10　Miramichi 河口湾沉积物类型分布图

湾顶向海而减低的。对底部沉积物的有机碳含量与黏土百分比进行比较，结果表明，存在着两个资料组合：一个是上部河口湾的，而另一个是海湾的。这些资料都反映出在河口湾内沉溺河段部分的污染较海湾部分严重。有机氮的分析也显示出海湾底部沉积物中的有机质富含氮，而沉溺河段的底质有机物中含氮较少。调查的资料表明，不仅底质中的有机物含量在外湾部分数量较少，而且有机物的类型在河口湾的内外两部分中也有所不同。对有机物的碳同位素组成分析表明，陆源有机物在河口湾沉溺河段的有机物中占有很大的比例。在这个区域内，$\delta^{13}C$ 值是－27.1%～－27.5‰，而其他资料表明，在湾口部分同位素比例更大，其 $\delta^{13}C$ 值是－22.8%～－23.9‰。这说明了湾口沉积物中 40％的有机物是从海洋生物中来的。

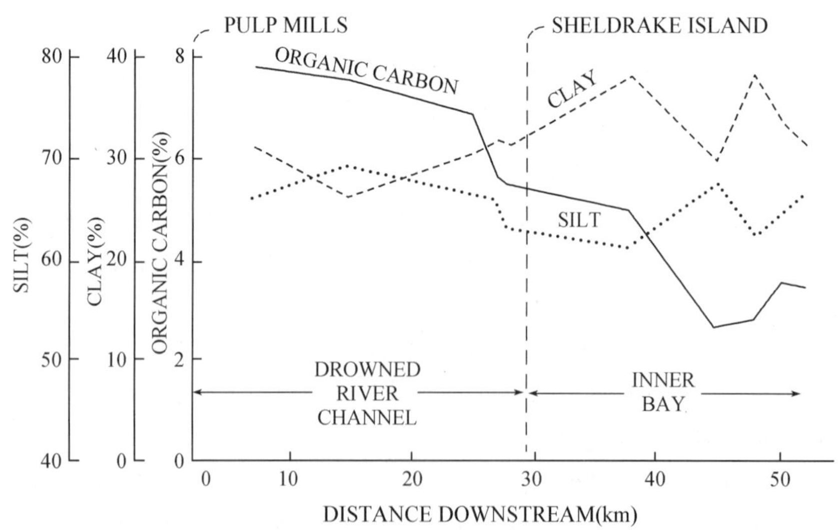

图 11　Miramichi 河口湾底质沉积物中，粉砂、黏土和有机碳含量变化剖面图

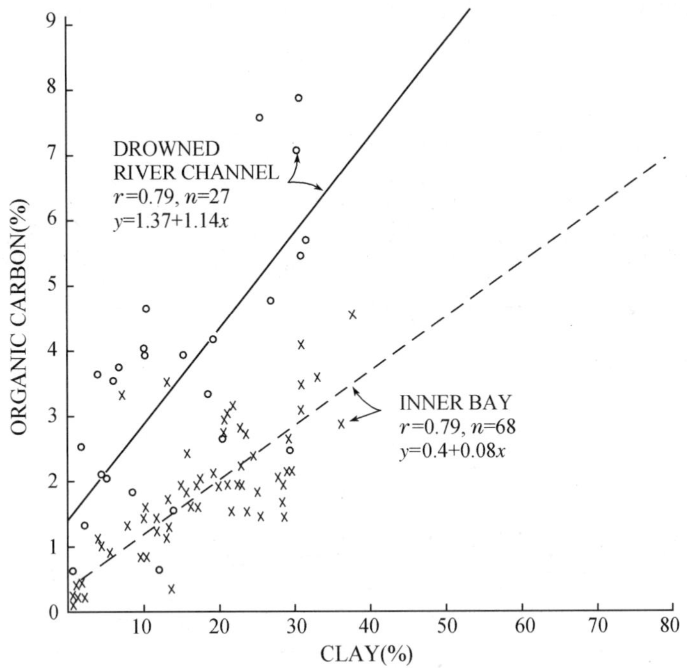

图 12　Miramichi 河口湾沉溺河道和内湾的底质沉积物中，有机碳含量和黏土百分数之间的线性回归分析

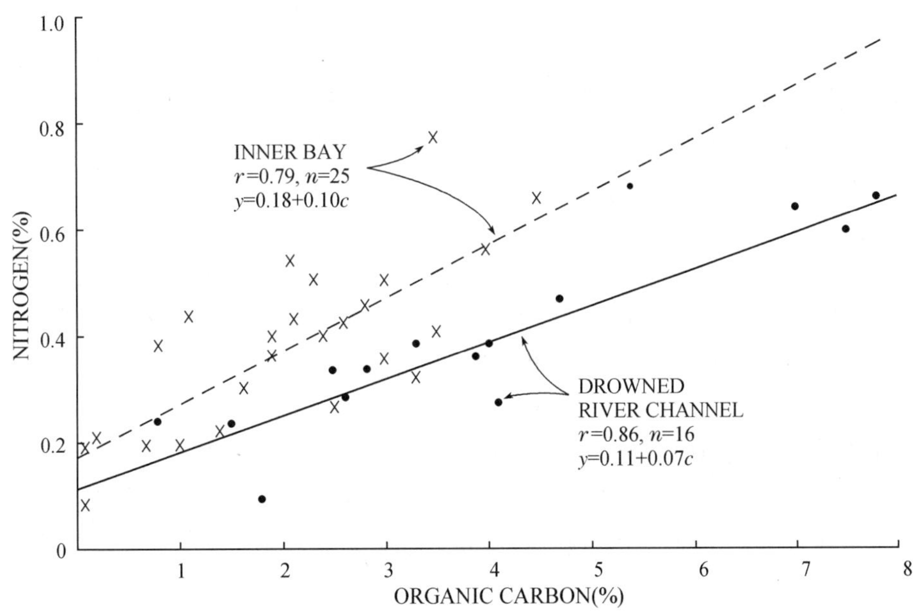

图 13　Miramichi 河口湾底质沉积物中，氮和有机碳含量间的线性回归分析

（分析结果表明，河口湾内湾段和沉溺河道段沉积物中有机物的形态不同）

在两年期间，对从河口湾中 57 个站位所采集来的水、悬浮质、沉积物样品进行了上万次的化学元素测定（图 14）。为了确定年周期的地球化学转移过程的特性，我们在一年的不同时间与不同河流流量条件下进行了采样与重复观测。由于 Miramichi 河的月平均流量变化大，可从小于 100 m^3/s 到大于 800 m^3/s，所以我们的测量和观测也不得不适应这一变化的广度。

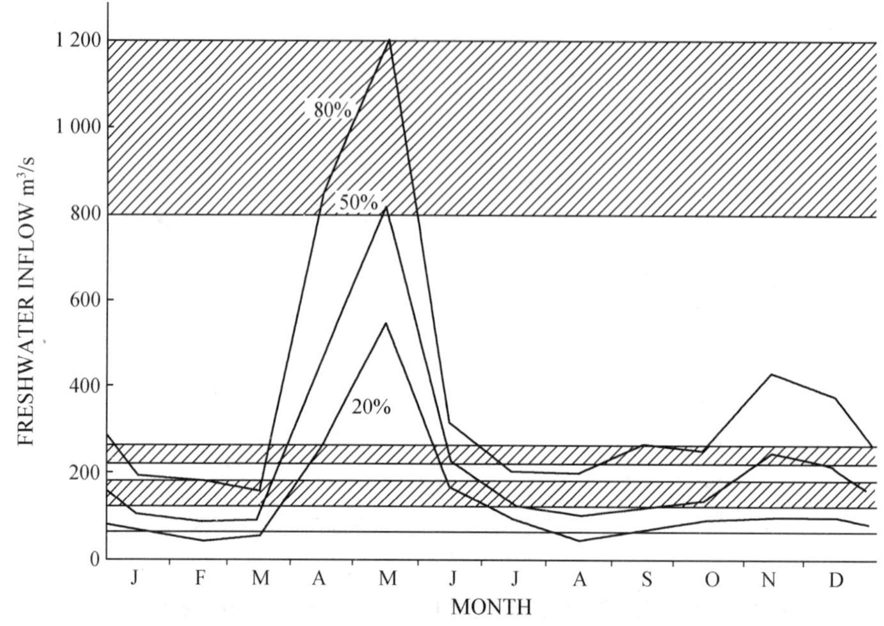

图 14　根据十年累积资料，Miramichi 河口湾平均月淡水流量图

（平行两斜线区表示已做了地球化学转移过程的研究）

悬浮质的浓度可从 100 mg/L 到少于 0.1 mg/L，这取决于在河口湾内的位置、水样的深度、潮流过程阶段以及河流的径流量。悬浮质的成分也有较大的差异，例如：悬浮质中有机碳的百分比变化可从 50%～2.5%，锰的浓度可从最大值的 7.5 mg/g 变到最小值的 0.5 mg/g。

水样中总的不稳定金属的浓度变化范围也较宽。在接近河口湾淡水终端处，不稳定 Fe（可溶性 Fe 在低 pH 值时可以在 MIBK 中被 APDC 和 DEDTC 萃取出来）的平均值是 200μg/L，对总的不稳定金属 Mn、Zn、Cu、Pb、Cd 中也有相同的趋势。

通过应用一些多变量的回归分析，我们发现，大多数金属的浓度与水体盐度具有负相关关系；但有时与悬浮质以及颗粒有机碳的浓度成正相关。这些关系随季节不同而变化，表明系悬浮质和可溶性金属间的地球化学相互作用程度不同。

各种地球化学测定值与河口湾水体中的盐度的关系采用沿着整个湾长而分布的相对浓度剖面的形式来表示。确定相对浓度因子应考虑到在河口湾任何一点的淡水分量。这个分量是由河口湾外处的本底盐度与取样点处的平均盐度之差与本底盐度之值来决定的，即

$$F_w=\frac{S_b-\bar{S}}{S_b}$$

在河水中这个比值是 1。

因此，任何地球化学的性质可以与淡水分量相比较（图 15 和图 16）。如果一个金属浓度的变化确实是与淡水分量的变化有关，那么，浓度因子的变化必定是零。同时还可以推想，绝对浓度的变化则是由于淡水和咸水间的简单混合而引起。

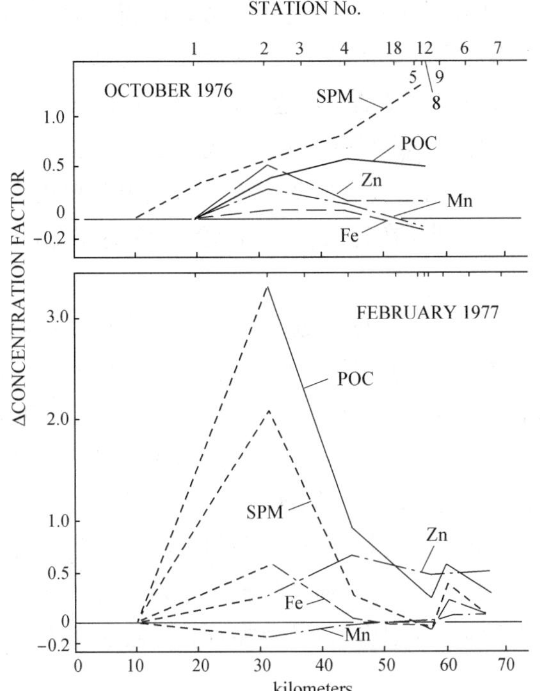

图 15　通过 Miramichi 河口湾每一观测站，各种地球化学性质浓度因子的变化与所测的淡水分量的关系
（Δ浓度因子代表变量的平均浓度因子和淡水分量的差值，因此，曲线可以描述河口湾内相对于恒定混合的偏离程度。数据表示两种低淡水流量的情况）

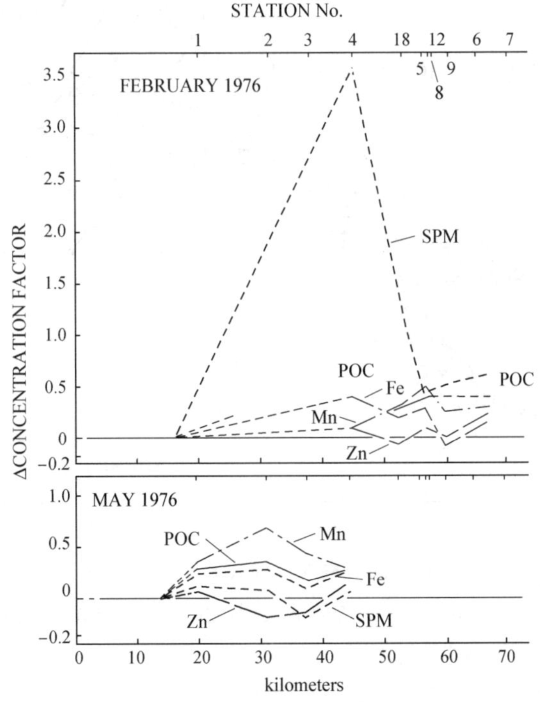

图 16　通过 Miramichi 河口湾每一观测站，各种地球化学性质浓度因子的变化与所测的淡水分量的关系
（Δ浓度因子代表变量的平均浓度因子和淡水分量的差值，因此，曲线可以描述河口湾内相对于恒定混合的偏离程度。数据表示两种高淡水流量的情况）

可以看出，悬浮质和颗粒有机碳总浓度等参数，其行径不如我们所预期的那样恒定。其恒定混合的最大偏离出现在河口湾的上部，该处由于底部沉积物被再悬浮而形成了最大的混浊度。可溶性金属恒定浓度的偏离也是出现在河口湾的上部，而这种偏离表明，在悬浮质与水体间有一些地球化学上的相互作用，或者是加入了来自另一源地的金属成分。少数样品中具有这样的迹象：可溶性金属的浓度降低到由简单混合而成的标准以下。这就指明了像 Zn 这类金属，特别是在特大春汛和高混浊度期间，由于存在某些作用而从溶液中被转移掉了。

我们采用潮棱柱充溢模式来计算通过河口湾的化学元素搬运速率(图 17～图 19)。该模式是根据这样的一个假定："进潮量取决于进入河口湾的淡水流入量，而最终咸淡水发生彻底地交换"而作出的。它把河口湾划分为一系列段落。位于湾顶的第一段确定为，在每一潮汐周期内流入湾中的淡水量都是相等的。向海侧的每一个相邻段落所特有的高潮量相等于相邻的下一段落的低潮量。因此，在低河水径流时期，模式中有 13 个段落，而各段皆有预报的盐度。根据实测资料，我们把预报的淡水分量与经实测计算出的淡水分量进行比较，结果发现实测的淡水分量通常比低淡水流情况下预报出的淡水分量高。这是因为在河口湾上段缺乏彻底地混合，以及存在垂直分层的缘故。

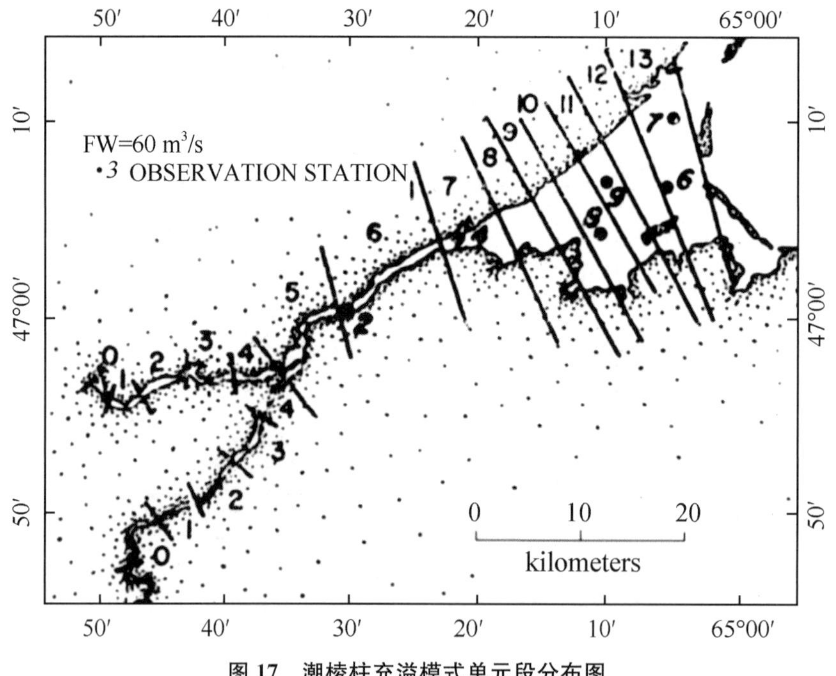

图 17 潮棱柱充溢模式单元段分布图

(淡化流量为 60 m^3/s，黑点外侧的观测站位用来获得在一个或多个潮周期内所测得的水质数据)

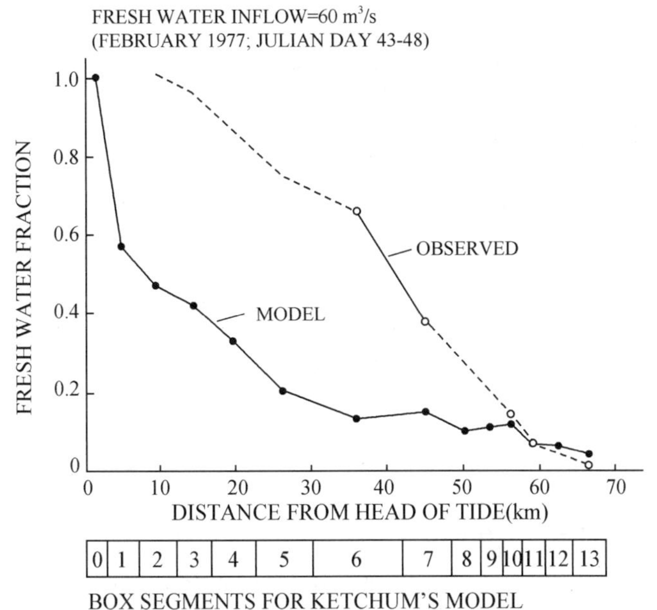

图 18　Miramichi 河口湾各站位淡水分量的测定值和推算值(根据潮棱柱充溢模式)比较图

(淡水流量 60 m^3/s)

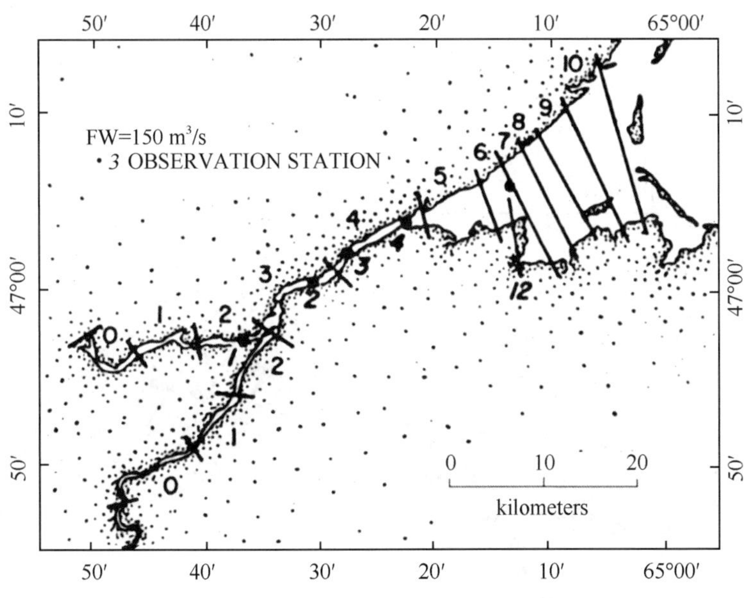

图 19　淡水流量为 150 m^3/s 时潮棱柱模式各单元段分布图

当淡水流入量增加时,模式中的段落数目减少了,而预报的淡水分量与实测的淡水分量间的相似程度增加了(图 20～图 25)。在模拟的最高淡水水流条件下,段落的数目减少到 4 个,此时预报的盐度比实测的盐度低。其次是由于在河口湾上段缺乏彻底的混合,虽然有着大流量的淡水沿着表层向海流去,但是盐水还是沿着底部而入侵到河口湾的上段。

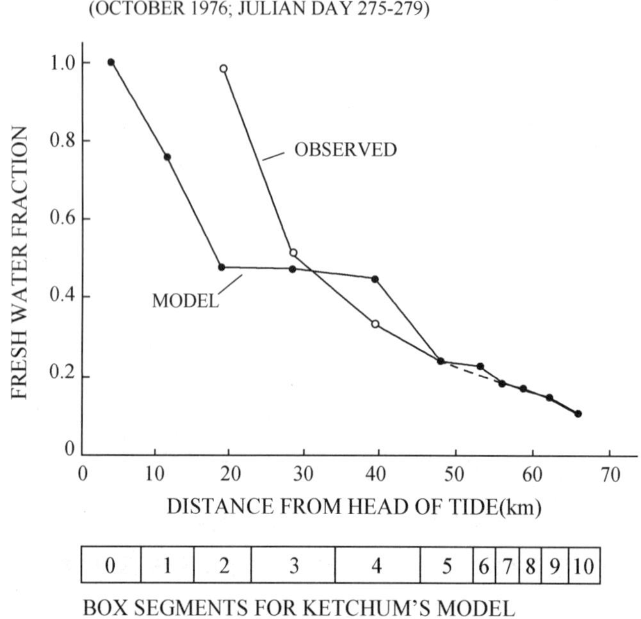

图 20 淡水流量为 150 m^3/s 时淡水分量的测定值和推算值比较图

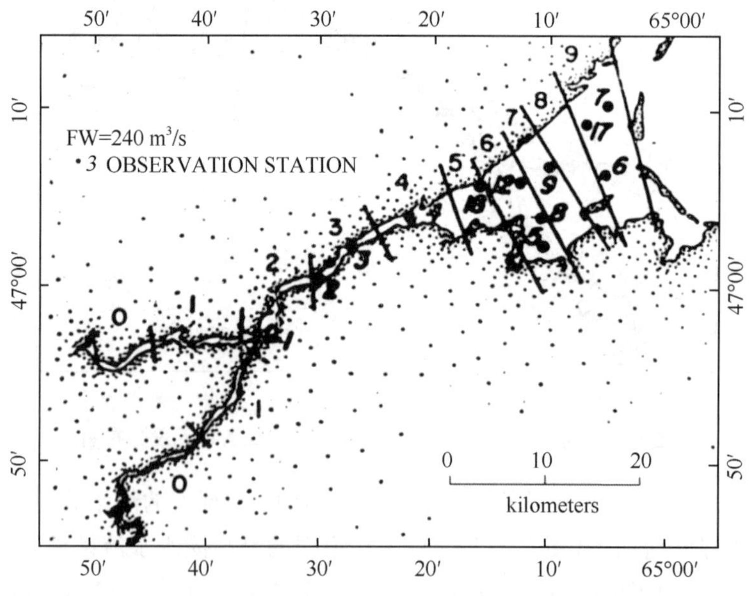

图 21 淡水流量为 240 m^3/s 时潮棱柱模式各单元段分布图

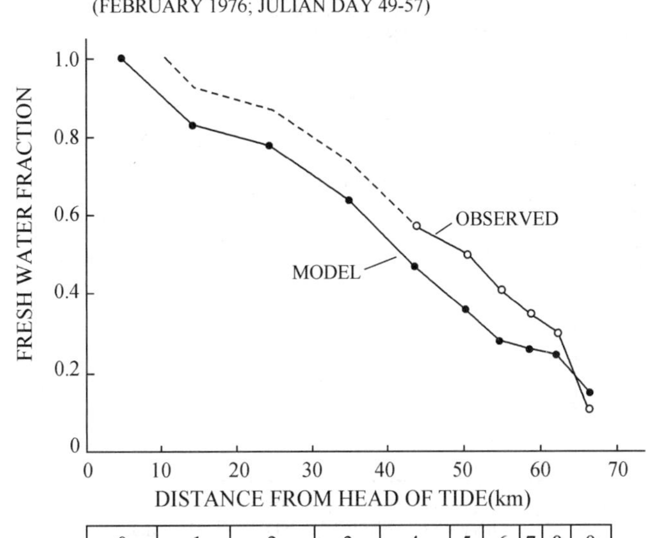

图 22　淡水流量为 240 m^3/s 时淡水分量的测定值和推算值比较图

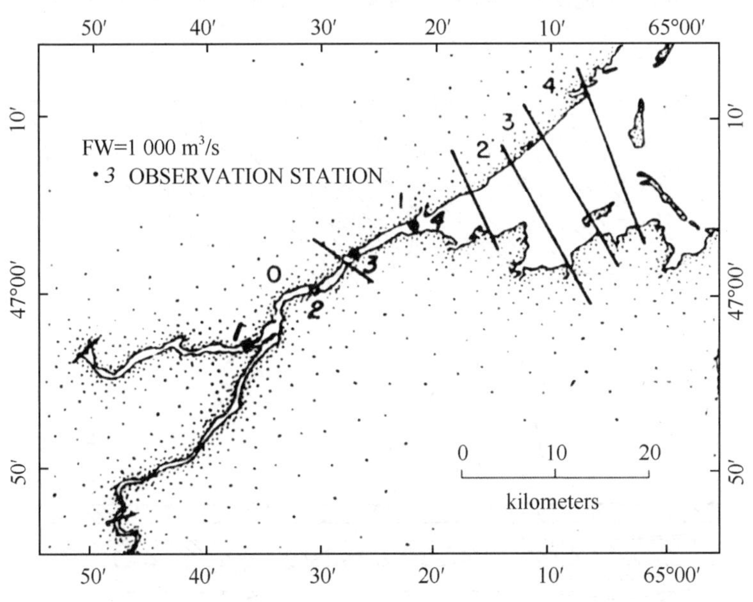

图 23　淡水流量为 1 000 m^3/s 时潮棱柱模式各单元段分布图

我们已经能够运用这些模式去确定通过河口湾各段落的任何物质的搬运速率。观测到的淡水分量乘以模式每一段落所观测到的浓度，就可求得通过任一段面的搬运量。一年内每一季节的每一种模式的应用范围，是与有着不同的优势平均流量条件的当年天数相应的。根据十年河流流量累积概率资料我们确定了：淡水流量为 60 m^3/s，平均一年出现 14 天；流量为 150 m^3/s 一年出现 101 天，而当流量增大至 1 000 m^3/s 时，一年的出现率为 39 天。

如果以通过该模式的临界段落的日流量乘以一年内可以应用模式的天数，那么我们就可以确定通过该地段的悬浮质数量和可溶性金属量。临界段面是指：河流进入河口湾

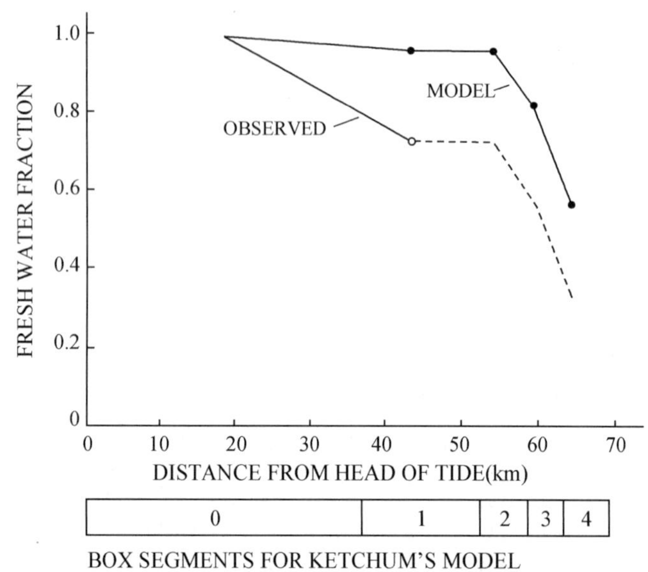

图 24　淡水流量为 1 000 m^3/s 时淡水分量的测定值和推算值比较图

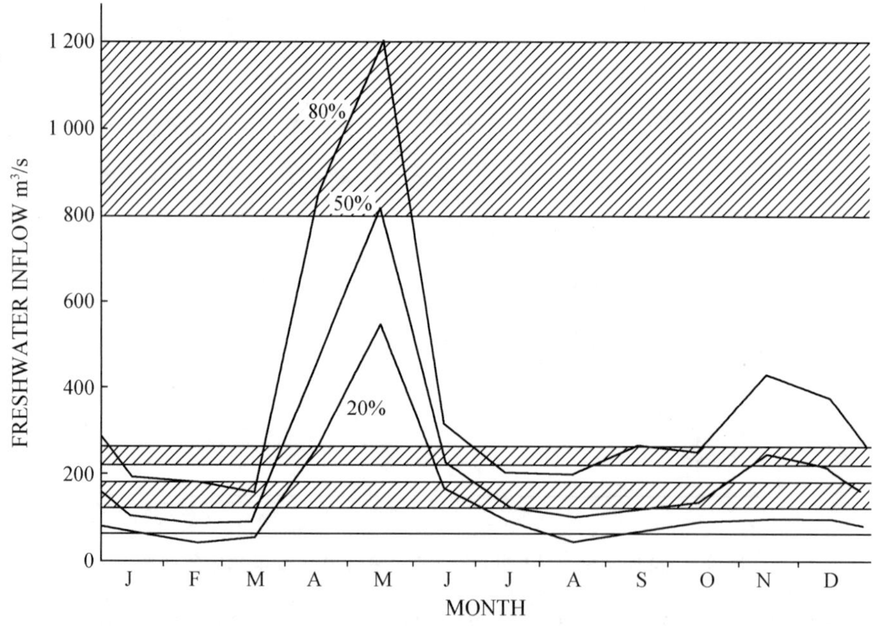

图 25　根据十年累积资料 Miramichi 河口湾平均月淡水流量图

（由所得数据可以确定每一种潮棱柱充溢模式所代表的每年的天数为：60 m^3/s 模式用 14 天；150 m^3/s 模式用 101 天；240 m^3/s 模式用 211 天；1 000 m^3/s 模式用 39 天）

处的断面；河口湾内的沉溺河段与海湾之间的断面；最后一个断面是在河口湾进入 St. Lawrence 湾处。

根据上述的运算结果，我们了解到每年有 95 600 t(10^3 kg)的悬浮物质进入河口湾，而有 93 900 t 悬浮质输送到海湾中，其中 70 750 t 是来自河口湾的(表 1)。这意味着基本上每年只有 1 700 t 悬浮质泥沙沉积在河口湾中的沉溺河段内，而 23 150 t 悬浮质泥沙沉积于海湾内。在河口湾与海湾之间，约占年输沙量的 46％是在春汛的 39 天中进行的。模式计算还表明，在春汛期间几乎与全年泥沙堆积量相当的沉积会发生在河口湾上段。在一年中的其他时期，由于底质沉积物不断地被再悬浮，从而影响了泥沙在底部的堆积。

表 1　通过 Miramichi 河口湾悬浮质和活性金属年输送量　　（公吨）

		Fe	Mn	Zn	Cu	Pb	Cd
	SPM						
RIVER	95 600	800	260	40	7.7	6.1	0.43
	PARTICULATE	4200	138	40			
RIVER ESTUARY	93 900	770	270	35	11.7		
	PARTICULATE	3 675	208	28			
INNER BAY	70 750	980	380	50	8.5		
	PARTICULATE	2 020	221	40			

当我们检查被悬浮质携带的金属搬运量与以溶液状态存在的金属搬运量时，就会注意到，进入河口湾中的悬浮质 Fe 5 倍于可溶性 Fe 的搬运量，甚至比 5 倍还大。然而，当搬运到河口湾终端时，悬浮质 Fe 的搬运量仅相当于可溶性 Fe 量的 2 倍多一点。这是由于有些悬浮质 Fe 沉积于湾底，而有些可溶性 Fe 加入了搬运系列。相反，可溶性 Mn 的搬运量却比悬浮质 Mn 的搬运量大。值得注意的是，从河流到河口湾口，悬浮质 Mn 的搬运量是增加的。在河流与河口湾的上界，悬浮质 Zn 的搬运量几乎与可溶性 Zn 的搬运量相等。但是当 Zn 元素继续向海搬运时，可溶性形态被搬运的比例就逐渐加大。因为没有定量测定悬浮质上的 Cu、Pb、Cd，所以未能算出这些金属以悬浮质形式被搬运的比例。

根据搬运量计算结果，加上底质与悬浮质组成的资料，我们做出一个关于 Fe、Mn、Zn 三种金属在河口湾内地球化学转移和相互作用的模式。我们确定进入河口湾的悬浮质 Fe 的平均浓度是 4.93％，而其总 Fe 的量的 17.6％则是活性的或者是能被弱酸提取的。当悬浮质 Fe 搬运到河口湾以外时，其浓度已减少为 4.01％，并且有 22.5％的这种金属 Fe 是活性的。最后，从河口湾输送出的悬浮质中仅含有 2.85％的 Fe，而其活性部分减少到总量的 13.2％。应用这些数据我计算了通过河口湾临界段落悬浮质上的总 Fe、活性 Fe 和残留 Fe 的比例，然后又将这些 Fe 的流通量与可溶性形态被输运的 Fe 进行比较。我们的模式表明，每年有 5 000 t Fe 自河流输送到河口湾中，其中有 800 t Fe 是可溶性的，4 200 t 是呈悬浮质形态的。而悬浮质 Fe 是由 726 t 活性 Fe 与 3 400 t 残余 Fe 组成的。在河口湾的上段，每年有 75 t 悬浮质 Fe 随着悬浮质泥沙而沉积下来。我们确定出从河口湾进入到海湾中的悬浮质中包含着 0.38％Fe，这个总 Fe 量少于汇入到河口湾中的量。我们估计是从悬浮质中的活性部分损失的，有时这些 Fe 被加入到可溶性浓度中去了。这与我前面所提到的观察结果是一致的，我们发现在河口湾上段 Fe 浓度通常大于从恒定混合所预测的浓度。在一年循环的部分期间，一些添加在溶液中的不稳定金属也被吸附于底部沉积物上。如果吸附仅仅发生在底质表层的 0.1 cm 处，就已足够算出，如前所述的从河流到河口湾底质沉积物中的 Fe 从 2.28％增加到 2.72％。

还对从河口湾输送到海湾的 4 535 t Fe 进行了类似的比较计算(图 26)。在海湾环境中,将 821 t 活性 Fe 加到可溶性金属的浓度中,并且将 611 t 可溶性 Fe 加入到底部沉积物中去时,全部 928 t 颗粒 Fe 沉积到海湾淤泥底上。因此,底部的总浓度增加到 3.35%。最后,总数为 3 000 t 的 Fe 从河口湾输出到海湾,其中有 980 t Fe 以可溶性形式输运,而仅有 266 t Fe 是以活性形式附于悬浮质颗粒,并随之向海湾搬运。

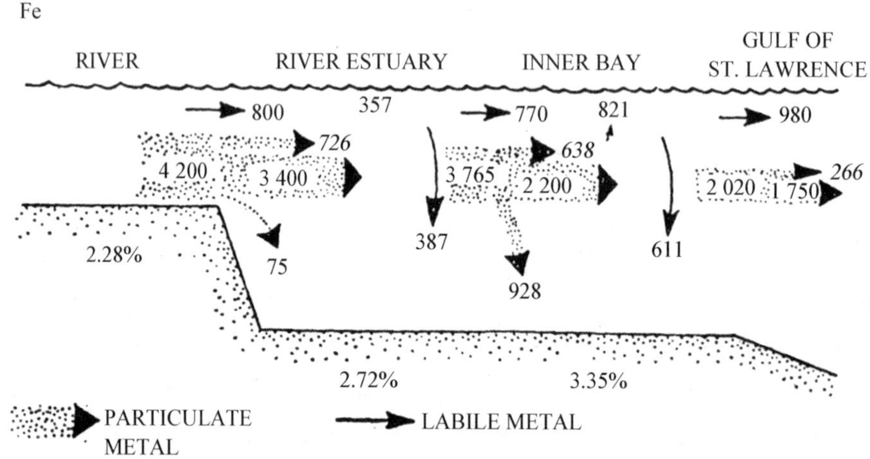

图 26　Miramichi 河口湾内颗粒 Fe 与活性 Fe 相互作用的图解模式

(输送量以每年的吨数表示,百分数表示细颗粒底部淤泥中 Fe 的浓度。颗粒物的斜体数字表示有反应倾向金属的吨数)

这个模式表明,河口湾底部沉积物是来自河流中 Fe 的沉降池。在 Miramichi 河及河口湾系统中,每年有 2 000 t Fe 被留在河口湾的沉积物中。这样大的堆积量可能是因为大量的有机物质在河口湾一般性污染条件下而加积于底部的。

用同样的程序计算的颗粒锰,我们注意到,在地球化学的相互作用间与 Fe 有明显的相反趋势。从河口湾净输出的不稳定 Mn 以及颗粒 Mn 较从河流流来的多。这说明了在河口湾内 Mn 有着重要的外加来源。

与颗粒 Fe 的显著对照是,呈悬浮质状态的 Mn 浓度较底部沉积物中的 Mn 浓度高出很多(图 27)。在河流内有 0.144%的悬浮质是由 Mn 所组成的,而在底质中 Mn 的成分是 0.07%。在河口湾内,悬浮质中的 Mn 增加为 0.222%,而底部沉积物中的 Mn 浓度减少为 0.042%。当悬浮质到达海湾时,悬浮质中的 Mn 浓度增加到 0.313%,而湾底淤泥中的 Mn 浓度减少到 0.04%。另外一个与 Fe 模式不同之处是,有更多的颗粒 Mn 为活性的。在河流的悬浮质中,总颗粒 Mn 中有 96%是活性的,而在河口湾中占 89%;到海湾时减少到 77.3%。

当悬浮质泥沙从河流中被输送堆积到河口湾时,仅有 2 t 颗粒 Mn 伴随泥沙而沉积。其余的 136 t 颗粒 Mn 又增加,从河口湾汇入到海湾中的 72 t 颗粒 Mn,必然还有 10 t 不稳定 Mn 被加入到水体中,这样才有 270 t Mn 以可溶性形式输出到海湾中。从底部沉积物中释放出的 Mn,部分是从新沉积的悬浮质泥沙中释放出的,部分则是从上部扩散到底部沉积物中去的。因为河口湾底部的淤泥大部分处于还原状态,因此氧化锰的覆盖层会被还原而使 Mn 再行活动。

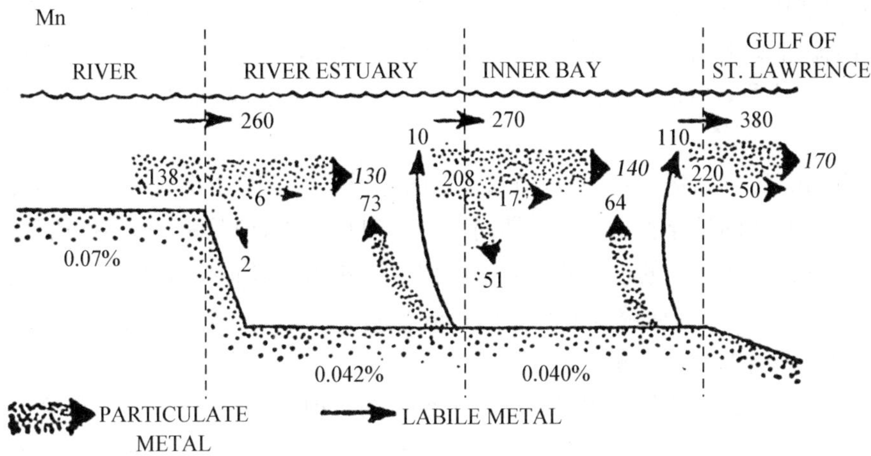

图 27　Miramichi 河口湾内颗粒 Mn 与活性 Mn 相互作用的图解模式

（输送量以每年的吨数表示，百分数表示细颗粒底部淤泥中 Mn 的浓度。颗粒物的斜体数字表示有反应倾向金属的吨数）

我们了解到有 51 t Mn 沉积到这个海湾环境中，这些 Mn 是以悬浮质形态从河口湾搬运到海湾中的。其余搬运出海湾剩下的 157 t 颗粒 Mn 必定又补充了 64 t 颗粒 Mn 才最后达到总输出量为 220 t，不稳定金属的输出量也达到了 380 t。因此，必然有 110 t 的 Mn 以可溶性形式加入到年输送量中。我们再一次估计到有净重达 150 t 的 Mn 是从海湾底部的淤泥中释放回来的，这样才达到了从河口湾输出 Mn 的总量。这就意味着最初被带入到河口湾中的 398 t Mn 又加入了 202 t 主要是自湾底淤泥所供应的 Mn。这后一种情况只有当河口湾受到污染，湾底淤泥处于还原状态时才可发生。

Zn 的相互作用模式与上述的 Fe、Mn 模式是类似的（图 28）。在河口湾内，Zn 的行径和 Fe 一样，每年以可溶性以及悬浮质的形式被运走而减少；有些可能被吸附在底质沉积

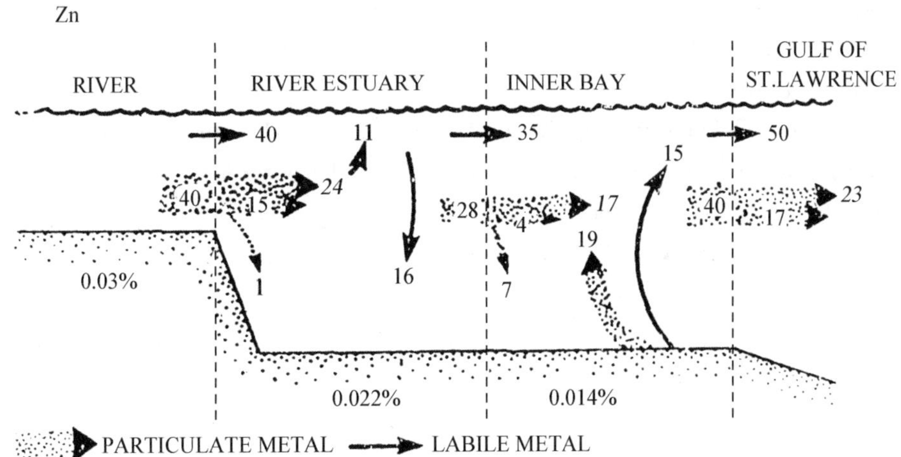

图 28　Miramichi 河口湾内颗粒 Zn 与活性 Zn 相互作用的图解模式

（输送量以每年的吨数表示。百分数表示细颗粒底部淤泥中 Zn 的浓度。颗粒物的斜体数字表示有反应倾向金属的吨数）

物上或者直接沉淀了。在海湾中，Zn 的相互作用与从底部淤泥中倒流回来的 Mn 类似。总之，从河流向入海的河口湾终端，Zn 量稍微有所增长，占增加量的 56%看来是在悬浮质上的。在河流段，活性颗粒 Zn 占 62%；河口湾中占 82%；从河口湾输出的物质中，Zn 占 58%。

在 Miramichi 河口湾体系中，从河流带下的悬浮质的 84%是发生在春汛的 39 天中。然后，大约有 45%的悬浮质是在春汛期间沉积于河口湾内，而从整个体系中输出的悬浮质泥沙，有 55%是在一年的这个期间内进行的。一年内的其他时间中，春季已沉积的悬浮质，很多又被再悬浮起而被带出河口湾上段。这种年周期的作用过程为颗粒的和不稳定金属的重新分布提供了足够的机会。

从 Miramichi 河口湾研究中得到的一个重要的教训是，如果要准确的估价一个河口湾的年流量，进行一年以上的观测工作是非常重要的。错误估算了净年流量可能是由于没有在恰当地、具有代表性的段落观测物质浓度而造成的。尤其是，当我们的观测对象是要对从一条已污染了的河流或河口湾来的污染所产生的影响作出评价时，这方面的考虑更具有特殊的意义。

主要参考文献

[1] Buckley, D. E. 1972. Geochemical interaction of suspended silicates with river and marine estuarine water. 24th International Geological Congress, Section 10, 282 - 290.

[2] Buckley, D. E. and Winters, G. V. Inpress. Geochemical transport through the Miramichi Estuary. *Canadian Journal of Fisheries and Aquatic Sciences*.

[3] Rashid, M. A., and Reinson, G. E. 1979. Organic matter in Surficial Sediments of the Miramichi Estuary, New Brunswick, Canada. *Estuarine and Coastal Marine Science*, 8: 23 - 36.

[4] Vilks, G. and Krauel, D. P. 1982. Environmental geology of the Miramichi Estuary: Physical Oceanography. Geological Survey of Canada, Paper 81 - 24, 53.

[5] Wagner, F. J. E. and Schafer, C. T. 1979. Upper Holocene Paleoceanography of inner Miramichi Bay. *Maritime Sediments*, 16(2): 5 - 10.

[6] Willey, J. D. and Fitzgerald, R. A. 1980. Trace metal geochemistry in sediments from the Miramichi estuary, New Brunswick. *Canadian Journal of Earth Sciences*, 17(2): 254 - 265.

[7] Winters, G. V. 1981. Environmental geology of the Miramichi Estuary: Suspended Sediment transport. Geological Survey of Canada, Paper 81 - 16, 12.

海洋环境研究*

近二十年来，海洋学各个领域都有显著的进展。其原因是人们越来越认识到海洋蕴藏着巨大的矿产和能量资源，而渔业资源却是有限的。与此同时还认识到海洋环境在容纳由于人类活动所加予的变化能力也是有限的。

海洋环境研究是在50年代从海岸带开始的，当时开始意识到，海岸带的许多侵蚀和淤积问题是由于在海岸和港口设施的设计与管理中缺乏经验而造成的。同时还认识到，由于土地侵蚀以及污染物被输送到海湾和河口，严重地影响到近海地域的环境质量。人们还逐渐地懂得，由于不断地向海中倾倒城市和工业废物，可能污染着整个大陆架环境。

时至今日，我们已经认识到，世界大洋可能被持续增加的工业产物和废品所改变或污染。长期以来，已经知道大气中的二氧化碳含量正在增加，目前海洋表面还能容纳这一增量，但它对于海洋最终的衰落或者说可能产生什么结果，所有的海洋学家们还没有一致的意见。广泛散布的塑料或其他合成产品，现已可在外海海面上观察到。由于海事所造成的原油溢出，对海滨造成的明显影响也为人所共知了。

或许我们已经进入一个研究海洋环境问题的新阶段。到目前为止，我们已进行了大量工作，以求确定出由于经验不足或意外事故发生后对海洋环境所产生的影响。现在我们正进行研究，即试图估计现在与将来的废物对于环境的危害性。下面的一些例子将说明这一研究课题的新进展。

现在，通常是在计划新的石油勘探与开始钻井前先进行环境效果研究，然后才能允许进行钻探活动。对于环境影响的研究，必须考虑到一些突然事故发生的情况，以确定石油可能被释放出来的量，以及石油会流过哪些地方，最后可能在何处沉积下来。危害性的估计应该考虑到，对生物资源的危害程度以及采取什么措施可以限制与排除石油。

关于向海洋倾倒废物，一些国家已经有了限制废物的数量与形式的国际协定。此外也有一些国家迫使他们自己的政府订出向海洋倾倒废物的规定。这些规定要求，废物对海洋生物必须没有毒害，并且大量的废物必须限制于预定的堆放处。

还有很多问题关系到非故意的弄脏和污染，因此要求大量的研究去估计和解决。相对于开阔外海，很多港湾和港口已被严重污染，甚至在原来的污染源已停止以后，这些地区还会在很多年内继续被污染。长效的废物例如放射物质就给我们提出了特殊的问题，因为我们对于它在海洋环境中的活性情况了解得很少。

为了说明我们所面临的这类海洋环境研究，我将介绍加拿大 Nova Scotia 所进行的两项研究。这两项研究最初都是从小河口湾或海湾开始的，但是具有不同的目标。一种情况是，海岸带未被人类活动所污染，但它却是沿岸海洋环境重金属污染源的所在地。另

---

* 王颖译：加拿大 Bedford 海洋研究所 Dale Buckley 研究员在国家海洋局第二、三海洋研究所的讲演，1982年。

一种情况是已经了解到港湾区域可能会被污染，但是不了解这种污染的范围有多大，也不知道这种污染对生物的效果如何。

多年来，已经很好地了解到海洋生物食物链的构成像个金字塔(图 1)，金字塔的基础是浮游植物，它为浮游动物提供食料，而后者又是海洋大鱼的饵料，鱼类被捕获供人类消费。在食物链的每一层生物量的比率大约是 10：1，即在低层的生物量是 10 个单位，则其邻接的上一层就是一个单位。假定有一个污染物出现于食物链的最低层，那么这个污染物的浓度可以在食物链的上下各层中以 10 倍的速度增加。浮游植物中汞污染可以说明这种情况。

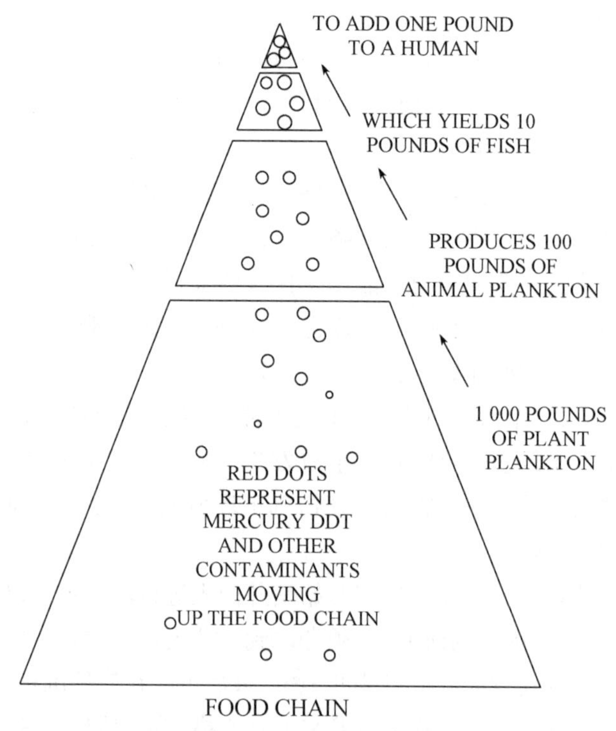

图 1　海洋生物食物链呈金字塔状

(随着食物链的消耗，污染物浓度在较高营养阶层上的图解)

1969 年，由于在两种远洋鱼类——金枪鱼和箭鱼中发现含有高剂量的汞，所以这两种鱼从加拿大鱼品市场上撤掉了。通常这些鱼类是从 Nova Scotia 大陆架边缘捕捞到的。人们不了解为什么在如此远离主要工业区和汞污染源的地方，这些鱼类含汞量这么高。在生物食物链的其他层中若含有汞浓度，就可以解释大鱼中汞含量的水平。

在贝德福海洋研究所，我们进行了多项调查，测定了从加拿大东岸一些工厂流出物中的汞浓度(图 2)。在某些情况下，我们发现在流出物的悬浮质中汞含量增高了；同时还发现，不仅在工厂流出物的邻近地区，底质沉积物中的汞含量高，而且大部分流出的水中汞量也稍有提高。然而，这些流出物在海域中的扩散似乎有限，仅达到距源地数公里的地方。几乎所有的工业源与发现有污染的远洋鱼类之处的距离都有数千公里。

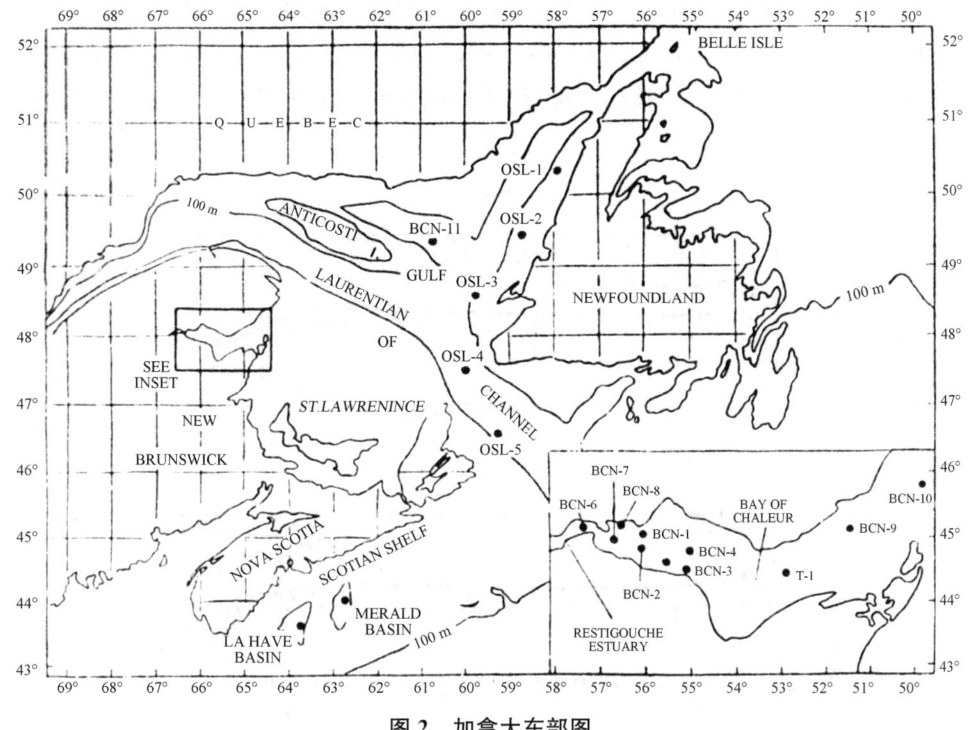

图 2　加拿大东部图

（在远离 Nova Scotia 大陆架边缘处的沿 Restigouche 河口湾有一些汞的工业污染源。La Have 河口湾位于 La Have 海盆北部 Nova Scotia 海岸）

然后我们开始研究 Nova Scotia 东南岸的 La Have 河以及河口湾（图 3），因为这一带是距离 Nova Scotia 大陆架边缘最近的陆地。这条小河流经花岗岩和前寒武纪与寒武纪变质页岩区，在页岩中有一些硫化矿物。在这条小河流域及河口湾，人口很少而且几乎没有工业发展。

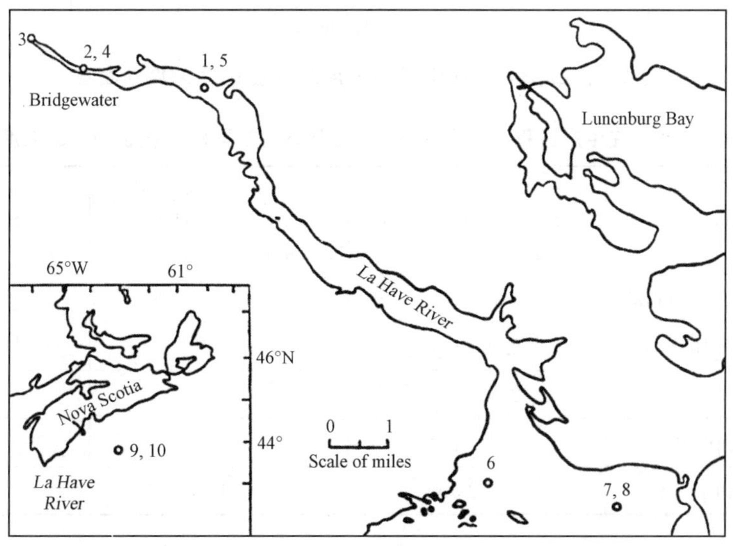

图 3　位于加拿大 Nova Scotia 东南岸的 La Have 河及其河口湾

（为分析汞所采集的地球化学样品的站位图）

小河河口湾长约 15 km，宽度小于 1 km，它面向并汇入内陆架上的一个浅海湾。从河流、河口湾以及海湾中采取了水样、悬浮质以及底质样品进行了化学分析。从经过滤的水样、浓缩的悬浮质样以及经冲洗、烘干的底质样中分别进行了总汞的分析(图 4，表 1，表 2)。

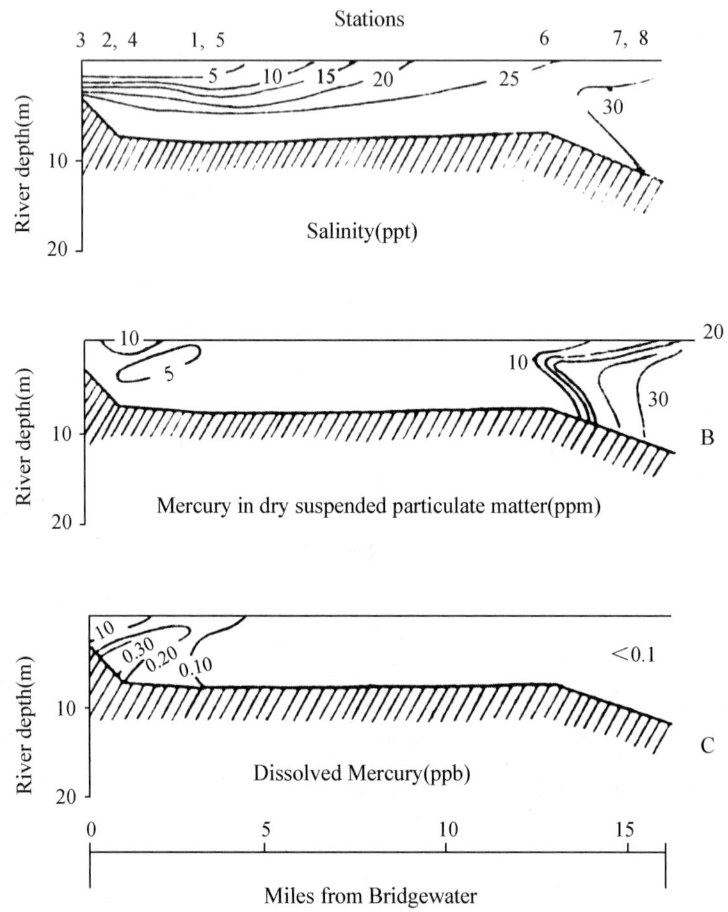

图 4　沿 La Have 河口湾水及悬浮质中盐度、汞浓度剖面图

表 1　邻近在生产过程中用汞的一些工业区水、悬浮物、沉积物中汞的浓度

Sample	Dissolved Hg(ppb)	Hg in Suspended Particulate Matter (ppm, dry weight)	Hg in Bottom Sediments (ppm, dry weight)
Paper mill (settling pond)	0.080	10.0	—
Paper mill (effluent)	2.0～3.4	5.6	1.16
Pertilizer plant	2.6～4.0	32.0	0.56
Smelting plant	2.0～4.0	—	0.71
Chlor-alkali plant	80～2 000	14.0	2.02

表 2 La Have 河从河口湾中水、悬浮质、沉积物中汞的浓度

Station	Depth (meters)	Salinity (ppt)	Dissolved Hg (ppb)	Total Particulate Matter (g/l)	Hg in Suspended Particulate Matter (ppm, dry weight)	Particulate Hg (ppb)
3(mean)	0	0.0	0.036	0.001	8.05	0.008
2, 4	0	1.9	0.090	0.001	10.9	0.01
2, 4	1.5	16.2	0.380	0.003	3.93	0.01
2, 4	6.5	26.6	0.134	0.007	7.12	0.05
2, 4(mean)		14.9	0.201	0.004	7.32	0.03
2, 4	Bottom	Sediments	0.27ppm			
1, 5	0	5.4	0.175	0.002	3.59	0.01
1, 5	1.5	8.8	0.057	0.002	6.39	0.01
1, 5	7.5	28.4	0.091	0.017	7.58	0.13
1, 5(mean)		14.2	0.108	0.007	5.85	0.04
1, 5	Bottom	Sediments	1.06ppm			
6	0	24.9	0.057	0.003	6.32	0.02
6	2.0	29.0	0.056	0.003	20.2	0.06
6	6.0	28.4	0.058	0.003	7.05	0.02
6(mean)		27.4	0.057	0.003	11.2	0.03
6	Bottom	Sediments	0.12ppm			
7, 8	0	26.5	0.054	0.002	14.0	0.03
7, 8	2.0	28.7	0.076	0.002	30.2	0.06
7, 8	10.5	29.6	0.046	0.002	34.4	0.07
7, 8(mean)		28.3	0.059	0.002	26.2	0.05
7, 8	Bottom	Sediments	0.09ppm			
9, 10	2.0	30.0	0.071	0.002	20.2	0.04
9, 10	20.0	30.3	0.088	0.002	5.12	0.01
9, 10	240	34.0	0.058	0.003	2.04	0.01
9, 10	Bottom	Sediments	0.14ppm			

河水中的汞浓度比较低，大约 0.04 ppb，我们认为是在这种地质地形河流含汞的平均数量。在半咸水中的浓度(指含盐量为 15.0‰的水体)相当高，已经达到 0.3 ppb。在河口湾向海端，可溶性浓度减少到 0.06 ppb，发现在大陆架的开阔水域海水中也具有相同的浓度(图 5)。

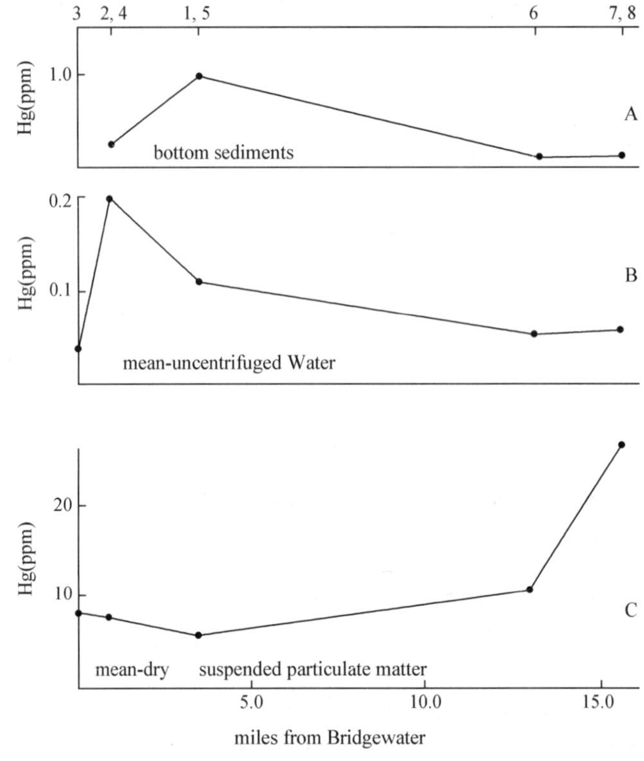

图 5　沿 La Have 河口湾在沉积物、未经离心的水及悬浮质中汞的平均浓度

被河水携带的悬浮质中汞浓度比较高，平均为 8.04 ppm。这个数量远远高于一般河流中的汞含量，甚至比污染河流中的含量还高。从河口湾上段采取的悬浮质样品中汞浓度仍然很高，仅在河口湾中部略有减低。在河口湾的含盐量达到 30‰，而悬浮质中汞浓度较湾中部超过 4 倍。这样高的含量一直延续甚至越过大陆架。

将采自河流、河口湾以及大陆架的底部沉积物进行了汞的分析，发现其含量亦较预期的要高。最高的汞含量出现在河口湾上段、河口湾的半咸水末端，浓度超过 1.0 ppm。

在检验这些数据时，我们注意到在整个区域中悬浮质的总含量都是相当低的，大约是 2～20 mg/L。这样少的悬浮质却浓集了略低于 20%汞的总输量，以致在悬浮质上的汞浓度就很高了。

为了确定为什么在悬浮质上会有这么高的汞浓度，我们对一些样品进行了粒径分析以了解在每一粒级上的汞浓度(表 3)。当我们分别分析每一部分时，我们发现 87%的汞

表 3　由 La Have 河口湾所得沉积物样的不同粒级中的汞的浓度

Diameter (microns)	Mean Specified Surface Area (cm^2/g)	Percent of Size Fraction in Sample (by weight)	Hg (ppm)	Percent of Total Concentration
>60	<400	2%	0.51	1%
20～60	600	20%	0.44	12%
4～20	2 000	49%	0.68	44%
<4	12 000	29%	1.12	43%

集中在黏土质粉砂沉积物的黏土与粉砂级上。细黏土部分含有汞总量的 43%，相当于浓度 1.12 ppm。

这个发现表明始终处于悬浮状态的细粒黏土携带着高浓度的汞进行了长距离的搬运。因此假定，La Have 河搬运的粉砂和黏土富集了汞，但当这些悬浮质进入到河口湾后，有一些粉砂和黏土沉积于半咸水的淤泥中，而细黏土仍呈悬浮状态，并吸附了水体中的汞使具有高浓度汞的悬浮物被带出河口湾而至大陆架上(图 6)。

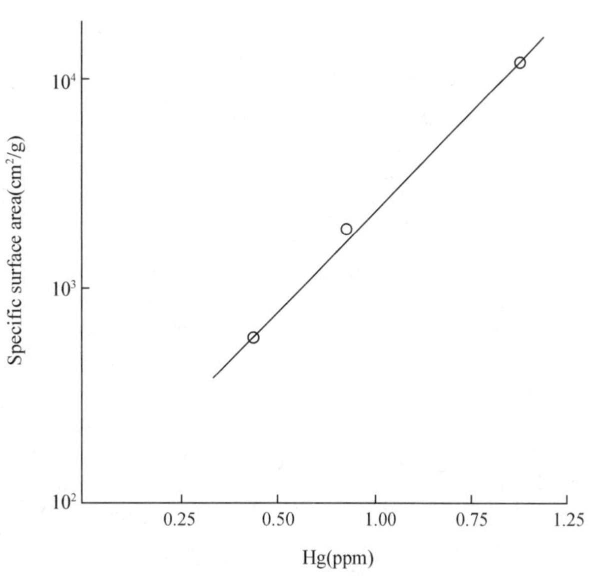

图 6　悬浮质颗粒表面积与汞浓度的关系

现在再回过头来讨论最初谈及的污染鱼的问题。后来我们发现从河口湾流出的细粒悬浮质越过大陆架，主要集中在沿着陆架边缘盆地中的雾状层内。该处是远洋鱼取食的饵料场所，当鱼捕食浮游生物时也就渗入了大量富含汞的悬浮质。结果我们就有了被污染的鱼。

这个环境问题虽然可能有些间接联系，但不是直接与现代的人类活动有关。汞源来自风化的自然矿化岩石，过去一百多年由于人类砍伐林木而使得这些岩石暴露在外。

第二个海域环境研究的例子毫无疑问地受到人类活动的影响。在这种情况下，我们进行研究以了解这一影响的严重程度。

研究区域是介于 Nova Scotia 大陆与 Cape Breton 角岛之间的 Canso 海峡。此海峡长达 40 km，宽 2 km，其中部的平均深度是 45 m。1952 年此海峡由于建立了一条公路堤坝而被阻隔。这个建筑物有效地把海峡分为两个段落。北段伸入到 St. Lawrence 湾，冬季表面开始结冰。而南半部面临的 Nova Scotia 的大西洋海岸并保持着全年不冰冻。由于人力对海洋环境改造的结果，在堤坝的南部就形成了一个理想的深水海港。60 年代和 70 年代初期，在 Canso 海峡南半部沿岸地带建了若干新工业。这些工业包括一个石膏厂、纸浆与造纸厂、重水制造厂、热电厂和一个炼油厂(图 7)。其中一些工厂生产时实际上并未对废物排入邻近的海洋加以限制，结果，从各工厂中排出的废物堆积于海岸地带以致于使得航道发生堵塞。

后来我们开始了全海峡地域的研究，测定水层与水循环的海洋条件(图 8～图 12)。观测了季节水温变化与当地天气的关系以及风况对表层水循环的影响。我们测定了海水的浑浊度，采集了水样与悬浮质样以分析这些样品的有机质、痕量元素与细菌。我们也用表层咬合取样器和取样管采集了底质样。对这些沉积样品进行了底栖生物和残留化石的研究，同样还进行了沉积学与地球化学分析。

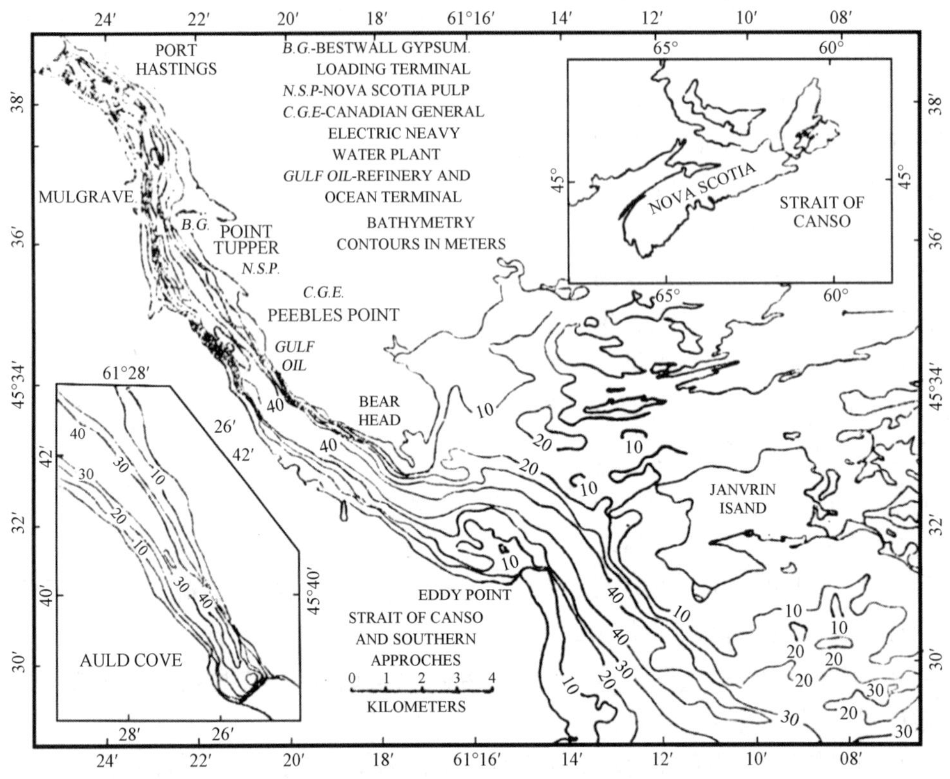

图 7 在 Canso 海峡区研究新的工业污染效应

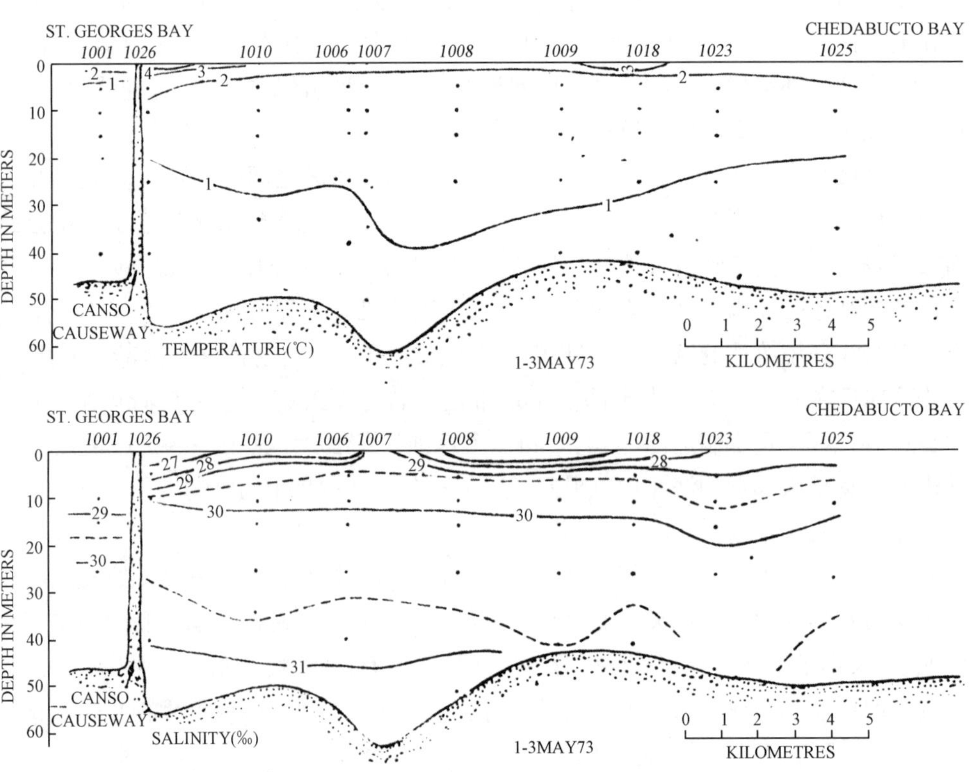

图 8 1973 年 5 月 1 日至 3 日 Canso 海峡盐度、温度剖面图

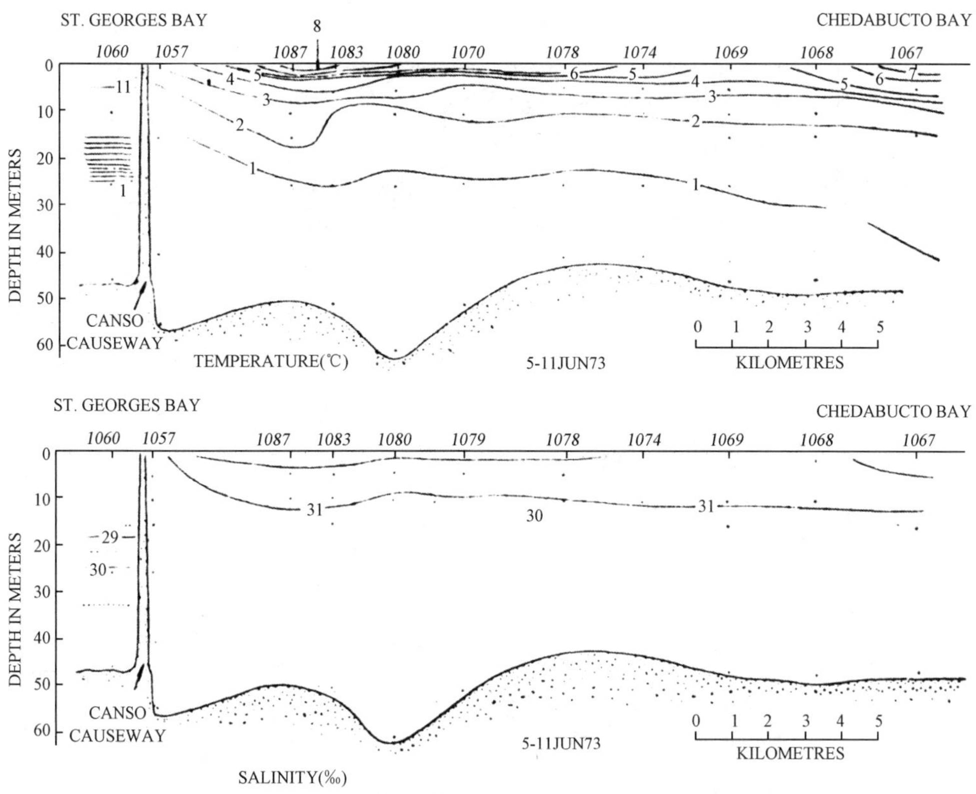

图 9　1973 年 6 月 5 日至 11 日 Canso 海峡盐度、温度剖面图

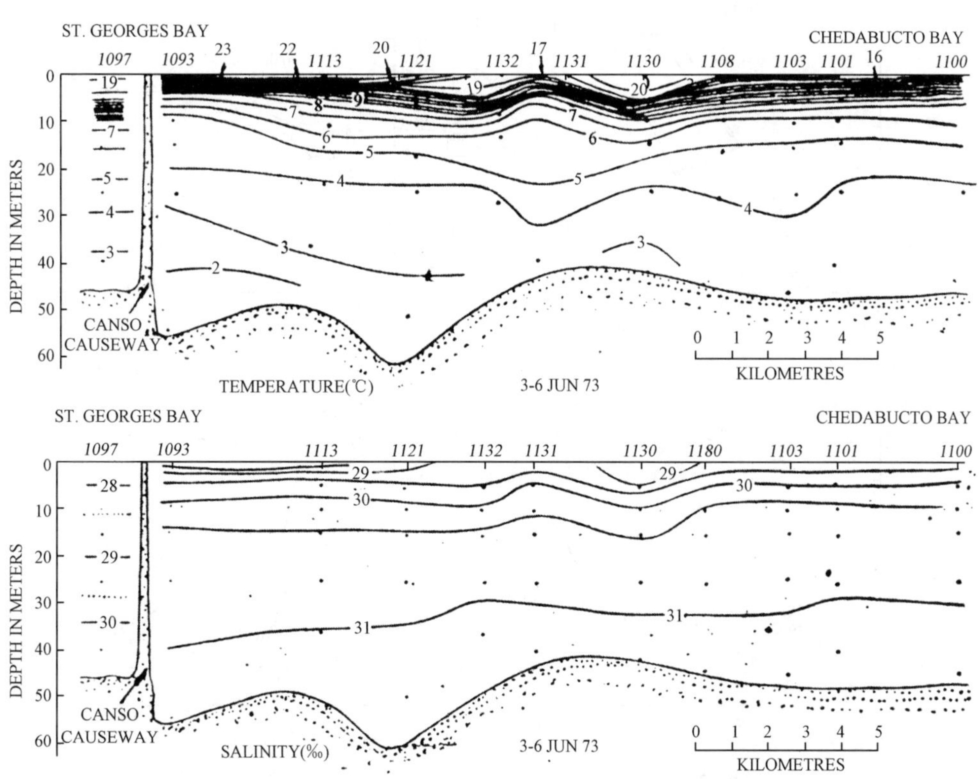

图 10　1973 年 7 月 3 日至 6 日 Canso 海峡盐度、温度剖面图

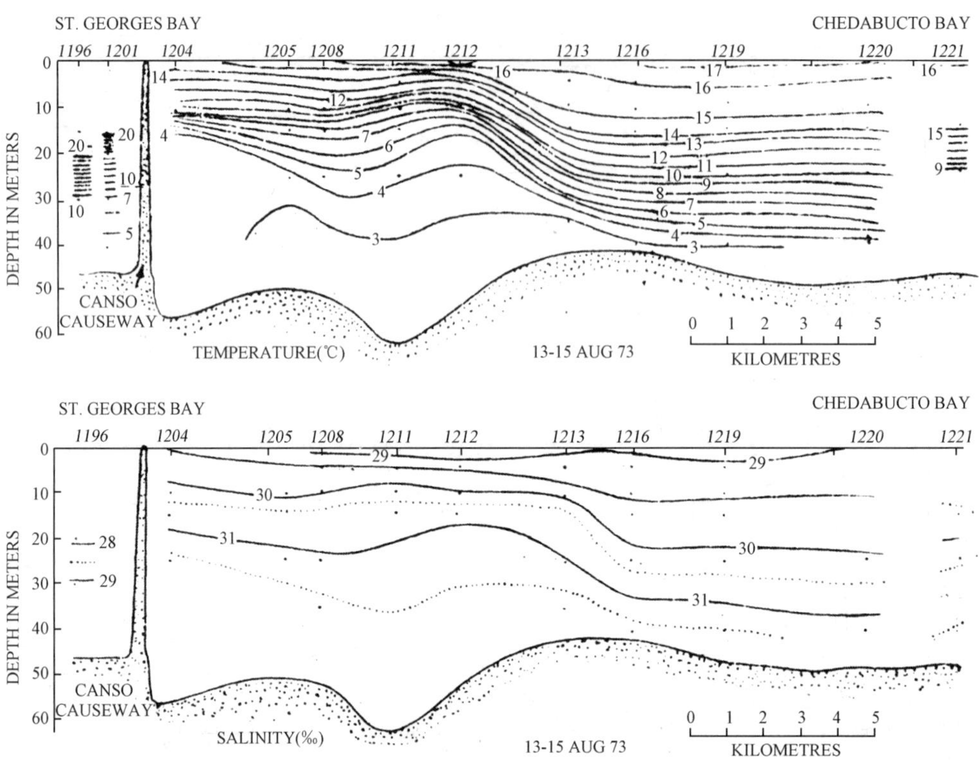

图 11　1973 年 8 月 13 日至 15 日 Canso 海峡盐度、温度剖面图

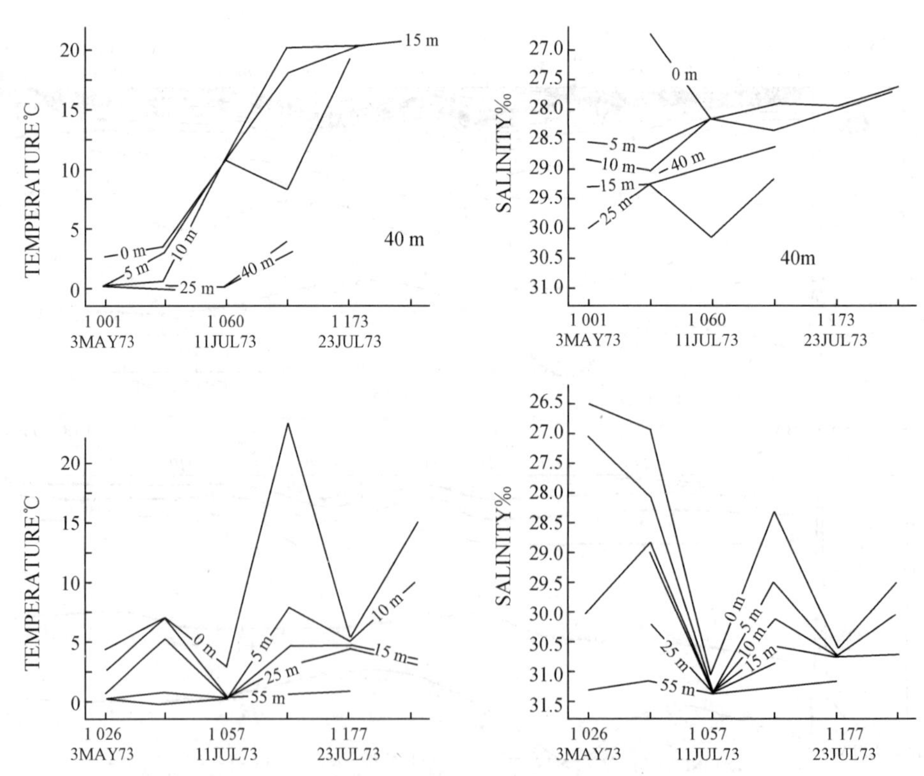

图 12　Canso 海峡堤坝两侧温度和盐度的季节变化

（图的上部指堤坝北部区域，下部指堤坝南部区域）

用激光衰减仪测定的水混浊度表明从工业化地区出来的浑浊水流，其范围沿海峡长达 4～10 km(由于风况与流况不同而致)(图 13、图 14，表 4、表 5)。悬浮质中表明有高达

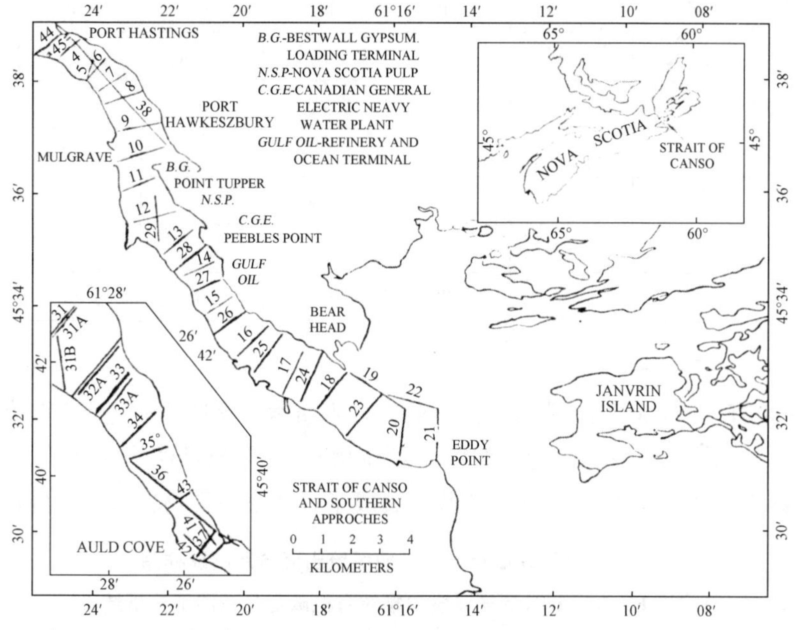

图 13　Canso 海峡多学科研究调查区域

(工业区位于海峡的东北部)

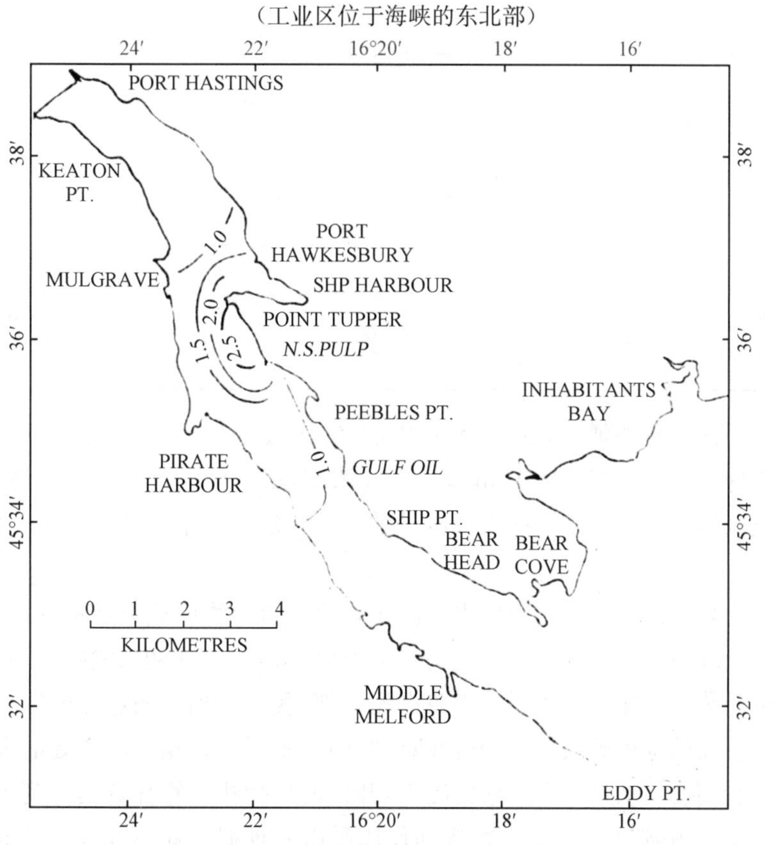

图 14　Canso 海峡工业区的悬浮质浓度(用激光衰减仪测定)

(等值线表示悬浮质浓度 mg/L)

表 4 取样点与工业区流出液的距离随样品中可溶性 Mn、Fe、Zn、Cu 浓度以及总悬浮质和颗粒有机碳浓度变化的相关矩阵

	XKm	SPM	Mn	Fe	Zn	Cu	POC
XKm	1.000	—	—	—	—.60	—	—
SPM	—	1.00	.58	.70	.49	—	—
Mn	—	.58	1.00	—	.83	.62	—
Fe	—	.70	—	1.00	—	—	—
Zn	.60	.49	.83	—	1.00	.61	—
Cu	—	—	.62	—	.61	1.00	—
POC	—	—	—	—	—	—	1.00

90%的物质是有机质，从纸浆、造纸厂流出的富含木质素。这些物质还含有富集的过渡金属和重金属。Mn、Fe、Zn 和 Cu 也被吸附于颗粒物质上，但当悬浮物质从流出源移走时，Mn 和 Zn 易被溶解在海水中。因此，这些金属成了水柱中污染物扩散的良好示踪剂。

表 5 Canso 海峡取样点与工业区流出液的距离随样品中可溶性 Fe、Zn、Cu 以及悬浮质和颗粒有机碳变化的因子分析结果

Factor→	1	2	3
XKm	—	.97	—
SPM	.87	—	—
Mn	.65	—.50	—.40
Fe	.88	—	—
Zn	.48	—.64	—.50
Cu	.59	—	—.54
POC	—	—	—.89
Total Problem Variance	37%	24%	22%

在紧邻工厂区并再外延 0.5 km 处，水中的 Zn 浓度至少 10 倍于自污染海峡向外数十千米处的本底浓度。距工厂区 1 km 处的 Zn 浓度，虽然已经经过强力的海洋混合，但仍较本底浓度高出 5 倍。距工厂区 10 km 处的 Zn 浓度，仍然至少是通常本底浓度的 2.5 倍(图 15)。

鉴定了从海峡地区沉积岩芯中取出的沉积物表明流出物已造成严重的污染，至少在底质的上部 20 cm 中含有强还原的有机物与沉积物(图 16)。在初次检定时，这些沉积物皆发出强烈的硫化氢气味。在表层的黑色淤泥下部，我们见到了弱还原的绿色淤泥层与经过生物高度扰动的棕色砂层。人们曾注意到，距严重污染区相当距离处的大部分底部淤泥层全被生物高度扰动过。在某些情况下，开口的生物扰动管孔竟延伸到表层以下的 3 m 深处。后来了解到这么深的生物扰动作用是由一种掘穴虾 Axius Seratus 造成的。它适宜居于具有有机物碎屑的污染区边缘或在有大量细菌有机质自然降低的地区。

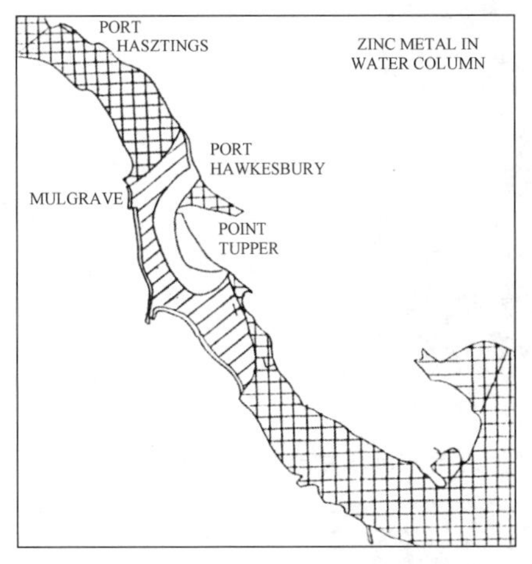

图 15　Canso 海峡工业流出液周围水体中可溶性 Zn 分布图

（紧接流出液区 Zn 为本底浓度的 10 倍，第二区 5 倍，第三区则为 2.5 倍）

图 16　Canso 海峡污染区沉积物岩芯样

（表层上部含还原性的有机物及黑色淤泥）

对沉积样品进行有孔虫含量鉴定时发现大部分严重污染物的沉积物几乎没有任何活的有孔虫品种。然而，在距离流出物区一定距离处发现了有一种活的有孔虫；在污染源外 2～5 km 处，发现有 15 种活的有孔虫生活在海底淤泥中。超过 15 km 以外，活的有孔虫种数增加到 50 种。由于整个海峡的海洋状况，在任何时候都是比较类似的，因此，活的有孔虫种属数目的不同基本上反映了当地的污染强度。

根据分析沉积岩芯的结果，我们能够确定这个海峡过去的环境状况。在有些情况下，较长的沉积岩芯（岩芯长可达 1.5 m）的沉积物是在数百年或是数千年中形成的。确定沉积物年龄可以根据保存于岩芯中的孢粉，或者是原子武器的裂变产物（^{90}Sr、^{137}Cs），有的还用碳酸盐贝壳进行^{14}C 定年。根据这些岩芯样，可以证实，这个海峡曾供养了多品种的动物群落一直到 1950 年，从 1950 年到现在，很多品种急剧减少，有些底栖生物甚至完全消失。在一些沉积岩芯的表层，我们注意到有机物和过渡金属的浓度显著增加，这种情况明显地与海峡的污染状况有关。

除了对污染反应敏感的底栖有孔虫的数量进行了分析外，我们还分析了介形虫与软体动物数量，结果发现这两种底栖无脊椎动物对污染的敏感程度较有孔虫更高；介形虫的缺失带较有孔虫与软体动物更大；而软体动物的缺失带则较介形虫小而较有孔虫大。因此，根据这些底栖动物生活在污染的底质沉积中的能力就可以建立对环境污染敏感性的指标。根据底栖动物的生存能力所反映出的污染程度，给出了一个分布图示，此图示表明在海峡内污染物分布的综合结果。底质中高于平均浓度的 Zn 的分布也能表明污染强度。因此，事实上含 Zn 浓度大于通常本底浓度 2.5 倍的区域也正是没有活的软体动物的区域（图 17～图 25）。

图 17　被掘穴虾高度扰动过的绿色和棕色淤泥混合的沉积物岩芯样

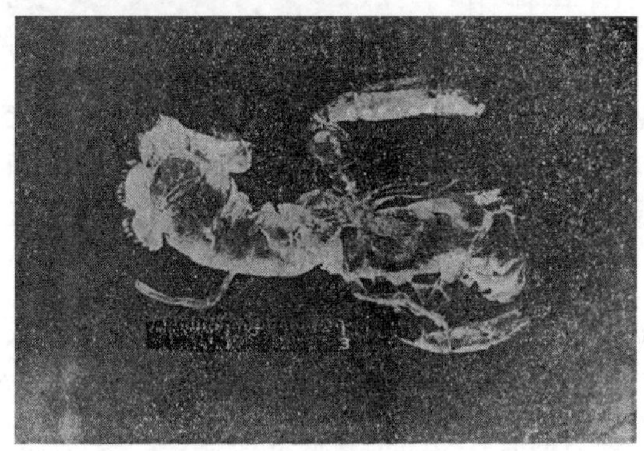

图 18　掘穴的“超级虾”Axius Seratus. 它能制造开口的穴孔深至沉积物下部 3 m 处

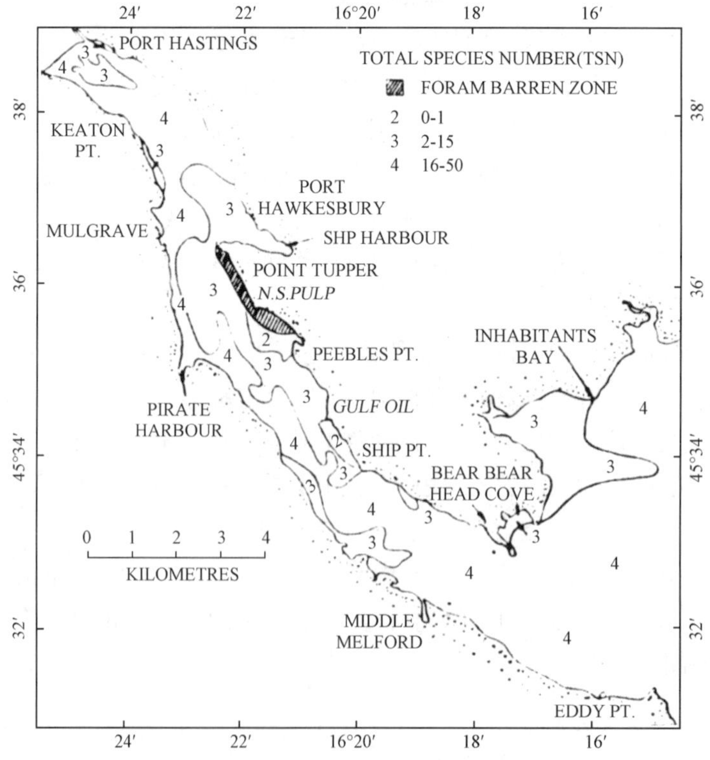

图 19　Canso 海峡底部淤泥中活有孔虫总的种数

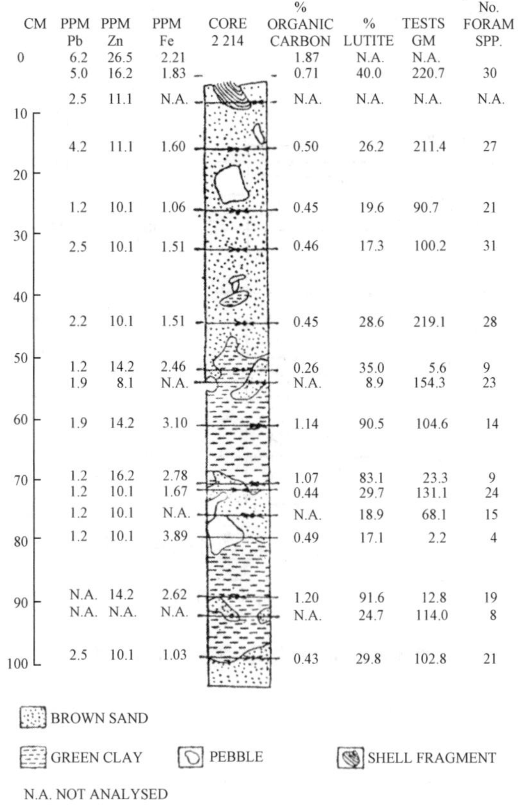

图 20　在 Canso 海峡邻近污染区所取的重力岩芯样

（注意到表层中金属含量较高但无有孔虫。表层以下几厘米古环境条件即趋正常）

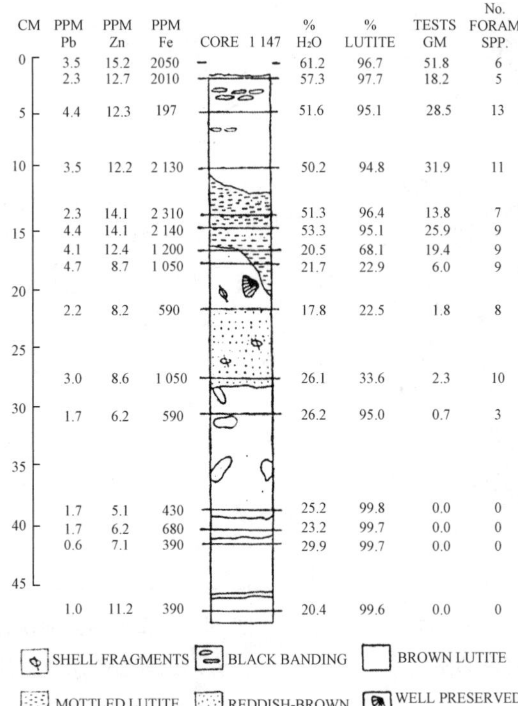

图 21　在 Canso 海峡邻近污染区所取重力岩芯样

（在被污染的表层淤泥中金属浓度稍高，而有孔虫种数稍有减少。深 37 cm 处是 13 000 年老的冰期浅水沉积）

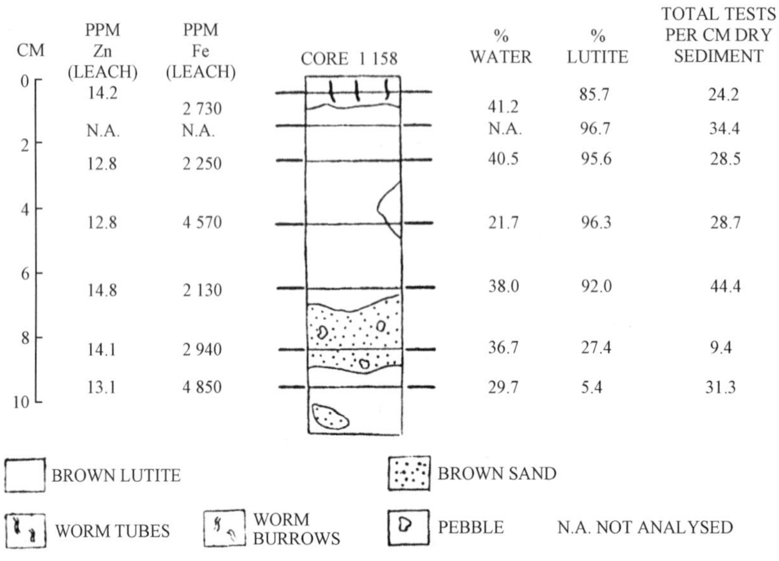

图 22　在 Canso 海峡未受污染区所取重力岩芯样

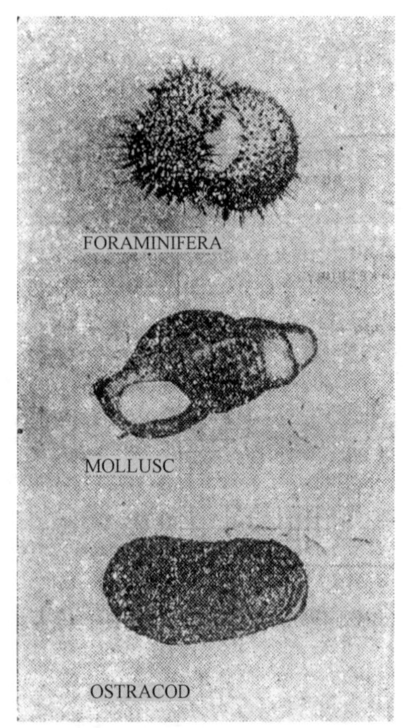

图 23　可指示受污染强度的动物。有孔虫、软体动物和介形虫的照相

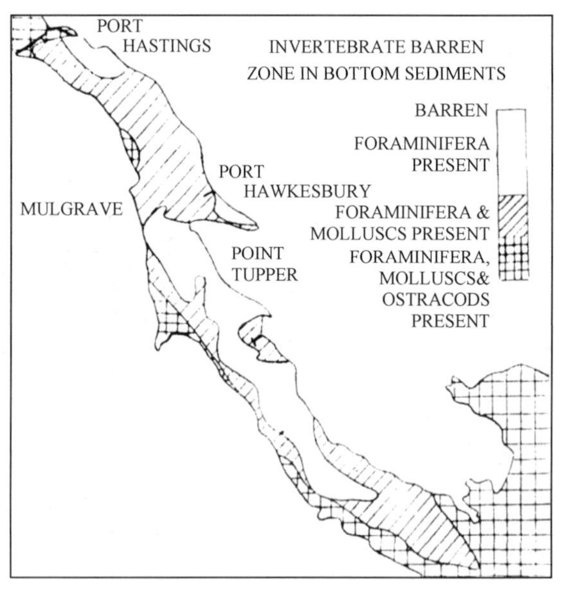

图 24　因工业污染而使有孔虫、软体动物、介形虫等活种无法生存的区域分布图

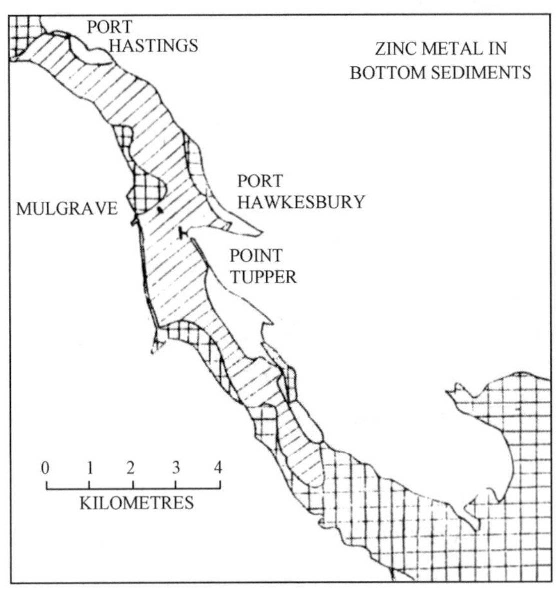

图 25　底质沉积物中 Zn 的分布图

（邻近工业流出液处 Zn 的浓度高于本底 2.5 倍，该处没有发现活的无脊椎动物）

从这个滨岸海洋环境的研究中可以得到一些重要的教训。即使工业的集中程度很小，开办的时间也很短，但是由于缺少对废物排入海洋中的控制，结果还是产生了明显的影响。在特别的污染物如过渡金属或有机质与造成了底栖海洋动物种属减少或消失之间还没有建立起直接的相关联系，但是，我们能够运用这些数量上的动态特征来作一个污染强度的指标。通过较多的污染毒性效果的研究以便能有一个标志污染最佳效果的更好办法。

这类环境研究的另一方面是我们要正确地了解和预期海洋污染的效果。我们必须了解那些把污染物从源地搬运出去或扩散开来以致对自然资源产生危害的一些作用过程。当我们正确地了解这些作用时，我们就可以用模式来表现出这些有毒物质是怎样移动的、怎样堆积的以及怎样与沉积物和生物界进行相互作用。这些模式可以考虑到水力循环、潮流动力以及天气影响等等变化状况。把这些作用包括到海洋模式中，并与地球化学以及沉积作用过程结合在一起，我们就可在污染发生前给以有效的制止，以避免恶劣的后果。目前这一方法已被发展运用于许多海洋环境以及需要解决的各种问题的研究之中。

参考文献

[1] Buckley, D. E., Owens, E. H., Schafer, C. T., Vilks, G., Cranston, R. E., Rashid, M. A., Wagner, F. J. E. and Walker, D. A. 1974. Canso Strait and Chedabucta Bay: A multidisciplinary study of the impact of man on the marine environment. In Offshore Geology of eastern Canada, Pelletier, B. R. (ed), Geological Survey of Canada, Paper 74 - 30, 1:133 - 160.

[2] Vandermeulen, J. H., Buckley, D. E., Levy, E. M., Long, B. F. N., McLaren, P. and Wells, P. G. 1978. Preliminary report-Bedford Institute of Oceanography scientific visit to AMOCO CADIZ

spill site. Spill Technology Newsletter, Environmental Emergency Branch, Department of Fisheries and Environment, 3(2): 7 - 13.

[3] Vandermeulen, J. H., Buckly, D. E. [Eds.]. 1980. The Kurdistan oil spill March 15, 1978. Activities and observations of the Bedford Institute of Oceanography Scientific Response Team. BIO Report Series, BI-R - 80 - 2.

[4] Cranston, R. E. and Buckley, D. E. 1972. Mercury pathways in a river and estuary. *Environment Science and Technology*, 6: 274 - 278.

[5] Pemberton, G. S., Risk, M. J., and Buckley, D. E. 1975. Supershrimp: Deep bioturbation in the Strait of Canso, Nova Scotia. *Science*, 192: 790 - 791.

[6] Risk, M. J., Venter, R. D., Pemberton, S. G. and Buckley, D. E. 1978. Computer simulation and sedimentological implications, burrowing by *Axius serratus*. *Canadian Journal of Earth Sciences*, 15(8): 1370 - 1374.

[7] Vandermeulen, J. H., Buckley, D. E., Levy, E. M., Long, B., McLaren, P. and Wells, P. G. 1978. Immediate impact of AMOCO CADIZ environmental oiling: Oil behavior and burial and biological aspects. *Centre National pourl' Exploitation des Oceans*, *Actes de Colloques*, (6): 159 - 174.

[8] Vandermeulen, J. H., Buckley, D. E., Levy, E. M., Long, B. F. N., McLaren, P. and Wells, P. G. 1979. Sediment penetration of AMOCO CADIZ oil, potential for future release, and toxicity. *Marine Pollution Bulletin*, 10: 222 - 227.

[9] Cranston, R. E. 1974. Geochemical interaction in the recently industrialized strait of Canso, Proceedings of the International Conference on Transport of Persistent Chemicals in Aquatic Ecosystems, Ottawa, Canada, I - 59 - I67.

[10] Cranston, R. E. 1975. Accumulation and distribution of total mercury in estuarine Sediment. *Estuarine and Coastal Marine Science*, 4: 695 - 700.

[11] Owens, E. H., and Rashid, M. A. 1976. Coastal environments and oil spill residues in Chedabucto Bay, Nova Scotia. *Canada Journal of Earth Sciences*, 13(7): 908 - 928.

[12] Schafer, C. T., Wagner, F. J. E., and Ferguson, C. 1975. Occurenc of foraminifera, molluscs, and ostracods adjacent to the industrialized shoreline of Canso Strait, Nova Scotia. *Water, Air, and Soil Pollution*, 5: 79 - 96.

[13] Vilks, G., Schafer, C. T. and Walker, D. A. 1975. The influence of a causeway on oceanography and foraminfera in the Strait of Canso, Nova Scotia. *Canada Journal of Earth Sciences*, 12(12): 2086 - 2102.

The Construction of Advanced Coast Laboratory Series Through International Cooperation and the Laboratory Own Efforts*

The establishment of the laboratory has been through three stages from initial stage of Coastal Geomorphology & Sedimentology Laboratory during 1963 to 1984; secondary stage with significant support from International Development Research Centre of Canada (IDRC) from 1985 to 1991; the third stage completely formed a new laboratory series by World Bank Loan during 1990 to 1994. It was officially named as State Pilot Laboratory of Coast & Island Exploitation by the State Planning Commission of China, State Education Commission of China and State Scientific & Technology Commission of China in 1990 as one of the seven State Pilot Labs in "Key Studies Development Project" founded by World Bank.

There are 20 formal members in the lab, including a member of Academic Science of China, 3 professors, 5 associate professors, 1 senior engineer, 4 of lecturer and engineer and 6 technicians. 34 post graduate students including 16 students study for Ph. D, 17 students for Master Degree and one Post Doctorial. Under the philosophy "Opening, Flowing, Uniting", the SCIEL has 26 guest or visiting Professors, seven of them are from overseas.

First point of our operating experiences, through the pilot project, is with State-of-the-art equipment and the organizational pattern, which makes the lab to be one of the most advanced labs in the world on coastal science. The establishment of SCIEL has been proved to be a great success in practice through self-support and international contribution, with little Chinese government financial component.

The State Pilot Laboratory of Coast & Island Exploitation is a new type of laboratory series in Geo and Ocean Sciences. It studies the natures of environment and resources of coast, island and continental shelf for improving regional sustainable development and advanced education on Global Changes. It is an institution of applied fundamental sciences.

The present conception of Coastal Ocean is in the wide meaning including coast, continental shelf, continental slope and the continental rise, which covers whole area of the land and ocean interaction. Dynamic processes of the interaction zone are active,

* Ying Wang:撰写于 1995 年 3 月 31 日。

changing rapidly, with the close relationship to the surfaces of lithosphere, hydrosphere, atmosphere and biosphere. Coastal zone has abounded and regeneratable resources: wind and marine energy, sand placer and mineral resources, tidal flat, land and embayment resources, food and tourist resources. To exploit the coastal environment and resources, it needs sufficient work on survey, study and planning, according to the nature of coastal evolution with the view point of systematical science to provide technical direction and assisting for strategy and regulation making. The reason is avoiding over exploitation to destroy natural ecological balance and routine a sustainable development of coastal zone. Sea island has the same nature as sea coast. There are 6 000 islands in China with the total length of coastline 14 000 km. It is valuable and important resources of China for the economic and transportation development, oceanic weathering forecast, and marine territory defence. The specific study object, purpose and functions have determined the laboratory with multidisciplinary features of marine, geology, geography, biology and technology sciences. It is a new type of laboratory of Geo and Ocean sciences.

The laboratory consists of three parts, the structure can be illustrated as figure 1.

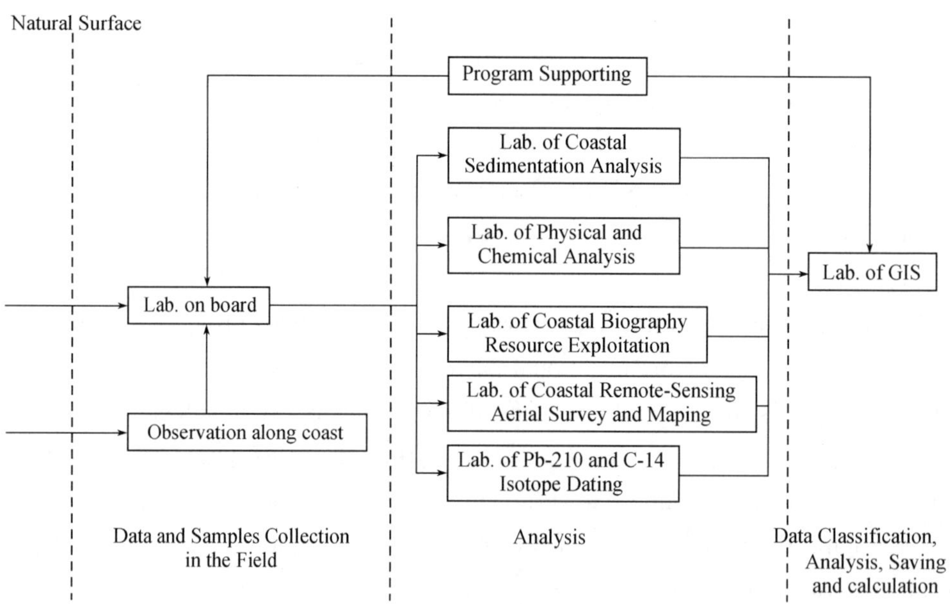

Figure 1 The structure of SCIEL

The first part is **the lab of Coast Ocean Survey.** All equipments are high-tech of acoustic, optical and electronic systems, larger in size and very expensive. Such as *Geopulse Seismic Profiler*, *Side Scan Sonar*, *Dual Frequency Echo Sounder*, *Global Position System*, *C. T. D. System*, etc. These are key instruments for surveying the environment of coastal and continental shelf. The lab of coast ocean survey has been filled up the gap of marine labs in China. It was provided by International Development

Research Centre of Canada (IDRC), through Sino-Canada joint research by SCIEL and Atlantic Geoscience Centre of Bedford Institute of Oceanography Canada. The new instruments and an isotopic dating laboratory Pb^{210}, Po^{210} and $Pu^{238-239}$ for precisely estimating deposition rates, the total value of the Canadian contribution is 400 000 U. S. dollars. With the additional financial source by World Bank Loan, the value of coastal ocean survey equipment in SCIEL reaches 430 000 U. S. dollars, which makes the SCIEL jump to the frontier rank of coastal ocean labs in the world.

The second part of the SCIEL is **the Analyses Series** for water, sediment and biological samples collected from field, including:

(1) Sedimentary labs (size, mineral, microfossils, sedimentary structure, etc.);

(2) Marine physical and chemical analysis (water quality, temperature, salinity, density, wave and current calculation);

(3) Coastal biological analyses;

(4) Isotopic dating laboratory (Pb^{210}, C^{14}, etc);

(5) Lab of remote-sensing serial survey and mapping.

The sedimentary laboratory was set up in 1980's through our own efforts to accumulate the money from contract projects. Then, it was reformed by using computer system in 1993 with the help from our Canadian colleagues. All of the value is equal to 400 000 U. S. dollars. The isotopic dating lab of Pb^{210} was set up completely by Canada; the computer system in coastal biological lab was set up from the World Bank Loan, and the water quality analyses equipment, of which the total value is 240 000 U. S. dollars, was contributed by Mr. Wang, Chao-liang, an alum from Hongkong.

The third part of the SCIEL is **the lab of Marine-Geographical Information System** provided by the World Bank Loan of total 130 000 U. S. dollars, for all data classification, saving, calculation, analysis and modelling. It is the data base, and also the part of strategy and decision making.

Three parts are closely combined to be a new type of related laboratory series, fully equipped, and is highly controlled by computer systems, which can improve the coastal research on environment, resources and disaster to a new step of science.

The original laboratory equipment from our own finance is about 300 000 U. S. dollars, but now the total hardware value of SCIEL is more than 1 300 000 U. S. dollars, which is more than or equal to the amount from the World Bank Loan to each Key Lab. Why don't you accept the SCIEL as one of the Key Labs in China, SCIEL is a high rank lab of earth science and it is even better as only using 420 000 U. S. dollars from the World Bank Loan to set up a lab series. The reason for us having the Canadian and Hongkong alum supporting is that we have good reputation through the joint studies, they trust our scientific ability. With the new equipment supporting, we can get even higher level scientific result, they would like to improve the ability for us to be one of the most advanced labs in the world coastal science.

Thus, through international support and our own effort to set up the advanced laboratory series is one of the most important experiences for Pilot Project of Coast and Island Labs.

The head of SCIEL has the working experiences in North America, she knows exactly what kind and the qualities of instruments we need for coastal ocean study. Thus, right from start of the instrument selecting, we pay attention to the quality of high standard instruments, light one for working on shallow water and easy to equip on the small boat. Surely, the reasonable price to buy and choosing the companies which have the sub-branch in China for handling the repair service are important. An enthusiastic and capable chief scientist has improved the lab open to the public early during its constructive stage.

Secondly, International Joint Research Produces High Level Research, and Training New Generation Scholar Through the Research Combined with Education.

The Coastal and Island Lab is located in the Nanjing University, occupied 800 m^2 in a 3 000 m^2 building joined using with Department of Geo & Ocean Sciences. The lab is relatively independent in the research project, management and finance, but joint library and service facilities with the department. Part of the laboratory facilities, such as Remote Sensing, Mapping, Environment Geography and Marine Biological Lab, are combined with departments for strengthening their capabilities.

Education combines with scientific research is important for post graduate program, SCIEL covers one position of post doctorial, four of Ph. D study fields (including Physical Geography, Geomorphology & Quaternary Geology, Human Geography, Botanic-Biology), and six of Master degree fields (Remote Sensing & Cartography, Environment Geography, and four others area as same as Ph. D field).

Scientists are the soul of the laboratory, and post graduate students are the new blood of the scientific "body". SCIEL needs a group of people to survey on the boat, to do analyses and computing the results in the lab. Even though, students graduated and leave the lab year by year, however, we are benefited by the continuing support by younger generation to improve and to deepen our studies. It might be better than independent academic institute.

SCIEL has set up academic links with Canadian, British and American Institutes or Departments on Coast and Continental Shelf Studies. We have invited several foreign scholars to be the concurrent professors. Through international collaboration, a number of technicians are also trained to install, test and operate equipment according to their design specifications. It is important for technicians to turn hardware according to international standards, for instance, a technician working in an isotopic dating laboratory who had used ordinary water in place of distilled water by mistake. Having compared the analysis results with that obtained in foreign sister laboratories, we found the mistake and reanalysed about 1 000 samples until we got the correct results. Through much effort and practice, the Pb^{210} isotopic laboratory has achieved the standards set by the International Atomic Energy Commission (IAEA). In another example, a technician had thrown away nuclear filter paper thinking it was ordinary one. So both the training of technicians and the standardization of equipment are very important issues for SCIEL management. Besides, the safety regulation especially working on the boat drilling on sea bottom our foreign colleagues are always pay attention to the safety aspect. We are learning and learned a lot from our international colleagues. On the other hand, foreign scientists are also satisfied to learn the knowledge of coast environment, resources and the method for regional comprehensive analyses method from Chinese scientists.

Sino-Canada joint project had surveyed 300 km long coast ocean around the south, west and north side of Hainan Island. It concluded that three harbours: Yangpu, Sanya and Haikou can be opened and expended as the large size harbours of 50 000, 30 000 and 20 000 tones. Soon after, the Yangpu harbour has constructed the deep navigation channel and two docks for 20 000 tone ships landing in 1990, and the result is that there is no siltation problem as we planned. During the period of joint studies, 12 Chinese scientists went to Canada either for training or doing cooperative studies, 15 Canadian scientists and senior technicians worked in China for exchanges. 58 papers have been published and joint held *the International Symposium on Exploitation and Management Island Coast and Embayment Resources* (Nov. 1990), published proceedings of Island Environment and Coast Development (1993). Canadian IDRC evaluated the joint project is one of the ten most successful projects since IDRC set up for 20 years, the joint Evaluation Committee by Chinese, Canadian and American scientists reviewed the whole program and working results, have concluded that the study is "at the international leading level of coastal science".

SCIEL had also organized "*the 5th Symposium on Man's Impact on Coastal Environments for Asia and Pacific*", and published "*Ecosystem and Environment Activities*" (1990, 10). In 1993, SCIEL organized and helped Chinese Society of Oceanography to hold the 1993 "*International Pacific Conference on Marine Science and Technology, Regional Conference, Beijing*". It was the first time PACON International have a conference in the continent of Asia, and outside of island countries. Once again, Professor Wang, Ying as the chief editor joint with Professor David Hopley from Australia published the proceedings of 1993 PACON (1994).

The cooperative study with Department of Oceanography, University of Southampton conducted the study on sedimentary dynamics of tidal flat and four Ph. D students had been trained by joint supervision. Professor Wang, Ying, the director of the SCIEL, has been pointed as the chairman of the Working Group: "*Relative Sea Level and World Muddy Coast Studies*" by Scientific Commission on Ocean Research (SCOR) during Oct. 1994 general conference, which means that international academy has confirmed the SCIEL's research achievement in the area. Wang has been invited by UNESCO to give a talk on "Sediment Flux to the Continental Shelf" during the Coastal Ocean Workshop in 1994, she has also edited the same chapter for publishing a book of "The Sea" by UNESCO.

With the modernized equipment, quite good scientific team and international academic cooperation, SCIEL had conducted a number of research projects such as: "*The Comprehensive Survey on Island Environment and Resources of Jiangsu Province*", "*The State Holiday Resources—Yalong Bay Submarine Tourist Planning*", "*The Shimei Bay Sea Shore Tourist Planning*", "*Beach Nourishment Project of Sanya*" etc. The characteristics of the studies are with multi-items data of the coastal environment including wave, current, sediment, biology, submarine geology etc. Based on the detail studies, an high level comprehensive planning can be made and all with excellent evaluation remarks.

SCIEL has received a key study project from Natural Science Foundation of China: **"Study on the Evolution of Sandy Ridge Field of Yellow Sea"**. It is an unique submarine sandy ridge field located at: 31°55′ to 34°24′ North latitude, and 120°40′ to 122°32′ East longitude. It consists of 70 sandy bodies and extends 200 km long from north to south, 90 km wide from west side of coast to the east. In 1994, with the new equipments of GPS, Geopulse Profiler and Double Frequency Echo Sounding, SCIEL conducted a comprehensive survey on the sandy ridge field with more than 600 km long seismic profiling data. Through the analysis of sedimentary structures, it is found that the sandy ridge field is a combination pattern including ancient Yangtze River delta and continental shelf sandy bodies. The sandy ridge field may be formed during late pleistocene. One largest tidal channel, Huangsha Yang, was an old Yangtze River valley, with the stable water depth, the channel can be used as navigation route for 200 000 tone ships. In 1995, the Commercial Department of United States of America decided to set up a natural gas generated power station there, with the total capital of 3 400 000 000 U. S. dollars. The theoretic study has improved the coastal construction. As a result, SCIEL has a new project of studying the stability of Huangsha Yang channel for the engineering construction. From this kind of contract project, SCIEL can get 600 000 to 1 000 000 yuan each year as their own financial sources.

Third, to strengthen the laboratory organization and emphasize the frontier studies of coastal science.

With the concern and warm help from State Planning Commission of China, Natural Science Foundation of China, State Education Commission of China, and the

World Bank, through their delegates to check the progress of pilot project in SCIEL, to exchange the information and experiences between each pilot laboratory, to summary and direct the lab's construction, management, training, etc, the SCIEL develops healthily and rapidly. It gradually forms a systematic team for management or operating the laboratory. Two senior professors are in charge of academic studies, one is the director of the lab, being responsible for the guiding principles, fundamental policy and public affairs; the other professor, the deputy director of the lab, takes responsibility for training the post graduate students and the research programs. One assistant is in charge of the administrative work and field survey activities, one secretary is response to all of the paper work and as the scientific assistant of the director. Another deputy director is in charge of the equipment affairs and the daily operating of the lab.

The major step for SCIEL to carry out its "Opening, Flowing, Uniting" policy is establishing the Academic Committee. With the permission of the State Education Committee of China, the Academic Committee of SCIEL was established in May of 1992. It consists of 19 members and one secretary. All the 19 members are scientists from marine and geoscience, including 4 members of Chinese Science Academy, 6 professors from Nanjing University, other 11 members from all over the coastal region of China. Besides, one is from U. S. A. and one is from Canada. Professor Ren, Mei-e, the member of Chinese Science Academy, is the chairman of the committee, Professor Shi, Yafeng and Professor Ye, Zhizheng, the members of Chinese Science Academy, are vice chairman of the committee, Professor Wang, Ying, the director of SCIEL, takes the vice chair of the committee too.

The Academic Committee approved the foundation of SCIEL and the scientific team organization, highly identified the academic exchanges and scholar training. Through the fully discussion, the committee has defined the following topics as major research topics for application in a three year period.

(1) Trend analysis of climate, sea level changes and coastal response.

(2) Embayment Sedimentation, its structure and sedimentary rate.

(3) Geomorphological and dynamic system of tidal inlets and it's stability studies.

(4) Sedimentary dynamics and geomorphological evolution of tidal flat, and its exploitation.

(5) River-Sea sedimentary dynamic system and continental shelf sedimentation.

(6) Shallow sea submarine mapping and the new technic application.

(7) Study on tidal flat and shallow sea ecosystem, resources and exploitation.

(8) Island resources, environment and territory history studies.

(9) Study on the coastal disaster development and prevention.

(10) R&D on remote sensing and GIS application of environment and resources of coast and island.

Academic Committee evaluated research applications and approved a number of candidates with 2 000 to 10 000 yuan supporting for each topic, mainly for the using of our experimental equipment and local travel. Financial resources for supporting candidates are from SCIEL own project through the industry contract, and the State Education Committee of China give supports for the opening research twice. Total 18 research topics have been approved and supported for total 155 000 yuan since 1992.

The Academic Committee had two meetings in the year of 1992 and 1993. It had also organized seminars for one day or two during the meeting period. Several topics of the study supported by SCIEL have been finished with quite good results, such as "*Micro Sea Algae Study*" by Professor Zhen, Zhaoqi, the micro algae have been adopted widely by food industry. "*The Coastal Geomorphology of Taiwan*", by Professor Zhen, Zhaoxuan from South China Normal University, finished the study and publishing it as a book. The topic of "*Pb^{210} Profile and its Sedimentation Meaning in the High Turbidity Environment*" finished by Second Institute of Oceanography of State Oceanic Administration with SCIEL, has got excellent thesis prize from Nanjing University. Research on sea level rising and the coast response, Professor Ren, Mei-e published the paper: "*The influences on Submarine Sandy Ridges and North Jiangsu Tidal Flat by the rising of the Sea Level*", Prof. Wang, Ying's paper on "*Sea Level Rising and the Beach Response*" estimates the major tourist beaches in China will loss 15%～60% of its present area while the sea level rising to 0.5m during 21st century.

The third part of the capability building is to training the technicians, to let them mastering the fundamental principle of experiment, instrument function and manipulation, and through the practice to standardize the working procedure.

Through the Pilot Laboratory Project, SCIEL has improved their survey and research ability, expended their service scope, thus improving the opening, cooperating and accumulating more income. As a result, SCIEL has improved their capability building again. It is believed that the policy of "Opening, Flowing and Uniting" has forced the laboratory to be in the healthy circulation.

At the present stage, SCIEL has realized that the most important aspect is to study the frontier area of coastal marine science. Several topics have been carried out, which study on global sea level changes and the coast response in the lowland area, such as the coasts along the North China plain, North Jiangsu plain and the Changjiang River delta. These topics have summarized the post glacial sea level changes and the local tectonic influence. More attention will be paid to the study on coasts and islands environment, resources, human impacts and sustainable development, more theoretic study on marine territorial right, and through the service of research to train scholars for 21 century.

The summary for public evaluation and for improving the laboratory studies in the future.

Brief Introuction to State Pilot Laboratory of Coast & Island Exploitation*

The State Pilot Laboratory of Coast & Island Exploitation was selected as one of the seven state pilot labs in "key Studies Development Project" founded by World Bank. The establishment of the laboratory has been through three stages, from initial stage of a single sedimentary lab during the early of 1980's; Secondary stage with significant support from International Development Research Centre of Canada from 1985 to 1991, the third stage completely formed a new laboratory series by World Bank in 1990's. It was officially named as State Pilot Laboratory of Coast Island Exploitation (SCIEL). The total value of the labs is about 2 million U. S. dollars. The establishment of SCIEL has been proved to be a great success in practice, and has contributed significantly to the development of Coastal Ocean Sciences at Nanjing University. Especially, through Sino-Canada international joint research, it was provided by IDRC (International Development Research Centre) with new instruments for inner continental shelf surveys, SCIEL has set up an isotopic dating laboratory Pb^{210} which can be used to determine the activities of Pb^{210}, Po^{210} and Pu for precisely estimating deposition rates. The Pb^{210} isotopic laboratory has achieved the standards set by the International Atomic Energy Commission (IAEA). All the equipment is controlled by computer systems. It is the unique part of the lab than other labs in the universities and hardware upgrading with State-of-the-art equipment, which makes the lab to be one of the most advanced labs in the world on coastal science.

The State Pilot Laboratory of Coast & Island Exploitation is a new type of Laboratory Series in Geo & Ocean Sciences-If studies the natural processes of land-sea interaction, the natures of environment and resources of coast, island and continental shelf for improving regional sustainable development and advanced education on global changes. It is an institution for applied fundamental sciences. There are 16 permanent 25 visiting professors and about 10 post graduate students each year in SCIEL. Professor Wang Ying, a famous female scientist on marine studies, acts as the director of the Lab. The academic committee of SCIEL has been organized with Professor Ren Mei-e, a member of Academic Sinica of China, as the chairman. In order to promote

* Ying Wang: *Annual Science Report-Supplement of Journal of Nanjing University*, 1995, Vol. 31, (English Series 3): pp. 98－99.

domestic and international research and education in fields of coast & island exploitation, SCIEL has been opened to a larger pool of Chinese and foreign scientists and scholars with the approval of State Planning Commission of China and the State Education Commission of China under its philosophy "Opening, Flowing, Uniting".

1　Main Research topics

(1) Trend analyses of climate, sea level changes and coastal response.

(2) Embayment sedimentation, its structure and sedimentary rate.

(3) Geomorphological and dynamic system of tidal inlets and the stability studies.

(4) Sedimentary dynamics and geomorphological evolution of tidal flat, and its exploitation.

(5) River—Sea dynamics and sedimentary system.

(6) Shallow Sea submarine mapping and the new technic application.

(7) Study on tidal flat and shallow sea ecosystem, resources and exploitation.

(8) Island resources, environment and territory history studies.

(9) Study on the coastal disaster development and prevention.

(10) R & D on remote sensing and GIS application of coastal & island environment and resources studies.

2　Main Equipment

(1) Marine observatory

Microwave Trisponder Positioning System (Model: 542)

Subbottom Profiling system (Model: 5210A)

SCUBA Diving Equipment

Nikon Underwater Camera

Dual Frequency Depth Sounder System

Global Positioning System (G. P. S.)

Side Scan Sonar

C. T. D. System

Direct Reading Current Meter

4. 5 in Inflatable Rubber Boat

Tide and Wave Recorder

Lehigh Core System

(2) Analytical Experiment

Pb^{210} Dating System

Photoelectic Grain Size Analyser (Model SKC - 2000)

Automatic Grain Size Analyser (Model: LDY-1)

Normal Upright Transmitted Light Microscope (Model: AX10MAT)

Stereomicroscope (Model: OPTON SV, DR,DRC, DV4)

Water Quality Analysis System

Automatic Temperature controlling Chamber

(3) Marine—geographic Information System

Graphic Workstation

Pentium Computers

Colour Inkjet Plotter

Draft master Plotter

Software of ARC/TNFO, ERDAS

Rewritable Optical Disk Drive

Optical Disk

X Window Base Unit

Laser Printer

Digitizer Ao size

Computer Aided Cartographic system

Scanget

State Pilot Laboratory of Coast & Island Exploitation (SCIEL)*

The establishment of the laboratory has occurred in three stages. It started as the Coastal Geomorphology & Sedimentology Laboratory between 1963 and 1988. Its secondary stage evolved with significant support from the International Development Research Center of Canada (1988 to 1992). The third stage allowed the organization of a completely new laboratory structure that was funded by World Bank Loan (1990 to 1994). The new laboratory was officially named as the State Pilot Laboratory of Coast & Island Exploitation (SCIEL) by the State Planning Commission of China, State Education Commission of China, and State Scientific & Technology Commission of China in 1990. SCIEL is one of seven State Pilot Labs in the "Key Studies Development Project" funded by the World Bank. Recently, it passed an evaluation by the State Education Commission of China (1995).

SCIEL is an institution of applied fundamental sciences. It studies the natures of resources, environment and ecosystems of coast, tidal flat, island and continental shelf for improving the effectiveness of regional sustainable development, and for advanced study of Global Change issues.

The present conception of "coastal zone" covers the entire field of land and ocean interaction. Dynamic processes in this environment are active, changing rapidly, and have a close relationship to the surfaces of lithosphere, hydrosphere, atmosphere, and biosphere. The coastal zone abounds in renewable resources, and has the highest density of population and an exciting velocity of economic development. To effectively exploit coastal environments and resources in a sustainable way, work is needed on survey, study and planning, according to the nature of coastal evolution, with the view point of systematical science to provide technical direction and assistance for strategy and regulation formulation. Specific study objectives and functions have produced a laboratory profile with multidisciplinary features (marine geology, geography, biology and technology sciences).

The laboratory is composed of three main parts (marine observatory, experiment analysis, and Marine-geographic information systems) and also includes sample archiving, instrument repair and support sections.

---

* Ying Wang, Charlie T. Schafer, Yongzhan Zhang: *Annual Science Report—Supplement of Journal of Nanjing University*, 1996, Vol. 32 (English Series 4): pp. 20 - 22.

The first section of SCIEL is the Laboratory of Coast Ocean Survey. All equipment is high-tech and includes acoustic，optical and electronic systems：

Global Positioning System

Microwave Transponder Positioning System

Geopulse Seismic Profiler

Side Scan Sonar

Dual Frequency Echo Sounder

Acoustic Doppler Current Profiles

C. T. D. System

SCUBA Diving Equipment

Underwater Video

Nikon Underwater Camera

Lehigh Coring System

Current Meters，Wave and Tide Recorders

Small Boats and Communication Systems

The second section of SCIEL is the Analyses Section. Equipment in this part includes：

Photoelectric Grain Size Analyzer (Model：SKC－2000)

Automatic Grain Size Analyzer (Model：LDY－1)

Automatic Grain Size Analysis System (For Sand)

Normal Upright Transmitted Microscope

Stereomicroscope

High Resolution Power Microscope

Water Quality Analysis System

Stable Isotope Pb－210 Dating System

Automatic Temperature Control Chamber

The third section of SCIEL is the Laboratory of Marine-Geographical Information Systems. Software in this section is used for data classification，storage，calculation，analysis and modelling. The key equipment in this section includes：

HP Graphic Workstation

Pentium Computer

Digitizer (A. Size)

Scanget

Colour Inkjet Plotter

Draftmaster Plotter

Laser Printer

X Window Base Unit

Rewritable Optical Disk Driver

Optical Disk

Software ARC/INFO, ERDAS

Computer-Aided Cartographic System

The three laboratory sections are closely interrelated into a new type of "integrated" laboratory infrastructure, which can be employed to improve the output and quality of coastal research on environment, resources and disaster prevention.

The famous geoscientist, a member of the Chinese Academic Science, Professor Ren Mei-e is the honour Chairman of the Academic Committee of SCIEL. Professor Wang Ying, a famous female scientist in marine studies, is the current director of SCIEL. The Academic Committee of SCIEL includes Professor Liu Guangding, member of the Chinese Academic Science, who is the Chairman, Professor Wang Pingxian, a member of the Chinese Academic Science, Professor Ye Zhizhen, a member of the Chinese Academic Science, and Professor Zhu Dakui are the vice Chairmen of the Academic Committee of SCIEL. There are 20 formal members of the SCIEL organization including 3 professors, 8 associate professors and senior engineers, 3 lecturers and engineers and 6 technicians. There are also some concurrent professors and foreign members.

The major study fields defined by the Academic Committee of the SCIEL include:

◆ Global Change, Sea Level Change and Coastal Response

◆ Embayment Sedimentation, Its Structure and Sedimentary Rate

◆ Geomorphological and Dynamic System of Tidal Inlets and Inlet Stability Studies

◆ Sedimentary Dynamics and Geomorphological Evolution of Tidal Flat

◆ River-Sea Sedimentary Dynamic Systems and Continental Shelf Sedimentation

◆ Shallow Sea Submarine Mapping and New Techniques

◆ Tidal Flat and Shallow Sea Ecosystem, Resources and Exploitation

◆ Island Resources, Environment and Territory History Studies

◆ Coastal Disaster Development and Prevention

◆ R & D on Marine Remote Sensing and Geographic Information System Applications of Environment and Resources of Coast and Island

◆ Coastal Exploitation, Survey, Planning and Shallow Sea Engineering, including Harbour Sitting and Navigation Channel Selection, Tidal Flat Reclamation, Beach Nourishment, Coastal Protection, Coastal Sewage Project, Sea Bottom Stability Study for Marine Platforms and Pipe Lines.

Because the SCIEL includes Physical Geography as one of the State Key Studies in China, it also serves as a base to provide advanced qualified scientists and technicians to the State. Education combined with scientific research is important for post graduate programs. SCIEL supervises one post doctorial position, four of Ph. D study fields (including Physical Geography, Geomorphology & Quaternary Geology, Human

Geography, Botanic-Biology), and six master's degree fields (Remote Sensing & Cartography, Environment Geography, and four others area as same as Ph. D field). Since 1990, about 50 post graduate students have obtained their degrees and left the lab, but also there are 13 post graduate students currently enrolled in the SCIEL program (including 2 post doctorial students, 8 Ph. D. students and 3 master's degree students).

With advanced equipment and strong scientific leadership, the SCIEL has already conducted 20 projects through international and national cooperation. Among these, there is 1 item of National Key Studies, 6 items of National Scientific Foundation Commission of China Project, 4 topics from State Department, 2 items of International Joint Program, and 7 projects arising from industry contracts. Several topics have reached high levels of achievement, such as the comprehensive survey on island environment and resources, Jiangsu province, China-Canada joint study on embayment sedimentation rate and deep water harbour development, the cooperative study on sedimentary dynamics of tidal flats, recent sea level change and its effects on the deltas of the Yellow, Changjiang and Pearl Rivers, coral reef coast and global change study, study on formation and evolution of submarine radiative sand ridges off Jiangsu coast and the navigation channel stability of Yangkou harbour, the state holiday resources-Yalong Bay submarine tourist planning, beach nourishment studies of Sanya, the geophysical survey and study for artificial island construction and sand sources in Shenzhen Bay etc. The SCIEL also organized "the International Symposium on Exploitation and Management of Island Coast and Embayment Rescourcs", "the 5th Symposium on Man's Impact on Coastal Environments for Asia and Pacific", and the "International Pacific Conference on Marine Science and Technology, Regional Conference, Beijing."

Now SCIEL staff are also conducting projects on global sea level changes and coastal responses, combined with SCOR WG 106: "Relative sea level and world muddy coast", study on the evolution of a submarine sand ridge field in the Yellow Sea, study on environment, resources and sustainable development of the Changjiang River delta, study on island resource and environment exploitation and sustainable development of ecosystems, and integrated coastal zone management of China. Many of these have led to significant economic benefits.

According to its "Opening, Flowing, Uniting" policy, and the situation of the scientific development and characteristics of the lab, SCIEL established its own Academic Committee. The Committee has evaluated research applications and approved 18 research items since 1990. 10 items have already been completed. All of these topics extend the research field of SCIEL.

The International Conference on Relative Sea Level and Muddy Coasts of the World by Working Group 106 of SCOR, ICSU Held in Nanjing University*

The Scientific Committee on Oceanic Research (SCOR) as a leading non-governmental organization for the promotion and coordination of international oceanographic activities, reflects organization for the promotion and coordination of international oceanographic activities, reflects the multidisciplinary nature of oceanography. SCOR has related closely with UNESCO as a unit of the International Council of Scientific Unions (ICSU). During 39 years operation, SCOR has organized 108 working groups and numerous conferences on international cooperation, made contributions to the world wide ocean research, such a JGOFS (Joint Global Ocean Flux Study), Global Climate Change, Oceanic Circulation etc. Presently, the impact of Global Change has become a more and more important issue, as the progressive sea level rising relates to densely populated coast zone; as a result, the SCOR working group of 89 has studied on "Sea Level Rise and the Beach Erosion" since 1989, in which Professor Wang Ying from Nanjing University is the corresponding member from China, and the SCOR working group 106 on "Relative Sea Level & Muddy Coasts of the World" was set up during the 32nd Congress of SCOR in Canada in October. 1994. The WG 106 consists of 9 members and 5 corresponding members from the major muddy coast countries of the world, including Australia, Brazil, Canada, China, Germany, India, the Netherlands, the Philippines and the U. S. A (Table 1 Name list). Professor Wang Ying has been appointed as the Chairman of the group. Professor Terry Healy, the vice president of SCOR, takes the part as the reporter of WG to SCOR.

As the first meeting on "Relative Sea Level and Muddy Coasts of the World" of SCOR WG106, it was successfully held in Nanjing University during 3 - 9 October 1995. The conference was cosponsored by the Headquarters of SCOR, National Science Foundation of China, State Education Commission of China, State Oceanic Administration of China through China SCOR, and State Pilot Laboratory of Coast & Island Exploitation, Nanjing University; all gave their financial support and helpful

* Ying Wang: *Annual Science Report—Supplement of Journal of Nanjing University*, 1996, Vol. 32 (English Series 4): pp. 28 - 30.

suggestions. 60 participants including scientists, officers and post graduates students from China and 9 countries attended the conference. The purpose of the conference was to provide a forum for exchanging previous study results, to identify the problems which would benefit from coordinate international attention, and to develop a scientific plan to focus research on the topic of relative sea level and world muddy coasts, and to assist with their implementation.

It was a high level conference and consisted of two parts. The first part was three days for paper presentation and discussion; 16 papers were chosen from 25 abstracts for oral presentation. The papers were given by members of the WG and invite speakers, such as Professor Ren Mei-e, an esteemed member of the Chinese Academy Sciences, Professor Terry Healy of University of Waikato, New Zealand, vice president of SCOR, Professor Chen Jiyu, the leading scientist of estuary studies in China, and Mr. Huang Jing from the Administrative Center for China's Agenda 21. The second part was 3 days for an excursion to typical muddy coast in China along the Yellow Sea for a comparative study on muddy coasts to the other countries. The muddy tidal flat coasts in China are under control by the river-sea systems and through the influence of human activities. The working group was also exposed to some economic consequences of coast developments on a muddy coast of a developing country (China). The conference spent several days in discussions on the classification, meaning and dynamic processes effected by the rising of sea level. Arising from the discussion, the SCOR WG106 decided on a project to create a book in muddy coasts including ten sections. Submissions will be made by various members to sections with which they have special expertise; and all the members are expected to contribute to the section of human impacts, gaps in knowledge, recommendations and literature cited on muddy coasts. Submissions and contributions are expected to reach the section leaders by February 1996, and after initial editing, they are to be forwarded to Professor Wang Ying, the Chair, for the future editing. The next meeting for the final review of the manuscripts for publication will be held in South Carolina of U. S. in the autumn of 1996.

Table1 Final name list of WG 106

RELATIVE SEA LEVEL AND MUDDY COASTS OF THE WORLD

Prof. Wang Ying
Chairperson of the WG106, and the Chair of the conference
Department of Geo & Ocean Sciences
State Pilot Laboratory of Coast & Island Exploitation
Nanjing University, Nanjing, China

Dr. Pieter G. E. F. Augustinus
Institute for Marine and Atmospheric Research (IMAU)

University of Utrecht, Utrecht, The Netherlands.

Dr. Ashish Metha
Coastal and Ocean Engineering
University of Florida
Florida, U. S. A.

Prof. Bjorn Kjerfve
Marine Science Program, Department of Geological Sciences
University of South Carolina, Columbia, SC, U. S. A.

Dr. M. Baba
Center for Earth Sciences Studies
P. O. Akkulam, Trivandrum - 695 031, India

Dr. Carl Amos
Geological Survey of Canada Atlantic
Bedford Institute of Oceanography, Dartmouth, N. S. M Canada

Dr. Eric Wolanski
Australian Institute of Marine Science
Townsville M. C. , Queensland, Australia

Dr. Eduardo Marone
Centro de Estudos do Mar-UFPR
Pontal do Sul. Paranagua, PR, Brazil

Dr. Miguel D. Fortes
Marine Science Institute CS
University of the Philippines, Quezon City, The Philippines
Corresponding Members

Prof. Wang, Ji
National Marine Data Information Service
State Oceanic Administration of China, Tianjin, China

Prof. Han Mukang
Department of Geography
Beijing University, Beijing, China.

Dr. Tad Marty

The National Tidal Facility
The Flinders University, of South Australia
SA, Australia

Prof. Burg Flemming
Senckenburg Institute, Division of Marine Science
Schleusenstrasse, W. lhelmshaven, Germany

Dr. Yara Schaeffer-Novelii
Department of Biological Oceanography
University of Sao Paulo, Sao Paulo, SP Brazil

Executive Committee Reporter for WG 106

Prof. Terry Healy
Vice-President of SCOR
Department of Earth Sciences
University of Waikato, Hamilton, New Zealand

人类与世界海洋*

1 会议概况

太平洋海洋科技协会(The Pacific Congresses on Marine Science and Technology,简称 PACON)1999 年会议是于 1999 年 6 月 23～25 日在莫斯科举行。会议主题是"人类与世界海洋"。会议代表有 205 人,其中中国代表有 24 位(含家属 4 名),分别来自国家海洋局、南京大学、上海交通大学、厦门大学、香港城市大学、香港中文大学和台湾大学等单位。

PACON 中国分会主席国家海洋局副局长杨文鹤、副主席王颖参加会议,作报告,并出席理事会。台湾代表陈汝勤教授未出席,而由台大海洋所所长梁乃匡教授参会并代陈出席理事会,提交正式年度报告。在理事会上他表示要扩大台湾在 PACON 的团体会员名额。中国分会在 1998 国际海洋年中做了许多工作,但未提交报告。这次杨文鹤副局长及时做了口头发言,并提交了正式年度报告。杨认为,海洋局和南京大学均系中国 PACON 发起单位,应以团体会员名义参加 PACON,扩大交流与传播影响。

南京大学海岸与海岛开发国家试点实验室及城市资源系共有 6 位代表参加,组成一个小组参加 PACON 学术活动,扩大了南京大学海岸海洋学术工作的影响。会议中组织了对莫斯科城市的考察,会后考察了俄罗斯西部海岸和圣彼得堡。这是一次地理学考察活动,也是一次中、俄文化活动。

中国由杨文鹤代表严宏谟任大会名誉主席,其他 4 位名誉主席分别是俄科学院院士 NiKolai P. Laverou,日 Kenji Hotta,韩 Hyung Tack Huk,美 James J. Sullivan,王颖为大会的学术项目协调人之一。

PACON 99 由 4 次大会报告、26 个分会、4 个研讨会、展贴与科技展览等 5 个部分组成,演讲文章约 300 篇。

大会报告有俄国科学院 Shirshov 海洋研究所 Sergei. S. Lappo 教授的《展望第 3 个千年的海洋学》,莫斯科石油天然气研究所、俄国科学院院士 Anotoly N. Dmitriesky 的《海洋油、气研究》,中国海洋环境预报中心李允武研究员的《海洋科技在中国之发展》,俄国 Shirshov 海洋研究所 Anatoly M. Sagalevitch 的《深海人控潜水器"Mir - 1"及"Mir - 2"运作的不同用途》。大会报告以俄国科学院院士居多。

按照 PACON 传统,在欢迎晚宴上讲演的是美籍俄裔学者、夏威夷大学海洋系教授 Alexandar Malahoff,题目是《海洋研究的前沿领域:技术与资源》。他着重谈到磷、硅在生命发展中的作用,属前沿课题。

26 个分会,分为大洋科学技术(OST)与海洋资源管理与发展(MRMD)两大部分。

---

* 王颖:《国际学术动态》,2000 年第 4 卷,第 23 - 26,38 页。

PACON 具有海洋技术的优势，所以 OST 的分会多，参加的人数与报告也多。各分会皆有一名俄国学者任分会主席，这是历次 PACON 会议所未有的特点。

2 大洋科学技术(OST)

OST－1，海洋遥感。两位俄国主席，共 18 篇报告，其中俄国学者 12 篇，日本、美国各 1 篇，中国大陆和台北共 4 篇。报告数量既反映出热点，也反映出有关国家的专长领域。但实际报告仅 10 篇，中方缺席较多。

OST－2，海气交换和气候问题。俄国 2 位主席，加拿大 1 位主席，涉及全球气候变暖、南方涛动等，19 篇报告。以俄为主，韩、美次之，各 2 篇，实际报告为 12 篇。

OST－3，海洋物理。由俄、德代表任主席，共有 16 篇报告，俄国 10 篇，包括与美、加合作各 1 篇，法国及中国台湾各 1 篇。实际报告 11 篇，还有 18 篇俄国学者的展贴报告。

OST－4，海洋仪器、技术和水下机器人。由 2 位俄国学者主持，共 7 篇报告，除 1 篇德国学者报告外，均为俄国学者所作。

OST－5，海底地震与海啸。由俄国学者和日本学者主持，共 8 篇报告，研究成果均佳，俄、日、加、美各国学者均有。本专题有 9 篇俄国学者的展贴。

OST－6A，海洋地质、地球化学与物理。由俄、美学者主持，共 21 篇报告，除 1 篇美国报告外，其余 20 篇均为俄学者提交，且用俄文报告，当我去听时，主持人惊慌，为我提供英文报告文本。此分会以地球动力与大构造体系为主，会议讨论热烈，多为俄国年老学者，争论相持不让。本议题有 24 篇展贴，以区域海洋地质——火山、地震、构造与沉积特点等为主。

OST－6B，共 6 篇报告，均俄国学者所作，有 11 篇展贴。

OST－6C，专题同上，共 6 篇报告，11 篇由俄学者所作，以及 1 篇俄美合作展贴。

海洋地质、地球物理与地球化学为本次会议中心，共 32 篇报告，23 篇展贴，而且参加人数均多，反映了俄国在海洋领域的热点与强项。往届 PACON 会议却未设过海洋地质分会场。

OST－7，海洋生物成就。俄学者主持。中国厦大、香港中文大学各有 1～2 篇报告，余为俄国学者所作，共 11 篇报告。

OST－8，海洋环境问题。由俄、日学者主持，共 7 篇报告。中国海洋局学者作了 1 篇有关海洋沉积容量的报告；香港城市大学黄玉山、谭凤玉作了红树林区污染报告；一位日本学者作了白令海生态系动力和关于贻贝污染灾害的报告；一位加拿大专家作了太平洋河口湾的环境临界状况报告，其余均为俄国人报告。

OST－9，海洋灾害事件。由俄、西班牙和韩国代表主持，共 9 篇报告，其中有 2 篇西班牙与俄国合作报告，主要为大气作用之长期影响。除中国台湾梁乃匡的台风浪高的预报外，均为俄国的有关大幅度波高风波研究。本专题有 3 篇俄国学者的展贴：海啸灾害、风暴潮和海啸观测资料。

大洋科技是 PACON 学术交流的主要领域，本次又以海洋地质、海洋地震与海啸等为特色。

3 海洋资源管理与发展(MRMD)

MRMD-1,大洋外海与海岸矿物资源。由俄、美学者主持,共7篇报告,除1篇美国学者,1篇德、瑞与日合作的锰结核研究,余为俄国学者所作,涉及大洋锰结核矿物资源及其成因机制讨论。

MRMD-2,海岸海洋油气资源(有关方面及环境问题)。由俄、美学者联合主持。报告6篇,均由俄学者所作。

MRMD-3,海洋食物资源。由俄、澳学者主持,涉及贝、藻养殖和新的生物技术,报告6篇,其中2篇为乌克兰代表所作,余为俄国学者完成。

MRMD-4,海洋能。美、俄学者主持,共11篇报告,除1篇美国报告外,余为俄国报告。

MRMD-5,海岸带持续发展(经济、社会、环境)。俄、美学者主持,共16个报告。其中有5个中国报告,2个日本报告,余为俄国报告。南京大学作了4个报告,均有好的效果。杨达源教授研究废黄河口海岸变化,幻灯显示清晰的研究内容及成果;王颖教授讲解黄河断流与海岸环境效应,提出气候变暖是人为引水过量之结果,给出引水临界值与例证海岸之变化;朱晓东教授报告辐射沙洲演化及其对持续发展的影响;邹欣庆副教授用多媒体显示讲授了海岸环境管理与人群需求分析。香港中文大学生物与化学系教授黄玉山副校长讲解了潮间带红树林生态系与城市污水处理渗积问题,主要分析了5年的实验效果,认为红树林生态无变化,表层沉积中重金属增多,而水质得以清滤。

MRMD-6,海洋运输及港口。

MRMD-7,海洋政策与国际合作。由美、俄学者主持。印度海洋研究学会的S. Z. Qasin的报告为《21世纪的印度洋》,美国夏威夷东西方中心Mark J. Valencai报告了《东北亚潜在的海滨热点与可能性分析》,香港城市大学陈梁博士有一篇报告,余为俄国6个报告,涉及海洋,俄国对海洋服务、考察与海洋资源系统性管理之贡献,21世纪国际合作等。

4 Workshop

Workshop-1,放射物质对世界海洋的污染,围绕着俄国的10篇展贴进行。

Workshop-2,NEUTRINO ASTRONOMY From the Deep Ocean 深海中微子天文学,Deep Sea Optical NEUTRINO Detection and Related Oceanological Problem & Possibilities for Hydrocoustical Detection of Super High Energy Cosmic NEUTRINO.

有12个报告由俄国学者提出,围绕上述三方面讨论。有1篇为日本学者提出。我国尚未对这方面予以关注。

Workshop-3,妇女在海洋科学/商业及教育中的作用。由俄、澳学者主持,内容涉及俄国在EWISH篇章中的关注与活动,俄-美相互联系对妇女从事科学的支持,俄国地质学的女学者——有前途吗,妇女科学家信息与交流,俄国在妇女与性别方面的研究,科学界的妇女与男人——两个侧面等。

Workshop-4，旅游与娱乐，由俄、澳、韩国学者主持，共7篇文章，由俄、美、韩、日、澳学者提供讨论。

5　参观访问

在本次常务理事会上确定王颖为PACON International奖励委员会成员。2001年将在中国海口举行PACON专题会，主题为海洋世纪，管理/持续发展与合作。莫斯科会议的特点与不足可为海口会议提供借鉴。

我这次是生平第一次访问莫斯科，总体感到莫斯科市建筑整齐，空气污染程度较北京低，仍蓝天在望，树木丛生，街上小汽车多，私车多。莫斯科号称800万人口，有250万辆车，且是近6年增长的，反映了一定的经济实力。俄国科学院建筑宏伟高大，有空调，但维护与开放效率欠佳，PACON会议在该处举行。

代表住CnymHuke卫星旅店，房内设施尚整洁，无空调，而俄式建筑厚墙，半开窗，通风不好。室内温度28℃。感觉极热，近两年俄国气温增高明显，据说1999年4月莫斯科气温达20℃～25℃，而5月却降雪。开会这几天，住宿处干热、流汗，使人无法入睡。在俄几日体会到全球变化在高纬地带增温明显，加拿大亦如此。考察汽车陈旧，无空调，且不停在科学院大门前。原以为不能停，后见一日本团体的大巴停靠，才知是跟我们开车的司机为己方便而不愿拐入街内停靠。

莫斯科运河，克里姆林宫建筑，Amory博物馆，大教堂，莫斯科大学等建筑均好，但旅游票太贵，均以数十美元计。一城市览介加上克里姆林宫参观门票高达49美元，还不包括参观钻石馆等。类似20世纪80年代我国开放初期，对外国人收价特高。

食物多油脂，淀粉，俄人多胖，但莫市较边缘地区的人略瘦些，街上行人衣着鲜艳，据导游介绍普通人月收入仅200卢布(不足80元人民币)，教师与工程师为500卢布，月房租至少需200卢布。

中国香港代表4人到达，当日乘地铁外出即遭人抢劫与从头浇水等侮辱。旅馆中，外门入夜有2名带手枪的军人把守，内门有2名壮汉站立铁栅栏门两侧，以维护安全。所以，我在莫市，除开会与集体参观活动外，均不外出。室内燥热，朝阳东晒，只能开门通风，倚门手持本子写字，以取凉意。

会后，中美学者一行14人参观了圣彼得堡，在1924—1991年该城名为列宁格勒，是纪念列宁领导十月革命，于其逝世时所命名的。该市位于波罗的海的芬兰湾畔，现有人口450万，约相当于莫斯科城市人口的1/2。1941年德国纳粹入侵时，该市设两道防线，第一防线被攻破，沙皇村夏宫被侵占，遭掠夺破坏。第二道防线没攻破，纳粹围城900个日夜，城中粮食断绝，苏联人喜爱的狗、猫也被吃掉，后来每人每日配给面包160克(如火柴盒大小)，而其中1/2为面粉，其他多为树叶、木屑甚至黏土。因此，每日城中有40 000人被饿死，导游Lena的外婆于1943年饿死。冬日来临时，未被德军包围的唯一的湖(Lada湖)上结冰，成为运粮、运药和送妇女儿童离开列宁格勒市的“生命之路”。有些人是冻死的，连士兵在内，累计约死200万人。Lena讲到这段历史，不禁凄然、沉默。

圣彼得堡街市整齐，所住之 Pukofckaq 旅馆为四星级，位于近郊，距沙皇村夏宫约 45 分钟车程。圣彼得堡建筑与居民习俗明显受德国文化及西方影响，旅馆中服务设施均具英文说明，售货员全讲英文，比较便利。旅馆大厅内有小卖摊，出售 Matlещка 木偶，大的为 40 dollars，十多个木偶装一套的，小的几个 dollars，以及琥珀制品，价格昂贵，以美元计，但比莫斯科便宜。10 美元相当 219～243 卢布，但店中多以 250 卢布为 10 美元起价，收费为美元，但发票写成卢布。

参观了涅瓦河、冬宫等处，了解到彼得大帝之妻生一子，被 Peter 杀死，后由其妻生的女儿即位，不久又由彼德的次子尼古拉一世即位。但继彼得大帝之后的主要沙皇是叶卡捷琳娜，德裔皇后，后为女王 30 年，神采风姿颇丰，对俄国建树多，所以冬宫、夏宫均具欧洲文化特色。

夏宫——沙皇村，为一大型皇家林苑，位于芬兰湾畔，黄色建筑，约两层楼，但高大、凉爽，屋后有喷泉，金色塑像，格式仿凡尔赛宫，室内金碧辉煌，多用金装饰与玻璃灯具。设施多系二战后及近期恢复，宫廷后宅奢靡程度超过中国历代封建王朝。

彼得堡市容整齐，已陈旧，教堂之圆型金顶未若克里姆林宫灿烂，Lena 说："斯大林、勃列日涅夫领导不喜欢列宁格勒，没来过，但戈尔巴乔夫一上台就来列市……"这种情况若非访问此市，不能得知。

第10届太平洋科学技术协会科学大会述评*

第10届国际太平洋海洋科学技术协会(PACON International)于2002年7月21～26日在日本千叶县慕张的国际会议厅举行。参加会议注册的共266人,在5天的正式会议期间,每日约100人参会,以日本人居多。其中研究生及企业界代表多参加一天或两天会,这是在日本召开PACON会议的特点,每天均办理注册,受益于学术会议的人数远超过开幕式出席人数,对此我颇有感受。与会的日本代表多为青年研究生,其中又以海洋工程组的参与者为多,其次为美国、韩国。中国有9人出席(南京大学王颖、杨达源;华东师范大学刘敏;南海海洋研究所张乔民;台湾高雄市副市长侯和雄及其子,及庄甲子、郭义雄;香港中文大学1人)。此外还有泰国、菲律宾(4人)、越南(2人)、俄国及欧洲国家代表(包括PACON所有的13个Chapter代表在内)。PACON因涉及海洋资源环境及科技,所以参会代表包括学者、学生、管理者、工程技术人员及退役海军等。

一、大会报告

(1) 7月22日上午的开幕式仅30分钟,由美国NOAA原负责人John Carey主持,接着为本届会议日本与美国的荣誉主席发言(Mr. Hajime Sako及Ms. Magarat Davidson),以及PACON主席Narendra Saxena作工作报告。参加开幕式人员约100人。

(2) 大会报告共6次。开幕式后,由美国National Science Foundation的副主任、女海洋生物学家Margaret Leinen博士作大会报告,报告题目是"The Ocean Century",她的报告内容丰富,涉及观测海洋的挑战、研究海洋内层动力学、海岸带多点监测、新仪器的研制;海底火山喷发与海洋物理化学、环境变化及所形成之海洋生物量增长;甲烷天然气可燃冰;海洋世纪与生命世纪之密切关联;多学科、多国结合对全球变化研究;咨询网络;海洋与人类活动,这是在海洋国际会议中首次强调关注的。她又特别指出:海洋在人的生活中,更需关注海岸城市的发展,比郊野更重要。她的报告,反映出美国在海洋方面的关注领域。

(3) 7月23日印裔加籍学者Dr. Tad Murty作了"Climate Change and Global Marine Hazards"的报告,他是海洋气象学家,报告所引用的资料较老,多为1970—1980年代资料,并以区域局限性的北美海平面及大洋资料为主,他认为全球将变冷。

(4) 7月24日的报告是Dr. Hitoshi Hotta(Japan Marine Science and Technology Centre)的"Challenge for unknown in the ocean"。他是私人公司代表,年纪40岁左右,报告很精彩。他将海洋工程技术与地球科学内容紧密结合,理论与应用价值高。个人感觉,日本私立大学与公司的活力大!

---

* 王颖:《国际学术动态》,2002年第6期,第22-24页。

(5) 7月25日大会报告者是 Dr. Young C. Kim 韩裔美国人,现任 California State University 土木工程系主任,他报告的题目为“Progress and Challenges of Coastal Engineering”。他指出应将传统海岸工程(防浪堤、丁坝),发展为“海岸管理”、“可持续发展”等,他还报告了海岸工程界的代表人物以美、日、韩为主。

(6) 7月26日的大会报告题目是“Global Ocean Observations for Climate Services”,由 Dr. Sidery W. Thurston 所作,他是美国 NOAA 的成员,负责 Global Program。这个报告反映了海洋观测之重要,研究大洋内部变化机制,将海平面、南方涛动、ELNINO、LANINO 等变化联系在一起分析,涉及太平洋及大西洋的涛动等研究新进展。

另外,在7月23日晚 PACON 表彰晚宴上亦有大会学术报告,由东京大学的 Masayuki MacTakahashi 教授宣讲,题为“Deep Ocean Water Utilization and Future Challenges Toward sustainable Society”。日本的研究工作密切结合大洋、岛屿的特点,利用深层冷水作冷却、空调、人工做深层水上升流富集营养盐以养殖鱼虾、冷热差动力发电,其研究选题及理论与应用并重,值得学习。

二、分会报告

学术分会大体上可分为四部分:

(1) 海洋科学技术(OST),据内容不同分21个专题,报告达350份,反映出 PACON 特点以海洋科技为主。另外是一个人可作多题报告,每个报告时间均有20分钟,可充分表达。其中又以气候变化研究、海底扩张与大洋观测技术为特色。

(2) 海岸科技(CST),划分了10个 Session,共作了60个报告。日本海岸被利用充分,沿岸皆有人工填筑滩地成陆,我所住的千叶县 Prince Hotel 就是由潮滩筑陆,而潮滩是发育在海湾的基础上。坝、路连筑的小岛等工程,尤以东京湾自房总半岛至东京的大桥为代表。人工潟湖,变海湾为潟湖,进一步利用,这些我认为值得派遣中国学者去日本学习。留学日本,可能还会促使我国青年学者学会彬彬有礼与尊老敬业。这些原是中华文明的传统,可惜经“文革”破坏几尽。这次我从南大的留日学生孟昭武身上又见到有礼貌与敬业之风貌。

(3) 海洋资源管理与发展,Marine Resource Management and Development (MRMD),22个 Session,涉及珊瑚礁,海洋保护区,休闲与旅游,红树林,有害藻类暴发(赤潮),深海水利用,海洋废弃物,一体化管理等。

(4) PACON 第一次举办5个专题论坛:

Forum1:渔业(Fisheries);

Forum2:一体化海岸管理与资源管理(ICZM & Resource Management Issues);

Forum3:南沙群岛海洋环境(Spartley Islands-Environmental Issues);

Forum4:海岸与外海工程(Coastal & Offshore Engineering);

Forum5:海洋教育与科技(Education,Marine Science and Technology)。

(5) Worshop。① 溢油与萨哈林石油发展项目(Oil Spill and Sakhalin Oil Developing Project);② 妇女在海洋科学、技术与政策(Women in Marine Science,Technology and

Policy)。

上述报告分会我虽未一一列出名称，已可看出 PACON 包含了广泛的海洋科学技术内容，而学术讨论自始至终认真组织，内容广泛，是值得我国学者参加的海洋科技学术活动。它自 1982 年以来，每两年一次大会，分别在日本、夏威夷或澳洲等岛屿举行。1990 年在北京举行的是 PACON 在大陆上举行的第一次会议，嗣后在每两次大会间均插入一次会议，成为每年举行一次会议，每两年一次大会，4 年增改选一次 PACON 常委。

目前 PACON 已有 13 个 Chapter，包括美国的夏威夷、加利福尼亚与华盛顿，日本，韩国，澳大利亚，中国北京与台湾，俄罗斯，泰国，菲律宾，印尼及越南(按成立先后顺)。会员约 100 人。往常大会的与会人数达 600 人，而后减少，原因可能在于东京，夏威夷两地消费费用高，PACON 的注册费高(500 美元左右)。

我应 PACON International 新任主席日本大学教授 Dr. Kenji Hotta 邀请参加南沙群岛论坛，原主席 Narendra Saxena 教授邀请安排我参加“Coastal & Offshore Engineerning”论坛，“一体化海洋管理”论坛，妇女论坛主席亦邀我参加，加上我自己原拟参加海岛规划，我共作了 4 个报告:《南海海底地貌》,《介绍中国的海岸海洋工程》,《妇女在海洋科技》和《亚龙湾海岸规划》。但没有参加一体化管理，各报告均 20 分钟，用 Powerpoint 报告，均获成功，尤以 3 个论坛报告，内容翔实，图表新颖，影响深远。

三、小结

(1) 必须说明，在南沙群岛论坛中，我有意识地介绍整个南海海底地质地貌特点，突出反映南沙群岛的基底岩层曾是中国古大陆架的一部分，后经折断下沉成为阶梯状大陆坡，与西沙群岛情况相同，但由于南海中部张裂，为形成的深海盆所分开，但南侧也有断裂谷将南沙群岛与其他大陆架分开。其次，我指出断续线标明中国的海域疆界，是历史传统，并由 1948 年出版的中国疆界图划定，是第二次世界大战后被同盟国波茨坦公告所肯定的，南沙群岛归属我国无可置疑。

张乔民(南海海洋所)介绍了南海水质底质环境特征，反映了中国对该区的深入研究与了解。

国外代表称南沙群岛为 Spratley Islands，台湾侯和雄仍称其为南沙群岛，并讲了在中沙与南沙群岛中国所做的工作。

该日上午报告在我的发言后结束。下午，我参加“海岸与海洋工程论坛”，第一个发言，因而不知南沙群岛论坛上越南与菲律宾代表发言。据张乔民讲，越南代表 BUI Hong Long 是海洋研究所副所长及海洋物理学博士，他谈了他们的很多工作，后由主持论坛的菲律宾代表 Dr. Chua，Thia Eng. 讲:“越南做了很多工作，今天中国代表介绍的情况反映他们的工作更多，而且所报告的仅是他们工作的一页，他们发表过 67 本著作……”，打断了 Long 的发言。

这次论坛因为涉及南沙群岛，高度政治敏感区，所以，会前，我曾向 Kenji Hotta 提出，是否适宜讨论，Kenji 回信:“仅谈环境方面的问题”。在论坛上未出现关于疆界权益争论，仅我一人指明断续线是中国的海疆界线。

美国代表 Michael P. Crosby 为南沙群岛论坛的 Co-chair，他建议"在 PACON 建立执行委员会，在每次 PACON 会上讨论南沙海域问题"，遭到多国代表反对，他又改为建科学委员会，后来在常务理事会上，未被通过。我提出不宜建立任何形式的委员会，美国的 John Carey 与其他美国常委均反对建立任何委员会，因此，本次南沙群岛论坛只是这么结场。

(2) PACON 新主席为日本大学的 Kenji Hotta 教授。美国夏威夷大学教授 N. Saxena 任主席 22 年后，正式辞职，理事会推其为荣誉终身常务理事。

新一届的理事会中，中国有杨文鹤及王颖(见附表)。Saxena 表达了对中国严宏谟教授的敬意，并提出"会上正式通过书面感谢他为建立 PACON 中国分部的功绩，严是好友……"Saxena 愿访问中国，阐明他是支持只有一个中国 PACON，台湾是以省参加。

(3) 通过 2003 年在高雄举行 PACON 年会 3 天。高雄市侯和雄代表，很活跃，在闭幕式大会上一再邀诸大陆代表参加，特别提出邀请王颖。我在理事会上提出，中国代表这次不能有很多人到日本开会，因为对中国人签证手续很麻烦。大陆代表到台湾开会更麻烦，需要在香港金钟大厦中挤站 4～5 个小时取入境证。2002 年 3 月我刚从台湾回来，所以下次会不参加了。

我认为，PACON 在海洋科技有广泛交流，且以太平洋地区为主，但不能因台湾(4 个人参加)而我们不去，我们从 1990 年参加至今，比台湾省早，比俄、越、泰等国都早，这个国际学术论坛不能放弃。我们仍应积极参加学术活动，政治立场坚定。这次我在南沙群岛论坛的报告，以及介绍中国妇女在海洋科技、中国海岸与近海工程等报告均以高度的科学分析、事实明晰地表达，赢得听会者肯定，反映了中国海洋工作的力量。今后日本 Kenji Hotta 做主席，他要增加会员数量，中国应继续参加与会。这次我了解到：所谓 Chapter 年度报告，仅是写 Chapter 主席、副主席的名字，以及会员的总数(是终身会员还是年度会员)即可，这方面工作，我认为可以做。

Name List of Board of Directors	
Chapter	Director
Australia	John Benzie
California	Young C. Kim
China	Wenhe Yang, Ying Wang
Europe	H. D. Knauth
Hawaii	Lorenz Magaard
India	S. Z. Qasim
Indonesia	Dietriech Bengen
Japan	Kenji Hotta Hajime Sake
Korea	Hyung Tack Huh
Phillippines	Eduardo Santos
Russia	Sergei Shapovalov
Taiwan	Ju-Chin Chen
Washington DC	John Carey

访问美国汇报*

一、概况

南京大学地理与海洋学院王颖院士与朱大奎退休教授于2009年1月16日～2月26日，利用寒假时间访问美国。

在美40天，主要有三个内容。

(一) 访问

(1) 应加州大学(University of California, San Diego)Scripps海洋研究所Gustaf Arrhenius教授邀请赴美，访问Scripps了解当前海洋地质学进展，与Gustaf Arrhenius讨论古海洋、海洋演变问题，撰写论文。

(2) 应Dr. Charles T Schafer之邀请，访问考察Florida海岸。Dr. C. Schafer是王颖在Bedford的同事，一同做大西洋深海调查的同行，是南京大学王颖教授获得的第一个中加海洋科技合作项目的合作者。该项目(1986—1989年)南京大学海洋室获加拿大IDRC 46万美元的科学援助，建成南京大学海岸重点实验室第一批现代化海洋勘测实验设备及同位素实验室，该项目合作中南大教师有30多人次赴Bedford海洋所合作研究、进修培训。

(3) 访问South Florida大学海洋学院(University of South Florida, College of Marine Science)，参加Dr. James Luglea的讲演及会见交谈，James曾为Woolshole海洋所所长，王颖与Woolshole多位学者有长期的合作交往(K. O. Emery, D. G. Aubrey, J. D. Milliman)。会见Dr. Chuanmin Wu，吴传民副教授，中国科技大学物理系毕业，获Florida大学PhD学位，现为University of South Florida, College of Marine Science海洋遥感研究所主任。王颖院士的博士生于堃公派出国，在Dr. Chuanmin Hu(其名片上写成Hu，实应为Wu)的指导下作海洋遥感与海岸变化的研究。南佛罗里达大学海洋学院是个独立的学院，实验室设备完善，有自备的船坞与考察船。Florida海岸有其环境特色，教学与研究并重。学术气氛浓，副院长为海洋地质学家。吴传民副教授从物理、遥感方向研究海洋，学风严谨。他认识高抒教授，我建议南大海洋室宜与南佛州大学海洋科学院建立合作联系。

(二) 考察

有三周时间野外考察。南加州海岸，Florida西海岸及专门租车去了Arizona沙漠与

* 王颖外出开会与合作研究，均有访问总结，现摘录一、二，作为她写作的一个方面的反映。

红岩盆地——古海洋考察，这是十九世纪末美国开发西部，诞生地貌学的地方，不少地貌学经典理论出在美国西部。

（三）度假

在美国过春节，与女儿家人团聚。

二、野外考察记录[①]

（一）南加州海岸

从 Los Angeles—Long Beach—San Diego 是北美海岸山脉紧邻太平洋海岸的绵延 300～400 km 的砂质海岸，粗砂质海滩宽 200～500 m，岸外大陆架甚窄仅 1～2 km 即进入大陆坡，太平洋深海底，故整个海岸侵蚀作用明显。

Scripps 海洋研究所在 San Diego 的 La jolla 海岸，这段海岸研究最多，有不少经典理论出于此处，如海岸泥沙运动，沉积物的横向运动，纵向运移，不少观测试验及计算出于此处。特别是 F. P. Shepard 海滩循环的理论影响深远。Shepard 从 1937—1945 年对 La jolla 海滩剖面逐月测量从高潮线至水下岸坡基部，得出海滩剖面的季节变化规律。冬季海滩受侵蚀，冬季风暴浪强烈冲刷侵蚀海滩，将海滩沙子搬到水下海底堆积，而夏季平缓的波浪，冲刷水下沙坝，将泥沙又搬回岸上，使海滩堆积，形成海岸沙坝。这一经典性实测结论，在中国，在世界各地均有相应实例，而在海岸工程上很有实用意义。故在考察南加州海岸时，总会联想到这些事。Scripps 海洋研究所拥有海岸研究部门，但海洋地质似不及海洋生物、物理与化学海洋研究影响突出。

南加州海岸紧邻太平洋深水海底，风浪大，是激浪卷浪最强烈的海岸，故南加州海滩是冲浪运动最佳区域，有暖流影响，终年温暖，海滩冲浪运动带来大量游客。加州海岸海滨旅游设施、海洋公园、度假村、海滨酒店众多，大多设计科学新颖。考察南加州海岸也为我们在中国做海岸带调查、规划、设计时，提供许多启发、参考，是一种有效的学习。

（二）Florida 西海岸

Dr. Charles T Schafer 是加拿大 Bedford 海洋研究所高级研究员，原籍美国，冬季回 Florida 过冬。这次邀请我们一起考察 Florida 西海岸，Tampa 湾沿岸是低平的沙坝、潟湖海岸区域，湿地保护区范围大，湿地水面约占 Florida 南部 1/3 的面积，多海洋生物与水禽。海岸沙坝-潟湖规模巨大，Tampa 西海岸（沿墨西哥湾）有绵延上百公里的海岸沙坝，其上叠加着宽广的、白色的贝壳海滩，成为该海岸特色，沙坝与陆地之间隔以宽大的潟湖。现海岸沙坝已开发为旅游度假区，有众多的热带度假村、酒店及私人度假别墅。

海岸大沙坝有许多 Tidal Inlet 通道，构成 Tampa Bay 与墨西哥湾的通道。当地将此一段段的沙坝（大沙坝的某一段）称为 Sand Key，在地图上有许多 Key 的地名。这系列大

---

① 摘自王颖的野外记录。

沙坝是 W、NW 向强风浪将墨西哥湾沿岸浅水区的沙坝向岸堆积，围封 Tampa Bay。Florida 南端著名的 The Key West 是沿大陆架边缘的沙坝链，可能亦是同 NW 向来自陆地方向风浪与海洋向(SE, SW)风浪相互交叠作用有关。可以判断 Florida 西海岸受 W、NW 向强风浪作用明显。

沿海大沙坝建有一些国家公园，均按其自然环境特点而建(Natural Park)，在原来大面积的海岸沙坝上的树林植被，稍加人工整理，开辟林间与沙丘、沙岛间小路，补种一些适合当地环境的亚热带树木花草，供居民活动。这些公园均有游客中心、停车场、野餐休息场所，其他都是天然的海滩、树丛、草地、湖沼水面。我们所到之处收集了许多图件资料，也购买了一些书刊音响，以作教学与国内海岸规划工作之参考。

(三) Arizona 山区的考察

现代地貌学的产生，同 19 世纪末美国开发西部有关，当时大批地质学家去西部工作，西部干旱荒漠，植被稀少，地质构造在地形上显露清晰，像科罗拉多河的研究，提出河流发源与地质构造关系的新概念—先成河、叠置河等；山地夷平面、准平原的概念，特别是时任美国地质学会会长的 W. M. Davis 提出侵蚀循环学说，把地形看成有幼年、壮年、老年的发育阶段的，促进地貌学的产生。这些故事在我们年轻时、大学二年级读书时知道的，如今能去看看，的确很高兴。

在 Arizona 可以看到大规模的层状地貌；山顶夷平面、山地剥蚀面，山坡有 2～3 阶地，山麓是大规模的洪积扇，准平原，这种境象，与我国新疆相似，只是规模更大些，地壳抬升大而下切量更大。

在 Sedona 地区，是红砂岩的层砂岩始形成于 330 百万年前(MYA)。红墙岩层之上有四层红色沉积层：

(1) Supai 组(600 呎，200 m)厚，形成于 320～285 MYA(million years ago)，该时为一位居两海之间的海岸平原，地壳上升与海洋被充填，形成砂岩与泥岩间层。Supai 组可于 Midgely 桥下的 Wilson 峡谷见到。

(2) 自 285～275 百万年前之间，河流从山地(即今日的 Rockies)流出，改道并形成一个大片的洪积平原，即 Hermit 建造(Formation)，是坡积的泥岩、砂岩及砾岩。Sedona 即建立于此建造的深红色砂岩之上。在一些道路切出处及脚下可见此层。

(3) 当向上看时，一些亮桔红色的尖塔、与尖山堡，曾是海岸沙丘层，形成于 275～270 百万年前，大约有 700 呎(250 m)厚，此为 Schnebly Hill 建造。表现为薄层的、色彩浓淡的水平地层。这一层包括四个不同的次层。易于分辨出是近灰色的 Fort Apache 石灰岩，位于本建造的底部，常会形成抗蚀的帽岩(Caprock，如 Snoopy 的鼻子)，多处均为可见此层为 10～12 呎(2.5～4 m)厚块状的带分布在红砂岩中。

(4) 在邻近 Bognton 峡谷处，石灰岩减薄，表明该处为古海洋的边缘。海洋退去(Receded)后，北 Arizona 是一片广阔的沙漠，为风力作用的内陆，在 270～265 MYA。沙丘层为淡金色，具有完美的交错层理，是为 Coconino 砂岩，因为沙丘层是气候变化与海岸隐退的继续(沉积)，所以，Coconino 层插入(穿入)偏红色的 Schnebly Hill 砂岩层中，所以，很难以分辨何处为 Coconino 层的起始点。

红砂岩系海洋沉积之灰岩，后为富含氧化铁沉积淋染成红色，海洋干枯后为风成沙漠活跃场所与沉积，嗣后，又有火山熔岩流喷溢，其地貌可与华南丹霞地貌相比较，红砂岩地貌分布具有世界对比之广泛性。

Sedona 多彩的尖山，锥形体，尖塔与峡谷是从 Colorado Plateau 切割下来的，科罗拉多高原是一个广阔的高地，延展环绕着西南部的 Four Corners 地区。在高原的南部边缘，由 Mogollon 边缘（Muggy-own）形成了 Sedona 区的北部与东部急斜的悬崖。

Sedona 东北部 Sunset Crate 是一片火山喷发区域，有 500 多座火山锥，1000 年前喷发，为红黑色火山灰、火山砾石及 Lava 熔岩流。熔岩流使地面起伏不平，在 Wupatki 处为火山岩砾、渣、灰沉积的地面，火山喷发堵塞了红砂岩河道的上游，使河道干涸。红砂岩层被印第安人建为民居小屋（400AD—1700），形成村落。

Florida 及 Arizona 两处考察的确丰富了我们的教学内容，并使古海洋与沙漠相关的研究获得力证。

三、一点建议

我们常向海外留学生、学者谈及国内情况，欢迎他们回国、来南大。现状是已有经验、高职称、有稳定工作的学者，大多是不会动的，而刚得 PHD 学位的年轻学者，常常不合国内要人的标准。最好南大能多吸引年轻学者来南大工作。

加拿大总督戴维约翰斯顿阁下访问南京大学时的发言*

Dear Professor David Johnston

your honor

Thanks again for your long period concerning and warm friendship to Nanjing University.

Learning and Creation is always the major topics of universities, and the diplomacy of knowledge is closed contact to set up scientific & higher education cooperation with International University or Institutions. Personally, I do believe so!

I was one of first scientists, supported by Chinese government, send to Canada for refresh study and joint research during early 1979, while China just to reopen door to the world. After 3 years study, Nanjing University set up cooperation with Bedford Institute of Oceanography and Dalhousie University during 1985 to 1989. The cooperation towards to support Nanjing University to create a key laboratory of Coast & Island Development, and both side has cooperated to hold the 1st International Conference of Hainan Island during Nov. 1990. Topic was concerned: Island Environment, Resources and its application.

Then, with President David Johnston's support, Nanjing University cooperated with University of Waterloo, more than 20 years cooperation. As a result, to create a Sino-Canadian College in N. U. to carry on the scientific exchanges and young scientists training more than hundred persons, especially to train 51 persons, then working in different level of Hainan government.

At this stage, as a Marine Geologist, myself, deeply hope to have your honor's consideration to strengthen the cooperation between Nanjing University with Dalhousie University and Bedford Institute of Oceanography. To improve scientific cooperation on Marine Science, including Oceanography, Marine Geology & International Law of Sea. Which will give us great help and also benefit the both side along the Pacific & Atlantic Ocean. It will improve the regional peaceful life greatly!

Thanks!

* 王颖：加拿大总督戴维约翰斯顿阁下访问南京大学时的发言，2013. 10. 21。

中加合作、推动海疆研究新拓展*

一、访问目的与意义

响应中央在南海问题上提出的“主权在我，搁置争议，共同开发”的争端解决模式与通过谈判解决问题的精神，及加拿大国家总督、南京大学兼职教授 David Johnston 于 2013 年 10 月访问南京大学讲演时，“希望南大发挥知识外交优势与加拿大大学扩大合作联系”的建议，想到加拿大是兼具广袤陆地与周边海洋的大国，地球环境资源具有特色，地质与海洋科学、教育发达；无论与我国建交前或建交后，均无领土争端；而且具有与美国在大西洋海域疆界纷争获得良好解决之前例；这些均值得我们与之学习交往，以期从中获取良知而受益。开放与国际交流，扩大我国在南海开发经营与历史传统，同样是重要的方面。

从 1980 年我国“改革开放”以来，首批派赴北美的 53 位学者之中，有 3 位赴加拿大研学（复旦大学生物学家苏德明，上海第二医科大学内科医生荣烨之以及南京大学海洋地质学王颖），学成后均回国效力。南京大学正是在加拿大国际发展机构（IDRC 与 CIDA）的资助下，于 20 世纪 90 年代创立了海岸与海岛开发国家试点实验室（State Pilot Laboratory，于 2000 年后改为教育部重点实验室），与加拿大有关大学及研究机构有着 30 多年的交往传统。故此，在南海协同研究创新中心初建之际，访问加拿大相关研学机构成为首选。本次出访主要访问加拿大达尔豪谢（Dalhousie University）、贝德福德海洋研究所（Bedford Institute of Oceanography）及滑铁卢大学（Waterloo University），目的在于学习、交流、探讨与建立正式的合作研究关系，通过学习、交流与传播，以促进对我国东南海洋疆域的研究、人才培养以及提出解决海疆权益争端相宜途径之咨询建议，提供决策参考。

二、行程与出访单位人员

（1）出访七天，包括旅途往返及日期转换共八日（2013 年 12 月 8 日～15 日）。具体行程如下：

12 月 8 日，凌晨抵达哈利法克斯市；

12 月 9 日，赴达尔豪谢大学；

9：00～9：40，会见副校长 Martha 博士与国际关系执行主任 Alain Boutet 博士；

9：45～11：45，会见理学院院长 Chris Moore 博士，与海洋系教授见面；

13：30～15：15，与地球科学系主任 Rebecca Jamieson 博士及教授会谈；

* 王颖：《南海经纬》，173－184 页，南京：南京大学出版社，2016 年。

15:30～16:30，访问海洋与环境法研究院(MELAW)，与 David Vander Zwaag 教授、Phillip M. Saunders 副教授会谈。

12 月 10 日：

9:00～10:00，访问 MEOAPR 的实验室，与 Douglas W. R. Wallace 博士见面；

10:00～10:30，与研究生院院长 Bernard Boudrea 博士会谈，讨论联合培养博士生的事宜(Joint Doctoral Degree Programming)；

10:30～12:00，与海洋系青年学者和研究生见面交流；

下午参观海洋系生物学系列实验室。

12 月 11 日：

9:00～12:00，王颖、高抒、肖冰与申锦瑜讨论议定合作协议，倪培、殷勇、杨競红参观地质系实验室；

13:30～16:30，赴 Bedford 海洋研究所参观访问。

12 月 12 日，上午与达尔豪谢大学相关负责人商定协议书；中午代表团离开哈利法克斯赴多伦多。

12 月 13 日，上午赴滑铁卢大学，与地理、水科学院系教授会谈；午餐与副校长 Dr. Nello Angerilli 会面。

12 月 14 日，星期六考察美加边境大瀑布。

12 月 15 日，由多伦多返回上海。

(2) 访问会面交流人员如下：

1. Dalhousie University

Dr. Martha Crago, Vice President Research

Dr. Alain Boutet, Executive Director, International relations

Dr. Chris Moore, Dean of Science, Professor of Psychology

Dr. Bernard Boudreau, Dean of Graduate Studies

Dr. Robert Moore, Assistant Dean of Science, Professor of Oceanography

Dr. Marlon Lewis, Professor and Chair of Department of Oceanography

Dr. Douglas W. R. Wallace, Canada Excellence Research Chair in Ocean Science and Technology

Dr. Keith Louden, Professor of Oceanography

Dr. Keith Thompson, Professor of Oceanography

Dr. Barry Ruddick, Professor of Oceanography

Dr. Jinyu Sheng, Professor of Oceanography

Dr. Tetjana Ross, Professor of Oceanography

Dr. Rebecca Jamieson, Professor and Chair of Department of Earth Science

Dr. Grant Wach, Professor, Petroleum Geoscience

Dr. David Scott, Professor of Environment and Micropaleontology

Dr. Nicholas Culshaw, Professor of Geology

Dr. Jon Grant, Professor, Benthic Ecology, Aquaculture, Conservation Biology

Dr. David Vander Zwaag, Canada Research Chair in Ocean Law & Governance, Professor of Law

Dr. Phillip M. Saunders, Associate Professor of Law

Dr. Neil Gall, Executive Director, MEOPAR

Dr. Stephen Harlten, Assistant Vice President Industry Relations

2. Bedford Institue of Oceanography

Dr. Charles Schafer, Marine Geologist, Professor Emeritus

Dr. John Smith, Director of Marine Chemistry Laboratory

3. Waterloo University

Dr. Nello Angerilli, Associate Vice President, International

Dr. André Roy, Professor and Dean, Faculty of Environment (ENV)

Dr. Peter Deadman, Chair and Associate Professor, Department of Geography and Environmental Management, ENV

Dr. Bob Gillham, Executive Director of Water Institute, Distinguished Professor Emeritus

Dr. Kevin Boehmer, Managing Director of Water Institute

Mr. Drew Knight, Director of Global Alliances, Waterloo International/Office of Research

Ms. Suping Zhao, International Relations Specialist, Waterloo International

三、出访单位与交流情况

(一) 达尔豪谢大学

达尔豪谢大学，位于加拿大东北部 Nova Scotia 省首府 Halifax 市，加东是欧洲人乘五月花号(Mayflower)至美洲大陆的登陆点之一。大学建于 1818 年，具有悠久的文化历史传统，也是相对于中部大湖区较为保守的地区。达尔豪谢大学含文学、理学、农学、医学与牙医科学、建筑与规划科学、艺术与社会科学、法学与管理科学，以及计算机技术科学等，各学科皆设本科与研究生教育，本科教育与行政管理集中于 Halifax 市南部海湾区的校本部，其他分设于 Carleton 及 Sexton 校园中。在 Halifax 市前后 4 天(12 月 9 日至 12 日上午)，主要访问达尔豪谢大学，访问成功既缘于代表团在出访前的认真准备与及时传达目的与要求信息，亦与达大校系领导对此访问的重视密切相关。

出访首日，副校长 Martha Crago 博士(多次访华到达南大)与国际关系执行主任 Alain Boutet 博士会见了全体代表，详细介绍了达尔豪谢大学系科特点、人才培养与国际合作交流情况。该校共有 18 220 名学生，教学与研究并重，是加拿大东部注重研究且经费最多的学校，以海洋科学、医学与保健、材料科学、海洋与环境法、文化与管理科学著称，新领域有农业、能源、计算机科学。达大与毗邻的贝德福德海洋研究所(B. I. O.)结合成为加拿大海洋研究的中心,2013年达大成为加拿大第一个开始招收海洋学本科生的高校。在达尔豪谢大学,海洋科学与工程系科计算机系、社会与管理系科联系密切。该校有超过120名教授在从事与海洋有关的研究工作,多学科交叉,协同合作,如海洋系、计算机系、生命科学院、海洋与环境法学院等,在海洋地质、海洋新技术开发、海洋生物、海洋法、海岸带管理等方面开展研究。海岸管理即是跨学科组成的中心与所。"MEOPAR"是杰出的网络组织:由四个研究所联合建立"Halifax Marine Institute",并联合加拿大东部三省(新斯科舍、爱德华太子岛与纽芬兰)工业与海洋,结合450家研究单位从事与海洋相关的研究。达大海洋研究与贝德福研究所关系密切。会见时,中科院院士王颖教授阐明了与达大交往的传统关系、本次出访的目的与人员组成,希望能进行合作研究项目申请"The Cooperative Study on Subtropic South China Sea and Subarctic Atlantic Ocean: Environment, Resources and Marine Boundary Dispute"。项目内容包括共同研讨,交流讲学,交换研究生,合作举办国际研讨会,出版论文等。在2014年开始阶段,由南海中心邀请加方教授访华讲学,举办海洋法培训班等;继以双方各自申请国家资助等。Martha Crago副校长表示支持,将征求院系人员意见以具体落实,并于10日晚宴请了代表团全体,表达愿意再次访问南大落实合作事宜。

达大助理副校长、负责工业关系的Stephen P. Hartlen先生访问过我国,午餐时会见了代表团全体,表达了也希望在应用学科方面加强合作交流的意向。代表团访问三个院系时,分别由理学院院长Chris Moore博士(心理学教授)、副院长Robert Moore(海洋学教授)和研究生院院长Bernard Boudreau博士接待并交谈具体合作事宜。嗣后,由理学院院长、国际关系执行主任向南大代表团提交Chris与南大戴者华研究撰写的合作协议初稿,请南大代表提出修改意见。

12月9日上午研究生院院长Bernard Boudreau博士会见南大代表时,介绍该校有3 500名研究生,其中6%从事研究论文工作,4%为职业学位,700～800名博士生,900名来自国外的研究生中27%为中国学生,以学习计算机科学为主,主要模式为"4+2"模式。来达大学习的本科学生是三年半在中国、半年在达大、不需考TOEFL,但要学习英文,可学计算机科学及其他学科。

本次出访主要集中于理学院与法学院系科。达尔豪谢大学的理学院包含生物学系(生物化学与分子生物学)、微生物与免疫学系、化学系、地球科学系、环境科学系、数学与统计学系、海洋科学系、海洋生物学系、物理与大气科学系、心理学与神经系统科学系、经济系等。理学院所含的专业具有学科交叉与发展进步快速的特点。相对地说,南大的系科仍具传统的特色。理学院院长Chris Moore教授访问过南大,与南大外办副主任戴者华拟定合作协议文本初稿。

(1) 海洋系教学研究工作具有学科交叉特色。该系包括物理海洋关于潮汐和海流、海水化学、海洋动物与植物学,以及海底沉积等专业。学者的研究领域从极地到热带海洋,重视海-气相互作用、全球变化与人类活动影响。由加拿大10所大学、40名研究人员与研究生组成的"全球海洋水圈与大气圈预报及可预报网络(COAPP)"的总部及海洋环境预报中心设在达大海洋系。我们在达大会见了海洋系教授Robert Moore,他是达大主管科研的理学院副院长，多年从事海洋科学教育。Douglas W. R. Wallace 教授是“加拿大杰出教授组织”的主席，全加拿大资助 14 个有潜力冲诺贝尔奖的优秀科学家，他是其中唯一的海洋科学家。该组织是由非政府组织(NGO)支持，是海洋科学的三个顶级组织之一，包括海洋生物与微生物、物理海洋、化学海洋、地球科学以及海洋法等。他表示了与南大合作的诚挚愿望，此反应重要！我们拟邀请他到南大讲学(海洋学与海洋化学)。在海洋系还会见了 Keith Louden 博士，他曾与地球科学系的 Mladen Nedimović教授合作从事海底地球物理勘探地壳与洋壳、研究水合物。“MEOPAR”即 Marine Environment Observatory Prediction and Response，是加拿大设立的 14 个国立研究中心之一，而海洋中心实验室放在达尔豪谢大学，并从海洋系所在的 Life Science Building 后侧向外扩建了透亮的以玻璃幕墙相隔的实验室，目前有 2 个雇员，将招聘至 5 人，主要从事海岸海洋环境观测及海洋紧急情况预报。2012—2017 年，加拿大政府资助该中心实验室约 2 500 万加元。

(2) 地球科学系(Dept. of Earth Science)是由建于 1879 年的地质系发展而成的，当时是加拿大大学中的第一个地质系。该系现有 12 位教授及 5 位教员(“Instructor”指讲师及助教)；有本科生 120 名，15～17 名修 honours 学位(大学毕业后学习 2 年，介于本科与研究生之间，具有双学位性质)的学生；30～40 名研究生。现任系主任 Rebecca Jamieson 博士是 1980 年进入达大的，研究前寒武纪变质岩、适应于加拿大地质的环境背景。该系具有地质与环境结合的地学特点，拥有 Coastal Geology & Geomorphology, Orogenic Tectonics & Geodynamics, Geodynamic model, Geochronology & thermochronology, Petrology, Marine Geophysics, Geochemistry & Mineral Deposit 等方面的教学与研究专业组织。她谈道：“因业务领域不宽，而不提供专业职位工作。”笔者认为他们将地球作为一个整体研究，从表层到深部，将地貌、地球动力与岩层构造、矿产等联系一体，反映出地球科学(Earth Science)全貌。这点令人惊讶：因为一直感到加拿大东部较保守，但从其地球科学所包括之内涵来看，却比我们先进。该系在地貌与地层结构成像、结合加拿大地盾与玄武岩流活动之地区特点，以及在钻石矿研究方面有独到之处。Rebecca Jamieson 教授认真接待代表团并介绍该系情况，对合作研究表示“需征求相关教授的意见后定”，似较其他系有保留。在返回南京大学后，笔者收到 Robert Moore 的电子邮件，他介绍 Mladen Nedimović博士因出差在外，未与南大代表会见，但表达了与南大合作的迫切愿望。随即笔者又收到 Mladen Nedimović博士的电子邮件：介绍自己是地球物理学家，应用地震剖面研究地质结构、地震灾害等，深望与南大合作……地球物理与地震地质正为我们学校所需，却是 Rebecca 未曾表态的。笔者当即于 12 月 20 日回信表示，肯定与其合作。

在地质系，倪培教授与杨競红副教授参观了相关的实验室，认为南大与之合作将会扩展领域，收益也会很丰硕。我国地学人才培养需探求从地球圈层系统交互作用融为一体的效应，应改变现有的以圈层分设系科的状况，至少应有一所大学涉及这方面的综合研究与教学。

目前在达大约有 1 000 名中国留学生(1979 年 2 月时仅王颖一人来自中国大陆)，其中研究生 200 名，主要学习工商贸易、医学护士、管理工程与计算机科学，以应用为主。我们会见了8位在海洋系与地质系留学的研究生及访问学者。他们分别来自中国海洋大学、中国地质大学,厦门大学及中山大学,以学习物理海洋、海洋地质及遥感模型为主,其中来自中国地质大学、在达大攻读博士学位的陶静是研究海洋悬浮体及外滨沉积的,她表示愿在毕业后到南大工作,我们也表达了诚挚的欢迎。

在海洋系学习的中国留学生主要师从申锦瑜教授。申教授毕业于河海大学海洋水文专业,后留学加拿大的纽芬兰纪念大学,获博士学位,在达大海洋系从事博士后研究,留校工作。由于他重视多学科交叉,教学研究成果优异,发表成果多,已升至正教授。目前他有专项研究课题,热心服务于中加交流,为达大所重视。我们这次访问,他给予了鼎力支持与热心帮助,可以说是中国留学生在加成功建业的楷模。

(3) 达大的海洋与环境法学院(The Marine & Environmental Law Institute),位于自校本部行政楼向东沿大学街(University Ave.)的独栋楼。大学街两侧分别为达大的院系、艺术中心,直至医学院,没有围墙分隔,却组成了实质性的 University Avenue。该学院以海洋法与管理、环境法与管理、特殊区关注(大西洋,北极,加勒比及东南亚)以及政体建设与维护等四个方面的教学研究与成就,在全球具有领先地位。

12月9日在当天访问达大的一系列活动十分紧凑,至海洋与环境法学院时,已届黄昏,但是达大的国际关系执行主任 Alain Boutet 博士、加拿大海洋法与管理研究首席科学家、达大海洋法教授 David Vander Zwaag 博士与 Phillip M. Saunders 副教授均等候在学院门前迎接,热情地介绍了该院海洋法方面有5位教授,执教海洋与环境法,学生毕业时发予法学证书。我方代表提出:希望2014年达尔豪谢大学法学院教授赴南大讲学,举办海洋法培训班,介绍海疆争端实例、解决途径与结果等。David 提出对短期培训最感兴趣,已与中国政法大学合办过,他们对越南与中国在南海的争论、印度与巴基斯坦争论等十分关注,且在北京时与中国海洋法大法官高之国研究员曾进行了会谈交流等。Phillip 教授介绍了达大举办过多次区域海洋与环境法培训,如南太平洋区及边界项目等。他们均表示愿访问南京大学,进行合作项目研讨。他们介绍了由达大和马耳他国际海洋研究所(IOI)共同主编出版的刊物 Ocean Year Book(海洋年鉴),每年一期,汇集全球海洋事务,迄今已出版27期。

总之,对海洋科学系、地球科学系以及海洋与环境学院的参观访问是达到了初建联系、互通情况与表达双方合作愿望的目的。在飞雪冰冻的当天,访问与交流却是紧张而热烈的。

(二) Bedford Institute of Oceanography

Bedford Institute of Oceanography (B. I. O.)是加拿大于1962年建立的研究海洋的国立机构,位于 Halifax 市对面与之相隔 Bedford Basin 的 Dartmouth 市(类似与南京隔江相望的浦口区)。B. I. O. 与美国的 Scripps 海洋研究所、Woods Hole 海洋研究所一起被誉为世界三大海洋研究所。由于它建立较新近,加上加拿大土地广博,所以研究所设备较全,是以海洋地质(Atlantic Geoscience Centre, A. G. C.)、物理海洋、海洋化学以及海洋生物与生态学为特长的组合,并拥有调查船队、码头、技术支援车间等仓储设施,规模与设备堪属国际一流。王颖于1979年参加 A. G. C. 调查研究 Labrador 海中部及大西洋,是第一位从中国大陆赴 B. I. O. 的访问学者。至今该所已有数位来自中国的正式研究人员与多位进修人员。基于 20 世纪 80 至 90 年代参加南京大学海岸研究室在南海的调查研究，并帮助南京大学建立海岸与海岛实验室的 Charlie Schafer 博士(海洋地质研究员，已退休)及 John Smith 博士(海洋化学研究员)的支持，代表团成员于 12 月 11 日下午参观了海洋化学实验室系列，海洋沉积实验室的粒度、钻孔分析与沉积柱仓储室，以及礼堂、图书室等，获得了深刻的印象，认为可为南大海洋科学院的建设发展所效法。代表们还观察了位于 B. I. O. 大门外的冰蚀基岩磨光面与擦痕，领略了最后冰川活动所造成的加拿大大地风貌。11 日晚代表团成员回请了达尔豪谢大学接待的有关教授，交谈更加深入。

代表团一行于 12 日上午与达尔豪谢理学院院长、国际关系执行主任进行最后会谈，研定了由高抒、倪培、肖冰三教授修订的两校协议原稿(附件 1)，双方达成基本共识，商定将进一步征求意见后经通讯交流定稿。双方还商定：邀请南大校长陈骏教授于 2014 年 5 月访加，签订经双方审定的协议；南大邀请加方副校长等一行于 2014 年 10 月访问南京大学，进一步商定合作计划。

代表团一行于下午 1 时半离开 Nova Scotia 赴多伦多市，以备从多伦多市乘国际航班回国。

(三) 滑铁卢大学

12 月 13 日(星期五)，代表团成员出访距多市约 60km 的滑铁卢大学，拟商讨进一步巩固与扩大和滑铁卢大学的合作交流。当日多伦多地区大雪，包租的汽车因零件冰冻，难以发动，司机延迟到 11 时 20 分才到达滑铁卢大学。代表团会见了等候在该处的六位成员：André Roy 教授，环境学院院长；Dr. Peter Deadman 副教授，地理与环境管理系主任；Robert W. Gillham 博士，杰出的退休教授，现任滑铁卢大学水科学研究所执行主任；Kevin Boehmer 博士，水科学研究所管理主任；Drew Knight 先生，滑铁卢国际研究室全球联盟主任，1990—2000 年组织与南大合作的负责人；赵素平女士，滑铁卢外办、国际关系专家。

因为迟到，王颖开门见山地直述访问目的：此次主要是探讨南京大学南海研究协同创新中心与滑铁卢大学有关院系建立继续合作关系，尤其是在水科学与地理信息科学领域的合作；希望进一步加强在淡水资源环境等方面的交流讲学与研究生培训；希望中加教授双方互访与讲学，中、加双方交换研究生等。加方水科学研究所成员表达了与南大合作的诚挚愿望，谈到目前来校的中国自费研究生每年约有 6 名，攻读硕士学位或工程师证书。录取的关键是专业教授接受与否。中国学生每学年学费约 10 000～11 000 加元。环境学院院长表示：加拿大学生也愿赴中国学习，但主要通过教授与中国的合作项目去；通过“2+2”的研究生培训模式每年接收 5～10 名中国学生，而其有 2～5 名来自南大，以学习物理、经济、历史及地球科学为多。地理与环境专业的中国留学生每年有 2～4 名。Peter Deadman 介绍：学习人文或自然地理学的研究生多与水资源及湿地水文学相联系；该校有一个气候变化研究中心(Climatic Change Centre)，以教师为主体，也关注海岸资源、海平面变化以及加勒比海区的旅游资源；“海平面变化”是研究大湖区及北冰洋冰层的融解效应、蒸发效应及在圣劳伦斯河建坝对海面的影响效应，具体而实际；在淡水资源方面，关注水资源管理、水质与生态、营养物循环等。加方表示以上各方面均愿与南大积极合作。

滑铁卢大学主持国际合作的副校长 Nello Angerilli 教授中午招待代表团午餐时向王颖表示：应当建立正式的南京大学与滑铁卢大学合作研究所（他再次强调是 Nanjing University and Waterloo University Research Institute）。王颖表示支持此建议，先从地球科学着手。Nello 提出要与王颖就此事以电子邮件的方式联系讨论，并将于 2014 年 4 月访问南大时与校领导研定此机构事宜。

王颖希望 2014 年 5 月陈骏校长出访加拿大两处院校，签订正式合作协议。会谈中笔者感到：本次虽在滑铁卢大学停留的时间短，但因双方有较长的合作关系，相互了解深入，更可能是多伦多地区院校较加东更为开放，所以开门见山地直接讨论合作项目内容，更见成效，尤其是建立正式合作机构的建议，令人鼓舞。

综上所述，以南大七位教师代表南海研究资源环境与海疆权益争端平台出访加拿大院校，获得了积极圆满之结果。通过访加，代表团成员亦深受鼓舞。

利用到达 Halifax 市的星期日下午及返回前的星期六休假日，首次访加成员特别是地球科学教师考察了 Peggy's Cove 花岗岩海蚀岸及美、加边界的 Niagara 大瀑布。前者为天寒地冻的严峻的海岸景象，后者因暴雪纷飞、水雾迷漫，难观全貌。但五位代表仍冒严寒、临河观察，以致外衣结冰，才看到了偶尔显现的马蹄形大瀑布，为其雄伟的地貌与水力资源而欣服。

总之，南大代表通过积极的、紧张的活动达到预期目的，获得建立合作项目和加强对加拿大环境、资源与人文教育深入认识之双丰收。

四、结果与建议

(1) 制定与达尔豪谢大学、滑铁卢大学合作协议，建立双方教、研合作交流的正式关系。

(2) 2014 年举办两校海洋法培训班，开始教授讲学交流。

(3) 研讨与筹建南京大学与滑铁卢大学正式研究机构。

(4) 平台拟以 2013 年 30 万经费支持，即时开展上述交流活动。

(5) 建议组织以陈骏校长为首的小型代表团于 2014 年 5 月访加，签订合作协议。

附件 1

Agreement of Collaboration
between
Dalhousie University
Halifax, Nova Scotia, Canada
and
Nanjing University
Nanjing, Jiangsu, China

In order to facilitate the broadest academic relations and research collaborations, Dalhousie University and Nanjing University agreed to formally establish their cooperative relationship and in doing so provide a framework for further collaborative activities between the two institutions.

1. Areas of Academic Interest

This agreement initially will reflect the current informal collaboration in the field of ocean studies between the two institutions. Depending on the availability of resources, every effort will be made to maximize productive relationships and exchanges as widely as possible between interested members of both institutions. As a first step, the agreement focuses on the ocean studies including (a) Oceanography, (b) Earth Sciences, and (c) Law of Sea. Financial support to these three areas will be provided by the recently established research network known as "Collaborative Innovation Center of South China Sea" (CICSCS) funded by the Ministry of Education, the People's Republic of China.

2. Cooperation in Ocean Studies

It is agreed that the two institutions will cooperate in the development of cooperative activities, subject to available resources. Cooperative activities under this agreement include or may be expanded to include faculty and researcher exchanges; student exchanges and mutual recognition and accreditation of courses; study abroad or visiting student programs; joint academic programs; graduate student advising;

development and/or delivery of courses; joint project development for external funding support; and any other area that may be mutually agreed upon by the two institutions.

3. Cooperative Research

It is recognized that significant opportunities exist for cooperative research between the two institutions. Such clearly beneficial activities may require specific arrangements, in particular with respect to financing. In view of the importance of cooperative research, both institutions agree to give high priority to this within their overall relationship, and to make every effort to develop programs within this area.

4. Exchange of Faculty Members and Researchers

Both institutions will endeavor to facilitate visits by faculty members and researchers from the other. Visiting faculty members and researchers will comply with administrative procedures required by and the regulations of the host institution.

Both institutions may provide economic support for visiting faculty members or researchers from the other institution, but are not required to do so.

Both institutions will accord visiting faculty members and researchers from the other institution the use of research space (which must be negotiated and agreed upon prior to the visit), libraries and other facilities, opportunities to audit lectures free of charge, and to the extent possible, other common courtesies generally granted to guest scholars.

5. Student Exchanges

Both institutions will consider establishing exchange programs for graduate and undergraduate students. The detailed conditions governing any such student exchange will be set forth in a separate agreement.

6. Exchange of Scientific and Information Materials

Both institutions agree to exchange as widely as is practicable such items as university calendars, prospectuses, course outlines, teaching materials, reference materials, scientific publications, journals and databases.

7. Operational Constraints

It is recognized that the implementation of program activities under this agreement will in every case be dependent upon the availability of necessary resources, either from within the institutions concerned or from external sources. In the case of externally funded projects, cooperative activities will be subject to the terms of the project of which they are a part.

8. Specific Activities

Both institutions agree that specific activities that may take place as a result of this agreement may be documented in subsequent arrangements. In particular, the two institutions agree to work towards the activities in three areas of (a) Oceanography, (b) Earth Sciences and (c) Law of Sea. Specific tasks include

(1) An international symposium to be held in October 2014;

(2) One-day lectures at Nanjing University to be given by professors at Dalhousie University (travel expenses of ten DAL professors will partially be covered by Nanjing University);

(3) The Joint Doctoral Program (JDP) in Oceanography, Earth Sciences and Law of Sea;

(4) Selection of supervisors for PhD students registered in the JDP;

(5) Collaborative research projects (proposals to be submitted jointly by researchers at the two institutions to funding agencies such as the Ministry of Science and Technology of China, the Natural Science Foundation of China, and the Natural Science and Engineering Research Council of Canada);

(6) Short-term training courses in the Law of Sea.

9. Institutional Contacts

The contact at Dalhousie University for this agreement will be Dr. Alain Boutet, Executive Director International Relations. The contact at Nanjing University will be DAI Zhehua, Deputy Director, Office of International Cooperation and Exchanges.

10. Duration of this agreement

This agreement shall come into effect on (1st April 2014), and shall have a duration of five years, after which extension, amendment or other changes may be made as agreed by both institutions. Within the initial five year period, the terms of this agreement may be amended as required, subject to approval by each institution.

11. Approval Procedure

Should this agreement require the approval of the Senate or other bodies within either institution, the signatories agree to initiate action for such approval at their respective institutions.

Signed by:

Dalhousie University	Nanjing University
__________________	__________________
Dr. Richard Florizone	Professor Jun Chen
President & Vice Chancellor	President
Date: __________________	

王颖教授在中国地理学会海洋地理专业委员会成立大会的报告*

主席、各位代表、各位来宾：

中国地理学会海洋地理专业委员会成立大会暨第一次专业委员会会议现在开幕了，我代表筹备组向大会汇报筹备经过。

一

1987年4月14日—18日，在中国地理学会五届二次理事会上通过成立《海洋地理专业委员会》，挂靠在南京大学大地海洋科学系，由本人负责筹备。经发函向全国27个沿海单位征求意见，到今年为止，共收到22个单位的来函来信，表示支持海洋地理专业委员会的成立，表示愿意参加专业委员会.其中有12个大学的系所，它们是：北京大学地理系、华东师范大学地理系、华东师范大学河口海岸研究所、杭州大学地理系、中山大学地理系、华南师范大学地理系、福建师范大学地理系、辽宁师范大学地理系海洋资源所、同济大学海洋地质系、青岛海洋大学海洋地质系、厦门大学海洋系热带海洋研究所及南京大学大地海洋科学系。还有10个科学院、海洋局、能源部、交通部和地矿部等系统的有关研究所，它们是：中国科学院地理研究所、南京地理与湖泊研究所、广州地理研究所、南海海洋研究所、国家海洋局第二海洋研究所、海洋环境保护研究所海洋地质室、海洋科技情报研究所、南京水利科学研究院河港所及地矿部海洋地质研究所。

这些单位可以代表我国沿海各省的海洋教育、研究以及事业机构，因此建议上述各单位至少出任委员一名，经各单位推荐，报请中国地理学会理事会研究审定，共推选出委员33名、主任委员1名、副主任委员7名以及顾问3名、秘书长1名。人选主要是考虑到海洋地理的各个专业以及各大单位与区域的代表性。其中可能高校的同志稍多了些，这反映了我国目前已开展海洋地理研究工作单位之实际情况。愿海洋地理专业委员会在中国科协、中国地理学会的关怀与领导下，能为推动海洋工作做出一定的贡献。

二

成立海洋地理专业委员会既作为国际地理学联合会海洋地理研究组的对口组织（IGU Study Group on Marine Geography），亦是符合当前地理科学发展的需要。

* 王颖：《南京大学学报》（自然科学，地理学专辑），1988年总第9期，第201-203页。

(一)

IGU的海洋地理研究组于1986年成立,经中国地理学会吴传钧副理事长推荐,朱大奎教授为通讯成员,以后由于成立海洋地理专业委员会,王颖被聘为成员(corresponding member)。而在该成员组织中,青岛海洋大学杨作升副教授为成员之一。

第一次IGU的海洋地理研究组会议是1987年7月3日—6日,在英国Cardiff的Wales大学科技研究所(Institute of Science and Technology,University of Wales)举行。会议据发展趋势,确定三个研究题目。这三个题目主要涉及1982年的国际海洋法讨论,将问题集中于:人类活动对海洋管理之相互关系。这是涉及各方面的广泛关注的大问题。在这个课题下,提出了三方面问题。在第一次讨论会上,大家兴趣大,并产生了下属课题。

(1) 海洋法中的地理牵连(Implication),即涉及海洋法的地理问题

① 海洋的法律与政治地理;

② 海洋边界的确定;

③ 海洋资料基础和技术管理。

(2) 海洋的利用与管理

① 以发展中国家为重点的国际组织与发展;

② 与海岸相接的海滨水域和外滨以外海域的管理问题;

③ 关于海洋各方面利用的管理,如:航运、渔业、油气开发等。以及由此产生的区域课题。

(3) 海洋地理学的国际合作

海洋地理学发展中的国际合作,包括研究项目、海洋地理之训练与应用,以及各个不同国家的海洋地理发展问题等。

今年8月22日—23日,将在悉尼召开的第26届地理学大会上召开海洋地理研究组第二次会议。我想,我们这次成立海洋地理专业委员会之事,将要在会上有所报道与反映。同时,欢迎大家写稿,我可代为综合发言。

(二)

众所周知,地理科学是以地球表层为对象,研究它的资源、环境、以及人与环境之关系与影响等。地理科学在战略决策、规划、设计以及政治、经济等各方面发挥越来越重要的地位。对于陆地我们尚需投入大量的工作,而面临的世界性的人口、能源、水资源、矿产、食物、土地和环境,以及边缘政治等方面问题摆在我们面前,使我们不得不向占地球表面2/3的海洋进军,开发研究海洋。海洋既是地理环境中的自然要素之一,亦是资源的巨大宝库一潜在的源地。世界各先进国家在20世纪50年代就已掀起"向海洋进军"的热潮。现今全世界海洋产业的年收益约计为2 800亿美元。21世纪是"海洋经济"的世纪。经济开发又涉及领海、大陆架、专属经济区、公海开发等项政治、外交与国际问题。这些是世界各国所面临的问题,也是我们"海洋地理"从事研究的任务与领域。我国作为一个既是大陆又是海洋的国家,需要急起直追,积极开展海洋地理研究工作。同时,考虑到当前世界经济开发与政治活动有自西向东,向亚洲、大洋洲太平洋地区转移的趋势,这些地区将是21世纪最有活力的区域。我国实行开放政策,十三大及正在召开的七届一次人大会上,

提出了建立沿海经济区、发展外向型经济等。这些,都将促进海洋地理工作之开展。基于这样的国内外形势的推动,中国地理学会五届二次理事会决定成立海洋地理专业委员会,来适应国家建设的需要。这给我们地理科学工作者开辟了一个发展的新领域,提出了一项新任务。我们应该注意从自然、经济、技术、政治、国际、外交等各方面交叉地进行海洋研究。我想这也是海洋地理区别于一般海洋学的一个特点,是社会科学与自然科学、技术科学的新交叉,是国内与国际问题的新交叉与渗透。

三

参加这次会议的代表中,既有德高望重的老一辈科学家,也有占绝大多数的中青年学者。代表们来自沿海各单位,专业涉及海洋水文、海气交换、海洋动力、工程、地质地貌、海洋沉积、生物、环保、海洋遥感、信息系统与海图制作、海岸与海洋经济规划、海岸带立法、管理、海洋法等诸方面,人数不多,但代表性广泛。我想这反映了海洋地理专业委员会是中国地理学会领导下的一级团体,以海洋地理为主要课题,涉及海洋、海岸的资源开发、环境利用、立法管理以及国际交流等诸方面,是一个多学科交叉的学术团体。希望专业委员会的建立能加强我们各单位之间合作交流。加强对海洋资源、环境的综合研究与区域开发规划,为政府提供规划决策的前期研究与决策依据。同时,促进国内各单位的学术交流以及与国际地理学联合会有关组织对口,积极从事国际学术交流活动。

这次大会希望能开始这种交流与协作,希望各位代表能对专业委员会的性质、任务、工作目的、联络办法等进行讨论,取得明确意见,并请对专业委员会组成提出宝贵意见,以便使海洋地理专业委员会真正地推动这方面工作的开展并做出贡献。

四

最后,我想说:专业委员会设在南京大学大地海洋科学系,这是地理学会对我系建设的热情鼓励与大力推动。同时,因为它是第一个挂靠在我系的学术团体,使得我们感到由衷的高兴。我们非常感谢各兄弟单位对我们的支持与帮助。南京地处东部沿海之中并属于内外联通的黄金水道之滨。所以从地理位置看,南京是一个南、北、河、海之中心。从单位设置看,这里是地学的一个中心,有中国科学院南京地理与湖泊研究所、南京地质古生物研究所、南京地质土壤研究所、河海大学、南京水利科学院、南京航务工程专科学校、南京师范大学地理系等各单位。有理、工、文、法等各学科在这里交叉。

我们系从事海岸地貌、沉积、海图等方面研究与教学多年,近年来又增加了遥感、港市海洋经济等项专业工作。1986 年成立了跨系的海洋科学中心,挂靠在我系。又增加了台风与海气交换、海藻、大米草、构造、沉积岩等方面。加之,综合性大学有贸易、经济、海洋法、现代分析中心、计算中心等方面配合,因此,为开展综合性海洋工作提供了一定的基础。海洋地理专业委员会的成立必然促进南京大学海洋工作之开展。我们愿尽自己微薄之力,做好这方面的学术交流、组织服务与联络工作。谢谢大家在百忙中来宁参加会议,希望诸位不吝提出意见,以帮助我们搞好会议与今后工作。祝代表们在宁愉快。谢谢!

海洋研究倍受重视*

第7次太平洋海洋科技会议于1996年6月16日—22日在美国夏威夷州的檀香山举行。会议出席人员约400人左右，主要来自美、日、朝、中。

会议设有40个分会。在海洋科学与技术方面有20个：海洋遥感；海洋声学系统；专属经济区制图与电子海图；海洋地质制图新发展；海啸（系太平洋地区重要专题，日本人研究多）；水下汽车与海洋机器人；高性能船只；全球定位系统在遥感中的应用；海洋海平面变化（主要是澳大利亚人研究）；海洋与海岛工程；飓风与风暴潮；化学海洋学；海洋光学；石油漏溢处理；海洋仪器的进展；海洋潮汐模拟/数字化制图；珊瑚礁；第四纪大陆架，等等。在海洋资源管理方面有15个：海事与政策；海洋采矿；海洋能源；渔业资源演化与管理；海洋生物技术发展；海洋环境管理；海洋空间利用；海洋科学与技术教育；海洋资源管理；太平洋持续发展原则；环境信息资料库管理；河口湾与海岸水域的水质管理；太平洋海洋科技事业组织，等等。另5个是：妇女与科技；工业论坛；海洋安全体系的发展趋势；海洋声束；太平洋灾害中心；返回海洋——教育内容。

各分会主题全面反映出海洋科技重点和太平洋地区特点，当前重要项目有海底制图（涉及到海洋权益），海洋科技发展相关的组织、教育、仪器，以及立法管理等。

大陆架分组是新设立的，在这一方面，澳大利亚学者着眼于对大陆架海域资源的管理；香港大学严维新博士着眼于沉积孔取样分析、定年分析与气候变化，属于传统学科内容，不过将过去深海研究的一套用于陆架浅海区；笔者与朱大奎教授则从河海体系来分析陆架沉积作用与堆积过程，属于动力沉积学研究。

讨论内容偏重于海洋高新技术。近年来，海底制图越来越受到重视，美国NOAA（海洋与大气局）有大量经费支持海底制图。联合国海洋法将毗连海域划归沿岸国，扩大了海洋管辖权益，各国开展了积极行动，以使划界有益本国。开展此项工作，首先需摸清海底的地形环境和蕴藏的资源，美国进行此项目已近5年，主要海域已测制详细地图。我国“九五计划”已在海底制图方面有1.5亿元的投资，主要项目由海洋局与地矿部承担。

* 王颖：《国际学术动态》，1997年第3期，第74页。

在南京大学海洋科学研究院扩建与更名典礼上的讲话*

尊敬的陈校长，各位院士，各位教授与代表：

酝酿已久的"海洋科学研究院"扩建与揭牌仪式今天举行，十分高兴。

一、"21世纪是海洋世纪"，在我国揭开了新的一页：继发射"天宫"航天器，我国的"蛟龙号"深潜器与第一艘航母正式下海，在东海钓鱼岛及南海岛礁海域进行渔政定期巡视，在黄岩岛海域对峙，以及"海洋科学"列入"十二五"重点科技发展规划等。说明我国对海洋科学、海洋开发及维护海疆权益之重视。

我国是四海一洋，除台湾以东直临太平洋外，渤、黄、东、南海外围均有一系列岛屿环绕，是半封闭的陆缘海域(图1)。海疆问题与周边邻国均有争执，我国海洋事务所面临的自然环境与政治形势均十分严峻！

二、南京大学海洋科学研究与教育始于1958年周总理号召"三年改变港口面貌"，地理系师生参加"天津新港回淤来源研究"，开始了南大以动力、地貌、沉积相结合进行选港的科学研究特色与人才培养。1962年建立先后有5名专职科研人员的海岸研究组(王颖讲师，陈万里、何浩明、张忍顺、张连选)。

1963年　任美锷教授下海，担任国家科委海洋组成员。

1984年　扩建为"海岸地貌与沉积研究室"，14名编制。

1986年　联合气象系、生物系、地质系及法学方面力量建立"南京大学海洋科学中心"，多学科交叉、联合从事海洋研究。

1988年　由加拿大IDRC资助41万美元，扩充购置海洋设备。

1990年　世行贷款40万美元，建成由海陆带勘测、实验分析、信息集成与模拟三个系列的"海岸与海岛开发国家试点实验室"(State Pilot Lab of Coast & Island Exploitation)属国家计委、教育部及自然科学基金会领导，试点成功。后改制由国家科委领导。

2000年定为教育部重点实验室至今。以研究海陆交互作用带——海岸带、大陆架、大陆坡及大陆隆为特色，目前实验室有实虚并举的(不含研究生)编制42名，从事科研教学与应用开发，在选建海港(曹妃甸深水港、洋口深水港)，数字南海，海南岛研究开发与干部培养方面，成果卓著。并承担海洋地质学省重点学科与海岸海洋科学新兴学科点的教学任务，培养本科、硕士与博士研究生及博士后工作站。

* 王颖：在南京大学海洋科学研究院扩建与更名典礼上的讲话，2012.6.5。

图 1 岛屿环绕的边缘海(附彩图)

三、南京大学地学学科强，加强海洋学科建设，是适应国家需求，增强南大学科群体力量，为国贡献的重要举措。

海洋科学不仅是自然科学，也包含社会、人文与工程科学，扩建更名为南大“海洋科学研究院”，可联合生机勃发的生命科学、经济学、法学、历史与管理等学科力量，通过合作承担国家研究项目及大型应用项目(如 2011 南海疆域综合项目)，培养复合型海洋人才，发展具有特色的海岸海洋学科，逐步在南大建立一级学科点，为海洋经济、军事、文化发展，为维护海洋权益贡献力量。

四、经酝酿后建议的院组织机构

“研究院”实行院长与副院长协商负责制。

设有学术委员会和学科建设咨询委员会，目的是加强老中青科学家结合、促进校内外密切合作，把握研究院学科发展方向，为研究院设立战略发展目标，支持海洋学科的发展，并最终为在南京大学海洋学一级学科点提供指导与咨询。

研究院以“海岸海洋科学系”及海岸与海岛开发教育部重点实验室现有的研究力量为

基础，充分吸收南京大学其他院系与海洋学相关的研究人员，实行固定研究人员与兼职研究人员相结合的科研团队建设机制，各研究团队之间通过综合性项目进行有机协作。

初步建议的组织构架组成人员如下：

1. 海洋科学研究院行政机构建议组成：

院　长：高抒教授

副院长：邹欣庆教授（常务），王元教授（大气），
　　倪培教授（地科），安树青教授（生物），
　　肖冰教授（法学），朱晓东教授（环科）

院长助理：杨旸，许叶华

院长办公室主任：蒋松柳

2. 海洋科学研究院学术委员会建议组成：

学术委员会主任：胡敦欣院士（中科院院士、海洋研究所）

学术委员会副主任：蒋少涌教授（南京大学）
　　鹿化煜教授（南京大学）
　　王晓蓉教授（南京大学）
　　范从来教授（南京大学）

学术委员：赵美训教授（中国海洋大学）
　　左军成教授（中国海洋大学）
　　于仁成研究员（中科院海洋所）
　　李铁钢研究员（中科院海洋所）
　　印萍研究员（青岛海洋地质研究所）
　　业渝光研究员（青岛海洋地质研究所）
　　戴民汉教授（厦门大学）
　　余克服研究员（中科院南海研究所）
　　李满春教授（南京大学）
　　侯书贵教授（南京大学）
　　凌洪飞教授（南京大学）
　　葛晨东教授（南京大学）
　　……（待续，待定）

3. 海洋科学研究院学科建设咨询委员会建议组成：

学科建设咨询委员会主任：王颖院士

学科建设咨询委员会副主任：符淙斌院士，李成教授

学科建设咨询委员：王德滋院士
　　薛禹群院士
　　伍荣生院士
　　张峻峰教授……

五、初步计划形成两个大的研究方向及十一个研究团队包括：

海洋自然过程研究方向：

对海岸海洋地质历史时间的演化过程进行反演，特别是晚更新世以来沉积物所揭示的全球变化信息提取；

研究陆海、海气相互作用机理的海岸海洋环境效应；

揭示海岸海洋沉积动力过程、海洋地球化学过程；

探讨海岸海洋生态系统，特别是湿地生态系统过程及其对地球系统的影响等。

海洋资源、环境与管理研究方向：

研究海岸带湿地资源、空间资源的分布规律以及开发利用模式；

研究中国边缘海矿产资源的形成与分布规律；

根据海岸海洋的自然过程特点，研究海岸海洋环境及保护策略；

依据海岸海洋资源的特点，探索我国特色的海洋经济发展模式；

从资源与环境特点出发，结合海洋法的研究，探索一体化管理新思路；

从地学视角，加强海洋疆界理论的研究，服务于维护国家海洋权益。

可形成学术梯队的学科与核心成员，建议如下：

1. 晚更新世以来海陆交互作用、地貌特征与沉积相

团队学术带头人：王颖院士

团队成员：殷勇，张永战，李海宇，何华春，蒋松柳

2. 海洋沉积动力学

团队学术带头人：高抒教授

团队成员：汪亚平，高建华，张继才，杨旸

3. 海洋地球化学

团队学术带头人：蒋少涌教授

团队成员：凌洪飞，葛晨东，杨竞红，杨涛，魏海珍，李艳平……

4. 极地海洋

团队学术带头人：侯书贵教授

团队成员：庞洪喜，周丽娅……

5. 海洋气象学

团队学术带头人：王元教授

团队成员：张录军……

6. 海洋生物

团队学术带头人：安树青教授

团队成员：……

7. 海洋经济

团队学术带头人：范从来教授

团队成员：路瑶……

8. 海洋法

团队学术带头人：肖冰教授

团队成员：税兵，李斌，彭岳，孙雯，张华

9. 海洋矿产

团队学术带头人：倪培教授

团队成员：蔡元峰、赵奎东、吴昌志、丁俊英……

10. 海岸海洋一体化管理

团队学术带头人：邹欣庆教授

团队成员：于文金，左平，杜永芬……

11. 海洋环境保护

团队学术带头人：王晓蓉教授

团队成员

六、研究院三年建设目标

2012 年：

1. 完成研究院学术委员会的组建，并召开第一届学术指导委员会，讨论研究院章程。

2. 完成研究院组织工作，计划申请固定人员编制。

3. 组织 2012 年"973"项目的申报工作。

4. 整合南京大学相关研究力量，申请海洋学一级学科博士点并争取通过评审。

5. 申请学校"研究院"首期建设费用。

6. 申请"研究院"在仙林校区的办公用房。

2013 年：

1. 围绕研究院设定的重点研究领域，在海内外聘请高层次研究人员，合作研究。

2. 开展"海洋学"一级学科点的研究生招生工作。

3. 建立与国家海洋局以及江苏省地方政府的有机互动机制，积极参与到国家和地方的涉海建设及决策建议中。

2014 年：

1. 争取吸纳"千人计划"层次的海外高层次学术人才，并围绕其进行学术团队的建设。

2. 研究院实体化。

3. 在研究院的基础上，建立"海洋科学学院"，为南大增强学科群力量。

希望得到校领导、有关院系领导及各位教授的大力支持。

谢谢！

在旅游与全球变化国际学术研讨会上的讲话*

各位代表、各位来宾：

早上好！在这繁花似锦的五月，在喜庆南京大学建校110周年之际，迎来"旅游与全球变化"国际学术讨论会，真是双喜临门，请允许我代表学术委员会全体向诸位表示热烈欢迎！

旅游是当代叩击人心弦的热门活动。可令人开阔视野、陶冶心情、焕发精神、增强知识、促进健康、锻炼意志与促进文化、风俗、物品与友谊交流，是一项全民喜欢的有益活动。

将旅游与全球变化结合在一起，是一项有新意的课题。全球变化含意很广，它可以是全球气候变化：气温变暖、冰川融化、海平面上升，提供更多的旅游空间——极地、荒野以至沉溺的威尼斯或新西兰南岛大U型谷峡湾"fjord"，深长的海湾(图1和图2：新西兰、澳大利亚)。

图1　新西兰 Milford Sound 沿冰川谷发育的峡湾(Fjord)海岸

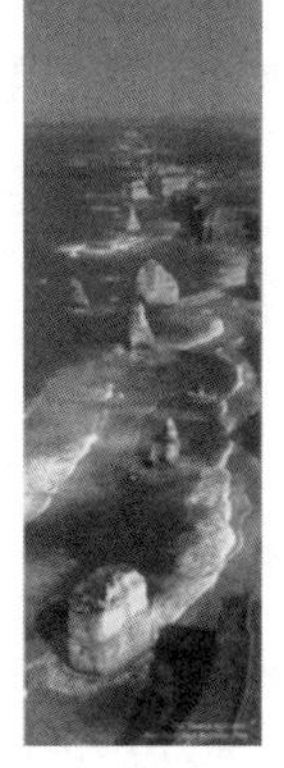

图2　澳大利亚 The Twelve Apostles 海岸"十二信徒"

* 王颖：在旅游与全球变化国际学术研讨会上的讲话，2012.5.23。

但是全球变化≠全球气候变化，不仅仅是指气温变化，而且其中与每人关系最直接的是当代21世纪初，政治形势与全球经济变化。我不想深入展开此课题，仅举一例，当年郑和下西洋(图3)，浩浩荡荡的船队到东南亚，是交流，也是一项组团的大型旅游活动，虽其原本目的是永乐帝要寻找“正统”皇帝后裔侄子朱允炆的下落。从中国经东南亚到东非，所到之处也传播了军威，交流了文化、民俗、丝绸与工艺产品，这一壮举是当时中国封建社会经济发展到一定高度的产物。郑和率领的宝船百艘，最大的长达44丈(145.2 m)、宽18丈(约60 m)，载重量数千吨，远远超过哥伦布(最大的船250 t)或一个世纪后麦哲伦所率领的船队(表1)。李约瑟估计：“约在1420年，明代的水师在历史上可能比任何其他亚洲国家的任何时代都出色，甚至较同时代的任何欧洲国家，乃至于当时所有欧洲国家联合起来非他对手”。

图3　郑和像(据 http://baike.baidu.com/view/1988.htm)

表1　15—16世纪时船队比较

郑和船队(1430—1431年)	哥伦布船队(1490年)	麦哲伦船队(1591年)
3 800艘(千吨级) 1 350艘巡船 1 350艘战船 400艘大船 400艘漕船 250艘远洋宝船及数千商船 每艘船450～690人	最大旋舰“圣玛利亚” 250吨，90人 达伽马船队 主力舰120吨 船长25 m	5艘船265人 两艘130吨 两艘90吨 一艘60吨

中国数千年不衰的传承文化，吸引着外宾，也吸引着国人。据国家旅游局2012.5.17发布的信息：2011年完成旅游投资2 064.26亿元。目前旅游景区主要为自然类和历史、文化类(占41.82%)，乡村旅游和度假旅游类异军突起。科技教育类、工业类和主体游乐类旅游景区数量少。海洋旅游、高山滑雪、温泉康体、邮轮游艇、自驾车游等深度体验享乐

式为新领域投入项目*。当代更为突出的是国人外出旅游，成为一全球现象。一团又一团，成批的中国游客蜂拥而至到国外或出境参观。初为新、马、泰；近而日、韩、西欧；当前最多的是去北美，美国、加拿大，不仅成人旅游，而且小学生组团旅游。境内游不仅港澳而且台澎金马成旅游新热点。旅游促进了对中国的认识，对中国人民的了解——古老文化、朴素上进、友好的人民，一改过去被歪曲的形象（20 世纪 80 年代末在北京召开的 NGO 世妇会时还有不少外国人奇怪为什么中国女人不是裹小脚的?），更明显的是：大至法国的巴黎，小至近邻济州岛，各地商店、商品均有中文标注，都重视中国旅客，中国游客肯花钱，喜欢采购纪念品，往往是一包包、一批批的带回内地，馈赠亲友，现在是购置化妆品、名牌箱包等奢侈品，因为价格比国内便宜、合理。个人认为这也是“全球变化”的一个重要现象，研究人的交流新趋势，关注到成批成千上万的人潮安全出访，并处理好环境保护，其意义大矣！所以我想说“全球变化与旅游”这一课题中最大的内容是人员的交往，是东西方、南北方的交流。21 世纪，中国这一人口大国，经济复甦，在东方兴起，不仅迎来一批又一批来自东南西北的来宾，不仅仅是参观故宫、登上长城，不仅仅是去黄土高原拜访周、秦、汉皇陵与始皇兵马俑，而且赴云南采风，越虎跳峡，攀雪山，驾车远赴西藏与帕米尔高原，一年年，一次次！经济发达，提高了人民生活素质，旅游是一项重要活动。我国政府配合设定了一个又一个长短假，旅游业兴起发达。中国的经济发达，城乡规划建设成规模地开展，多彩的衣、食、住、行与民俗风情，五千年人与自然的和谐相关遗迹等等……吸引世人注目。旅游开发增进保护自然与社会环境的教育，既加强文化、民俗以至人民生活之交流，又能促进建立健康、持久的友谊关系，这可能是一个重要的课题。大学、政府机关、社会团体要以先见之明迎接这一发展趋势，抓住新机遇，及早研发处理新问题的方式与途径，这是我对本次会议的一点设想，提出个建议，供大家参考，期待以学习。

谢谢！

---

* 人民日报海外版，2012.5.17，04 版。

在庆祝南京大学建地理海洋学院 90 周年大会上讲话*

大会主席：

各位院士、各位老师、各位先后同学们、各位来宾，看到如此多的校友返校，十分感动。

这是我平生第一次参加地理学系科 90 周年庆典，也是唯一的一次，太重要以致难于启齿“讲演”，只能作为 1956 届毕业生谈谈三点感受。临开会前陈毓川院士让我也代表他讲。

一、实际上是建立地理学系九十周年。

南京大学地学力量强，最早是 1921 年建立的地学系，继之，于 1930 年分成地理学系与地质学系，后建气象学系（1945 年）。1987 年改名为大地海洋科学系，1995 年改为城市与资源学系。于 1984 年由地理、地质、大气与化学等系组建环境科学系。地理始于 1921 年，海洋始于 1962 年，由任美锷教授、许廷官书记大力支持建设海洋组，设有 5 名专职编制，及购置一批海洋仪器。1990 年为海岸与海岛开发国家试点实验室，试点成功。2006 年建立地理与海洋科学学院。（下设城市与区域规划系、海岸海洋科学系、国土资源与旅游系、地图与地理信息系统系），建学系 90 年及学科 50 年，均已成年。南京大学可以说是国内地球科学历史悠久、系科最全的，地质、地理、大气、环境、海洋。加上天文、数理以及文、史、法，力量雄厚。回顾历史反映出地学是个伟大的“母亲”，伴随着时代发展与科学进步而派生出一个个的新的交叉学科系。当前，可能是因为“还在生产的时期，新生子女多”，妈妈一时力弱，但后劲居上，校内外团结，必强大发展。

二、庆祝地理海洋学院（南大是 1962 年正式建立有专职科研的海洋组，以海陆交互作用为特色，至今已 50 年，约为地理学系年龄的 1/2）。建院 90 年缅怀先师的业绩：

竺可桢教授创建地理学系，先后由张其昀、胡焕庸两教授任系主任，继之，李旭旦教授、任美锷教授任系主任；著名教授有孙本文、黄国璋、李海晨、杨怀仁、宋家泰、沈汝生、苏永暄、杨纫章、张同铸、黄瑞农、刘振中等任教。当年地理学教育具有很强的地质与气象气候学教育特色，先后有李学清、陈旭、孙鼐、姚文光、郭令智、肖楠森、杨文达、俞剑华、王德滋（教务处长）等地质学前辈教授，与幺枕生、徐尔灏气象学教授来系任教。他们兢兢业业，热心教书育人，终生不渝。地理系后继教师尚有：杨戊、杨森源、余序君、卢友裕、吴友仁、胡友元、金瑾乐、曾尊固、陈丙咸、雍万里等一批于 1952 年、1953 年毕业后任教。前后有三位政治辅导教师：李乾亨、白秀珍、毛德馨，循循善诱，耐心指导我们走红专道路。师恩难忘，饮水不忘掘井人，在此九十年庆典之际，怀念与感谢师长，期以榜样励志，承前启后的进步！在南京大学地学历届老师教导下，为国输送了大量人才，在各条战线建功立业。今日济济一堂，显示了科学教育生命力。

---

* 王颖：在庆祝南京大学建地理海洋学院 90 周年大会上讲话，2012.5.19。

三、地理科学、海洋科学都是综合性大科学体系，涉及到自然环境、资源因素与人类生存、生产活动两大方面，同时需要技术科学支撑。均应具有数、理、化、生以及文、史、哲的基础。大学本科只是基础，青年时期，打好坚实的基础很重要，但不以本科专业定终身。还应据各人基础与兴趣，在大学选修课程；尚有不同阶段的研究生学习，学科交叉；并为适应国家与社会需求，通过实践学习，总结与成长更重要。大千世界需要各种人才，不断学习与进步，形成特色，十分重要。

回顾我们自己，不也是这么一步又一步，通过本科阶段的基础学习，工作实践或研究生阶段增强科学理论与实践的学习与锻炼，逐渐成长，适应国家发展需求，在完成各项任务过程中又获得实践，总结成长工作能力与科学品质。在我们 1952—1956 年同一个班级中，先后有三位当选为中国科学院、中国工程院的地学院士，不能不说是南京大学地学教育之成功，这个纪录可能还是空前的。可以说我们每前进一步又不断有吸引力，有兴趣吸引你学习、钻研与获取工作成果。确实是处处有学问、时时要学习，终生在进步！这可能是我们在享受国家社会进步成果之同时，又为国家社会做出贡献的必由之路。

谢谢！

在南通市人民政府的发言*

陆海统筹、江海联动，建设以大上海为中心的长三角北翼港口群，发展为国际航运、海工制造、轻纺、商贸与生态人居的现代化江海港城。

一、感谢丁大卫书记的盛情邀请，南通市委、市政府及地县各级领导的热情接待，尤其是夏明文院长的细心组织、全程陪同考察。我本人、南京大学地理与海洋科学学院院长高抒教授、海岸海洋科学系主任葛晨东教授、南京大学海洋科学研究院常务副院长邹欣庆教授、江苏海岸港口研究专家朱大奎教授及海洋地质学研究生王敏京、唐盟，在2～3天内，看到、听到、想到的，深感启发学习收获大。南通市已由老概念中的围海农田与轻纺业城镇跃为航海造船、重型海工制造、小飞机制造、滨海新区等具有海洋经济特色的现代化工业、商贸与生态人居城，令人叹服。尤其是近几年发展之快，非眼见难以想象：启东在兴起、通州湾在着力规划、洋口港液化天然气(LNG)33 km^2 围地将建成，年输350亿～500亿 m^3 液化天然气，国内第一。高耸的沿海风机，已并网送电。33 km^2 围地在引来石化工业，与12.5 km黄海大桥平行建立管道栈桥现用与发展立体式利用，西太阳沙建立冷却制N循环利用经济系统。我深深感到如东展现了深水大港与新能源示范基地的风貌，建港十多年后，才成气候。这些都是南通市的巨大进步与成就，现代化的港口群与工业带在成长！

二、讲讲对南通市区位优势与发展建议，实际上南通市各部门均已有详细规划，此处一孔之见提出，为补充性意见。

1. 南通扼长江与黄海之交汇点，区位优势独特，全新世(近10 000a)来发展的江海平原，土地平坦，滩涂广袤，降水与长江河网交叉而淡水丰富，四季分明，冷暖适宜，人民教育与劳动素质高，这些有利因素结合为国内仅有。渤海湾平原、京津大地人气积聚，但缺乏淡水，土地盐碱化；浙闽粤沿海淡水丰富，但却缺乏大片连绵的平原，多为港湾平原。而江苏平原西起大运河，东至黄海之滨，有江、淮、河之利，开发悠久，有6 000～8 000年历史，非其他处可比，而南通又是“皇冠平原”上的宝石，紧临长江，扼守江海之交。

2. 平原海岸除大河口之外，传统上缺乏深水海域，水浅沙多，建港不利。而南侧的长江口有深水航道，却在江口有拦门沙之弊，疏浚维持12.5 m，10万t级大船需趁潮入港。

但是，古长江它在34 000年前从弶港入海，大量泥沙在黄海堆积了巨大的辐射沙脊群。数十条沙脊平行呈放射状向海展开，沙脊间为潮流通道，苏北潮差大(大于杭州湾)，平均潮差4.5 m±，最大潮差9.26 m，南京大学与河海大学在黄沙洋测得。而黄沙洋(－15 m)与烂沙洋是沿长江古河道发育的现代潮流通道，地居北半球，古长江受地转力至今具有自北向南移的趋势，已知最早从李堡出海，后移至黄沙洋，再移至烂沙洋，再移至小庙泓，再移至长江北支，主泓又南移至长江口南支，而现在又移至南航道了！那么，留在

---

* 王颖：在南通市人民政府的发言，2013.8.7。

南通市管辖境内的烂沙洋，因停留时间久(弶港处为突出三角岬)，河道宽，现在有强大潮流冲刷，却无河沙下泄，不会形成拦门沙，所以，保留了一条天然深水航道，南京大学于1980年代测辐射沙脊群，了解到利用古长江发育的烂沙洋深水通道，提出可建深水港，几经波折，最后，在南京古南都饭店，努力说服交通部港航部门的专家(大卫书记帮我一起说服)，－17 m的水深可建10万～20万t级的航道，如全程浚深，以两侧沙脊为掩护可建30万t级深水大港。在烂沙洋建洋口港，未经浚深即建成10万t级港口。这是南通市，也是江苏沿海最佳港口，建港至今未形成骤淤(而海南新村建军港，5月航道打通，经一次台风暴浪即淤没了)。洋口港没形成突变性淤积，证实了20年前我向交通部专家的诺言。因为，我们是通过调查研究的科学论证！洋口港建成后，现在各个地方均要建港，当年洋口港我们是多年调查、实验之结果。现我已老矣，而当年新兵邹欣庆他是现代908海岸调查的队长，接班调研了东台苦水洋！

洋口港是南通市自力更生建港，获得中石油等部门的支持，现已宏伟规模初现，但是没发挥其潜力，它应获得省与国家给予经费与政策上的大力支持，疏通航道建成30万t级深水大港，形成以上海港为中心，南有北仑—洋山港，北有洋口港的东方大港港口群。这次我赴洋口港很高兴，因为詹立风书记很明确，集中力量建洋口港，而不似前些年，洋口刚建成就要建金牛港，我听了后头大了，几年不愿去洋口。今年好，集中力量做深、做大洋口港，今后有力量可建黄沙洋，可建其他规划中港口。我认为南通在建港方针上，应集中力量明确重点支持，做好大、中、小港口规划，有分工与互补，可以吃一个、拿一个、看一个，这是有规划远见，而不是熊瞎子掰苞米，吃一个、丢一个，恕我直言。作为南京大学教师的一员，我是站在全国的角度，站在江苏省角度，也站在居江海优势的南通市立场上说这句话，多年来与南通市与洋口建立战斗友谊，“士为知己者死”，我要为地方老友讲实话。

其次，吕泗港是沿古长江南移的支汊小庙泓建港，历史上的吕四洋是指辐射沙脊群中的小庙泓外海。此传统渔港可发展为现代化的渔业基地，前日已见做四个挖入式港池，3 000艘渔船之港口，此规划好，实际上小庙泓可进一步浚深，发展为10万t级港口，并且加宽通吕运河，使从吕泗到南通江海联通。我想，应规划长远些，为1 000 t级的船只不够，宜设计为3 000 t～5 000 t级船只，要吸取南京机场高速、江苏高速公路，从双向各双道，改为3道又改为4道的折腾，有远见些，一次挖深，加宽些。可否改为90 m宽，7 m水深江海运河通道！南京公路多次加宽造成交通不便、事故与污染增多之弊应引以为戒。

3. 讲讲港口发展，港口选址要求水深、浪静、少淤。这是一对矛盾，水深处浪大，浪静处易淤，所以要科学选港。我国初始的港口多沿河，20世纪30—50年代建港口多沿海岸为凹入式与突堤当风浪，大连、天津、青岛、秦皇岛港均如此。我本人参与60年代秦皇岛南李庄5万t级油码头(1961—1964年选建)、70年代青岛黄岛5万t油码头、1959年天津新港回淤研究等，以及海南洋浦港、马村港选址，均为海岸港口，这是第一代港口。

第二代是人工岛式。我组织进行了河北曹妃甸利用26 km远之岸外沙岛与其临海侧之渤海潮流通道组合为港址，30万t级成功；江苏洋口港利用12.5 km远水下沙脊与烂沙洋深水通道选建深水港址，建成之例证。洋山港距岸30多km，这是深水大港发展趋势。

科学进步，油输需大吨位船只，多需20万～30万t大船，运费降低，而其抗风浪力大，在外海建人工岛(选有沙脊处),是第二代建港成功之发展趋势。

第三代,应发展港口群、人工岛大港与运河联通之河港结合,进一步繁荣河港,使物资输运从港→港,进一步发展为港→港→户！近大都市与工业带便利使用。所以,南通应据自己扼江临海已有大小港口的基础发展江海联运的港口群。

南通紧靠上海,以上海为中心,发展长三角北翼港口群,十分重要,利国利民,亦利省利市!

长三角范围北起洋口港,南止洋山港——扩及北仑港(地理上现代长三角不到此,但现代经济发展到北仑宁波;但古扬子大三角洲却北起连云港、南至台湾海峡北端,那是将来发展的),所以,可以北翼到洋口港,南翼至北仑、宁波港,以上海港为中心,组建东方国际航运、商贸金融大都市带,在21世纪发展辉煌。南通市作为长三角北翼中心城市,港口发展宜在已有基础上有重点、有步骤的发展稳进。

(1) 南通市域为江海联运之枢纽。可以将规划中的通州湾定位为商贸港:沿长江港口为造船与海工基地;古城为生态旅游与传统文化中心。

(2) 吕泗港为现代化中心渔港,宜发展为远洋捕捞渔业基地,通吕运河扩展为江海联通起点之一,达海通江与水产食品渔港风情旅游度假区。

(3) 洋口港发展为长三角北翼第一深水大港,国际航运与新能源示范基地以及打造江海联运第二出海口大通道。

江海联运发展壮大原有港口形成拳头的港口群,然后再发展新港口。建一个港要十年,发展成气候又需十年,宜抓紧时间,早为民创福祉为根本。今后,是子孙创业。现在奠定基础、勾画发展规划蓝图。

4. 运河开挖利用古河道

两条长江古河道仍展现卫星图上,利用古河道,浚深促进航运可事半功倍。运河应顺直、选捷径、不易淤,30～80 km长度,河宽90～130 m,水深12～20 m,北起黄海,南达长江南通内岸,为辅助长江口的第二通海口大通道。

先例:北欧荷、比,运河与海港系统,打造东方鹿特丹江海港口联运。

在南海论坛上发言稿*

按:中国南海研究协同创新中心,每年举行一次年度论坛,主要围绕南海权益维护,资源能源开发,区域和平发展三大战略需求,从南海司法化应对、周边国家南海政策研判、南海共同开发、南海情势推演、南海舆情有效应对等多个方面,邀请国内知名专家学者,研讨南海研究战略需求及重大任务,深化南海研究的智库建设,力求高规格、高水平为国家南海战略建言献策。

2016年论坛设立特别演讲、大会报告、分论坛,邀请国内知名学者、涉海部委领导、国内涉海机构代表与会,针对菲律宾南海仲裁案、南海争议化解思路、南海行为准则(COC)磋商、亚太海洋秩序建设与中国的主张等核心议题展开研讨。2016年12月7日与8日在江苏省人大常委会会议厅(南京市中山北路32号)举行了开幕式及特别讲演。开幕式上已有中国南海研究协同创新中心管委会主任杨忠代表南京大学,江苏省对外友好交流促进会会长林祥国,教育部高校社会科学评价中心主任李建平致辞。

王颖作为南海中心主任致欢迎词,其发言如下:

各位领导、各位代表:值此建设海洋强国与一带一路国家大政实施之际,召开了2016年度的"南海论坛",意义非凡。感谢外交部边海司,条法司领导,国际海洋法法庭大法官,教育部、南京军区、中国科学院、南京大学、厦门大学、中山大学、四川大学、解放军理工大学等高校以及军队有关部门领导与代表参加本次论坛,规模增大,内容紧扣南海问题,深表感谢,深受鼓舞。我原不准备发言,但南海中心执行主任朱锋教授安排我发言,请原谅我随意说一些。

最近邀请一位新当选的中科院院士来访,谈话中,他问我:"怎么会在南京大学建立南海中心?"我一愣,因为在我认为是顺理成章之事,一下子却没回答,我想:为什么不能在南大! Why not?

一、南京大学地处:南京通过长江直达黄、东海,当年郑和所乘用的大海船与船队舰艇均是南京或南京主持建造,船队从浏河经长江出海下南洋;甲午海战清廷惨败,与英、法等不平等条约是在南京下关海宁寺(静海寺)与停泊于长江的船上签约的;鲁迅就读的水师学堂设在南京,新中国建立的海军学院是在南京(下关);至今,海军指挥学院仍在南京……。南京,一直是通海,是与海洋密切联系的。

二、为什么在南京大学建立南海中心?!

1. 南京大学是百年老校(1888年创建),地球科学系科齐全,在国内综合大学教育中,地学教育的力量最强。

2. 1921年著名的科学家竺可桢教授创建地学系,继之,发展为地理学系(1930年)、地质学系(1930年)、气象学系(1944年)。1987年地理系扩建为大地海洋科学系,2006年

* 王颖:2016年度南海论坛上的发言稿,2016年12月7日。

5 月建设为地理与海洋科学学院。

3. 1958 年，南京大学地理系（当年还有华东师大，北京大学，中山大学，华南师范学院与山东师范学院的地理系）参加了交通部会同中国科学院与教育部组织的“天津新港回淤研究”项目，探求港口回淤之泥沙来源与解决港口淤积之措施，当年的院校成员多为后日我国海岸带研究的骨干。

4. 20 世纪 60 年代号召老专家下海，南京大学地理系主任任美锷教授开始研究海岸，并继续从事西南地区喀斯特研究。1963 年春季，海军邀请一批专家在青岛东海饭店开会，讨论建立国家海洋局*。王颖随同任美锷教授参加会议，亦举手支持建立海洋局。南京大学地理系于 1963 年夏季建立海岸研究组，设专职科研：王颖任组长，组员：陈万里、张忍顺、何浩明、马仲荃、尤坤元及张连选，王秀娟等。

5. 继 1958 年参加天津新港回淤调研后，王颖参加了北京大学进行的选港与海滨砂矿勘探。王颖带领两名勘探工（勇玉林师傅与小王）参加冶金部第五勘探队在山东半岛探查为航天器特种金属冶炼所需的锆英石工作，勘察了山东半岛东部（从蓬莱→烟台→威海→成山头→石岛→乳山县）。完成了 C1 级矿床选定任务；1961 年后，王颖领导南京大学海岸组，先后完成秦皇岛油港、南李庄军港、山海关船厂、北海地角海军港、广东镇海港、湛江海军码头、海南铁炉港、三亚港及海口港扩建与新码头选址、洋浦港选建以及渤海曹妃甸深水港、黄海洋口深水港选建及南黄海辐射沙脊群的成因及资源环境与开发利用研究等。长期实践研究，提高了研究团队对海岸发生与发展研究的科学水平。

6. 20 世纪 80—90 年代，南京大学与加拿大 Bedford 海洋研究所，先后获加拿大国际发展研究中心（IDRC）及加拿大国际发展署（CIDA）资助，开展中加海洋合作研究海南岛海岸，建设海洋勘测实验室、^{210}Pb 同位素定年实验室。因而奠定坚实基础，推动了由国家计委、科委与教委批准，以世界银行贷款建设了实验室系列，组成“海岸与海岛开发国家试点实验室”（State Pilot Laboratory of Coast & Island Exploitation）。后归为教育部重点实验室（The Key Laboratory of Coast & Island Development, Ministry of Education, China），扩大研究领域至大陆架、大陆坡与深海盆，组成海陆过渡带研究的系列实验室专门从事海岸海洋（Coastal Ocean）研究与人才培养。开展国际海洋科学合作，专题研究河—海交互作用海岸海洋发育、海平面变化与效应等。仅在海南省一处，南京大学与加拿大 Waterloo 大学等合作培养了 50 名硕士研究生水平的干部。嗣后，大力支持在海口建设的中国南海研究院开展合作研究。

7. 21 世纪发展多学科交叉合作的海洋疆界研究。中共南大党委书记洪银兴教授支

---

* 记得当时是由海军袁也烈将军、于笑虹将军（国防部、六机部第 7 研究院院长）、律巍（航保部）及刘志平大校组织会议；哈尔滨军事工程学院刘恩兰教授（将军待遇）与秘书万延森中尉参加；国家科委王遵伋主任（女）；地质部郦治铮教授、刘光鼎研究员；中国科学院海洋研究员童第周所长、曾呈奎与张玺副所长，毛汉礼研究员，及年青的助研秦蕴珊；南京大学任美锷教授、王颖讲师；青岛海洋学院的教务长赫崇本教授；华东水利学院的严恺校长等人参加会议。犹记得在会议过程中，赫教务长问我：“你不是研究生毕业后由北京大学分配到海洋学院的吗？怎么到了南大……”？真的很难一言说清楚。会议期间看地道战电影，非达六级不能看，而我是由王遵伋主任证明为会议成员而进入放映厅的，至今我仍记得这一情节。当今，当年参加会议的成员，可能只有刘光鼎、万延森与我在世。

持创建，经教育部批准的“中国南海研究协同创新中心”。开展南海调查研究，通过调研，论证了南海断续线为中国南海海疆国界线(2014)，研定1947年划定的南海海洋疆界不仅依据历史传统与长期开发经营，而且具有海底地形区别的自然科学依据。

2014—2015年，开展西沙群岛调查；2016年4月考察了曾母暗沙；12月在九章群礁海域调查了12个无人岛礁，抗风浪，耐晕船，吃、住在渔船，南海中心15位师生在遥远的南沙海域完成了调查任务。期待着2017年能允许我们登陆人工岛，进行岛礁稳定性科研工作。

总之，我们有基础，有队伍，有信心，为南海与中国边缘海工作努力奉献。谢谢大家支持。

第八篇
回忆与纪念

时代发展、科学进步与导师培育。

中国"海洋地理"记事*

海洋地理学是地理科学的新分支,具有自然科学、社会科学、技术科学相互交叉渗透的特点,它是地理学与海洋学之间新的结合点学科。海洋地理学在中国的发展始于20世纪70年代,随着海洋国土概念在国际的兴起,近十年来,海洋地理学发展迅速。

1973年,中国科学院贯彻周恩来总理"重视基础研究和加强基础理论研究"的指示,成立了以竺可桢副院长为主任,黄秉维教授与郭敬辉教授为副主任的"中国自然地理"编辑委员会,将编著"中国自然地理"列入中国科学院1973—1980年重点学科规划之中,海洋地理作为"中国自然地理"专著系列的一个分篇组织编著与出版。这可能是"海洋地理学"正式出现于近代中国地球科学领域的肇始。

记得是在1973年,郭敬辉教授在南京华东水利学院开会,我去看他时,他谈到了"中国自然地理"专著计划,嘱我参加海洋地理篇编著工作,并让我推荐人选,当时我提出南京大学的朱大奎编写过地理小丛书(侯仁之主编)的《海底世界》,青岛海洋研究所的金翔龙从事海底构造工作,适合于编写中国海海底地质地貌,他很高兴地同意了。为此,我参加了中国自然地理专著编写的兰州会议,在那次会议上,结识了瞿宁淑先生,以后是她代表郭教授具体组织海洋地理篇工作。

1974年,在大连黑石礁宾馆召开了海洋地理编写工作会议,是由海洋局主持的,郭敬辉、瞿宁淑出席了会议。出席会议的还有外交部条法司的代表厉声教,中科院海洋研究所的曾呈奎所长和刘瑞玉研究员,南京大学任美锷教授、王颖讲师,海洋局的窦振兴助研等代表以及海洋地质方面的代表。因岁月久远,我已记不清人名与职务了,印象很深的是海道测量部的律巍部长,海洋局的肖卓能参谋与姜富参谋,身着海军装,神采奕奕。海军给予了大连会议热情的支持与安排,组织参观旅顺口海军基地与白玉山炮台。那是在"文革"岁月中的一次集会,周总理的指示如和煦的春风复苏着知识分子的心灵,推动着海洋地理的编写工作,代表们就海洋地理编写内容,大陆架定义的提法等进行了广泛深入的讨论,并就章节内容,分工,达成共识。可以说,大连会议是一次"动员大会",是中国海洋地理学启动的里程碑,从一开始,海洋地理即具有地理、海洋、大气、地质及军事、外交等多方面合作与关注的特性。

《海洋地理篇》是由海洋局负责主持,组织各单位共同编写的,是全国范围协作编写的第一本海洋地理学著作。初稿完成后,于1975年在杭州西子湖畔的华山饭店进行了通稿与讨论修改,先后邀请有关专家和教授审稿。

1976年1月于广州三元里矿泉别墅,由国家海洋局副局长律巍主持,郭敬辉与瞿宁淑指导,集中了编写组人员对全稿进行了认真的修改。矿泉别墅的环境设施很好,与会工

---

* 王颖:吴传钧,施雅风主编,《中国地理学90年发展回忆录》,470-474页,北京:学苑出版社,1999年。

作人员专注于写稿，很少外出。但这次会议却在香港地区报纸上有报道："中国集中一批专家于广州矿泉别墅，从事海洋战略研究……"在这次会议上，会议负责人决定："书稿以内部资料发表，不外传。"也正是这次会议上，传来了周总理逝世的噩耗，与会同志泣不成声，不顾当时"通知"的规定，主动地带上了白花黑纱，向总理遗像默哀，心底的一线光亮与温暖又一次被掐灭了，再次地陷入了压抑与失望！《海洋地理篇》是在惊涛骇浪中孕育成型的！

1976年春，由郭敬辉副主编主持，组织了海洋地理各章的负责人刘瑞玉、王颖、窦振兴等，在北京中关村科学院招待所进行交付出版前的终稿审定工作。虽早春天气多变，工作环境拥挤，但几位同志和谐相处，有序地工作，展示着祖国的辽阔海空。清明节临近，去天安门悼念周恩来总理的人潮情思奔腾。"四人帮"加强了管制，连续两次到中科院招待所盘查。郭敬辉先生很惦念，连忙召我去他家，谆谆嘱托："你从南京来的，那次列车上贴有大标语，对南京来人多次盘问，我都挡住了，要千万小心。"次日在去天安门广场后赴王府井街口的国家海洋局，肖卓能与姜富两同志告嘱："要小心，天安门广场不能再去了……"其实，我是一个普通的教师，只是凭吊烈士，缅怀总理，而民众的激情的确给压抑的心情以激励、以鼓舞。即使如此，也是被禁止的，心情更加悲愤。接着又传来姐姐在西安病逝的消息。为了共同统稿，没去悼别，直到《海洋地理篇》书稿完成后，各人才分别离京返家。经历春寒与悲愤交加，我回宁后不久，全身关节肿胀而卧病不起，直至夏季，由省人民医院的万锦堂老中医诊治，才重新站立活动。

《海洋地理篇》全书约30万字，以内部形式发行刊出(科学出版社，1979)。这本著作凝聚着两代知识分子的辛勤劳动，经历了20世纪70年代人民生活中悲痛的一页。中国自然地理专著系列出版后，于1986年获中国科学院科技进步一等奖，使人深受鼓舞。

1984年，中国地理学会第四届理事会召开时，据国际地理学的发展趋势，著名的地理学家吴传钧教授，倡议在中国发展海洋地理。1987年4月，中国地理学会五届二次理事会在南京大学召开，出席会议的有黄秉维院士、任美锷院士、施雅风院士、陈述彭院士、吴传钧院士、王乃樑教授、陈吉余教授、曾昭璇教授、张兰生教授等多位地理学界的著名专家。与会者讨论决定成立"海洋地理专业委员会"，挂靠在南京大学大地海洋科学系。由系主任王颖负责筹备。海洋地理专业委员会获得了沿海12所大学、4个科研单位及海洋科技管理与产业部门的支持，如华东师大、同济大学、华南师大、中山大学、海洋大学、厦门大学、杭州大学、福建师大、中国科学院地理研究所、海洋研究所、南海海洋所、国家海洋局、海洋局第二海洋研究所、国家海洋环境监测中心(大连)、海洋科技情报研究所(天津)、南京水利科学研究所及中国海洋学会等单位，以及多专业方面的33位代表的支持。特别是地理学会与瞿宁淑秘书长的关怀与指导，海洋地理专业委员会于1988年春正式成立，由王颖任主任，苏纪兰、刘南威、刘以宣、李克让、罗章仁、张耀光、恽才兴、杨作升任副主任(出席国际海洋地理委员会的理事、中国国家代表)，委员22名。第二届学术委员会于1991年成立，副主任中增加了深圳赤湾南山开发公司的田汝耕，增加了海洋战略研究与海洋经济研究学者，以及港、澳、台地区的代表，共46位委员。第三届委员会于1996年夏成立，选朱大奎教授任主任，副主任9名，增加了中青年科学家的比例，委员有变动，人数减为41名，任美锷、吴传钧、陈吉余、曾昭璇连任顾问委员。海洋地理从成立起即与国际地理学会的海洋地理专业委员会联系合作，首任中国理事为杨作升教授，后二届为王颖教授。

海洋专业委员会是自筹经费无营利的学术团体，发挥多学科交叉的优势，先后组织了在华南珠江三角洲、华东连云港地区以及东北旅大地区的海岸带考察，组织了专题研讨：海平面变化、海岸侵蚀、沿海灾害、海洋资源开发、海洋经济以及海洋地理当代研究内容与发展等。1990年"海洋地理专业委员会"第二次会议上，王颖建议，经各位委员讨论，决定将各单位、各委员的多年研究心得归纳著作，出版《中国海洋地理》专著，为海洋资源开发、环境保护以及海洋权益服务。初稿于1991年完成，在第三次会议(广州)时进行了充分讨论及逐章修改。后由张耀光、李从先、恽才兴、朱大奎负责分篇统稿，专业委员会秘书长尤坤元联络与编辑，最后由朱大奎、王颖校阅定稿，经过43位委员共同努力，经历四年，在科学出版社第三编辑姚岁寒与朱咸滨的大力支持下，编写出版了《中国海洋地理》。这本专著，内容充实，涵盖范围广，自然科学与社会科学结合，由多方面专家联合著作完成，有理论、有数据、图文并茂，由科学出版社正式出版发行(1996年)。该书反映出中国地学界推动学科交叉，为国家服务的努力，是改革开放新天地的一项科学新成果。该书与一批海洋地理有关论文，获1998年教育部科技进步一等奖。此外，专业委员会还组织出版了《海平面变化与海岸侵蚀专辑》(1995年，南京大学出版社)。为迎接1998国际海洋年出版了《海洋在召唤》系列丛书共十本(1998年，广西教育出版社)。

联合国《21世纪议程》指出：海洋是生命支持系统的一个基本组成部分，也是一种有助于实现可持续发展的宝贵财富。海洋地理专业委员会进一步明确任务有三：促进海洋资源的合理开发利用与生态环境的可持续发展；促进海洋与地理交叉学科的发展与人才培养；促进海洋国土权益的维护、新技术的应用与一体化的海洋管理。这些任务必将推动我国海洋地理学的发展。

大学生活回忆*

时光飞逝，转瞬已是50年。犹记1952年初夏，我从北京慕贞女中考取南大地理系，初到南京，十分惊奇：偌大个城市，人烟稀少；城中大道即使是中山路也仅是马路中间铺沥青，两侧均为石子路，南京大学附近的天津路，百步坡均为方块石子铺就，下雨时，经过冲洗的青白石块相间，十分干净；1954年冬大雪奇寒，石子路冰冻，学生滑倒无数，很多竹壳热水瓶胆打破，一时都难以配到；南园农学系圆顶铁皮活动房周围桃林丛丛；大校门在今教学楼东侧，大屋顶的庙门式建筑，设有门房与信箱，一下课学生纷纷拥至，找值班的老陶询取信件；校门以西是陶家菜园，果木花树相间，颇富田园情趣；整个校园以枸骨绿篱相隔，看不清边界。迷蒙细雨江南，北大楼前雪松挺拔，溲疏白花颤颤，春日的杜鹃花、夏日的蔷薇、秋菊冬梅、清香四溢，的确令我这来自北国苍茫大地的青年，感到新奇、陶醉于斯。

当时院系调整不久，在原来金陵大学的校址上，1952年全国统一招生，招来了大批青年学生，尤其以上海籍学生为多，处处均可听到"阿拉、阿拉"瓜拉松脆的上海话。放假时，"阿拉"们回沪，校园都变得安静无声。学生宿舍大部分在校本部，高年级男生住甲、乙、丙、丁、戊、己楼，女生全部住庚、辛、壬楼，每日专门由工友老金挑开水，一天两瓶定量供应，多用热水需自己去打。女生厕所仅有四个茅坑，清晨人多如厕需排队，宿舍楼专门有一陶姓老妈妈，入夜放置马桶，清晨提走。一年级新生男同学住西园大房间，楼旁有个大池塘，水面青苔，塘畔垂柳，鸣蝉与蛙声相伴，颇令学子思乡。

开学不久，新生闹情绪，我清楚地记得政治辅导员李乾亨先生为众生一一排解，青年团书记杨世杰在大礼堂作报告："希望大家克服暂时性的、前进中的困难……"师长们的亲切关怀与真诚话语，安定了我们的情绪。由于一下子来了许多新生，在平仓巷加盖了大草棚食堂，伙食很好而且吃饭不要钱，顿顿两菜一汤，有萝卜烧肉，茭瓜毛豆烧鸭丁……中秋节发月饼，餐后备有大桶豆浆供饮用，以补充营养。夏日晚会时，伙房师傅还在大操场供应白糖绿豆汤。记得当时一位女同学平时仅90多斤，一次吃蛋炒饭，十分可口，猛吃几碗，饭后到体育馆称重，达到100斤，有位女生用豆浆洗碗，还有些男生吃馒头剥皮，剩菜桶中浮满馒头，令人不忍。当有人提出规劝时，被批评者还反"哼"几声！如今她(他)们已是德高望重的教授或研究员，儿孙绕膝，忆及往事，大笑不止！当年旧事印象很深，至今我仍忘不了那些大师傅身着白围裙，脚踏木屐，健步快行地抬菜、送汤供应大家的情景。50年代的南京可能每人均有一双木屐、北方人称"塌拉板"，大街小巷塌拉之声不绝，伴以高轮的四人马车的节奏声，形成一道城市特色景象，当年公共汽车不多，外出多乘马车，从鼓楼到中山陵，每人3分钱，后涨到6分。现今，载客马车没有了，木屐已销声匿迹，设置于小巷口的老虎灶，与街头巷尾的小便池也消失了，时代在发展，人们的衣、食、住、行习惯也

---

* 王颖：《南大校友通讯》，2004年第24期，第49－51页。
2002年5月曾载于：《南京大学地理学系建系八十周年纪念》，48－50页。

在变化。

院系调整后的地理系设在西大楼，楼内光线很暗，系内行政人员很少。系主任是任美锷教授，并兼中国科学院地理研究所所长，系秘书是白秀珍老师，政治辅导员是李乾亨老师（后来他教授政治经济学，由毛德馨老师接任辅导员）。1954 年系办公室增加了一位年青的教务员管素莲，许培生先生是系绘图员，还有一位工友孟朝周师傅，负责清洁与管理系图书室，当时，刘振中老师是助教兼管图书室，全系就这么六位行政与管理人员。而我们这帮一年级新生不懂规矩，常在系图书室说说笑笑，声音较大，老同学罗贻德忍不住劝说："你们怎么像小学生一样吵吵闹闹。"没想到这位一向帮助我们的老同学，立遭围攻，沈道齐说："你是幼儿园小朋友！"王飞燕哼了一声，王颖等立即附和："你是幼儿园！"面对这一帮天真无知的新生，罗贻德只能无奈作罢！想想也奇怪，若干年过去了，很多事已记不清了，但初进大学的这些情景，却仍历历在目，可能是富有生活情趣吧！

大学生活中最令人深铭于心的是，南大对我们的教育，师恩难忘，至今受益！

1952—1956 年的大学生是很幸运的一代，教授基础课的老师都很强：孙鼐教授讲授"普通地质学"，他讲课深入浅出、引发兴趣，至今我们仍记得他教的硬度口诀："滑石方、氟磷长，石黄刚金刚"。当年地理系一年级有 80 个学生，分甲、乙两班，地质实习由肖楠森先生指导我们乙班，郭令智先生指导甲班，从方山玄武岩、洞玄观砾石层、石炭二叠纪灰岩到头台洞仑山灰岩，至今仍记忆清晰，甚至下山宜侧身横足也是肖老师教我们的。"地貌学""第四纪地质学"是杨怀仁教授讲授，亲自指导我们在九华山与黄土高原实习，言传身授地教我们素描地形与分析地层剖面；教"岩石学"的是李学清教授，他一丝不苟地批改我们作业；教"构造地质学"的是姚文光教授、俞鸿年与施央申任助教；"古生物学"是陈旭教授讲授；"地史学"是俞剑华先生教授；讲"中国地质学"的是杨鸿达教授；教"气象气候学"的是幺枕生教授（乙班）、徐尔灏教甲班，季风、冷暖气流锋面交绥，至今仍为我们所应用；"高等数学"由徐曼英先生讲授；"普通化学"教师是陆礼光；"大学物理"是吴汝麟教授；"土壤地理学"是马溶之先生教；"土壤学"是位年轻的周湘泉先生教，教学非常风趣；"普通测量学"由南工的刘海清先生讲授，实习是一位汪姓助教；"水文学与中国水文地理"是杨纫章老师教的，她口齿清晰，讲课有条理，循循善诱，学生获益很大。此外我还选修了耿伯介先生的"植物基础"，仲崇信先生的"植物地理学"。记得在中山陵实习时，由于我提醒另一位同学别讲话，被仲先生以为是我讲话，让我俩都站出队列，十分尴尬！当年地理系教学计划有很强的地质地貌与自然科学为基础，同时，也学了大量人文与区域地理学，系主任任美锷教授亲自讲授"经济地理学概论"，他每次自带饮水，风度翩翩地到西园大教室讲课，他讲课原理清晰，数据准确，启发兴趣，引人深思，从不照讲稿宣读。沈汝生教授与宋家泰先生先后为我们讲授"中国经济地理"，苏永煊教授讲"世界自然地理"，他讲课时表情丰富，意趣盎然。记得当时讲新民主主义论的是一位军事学院政治理论教研室主任孙姓教师，他身材高大，军装整齐，但讲课十分生动，听说他参加过一二·九学生运动，1955 年升为少将，可惜我记不得名字了。把政治课讲得很生动，关键在老师。李乾亨先生教的政治经济学，吸引着同学们学习、钻研的兴趣，赢得同学们的尊重。还有我们敬爱的孙叔平副校长讲授哲学"辩证唯物主义与历史唯物主义"，句句落地有声，引发学习兴趣。孙副校长每学期每阶段结合形势讲学校的任务、学生成长的道路，感情真挚，语言朴素，大处着眼，细处注意，对我们做人教诲至深。当年的基础课与涉及地球科学主要方面多学科交叉的专业课，十分扎实。四年期间，学习的课程可以自选，门类丰富，名师教导印象深，应用于实践，至今受益无穷。学习地球科学，开阔了视野，启迪思维，增长知识才干，锻炼了健康的身体，练就坚强意志，我越来越热爱地理学，热爱祖国大地河山，至今无悔。回忆至此，不能不深深感谢南京大学，感谢我们敬爱的各位师长辛勤的谆谆教导，一代宗师，为新中国培养了一批又一批的地理人才，至今仍分布于祖国的东西南北，扎根发芽，立功建勋。回想当年教育，任美锷教授教导：要成为新中国的一代建设者，促进祖国的繁荣富强！当年的学习很紧张，尤其是有一段时期，学习苏联的六小时一贯制，每天从清晨 8 时到下午 2 时连上六小时课，中间在北大楼草坪供应馒头。由于年轻，六小时听课并不感到紧张，但常常想吃东西，课间或自习期间，会跑到校门口买肉松烧饼、花生牛扎糖，或者去平仓巷后门买老头小铺的“奶油锅巴”。

学生生活很有意义，每日课后有体育锻炼，进行劳卫制二级标准测验，我因百米跑步 16.1 秒，仅因此一项的 1 秒之差而未得劳卫制奖章，失去了优秀生的称号。当时优秀生要门门考试优，体育达劳卫制，德智体全面发展，我们班仅李吉均同学获此称号。但全班体育锻炼好，由陈治平同学组织了“国防战士队”，由牟昀智同学指挥参加全校歌咏比赛获团体第一名，最后全班获“红色地貌班优秀集体”。当时的学生生活很丰富，尤其是在二年级时，四川大学地理系的数十位同学合并入南大，班级增加了新的生力军，川江号子，诗歌文章以及四川青年的勤劳勇敢，刻苦朴素，为江南学子也为我这为数极少的北方学生树立了榜样。当年李吉均每日以冷水洗澡，令人惊叹，可能青年时的锻炼，为他以后在青藏高原的长期辛勤调研打下了基础。南北方结合，四川女力士牟奇俊的铁饼，王颖的标枪，沈道齐的 60m 跳绳跑与 1500m 自行车比速（南京市第一名），高曼娜的 200m 中跑，地理系女生 400m 接力及女子拔河均获校运会第一名，为全校女子团体总分亚军，仅次于女生众多的化学系。加之，王飞燕的排球，高曼娜篮球均为校队主力，因此，地理系女生的“牟奇俊锻炼小组”全校知名。四年的大学生活丰富有意义，尤其是野外实习时，男生帮女生，年长帮年幼，凝结了真挚的友谊。我们曾冒雨登上九华山，黑夜翻越天目山，大雨中同学们唱起勘探队员之歌，充满着火热的青春激情；1954 年长江大水，我们在大通港趸船甲板上草席单衣过夜，仅凭一张报纸御风，清晨下船梯取水，一踏草垫发现很多水蛇，发大水连蛇也依船避难。而山西离石黄土高原却是滴水如银，地窖中存储的水都长出红色絮状体，视情动心，同学们都自觉地节水、惜水，并立志在黄土高原做好找水调查。名师执教，基础课扎实，专业课多学科交叉，涵盖地球各圈层，理论与实践教学并重，四年学习不间断……，所以毕业后，同学们遍布祖国各地，可适应如地质勘探、水利、铁道、外交、科研与高校各条战线工作。数十年不断地辛勤工作，为祖国建设，为地理学发展添砖加瓦，一步一步地前进，其中，李吉均、王颖、陈毓川三人已先后成为中国科学院与中国工程院院士，教授、总工程师的比例更大，但也有少数同学早亡夭折。同学四年，友情永志！

时至 21 世纪，南大有了巨大的进步发展，学科建设齐全，教学研究成果累累，名列国内高校前茅，声誉蒸蒸日上。校园楼群增加，设施日臻完善，即使是位置依旧的大操场、北大楼前草坪、东大楼与西大楼也更新了面貌。原甲、乙、丙、丁宿舍楼群，外貌依旧，“调卿宿舍”，铭石仍在，而楼内已更换了新内容，这些均令人振奋。只是忘不了记忆中的绿篱、垂柳、桃林与池塘，那些富有江南风貌的情景，也许已在浦苑新校区获得了重现。看今日之进步，忆当年的情景，点滴回忆，藉以表达赤子心情，以期在母校发展的历史长河中作一刹那的回响吧！

时代的召唤 导师的教育*

时光飞逝，转瞬间我已68岁，可自感仍很幼稚、蒙蒙眬眬地不懂世事，回顾一生似乎还早，也还不知如何落笔。认真地思索，感到自己成长的每个阶段，皆反映着所经历的时代特色。

1935年2月，我生于东北一户军人的家庭，当年父亲在河南潢川驻防。抗日战争爆发，国共合作东北骑兵师归朱德将军领导，父亲转战在太行山东西与黄河南北，1938年冬父亲因肺病，从前方返至西安病逝。从"白山黑水到关内流亡"，日机轰炸，生活动荡，使年幼的我深感"国家兴亡，匹夫有责"，精忠报国的岳飞是铭记于心的英雄形象。在西安师范附属小学学习时，老师们敦厚慈爱的教导，启发我立志勤奋学习，长大报国。我每日随由新城内向南外飞的晨鸦鸣叫声起床，背起书包上学，黄昏又随群鸦返巢而归家，从不迟到早退，当年的王肃波(父亲为我起的名字)，年年名列前茅，是初秋三八级班的"模范生"。至今我仍怀念小学校长王汇百，老师王敬斋、何纫秋、马淑箴、王庭悟等。新中国成立后王汇百与何纫秋曾被评为模范教师。成人后，每当回到西安，必然回母校看看，可是当年西安新城的成千上万只乌鸦却渺无踪影了。抗日战争胜利，人们热泪盈眶，企盼返乡团聚。却因遭遇母亡家散，姐弟三人分别寄宿学校，我就学于西安东关外汇文中学，学习成绩平平。新中国成立后，交通恢复，生活安定，在亲友安排下，我得以返回北京考入女十三中(原慕贞女中)。高中三年学习扎实，国家稳定，回忆那时，感到真是"解放区是明朗的天，解放区的人民好喜欢!"同学互助友爱，《钢铁是怎样炼成的》《卓娅与叔拉的故事》《远离莫斯科的地方》，开阔了我的视野，陶冶了我的情操。忆当年刘植莲老师常给我的作文上留下点点"红瓜子"，意味着这些是好文句；温文儒雅的王文漪老师每次都是从清华乘校车赶到崇文门来教我们英语；教化学、物理与代数的老师们均耐心启发，引发兴趣，教导学生努力学习与钻研。有位代课老师在沙鸥照相馆摄影，他把美学、艺术融合至数学教学中，启发学生，妙趣横生。中学的伙食很平常，每天两顿高粱米饭，吃得饱饱的。团总支张惠如老师住校辅导，使我们感到生活在一个友爱的大家庭中，朝气向上。我加入了新民主主义青年团，参加了天安门前接受检阅的红五星行列。高中毕业我被通知留在北京市东城区，参加"亚洲与太平洋和平会议"的宣传工作，我十分高兴地接受了。但不久又通知我，1952年报考大学的生源不满，已录取入学的不得留下工作。当年我并不懂如何填报志愿，向老师请教。"你喜欢什么?""我喜欢森林，向往海洋……"地理老师贾懋谦说："你学地理吧!""上哪学?""到南京大学，那里有位著名的女地理学家刘恩兰。"就这样我报考了南京大学，并以第一志愿高分录取。

1952年的南京市，人口少，街巷多由方石块铺就，仅马路中央铺有柏油，雨季时路面被冲洗得干干净净，时闻马车驶过的轮蹄声。以绿篱作围墙的原金陵大学校园中有五六

---

* 王颖：《科学的道路》，973－977页，上海教育出版社，2005年。

个池塘，池塘间有菜园田地，垂柳蛙声，别有一番情趣。初到南京，十分陌生，不由得凄然泪下，企盼着寒假返京与中学好友相聚。随着学习的进行，逐渐适应了新环境。1952 年大学是全国统一招生，学生数量猛增，宿舍用房不足，食堂是草棚餐厅。当年上大学不交学费，吃饭不要钱。直到大学三年级，每日三餐，八人一桌，济济一堂，荤素搭配，米、面俱有，餐餐有豆浆，夏日供应绿豆汤，中秋发月饼，学习生活紧张而健康。南大是老大学，注重基础教学，而且院系调整后，地学精英集中南大，地质、地理、气象气候系科齐全，为学生打下坚实的地球科学基础。当年教师年富力强，敬业治教，循循善诱，同学们满怀热情地努力吸收。在南京大学，深受名师教导，终身受益。至今教师们的形象仍历历在目：徐曼英讲授高等数学，熊子敬讲授物理学，幺枕生讲授气象气候学，孙叔平讲授新民主主义，李乾亨讲授政治经济学，孙鼐讲授普通地质学，肖楠森、郭令智指导实习，姚文光讲构造地质，李学清讲沉积岩石学，陈旭讲古生物学，俞剑华讲地史学，杨文达讲中国地质，杨怀仁讲地貌学、第四纪地质学，马溶之讲土壤地理学，耿伯介与仲崇信讲植物基础与植物地理，任美锷讲授经济地理概论，杨纫章讲中国气候与水文地理。尤其是党支部白秀珍老师，她毕业于金陵女大，一生忠于工作，勤恳育人，每每细致谈心，解脱学生的思想困惑，引导我们走上革命的道路。一代名师致力教育，为祖国培育了几代德、智、体全面发展的科学人才，仅我们一个班级后来就有三位同学先后当选为院士。课堂与野外实践结合是地学教育的特点，四年学习，我们攀登了宁镇与皖南山脉，远赴太行、吕梁与黄土高原参加水土保持调查，野外实习增长了能力、锻炼了身体，培养坚韧刻苦的意志。青年时代的我适逢新中国建立，沐浴着党的阳光，在一个朝气蓬勃、日新月异、充满关爱的大学环境中得以健康地成长，真是生长逢时！

1956 年初，国际地理学大会在印度阿里迦(Aligah)穆斯林大学召开，我国派出了以团中央谢邦定为团长的中国青年代表团，两位地理学家：孙敬之教授(经济地理)与郭敬辉教授(水文地理)，岑悦芳翻译及两位学生：北京师范大学地理系研究生陈冠云、南京大学地理系王颖(我)。出国的任务是很紧张的，团组织亲切嘱咐，女同学借给我花衣裙，准备小礼品及带给印度大学生的祝愿。在北京，大家齐住团中央，准备大会报告与介绍中国地理学进展的文稿，反反复复地改写了若干次，然后译成英文，打印成册。外交部亚洲司林林同志会见了我们，介绍了印度的风俗、外交礼仪。当年 1 月，从北京出发时，大雪纷飞，我们刚登上火车即驶动了，在火车的包厢中仍在准备文稿。第二日至广州时，却是春意盎然，我们未作停留，在火车站换了服装即登车赴中国香港，住在新华社在香港浅水湾的一幢别墅中。当时香港淡水紧缺，每日仅是半日供水。我们到达已是午后，因别墅用游泳池蓄水，我们六人，得以洗发、沐浴，洗去了北国的风尘。记得我是最后洗，头发长洗得慢，出浴后，团长与教授们已等我开会了，我十分不安，暗下决心："行动要迅速，绝不能拖后腿。"那次会议很严肃，因为正逢万隆会议克什米尔公主号飞机失事之后，我们的行程是乘英国海外航空公司飞机经仰光飞新德里，分析情况的利弊后，个个表态做好遇险准备。我现在已记不清乘机的情况，仅记得到仰光时，阳光耀眼，热风扑面，休息在一个外敞式的帐篷中，饮料很甜腻，招来很多苍蝇。出席国际地理学大会人员很多，英国的斯坦普、苏联的格拉西莫夫等都是大师级教授。孙敬之的报告——《人口增长与粮食问题》，引起热烈反响。在新德里参加了印度国庆观礼，远远地见到尼赫鲁总理。当年曾为中印友好的欢情所感动，也深为自己“有口难开”而自惭：英文能听（因为中学学英文），但不能讲（当时大学改学俄语），因而立下志愿，一定要掌握英文！中国驻印度大使袁仲贤将军接见了我们，深情地关注、安排代表团在印度多访问些地方，领略地理风光，多与人民接触。我们在新德里参观了红堡、阿格拉的塔姬·玛哈陵、喜马拉雅山麓的西瓦里克沉积层、拉贾斯坦的古城、德干高原的铜矿，直到孟买海港，增长知识，交接朋友，也曾惊险遇虎，汽车抛锚。记得在德干高原的火车上，我打瞌睡，郭敬辉教授狠狠地批评我：“没出息！不珍惜这么好的时光，努力学习。”他几乎是咬牙切齿地说。我很震惊，因为他一向对我慈爱有加，这么严厉地批评我仅此一次。但从此，我外出考察再不打瞌睡，沿途观察学习与思考，获益至大。

1956 年大学毕业时，我国提出“向科学进军”的号召，我和韩同春同学被分配到北京大学投考副博士研究生，韩同春因被李海晨教授留校而未去北大，我考入地质地理系地貌学专业，导师是王乃樑教授。北京大学学术气氛浓，研究生教育规范：全校统修自然辩证法、外语，各系皆有专业课程，我学习的课程有：现代地貌学基本理论与问题、砂矿地质学、沉积岩石学等，均由苏联专家讲学。B·Γ·列别杰夫副教授讲授地貌与砂矿两课（韩慕康任翻译），他理论与实践均强，讲课句句如珠玑，吸引着研究生及来自全国的进修生学习。王乃樑教授与列别杰夫密切合作，指导我们进行了西山寨口、大同盆地、聚乐堡火山群及辽东半岛砂矿地质实习。当年王乃樑教授刚从法国留学归来，英语与法语流利，还能与列别杰夫用俄语交谈，令我敬佩。两位教授教导我们研究生：必须对大学生开设讲座与指导野外实习，毕业时要能独立讲授基础与专业两门课程。在导师指导下，进行了山东半岛海滨砂矿分布规律的研究，我带领一个小组，从烟台到威海，至成山头经荣城，到达莫耶岛、石岛，踏遍了每一个海湾，在每一个沙体上采样分析，追寻原生矿源及砂矿富集带。风餐露宿，旧衣草帽淘沙盆，漫山遍野地探索，几次被觉悟高的老区人民报告民兵而遭围困，释然后，捧出大蒜让我们做“就”（胶东土语：下饭的菜）。我们终于完成任务为圈定 C1 级矿物量做出贡献。通过砂矿工作，使我提高了地质基础工作能力，将营力、海岸地貌与沉积密切联系，探索其内在联系与发展规律，完成了一篇学年论文。2000 年，我在北大评审重点学科时，系图书室的老师还将这篇文章拿给我看，仍保存完好，深为感动。最近，还收到当年在山东一起工作时，冶金部第 5 勘探队工人勇玉林同志来信，那是一段共同战斗难忘的岁月呵！继之而来的苏联海岸学权威——曾科维奇院士与列昂杰夫教授给我们开设了海岸讲座，指导我们进行了胶辽半岛海岸考察。1957 年提出“教育为无产阶级政治服务，与生产劳动相结合”，结合天津新港回淤来源与整治研究，我被导师指定以“海岸地貌与沉积”为专门方向。为追溯新港泥沙来源，在中科院海洋所与天津新港回淤研究站徐选总工程师组织支持下，我带领北队，从滦河口考察到天津新港，以后又从新港到黄河口从事艰苦的淤泥海岸断面调查与重复观测。实践出真知，在一步一陷（至 40 cm 深）的淤泥滩上，背着仪器在一次次 6 个小时的落潮间隙中测量、采样；在风浪颠簸的舢板中测流采样，浅滩上浪大船荡，学生们晕船了，我身为队长必须坚持定时观测，记录与采水均是在呕吐的间隙中完成的，有次实在晕得受不住了，我跳下船去站在齐肩的水中，希望能喘息一下，但跳下船仍晕，而且更冷，只有上船再记。如此艰苦地实践，使我认识到淤泥质潮滩的动力、沉积与地貌的分带性规律，总结出潮滩与波浪动力环境下砂质海滩的不同特性，获得国内外的肯定。以后我们又从事了秦皇岛油港、张庄渔港、山海关船厂、黄岛油码头、龙口新港、北海港、三亚港、铁炉港等港口与航道选址等项研究，在完成生产项目的同时，增长了知识，积累了经验，提高了科学能力，总结出科学成果，同时锻炼了意志，增强了体力。这种锻炼使我经得起“文革”的浩劫：身患肾盂肾炎，参加修建长江大桥，我担任背水泥（一包 50 斤）与铲装石子工作，一车挨一车地装，得不到喘息的机会。分配我干重体力活的原因在于：出身“军阀”家庭，是走“白专道路”的修正主义苗子；抱病下溧阳背包行军 100 里；在农场修猪圈、插秧、割稻，因割得慢，总是接下别人的稻行，因而割的最多，最后连小指也割下一半！这些我都没在乎，挺一下就过去了，但被冤枉时心灵创伤却很沉重。无可奈何，只有不去想为什么。只能自得其乐：农场中空气清新，工余可以看报纸，听听样板戏，每周赶集一次，赶早洗第一批“浴锅”，平息着内心的忧困苦闷。当时流行自己做敲煤油炉，我不会做，买了一个，有了炉子又带来一些乐趣：可以不再受打开水时间与数量的限制，赶集后可炖一小锅红烧肉，农家刚宰的猪十分新鲜，肉都会跳！实际上，“运动”几年了，就那么些人，那么些事，批来批去，批人的人可能也厌烦了，或许也明白了是“莫须有”，因为原先斗别人的人，后来也挨斗了。

邓小平同志的改革春风，确实是扭转乾坤，复苏中国大地，拯救中国人民。国家复兴，我也被召回集中住校学习，学习间隙，常常听任美锷教授讲解 *National Geography* 杂志的文章，高加索、洪昌、秘鲁山地，长寿之三极……我也经常阅读该杂志，提高了英文阅读能力。没想到 1978 年选派访问学者：我因身体健康，“文革”中未沾劣迹，英文考试（系、校、北京）三关通过，而为第一批入选者。赴北京语言学院学习，又受益于几位好老师的教导，英语会话能力提高快，经加拿大大使面试宴请后，于 1979 年 2 月初，我与复旦大学苏德明，上海的荣医生，三人一起乘机赴加拿大。第一批留学经历也不容易，三个人的旅途零用是 1 美元，舍不得花，在东京成田机场等候转机，从上午至晚上 8 时没进食物，我观察到有免费的饮水机，脚踏就可饮水，三人解了饥渴。多少年过去了，这事仍历历在目。后来，每经过成田机场，我都会吃碗汤面以资纪念。

当时，我国驻加拿大大使王栋（据说他在白求恩大夫小组工作过），教育处杨靖华同志，对我们关怀备至，介绍加国科学、文化与民俗，购置冬装，安排我们分赴学校：苏在 Montral，荣在 Toronto，我在遥远的大西洋岸的 Halifax。最初我在 Dalhousie 大学地质学系，作为预备学员，因为系里不知道如何安排我，经过半年考察后，定为 Research Fellow，这在当时是不多的，得益于我在国内的多年研究工作，在海岸科学领域确有真知实力。后来，Dalhousie 大学收了一批中国学者，他们认为中国学者初来时虽有语言交流的障碍，但专业工作与人品绝对优秀。最近送我一本个人诗集的张汉飞，就是当年在 Dalhousie 大学化学系。我改为研究员后，省下了使馆为我交的学费，系主任 H. B. S. Cooke 教授安排用此款为我购买野外装备，请学生助手，以及后来购车，从事芬地湾淤泥海岸调查。加拿大国土辽阔，具有高纬、寒温带与极地气候和海洋环境，与我国海洋环境互为补充，丰富了科学视野。加拿大人口少，人力资源珍贵，时间珍贵，重视生命与个人生活，在加工作尽量少求问于人，促使我勤于用脑与自己动手。学会驾车取得驾照，获得了野外考察的自由，我是开放后 Dalhousie 大学的第一位中国学者，使馆让我负责接待后来者，每来一人我均帮他们熟悉校园环境，见导师，租房安居，购物开伙。一起读报谈心，假日开车去 Acadia 参加苹果花节或采摘苹果，或去海滨或新年聚会，那辆汽车的确发挥了作用。青岛海洋研究所的曾呈奎院士与吴教授夫妇等来 Halifax 从事海洋研究，也由我接机与协助组织中加学者聚餐。当三年学习期满时，大使馆教育处希望我能多留些时日，协助工作，我婉言谢绝了。后来，是大使馆关心支持使我开始了中加合作项目。我体会：没有天才，只有勤奋，只要认真努力，完成所托的事，工作的机会会很多。在 Dalhousie 大学一年后，我又被介绍参加 Bedford 海洋研究所（B. I. O. ）工作，一流的国家研究机构，经费足，设备现代化，有大小不同型号考察船，专用码头与配套实验室，研究工作有序、有节奏，环境优美安静。在 Bedford 两年多的时间，我完成了 Cape Breton 岛鼓丘海岸研究；参加了 Nova Scotia、Labrador 大陆架、纽芬兰峡湾海岸考察，大西洋洋中脊钻探、SOHM 深海平原勘测钻探与 Porto Rico 海沟考察；更获得乘 PISCES Ⅳ号深潜器之珍贵机遇，下潜考察 St. Lowrence 海底，扩展了自己在海洋地质学的科学能力与研究成果。珍惜工作良机，竭尽全力完成工作，绝不辜负祖国人民的期望。北大西洋浪高流急，船只颠簸，克服晕船之苦，摄下了座座漂移的冰山与低垂的北极光帷幕。热带风光的 Bermuda，湛蓝的海水与白色的石灰岩岛相映生辉，但当 CSS Hudson 号考察船开航向 Porto Rico 海沟方向行驶时，突遇 10 级风暴，大海怒涛汹涌，船只抛上掷下，甲板的泳池也击碎了，这时的我已经受了多日海上磨炼，却能坚持工作，获得了百慕大海域的实况资料。出海后，抓紧回国前的两月时间，应用 B. I. O 先进的扫描电子显微镜完成了从山地到深海，从亚洲至北美所搜集到的各种环境石英砂样品的观察、分析与鉴定工作，出版了《石英砂表面结构与沉积环境的模式图集》，供沉积学研究工作参考应用。加拿大三年，使我在海洋科学领域获得全面锻炼，掌握了从外业到实验室分析工作与总结成果的研究能力，推动我发挥在国际海岸海洋科学领域的合作研究与影响。

随着《联合国海洋法》于 1994 年正式实施，全球海洋权益的重新划分，推动着“海岸海洋”（Coastal Ocean）科学概念于 1994 年被 UNESCO 政府间海洋委员会（IOC），1996 年被国际地理联合会（IGU）正式确定。我国政治经济形势的繁荣发展，推动着我们学校、重点实验室及新型交叉学科的建设发展，我自己也在努力工作与继续前进。又一个十年历程，负责完成了中、加合作两个项目：海南深水港建设（IDRC）、海岸生态环境监测与高等教育（CCHEP），于 2001 年 6 月 Waterloo 大学毕业典礼上，正式获得环境科学荣誉博士学位。最为鼓舞的是，在 2001 年 11 月当选为中国科学院院士。

回顾自己的成长，离不开祖国的需要与时代的召唤，在党的温暖阳光照耀下，受益于师长们的教导，培养我一步一步地前进。大学从事地理与地质学习，奠定了我从事地球科学教学与研究的基础；研究生阶段获得地貌学与沉积学科学能力的提高，奠定了海岸科学的基础；毕业后，在我国“三年改变港口面貌”的导向下，通过十多年海港选建工作，大大充实了对海岸发育演变规律的认识与服务于海岸建设的科学能力；改革开放的东风把我送出国外深造，使我在海洋领域攀登，增加海洋地质学物探与海洋沉积分析的实验工作能力，参加了从极地海洋到热带边缘海，跨越洋中脊到深海平原的工作实践。在研究工作中学习，在生产实践中学习，一步一步前进，一层一层地攀登科学高峰。回顾历程，使我体会到：学无止境，不要囿于大学学习的专业基础，应随着时代前进的步伐，不断地学习、扩展，交叉与融合，使科学能力真正服务于祖国发展的需要。饮水思源，立志不懈，在教育科研岗位，为培养品学兼优的人才，为祖国的现代化与科学发展贡献力量。

忆导师王乃樑教授*
——九十诞辰纪念

1956 年 8 月，我从南京大学地理系地貌学专业毕业，被分到北京大学地质地理系准备入考副博士研究生。到北大报到后，被告知暂定为见习助教，若考取了就当研究生，若考不取则留为助教。记得去见王乃樑教授(他当时是地貌教研室主任)时，王先生安排我使用办公室靠东墙的一张桌子，并很和善地对我说："来了就准备考试吧!"1957 年 2 月，通过入学考试，我被录取为四年制的副博士研究生，师从王乃樑教授，开始在北京大学的学习生活。

1957 年春，苏联利沃夫大学的地貌学家 B. Г. 列别杰夫副教授应聘到北大担任校长顾问并讲授"地貌学基本理论问题"与"砂矿地质学"。先我一年进入北大的崔之久、周慧祥与钱宗麟三位都是列别杰夫的副博士研究生。我们四位，还有地貌教研室的曹家欣老师、欧阳青老师亦参加了专家讲学班的学习，参加专家讲学班的还有来自全国各地的 40 多位青年教师，其中由专家亲自指导的有长春地质学院刁正清、北京地质学院陈华慧、中南矿业学院沈永欢及中山大学李见贤。讲学班由韩慕康和江美球两位先生分别担任口译和笔译。我们共同组成了一个富有活力的年轻集体，由王乃樑教授与 B. Г. 列别杰夫副教授共同指导。

当时王先生约四十岁，风华正茂，文静而又帅气。从法国返国后，一直骑一辆轻型单车上班，彬彬有礼，对待师生一视同仁，和蔼相处。列别杰夫讲课时，常就一些术语的翻译，征询王先生的意见，他都认真给予回答。王先生能流利地应用英语、法语和俄语。20 世纪 50～60 年代，我的外语还是"一瓶不满，半瓶晃荡"，原因是中学学英语，而入大学后，全部改学俄文，两种外语均未学好。先生的外语水平使我很惊异，见他每日均在阅读一本厚厚的俄文专业书，忍不住去问，"您的俄文怎么也这么好?"他说："俄文与法文有相似之处，也曾短期突击学习过俄语……"。90 年代他病重住在中日医院时，仍以阅读英文小说来度过漫漫的长夜。王乃樑先生的外语水平是我结识到的所有人当中的最佳者。

1957 年春季，王乃樑教授带领我们去大同盆地野外实习，由列别杰夫副教授教我们绘制地貌综合剖面，分析地貌发育阶段，王乃樑教授指导我们从事沉积物与沉积层分析，分辨沉积相，将地貌与沉积结合研究，以探索地貌发育过程之真谛。随后，让我们指导自然地理专业与地貌专业大学生的实习(寨口与大同盆地)。通过教学实践，加深了我们对河流地貌，对火山地貌，对区域环境特点的认识与理解，学会了运用多学科的综合，学会了应用对比研究的方法，将地貌特征与沉积组成相结合，分析动力成因与动态变化，终身受益。

王先生与列别杰夫还指导了我们赴辽东半岛与山东半岛从事砂矿地质调查研究，这

---

* 王颖：北京大学环境学院王乃樑文集编辑组，《王乃樑文集》，334－337 页，学苑出版社，2006 年。

种实践与应用工作，是非常重要的学习，的确可以培养学生实际才干，增强智力与健康体魄。我和两位工人师傅一起采样淘砂，早出晚归，步行千里，终于摸清了山东半岛东部海滨砂矿的分布情况，为探求C1级锆英石砂矿床做出了应有的贡献。50多年后，收到当年在山东半岛共同工作的勇玉林师傅的来信，往事历历在目，值得回顾！

20世纪50年代末期，国家发出“教育为无产阶级政治服务”“教育与生产劳动相结合”的号召，王先生亲自带领我聆听苏联海岸研究权威B. П. 曾柯维奇院士等三位专家开设的海岸科学讲座。为了使海岸地貌与沉积的科学研究直接为港口选址服务，他指派我带领徐海鹏、任明达、李容全、高善明、濮静娟等北大地貌专业师生，先后参加了“新港回淤泥沙来源研究——滦河三角洲研究”“苏北黄沙港选址研究”等与天津新港回淤研究有关的生产项目。通过深入实地调查研究，理论联系实际，我们查清了新港泥沙的主要来源，提出无须沿岸建堤来拦截滦河与黄河泥沙的科学论断，为我国首次在淤泥质平原海岸建造深水港做出了贡献。

王乃樑教授不拘泥于法国留学时所学的专业，回国后将沉积学研究应用于地貌发育与动态分析，通过沉积将动力、地貌结合为一体，不断地探索新的科学理论，在火山地貌与新构造运动、河湖地貌与沉积相、喀斯特地貌、海岸地貌等方面，培养了一批年轻的地貌学者。他指导我们将科学研究与生产任务相结合，向地貌学教育与研究的纵深发展。他真正是实实在在地，一生不间断地，虚怀若谷，在地貌学教育与科学研究方面做出重要贡献。

王乃樑先生文学造诣深，他发表的文章都是经过深入思考严密分析写成的。但在日常生活中先生却颇有风趣。记得一次在寨口实习时，田野中惊散了一群胖嘟嘟的小黑猪，我们都停步以待，王先生说：“什么东西小的时候都很可爱！”一句不惊人的话，风趣横生，使大家忘却了疲劳。1983年在厦门进行海岸考察时，我发现新相机的镜头盖失落了，十分懊恼，王先生嘱我再回头找找。果然，在前一站点找到了，王先生风趣地说：“找到了比没丢还高兴！”他就是这样顺其自然地处事，真真实实地待人，他这种心平气和的精神也感染了周围，提携我们以乐观的态度去处事。在登上厦门曾山考察时，看到在120 m高处有巨型海蚀穴、残留的海蚀柱及海蚀阶地遗迹，十分惊奇。当时一位教授认为“这么高的位置不应该是海蚀，而是风蚀。新疆山地有很多吹蚀凹穴”，的确，在干燥区的花岗岩山地上，常存在着类似于海蚀穴的风蚀凹穴，但是却不会有磨蚀阶地与残留柱体的组合形态。当热烈争论之时，王乃樑教授说“我支持以组合地貌判断其为海蚀成因的意见”，坚定了我的认识。

1961年春季从北大毕业后，我回到南京大学工作，常常在相关会议上见到王先生。记得一次在南京召开地貌学会议，王乃樑与潘德扬老师均到会。王乃樑老师在考察到雨花台砾石层时，拿了几块砾石问我：“你看它们有何特征？”我从未注意到雨花台砾石的形态特点，王老师一句话使我茅塞顿开，我很快意识到“扁平度高”，他指出：“应当系统地做些量计与研究。”这个叮嘱，一直在我心头萦绕，后来我也曾量计了数千块砾石，但始终没完成研究成果的总结，至今遗憾未能答复老师的关注。

王先生真诚爱国，一心跟党走，大事小事都听从。记得在1958年时，他身为教授也曾手执弹弓在北大校园中转悠，打麻雀，除“四害”。现在想起那段往事，实感荒谬。在“文革”动乱初期，王先生到南京来过我家，他送给两个孩子一人一个吹气娃娃。我的女儿们很高兴，一再爬上爬下在他身边玩耍。他很感慨地说："在北京，已经没人敢与我谈话了，在南京却受到孩子们的稚情欢迎！"70 年代末，王先生一再嘱咐我将两个孩子暑假中带到北京"增长知识与见地"，是前情不忘，深情关注。他出资购买火车票让她们赴京，两个孩子真是高兴啊，到了天安门、故宫、香山，领略了首都风光，受到了文化熏陶、启蒙教育，终身受益。这件事，我一直铭记于心。王先生子女多，家庭经费开支大，他对自己一直是省吃俭用，很少买新衣，一件卡其布反毛领夹克，仍是从法国带回的，但穿上总是风采奕奕，与众不同。我当时即下定心愿：一定要报答老师关怀，对人也要主动关心，要"雪中送炭"。

80 年代后，每逢赴京，我总去拜访王先生与何先生一家。何申先生总是全身心在卫生、健康教育工作上，在城里协和医学院工作的时间多，出城在家的时间少。那时，电话联系不像现在，我每次去事先很少打招呼，直接就到燕东园 21 号王先生家中。当时王先生住房紧张，家里不仅有子女，还要赡养老人，但他总是娓娓地与我们交谈，互道别情与对时事的看法，平静而舒畅，对我们时有勉励，时有启发，内心不平也可以互有分析与排解。至今我仍忘不了窗外的竹影斜阳与花叶飘香和那一次次谈话所绵延的真挚的师生情谊。去京必去老师家，已成为我多年的习惯。碰到吃饭时，大家共同用餐——饭菜均是从食堂买来的，但吃得非常惬意。

在老师多年的关心下，我有所进步，有所成长，尤其是在申请院士未通过时，老师总是指出我的不足，鼓励我继续努力。多年后，一些资深院士还提起当年王先生多次向他们表达了对我的关切与介绍的事，谆谆勉励我。

90 年代一个初冬的夜晚，我趁在赴清华大学开会之机，到老师家看望。那时老师已病重卧床，多日不会客。我去了，只见病榻旁堆满着报纸与书籍。何先生让我稍等，老师正在灌肠。见到他病重，我心乱如麻，为照顾他休息，我只能探视几分钟。告别时，王先生知道我要走了，他清楚地说："王颖，你惦记着来看我。灌肠很麻烦，裤子都是冰凉的，很难受。我心里很明白，你别忘了我啊！"我当时心如刀割，但又不能放声大哭，只能强忍地说："老师，我永远不会忘记您对我的教导，对我们的爱护。让我来磕头答谢师恩吧。"于是我恭恭敬敬地磕了三个头，然后又深深地一鞠躬，含悲告别了我尊敬的导师——王乃樑教授。不久后，王先生逝世了。这次见面竟成永诀！但我永远忘不了师恩如山，北大燕东园 21 号小园的往事也永存我记忆！

时光飞逝，至今我已年过七十，我依然铭记导师的音容，活着就要像王先生等师辈那样：真挚、淳朴、爱国、奉献、和爱众生！

2006 年 9 月 2 日于南京大学

科学建树　贡献终生*
——纪念刘东生院士

我认识刘东生教授是在20世纪50年代后期，当时我是北大的研究生，因为不是同辈，很少接触，但我很敬佩他。刘先生和我的导师王乃樑教授曾是西南联大的同学，他们千里跋涉，从天津南开中学到大后方求学，并满怀热情地担任美军在华电台的英语播音员，对外传播抗日战争的实况。我最早的印象就是他们热诚爱国，地学与英语双优。刘先生和王乃樑教授的友谊长达数十年，共同从事第四纪研究。刘先生一直关心地貌学与第四纪地质学的教育，对北京大学的青年教师与学生，如曹家欣、崔之久、钱宗麟等关怀备至；曾多次亲临南京大学地貌学与第四纪地质学专业讲学、指导。在黄土高原考察中，吸收了当时的青年教师与大学生参加考察实践，培养、指导他们，这些人日后成为了学术骨干。

刘东生院士一生不断地探索、研究，从古脊椎动物工作起始，奠定了第四纪研究的坚实基础。50年代，组织了北京大学、北京地质学院、南京大学等青年师生数十人，参加了黄土高原十条大剖面调查，从山西太行山、吕梁山，经子午岭到甘肃临洮和宁夏的固原等地，从大青山麓到秦岭北坡，实地步行调查黄土的分布，从基岩到顶层的黄土剖面物质组成、沉积结构以及从东到西各地点的午城黄土、离石黄土与马兰黄土的分布变化等。首次全面大规模的黄土地质调查，汇集大量实际资料与综合研究的成果，撰写了《黄河中游黄土》《中国的黄土堆积》和《黄土的物质成分与结构》三部论著，有力地向国内外展示了中国地球科学家在黄土风成沉积研究的重要进步。刘先生没有停滞，进而深入探索、研究黄土-古土壤互层结构，揭示了冰期-间冰期的多旋回特点，质疑居正统地位的“第四纪四次大冰期”的理论，以事实论证了第四纪气候变化的多旋回理论。这是继黄土新风成学说之后，在地学理论方面的又一项重要进步。嗣后，刘先生继续深入剖析系统的黄土地层，建立了近250万年来古气候变化的陆相沉积记录，与深海沉积研究媲美，共同谱写了地球环境变化——第四纪历史的新篇章。巨大的工程，凝聚了世界科学家努力探索的结晶，刘东生院士代表中国学者作出了为世人瞩目的重要贡献。

人的一生，即使活100岁，与地球年龄46亿年相比，仍是瞬间的分秒。刘东生院士孜孜不倦地追求，为后辈树立了光辉的榜样：他致力于科学研究，建设第四纪地质科学；发展地球环境系统科学；培养了一代又一代的地球科学家，从不同地学分支领域，从不同地区特性研究，培养人才。真正做到“十年立著，百年树人”，展现了地学大师的风范，冀望人们一代又一代地阅读、探索与详解“地球巨书”。

近年，我参加《第四纪研究》编委会工作，亲身感受到刘东生院士的大师品格，他年过80，仍精力充沛地关注《第四纪研究》的每期内容，文章的水平与组成特点。他工作深入细

* 王颖：中国第四纪科学研究会编，《纪念刘东生院士》，47－48页，商务印书馆，2009年。

致，心平气和地处事，和蔼地待人。他因生病住院，不能到南京参加最近一届的全国第四纪学术大会，为此，专门书信给我，说明与嘱托。2007 年底，他亲自撰写纪念孙殿卿院士的文章，并主编文集。这一件件的事例均使我深为感动，我虽晚于他两代之多，但却没有他那样的精力与精神不断地探索与关览全局，也不曾像他那样耐心细致地待人接物。刘先生的风范真是我学习的榜样。

刘东生院士的科学成就为国内外所肯定，他先后获得国内外科学奖励与最高科学奖：国家自然科学一等奖、二等奖，中国绿色科技奖特别金奖，中国科学院自然科学一等奖，中国科学院科学技术进步特等奖，陈嘉庚奖，何梁何利奖，2002 年国际环境科学成就奖——泰勒奖，2003 年国家最高科学技术奖。荣誉高、责任重，但刘先生从不间断地、持之以恒地努力工作，前后担任一系列的国内外学术职务，如：国际第四纪研究联合会副主席(1982—1991)、主席(1991—1995)，中国第四纪研究会主席(1979—1999)，以及全球环境大断面(PEP Ⅱ)项目首席科学家(1994—2000)等。与此同时还担任《第四纪研究》《环境科学学报》和《极地研究》主编。我想，刘先生始终保持高昂的工作热忱，其精力来源于对祖国、对人民的赤子热诚。伟大的科学目标赋予持久的动力，科学团队力量的和谐配合，造就了最高科学成就。

刘东生院士为我们树立了永铭于心的榜样。

遥远的记忆*
——大学教育 受益无穷

人生有很多很多机遇，未容细想，也就走过来了。我选学地球科学，既不是父母之教诲(他们已早亡)，也不是个人的向往，但却是机遇。1952 年我高中毕业于北京慕贞女子中学(后改名育新女中、女十三中等)，要填报高校志愿，不知如何填。请教当时辅导我们班的老师贾懋谦，他毕业于北京师范大学地理学系。他问我，“你喜欢自然吗?”“喜欢。”“喜欢什么?”“森林、海洋……”因为我当时刚读完苏联小说《远离莫斯科的地方》。“那么，你报考地理学系吧。”“去哪?”“去南京大学，有位著名的女地理学家刘恩兰，她也是海洋学家。”就这样，我报考了南京大学地理学系。发榜后才知道，华北区报考到南京大学地理学系的就两名：王颖、袁广本，这就是我学习地理学的始末。而当时录取的 60 多人，作为第一志愿的只有几个人。有人闹情绪，而我却仍是浑浑噩噩的，没太多想法。

1952 年，南京大学招收了两年制金属与非金属矿产地质及勘探专科班，地理系学生与地质学专科的学生一起上课，学了很多地质基础课，而且均是名师所授，我愿写下来以志铭记：教授普通地质学的是孙鼐老师，两个班级的助教，分别是俞鸿年与施央申老师；讲授构造地质学的是姚文光老师；讲授岩石学的是李学清老师；教授古生物学的是陈旭老师；地史学是俞剑华老师讲授；中国地质学是杨鸿达老师。野外地质实习指导老师是郭令智和萧楠森老师。这些老师的教育，使我们打下了坚实的地质学基础，对嗣后工作受益一生。

在这些课程中，孙鼐老师是我们在地质学方面的启蒙老师。他当年约 40 多岁，讲话南京口音重，但讲课时铿锵有力，如掷金石于地，使我得以牢记于心。他讲课中常用通俗的道理，讲解当时对我们来说仍很抽象的地质学内容：地层、褶皱、断层及地质年表，使学生能清晰了解，而且兴趣盎然，至今仍清晰印刻于脑中。最感兴趣的是记矿物硬度表，孙鼐老师念了一个口诀：“滑石方，氟磷长，石黄刚金刚”，尤其最后一句“刚金刚”，老师的南京口音重，鼻音也重，至今仍回萦于耳。50 年后，我曾问当时同班的陈毓川同学：“你还记得孙鼐老师教的硬度表吗?”他笑眯眯地念出：“滑、石、方，氟、磷、长，石、黄、刚、金刚”，老同学都哈哈大笑起来。孙鼐教授讲课给人印象之深由是而知。

老师对学生的启蒙教育、讲课与实践教学都很重要。当年我并没挑选学什么，也不知道该学什么。进了南大后受教于一大批名师，却受益匪浅！记得李学清老师教学耐心细致，亲自修改我们的岩石标本描述记录。姚文光老师用手势比拟构造变动，启发学生深思。萧楠森老师指导我所在的乙班地质实习，是从方山开始向老地层走，甲班由郭令智老师指导的是按从老到新的地层学习。他们以矫健的步伐，上下奔走，循循善诱地教导，唤起了我们学习的热情，不顾炎热夏季跟随老师走遍宁镇山脉地区。记得还有一位王通助

---

* 王颖：《孙鼐纪念文集》，19－20 页，南京大学出版社，2010 年。

教带我们爬紫金山时，从阳坡上，在森林中走失了道路，湿热、蒸腾、钻荆棘丛，确实是一番攀登的艰辛，但年轻人却互相嬉笑鼓劲，终于登上山顶。当时那份高兴之心情似乎至今仍有笑声荡漾于淋漓的汗水中。

在1952—1956年间，我们上大学时，出版的专业书籍很少，记得只有两本书。一本是孙鼐老师的《普通地质学》，一本是南工刘海清讲师的《普通测量学》。那时，在我们学生心中，感到能出版专业书，真是了不起！两本书均不厚，普通测量学比普通地质学还稍厚些，但是，字字如珠玑，我们学起来如获至宝般地钻研书中的原理与认知方法。不像后来，书多了厚了，但不易通读学习了。这使我反思：教科书究竟如何写？应写多少为宜？实际上，学习基本理论与思维方法，可拟在实用中举一反三，积累能力，而不一定很多叙述，照搬实例是不适宜的。

大学学习时，我们不了解也不探求了解老师的职称，至今，我仍不知那些老师当年的职称如何。1963年当我被提升为讲师时，真是有“倍受鼓舞之感”。当年，讲师被列为高级知识分子，而现在，大学毕业尤其是研究生毕业后对担任讲师或工程师似乎已不在意，至少要是副教授，升至教授还不够，还要是博导，博导不够还应上升……所幸现在还有CEO，有“总”这些为人所在乎的称谓，否则，全都奔教授，教师队伍越来越单一，没有梯队，没有分工，如何提高教学内容与质量呢?！我是1984年被特批为教授，1991年评为博导。当年是情况特殊，现在不必那么长，但也不能以连升职称为目的！我不知道20世纪50年代时，我的老师们职称如何，但在我心目中，不，在我们班同学的心目中，至今仍认为老师们的学术水平、教学内容与讲授水平都是很高的，都是尽职尽责，教书育人，为人师表。最后我想说，我曾在北京西路二号新村宿舍住过20年，与孙鼐教授是前后排楼舍之邻居。许多年过去了，他的容貌没多大变化，身体也没发福。常见他从宿舍大院门房一侧的石阶上下进出，精神好，步履不困难，只是向他行礼、打招呼时，他会愣一下，想一想，或许是听力稍弱之故吧。我常想，这是从事地球科学工作的优越性，经常出野外，体质较一般人增强很多。仔细想想，老一辈地质、地貌学家多长寿，一般都是享年在八十岁以上。孙鼐教授的健康身体是在长年累月的地质工作中锻炼和培养出来的。它给人启示：体力活动之重要，地球科学有益于人生。

人类生活于地球之上，衣食住行，生活质量(温暖、炎热、寒冷、适宜)均取之于地球，人类生命的健康延续，依赖于地球，必须了解地球，了解其环境与变化特征。地球已有46亿年的发展历史，人们的生命有限，要能尽本分地对所生存的地球环境有所了解与有所贡献，必须承上启下地学习、了解与传介。真正的知识、能力与健康的身体都是重要的。回忆以往，孙鼐教授、老一辈师长的教导与榜样，铭刻于心，永远常青。

缅怀江上舟博士*

人生数十年，感觉似乎很长。但实际上很短、很快。真如刹那般的一日又一日地飞过，即使是 3 万多天也令人感觉很快。真正体会到珍惜时间的重要。回顾成长的一生，似乎是漫长的经历，世事繁多：父母的抚育，从婴儿哺乳、牙牙学语到学会行走独立生活；学校的教育，经历一个又一个的考试、一级又一级的迈进；社会的关怀、支持与实际锻炼，一步又一步地经历过来；家庭、社会、国家乃至世界均给予扶植与付出。在有限的生命内，为国家与社会服务，做出一定的贡献，即使是回馈，也是做人的必备品格与基本任务。大千世界正是在一代又一代人的辛勤努力工作中发展前进的。回想江上舟同志的一生是实现了他的人生价值与回报了国家与社会的培养。

初识江上舟同志是在 1988 年 11 月，当南京大学海洋地貌与沉积研究室王颖与朱大奎教授一行与加拿大贝德福海洋研究所(Bedford Institute of Oceanography)Charlie T. Schafer 博士与 John Smith 博士等一行开展“海南岛港湾沉积速率与深水港建设研究”合作项目时，该项目是在我国国家科委支持下开展的，获得加拿大国际发展研究中心(IDRC International Developing Research Centre)及中国国家教委的资助。当时，我们正在三亚海岸工作，因出海需要，我与 Charlie T. Schafer 博士到三亚市政府联系工作，市长接见后，由江上舟同志负责接待安排我们工作。当时，他担负副市长的工作，却尚无命名。工作交往中，我和 Charlie 均很惊讶：在边远的海岛城镇(当年三亚尚在发展初期)，有这么一位年青帅气的市领导，真挚的笑容，明亮的大眼睛，能说一口流利的外语，对我们的工作给予了认真细致的帮助。Charlie 是德裔美籍的加拿大人，他与江两人以德语交谈，十分融洽，双方均有他乡遇故知之感。从此，我们开始了交往。后来得知：他毕业于瑞士苏黎世理工学院，获自动化工程博士学位；他和夫人吴启迪博士从瑞士学成归国后，吴在上海同济大学从事高等教育工作，江到边远的海南岛从事开拓建设，一干就是十数年。江的父亲江一真曾任首届福州市委书记，后任卫生部、农林部部长，其大名是常见于报道，但我们从未将父、子联系考虑过。

当中加合作第一个项目结束时，中加联合于 1990 年 11 月 1 日至 10 日，在海南岛召开了首次国际会议：“国际海岛海岸与港湾资源开发管理学术会议”，前后衔接地于海口及三亚召开，参加会议的有 11 个国家(中、加、美、澳、苏、德、丹、印、韩、孟加拉与斯里兰卡等)约 80 位代表，这在当时是空前的。会后选出 52 篇文章，发表了“Island Environment and Coast Development”英文专著(1993.1 南京大学出版社)。在海口开会时，获得当时海南省领导大力支持，邓鸿勋书记与省长刘建峰均出席开幕式，刘省长与孟庆平副省长致开幕词与讲话，表达对代表的欢迎。记得当时在大会与分会发言的有国家海洋局葛有信副局长，南京大学党委书记陆渝蓉教授，海南省交通运输局马文金局长，海南港务局欧阳

* 王颖：2013 年 2 月供稿于《江上舟在海南》。

宝奎局长和海口市市长李金云及三亚市副市长刘明哲。会议开得十分成功，中加合作的“海南岛港湾沉积速率与深水港开发”系列报告，涉及三亚港扩建、洋浦港选址、东水港选址及亚龙湾开发规划等，形成海南海岸环境工程的特色与论文核心。记得在晚会上，孟庆平副省长欢唱了民歌“回娘家”，王颖即兴表演，获得代表们的欢迎。

在三亚开会时，江上舟博士给予大力支持并做了“三亚市资源的综合开发和利用”学术报告。他在报告中介绍了“三亚的地理位置、气候、市辖区。1988年时，人口为34万，其中44.4%为黎、苗与回族；当年74%为农业生产，年净收入为4亿元。所以，当时此遥远的南陲小城是不富裕的，但发展潜力巨大：长年无冬的气候条件与青山、河流、海滨景观独特，纯净的石英、珊瑚质沙滩总长达100 km，与多彩的民族文化组合形成“不是夏威夷，胜似夏威夷”的旅游资源；加之，13 000 km^2 外海渔场，多样的小岛、生物与30多种矿产等，三亚市发展的潜力巨大”。江在报告中指出：“海南建省三年来，三亚有了巨大的进步。在发展中要注意各部分的协调与综合利用资源，宜将当前的需求与长远的利益结合，提高生活质量，在短期发展致富过程中，建设经济特区。”他进一步谈到“三亚从传统的海港、渔业与制盐业向现代化国际海滨旅游城发展的设想，传统产业应赋有新内容等”。(附：报告原文)

江上舟的报告代表了当年三亚市领导的开拓性工作，获得代表们的热情关注。海南首届国际会议是成功的。在总结时，中加合作双方一致对海南省各级领导及人民表示衷心感谢。尤其是对在会议中与我们亲密合作的江上舟和欧阳宝奎两位表达了深切的谢意，为此我专门向邓书记和刘省长汇报了对海南省年青一代新型领导工作的敬佩，从他们身上反映党领导下，中国发展中的巨大生命力。

江上舟博士正式担任三亚副市长后，为三亚市整体规划与分期建设投入了大量精力。我亲临感受着三亚面貌的改变：当1964年我第一次赴三亚，是在榆林海军基地参加铁炉港建设方案。当时，仅榆林海、陆两军基地设施较全，其他地方是一边远的小镇。至1966年从事三亚港扩建调查时，三亚仅有一条红砂大道，贯通东西到莺歌海。我们住在港务局刚建的宿舍楼中，为开展实验过滤砂样时，只有三亚医院有天平可借用称含砂量。港内渔船多，渔民杂居在以长柱支架的椰寮中的木板上。街市上仅有两家小吃店出售“面包仔”“咕咕奶”(后知为可可奶)与“伊面”，当时，三亚地产西瓜是白色无甜味的，被用来炒菜吃。1966年“文革”风波波及三亚市，当时的中学生红卫兵针对在绿色植被草坛标语“毛主席万岁”上洒清水时，竟抗议：你们怎敢对伟大领袖浇凉水？真令人难以理解。但当90年代初，再次赴三亚市，三亚变了，市容整齐，滨海椰林绿地在发展，街市繁荣了，正在兴建凤凰机场……我对在天涯海角海滨骑马游憩活动，马粪污染沙滩，小贩阻拉游人兜售珍珠、贝壳，以及机场修建时要注意保留古海湾自然遗迹，海滩应留出一定宽度而不能建筑等，向江副市长提出改进建议，他均十分支持，并说：“我们在规划、整理与建设中……”后来，天涯海角围管起来了，沙滩洁净、海风宜人，商业小街建设起来了，既保留了海岸自然风光，又照顾了当地民众的商业生路，满足了游客需求。三亚一天天在发展变化，凤凰机场通航，道路扩展平整，南山寺兴建富含文化韵律，回辉镇清真礼拜寺新貌展现民族风情。三亚在建设中发展，昔日的海滨民族村镇成长为热带海岛的旅游中心，深深体会到，江上舟、王永春、王万超等领导及各民族群众以他们的勤劳智慧，实现着现代化的建设蓝图。

发展三亚的理想志愿在实践中前进，当三亚市中心建设至一定规模时，1992 年三亚市政府在海口市召开“三亚亚龙湾规划论证会”。主题内容由清华大学规划系教授报告亚龙湾总体规划，三亚市副市长江上舟博士专门邀请南京大学海岸与海岛开发国家试点实验室的朱大奎教授报告发展三亚海洋旅游的设想。会后，三亚市政府委托王颖教授等人进行了亚龙湾海洋旅游调查、勘测与规划研究，经过两年工作，完成了亚龙湾海洋旅游规划，并经过评审，明确该项海洋规划可与清华大学所做的岸陆规划相衔接。嗣后，江上舟副市长又专门到南京大学与王、朱两位教授商谈建设白排人工岛，在岛上建设类似悉尼歌剧院般的标志性建筑，为三亚树立国际热带海滨城市形象等，虽经过多次研讨酝酿，可能由于经费或因江上舟博士调走而未能实现。

亚龙湾建设使三亚市发展更上一层楼，亚龙湾与三亚湾发展形成三亚城市核心的东、西两翼，亚龙湾发展具有国际品位，兼有碧海绿野风光的旅游胜地；西部三亚湾则具有国内假日民居休闲之特点；三处结合，形成环海岛南岸的熠熠闪烁的明珠项链。

与江上舟的另一次合作是在 1992 年，国际海洋年期间于 6 月 22 日至 26 日在意大利热内亚(Genoa)召开“全球变化中的海洋管理国际学术大会”以及相伴的国际海洋博览会活动。意大利热内亚大学教授 Adalberto Vallega，时任国际地理学会副主席与海洋地理专业委员会主任委员负责该学术大会。他邀请我赴会并在大会上做“中国海岸管理”专题发言。同时，让我代邀 5 位中国代表参加海洋年大会活动，Vallega 教授负责安排由大会提供 5 人的国际旅费与会议期间的全部费用。当年我国参加国际学术活动未若现在这么频繁与经常，所以，对这么个盛情相邀，我认真选择了在北京的国家海洋局的杨锦森研究员，海南三亚市江上舟博士，中国科学院院士任美锷教授，及华东师范大学河口海岸室水文动力学沈焕庭教授共四位赴会。其实是应邀请 5 位，而我理解是包括我为 5 位。江上舟博士愉快地接受邀请，我希望他能在会上介绍海岛海洋建设的体会与展望国际合作前景，通过中外交流有助于发展海洋文化与生态旅游事业。我们一行 5 人参加了会议交流，也开阔了视野。但是，他因在国内临行前工作繁忙，第二天未能早起而迟到了，失去了一次发言机会。但整个会议活动是有成效的。事隔多年，我记得同济大学吴启迪校长对我还讲起该事：“当年出国不容易，他全额资助出国还迟到了。”实际上，海洋年年会规模大，个别代表迟赴会未被注意。

90 年代，我赴洋浦港开会，洋浦港已批准为经济开发区，具有“小特区”的地位，洋浦港口区设立了门卡与围栏，出入凭证。见到江上舟博士，已调任为开发区主任，他仍是那么热情、精力饱满但却略显风尘仆仆地迎接我们。故人相遇，十分高兴，但我也略有惋惜，心中略有不忍：三亚市刚具规模，又调任了，年青一代的共产党人保持了光荣传统，迎接一个又一个新的挑战，勇担重担！但是，发展中的三亚也需要熟悉三亚发展的领导针对发展趋势中出现的新问题，提出新一轮、更切合实际，或说更符合人地和谐相关发展的规划与建设大计！调动岗位是可免“地方主义”弊病，但重新学习与认识一个地方的特点与相宜发展途径，却的确需要通过一定时间的实践总结与提高啊！什么事都不能千篇一律，而应因地、因时、因人制宜。我佩服江的实干热情，但却感到他多年连续苦干，已略显憔悴！

初建时的洋浦经济区已不见绿色田野与黎族民居，崎岖的玄武岩岗丘已推成平坦大地，街道初成网格，仅港口宾馆小楼园中尚见热带树丛，港口内船只不多，港区堆放着桉树木片，输出为造纸原料。大约是2008年前后，又一次访问洋浦区，港口在扩建，船只穿梭往返，打桩机轰鸣，景象繁荣，发挥着海南西岸石化与农产品出入港基地作用。港市在发展，我联想到母鸡神——神尖岬海底抬升的古海岸遗迹能否还被保存下来？地区的发展宜前后衔接，宜将当代与未来结合规划。但江上舟博士已调离洋浦，据说调至上海，而正在北京协助工作！我未能在洋浦见到他反映意见，却能理解：学成归国建设南疆海岛，廿年如一日地努力，家庭却分居三处，父母在北京，夫人在沪担负高校重任，谁也不愿离开工作岗位，年事渐长，人到中年返回内陆邻近家人，相互关照也是应当的。江上舟博士就是这么毫无怨言，奋不顾身地一次又一次肩负新任务，迎接新挑战，满怀热情地建设祖国！他的精神感人，我们虽未再合作，也很少联系，但他的形象在我心目中，始终是热情、乐观与坚毅的，是新中国培养的一代知识分子的代表。他身患重病仍坚持开拓，坚持实践前进，英年早逝，令人实难相信，难以面对事实。

2012年10月，赴加拿大Halifax参加Bedford海洋研究所建所50周年庆典时，当年在海南建港调查的老友：C. T. Schafer，J. Smith，K. Roberson，K. Asprey等又会面了，昔日中年人均已步入白发行列，年青的K. Ellis女士早逝。回忆当年三亚-洋浦海上工作，几乎不约而同地悼念我们共同的好友——江上舟博士！敬爱的江，安息吧！你虽离开人世，但你开启建立的事业，你的工作精神与伟大人格魅力将永驻友人之心，永载海南发展建设的史册。

2013.2.14完稿

附：江上舟报告原文（据 Wang Ying，Schafer Charles T. Island Environment and Coast Development. Nanjing：Nanjing University Press，1992.）

Comprehensive Exploitation and Utilization of the Resources of Sanya City*

JIANG Shangzhou
Deputy Mayor，Sanya City Government

1. INTRODUCTION

Sanya City is situated in the southern tip of Hainan Island，18°09′34″～18°37′27″N. It is the southernmost seaside city of China. The well-known " Tianya Haijiao" (the end of the earth) is in the city.

The total area of Sanya City is 1 887.35 km^2，with 179.25 km of coastline including 19 embayments. Its six main islands are：Wugongzhou，Ximaodao，Dongmaodao，Xizhou，Dongzhou and Wild Boar Island. There are seven important embayments：Yulingang，Sanya Harbour，Yalong Bay，Tielugang，Anyougang，Jiaotougang and Gangmengang. There are more than 200 hills，among which the highest one is Jianlin，1 091 m above sea level. The three main rivers are：Ninyuan River，Tengqiao River and Sanya River.

Climatologically，the annual mean temperature is 25.4 ℃，the highest monthly mean temperature is 26.4 ℃～28.5 ℃，the lowest monthly mean temperature is 17.4 ℃～21.0 ℃. The highest temperature is 35.8 ℃ and the lowest is 6 ℃. The annual rainfall is 1 279.5 mm. The relative humidity is 79%. Annual mean typhoon visiting frequency is 3.7.

The City governs 2 districts，13 towns and 55 farms. Its population of over 340 thousand consists of Han，Li，Miao and Hui nationalities，(datum from 1988，same year for the data in the following text)，of which 44.4% are the minority nationalities，and 74% (over 250 thousand) are agriculturally based.

The gross national product in 1988 was approximately 0.4 billion yuan，the national

* 为纪念江上舟——一位20世纪80年代早期从瑞士归国的博士在海南从事开拓工作时，支持我们中加合作主办的国际会议上的讲话，愿藉此文集留下对他深深的敬佩与怀念。（王颖 2016.12.23）

income 0. 3 billion yuan, and the gross value industrial output 0. 32 billion yuan.

Briefly, Sanya city is in a remote area with poor economic and cultural background. However, it is a city with great potential for development and a prosperous future with reform thanks to its advantageous geographic position, special climatic conditions, beautiful natural scenery as well as various rich resources.

II. RESOURCE ADVANTAGE AND DEVELOPMENT PLANNING

An area should be developed by taking advantaged of its own special conditions and resources. Sanya city is ripe for development with its rich resources such as land, tropical crops, aquatic products, minerals and tourism.

At present, Sanya City still has 500 km^2 of agricultural land and 200 km^2 of scenic areas to be developed.

Owing to its year around warm climate of long summer and without winter, Sanya has long been called "the natural great green house" of China, and has become a crop breeding base that provides seeds for the mainland of China. Sanya is in a favourable position for growing season free vegetables, tropical crops and fruits, Southern Chinese medicines and rubber.

Sanya city has a 13 000 km^2 fishing sea area with an annual catch of 18 000 tonnes. This area is rich in sea-food especially "the three rare delicacies of Sanya"—shark's fin, sea cucumber and abalone, because of its short distance from the South China deepwater fishing ground. Sanya has good conditions for aquatic breeding, as there is no winter season. We have begun to breed rock fish, pearl shell and shrimp in the tidal inlets (340 ha of the total 1 900 ha). At present, we have two salterns, the Yuya Saltern and the Tielu Saltern, producing 36 000 tonnes of salt yearly (70 000 tonnes maximum).

Sanya is also rich in over 30 mineral resources, such as phosphorus, iron, gold, titanium, crystal, granite, quartz, marble, and limestone etc. , among which granite limestone and quartz all have reserves of more than 100 million tonnes. Especially important, there are 140 billion m^3 of proved natural gas reserves in the Yingge sea area of the South China Sea, only 90 km from Sanya City.

Since the founding of Hainan as a province and the development of a large special economic zone, Sanya, once a remote small city with poor communication with the outside world, has now become an important port due to its advantageous geographic position. Sanya port is a tidal inlet with favourable natural conditions: well sheltered and free of siltation. At present, there are seven berths of different sizes (the largest, 5 000 tonnes). There have been regular liners linking Sanya with Hongkong, Guangzhou, Shantou and Haikou. The existing Sanya airport has now provided civilian service: linking Sanya and Guangzhou. The Phoenix Civilian Airport, under construction, will be put into use in 1992 and will be able to handle today's largest jets. This new airport will greatly improve the city's transport communication capabilities.

The favourable geographic position, nice tropical marine climate and typical tropical scenery are all precious tourism resources. We have more than 100 km of wonderful sand beaches attracting modern tourists with bright sunlight, clear seawater, shine and soft sand, green forests and fresh air. The famous scenery spots include: Yalong Bay, Dadonghai, Luhuitou, Sanya Bay and Tianya Hajao. Yalong Bay, maybe the most beautiful beach in the world, is called "not Hawaii, but more beautiful than Hawaii".

The Overall Plan of Sanya City (OPSC) was carried out by the Hainan Branch of the Institute of China Urban Planning and Design in 1988 when Hainan Province was founded. OPSC determined the final goal of Sanya City to be a modern international city with a prosperous economy, culture, and fine scenery, mainly through the development of tourism and high technology (hitch) industry. The main recommendations of OPSC are: (1) make Sanya become an international tourist city and winter holiday resort as well as an important scenic area for our own citizens by promoting its tourism and scenic spots; (2) centre on hitch and technology-dominated industries such as electronics, food, light and textile industry while properly developing the construction material industry, then, on a small scale, other industries; (3) develop Sanya into a central city and foreign trade base of south Hainan Island. On the basis of OPSC, Sanya Municipal Government invited the Institute of Architecture Planning of Tsing Hua University to carry out an aesthetically pleasing urban design for the 10 km^2 of the downtown area of Sanya City this year. The experts agreed that Sanya was beautiful with its hills, streams and seas, which is rare not only in China, but also in the world. The downtown area itself is a beautiful scenic spot, as if it was a large wonderful park. What is more, there are famous suburban scenic spots outside the downtown.

III. COMPREHENSIVE EXPLOITATION AND UTILIZATION OF RESOURCES ACCORDING TO OPSC

As a newly-developed seaside city, Sanya is just in its infancy and far from mature. Its urban facilities have not been completed, and we lack experience in construction and management. The municipal government's main functions should be making plans, and directing the development and construction. The overall plan that is a basis for the whole development and construction programme has been created and during the three years since the foundation of Hainan Province, Sanya city has made great progress and the situation is good. However, we felt that in this period we still had some problems and difficulties on how to coordinate the development of different sectors, and utilize various resources rationally and comprehensively. Sanya was a poor remote region of minority nationalities. We have another problems in dealing with the balance between immediate interests and long-term ones. We have been somewhat overanxious for quick results and even neglected long-term interests in order to develop our economy, improve our standard of living and become rich in a short time by founding the province and creating a special economic zone.

We have already introduced Sanya's resources briefly. All of these resources still have great potential for development except for the relatively mature nearshore fishery and saltern industry. At one time, resources such as quarries, titanium mining and coral weathering were abused and there was a problem with illegal landuse for home construction. These difficulties have been controlled by taking powerful measures. However, conflicts between different sectors among construction, tourism, fishing, harbour transportation and salterns may be a long-standing problem. Problems between urban land use and harbour development remain unsolved.

The historical development of Sanya city can be attributed to its port, fishery and saltern. Today, Sanya has evolved from a barren land into a prosperous city replacing the old Yazhou city to become the economic and political centre of South Hainan Province due to the existence of its excellent harbour. This just follows the saying "the city develops as the port develops". Sanya harbour will continue to be developed in commerce and trade in accordance to OPSC but it should be controlled because of its close interaction with the Sanya River mouth. Situated at the Sanya River mouth, Sanya harbour is a tidal inlet with good physical conditions due to the 3. 9 million cubic tidal prism of Sanya River. The ebb current velocity is obviously greater than that of flood current, and keeps Sanya harbour navigation channel free from siltation. Scientists from Nanjing University conducted several field investigation during 1970's—1980's and calculated the siltation rate to be only 1. 9 cm/yr for the harbour , and 0. 1 cm/yr for the navigation channel—negligible siltation. However, with development, Sanya City is being constructed at high speed, which causes an escalation in reclamation of the Sanya estuary for land. It is well-known that a decrease of the tidal prism will cause a serious increase of siltation at a rapid rate.

It is difficult to calculate the quantitative rate of siltation of the harbour because of the complex environmental factors forming the embayment. So we should be a little more pessimistic in judging the problem. Once the environmental balance is damaged, the siltation will become an avalance, seriously affecting the existence and development of Sanya Harbour. The reclamation of the estuary will hamper the drainage from floods and will be harmful to agriculture. According to the report from Sanya Harbour Bureau, siltation has recently begun to increase obviously a problem, which should be paid attention to immediately. Also, new reclamation is being considered for building a wind shelter for fishing boats. I think that such a project is not proper for this area and should be carefully monitored or preferably conducted at another site.

This "International Symposium on Exploitation and Management of Island Coast and Embayment Resources" is very timely. The scientific work produced by experts from different parts of the world provide a basis for our construction. I believe that the municipal government will make policy on a scientific basis. To exploit the resources of Sanya City rationally and comprehensively, the following basic principles should be kept in mind: first, the construction must be acted on in accordance with the OPSC and with high standards, second, environmental protection and development must be well coordinated with immediate measures taken to control pollution problems and stop environmental damages, and third, urban construction and harbour development should also be well coordinated. It is wrong to sacrifice one thing's development at the expense of another. In conclusion, we should no longer reclaim the estuary for land use in principle.

王颖　读后感

多么好的思想与文采！江，我深深地感到你早应发挥更大的作用！英年早逝，是国家人才的损失！

海南洋浦港选建研究记实*

南京大学海岸海洋研究始自1958年从事天津新港回淤研究，1962—1963年北海地角海军码头回淤调研。至1964年开展了一系列的海岸海洋动力、地貌与沉积研究，确定海岸的冲、淤动态，为海港选址服务。海港选建初始于河北秦皇岛、辽宁盘锦止锚湾与营口鲅鱼圈。同期，调查研究工作大量投入海南岛，先后从事了铁炉港航道整治、三亚港扩建与万吨码头选址、洋浦港选建、秀英港扩建与东水港选址等。多次深入地调查研究，完成了港口选址任务，在实践中锻炼培养了人才，促进了中加合作研究与实验室发展壮大。海岸动力地貌与沉积研究组至1990年扩大为海岸与海岛开发国家试点实验室，试点以优异成绩通过；2000年被批准建为教育部重点实验室。期间，曾三次在洋浦地区从事海港选址与建设规划研究，这些项目中，我均参加实地调查并担任学术负责工作。现追忆记之。

(1) 1983年，基于“加快海南岛的开发建设”要求，当年12月交通部水运规划设计院委托南京大学进行“洋浦深水港建设的泥沙回淤研究”，着重解决拦门沙浅滩泥沙来源与试挖航道的可能性。缘由是因有研究单位认为洋浦湾大浅滩不宜建港。南京大学于1983年12月20日至1984年1月20日进行了洋浦湾与内港新英湾地区的地貌调查与制图，外湾拦门沙浅滩与通向内港的沉溺河谷(预选航道)的全潮水文测验及采集了160个海湾底质及沿岸沉积物样品。于1984年2月中旬开展了实验室分析：底质及钻孔沉积物粒度分析175个，岩芯沉积层分析，碳酸钙含量测定，沉积物^{14}C年代测定，海图对比分析以及波浪、潮流水文计算等。在上述外业及室内计算分析的基础上，完成了“海南岛洋浦港动力地貌调查报告”，阐明该区海岸与海底的地貌结构与沉积物分布特性，以及海洋水文动力条件(王颖等著：《海南潮汐汉道港湾海岸》，P2，北京：中国环境科学出版社，1998)。

(2) 1984年9月3日—11月4日，1985年1月18日—2月19日，南京大学于秋、冬季风暴频繁时期对洋浦湾地区进行了海洋气象水文观测，并向北扩展至神尖、母鸡神与龙门海岸，南到大铲与排浦海岸进行了深入的地质构造、岩层特性与地貌调查。于1985年4—6月进行室内总结、计算与实验分析：完成地质地貌图绘制，底质及浅钻沉积柱样品分析298个，12个断面与30个测站的全潮水文验算，沉积物粒度分析267个，重矿鉴定30个，石英砂样品表面结构分析20个，^{14}C年代测定样品6个，化学分析样品24个，微量元素样品分析25个，岩石样品鉴定3个，同时进行了航片与相关卫片的判读分析，河流输沙量分析等。总结完成了“洋浦港泥沙来源与回淤问题研究报告及图集”。历经两年的深入调查研究获知：洋浦湾与新英湾位于北部龙门至神尖处晚更新世及全新世火山熔岩喷溢活动区的南缘末端；母鸡神为火山口多次喷发的岩柱，神尖为抬升的海底沉积层，顶部为

---

* 王颖：2013年2月供稿于《江上舟在海南》。

凝灰岩覆盖，均为火山喷发与构造抬升的自然遗证，很珍贵；洋浦鼻、西浦与洋浦村为由喷溢的玄武岩流组成的熔岩基座台地，上覆抬升的海相沉积，现代构造活动趋于平稳；洋浦湾北部濒临西浦与洋浦台地，南邻洋浦大浅滩，自西向东延伸至新英湾内口与白马井之间，为一条沉溺谷地，谷地长达 10 km，宽 400～500 m，水深 5～25 m，具明显的蜿蜒曲流与下沉深切的河谷形态，基底为湛江系砂、砾层；沉溺河谷东自白马井与新英湾口向西至洋浦鼻末端，逐次降低，平均坡降 0.058%，深槽与 NE－SW 构造断裂带相近，是沿差异升降断裂带（北部抬升，南部下沉）发育的河谷，在全新世海面上升过程中淹没为溺谷，水道中沉积细砂、淤泥；目前东段河道已被熔岩阻隔断源，无泥沙下泄；拦门沙发育于深槽西段通向海域的入口处，水深仅 5 m，近东西长 400 m，近南北宽 80～150 m，砂砾质组成，层厚约 7 m，据 ^{14}C 定年为 8 500～6 000 aB. P.，其沉积速率 8 500 年以来平均为 0.16 cm/a，其中，8 500～6 350 年前为 0.35 cm/a，6 350～3 200 年前为 0.1 cm/a，沉积缓慢（王颖等著：《海南潮汐汊道港湾海岸》，P240，北京：中国环境科学出版社，1998）。研究表明：周边玄武岩台地岸稳定；沉溺河谷可辟为深水航道，河谷中为细砾粗砂质沉积，沉积层薄，表明无大量活沙供应，可开挖而不致形成骤淤或严重回淤。科学研究成果论证利用洋浦湾沉溺河谷可建深水港。该研究通过鉴定而作为洋浦港建设与海岸冲淤动态研究的科学依据，洋浦港于 1990 年建成 2 万 t 级深水码头 2 座，于 20 世纪末建成 5 万 t 级码头，航道已经开挖利用，而未形成骤淤，成为海南岛西岸与北部湾区最大深水港。

(3) 1988 年 11 月 21 日—12 月 30 日，获加拿大国际发展研究中心（IDRC）与中国教委支持，南京大学海岸与海岛开发国家试点实验室与贝德福海洋研究所的学者合作，对洋浦港所在的洋浦湾进行大面积补充性与重复监测调查，工作量大，深入调查了沉溺谷地，其工作包括：沿深水航道流向及垂直航道方向进行地震地层剖面调查，获得 32.9km 长的航道地层结构剖面，深水谷地剖面昭然显现（图 1），确切地反映出该处系由沉溺河谷发育的潮流通道，被选定为深水航道是十分恰当的；在深水航道进行了潜水观察与钻孔取样，该项工作是由加拿大 C. T. Schafer 博士，K. Ellis 女士及 Paul Macpherson 先生，以及中方博士研究生邓伟东共同进行的，获取 2 个潜水孔，柱长 1.29 m，证实该通道沉积量不大，具有一定的风暴作用影响；沿航道自新英湾内港向外港口拦门沙以及洋浦湾外浅海取得 7 个沉积孔，沉积柱总长 9.56m，采取 900 个沉积样品；由潘少明教授、施晓冬技术员及柯贤坤教授进行了 ^{210}Pb 定量分析沉积速率；沉积物粒径，矿物组成，化学元素，有孔虫与孢粉等实验分析鉴定分别由王雪瑜高工与许叶华助工，周旅复副教授与施炳文技术员，尤坤元副教授与朱晓东教授，王晓蓉教授，韩辉友副教授与于革博士（南京农业大学）等分工负责完成。^{14}C 测定由曹琼英高工、李红技术员、盛莉莉工程师完成。此外，在洋浦海域进行了三船同步全潮水文与含沙量测验，由朱大奎教授负责组织领导，罗哲文博士、柯贤坤博士、戴军博士、杨巨海副教授、蒋松柳工程师、许叶华助工进行观测，由罗哲文、柯贤坤、戴军完成分析计算，许叶华完成含沙量分析。地震剖面与 Lehigh 柱状样品由王颖教授、John Smith 博士与朱大奎教授负责，杨巨海副教授、朱晓东、施炳文以及当年的硕士研究生曹桂云、邓伟东、李德贤合作进行，完成现场观测、记录描述、分样，区域成果鉴定与综合分析（王颖等著：《海南潮汐汊道港湾海岸》，P3，北京：中国环境科学出版社，1998）。调查研究工作量是巨大的，海上天气变化多端，全潮测验 25 小时（半潮 13 小时），涌浪、日晒、风暴袭击等，参加工作的人员克服千难万苦，完成调研任务，获得扎实资料与数据，为洋浦港口选址建设奠定坚实的科学基础，确保港口的有效利用。在洋浦港选址与重复观测工作中，是交通部水规院提出任务，支持前期调查费用，复查是由中加合作项目出资，中、加人员克服困难，共同深入研究努力完成。在整个调查工作中，洋浦地方部门与人民对调查工作中所需的食、宿、交通，出海船只与临时工协助以及海上安全保证等，均给予大力支持。众志成城，前人努力，功不可没，应略加笔墨以铭记之。

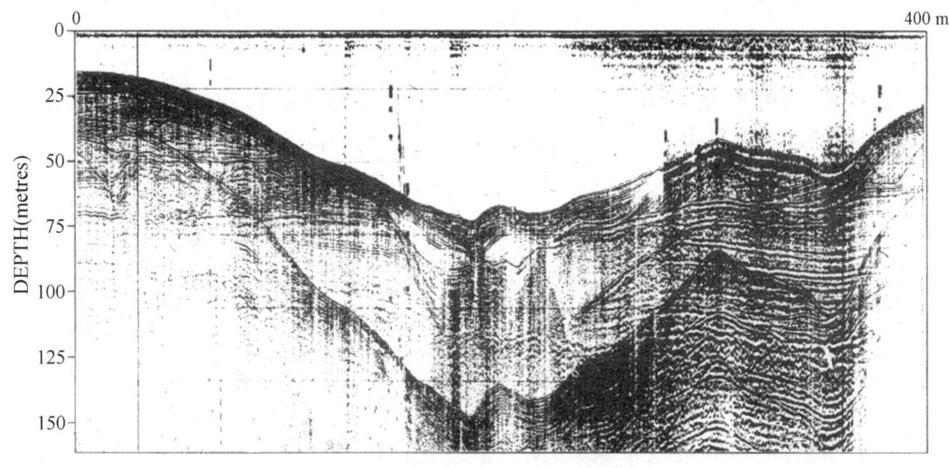

图 1　洋浦沉溺河谷地震剖面

(Wang Ying, Schafer Charles T. Island Environment and Coast Development. P171, Nanjing: Nanjing University Press, 1992.)

忆当年，看今朝，心潮澎湃，犹记初赴洋浦实属不易，火山喷溢的玄武岩地崎岖不平，唯一的"现代化"交通工具是手扶拖拉机，雇佣很方便，村民熟悉道路与村镇，但驶过一道道的玄武岩陇岗，起伏颠簸，有如海浪般地上下起伏(但却无海面平滑)，乘车者很不舒服，但这条路跑多了，也就习惯而不为难了。进入半岛终端干冲公社洋浦村时，接近黎族聚居的村寨，坟茔座座，墓碑有小檐蔽雨，家族坟墓围聚一圈，地景奇特。村外有一口大井，甘泉充盈，村民汲水，并环井建围渠，供村民洗濯排水。番薯、稻田、农田葱葱，多甘蔗、木瓜与丛林；一派田园景色、生气勃勃。洋浦村内有红砂土路近东西贯通，村中有一百年榕树合围，绿荫如伞敝阳光，村老围树台憩坐，抽竹筒水烟聊天，村童环树嬉戏！此情此景，曾令我们顿忘旅途奔波而怡然自得。

我们最初居住在洋浦建港办事处，随同首次考察的有美国 Woods Hole 海洋研究所的 David Aubrey 博士，Syracuse 大学的 Marwyn S. Samuels 教授、夫人及两个孩子。犹记院中有大井圈，围绕一深水洋井，当年两个洋孩子被地产黑毛的塌肚小猪追赶，围井圈笑叫跑动，此情景仍历历在目，院南边有一排房子数间，供我们居住。因有外宾初访，客房中新蚊帐环绕老式木床、新床单、花枕巾，别有乡趣。冲凉用井水，一排木房三间，中隔板栏，记得 Samuels 夫人(菲律宾人)，高兴地对也在淋浴的 Aubrey 开玩笑说："David，不要左顾右看啊！"村居虽简陋但基本生活设施仍方便，每日鲜蔬鱼虾，众人生活得很愉快。

当年洋浦湾以北，火山喷溢海岸地貌十分奇特：德义岭火山锥崎岖不平，符上村南侧被火山熔岩流阻断的河道，仍"不屈不挠"地下切，形成无头无尾上下游均截断的深切河曲；莲花山下神尖是因火山喷发抬升的海底沉积层，颗粒浑圆的笠贝、有孔虫、扇贝，高悬在角锥般的山崖地层中，上覆凝灰岩保护层，使黏土粉砂质地层得以残留；母鸡神是多次抬升从地底拱出的火山颈，弯曲扭折，似依陆面海的抱窝母鸡，附近海滩上残留着一个个蚀低的圆圈形小火口与飞簷状的海砂—贝壳沉积岩层；龙门有抬升的海盆状的崖墙，烟囱状的喷火孔以及似乎仍在沸腾的熔岩流表层。我曾很想将烟囱状小喷火口搬回南大实验室，但最终决定仍保留在当地为一自然历史遗迹。罕有人迹的火山海岸保留着大自然变化绝妙的奇迹，使我们扩大眼界，习读了地质历史记录的“天书”，增长知识，提高科学分析能力，真实体会到，一份努力的汗水确有十份的科学收获！

再次访问洋浦港是20世纪90年代后期，该港已被批准建设为洋浦经济开发区。从儋州乘车到达洋浦一马平川，道路开宽已无起伏的玄武岩陇岗，入区大门建为设卡通道，开发区周边有铁丝网围隔。入内只见平整的大地，已不见绿色的田野、村口大井、黎族村寨与人群欢笑之声。区内道路初具形态，空旷的码头上散堆着桉树木材屑片（出口造纸的原料）。港口船只不多，沿新英湾与洋浦湾交界的岸上，建有几幢洋房式宾馆，园中植有芭蕉与丛林。建设在发展，我几乎不敢相信这就是我曾数次到来工作的驻地！洋浦鼻已近在眼前，建有灯标，港区北侧的大浅滩及岸上的木麻黄林仍在。

又一次访问是在21世纪头十年期间，港口已现繁荣忙碌景象，码头前船只停泊装卸货物，堆场上囤积货箱，人员往来频繁，出入港口要检查通行证件……故地重游，似曾相识又不识！神尖角崖尚在，但与岸相联，四周以矮岩围栏保护，与相邻的跨越式码头和栈桥相比，它似乎变小了，已非当年遥立浅海的伟姿。在洋浦港区唯一见到的友人是洋浦区原土地局局长夏明文同志，他来自长三角经济发达区，原籍无锡，建港初期到洋浦，数十年坚持在洋浦，满腔热忱、开拓奉献。当年南大曾分配一位获博士学位研究生到洋浦，工作约八年，转战深圳了，明文却始终坚守在洋浦。

时光飞逝，历史在续进，故人已初见白发，而中华民族勤劳勇敢、艰苦卓绝的光荣传统继续发扬，这是祖国大地变换新颜的动力源泉！

2013年2月15日稿改写于2月18日

纪念伟大的地理学家任美锷院士*
——任美锷院士诞辰 100 周年

我接受的地理科学教育，启蒙于 1952 年 9 月—1956 年 7 月在南京大学地理学系学习期间，身受任美锷教授、杨怀仁教授及地质、地理两系老师的教导，打下坚实的地学基础，受益终生。1961 年 2 月，在北京大学地质地理系研究生学习毕业后，又承任美锷教授接受，支持我继续从事海岸地貌与海洋沉积学工作，以及组建实验室。通过任美锷教授的言传身教与合作研究中的理解、默契，深感他爱国至诚，专业精深，科研与教育贡献重大，是一位伟大的地理学家。值此纪念任师 100 周年诞辰之际，同事们让我介绍任美锷院士的学术思想与成就，我深感惶恐，并非谦虚，我认为：能承担这项任务的最佳人选是学长包浩生教授。包浩生教授师承任美锷院士，数十年的教导与合作，其感受之深，认识之精辟，非旁人可比。可惜浩生英年早逝，我深为追悔莫及。我所能记述与表达的，只是自己的切身感受，确实有时间、经历之局限。在参考经任美锷院士审阅过的包浩生教授专文①，结合自己感受，做此发言，我想，这次会议却具有集腋成裘，点滴汇集而成江河之势的意义。

（1）中国科学院院士任美锷教授最大的特点是他以赤子之心，终生为祖国奉献，是一位热爱祖国的科学家。评介一个人，要依据其所生长的社会环境，以历史事实为凭证，纵观任美锷院士的一生是以爱国奉献为主线，以其经历的事实可证明。他学习与获得博士学位的专业是“地貌学”，毕业论文为“Clyde 河（英国）流域地貌发育”，学成于 1939 年回国报效，正逢战火连连，应竺可桢教授邀请，他历经战争烽火艰险到广西宜山与贵州遵义任职于浙江大学史地系，讲授“地形学”。结合地形与水系研究“遵义附近的地形”“贵阳附近地面与水系发育”，绘制了我国第一幅土地利用图，以应当时国家形势所需。40 年代初，结合着当时，“抗战胜利、建国必成”的形势发展，他从事川甘之间的江、河分水岭调查，研究地形与人生活动，提倡“建设地理学”，将地理研究与经济建设结合，应用于工业区域配置等，是适应当时的国家需要。积极学研发表“甘南川北之地形与人生活动”“地理研究与经济建设”“工业区位的理论与中国工业区域”，为事有成就的例证。

我是 1952 年考入南京大学地理系，当时，正逢院系调整后，地理系新生有 80 多名，后不少人因非本人志愿而退学，但因浙大与川大地理系师生并入，生员仍多。当时，系内分两个专业：经济地理专业与自然地理专业，但是自然地理学专业并未实建，而是办了地貌学与水文学两个专门化，一直至今仍以地貌学见长。当时，由于需要恢复新中国成立后曾经停滞过的经济地理学，任美锷教授开设了“经济地理学概论”，至今我仍记得他穿着白色夏威夷衫，灰绸长裤，自带一瓶白水，在南大西平房教室中孜孜不倦地讲学情景，他尽量讲普通话，但是宁波口音重，我们都能听懂，而且全神注入。其自然资源与开发利用结合，工

* 王颖：纪念任美锷院士诞辰 100 周年，2013. 11. 1.

① 包浩生，《任美锷教授八十华诞地理论文集》前言，南京大学出版社，1993.

业配置与城乡结构等始终贯穿讲课中，有理论有实际，学生听得入神。同时，他亲自编译出版了《台维斯地貌学论文选》(1958 年，科学出版社)，为学生传介经典的地貌学理论。至今我常想：当年的地理系学生，兼具自然与经济地理的课程，并有坚实的地质、地貌理论学习与科学实践，为以后的工作打下了雄厚坚实的基础，毕业后，在不同岗位上发挥了应有的作用；地球科学应是圈层系统的自然结合，系、科划分太细，是不易了解客观规律与人类生存发展建设之所需。

(2) 爱国主义思想贯彻于科学教研之始终，实践出真知，总结规律，成功地应用于实践，业绩印证任师是伟大的地理学家。

20 世纪 50 年代，任美锷教授作为南京地理研究所第一任所长，及中国地理学会副理事长，他应国务院邀请参加了“全国十二年科学技术规划”制定工作，受到周总理接待，曾与毛主席等中央领导合影，他倍受鼓舞。为我国经济发展需要，他带领师生积极赴西南山原密林，考察热带、亚热带生物资源，特别是深入橡胶种植园，调查分析自然地理环境条件，总结群众种植发展橡胶林的经验，提出橡胶种植北限可达 25°N，以至海拔 900～1 000 m 高的地带，为扩大橡胶种植基地，解决短缺的战略资源，提供科学依据。在实地调查基础上发表了论文“丽江和玉龙山地貌初步研究”“金沙江河谷地貌与河流袭夺问题”，以及《云南南部地貌区划》《中国热带亚热带地貌区划》专著。任美锷教授认为：中国自然条件复杂，进行自然区划在于“深入分析和研究各地区自然综合体内各种自然因素间的矛盾，以及自然界与农业生产间的矛盾，从而找出其中的主要矛盾……为各级自然区划的依据”“热量条件作为全国第一级自然区划依据，对我国东部地区是合适的，……西北干旱区，自然界与农业生产的主要矛盾是水分……”其总结是十分精辟的，适用于我国自然条件的复杂状况，他提出：“区划单位必须代表地域自然界全部差异，而不是某一个因素的差异，纯粹的热量带不宜作为区划的标准，应结合水分及受热量制约着的区域内一定土壤和植被类型的发展为依据，划分自然区、自然地区和自然省等区划单位等级系统。”在上述研究与区域分析的基础上，任美锷教授提出“中国准热带”理论，据云南独特的景观，认为：云南和四川西南部终年受热带气团控制，热带范围可达 28°N，由于是一个地形复杂的、经山地在上新世末与更新世期抬升的山原，随着海拔高度及地貌结构不同，出现热带(800 m 以下)、准热带(800～1 000 m，或 1 100 m)、亚热带(1 000～1 600 m)和温带景观。准热带与热带在热量之间存在差异，但热量与景观与亚热带有本质之区别。继云南山原之垂直地带分布后，他进一步分析厦门、广州、南宁一线东南沿海处于东亚季风区热带北部边缘，冬季虽受寒潮影响气温偏低，个别年份热带作物受影响，但从自然景观要素与农作物的特征分析，均与热带性质相似，与亚热带有本质区别，故属热带范畴，称此为“准热带”。准热带理论，为扩大热带作物生产需加以人工措施辅助，提供了科学依据，同时也是从实际调查到自然区划界定，进而为农业生产服务的重大进步。科学研究应用于实践，实践又推动了科学进步。

1972 年底，中国科学院竺可桢副院长会见任美锷教授时谈道：“至今还没有一本篇幅适当的中国自然地理著作问世，与我国的国际地位很不相称。”受此启发，任美锷教授以他讲授“中国自然地理”的原稿为基础，综合后续的研究成果，论述完成了《中国自然地理纲要》，由商务印书馆于 1979 年正式出版，获海内外热烈反响。在 1979 年 12 月召开的全国宣传会议上，时任总书记的胡耀邦对此书给予很高的评价，号召大家研读；大学地理学系以此书为教材。后经多次修订再版，中国外文图书出版社将“纲要”译成英文与西班牙文出版，向全世界发行。日本东京大学出版社以日文出版。1988 年，该书获国家教委颁发的“首届全国高等学校优秀教材特等奖”。

继之，在贯彻 12 年科学规划中，通过在云、贵地区的研究，深感西南地区广大石灰岩山地区，土地贫瘠，地表水渗漏，制约着农业发展与广大农民生活的改善。因此，他深入对喀斯特地貌调查与开发应用研究。通过实地调查与大量钻孔资料分析，发现在现代喀斯特作用带以下仍有深部喀斯特发育，从而提出，“喀斯特作用不是以河流下切之基面，不以海平面为下限，而是以石灰岩层的基底为下限。”这项理论贡献为三峡水库工程、岩溶区道路建设等提供了重要的参考依据。他进而总结发表了论文：“中国岩溶发育规律的若干问题”，明确地划分出中国喀斯特地貌的三种气候地带性类型。在此基础上，任美锷教授与其梯队，研究了周口店洞穴演化与古人类生存关系，总结出中国猿人所居之龙骨山洞穴受地质构造与岩性控制。据洞穴沉积与化石，他提出猿人居住约始自 50 万年前，而在距今 23 万年时迁移，这期间至今猿人洞穴发育经历 5 个阶段。此文在“中国科学”刊登后，获得热烈肯定，被认为是在洞穴演化与古人类生活关系方面具有重要进步的研究。任美锷教授在中国喀斯特发育理论研究，于 1985 年获国家教委科技进步二等奖。上述表明任美锷教授始终以国家需要为己任，从西北、西南，到华北，从干旱区研究到喀斯特石山，从今至古，一步一个脚印，深入实际调查研究，做出了一系列重要的地学理论贡献，成功地应用于实际建设。

1963 年后，随着我国建设开展，任美锷教授响应“国家三年改变港口面貌”的号召，积极参加海洋科学工作。1965 年参加海南铁炉港航道开挖整治工作，70 年代开始从事海洋沉积与古海洋工作，成为找寻海底油气藏从事区域探测的导向研究。先后评介与发表了“潮汐汊道与纳潮量”“古海洋学回顾与前瞻”等论文，并为铁炉港航道开挖，黄河三角洲建港提供了重要的依据。1979 年他担任江苏省海岸带与海涂资源调查队队长，虽于 1980 年 7 月患眼球视网膜剥离先后三次手术，左眼几乎失明，但仍然完成了江苏海岸带作为全国试点的先导调查任务，于 1986 年获江苏省人民政府“江苏省科技进步特等奖”。同时，完成“海平面上升对我国三大三角洲影响”的研究专题，出版专著并总结发表了“近 80 年我国相对海平面变化”论文。任美锷教授于 1980 年被评为中国科学院学部委员(院士前身)。鉴于他在地貌学研究中的成就，于 1988 年 6 月被英国皇家地理学会授予“维多利亚奖”最高荣誉，为我国乃至东亚地区获此殊荣的第一人。他的学术贡献与所建业绩足以说明他是一位伟大的中国地理学家。

(3) 任美锷院士爱国奉献，响应祖国在各个阶段的需要与号召，亲历调查，深入研究，勤奋总结，一步一步地前进与探索，留下了实实在在的业绩与学科建树。在他的一生中，始终寓教学于研究中，孜孜不倦、言传身教，为我国培养出四代地学人才，其间有六、七位院士。后辈们亦谨承师训，孜孜不倦地为祖国奉献。任美锷院士教学与研究十分严谨，但生活作风却十分朴实，平易待人。我与他相处近 50 年，却从未见他发过脾气，即使在“文革”动乱中，被批斗，被下放农场，却始终未听到他有任何抱怨，始终是平和、严肃，勤恳于教研，一直到安静地离世。每每回忆，更加深对他的敬佩与怀念！学习他爱国与科学贡献，始终不渝地是我学习与努力的方向！

乐亭情怀*

滦河是冀东平原的母亲河。它源自河北省巴彦古尔图山麓，流经内蒙古高原、燕山山脉至乐亭县的兜网铺入海，是一条长 877 km 的多沙性河流。滦河在潘家口以北流出燕山后顿失落差，长期的泥沙堆积发育形成了水土肥美的冀东平原，平原上的滦河却成为曲流摆动的游荡河。河道沿平原倾斜从西向东迁移，尚可追溯古河道遗迹：最西侧是从北部山地向南，大体上在南堡一带，现今的咀东河（双龙河）是在古滦河的遗道上发育的；嗣后，向东迁至西河口（溯河口）及大庄河（小清河口）入海，当时，滦河的水沙量大，汇入海中的泥沙，又被波浪在岸外推积为曹妃甸沙岛、腰坨、草木坨与蛤坨等沙坝；继之，滦河又东迁至大清河、湖林河口入海，泥沙被波浪推积成石臼坨、月坨与打网岗等小沙岛；继而东移至老米沟、浪窝口入海，泥沙又被激浪推积为一系列环绕海岸的沙坝；仍继续东移经姜各庄稻子沟、新开口入海，形成蛇岗与现代河口的一系列沙坝。滦河也曾在北部七里海入海，但为时不长。河流摆荡，形成岗洼起伏的平原与沙岛环绕的河口，反映出冀东平原具有西部抬升向东倾斜的趋势，海岸带盛行偏东向风、浪，这种状况自人类历史以来都变化很小。我曾从 1958—1961 年与相关院校师生在冀东平原海岸工作，从现代滦河口到天津新港，在海岸与沿海平原步行观察与填图测量，走遍了平原的港湾与村落。经过分析，明确认识到滦河泥沙曾西达南堡，但对于天津新港回淤无影响。二十世纪八十年代起，又在曹妃甸深水海港选址工作中，认识到渤海潮流主通道沿曹妃甸沙岛外缘流过，沙岛基础沙层厚实，位置较稳定，故依托曹妃甸沙岛，将外侧深水潮流通道选建为 30 万 t 级的深水大港，建港至今码头无回淤。21 世纪又继续关注该港域的发展规划。长期的累积，使我对滦河三角洲平原了解增多，热爱加深，却怎知，自家与滦河与乐亭却有着深厚的历史渊源。

谢谢乐亭县领导与父老乡亲，承寄《读乐亭》给我，每每使我怀着浓厚的兴趣去读、去遐想。乐亭县在我脑海中似熟悉又不解，《读乐亭》每次均给了我点点滴滴而汇聚的乡情与亲情。我的父亲叫王奇峰，号峙亭，是抗日将领。他青年时从保定军官学校毕业，服务于东北军，转战冀辽热，因在通辽抗日，歼灭敌军有功，于 1933 年升任骑兵第三师师长。1935 年被授予陆军少将，1937 年被授予陆军中将，1938 年逝世时任国民革命军骑兵第四师中将师长。国共合作时，他与刘伯承、徐向前两将军一起，归朱德司令指挥，在山西军寨曾与刘、徐在一原姓宅中合影，此照片"文革"中被毁。后因在太行山抗日，风雨淋体，积劳成疾，自前线送回西安仅数日而逝世。当年我仅三岁，随母亲像其他东北同胞一样地流落在陕西西安市。我的籍贯填写是辽宁省康平县人，却从未到过康平，只是小时听家人讲所知：我家是祖父一辈从河北省乐亭县汤家河村大王庄迁至关外，祖父仍讲地道的乐亭呔音，年轻时下关东至康平县哈啦呼哨村娶妻定居。父亲是康平人，而王家先祖却是从山西省洪洞县大槐树迁至河北的。具体事迹我一点不知，但近年从报纸知王氏宗亲在山西，从

* 王颖：乐亭文化研究会主编《读乐亭》，2004 年第 38 辑，第 43－45 页。

《读乐亭》确知有王姓先祖是明朝永乐二年由山西洪洞县迁至乐亭的，文籍与传说相合，而且常听说："小脚指甲分瓣是大槐树人的遗证"！确合实情但又不能肯定此语是否无误？

1958—1959年参加天津新港回淤研究，我负责北岸从滦河口到新港确定滦河泥沙对新港有否影响。由北京大学、北京师大、南京大学、福建师院与华南师院的师生共同组队，足迹遍及海岸带陆上与浅水部分。考察去过昌黎、滦县及乐亭，还特别去了大黑坨以了解它是否古海岸贝壳沙堤。这是我平生第一次到乐亭，主要见沿海平原、沙坨、海滩、苇塘、盐田及农场。

20世纪80年代考察古滦河三角洲，从山地向海穿插调查，与海上测量地层断面及泥沙采样结合，数年工作的成果选定在曹妃甸沙岛外侧、沿渤海主要潮流通道可建30万t级大港。经历多次汇报、评审、现场考察与质疑评定，唐山市委大力支持，又组织规划设计，终于批准投建。大港建成至今无回淤，选建深水大港过程中，结交了不少地方领导与同志。如：我深深敬佩的薛渤珣同志，30年如一日在曹妃甸海港建设中孜孜不倦地努力，继承了老一辈革命传统又为后继树立榜样。21世纪初，承他与县委联系安排我两次赴乐亭寻根，于2008年找到了原汤家河大王庄，现在地名已改为柴庄村大王庄，王家仍有老人在。并奇迹般地结识到李印亭老人(当年91岁)，仍记得在"他16岁下地干活时见到王奇峰曾到乐亭大王庄祭祖"(后来我知道大约是1933年，父亲提升为骑兵第三师师长，在换防时曾去乐亭三天)，"张学良部队的师长军人骑马，在祖坟前烧纸……"那一次，我还了解到父亲同族的堂兄弟辈王雨亭、王任亭、王秀亭、王翠亭的后人王同辉尚在大王庄住。真是奇迹，似乎我与乐亭有着千丝万缕般的维系，怎么那么巧，我前后数次考察滦河三角洲平原海岸，选建深水港，又寻祖迹到汤家河大王庄！当天，仍住在村中的一位王姓远房侄孙带我到祖坟的遗址，我按照乡俗祭酒与烧纸以示悼念。此行又了解到在祖籍乐亭还有王家"亭"字辈的后人。世事忙碌又是几年过去了，再未回访过乐亭，更未曾去洪洞县！何况，父亲生长之地籍贯是在法库县，该地现今为辽宁省法库县卧牛石乡哈啦呼哨村。但他填写自己籍贯为康平县，我也跟着填，后知他为上中学必须填写是康平县人才能被接收，所以改祖籍法库县为康平县。我是2006年后才知，无怪乎20世纪50年代末时在北京大学审干时，曾有人到康平县查访却找不到哈啦呼哨村？而且有意思的是，我最近翻阅故人书信时，才知外公张某某原籍也是唐山乐亭人，年轻时下关东，在辽宁省通辽县与姥姥(旗人女子)结婚，后去通辽成亲，生有一子二女，后仅第二个女儿(我母亲)成长，结婚后将他们迁至北京，住后海南官房7号，40年代初外公外婆均逝世，葬北京香山地区。我因父母早逝，不知爷爷、外公与姥姥的名字！我冀东平原情系来自选建海港，可谁又知与乐亭情怀来自父、母双方！

乐亭所在是山清水秀，水土肥美的沿海平原，农作物生长茂盛，白浪淘沙的渤海鱼虾丰富，自然景观人文与历史遗迹引人，更有先行发达的工矿业，现今经济繁荣人居康乐，《读乐亭》又将各方面人脉相联，滨海沙地平原人才辈出，谢谢《读乐亭》，谢谢乐亭的父老乡亲。

有感于渤海湾工作*
——致王宏研究员的信

王宏　研究员：

你好，正逢南大暑假，略有空暇，愿和你聊天。为什么我说对渤海湾有深厚感情？来源于我在该区延续工作先后约 60 年。

① 我于 1958 年—1963 年期间，参加天津新港泥沙来源与回淤研究，在渤海湾西岸从事调查与定位观测，从未想到却在“苦海盐边”之地，认识了解到贝壳堤与潮滩分带性的自然发展变化反映近岸海蚀-海积演变规律，而且受制于岸坡、动力、泥沙物质成分的内在联系；② 1964—1966 年主持秦皇岛到山海关的海港选址，深入调查了秦皇岛、南李庄、山海关、止锚湾与长山寺预选海岸段，作出各段海岸环境评价，为交通部第一航务工程局选定秦皇岛新港，南李庄海军码头及山海关船厂港址，均投建成功；③ 90 年代负责选定曹妃甸沙岛与岛后水域之组合为 30 万 t 级深水港址，经交通部审批，投建成功，为河北省最大深水港，未回淤，2000 年后又从事该港域的生态环境变化调研；④ 2008 年承天津地质调查中心王宏研究员的指导，踏勘了渤海湾西岸牡蛎礁，深感渤海西海岸平原海岸环境在海陆交互作用过程的丰富内涵——贝壳堤反映潮滩海岸的蚀积过程，而牡蛎礁是粉砂、淤泥质浅海湾环境的生物指标，深赋科学意义；⑤ 2012—2014 年，负责实施中科院地学部发展战略项目“中国海陆交互作用带环境特征与相宜利用”考察渤海湾与黄河口海岸带的现状新貌与问题。近 60 年的亲历见证，深感渤海湾地区的平原环境貌似单调的苦海盐边，内涵却十分丰富：包括多次河海交互作用过程的遗迹；海岸带泥沙组分粗细与生物种属结合的特点环境；沉积组成与岸坡及对海浪作用的影响效应等；加之，位于京津外围，社会、经济与科学的发展快，人类活动对当代海岸环境影响变化大，所以，在中国沿海平原中是最具自然与人类双重影响结合的代表区。我为李凤林先生和你们的工作以及这本图集的出版而欢欣，但我和你谈的这段文字仅为学者间的交流而不愿写于序中，否则本末倒置而篇幅太大。又因在你的序稿中提到过（你所了解到的）我在渤海湾的工作，藉此我愿与你回顾这段涉及我青年与中年的工作历史。我们，均是历史长河中的一颗沙粒。但是，颗颗沙粒坚实，脚踏实地地为祖国工作，这是我们共同点，是可贵的品质，愿共勉之。

王　颖

2015.8.7

* 王颖：致王宏研究员的信，2015.8.7。

北京大学地貌专业建立60周年、地表过程和第四纪环境学术论坛暨王乃樑先生诞辰100周年纪念活动发言稿*

导师王乃樑教授，忠贞爱国，1950年新中国刚成立不久，不等即将举行的学位论文答辩，义无反顾地从法国返回祖国，在北京大学教育奉献。他教书育人，创建地貌学专业。将地貌与动力，特别是与沉积学研究结合，沉积相是地貌发育的自然记录，将形态研究与形成动力结合，以沉积为证，推动了地貌学发展。他重视调查研究与基础理论结合推动地貌学的应用，密切为国家建设服务(海港选建，砂矿勘探，地震、泥石流等地质灾害防治)；致力于砂矿地貌学研究，海岸动力地貌学研究，新构造运动与地貌，以及沉积学研究，促进部门地貌学深入，拓宽发展交叉型新学科；培育了一代，又一代的新型地貌-沉积学家，对地学贡献是千秋万世。

他深具人格魅力，热心教育，和蔼细致，谆谆善诱，作风朴实，业务精湛。他的教学，研究写作是字句斟酌，“每个字都是‘镚儿’出来的”，治学极为认真严谨。

他外语好。英语可为广播口译，法文经留学而熟练运用，后来又突击学俄文，他可与苏联专家B.r.列别杰夫自由交换学术意见，当韩慕康同志为列别杰夫讲课翻译时，他常给予专业译名的补充帮助。而我认为韩慕康是俄语口译与地貌学结合的第一位中国专家。我这个学生却是自愧不如，仅拥一门英语，而学过两年的俄语全忘了！

他孝顺父母，爱护子女，甚至幼小的动物，仍记得他指导我们在寨口实习时，他为沿坡奔跑的小猪让路，并讲：“什么东西小的都可爱！”年长的国际知名地貌学家，依然怀稚子之情。

我深深地怀念王乃樑老师，我记住他在临终前，我最后一次探望分别时他的话：“灌肠很难受！王颖，不要忘记我呀。”今天这个学术会议是怀念、继承王乃樑教授嘱托的最好证明。我永远铭记，虽然我没学好，但我时时记住老师教导，做一个合格的教师与忠实的地貌-沉积学工作者！

---

* 王颖：在北京大学地貌专业建立60周年、地表过程和第四纪环境学术论坛暨王乃樑先生诞辰100周年纪念活动上的发言，2016.8.20。

怀念老师郭敬辉教授*

(1916.10—1985.4.5)

郭敬辉研究员是我敬仰的导师,他热爱地理科学,终生不渝。在抗日烽火燃烧的年代,于1937年毅然投笔从戎,参加了华北平原游击与八路军组织的抗日战斗。忍饥耐劳冒着枪林弹雨与日寇斗争,立下了不朽的战功,于1939年批准为中国共产党党员。在烽火战斗的年代,结识了双枪女八路李淑杰,结为夫妇,她比郭敬辉小5岁,但参加革命早半年。新中国成立后,双双从事科学与教育工作,郭先生专长于水文地理学,主持创建河北北京师范专科学校。1952年任职于河北北京师范学院地理系主任。1955年应聘于中国科学院地理研究所,先后担任副研究员、研究员,兼任过副所长与水文研究室主任以及西南地理研究所、石家庄农业现代化研究所所长。他专注研究中国与区域水文地理环境,1954年完成了第一张"中国径流模数图",1955年在地理学报发表了"中国地表径流",引起地理界与水利界的重视。继之,于1957年在"科学通报"上发表了中国河川径流资源的估算值(27 841亿m^3),与1987年经过精确量算的结果27 115亿m^3仅差2.6%(汤奇成,2010)。1958年在地理学报上发表了"中国地表径流形成的自然地理因素"。在20世纪50年代,能在三年中发表三篇系列的重要科学论文,实为稀少,这是他孜孜钻研并组织团队工作的重要成就。继之,从事新疆综合考察、南水北调考察等项工作,为中国区域水文与江河治理奠定了坚实的河川水文学基础。他重视实践调查与总结分析,重视科学活动与支持交叉学科河口学与海岸科学发展,担任过中国地理学会副理事长、水文专业委员会主任,曾任北京市地理学会理事长、中国海洋学会副理事长、中国水利学会理事、中国地名委员会委员等职。他以赤诚的热心,长期不懈地关爱与培植年青的学者,一贯主张"依靠青年使后来者居上",培养了李涛、刘昌明、李德美等水文地理学家,也关怀海洋科学的张利丰、王颖等人。我是他的院外学生,他是我的校外导师与科教事业的培育人。

郭敬辉老师与李淑杰先生于1955年期间定居于北京,初租住于前门外西打磨厂51号民房,后移居中关村科学院宿舍,最后定居于条件较好的南沙沟中科院单元住房。两位都是三·八级干部,却都深入基层。印象中,李先生更忙,常年在百花山天文台值班,至离休,才长居南沙沟。他们育有一子(郭鹏武),二女(郭玲玲、郭莉莉),除莉莉幼时居中关村家中外,鹏武与玲玲均在香山慈幼院寄读长大。

怀念郭敬辉老师,藉以我个人与郭老师有关的两件事例反映其为人。

第一件事:我初识郭敬辉老师是在1955年底与1956年1月间,为参加在印度阿里迦穆斯林大学召开的国际地理学大会,我国首次派出了一个六人的代表团:团长是团中央国际联络部的谢邦定,翻译是岑悦芳(男),团员为中国人民大学的经济地理学孙敬之教授,中科院地理研究所的郭敬辉研究员,北京师范大学地理学系研究生陈冠云以及南京大学

---

* 王颖:原为在"纪念郭敬辉先生诞辰100周年学术座谈会"上的发言,2016.10.11。

地理系四年级的大学生王颖。出国前集中于团中央，准备科学报告稿："食物来源与人口增长"与"中国水文地理"，我配合郭老师记录他口授修改的"中国水文地理"文稿。确实为他出口成章，对数据记忆的能力而惊讶！首次外出，团中央领导还组织报告，让我们学习有关印度国情与风俗，国际交往礼仪以及制装与准备礼品。我记得男同志是量制两套西装(正装与便装)，配购衬衫、领带与皮鞋。我是唯一的女代表，规定制4件旗袍，自选衣料，由"造寸"服装店裁缝来量身定制(咖啡色织锦缎蝴蝶彩绣旗袍与浅绿薄呢旗袍两件为正装，两件绸布料旗袍为便装)，在百货大楼购买的皮鞋。郭老师讲："你怎么买了一双猪皮皮鞋?"当时刚采用猪皮制鞋，质地很粗！因我初次出国不懂规矩，这些细节亦被关照。郭老师自始至终对我们两个学生给予更多细致的关照。从北京出发时，每人除带自己行装外，还需分带代表团的书刊与礼品，只有一个箱子怎么装？郭老师与谢邦定分别示范分类装箱扩大容量的方法，使我日后的外出都受益。出发时，谢邦定同志仍忙于在团中央开会，直到最后一分钟登上火车时，火车即缓缓地驶动，我感到惊愕，可常年外事工作忙碌的谢、岑两位却习以为常，安然处之。

离站时北京飞雪，到广州时却很炎热，在车站休息室换下旧时衣服，穿上新装，至中国香港时已满头大汗。接站的是驻香港办事处代表，外表像位大亨：体胖、西服领带正装，口含雪茄，实质却是一位在白区工作的老外交。他向我们介绍："前次，赴印尼参加万隆会议包乘克什米尔公主号飞机的代表人员殉难，万幸周恩来总理不在机上！这次，乘英航，中国人少，会安全。"到香港当日即学习，我们均表态："忠于祖国、坚定地应对突发事件。"当时香港本岛淡水少，用水困难，仅半日供水。但驻港办事处为一花园洋房，有一游泳池蓄水，我们得以外出前洗头净身，轮流使用浴室，至我为最后一个洗，洗时，众人已催等我开会，使我很忐忑。跟着两位老师发言，我的表态很痛快，郭老师讲："看来你既无父母，又无男朋友，无后顾牵挂"，使我稍安心。实际上大家均表明心志，一定完成祖国交予的出访任务，不畏难、不叛国！第二日乘英国海外航空公司(BOAC)的飞机赴新德里，途经仰光稍停，见到阳光照耀下的大金塔，简易的帐篷式候机楼，饮品周围苍蝇飞舞，只觉得昏昏沉沉地难以睁眼。在新德里住在"大使饭店"，清洁明亮的西式建筑，在旅馆我第一次接触到印度的等级制度，总感觉室内有个黑影子，我一进门就闪掉了，最后终于给我发现一黑瘦的仆役在清整房间，因属贱民阶层，不能见客人！

阿里迦穆斯林大学在新德里南边的阿里迦小镇上，会议期间我住在女子学院院长家中，布满玫瑰花的卧室，清洁，舒适，但我不习惯入夜的报更声音，在院中巡廻，一小时喊一次，音似："What am I know"？颤抖凄凉以发瘆！另外，印人如厕是座在一板床上，需对准一个不大的葫芦状的洞孔通粪池，非常不方便，而每日外事活动多，时间很紧。孙敬之老师做的大会报告："食物来源与人口增长"，效果好，反映热烈，粉碎了中国人多缺粮饿死人之谣传(后来"大跃进"弄虚作假，却确实有人饿死)。郭敬辉解答了其他有关中国地理学问题，我当时英文差(中学学英文，大学学俄文，结果，英文与俄文都不过关)，解答问题只能半语半手势交流，一位印度学生说："中国学生英文少少的?!"大使馆的李翻译讲："但她中文多多的!"应对印度尚无统一语言的状况。印度多民族各用其语：印度语、乌尔都语等，大学中以印式英语为主！自那次，促使我一定要学好英语，要流利应用。在会上，我见到了苏联科学院士格拉西莫夫，他以英语做学术报告、解答问题与交流，获得与会者尊重，显示了科学力量。水文地理学家列宁格勒大学教授李渥维奇，以科学交流促进了友谊与了解，他很友善，送给我一个胸针：打开蓝色金属封面内为一套连接的风景照片，我十分喜爱。这次会议对我日后学习有很大影响：要努力攀登科学高峰。由于交流成功，宣传介绍了新中国科教之成就，驻印大使袁仲贤将军要求安排代表团在印停留多做些访问交流活动。所以会后，东访加尔各答大学；向西，经喜马拉雅山南麓，到拉贾斯坦，再经德干高原到孟买，访问有关院校一个月后，乘船回国。沿途访问的确增长知识，同时宣传中印友好兄弟情谊，帮助使馆运回一批国内需要的物品及带回一名返国留学生。因此，不仅乘船航行于印度洋，经马六甲海峡经南海返抵香港，而且也得以乘火车而东西横贯印度次大陆。印度火车的头等车厢是一个一个包厢与其他车厢分隔的，卧铺式车厢中只有我们六人。仅到站时，才有专人送饮食。一次乘车途中我打瞌睡，睡着了，郭老师叫醒我狠狠地批评："没出息！这么难得的机会不好好观察学习。"我既汗颜，又感动，立刻起坐，沿途认真观察。幸亏郭老师提醒，否则错过了观察学习的机会，那次是我平生唯一的一次横贯德干高原。也是从那时起，我外出一定会留神观察，不打盹，受益不少！在印度，曾于喜马拉雅山麓汽车行驶时，夜遇白额老虎拦车跳跃越过，无惊险，却令人兴奋而驱走瞌睡。在印度参观一寺庙时，我们鼓动郭老师在女神脚上拴个红绳以求子，莉莉是 1956 年生人，但却不是因红绳而生，因月份不符！在印度经月，实不惯咖喱与辣椒饭菜，幸亏郭老师提议并坚持，使得我们可在使领馆中吃几次中餐，改善口腹！孟买是印度西部的港口城市，棕榈树掩映的下的小楼，热带气候下的西欧式建筑，别具风情。我们考察了阿拉伯海沿岸，并参观了电影摄制厂，结识了著名的电影演员——二亩地的男主角等，的确广泛开展了友谊活动，没辜负袁大使高瞻远瞩的期望。访问印度结束后，返程乘意大利邮船自阿拉伯海，驶经印度洋、南海，驶往香港。印度洋风浪大，多晕船，我久坐舱外甲板上不回房，天渐渐黑了，郭、谢两位几次催叫我进船，"一个人在甲板，被人扔下海怎办？"当时，外轮中乘客情况还是很复杂的。

印度之行后，我一直得到郭老师的亲切教导，尤其 1956 年 9 月分配到北京大学地质地理系工作与学习时，经常向他请教水文学与自然地理学问题，他常说："你为何不到地理所工作，否则我也会送你到苏联学习！"在南大学习时，我曾入选为留苏预备生，可惜体检时健康检查未通过。后来，我常想，当年若去苏联留学，就不会专注英语，也不会在 1979 年被选为首批赴加拿大的访问学者而在加拿大研学三年了！是命是天意？均不是！而是我们是在国家发展过程中成长起来的，每一步都得益于党与国家的进步与发展。

20 世纪 60 年代初期，正逢国家"困难时期"，粮、棉、油匮乏，郭敬辉教授虽然有高知补贴，但他家孩子多，粮食也不富余，可是他不时叫我去中关村家中，了解我学习情况，还常包饺子使我饱餐一顿。他讲："打游击时能包顿饺子像过年一样高兴，我们都会包，两手一捏就一个饺子。"虽位居高研，却仍平易近人，保持了老八路作风。我 1959 年元旦在北大结婚的，不懂又不好意思请人，仅在和地貌专业大学生的元旦联欢会上被加了个婚庆祝贺，没请任何人。事后他和李先生埋怨："你这个孩子，结婚怎么不告诉我们，让我们主婚？我还是备了一份与给刘昌明的礼物一样，送你。"当时，我还不认识刘昌明，但我知道郭老师送他一对黑色描金的福建漆器花瓶，而我是一对蓝色描花的漆器花瓶，比刘的略小一些。一对漆器花瓶是一份珍贵的秀丽的礼品，倾注着郭老师与李先生对年轻学生之关爱！①

我3岁时丧父，12岁丧母，是个孤儿，能收到这些礼品十分感恩与珍视。1959年北京大学地质地理系地貌专业同学在庆新年晚会上还对我们新婚祝贺，至今，我仍萦记得当时薄显耀、李容全、周儒忠与杨连山诸同学在会上的祝贺之情。

第二件事：1972年，在周恩来总理指示："中国科学院应重视基础研究和加强基础理论研究"后，"中国自然地理"专著系列被纳入科学院1973—1980年重点科学规划之中，竺可桢院士欣然受命，提出了工作与组织建议，出任编辑委员会主任，黄秉维、郭敬辉为副主任。中国自然地理全书分12册，包括：总论、地貌、气候、地表水、地下水、土壤地理、植物地理、动物地理、古地理、历史自然地理、自然条件与农业生产、海洋地理。海洋地理由中国海洋学会副理事长郭敬辉与编委会委员瞿宁淑及国家海洋局副局长律巍组织领导。第一版的《海洋地理》一书分工编写是：第一、六章（绪论、我国海洋事业的发展概况）由国家海洋局东北海洋工作站赵叔松执笔，第二章（海底地质地貌）由南京大学地理系王颖、朱大奎及中国科学院海洋研究所金翔龙执笔，第三章（海洋气候）和第四章的海浪一节由中国科学院地理研究所张丕远、李克让执笔；第四章（海洋水文）由东北海洋工作站窦振兴、王仁树、董须瑜、张立琨执笔；第五章（海洋生物）由中国科学院海洋研究所曾呈奎、刘瑞玉、成庆太、王存信、陈清潮、唐质灿、陆保仁执笔。初稿编写后，由每章派出一位代表在广州市矿泉别墅统稿与整编。郭敬辉与瞿宁淑两位老师亲临组织，即使是毫不张扬的撰写工作，在当时，还是被香港新闻报道："在矿泉别墅集中了一批专家在研究海洋规划……"约月余工作，初稿统成后，又经全国各有关单位的专家审核与补充修订，当时请到的有陈吉余、王宝灿、谢钦春、夏勘沅、陈则实、任允武、林之光、秦曾灏、费鸿年及张立政等人。该书共298 000字，于1979年10月由科学出版社出版。该书是我国第一部《海洋地理》专著，是由交叉学科的多位中青年研究人员共同完成，在编辑与出版过程中，始终贯穿着郭敬辉老师与瞿宁淑同志的指导与帮助，推动着多学科交叉的我国第一部《海洋地理》出版诞生。

郭敬辉教授重视与支持发展交叉学科——水文地理与海洋地理，郭敬辉教授重视调查研究，集各家之长组织编写与出版专著，从无到有，一步一个具有学科发展里程碑式的著作，承上启下，推动科学进步。在整个过程中重视对青年人的培养、提携，使之成长提高，从无单位之分与本位之界，培养了一批具交叉学科特点的人才。

他终生忠诚爱国奉献，不计名利，忠厚朴实，平易近人，亲切爱护青年学生，为我们树立了中国科学家崇高的品质、可贵形象与榜样。

郭老师，我们怀念您！您的教导与榜样永远铭志于心！

---

① 1959年我结婚时，共收到四份礼品：一份是朱大奎妈妈（凌恩佑）送我一件银灰织锦缎丝棉袄，何屏仪大姨与汝成大哥送的一只镶了颗珍珠的戒指，王维君大婶送的一只粉红乔其纱罩的台灯（路上摔坏了石头灯柱），以及郭老师与李先生送的一对花瓶。

挚念刘光鼎院士*

刘光鼎院士于 2018 年 8 月 7 日在北京逝世，消息传来，不胜悲痛！其生姿勃勃的形象却始终萦回于脑际：一代科学英才、创建中国海石油勘探巨业、功勋昭著，身教言传、力行楷模、培育人才、奉献继世，光明磊落、无私奉公，严于律己、忠厚待人，诗文并茂、酌酒促勤，宏才大略、太极国粹传人。

最早与刘先生相识，是 1963 年 3 月，在青岛东海饭店——一座屹立于黄海之滨的基岩海岸上的淡绿色船舰型建筑——举行的一次重要的海洋会议上。会议是由国家科学技术委员会海洋专业组织召开的，研讨我国海洋科学十年发展规划草案。与会专家一致认为，为加速发展我国海洋事业，建议成立“国家海洋局”。至今我仍记得会议代表的形象：国家科委出席的是王遵伋处长，一位和蔼可亲、文质彬彬的女处长；出席的海军代表：袁也烈少将，马靴戎装、一派将军气概，海洋组副组长于笑虹将军，身着挺括的海蓝色呢装，身着海军便服的刘志平大校主持会议，航保局的律巍、刘恩兰教授（海军将军级待遇的特聘专家）、秘书万延森中尉（南京大学地貌专业，1958 年毕业）；中国科学院海洋研究所所长童第周教授，副所长张玺及曾呈奎（均为海洋生物专家，张玺发言谈到贝类“招潮”，还以手势表达），物理海洋学家毛汉礼研究员及尤芳湖助理研究员；海洋地质界代表为长春地质学院的业治铮教授，北京地质学院地球物理勘探教研室主任刘光鼎先生；高校代表有：山东海洋学院教务长赫崇本教授，南京大学任美锷教授及助手王颖讲师。会议经过热烈讨论，一致建议成立国家海洋局。仍记得会议全过程十分保密，甚至晚间放映电影“地道战”，非一定级别不得入内，王颖进入被阻，经过同行的王遵伋处长解说，得以观看。若干年后，重游故地，东海饭店杂立于众多的房建之中，已不显当年雄伟之姿。当年的代表多故世，可能仅余万延森和我——两位代表的助手了!? 国家海洋局建立 55 年后，也已改变，世事沧桑，在缅怀故人之时，值得留记。

现今追忆，当年刘光鼎先生不过 34 岁，但已事业有成，发言时谈到在海上地震勘测，有时甚至租用木质渔船“放炮”，乘风破浪、工作艰苦，但却获得渤海油气构造，为打破国外对中国的石油封锁，继大庆之后，二次创业，开发海洋油气，功勋昭著。他与业治铮教授一起，师友相交、十分默契。业先生文质彬彬、和气而不苟言笑，刘先生身穿棕黄色皮夹克，具有海洋地质人气魄，会间很少与人交谈，我感到他俩均有些傲气。没承想，至 20 世纪 80 年代，却成为我的良师益友。当然，也缘于地质部海洋地质研究所在南京，同一地区同为海洋地质领域工作，联系交往密切。尚有一印象，大约是在 60 年代“文化大革命”中后期，我们均遭批判，偶然地在杨公井遇到刘先生，苦中作乐共同餐叙，他谈到被“革命群众”追捕躲藏，幸得一工友热诚相助……虽经颠沛流离之苦，但仍坦然小酌，热衷于海洋石油

---

* 王颖：朱日祥，郝天珧，江为为，刘少华主编，《刘光鼎先生追思集》，12 - 14 页，北京：科学出版社，2019 年。

地质工作，豪情不减。我深受鼓舞，深感这位年华正茂的老党员忠贞于中华、大气凛然！后来，形势好转，80年代以后，他出差南京，我们经常交流，邀请他给南大师生讲学，期间我常备洋河大曲，请他品尝，他也很高兴地交谈，但只喝不带，即使是贵州茅台，喝不完，也存放南京友人家中，以备下次再饮，他告诉我："老太太照顾我健康，劝戒酒。"我十分理解师母之深情，但也理解，海洋考察，风寒重，为赶潮时及在风浪平静时加班多做，饮酒却是驱寒、提神的重要手段，可以理解。我知道地学界中常以业治铮院士、刘光鼎院士及朱夏院士为"新生代"的酒仙。我不认识朱夏院士，而常听到人称刘先生为"海洋地质之父"，我认为：业先生是海洋地质之父，因此总会说："刘先生是海洋地质之叔"，每当此时，刘先生总是注视我而不表态。似乎是"不同意"？但不值得反驳！他们两位是我国海洋地质与油气勘探工作的开拓者，一位着重于沉积学研究与人才培养，另一位着重于海洋物探，均为调研与圈定油气藏构造与层储，各有所长，按中国习惯，十年为一代，所以，我认为：年长者是父，年轻者是叔。

刘先生大气豪爽，却又有深厚的文学与武学功底，他是吴氏太极拳一代传人，也爱看金庸武侠小说。我酷爱阅读文艺小说，但不热衷武侠小说。受刘先生身教启示，去书店购买了金庸全套，准备阅读，不知为何，灵机一动，也订购了一套给刘先生，结果惹了麻烦，成纸箱的书到后，刘先生自己搬不动，设法借了一个小铲车，推运回家。事后，我很内疚，检讨自己不动脑子的突发行为，那么多册的书籍，他怎么能搬得动呢？在学校工作有学生帮助，研究所下班后同事均回家，只能自己取！这些细节小事，却萦回于脑，至今不忘。

刘先生从北大物理系毕业，至工作岗位从事地球物理勘探与教学，应事业所需，专业能力不断进步与扩展，为探寻海洋石油，他又从事于发展海洋物探，圈定海洋油气藏，为发展祖国海洋石油工业，为解除以油气能源封杀新中国工业发展的危机，建功立业。时时进步，终生奉献。他的为人、为学、为业的爱国奉献精神，在海洋科学的建树，开发油气能源之成果，将永铭于史，永驻于我心，始终是我学习的榜样。值此年逾83岁之后，谨以片段追忆补白，以志纪念。

深切怀念孙枢院士*

孙枢院士病逝，令人不胜哀悼，他的音容笑貌仍生动地映现于我的脑海中。回忆我最初见到他，是20世纪80年代初期，在北美参加海洋科学会议时。当年印象深刻：他身材修长，始终温文尔雅，诚挚地答问交流，我深为敬佩这位北京学者的言行与科学造诣！

此后，约在90年代，得知孙枢到南京大学开会，我即赴南园宾馆探访，适逢他外出未遇，匆忙之中，留下一包在后校门刚买的糖炒热栗子，既表欢迎，也是回忆50年代时南大学生冬日相聚，热栗子是深为学子喜欢的稀罕食物。以后曾在国际沉积学大会时偶有相遇，但均匆匆招呼而未曾深谈，因均忙于所负责的会议事宜。2001年以后，开会时常与孙枢学长及夫人谢翠华相聚，友谊深厚，交流良多。

尤其当2013—2015年进行中国科学院院士咨询项目“南海资源环境与海疆权益”时，孙枢院士为项目组专家，共同讨论项目任务、咨询课题，并进行现场考察调研。在整个过程中，孙枢院士给予很多指导与帮助。仍记得2013年4月，在北京召开的咨询会议上，他提出：“在南海应重视中国人之间的合作……海洋油气开发可给予台湾学者一些工作，太平岛是中国在南海的一个立足点，是一个不沉的航空母舰。咨询工作宜从历史到现实，比较系统地了解南海情况，明确南海存在的问题。对三沙市今后发展提出的具有积极性的现实建议，实是我们项目成功的一部分。……我们已经公布了油气招标9个区块，意在南海油气有所突破，勘探工作需由北向南推进，要有一个基地……采取什么策略与对策对于南海的资源和维权是一个关键部分。”孙院士言简意赅，给予咨询工作十分重要的指导。

右起：3. 孙枢，4. 王颖，5. 苏纪兰，6. 张乔民

图1 在广西南宁“第十五届海峡两岸地貌学研讨会暨‘南海资源环境与海疆权益’咨询项目2014联合学术年会”会议(附彩图)

2014年在广西考察北仑河口后，于7月20日在南宁召开的第三次咨询会议上，孙枢院士总结会议发言，明确会议的正式意见：“断续线宜称为海疆国界线，不宜称九段线；通过中科院咨询委员会这一正式渠道向政府有关部门提出建议(补充国民政府证明资料和其他历史证明资料)；汇集咨询成员的研究文章形成文集送审；建议有关同志继续已安排的工作。”孙枢院士的发言，

* 王颖：《孙枢追思文集》编辑组编，《孙枢追思文集》，311-312页，北京：科学出版社，2019年。

有力地肯定了咨询工作组在明确南海海疆国界线的重要贡献。

孙枢院士的发言，进一步明确了在南海工作的重点：资源环境与海疆权益。对当前与今后的南海工作均有重要意义。

从20世纪90年代末开始，南京大学海岸与海岛开发教育部重点实验室与成都理工大学地质灾害防治与地质环境保护国家重点实验室联合与台湾大学地理环境资源学系合作，组织召开了"海峡两岸地貌学研讨会"，从每两年一次会议到每年一次会议，分别在大陆与台湾轮流进行，已召开了17届。

两岸地貌学术交流获得中国第四纪研究会韩家懋秘书长大力支持，孙枢院士给予帮助，他因忙碌与患病而未能参会，但支持夫人谢翠华同志参加学术会议，给予帮助。翠华同志多年从事秘书与学会组织工作，经验丰富，热心、细心、耐心，她本人也热爱野外考察，每次参加研讨会给予很多具体帮助并积极参加野外与海上考察。所以，友谊继续，至今不断。

与孙枢院士相识多年，从无私交往来，但工作与友谊绵长。我深深地敬仰他——诚挚、质朴，对国家、人民无私地默默奉献，一片丹心照汗青！谨以近期实例为文铭谢学长。

a. 左起：王颖、孙枢、李吉均

b. 左起：谢翠华、林俊全(台湾大学地理学院)、王颖、孙枢

图2　"海峡两岸地貌学研讨会"会后考察(2004)(附彩图)

教育树人 活力永续*
——铭谢北大母校的教育培养

我是1952年夏季毕业于北京市慕贞女子中学①,通过全国统一招考,录取入第一志愿的南京大学地理系学习,于1956年7月毕业于地理系地貌学专业。经学校分配,到北京大学地质地理系任助教,备考当年首设的副博士研究生。因此,于1956年7月告别了学习生活四年的南大母校,乘火车赴京进入北京大学地质地理系,入住26斋女生宿舍,室友是先我一年到北大工作的周慧祥校友。1957年2月,考试通过,入学为地貌学专业副博士研究生,学习四年,导师是王乃樑教授。当年研究生按助教要求,需担任教学工作(讲课与指导野外实习),学校照顾购买参考书与野外考察,每月助学金66.8元,稍高于约64元的助教月薪。

当年是新、旧时代交替的时期,破旧立新,建立新秩序,因此政治运动多:1957年反右,大字报铺天盖地;1958年"大跃进",大炼钢铁,亩产万斤,报刊上登着照片——小孩可站立于稻田密集的稻穗上;接着60年代困难时期,粮食匮乏,打麻雀除"四害"鼠口夺粮——校园中摩托车来回行驶,震荡得雀鸟停不下来,黄昏时纷纷落地死亡;食堂凭固定日期的早、中、晚餐券购饭,我因在王府井购书时钱夹被窃,顿失餐券,幸亏居住科学院中关村宿舍的老同学李安琦有粮米,供我一周饭食,在当时真是救命义举!一周后,公安局通知我领回钱包——小偷拿掉现金后将钱包扔于王府井大街垃圾筒中,被清洁工人捡出交到派出所,后转北大交我,饭票仍在,使我后半月得以有食。这是当年社会状况——小偷扒窃,清洁工义举,公安警察高效负责,同学间友爱支持,好人居多!即使在运动连连的动荡年代,北京大学仍坚持举办苏联专家讲学班,保证研究生与进修生的学习,我们没参加下乡挖地瓜,没去十三陵修水库。当时连王乃樑教授都拿个枝杈弹弓在燕园巡回打麻雀除"四害"!但研究生与进修生班学员仍坚持听课。B. Г. 列别杰夫副教授给我们讲授"地貌学基本理论与问题"和"砂矿地质学",他讲课深入浅出,理论与实践结合,凝聚了中外科学成就,使我们受益终身。他课上的学员后来在北京大学、南京大学、北京地质学院、长春地质学院、中南地质学院、中山大学与华东师范大学等院系开设了系列课程,对地学教育与开采航天需用的稀有元素砂矿发挥了重要作用。北京大学始终坚持教育领先、育人工作重要的理念与学风。

当年我也参加过群众运动,展示用子弹壳打成扁平的筒嘴喷出热水汽做"超声波"融冰化雪的试验,实为愚昧传讹。当年的同事中,有人因信中一句话或因几句牢骚,被批判而不认错,因此被定为右派,虽多年后被平反了,但因想不开而自尽的人却无法生还了。历史的教训应铭记。每个人处世均应客观地实事求是,尊重生命予人尊严。

---

* 王颖:蒋朗朗主编,《精神的魅力2018(一)》,121-125页,北京:北京大学出版社,2018年。

① 1951年改名为育新女中,1952年改名为北京女十三中,1972年更名为北京市一二五中学。

北京大学文、理、法、政学科齐全,具有基础雄厚、创新优先的学术氛围,持续有力地促进新型的交叉学科发展。以地学为例,20 世纪 50 年代,国家工农业建设发展急需地质学人才,除建立两年制的地质专科外,还从清华、北大等院校抽调了一批地学人才,组建成立北京地质学院(中国地质大学前身)。嗣后,北京大学又集中了岩矿、古生物、地貌、地图、自然地理及经济地理的教师,组建成新型的地质地理学系,系主任是历史地理学家侯仁之教授,地球化学教研室主任是王嘉荫教授,古生物教研室主任是乐森璕教授,地貌学教研室主任是王乃樑教授,经济地理学教研室主任是仇为之教授,土壤地理教研室主任是李孝芳教授,地图学教研室主任是刘迪生讲师;按专业招收本科生与研究生,聘请苏联专家协助培养研究生与青年教师;先后在山西大同阳源盆地、周口店、百花山及北戴河等处建立野外实习基地等。1958—1959 年聘请讲学的苏联专家有:沉积岩石学家 A. П. 列兹尼科夫教授,地貌学家 B. Г. 列别杰夫副教授,国际海岸学权威 B. П. 曾柯维奇教授及 O. K. 列昂杰夫教授等,他们重视地质基础、地貌结构与动力作用以及与沉积组成相结合,理论与实践相结合,以集中的专科讲学与短期的学术讲座的形式,传播了先进的科学理论与成果,提高了我国专业教师的科学认知与实践工作能力,加快发展了我国地球科学水平,为国家建设培养了交叉学科骨干队伍,在工、农、科、教领域发挥了重要作用。新兴的地貌学正是在该期间,首先在北大获得了迅速的成长与发展,发挥着人才培养与支持生产建设的领军作用。至今,在官厅水库的建设、山东半岛稀有元素砂矿开采、西南喀斯特地貌与青藏高原山地冰川发育规律等方面,仍发挥着生产建设效应。

北京大学拥有学科齐全的基础科学、浓厚的创新氛围,本科生与研究生教育在教学、实验、研究、实践,论文、设计中形成严密的科学体系与自由的学术氛围,孕育着强大的生命力,发展出新型的交叉学科,促进人才成长。以专业学科的组合与发展为例,有中文与英文,西语、东语与外国语;地质、地理与地貌,大气与环境,地球与空间科学,数学与力学,计算机与信息,软件与微电子科学,生物、生命与医学,基础医学、专科医学,药学与公共卫生学,经济、人口、管理与国家发展学科等,铭刻着学科发展的进步历程。当前历史、考古与文博,哲学与宗教等学科发展亦引起国人的广泛注意。北大学科的成长进步,展现出雄厚的基础,具有茂盛的新学科发展生命力。

北大燕园诗情画意,令人留恋。未名湖、石舫、博雅塔、临湖轩、钟亭、办公楼、华表、红漆大校门、高悬蓝底金字的北京大学校匾与烘托校门的双狮,集映中国文化传统之风貌。地学、物理、化学、生命、医学、文、史、哲、法、图书与管理等楼群寓学科教育之内涵。教室、场馆、斋舍、食堂,花前林下,汇聚着活跃的人群。即使是校园现代化发展的今天,在我脑海中仍时时映出当年北大生活之景:大食堂外,一群群捧着搪瓷饭碗,边吃边看通知或大字报的青年学生;教工食堂中师傅热情服务的身影,值班的校卫队以及 26 斋宿舍收集衣物洗涤的女工等,点点滴滴仍萦记心头!尤其是师恩难忘:导师王乃樑教授原籍福建,却说着一口地道的北京话;他有儒雅朴实的风貌,多年穿着在留学时购置的黄色咔叽外衣,骑单车代步;他有坚定的爱国信念,于 1950 年新中国刚成立即放弃论文答辩返回祖国任教;他经历了“文革”动乱年代,备受批斗侮辱,却仍坚持恢复地貌学教育,建设沉积相科学,调查、研究、讲学的足迹遍及西部山川与东部海洋原野;他对学生谆谆指导,对同仁友善,对家人关爱,尊重、照顾年老生病的父亲,侍奉父侧的继母,多年如一日,我甚至还记得他将父亲的骨灰盒置于他单身宿舍的书架上；他真正是品学兼优，谦虚谨慎，每篇文章、每次讲课均字斟句酌，掷地有声。正是他致力于将地貌形态研究与动力分析结合，以沉积相之天然记录为阐明地貌发育历史的依据，进而阐明其发展的动态。是他，推动了中国地貌学科学教育与人才培养。他科学功底深厚，学术贡献昭著，却淡泊名利，无私奉献。

侯仁之教授是地质地理系主任、历史地理学家。他致力于研究北京城市的历史沿革，对北京市的扩建发展——城市特性、格局，中轴线与四环，城建对历史遗迹的保护等方面有重要贡献。今日北京已扩展超出五环，超过了他所预料。他和善待人，学生受教，如沐春风。犹记得首次见到侯先生是在 1955 年 2 月，为参加在印度 Aligar 回教大学召开的国际地理学大会，专赴燕南园的小楼侯宅向侯先生请教。适逢他午休，但仍亲切地接待与指导我，让我受益深远。至今我仍为当年不懂事地扰他午休而内疚。侯先生的研究生王北辰，和我同届入学，但他略年长，文、史功底强，毕业后留北大任教直至逝世。王嘉荫教授的两位研究生冯钟燕与李应运是与我同届的研究生，当时冯已是讲师，温文儒雅，待人接物诚朴，李应运乐观好学。他们师从王嘉荫教授，专攻岩石地球化学，毕业后，冯钟燕留北大任教，李应运初分到南京大学地质系，后转至安徽，任地质总工程师。乐森璕教授任教于地史古生物专业，记忆中他精神饱满，走路步伐一致、每步计数，是多年野外工作久而成习惯。

这些名师正当中年，讲学精湛，野外教学指导细致精辟，让学生受益至深。新建的地质地理系学科力量茂盛成长：经济地理学仇为之教授、胡兆量讲师；自然地理学李孝芳教授、陈昌笃讲师、陈静生讲师；地图学刘迪生讲师；地貌专业力量强：王乃樑教授、潘德扬讲师、刘心务讲师、曹家欣讲师、欧阳青讲师、韩慕康与江美球两位专业翻译、新留助教林自立；研究生有崔之久、钱宗麟、周慧祥、王颖，以及一批进修讲师：陈华慧（北京地质学院）、刁正清（东北地质学院）、沈永欢（湖南矿冶学院）、李见贤（中山大学）、刘树人（华东师范大学）。全系在地学楼学习、钻研，出野外调查，充满着欣欣向荣的科学、友谊气氛。

50 年代成长的这一批新生的地学力量，长期地在我国地球科学战线努力奉献，为学科建设、人才培养以及区域工农业建设，做出了卓越的贡献。

回顾总结既往经历，有助于嗣后发展。当代地学教育需再重视加强专业体系教育：基础教育（文理科学基础与地球科学基础）——专业科学教育与实践——综合科学与实践能力的培养。

2017 年 3 月

王颖谈海洋地质学*
——海洋学中的一门新兴学科

[编者按]王颖同志是南京大学地理系的副教授，她从1957年开始，就从事我国海岸地貌的研究，二十多年来积累了丰富的经验，解决了生产上不少具体问题。特别是1979—1982年在加拿大从事海洋地质研究工作，取得了重大成绩。她写的有关鼓丘方面的论文，被国外科学家称为"鼓丘海岸的典型文献"。同志们称她为地理科学中"冲出亚洲，走向世界"的女科学家。为此，本刊特地采访了她，现报道如下：

记者：王颖老师，你从事海洋地质研究工作已经二十多年了，请你谈谈这方面的情况。

王颖：好，先从海洋学谈起。因为海洋地质学是属于海洋科学的。总的来看，现代海洋科学是从第二次世界大战后蓬勃发展起来的，海洋地质学亦是如此。由于军事活动、海上交通运输、海底矿产资源及领土等问题，都刺激了海洋学的发展，而二次大战后部分船舰与舰艇转为民用也促进了海洋地质学的发展。

海洋科学的特点是：综合性、全球性和技术复杂性。

为什么说是综合性和技术复杂性呢？因为海洋研究不像陆地上的研究是看得见、摸得着的，在国外甚至有人说，到今天为止我们对海底的了解还不如对月球表面了解得更清楚，因为通过拍照月球表面可以看得很清楚，而海底呢？厚层的海水覆盖就使对它的研究很困难了。而且往往对海底进行一个任何过程的研究都是互相影响、互相制约的，例如，海洋水圈、海里的生物、海水的物理特性、海浪和水流对海底地形的塑造与影响，而反过来内动力所影响的海底地形又对波浪有影响，所以它是综合性很强的。就是研究海洋地质的话，必须具备海流、波浪的知识，还必须有海水化学、物理特性的知识，同时又需要了解生物过程。这就是说水圈、岩石圈、生物圈、大气圈都是在这里相互交汇、相互作用的，所以它综合性特别强，需要有多种学科协同作战。另外，所以要协同作战，这是由于它技术上的复杂性有关。例如，进行一次海洋调查，必须要用船，船上要装备很多仪器，如测深、地震地层仪、水下摄影、重力、磁力测量，水下取样钻探等以及潜水器具等，出海一次花费很大，因此要求每次航程要相应的多学科，以便尽可能在一次出海中获得较多的成果。所以其综合性强也是由它技术复杂性所决定的。而由于技术复杂性，也决定了多学科作战，以便取得最大的经济效益。

还有一个特点是全球规模的影响。全球大洋是互相沟通的，如海洋环流，40°S的西风怒吼区所形成的波浪，海流能影响得很远。因此对海洋现象不能仅仅从一个地区孤立来研究。又如夏威夷、智利海底地震所产生的地震波，对日本海岸带来灾害性的影响。因此对海洋科学研究来说，它还有一个全球性的特点。

---

* 报道于：《世界科学》，1983年10月，第1-3,6页。

海洋科学的发展和生产需要与应用分不开的。以海洋学的发展为例，各国多半是从渔业捕捞开始，注意到海洋研究。因此生物海洋学和海洋生物学发展得最早，进一步是物理海洋学的发展。由于交通运输的发展，船只航行需要有风浪、水文与天气预报，这样就促使了物理海洋学发展。近年来，人们集中注意力于海底矿产，这不仅是在国外。就是我国的情况也是如此，紧接着海洋物理学的发展，海洋地质学以茁壮蓬勃的声势登上了海洋科学舞台。

海洋地质学的发展首先是从应用方面即生产活动的需要而促进的。最主要的是：

（1）海底矿产资源的开发利用。如：① 海底石油和天然气，蕴藏丰富。主要分布于：海湾、海岸潟湖沉积、三角洲沉积、大陆架沉积盆地以及大陆坡下的深海沉积扇或沉积盆地中等。勘探与开发海底油气资源如美国开发墨西哥湾油田，英国开发北海大陆架石油，我国开发渤海与南海油气储藏都有力地推动了海洋地质学的发展。② 多金属矿瘤，即锰结核。这是一种深海资源，分布在 2 000～6 000 m 深的海底，如在太平洋 6°～20°N、110°～180°W 之间有 600 万 km^2、水深为 3 200～5 900 m 的海底，锰结核在那里富积的程度每平方米在 10 kg 以上。现在世界上已经建立 8 个跨国财团约有 100 多家公司在从事勘探与试采。据海洋情报研究所的资料，太平洋底锰结核的金属含量为：镍 164 亿 t，铜 88 亿 t，比较一下镍在陆地上的蕴藏量为 0.6 亿 t，铜为 4.10 亿 t，而铜是非常需要的。钴为 58 亿 t，陆地上是 0.06 亿 t；而锰结核中锰的含量就更高了，达 4 000 亿 t，陆地上仅 60 亿 t。所以在锰结核中金属的含量都要比陆地上多得多，引起了人们的巨大兴趣。因此，石油的开发引起了对大陆架的研究，而锰结核的开发引起对深海的研究，特别是海底平原的研究。③ 海底的砂矿。海岸带有砂矿，海底也有砂矿，如金、金刚石、锆英石、金红石、独居石等一些稀有矿物，这些矿物在原岩中是分散的，不足以成为矿床开采，但岩石经过风化松散以后，这些重矿离解出来了，再经过流水搬运与波浪淘洗作用，最后堆积富聚起来，形成海滨砂矿。有些砂矿具有重要的开采价值，如山东半岛南部的锆英石砂矿，美国在阿拉斯加大陆架开采金砂矿，马来西亚的砂锡矿，印度、澳大利亚开采金红石、独居石砂矿，南非开采金刚石砂矿，这些都促使了对海岸堆积过程以及大陆架形成发展的研究。

（2）港口交通。事实上对海洋的开发利用是从海岸带开始的，即从港口与海上交通开始的。港口设置在海岸带，海岸是人们进出海洋的门户。过去我们了解海岸只是一条线，实际上不是这样。海岸应该是一个带，它是海水与陆地互相交界、相互作用的地带，是个“两栖”地带，既有沿岸陆地又有水下岸坡，水下大概达到 20～30 m 水深，也有十多米的，也是波浪开始扰动泥沙，其水深相当于当地平均波长的 1/2～1/3 的地方。不是所有的海岸带都适宜建港的。譬如，大连港是位于适宜建港的天然海湾，而天津新港建于淤泥质海滩，这是不适宜的。新建海港需要进行动力地貌调查，选择理想的港址。所以随着港口的发展促使了对海岸进行研究，研究海岸带的波浪、水流、泥沙运动，地貌的形成与变化情况，今后海岸带还要注意对生态环境的研究。

（3）通讯。如铺设海底电缆与管道等问题。要求在海底选择一条地形适宜、波浪、水流扰动强度较小及泥沙活动强度较小的地方，这就需要了解海底地貌与海底表面地质过程的情况。例如，在北海铺设电缆管道，洋底浑浊流切断了海底电缆，需要防止海底沙浪移动问题。又如在加拿大，浮动冰山切断海底石油管道，促进了研究冰刻痕，了解浮冰移动规律。

(4) 海洋环境问题。石油资源的开发,特别是海上平台、钻井的设置,涉及环境海洋地质问题。如沙浪移动规律影响海上平台的稳定性。海底沉积层如粉砂层、砂层、黏土层等不同的土层与海底工程地基的稳定性问题,这些工作与陆地上工程地质不太相同,国外称为环境海洋地质问题,就是要研究水流、泥沙、海底地貌的形成和变化以及沉积层稳定特性与分布等特点,所以不仅开采石油问题,还有一个环境的研究,这是很重要的。我参加过加拿大的大陆架与深海调查,其中涉及环境海洋地质问题,如深海原子能废物埋藏的可能性问题。随着核电站和核燃料的使用,产生了对高放射性的核废物处理问题。有些国家把核燃料废物投放于海底,但究竟人类共有的海床能否放置核废物,现在英、美、法、加拿大、荷、日等国参加的国际海床研究组织专门研究这个问题,我在加拿大也参加了,是作为海洋地貌和沉积方面参加了对大西洋梭木深海平原的调查研究工作,了解到洋底沉积与其上海水之间有着一个活动交换层,通过海水,泥沙,地球化学元素与生物之交换转移,核废物的放射性会辗转影响到水里,进一步会影响到生物,因此海底埋藏核废物需慎重加以研讨。这是个环境海洋地质问题,用沉积学的方法,用地球化学的方法研究确定物质来源,用浮游体、古生物,用石英砂表面结构确定物质来源,知道影响范围。

(5) 领土问题也促进了对大陆架的研究。

以上五个方面,由于应用问题提出了对海洋地质的要求,也促进了海洋地质学的发展。

但在另一方面不是应用问题,而是理论本身促进了它的发展。最明显的是海底构造。环太平洋的火山、地震带,喜马拉雅与地中海的地震带分布,进而提出海洋是如何起源与发展,以及海洋演变问题。在第二次世界大战以后,加强了海底调查,了解到地球是几个大的板块组成的,并且地球上地壳的厚度并不是一样的,大陆上地壳很厚,可达 60 多 km,海洋里地壳厚度仅 3 km,平均为 10～12 km,并了解到大洋的年龄很年轻,不超过 2 亿年。大西洋洋中脊中部有一个大断裂,炽热的熔岩就通过这个大断裂喷发出来。所以现在海洋地质学中一个重要方面是研究海底构造,研究洋壳与板块运动机制,这是全球性的问题。二十世纪五十年代末期板块运动的研究是震撼地质科学的革命,有力地推动了海洋地质学的发展。

记者:我国在海洋地质学的研究状况呢?

王:国外对海洋研究很重视。按研究实力与影响地位大致上可分为北美、欧洲、亚洲太平洋三大研究中心,当前以北美领先。

我国海洋科学开展较迟,五十年代开始。而海洋地质学更迟,五十年代末才开始。国内目前研究海洋地质的有五个系统:① 科学院;② 教育部;③ 海洋局;④ 地质部;⑤ 石油部。科学院海洋研究所在 1958 年海洋地质普查中进行了渤海湾及东海大陆架调查;教育部高校系统结合港口建设、河口治理及海滨砂矿等问题从事海岸和河口研究;五十年代末期地质部就开展了渤海与南海石油研究;六十年代石油部迅速开展海洋石油地质工作。总的来说,海洋地质与海洋科学一样,在我们国家是比较新兴的。最近钱学森、钱伟长等著名科学家提出要像当年发展核工业、航天工业那样重视、发展海洋事业,这是十分正确的。

国内研究情况是，海岸研究首先发展，现在也是重点，它联系我国生产的实际任务，工作进行得很深入。当时有苏联海岸专家曾科维奇指导。我们的海岸研究在世界上也是具相当水平的，把海岸作为陆上、海下的整体来看，研究它的动力作用，研究它的现代过程。但我们也有不足之处，就是手段落后。

大陆架调查二十世纪五十年代末开始，而主要发展在七十年代，如对渤海、南黄海、南海的石油普查勘探，现在已经形成较大的研究力量。在深海研究上还是零星的，如对太平洋导弹溅落区及锰结核的调查研究等。而国外情况跟我们不一样，它们海岸开发早，海岸工程手段强。所以，目前海岸带研究着重于海滩侵蚀防护、土地利用规划和管理，并发展到海岸生态系的研究，对海岸带研究的手段（船只定位、测深、取样、制图等）先进，省时高效。他们在大陆架的研究较深入，因为要开发石油。同时国外在深海地区的研究也较深入，如板块运动、深海浑浊流问题等，并广泛进行了深海钻探，在大西洋、地中海、太平洋都搞。这方面我国没有参加，任美锷教授为此呼吁了两次。

我国处于亚洲大陆东部与太平洋的边缘，因此我们研究板块理论比人家有独特的地方。有潮海、季风、波浪加上长江、黄河等大河的影响作用，因此海岸和海底都有特色。从海岸上来说，淤泥质平原海岸的研究、港湾海岸的研究，这个特色我们抓住了。还有堆积成因的大陆架，由于大河三角洲的堆积及很多沉积盆地，因此油气藏的远景在大陆架与海岸带堆积体内都是很丰富的。再有一个构造上的特点，一些边缘海，如白令海、日本海到东海、南海都呈现为一些菱形的海盆，我国位于岛弧海沟系的内侧，对于边缘海海盆成因的研究，洋壳运动在太平洋边缘发生的变化，能搞清的话，都具有全球性的重大意义。所以说我国海洋地质研究是很有特色的，只要我们花上一定的力量，是能取得突出成绩的。

总的来说，我国研究海岸带能结合实际问题，区域综合分析比较强，了解海岸地质过程比较深入，所以能解决很多生产实际问题，这样就进一步发展了海岸研究。我国海岸研究中还有一个特色是把海岸动力、地貌和沉积结合起来；同时我们海岸地貌又与港口工程结合起来。大陆架工作开展不久，但生命力强大发展迅速。远洋深海工作尚有待于发展，这就是我国在海洋地质方面简单的研究情况。

记者：这方面你已经谈了很多了，那么海洋地质学研究的内容和方法呢？

王：根据以上所讲，就可以知道海洋地质学是研究海岸和海底，就是被海水覆盖和作用这一地带岩石圈的形态、成因和变化的。具体包括：① 海岸；② 大陆架及其底下的大陆坡；③ 深海（包括深海平原，深海沟）。

海洋地质学的研究内容呢？我认为：① 海岸与海底地貌，过去没有单独提出来，实际上它很重要，海底山脉、海沟都是构造地貌，反映了它的成因。人们都是从海底地貌着手来研究海底构造的，所以它很受重视。② 研究海洋沉积和沉积层，包括微体古生物松散沉积。③ 海底构造。

这三方面研究方法不一样。虽然都需要应用船只，通过雷达、无线电、卫星或声学导航定位，但海底地貌通过回声测深仪、旁侧声纳、海底扫描摄像可以取得，主要取得它的形态结构与分布，而海底沉积要通过取样、打钻再进行分析研究。研究海底构造则是应用地球物理方法，如地震、重力、磁力、钻探。现在已有很多国家用小的潜水器深入到海底进行直接观测。所以说如果登月是一个重大成就的话，那么使用潜水器进入几千米深的海底也是一项重大成就。但它们两者在方法上不一样。

记者:你在加拿大工作了三年,取得了很大成绩。现在再请你谈谈今后工作的打算。

王:我今后打算有三个:① 就是自己本专业,继续从事海岸研究工作,尤其是淤泥质平原海岸和南、北方的砂砾质港湾式海岸。② 参加大陆架和深海海洋沉积的调查研究。③ 我在加拿大期间,向加拿大同行学习了沉积物的分析方法,深海的工作方法。还了解了加拿大科研组织机构情况、研究课题如何选定及他们的研究方法,以便借鉴。加拿大在海洋科学上是较先进的,而且它与美国随时沟通。因此我们可以通过加拿大,保持了一条稳定渠道,来了解北美海洋中心的研究状况,这对我们来说是很重要的。我愿意尽我一点力量,促进中国与加拿大、美、英在海洋地质科学方面的友好交流。

我另外还想说一点,就是我国应该进一步在海洋科学上加强对外交流,多订外文原版杂志,尽快了解信息。还要加强仪器建设。但更主要的是:科研政策稳定,对某一个科研项目不要轻易上马、下马,同时人才要集中使用,打破部门所有制。

记者:谢谢你,介绍了这么多的情况。预祝你在海洋地质科学的研究中取得新的成绩。

(本刊记者:蒋 平)

王颖院士建议：江苏应大力发展海洋经济*

今年7月，国务院正式批准了江苏省沿海开发规划，这标志着江苏沿海开发由此上升为国家战略。加快江苏沿海地区开发进程对长三角地区率先发展，乃至我国中部崛起和西部开发具有重大战略意义，也是江苏发展的一大机遇。

近日，为全面贯彻国务院江苏沿海地区发展纲要和江苏省沿海开发工作会议的精神，江苏省委研究室、省政府研究室和省科协等单位联合举办了江苏沿海开发高层论坛，主题为“江苏沿海开发与科技支撑”。江苏省有关委办厅局，沿海三市连云港、盐城、南通等市县有关领导，南京大学、河海大学和东南大学等高校和科研院所的专家学者与研究生200余人参加。论坛由江苏省委研究室副主任刘松汉主持。

中国科学院院士王颖应邀出席本次论坛，并作了题为《江苏省海岸海洋环境资源特点与开发建议》的报告。她以她领导的团队长期科研工作的成果为依据，精辟地阐述了三方面内容：当代国内外对海岸海洋的认识与关注、江苏海岸海洋环境资源的特点和对当前江苏沿海开发的建议。

王颖院士提出的六条建议是：一、在省委、省政府大力领导与关怀支持下，应用908新一轮海洋调查资料，由领导、专家与地方相结合制定具有前瞻性、与生态环境相和谐的江苏省海洋经济发展规划；二、以海港建设为龙头，扩展建设连云港深水大港，大力支持建设洋口深水港、吕四港与大丰港组成长三角北翼港口群；三、大力发展外向型海洋经济，重点发展海洋交通运输、海洋渔业、油气、能源、制造业与外贸为特色的临港产业带；四、建设现代化的大农业与大力开发利用可再生能源（太阳能、潮汐能与风能），发展为绿色有机现代化大农业基地；五、巩固丹顶鹤与麋鹿自然保护区，建设生态型的城镇体系，组建自然灾害预警减防灾体系，发展人与自然和谐持续健康发展的滨海江苏经济带；六、培养海洋科技人才，加强海岸海洋开发与管理力量。报告在地方基层领导中产生了强烈反响。大家一致认为，她的报告具有很高的理论价值和现实指导意义。

一些沿海市县的领导表示，听完王颖院士的报告，了解到在江苏沿海宽广的辐射沙脊群中深藏着可以建设深水大港优良的天然港航条件。认识到要推动沿海开发在高起点上加快进程，必须充分发挥科技的作用，需要科研院所和大学为沿海经济社会发展提供全方位的科技和人才支持。

一些代表表示，江苏沿海地区是长三角的重要组成部分，自然资源丰富、经济腹地广阔、区位优势突出，未来发展海洋经济的潜力巨大，是我国东部地区重要的经济增长极，战略地位非常重要。他们说，王颖院士的报告和研究成果为推进江苏沿海大开发提供了重要的科学依据和支撑。

---

* 朱小卫：《科学时报》，2009年11月9日，转载于：南京大学党委宣传部、南京大学新闻中心，《声誉》，2009年第6期，第63－64页。

江苏省政府研究室副主任包锦球在作论坛总结时指出:“此次论坛十分成功,七位专家和领导的演讲精彩纷呈,时间虽短,但议题集中、重点突出,探讨的问题层次高,研究的成果具有较强的操作性。”

王颖——海洋地貌与沉积学家*

王颖，思想敏捷，基础宽广，在科学面前大胆无畏。这样的人，我愿意在科学上与她交往。

——柯林斯（英国 Southampton 大学教授）

王颖，1935 年生，原籍辽宁省康平县。海洋地貌与沉积学家，中国科学院院士，中国共产党党员。1956 年毕业于南京大学地理学系。1957 年 2 月入北京大学地质地理学系攻读地貌学专业副博士研究生，1961 年 2 月毕业。现任南京大学教授，地学院院长；中国海洋学会名誉理事长，中国第四纪研究委员会副理事长，中国海洋湖沼学会常务理事，中国地理学会理事；国际海洋学研究委员会（SCOR）世界淤泥质海岸与海平面变化工作组主席，国际太平洋海洋科学技术协会（PACON）常务理事兼中国分部副主席，国际地理学联合会海洋地理学专业委员会常务理事，国际地貌学家联合会（IAG）执行委员会委员；"全国三八红旗手"。主要著作有《中国海洋地理》《海岸地貌学》《黄海陆架辐射沙脊群》《海南潮汐汉道港湾海岸》等。

图 1　2001 年 6 月，王颖获加拿大 Waterloo 大学环境科学荣誉博士后与该校校长 Prof. David Johnston（左 2）、校董主席（右 1）及爱人朱大奎教授（左 1）合影（附彩图）

* 于洸：郭建荣主编，《北大的才女们》，179－193 页，北京大学出版社，2009 年。

1　理想的航船从这里扬帆

王颖，这位大海的女儿，从小就有一个漂洋过海的美梦。幼年时，地理课及美妙的童话中，关于美丽大海的描述，深深地吸引着她。她多么想到大海上去航行啊！在北京读中学时，听说南京大学有位女地理学家，一口气能登上峨眉山，她佩服不已。她想，我要成为这样的人。高中毕业时，如何填写报考大学的志愿，她向老师请教。老师问："你喜欢什么？"她说："我喜欢森林，向往海洋……"老师说："你学地理吧！"这样，1952 年夏，她第一志愿报了南京大学地理学系，并以高分被录取。

南京大学注重基础学科的教学，地质、地理、气象、气候学科齐全，为学生打下了坚实的地球科学基础。当年，教师年富力强，敬业治教，循循善诱，同学们满腔热情地努力学习。课堂教学与野外实践结合，四年学习期间，王颖所在的班级攀登过宁镇山脉与皖南山脉，远赴太行山、吕梁山及黄土高原参加水土保持调查。野外实习使学生增长了能力，锻炼了身体，培养了坚韧刻苦的意志。王颖回忆说："青年时代的我，适逢新中国建立，沐浴着党的阳光，在一个朝气蓬勃、日新月异、充满关爱的环境中，得以健康地成长，真是生长逢时！"

1956 年初，国际地理学大会在印度阿里迦(Aligah)穆斯林大学召开，我国派出以一位团中央负责人为团长的中国青年代表团与会，这是新中国成立后派出参加国际学术会议的第一个地理学代表团，除团长、两位地理学家、一位翻译外，有一名地理学研究生，一名地理学本科生。刚刚 20 岁的王颖有幸是这个代表团的成员之一。他们一起准备大会报告与介绍中国地理学进展的文稿。中国代表团的报告在会上引起了强烈反响。但同时，王颖也亲眼见到有的外国学者趾高气扬的样子，亲耳听到有的外国学生嘲笑我们"英语少少的"。她心里憋了一口气，在日记上写道："我们政治上翻了身，科学上也要翻身啊！……我们有几千年的文明史，有得天独厚的地理环境，为什么不能做出具有世界先进水平的成果呢？"代表团长对王颖和另一位研究生说："新中国的政治影响，中国的自然环境和悠久的历史，使我们的报告获得了成功，赢得了尊重。但今后我们不能光靠政治影响，老讲五千年的文明史。下次会议，就要靠你们这一代了。要又红又专啊！"王颖的心潮像海洋一样翻滚，几天的国际学术会议，把她的眼界由小天地扩展到全世界，她追求的目标，不仅是要成为一名我国的地理学家，而且上升到要为国家、民族争光。

1956 年，党中央发出"向科学进军"的号召。这年夏天，王颖大学毕业后，被分配到北京大学地质地理学系任见习助教，并准备报考我国在 1956 年招生、1957 年始业的四年制副博士研究生。她被录取为地貌学专业的研究生，导师是著名地貌学家王乃樑教授。北京大学学术气氛浓厚，研究生教育规范。除全校统修自然辩证法、外语之外，她学习的专业课程有"沉积岩石学""沉积相与建造""现代地貌学基本理论与问题""砂矿地质学原理"等，均由苏联专家讲学。王乃樑与苏联专家 B. Г. 列别杰夫两位教授密切合作，指导研究生进行了北京西山寨口、大同盆地、聚乐堡火山群及辽东半岛砂矿地质实习。两位教授定期(约两周一次)共同与研究生个别交谈，帮助每位研究生明确其学习方向、发展道路，指导学习方法。读研究生期间，在专业上，必须通过专业基础考试，专业外语考试，参加专业基础课的部分讲授，指导本科生的野外实习，完成一篇学年论文，一篇毕业论文。

那时，提倡“教育与生产劳动相结合”。王颖的学年论文做“山东半岛海滨砂矿”方面的课题。她带领一个小组，从烟台到威海，从成山头，经荣城，到莫耶岛、石岛，踏遍了山东半岛的每个港湾，在每个砂体上采样、分析，找寻砂矿富集带，追寻原生矿源。他们风餐露宿，漫山遍野地奔波、探索，终于完成任务，为圈定C1级矿物量做出贡献。通过做砂矿课题的研究工作，使她提高了基础地质工作的能力，将地质营力、海岸地貌与沉积密切联系起来，探索其内在联系与发展规律，写出《山东半岛滨海砂矿形成与开发》的学年论文。

接着，苏联海岸学权威曾科维奇院士与列昂杰夫教授给研究生开设了“海岸讲座”，指导研究生进行了胶辽半岛海岸考察。导师为王颖确定以“海岸地貌与沉积”为专门方向，结合“天津新港回淤来源与整治的研究”作为毕业论文课题。天津新港涨潮时一片汪洋，落潮时就成为一个淤泥滩。他们要研究这些淤泥从何而来？怎么控制它？当时苏联专家认为，新港的淤泥是从150km以外的黄河入海口运移过来的。为了查明泥沙的来源，王颖带领一个小分队，在中科院海洋所与天津新港回淤研究站的支持下，从滦河口到天津新港，从天津新港到黄河口，从事艰苦的淤泥海岸断面调查与重复观测，在一步一陷（深至40cm）的淤泥滩上，背着仪器一次次在6个小时的落潮间隙中测量、采样，在风浪颠簸的舢板上测流、采样。浅滩上浪大船荡，学生们晕船了，身为队长的她必须坚持定期观测，记录与采水样都是在呕吐的间隙中完成的。有一次她也晕得实在受不住了，便跳下船，站在齐肩的水中，希望能喘息一下，但仍然晕，而且更冷，只有上船再记。通过艰苦的工作，终于搞清楚：港口的淤泥不是从黄河口来的，而是当地的浅滩造成的。通过研究，使她认识到淤泥质潮滩的动力、沉积与地貌的分带性规律，总结出潮滩与波浪动力环境下砂质海滩的不同特性，为治理港口淤泥提供了科学根据。在这次考察的基础上，写出题为《中国淤泥质海岸特征与发育》的研究生毕业论文。

四年的大学生活和四年的研究生经历，滨海砂矿和淤泥质海岸的研究，为她攀登科学高峰打下了坚实的基础，训练了基本技能，这位大海女儿理想的航船从这里扬帆。

2　在海浪与风暴中搏击

王颖是改革开放后较早出国的访问学者，1979年2月上飞机的那天，大雪漫天，望着机场上风雪飘舞中的五星红旗，她心中默默地念叨：“祖国！您的女儿决不辜负您的期望。”王颖赴加拿大学习深造，到位于哈立法克斯（Halifax）市的达尔豪谢（Dalhousie）大学，进修海洋地质与沉积学。这里是加拿大东海岸的海洋学中心。起初，学校还不了解她，把她作为“预备学员”进行安排。经过半年考察后，决定让她做研究员（Research Fellow），发给她研究员证书，还给她配备了助手。这在当时访问学者中是不多见的，这得益于她在国内多年的研究工作，在海洋科学领域确有真知实力。

一天，系主任柯克（H. B. S. Cooke）教授与她讨论确定研究课题。这位白鬓红颜的老人问她：“你准备做什么工作？”答：“鼓丘海岸研究。”教授点头说：“很好！那你就在哈立法克斯地区做些工作吧！比较近，资料也比较多。”教授是一片好心，想照顾这位四十多岁的中国妇女。但王颖却十分坚决地说：“不，我选择了开普不列颠（Cape Breton）岛东南海岸中一段典型的鼓丘海岸。”“为什么？那儿很远呀！”教授惊愕了。他知道，那一带荒无人烟，条件艰苦。鼓丘海岸发育在高纬度地区，是经过大陆的冰流作用形成的海岸。这种海岸，港湾曲折，岛屿众多，主要分布在美洲大陆东部的丘陵和平原地区，中国是没有的，文献上也没有系统的记述，理论上还是空白。王颖暗下决心，要用加拿大的先进设备，用自己在中国海岸工作的经验，做出成果，由中国人来填补这个空白。因此，她查阅了大量资料，用两个月的时间踏勘了这一带近千公里的海岸，才选定了这段近 50km 长的典型鼓丘海岸。王颖从地质上、海岸特点上详细论证了这个地区是典型的鼓丘海岸。听完王颖的陈述，教授被这位中国女学者的勇气和踏实的科学态度深深地感动了，连连说："很好！很好！赶快把研究计划和方案拿出来。"

王颖的研究计划引起了加拿大同行的极大兴趣，不仅她所在的达尔豪谢大学愿意提供经费，贝德福德(Bedford)海洋研究所大西洋地质中心也愿意提供经费、助手和设备，并聘请王颖做兼职研究人员。由于工作需要，中、加两国有关部门研究确定，将王颖的进修期从两年改为三年。

在开普不列颠岛进行鼓丘海岸考察时，他们从南往北沿着大海走，只听到自己走路的声音。遇上风暴天气，天是铅灰色的，海是黑沉沉的，大西洋的海浪一个接一个地扑过来。但无论风浪多么大，她都不能停止观测，都要在预定的时间内拿出成果来。功夫不负有心人。经过两年多的艰苦磨难，王颖写出《开普不列颠岛东南部鼓丘海岸动力地貌学》(与 D. J. W. Piper 合著)的学术论文，在加拿大《海洋沉积与大西洋地质学》杂志 1982 年第 18 期上，排在第一篇的位置发表，被地质学界评论为"把中国经验应用于加拿大区域海岸研究中，成功地为加拿大海岸研究开拓了一个新领域，是鼓丘海岸的典型文献"。

图 2　王颖在美国讲学并与同行交流

在加拿大三年期间，王颖珍惜工作良机，竭尽全力完成工作，参加了 Nova Scotia 大陆边缘、Labrador 大陆架和纽芬兰峡湾海岸考察，进行中、加淤泥质海岸和大陆架的比较研究；参加了大西洋洋中脊钻探、SOHM 深海平原勘测钻探及 Porto Rico 海沟考察，了解洋中脊和深海平原，进行海洋地质与深海环境调查，研究海底埋藏高品位核废料的可能性。大西洋百慕大魔鬼三角海域，不知覆没了多少船只，吞噬了多少航海的英雄好汉。王颖曾三次到百慕大海域考察。北大西洋流急浪高，船只颠簸。一次遇到 10 级风暴，大海怒涛汹涌，船只被抛上掷下，一只法国船只沉没了，他们船上很多人也倒下了，王颖一边吃药，一边坚持摄影,获得了百慕大海域许多珍贵的实况资料。她还主动要求乘“派塞斯”(PISCES)Ⅳ号深潜器,下潜到216 m深的海底,对圣劳伦斯(St. Lawrence)海湾进行海底调查。深潜器里,既不能坐,也不能站,她伏在里面整整工作了两个小时。海浪扰动,呼吸不适,上岸后,她的嘴唇颜色都变深了,加拿大同行们都敬佩不已。他们说:“很少有妇女到海底工作,你真不简单啊!”他们著文称王颖是“中国第一个用深潜器从事海底地质调查的科学家”。

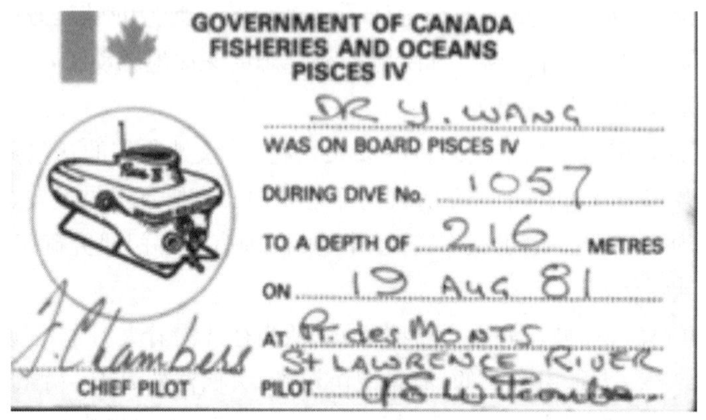
GOVERNMENT OF CANADA
FISHERIES AND OCEANS
PISCES IV

DR Y. WANG
WAS ON BOARD PISCES IV
DURING DIVE No. 1057
TO A DEPTH OF 216 METRES
ON 19 AUG 81
AT Pt. des MONTS
St LAWRENCE RIVER

CHIEF PILOT
PILOT

图3 王颖1981年下潜大西洋圣劳伦斯海湾海底216 m考察证书(附彩图)

王颖通过国内外朋友搜集到黄河、长江、珠江的砂样,西藏的冰川砂,甘肃的黄土粉砂,美国、加拿大、大西洋及大陆架的砂样,在回国前的几个月里,抓紧时间,利用贝德福德海洋研究所先进的扫描电子显微镜,对一百多个样品的数千颗砂粒进行了处理、观察和摄影,完成了1 300张照片,取得了几千个数据资料,完成了《石英砂表面结构与沉积环境的模式图集》(与B. 迪纳瑞尔合著),在王颖离开加拿大前两天,1982年2月12日,四位专家审定同意出版,1985年由科学出版社以中、英文出版。

图4 王颖院士与外国同行在一起

1980年4月,加拿大全国海洋会议在伯林顿举行,一位来自中华人民共和国的女学者庄重地登上讲坛,用英语作关于中国海岸研究的学术报告——她就是王颖。24年前在印度召开的那次国际地理学讨论会的情景,王颖永远铭记在心:“要又红又专!要又红又专!”她知道,要攀上科学高峰,必须从脚下做起。现在,她终于从淤泥滩走上国际讲坛。

在加拿大期间,王颖曾七次参加海洋地质方面的国际学术会议,在会上报告或展示她的学术论文。她的报告被赞为“精彩的报告”,论文被收入会议论文集出版。外国专家们对王颖评价很高。加拿大环境海岸地质研究所主任派帕教授曾给我国驻加拿大大使写了一封很长的信,高度赞扬王颖“为加拿大的海洋地质科学作出了四个独特的重要贡献”。

在王颖家里的墙壁上,挂着一幅金属版画,这是王颖离开加拿大时,贝德福德海洋研究所赠送给她的。版画上刻着王颖工作过的考察船,写着:“送给王颖,为了纪念她在中加海洋地质合作中开拓性的工作。”

是的,在王颖心中也刻着一艘理想的航船,几十年来,她驾驶着它,乘风破浪,闯过一道道难关,开拓出一个又一个新局面,获得一个又一个新成果。

3　让生命在海洋科学研究中闪光

王颖在地理科学、海洋科学的教学和科研方面做了大量工作,曾担任过南京大学大地海洋科学系系主任,先后讲授过“地貌学与第四纪地质学”“海洋动力地貌学”“海洋地质学”“砂矿地质学”“全球变化与海洋专题讲座”“海岸海洋环境、资源与一体化管理”“海岸海洋科学概论”等多种本科生和研究生课程。已培养硕士 20 余名,博士 15 名。由于在教学与科研工作中成绩显著,1983 年 12 月获江苏省妇女联合会授予的“三八红旗手标兵”称号;1984 年获国务院人事部授予的“中青年有突出贡献专家”称号;1994 年获南京大学第四届研究生导师教书育人奖。1991 年获国务院颁发的政府津贴;1985 年和 2002 年两次获全国妇女联合会授予的“全国三八红旗手”称号。

王颖长期从事具有地域特点的淤泥潮滩海岸、鼓丘海岸及河海体系与大陆架沉积等方面的研究,运用地质学、地理学、海洋学等多学科结合的知识与造诣,总结潮滩动力环境的沉积与生态模式,分析中、新生代泥沙、粉砂沉积环境,从中国主要河流对大陆架沉积作用的研究,深入到河海体系相互作用、沉积物搬运与陆源通量、黄海辐射沙洲的形成与演变等方面的研究,推动了具有学科交叉特色的海岸海洋科学和海洋地理学的发展,并将海陆相互作用与全球变化的研究相结合,应用于海岸、海港建设。1961 年以来,出版专著 16 部(包括主编与合著)、发表论文 130 多篇,其中有 40 多篇在国外学术刊物上发表。她是一位被国际公认的海洋学领域的著名科学家。

在海岸动力和地貌学研究方面,从渤海湾到古黄河口、长江口,一直到珠江三角洲、海南岛,绵延一万多公里的中国海岸,大都留下了她的足迹。淤泥质海岸是中国海岸的一个特色。从 20 世纪 60 年代起,她就对渤海、黄海淤泥质海岸进行广泛的研究,总结出淤泥质潮间带浅滩微地貌与沉积的分带性规律,及淤泥质海岸演化的动力过程与演化规律。她 60 年代初发表的关于潮滩海岸特征的学术论文,比英国著名学者发表的关于潮滩研究的学术论文还要早两年。她提出了“黄河改造、潮滩反馈与贝壳堤发育假说”。她对中、加、英三国的潮滩进行对比研究,总结出三种动力环境的淤泥潮滩沉积与生态模式,分析中、新生代淤泥粉砂岩沉积环境,把我国的潮滩研究推向国际先进水平。她还针对中国海岸的特点,着重研究季风波浪、潮流及大河河流对海岸的作用及海岸演变,研究海岸带海

图 5　王颖在杭州湾潮滩工作

洋与大陆的相互作用过程、机制，以及人类活动的影响，总结出中国海岸的类型及其发育演化的规律。华南广泛分布着港湾海岸，王颖多年对广东、广西、海南岛港湾海岸潮汐汊道进行研究，总结出纳潮量与港湾蚀积相关的机制及潮汐汊道港湾海岸演化过程的规律，并应用于海港选址与航道工程。其系列研究成果，发表在国内外学术刊物上，受到海洋学家的广泛关注，有关国际会议邀请她到会报告，并被推选为国际海洋学研究委员会世界淤泥质海岸与相对海平面变化工作组主席。

在海洋沉积研究方面，王颖几十年来一直在海洋现场考察，足迹遍及中国海、大西洋及太平洋边缘海，曾六次赴大西洋进行远洋调查，考察大西洋深海平原、波多黎谷海沟、大西洋洋中脊、北极圈拉不拉多海底等。她在《Sedimentology》杂志上发表的有关深海平原浊流砂的论文，是对深海底浊流动力作用的首次发现与论述，她论证了深海底仍有强大的动力环境，可将大陆架的砂带到大西洋中部的深海底，在理论上提出了深海沉积环境新的概念，在应用上对国际上有人提出要利用深海底埋藏核废料提出质疑，指出深海底仍具不稳定性，需做进一步的防护工程。王颖主持"南黄海海底沙脊群形成演变研究"的项目，这个大的研究集体，通过数年研究，查明了这 2 万多 km^2 海底沙脊沙滩的泥沙来源、形成过程与演变趋势，在大陆架沉积作用、堆积型大陆架的形成等方面作了理论上的阐述，并对这片沙脊沙洲的开发提出意见。她依据多次海洋考察积累的丰富资料和经验，结合先进技术手段的研究，在国内外发表了一系列高水平的论文。前国际沉积学会主席、牛津大学 H. G. Reading 教授认为王颖的论文《中国主要河流泥沙与大陆架沉积作用》，是"河海相互作用过程的最佳总结"。由王颖主编并撰写第一、二章的《中国海洋地理》，被《地理学报》评为我国第一部具有划时代意义的海洋地理学专著。

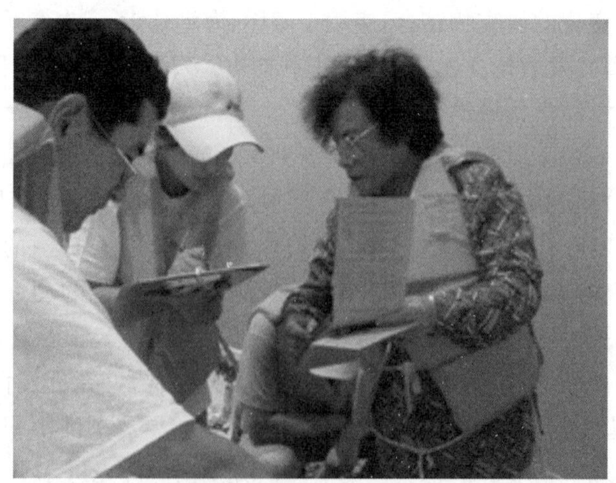
图 6　2003 年 6 月王颖在苏北洪泽湖考察并指导博士后研究生(附彩图)

在海岸海洋科学研究方面，从20世纪90年代起，王颖科研工作的主攻方向是对海岸海洋的研究。1994年联合国教科文组织在比利时列日大学召开的海岸海洋工作会议上，明确海岸海洋的范围包括海岸带、大陆架、大陆坡与大陆隆，即整个海陆过渡带。王颖应邀在会上作“陆源通量与海洋沉积”的报告，受到与会学者的高度评价，后被哈佛大学收入当代海洋的权威著作《The Sea》第10卷《Global Coastal Ocean》海洋专集，成为具有国际影响的科技文献。王颖十分注意科学研究工作与国家的经济建设相结合，将海岸动力、海岸演变、泥沙来源及运移规律、海岸泥沙回淤量、海岸冲淤变化预测、潮滩沉积作用及沉积相等研究成果，应用于海洋石油天然气勘查、海港与航道选址、海岸工程、航道回淤治理等。她从事过天津新港、秦皇岛油港、秦皇岛渔港、山海关船厂、黄岛油港、龙口海洋石油基地、北海港、三亚港、洋浦港等港口与航道选址研究，由她为负责人或主要执笔人撰写的关于海岸动力地貌、港址研究、回淤研究的调研报告35册，已有28项投入建设。如天津新港泥沙来源及减轻回淤的研究，成为天津新港扩建的科学依据。国家重点工程秦皇岛油港及秦皇岛煤港的一、二期工程，建成后多年证明预报正确，措施得当。在三亚港、洋浦港的建设中，运用和发展了潮汐汊道理论，成功地解决了港口建设中回淤的问题，洋浦港于1990年建成两万吨级泊位，十多年来没有回淤。在南黄海，论证其沙脊群间潮道为古河谷，地形及水深将长期保持稳定，可建20万t深水港，现已开工建设，将对江苏经济发展作出重要贡献。近年来，还将海岸研究应用于滨水城市规划、海洋旅游规划，建设人工海滩、人工岛等。

图7　2003年10月王颖与师生一起考察连云港古海岸遗迹(附彩图)

二十多年来，王颖的国际学术交流与合作相当广泛，与加拿大、英国、美国、澳大利亚有8项海岸海洋合作研究项目，组织并主持过7次有关海岸海洋的国际学术会议，利用国际合作的研究经费在南京大学建设了“海岸与海岛开发国家实验室”。国际太平洋海洋科技协会曾授予她“科学服务贡献奖”(1993年、2004年)。为了表彰王颖在中、加合作，特别是在海岸地貌学及海岸管理领域取得的突出成就，2001年6月13日，加拿大滑铁卢大学授予她“环境科学名誉博士”学位。滑铁卢大学还与加拿大邮政总局合作，出版了印有王颖工作照的纪念邮票，一版25枚，成为上了被誉为国家名片的加拿大邮票的第一位中国科学家。

王颖的科研论著和编著的教材多次获奖。“海岸动力地貌的研究(海港选址)”获全国科学大会重大贡献奖三个受奖提名人之一(1978年);“海岸发育与岸线变迁研究”获江苏省科技进步三等奖(1984年);中国自然地理专著“海洋地理篇”获中国科学院科技进步奖一等奖(王颖等,1986年);高科技知识丛书《海洋技术》(副主编)获中共中央宣传部“五个一工程”作品入选奖(1993年)及国家科技进步奖三等奖(1996年),“亚龙湾海洋旅游勘测研究与规划”(第一完成人)获国家教委科技进步奖三等奖(1995年);“江苏省海岛资源综合调查与开发试点研究”(完成人之一)获江苏省科技进步奖二等奖(1995年);《海岸地貌学》(王颖、朱大奎著)获南京大学优秀教材一等奖(1995年),国家教委第三届高等学校优秀教材一等奖(1996年),国家教委科技进步奖二等奖(1997年);“中国海洋地理研究”(第一完成人)获国家教育部科技进步奖一等奖(1998年);“地球系统科学创新人才的培养方式与实践”(第一完成人)获江苏省高等教育教学成果一等奖(2004年)。

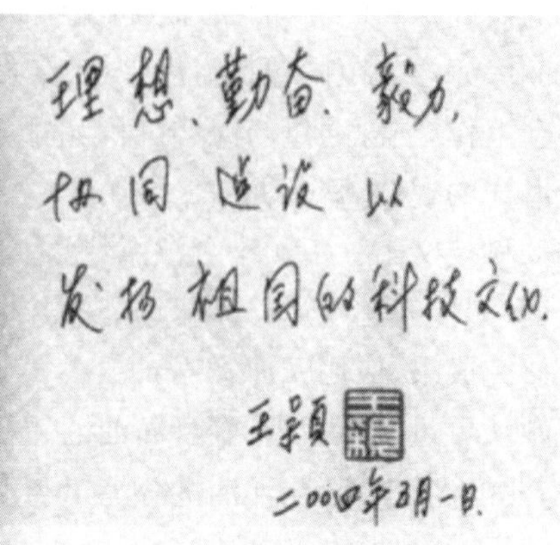

图8　王颖题词

王颖曾说:“人活着要有理想,为追求理想,还要有刻苦、实干与敢于面对挑战、不断追求的坚持精神。”她就是这样,非常高标准地从事教书育人与科学研究工作,一步一步地前进,一层一层地攀登科学高峰,在教书育人中学习,在研究工作中学习,在生产实践中学习。她体会到:学无止境,不要囿于大学学习的专业基础,应随着时代前进的步伐,不断地学习、扩展,交叉与融合,使科学能力真正服务于祖国发展的需要。饮水思源,立志不懈,在教育科研岗位上,为培养品学兼优的人才,为祖国的现代化与科学发展贡献力量。这就是王颖成功之道。

图9　幸福家庭(附彩图)

王颖从1956年大学毕业至今,工作了五十余载。王颖的工作成果都是在与大海的拼搏中获得的。是啊!热爱大海,构成她丰富的人格内涵;献身大海,体现了她不知疲倦的进取精神。这就是王颖!

附录

王颖主要著作目录

1. 王颖. 祖国的海岸[M]. 北京:科学出版社,1976.

2. 中国科学院《中国自然地理》编辑委员会(王颖为主要编写成员之一),中国自然地理——海洋地理[M]. 北京:科学出版社,1979.

3. 王颖,[加]B. 迪纳瑞尔著. 石英砂表面结构模式图集(中、英文版)[M]. 北京:科学出版社,1985.

4. 王颖,朱大奎等. 秦皇岛海岸研究[M]. 南京:南京大学出版社,1988.

5. Wang Ying. Proceedings of the Fifth MICE Symposium for Asia and Pacific: Ecosystem and Environment of Tidal Flat Coast Effected by Human Beings' Activities [M]. Nanjing:Nanjing University Press,1990.

6. Wang Ying and Schafer, Charles T. Island Environment and Coast Development [M]. Nanjing:Nanjing University Press,1992.

7. 朱大奎主编,薛鸿超,王颖副主编. 海洋技术[M]. 南京:江苏科技出版社,1993.

8. David Hopley and Wang Ying. Proceedings of the 1993 PACON China Symposium: Estuarine and Coastal Processes[M]. Published by the Sir George Fisher Centre for Tropical Marine Studies, James Cook University of North Queensland, Townsville, Australia,1994.

9. 王颖,朱大奎编著. 海岸地貌学[M]. 北京:高等教育出版社,1994.

10. 王颖主编. 中国海洋地理[M]. 北京:科学出版社,1996.

11. 王颖著. 海岸——通向海洋的虹桥[M]. 南宁:广西教育出版社,1998.

12. 王颖等著. 海南潮汐汊道港湾海岸[M]. 北京:中国环境科学出版社,1998.

13. 蔡明理,王颖编著. 黄河三角洲发育演变及对渤、黄海的影响[M]. 南京:河海大学出版社,1999.

14. 朱大奎,王颖,陈方编著. 环境地质学[M]. 北京:高等教育出版社,2000.

15. Yang Wencai, Paul Robinson, Fu Rongshan, Wang Ying. Geodynamic Processes and Our Living Environment[M]. Beijing: Geological Publishing House,2001.

16. Terry Healy, Ying Wang and Judy—Ann Healy (editors). Muddy Coasts of the World: Processes, Deposits and Function[M]. Published by ELSEVIER,2002.

17. 王颖主编. 黄海陆架辐射沙脊群[M]. 北京:中国环境科学出版社,2002.

18. 王颖主编. 中国区域海洋学——海洋地貌学[M]. 北京:海洋出版社,2012.

19. 张长宽主编，王建，王颖，王国祥等副主编. 江苏近海海洋综合调查与评价总报告[M]. 北京：科学出版社，2012.

20. 王颖主编. 中国海洋地理[M]. 北京：科学出版社，2013.

21. 王颖主编. 南黄海辐射沙脊群环境与资源[M]. 北京：海洋出版社，2014.

22. 朱大奎，王颖著. 工程海岸学[M]. 北京：科学出版社，2014.

23. 朱大奎，王颖，杨柳燕，邹欣庆著. 曹妃甸海洋生态研究——唐山曹妃甸工业区建设对海岸海洋生态影响与预测研究[M]. 南京：南京大学出版社，2015.

24. 中国海洋学会编著，苏纪兰主编，王颖副主编. 中国海洋学学科史[M]. 北京：中国科学技术出版社，2015.

25. 中国科学院编（王颖任撰写组组长）. 中国学科发展战略——海岸海洋科学[M]. 北京：科学出版社，2016.

26. 朱大奎，王颖编著. 环境地质学（第二版）[M]. 南京：南京大学出版社，2020.

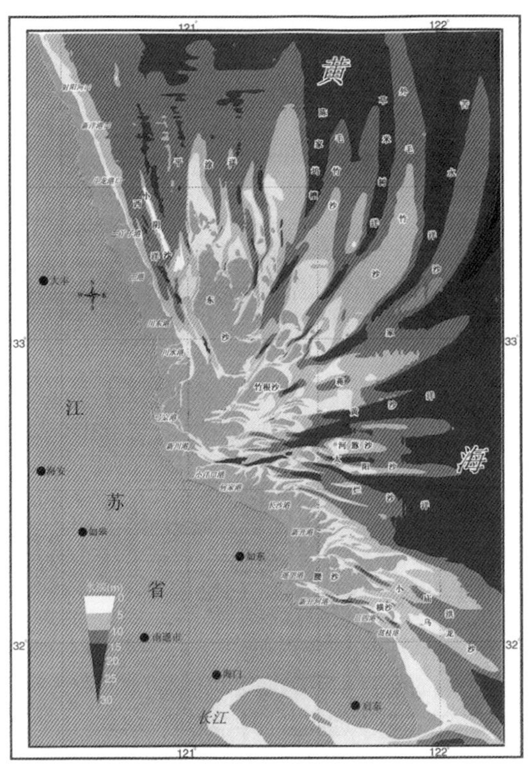

南黄海辐射沙脊群(83页　图2，112页　图9，131页　图5左，157页　图6，170页　图8，618页　图1，767页　图5)

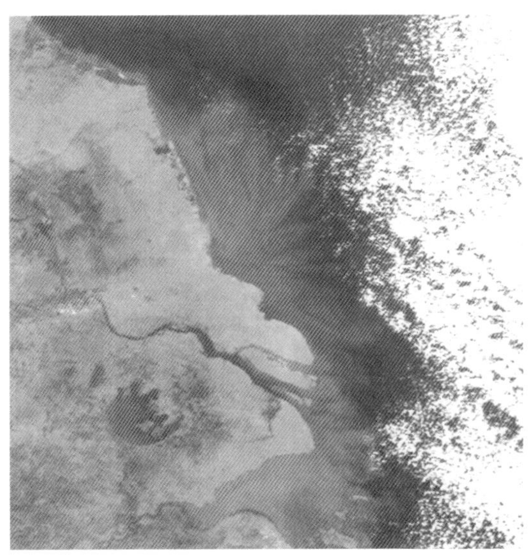

辐射沙脊群卫星图片(85页　图4，158页　图7，620页　图5)

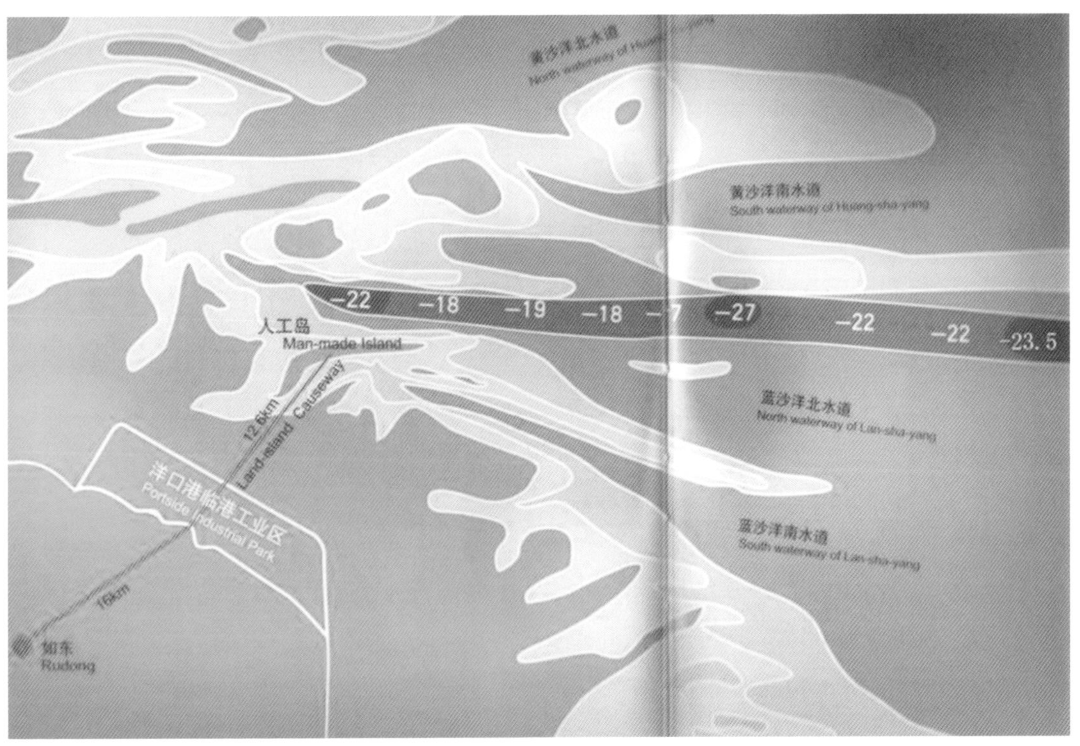

长三角北翼洋口港深水通道(113页　图10)

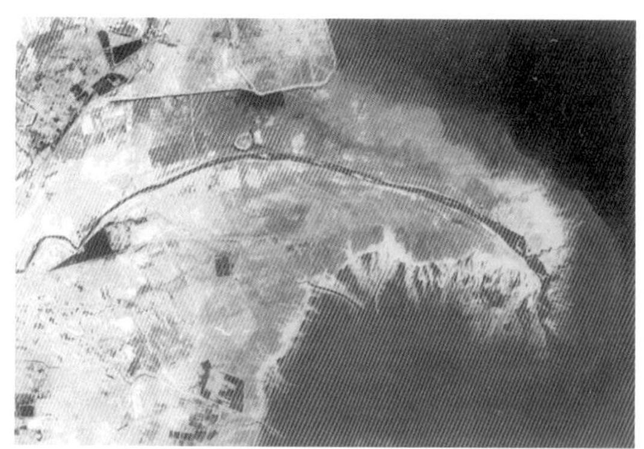

黄河三角洲（94页 图6，105页 Fig.8）

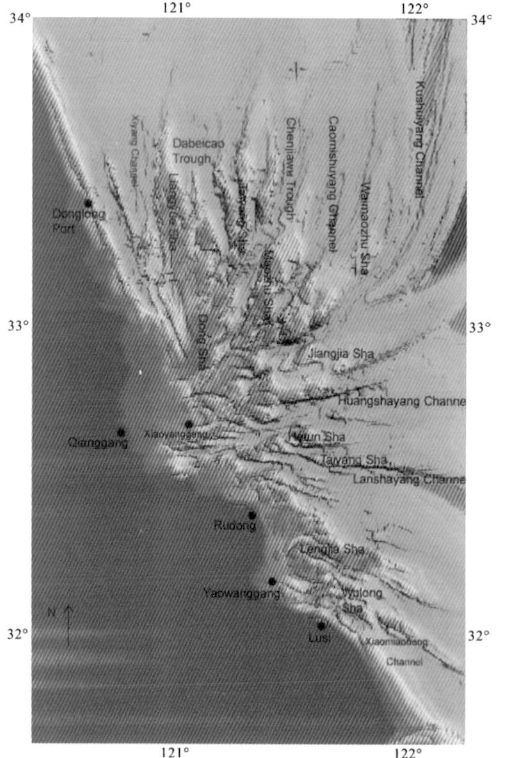

Radioactive Sandy Ridges（100页 Fig.3）

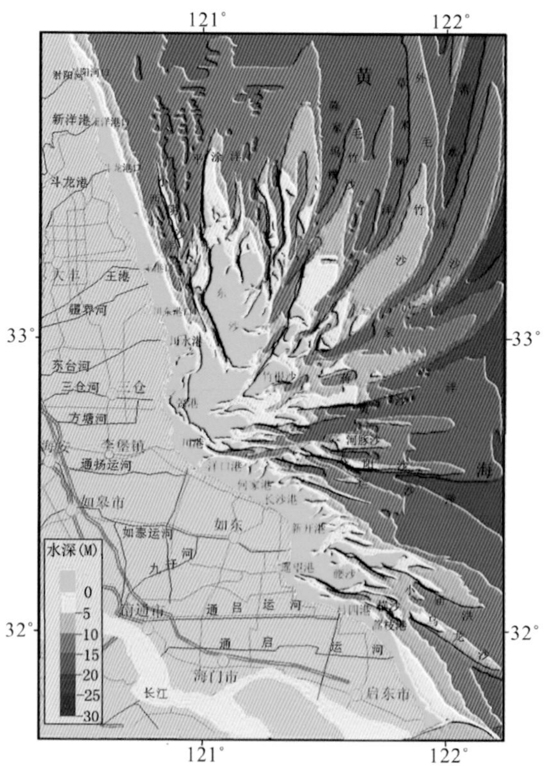

南黄海辐射沙脊群全貌（130页 图4）

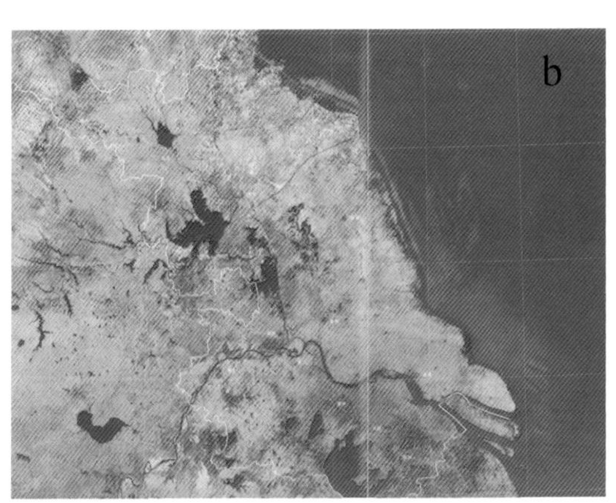

Remote Sensing of Jiangsu（104页 Fig.7右）

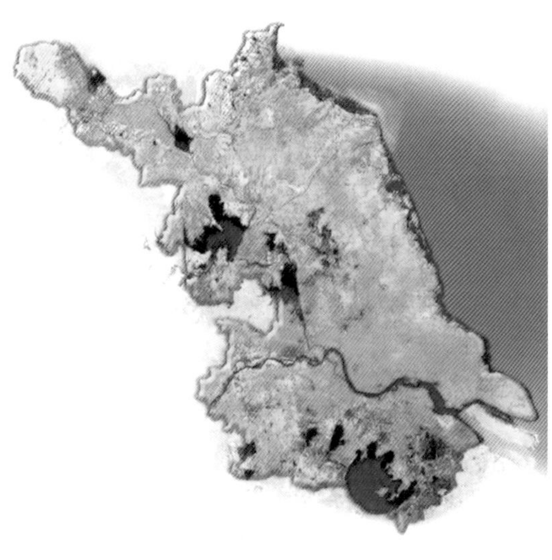

江苏省卫星图（110页 图1-2）

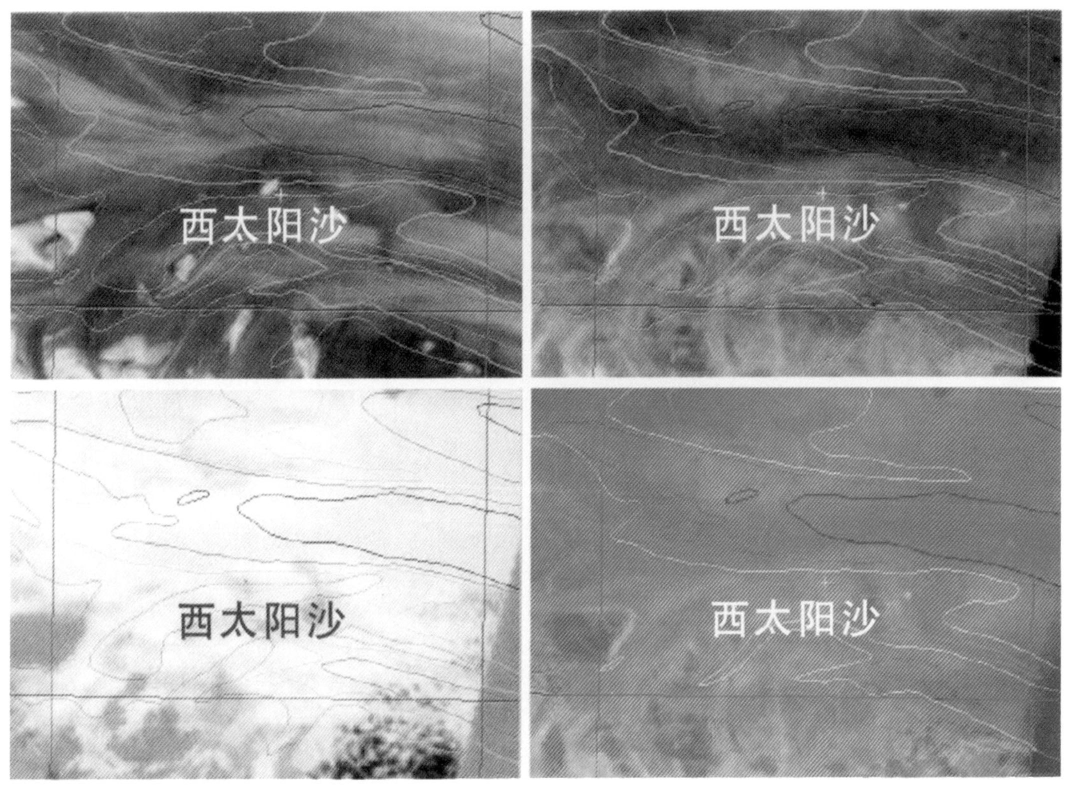

西太阳沙与烂沙洋遥感图像（86页 图6，161页 图14）

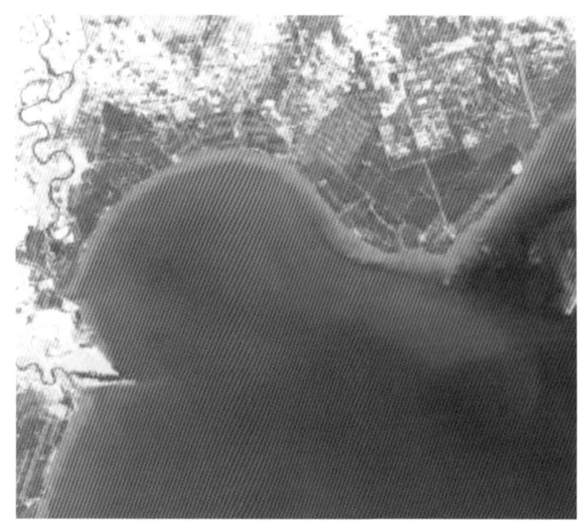

曹妃甸沙岛(128页 图1底图)

曹妃甸港口组合遥感影像(134页 图8)

曹妃甸港(129页 图3)

洋山港(133页 图7b)

平原海岸遥感影像(136页 图9底图，167页 图3)

鸭绿江口(219页 图1)

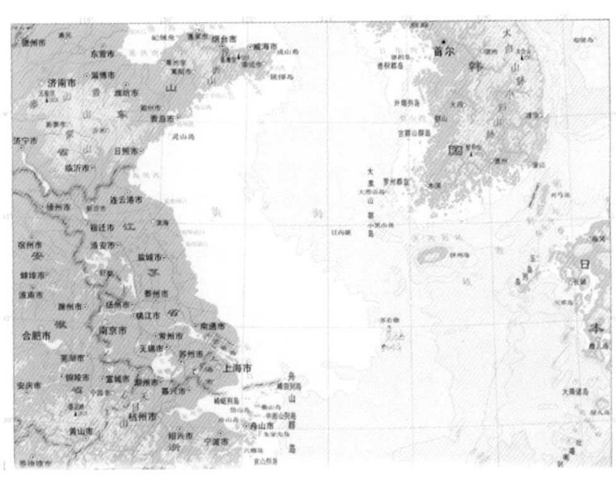

苏岩礁位置图(220页 图2)

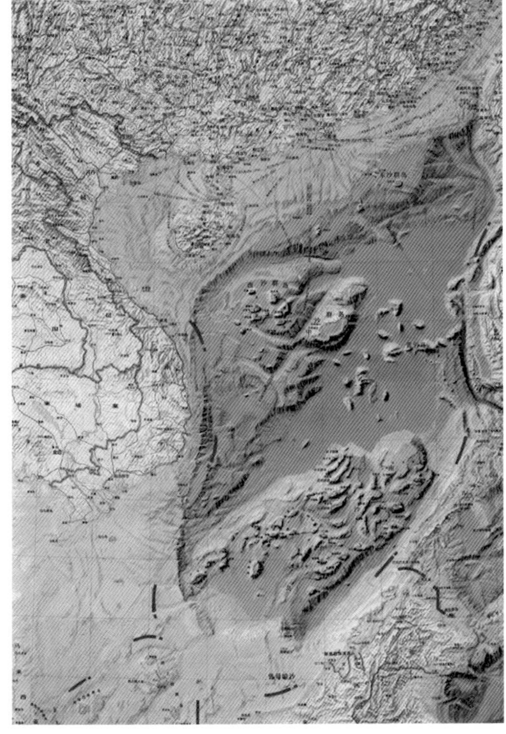

南海海底图(224页 图7)

1946年11月24日中国海军收复西沙群岛纪念碑(226页 图9)

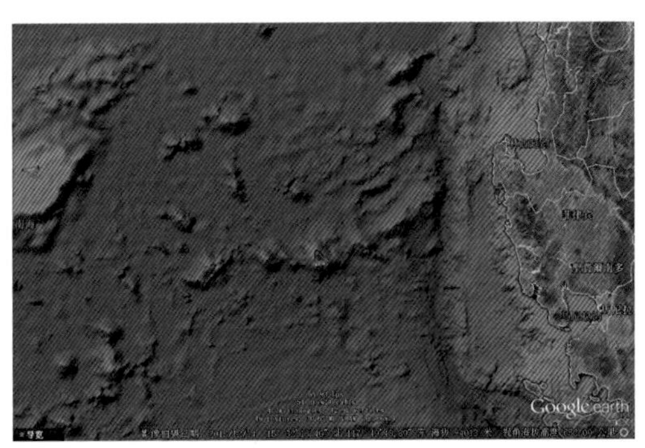

南海深海盆中近东西向海山上的黄岩岛(269页 图4)

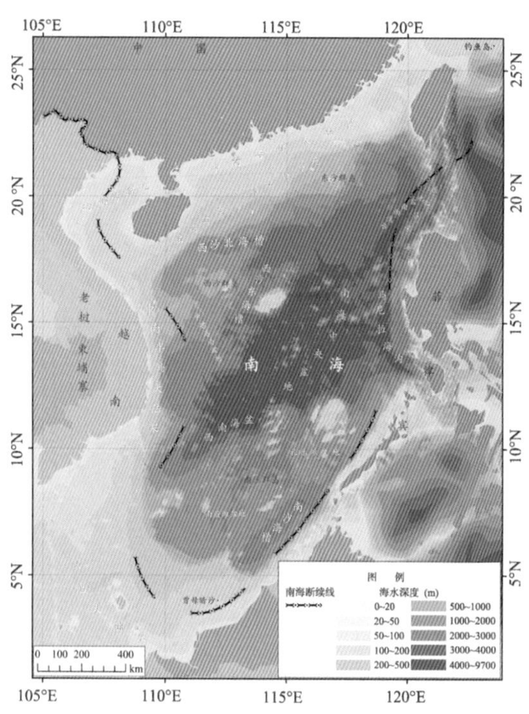

中国南海海底数字地形图(237页 图2)

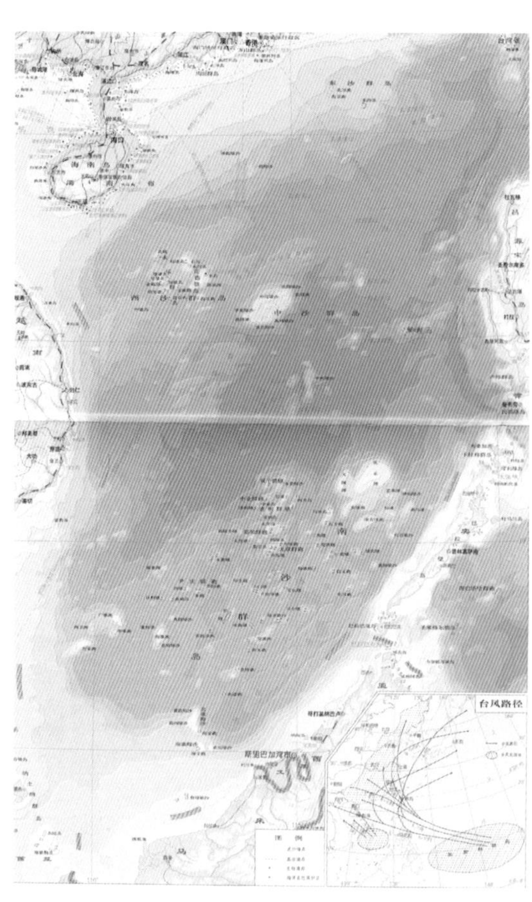

南海海域的中沙群岛与黄岩岛(266页 图1)

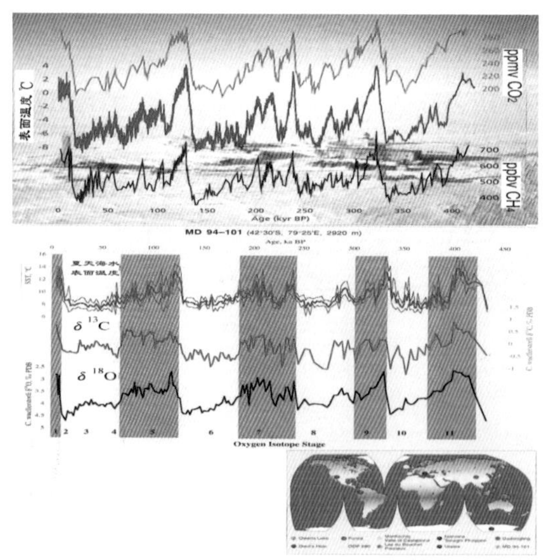

海平面、大气温度和海水温度三者间正相关示意(454页 图2，470页 图4，761页 图1)

太空看地球(465页 图1)

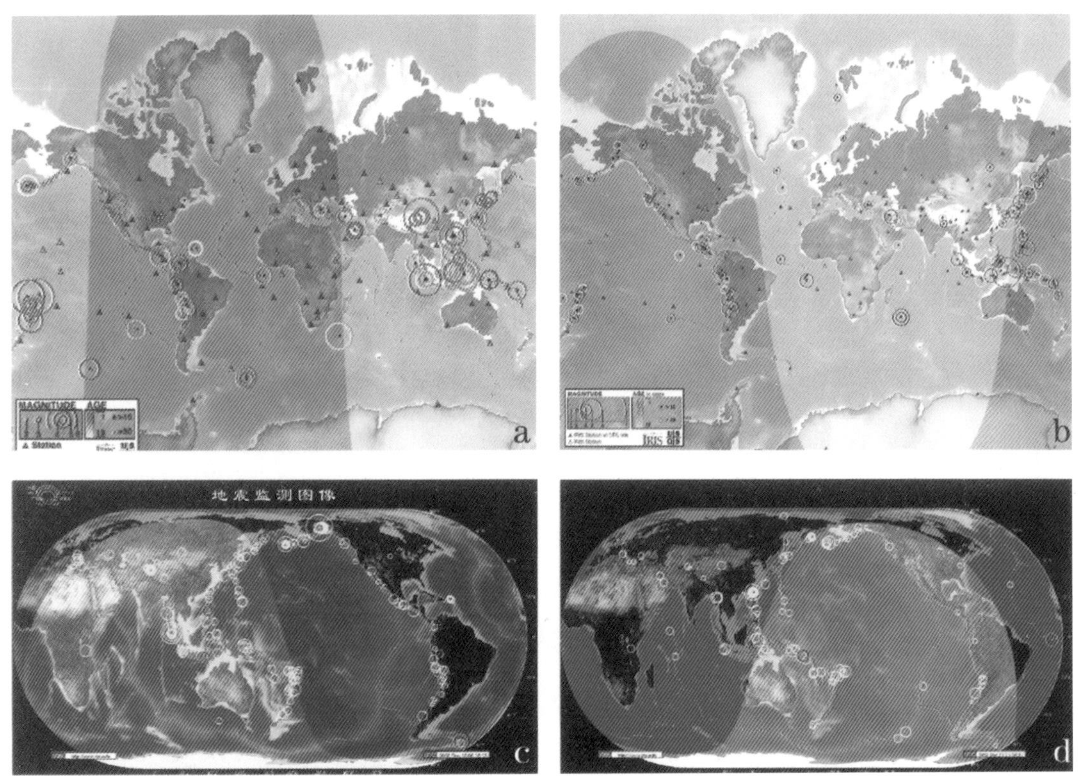

全球地震分布(455页 图3)

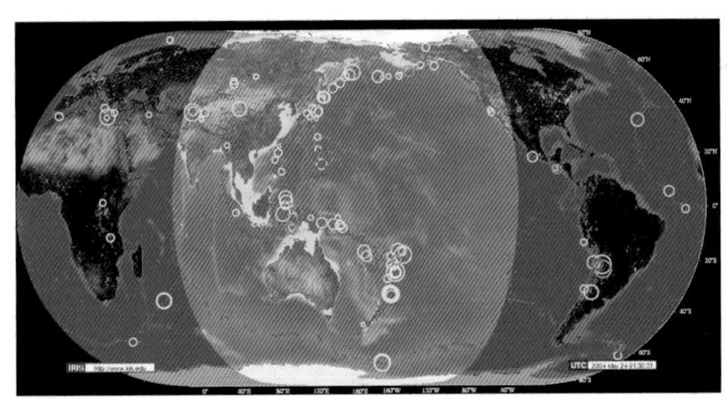

太平洋圈与全球火山地震分布图(2004年)(472页 图9)

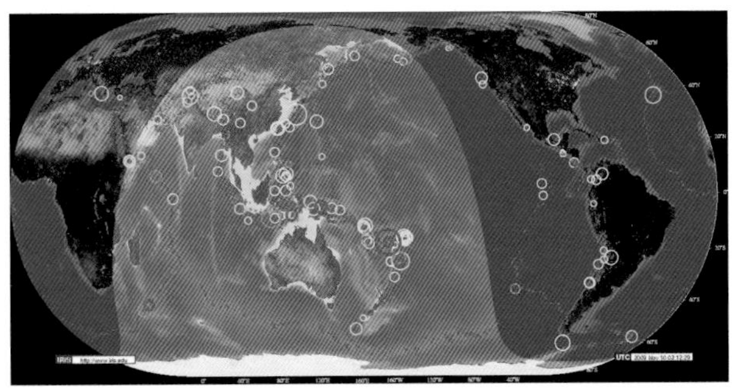

太平洋圈与全球火山地震分布图(2009年)(763页 图3)

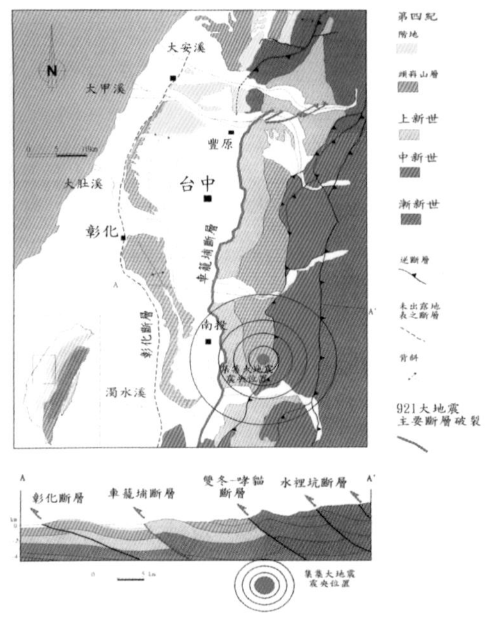

台湾岛集集镇地震
(449页 图7，456页 图4，474页 图13)

岛屿环绕的边缘海(923页 图1)

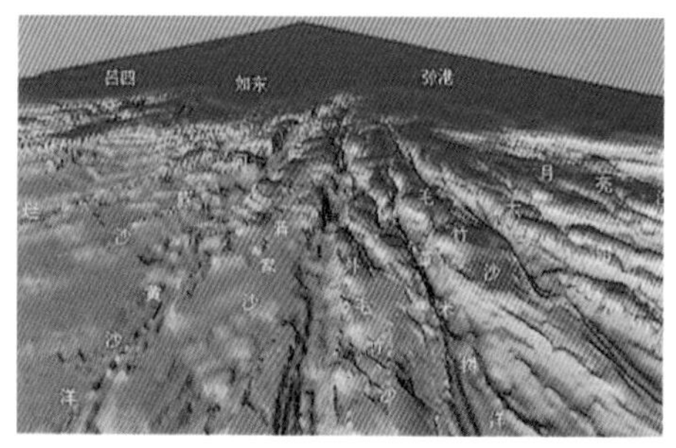

辐射沙脊群海底三维效果(618页 图2)

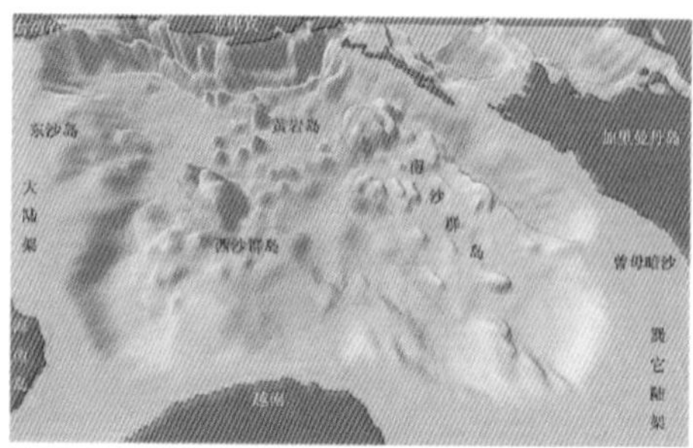

南海巽他陆架立体地形(621页 图6)

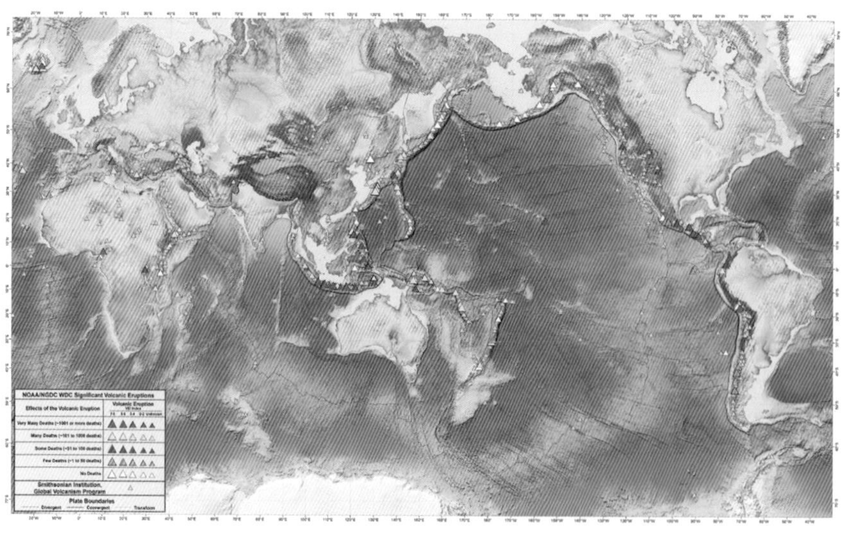

公元前4360年至公元2013年间主要火山喷发的分布图(635页 图1)

大同火山(636页 图3)

六合桂子山石柱林(638页 图4)

富士山火山口(641页 图8)

耸立于那不勒斯海湾平原的意大利维苏威火山
(647页 图16)

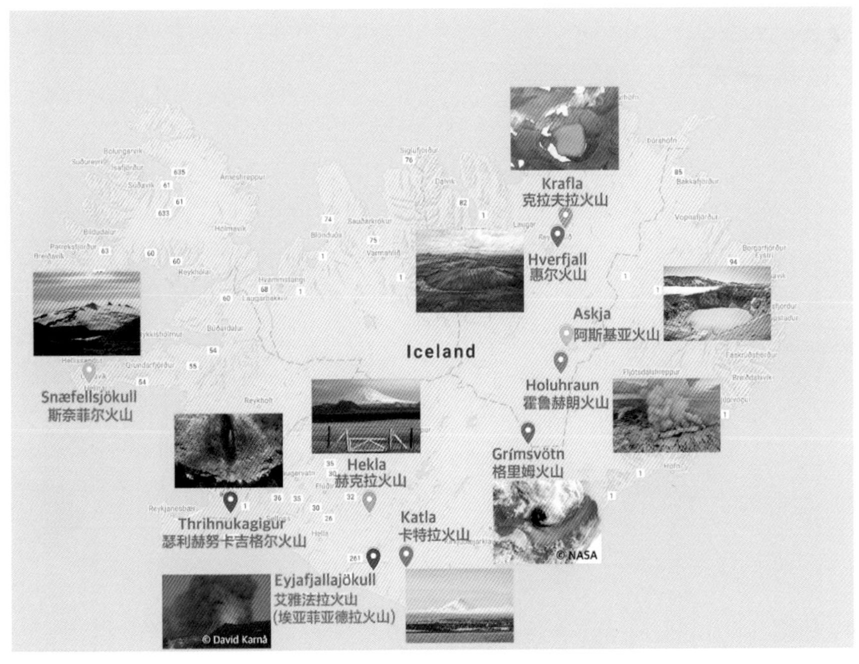

冰岛最著名的火山(654页 图25)

长白山天池火口湖(645页 图12)

冒纳罗亚火山口(650页 图20)

惠尔死火山口(656页 图29)

大屯火山群(646页 图14)

艾雅法拉火山喷发(655页 图26)

“海峡两岸地貌学研讨会”会后考察(2004)(987页 图2)

“第十五届海峡两岸地貌学研讨会”暨“南海资源环境与海疆权益”咨询项目2014年联合学术年会(986页 图1)

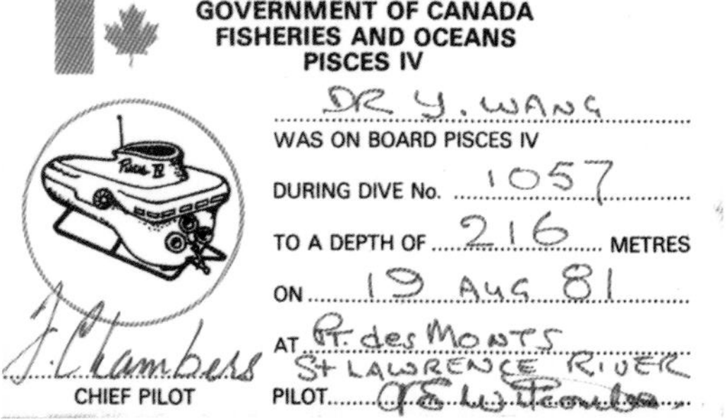

GOVERNMENT OF CANADA
FISHERIES AND OCEANS
PISCES IV

DR Y. WANG
WAS ON BOARD PISCES IV
DURING DIVE No. 1057
TO A DEPTH OF 216 METRES
ON 19 AUG 81
AT Pt. des MONTS ST LAWRENCE RIVER
PILOT
CHIEF PILOT

王颖1981年下潜大西洋圣劳伦斯海湾湾底216m下潜证书(1002页 图3)

2001年6月，王颖获加拿大Waterloo大学环境科学荣誉博士，与该校校长(时任)Prof. David Johnston(左2)、校董主席(右1)及爱人朱大奎教授(左1)合影(998页 图1)

2003年6月王颖在苏北洪泽湖考察并指导博士后研究生(1004页 图6)

2003年10月王颖与师生一起考察连云港古海岸遗迹(1005页 图7)

幸福家庭(1006页 图9)